The Chora of Metaponto 3
Archaeological Field Survey
Bradano to Basento

Vol. I

The Chora of Metaponto 3
Archaeological Field Survey
Bradano to Basento

Vol. I

Edited by Joseph Coleman Carter and Alberto Prieto

Institute of Classical Archaeology
Packard Humanities Institute
University of Texas Press, Austin

Printed in China

First edition, 2011

∞ The paper used in this book meets the minimum requirements of
ANSI/NISO Z39.48-1992 (R1997) (Permanence of Paper).

The chora of Metaponto 3 / edited by Joseph Coleman Carter and Alberto Prieto. — 1st ed.
p. cm.
Includes bibliographical references and index.
ISBN 978-0-292-72678-9 (cloth : alk. paper)
1. Metapontum (Extinct city) 2. Metapontum (Extinct city)—Rural life and customs. 3. Agricultural colonies—Italy—Metapontum (Extinct city). 4. Greeks—Italy—Metapontum (Extinct city). 5. Excavations (Archaeology)—Italy—Metapontum (Extinct city). 6. Metaponto Region (Italy)—Antiquities. I. Carter, Joseph Coleman. II. Prieto, Alberto. III. Title: Chora of Metaponto three.
DG70.M52C56 2011
937'.773—dc22

2011004131
978-0-292-73481-4 (E-book)

For reasons of economy and speed, this volume has been printed from camera-ready copy furnished by ICA, which assumes full responsibility for its content.

To the memory of Dinu Adamesteanu,
to Antonio De Siena,
to students of Metaponto, past, present, and future.

Special Acknowledgment

These volumes and others in this series could not have been conceived, much less carried through to publication, without the generous financial support of the Packard Humanities Institute.

The Director and staff of the Institute of Classical Archaeology and its collaborators express their profound gratitude.

Contents

Volume I

Acknowledgments xi

Introduction xvii
Joseph Coleman Carter

I. The Land

1. Geologic Background of the Metapontino 3
Robert L. Folk

2. Geomorphology and Geoarchaeology of the Metapontino 31
James T. Abbott

II. Methods and Analytical Tools

3. Survey Design and Field Methods 71
Alberto Prieto

4. GIS Methods and Considerations 93
Peter H. Dana

III. Survey Materials and Assemblages

5. The Survey Assemblages 129
Keith Swift

6. Archaic and Black-Gloss Fine Ware 143
Elisa Lanza Catti, Francesca Silvestrelli, Keith Swift, Amelia Tubelli, and Eloisa Vittoria

7. Grey Ware 271
Eloisa Vittoria

8. Figured Ware 303
Francesca Silvestrelli

9. Plain and Banded Ware 337
Eloisa Vittoria

10. Lamps 425
Emanuela Conoci with Eloisa Vittoria

11. Greek and Roman Republican Cooking Wares 439
Ruth Smadar Gabrieli

12. Archaic to Late-Republican Transport Amphorae 455
Keith Swift

13. Terracotta Objects 489
Rebecca Miller Ammerman

14. Loom Weights 539
Lin Foxhall

15. Coins 555
Anna Rita Parente

Volume II

IV. Prolegomena to the Settlement of the Chora
Joseph Coleman Carter

16. Introduction 559
17. The Prehistory of the Chora 569
18. Types of Structures in the Chora 591
Alberto Prieto with Joseph Coleman Carter
19. An Approach to the Chora 617

V. The Historic Development of the Chora, 625–25 BC
Joseph Coleman Carter

20. The Pioneers: Historical Development of the Chora, **600**. 641
21. Golden Harvests: The Expansion of the Mid-6th Century BC. 677
22. Crisis in the Chora, **500** 727
23. Reform and Rebirth: The Mid-5th Century BC 745
24. The Chora in **400** 785
25. The Refounding of Metaponto 809
26. A Century of Decline 869
27. The Last Two Centuries BC 893

Appendices

A. EAW of Dated Sites, **600–50** 923
B. Farmhouse Significance (FMCE), **600–50**. 951
C. Basic Graphs and Tables. 963
D. Exploratory Statistical Analyses 1007
Alberto Prieto

Volume III

VI. Division Lines: New Data and Analysis

28. The 1999 Pizzica Excavations: The Division Lines Reconsidered 1027
Joseph Coleman Carter

29. Catalog of the Objects from the 1999 Pizzica Excavations 1053
Joseph Coleman Carter, with contributions by Lucilla Burn, Anna Rita Parente, Francesca Silvestrelli, Keith Swift, and Eloisa Vittoria

30. The Human Remains 1107
Renata J. Henneberg and Maciej Henneberg

31. Aerial Photography and the Organization of the Chora 1115
Alberto Prieto

VII. The Roman Period

32. Imperial and Late Roman Settlement in the Metapontino 1137
Erminia Lapadula

33. Imperial and Late Roman Ceramics 1147
Erminia Lapadula

VIII. The Medieval and Post-Medieval Periods

34. Medieval and Post-Medieval Settlement in the Metapontino. 1161
Erminia Lapadula

35. Medieval and Post-Medieval Ceramics 1173
Erminia Lapadula

IX. Reference Materials

References 1205

Index 1255

Volume IV Atlas

I. Introduction vii
Jessica Trelogan

Map Key. ix

II. Overview Maps 1

Map 1. *Magna Grecia and Sicily, 7th to 1st centuries BC*

Map 2. *Survey Transects, Bradano to Cavone*

Map 3. *All Sites, Bradano to Basento*

III. Period Maps 7

Map 4. *Prehistoric Sites*

Maps 5–14. *Greek Sites by Fifty-Year Periods*

- Map 5. **600** (625–575 BC)
- Map 6. **550** (575–525 BC)
- Map 7. **500** (525–475 BC)
- Map 8. **450** (475–425 BC)
- Map 9. **400** (425–375 BC)
- Map 10. **350** (375–325 BC)
- Map 11. **300** (325–275 BC)
- Map 12. **250** (275–225 BC)
- Map 13. **200** (225–175 BC)
- Map 14. **150, 100, 50** (175–25 BC)

Map 15. *Roman Sites*

Map 16. *Medieval Sites*

IV. The Modern Landscape 35

Map 17. *1877 Map of Metaponto (Istituto Geografico Militare)*

Map 18. *Mosaic of Aerial Photographs (with Division Lines)*

Map 19. *Mosaic of Aerial Photographs (without Division Lines)*

Map 20. *Closeup of Map 19, Central Plateau*

Map 21. *Closeup of Map 19, Pantanello Region*

Map 22. *QuickBird Satellite Image, July 2006*

V. Localities. 47
Alberto Prieto

VI. Gazetteer of Sites 55
Alberto Prieto and Cesare D'Annibale, Jon Morter, and Steve Thompson with Allison Devereux

VII. Assemblage Tables 87
Keith Swift with Allison Devereux

VIII. Index of Place Names 121

Acknowledgments

This study of the Metapontine chora was three decades in the making: the first two devoted to fieldwork; the final to the analysis of the materials, the synthetic studies, and the production of these four volumes. An enterprise of such complexity and long duration could not have been realized without the continuing hospitality and encouragement of our hosts, the commitment of the team in the field, laboratory, and office, and the generous and steadfast backing of our sponsors and donors.

Over the past thirty years, archaeological field survey has developed enormously, incorporating new techniques and technologies, to become a leading discipline in archaeology. While maintaining its original focus and spirit, the Metaponto survey has contributed to that growth. The initial stimulus and enthusiasm of the first field campaign, though great, would not have sustained the project for so many years had it not been for the openness to innovation on the part of the participants. The sense of being on a new frontier of knowledge provided the essential spark to a collection of unusually motivated, talented, and supportive individuals, whose contributions it is here a pleasure to acknowledge.

Our Hosts

The Soprintendenza per i Beni Archeologici della Basilicata provided the ideal conditions for the survey, and promoted it from the beginning. The ICA project is a continuation of research begun by its first Superintendent, Professor Dinu Adamesteanu. His successors, Elena Lattanzi, Angelo Bottini, Maria Luisa Nava, and Antonio De Siena maintained the tradition. Special thanks are due De Siena. Throughout a long career, he has advanced the research agenda of Adamesteanu and made brilliant contributions to it. Our gratitude not only for the hospitality, but for a lifetime of intellectual stimulation and friendship is expressed in the dedication. The relations of trust and mutual understanding between ICA and the Soprintendenza has made it possible to carry forward a broad range of research projects, including excavation of farmhouses and necropoleis identified by the survey, an opportunity rarely if ever realized elsewhere. The 1999 excavation at Pizzica, the subject of three chapters in Volume III, in fact was a collaborative effort, with expenses and personnel shared.

In all the years of ICA's presence at Metaponto, which began in 1974, the Region of Basilicata has been a gracious host, providing workspace in the early years at the Azienda Sperimentale "Pantanello." The convention of 2000 between ICA and the Region established a home for the "Centro di Agroarcheologica," a unit of ICA devoted to the study of the history of agriculture in the region. The Region provided space for living, a library, and office quarters at Pantanello, where much of the research and writing of these volumes took place. Warm thanks are also due the President of the Region, Vito Di Filippo, the Dipartimento di Agricoltura e Foreste, its Assessori and Dirigente past and present, and in particular to Giuseppe D'Agrosa, Salvatore Martelli, and the Torraco family for many kindnesses.

The Survey Team

The first field survey teams were led by Cesare D'Annibale. The 1981 summer campaign of about six weeks was unexpectedly productive and exciting. Well over a hundred sites were discovered, and the momentum carried over into 1982, with an extended campaign in the summer and fall. D'Annibale was, in his early twenties, already experienced with a remarkable ability to "read" the landscape, a sixth sense about where sites should be located, and a relaxed style that belied his deep seriousness and conviction about survey archaeology. The teams responded with consistently outstanding results. It is no exaggeration to say that the survey project owes much of its success to his early leadership. He never lost his enthusiasm and in fact has returned periodically (on leave from his employment at Parks Canada) to lead teams, most recently in 2007.

Successive team leaders Jon Morter, Steve Thompson, and Alberto Prieto carried the project forward along the same lines, but with modifications and improvements. A list of all team members from 1981 to 2001, representing a wide range of institutions and nationalities, appears below. Two participants in the survey, Robert L. Folk and James Abbott, conducted geological and geomorphological research that proved to be of fundamental importance in the interpretation of the survey results.

The Study Teams

Much of the last decade has been devoted to the study of the materials recovered during the survey, which consisted of tens of thousands of pottery fragments and other largely ceramic objects. The study of black-gloss pottery was begun in 2000 by Vincenzo Cracolici and Eloisa Vittoria. They were joined by Marsha Robbins, who was responsible for the preliminary analysis of the amphorae, and Ruth Smadar Gabrieli, who studied the cooking ware. The study of the Roman and Medieval pottery was entrusted to Erminia Lapadula, a native of nearby Pisticci. Rebecca Miller Ammerman, known for her work on terracotta figurines from Poseidonia, the sister polis of Metaponto, undertook to organize and study the often small and puzzling fragments from the survey sites. Lucilla Burn contributed a study of the figured pottery from the 1999 survey and excavation at Pizzica, following her earlier work on that topic for the Pantanello necropoleis. Renata and Maciej Henneberg, who had also worked on the the Pantanello necropoleis materials, continued their research on the ancient population with the study of the human remains from the Pizzica excavations.

ICA had been heavily involved in research at Chersonesos in Crimea for a number of years, but began again in 2007 to concentrate its efforts on Italian projects. A fresh group of students and scholars, all highly trained and motivated, arrived at the Centro to join the above-mentioned researchers. They included Keith Swift, Elisa Lanza Catti, Francesca Silvestrelli, Amelia Tubelli, Emanuela Conoci, Margherita Nigro, and Anna Rita Parente. Lin Foxhall, who participated with John Bintliff in a review of early progress in the survey study in 2004, herself undertook the study of the loom weights. This talented team of established scholars from institutions in Italy, England, Australia, Canada, and the United States contributed excellent studies of their respective specialties, thereby providing a solid basis for interpretation. They worked closely with Tiziana Ria and project coordinator Esmeralda Moscatelli, improving the overall results. In a project of such dimensions, complexity, and duration, this degree of cooperation was as unexpected as it is welcome.

As able as the study team was and is, there needed to be overarching control and direction. From 2001 to 2007, while the undersigned was recovering from a long illness, that responsibility was undertaken by Alberto Prieto. Alberto's contribution to the project, like Cesare's, was fundamental on many fronts. He monitored the study of the ceramic materials and advised on the creation of databases. For the last two years, the logistical aspects have been very effectively handled by Esmeralda Moscatelli, the coordinator of Italian projects, while Keith Swift helped sort out questions about the ceramic data and the translation of the terminology. Alberto's major contribution to the project, however, was his development of the Metaponto Geographic Information System (GIS) and his mastery of the data that underlie the whole of this study. Alberto, together with Peter Dana, and the undersigned, developed the system of dating survey sites and evaluating their relative importance. A phenomenally quick learner, particularly adept at computer operations, and a tireless worker, Alberto produced results that would have taken others far longer to achieve. His contributions continue down to the present. Thanks also go to Erika Hale, from the Division of Statistics and Scientific Computation, at The University of Texas at Austin, for her help on the statistical analyses and chi-square tabulations.

The Staff and Production Team

As for all ICA publications, the editing of manuscripts, the preparation of illustrations, photographs, maps, tables, and graphs, the layout of the volumes, and the initial proofreading are carried out in ICA's offices in Austin. Accurate topographical maps, as will be immediately apparent, were essential to the analysis of settlement patterns in Volumes II and III. Created by Jessica Trelogan, these appear interspersed throughout the text, illuminating the discussion, and, on a larger scale, comprise the contents of Volume IV. Jessica was assisted by Dan Hudson, Jessica Nowlin, Allisa Stoimenoff, and Jonathan MacClellan, who spent many hours digitizing maps. Gordon Lobay and Dan Davis worked on early versions of the Gazetteer of Sites in the Atlas.

Many of the photographs of objects and a number of drawings are the work of Cesare Raho. Chris Williams has made other photographs and has shouldered the responsibility of preparing all for publication. Deena Berg, who joined the ICA staff in 2007, has been the driving force behind this production and others. She brought to the job a deep understanding of the subject matter, having previously earned a Ph.D. in Classical Archaeology. Her editorial acumen, attention to detail, and insistence on maintaining schedules has kept us all on track. The editing job tested her considerable talents as a diplomat and gentle taskmaster. Jessica, Chris, and Deena have contributed immeasurably to the physical appearance and readability of these volumes. All involved in book production at ICA have been greatly assisted by graduate student workers Dan Davis and Lauren Jackson. Lauren, as Dan, Pat Irwin, and Debbie Bartz before her, had the unenviable job of deciphering my handwritten manuscripts. I owe them all a deep, personal debt of gratitude. Thanks go as well to the other production team members, Estéban Hinojosa, and especially Allison Devereux, who patiently organized and formatted the many tables in Volumes II and IV, and Alison Parker who helped proof the manuscript for review. Dan Patrick, Eliza Walton, and Austin Parker helped prepare graphic images. For any remaining infelicities, the undersigned general editor must take responsibility. Carol Cook, our long-time administrator helped, as always, to keep ICA together and see us through the day-to-day bureaucratic hurdles and financial tasks.

After Deena's unexpected leave for reasons of health in March 2010, Jessica stepped in with Lauren, Allison, Chris, and Estéban for the final revisions and corrections, and to see the volumes to press.

We are all grateful to the staff at The University of Texas Press for the consideration and care they have taken with the production of these and previous volumes. Particular thanks go to Jim Burr, our editor at the Press, and to recently retired director, Joanna Hitchcock. The relations between the Press and ICA since the publication of the first volume of the Metaponto series in 1998 have been productive and cordial—truly ideal. We shall miss her very much. Comments from the external readers, John Cherry and David Ridgway, have been extremely helpful and encouraging.

The Donors

For the first quarter century of its existence, ICA depended almost entirely on the generosity of individual donors, many from Texas, for the funding to carry out its fieldwork. A list of these loyal donors appears at the end of these acknowledgments, following the list of team members. For many years the contributions were doubled by matching grants from the National Endowment for the Humanities, until 1996 when NEH and its enlightened programs in archaeology that promoted publication were eliminated by Congress.

Then, in 1997, as if in answer to a prayer, I received a telephone call from David Packard, who was then creating the Packard Humanities Institute and the rest is, very happily, history. Annual grants from PHI have made possible for the last dozen years not only major research and conservation projects at Chersonesos, but also the continuation of all aspects of ICA's work in southern Italy, with the emphasis on publishing the results of a quarter century of intensive fieldwork. Additional fieldwork for this publication included the survey and excavation at Pizzica in 1999 and carrying out revisits of sites in the chora.

Without the grants from PHI the detailed study of the survey pottery would have been inconceivable. It represents thousands of hours of painstaking examination and evaluation of pottery of all classes. The results are the true basis for the dating of the sites, and for the understanding of the chora's transformation. The catalogs and databases that resulted from these studies will be shared with the wider public not only with the publication of Volumes I and IV, but also online in a form that will invite ongoing study of these materials.

The process of seeing patterns and relationships in this mass of data required intense concentration over a long period. In a busy academic world, opportunities to engage in extended study, reflection, and deep reading are increasingly rare. For this unique privilege, the authors of these volumes are profoundly grateful to PHI.

The books that you hold in your hands are the result of years of trial, error, and discussion. We were pleased that PHI encouraged the production of Volume IV, the Atlas, and through an agreement with The University of Texas Press made it economically possible for scholars and students to add all these volumes to their libraries at a moderate cost.

To PHI and those donors who for the past four decades have participated in the research at Metaponto, a heartfelt thank you.

Joseph Coleman Carter
December 2010
Austin, Texas

Sponsor
Packard Humanities Institute

Donors
Government and Educational Institutions
National Endowment for the Humanities
College of Liberal Arts, The University of Texas at Austin

Foundations
Austin Commuity Foundation
The Brown Foundation, Houston, Texas
The Ewing Halsell Foundation, San Antonio, Texas
The James R. Dougherty, Jr. Foundation
The Liss Foundation
The Spiros Martel Foundation, Houston, Texas
The Rachael & Ben Vaughan Foundation
The David and Lucille Packard Foundation, Palo Alto, California
The Bernard and Audré Rapoport Foundation, Waco, Texas
The Potts Sibley Foundation, Austin, Texas
The Wenner-Gren Foundation for Anthropological Research, New York, New York

Corporate Matching Grants
Highland Resources, Houston, Texas
The Washington Post
Texas Tradewinds Travel
IBM through Project Quest

Non-Profit Organizations
Center for the Study of Ancient Territories
National Geographic Society

Individuals
Mr. and Mrs. Daniel Balz, Bethesda, Maryland
Ms. Jane Bedichek, Scarsdale, New York
Dr. Leo Bini, Rome
Signora Laura Bini Mazzi, Rome
Signor Stefano Bini, Rome
Col. and Mrs. Dayton Blanchard, Cloudcroft, New Mexico
Mr. James Buchanan, Vancouver, British Columbia
Mr. Joseph C. Carter III, Versailles, Kentucky
Mrs. Mary L. Carter, Missoula, Montana
Mr. John Breckinridge Carter, Missoula, Montana
Ms. Carlyle Carter, Evanston, Illinois
Dr. William L. Carter, Tampa, Florida
Ms. Judy D. Coker, Austin, Texas
Mrs. Mary Cusack, London, England
Prof. Ronald DeFord, Austin, Texas
Mrs. Marion Weir DeFord, Austin, Texas
Mr. Gilbert M. Denman
Prof. Myron Dorfman, Austin, Texas
Mr. Jack M. Englert, Jr., Littleton, Colorado
Mr. Michael Guarino, San Antonio, Texas
Mr. T. Hadrian Hale, Houston, Texas
Ms. Harriet Harris, Austin, Texas
Ms. Sandra S. Heinrichs, Austin, Texas
Mr. and Mrs. Marshall Johnson, St. Paul, Minnesota
Mr. and Mrs. Daniel Y. Kruger, San Antonio, Texas
Ms. Amy Liss, Summit, New Jersey
Dr. Henry Liss, Summit, New Jersey
Mrs. Lois Ashton Larson, Elmhurst, Illinois
Mr. Warren Larson, Elmhurst, Illinois
Ms. Mary Link Malone, Dallas, Texas
Mrs. Barbara McCluer, San Antonio, Texas
Ms. Etelka McCluer, San Antonio, Texas
Ms. Elizabeth McGrosso, Austin, Texas
Ms. Vera Massaro, Austin, Texas
Prof. Lucy Shoe Meritt, Austin, Texas
Prof. Benjamin Dean Meritt, Austin, Texas
Mr. Bert Henry Michelsen, Hinsdale, Illinois
Ms. Kathleen A. Monahan, Austin, Texas
Mr. George Nalle, Jr., Austin, Texas
Ms. Anne Byrd Nalle, Austin, Texas
Mr. and Mrs. John O'Connor, Elkins, New Hampshire
Ms. Maconda Brown O'Connor, Houston, Texas
Mr. Ralph S. O'Connor, Houston, Texas
Mr. and Mrs. Bernard Rapoport, Waco, Texas
Dr. and Mrs. David M. Renison, Higganum, Connecticut
Mr. Phillip Schafer, Darien, Connecticut

Ms. Johanna Smith, Oak Hill, Texas
Mrs. Elizabeth H. Tracey, Denver, Colorado
Mr. Ben F. Vaughan, Jr., Corpus Christi, Texas
Mr. and Mrs. Ben F. Vaughan III, Austin, Texas
Ms. Genevieve Vaughan, Rome, Italy
Mr. and Mrs. William Willis, Austin, Texas
Ms. Nancy C. York, Austin, Texas
Col. Haskell Zipperman, San Antonio, Texas

Survey Team Members

1981
Cesare D'Annibale (field leader)
Prof. R.L. Folk (geologist)
Pamela Dyson
Stephan Wolohojian
Giuseppe Di Taranto
Boyce Cabaniss
Lucilla Burn

1982
Cesare D'Annibale (field leader)
Prof. R.L. Folk (geologist)
Claes Petersson
Pontus Reimers
Gene Storie
Jennifer Brehob
Sandra Langston
Ann Patterson

1983
Cesare D'Annibale (field leader)
Helen Dunlop
Maria Christensen
Christine Renaud

1984
Cesare D'Annibale (field leader)
Helen Dunlop
Maria Christensen
Miranda Richardson

1985
Cesare D'Annibale (field leader)
Helen Dunlop
Dick Grove

1990
Cesare D'Annibale (field leader)
Brian Ross
Maria Christensen Elliot
Gert Burgers

1992
Jon Morter (field leader)
Gianna Ayala
Kathy Collins
Miranda Grieder
Judy Gwyn
Marina Markantonatos
Patrick MacMillan
Glenda Middleton
Forrest Swan
Anne Toxey

1993
Jon Morter (field leader)
Gianna Ayala
Michael Guarino
Kara Nicholas
Stephen Thompson
James Abbott
Stefan Lunden
Lynn Fotheringham
Christine Lee
Marina Markantonatos

1994
Jon Morter (field leader)
Brice Erickson

1999
Stephen Thompson (field leader)
Alberto Prieto
Vincent LeCornu
Christian Hartnett
Eric Dugdale
Mireille Lee
Steven Gavel
Jessica Trelogan
Evgeniy Tkachuk
David Lummus

2000
Alberto Prieto (field leader)
Christian Hartnett
Cindy Young
Alexander Alderman
Michael Fronda
Kevin Bain
Anna Cavallo

2001
Stephen Thomspon (field leader)
Alberto Prieto (field leader)
Mark Armstrong
Heather Harper
Caitlin McCracken
Kelley Russell
Kelly Gilbride
Kelly Williams

Contributing Authors

James T. Abbott, Ph.D.
Archaeology Branch, Environmental Affairs Division
Texas Department of Transportation

Professor Rebecca Ammerman
Department of the Classics
Colgate University

Dr. Lucilla Burn
Keeper of Antiquities
Fitzwilliam Museum, Cambridge

Emanuela Conoci, Laurea
Università di Lecce

Peter H. Dana, Ph.D.
Department of Geography and the Environment
University of Texas at Austin

Professor Robert L. Folk
Professor Emeritus in Geology
University of Texas at Austin

Professor Lin Foxhall
School of Archaeology and Ancient History
University of Leicester

Dr. Ruth Smadar Gabrieli
University of Sidney

Elisa Lanza Catti, Dottore di Ricerca
Università di Padova

Erminia Lapadula, Laurea
Università di Siena

Margherita Nigro, Dottore di Ricerca
Università degli Studi di Napoli Federico II

Anna Rita Parente, Professore di Numismatica
Università della Basilicata

Alberto Prieto, Ph.D.
University of Texas at Austin

Francesca Silvestrelli, Ricercatore
Università del Salento

Keith Swift, D.Phil.
Oxford University

Amelia Tubelli, Dottore di Ricerca,
Università degli Studi di Napoli Federico II

Eloisa Vittoria, Laurea
Università di Viterbo

Introduction

Joseph Coleman Carter

Figure i The chora of Metaponto, Bradano valley, the city of Ginosa in the distance. Pipe is part of modern irrigation system. (AP)

This four-volume study is the fruit of nearly forty years of research in the ancient countryside (chora) of Metaponto, and a major step forward in the comprehensive plan of publication that began with the *Chora of Metaponto: the Necropoleis* (1998).[1] Using evidence mainly from intensive field survey, but also from excavation at sites representative of its principal periods, it includes a detailed examination of the landscape and the artifacts recovered from it (approximately 30,000 from the first half of the survey alone). This publication is not, however, a simple presentation of the factual information gathered over the course of two decades of fieldwork. It has greater ambitions. Its larger aim is to recreate the patterns of ancient life in a single territory over its millennial history.

The work presented here has been perforce a multidisciplinary, collaborative effort requiring specialists not only in more established fields such as ancient ceramic studies, numismatics, and geomorphology, but also expertise with newer and constantly evolving technologies like GIS and remote sensing. This wide-ranging approach to the landscape was new when this project began in 1974, and the two developed in tandem: as technologies have evolved, so have the methods. Special attention is paid here to their clear presentation, which is accompanied by access to original datasets, so that the reader may test our conclusions and, we hope, expand upon them.

The approach, especially to the still relatively unknown Greek countryside, is original (*Fig. i*). Although the majority of the material presented here relates to the period of Greek occupation, much attention was paid, during fieldwork and study, to the prehistoric, Roman, Medieval, and Post-Medieval periods as well as the paleoenvironment and the physical anthropology of the occupants of the ancient countryside.[2] Moreover, the Greek chora is examined in this volume not in isolation, but in the context of

[1] The name of the ancient Greek polis was "Metapontion," subsequently known in Roman times as "Metapontum." The more common Italian name "Metaponto" is used throughout these four volumes to refer to the polis, and "Metapontino" in reference to the general territory.

[2] Excavation results and more in-depth presentation of the non-Greek Metapontino are forthcoming in future volumes of this series.

Figure ii Dinu Adamesteanu discussing the mosaic of aerial photos, Pantanello, 1979. (CW)

the ancient city (asty), as part of a single political, economic, and social unit: the polis. The polis is seen as one phase—albeit the one leaving by far the heaviest imprint—in the development of a territory over the span of seven or eight millennia of human interaction with this land, the main protagonist of our story.

Background

The genesis of this project can be traced back to 1967, and the 7th International Convegno di studi sulla Magna Grecia in Taranto, where Prof. Dinu Adamesteanu, then head of the Soprintendenza alle Antichità della Basilicata, revealed for the first time the widespread settlement of the Greek countryside at Metaponto (*Fig. ii*). Apart from important extra-urban sanctuaries and a few excavations of farmhouses, the countryside at Metaponto and elsewhere had been generally ignored by archaeologists and historians. With his pioneering study of aerial photographs, field survey, and the excavation of farmhouses, Adamesteanu's discoveries made a great impact on his audience and eventually on the field of Classical Archaeology. Like all fundamental studies, this required a paradigm change, and it has taken time and much work to begin to understand the full implications of these discoveries. The atmosphere at that memorable encounter of historians and archaeologists was electrifying and it left an impression on this author, a graduate student at the time, and an enthusiasm to learn more about the Greek rural past that remains undiminished to this day.

Scholarly interest in the ancient Mediterranean countryside grew in the second half of the 20th century, both in regard to Greek colonization, as well as to Etruscan and Roman settlement. The survey of southern Etruria by the British School at Rome under John Ward-Perkins began in the 1950s. In Greece, the pioneering work of Michael Jameson in the Argolid, and of John Bintliff and Anthony Snodgrass in Boeotia in the 1970s were the first, and among the most influential. By the 1980s field survey had become widespread throughout the Mediterranean, providing indispensable evidence for a growing interest in and study of the ancient Greek countryside, manifest in the seminal volume *Problèmes de la terre en Grèce ancienne* (1973). Archaeology had begun to fill the enormous gaps in the undocumented history of the land and ancient world outside ancient urban centers.

The southern coast of the modern region of Basilicata was relatively undeveloped before the Agrarian Reform of the early 1950s as a result of historical circumstances and environmental conditions, and thus, this remote area of southern Italy was ideally suited to this kind of study. The landscape was relatively unchanged since ancient times. Because of Adamesteanu's prodigious, visionary work, it soon became apparent that the area around Metaponto was potentially among the best preserved examples of an ancient territory. Similar circumstances existed in the territory of Croton in Calabria in the immediate Post-WWII era and in the territory of Chersonesos on the shores of the Black Sea in southern Crimea until relatively recently, but conditions in both these areas have not been so favorable to research as those at Metaponto. Though the excavated farmhouses in the chora of Chersonesos are numerous and impressive, they remain isolated monuments awaiting comprehensive study. And so, for now, the chora of Metaponto will stand as the most fully documented and completely preserved example of an ancient Greek countryside.

The investigation of the territory of Metaponto by the Institute of Classical Archaeology began in 1974 on a hilltop at Pantanello overlooking the Basento valley and the Ionian Sea (*Fig. iii*). Nothing more than a few fragments of ancient roof tile was visible—hardly a premonition of the long and productive role that this multi-phased and multifaceted site would play in the discovery of the ancient countryside. From this fortunate beginning the project developed over

Figure iii Pantanello on June 25, 1974, the beginning of excavation at this extraordinarily productive site.

the next decade into a wide-ranging exploration of the whole chora in all of its major historical phases. It was a deliberate process, with each advance developing out of an earlier discovery, and it was aided by two essential circumstances. The first was the goodwill and support of local authorities, of the Soprintendenza per i Beni Archeologici della Basilicata, which was led by Adamesteanu from 1974 to 1977, and later by his successors, Elena Lattanzi, Angelo Bottini, M.L. Nava, and Antonio De Siena. The constant presence of De Siena, first as inspector, then Director of the Museo Nazionale di Metaponto, and finally as Superintendent (2010) has ensured the Metaponto and the Metapontino remained in the forefront of research in southern Italy. The support and encouragement of the Regione di Basilicata, on whose experimental farm "Pantanello" a number of key sites were discovered, has been crucial. The Regione has been a host to ICA and the Centro di Agroarcheologia at Pantanello, since 1974. We had the good fortune to be in the right place at the right time. The survey began just as a new area in the heart of the chora was being opened up to irrigation and deep-plowing, bringing to the surface evidence of hundreds of previously invisible sites, and inevitably destroying them in the process. The sites are often less than half a meter below ground and the modern plow blade reaches down a meter or more. Had we arrived a few years later the harvest would have been far more meager.

The results of the twenty years of research would probably have been archived years ago had it not been for the Packard Humanities Institute. PHI's generous support for ten years of research leading up to the publication, and for the production of these volumes, has not only made this study possible, but has greatly advanced archaeological research in an area of the ancient world where important discoveries are being made almost daily, but where the resources to study and publish are largely lacking. PHI's enlightened support of the young Italian scholars studying the survey materials, and those from excavated sites in the chora of Metaponto and also of Croton, is Maecenism on a scale that has never before been witnessed in Magna Grecia.

The Metaponto Project

Over the course of twenty-five years of active fieldwork the ICA Metaponto Project excavated a wide variety of rural sites, ranging from farmhouses to necropoleis and sanctuaries, and engaged in research directly related to daily life in the countryside. This included paleobotanical and archaeozoological studies of excavated plants and animal remains, physical anthropological studies of the human skeletal remains from a major rural necropolis, and geologic and geomorphological analyses of the chora. The scope of these activities was outlined in depth in the Jerome Lectures at the University of Michigan (2000) and the American Academy in Rome (2001), subsequently published as *The Discovery of the Greek Countryside at Metaponto* (2006) by the University of Michigan Press. It was translated as *La scoperta del territorio rurale greco di Metaponto* (2008) by Elisa Lanza Catti, and published by Edizioni Osanna, Venosa.

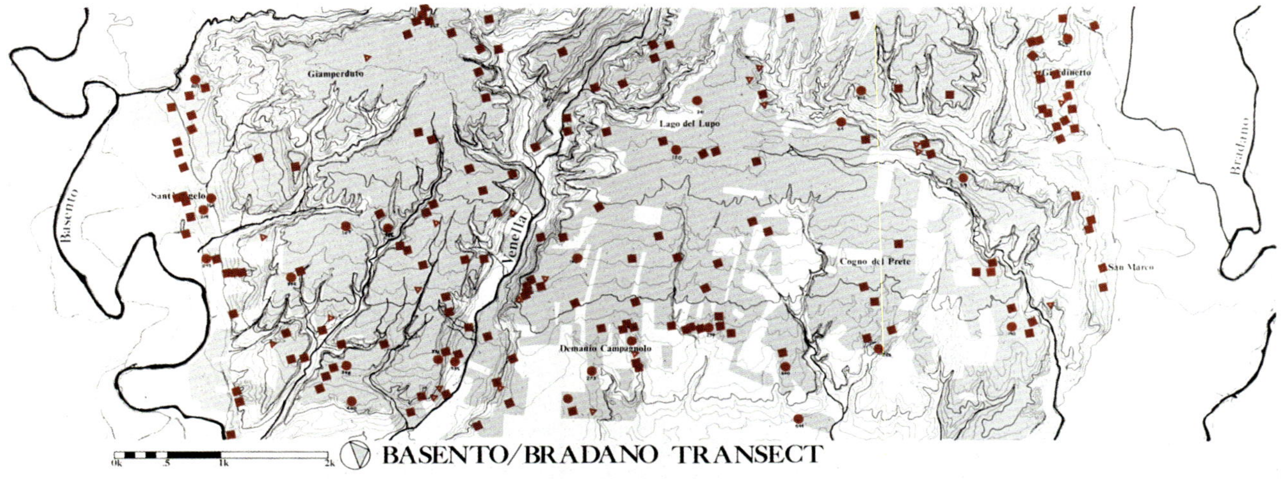

Figure iv Hand-drawn map of sites in the chora, ca. 1985.

In the early years of the project it was already obvious that intensified agriculture was rapidly destroying sites throughout the chora and beyond. ICA participated in rescue excavations at Saldone (1976) and witnessed firsthand the unprecedented removal of whole hillsides by huge earth-moving machines. The Pantanello necropoleis were discovered by the survey team as ancient olive trees were bulldozed to make way for an experimental field. All of this was part of a shift away from the traditional cultivation of cereals and tobacco that characterized the Metapontino in the immediate Post-WWII years, to more profitable fruit crops, such as grapes, oranges, kiwi, and avocados that were aimed at the markets of Northern Europe. The funding that enabled these programs came from various sources: the Regione Basilicata, the Italian government, and the European Union. The standard of living rose, but there would unfortunately also be regrettable consequences for the archaeology and for the natural beauty of the area. Thus almost from the beginning ICA's work was motivated also by a deep sense of an urgent mission: to save what could be saved, as both a record of the past, and of a Mediterranean landscape of quiet beauty that was fast disappearing.

This phenomenon was not limited to Metaponto. The alarm had been sounded decades earlier in the case of the Attic countryside, but it was too late there. In southern Italy the warning was simply ignored, and wonderful opportunities were allowed to slip away. This deteriorating environment made the original vision of Dinu Adamesteanu all the more remarkable and admirable, in its organized and implemented research, including, prominently, the initial survey of the Metapontino. It was international, multidisciplinary, non-political, intellectually exciting and, as this volume and others will testify, ultimately produced results that have fundamentally altered our understanding of the ancient countryside, and the ancient world.

The survey by the Soprintendenza under Adamesteanu in the 1960s provided valuable, irreplaceable early records of sites, especially those close to the coast and to the ancient city. This documentation was supplemented by Giovanni Uggeri's eyewitness account of features of the countryside that were no longer evident in the 1970s. Unfortunately, there was no immediate follow up. An initial campaign by several French teams, again in collaboration with the Soprintendenza, put many sites on the map, but a brief preliminary report had no sequel. The positions of these earlier survey sites have been noted and they often coincide with those recorded by the ICA survey in areas where the surveys overlap.

In the decade that elapsed between these early efforts and the systematic campaigns of the 1980s many sites were undoubtedly lost or seriously damaged, but this happened mainly in the area immediately outside of the ancient city, in the coastal plain and the low marine terraces overlooking it, where larger land holdings of the 19th and 20th centuries were located and where there was more capital to invest in improving agriculture. Further inland the farm plots tended to be smaller, many of them the result

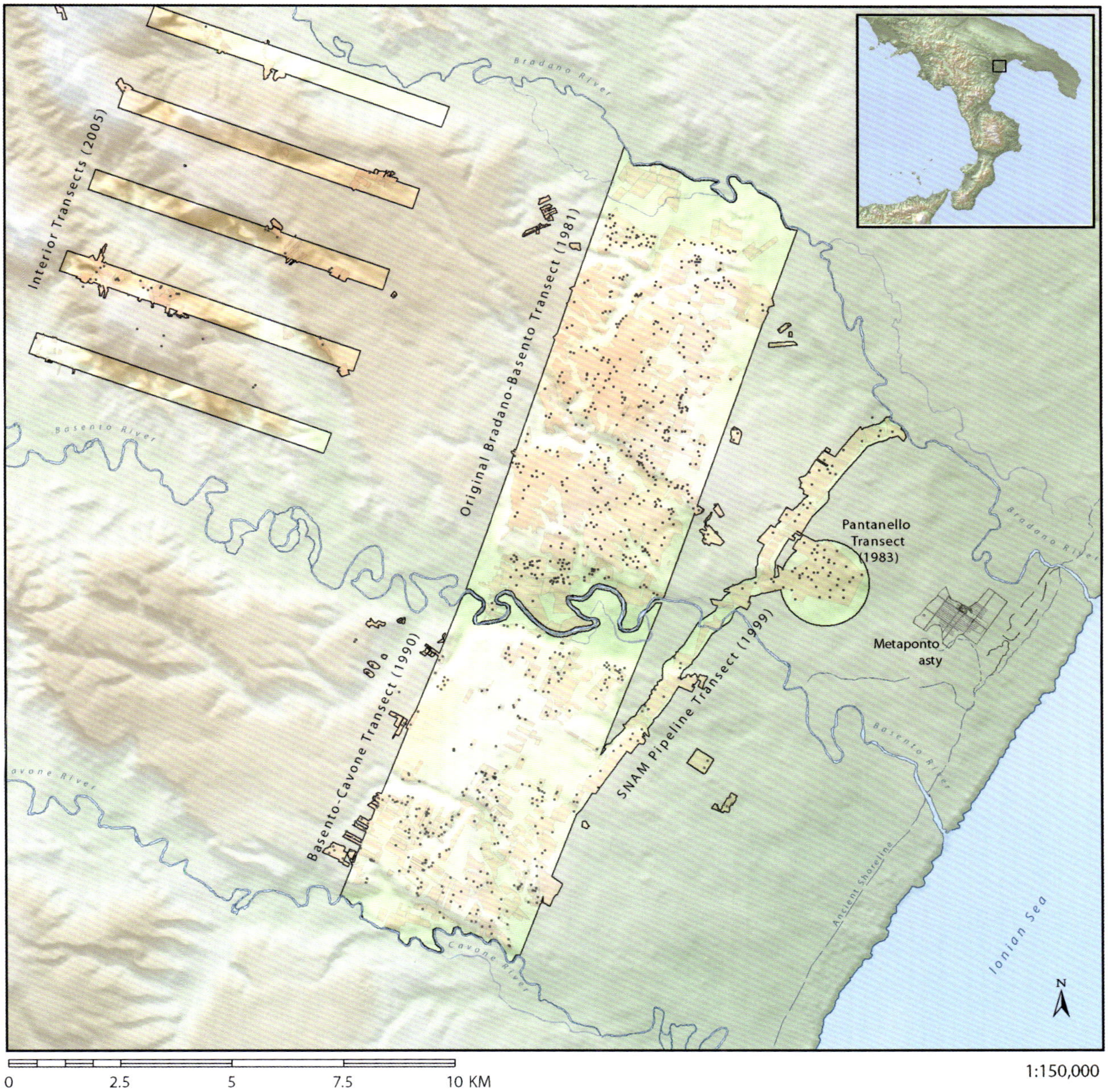

Figure v GIS generated map of the chora and ancient city (asty) of Metaponto with all transects indicated (1981–2005). (JT)

of the large-scale redistribution of property that accompanied the agrarian reforms of the 1950s. Here the agriculture tended to be traditional, with far less use of heavy equipment. This situation changed suddenly in 1980 as irrigation water carried by deeply buried pipes was introduced for the first time into a central area of the ancient chora, beginning 5 to 6 km from the coast and stretching farther inland. The area chosen for the first ICA intensive survey, the "Bradano-Basento Transect," measured 4.2 km wide by some 10 km in length, and stretched between the Bradano and Basento Rivers (*Figs. iv–v*). This transect, fortunately oriented to correspond with the area to be irrigated and deep-plowed, was designed to sample all the landscape typical of the chora. The years between 1970 and 1980 were fallow for survey. The results, however, of the renewed campaigns in 1981 and 1982 far surpassed our expectations.

Technology has greatly increased the effectiveness of field survey documentation and analysis. Databases and personal computers for recording, displaying, and relating the masses of information collected in the field became generally available in the early 1980s, just as the ICA survey began, and were the first step in the modernization of our methods. These innovations were followed by the introduction of the Global Positioning System (GPS), first used in ICA's 1999 campaign, and thereafter standard equipment in our toolkit, and finally the creation of a Geographic Information System (GIS) for the management of spatial information integrated with the databases. GIS technology, as will be immediately evident, has been fundamental to the analysis of survey results presented in this publication.

Asty and Chora

Large scale exploration of the ancient city's monumental center began in the mid-1960s with the creation of the Soprintendenza under the direction of Dinu Adamesteanu, and continued throughout the 1970s and early 1980s. The sanctuary and agora of the ancient city were excavated by the Soprintendenza, in collaboration with a team from the German Archaeological Institute in Rome under the direction of Dieter Mertens (see below). At the same time, with encouragement and collaboration from the Soprintendenza, ICA's program of farmhouse excavations (1979 to 1981) together with that of rural necropoleis and sanctuaries (1974 to 1999) defined the basic elements of the chora. The timing of these developments was also extremely fortunate. Interpretation of the field survey results benefited directly from the concurrent excavations in the chora, in which the range of rural site types and their chronology was revealed. The survey sites, in turn, created a spatial and temporal framework for the relatively few excavated sites scattered over a vast area. Information from the urban excavations helped establish dates for pottery and corroborated information regarding periods of prosperity and decline.

The original Bradano-Basento survey transect of some 40 km^2 sampled a large and representative area of the territory between the Bradano and Basento Rivers. When the decision was made in 1988 to survey Metaponto's chora in its entirety, the transect was extended from the Basento to the Cavone (*Fig. vi*), effectively doubling its size (the Basento-Cavone survey is the subject of another proposed study in this series). A smaller transect was added around Pantanello (1982), followed in 1999 by yet another transect along the path of oil and gas conduits that was to cut across the whole of the territory from north to south, passing close to the ancient city. Most recently a set of parallel transects was initiated in 2005, which run perpendicular to the coast and stretched inland to a distance of 27 km. Research in these areas is ongoing.

Figure vi Surface concentration at site 230, summer 1982.

Unique Aspects of the Metaponto Survey

The Metaponto survey is different in several respects from those carried out in other areas of the Mediterranean. The depth of the built-up topsoil and a sparse occupation after the Greek period to the present have resulted in remarkable preservation. While sites located on slopes are often partially eroded, heavy flooding has deeply buried the sites in the river valleys, protecting them and creating, in some cases, anaerobic conditions that favor especially good preservation of organic remains. Only a few sites in the valleys were found by the survey teams, though it is clear from excavations at Pantanello and during the construction of the oil and gas pipelines that such sites are present and may be numerous.

Sites situated on slopes, along with those in the elevated marine terraces that comprise two-thirds of the chora's area, are easily identified by distinct surface concentrations (*Fig. vi*). Finds include ceramics, pottery, roof tiles, as well as other building materials for habitations and rural structures; in the case of burials, sometimes human bone is found. These numerous surface remains, all recorded in the Gazetteer,

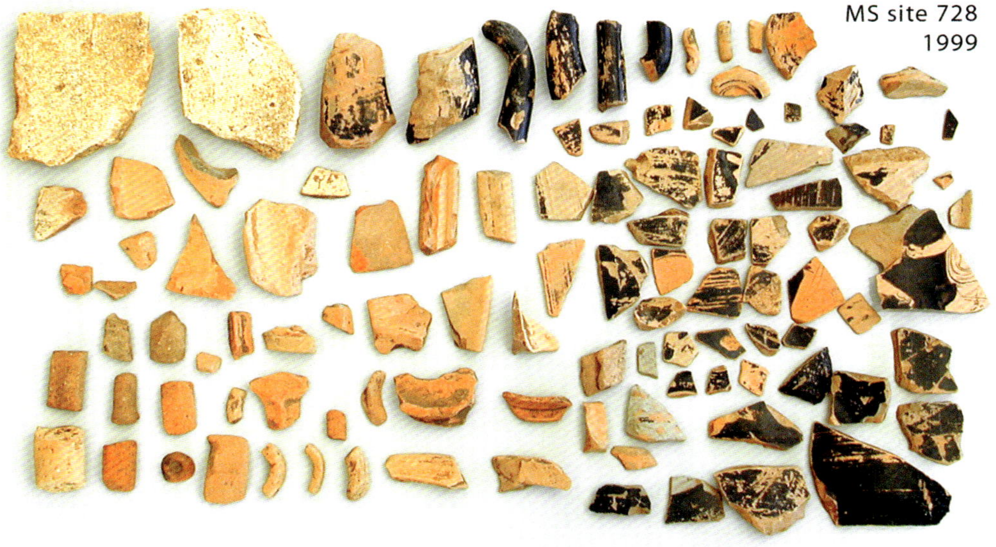

Figure vii Partial assemblage from site 728, 1999.

have greatly facilitated the identification of site types, which were regularly determined in the field. In the Metapontino there is not a continuous carpet of material, a condition which has often complicated locating and defining sites elsewhere.

Large quantities of ceramic materials were found at many, if not all, of the sites (*Fig. vii*). Fine wares, decorated and plain table wares, utilitarian and cooking wares, and amphorae have made it possible to date the sites with a high degree of accuracy (particularly in the Greek and Roman periods), and to describe a wide range of activities that took place on them. The average number of pottery fragments for the 560 sites of the Greek and Roman periods is 50, of which a high percentage can be identified by form and dated. Study of this material has confirmed the identification of site types made in the field. Much of the first volume of this publication deals with the pottery and other artifacts of the Greek and Roman periods that were recovered at sites between the Bradano and Basento. The Medieval pottery is discussed in the third volume; detailed analyses of the prehistoric pottery will appear in a separate study, which includes the Neolithic site at Pantanello.

Goals of this Study

The primary motivation for this comprehensive publication is to present, in as full and useful form as possible, the same data and analyses generated during comprehensive study of the survey sites between the Bradano and Basento that was available to the original researchers. This includes the following:

1. Detailed maps of the surveyed and surrounding areas of the chora (Vol. IV);
2. Complete catalogs and discussions for each class of material;
3. Data tables with explanatory graphs that summarize the data; and
4. Full discussion of the methods (and, where appropriate, algorithms) used in the spatial and statistical analyses of the sites and associated artifacts.

A similar level of documentation characterized the first study in this series on the necropoleis, *The Chora of Metaponto: The Necropoleis* (1998). The survey study will take a major step forward in usability: to supplement the print publication, the data presented here will be made available in digital form on the publication's companion website (metaponto.tacc.utexas.edu). Thus, it will be possible for other scholars and students not only to check the statements made in the text, but also to search for materials, formulate new hypotheses, test them, and, in effect, continue the research. These volumes are not intended to be the definitive word on the chora of Metaponto, but to act as a stimulus to further investigation. We hope that the studies of individual material classes will prove useful to all researchers

Figure viii View from the protohistoric and Greek site of Incoronata, overlooking the Basento valley.

working in Magna Grecia in Greek southern Italy and elsewhere in the Greek world. In some instances these will be among the first examples of a comprehensive typology and chronology of particular classes. These studies will need to be modified as new discoveries are made and science advances.

The present study of the chora, seen through the results of the survey, presents major advances over previous work on the subject by this writer, most recently in *Discovering the Greek Countryside* (2006). The studies of the nearly thirty thousand identifiable pottery fragments and other objects of all classes from the Greek and Roman periods has provided for the first time an accurate chronology and in addition a factual basis for describing the activities centered on the sites. They are placed at the beginning, in Volume I. The accurate positioning of sites on geo-referenced topographical maps made possible important insights into the development of the patterns of settlement in the chora (Volume IV). The recent discovery of sites deeply buried in the river valleys during the pipeline excavations of 2000–2002 and of the physical reality of the majority of the "division lines" visible in aerial photographs made during WWII and afterward has profoundly altered our perceptions of the chora as a whole (as discussed in Volume III of this study). These are just a few of the advances that appear for the first time in these pages.

This study has goals beyond the full presentation of data. It aims to create as complete a picture as possible of the ancient countryside with its inhabitants, animal as well as human, in historical terms. Thus, it will draw on previous studies of the necropoleis in *The Chora of Metaponto: The Necropoleis* (1998), and the animal populations of the chora in the recently published study, *The Chora of Metaponto 2: Archaeozoology of Pantanello and Five Other Sites in the Chora* (2010), as well as others nearing completion—on the farmhouses (five separate volumes), on the proto-historic village of Incoronata (*Fig. viii*), the Greek sanctuary and Roman tile factory at Pantanello, the Neolithic site at Pantanello, and long-anticipated studies of the paleobotany of the chora. Preliminary results from all of these can be found in *Discovering the Greek Countryside at Metaponto*, but should be considered only as rough approximations of what is to come.

The Survey in Its Context: Chronological Range

The chronology provided by the detailed studies of materials makes it possible to trace the settlement of the Metapontino in all its major prehistoric and historic periods. Volume II is almost entirely devoted to the Greek polis and the early years of Roman control in southern Italy. All but 5% of the 595 sites included in this study have at least one—and often the major—component in this period. Chapter 17, "The Prehistory of the Chora," summarizes the study results of the prehistoric settlement and pottery that will appear separately in the series. It does include a longer section on the Iron Age and protohistoric period immediately preceding the Greek settlement.

Primarily on the basis of pottery (Archaic fine wares, black gloss, and grey ware) the development of the chora has been divided into twelve uniform 50-year periods that begin in 625 BC and end in 25 BC. This is an artificial division made necessary by the nature of the evidence, as described in Chapter 4 of Volume I, but it does correspond remarkably closely to major turning points in the history of Metaponto and Magna Grecia.

The nature of the evidence near the end of these six centuries changes. There are far fewer sites and much less ceramic material. The subsequent Early Imperial (25 BC–AD 50), Imperial (AD 50–300), and Late Roman (AD 300–600) periods are dealt with separately in Volume III, together with the Medieval (7th–15th centuries) and Post-Medieval (16th–20th centuries) periods.

Man and Nature

Among the important themes illuminated by the study as it developed are the interaction of man and nature, the impact of settlers on the land in particular periods, and its future consequences. The agricultural practices of Neolithic farmers and the deforestation that resulted helped create the conditions of life in the heyday of the chora in the 6th century BC and later, as Abbott's work (Ch. 2) shows. Water was the crucial factor in shaping the patterns of settlement. Sites in all periods are located beside springs and drainages, as Folk (Ch. 1) points out, and as the detailed topographical maps in Volume IV make clear. No farmhouse is far from a water source. Rural sanctuaries were centered on all-weather springs that had come to be recognized as particularly beneficial, and where good sources of clay were located beside them, local ceramic production centers often grew. The destructive side of the relationship—excessive water in the fields and the flooding of the valleys—was evident from the start, beginning with the early attempts to drain the site of the asty, then followed by the massive public works projects of the 5th century BC and later in both the chora and the asty. In the end water was a considerable factor in the decline of the ancient city and its territory.

The Political World

The relations between the Greek settlers of the territory and the non-Greek peoples, the indigenous population of the proto-historic phase (8th and 7th century BC), and the very dynamic interrelationships of the chora and the urban center or asty in successive periods are themes that animate this study. Nowhere else in the Greek world can the reciprocal dependence of city and countryside be more clearly observed, because the asty, for the same reasons, was as well preserved since its abandonment in the 6th century AD as the chora. The excavations of the asty and the investigations of the chora have gone on contemporaneously now for many years. This study is one of the fruits of that close and harmonious collaboration and aims to shed light on the important relationships between these two interconnected spheres of the polis.

A concerted effort has been made throughout to view the chora and these relationships against the background of the major contemporary historical developments in Magna Grecia as a whole in all periods. This aspect of the study, especially in regards to the Greek and Roman periods, has benefitted greatly from the stimuli offered by the scholarly debates and discussions of the Convegni di Studi sulla Magna Grecia, held annually in Taranto since 1961. Though some aspects of the chora were particular to Metaponto, similar phenomena can be found in other poleis. The chora of Croton has been the subject of a parallel investigation by ICA that has revealed many similarities as well as a number of differences, but that is another story and the subject of a separate forthcoming volume, in the publication series devoted to that territory.

This study is, of necessity, the work of many hands and the fruit of decades of labor. It has been, we hope, worth the painstaking effort. It would be impossible to acknowledge adequately the contributions, large and small, made by generations of colleagues and students, donors and foundations. An attempt has been made in the annual reports that appear on ICA's website. One debt that all share, to Dinu Adamesteanu and to Antonio De Siena, is in part repaid by the publication of these volumes. Without their enormous contributions and without their constant support, this study would have been unimaginable.

Abbreviations

Site Type and Occupation

S	Sanctuary
F	Farmhouse
N	Necropolis
T	Tomb
K	Kiln
F/T	Farmhouse/Tomb
F/K	Farmhouse/Kiln
GAS	Greek agricultural structure
UGF	Unidentified Greek feature
NS	Nucleated settlement
Act	Activity
NR	Not recorded

Periods

Imp	Imperial
Gk	Greek
600	625–575 BC
550	575–525 BC
500	525–475 BC
450	475–425 BC
400	425–375 BC
350	375–325 BC
300	325–275 BC
250	275–225 BC
200	225–175 BC
150	175–125 BC
100	125–75 BC
50	75–25 BC

Bold numbers indicate period bins

Archaeological Material and Ceramic Classes

ACk	African cooking wares
AE	Bronze (coin)
AFW	Archaic fine ware
Amp	Amphora
ARS	African red slip wares
BF	Black-figure wares
BG	Black gloss
BoB	Black-on-buff ware
BT	Bricks and tiles
BW	Banded ware
Ck	Cooking ware
CW	Coarse ware
ETS	Eastern terra sigillata
Faunal	Faunal remains (bone, shell)
Gna	Gnathia wares
Grind	Grinding stone/quern
GW	Grey ware
Human	Human remains (bone)
IFW	Iron Age fine ware
Imp	Impasto pottery
ITS	Italian terra sigillata
La	Lamp
Li	Lithics
Lo	Loom weight
LRA	Late Roman transport amphora
LRC	Late Roman "C" ware
MED	Medieval wares
Min	Miniature
NFW	Neolithic fine ware
Pa	Painted wares (Medieval)
Pest	Pestle
PMED	Post-Medieval wares
PGlW	Post-Medieval glazed ware
PW	Plain ware
PWBW	Plain and banded ware
RF	Red-figure wares
ROM	Roman ware
RPW	Late Roman painted wares
Slag	Industrial waste (iron, ceramic)
TC	Terracotta plaque/figurine
UPW	Undecorated plain wares
Vot	Other ceramic votive object

General

Study area	Survey transects + 50 m buffer
MetSur	Metaponto Survey
NR	not recorded

Geospatial and Statistical Terms

GIS	Geographic Information System
GPS	Global Positioning System
UTM	Universal Transverse Mercator
DEM	Digital Elevation Model
SRTM	Shuttle Radar Topography Mission
IGM	Istituto Geografico Militare
FMCE	Final Multiple Criteria Evaluation
EAW	Equivalent Artifact Weight

Scatter

H	heavy
M	moderate
L	light
S	sparse
v.	very

Excavated Sites

CR	Crucinia
FF	Fattoria Fabrizio
FS	Fattoria Stephan
IC	Incoronata
SAG	Sant'Angelo Grieco
SAV	Sant'Angelo Vecchio
SB	San Biagio
SG	San Giovanni
SS	Saldone

I. The Land

1
Geologic Background of the Metapontino

Robert L. Folk

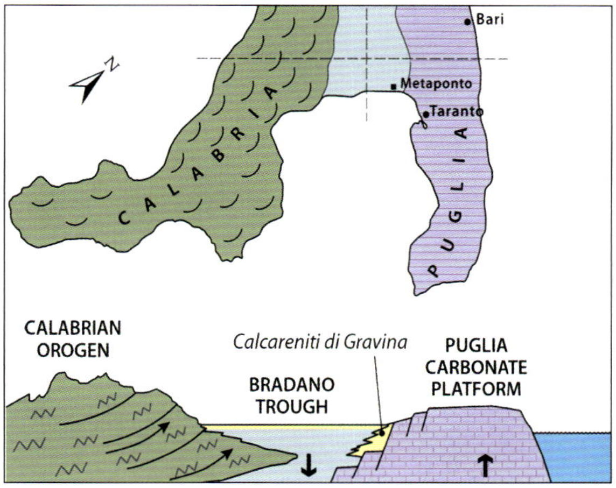

Figure 1.1 The three major crustal blocks comprising southern Italy: 1) the stable platform of undeformed Cretaceous limestone of Puglia; 2) the central down-dropped block filled by Plio-Pleistocene clays and sands, the "Bradano Trough"; and 3) the orogenic, actively deforming mountain arc of Calabria, slowly being thrust eastward (note the leading edge, the "Metaponto Overthrust," protruding into the Bradano Trough). The step-like series of faults marking the boundary between the Puglian block and the Bradano Trough is flanked by calcareous beach sediments (the calcarenita di Gravina) upon which are built troglodytic towns like Ginosa and Matera. (RLF/KMDL)

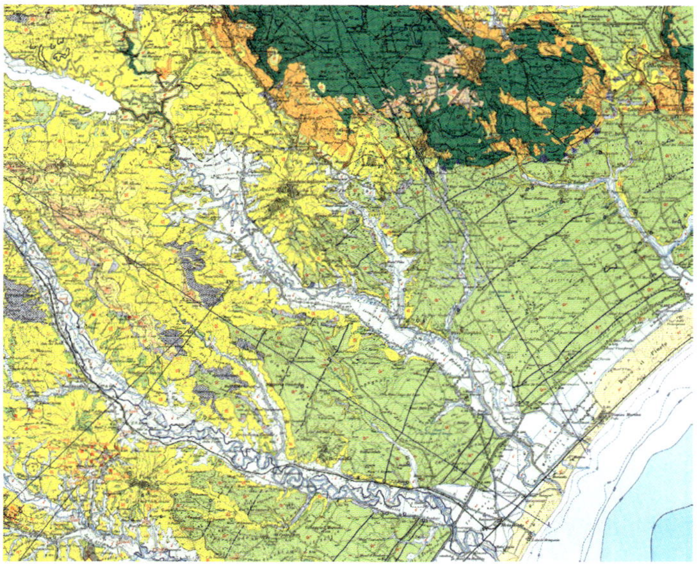

Figure 1.2 Foglio 201/Matera of the *Carta Geologica d'Italia* (scale 1:450,000) showing the geological formations south and southeast of Matera (located along the top edge, at the interface between the green and yellow areas), at the boundary between the Puglia carbonate block and the Bradano Trough. Key: dark green = limestone; buff = conglomerate; orange = calcarenite; yellow = clay; pale green = marine terrace. (Reproduced with the permission of APAT—Dip. Dif. del Suolo—Servizio Geologico d'Italia, Rome, prot. 45230. Orig. scale 1:100,000)

Geologic Structure

Southern Italy consists of three crustal blocks, their boundaries following the general northwest to southeast orientation of the peninsula as a whole (*Fig. 1.1*): (1) in the northeast, a gently-uplifted block of mainly horizontal and un-deformed, hard, light gray Cretaceous limestones approximately 90–110 m.y. (millions of years) in age, forming the low plateau of the Murge and extending from the tip of the Italian "heel" as far north as Foggia (Puglia); (2) in the center, a depressed crustal block called the *fossa bradanica* ("Bradano Trough"[1]), forming a lowland extending from the Ionian seacoast northward through Basilicata to Pescara (Abruzzo) on the Adriatic coast, underlaid mainly by Late Pliocene and Pleistocene marine clays 1–3 m.y. in age; (3) in the southwest, a strongly-uplifted orogenic belt of highly deformed rocks of various ages (the Calabrian arc or "toe"), a southern continuation of the Apennine mountain chain.[2]

The boundaries of these crustal blocks are marked by faults trending approximately N 45° W.[3] The eastern boundary of the *fossa bradanica* (which bounds the Apulian uplifted block), passing through Taranto and Matera, consists of normal faults along which blocks of crust have been down-dropped successively to the west, so that it now resembles a

[1] By Migliorini 1937; 1952. In the terminology of modern plate tectonics, this is known as the "Bradano Foredeep" (D'Argenio/Pescatore/Scandone 1975), a depressed "wrinkle" between the stable Apulian foreland and the Calabrian orogenic arc, which is being thrust eastward. See Roda 1967 for a massive reference list on early geologic work; see Van Bemmelen 1975, Channell/D'Argenio/Horvath 1979, and Boccaletti/Nicolich/Tortorici 1984 for summaries of plate tectonics and crustal movement in Italy.

[2] For recent data on the structure of southern Italy, see Tramutoli et al. 1984; Doglioni/Mongelli/Pieri 1994; Pieri/Sabato/Tropeano 1996.

[3] Jaboli and Roger 1954.

Figure 1.3 View of the Bradano Trough (foreground) and the Apulian carbonate block (Murge), looking east from Pomarico. (AP)

flight of steps (*Figs. 1.2–3*).[4] Here local fault trends run in two directions, N 35–50° W and N 80° W.[5] This faulting probably occurred in Middle to Late Pliocene time (2–4 m.y.). The western boundary of the *fossa bradanica* is marked by overthrust faults (nappes) representing great sheets of jumbled rock of Miocene to Cretaceous age (5–80 m.y.) that have slid eastward almost horizontally from the heights of the Calabrian arc (Apennine belt) into the soft and yielding Pleistocene clays of the Bradano Trough (*Fig. 1.4*). Such thrusting has happened repeatedly during the last 1–3 m.y. and probably continues in the present period. The northeastern margin of the buried thrust sheet, known as the "Metaponto Overthrust,"[6] passes in the subsurface ca. 750 m beneath the hilltop town of Pisticci (*Fig. 1.5*);[7] the sheet is 1–2 km thick.[8] The trend of the margin of this overthrust varies between N 45° W and N 55° W, and it extends almost half the length of the Italian peninsula, as far as Pescara (*Fig. 1.6*).[9] The overthrust also continues southward to split the Gulf of Taranto, forming a submarine valley.[10]

Faults of orthogonal northwest and northeast trend are present throughout the Gulf of Taranto.[11] A great many large and important examples lie inside the chora of Metaponto. Faulting activity began in Pliocene times (approximately 5 m.y. ago) and continues today: since 1900, there have been eleven earthquakes of magnitude 3 or 4 (dish-rattlers or better) with epicenters in the quadrilateral between the Matera-Ferrandina line and the Ionian coast.[12] The horrendous earthquake of November 23, 1980, leveled whole sections of the regional capital, Potenza, ca. 100 km northwest of Metaponto, but produced only minor effects (such as cracking of old churches) at Matera, Pisticci, Montescaglioso, and Pomarico, 20–40 km inland from Metaponto.[13]

Stratigraphy and Geologic History of the Bradano Trough

When the *fossa bradanica* began to subside by faulting in Pliocene times, polar glaciation was minimal and sea level was at least 200 m higher than at present. Marine waters flooded into the down-dropped trough as far north as Foggia and Ancona (Le Marche), cutting off the plateau of Puglia as an island.[14] A thick

[4] Carissimo et al. 1963; Desio 1973, 693; D'Argenio et al. 1975; Beneduce et al. 2004.

[5] Cocco et al. 1975, Fig. 3; Praturlon 1980, Fig. 1.4; Ciaranfi et al. 1983, Fig. 6.

[6] Ogniben 1969.

[7] See Carissimo et al. 1963, Fig. 20; Pieri 1966; D'Argenio et al. 1975.

[8] Ogniben 1969, Fig. 129; section on Foglio 201/Matera of the Geological Map of Italy.

[9] Maps in Crescenti 1975 and Praturlon 1980, Fig. 1.4, here simplified as Fig. 1.6.

[10] Rossi/Auroux/Mascle 1983; Tramutoli et al. 1984, Fig. 1.

[11] Tramutoli et al. 1984, Fig. 10.

[12] Caputo and Postpischl 1975. Ciaranfi et al. (1983, Fig. 1) show a map of epicenters for earthquakes during the last 1,000 years.

[13] Alexander 1981a, b.

[14] Desio 1973, 692; Ambrosetti and Carraro 1980; Pieri/Sabato/Tropeano 1996.

Figure 1.4 View of the Bradano Trough (foreground) and the Calabrian orogen (Apennines), looking southwest from Pomarico. (AP)

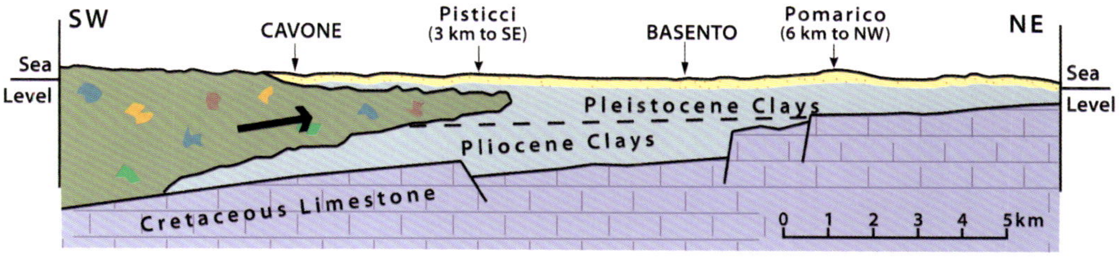

Figure 1.5 Cross section oriented NE across the western part of the Bradano Trough revealing the block-faulted Cretaceous basement, the capping Pleistocene gravels (in yellow), the Plio-Pleistocene marine clays filling the Bradano Trough, and the great Metaponto Overthrust sheet. (RLF/KMDL, after cross section #4 of Foglio 201/Matera, Fig. 1.2)

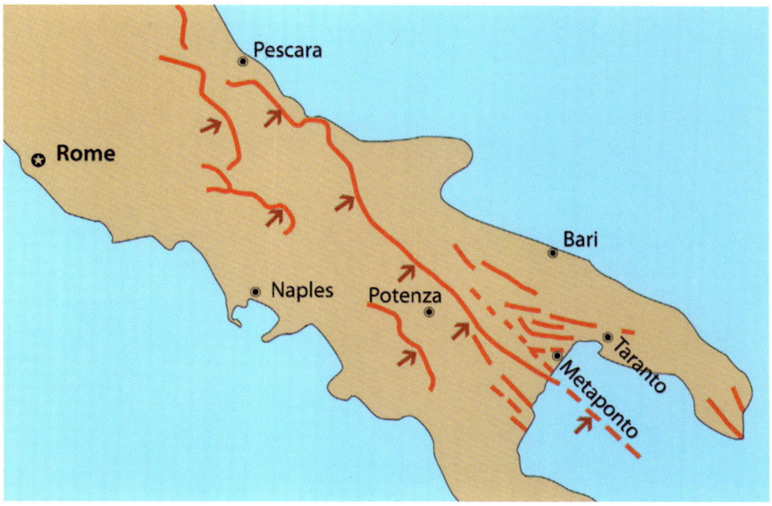

Figure 1.6 Fault pattern map of southern Italy. The edge of the Metaponto Overthrust (movement direction shown by arrows) passes all the way from the Ionian Sea to Pescara. Other major faults are also predominantly oriented NW. (RLF/CW, after Praturlon,1980)

Figure 1.7 Pliocene oceanic clays *(argille azzurre)* exposed along the north bank of the Cavone valley below Pisticci. (RLF)

Figure 1.8 Characteristic ribbed pattern on eroded "knuckles" of Pliocene clays caused by rilling. (RLF)

Figure 1.9 Exposures of Pliocene clays on slopes below Pisticci. (RLF)

sequence of blue-gray open-marine clays *(argille subappennine* or *argille azzurre),* derived from erosion of the burgeoning Apennine chain to the west, was deposited upon the down-dropped basement of Cretaceous limestone;[15] in places the accumulation of clays exceeded 500 m in thickness. These clays are of Upper Pliocene and Calabrian (earliest Pleistocene) age, 1–4 m.y. old,[16] and were laid down during the period of repeated glaciation with fluctuating and falling sea level. The open-marine clays contain cold-water marine fossils such as the microscopic foraminifer *Hyalinea balthica*[17] and the large clam *Arctica islandica*.[18] They

[15] Crescenti 1975.

[16] Nikiforova 1978.

[17] Bayliss 1969, Fig. 4.

[18] Raffi 1986.

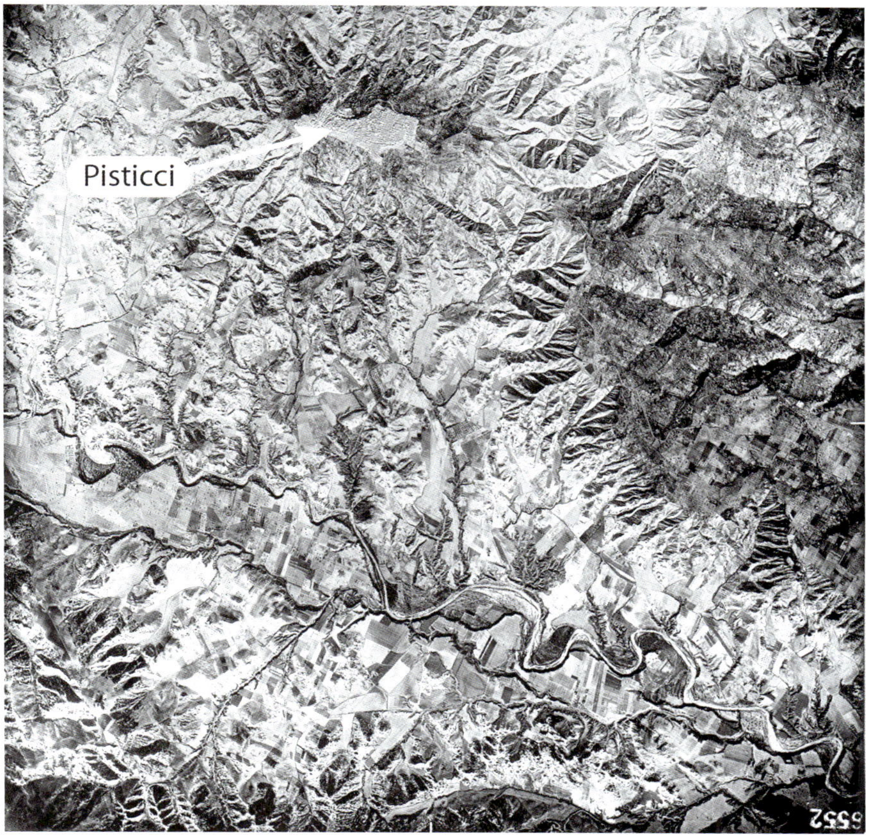

Figure 1.10 Vertical aerial photograph of the Cavone valley and southern Metapontino. Pisticci, on its flat, gravel terrace, and the surrounding eroded clay slopes are clearly visible. (GAI Neg. 109591/Pos. 6552, 1954-1955, approximate scale 1:72,000. Reproduced with the permission of the Aerofototeca, Rome, 2004.)

Figure 1.11 Structural damage at the edge of Pisticci caused by landsliding, 1980. (RLF)

Figure 1.12 Large landslide scar on a slope near Pisticci, 1980. (PD)

weather into steep, barren, yellowish-gray (Munsell Soil Color Chart 5Y 8/2) slopes often marked by severe gullying, badland topography, and mud-cracked surfaces (*Figs. 1.7–8*). They are easily observed on the slopes leading up to hill towns such as Pisticci, Pomarico, and Montescaglioso (*Figs. 1.9–10*) and are prominently visible along the Rome–Taranto railroad line as it passes through the Basento valley. (Locally, these strange erosional features are known as *calanchi*.[19]) The clays are responsible for serious land sliding problems, most notably at Montescaglioso and Pisticci (*Figs. 1.11–12*).[20] Near Metaponto these clays are exposed along the stream banks at Sant'Angelo Vecchio.

Deposition of marine clays gradually filled the Bradano Trough, while simultaneously sea level fell due to the growth of the ice sheets in northern Europe and North America. Consequently, the area changed from a deep-water sedimentary zone to a much shallower near-shore zone in which marine sand was deposited, culminating in thick shoreline gravels with marine pelecypods.[21] There ensued a period of widely-fluctuating sea level, as glaciers alternately waxed and waned in the northern parts of the world. When the glaciers were at their maximum, a large portion of the Earth's water was locked in ice, so that sea level

[19] Boenzi/Palmentola/Valduga 1976; Alexander 1980.

[20] Kayser 1958. Alexander (1982b) discusses erosion rates of the *argille azzurre*, while their technical properties are described by Del Prete and Valentini (1971).

[21] Sections in Ricchetti 1967; Carissimo et al. 1963; Desio 1973, 695.

Figure 1.13 View of the lowest marine terraces on the right (south) bank of the Bradano River, looking W from the vicinity of Ginosa Marina. (AP)

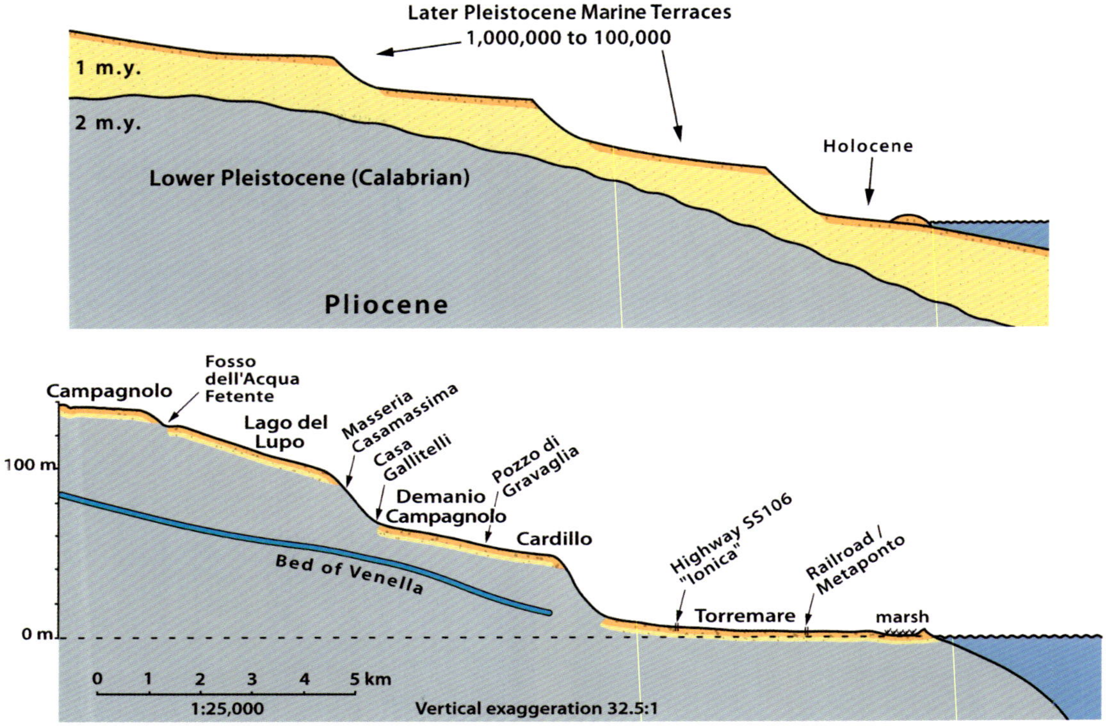

Figure 1.14 Pleistocene marine terraces and their mode of development:
(Upper) Schematic profile showing the Pleistocene marine terrace sands and gravels unconformably overlaying the Pliccene to Lower Pleistocene oceanic clays. *(Lower)* Accurate topographic profile from the Campagnolo locality southeast to the Ionian Sea, showing prominent Pleistocene terrace levels and scarps. (RLF/ KMDL)

was low; when the glaciers began to melt, sea level rose accordingly. The fluctuations covered an overall range of approximately 130 m, superimposed upon a generally falling long-term sea-level trend.[22] When sea level was high, erosion occurred on what was then the coastline, and a sea cliff was produced, marked by sandy or gravelly beaches and a smooth, gently sloping bottom offshore. When sea level dropped, the old, flat sea bottom was exposed to weathering, and gullies were cut. Then sea level would rise again, but not as far as before, and a new sea cliff would be cut; a sandy shore would develop at a lower elevation, and so on.

Thus, viewed in profile, the Ionian coast of Basilicata southeast of Pisticci exhibits a step-like profile in which short, steeper slopes (old sea cliffs) or beaches

[22] Fairbridge 1972.

Figure 1.15 Pleistocene terraces facing NE from Cozzo Presepe, overlooking the Bradano valley. The smooth terrace sloping gently seaward (to right) in the middle distance represents sea floor ca. 700,000 years ago. (RLF)

Figure 1.16 Flat marine terrace, 900,000 years old, at the edge of the Cavone valley, southwest of Marconia. Note the gullied Pliocene clays in the foreground. (PD)

Figure 1.17 View from the Sant'Angelo locality southwest across the Basento valley, showing in the distance a shoreline formed about 500,000 years ago: the flat terrace level bounded by a gentle escarpment that descends to the next lower terrace level. (RLF)

Figure 1.18 Minor Pleistocene shoreline (behind the modern farmhouse) of sand and gravel extending northeast, parallel to the present coast. View looking NW from a point between Marconia and Pisticci. (PD)

alternate with broad, gently seaward-sloping flats (old sea floors; *Figs. 1.13–18*). Shallow-water sands and gravels accumulated on the temporary sea bottoms, so that the "bedrock" Calabrian open-marine clays are covered by a 10–40 m thick blanket of sands and gravels deposited on the shallower Late Pleistocene sea bottom (*Fig. 1.19*).[23] That these are ancient shoreline conglomerates is demonstrated by the locally abundant fossil presence of the marine clam *Arctica islandica* (*Fig. 1.20*). Furthermore, the pebbles in the gravel are extremely flat and discoidal, characteristic of beaches (*Table 1.1; Figs. 1.21–22*). The ancient shorelines, which run parallel to the present coast, extend inland as far northwest as Pisticci; they are parallel with the NE-trending set of the Greek division lines.

Approximately 1 m.y. ago (Sicilian geologic time) the sea lapped at the foot of what is now the terrace upon which rest Pisticci and Montescaglioso, 20 km inland of its present position and at an elevation of 392 masl (at Pisticci).[24] In the intervening million years, sea level has fallen in an oscillating manner, while

Figure 1.19 Thick deposits of Pleistocene shoreline sands and gravels capping a marine terrace near Marconia. (RLF)

[23] Vezzani 1967.

[24] Vezzani 1967; Brückner 1982.

Modern beach pebbles, mouth of Cavone	$\psi = 0.42$
Pleistocene beach pebbles, Fosso Lavandaio	$\psi = 0.51$
Pleistocene beach pebbles, Fontana Cappa d'Amore	$\psi = 0.54$
Pleistocene river pebbles, Fosso Lavandaio	$\psi = 0.74$

Table 1.1 Sphericity data for pebbles in the Metapontino, method of Dobkins and Folk (1970); 50 pebbles measured at each location. The boundary between beach and river pebbles is $\psi = 0.65$, beaches having lower values.

Geologic time	Age (m. y.)	Level of Mediterranean Sea	Terrace formation, location, and present elevation
	4		
Pliocene			
	2.0 ± 0.3		
Calabrian		+130–210 m	
	1.2 ± 0.2		
Emilian			
	1.0 ± 0.2		
Sicilian		+80–90 m	Pisticci (340–390 m)
Gunz (glacial)	0.8 ± 0.2	-30 m	
Milazzian		+55–60 m	Lago del Lupo (90–150 m) Gravaglione (40–60 m)
Mindel (glacial)	0.5 ± 0.2	-60 m	
Tyrrhenian		+35–40 m	Pizzica (25–50 m)
Riss (glacial)	0.3 ± 0.1	-90 m	
Monastirian		+15–18 m	
Würm (glacial)	100,000 to 15,000 yrs.	-140 m	
Present		0 m	Metaponto (1 m)

Table 1.2 Sea level positions at various times, and periods of formation of the various terraces in the Metapontine hinterland. Data from various sources (see text).

Figure 1.20 Sample of Pleistocene shoreline conglomerate from a cliff behind Sant'Angelo Vecchio. The large fossil clams indicate a marine origin. (RLF)

Figure 1.21 Disk-shaped beach pebbles exposed near Ponte Fabrizio along the Venella. (RLF)

Figure 1.22 Comparison of pebbles. (See Table 1.1.)
a) Modern beach pebbles from near the mouth of the Cavone.
b) Pleistocene beach pebbles from the Fosso Lavandaio (note disk-like shapes resembling modern beach pebbles).
c) Pleistocene river pebbles from the Fosso Lavandaio—note the equidimensional shapes. (RLF)

southern Italy has been uplifted, forming a series of terraces (*Table 1.2; Fig. 1.13*).[25]

By correlating sea level elevations around the Mediterranean, the present elevations of these terraces on the Lucanian coast, and the geologic age, it can be demonstrated that southern Italy has been rising rapidly in an absolute sense. The hilltop now occupied by Pisticci has been uplifted some 250 m in the past million years, while at the southwestern edge of the terraces, near the foot of the Apennines, the uplift of the same terrace has been some 350 m, so that the coastal plain has not only been generally lifted but also tilted towards the northeast.[26] However, the rate of uplift has decelerated in the last half million years, as demonstrated by the mere 50 m rise in the terraces around the Lago del Lupo locality (*Fig. 1.13*).

Approximately 14,000 years ago, at the height of the Würm glaciation (roughly equivalent to the Wisconsin glaciation in North America), sea level was ca. 140 m lower than at present.[27] Rivers such as the Bradano and Basento cut deep valleys as they attempted to erode down to the new, lower base level. Then the glaciers began to melt rapidly, and sea level rose to approximately 2 m below its present level by approximately 3000 BC. By the time of Christ sea level was within ±0.5 m of its present position.[28] Thus, during the last 5000 years, sea level has risen only slightly, while vast amounts of sediment have been carried down to the coast by the Bradano, Basento, and other rivers, especially during the winter rainy season.[29] As a result, the shoreline, overloaded with sediment, has

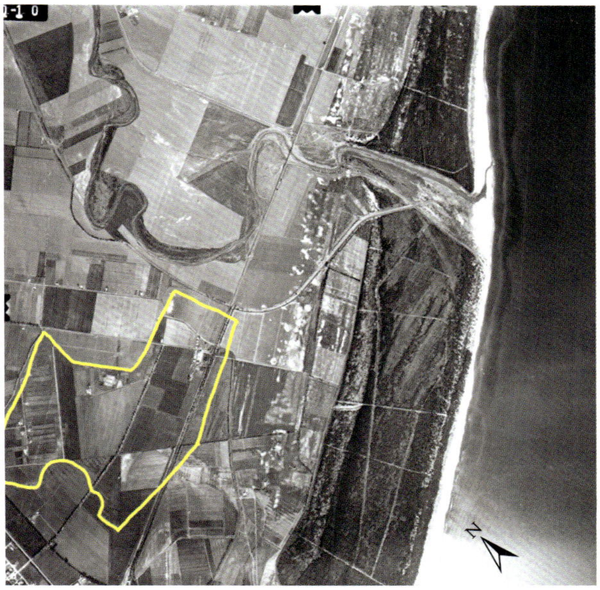

Figure 1.23 Vertical aerial photograph of the coastal plain around the mouth of the Bradano. The ancient shoreline follows the northwestern boundary of the dark, macchia-covered coastal strip. The ancient urban center is outlined in yellow. (Unknown series Neg. 79408/Pos. 4362, 1970, approximate scale 1:56,000. Reproduced with the permission of the Aerofototeca, Rome, 2004)

been built out to form a flat surface, often marshy, very near sea level.[30] During the Greco-Roman period flooding was frequent and severe, partly the result of extensive de-vegetation,[31] and the alluvial valleys were built up with flood-laid sediments.[32]

According to Kayser,[33] a strip of shoreline land ca. 2 km wide has been added since Metaponto was founded in the late 7th century BC, and the silting of the port was in part responsible for the decline of the city (*Fig. 1.23*). Even during the 19th century the shore continued to build out hundreds of meters further into the sea.[34] But since the construction of dams began in 1944, diminishing the frequency of flooding and choking off the sediment supply to less than a tenth of its natural amount,[35] the shore has begun to erode at an average of 2–3 m/year, occasionally as much as 10 m/year.[36] Present longshore drift in the area is from southwest to northeast,[37] so that sand

[25] The terraces are mapped in detail on Foglio 201/Matera of the Geological Map of Italy, and are described in detail by Vezzani (1967, 1975). Absolute Pleistocene dates are still in a state of flux, as they are based on several different criteria; dates in the table are a compromise from several sources. The oldest dates are given by Horowitz (1979, figs. 1.1–2); next by Fairbridge (1972), then Kukla (1977), then Berggren and Van Couvering (1974) and Nikiforova (1978); and the youngest dates by Haq/Berggren/Van Couvering (1977), Hsü et al. (1984), and Berggren/Kent/Van Couvering (1985). For example, the beginning of the Calabrian is placed at 1.6 m.y. by Haq/Berggren/Van Couvering, but at approximately 2.8 m.y. by Horowitz.

Selli (1962; 1977) discusses the definitions of these subdivisions of Pleistocene time in Italy, but sea levels and stage names have been in a state of revision (see, for example, Raffi 1986).

Elevations given are for the general Mediterranean sea level (data from Horowitz 1979). Localities in the Metaponto hinterland are identified from topographic maps using the terrace elevations and identifications of Vezzani 1967; 1975.

[26] For a general summary, see Brogan/Cluff/Taylor 1975; Ciaranfi et al. 1983, Fig. 7; Dai Pra and Hearty 1988.

[27] De Maio et al. 1979, 1340.

[28] Flemming 1969; Pirazzoli 1976; 1991, 89–93; Mörner 1979–1980.

[29] Vezzani 1967.

[30] Cotecchia/Dai Pra/Magri 1971; Boenzi/Caldara/Pennetta 1995.

[31] Neboit 1975; 1977.

[32] Terracing in the Basento valley has been discussed by Boenzi/di Gennaro/Pennetta (1978).

[33] Kayser 1958.

[34] Cocco 1976.

[35] Cotecchia/Dai Pra/Magri 1971.

[36] Cocco et al. 1975; Cocco 1976.

[37] Cocco 1976; Dell'Anna and De Marco 1977.

derived from beach erosion is swept toward Taranto and the "heel" of the Italian peninsula.

A sharp rise in sea level has been recognized between approximately 500 and 200 BC,[38] leading to extensive alluviation and marsh development. Boenzi et al.[39] documented a rapid rise in the water table in the asty of Metaponto, such that a drainage channel had to be excavated to prevent damage, but not before Temple B (Apollo Lykeios) collapsed; subsequently, by approximately 300 BC, the asty was largely depopulated. By Medieval times the population had left the lowlands and was concentrated in hill-towns such as Pisticci.[40] There is no doubt that, prior to 3000 BC, sea level was much lower (*Table 1.2*), streams had steeper gradients, and there was probably no coastal marshy belt. As sea level rose to approach its preglaciation level of approximately 130,000 years ago, it encountered the new, flat terrace level formed at that time and began to inundate it.

Such a rise may have created ideal conditions for malaria, causing the decline of the Greek coastal colonies in Magna Grecia. The inundation of the Ionian coast certainly led to an extensive development of swampy land in areas such as the Metapontino.[41] The presence of swampy land almost inevitably promotes malaria, since the mosquitoes that carry the disease need still, fresh water in which to lay their eggs. There is argument over when malaria was first introduced into southern Europe. Jones,[42] in an exhaustive review of classical texts, proposed that malaria entered coastal Greece and Italy by approximately 500 BC, was well-recognized and endemic in Italy by approximately 200 BC, and by the time of Christ was a deadly scourge known by the Romans to be associated with standing water and swampy lowlands.[43] A relatively high incidence of anemia in the population buried in the Pantanello necropolis outside of ancient Metaponto, spanning the entire period of the necropolis' use (575–275 BC) but most apparent during the 5th and early 4th centuries BC, may have served as a biological response to—and defense against—malaria.[44] Whether sea-level rise was the prime cause can probably never be determined, but the coincidence is suggestive.

Drainage Patterns

There are three main geologic factors controlling orientation of natural drainage channels in the Metapontino: (1) a deep-seated fracture system which imposes a strong northwest–southeast grain on the surface topographic features; (2) the presence of northwest to southeast oriented concretions (locally-cemented zones) in the Pleistocene sandstones; and (3) the presence of northeast to southwest trending scarps created by Pleistocene shorelines, with streams consequently running downslope in an orientation perpendicular to them. In the chora of Metaponto, these three orienting tendencies have acted congruently to produce a remarkable example of aligned drainage patterns.

(1) Fracture orientation

Evidence of a deep-seated fracture system (the Metaponto Overthrust and other related northwest-trending faults) is visible in many road-cuts and stream valleys in the Bradano–Basento area. Surface outcrops of sandstones and conglomerates demonstrate a fracture orientation parallel to the subsurface fault system: 9 of 16 fractures measured lie within ±10° of the N 50° W trend (*Figs. 1.24–25*). The thin ledges of sandstone are usually fractured in a northwest/southeast orientation parallel to the "grain" of the cemented concretions (*Fig. 1.26*; and "Oriented Concretions," below) because this is the easiest orientation for breakage, and it coincides with the major deep structural fracture lines beneath the Italian peninsula (*Fig. 1.6*). Occasionally other fracture directions are present as well, giving a sawtooth appearance to the edge of the cemented ledges.

In places the massive later Pleistocene conglomerates, too, are cut by large joints (fractures) running predominantly in the same northwest to southeast direction (*Fig. 1.27*). The most outstanding such locality is on the flank of the canyon that runs southwest of Masseria San Marco near Cozzo Presepe. Here, open joints dozens of centimeters wide cut vertically through several meters of conglomerate (*Fig. 1.28*), clearly demonstrating that the tectonic forces which produced the Metaponto Overthrust and other faulting in the Pleistocene have been—and still are—

[38] Cotecchia/Dai Pra/Magri 1969.

[39] Boenzi et al. 1988.

[40] Schmiedt and Chevallier 1959b, 23.

[41] Angel 1972.

[42] W.H.S. Jones 1907; 1909.

[43] For evaluation of this complex problem, including medical detective work, Classical texts, ancient climates, and sea levels, see Ackerknecht 1965, Bruce-Chwatt 1965, Brunt 1971, Angel 1972, Borza 1979, and Burke 1981. Most of these authors are skeptical of Jones' theory and favor the idea that malaria may have been present several centuries earlier.

[44] Henneberg and Henneberg 1998.

Figure 1.24 Orientation of fractures (red lines) and linear concretions (green lines) in the Metapontino. (RLF/DD)

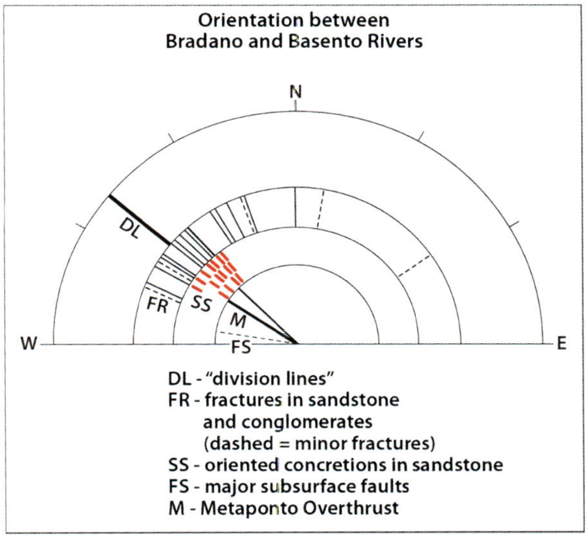

Figure 1.25 Orientation diagram showing parallelism of features between Bradano and Basento. (RLF/DD)

Figure 1.26 Fractures in Pleistocene conglomerate oriented N 50° W in a scarp just northwest of San Biagio. (RLF)

Figure 1.27 Fractures/joints in Pleistocene conglomerate near San Marco oriented N 50° W. (RLF)

Figure 1.28 Large NW fracture/joint near San Marco extending for several meters through the Pleistocene conglomerate. (RLF)

active in the region. Such large fractures easily orient any drainage that develops above them.[45]

(2) Oriented concretions and sandstone "cigars"

Road-cuts descending from the level of the Pleistocene marine terraces into modern stream valleys, and stream valley walls themselves, often display good outcrops of the later Pleistocene shoreline and sea-bottom sands and gravels (mainly Milazzian, 500,000–800,000 years old), deposited as sea level fell and the shoreline retreated across the area. In

[45] Otvos (1981) has documented a similar example of coastal drainage lines oriented by fractures in the United States along the coast of Mississippi. See Melton 1956, Mollard 1957, Lattman 1958, and Osmaston 1980 for other examples of naturally-formed lineations; Boccaletti/Coli/Napoleone 1980 for lineations in southern Italy northwest of Potenza; and Wise et al. 1981 and 1985 for other Italian lineations.

Figure 1.29 Cigar-shaped oriented sandstone concretions in conglomerate near Ponte Fabrizio protruding from the outcrop (center) and oriented N 50° W. (RLF)

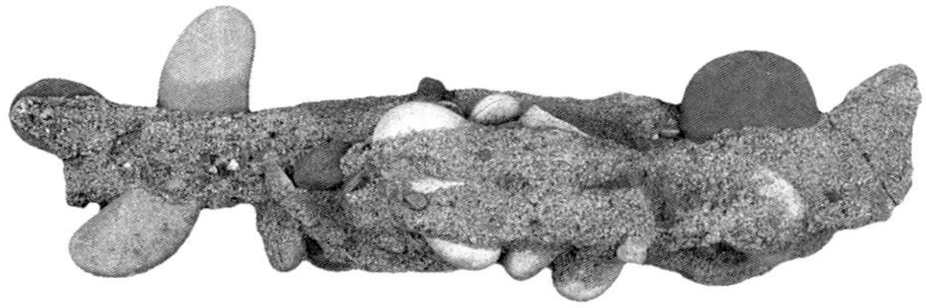

Figure 1.30 Hand sample of linearly-cemented conglomerate taken near Ponte Fabrizio. L: 25 cm. (DS)

Figure 1.31 Outcrop of cigar-shaped oriented sandstone concretions along the Gravaglione stream. (CW)

Figure 1.32 Loose sand with oriented sandstone concretions between Tinchi and Marconia (Basento-Cavone). The photograph, taken at noon, shows the parallel orientation of the "cigars" at N 70° W. (PD)

places the sands and gravels are patchily cemented with calcite, and in these cemented sandstones and conglomerates lies additional evidence for geologic controls on topographic lineation.

The most spectacular features are cigar-shaped cemented zones in otherwise loose sands or gravels. These "cigars" project from the outcrop in parallel sets oriented northwest-southeast, each protruding approximately 10–20 cm and having an average diameter of just a few centimeters (*Figs. 1.29–31*). Excellent examples are found along the Gravaglione in the Pizzica locality and along the paved road crossing the Venella stream valley at Ponte Fabrizio. The second type of feature are outcrops of sand, frequently in beds several centimeters thick, that are cemented by calcite and protrude as thin ledges (*Figs. 1.32–33*). The surfaces of these ledges often contain small ridges 1–3 cm wide and a few millimeters high; the ridges run in parallel northwest-southeast sets across the ledges, consistent in orientation with the isolated "cigars" (*Fig. 1.34*). Again, these features are the result of localized cementation.

Such oriented concretions are found in perhaps half of the sand outcrops in the Metapontino. At all of the 15 localities measured, the concretions and lineations

Figure 1.33 Loose sand with oriented sandstone concretions between Tinchi and Marconia (Basento-Cavone). The photograph, taken from the base of the outcrop looking upward, clearly shows the parallelism of the features. (RLF)

Figure 1.34 Cemented sandstone ledges detached from cliffs behind Sant'Angelo showing the same direction of lineation as the oriented sandstone concretions. (RLF)

follow an orientation between N 40° W and N 60° W, a remarkable consistency of trend caused by the downslope flow of groundwater towards the coast (*Figs. 1.24–25*).[46]

(3) Orientation by shoreline processes

As the shoreline of the Ionian Sea retreated irregularly from its high stand near Pisticci, it left a step-like surface on which the main flat areas slope towards the present coast at approximately 7 m/km (⅓°). Thus the major drainage courses tend to run naturally straight down the slope, i.e., in a direction of S 60° E. Fluctuations in the rate of shoreline retreat and normal shoreline processes at time of stillstand (surf activity, longshore currents) left a series of beaches, cliffs, and offshore bars that run parallel to the shoreline, i.e., in a N 30° E orientation. Ancient sea cliffs are represented by slightly steepened slopes, and old barrier bars or beaches survive today as long, narrow, gentle ridges of sand or gravel (*gi[o]voni* in the local dialect[47]). These features all tend to be oriented parallel to the coast and therefore serve as obstacles to the natural drainage pattern: upon meeting the *givoni,* water is forced to move in a direction perpendicular to the general trend of the peninsula (i.e., southwest/northeast). The result of downslope plus coast-parallel stream activity is a basically rectangular or "trellis" drainage pattern, which is often remarkably regular. In the cross-wise drainage all tributaries join from the northwest, as this is the primary orientation of downslope movement (*Fig. 1.10*).

Conglomerates, Sandstones, and Calcarenites: Summary of Petrology

The local conglomerates of the Metapontino are composed of pebbles, predominantly siltstone (often of very discoidal shapes) along with common limestone and some chert, vein quartz, metaquartzite, sandstone, and rare green plutonic rock, informally called "Calabrian jade" (*Fig. 1.20*). In petrologic terms these conglomerates are classed as "sedarenites,"[48] signifying that they consist primarily of fragments of older sedimentary rocks.

Some conglomerates on the level of the Gravaglione/Pizzica terrace are cemented with a purplish-black (Munsell Soil Color Chart 5R 4/1) manganese oxide. Chemical analysis of the sandy matrix of this conglomerate shows it to contain 1.6% MnO.[49] As the average sandstone contains only 0.2–0.7% MnO[50], these specimens represent a considerable enrichment in manganese, probably the result of weathering of manganese-bearing carbonate minerals in the limestone pebbles. This manganiferous material may have been used as a pigment in local ceramic production.

The local sandstones are good examples of "litharenites," sedimentary rocks composed of abundant sand-sized rock fragments (*Fig. 1.34*). Many of the coarser sandstones are more specifically designated as "calclithites" inasmuch as they consist predominantly of fragments of older limestones; they also contain abundant angular pieces of radiolarian chert and some metaquartzite. The finer sandstones are mainly phyllarenites, in which the main rock fragments are phyllite and slate, with common metaquartzite and feldspar and some serpentine. The composition of the rocks indicates derivation from the Calabrian orogenic arc, which contains a complicated assemblage of thrusted and deformed igneous, metamorphic, and sedimentary rocks. The sandstones also contain abundant fossil fragments, mainly coralline algae and mollusks. The sandstones are well-sorted but consist of angular grains, and thus they are texturally mature. Calcite cement is fine-to-medium crystalline spar (0.15–0.35 mm). The fossils and the well-sorted nature of the grains indicate a shallow, current-swept marine environmental origin.

Pleistocene calcarenites occur in an arc from Matera through Ginosa to Castellaneta and Massafra as a thick series of soft, white-to-yellowish rocks termed calcarenita di Gravina, attributed to Calabrian or Late Pliocene age (*Fig. 1.35*).[51] They are composed of sand-sized bits of eroded Cretaceous limestone and indigenous fossils and are known locally as *tufo.* Because of their superb workability they have been quarried for use as building stone from the colonial Greek period to the present day. (Many quarries are still active south of Laterza and around Castellaneta; *Fig. 1.36*). The temples and theater in the ancient urban center and the Temple of Hera on the Bradano

[46] McBride/Picard/Folk 1994. Similar parallel-oriented, cylindrical, log-like concretions have been noted elsewhere—Dakota (Todd 1896), Kansas (Elias 1931), and Nebraska (Schultz 1941)—and attributed to ancient ripple or beach trends, or to groundwater flow directions. Colton (1967) described concretions oriented by marine currents parallel to shore in Devonian shales, but these were not nearly as well-oriented as the concretions in southeastern Basilicata.

[47] Lacava 1891, 13; also Uggeri 1969, 53, for the etymology.

[48] Folk 1980.

[49] Analysis by K. Hoops.

[50] Pettijohn 1963.

[51] Beneduce et al. 2004; Sabato 1996; Pomar and Tropeano 2001.

Figure 1.35 Cut block of calcarenita di Gravina sandstone. W: 7.5 cm. (CW)

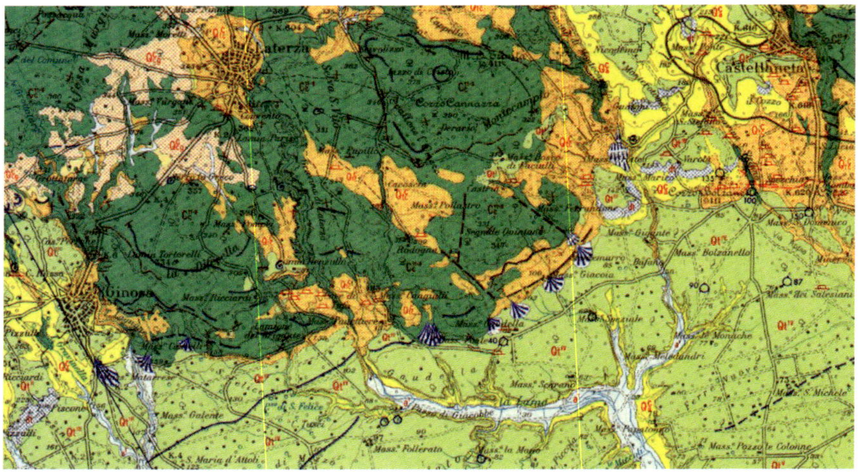

Figure 1.36 Detail of Foglio 201/Matera of the *Carta Geologica d'Italia* (scale 1:160,000) showing the geological formations around Laterza, Ginosa, and Castellaneta.
Key: dark green = limestone; buff = conglomerate; orange = calcarenite; yellow = clay; pale green = marine terrace. (Reproduced with the permission of APAT—Dip. Dif. del Suolo—Servizio Geologico d'Italia, Rome, prot. 45230. Orig. scale 1:100,000)

Figure 1.37 Architectural members in the ancient theater sculpted from calcarenita di Gravina. (RLF)

Figure 1.38 Fragment of a column capital in calcarenita di Gravina recovered during the excavation of the Pantanello sanctuary, 1981. Note the typical regional stratigraphy—sand-gravel cap over clay—in the background. (JCC)

Figure 1.39 *Left,* architectural blocks in calcarenita di Gravina strewn around the fortified outpost at Cozzo Presepe overlooking the Bradano River. (RLF)

(Tavole Palatine) are all constructed of this, or a very similar, stone (*Fig. 1.37*). Sarcophagi and monumental architectural blocks such as ashlars, column drums, and capitals, found scattered throughout the chora of Metaponto, are also commonly sculpted in these local calcarenites (*Figs. 1.38–39*).

Climate and Climatic Changes

Metaponto today has a warm, dry, Mediterranean climate. According to Kayser,[52] the annual rainfall is less than 600 mm (24 inches), and most of it falls in winter with torrential downpours and flooding of

[52] Kayser 1958.

the rivers.[53] The mean annual temperature is 17° C (63° F). There is not much available data on possible climatic changes in southern Italy for the past 3000 years. Brooks[54] obtained a climatic curve for Europe based mainly on data from the Alps. This shows (1) a very dry period about 1000 BC, (2) a sudden great increase in rainfall about 850 BC, and (3) a fluctuating period of lessening rainfall until about 350 BC. Lamb[55] showed a wetter period from 200 BC to AD 200, but then considered that warmth and dryness resumed and increased until about AD 400.[56] Kayser[57] gave evidence that even during historic times (1st to 13th centuries) the rivers were much larger than today and were navigable by large ships. Deforestation by humans during the Bronze Age[58] probably promoted silting of the river channels.

Lamb, and Imbrie and Imbrie, believed in a colder climate around 1000–500 BC, followed by a warming trend to AD 100.[59] Warmth climaxed during a glacial minimum about AD 500.[60] The Greek settlement at Metaponto would have been initiated during a cool and moist period (favorable to farming); it then experienced long-term warmer and drier conditions, with the most extensive drought about the 5th century AD.

The Division Lines, a Geologic Perspective

As pointed out by Carter and Prieto (Chs. 28 and 31), the identification of the path and extent of the division lines, which appear as light-colored anomalies in aerial photographs, is to a certain degree subjective, as all scientific observations made with the human eye must be: witness the very different schemes proposed by Uggeri[61] versus Adamesteanu and Vatin.[62] Identification of the lines on the ground is even more open to interpretation and personal bias. One uniform characteristic of all schemes is the N 50° W orientation of the main longitudinal lines—Prieto has measured slight variations—and their uniform spacing of between 195–220 m. The greatest discrepancy between authors is in the NE-trending transverse lines: Uggeri postulated a regular grid covering the entire area at right angles, while in great contrast Adamesteanu and Vatin saw only four lines all near the coast and running at a 78° angle to the longitudinal lines. Prieto recognized five or six transverse lines in the same area, but did not see the northwesterly two transverse lines of Adamesteanu and Vatin.

There are many logical questions that need to be considered for each hypothesis regarding the function of the lines, whether they be natural geologic lineations, farm-plot boundaries, canal/ditch systems, or ancient roads. All of the proposals have some favorable points, but none is "without sin." There appears to be a continuous gradation between (1) purely geologic drainage channels controlled by the N 50° W grain of the subsurface fracture system and oriented concretions, (2) anthropically-modified natural channels, (3) originally anthropic lines later modified by natural gullying and erosion, (4) entirely man-made lines, whether they be boundary lines for farm plots, roads, or drainage/irrigation channels. Disentangling the specific nature of these lines from air photos is a daunting task. I will here propose a few points that favor a geologic origin for at least some of the proposed lines; other lines, as shown by excavation, are clearly man-made features.

There is no doubt that the general orientation of the lines in the Bradano–Basento area (N 50° W) coincides with that of geologic features such as fractures and linear concretions (N 40–60° W). The Basento-Cavone sector provides an excellent confirmation of this relationship, as both geological features and division lines run N 70° W (*Figs. 1.40–1.41*).[63] Surely this double coincidence is no chance feature. The drainage of many present-day streams is parallel with these orientations, therefore man-made features such as roads, canals, farm boundaries, tombs, and houses naturally would follow the same drainage direction. It is probable that some of the line segments seen on air photos are natural geologic features, e.g., the sinuous channel in Figure 1.42; some are purely anthropic, such as those in the excavated road complex near Pizzica (Ch. 28). These latter are too wide, too straight, and too evenly spaced to be geologic in origin.

53 Vezzani 1967.
54 Brooks 1949, Fig. 32.
55 Lamb 1982b.
56 Lamb 1982a. In contrast, Boenzi/Caldara/Pennetta (1995) have proposed a hot climate for the area around Monte Gargano during the last two centuries BC, and minimal precipitation around AD 700.
57 Kayser 1958, 13.
58 Neboit 1975.
59 Lamb 1966, 63. Imbrie and Imbrie 1979, Fig. 43.
60 Denton and Porter 1970.
61 Uggeri 1969.
62 Adamesteanu and Vatin 1976.
63 Adamesteanu and Vatin 1976

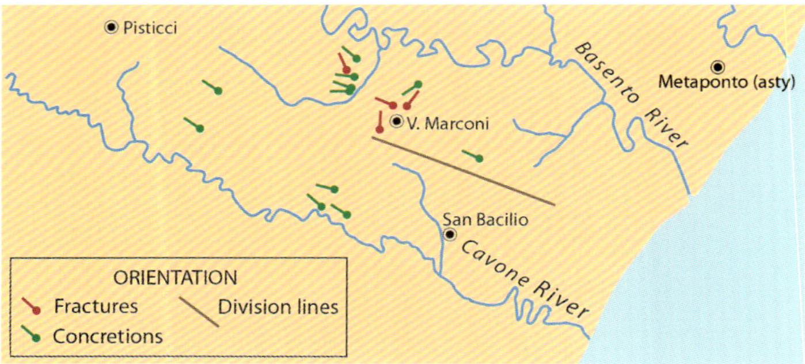

Figure 1.40 Orientation of concretions (green), fractures (red), and division lines between the Basento and Cavone Rivers. (RLF/DD)

I will first discuss some features that argue for a natural origin for some of the division lines. On the ground there are two areas in which there appeared to be the clearest visible evidence of a grid on our air photos: (1) around Pozzo di Gravaglia, Metaponto topographic sheet, Long. 4° 20' E of Rome, Lat. 40° 24' 30" N; (2) a more important region, the gently inclined plain designated Lago del Lupo, on the Masseria Girifalco topographic sheet, Long. 4° 17' E of Rome, Lat. 40° 26' N.

At Pozzo di Gravaglia, the NW-trending branches supposed by Adamesteanu and Vatin to be grid lines,[64] were in fact by 1980 concrete-lined straight drainage ditches which lie in a natural stream drainage system with normal, wide, gently-sloping valley sides. In the vicinity of Lago del Lupo (a triangle extending from Masseria Lago del Lupo to Masseria Appio and Casa Gallitelli) occur the lineations that are best visible on air photos and from the ground. The supposed division lines coincide with a wide variation in ground conditions: from proposed "lines" that are invisible on the ground, to narrow ditches 1 m or so wide (*Fig. 1.42b*), to very large valleys hundreds of meters wide and as much as 10 m deep (*Figs. 1.42a, c*). The wide variation in present-day visibility implies that they were not dug as a uniform series of boundary lines at one time in antiquity. In July 1980, J. Carter, C. Lyons, and I inspected a modern irrigation ditch 1–2 m deep that crossed the division lines near Casamassima at nearly right angles, and no difference in soil could be seen at the supposed positions of the lines.

The largest and supposedly the main central line (L15 of Prieto and others) that divided the Metapontine territory into two equal halves, and passes 450 m NE of Casamassima (*Fig. 1.42a*), shows no sign whatsoever on either the air photo or the topographic map of continuing south of its juncture with a large natural stream valley at about Long. 4° 17' 50" E, Lat. 40° 25' 30" N. If it were the enormous master division line, it should surely not disappear completely on the other side of a transverse drainage line. But this would be a normal characteristic of natural channels developed on a coastward-sloping surface (*Fig. 1.43*), since water does not run up-slope. Most of the present-day ditches or drainage lines that are oriented parallel with the main division lines (i.e., N 50° W) occur only on the north sides of cross-cutting drainage lines (e.g., lines 15e, 18c, and 19d): i.e., all the visible ditches flow SE down the slope of the old marine plain (this is true also of the oriented trellis drainage patterns seen outside the Metaponto area along the Ionian coast). If artificial ditches had been laid out for centuriation (farm boundaries), they should extend across transverse drainage lines, and we should today still find traces of ditches coming in from the south side as well (*Fig. 1.43*). This morphology would be acceptable for drainage ditches, but not for irrigation systems.

The sinuous course of the visible division line track in Figure 1.42c is certainly a characteristic of natural erosion. Furthermore, all the ditches of whatever size show gradually-sloping valley slopes adjacent to them, as in natural drainage systems (see Dana's analysis of slope in Ch. 4). Erosion in the past 2,500 years could not possibly have started with V- or U-shaped artificial cuts and gradually broadened them out into

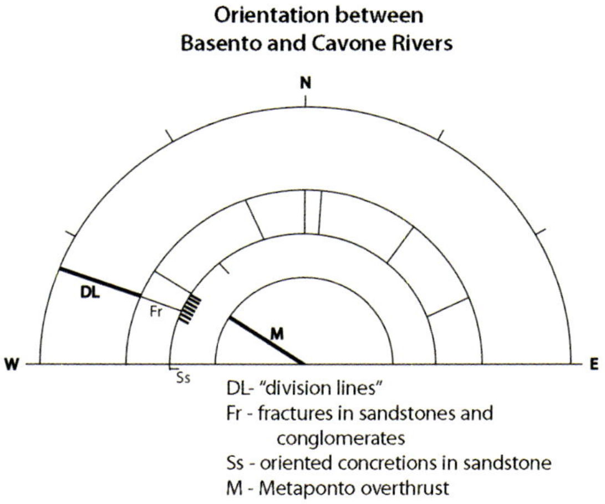

Figure 1.41 Orientation diagram for features between the Basento and Cavone. Fractures do not show any preferred orientation, but oriented concretions are parallel with the division lines. (RLF/DD)

64 Adamesteanu and Vatin 1976.

a

b

Figure 1.42 Division lines as they exist today. (RLF)

a) What has been interpreted as the main, central division line by archaeologists (photo taken from a point .5 km NE of Masseria Clemente, looking SE). Observe the large size and gentle slopes of the valley sides, a characteristic of natural drainage lines.

b) A small division line near Lago del Lupo. (CW)

c) A large division line near Lago del Lupo. Note the broad valley, lack of any rimming lines of excavated earth, and sinuous nature. (CW)

c

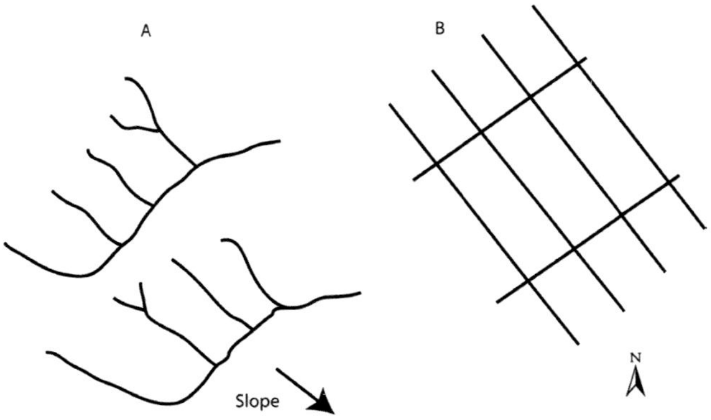

Figure 1.43 In a naturally developed drainage system (A) all the smaller tributaries flow down the sloping surface to the sea, thus all join the larger crosswise streams from the northwest side. In a system of ditches dug for division lines (B), junctures should occur on both the NW and SE sides of the crosswise ditches; the two sets should intersect in a grid. In the chora of Metaponto, the situation is as in A. (RLF)

an entire sloping valley profile hundreds of meters wide. The huge gullies (50–100 m deep) in the area of Campagnolo show the same rectilinear pattern and N 50° W orientation, and these are far too large to have originated as man-made ditches only 2–3 millennia ago (*Fig. 1.44*).

The two southeasternmost transverse division lines recognized by Adamesteanu and Vatin are clearly parallel to Pleistocene beach lines formed as sea level dropped sporadically (geologic map #201).[65] This is why they are not at right angles to the N 50° W lines. Their two northwesternmost lines are nearly parallel with two other old major beach lines, though not "on top" of them. These two lines were not recognized by Prieto, so their very existence is dubious.

The trellis drainage systems (streams intersecting at close to right angles) are found not only in the chora of Metaponto, but also along the entire instep of the Italian "arch," from the Agri River to the Lama stream. These are natural features, which control the orientation of present-day roads, but no one has yet proposed that these were Greek lines like those in the chora of Metaponto.

I have suggested that some of the lines may be geologic in origin, but there are clearly many that are man-made. For the lines that are definitely anthropic, what was their specific purpose? Property boundaries, water-management canals, or roads? Originally they were thought to be division lines for individual farm plots, along with other dividing features in the Mediterranean[66] and in Chersonesos. That they were not originally laid out as "centuriation" lines is now evident. Had they been farm plot division lines, they should have formed a right angle grid, and grid lines would have extended over the whole of the chora, not just occurring in the southernmost portion of it. Even in the portion where the transverse lines exist, they are unequally spaced. Surely this is not a way to divide rural territory into equal-sized plots! It is much easier to plant trees, hedges, drive stakes, or erect rock cairns as boundaries, rather than digging trenches several meters wide as boundaries, a staggering amount of unnecessary labor—unless the ditches were necessary for drainage, made to do double duty (see Ch. 28). Moreover, if they were grid lines, they should have been of uniform width and depth. In fact, most of them now show only discontinuous traces on the ground or in air photos, yet other purported ones are 5–10 m deep and tens of meters wide (*Fig. 1.42*) If they were dug as ditches, where did workers pile the dirt? There is no uniform evidence of ridges of dumped dirt running alongside the "ditches," even the largest ones (*Fig. 1.42*), although Uggeri observed dirt along the edges of ditches, discolored and mounding in the 1960s. In profile, the adjacent valleys gradually slope downward toward the ditches as is the case in natural drainage systems (*Fig. 1.42*); on the other hand, artificial ditches dug as boundaries would have been cut in straight lines across the topography, passing indiscriminately across valleys, uplands, and slopes, like the track of a pipeline. In fact, the "ditches" now visible all lie in normal valleys. On the air photos, the few extant division lines do appear to be straight; but on the ground they are definitely slightly sinuous (*Fig. 1.42*) with overlapping spurs, characteristic of natural drainage systems. If erosion had modified the originally straight drainage channels, a sinuous course might eventually develop, but it would not have overlapping spurs; instead, it would have low flat eroded areas where the channel had meandered and eroded the beds (*Fig. 1.45*).

The lines, whatever their first purpose, may have been subsequently used in places as boundary lines between farm plots, but it is not obvious that they were all laid out in a process of centuriation, although the regular 180–220 m spacing between them is a point in favor. Another possibility is that the lines represent

[65] Adamesteanu and Vatin 1976.

[66] Bradford 1957.

Figure 1.44 Deeply incised stream valleys, the result of natural erosional processes; valleys trend in the same direction as the division lines. *Above*, canyon NE of Lago del Lupo (PD); *below*, near Masseria Montesano, between Pisticci and Le Tinchi. (RLF)

a Greek system for water management, either to carry off excess water during the rainy season, or to use as irrigation canals during the dry summers. The area of most of the chora slopes seaward at about one-half degree (*Fig. 1.14*), thus it seems it should be naturally well drained with little chance of flooding. Furthermore, the surface material consists of several meters thick of sand and gravel which should be permeable enough to absorb water rather than shedding a large amount of damaging runoff. As pointed out already, why would the Greeks have the main flood runoff canal (line L15) directing water right at the asty of Metaponto? Although some lines stop at the nearest natural stream valley that flows laterally toward the rivers, others graze (sideswipe) them, as shown particularly by line 20b, which is 2.6 km long and nearly parallel to a large valley only 0.2 km to the west. The swarm of lines east of Masseria Lago del Lupo also runs parallel to existing stream channels, without leading into them.

If one considers the putative use as irrigation canals during the dry season, generally in other areas effective irrigation systems are constructed on flat floodplains with essentially zero slope, elevated only a few feet above a large river that feeds them. In the chora of Metaponto, one would have to be able to lift water from the Bradano or Basento up 10–30 m to the low plateaus of the chora. Furthermore, one would have to show that the base flow of these rivers would be adequate during the dry season, and there should have been a series of dams built to retain water for such a system to be operative. Rather than having a swarm of parallel longitudinal lines that would have to be fed at all of their top ends by a parade of water carts, one would have constructed one or two main channels to receive the lifted water, then branch out dendritically into distributive channels. Trying to get irrigation water out of the bed of the small transverse streams, such as the Fosso Appio, to feed this swarm of channels seems impossibly difficult. Water dumped into the channels would either soak rapidly into the permeable gravels, or run swiftly down the half-degree slope towards the sea. There should also have been lateral lines in the northern part of the chora to lead water into the farmers' fields. Furthermore, what was the source of water for the supposed irrigation lines around Pizzica? All in all, the hypothesis of an irrigation system seems to have many logical difficulties.

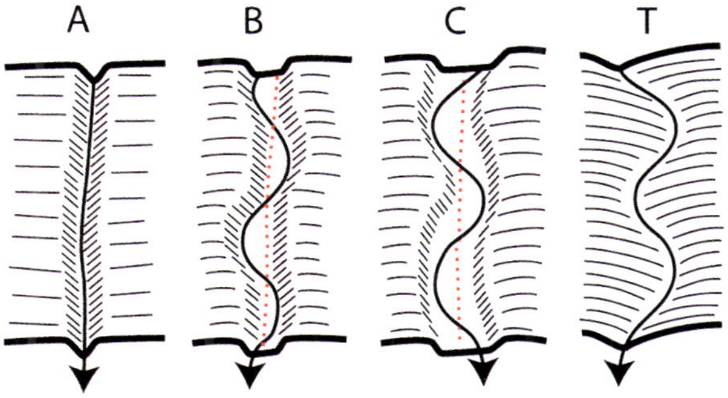

Figure 1.45 Erosional evolution of channels. Drainage ditches (A) should have been dug as straight lines. As erosion proceeded over the last 2,500 years, a ditch might begin to meander slightly, but if so it should leave truncated spurs (B, C). In a natural drainage system (T) valley sides descend gradually to the stream channel, with spurs that overlap as one sights along the stream course. At Metaponto channels look like the situation at T. (RLF)

Another hypothesis was that the lines represent an ancient road system. There is no doubt that some of the lines in the Pizzica area are in fact roads as shown by excavation (Ch. 28), a fact confirmed by the orientation of tombs found along them. But what about the rest of the chora at elevations greater than 50 m? If the longitudinal, N 50° W lines are roads, then this seems like an awfully dense array for a rural area; one would expect random paths for cart trails, not meters-wide excavated roads every 200 m for chariot traffic. Why is there such a dense array of NW roads and not a single trace of a crossing NE road in the chora in ancient times, again excepting the area around Pizzica? Roads should converge in direction as they neared an easy stream crossing, not shoot straight across like an arrow. They should at least go diagonally down stream valley sides so as to lessen the stress of driving on a steep slope. They do not do this. It also seems logical that rural tracks ought to converge towards the asty of Metaponto or towards important springs, instead of being a unidirectional symmetrical array. One road clearly visible on air photos is PT 02 near Pizzica, which appears as a wide swath of darker soil with straight sides and a constant width, as roads should be. Even so, its trace seems wider even than modern roads.

I have attempted to consider the strengths and weaknesses of all the hypotheses concerning the nature and purpose of the division lines. Unfortunately, all hypotheses so far proposed have serious logical difficulties, even if some evidence supports each of them. Perhaps their weaknesses may direct future research to give a final definitive answer.

Geology and Population Distribution

Settlement patterns: rural areas

The Metapontino is composed of dense, impermeable blue-gray Pliocene/Early Pleistocene pelagic clays overlain unconformably by a terraced series of loose sands and gravels, several tens of meters thick, which represent shallow marine sediments deposited as the sea retreated across eastern Basilicata to its present position during the later Pleistocene (*Fig. 1.14*). The sands and gravels are generally unconsolidated and consequently highly permeable and porous, but they are locally cemented by calcite into ledges varying in thickness from a few centimeters to several meters. The sand and gravel deposits which cap the terraces form level surfaces of loose, easily tillable material ideal for agriculture, which are today cultivated over almost their entire surface. They were presumably just as productive in antiquity.

The blue clays (*argille azzurre*) in the Metapontino tend to form gullied slopes that are useless for agriculture. These slopes are usually precipitously steep (*Figs. 1.7, 1.9*) and very prone to catastrophic landsliding (*Fig. 1.11*), especially during the wet winter season. Rainfall runs over the surface instead of infiltrating it, and the material is so viscous when wet, and so hard and compact when dry, that tillage is impossible. Franciosa relates that the blue clays confer a sad and monotonous aspect upon the region and are responsible for the scarcity of rural habitations; the hills appear as a series of pallid pyramids (*Fig. 1.8*) lacerated by gullies and deprived of vegetation.[67] Of the ten main types of lithologic structure in Basilicata, the *argille azzurre* class has the lowest population density: in 1931, it averaged only five persons per square kilometer. By comparison, the overlying yellow sands and gravels were characterized in the same year by a population density of 113 persons per square kilometer.[68]

Carlo Levi[69] evocatively described the Pleistocene clay outcrops around Aliano, approximately 55 km southwest of Metaponto: "*. . . da ogni parte non c'erano che precipizi di argilla bianca, su cui le case stavano come librate nell'aria; e d'ognintorno altra argilla bianca, senz'alberi e senz'erba, scavata dalle acque in buche, coni, piagge di aspetto maligno, come un paesaggio lunare.*" ("On every side there was nothing save precipices of white clay, on which the houses stood as if hovering in the air; and everywhere more white clay, without trees and without grass, scoured by the waters into cavities, cones, and declivities of malign aspect, like a lunar landscape.") Since the 1980s these differences have become less pronounced, as powerful land-leveling equipment and modern fertilizers have rendered areas of exposed *argille azzurre*—even slopes—amenable to cultivation.

Most of the rural habitations in the ancient Metapontine hinterland are found on the edges of sandy tablelands, part way up the sides of the stream valleys near the contact between the blue clays and the overlying permeable sands and gravels (*Fig. 1.46*). Houses would not have been built in the stream floodplains because of the danger of flooding. Adamesteanu

[67] Franciosa 1946, 72.

[68] Ibid.

[69] C. Levi 1947, 13 (translation by R.L. Folk).

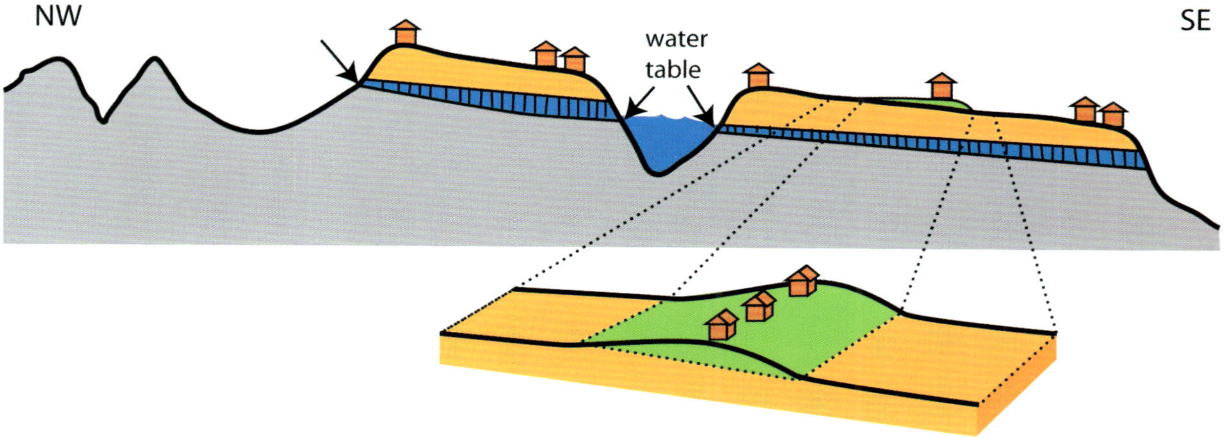

Figure 1.46 Reason for the siting of farmhouses. Impermeable Plio-Pleistocene oceanic clays are overlain by loose, permeable later Pleistocene gravels and sands laid down on the shallow sea bottom (see *Fig. 1.14*). Rainfall seeps down into the loose sand until it hits the impermeable clay, then runs horizontally until it emerges as springs below the brow of the hill. These became favored sites for early settlement. Later, wiser colonists realized that water could be obtained by digging shallow wells into the sand where the clay acts as a plug to keep the water from draining away too deeply, and holds up a shallow water table. Some settlements were also made along gravelly beach ridges (in lower view) which played a similar role as good water sources. (RLF)

and Vatin noted the concentration of farmsites on the flanks of stream valleys, but did not mention the geologic basis for this.[70] The reason they were built near the sand/clay contact is that the blue clay in this situation plays a beneficent role: rainwater percolates through the overlying sands, seeps down until it hits the impermeable clay, then travels laterally along the contact to emerge along the valley sides as a series of springs (*Figs. 1.47–49*). This is analogous to putting big piles of sand on a table and pouring water on it; the water will eventually emerge at the sand/table contact. Lacava (1891) and Franciosa (1946) mention the occurrence of springs at this contact. Many of these are still active today, e.g., at Ponte Fabrizio (*Fig. 1.48*). Farther up the Bradano valley, the same situation applies, because most of the springs there are also at the sand/clay contact.[71]

Inasmuch as the Pleistocene cap of sandy gravel slopes southeastward, the best place to look for spring sites is at the southeastern margin of these gravel sheets, because the groundwater moves in this direction, i.e. downslope (*Fig. 1.46*). Early Greeks would have taken advantage of this situation—using the level, sandy, well-drained, and tillable tablelands for farming and building their houses on or near the scarps to take advantage of the springs. Later, they might have gotten the idea that they could settle on the tablelands away from the scarp edge, and dig to the base of the sand where again water would be found, retained there by the clay beds. Today, water can be obtained at depths of 5–7 m in this sandy region;[72] however, where the sand cover is very thick (such as Incoronata), a hole would have to be dug too deeply in soft sand for such a well to be practicable.

There are two other types of areas where water conditions are favorable. One is along temporary old beach lines where a minor ridge of sand or gravel was left behind by the retreating shore (*Figs. 1.18* and *1.46*). These beach lines trend about N 30° E parallel to shore; in 1981 we found rows of habitations following the trends.

It is interesting that a similar geologic control apparently determines the location of ancient brick or tile kilns. All kilns so far discovered (Pizzica, Sant'Angelo, San Biagio, Sant'Angelo Grieco, and San Vito) are excavated in the Pliocene clays (*Fig. 1.50*) just below the contact with Pleistocene gravels. Thus the sites provide both ceramic raw material and the water for its manufacture. The same clays are being quarried today for brick manufacture, along the Basentana highway near Sant'Angelo.

I propose that the reason for the inland limitation of the Greek territory of Metaponto was also fundamentally geologic. The aggressive colonizers seized all the good land on the sandy marine terraces

[70] Adamesteanu and Vatin 1976.

[71] Baldassare and Radina 1971

[72] Sig. Casamassima, pers. comm. 1982.

Figure 1.47 Sharp contacts between overlying permeable Pleistocene gravelly sand and impermeable Pliocene clay. These are usually the sites for springs. *Above,* a roadcut just west of Ponte Fabrizio (RLF); *below,* near San Vito. (PD)

as far back as the terraces extended (e.g., to Pisticci and Montescaglioso, *Fig. 1.2*) and forced the indigenous Italians off onto the desolate blue clays in the rugged hinterland (much as the white man forced the American Indian onto less desirable territory). Thus Pisticci and Montescaglioso remained centers for the Italianate "aborigines."[73] In this regard geologic maps provide a good paleo-ethnic map: Greeks are on the Pleistocene sand, Italians on the Pliocene clays.

Settlement patterns: urban areas

There are three major factors determining location of larger modern towns in southern Basilicata and bordering Puglia. The towns had precursors in Classical times, and most of them had antecedent settlements on the same sites even back to prehistoric times; for geologic reasons these have remained optimal sites of habitation for several thousand years. The three major types of sites are: (1) seaports, (2) ancient channel

[73] Adamesteanu 1973, 61; 1974a, 207.

Figure 1.48 Springs active at the gravelly sand/clay contact. *Left,* Ponte Fabrizio (RLF); *right,* near Montescaglioso. (C. D'A)

Figure 1.49 Archaeological site at Ponte Fabrizio, located at the contact between gravelly sand and clay, across the valley from the spring in Fig. 1.48. (RLF)

Figure 1.50 Ancient kiln near San Vito. Kilns are systematically located within the Pliocene clays, just below the contact with the Pleistocene sandy gravels. (RLF)

Figure 1.51 Towns built on Calabrian fan-channel gravel ridges. *Above,* Pisticci, viewed from west of Pomarico; *below,* the linear ridge upon which Pomarico Vecchio lies, viewed from the present town of Pomarico. (RLF)

gravels, and (3) Pleistocene calcarenites. A fourth type of town (trulli area) is found on Cretaceous limestone.

(1) Seaports sprang up during the Greek occupation in about the 8th century BC; the shoreline of that time was as much as several kilometers farther inland than it is today. These port cities included Taras (modern Taranto), Metapontion, Herakleia, and Siris.

(2) Channel gravels of Early Pleistocene (Calabrian) age served as important town sites. These gravels are surviving remnants of ancient river channels. A striking feature of southern Basilicata, even to the layman, is the occurrence of these hilltop towns which rise dream-like from the forbidding landscape bordering the transportation routes (*Figs. 1.9, 1.51*). Guidebooks attribute the elevated location of these sites to "defense against Moors" and "avoidance of malaria"; these factors certainly operated, but the primary reason for settlement is more likely that they afforded a good source of groundwater. Romantic towns like Pisticci (*Figs. 1.9, 1.51, above*), Montescaglioso, and Pomarico (*Fig. 1.51, below*) loom high above the river valleys atop their scarified cliffs of gray clay. Each town is built on its isolated cap or ridge of gravel. Today they are accessible only by tortuous roads leading from the main highways built far below along the flat valleys that flank the Basento and Bradano Rivers.

What is the reason for the location of these gravel cappings? Examination of the geologic map (Foglio 201; *Fig. 1.2*) reveals a pattern to the occurrence of the gravel remnants. It suggests that they are a classic example of topographic inversion (*Fig. 1.52*).

During Calabrian times, the southern Apennines were rising and shed a vast apron of detrital sediment as a series of coalescing fans.[74] The shoreline was about 30 km inland of its present position, and sea level was some 200 m higher. This shoreline would have been approximately along the line of Pisticci, Pomarico Vecchio, Montescaglioso, and Laterza. Overloaded with gravel, streams came down from the Apennine front, apparently forming a series of anastomosing channels that divided outwards like fingers (cf. the modern mouth of the Po and other large deltas). At this time the coastal plain would have consisted of channels filled with gravel, together with interchannel areas of slightly higher elevation, made up of finer materials, mainly sands. This old coastal plain material is designated on Foglio 201 as the *conglomerato di Irsina*.[75]

As glaciation began and sea level accordingly fell, erosion began to act on the old gravel fans and deltas. The finer interchannel sands were eroded rapidly, but the channel gravels were more resistant; what was originally formed as a topographic low (the channels) subsequently became a linear topographic high, and inverse topography resulted. This is why the Calabrian gravels on the geologic map form a branching pattern, with fingers opening towards the southeast (*Fig. 1.52*). Further erosion dissected these fingers, leaving isolated "knuckles" where the gravel may have been thicker or locally cemented.

When human habitation began, these high gravels formed backbone-like ridges and isolated

[74] Segre 1978.

[75] Described by Ricchetti (1967).

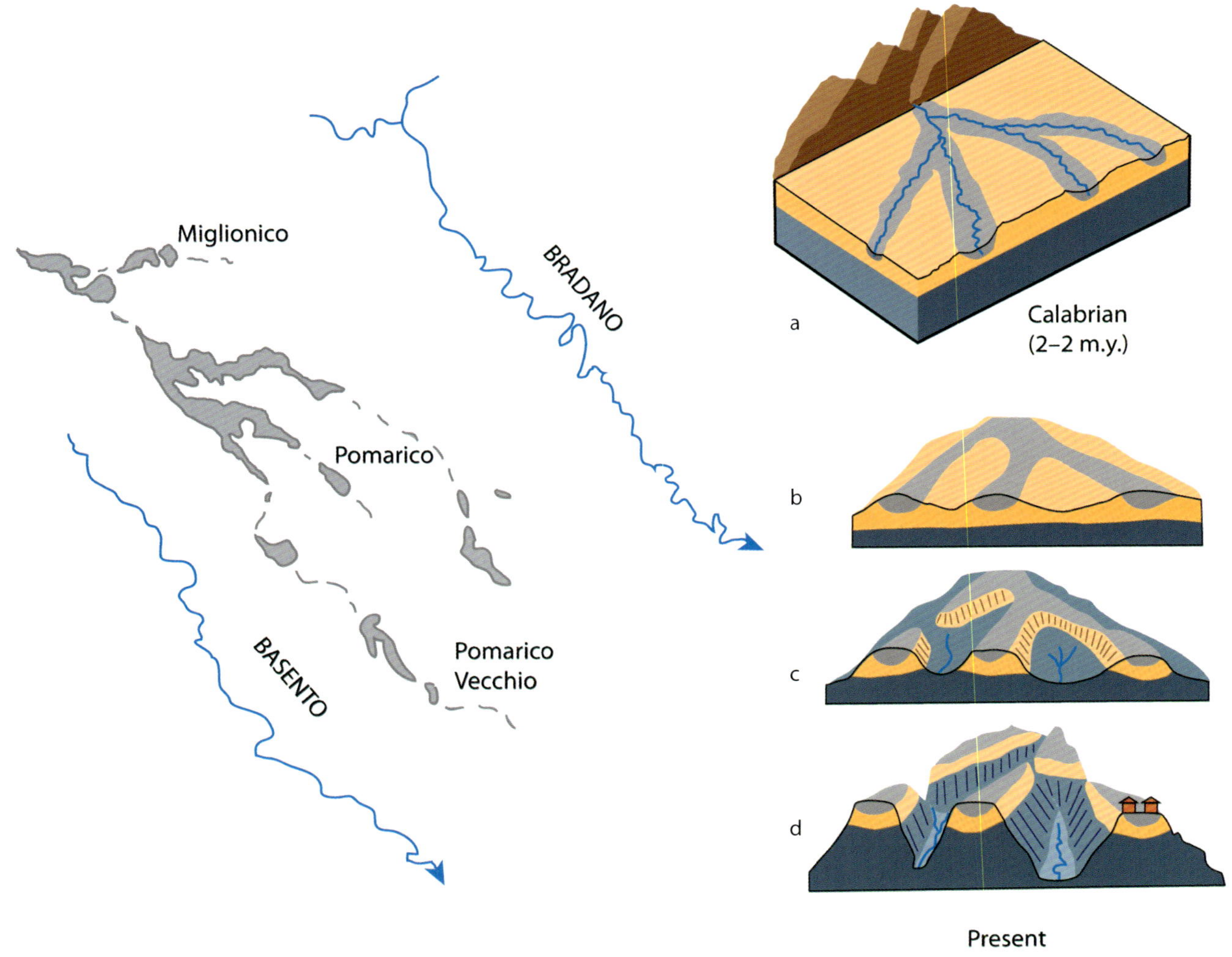

Figure 1.52 Diagrams showing the evolution of inverted topography and the geologic reason for the hill towns of Basilicata. *Left,* traced off the geologic map, Foglio 201, are the outcrops (in grey) of the Calabrian (Early Pleistocene) "conglomerato d'Irsina." They show a southeastward-branching pattern of ridges, upon which lie several towns. *Right,* the evolution of the topography. a) During Calabrian time, the rising Calabrian orogenic arc shed vast quantities of sediment, with gravel-filled channels branching SE, forming a series of alluvial fans. Between the channels, finer sands were deposited. b) As sea level dropped, the channels were abandoned and the sequence of sediments began to be eroded. c) The old gravelly channels were more difficult to erode than the softer sands, so eventually the channels became linear ridges. d) The gravel channels, overlying impermeable clays, acted as good traps for water. (RLF/DB)

pimples, resting atop older Pliocene clays. Thus the gravels have formed ideal traps for subsurface water, as rainfall percolates into them, seeps to the bottom of the gravel, and is held there by the impermeable clay. Springs would have been abundant at the gravel-clay contact. The channel gravels were ideal sites for early settlements and remain so today. Adamesteanu and Vatin point out that the earliest villages here were Iron Age Italic settlements, many of them later converted to fortresses by the Greeks because of their commanding positions.[76]

The area that stimulated the idea of topographic inversion is the finger-like pattern of gravels upon which Miglionico, Pomarico, and Pomarico Vecchio (as well as the hill of Cozzo Parlante) are located. I propose that another long gravel channel once extended from NW of Ferrandina (Monte Finesci) to Pisticci, and yet another between Timmari and Montescaglioso (though the latter is now all but obliterated).

(3) Pleistocene calcarenites occur in an arc from Matera through Ginosa to Castellaneta and Massafra as a thick series of soft, white to yellowish calcarenites termed calcarenita di Gravina and attributed to

[76] Adamesteanu and Vatin 1976.

Figure 1.53 Calcarenita di Gravina used for construction in the 15th century castle at Bernalda. Many of the modern buildings in Bernalda, Matera, Ginosa, Laterza, etc., are made of this soft stone, and quarries are still active. (PD).

Figure 1.54 The troglodytic portion of the town of Ginosa. Inhabitants excavated rooms and entire houses into outcrops of the calcarenita di Gravina because of its easy workability. (RLF)

Calabrian or Late Pliocene age. These soft rocks are made of sand-sized bits of fossils, locally known as *tufo*; because of their superb workability the rocks have been quarried for use as building stone from ancient Greek times to the present day. Many quarries are still active south of Laterza and around Castellaneta. The temples at Metaponto, the Tavole Palatine, the blocks from ruined temples at Pantanello, and the coffins of the Pantanello necropolis all appear to be made of this or a very similar stone. Monumental architectural blocks such as ashlars, columns, and capitals—found scattered and isolated throughout the hinterland of Metaponto—are also commonly made of this stone (*Figs. 1.39; 1.53*).

The calcarenita di Gravina are so easily carved that, for 2000 years, inhabitants have excavated homes into the cliffs made of this rock where it outcrops along river valleys. The curving arc of the calcarenita di Gravina is the explanation for the belt of trogloditic towns, among the most curious types of settlement in all Italy. The most famous of these towns is Matera, but Massafra,[77] Ginosa (*Fig. 1.54*), and Laterza are also excellent examples.

[77] Described by Allen 1969.

Figure 1.55 *Trulli* houses made out of slabs of Cretaceous limestone outcropping on the Puglian platform. *Left,* a typical street in Alberobello; *right,* an old farmhouse in trulli form. Only where thin-bedded limestone occurs does one find this quaint type of housing. (RLF)

The calcarenite formed during Plio-Pleistocene times, when sea level was much higher, and Puglia was an island, its shoreline running along the Matera-Ginosa-Castellaneta-Massafra arc. The island of Puglia was made of Cretaceous limestone, and streams draining this area carried much dissolved $CaCO_3$ to the sea. This supported an abundant calcareous biota, resulting in the coastal belt of $CaCO_3$ shell-fragment sand that eventually became weakly cemented into beachrock. Modern analogues of these Pleistocene carbonate beaches occur in the resort areas of the west coast of Puglia, such as at Porto Cesareo.

Trulli towns are composed of conical houses atop the Cretaceous rocks forming the plateau of Puglia (*Figs. 1.1–2*) in the trulli region—these houses are especially notable in towns such as Alberobello (*Fig. 1.55*), Locorotondo, and Martina Franca.[78] The houses are made of slabs of thin-bedded Cretaceous limestone, and this gnome-like style of architecture is confined to areas where that type of limestone crops out.

Thus we see that in southern Basilicata geologic factors have played an important role in determining locations of major townsites and also in influencing their typical architecture.

Acknowledgements

It is a pleasure to acknowledge Dr. Joseph C. Carter, expedition chief, for arranging my trips to Metaponto, making the stays there rewarding, and providing the proper accessories such as vino da tavola; Dr. Ingrid Edlund for support during my first field season there (1980); Pam Dyson and Cesare D'Annibale, my field partners on survey during 1981; Chris Williams for photography during the 1980 season; Michael Guarino, Deena Berg, and Boyce Cabaniss for discussions; David Stephens (Geology Department) for doing the splendid job of converting color slides to black-and-white prints; and Martha Ladyman and later Rosemary Bryant (Geology Department) for typing various drafts. Tullio Pescatore (University of Naples) and Keith A.W. Crook (Australian National University) made helpful comments on the text.

Appendix I

Pantanello necropolis, 1986

In 1986 I visited the necropolis very briefly, accompanied by graduate student James Miller. Two features were examined: (1) several deep trenches that had been dug on the slopes along the eastern border of the necropolis, and (2) the necropolis itself. The purpose was to see evidence for the existence of division lines or ancient roads.

The trenches were cut in order to intersect the trace of one of the division lines seen by Adamesteanu on the older air photographs. The trenches showed irregular layers of gravel, sand, and mud, typical of sediments deposited by stream or gully wash in natural situations. The section suspected of being an ancient division line was an asymmetrical channel, with steeper slope on the west side: a cross section fully expectable in a meandering gully. If there had been at one time a road, it was certainly very crude and irregular. To me the section was something one would normally see in the valley between two gentle hills. What should be done is to dig a similar trench in areas where no division lines are predicted, to see if those areas would show equivalent sediments.

[78] See Allen 1969.

The cross section of the "road" encountered in the necropolis showed a flat-topped gravel mound a few decimeters high, with steeply sloping sides. I am certain that this is a natural gravel deposit and not an ancient road. The material is composed of clean, sandy gravel without any clay binder, surely terribly loose footing for either humans or horse-drawn chariots. And a few days' wear would have stomped down the steep slopes of gravel on the side of the "roadbed." Furthermore, pebbles in this gravel are imbricated, sloping consistently to show a current from the NW—the same direction as the Bradano flows today. Pebbles have caliche crusts on their lower sides, showing that they are lying in place. Gravel roads built up by shoveling workmen should have pebbles that are randomly oriented and with caliche crusts indiscriminately on top or bottom.

This gravel deposit is a natural bar deposited by a braided stream. Similar bars can be seen on the river bed along the train tracks northwest of Bovino (line from Foggia to Naples). These show flat-topped gravel bars with steep sides, and the bars are linear, parallel with the river current. If buried, they would mimic very closely the gravel feature in the necropolis.

Appendix II

Metaponto necropoleis—sarcophagi and cist tombs (2/12/92)

All the samples given to me except one are made of very similar soft, porous rock. They are made of well-sorted and abraded, medium-sand-sized (0.5 mm) shell fragments with a variable quantity of larger shells such as mollusk fragments. The high porosity is due to an almost total lack of cementation by calcium carbonate. Minute crystals of calcite do occur, and they indicate that the slight amount of cement was probably precipitated in an ancient freshwater environment; thus the rock probably formed in an environment high on a beach. My thin section number TC-1512a-TXEu (collected from a sarcophagus in 1986) shows minute needles of lublinite, a form of calcite precipitated when water freezes. All samples are a pale creamy yellow due to infiltration of hydrous iron oxide.

The samples are essentially identical to those I already had collected from the Plio-Pleistocene age calcarenita di Gravina, which outcrop in a broad arc at the SW edge of the Puglia uplifted carbonate plateau; excellent outcrops of this are found at all the troglodyte towns such as Ginosa, Laterza, Castellaneta, and Massafra. This band of rock originally formed on a beach, similar to that at Gallipoli (Puglia) today.

One of the samples, labelled simply "sarcophagus," is totally different from the above. The shell fragments are much finer, "very fine sand" size, about 0.1 mm. Many of them have been dissolved out to leave holes. There is pervasive cementation by calcite. This rock either (a) is older or (b) saw a much more pervasive flux of freshwater through it. I would suggest that this much harder rock would be ideal for making a sturdy structure such as a sarcophagus.

The *pietra leccese* outcrops in a large territory around Lecce. It is of Miocene age and, from the descriptions, it must be very similar in megascopic appearance to the calcarenita di Gravina, the latter, however, being younger by a few million years. I saw the town of Lecce in 1981 and do remember that it is made of a soft, warm brown stone, but I do not recall looking at the stone with a hand lens.

I suspect from a few samples that it would be very difficult to tell calcarenita di Gravina from pietra leccese. Since both would appear to be of substantially equal appearance, ease of workmanship, color, and durability, each unit could be quarried with equal facility. It would seem to me to be more economical to quarry the stone at Laterza or in the vicinity of the calcarenita di Gravina, only some 30 km away, rather than from the area around Lecce, over 100 km distant.

If the Germans have some specific reason for deciding on the pietra leccese rather than the calcarenita Gravina, I would like to see their evidence.

The proper specific geologic name for the samples given to me would be "calcarenitic limestone." The rock could be correctly termed, in a broader sense, either "limestone" or "sandstone," the first term referring to the component composition (almost pure $CaCO_3$), the latter to the sandy texture.

2
Geomorphology and Geoarchaeology of the Metapontino

James T. Abbott

While geoarchaeology[1] has been defined in a number of different ways,[2] a succinct and useful definition is presented by Butzer, who describes it as archaeological inquiry using the methods and concepts of the earth sciences.[3] Geoarchaeology draws upon a number of geographic and earth-science disciplines, including sedimentology, stratigraphy, pedology, geomorphology, geochronology, and paleobotany, to address culturally relevant issues and to provide a context in which to interpret archaeological data. The aim of this investigation is to present the evolution of the Metapontine landscape during the last six millennia and thereby better understand the relationship between this landscape and the cultural history of the region. An examination of the alluvial stratigraphy of the three principal stream systems in the vicinity of Metaponto—the Bradano, Basento, and Cavone Rivers—was the basis of this investigation.

Landscapes are dynamic systems, evolving in response to natural and cultural processes. If a landscape is evolving rapidly, erosion in the uplands tends to produce sediments that accumulate in the lowlands and on the coast, or erosion in the valleys cuts into and flushes older sediments downstream and, ultimately, out of the system. Conversely, if a landscape is relatively stable, little erosion or sedimentation occurs in the valleys, and pedogenic processes gradually convert the surficial sediments into soils. Although examination of the upland sources is instructive, the majority of temporal information is likely to be preserved in the lowlands, where the sediments shed from higher portions of the landscape are deposited and preserved. Consequently, the geoarchaeological research in the Metapontino concentrated on the character and archeological implications of the sequence of Holocene sediments and soils preserved in the river valleys, the landscape features most accessible to study.

Following an overview of the geomorphic setting of the study area, this chapter summarizes the record of Late Quaternary landscape change preserved in the alluvial systems. It then addresses the mechanisms responsible for driving these landscape changes and closes with a consideration of the implications of the sequence for archaeological site preservation and visibility.[4]

Study Methods

With the exception of a small number of backhoe trenches excavated on Pleistocene and Holocene surfaces near the Pantanello archaeological complex,[5] all of the investigations reported in this study were conducted without the benefit of mechanical excavation equipment. The lack of means to obtain and examine exposures at will necessarily left several important questions unresolved, but the extensive cutbanks exposed by the Metapontine streams proved to be extraordinarily useful sources of information. After thorough reconnaissance, localities that exhibited relatively well-exposed and complete sequences were selected and documented.

Sampling localities were documented in the field using established criteria.[6] Field descriptions included observations on stratigraphic architecture and characteristics of bounding surfaces between stratigraphic units, as well as descriptions of textural trends, soil structure and consistency, sedimentary structures, Munsell color, soil features, and artifacts contained in each sedimentary unit and paleosol. Because of the height of the exposures and the rapid sedimentation rate indicated, measured sections were typically sampled at the relatively coarse interval of 50 cm,

---

[1] This study was conducted between 1990 and 1993 at the invitation of the Institute of Classical Archaeology, and was supported by ICA, The University of Texas at Austin Department of Geography, and National Science Foundation Dissertation Improvement Grant SES-9206033. Accepted for publication by ICA in 2002.

[2] See, for example, Gladfelter 1977; Waters 1992.

[3] Butzer 1982.

[4] The current presentation is necessarily synthetic; the evidence is presented in detail in the author's doctoral dissertation at The University of Texas at Austin (Abbott 1997). This geoarchaeological study was conducted on a relatively restricted budget and predated the remote-sensing based investigations conducted by ICA in subsequent years.

[5] See Abbott 1998.

[6] E.g., Olson 1976; Lindholm 1987.

Figure 2.1 Sampling the alluvial sequence at the La Canala Bridge locality in the Basento watershed. The "Post-Classic paleosol" is clearly visible halfway down the visible section. (SVS)

although a few sections were sampled at 25 cm intervals. Ped samples for micromorphological analysis were taken only from intercalated paleosols. Many alluvial sections had to be accessed by rappelling down vertical cutbanks up to 30 m high (*Fig. 2.1*); sampling localities were therefore further restricted to localities where a rope could be anchored near the edge of a given field, usually to a tree or an automobile. Because the majority of Holocene alluvial surfaces in the chora are occupied by active agricultural fields, these requirements were frequently impossible to satisfy, and several important sections were therefore described only in the field and sampled for radiocarbon.

Beyond examination of the alluvial stratigraphy, extensive geomorphic reconnaissance was conducted in the uplands to facilitate characterization of the overall landscape. Soils and exposures on the slopes and uplands were examined, and representative profiles were described and sampled. In the large Metapontine catchments, this task entailed a number of reconnaissance trips penetrating up to 100 km inland from the coast. However, little detailed work was conducted more than circa 40 km inland, primarily because exposed deposits were scarce in the interior. Nevertheless, this reconnaissance was invaluable for characterizing the Metapontine stream systems as a whole.

Sediment and soil analyses were conducted by the author in the Applied Geomorphology and Geoarchaeology Laboratory in the Department of Geography, The University of Texas at Austin. Geomorphic mapping of the chora (*Fig. 2.2*) was performed using stereopair aerial photographs with subsequent ground-truthing and revision in the field. The extent of geomorphic mapping was limited to the chora proper by the availability of relevant air photos.

Geological Setting

The geology of the Metapontino has been described in detail,[7] and it is therefore summarized only briefly here. Southern Italy is situated near the boundary between the African and European crustal plates and has therefore been subjected to relatively intense

[7] See Ch. 1.

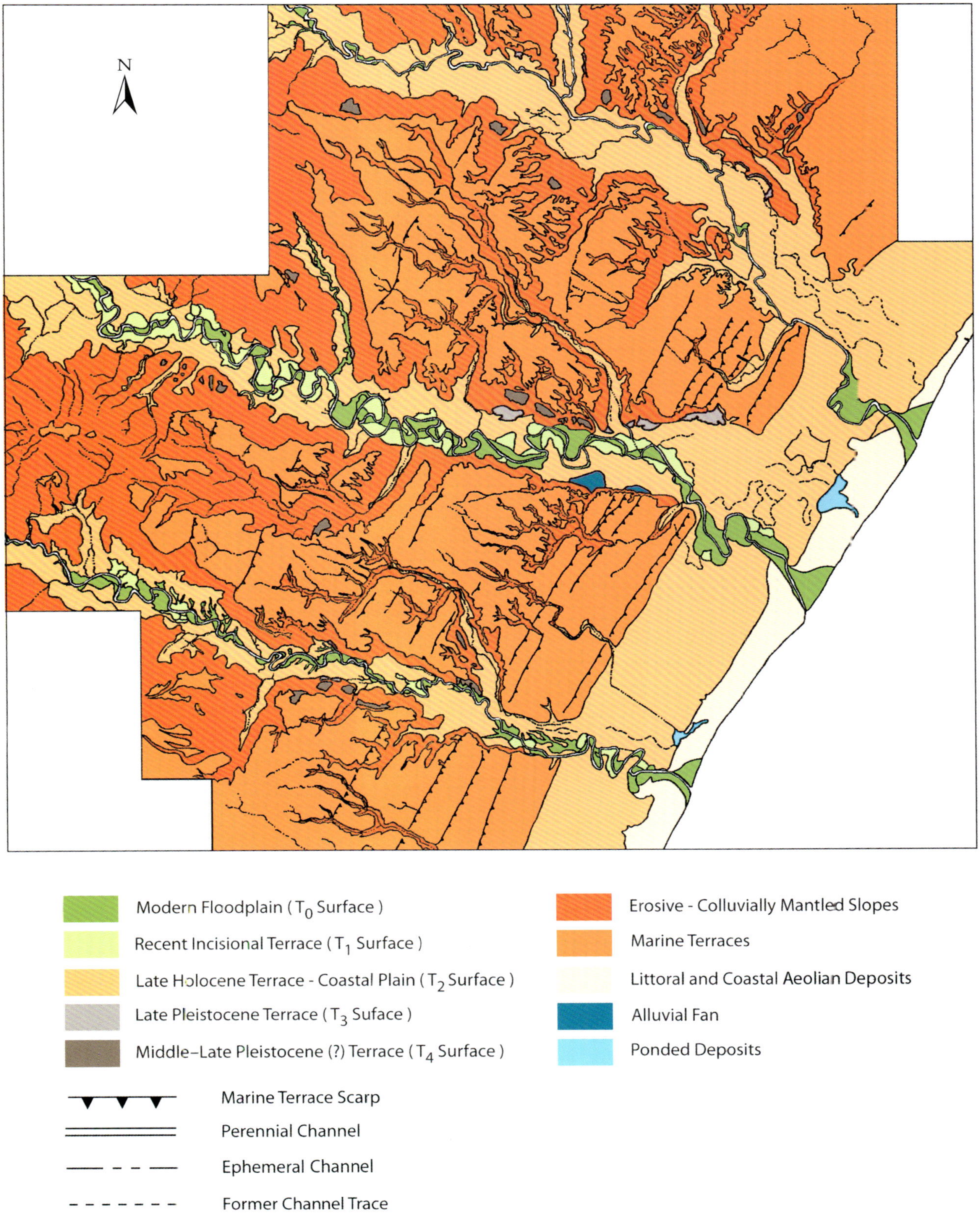

Figure 2.2 Generalized geomorphic map of the Metapontino. (JTA)

tectonic influences.[8] Beginning several million years ago, during the Late Pliocene and Early Pleistocene, a deep, linear, north–south depression in the earth's crust gradually developed behind the southern Apennines on the eastern side of the Italian peninsula. This depression, named the Bradano Trough, lies between the strongly deformed rocks of the southern Apennines and the flat-lying limestones of Apulia. It is termed a fore-arc basin in the language of structural geology. As this zone of downward folding and fracturing developed, it was invaded by the Ionian Sea and gradually filled with thousands of meters of silt and clay deposits shed from the adjacent peninsula.[9] Approximately one million years ago, these deposits began to be uplifted, causing the sea to withdraw to the southeast. As the sea gradually retreated during the Middle to Late Pleistocene, a succession of shorelines were constructed and abandoned, leaving a series of steplike, wave-cut surfaces (marine terraces). The terraces, which now rise to more than 400 m above the modern sea level because of tectonic uplift, are mantled with relatively thin deposits of beach sands and gravels. As the sea retreated, ancestors of the Bradano, Basento, Cavone, and Agri Rivers carved deep, parallel valleys across the terraces to the sea.[10]

Geological factors create an extremely high potential for erosion in the Metapontino. In most areas of the Mediterranean, the landscape is composed of hard bedrock covered by a thin veneer of soil. When fragile slopes are disrupted by vegetation clearance, agricultural activity, or animal grazing, soil erosion sets in quickly. Because the soil is thin, however, the potential for erosion is limited; once the supply of soil has been exhausted, the residual landscape is barren and rocky. Such a landscape is the bane of the farmer: crops are hard pressed to find a sufficient depth of soil in which to take root. The chora of Metaponto differs from the typical Mediterranean landscape in that its "bedrock" is unconsolidated, base-rich silty clay, often layered hundreds of meters thick. This practically infinite supply of relatively fertile soil (in an agricultural sense; true soil development is limited by erosion) at first seems an ideal situation for farming. Yet the same inexhaustible soil cover also carries an unlimited potential for erosion, a reality repeatedly experienced by the peoples inhabiting the area from the Neolithic period until today.

Generalized Geomorphology of Metaponto and Environs

The chora of Metaponto, *sensu stricto,* covers an area of approximately 500 km^2 stretching from the uplands northeast of the Bradano valley to the uplands southwest of the Cavone valley, and inland from the coast roughly 25 km to the boundary of the extant marine terraces with the lowest Lucanian mountains. It is in this coastal zone that the thick Holocene deposits of the study area are most prevalent, and this area therefore served as the primary focus of field investigations. Nevertheless, reconnaissance evaluation was conducted throughout the watersheds of the Bradano, Basento, and Cavone Rivers, and a few particularly informative inland sections were also sampled for radiocarbon and documented in relative detail.

The chora of Metaponto proper can be divided into distinct landform assemblages (*Fig. 2.3*). The first of these is the littoral zone extending along the modern coast, which is the locus of extensive deposition under the action of coastal, fluvial, and aeolian processes. The second major assemblage is the series of gently sloping marine terraces, separated by distinct scarps, which represent sculpting of the Plio-Pleistocene marine clays and deposition of littoral sediments along a sequence of successively lower paleocoastlines through the Middle to Late Pleistocene. The third major assemblage consists of active slopes; as far as 90 km inland from the coast, slopes in the Bradano valley are underlain by easily eroded Plio-Pleistocene muds, while the upper Cavone and Basento watersheds are dominated by harder rocks of Late Mesozoic and Tertiary age. In the Metapontino, the soft, readily erodible clays are exposed on the valley slopes along the margins of the marine terraces. Further inland, where the mantling marine terraces are almost completely dissected, these deposits form the primary substrate of the rugged Lucanian hill country. The fourth and last major assemblage in the chora proper consists of thick, fine-grained fluviatile deposits in the lower valleys of the Bradano, Basento, and Cavone Rivers and their respective tributaries. Thus Figure 2.3 is a typical scene in the Metapontine landscape.

[8] Caire 1975; Hsü 1977; Smith and Woodcock 1982; Embleton 1984.

[9] Valduga 1973; De Castro Coppa et al. 1979; Valentini/Cherubini/Guadagno 1979.

[10] Boenzi/Palmentola/Valduga 1976; Brückner 1982; see also Ch. 1.

Figure 2.3 View looking from a marine terrace down the Cavone valley to the coastal plain and the sea, illustrating the major landscape settings in the chora. (JTA)

The landscape that developed within the watersheds of the three principal streams changes markedly with increasing distance from the coast. The strong similarities of the Bradano, Basento, and Cavone systems in the coastal reach are replaced by diversity in their middle and upper reaches primarily because of differences in bedrock and relief. The biota of the upper drainages also differs, but all of the flora and fauna are relatively more lush than in the lowlands, reflecting the greater effective moisture received at elevation. The middle reaches of the Basento and Cavone systems consist of relatively narrow, straight valleys separated by high, thin interfluves capped with gravels and sands; significant tributaries are rare. The Bradano system, in contrast, is characterized by a number of large dendritic tributary valleys. The Plio-Pleistocene clays are still the dominant rock unit in the middle reaches of all three systems, but interbedded coarser clastic units representing submarine gravity slide deposits (turbidites) are also common, particularly in the more westerly part of the study area, and the rocks capping the upland surfaces are typically of fluvial and fluviodeltaic, rather than littoral, origin.

The steep valley slopes in the middle reaches of the streams exhibit traces of overlapping mass-movement scars at a variety of scales. They are mantled by landslide and debris flow deposits as well as thin deposits resulting from creep and surficial weathering. The smaller features are typically slides and flows, but there are also several apparent large rotational failures that form distinct, terracelike breaks in slope on the valley walls. These features, which are particularly common in the narrow middle Basento valley, are often difficult to distinguish on brief inspection from the occasional poorly preserved high fluvial strath terraces (i.e., terraces formed primarily by cutting rather than filling) inset against the valley wall. In contrast to these discontinuous remnants, there is a series of remarkably well-preserved, albeit discontinuous, broad terrace surfaces at elevations up to 200 m above the

Figure 2.4 High strath surfaces in the middle Cavone watershed. (JTA)

modern streambed in all three systems (*Fig. 2.4*). In the Cavone system, which is closest to the Apennine tectonic arc, these high terraces clearly represent a network of broad, smooth-floored dendritic tributary valleys, most lacking any defined channel at all, that were abandoned by rapid incision of the stream network.

While the coastal reaches of all three systems are dominated by narrow, deeply incised, meandering channels, the character of the modern streams changes completely 25–30 km inland from the coast. In the Basento and Cavone valleys, the streams shift abruptly to a broad, shallow, braided channel pattern, and the high Holocene terraces that dominate the lower valleys narrow and disappear as the floodplain expands (*Fig. 2.5*). The Bradano and its tributaries, in contrast, do not adopt braided patterns: at roughly the same distance inland, the Bradano has incised a very deep (20–35 m), narrow channel into the calcareous conglomerate underlying the broad valley bottom. This deep trench is confined to the modern channel; the surrounding valley is broad and open, floored with a thin (typically 2–3 m) veneer of alluvium and slope wash representing a continuation of the T_2 terrace surface.

The upper watersheds of the Basento and Cavone penetrate into harder rocks, changing the morphology of the landscape considerably. The transition is reflected in an abrupt change from broad, gravelly, braided systems to relatively straight channels flowing on bedrock or thin gravelly deposits over bedrock. In the Basento, the change occurs where the stream passes through a narrow, deep gorge cut through the "Lucanian Dolomites." This feature, named for its resemblance to the jagged topography of the Dolomite range in northern Italy, represents a roughly north–south trending outcrop of Miocene-age sedimentary rock. Inland of this constriction, the Basento watershed opens up into a moderately steep dendritic stream net draining a landscape that retains significant soil cover and is relatively densely vegetated with grasses, scattered maquis/scrub, and some trees. In the Cavone system, the stream heads in the hard Miocene clastics, where a steep, bedrock-confined channel controlled by joints and faults is developed. No appreciable accumulations of older fluvial deposits were observed in the upper reaches of the Basento or Cavone.

The vast majority of the Bradano system, in contrast, is underlain primarily by relatively soft Plio-Pleistocene clays (termed blue clays or *argille azzurre*) and contains the most heavily utilized agricultural land in the study area. Miocene-age clastics are confined to the uppermost part of the watershed. Unlike the other two systems, the Bradano and its tributaries exhibit fine-grained alluvial fills throughout. Exposure of the Holocene alluvial deposits upstream of the reservoir

Figure 2.5 Sampling of vertical exposure of Holocene fill in the upper Cavone drainage. Note the width and gravelly character of the T_0 surface and the remnant of T_2 surface in background. (JTA)

at San Giuliano is limited by a lack of deep incision. Although large areas of the upper Bradano watershed are forested, a greater proportion of the landscape is devoted to farming of the exposed, rolling substrate. Largely as a result of this agricultural use, contemporary erosion is considerable; in one field near Matera, concrete telephone poles crossing an agricultural field have been marooned on residual soil pedestals up to 1 m high (*Fig. 2.6*), implying a minimum erosion rate of 2–3 cm/yr.

The preceding summary briefly illustrates the variability of the landscape drained by the Bradano, Basento, and Cavone Rivers. The following sections describe each of the major physiographic components in more detail. The primary emphasis is on the coastal reach (that is, within approximately 35 km of the coast), where the majority of the record is preserved.

Figure 2.6 Concrete telephone pole stranded on a residual island of sediment by agriculturally stimulated slope erosion in the Bradano watershed. (JTA)

The Coastal Plain

The Metapontine coastal plain consists of a very gently sloping accumulation of sandy, silty, and gravelly sediments approximately 4 km wide, stretching between the scarp of the lowest marine terrace and the present shoreline. It is the setting of the ancient city of Metapontion and the modern towns of Metaponto Borgo and Metaponto Lido. As in many locations in the northern Mediterranean,[11] Late Holocene sedimentation on the Metapontine coastal plain has been extensive: the coastline has prograded as much

[11] Delano Smith 1978; Kraft/Rapp/Aschenbrenner 1975; Kraft/Aschenbrenner/Rapp 1977; van Andel et al. 1980.

as 1.5–2 km since the colonial Greek period, infilling and choking the ancient harbor.[12] Because of the recent origin of the parent material, soils are poorly developed. The coastal plain is underlain by relatively unweathered sands, gravels, and silty clays, with the finer material more typical of the outer plain. This assemblage is crosscut by a number of paleochannel traces, demonstrating that avulsion of the major streams has occurred repeatedly through the Late Holocene (*Fig. 2.2*).

Overall, the Metapontine coastal plain is the product of aggradation and outbuilding of the shoreline as sediment delivered to the coast by the streams is distributed by the northeastwardly flowing longshore drift. Cocco differentiates two interdigitated coastal zones along the coastline between the Sinni River and Taranto: (1) areas dominated by the major river deltas, characterized by sandy and gravelly beaches backed by low dune ridges and frequently swampy swales; and (2) areas away from the major deltas, characterized by chains of partially stabilized dunes reaching heights of up to 18 m.[13] The latter environment is particularly prominent northeast of the Bradano mouth, extending roughly from the town of Ginosa Marina to the outskirts of Taranto.

Archaeological exploration of the coastal zone, particularly between the Bradano and Basento in the vicinity of ancient Metapontion, indicates that the coastline has experienced pronounced progradation and moderate aggradation during the past several millennia as a result of dramatic increases in the volume of sediments delivered by the Bradano, Basento, Cavone, Agri, and Sinni Rivers.[14] Modifications to the rivers in the last century, particularly the construction of dams in the middle and upper reaches, have drastically diminished the supply of alluvial sediments delivered to the sea. As a result, the coastline is now experiencing active erosion, receding at rates that can exceed 10 m/yr locally.[15] Occasional high-magnitude precipitation events can, however, provoke widespread flooding and the deposition of significant volumes of sediments on the coastal plain. For example, large floods resulting from a series of heavy rains in November 1959 inundated hundreds of hectares in the coastal area and deposited several decimeters of fresh sediments across broad areas.[16] Although the full impact of such infrequent, extreme events over the last few millennia is difficult to calculate, it is telling that the ruins of the city of Metapontion, situated on the inner coastal plain, lie buried under approximately 1.5–2 m of aggraded overbank sediment. Clearly such flood events have occurred repeatedly during the Late Holocene, and they are probably responsible for inhibiting the formation of large coastal dunes near the river outlets.

Several ancient literary sources suggest that Italian coastal plains in general developed poor drainage and marshy conditions during the Late Classical, Hellenistic, and Early Republican Roman periods,[17] resulting in the proliferation of mosquitoes and a concomitant rise in the incidence of malaria among the population.[18] Geological evidence of an elevated water table during the colonial Greek period has been noted in near-surface sediments in the city[19] and at the Pantanello sanctuary,[20] and some pathogenic indications of malaria are observable in skeletons of several individuals recovered from the Pantanello necropolis, immediately inland from the coastal plain in the lowest part of the Basento valley.[21] Although Boenzi et al. attribute the rise in groundwater to climate change,[22] this study suggests that the rise in the water table late in the Greek occupation was primarily the result of cultural influences: specifically, agricultural activity that introduced massive influxes of sediments into the river systems, obstructing drainage and causing marshy conditions to develop in the lowlands along the coast. This argument is developed more fully in subsequent sections.

The Marine Terraces

Approximately 4 km inland from the modern coastline, the Metapontine landscape begins to rise in a series of stepped marine terraces, which represent the most extensive landscape assemblage in the chora proper. These subparallel terraces are cut into the Plio-Pleistocene marine clays and extend 16–20 km

[12] Boenzi/Cherubini/Giasi 1987; Kayser 1958; Caputo et al. 1991; Cocco/De Magistris/De Pippo 1978.

[13] Cocco 1975.

[14] Boenzi/Cherubini/Giasi 1987.

[15] Caputo et al. 1991; Cocco 1975; Cocco/De Magistris/De Pippo 1978.

[16] Ministero dei Lavori Pubblici 1961.

[17] E.g., Cic. *Lege agr.* 2.27.71; Caes. *B Civ.* 3.2; Strabo 5.6.1, 5.1.7.

[18] See Delano Smith 1979, 384–88.

[19] Boenzi/Cherubini/Giasi 1987.

[20] Abbott 1997.

[21] Henneberg and Henneberg 1998; Henneberg/Henneberg/Carter 1992.

[22] Boenzi/Cherubini/Giasi 1987; Boenzi et al. 1988.

Figure 2.7 Exposure of thick marine terrace gravels on the margin of the Bradano valley. (JTA)

inland from the coast. They are mantled with sands and gravels deposited along the succession of paleo-coastlines, varying in thickness from less than 2 m to more than 10 m; the thickest of these littoral deposits observed consist of stacked packets of moderately dipping discoidal gravels. The gravels underlie the lower, more complete marine terraces on the flanks of the Bradano valley, where they are well exposed in several large gravel pits (*Fig. 2.7*). Inland, the littoral deposits capping the higher terraces are typically much thinner and sand dominated (*Fig. 2.8*). Marine fossils are common in the sandier terrace mantles.

The broad, gently coastal-dipping terrace treads are separated by scarps that vary morphologically from extremely subtle linear rises to distinct topographic breaks and range in elevation from approximately 25 to 425 m above sea level. The lowest flights of terraces are well preserved, stretching with few interruptions from valley to valley, but with distance from the coast the terraces are increasingly dissected by gully systems and tributary streams, so that the highest terraces consist of small remnant platforms surrounded by extensive badland topography developed on the *argille azzurre*. The Medieval town of Pisticci occupies the entirety of one of these small inland remnant platforms. Like many other hilltop towns of Basilicata, Pisticci is beset with frequent and disastrous landslides,[23] indicating that spring-sapping and mass movements continue to actively whittle away the margins of the terrace.

Estimates of the number of marine terraces present, ranging from seven[24] to eleven,[25] are complicated by the extreme dissection of the older, higher examples. During geomorphic mapping of the study area, this author identified probable terrace scarps defining nine distinct treads (*Fig. 2.2*); several additional terraces, now completely dissected, were probably once present farther inland and at even higher elevations. Using fossil inclusions and a combination of paleomagnetic, tephrochronologic, and uranium series dates as evidence, Brückner argues that his eleven identified terraces were constructed between 775,000 and 63,000 years BC.[26] Thus the basic elements of the upland landscape appear to have been in place since at least the Late Paleolithic.

Soils on the terraces are typically deep red Alfisols developed in sandy or gravelly littoral deposits.[27] The

[23] Catenacci 1992.

[24] Boenzi/Palmentola/Valduga 1976.

[25] Brückner 1983.

[26] Ibid.

[27] USDA Soil Taxonomy system.

Figure 2.8 Exposure of the littoral cap on the margin of a marine terrace in the Cavone valley. (JTA)

persistence of such strongly developed soils on the flat or gently sloping terrace treads suggests that significant erosion from the marine terraces is localized around the headward-cutting gullies that are gradually dissecting the lower surfaces. Because of their high permeability and low slope, the terrace treads appear to be relatively resistant to erosion by sheet flow. However, cultivation of the surface is steadily degrading soil integrity, and contemporary erosion from agricultural fields is evident.

The elevation of the higher marine terraces is well above the highest sea level stands documented in the Mediterranean,[28] again indicating the likelihood of successive periods of tectonic uplift during the formation of the terraces. Similar sets of Pleistocene marine terraces have been reported at elevations ranging from 5–800 m above present sea level along many sections of the Mediterranean coast.[29] Although these terraces are often correlated with, and occasionally dated to, glacioeustatic high stands,[30] such correlations are not necessarily valid because local tectonic uplift can cause terraces to form at virtually any time. The extreme variation in the elevations of these relic shorelines clearly indicates the pervasive influence of neotectonic activity in the Mediterranean basin.

The Slopes and High Alluvial Terraces

Erosional slopes characterize the major and tributary valleys and some of the marine terrace risers, where the muds underlying the marine terraces are exposed at the surface. While many of the slopes are protected by a dense maquis vegetation, others are markedly erosional, so that a moderate- to high-density network of rills and small channels has developed, frequently achieving a true badland topography where poorly vegetated (*Fig. 2.9*). Erosion of the slopes is the combined result of sheetflow, channelized (rill) flow, subsurface sapping, and various forms of mass movement processes, including mudflow, earthflow, landslide, rotational failure, and block failure, occurring on a variety of scales. The slopes provide the environment most susceptible to erosion and therefore are responsible for the majority of locally derived sediments introduced into the drainages. The margins of the lower valleys exhibit a smooth, concave transition from the lower slopes to the principal alluvial surface, suggesting that a veneer of sheet-wash sediments commonly mantles the contact between the valley fill and the slopes and probably interfingers with stream alluvium at depth.

All three river systems are marked by relic fluvial strath terraces at elevations ranging from 40 m to more than 150 m above the level of the modern beds. These surfaces occur only along limited reaches of the valleys; they are almost entirely absent in the more constricted

[28] Vail/Mitchum/Thompson 1977; Horowitz 1979.

[29] E.g., Itzhaki 1961; Butzer 1975; Fuchs 1980; Bonfiglio 1981; Raffy et al. 1981; Ghisetti 1981; Zamani and Maroukian 1981.

[30] E.g., Malatesta and Zarlenga 1988; Dumas et al. 1988.

Figure 2.9 Badland topography developed in the Cavone watershed. (JTA)

reaches, making them very hard to correlate along the individual valleys as well as between different valleys. Soils developed on the terraces are relatively thin (often less than 1 m), although they are frequently highly rubified and contain both argillic and calcic or petrocalcic horizons. In the constricted parts of the lower and middle Basento and Cavone valleys, in particular, high terraces are poorly preserved, although beveled and colluvially overridden strath (cut bedrock) surfaces are present. Large areas of mass failure are common on the valley slopes and dissected parts of the interfluves; in several cases it was impossible to determine whether poorly defined benches on parts of the valley walls represent terrace straths or large rotational block failures. Both types of features are apparent elsewhere, however, and there are a few relatively well-preserved high terrace remnants (*Fig. 2.10*).

The readily erodible character of the local bedrock makes landslides and mudflows extremely common phenomena, exacting a heavy toll in lives and property in Basilicata.[31] Although rashes of penecontemporaneous mass movements typically occur in conjunction with large storms and the relatively frequent seismic shocks common throughout Lucania, isolated mass movements are also common because of incremental changes in slope, particularly along lithologic discontinuities where groundwater seeps lubricate shear planes and slowly undermine support. The scale of mass-movement processes varies from low-magnitude creep and minor slumps on devegetated slopes to occasional catastrophic landslides, in which up to several square kilometers of hillside can detach and slide rapidly downslope. Catenacci documents a long series of disastrous landslides and debris flows in Basilicata since World War II, including more than 100 that caused extensive property damage.[32] A number of these slides were provoked by a magnitude 6.5 earthquake on November 23, 1980, which killed more than 150 people and destroyed or heavily damaged more than 39,000 dwellings in the province of Potenza alone. Although this was a clearly catastrophic tectonic event, most of the slides and flows documented

[31] Catenacci 1992; Cavazza 1962; Guerricchio and Melidoro 1988.

[32] Catenacci 1992.

Figure 2.10 Relatively well-preserved high fluvial terraces in the Basento valley. (JTA)

by Catenacci were stimulated by intense or persistent fall, winter, and spring rainstorms that triggered dozens of discrete slope failures throughout the region. In the most extreme example, cumulative damage from a series of slides led to the condemnation of the entire hilltop town of Craco, in the Cavone watershed west of Pisticci, and the mandatory relocation of its residents to a new community in the lowlands.

The frequency and dramatic effects of rapid mass movements overshadow the equally important—and equally pervasive—rill and sheet erosion of the clay hillslopes: typical erosion rates on sparsely vegetated slopes approach 30 mm/yr.[33] This second type of erosion rapidly leads to badland development, a phenomenon particularly pronounced in the area 15–30 km from the coast, where dissection and lateral erosion of the older marine terraces have exposed extensive expanses of the underlying clays. Despite this exposure, lateral dissection of the terraces has been a long and slow process, and grass and/or maquis vegetation is well established on the clays in many areas. Badland development, however, can be quickly initiated when a slope is cleared or burned, or when mass movements expose unvegetated sediments. Once initiated, these erosional patches frequently destabilize larger areas by stimulating lateral rill erosion and additional mass movements, causing the badland to grow at the expense of adjacent vegetated slopes.

The character of the extensive badlands in the Metapontino is relatively complex. Alexander identifies two distinct, interdigitated types of badland development in Basilicata,[34] both of which differ in important details from classic badlands in other parts of the world:[35]

1. Calanchi badlands are characterized by extensive dendritic rill and gully development and are generally characteristic of the steeper slopes.
2. Biancane deposits, in contrast, are characterized by hummocks and cones of residual primary deposits separated by relatively smooth wash surfaces covered with very shallow, reticulated rills; they may occur both on relatively steep slopes and in areas with much gentler gradients.

Both types of surface are commonly underlain by complex networks of reticulate cracks and pipes whose form is determined primarily by patterns of desiccation cracking in the shallow subsurface.

[33] Rendell 1982, 1986; Alexander 1982b.

[34] Alexander 1982b.

[35] E.g., Schumm 1956; Selby 1993.

Figure 2.11 View across T_1 surface in the Cavone valley at Feroleto, illustrating thick sequence underlying the T_2 terrace in the lower valleys. (JTA)

Late Quaternary Alluvial Systems

The fourth major landform assemblage in the study area consists of stream deposits of varying ages. The Bradano Trough is drained by five rivers that are quite large by south Italian standards. These streams—(from southwest to northeast) the Sinni, Agri, Cavone, Basento, and Bradano—converge from relatively large watersheds in the interior, traverse roughly parallel valleys cut through the marine terraces along the trend of the trough, and drain into the Gulf of Taranto along a 30-km stretch of the coast. Thick Late Holocene alluvium is present in the valleys of these streams and in many of the small-scale tributaries that cut into and drain the terrace surfaces (*Fig. 2.11*). Only the three northeastern streams (Cavone, Basento, and Bradano) cross the chora of Metaponto; the Agri and Sinni cross the chora of ancient Herakleia farther south and are therefore not addressed in this study.

Remnants of at least five, and possibly as many as seven, distinct alluvial surfaces are detectable in the coastal reach of the three Metapontine streams. The principal alluvial surface within 40 km of the coast is the broad T_2 alluvial terrace. Rising from approximately 4–7 m above the modern Basento channel at the outer limit of the marine terraces (4 km inland) to 16–20 m above the channel in the vicinity of Pisticci Scalo (25 km inland), this terrace merges imperceptibly into the coastal plain downstream and typically grades smoothly into the lower slopes on the valley margins. Two inset alluvial surfaces, designated T_0 and T_1, are present in the relatively narrow channel trench cut through this broad surface. The discontinuous T_1 surface is incised 2–6 m below the T_2 surface and occupies abandoned meander scrolls cut during the last phase of stream incision. The continuous modern floodplain (T_0) is inset 2–5 m below the T_1 surface. In many cases, the T_1 terrace clearly represents an alluvial strath and is underlain by relatively thin (typically 1–3 m) sandy or silty sediments resting on a bench cut as the river incised older Holocene alluvium. The active T_0 surface, which floods regularly but is still utilized for agriculture in many parts of the lower valleys, is distinctly sandy and exhibits an irregular scrolled surface except where it has been plowed.

The Metapontine streams share a similar hydrologic pattern. Rainfall in southern Italy is relatively rare but frequently occurs in the form of intense storms between October and April. Discharge varies considerably throughout the year and on an annual basis, reflecting the influence of occasional heavy rains during autumn/winter and, to a lesser extent, spring. In general, peak discharge during the low-sun season is between 10 and 20 times the monthly mean, an indication of the relatively short duration of winter storms. In contrast, peak discharge during the summer months is typically only 1 to 2 times the mean. The largest floods on record occurred in November 1959,[36] the result of a very intense series of storms; peak discharge was 1,930 m^3/sec at the Tavole Palatine gauging station on the lower Bradano. This very high flow overwhelmed the Bradano valley system and coastal plain, submerging much of the lower valley under more than 1 m of water.

Because the landscape frequently receives no rain for months at a time, and because vast expanses have been cleared for agriculture, the occasional intense rains fall on a highly erodible substrate that has often been stripped of all protective cover. Consequently, the majority of geomorphic work happens relatively rapidly, and significant erosion can occur in a brief span of time.

Alluvial Stratigraphy

Previous research

The region surrounding Metaponto has been the subject of many studies of Late Quaternary problems. Some studies have focused on the coastal plain and/or marine terraces, the character and erosion of the clay substrate, or the generalized morphology of the landscape.[37] Others have examined aspects of the alluvial sequences in the vicinity of the Metapontino. One of the earliest and most perfunctory descriptions of the alluvium in the region was presented by Kayser, who recognized the Holocene age of the thick valley fills but did not address the alluvial sequence in any detail.[38] High terrace surfaces were first addressed by Ricchetti, who described a series of four undated, tectonically produced alluvial terrace surfaces in the middle valley of the Bradano and its tributaries.[39] Several investigators, including Vita-Finzi,[40] Boenzi et al.,[41] Neboit,[42] and Brückner,[43] have previously described the thick alluvial deposits in the study area in varying levels of detail. While Neboit was the first to explicitly link alluvial deposits in the Metapontino to soil erosion during Classical antiquity, the most comprehensive previous work on this aspect of the record is that of Brückner, who describes a sequence of four Holocene units in the region based on sections documented in the Bradano and Cavone drainages. As in the present study, dating was obtained through critical examination of the available suite of ages, and many anomalous humate assays were disregarded.

Brückner's basal unit (*sediment 1*) is approximately 3–6 m thick and grades up from basal gravelly sands into loams capped by a thick, very dark grey-brown cumulic paleosol (*paleosol A*). An age of 7250 BC was obtained from a snail shell contained in the fill, while ages between 6100 and 5700 BC were obtained from the soil. Brückner observed that a second cluster of ages, ranging between 3910 and 3350 BC, were also obtained by Neboit from deposits stratigraphically consistent with the sediment 1/paleosol A complex. Although he cites a lack of topographic continuity as a source of uncertainty, Brückner suggests that his sediment 1 may subsume two different stratigraphic units.

Above paleosol A, sediment 2 consists of 6–8 m of poorly bedded, silty clays containing muddy interbeds (which Brückner terms *paleosol-sediments*) and Greek- and Roman-era cultural materials. It is topped by a second, more weakly developed paleosol (*paleosol B*) dated AD 72–400 (calibrated). Brückner draws a strong contrast between the graded character of sediment 1 and the relatively uniform laminated silts and sands of sediment 2; he suggests that the former represents a "natural" deposit formed by climatic stimuli, while the latter represents a deposit resulting from pronounced anthropic disturbance of the ancient landscape. He also equates the paleosol capping the unit with a decrease in anthropic disturbance during the Late Roman period.

Brückner's sediment 3 is typically 1.5–3.5 m thick, occasionally up to 9 m thick. It is described as

[36] Ministero dei Lavori Pubblici 1961.

[37] E.g., Boenzi/Palmentola/Valduga 1976; Brückner 1982.

[38] Kayser 1958.

[39] Ricchetti 1967.

[40] Vita-Finzi 1975.

[41] Boenzi et al. 1989.

[42] Neboit 1977, 1983.

[43] Brückner 1980, 1983, 1986, 1990.

similar to sediment 2 and tentatively dated to the Medieval period.[44] Once again, Brückner attributes the increase in sediment yield to anthropic disturbance. This phase of alluviation was terminated by trenching of the stream sometime between the 11th and 15th centuries AD, after which time the fine-grained deposits of an inset terrace (*sediment 4*) accreted within the confines of the channel trench.

Stratigraphic Architecture

The present study defines a stratigraphic sequence that is broadly similar to the sequences described by Brückner[45] and Neboit[46] but differs from them in a number of important details. The sequence is based on the examination of a suite of 27 new radiocarbon ages (*Table 2.1*) and the ages obtained previously by Neboit[47] and Brückner[48] (*Graph 2.1*). In their lowest 25–35 km, all three of the principal valleys are infilled with thick, fine-grained accumulations of alluvium containing intercalated paleosols and stratified archaeological materials exposed by a low-gradient incised stream. Although some differences are apparent in the character, thickness, and geometry of the respective valley fills, broad similarities exist in the stratigraphy and alluvial architecture of the lower reaches of the three streams, suggesting that similar depositional controls affected each valley.

Nine informal allostratigraphic units, designated *unit A* through *unit I*, were identified in the chora of Metaponto during this study. These units were differentiated and correlated using a number of criteria, including stratigraphic position, granulometry, color, sedimentary structures and facies relationships within sediment bodies, soil morphology, radiocarbon data, and cultural inclusions. As used here, the term *allostratigraphic unit* refers to a regionally correlatable, three-dimensional body of sediments defined by high-order bounding surfaces (i.e., unconformities, including major erosional surfaces and cumulic soils). In this context, a given allostratigraphic unit represents the sediments that accreted in the valley systems during a major cycle of aggradation and may be composed of a wide variety of depositional facies. Although identification of such units in the stacked Holocene sequence is relatively straightforward, it is much more difficult in the discontinuous Pleistocene deposits inset into the slopes, because clear bounding relationships between the majority of spatially discrete strath terraces do not exist. Discussion will focus on units D through I, which form the contiguous valley fill. These Holocene units are relatively continuous, typically demarcated by well-defined bounding surfaces, and can be correlated along the valley systems and between valleys with a high degree of confidence. The remaining older Pleistocene units (A–C) will be addressed only in passing.

Figure 2.12 presents schematic composite cross-sections of the architectural relationships between the allostratigraphic units in the coastal reach of the Bradano, Basento, and Cavone Rivers. Units D–G, which span the Early/Middle Holocene through approximately the Medieval period, underlie the T_2 terrace in each valley. As the figure indicates, these fills represent a sequence of roughly tabular units, stacked in layers within the valley confines and separated by cumulic alluvial paleosols. Axial channel complexes associated with units F and G were also identified at several locations, as were a number of smaller tributary gully fills associated with the same units. In each case, the channels and gullies are relatively narrow and deep, suggesting that they incised and then filled rapidly with minimal lateral channel migration. Units H and I represent relatively restricted deposits occupying the modern channel trench, where they form the T_1 terrace and modern floodplain (T_0) and channel, respectively. These relatively recent deposits were not examined in detail.

Upstream, the three systems exhibit markedly divergent characteristics, so that correlation of deposits between valleys becomes much more problematic. Figure 2.13 represents schematic cross sections of the Cavone/Salandrella system near the upper Cavone section (28 km inland), the middle Basento system near the Grottole section (46 km inland), and the Fiumello Trove, a large tributary that enters the Bradano from the west some 67 km inland. These three cross sections illustrate the diversity of the inland deposits, but by no means represent the full range of architectural variability.

[44] Brückner 1986.
[45] Brückner 1982, 1983, 1986, 1990.
[46] Neboit 1977, 1983.
[47] Neboit 1983.
[48] Brückner 1986.

Locality	Early–Middle Holocene soil	Bronze Age soil	Post-Classic soil	Medieval fill–modern soil
Bradano valley system	6370 ±70 BP	4110 ± 80 BP	1540 ± 40 BP	—
Basento valley system	5990 ± 70 BP* 5990 ± 70 BP 6210 ± 80 BP	3820 ± 50 BP 4070 ± 100 BP 3140 ± 50 BP 3580 ± 50 BP	3760 ± 60 BP** 2050 ± 80 BP 1730 ± 60 BP* 2340 ± 40 BP 2610 ± 60 BP** 3130 ± 50 BP**	280 ± 50 BP 3630 ± 60 BP**
Cavone valley system	< ? 7790 ± 80 BP*** ? > 8650 ± 110 BP*** 8520 ± 90 BP***	4450 ± 70 BP	3480 ± 60 BP** 2040 ± 60 BP 1580 ± 50 BP	1840 ± 70 BP*** 5220 ± 70 BP**

*No $\partial^{13}C$ correction
** Age is too old due to incorportation of allogenic organic matter
*** Age is probably too old due to incorporation of allogenic matter

Table 2.1 Summary of radiocarbon ages obtained during this study, arranged by stratigraphic unit and drainage.

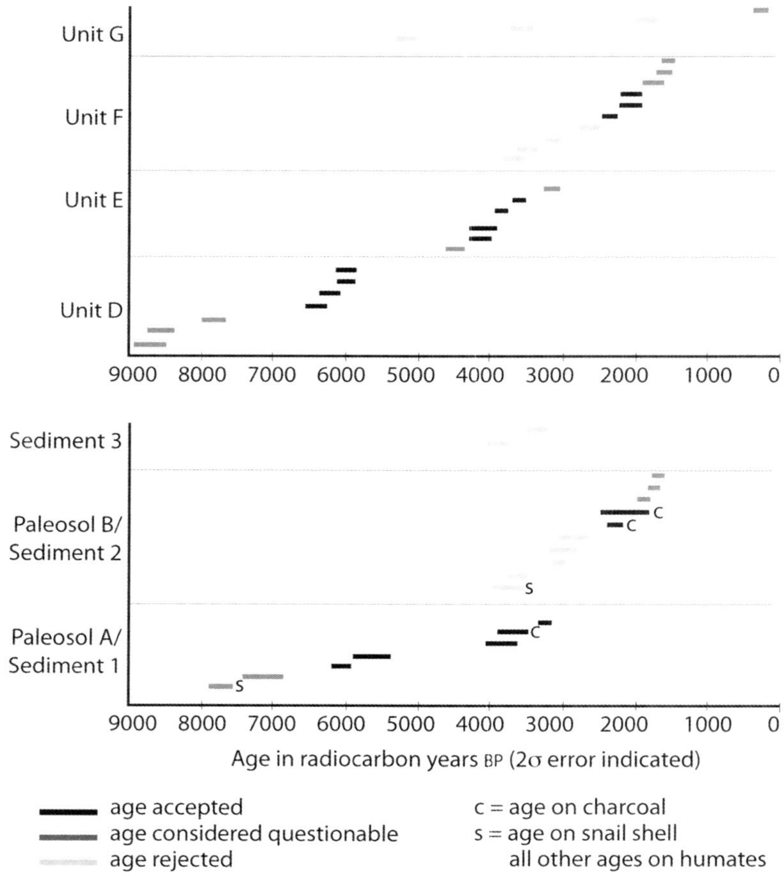

Graph 2.1 Summary of radiocarbon ages obtained from Metapontine alluvium during this study and previous studies.

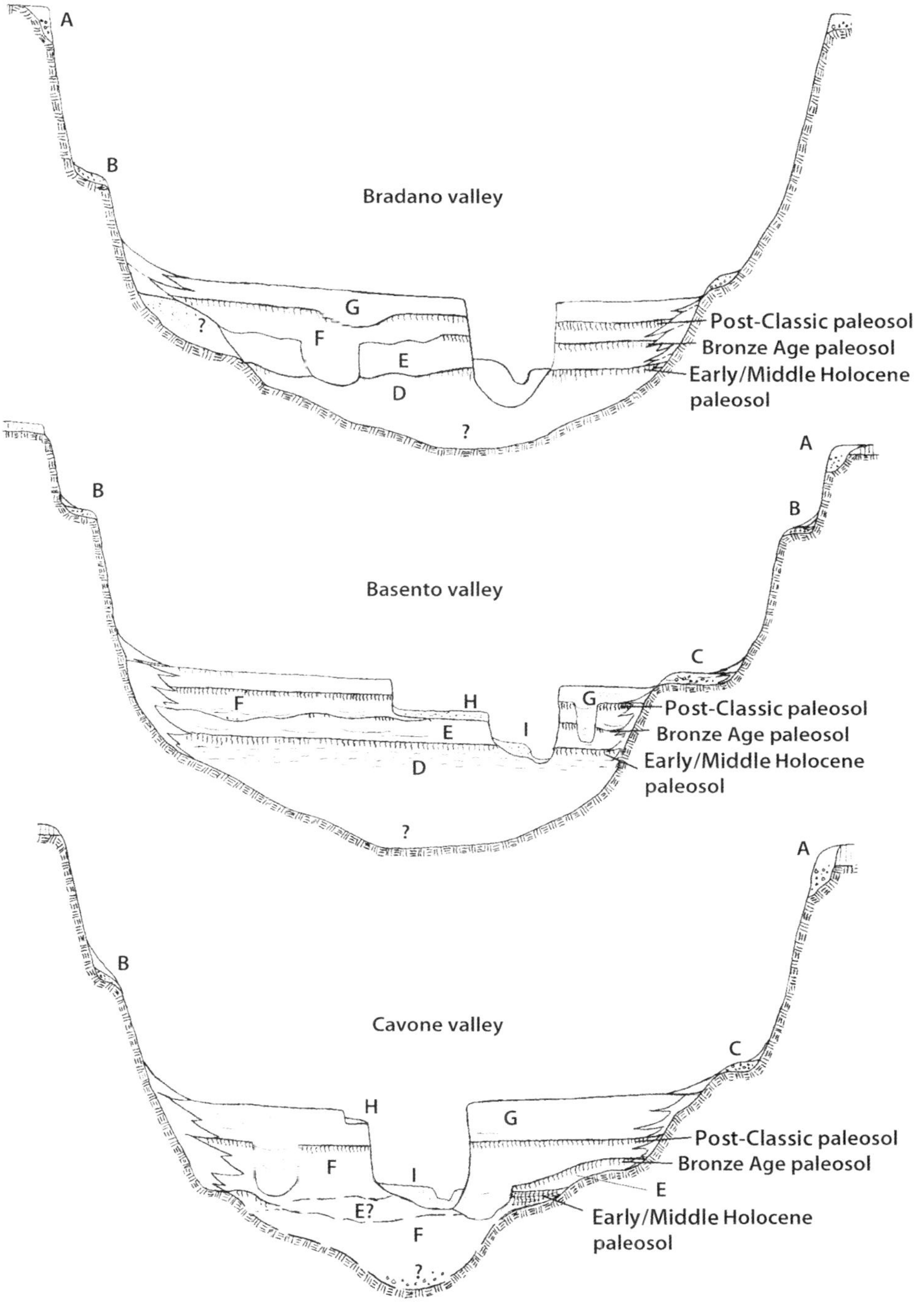

Figure 2.12 Generalized schematic cross sections of the three principal Metapontine streams in the coastal zone. (JTA)

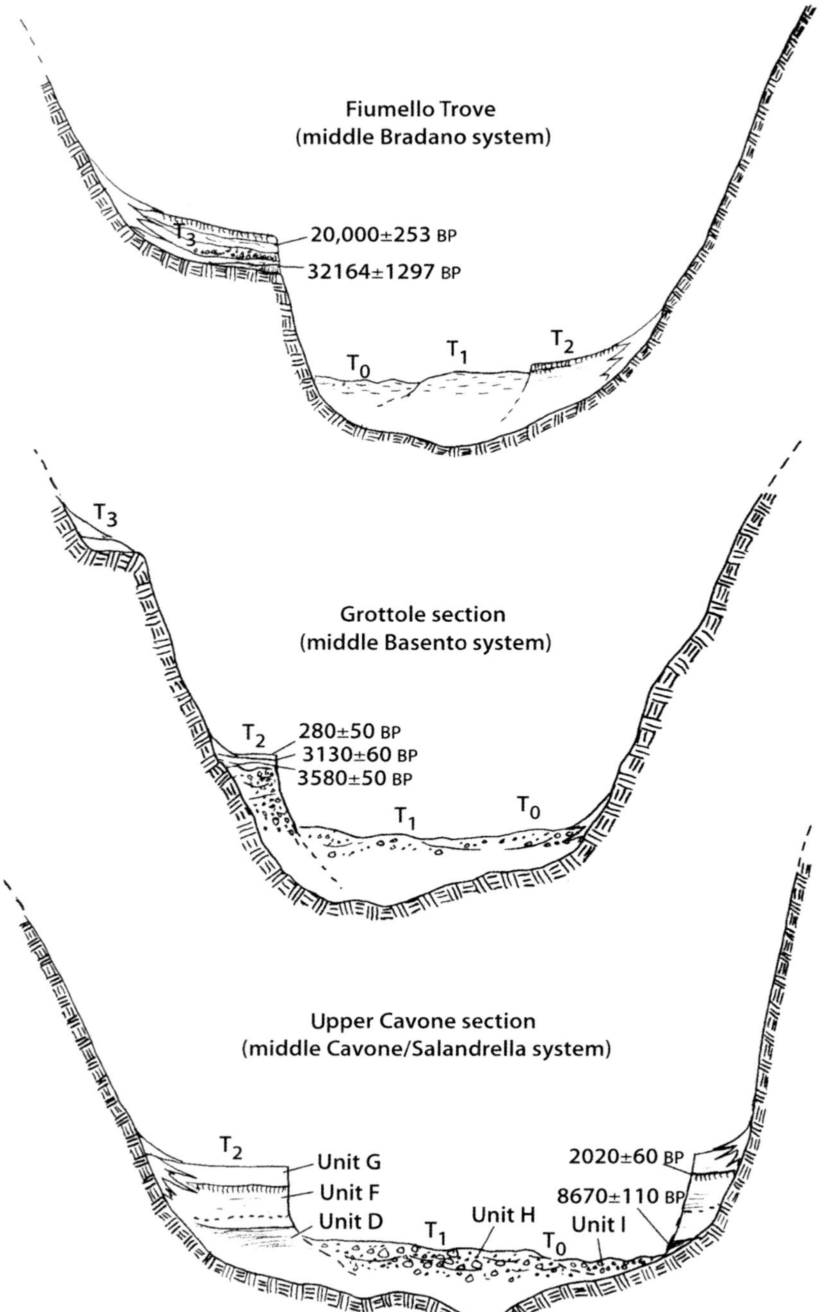

Figure 2.13 Generalized schematic cross sections of the middle reaches of the three principal Metapontine streams. (JTA)

Sedimentology

In general terms, the size and uniformity of grains transported and deposited by a sedimentary system reflect the energy characteristics of the system: rapid, energetic flow results in the transport of larger grains (clasts), while sustained energy conditions result in more uniformly sized (better-sorted) deposits. That said, the sedimentology of the Holocene alluvial deposits in the Metapontino is relatively complex. The grain size of Metapontine alluvial deposits generally indicates more about the types of sediments available in the system than about flow competence. Gravels in particular appear to have been relatively unavailable during most of the Holocene. The major exceptions to this generalization are found in areas where gravelly Pleistocene alluvial terraces on the valley margins were degraded, resulting in the introduction of minor amounts of gravel into the Holocene system. For the most part, both channel and overbank environments are characterized by sediments ranging from silts and silty clays to fine loamy sands. The preserved Pleistocene units, in contrast, are distinctly gravelly, reflecting more effective mobilization and transport of materials from areas in the upper drainages underlain by harder rocks.

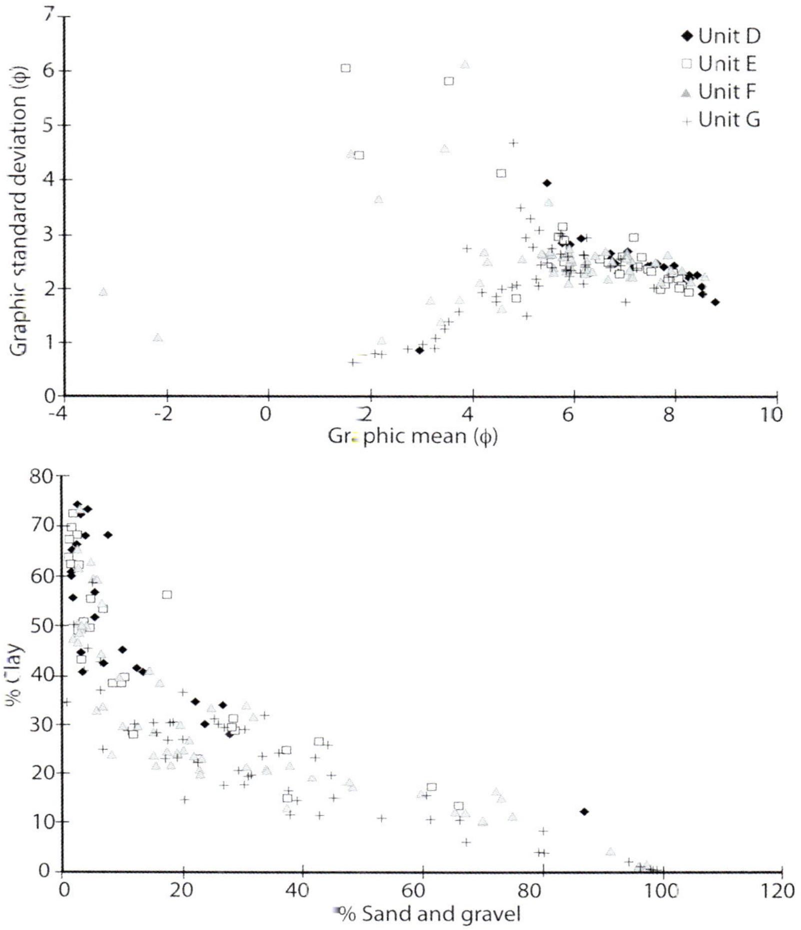

Graph 2.2 Textural characteristics of samples from deposits underlying the T_2 terraces.

Graph 2.2 illustrates the textural (grain size) characteristics of the four units underlying the T_2 terrace. As both plots in this figure indicate, a very high degree of textural overlap exists between the units, although unit D has a tendency to be relatively finer, and unit G to be slightly coarser, than the overall mean. The upper plot is a scatterplot of mean grain size vs. graphic standard deviation determined for approximately 160 samples from units D–G: one dominant and two subordinate clusters of points are apparent. The dominant cluster, in which 75–80% of the samples fall, exhibits graphic means of approximately 5–9ϕ and graphic standard deviations of 1.75–3.00ϕ. These samples represent the relatively poorly sorted silty loams and clays that make up the majority of the valley fill. One of the subordinate clusters consists of a series of quasi-linear points that slope down and away from the principal cluster. This group of samples, which are characterized by means of approximately 2–5ϕ and standard deviations of approximately 0.5–2.0ϕ, represents relatively well-sorted medium or fine sands deposited in paleochannels and thin, tabular beds interstratified in overbank deposits; more than half represent a deep channel fill associated with unit G at a single sample locality. The other subordinate "cluster" is a widely spaced cloud of points representing samples that are typically coarser and more poorly sorted than the majority of samples. These points represent silty clay deposits containing thin gravel stringers and dispersed, matrix-supported gravels derived from an adjacent Pleistocene alluvial terrace.

Despite their textural similarity, the four units underlying the T_2 terrace demonstrate several important differences, including varying facies assemblages

Locality	Late Pleistocene (unit C)	Middle Holocene (unit D)	Bronze Age fill (unit E)	Classical fill (unit F)	Medieval fill (unit G)
Pantanello necropolis	**Gm, Gp, St, Sp, Sh, Ss**	—	—	(gully fill) **Sr, Sh, Fl, Slm**; (colluvium) **Sh; Slm**	(poorly defined, minor colluvial unit) **Sh, Slm**
Pantanello sanctuary	**Gm, Gp, Sh**	—	—	(sheet wash/colluvium) **Fgs** (Fl); (alluvium) **Fm, Fgs;** *Fsc*	(sheet wash, alluvium) **Fgs,** *Fl*
Duck's Head no. 1 and no. 2	—	**Fss, Fsc,** *Fm*	**Fsc, Fm,** *Fl, Sp*	(alluvium) **Fsc;** (gully fill) **Sp, Sr, Ss**	**Fm, Fsm,** *Fsc*
La Canala confluence	**Gm**	**Fss**	**Fm**	**Fm**	**Fss, Fl, Sh**
La Canala Bridge	—	**Fsc, Fm,** *Sl*	**Fm, Fl,** *St, Sp, Sr*	**Fl, Fsc, Fss,** *St, Sp*	**Fl, Fss,** *Fm*
Basento factory locality	—	—	**Sp, Ss, Fl, Fgs**	**Fsc, Fl,** *Fm*	**Fsl, Sh**
Grottole locality	**Gm, Gp,** *St, Sp*	—	—	**Fsc, Fm**	**Fsc, Fm**
Lower Bradano	**Sh?**	—	**Sh; Fm?**	**Sp, Sr,** *Fgs*	**Fsc, Sl, Fm**
Bradano no. 1	—	**Fsc**	**Fsc, Fm,** thin interbeds of *Gm*	**Fsc, Gm, Fm**	**Fss**
Bradano Montescaglioso	—	—	**Fm**	**Slm, Fm**	**Sp, Sr, Sh, Fsc, Fm**
Calle Strada locality	**Fm, Sp, Ss, Gm, Fl, Fsc**	—	—	—	—
Cavone Tubazione	—	**Fm, Sl;** age unclear		**Sp, St, Sr, Sl,** *Fm*	**Sp, St, Sr, Sl, Fm**
Cavone no. 1 (Feroleto)	—	**Fsc, Sp, Sr, St,** *Gp, Gm, Fgs;* age unclear		**Fgs, Fss, Fm**	**Fss, Fsc, Fm, Fgs,** *Sl*
Upper Cavone	—	**Fss**	Presence not established	**Sp, St, Sr, Sl, Fss, Fsc, Fm**	**Sp, Sr, Sl, Fsc**
Pisticci Bridge no. 1	—	**Fm,** *Sp*	**Fm**	**Fm, Fsc, Fss, Fl, Sp,** Sr	**Fss,** *Fsc, Fm*
Pisticci Bridge no. 2	—	**Slm**	**Sr, Sp, Fl, Fsc, Fm**	**Gm, Fgs, Fm**	**Sh, Sl, Fsc, Fss, Fm**

bold type = dominant facies
italic type = secondary facies

Table 2.2 (above and opposite) Sedimentary character of each unit at the principal witness sections described during this study. Key on facing page. (See Abbott 1997 for locality, locations, and descriptions.)

Facies code	Lithofacies	Sedimentary structures and characteristics	Interpretation
Gm	Massive or crudely bedded gravel; may contain sand in interstices	Weak horizontal bedding; imbrication; abrupt basal contact	Longitudinal bar or channel lag; braided system
Gp	Gravel, planar cross beds	Clear horizontal bedding; imbrication; abrupt basal contact	Longitudinal bar; braided system
St	Trough-cross-bedded, medium to coarse sand or pebbly sand	Solitary or grouped trough cross beds; frequently associated with ripple bedding; abrupt basal contact	Dunes; associated with secondary channels in braided system and tributary and gully fills in meandering system
Sp	Planar-cross-bedded, medium to coarse sand or pebbly sand	Solitary or grouped planar cross beds; frequently associated with ripple bedding; may exhibit broad (3 m+) concave-up morphology; abrupt basal contact	Bars; sand waves; associated with braided systems and deep major channel fills in meandering system
Sr	Ripple-bedded, fine to very coarse sand	Ripple bedding, particularly climbing ripples; typically transitional basal contact	Associated with channel fills, splays, and channel-proximal overbank deposits; climbing ripples indicate very high sediment concentration
Sh	Horizontally laminated very fine to very coarse sand; may contain pebbles	Horizontal lamination; abrupt to transitional basal contact	Planar flow in channels and overbank environments in the meandering system; may be associated with energetic sheet wash from slopes and high terraces
Sl	Fine sand; low-angle cross beds	Low-angle planar cross beds; typically abrupt basal contact	Scour fills, splays in meandering system
Ss	Fine to coarse sand or pebbly sand infilling broad scours	Occasional ripples and cross beds; common scour surfaces; sub-horizontal gravel stringers; typically abrupt basal contact	Scour fills in braided system and gullies feeding into meandering system
Slm	Massive fine sandy loam	None; typically transitional basal contact	Overbank deposition; lack of structure probably due to gradual bioturbation of slowly aggrading floodplain
Fss	Strongly laminated silt and/or mud	Lamination; parts readily along bedding planes	Ponding on floodplain; fluctuating overbank flow; sheet-wash deposition; very rapid aggradation
Fsc	Weakly laminated to massive silt and/or mud	Visible textural or color lamination; does not part readily along bedding planes	Ponding on floodplain; overbank flow; sheet wash; pedogenic or biogenic turbation possible
Fm	Massive mud	None; typically transitional basal contact	Channel plug; ponding on floodplain; pedogenic turbation
Fgs	Massive, gravelly/sandy mud; coarse clasts matrix supported and/or in thin stringers	Occasional ripples and cross beds; common scour surfaces; dipping gravel stringers	Sheetwash; minor debris flow; bioturbation of stratified deposits
Fl	Sand, silt, and/or mud with fine lamination or very small ripples	Lamination and/or faint fine ripple bedding	Overbank flow; waning channel flow; sheet-wash deposition

Key for Table 2.2 (opposite).

and diagenetic traits imparted by groundwater. To facilitate description, the alluvial deposits in the chora were subdivided into 14 different facies using a system adapted from Miall,[49] but the facies types were modified to better represent the variability in the Metapontine systems. Table 2.2 illustrates the character and facies assemblages recorded at witness sections during the study. The characteristics reflect the combined effects of depositional conditions and post-depositional alterations, tempered by sediment supply.

Stratigraphic Sequence

Unit D

Unit D is the oldest Holocene unit exposed beneath the T_2 terrace tread. It is deeply buried in the valleys and typically does not crop out except on the outside of meander bends or in other positions in which circumstances have prevented the lowest part of the section from being covered by colluvium, bank collapse deposits, or inset alluvium. The upper, exposed part of unit D typically consists of thick accumulations of light grey to very dark grey-brown, laminated or massive organic silty clays; the clays often contain abundant iron-stained casts of interbedded vegetal matter, thick iron-stained root casts, small manganese nodules, and very intense orange and black redox mottling. This mud represents a poorly developed cumulic paleosol informally termed the "Early/Middle Holocene soil," corresponding to the bounding surface demarcating the top of the unit.

Only the uppermost few meters of unit D are typically exposed above the stream bed; the total thickness of the fill is therefore unknown. In a few places, fine laminated sandy or silty deposits are exposed beneath, or interbedded in, the muddy soil. In one locality, light yellow-brown laminated silts are present beneath approximately 2 m of the laminated unit D muds, suggesting that a more complete Late Pleistocene/Early Holocene sequence is present beneath the modern depth of exposure. Because the lower valleys would have been more deeply entrenched than at present in response to the substantially lower Late Pleistocene sea level, deposits dating to the Late Pleistocene full glacial period are probably present below the modern depth of incision in all three valleys.

The upper part of unit D reflects a period characterized by restricted drainage, in which ponding and paludal sedimentation dominated the lower valley systems. However, interbedded sandy deposits indicate that the valleys occasionally witnessed sustained flows that distributed coarser sediments. This marshy episode dates to the Early/Middle Holocene, terminating approximately 6,000 years ago, and probably relates to the last stages of rapid sea-level rise in the Mediterranean. Although the unit is identified in, and correlated among, all three valleys, the correlation of deposits in the Cavone drainage is tentative. In the Bradano and Basento valleys, the deposits consist of 1–2 m of greyish laminated or massive silty clays at the base of the exposed sections. Radiocarbon assays from these sections yielded four ages covering the range 5990–6370 BC (*Table 2.1*). The tentatively correlated unit D deposits in the Cavone valley, in contrast, show greater morphologic variability and date several thousand years earlier. Although these results may indicate somewhat earlier initiation and termination of paludal conditions in the Cavone in relation to the Bradano and Basento, it is also possible that the early ages are an anomaly introduced by the incorporation in the drainage of redeposited organic matter derived from older eroded soils. The Cavone system is situated closest to the Apennine front and exhibits the most pronounced evidence of tectonic uplift; the higher level of tectonism may have served to stimulate more intense soil erosion here than in the systems to the east, adding a more pervasive component of detrital organics during the Early/Middle Holocene.

Unit E

Unit E consists of a complex series of silty to sandy deposits, predominantly yellow-brown to pale brown, that are 1–5 m in thickness and grade upwards from cross-bedded to massive. As in unit D, they exhibit common evidence of intermittent saturation in the form of red-brown iron staining and mottling and small nodular accumulations of manganese. The deposits are sometimes capped by a relatively weak, brown to grey-brown cumulic paleosol, while in other cases the paleosol was erosively truncated prior to renewed aggradation. Six radiocarbon ages in the range 3140–4450 BC were obtained from this soil. Half of these ages fall into a 300-year span between 3820 and 4110 BC, while two are several hundred years younger, and one is several hundred years older (*Table 2.1*). Although a discrete fill event was not identified, additional corroborating ages of 3350, 3760, and 391

[49] Miall 1978, 1985.

BC were obtained previously by Neboit and Brückner from deposits identified by Brückner as sediment 1 and paleosol A. Although Brückner tentatively interpreted these three ages as anomalously young assays of an older soil, he did acknowledge the possibility that they might indicate the existence of two temporally discrete pulses of sediment in his sediment 1. That hypothesis is supported by the results of this study.

Unit E represents a period of relatively intense clastic sedimentation in the lower valley systems beginning sometime after approximately 6000 BC and ending ca. 4500–3500 BC. It therefore corresponds to the Eneolithic through Early Bronze Age cultural periods and is accordingly termed "the Bronze Age fill." The unit is typically graded, fining upward from relatively sandy, cross-bedded basal deposits into massive or weakly laminated silts and silty loams higher in the section. Where preserved, the sequence is relatively consistent across the valleys; it appears to represent a transition from a sandy, bed-load dominated system characterized by one or more wide, shallow channels to a system carrying predominantly suspended loads in a deeper, more restricted channel. Occasional beds of laminated or ripple-bedded silty sands occur throughout the column, indicating brief periods of higher energy flow on the aggrading floodplain surface, but massive deposits probably representing suspended load sedimentation on a vegetated floodplain are much more common. Gravels are present only where there is a local source in the form of a degrading Pleistocene marine terrace. Thin, localized muds resembling weak paleosols are also present in a number of sections but cannot be correlated between different exposures; they appear to be the product of localized ephemeral ponding on the aggrading floodplain and suggest that the system frequently developed a scrolled topography, facilitating the formation of short-lived marshy environments where muds and organics were concentrated. No evidence of significant lateral accretion deposits was detected, and it is unlikely that the system saw pronounced meandering of the channel, and therefore extensive cannibalization of its own deposits, during the latter stages of accretion

The paleosol at the top of unit E, termed "the Bronze Age soil," is frequently absent because of widespread scouring of the surface during the initial stages of the subsequent phase of aggradation. For this reason, and because unit E is very similar in character to the subsequent unit, it is often difficult to discriminate between the unit E fill and the subsequent unit F fill in individual sections. This situation is particularly applicable in the Cavone valley, where unit E can be positively identified at only one documented locality but is suspected to be present in several other sections. Whether these sections actually contain unit E sediments or not, subsequent scour has clearly removed the Bronze Age soil and portions of the underlying fill from most exposures in the Cavone system. Preservation of the soil is more extensive in the Bradano and Basento systems but is nevertheless still far from ubiquitous.

Unit F

Unit F consists of another massive series of sediment pulses introduced into the valley systems between approximately 3000 and 1800 BC. Unit F varies from circa 2–8 m in thickness and is capped by a relatively thin, muddy paleosol that exhibits different morphologies in various parts of the valley systems. The fill is very similar to unit E in character. In many cases, it grades up from yellow-brown to grey-brown cross-bedded and ripple-bedded sandy and silty loams at the base, to yellow-brown to pale brown massive or laminated silts, silt loams, and silty clays at the top; other sections are composed entirely of massive to laminated silty loams and silty clay loams. As in the underlying units, iron and manganese staining and red-brown mottling are common, suggesting that the fill was intermittently saturated by a fluctuating water table. This sediment pulse is roughly coincident with the Greek and Early Roman occupations of the Metapontino and is thus termed "the Classical fill." The soil capping the unit appears to represent a period of relative stability beginning about 1800–2000 BC and lasting for several hundred years. This "Post-Classic paleosol" is almost always present in sections in all three systems and is the most consistent and useful stratigraphic marker in the sequence.

At first glance, radiocarbon ages from the unit appear somewhat ambiguous. Humate samples collected from the paleosol during the current study yielded 10 radiocarbon ages between 3760 and 1540 BC (*Table 2.1*). A number of the older radiocarbon ages, however, are stratigraphically reversed with other dates and therefore demonstrably too early. Moreover, the consistent presence of Greek and Roman cultural material in the upper 1–2 m of the fill demonstrates that the pre-Classical ages are anomalies resulting from the incorporation of allochthonous organic matter. Accepted ages from the soil obtained during this study range between 2340 and 1540 BC, the approximate interval of soil formation. Brückner and Neboit also obtained a number of radiocarbon ages from the fill and capping soil, several of which are also obviously too old because of organic sedimentation. Ages from the body of the fill (Brückner's sediment 2) range from 3760 to 1730 BC, while the paleosol yielded ages between 3000 and 1690 BC. Altogether, 12 acceptable ages spanning 2340 to 1540 BC have been obtained from unit F. All of these ages were obtained from the upper 1–2 m of the unit and are associated with either interstratified cultural material or the capping paleosol. Consequently, the timing of the initial shift from the stability marking the Late Bronze Age to renewed rapid aggradation is poorly constrained, but the change appears to have occurred at some point in the first half of the first millennium BC.

As in unit E, the initial stages of aggradation in unit F are frequently characterized by diverse cross-bedded and ripple-bedded fine sands and silts indicating a broad, shallowly incised, and braided system. Cross-bedded deposits, however, are absent at the base of approximately half of the documented sections, indicating that braided channels did not cover the entire valley floor. Although cross-bedded basal deposits might be expected to occur preferentially in sections where the Bronze Age soil was erosively truncated during the initial stages of unit F aggradation, there is little relationship between preservation of the soil and the presence of cross-bedded deposits at the contact. Of the documented sections in which unit E is overlain by deposits representing the early stages of unit F aggradation, half exhibit either cross-bedded deposits resting on the preserved soil or massive deposits resting on a truncation surface cut below the soil. These sections indicate that initial scour of the relatively stable "Bronze Age" valley floor was localized and that this scour was followed by the establishment of broad, shallow channels in some places and of overbank environments in others.

The frequent presence of climbing ripples in the lower deposits suggests that floods during the initial stages of unit F aggradation were marked by very high concentrations of sediment. As the unit aggraded, the system shifted from a mixed-load system to a predominantly suspended-load system, probably indicating that the channels were deepening as the floodplain aggraded. The majority of the fill consists of massive silty loams and silty clays probably representing gradual deposition on a vegetated floodplain, with the action of flora and fauna serving to homogenize the deposits. It is also likely that human activity contributed to the mixing of the deposits in the upper part of the fill, although conclusive evidence of cultivation was not observed.

All identified Greek cultural strata are associated with the massive silts and silty clays in the upper part of the unit. Despite the rarity of bed-load deposition during the latter stages of accumulation, sedimentation was rapid enough to prevent organic matter from building up in significant concentrations and forming a cumulic soil until the final stages of aggradation. It is also likely that the massive silts and silty clays include thick deposits attributable to individual high-magnitude flood events, in addition to thin (1–3 cm) strata laid down by seasonal flooding. In one study locality, two discrete strata of Greek cultural material were observed to be separated by as much as 40–50 cm of sediment, suggesting that one or more large flood events deposited a significant volume of sediments in a relatively short period of time, destroying a settlement that was subsequently rebuilt (*Fig. 2.14*).

Although several documented deep, narrow paleochannels appear to be roughly contemporaneous with the formation of the Post-Classic soil, there are few documented examples of channels associated with the preceeding period of rapid aggradation. One such channel, however, was observed at the Pisticci Bridge locality. Unlike the later channel fills, all of which are composed almost exclusively of cross-bedded sands and silts, the channel at Pisticci Bridge contains a diverse fill sequence including cycles of cross-bedded sands, massive to laminated silts, and massive muds. The Post-Classic soil is developed across the top of this channel, indicating that it incised and then filled in completely during the unit F aggradational cycle. This channel deposit demonstrates two significant features: (1) the formation and subsequent infilling of relatively deep paleochannels during the accumulation of unit F; and (2) the prevalence of relatively low-energy conditions at various times, during which thick accumulations of mud were deposited in the channel. The latter phenomenon indicates the streams were

Figure 2.14 Outcrop of stratified Greek remains in the Ferçleto locality. (AL)

and at depths of up to 1.5 m below the soil in localities farther inland. Another intriguing but inconclusive line of evidence is the degree of soil development exhibited by the Post-Classic paleosol at various sections: in general, the documented exposures of unit F closest to the coast exhibit stronger, more horizonated profiles relative to the majority of the Post-Classic soil exposures farther inland. Thus the surface near the coast may have been stable for a relatively longer period and accordingly experienced a somewhat longer period of pedogenesis.

In addition to the silty loams constituting the bulk of the fill, unit F also subsumes a variety of gully fill deposits inset into the higher terraces and the valley-bottom alluvium. These gully deposits, noted at several localities, were formed by lateral feeder channels that rapidly incised the upper part of unit F very late in its development and then infilled just as rapidly with relatively coarse, rubified sediments derived from the marine terraces. Where they were well documented, the gullies were cut after the fill had stabilized, probably in response to incision of the trunk stream. Nevertheless, in all cases the sedimentary structures reflect rapid aggradation. In one locality, the top of the gully fill is marked by a weak soil that welds into the Post-Classic soil at the gully margin, indicating that the gully incised and refilled rapidly enough to allow some soil development to occur prior to the renewed aggradation of unit G.

not simply flashy, intermittent channels during the entire period; in fact, historical records suggest that the Bradano channel was permanent and navigable during at least one part of the later Greek period.[50] Thus there is evidence of significant dynamism in the basic character of the systems during the accumulation of unit F.

There is some evidence that the onset of stability, as well as the formation of the Post-Classic soil, was time-transgressive, beginning earlier in the lower valleys and gradually moving inland over a span of several hundred years. Suggestive of this possibility is the depth of the Greek strata, which are contained within the Post-Classic soil close to the coastal plain

Unit G

Unit G constitutes the last period of significant valley-wide aggradation in the three systems. It consists primarily of distinctly laminated, yellow-brown to brown-grey silty loam and clay; thickness varies from circa 2–10 m. Redox phenomena such as iron and manganese staining and mottling occur in the lower part of the fill but in far less frequency and intensity than in the underlying fills. While it also tends to be slightly coarser than the upper part of unit F, unit G is also composed primarily of silty loams and clays reflecting the sediments available to the system.

The unit is capped by the modern terrace soil, which typically exhibits a relatively weak A-C or Ap-C profile. On the margins of the valleys, the T_2

[50] Delano Smith 1978.

surface slopes up in a smooth, gentle arc, grading almost imperceptibly into the lower toeslopes. This structure suggests that slope wash interfingers extensively with the alluvium and contributes a significant fraction of the laminated deposits, particularly towards the valley margins. As in the underlying units, gravels are entirely absent except where they have been introduced from an adjacent Pleistocene marine terrace or a tributary draining the marine terraces. Thin, localized mud drapes, on the other hand, are quite common throughout the fill. In most cases the muds exhibit a weak structure, causing them to resemble weak cumulic paleosols. They are typically less than 40 cm thick, but a deposit of muds approximately 3 m thick was documented in one locality, indicative of ponding at the mouth of a small tributary.

Although the sediments forming unit G typically exhibit little textural variation from the underlying fills (*Graph 2.2*), there is one relatively striking difference between units F and G not reflected in granulometric data: a clear tendency (but not an entirely consistent trend) towards more distinct lamination in unit G than in unit F, indicating that the turbation processes that homogenized the upper part of unit F were much less active during the aggradation of unit G. As outlined above, it is unclear whether the homogenization of the upper part of unit F was anthropically induced (i.e., by plowing) or produced by a variety of natural bioturbation mechanisms. If the latter is the cause, then the strongly laminated deposits of unit G reflect either a more rapid rate of aggradation than occurred in the upper part of unit F, limiting the potential for bioturbation, or a very different (i.e., sparser) floral assemblage on the floodplain. In either case, however, the widespread preservation of primary depositional laminae indicates that agricultural utilization of the floodplain during the accumulation of unit G was minimal, for such laminae could not have survived cultivation.

Another noteworthy aspect of unit G is the presence of narrow, deeply incised channels at its base. These deep, U-shaped channels laterally truncate the Post-Classic soil, penetrating 4–10 m into the underlying deposits. At no point were channel walls definable above the level of the Post-Classic soil, indicating that the channels were roughly contemporaneous with the formation of the soil. The channels are infilled with thick, gently concave-up packets of cross-bedded fine sands and silts that grade up into the laminated deposits making up the body of the fill, demonstrating that the channels simply aggraded and spilled out across the floodplain, thereby initiating the aggradation of the unit as a whole. No channel deposits contemporary with the body of the fill were noted, and it is unlikely that a deep channel was ever present as the unit aggraded.

Few radiocarbon ages are available from unit G, and none of those determined are reliable indicators of the age of deposition. Ages from the body of the unit G fill obtained during the current study range from 5220 to 1840 BC; Brückner obtained ages of 3940 and 3340 BC (*Graph 2.1*). Again, none of these ages are reliable, although the youngest probably represents a deviation of only a few hundred years from the true age, the oldest a deviation of up to 4,000 years. In addition, an age of 280 ± 50 BC was calculated on the basis of a humate age obtained from the active soil on an inland terrace remnant tentatively correlated with the T_2 terrace downstream. Although this age is too young to represent the age of deposition, it is a reasonable estimate for the mean residence age of a soil developed on a surface that stabilized 500–600 years ago.

Units H and I

Units H and I are recent deposits situated in the channel trench incised in the broad valley bottoms. Both units consist of laminated, cross-bedded, and massive sands and silts with occasional localized massive muds, exhibiting little or no pedogenic alteration. Unit H comprises a series of discontinuous inset surfaces lying 2–6 m below the principal (T_2) surface and 2–8 m above the modern stream. These terrace remnants are collectively designated the T_1 surface, although they actually correspond to several different stages of incision. Although exposure on the margins of these surfaces is poor, in several cases the terraces are clearly underlain by only a thin (1–2 m) mantle of deposits resting on an alluvial strath. Thus the surfaces are probably remnants of incised former floodplains created and then quickly abandoned as the stream cut down, rather than thick fill terraces. Unit I is composed of similar material underlying the modern channel/floodplain complex. Its thickness is unknown, but the streams appear to be actively incising, and therefore the thickness of the modern fill probably does not exceed a few meters.

No ages are available from either unit, and no reliable age is available from the upper part of unit G to constrain the timing of initial incision and abandonment of the T_2 surface. The lack of pedogenic modification, however, suggests that incision probably began no earlier than approximately 500 years ago. Another clue to the approximate timing of incision is provided by a brick pavement, attributed (by a local informant) to an 18th-century structure, which is exposed on the T_2 surface at the confluence of the Basento and the La Canala tributary northwest of Marconia. Although the reported age is unconfirmed, the remains clearly have some antiquity and demonstrate that the T_2 surface has not seen fresh deposition for at least several hundred years at this locality.

Summary of Late Quaternary Landscape Evolution in the Metapontino

This section synthesizes the preceding information into a chronological model of the evolution of the Metapontine landscape under the influence of the various controls. The Pleistocene alluvial deposits of the Metapontino and areas further inland consist of gravelly and sandy sediments deposited by broad, braided channel systems. These deposits cap the inland interfluves and form a series of high, discontinuous terraces inset into the valley walls. The high terrace deposits are termed units A–C in this study for purposes of discussion, although it is likely that these designations subsume more than three discrete fills. With the exception of one terrace system in the middle Bradano drainage, which dates to approximately 35,000–20,000 BC, these older deposits are undated. However, they interdigitate with other elements of the landscape in a manner suggesting a date as early as the Middle Pleistocene, although, because of rapid uplift, many of the high strath terraces are probably much younger than their elevation implies The most noteworthy aspect of the Pleistocene sequences is the prominent component of coarse clastics, which represent material ultimately derived from hard rocks in the highest reaches of the drainages. These gravels and sands are the product of physical weathering at altitude during the cold phases of the Pleistocene and were transported toward the sea in broad, gravelly systems that filled the narrow valleys. The stream systems were considerably steeper than the modern drainages and created deep gravelly fills in the middle valleys which have since been largely cannibalized as the streams adjusted to Holocene conditions.

During the last full glacial period, the Mediterranean Sea lay as much as 180 m below its modern elevation. Such a lowered base level had a tremendous effect on the coastal reaches of the fluvial systems, causing them to entrench deeply into the nonresistant substrate. As sea level rose through the Early Holocene, the sea invaded the entrenched valleys, forming a Ria-type coastline and fostering the deposition of thick plugs of laminated to massive, organic-rich, muddy sediments in the swampy lower valleys. These sediments (unit D) are the oldest exposed at the base of the modern cutbanks. By about 5500 BC, deposition of these paludal muds gave way to terrestrial clastic sedimentation (unit E) as sea level stabilized and the coastal plain began to prograde.

The aggradation of unit E at Metaponto is interpreted as at least partially the result of land clearance and agricultural activity in the Late Neolithic through Early Bronze Age periods, but it may also reflect an increase in the frequency or severity of winter rains. The unit was laid down as a tabular deposit 1–5 m thick across the older laminated muds (unit D) and apparently did not incise a deep channel. By about 4000 BC, principal aggradation of unit E had ceased, and the cumulic Bronze Age soil began to form, possibly in response to incision of the principal channels.

After approximately 700–1100 years of pedogenesis, the systems began to aggrade again, forming unit F. The deep, narrow channel trenches that incised unit E during the formation of the Bronze Age soil were quickly infilled, and the streams began to erode the upper part of unit E, stripping the weakly developed Bronze Age soil in many parts of the valleys. This relatively abrupt shift is interpreted as largely a result of human disruption of the lower watersheds during the Iron Age and Greek periods, but the truncation of the Bronze Age soil clearly indicates that intense storms generating powerful erosive floods were also a factor. As the fill aggraded, it shifted from ripple-bedded and cross-bedded to massive overbank deposition; the change is probably an indication that the channels were deepening, finally resulting in a fill that varies from approximately 2–8 m in thickness. It is in this upper part of the fill that interstratified Greek cultural material is most prominent. Although mostly fine-grained and massive, aggradation of the fill appears to have been stimulated primarily by anthropic disturbance of the uplands and slopes and controlled by complex hydrologic dynamics in the lower streams, leading to fluctuations of the water table and obstruction of drainage.

As the deposition of unit F slowed, the Post-Classic soil began to form. This shift is interpreted as a result of decreased sediment yield from the slopes following depopulation and change in land use, but it also probably reflects the dominance of a relatively high water table in the lower valleys caused by restriction of flow in the channels. There is some evidence that the initial formation of the Post-Classic soil was time-transgressive, developing first in the lower valleys and extending gradually upstream. Such an interpretation is consistent with the growth of coastal swamps, probably one of the principal factors contributing to the decline of the Greek occupation due to the loss of arable land, and the health problems that they introduced.[51] The development of marshy areas was clearly not uniform, however, as relatively deep gullies were also cut and filled during the same relatively brief period, suggesting that the elevation of the water table and the extent of marshy areas fluctuated as the streams struggled to accommodate the unrelentingly high sediment loads.

By the Republican Roman period (roughly 200–100 BC), aggradation of unit F had ceased, indicating that the volume of introduced sediment was sharply diminished. However, the distribution of anomalous radiocarbon ages resulting from localized soil erosion suggests that land use expanded to include the steeper slopes and other marginal areas during this period, a probable indication of increased pastoral activity.[52] As the sediment load decreased, the streams appear to have incised somewhat, stabilizing the surface and allowing the Post-Classic soil to develop.

At some point, estimated at 500–800 BC, the Metapontine streams again incised the floodplain, forming the modern T_2 terrace. Although this incision has been attributed to climatic or political changes,[53] a more likely cause is that changes occurred in the hydrology of the lower valleys primarily because of intentional reclamation of the lowlands as the population again expanded coastward from the hilltop settlements of the Medieval period. Deposits associated with this phase of activity are thin (typically less than 2 m), suggesting that most of the sediments produced in the inland areas during this period were transported to the sea. The last phase of incision, which created the narrow, discontinuous T_1 terrace as the streams incised again in the narrow channel trench, is tentatively attributed to more efficient reclamation of the coastal zone in the early 20th century.

Graph 2.3 illustrates the timing of the Metapontine sequences in comparison to cultural sequences and selected indices of environmental change elsewhere in Italy and southern Europe. Two aspects of the data are notable: 1) there is very little synchronism apparent among the environmental indices and between them and the alluvial sequences at Metaponto and Croton; 2) nevertheless, there is a crude correspondence between the two sequences.

As described previously, the synchronism can be attributed not to any single factor, but rather to the coincidence of climatic, anthropic, and systemic influences in the development of the landscape.

Mechanisms Driving Landscape Change

Although the Late Quaternary alluvial stratigraphic record of the Mediterranean had been addressed previously,[54] the publication of Vita-Finzi's seminal study *The Mediterranean Valleys* in 1969 was a pivotal event in the field. Vita-Finzi identified two phases of synchronous, pan-Mediterranean valley aggradation, which he named the Older Fill and Younger Fill and attributed to changes in climate. The concept was immediately criticized as simplistic and for ignoring the considerable and disparate impact of human activity on the landscape,[55] but it nevertheless served as a catalyst for a series of detailed regional studies throughout the northern Mediterranean. Now, 40 years after its appearance, the basic thesis of Vita-Finzi's study has been largely discredited, as study after study has demonstrated unique, asynchronous alluvial stratigraphic sequences throughout the region that often correlate well with local variations in the archaeological record, and in particular with changes in settlement density.[56] Nonetheless, the

[51] Henneberg and Henneberg 1998; Henneberg/Henneberg/Carter 1992.

[52] Abbott 1997.

[53] Brückner 1990.

[54] E.g., Judson 1963; Eisma 1964.

[55] E.g., Butzer 1969.

[56] E.g., Davidson 1980; Delano Smith 1981; Brückner 1982, 1986; Boenzi 1984; Pope and Van Andel 1984; Van Andel/Zangger/Demitrack 1990; Van Andel and Zangger 1990; Chester and James 1991; Barker

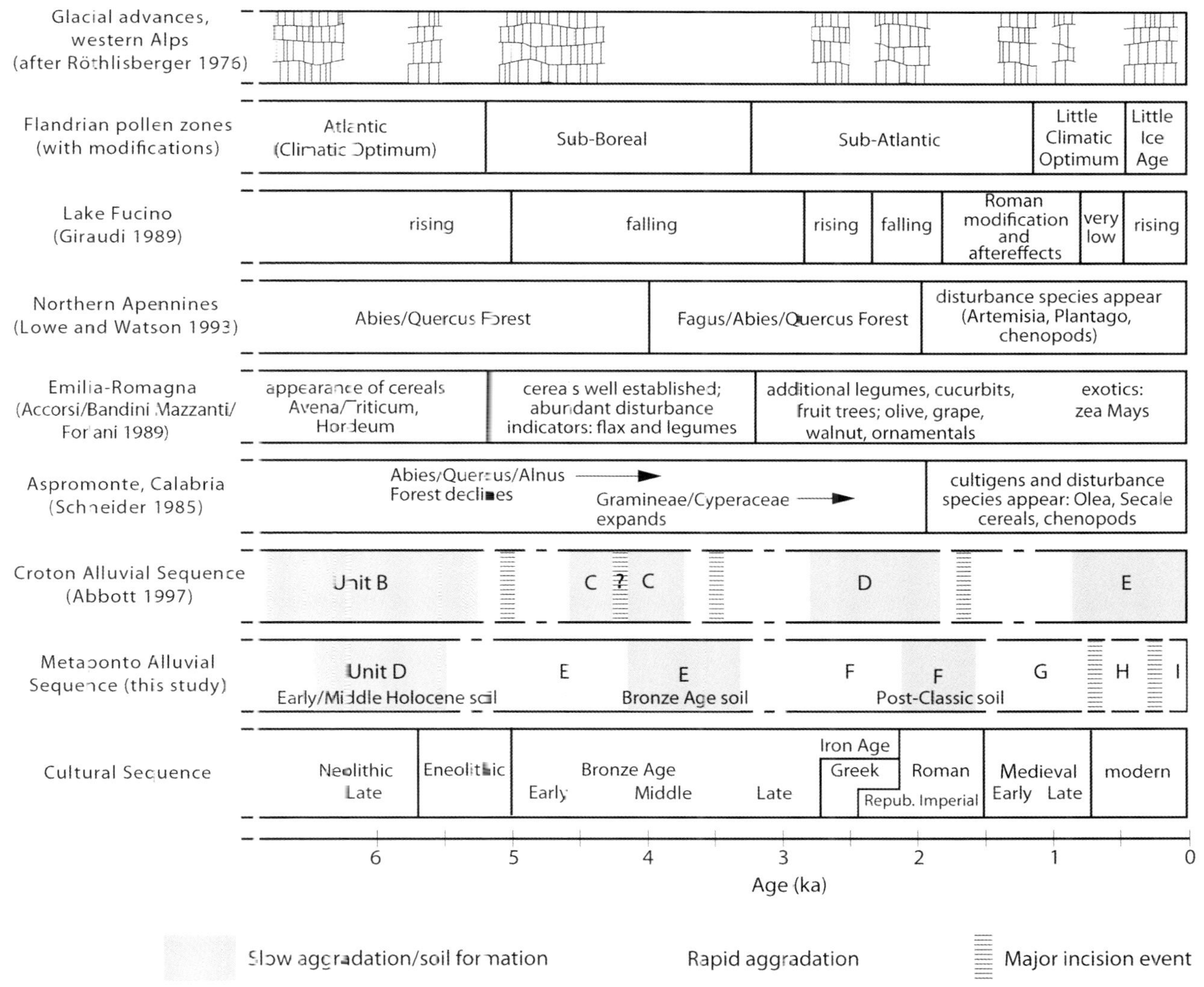

Graph 2.3 Comparison of the Metapontine sequence with other disparate evidence for Holocene environmental change in Italy.

basic question of underlying causality often continues to be posed in comparatively simple terms, with a great deal of emphasis on the relative roles of climate and human activity.[57] While human impacts have arguably played an important role in the trajectory of Late Holocene landscape evolution in the chora of Metaponto, the sequence belies a complexity that undermines the simplistic conclusion that human impacts were more important than climatic impacts. On the contrary, a realistic appreciation of the processes driving the evolution of these landscapes requires consideration of a number of interacting controls, including lithology, structure, tectonics, climate change, volcanism, human activity, and complex response of the landscape system.[58] This section examines the controls that interacted to influence the trajectory of landscape evolution in the Metapontino.

The Influence of Lithology and Structure

The bedrock underlying the Metapontino is unlike the hard rocks that underlie most areas of the northern Mediterranean. This difference explains much of the unusual character of the Metapontine alluvial record. In general, lithology influences the development of a given landscape through the exposure of rocks with differing resistance to erosion: rocks with relatively high resistance erode relatively slowly and develop

and Hunt 1995; Hunt and Gilbertson 1995.

57 E.g., Delano Smith 1981; Devereaux 1982; Abbott and Valastro 1995.

58 Barker and Hunt 1995; Frederick 2001.

positive relief, while rocks with low resistance erode relatively rapidly and develop negative relief. The pattern of outcrop exhibited by the rocks in a given region is a function of the underlying geologic structure and the suite of denudation processes in operation. Although many factors are involved, erosion resistance is principally a function of the hardness of the rock, its permeability, and its susceptibility to chemical attack under the prevailing environmental conditions. These factors are tempered by the density of vegetative cover, which serves to consolidate the sediment and reduce runoff, and the prevailing slope, which provides the energy available for water erosion.

Once its protective soil and vegetation cover is removed, the weakly lithified or unlithified clay substrate of the Bradano Trough has poor resistance to erosion. Consequently, the region is a young, active landscape that has developed entirely within the Quaternary period. It has evolved rapidly within this geologically brief time span, experiencing large-scale topographic inversion of much of the landscape inland of the chora proper. The rate and character of landscape development in the study area is therefore attributable above all to the low erosion resistance of the unconsolidated Plio-Pleistocene clay bedrock, coupled with the much higher resistance afforded by the poorly consolidated, but highly porous, coarse clastics that mantle the marine terraces and inland interfluves.

The long-term trajectory of landscape evolution within the areas mantled by littoral clastics is dictated by these two superimposed deposits. Although they are relatively thin, the coarse, permeable clastics mantling the marine terraces provide an effective armor protecting the surfaces from erosional downwearing. Rather than running off, most of the incident precipitation infiltrates into the coarse clastic cap, limiting the amount of sheet erosion on the surface. The infiltrating moisture, however, quickly encounters a permeability barrier at the top of the underlying clays and emerges as springs around the margins of the terraces. These springs result in enhanced erosion of the scarps and upper slopes through sapping and mass movement, causing the terrace margins to retreat laterally.

In addition to lateral retreat of the upland surfaces overlooking the major valleys, the marine terraces also exhibit a tendency to develop networks of headward-cutting gullies that progressively dissect the clastic cap. Once these headward cutting channels expose and incise the underlying clays, the channels broaden and deepen through the same processes affecting the terrace margins. As the littoral cap is removed and the underlying clays are exposed to attack, they erode rapidly. In some cases, the erosion serves to oversteepen the slopes and facilitate further retreat of the scarp through mass wasting. Over the long term, the development of badlands decreases the angle of the long slope profile from the marine terrace rim to the valley floor; however, badland dissection of the higher valley walls results in complex local slopes that are generally quite steep. The overall pattern is clearly illustrated in Figure 2.1, in which the extent of the marine terraces shrinks systematically at the expense of clay slopes and badlands with increasing distance from the coast, until the highest terrace consists of a few small remnants surrounded by an extensive erosional landscape.

The rapid erosion characteristic of the Plio-Pleistocene clays of southern Italy has been documented by several investigators.[59] To understand landscape evolution in the region, one must be aware of the fragility of the landscape underlain by these soft sediments. Because the clays of the Bradano Trough are hundreds of meters thick, erosion, once initiated, is difficult to check. In landscapes underlain by hard rocks, susceptible sediments are limited to a relatively thin weathered mantle; once these sediments are depleted, erosion is necessarily checked until weathering can produce more sediments. No such limiting mechanism exists in the study area. Other than the amount and intensity of incident precipitation, the only real limiting factor is the prevailing base level. Vegetation cover is therefore crucial for stabilizing the slopes. Once removed, it is difficult to reestablish, even in the absence of grazing animals.

A crucial corollary to the erosion-prone slopes is the resulting high volumes of sediment that can be rapidly introduced into the stream systems. One of the most important controls on Late Quaternary alluvial behavior in the chora was periodic overloading of the streams with more sediments than they were competent to handle. Almost all of the Holocene sediments were derived from the silty clays that underlie the lower Metapontine drainages, but coarser, sandy deposits are also visible as relatively minor facies throughout the Holocene deposits. Rubified

[59] E.g., Kayser 1958; Cavazza 1962; Valentini/Cherubini/Guadagno 1979; Rendell 1982, 1986.

sandy tributary gully fills signify localized degradation of the terraces in the lower watersheds during the Late Holocene. Yet coarse clastics derived from the upper Metapontine watersheds, which form a dominant part of the Pleistocene fills, are almost entirely absent in the Holocene sequence in the lower valleys, indicating that sediment yield from the upper drainages was dramatically reduced in the Late Holocene.

The Influence of Tectonism and Volcanism

Like structural features resulting from tectonic activity, active tectonism strongly influences the evolution of a landscape. The chora of Metaponto is characterized by a broad coastal plain, deeply incised river channels, elevated Pleistocene marine terrace remnants, and subparallel sets of marine terrace scarps. The region has been uplifted in stages throughout the Late Quaternary as a result of compression and nappe intrusion from the Apennine arc to the west. Direct effects of tectonic activity, such as uplift, subsidence, or faulting, are the result of active controls. These types of active influences not only affect the trajectory and rate of landscape development but are also capable of abruptly altering system behavior.

The influence of Quaternary volcanism on the development of the landscape is limited to the effects of ashfalls within the watersheds. While no volcanic ash beds were noted in the sequence by the author, one has been described previously in the Basento valley.[60] The effect of the aerosolic influx of volcanic ash into a drainage system can influence the behavior in ways that range from relatively minor to profound, depending on a number of factors, including the amount and temperature of the ash and the size and initial state of the system. Although ash tends to settle across a drainage net in a relatively uniform manner, fluvial and aeolian processes rapidly redistribute the ash in the system, concentrating it in drainages and aeolian traps. Because ash is typically silt-sized and poorly consolidated, even a thin blanket spread across a watershed can result in a considerable introduction of sediments into the drainages with the next rain, potentially causing a temporary but substantial alteration of system hydrology. Appreciable accumulations of ash can also have an adverse impact on vegetation in the drainage, particularly on vulnerable annuals such as grasses and forbs, through burial and also possibly through the instantaneous creation of an inhospitable chemical environment. Hot ashfalls are often even more devastating to vegetation, but such falls typically occur only very close to the source.

[60] Boenzi et al. 1989.

The Influence of Climate Change

The influence of patterns of precipitation, temperature, humidity, and wind is one of the most widely recognized controls on geomorphic behavior, and has been used as a basis for organizing the study of landforms.[61] Similarly, changes in climatic parameters are one of the most common explanations for changes in the behavior of geomorphic systems during the Late Quaternary.[62] The character of systemic response to such changes in climate is far from simple, however. Landscape responses to climate change, like responses to change in other systemic inputs, may vary considerably depending on the magnitude, direction, and character of the change, preexisting system state, and character of intrinsic geomorphic thresholds.[63]

Climate change operates at many scales. At the broadest relevant scale, global climate has fluctuated between much colder and somewhat warmer conditions than at present during the last few million years, profoundly influencing the evolution of the surface of the earth. The most obvious effects are associated with the massive continental ice sheets, but vestiges of colder Pleistocene climates are found at all latitudes.[64] The oscillations in global climate have had several significant effects in the study area, including the production of large quantities of coarse clastics in the high parts of the Metapontine drainages and the incision of the lower valleys as they adjusted their gradients to the dramatically lowered sea level during the cold phases, followed by aggradation of the lower reaches as the continental glaciers retreated and sea level rose again to the modern level.

The impact of climate change is more difficult to gauge at relatively refined scales, particularly in the Late Holocene, because the climate record of the last 5,000 years in the northern Mediterranean is largely obscured by the effects of human activity. Most types of proxy climate data (e.g., paleovegetation records derived from pollen or macrofossils, sedimentation rates

[61] E.g., Büdel 1982; Tricart and Cailleux 1972.

[62] E.g., Knox 1983, 1984; Wells/McFadden/Dohrenwend 1987; S.A. Hall 1990; Bull 1991.

[63] Knox 1984; Bull 1991.

[64] Büdel 1982; Tricart and Cailleux 1972.

and shifts in geomorphic trajectory, indices provided by soil development and morphology) are also affected by human activity. It is very difficult, for example, to determine in isolation whether a loss of oak witnessed by a dramatic decrease in the concentration of oak pollen in a stratigraphic column was due to climate change or anthropic deforestation. Better evidence is provided by some types of proxy data, such as variations in lake levels or alpine glaciers, and oxygen isotope analysis of sediments and biotic remains such as shells; unfortunately, these data sources often disagree and rarely provide a coherent picture.[65] Nevertheless, while the specifics of Late Holocene climatic variation in the Mediterranean region are far from clear, there is little evidence to indicate that severe changes in climate occurred. The picture, though fuzzy, is essentially characterized by relative constancy in annual temperature and precipitation.

While major climatic changes are not apparent in the Late Holocene paleoenvironmental record, it would almost certainly be erroneous to assume that Late Holocene climates in the northern Mediterranean have been constant, or that changes in climate have not affected the geomorphic evolution of the area studied. The character and timing of such changes are obscure, but there is incontrovertible evidence of shifts in climatic parameters in the study area during the past 5,000 years. Many of these Late Holocene changes were probably simple shifts in the timing, frequency, and intensity of precipitation, and particularly in the frequency and timing of strong storms, rather than significant changes in annual temperature or precipitation having strong impacts on the density and character of vegetative cover. Because the landscape is so vulnerable to erosion by intense rain, relatively minor increases or decreases in the frequency and intensity of storms would induce significant changes in the rates of sediment yield, causing the systems to adjust their gradients and depositional habits.

As the following section argues in more detail, the age and character of the last three major fill events in the Metapontino suggest that anthropic disturbance of the slopes played a very important role in stimulating aggradation in the valleys. Nevertheless, it is important to realize that, while human disruption can directly stimulate relatively low-magnitude gravity-driven erosion in a given ecosystem, the principal role of human activity in a semi-arid setting is to facilitate erosion by creating the conditions in which winds and runoff from precipitation events can mobilize and transport elevated volumes of sediments from the slopes. Changes in mean annual precipitation are therefore probably less important overall than changes in the frequency, duration, and intensity of storms.

The Influence of Human Action

The effects of cultural activity pervade the modern Metapontine landscape. Severe slope erosion is observable in contemporary agricultural fields, and this erosion is introducing prodigious quantities of sediments into the streams, forcing them to adjust their hydrologic character. The influence of prehistoric and historic human activity is less clear, because the morphological features produced by that activity have been largely obscured by subsequent erosion and deposition. Interpretation of former landscape impacts must therefore be made using indirect evidence. In this study, the principal line of evidence examined is the thickness and character of the resulting deposits.

Human beings can affect the landscape in many different ways.[66] Arguably, the three most important mechanisms of landscape disturbance by preindustrial civilizations are agriculture, pastoralism, and clearance of native vegetation to prepare land for settlement and use. Each of these activities tends to stimulate erosion, but it is frequently difficult to correlate the resulting deposits with specific types of land impacts, or even to demonstrate a causal linkage between cultural disruption of a landscape and subsequent changes in landscape activity.[67] Rather, the causal link must be suggested by demonstrating synchronism between cultural episodes and periods of landscape change. If it can be demonstrated that cycles of landscape activity parallel cultural changes closely enough to suggest a stimulus-response relationship, then the argument can be made that cultural activity was in fact the impetus for a given change.

Generally accepted models of cultural development in the region indicate three major periods of the archaeological record in which human activity had the greatest potential to substantially disrupt the equilibrium of the south Italian environment: the Neolithic, when agriculture, stock raising, and sedentism replaced

[65] Bradley 1985, 240.

[66] Nir 1983.

[67] Frederick 2001.

the mobile hunter-gatherer existence of the Mesolithic period; the Late Iron Age and Classical (Greco-Roman) periods, when population swelled dramatically and a market economy was established; and the Late Medieval through modern periods, when population once again grew following a near collapse in the Late Classical period.[68] Significantly, the last two periods exhibit high dynamism in the sedimentary record of the Metapontino, corresponding to units F and G–I, respectively, while the intervening periods witnessed formation of the Bronze Age and Post-Classic paleosols. This synchronism between geomorphic activity and cultural presence strongly suggests that landscape evolution during the last few thousand years has been driven largely by cycles of cultural impact. The data furthermore intimate that the preceding episode of aggradation (unit E), corresponding to the Eneolithic and Early Bronze Age, may also have been largely stimulated by human activity.

Investigators have argued that human disruption of the landscape in the Classical period resulted in dramatic sedimentation of the valley systems, while previous episodes of sedimentation were stimulated by climate change.[69] Brückner supports this argument by characterizing the preceding fill (sediment 1) as texturally graded, hence representative of a "natural" fining-upward sequence, in contrast to a massive, fine-grained Classical fill (sediment 2). Textural analyses and sedimentary structure descriptions conducted during this study do not support the contrast that Brückner draws between his sediments 1 and 2. Rather, the character of the preceding unit E fill is quite similar to the Classical fill of unit F; they share textural attributes and both tend to grade up from ripple-bedded and cross-bedded sandy silts near the base into massive silts near the top. In fact, the principal difference between units E and F is that the Bronze Age soil, which caps unit E, was frequently truncated by scour at the onset of unit F sedimentation, while the Post-Classic soil capping unit F is almost always present. The discontinuous presence of the Bronze Age soil may have caused Brückner to confuse and conflate the two units. Where the Bronze Age soil is well-preserved, he correlated it with his paleosol A, grouping the unit E fill in his sediment 1, while in other cases he correlated the Early/Middle Holocene soil with paleosol A, grouping unit E in his sediment 2.

Despite the lack of firm evidence for or against Brückner's argument denying human stimulus as the cause of pre-Classical erosional episodes, two aspects of the record suggest that human agency was, in fact, responsible. First, the similarity of the unit E and F fills suggests that the factors responsible for these two episodes of increased sedimentation were similar. Because the impact of the rural Greek (and possibly Lucanian) agricultural system on the landscape is arguably the principal cause of the unit F fill, and because agriculture and land clearance were initiated in the study area several thousand years before the Greeks arrived, it is reasonable to postulate a similar cause-effect relationship active in this earlier period. Second, and perhaps most compelling, is the presence of finely divided charcoal in the intervening soils. Three of the four examples (75%) of the Bronze Age soil examined micromorphologically for this study contain charcoal, a common indicator of anthropic disturbance.[70] Finely divided charcoal is also characteristic of the Post-Classic soil (present in 83% of the samples), but only one (33%) of the Early/Middle Holocene soil thin sections, and none of the Pleistocene thin sections, contain charcoal. More evidence is required to resolve the issue, but it is clear that units E and F represent similar massive influxes of sediments, which are interpreted here as largely the result of vegetation clearance and agricultural activity in the Eneolithic/Early Bronze Age and the Late Iron Age/Greek periods, respectively.

Although the textural characters of units E, F, and G are practically indistinguishable because of lithologic control exerted by the source "rocks," the sedimentary structures characteristic of unit G differ slightly from those typical of units E and F. Unit G typically exhibits distinct lamination throughout, while units E and F do not. The reason for this difference is not clear, and there are two possible explanations for it:

1. The upper part of unit F, and possibly unit E, was regularly turbated by plowing as it aggraded. The valley fills contain interstratified remains of Greek rural settlements, so that there is every reason to believe the valleys were intensively cultivated during the Classical period. In contrast,

[68] E.g., Stevenson 1947; Dunbabin 1948; Piggot 1965; Trump 1966; Toynbee 1965; Braudel 1972; R.D. Whitehouse 1968, 1981; Barker 1972, 1981, 1985; Coles and Harding 1979; De La Geniere 1979; Holloway 1981; Bietti Sestieri 1984; Champion et al. 1984.

[69] E.g., Brückner 1986, 1990; Neboit 1977, 1983.

[70] Courty/Goldberg/MacPhail 1989; MacPhail 1992.

the preserved sedimentary structures of unit G demonstrate an absence of agricultural activity as it accreted in the coastal zone, probably because Medieval settlements had been relocated to elevated, defensible sites in the interior.

2. The speed or style of aggradation was different, and sedimentary laminae were more prone to form—or less likely to be obliterated—by the action of floodplain vegetation. Possible explanations for a shift in depositional style during aggradation of unit G include (a) an inland shift of the principal source of sediments introduced into the system, following the population, which resulted in longer transport and more organized deposition of trunk stream sediments in the valleys (i.e., better sorting of sediments delivered by overbank pulses); and (b) increased sediment yields from the slopes of the coastal zone due to grazing or clearance, creating vast sheets of interdigitated slope-wash sediments. However, even if laminated sediments in the upper part of unit G can occasionally be found extending well up onto the colluvial toeslopes, suggesting that sheet erosion and deposition were essential mechanisms in the aggradation of the fill, it is unclear why this phenomenon is not present in the fills associated with the preceding periods, particularly the intensive Greek occupation.

The chora of Metaponto is an atypical northern Mediterranean landscape in both its physical characteristics and its anthropic overprint. The most striking abnormality is the almost total absence of agricultural terraces. Hillslope terracing slows or stops the loss of vital agricultural topsoil, but construction and maintenance of stone walls require considerable effort and expense. The lack of terracing in the Metapontino may reflect the paucity of suitable stone for the construction of retaining walls, as terraces would have been constructed if suitable stone had been available. However, it is also true that the inhabitants had little direct incentive to invest in such systems. The Metapontino simply does not fit the typical northern Mediterranean mold of steep, rocky terrain mantled with thin, patchy vestiges of ancient *terra rossa* soils. Unlike areas characterized by thin soils developed on hard bedrock, soil loss in the study area is not attended by significant decline in productivity. After the initial loss of Pleistocene soil cover, no matter how severe the surficial erosion, the freshly exposed sediments consist of the same base-rich clay and are almost as productive as the lost sediment. Consequently, adaptive decisions made by the ancient populations fell within a different ecological context and were unlike economic strategies employed elsewhere in the northern Mediterranean. For the individual farmer attempting to make a living on a given tract of land, excessive soil loss was probably not considered a serious problem, particularly since the construction of agricultural terraces involved importation of massive quantities of stone. Over the longer term, however, this assumption had repercussions for the sustainability of the agricultural system as sediment cascaded into the stream systems, overwhelming their hydrological capacity and altering the character of the landscape.

Interaction of Controls: Complex Systemic Response

Complex response, a concept originating in the application of systems theory to geomorphology, is a powerful and elegant tool for relating stimulus to resultant activity in a manner that acknowledges the complexity of natural systems.[71] In essence, complex response holds that, as a result of a given stimulus, a geomorphic system may respond with a series of temporally and/or spatially disparate adjustments as the system seeks to regain equilibrium.

As an example, a hypothetical stream catchment is subjected to massive devegetation in its upper reaches by a catastrophic fire. Sediments on the hillsides are suddenly rendered highly susceptible to erosion, and a large influx of sediments is introduced into the upper catchment, clogging the system and causing the channel to aggrade and steepen. With time, the influx of material lessens as the stored sediments on the slopes are depleted; the stream then begins to incise the oversteepened bed, eroding its deposits and shunting them gradually downstream. As these deposits reach the lower catchment, the declining gradient once again reduces competence to the point that the lower channel begins to aggrade even as the upper part of the system continues to degrade. The sediments are gradually shunted through the system until the lower reaches, deprived of the formerly high influx of clastics, incise again.

[71] E.g., Schumm 1977; Chorley/Schumm/Sugden 1984; Bull 1991.

This simple example illustrates two important points about complex response. First, it has a temporal scale: in the example, a given point in the stream will experience a series of disparate adjustments (e.g., aggrading and incising) until quasi-equilibrium conditions are reestablished. Moreover, if the affected point is distant from the locus of disturbance (thus the lower reach of the stream in the above example), the initial response may occur long after the stimulus, as disequilibrium is gradually transmitted through the system. Second, complex response also has a spatial scale, in that different parts of the system may be affected in entirely different ways at the same time (as where, in the above example, the upper basin incises while the lower basin aggrades).

The true role of complex response in the evolution of the Metapontino is impossible to evaluate, but nevertheless there are two major aspects of the record in which complex response appears to have played an important part. The first aspect is the long-term movement of the coarse clasts out of the upper Metapontine drainages, where they were produced by physical weathering during the Pleistocene stadials. During the Pleistocene, these clasts were steadily shunted downstream towards the coast, forming thick (30+ m) accumulations in the middle valley of the Basento and a relatively steep channel graded to the lowered base level. Although identification of specific related deposits in the lower drainages is difficult, much of the Holocene activity in the system clearly represents adjustment following the elimination of these coarse clastics from the sediment yield and the change in base level that occurred as the Mediterranean rose to its modern elevation. In the middle valleys, intact remnants of the coarse Pleistocene clastic load are sporadically preserved as terraces along the middle reaches of the streams: at the Grottole locality in the Basento valley, for example, the section consists of approximately 5 m of relatively fine-grained Holocene deposits resting on at least 15 m of coarse bedded gravels and sands of the braided Pleistocene system. The relatively thin capping Holocene sequence appears to be a compressed analog of the thick Holocene fills in the coastal reach, yet it is stranded on a terrace tread 30 m above the modern braided system. This overbank sequence could not have been deposited unless the channel was considerably higher than it is today. The clear conclusions are that incision in the middle valley is comparatively recent and that tremendous volumes of gravelly sediment have been removed in the past few thousand years.

One of the most vexing problems persisting in the Metapontine sequence is the question of the disposition of these gravels. Although the primary late Pleistocene sequence is certainly preserved at depth in the valleys, the exposed sections in the coastal reach contain essentially no gravels. The most extreme case is the Basento system, in which several square kilometers of gravelly fill appear to have been eroded and shunted to the sea through a narrow, low-gradient trench cut through the Holocene fill, leaving almost no trace in the process. One possible explanation for the absence of the gravel is previous erosion of the bulk of the Pleistocene fill (i.e., in the Early/Middle Holocene), leaving terrace remnants at intervals along the valley, and subsequent burial of these terraces by fine-grained aggradation in the Middle/Late Holocene. Thus the bulk of the material eroded from the middle valleys in the last few thousand years may not have been as gravel-laden as the preserved terraces suggest. Although there is no firm evidence to support it, this hypothesis would help explain the absence of gravel in the coastal system and may account for the pulse of sediments in the coastal reach during the Medieval period. (Also noteworthy is the apparent coincidence of this last phase of incision with the relocation of the population into the interior.) Whatever the explanation, it is clear that the Metapontine alluvial system is still striving to achieve equilibrium, as demonstrated by the abrupt change in gradient and depositional style observable 30–40 km inland from the coast.

The second aspect of the overall sequence in which complex response has arguably played an important role in the Metapontino is control of the sequence of aggradation and incision in the coastal reach. The response was not the result of long-term shifts in climate, but rather of fluctuations in sediment yields induced primarily by anthropic disturbance. As these sediments periodically flooded into the systems, they overwhelmed the streams' capacity to transport them, causing the channels to aggrade until bars and dunes blocked the exit to the sea; the blockage in turn caused localized ponding and a severe flooding hazard in the lower reaches of the systems, elevating the water table as far as several kilometers upstream. In time, as the sediment influx waned, the systems would adjust (typically by avulsion, thus bypassing the sediment plugs), the channels would be reestablished, and the water table would drop briefly until another large plug of sediments was introduced. Such complex interplay between sediment yield and sediment transport, and not external factors such as tectonic uplift or pronounced climate change, was the principal control on patterns of sedimentation in the coastal reach during the Late Holocene and almost certainly placed considerable stress on the agrosystem established in the lowlands, in addition to creating abundant habitat for mosquitoes and thus facilitating the spread of malaria. Physical evidence supporting this model is found in the intense redox alterations typical of buried paleosols and sediments in the coastal zone, rapidly forming and filling tributary gullies in the lower valleys; the traces of multiple avulsion channels on the floodplain; and, by extension, the evidence for malaria in the physical remains from the Pantanello necropolis.[72] All of these features are best explained by the model outlined above, for they occurred much too rapidly to be explained by changes in climate or by tectonic uplift.

Brückner,[73] Boenzi et al.,[74] and Neboit[75] suggest that the final, two-stage phase of incision that created the T_2 and T_1 terraces in the Metapontino was related to renewed tectonic uplift. In his 1990 publication, however, Brückner hedges his bet, attributing the incision and subsequent aggradation in the channel trench to either tectonic uplift in the 11th or 12th century AD (why tectonic uplift should occur within this particular time frame is not clear) or "political changes" affecting the rate of sediment delivery in the 14th–15th centuries AD. While tectonics—and indeed climate changes associated with the "Little Ice Age"—probably played some role,[76] it is unlikely that there has been sufficient regional uplift or climate change in the past 400–500 years to stimulate 20 m of downcutting. Rather, the two-stage incision is probably the result of the initial reclamation of the swampy lowlands following the end of the Medieval period, which caused the streams to incise somewhat and form the modern T_1 surface, followed by a second drop after the land reclamation projects initiated under Mussolini successfully drained the coastal plain and lower valleys.

[72] Henneberg and Henneberg 1998; Henneberg/Henneberg/Carter 1992.
[73] Brückner 1986.
[74] Boenzi/Palmentola/Valduga 1976; Boenzi/Cherubini/Giasi 1987.
[75] Neboit 1983.
[76] Cf. Dufaure 1976.

Geoarchaeological Implications

This final section addresses the implications of the Metapontine landscape record for archaeological site visibility and preservation. Where appropriate, the conclusions are accompanied by suggestions for additional geoarchaeological work.

Lowland coastal settings

Sites of the Medieval or earlier periods in lowland coastal settings are typically buried beneath 1 m or more of recent alluvium and therefore have extremely low visibility on the principal (T_2) terrace. In contrast, the inset (T_1 and possibly T_0) terraces are frequently strath surfaces exposing cultural material of Neolithic to Roman age, depending on the depth of incision and the thickness of associated veneer deposits.

The nearly ubiquitous preservation of the Post-Classic soil indicates that lateral cutting in the lower streams has been minimal during the last few millennia, and that the sedimentary character of the sequence depositional energy has been relatively low. The potential of site preservation and integrity in the alluvial environment is therefore relatively high. However, the much poorer preservation of the Bronze Age soil implies that lateral incision and surface scouring of the Bronze Age fill during the Classical reactivation was common, so that the potential for widespread preservation of interstratified Neolithic and Bronze Age sites is much lower. Nevertheless, while they are more difficult to find, there is great potential for the local preservation of Neolithic and Bronze Age sites in the alluvial fill.

In many cases, sites in alluvial settings were probably buried relatively quickly and therefore have the potential to provide a more detailed, temporally restricted "snapshot" of site structure and composition than sites in other contexts. The thickness and rapidity of aggradation also limit the potential for palimpsest mixing of unrelated components; in fact, several Greek sites were observed in section exhibiting multiple distinct occupation surfaces separated by 10–20 cm of flood sediments (*Fig. 2.14*), perhaps allowing ready characterization of the temporal trends of individual occupations.

The unit G fill is sufficiently thick to have protected the majority of buried Greek and Roman sites from modern deep plowing, although some sites may have been affected, particularly in the Bradano drainage, where the Medieval fill is relatively thin and is occasionally removed by bulldozer to expose the more fertile Post-Classic paleosol. The laminated character of unit G indicates that intensive, concomitant plowing did not occur as the unit aggraded from approximately the Late Roman to the Medieval periods.

The distribution of identified sites in the intensively surveyed transect between the Bradano and Basento Rivers indicates that some Greek and Roman sites are visible on the surface of the Basento valley.[77] While it is possible that some of these sites represent archaeological material washed down from the slopes and/or exposed in strath terraces, the presence of these sites suggests that there may be remnants of an undetected Late Pleistocene/Early Holocene terrace buried beneath the modern T_2 surface. If this is true, the older terrace would have been overridden and buried in the last few thousand years; prior to the Medieval period, it would have been a prominent feature rising above the surrounding floodplain. Such a feature would have been highly attractive for settlement, particularly as the lower alluvial surfaces became increasingly marshy. Despite the improbability that a widely preserved terrace lies buried in the valley, the persistence of the channel location inland of the marine terrace margin indicates that such a fill is widely preserved without being significantly exposed. Therefore machine prospection of the T_2 terrace in areas away from the modern channel, and of a sample of the sites identified during survey, is recommended to clarify the stratigraphic picture.

Although the potential for organic preservation in the more recent units exists, fluctuating groundwater levels have probably adversely affected the preservation of organic materials in many contexts. Pollen preservation in the alluvial deposits is also apparently quite poor.[78] Nevertheless, outstanding organic preservation was discovered in several investigated contexts, notably the Greek material recovered from the trench on the T_2 terrace in front of the Pantanello sanctuary site. In fact, the preservation of ovicaprid bone exposed in the mud plug of this trench is the best that this author has seen from any archaeological context; it compared favorably to bones exposed on the surface for a period of a few weeks. Furthermore, despite having clearly been subjected to repeated water table fluctuations, traces of organic materials and iron-stained casts of vegetal material were frequently recovered in a number of sections through the laminated muds of unit D. It is therefore recommended that future investigations of sites in such muddy settings include procedures to recover faunal and paleobotanical information.

Marine terraces

Most sites on the marine terraces are likely to be either exposed or buried under a thin veneer of sheetwash sediments. In general, these sites were probably exposed for a long period after abandonment and are thus highly vulnerable to disturbance by the plow, which is by now a nearly ubiquitous phenomenon on the terraces. The vulnerability of these sites is clearly indicated by the large number located during surface survey of agricultural fields. Preservation potential is slightly higher on the downslope side of the marine terrace scarps, where somewhat thicker sheet wash and colluvial sediments tend to accumulate. As the geological study has revealed, such scarps are also likely to host springs and seeps and thus were favored loci for settlements.

As discussed elsewhere in this volume, recent mechanical excavations on the marine terraces have exposed several examples of cultural features believed to belong to the network of division lines previously identified from aerial photographs. These features are broad, linear excavations into the substrate of the terraces, filled with soil presumably eroded from the surrounding landscape. While their function remains unclear at this stage of investigation, the considerable effort and expense invested in their creation by the Greek population are obvious. Because the physical and chemical character of the infilling sediments has the potential to shed considerable light on these features, detailed geoarchaeological examination of several cross sections is recommended.

[77] D'Annibale 1983a; Carter 1992; Carter and Costantini 1994; *Ancient Territories* 2000, 18.

[78] J. Ibañez, pers. comm., 1992.

Slopes

The preservation and visibility of sites on the slopes, both in the chora proper and in areas further inland, are generally erratic and unpredictable. Although particularly large or recent mass movements are often identifiable on aerial photographs, examination of the few roadcuts in the chora that reveal exposures of the slope deposits indicates that both sudden and incremental slope movements have been widespread in the chora and the inland areas for thousands of years. Therefore buried and surficial sites of a variety of ages in many different contexts are to be expected. This situation is particularly true in toeslope contexts, where incremental additions of slope wash could have gradually buried a site even during the span of occupation.

The slope environment is extremely active, so that long-term preservation and visibility of sites is unlikely. Because of the progressive nature of slope evolution, the potential for preservation of sites, particularly in surficial contexts, decreases with age. For this reason, estimates of the relative intensity of settlement based on opportunistic discovery of sites are ill founded, and comparative statements on the intensity of ancient settlement in the Bradano Trough and adjacent areas such as the relatively stable Tavoliere[79] are clearly spurious.

Middle and upper valleys

Preservation of Holocene-age alluvium is rare in the middle and upper valleys of the Basento and Cavone systems; the majority of the valley fill has been eroded in the last few millennia, taking with it most of the archaeological record formerly present in the valleys. Consequently, the archaeological potential in these areas is very low, although there are clearly several sites preserved in favorable settings.

[79] E.g., R.D. Whitehouse 1981.

II. Methods and Analytical Tools

3
Survey Design and Field Methods

Alberto Prieto

This chapter presents the historical background and methodological framework of the ICA survey in the chora of Metaponto. It begins with a brief review of early research in the Metapontino and of the transformation of the landscape with the arrival of mechanized agriculture and industrial development, both of which served as catalysts for ICA's work. Then the development of the survey project is traced in the three main phases of its activities between the Bradano and Basento Rivers in the period 1981–2001, culminating in the creation of a Geographic Information System (GIS) for the territory; each phase is described in detail and compared with the others, in terms of its aims and methods. The final section evaluates the results of the Metaponto survey in the light of the ground conditions, research design, and field methodologies.

1964–1980: Context for the ICA Survey

Prior research

Landowner Michele Lacava began conducting dilettante archaeological-topographic research in the late 19th century.[1] Further study followed with the inception of the Soprintendenza per i Beni Archeologici della Basilicata in 1964 under the direction of Dinu Adamesteanu. In his first years as superintendent, Adamesteanu studied intensively the aerophotographic coverages of the Metapontino (see Ch. 31). He identified several dozen concentrations of stone, roof tile, and pottery—visible as discolorations of the surface soil—which he believed to be mostly the plow-disturbed remains of Greco-Roman farmhouses.[2] Adamesteanu encouraged his multinational team of excavators and topographers, including Giovanni Uggeri and a French contingent,[3] to explore the territory between the Bradano and Basento Rivers. The aim was to locate more sites and to position precisely as many of them as possible on large-scale maps.[4] In 1979–1980, ICA teams excavated a complex of Greco-Roman sites in the Sant'Angelo locality beside the Basento; at this time they also experimented with small pilot surveys of adjacent fields to create a local land-use context.

In these pioneering efforts, study areas were selected opportunistically and explored. They were typically fields that were either adjacent to roads or excavations, or were under intensive cultivation. Selection of study areas was often influenced by the indications of cultivators or laborers employed in the various infrastructural engineering projects that were quickly transforming the landscape (see below); these people were intimately familiar with individual fields and could lead researchers directly to sites that they had helped disturb. The intervals between field walkers and collection strategies were not standardized, and little note was taken of the topographical and geological variability of the landscape.

By late 1980 it became clear that the extensive, unsystematic approach of these early surveys could not provide satisfactory answers to some of the basic questions that were motivating research in the Metapontino and most field-survey projects active in the Mediterranean basin at this time:

1. What activities occurred in the landscape in a given period?
2. Where (on what landforms, soil types, or geological-geomorphological formations) did ancient populations perform these activities, and why did they choose to perform the relevant activities there?
3. Did use of the landscape change through time, as measured, e.g., by the relative proportions of land used for the various activities?

[1] Lacava 1891.

[2] Adamesteanu 1970a, 235–36; 1973, 52; 1974a, 80. See also Adamesteanu 1968, 253.

[3] Chevallier 1971.

[4] Uggeri 1969; Adamesteanu 1967b, 46; 1968, 253–54; 1973, 51–52. Several hundred sites were ultimately mapped between the Bradano and Cavone Rivers: see Adamesteanu and Vatin 1976, 116–17, as well as Chevallier 1971. Schmiedt and Chevallier (1959) encountered several ancient sites in their brief ground-truthing mission to identify division lines observed in the aerial photographs.

Figure 3.1 Irrigation pipeline trench, Lago del Lupo, 1981. (PD)

Figure 3.2 Newly-planted peach orchard with irrigation pumps at site 325, 1982. (CP)

Figure 3.3 Part of a plowed field recently cut from a sloping alluvial terrace surface at site 380, 1982. (CP)

An intensive, systematic coverage of a defined study area was needed, along with standardized documentation and artifact-collection procedures. ICA engaged Cesare D'Annibale as project and field director (a position he has held at various times down to the present) to define these parameters and direct a team in the field.

Agricultural intensification

A significant impetus behind the shift in research strategy was the major changes that were occurring in the landscape. These provided a unique methodological window of opportunity for intensive survey, while simultaneously creating a sense of urgency. The introduction of mechanized agriculture and the successful completion of the reclamation of the Metapontino in the 1950s by the Italian government's *riforma agraria* had caused the disturbance or destruction by deep plowing and excavation of dozens of archaeological sites in the coastal plain and on the lowest marine terraces.[5] The development of the basic rural infrastructure in the 1960s and 1970s damaged more sites,[6] but the destruction was limited because the work of creating roads, raised irrigation conduits, and drainage canals involved relatively shallow excavations along long, thin corridors of relatively low-lying land.

By 1980, however, the rate of landscape change in the Metapontino began to accelerate rapidly. Deep ditches were excavated far and wide across the marine terraces for the installation of the first subterranean irrigation pipelines (*Fig. 3.1*), a capillary system essential for the large-scale cultivation of the high-value orchard and garden crops —oranges, melons, peaches, vines, nuts, etc. (*Fig. 3.2*). These crops were to revive the regional economy through sale to other parts of Italy and the European Common Market, largely at the expense of the pervasive "traditional"

[5] Schmiedt and Chevallier 1959, 994; Adamesteanu 1968, 252; 1972, 452; 1973, 52; Chevallier 1971, 311–12.

[6] De Siena 2002, 25.

Figure 3.4 Recently planted vineyard, Basento valley, 1982. (CP)

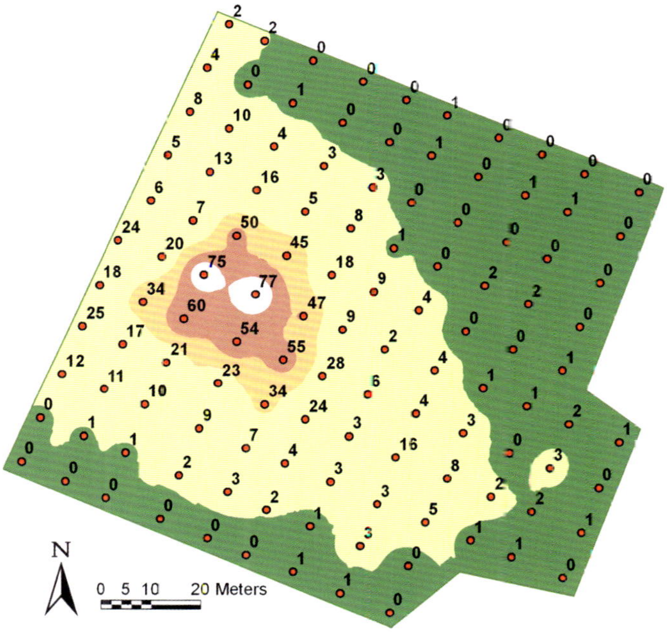

Figure 3.5 Map of the distribution of artifact density at site 16 in 2001. Red dots indicate 10-m intervals. The green area of the continuous (raster) inverse-distance–weighted surface indicates a range of 0–2 artifacts counted per 10-m interval. (AP)

crops, particularly wheat and olives. Local cultivators made greater use of bulldozers, backhoes, and hydraulic excavators to level and smooth the terrain (*Fig. 3.3*) and increase the amount of arable land; in addition, they used deeper-digging—and more destructive—plows to prepare the soil for the implantation of orchards and vineyards that contained hundreds of young plants, with feeder pipes bringing water to each root system (*Fig. 3.4*). This new activity accelerated the rate of site exposure across the heart of the Greek chora, and it was clear that the same intensive agriculture responsible for their discovery would damage the sites beyond recognition within a few years—the same fate that had befallen many of the sites nearer to the asty.

1981–90: The Intensive Survey

Problem orientation

The goals of the ICA survey in 1981 were consistent with those of the majority of intensive Mediterranean survey projects active at that time. The articulation of those goals began with the definition of the basic units of research, that is, the concentrations of artifacts and ecofacts that appeared on the surface of cultivated fields, unambiguously identified as foci of various activities, or "sites." Thanks to the concerted program of excavations across the Metapontino supervised by Adamesteanu between 1964 and 1980, there existed already a substantial corpus of knowledge about the various types of sites from all major historical periods (see Ch. 18). This corpus has continued to be enriched through regular excavations in the territory, mostly salvage/rescue work, under the direction of Antonio De Siena, the *ispettore* at Metaponto. The ICA survey team has, therefore, always been in the relatively rare and enviable position of having the following information in advance:

1. that various classes or types of sites were certainly present in the landscape;
2. what most of these classes/types are; and
3. what kinds of artifacts, ecofacts, and other remains are likely to appear on the surface as expressions of the buried remains.[7]

The primacy of sites in the survey methodology, particularly historical sites, was further reinforced and justified by their typical appearance on the surface as well-defined, discrete, but variably dense scatters of artifacts and stone separated by considerable expanses of archaeologically sterile (i.e., artifact-/ecofact-free) plow soil, no matter to which period they belonged (*Figs. 3.4–5*). Prehistoric activity, too, was considered mostly as centered on specific locations (sites) on the

[7] On the contrary situation in the Aegean basin generally, see, e.g., Cherry 1994; Osborne 1985a, 1987, 1992, and 1996.

basis of the prior discovery of several major cemeteries and settlements.

Unlike other Mediterranean landscapes—mainland Greece in particular—the Metapontino is relatively free of "background noise," that is, the nearly ubiquitous, continuous light scatter of artifacts that increases the difficulty of identifying sites. Therefore the monitoring of off-site artifact distributions was initially not considered necessary.[8] The effects of this choice on the recovery of the prehistoric record are discussed in a subsequent section.

This absence of pervasive artifact scatter in the Metapontino is likely due to its land-use history. Following the collapse of Late Roman Metapontine society in the 6th century AD, the landscape remained practically unoccupied and unexploited. Only in the early centuries of the second millennium did a very modest population reinstall itself on the land, which by then became the property of ecclesiastical entities or private agro-pastoral estates (*masserie*) attached to the Norman kingdom and its successors. The "feudal" phase of the Metapontino's history—which in reality ended only with the expropriation and redistribution of land in the agrarian reform of the 1950s—saw most of the population concentrated in the large hilltop settlements (Pisticci, Montescaglioso, and, after 1498, Bernalda). Such a situation curtails the capillary construction of towns/villages, farmhouses, silos, cult complexes, and other such structures that imprint their own archaeological "signatures" in the landscape on top of the ancient one, and even obscure it altogether if the activities associated with these later features are sufficiently intense and well-distributed (as is the case, for example, in much of Tuscany, Lazio, and Campania). In this period, the Metapontine landscape was subjected to relatively meager use, and the ancient sites were left largely undisturbed by agro-pastoral activity. Furthermore, until the arrival of deep plowing and other mechanical intrusions after the Second World War, ox-drawn plowing could generally penetrate only 20–40 cm; many ancient sites lay beneath more than 50 cm of accumulated deposits and therefore went untouched.

Centuries of geomorphological action (see Ch. 2) and half a century of modern mechanized agriculture have altered this situation, so that many formerly compact sites are now washing down slopes or lie spread across large areas and even entire fields. In 1982, the individual tombs of necropolis site 401, for example, were distinctively visible on the small promontory approximately 0.5 km northwest of, and above, the San Biagio sanctuary; the revisit in 2001 found only a weak, indiscriminate scatter of artifacts spread across the length and breadth of the vineyard (plot 2001.046) that had been planted in the meantime.[9] The wide range of forms/wares and large quantities of loose artifacts collected in several fields using the plot-based methodology of 1999–2001 (see below) strongly suggest the existence of Greek farmhouses in the relevant areas, but discernible concentrations that could be identified as sites were notably absent. Even in these worst cases, where artifact-rich sites have been severely degraded by highly intrusive agricultural work, the average distance between sites in the Metapontino is such that only very rarely do sites overlap physically. The absence of continuous off-site surface artifact scatter remains a characteristic feature of the Metapontino today, and sites remain the fundamental unit of landscape analysis.[10] This means that many of the theoretical models created in Aegean contexts to account for the spread of artifacts from historical periods across the landscape, principally premodern manuring,[11] do not apply at Metaponto, and so the agents responsible for the displacement of historical Metapontine artifacts must be of more recent and readily retraceable origin (see below).

Definitions of "site" in archaeological survey are notoriously problematic and contentious, but almost all agree on a basic requirement that a site be a locus of human activity. Many definitions assume, usually

[8] On off-site artifacts and site identification, see, e.g., Rupp et al. 1983, 28; Cherry 1983, 395; Bintliff and Snodgrass 1985, 130; Cherry et al. 1988, 159; Wright et al. 1990, 606; *Keos* 20–21, 45–48; Pettegrew 2001, 204–5; Bintliff and Howard 2004.

[9] When a survey returns to an area already investigated, a *site* documented in previous seasons is considered "revisited" because its existence is already known—there is an expectation of finding it; a *plot* is considered "re-examined" because the team looks at the entire field afresh in order to detect sites that may have gone unobserved in previous seasons.

[10] The phenomena closest to off-site artifact distributions in the Metapontino are generally encountered only at *masserie* (estate farms) active since the Medieval period, such as Sant'Angelo Vecchio. In these instances a series of historical factors have operated to create heavy artifact "haloes" immediately around the core farm structures (which are typically grouped closely together), principally construction of new buildings, repeated remodeling/reroofing of the oldest buildings, and waste-discard activities. Yet these haloes are not truly independent artifact distributions, but rather extensions of the *masserie* themselves.

[11] For explanations of off-site distributions, see, among others, Bintliff and Snodgrass 1988; Alcock/Cherry/Davis 1994; Fentress 2000; Given 2004.

implicitly, that "sites" are characterized by the presence of artifacts.[12] Such was the case at the outset of the Metaponto survey (see above), to judge from the archaeological record derived from excavations. But the picture quickly became more complex. In the second year the survey team encountered two quarries, recognized by the presence of partially cut blocks in exposed stone outcrops (sites 141, 171); although they offer no collectible artifacts, these quarries (along with site 736) are clearly loci of human activity and should be counted as sites. The insistence of other published definitions on a minimum density of artifacts and/or the presence of architectural remains is not satisfactory for Metaponto because it does not allow for the extreme variability in site preservation and surface manifestation resulting from agricultural activity.[13] Yet other definitions assume, again usually implicitly, that only loci of certain activities, in particular habitation, are fit to be considered "sites."[14] Such definitions apparently do not consider individual tombs to be sites, for example, even though tombs are without question loci of human activity. The difference, as far as tombs are concerned, is that the activity in question is generally limited to a single instance: by definition, activity at a tomb—except for acts of veneration—should not be repeated.

For the purpose of the discussions that follow, a *site* is defined as a locus of *deliberate* human activity that may or may not be characterized by the presence of durable remains (artifacts and ecofacts).

From the start, the strictly site-based orientation of the Metaponto survey, as well as the general absence of off-site material, entailed that the project would be essentially a census, intended to identify, locate, and document discrete vestigial loci of ancient land-use activities, in particular habitation/settlement (including domestic industry), burial, industry, and cult. From the numerical and physical distributions of the various types of loci would be derived models of their relationships to each other and to natural resources, as well as their relative importance and/or scale in the various historical periods of interest, from the Neolithic to recent times. The survey was designed to provide a broad history of the Metapontine landscape that would complement the localized pictures obtained from ICA's many excavation projects, and provide a spatial framework for them. Though all periods and types of sites would be sampled and studied, the greatest emphasis would inevitably be placed on settlements of the Greek period, by virtue both of ICA's research interests in Greek rural life and of the overwhelming importance—and equally overwhelming tangible record—of the Greek presence in the Metapontino. The intention was to reconstruct the social, economic, and even political history of the polis in Greek-occupied southern Italy through the expected variation in intensity of activity from period to period, as measured by changing numbers and arrangements of sites.

Study areas

The initial survey design considered the roles of topography, geology/geomorphology, and socio-political conditions in the history of land use in the Metapontino. One of the original goals was the reconstruction of settlement patterns in both the Greek chora and the eschatia, the "marginal" zone marking the transition to the indigenous Italic world, in order to provide further data on Greek-indigenous relations during the "precolonial" and colonial periods. This was intended to complement the data obtained from excavations at Incoronata and elsewhere.[15] Accordingly, a series of compact study areas—transects—were envisioned, placed roughly parallel to the trend of the coastline. These would begin some 6 km from the shore, stepping successively inland towards the exposed clay hills (*calanchi*) that mark the transition from the Pleistocene marine terraces to the Lucanian mountains, and would stretch from the Bradano to the Basento. The census created by these transects would include sites from most of the topographical and geological/geomorphological features of the Metapontino: fluvial floodplains and alluvial terraces, river and stream valley walls and related transitional features, and Pleistocene marine terraces. The fourth major geological/geomorphological unit, the coastal plain, was intentionally excluded because it had been disturbed by modern agricultural and building activities, as well as being covered regularly with alluvial deposits.

[12] In the Argolid Exploration Project, for example, "the term 'site' is . . . nothing more than a convenient way to designate a locality where cultural materials were found, apparently belonging together" (*Greek Countryside*, 221).

[13] See, e.g., Rupp et al. 1983, 28; the various ideas summarized in Schofield 1991, 3–4; and the statistically based concept in Keay and Millett 1991, 137.

[14] Keay and Millett 1991, 132, 133.

[15] See Carter 2006a, Ch. 2.

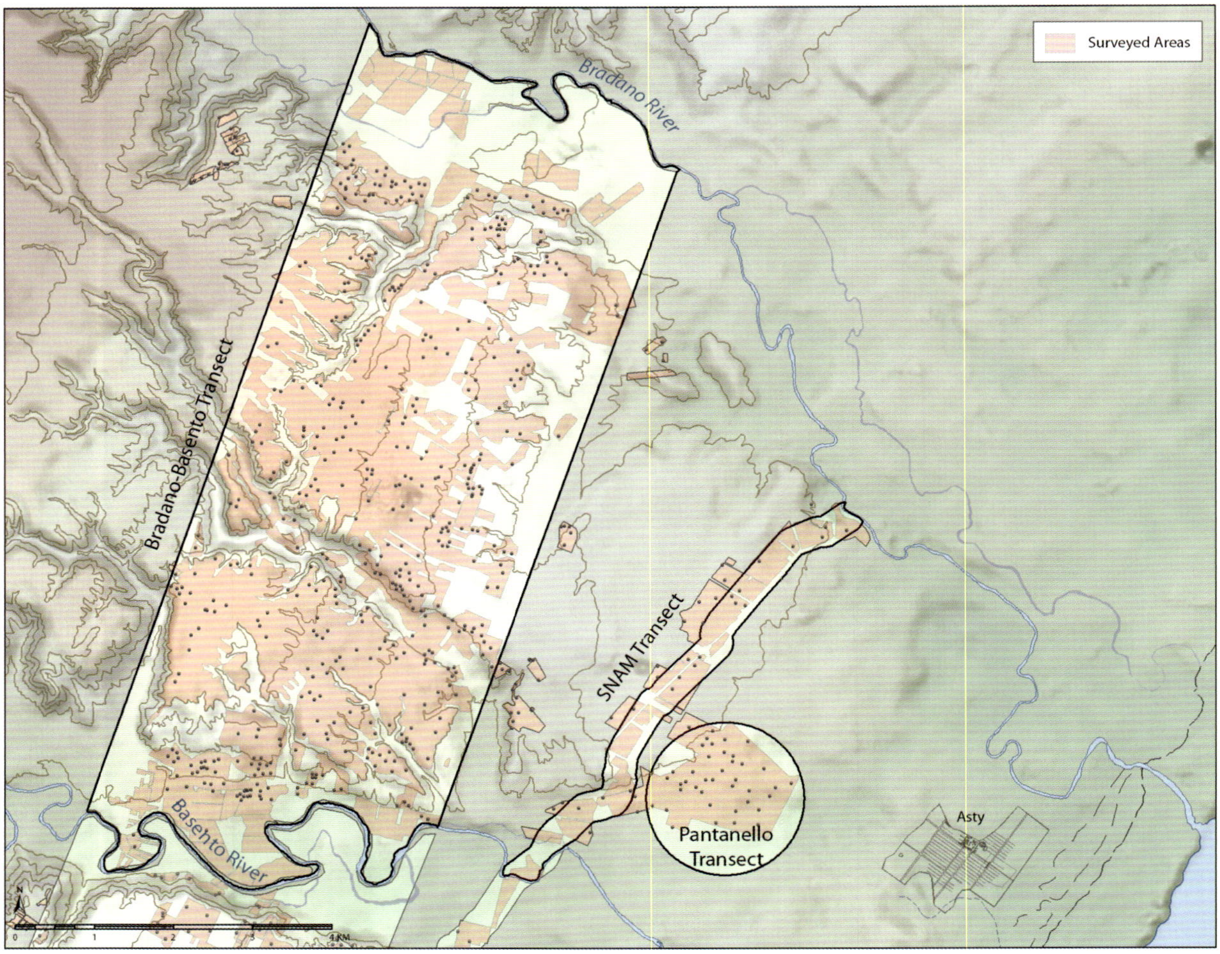

Figure 3.6 ICA Bradano-Basento study areas, showing transects imposed on elevation map. (JT)

The initial transect—hereafter referred to as the Bradano-Basento transect—was roughly 10 km long and 4 km wide and comprised 44.5 km^2. This quasi-rectangular area was demarcated approximately midway between the coast and the eroded badlands, some 9 km from the shore along its closest (southeastern) boundary (*Fig. 3.6*).[16] The transect was intentionally placed across the heart of the Greek chora, approximately 6–12 km from the asty. This area was known to contain numerous sites of Greek and Roman date, which had been documented in the excavation efforts and extensive walking surveys of the 1950s to 1960s, or noted in the aerial photographs, along with the greatest concentration of remains of the division lines.[17] This was also the area subjected to the most intensive mechanical modifications in 1980–81, as mentioned above. The distribution of geomorphological zones (*Fig. 3.7*) in the Bradano-Basento transect was dominated by the Pleistocene marine terraces (19.1 km^2/43%) but included significant samples of fluvial floodplains/alluvial terraces (11.9 km^2/27%) and river/stream valley walls and related transitional features (13.5 km^2 /30%).

The survey team occasionally worked outside the main rectangular transect in order to explore areas pertinent to larger research questions. A small area around the fontana at Cappa d'Amore, immediately

[16] The four approximate corners of this transect are defined by the Masseria D'Alessio/Giardinetto locality, the Masseria Panetto/Serra Marina locality, the sanctuary at San Biagio, and the Fosso della Guardia stream/Cozzo della Guardia locality immediately east-southeast of Bernalda.

[17] One of the principal goals of the survey project was, in fact, the fulfillment of Adamesteanu's ambition of documenting and analyzing the spatial relationships between the observable Greek sites and the network of division lines in order to determine the chronology and function of the latter. Adamesteanu's excavation program had concentrated on Greek farmhouses closely aligned to the central division-line feature in the Lago del Lupo and Demanio Campagnolo localities, and from the results he concluded that the lines were created in the Archaic period and had served to organize the territory into lots (Adamesteanu 1968, 253–54; 1970a, 234–36; 1973, 51–60).

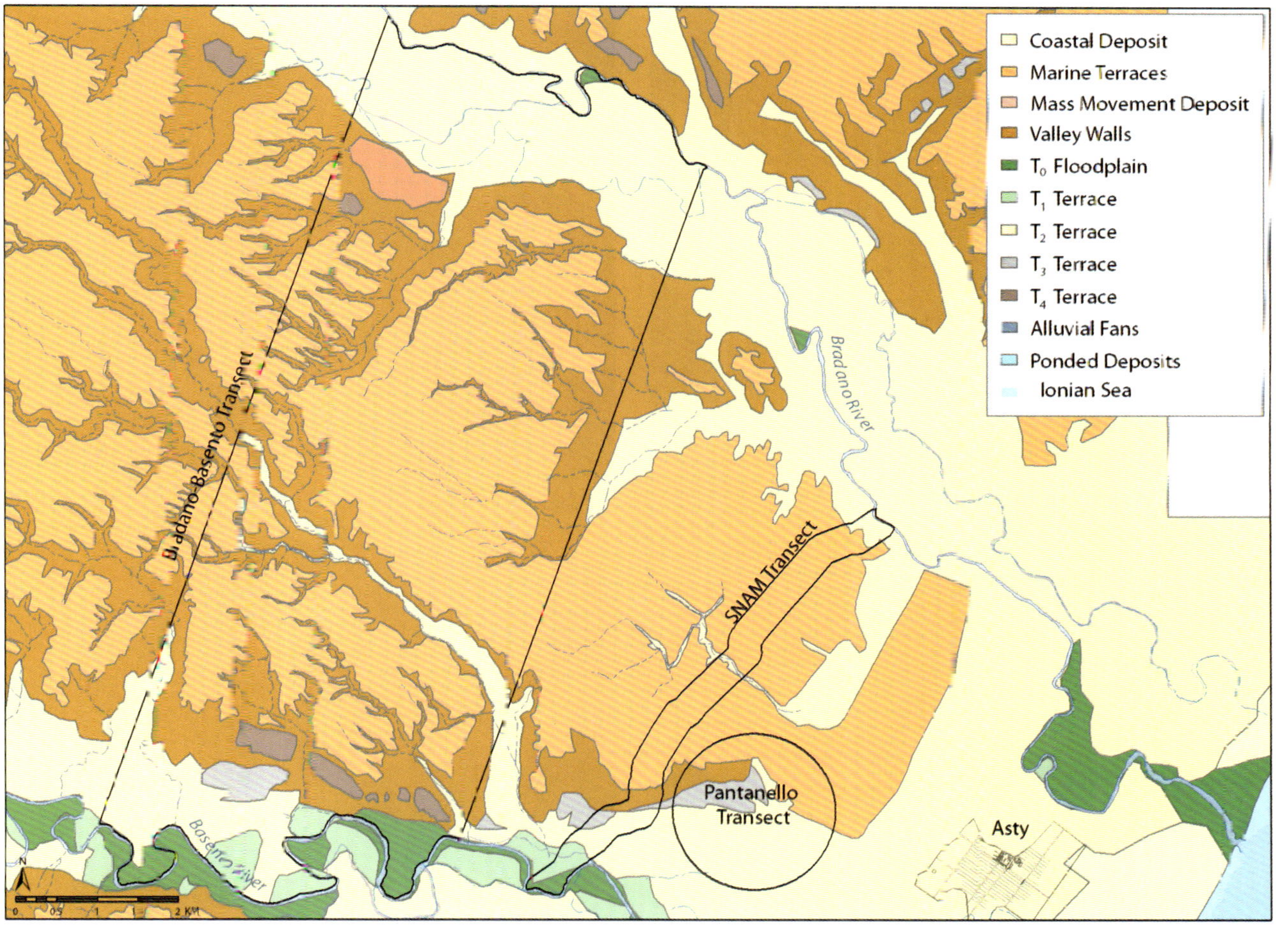

Figure 3.7 ICA Bradano-Basento study areas, showing transects imposed over geomorphology map. (AP/JT after J. Abbott)

southeast of the transect in the center of the chora, was surveyed to provide a land-use context for the Archaic Greek farmhouse and kilns excavated there in the 1960s (sites 33–42).[18] A short stint in the Serra Marina locality, also just southeast of the transect near its northern edge overlooking the Bradano River, produced three sites (197–99).

Field Procedures

Though the survey project aimed for total coverage of the designated study areas, the diversity of the modern agricultural landscape rendered many fields—even entire localities—inaccessible, or obscured large portions of the ground in a shroud of vegetation. In these cases examination of the surface was a futile exercise. Mechanical disturbances in the Pezza Grande locality near the northeastern edge of the main transect, for example, were so destructive that artifact clusters were displaced and dispersed over significant distances, and consequently site locations could not be assessed accurately. Topography and geomorphology also restricted survey coverage. The floors and low alluvial terraces of the Bradano and Basento valleys were quickly determined to be largely unproductive because of the repeated mass wasting and deposition events that had buried and obliterated the ancient land-use surfaces under meters of alluvial and colluvial fill from higher localities (see below and Ch. 1–2). The large expanses of eroded stream valleys and the unstable scarps forming the transition between the marine terraces and the fluvial systems were generally too precipitous for survey except along the upper and lower edges, where sites could be detected by trails of artifacts washing downslope (*Fig. 3.8*).

In practice, the team followed the standard procedure of many survey projects in examining only

[18] The plan is published in Adamesteanu 1974, 86.

Figure 3.8 Remains of a Post-Medieval kiln (site 379) washing out at the bottom of a valley wall, 1982. (CP)

Figure 3.9 An uncultivated and overgrown section of a vineyard at site 221, 1982. (CP)

fields offering uniform conditions of vegetation/land use and reasonably good visibility.[19] These were typically orchards, vineyards, and actively cultivated fields—i.e., any field in which large portions of the surface were exposed to harrowing or plowing—but even scrub-covered slopes and certain uncultivated or overgrown fields were examined if enough of the surface was exposed (*Fig. 3.9*). Certain types of ground cover, such as cut wheat and dense weed growth (in uncultivated fields), were frequently bypassed because they offered the poorest visibility conditions, often close to zero (*Fig. 3.10*). Thus, of the 44.5 km^2 area contained within the main rectangular transect, 21.3 km^2 were studied between 1981 and 1984. Although the actual coverage appears small when compared to the intended coverage, it should be noted that nearly one-third of the surface area contained within the limits of the main transect consists of stream valleys/walls and related unstable landforms (13.5 km^2), and of these topographic features only a little more than one-third (4.2 km^2) could be surveyed where the scrub was not impenetrably dense or the scarps too steep (*Fig. 3.11*).[20] In the end, nearly one-quarter of the entire transect was not amenable to survey. Similarly, although the circular transect around the Pantanello sanctuary encompassed an area of 3 km^2, only 1.7 km^2 of its surface was accessible, as well as an additional 0.2 km^2 of territory adjacent to it.

Figure 3.10 Wheat field after harvest at site 198, 1982. (CP)

The standard interval between field-walkers was 10 m, employed mainly in empty, plowed fields, harvested wheat fields, and uncultivated fields. In cultivated fields, where rows of vegetation or farming equipment (vine stays, irrigation tubes, etc.) were present, the team was compelled to walk in the unplanted spaces or between rows of equipment, at intervals of usually less than 10 m and often as little as 3 m, in order to avoid damaging the crops and

[19] Cherry et al. 1988, 162; Wright et al. 1990, 604.

[20] By contrast, 19% (2.3 km^2) of the fluvial floodplains and alluvial terraces within the transect boundaries were surveyed, and 77% (14.7 km^2) of the marine terraces.

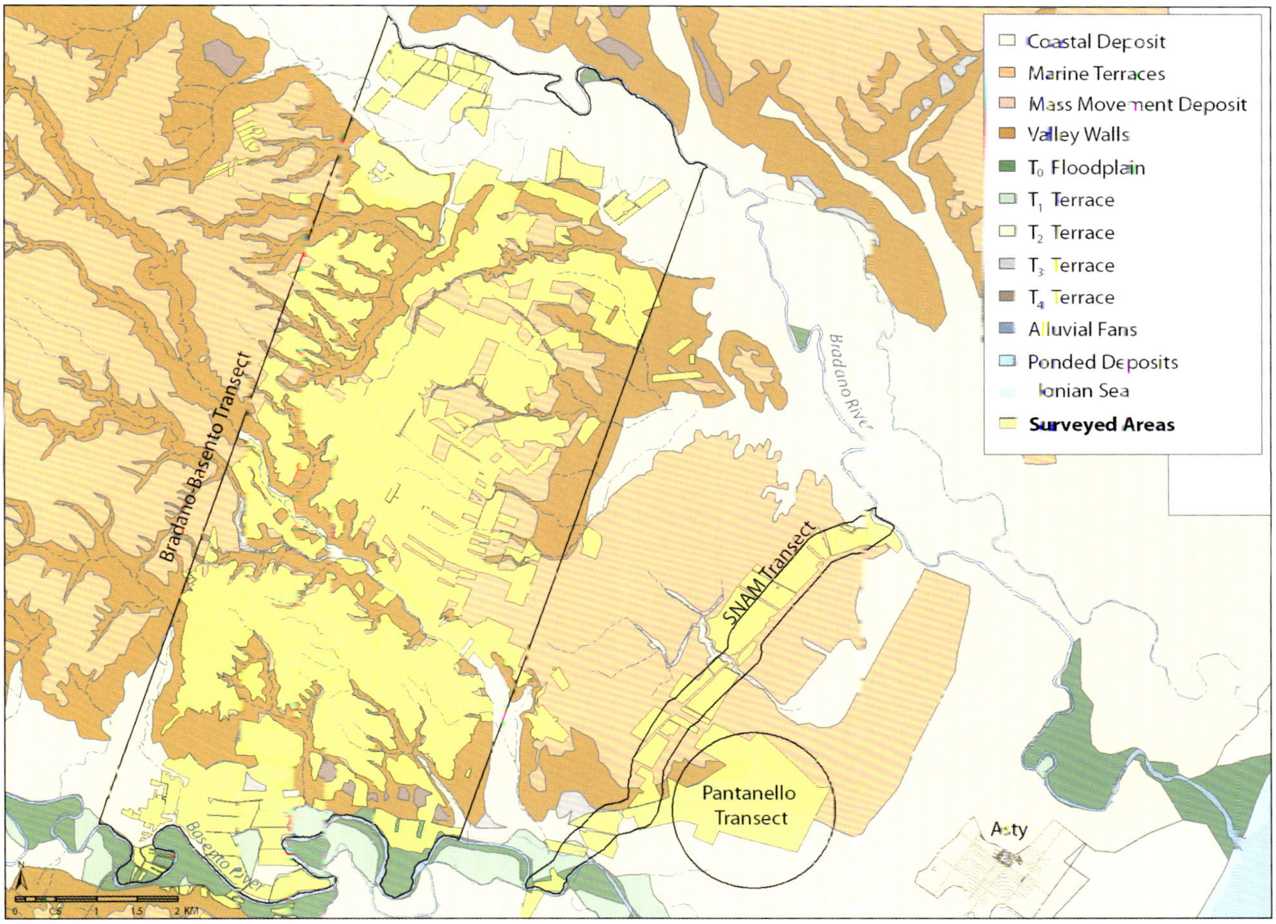

Figure 3.11 ICA survey coverage over geomorphology map. Bright yellow indicates surveyed areas. (AP/JT after J. Abbott)

equipment (*Fig. 3.12*). The maximum 10 m survey interval was a compromise between speed and intensity: while the crew could not cover ground as quickly as with a wider interval (15–20 m), the probability of perceiving even the smallest known sites (isolated tombs, for example the remains of which cover an area of just a few square meters), as well as those most difficult to distinguish from the surface (prehistoric sites, because of the typically dark colors of the lithic and ceramic artifacts), was significantly increased. The designation of an activity locus as a site was a purely subjective decision at the discretion of the field director, based on factors such as topographic/geomorphological setting, physical condition, artifact classes present (if any), and apparent function.

Site positions were marked on sheets of the black-and-white 1:10,000-scale topographic map prepared for the Soprintendenza per i Beni Archeologici della Basilicata in Potenza by ESACTA Srl, a Rome-based

Figure 3.12 Orchard, site 261, 1982. (CP)

Figure 3.13 Documentation of a site in progress, 1982. (CP)

photogrammetry and cartography company.[21] The positions were determined by counting measured paces and marking the obtained distances in reference to natural and artificial features (streams, road intersections, structures, plot boundaries) visible on the ESACTA sheets.

The "first generation" of the intensive field survey did not treat modern agricultural fields ("plots" in later Metaponto survey terminology) as archaeological entities, so that information on the land use/cover, boundaries, and other characteristics of individual fields was not recorded, but the boundaries of the cumulative survey coverage were sketched on the sheets of a 1:25,000-scale topographic map.

Figure 3.14 Site 199, located in multiple modern land-use contexts, 1982. (CP)

[21] Adamesteanu 1971, 483; 1972, 451; 1974a, 82; Adamesteanu and Vatin 1976, 111. The map was prepared in the early 1970s from low-altitude aerial photographs taken in 1967–68, in 16 sheets (60 x 55 cm/23.5 x 21.5 in, approximately equal to 5.5 x 4.5 km on the ground) covering the Metapontino from the Bradano to Cavone Rivers and from the coastline to an imaginary line passing immediately behind the Cozzo Presepe promontory and the San Vito locality. The sheets covering the various Bradano-Basento survey transects are "Masseria delle Palme," "Masseria Girifalco," "Lago del Lupo," "Serra Marina," "Bernalda," "Masseria S.Angelo," and "Case Teresa." All lie within *foglio* 201/Matera of the *Istituto Geografico Militare* (IGM) standard 1:100,000-scale *Carta d'Italia* and *fogli* 491/Ferrandina, 492/Ginosa, and 508/Policoro of the IGM's 1:50,000-scale M792 topographic series (corresponding to Sections II NW/"Masseria Girifalco," II SW/"Metaponto," III NE/"Masseria Gaudella," and III SE/"Bernalda" of the IGM's 1:25,000-scale topographic subseries). For the mission and history of ESACTA, see Boemi 2003b, 39.

Figure 3.15 Moderate pottery/tile scatter, site 113, 1981 (CP)

Figure 3.16 Very heavy pottery/tile scatter, site 213, 1982. (CP)

Figure 3.17 Worked stone blocks, site 160,1982. (CP)

Figure 3.18 Exposed section of stone wall, site 242 1982. (CP)

Upon discovery, a site was assigned a serial number (beginning with 1), and its description was recorded in small ruled notebooks (*Fig. 3.13*), the pages of which had been prepared with the following entries in a standard format:

1. Immediate topographic setting
2. Elevation in meters above sea level (masl)
3. Current land use/cover: one primary use and up to two secondary uses (*Fig. 3.14*)
4. Name of nearest potential source of water (spring, river, or stream)
5. Artifact scatter area, expressed as the product of maximum scatter length and maximum scatter width pace-measured to the nearest 5 m (e.g., 40 x 30 m)
6. Tile scatter density (ordinal scale: very light, light, moderate, heavy, very heavy) (*Figs. 3.15–16*)
7. Other artifact scatter density (ordinal scale: very light, light, moderate, heavy, very heavy) (*Figs. 3.15–16*)
8. Uncollected artifacts (e.g., pithos fragments, foundation stones, tomb coverings) (*Fig. 3.17;* see also Ch. 19)
9. Features of interest (e.g., changes in soil color, ash deposits, slight rises indicative of very localized densities of habitation debris, exposed structural remains, etc.) (*Fig. 3.18;* see also Ch. 19)
10. Types/classes of artifacts collected.

The artifact collection strategy was designed to capture the maximum information about site function and chronology. The team combed the entire area of the scatter at random, up to the very edges, and collected all visible diagnostic ceramic artifacts—rims, bases/toes, handles, and decorated or otherwise unusual body sherds—of all classes from all periods (with the exception of artifacts produced in the last 100 years, such as industrial roof tiles, glazed ceramics, and fragments of agricultural equipment), as well as all lithic artifacts, loom weights, terracotta figurines, architectural terracottas, a sample of roof tile fragments, exposed human bones, and any other artifacts and ecofacts that were functionally or chronologically determinant, or otherwise unusual (*Fig. 3.19*). Even if the sampling strategy eschewed accurate reconstruction of the relative proportions of the various artifact classes (typically accomplished by total collection of artifacts within defined sample areas), it did nevertheless guarantee the representation of all artifact classes present at the site in the collected assemblage. If, at the end of collec-

Figure 3.19 Artifact collection in progress in a typical plowed field, 1982. (CP)

tion, the assemblage appeared to contain no diagnostic fragments of a given artifact class noted on the surface (e.g., cooking ware), then one or more body fragments belonging to that class were collected as documentation of its presence in the site assemblage.[22]

In the first summer of activity (1981), the survey team began work on June 1 and finished on July 23, documenting 139 sites (1–139) in thirty-six working days. In 1982 the survey crew operated in the summer (June 15–July 29) and then again in the fall (September 6–October 21) in order to walk many fields inaccessible in the summer because of the presence of standing or cut wheat; 272 sites (140–411) were documented in fifty-eight working days (thirty-three in the summer, twenty-five in the fall). At the end of the fall campaign the survey effort was briefly redirected to develop a localized land-use context for the ICA excavations at the Pantanello sanctuary. A small transect was designed in the form of a circle with radius of 1 km, centered on the sanctuary, and a large amount of the landscape contained within it and immediately outside of it was surveyed, with 42 sites (412–53) located in six working days (*Fig. 3.6*). In 1983, an opportunity to survey the chora of Croton arose, and so the principal efforts and energies of the survey project were shifted there. As a consequence, the original survey design involving multiple parallel transects was suspended. A comparatively shorter survey campaign was undertaken at Metaponto, again divided between two seasons, late spring (May 23–June 4) and early fall (October 4–21); 83 sites (454–536) were documented in twenty working days (eleven in the summer, nine in the fall). The final season, in 1984, was the briefest of all (July 3–7) and identified 23 sites (538–60) in five working days.

Revisits

In the first four years of activity D'Annibale had noticed that certain activity phases of many sites, in particular the prehistoric and Medieval phases, were represented on the surface by very weak artifact signatures. His knowledge of the excavated sites and other sites discovered in the intensive survey led him to reason that this apparent weakness was likely caused by stochastic mechanisms of artifact exposure, which in many cases resulted in the proportional underrepresentation of prehistoric and Medieval artifacts within the great mass of Greco-Roman artifacts. Taking advantage of another major phase of agricultural intensification in 1988–90 that changed the crops and cultivation regimes of many fields, D'Annibale revisited many of the sites characterized by a weak prehistoric or Medieval artifact presence in order to compare the quantity and quality of available artifacts from both periods.[23] Where more of the artifacts in question appeared, he collected an additional sample to fill out the functional and chronological picture created by the original artifact collections.

Since this revisit effort was motivated entirely by gaps in artifact collection, no updated information about site conditions or characteristics was recorded.

[22] The background and methods of the 1981–84 survey project are described in D'Annibale 1983a, 5–8; 1983b, 191–92; Carter and D'Annibale 1985, 148–49; S.M. Thompson 2004, 69–70.

[23] On the reasons for resurvey, see Rupp and Keller 1983, 4; Stanley Price et al. 1983, 32–33.

1992–94: Changes in Orientation and Method

A change in the leadership of the survey effort in 1992 brought changes to the overall design and field methods. The new director, Jon Morter, maintained the survey's intensive character, with typical walking intervals of 5, 10, or 15 m,[24] but turned the project's focus away from the original plan of exploring more of the Bradano-Basento area in order to delineate the "boundary" between Greek and indigenous settlement. Morter instead wanted to create a sample of sites in the southern half of the chora between the Basento and Cavone Rivers to compare to the sample obtained between the Bradano and Basento in terms of site density, forms of land use, and chronology of land use. A new transect was designed as a direct continuation of the original main Bradano-Basento transect, stretching southwest from the south bank of the Basento to the north bank of the Cavone. Although survey efforts between 1992 and 1994 were concentrated in this new transect,[25] the team made several brief visits to the Bradano-Basento area in seven days between June 29 and July 15, 1993, in order to examine freshly disturbed fields. Thirteen new fields were surveyed on the lowest marine terraces, between the circular Pantanello transect and the original rectangular transect, adding 0.7 km^2 to the total coverage; inside and immediately adjoining these fields 18 new sites were documented (622–39). At the same time, several of the sites discovered in 1983 were revisited (372, 401, 413, 415, 451–52), and new artifact collections were made at most of them.

Under Morter's direction sites continued to be described as above, with several important modifications. The distinction between roof tile fragments and all other ceramic artifacts (see above) was no longer considered meaningful, and therefore the two original ordinal-scale approximations of artifact density became a single measure using the same scale, with the result that the reported artifact density was based on all visible artifacts. The Basento-Cavone transect was generally cultivated in more visually obstructive crops, which adversely affected site identification and artifact collection. Morter became aware of the problem and instituted the recording of visibility on an explicitly impressionistic, ordinal scale (very poor, poor, moderate, good, very good, excellent) as an additional consideration in the evaluation of artifact scatters and the data recovered from them. The greater intensity of mechanical landscape alteration in the Basento-Cavone area, and its powerful effects on site preservation, prompted an attempt at systematic recording of observations of work in progress or recently completed. Sites were represented as polygons of proportional dimensions, rather than as simple points, on the ESACTA topographic map. Artifact collection continued to target diagnostic items and all chronologically sensitive items through random, but complete, inspection of the entire site surface.

An important consequence of the 1992–94 methodology shift was the discontinued recording of the types, quantities, and conditions of uncollected artifacts and items associated with ancient land use. Notes were not kept regarding the large fluvial cobbles and blocks of conglomerate, sandstone, and calcarenita di Gravina employed in virtually every Greek and Roman structure, from tombs to farmhouses to temples, as well as the occasional fragments of pithoi, mill stones, or stones from presses, which were too large to be removed from the site or were reemployed in modern constructions. Notation of other significant features such as soil discolorations, slight rises, and ash deposits also ceased. Such observations, though nonsystematic and subjective, have proved an invaluable aid in determining the functional class of a given site.

A fundamental change in survey methodology between 1981–84 and 1992–94 was the introduction of the detailed documentation of entire fields as a complement to the documentation of the sites within them. Noting the serious disturbance of the archaeological record between Basento and Cavone caused by pervasive deep digging and soil transport, Morter anticipated frequent encounters with isolated artifacts and disembodied scatters. Beginning in 1992, therefore, surveyed fields (plots) were treated as independent archaeological entities, whether or not they contained sites, so that there would be a detailed record of the conditions of examined areas and off-site artifacts could be integrated into the survey corpus. A plot was defined as a territorial unit characterized by a single type of land use/cover, whether anthropic (agricultural) or natural (e.g., thicket or woods).

The plots were labeled with a serial numeric system parallel to the site numbering system, begin-

[24] Certain plowed fields and other types of cultivation (e.g., melons, fruit orchards) have recorded walking intervals of 1, 3, 6, 8, and up to 20 m.

[25] The results of this and subsequent survey seasons between Basento and Cavone will be presented in *Basento–Cavone* (forthcoming).

ning at 1 and continuing in sequence from season to season. Beyond the physical boundaries traced on the 1:10,000 map using paces and topographic landmarks, plots were described in terms of salient features such as the immediate topographic setting, land use/cover, and surface visibility (ordinal scale, as above). Nonsystematic observations on soil type, mechanical landscape alteration, surface treatment (harrowing, normal plowing, deep plowing), and the presence of intrusive anthropogenic features (reservoirs, subterranean pipelines) were also included in the plot documentation.

The introduction of plots allowed the survey to account for off-site artifacts found in small quantities in the Basento-Cavone area. All artifacts of chronological or functional interest (i.e., all prehistoric artifacts, all morphologically significant ceramic sherds, all decorated ceramic sherds) within a plot were collected and identified by the plot in which they were found. Most of the off-site artifacts attributable to historical periods (roof tile and vase fragments) clearly belonged to concentrated artifact scatters that had been disturbed by certain types of recent mechanical activity such as trenching, leveling, and soil replacement, the signs of which were readily visible in most plots.

The isolated prehistoric artifacts, however, could not be generally ascribed to displacement because of the rarity of prehistoric sites in the Metapontino relative to those of Greek and Roman date. In most cases these had to be the products of original discard activity. The organization of the survey by plots therefore revealed an important new aspect of the local prehistory—pervasive discard activity—that radically altered the perceived land-use patterns. This aspect was entirely absent from the previous research in the Bradano-Basento transect, either because the team was not attuned to isolated prehistoric artifacts and did not recognize them, or because the survey design did not consider isolated prehistoric artifacts to be meaningful and ignored them. (The problem was somewhat rectified by the revisit/resurvey project of 2000–2001; see below.)

1999: Rescue Archaeology, Artifact Mapping, and GIS

Problem orientation and study area

In 1999, significant petroleum and methane reserves were discovered by the Italian energy consortium Azienda Generale Italiana Petroli/Ente Nazionale Idrocarburi (Agip/Eni) at Monte Alpi in the mountains of southwestern Lucania. Another arm of the consortium, Società Nazionale Metanodotti (SNAM), was contracted to design and construct the subterranean pipeline that would transport petroleum from Monte Alpi to the Agip refinery located 120 km to the northeast at Taranto, passing through the entire length of the Metapontino. The projected course of the pipeline entered the Metapontino at a point 10 km from the coast in the Cavone valley, crossed the lowest marine terraces of the Basento-Cavone area (San Basilio and San Teodoro localities), entered the Basento valley at a point 500 m east of the protohistoric site of Incoronata, crossed the lowest marine terrace between the Bradano and Basento Rivers (Pizzica locality), and finally exited the Metapontino at a point circa 1 km northwest of the Temple of Hera (Tavole Palatine) and 4 km northwest of the ancient asty (*Fig. 3.6*). At a later stage a second pipeline was added to the project to carry the methane to Taranto. The second pipeline, though essentially parallel to the first, was located circa 100 m further north and began only in the Basento valley near Pantanello.

The Soprintendenza per i Beni Archeologici della Basilicata invited ICA to perform an intensive survey of the areas in the direct path of the two pipelines in order to identify and locate all sites potentially affected by the construction. The sites discovered would be ranked by likely significance and preservation, and then the most promising of them would become the objects of salvage excavations before and during construction of the pipelines.[26] The fieldwork in the 6 km of corridor between Bradano and Basento was completed in a campaign of seventeen days distributed over two seasons (June 23–August 4 and September 23–October 7) under the direction of Stephen M. Thompson. In total, 67 fields were examined, adding 1.9 km^2 to the overall Bradano–Basento survey coverage, and 26 new sites were documented (sites 726–45, 760–64, and 769). Additionally, one previously documented site (413) was revisited.

[26] See *Ancient Territories* 1999; Nava 2003b, 2000.

Figure 3.20 Creation of a baseline for density mapping, 2001. (AP)

The staging zones for the assembly of the pipe segments and the excavation of the trenches would occupy only about the first 20 m to either side of the actual trenches. The intrusion of the construction zone into most fields was therefore minimal (20 m), and the total width of the affected area, including the gap between the pipelines, was 150 m. Thompson conservatively defined the new survey transect as the area 100 m on both sides of both pipelines, but in practice he examined all accessible fields that were either fully or partially contained within this limit, up to circa 500 m from the pipeline paths, recognizing a unique opportunity to create another small but important intensive coverage in an area close to the asty. As in the previous two generations of the survey, fields that presented unusually poor visibility conditions or were otherwise inaccessible were not examined, and so occasionally the width of the coverage was less than the stipulated minimum of 100 m. The standard walking interval was fixed at 10 m, as it had been in 1981–84, although it was often collapsed (to 6 or 8 m) or expanded (to 12 m) to accommodate the row spacings of certain crops

The documentation of sites and plots in 1999 largely followed the system used in 1992–94, which in turn borrowed heavily from the original site recording system of 1981–84, with a few significant innovations. Plot boundaries continued to be traced on the 1:10,000 topographic map, using an orienteering compass and paces to determine distances and bearings from landmarks. A subjective evaluation on an ordinal scale (none, low, moderate, high) of the density of off-site artifacts in plots was introduced. The collection of chronologically or functionally significant off-site artifacts in plots continued. The notation of uncollected artifacts, building stones, and other objects associated with past land use remained sporadic, while visible changes in soil texture or color and similar variations in the terrain were noted once again.

Artifact counting/mapping

Thompson wanted to introduce to the Metaponto survey a more objective basis for calculating the sizes of sites, modeling the effects of surface visibility on the perceptibility/recognizability of sites and on the perceived density of on-site artifacts, and describing the spatial patterning of selected artifact classes. He consequently devised a system for counting and recording surface artifacts across sites and selected plots (see below) and instituted the recording of surface visibility for both plots and sites as a percentage (in increments of 5%) representing the visible fraction of the ground surface, rather than on an ordinal scale as previously. "Corrected" artifact counts and densities in plots and sites were to be obtained from the recorded raw counts by multiplying the latter by the reciprocal fractional value of the visibility rating (e.g., 10 artifacts in conditions of 20% visibility would become 50 artifacts in conditions of 100% visibility).

The standard procedure for counting and mapping artifacts at most sites, where visibility and/or physical access were not severely hindered (e.g., by dense vegetation, excavation, or the extensive presence of concrete), involved the creation of an orthogonal grid along the edges of the artifact scatter. A baseline was laid along one edge (and extending several meters beyond it) with a 100-m reel tape (*Fig. 3.20*); fixed intervals along that baseline (typically 10 m, but also 5 m) were then marked with wire-mounted colored plastic surveyor's flags as far as was necessary (*Fig. 3.21*). A transverse line was laid at a right angle to the baseline—measured by an orienteering compass—so that the ends of the two lines met to form an origin point for the grid, and fixed intervals were marked along this transverse line as far as was necessary. The reel tape was then used to form the missing two sides of a square or rectangle (depending on the size and shape of the scatter), which were marked at the same

Figure 3.21 Flagging a baseline, 2001. (AP)

Figure 3.22 Counting and collecting during density mapping, 1999. (AP)

fixed intervals with flags. The resulting grid could be extended to cover any size of scatter by projecting the baselines and transverse lines.

A scaled representation of the grid was created on one or more sheets of paper with a millimeter grid, on which were also recorded the positions and properties of nearby salient topographic landmarks (roads, electrical towers or poles, irrigation canals, bridges, terrace edges, gullies, etc.). The team then lined up along one side of the grid, starting from a corner, with each person standing at one of the fixed 5- or 10-m intervals behind a flag. The team walked forward in line as far as the next fixed interval (again, typically 10 m, but occasionally 5 m), visible in the colored flags planted along the edge line parallel to the direction of travel. As the team proceeded, the members used hand-held mechanical counters to record the number of artifacts in their immediate path, constrained to approximately 1 m to either side of the line of travel (total of about 2 m). The team counted separately all artifact classes of special functional or chronological interest: lithics, prehistoric pottery (both impasto and fine wares), black-gloss pottery, terracotta figurines, loom weights, Roman fine wares (grey ware, TSI, and ARS), and Medieval ceramics. Each walker's general and/or special artifact count was recorded in the appropriate position on the scaled map. As they walked and counted, the team members also collected all diagnostic artifacts in their immediate path (*Fig. 3.22*).

At most sites, upon completion of the counting, recording, and collecting, the team walked across the site a second time solely to collect any remaining diagnostic artifacts, still moving parallel to their counting/collecting traverses, but following fixed (though unmarked) intervals between the flagged positions.[27] The artifact collection strategy introduced in 1999 therefore continued to use the definition of diagnostic artifacts formulated in 1981, but abandoned total collection in favor of a partial collection that could be equated intuitively to total collection through multiplication by the reciprocal of the fraction of the surface actually examined.[28]

[27] For example, in a typical grid of 10 x 10 m intervals, one, two, or even three members of the team would be assigned to collect artifacts at regular intervals between two flagged positions. One person collecting artifacts between the two flagged positions meant that the effective collection interval of the site was 5 m. Two people collecting artifacts between the two flagged positions resulted in an effective collection interval of 3 m, and three people in an interval of 2.5 m. Sites mapped at intervals of 5 m were generally collected at intervals of only 2.5 m, i.e., by one person walking between mapped rows.

[28] If a site were collected at intervals of 5 m, then a total of 2 m of the surface would have been examined out of every 5 m, since the walker at each end of those 5 m has a field of view of 1 m (the other 1 m of the field of view falls in the next 5 m interval); this is 40% of the surface, and by extension 40% of the available diagnostic artifacts would have been collected. Therefore a collected artifact total of, e.g., 100 diagnostic artifacts would yield a hypothetical total of 250 diagnostic artifacts (100/0.4) for

Around the counts of artifacts recorded on these maps contour lines were drawn to represent the size and shape of a site. The outer contour was typically defined at circa five artifacts per 10 m, and additional contours were drawn at, for example, 20, 50, and 100-plus artifacts per 10 m. The shape and area of the site were then transferred by hand to the 1:10,000 topographic map.[29]

At the end of 1999 Thompson oversaw the creation of a Geographic Information System (GIS) of the Metaponto survey with ESRI's ArcView 3.x software as the central platform for organizing, storing, and analyzing the various types of data produced since 1981. Certain basic data from the ESACTA map (elevation contours, modern roads, and hydrologic features), digitized in CAD format by Morter, were converted to ESRI shapefile format. The positions of the 1981–84 sites along with the positions, sizes, and shapes of the 1992–99 sites and plots, all of which had been recorded on the ESACTA sheets, were digitized directly in ArcView from georeferenced high-resolution scans of the marked sheets. The attributes of the sites and plots recorded in the field documentation were imported from prepared databases and attached directly to the digital geographic features in the shapefiles. Other basic information, such as the shapes and positions of the various transects, the extent of the 1981–84 survey coverage, and the locations of salient features in the landscape, were also added by digitization from georeferenced source maps.

2000–2001: Resurvey, Revisits, and GPS

Following the 1999 season, Thompson organized the various Bradano–Basento survey data sources for final publication. He became concerned that the data sets created in 1999 with his analytically more rigorous methodology of data collection were basically incompatible with the early 1981–94 data and therefore could not be easily integrated into the bulk of the survey data. The numeric representations of site area, shape, and artifact density derived from the maps of artifact counts had no equivalent objective basis in the first-generation data. The partial artifact collections of 1999 could be equated to the complete collections of 1981–84 by mathematical manipulation using the collection interval (percentage of the surface examined); the 1981–84 data, however, had no equivalent for the approximate total surface artifact counts that the artifact mapping made possible (i.e., by dividing the sum of all artifact counts in the density map by the fraction of the overall surface observed during counting).

Thompson accordingly dedicated the 2000 field campaign to the resurvey of several large sections of the original main Bradano–Basento transect using the later methods in order to translate the early survey data into terms comparable to the 1999 data. He expected to be able to refine the original rectangular area estimates and ordinal-scale density estimates with artifact mapping and visibility correction, and to substitute the original total collected artifact counts with total surface artifact counts. If a sufficient sample of the 1981–84 sites were revisited, the new data would lead to general correction factors that could be applied to all or most of the sites. New artifact collections would contribute to the chronological and functional description of individual sites, and they could also be compared with the original collections as a measure of the utility of resurvey/revisits in the Metapontino. At the same time, the resurvey would provide a measure for evaluating the accuracy of the first-generation intensive survey results in the comparison of site densities and distributions obtained from two methodologically diverse examinations of the same areas.

The 2000 campaign was designed to include a small amount of new coverage as well. Previously unexamined fields adjacent to areas covered between 1981 and 1984 were examined, provided the conditions of visibility and accessibility were adequate. New coverage was also added in the river valleys. Abbott's 1990–93 geomorphological investigations in the three river valleys had concluded that the alluvial and colluvial accumulation over two millennia was so great as to have buried the Greco-Roman occupation levels in all three valleys under meters of fill (see Ch. 2), confirming D'Annibale's assumptions and appearing to validate his decision to avoid the valley bottoms in general (see above). Yet in 1999 the survey team had encountered several sites on the surface of the lower Basento valley floor. Their presence seemed to contradict Abbott's results and suggested a reevaluation of the valleys to determine whether the sites in the Basento valley were an isolated phenomenon

the entire site.

[29] The methodology of the 1999 survey season is described in *Ancient Territories* 1999, 13–15. See also Carter/Thompson/Trelogan 2004; S.M. Thompson 2004, 70–71.

Figure 3.23 GPS receiver with real-time differential correction used for plot and site mapping, 2001. (SMT)

or part of a wider trend. Therefore part of the 2000 survey campaign was devoted to an exploration of the Bradano valley and to further examination of the Basento valley, in the areas falling within the main Bradano-Basento transect.

In 26 working days (June 28–August 8) 1.8 km^2 of territory in two localities examined in 1981–84 were reexamined—equivalent to 65 plots—as well as two plots surveyed the previous year in a third locality; 27 sites were revisited (23 from 1981–184, four from 1999). In addition, 67 new plots and 14 new sites (771–83, 787) were surveyed in the Bradano and Basento River valleys and in the areas designated for resurvey, adding 2.4 km^2 to the coverage of the main transect and making a cumulative total of 23.7 km^2.

The only modification of field methods between 1999 and 2000 was the integration of Global Positioning System (GPS) receivers into the work flow as the primary navigation and mapping tool. The survey team had begun to experiment with GPS in 1999, recording the coordinates of the corners of artifact-count grids and the locations of the highest counts (considered to be the site-centers). In 2000 the survey team replaced the ESACTA topographic map, compass, and pacing with a Trimble Pathfinder Pro XL 8-channel receiver, which was carried around the perimeter of every plot. The recording of several solutions (positions) per minute allowed a much more accurate representation of both size and shape than was possible through the "traditional" method of pacing and triangulation.[30] Although the recording of coordinates of grid corners and site centers with GPS was discontinued in 2000, the precise recording of plot boundaries with GPS allowed for an improvement in the positioning of artifact scatters. The team plotted the location of all artifact-count maps relative to a plot's boundaries, according to which information any artifact scatter could be accurately positioned in the GIS environment with minimal error. Thus the resurvey/revisit project had the additional merit of providing a means of gauging the accuracy of the original map-derived site positions (at least in cases where sites were still recognizable on the surface).[31]

The results of the 2000 season immediately demonstrated that the passage of nearly 20 years had adversely affected the archaeological landscape: the known sites had been so badly degraded that the new data could never reasonably reflect the conditions of 1981–84. The first-generation results therefore could not be translated into the more objective standards of the most recent data. Disappointment dissolved in the realization that the resurvey would address a completely different but equally interesting and urgent issue: analysis of the changes in modern land-use habits and agricultural practices, and of their effects on the condition of archaeological sites over the span of two decades. Would the number of sites documented in a given area remain the same after nearly 20 years? How did the conditions of sites change over time in terms of artifact density and scatter extent, and were changes associated with specific agricultural regimes?

The 2000 resurvey/revisit results were not a large enough sample for valid analyses, and so the project was extended into the 2001 season in order to acquire more data. In 16 working days (June 25–July 19) 101 plots in four localities (0.9 km^2 of territory) that had

[30] All GPS solutions were calculated in the UTM projection (Zone 33T/33 North) on the WGS 1984 datum. The receiver subscribed to the LandStar satellite correction service, which broadcast differential correction calculations to the receiver in real time, so the ultimate accuracy of all solutions obtained in the field was sub-meter.

[31] The methodology of the 2000 survey season is described in Prieto 2000, 17–18. See also S.M. Thompson 2004, 70–71.

been examined in 1981–84 were reexamined, as well as three plots in a fifth locality that had been surveyed in 1999; 27 sites from the first generation were revisited, along with three sites originally documented in 1999. In addition, 15 new plots were surveyed, adding 0.1 km^2 to the accumulated coverage of the main transect. In these new plots eight new sites (791–93, 796–800) were discovered.[32]

The 2001 season introduced only geospatial methodological advances. The team's GPS capability was upgraded with a Trimble Pathfinder Pro XRS 12-channel receiver, which was again used to map plots with real-time differential correction from LandStar (*Fig. 3.23*). In evaluating the data from the 1999 campaign, Thompson had developed a method for georeferencing the artifact-count maps using the GPS-derived corner coordinates (see above) so that the maps themselves could be displayed in their true geographic position and orientation in the GIS, and the artifact counts could be analyzed directly in it.[33] Therefore in the 2001 season the corner coordinates of every artifact-count map were recorded with the GPS receiver for georeferencing of the maps (*Fig. 3.5*).

Comparison and Critique of Methods, 1981–2001

Despite its many merits, the artifact counting introduced in 1999 ultimately revealed less about sites in the Metapontino than anticipated, so it was abandoned in favor of less time-consuming methods. The improvement, relative to the 1981–84 methods, in the measurement of certain characteristics of sites is mostly marginal, while certain characteristics of sites made quantifiable for the first time either have little intrinsic meaning or simply find no corresponding characteristics in the early documentation. For these reasons much of the information in the artifact-count maps has been modified, for both the purposes of analysis and presentation in the Gazetteer, to correspond more closely to the 1981–94 data generally in the form of a reduction in resolution (i.e., translation from ratio to ordinal scale).

The original subjective evaluation of sites in 1981–84, largely continued in 1992–94, was more efficient than the artifact-count methodology introduced in 1999. Thousands of sites scattered over hundreds of square kilometers are constantly threatened by modern agriculture and development; the task of locating and documenting a significant fraction of them before destruction or dispersion, with the goal of creating reasonably accurate representations of ancient land use in the various historical phases, called—and still calls—for a methodology that strikes a balance between the speed of execution and the rate of coverage. The limited but important off-site archaeology, which consists mostly of prehistoric discard locations, is adequately handled by the plot system. Experience has demonstrated that artifact counting, on the contrary, is a slow and resource-intensive process best applied to a limited number of sites, rather than to all indiscriminately, and that it must be performed even more intensively if any substantive information about a buried structure is to be recovered.[34]

Evaluating the Results of the Metaponto Survey

The conclusion of the discussion of survey methodology is a good point at which to pause and broadly assess the overall reliability and representativeness of the results before the latter are presented in subsequent chapters. Simply stated, how accurately do the results of the Metaponto survey reflect the real distribution and intensity of the various rural activities in the various phases of the landscape's history?

The answer begins with consideration of the quantities of sites and other types of locations (for example, prehistoric discard locations) recovered. It is by now universally recognized that survey results, like excavation results, must not be treated as a direct sample of the original cumulative archaeological universe (consisting, in the case of the Metapontino, mostly of sites, with some off-site artifact distributions of recent origin), but rather as a sample of those features—and, more to the point, a sample of a sample. In the millennia between prehistory and the modern era, some portion of the remains left by

[32] The aims and results of the 2001 survey campaign are summarized in S.M. Thompson 2001, 74–79. The methods are described in S.M. Thompson 2004, 70–71, where the overall results of the resurvey/revisit project are examined in detail.

[33] The process began with conversion of the artifact counts to digital format. This was accomplished by first defining the shape and extent of a given map-grid as an artificial xy coordinate plane with an arbitrary 0,0 origin and as many rows/columns as traverses and samples. The artificial coordinates and artifact counts were then entered into a spreadsheet program in separate columns. The tables thus created were imported into the GIS via compatible database formats; once in the GIS, the artificial grid corners were assigned their true geographic coordinates as calculated by the GPS receiver, from which information the GIS calculated coordinates for each artifact-count location and converted all of the locations to points with artifact counts as z attributes.

[34] For a similar cautionary evaluation of intensive artifact counting, see Fentress 2000, 49–50.

the occupants of a given period was disturbed (damaged, displaced, destroyed, or reconstructed) by the activities of subsequent generations of occupants; at the same time, pervasive geomorphological processes such as soil deposition and mass wasting and changing patterns of vegetation (e.g., new tree growth) eventually obscured or disturbed most of the archaeological features of the Metapontino. By means of these two filters, the majority of the original universe of sites and off-site artifacts has been rendered invisible to surface detection. The recent agricultural and infrastructural work and erosional processes essential for revealing the archaeological features act as additional filters, in that they occur discontinuously across the landscape and therefore reveal only a portion of the features that are completely or partially intact, and they frequently destroy those features before they are discovered.[35]

This reduced sample of the original universe of sites and other features is subsequently sampled by the survey methodology. The quantity of the available features discovered in the study area of the survey depends directly on three principal factors:

1. The extent and topographic-geomorphological variety of the survey coverage, which are almost invariably discontinuous and imbalanced because of the necessary (but arbitrarily defined) minimum criteria of visibility and physical accessibility
2. Human error in recognizing the features of interest, which in turn depends largely (although not exclusively) on the intensity of the examination represented by the interval between field-walkers
3. The nature of the features themselves: sites shorter or narrower than the field-walker interval are more likely to go unobserved, artifact-discard locations are almost exclusively discovered by chance, and it is possible that many ancient structures were built and stocked with perishable materials, such as mud-brick-and-thatch huts/shacks filled with wooden, wicker, and skin utensils, which would generally leave no trace.[36]

The possible and probable effects of these three factors on the Metaponto survey results merit some consideration.

1. As mentioned earlier in this chapter, the original survey design aimed for a sample composed of 43% marine terraces, 30% river/stream valley walls and related transitional features, and 27% fluvial floodplains/alluvial terraces in the main (*Fig. 3.7*). These proportions are reasonably close to the proportions of the same categories measured across the entire Metapontino, according to Abbott's geomorphological map (37%, 27%, and 36%, respectively), but the combined coverage of all survey seasons between the Bradano and Basento Rivers, from 1981 through 2001, amounts to 62%, 16%, and 22%, respectively (*Fig. 3.11*). This overrepresentation of the broad, nearly flat areas is a product of their generally easier access and better visibility, while the underrepresentation of the narrow, low valleys and the slopes stems from a natural desire to avoid overgrown and unstable areas, where there is a greater likelihood that sites located on these landforms—whether intentionally or not (i.e., if they were displaced by slumping)—would be buried beyond reach or so fragmentary in appearance as to be illegible.

Here it can be said with reasonable certainty only that greater coverage of the valleys and slopes would not have been a good investment of time and effort because of the inherent uncertainty whether the results—inevitably meager—were to be attributed more to the valleys' peculiar history and conditions or to a real but limited presence of archaeological remains. The land-use history of the valleys is an unfortunate casualty of nature. The disproportionately greater coverage of the terraces in the survey may be translated into a significant oversampling of sites and consequently a misrepresentation of the real land-use patterns of antiquity, but this defect is arguably counterbalanced by the gain in knowledge of the overall scale of ancient land use (i.e., raw quantities of sites) that would not have been apparent in a proportionate distribution of coverage across geomorphological zones.

2. The results of the 2000–2001 resurvey project and the handful of revisits in 1993 and 1999 provide a limited glimpse into the influence of the field methods, as well as the extent and nature of coverage (see

[35] See also *Greek Countryside*, 228–46.

[36] See also *Laconia* I, 40–44; Cherry 1983, 400–401.

above). Only four new sites (772, 791, 797, 798) were discovered in areas that had been examined previously, a figure that compares very favorably with the 82 sites documented in the same areas between 1981 and 1984. Three of the new sites (772, 797, 798) were larger than 15 x 15 m and therefore would have been observed in the 1980s at the standard 10 m interval; the reason that they went unobserved is clearly that they had not yet been exposed by agricultural activity.[37] From these results one can cautiously conclude that the 10 m survey interval employed rigorously in 1981–84 and 1999–2001 has afforded significant compensation for any inherent deficiencies in the crew's ability to detect sites, and the obscuring effects of geomorphological action and agricultural activity in intensively cultivated areas are relatively weak, a combination of circumstances that has resulted in the discovery of the majority of available sites in intensively cultivated areas.[38]

3. The effects of the physical nature of the archaeological record are not amenable to estimation. The pressed-clay/straw (daub) fragments recovered at a very few sites with strong prehistoric components, such as site 270, are a salutary reminder that not all structural materials have survived the passage of time, but in the absence of excavation results from undamaged, desiccated, or anaerobic environments (such as the Pantanello sanctuary) it is impossible to develop a rational approach to determine how many sites built of nonperishable materials might have once existed in any period. Sites smaller than 10 m in length/width have been discovered in relative abundance (circa 35), as have prehistoric artifact discard locations, but in the absence of broad coverage at intervals of 5 m or less, there is no certainty that these results are an accurate representation of the true proportion and distribution of such small features across the ancient landscape. The smallest and most ephemeral of sites are therefore the subjects only of speculation.

In the Metapontino, at least, it is impossible to model with any reliability the effects of these two layers of sampling, or even the various contributing factors individually or collectively, on the distributions of sites and other features.[39] The number of features lost already in antiquity to decay, demolition, and dismantling defies mathematical estimation because of the absence of data that accompanies the problem by definition. The effects of recent agricultural work are no less difficult to gauge: in fields where the topsoil has been either removed or replaced, the number of original features (if any) is generally unrecoverable because of the absence or incoherence of remnant artifact patterns. One could attempt to model the loss of features to geomorphological processes (slumping, deposition) based on the discovery of sites exposed in stream cuts, clipped by the excavation of trenches for pipelines or roads, and perched precariously at the edges of the terraces, but the results would certainly either over- or understate the ancient reality because the same processes have not acted with equal intensity in all applicable areas, or necessarily with the same results. Even if one could determine that the same amount of land slumped in similar topographic-geomorphological conditions in two different localities, for example, how could one ascertain that the same quantity of features was likely destroyed in the process in both places? The discovery of sites in cuts and trenches, furthermore, is essentially random because of the extremely narrow view into the subsurface, so modeling is impossible. The valley problem—the recognized underrepresentation of archaeological features in the stream and river valleys caused by the region's massive erosion/deposition cycles—could realistically be resolved through the systematic, controlled mechanical clearing of one or more very large sample areas, but the conditions for such a solution are unlikely ever to arise, barring an extraordinary industrial or public-works project.

The question of the reliability and representativeness of the Metaponto survey data can also be

[37] The same three sites were discovered in vineyards planted after 1984.

[38] On the discontinuous appearance of sites on the surface over time, see, e.g., Barker 1995, 5.

[39] Attempts to develop models for these problems in other locations are laudable but still too primitive to be anything but discussion points. The GIS-based kernel approach for visibility "correction" of site patterns presented by Terrenato (2000), for example, appears to re-create site distributions degraded by low visibility with reasonable accuracy. The conditions of Terrenato's experiment are overly simplistic, however, and the logic is circular: the test distribution of sites was randomly generated, rather than controlled by any real-world factor such as soil type or topographic variation, and the success of the results is gauged against a pattern known in advance—no data from the field could ever be tested in this way because of the interference of topography, geomorphological processes, and land-use history.

approached through consideration of how closely the data collected at individual locations in the field—particularly sites—reflects the reality beneath the surface. Such considerations, as well as attempts to model subsurface remains from surface data, are complicated by the infinite variety of combined effects of geomorphological processes, topography, land use, and site condition.[40] Every site has endured a unique history at the mercy of nature and man, which only excavation can reconstruct with any plausible accuracy. The signature of area, density, and functional-chronological representation of any given artifact scatter is therefore a unique product, despite the fact that sites with very different histories may produce similar signatures, and so statistical methods for identifying sites functionally by their signatures are misleading in the absence of concrete data on historical development from excavation. A sobering example of how even the original total diagnostic collection strategy could not overcome the gulf between surface and subsurface is provided by site 1, the Greco-Roman farmhouse at Sant'Angelo Grieco/Nuovo excavated in 1981. The meager quantities of grey ware and terra sigillata italica available on the surface—7 and 1 sherds, respectively—suggested a weak Roman presence in an abandoned Greek farmhouse. The excavation, however, produced approximately 517 sherds of grey ware and 36 sherds of terra sigillata in a Greek farmhouse that had been extensively restructured in the Late Republic.

The results of the revisit/resurvey project suggest, finally, that in general the Metaponto survey has been highly successful at recovering information about all phases of occupation or activity at a given site. At only 9 of the 82 sites discovered in 1981–84 and revisited in 1993–2001 was a new historic component discovered that had not been identified in the first visit (sites 174, 327, 353, 372, 375, 395, 397, 465, 528), and in most cases the new component was lithic discard involving a single artifact, the most difficult phenomenon to detect.

Although a large percentage of the sites in the Metapontino are now lost through natural or anthropic action, the results of the ICA survey create a detailed picture of the ancient Metapontino, one far richer than could ever otherwise have been realized. It is unlikely that any other field survey methodology, no matter how rigorous and intensive, would have created a significantly fuller picture. The ICA survey, carried out with rigorous methodology and enthusiasm by dedicated teams working for two decades, happened at a critical time in the Metapontino's history. Both the only record and the best record possible, the survey is a good approximation that sheds light on this important ancient landscape.

[40] The latter includes the full range of processes active in all phases of a site's existence—habitation, abandonment, and postabandonment—described by Pettegrew (2001).

4
GIS Methods and Considerations

Peter H. Dana

This chapter describes the processes used to estimate the spatial and temporal distribution of settlement sites at Metaponto in a Geographic Information Systems (GIS) environment. An artifact database was linked to layers of spatially referenced information and subjected to various spatial analyses, resulting in a series of displays of settlement patterns by period over the study area. The materials available for analysis included a relational artifact database containing almost 11,000 records, a set of point and polygon GIS layers describing the positions of sites and plots, and a set of supporting GIS layers consisting of streams, elevations, geomorphological features, and scanned maps and aerial photographs. The study area in this volume, that part of the chora of Metaponto between the Bradano and Basento Rivers in southern Italy, covers an area of ca. 10,000 hectares at approximately 16° 45′ east longitude and 40° 25′ north latitude. Microsoft Access and Excel, ESRI's ArcGIS, Golden Software's Surfer, The MathWorks' MatLab, and the SPSS Graduate Pack were used for the analyses. Most processing was done within the Universal Transverse Mercator (UTM) coordinate system. Other coordinate systems were tested and used where appropriate. The GIS platform was tested for its ability to handle the coordinate systems and geographic operators.

Site locations were dated using fine-ware artifacts. Date periods were established, and a series of dates representing periods of maximum site occupation were computed. Sites were categorized into types, such as farmhouses, tombs, sanctuaries, and others. Error analysis and quality control were monitored throughout the project. The results of this process consist of estimates of spatial and temporal change in settlement patterns based on the ensemble of artifacts for sites identified as farm sites.

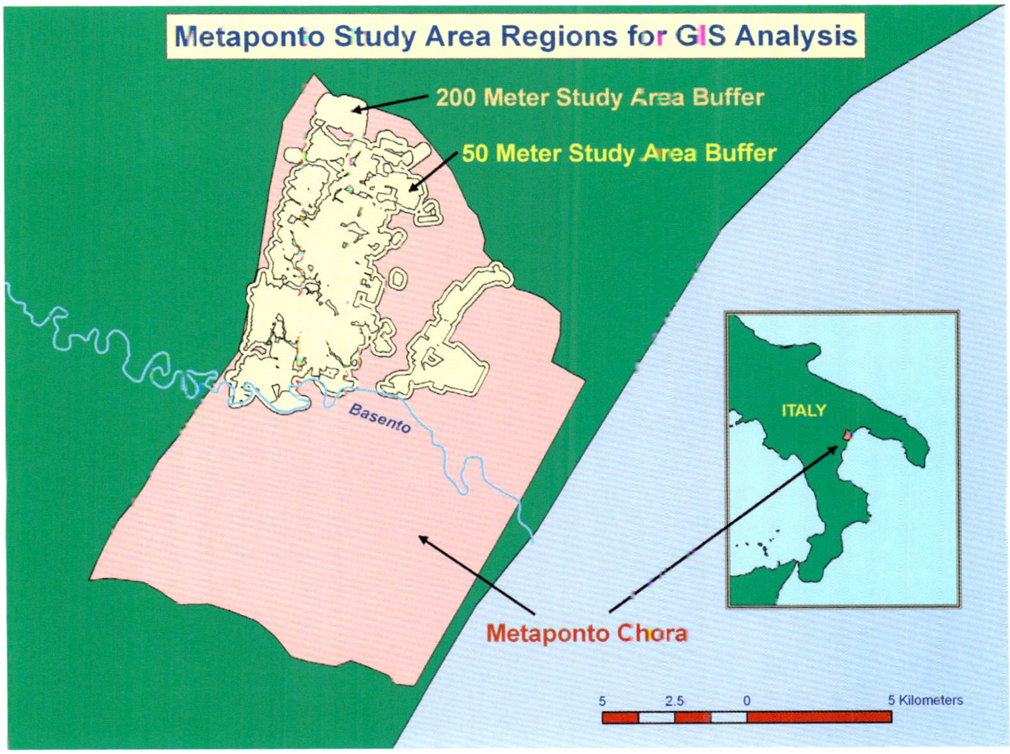

Figure 4.1 Metaponto study area regions for GIS analysis. (PD)

Source Materials

A Microsoft Access database table consisting of 10,995 records, created by the various artifact specialists who contributed to the project, forms the basis for this study. Each record is labeled with the number of the site or survey plot where each artifact was found, with chronological information contained in other fields.

The geographic attributes of the sites and plots are derived from several sources. Sites 1–787 were imported into the GIS as an event theme from a database of site attributes including northing and easting positions, which were read from the original 1:10,000 topographic sheets employed in the field campaigns. Sites 793–800 were located using a Global Positioning System receiver with differential correction capability. Plot boundaries were either digitized from scanned field maps (1993–1999) or traced with the GPS receiver (2000–2001).

Elevation data is recorded in at least one of four sources: a Digital Elevation Model (DEM) of the Metaponto study area generated from Shuttle Radar Topography Mission (SRTM) data, elevation contour polygons of the study area digitized from the 1:10,000 topographic maps, individual site location elevations read from the 1:10,000 topographic maps, and GPS readings for sites documented in 2000–2001.

Hydrology data, consisting of natural streams and man-made canals, were digitized from the 1:10,000 topographic maps and from the Istituto Geografico Militare (IGM)'s 1:50,000 topographic maps (sheets 491/Ferrandina, 492/Ginosa, and 508/Policoro). James T. Abbott's geomorphological map of the chora prepared from aerial photographs was digitized to obtain the geomorphological data.

Study Area

We defined the study area at its largest extent by the chora of Metaponto. This region contains all the surveyed plots and sites in the Metapontino from the Bradano to Cavone Rivers. For this volume we confined the study area to that part of the chora north of the Basento River. We then decided to restrict analysis further through the creation of a 200-m buffer around all the sites and plots north of the Basento that can be dated with fine-ware artifacts. Our most restrictive definition of the study area is defined by a 50-m buffer around dateable locations north of the Basento. These various study area polygons are depicted in Fig. 4.1.

Analysis Platforms

We began the project with ESRI's ArcGIS version 8.3 and finished with version 9.3. Most spatial analysis and all graphic display of analysis results were performed in ArcMap, the kernel ArcGIS module. We used The MathWorks' MatLab programming environment to produce custom data files and to perform cluster analysis using algorithms developed by the author. We employed SPSS as a platform for several statistical analysis procedures that were essentially non-spatial.

Coordinate Systems and Spatial Analysis

The coordinate system used for GIS analysis can have profound effects on that analysis. For the definition of single points on the surface of the earth any coordinate system that allows for the conversion to and from longitude, latitude, and height with reference to a specific geodetic datum can be used for accurate portrayal of position. Surveyors and geodesists make use of numerical methods for computing distance and azimuths along specified paths between points with reference to an ellipsoidal earth surface. GIS platforms, however, use a variety of methods for the computation of distances and angles. Distances and angles, as well as quantities derived from them such as area, aspect, slope, and stream sinuosity, are often used within GIS spatial analysis processes. GIS platforms rarely use geodetic computational methods (I know of none that do); rather, they use simple and fast operations, often based on Pythagorean methods and simple trigonometric functions on the surface of a plane. The plane on which these calculations are performed varies between platforms, databases, and specific functions.

ESRI's ArcMap is a typical GIS platform. Under different circumstances ArcMap computes the distance between points and angles from one point to another in significantly different ways. If one takes a database of points such as the major cities of Italy and produces a proximity (allocation) grid, the computations take place in the coordinate system of the original database, irrespective of the display projection of the data frame within which the allocation is performed. Allocation assigns to each cell in a grid the value of the point closest to it. Different allocation grids are produced depending on the coordinate system of the database representing the points. Fig. 4.2 shows two sets of allocation polygons, one produced from a database of Italian cities stored in longitude

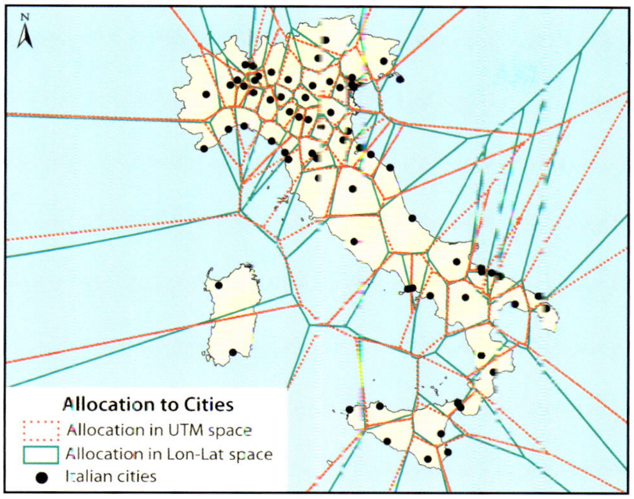

Figure 4.2 Allocation in UTM and longitude/latitude space.

and latitude, the other produced from a database stored in UTM coordinates.

Another important consideration in the evaluation of the project coordinate system is the difference between distance computations in different layer projections. Unlike allocation, which takes place in the coordinate system of the database, the select-by-distance function in ArcMap computes distances in the data frame projection. Fig. 4.3 shows the difference between selected sets of cities within 500 km of Rome as computed in longitude/latitude space and in UTM space.

Buffering is a common operation in GIS analysis. In ArcMap 8.3 buffer polygons produced via the Buffer Wizard are computed in a local Hotine Oblique Mercator projection; the entities selected for buffering determine the internal parameters used to produce the buffer. The resulting shape, then, is a function of this special projection and the shapes selected for buffering. This means that even a simple circular buffer created in ArcMap will be different from the circular buffer created in any other GIS platform on a projection surface that does not exactly match the unique internally generated Hotine parameters that are not made available to the ArcMap user. In ArcMap 9.0 buffer polygons generated using the Toolbox Buffer tool are calculated using Euclidean distance in whatever coordinate system is defined for the buffer's source file. When an ESRI geodatabase feature class is the basis for a buffer, the buffer polygon is defined as a single arc with a center and a single edge point; this buffer represents different spaces when the

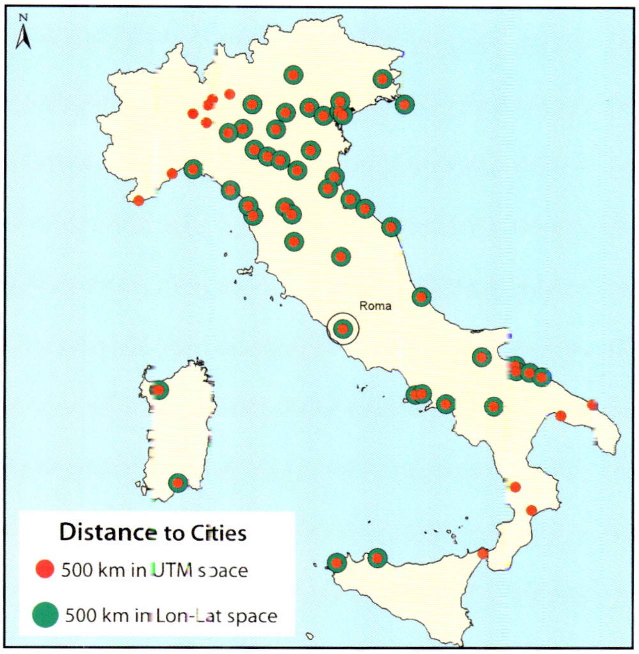

Figure 4.3 Selection by distance in UTM and longitude/latitude space.

data frame is projected in different systems; a buffer's graphic display can be quite different from what the buffer actually encompasses. Fig. 4.4 shows the graphic display of a variety of buffers, all based on a distance from Rome of 414 km. All but the "geodesic" buffer were produced using either the Buffer Wizard or the Toolbox Buffer tool in ArcMap 9.0. The geodesic buffer was produced using an ellipsoidal geodesic buffer generator coded in MatLab.

In effect, an analyst selecting points within even a simple circular buffer might select a different set of points from what an analyst who uses another method to compute the distance between the center point and each candidate point might select. Fig. 4.5 shows the difference between points selected using a variety of 414-km buffers.

Description of the Current Metaponto Coordinate System

The primary coordinate system used in this project is the Universal Transverse Mercator (UTM) zone 33, referenced to the World Geodetic System 1984 (WGS-84) geodetic datum. Data originally referenced to the Gauss-Boaga coordinate system of the ESACTA 1:10,000 topographic maps were converted to UTM/WGS-84 eastings and northings in meters using ESRI's Data Automation Kit (DAK) program.

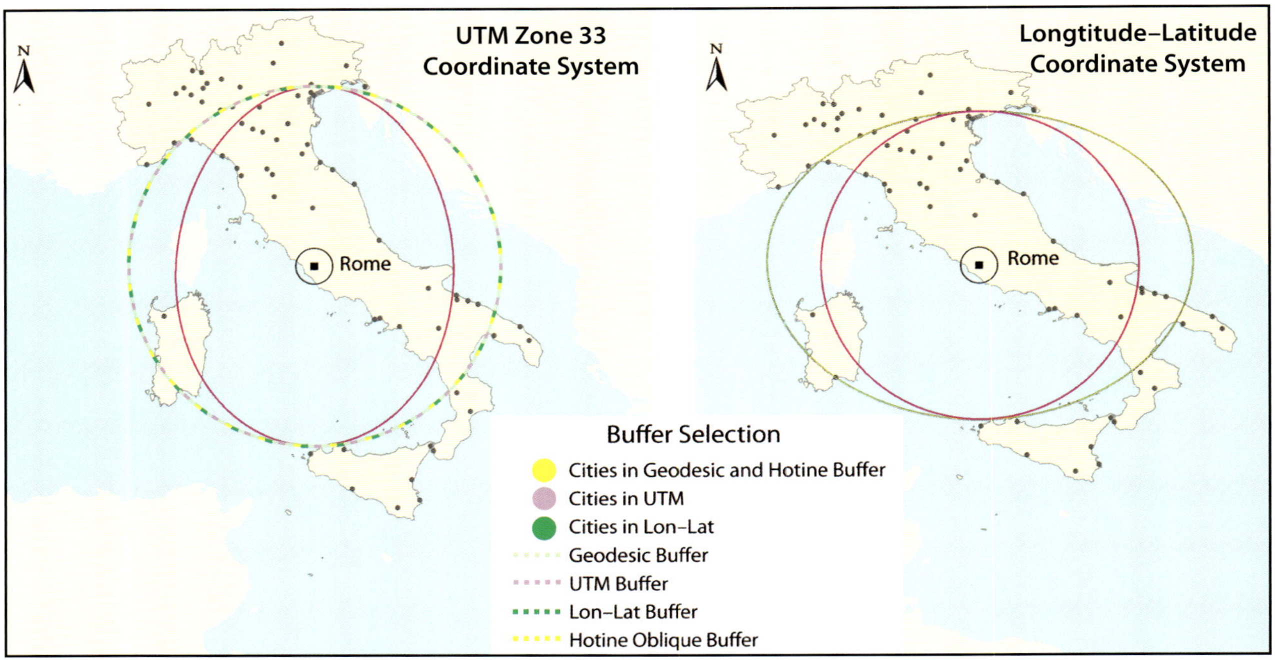

Figure 4.4 Various buffer types.

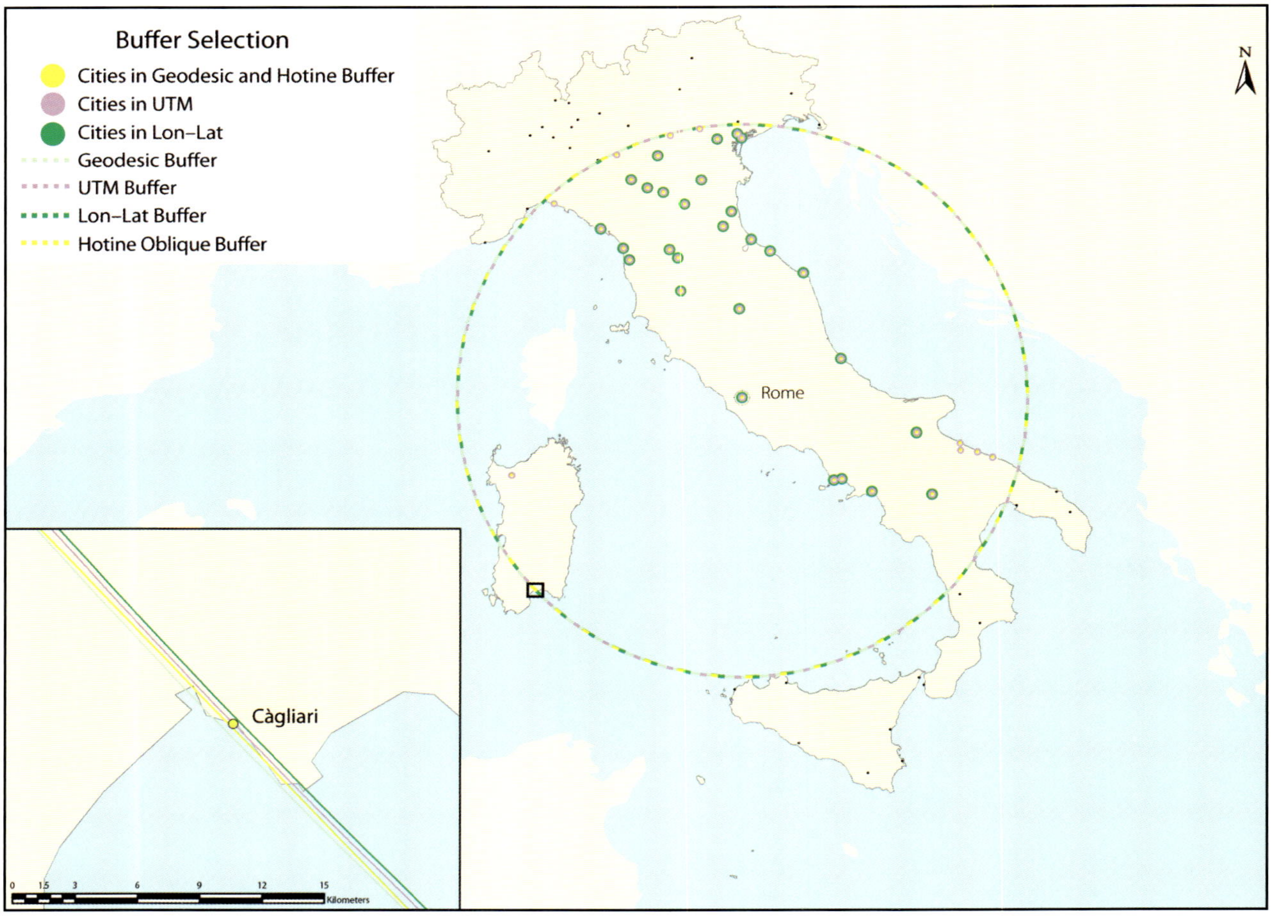

Figure 4.5 Selection within various 414-km buffers.

The complete coordinate system specifications now used for the Metaponto GIS are described within ArcMap as "WGS-84 UTM Zone 33N." This system has components consisting of a geodetic datum, a map projection type, map projection parameters, and distance units. The geodetic datum is the World Geodetic System 1984 designed by the United States Department of Defense (described in NIMA 1997). The map projection used in the coordinate system is the Transverse Mercator as defined for the Universal Transverse Mercator (UTM) segment of the Military Grid Reference System (MGRS, described in NIMA 1989 and NIMA 1990). The projection parameters for UTM Zone 33N are central meridian (CM) at 15° east longitude, scale at CM of 0.9996, units in meters, latitude of origin at 0°, false easting of 500,000 m, and false northing of 0 m.

The MGRS was designed by the United States Army to provide mapping, navigation, and coordinate systems for global operations in the 1940s. North of 84° north latitude, and south of 80° south latitude, MGRS specifies the use of the Universal Polar Stereographic (UPS) grid based on the stereographic projection. Between 80° south latitude and 84° north latitude MGRS specifies use of the Universal Transverse Mercator (UTM) system. This system defines 60 longitudinal zones, each six degrees in width, starting at 180° west longitude with easting zone 1 and ending at 180° east in easting zone 60. Each zone has an implicit CM in the center of the zone. Easting Zone 1 has a CM of 177° west longitude. The CM for Easting Zone 30 lies at 3° west longitude. The CM for Zone 31 is 3° east longitude. The CM for Zone 60 is 177° east longitude.

20 northing zones are defined for UTM. Each spans eight degrees of latitude and is designated by an alphabetic character, starting with Zone C between 80 and 72° south latitude and ending at Zone X between 72 and 84° north latitude. Zone X spans 12 degrees rather than the normal eight and is used to provide a single zone between 72 and 84° north latitude for operations within the land masses up to 84° north latitude. The alphabetic characters start at C and end at X, ignoring characters I and O, which might be mistaken for numerical values 1 and 0.

Defined by easting zones six degrees in width, the UTM system develops coordinates in an orthogonal plane defined by the Transverse Mercator projection. Positions are described by eastings and northings, distance respectively from the CM and from the latitude of origin on the surface of the TM projection plane. Because zones are six degrees wide, their width is ca. 668 km at the equator, 116 km at 80° south latitude, and 70 km at 84° north latitude. To avoid the use of negative numbers, all eastings are offset by a false value of 500,000 m, allowing one zone to be used as far as 500 km west of the CM before negative numbers would be required. With a nominal equatorial width of 667 km, and a distance from the CM to the zone boundary of 334 km, a false easting of 500 km allows operations to span zone boundaries. Northern latitudes do not employ a false northing, while operations in the southern hemisphere use a false northing of 10,000,000 m to provide positive northing values.

In MGRS, for both UTM and UPS, numeric values of 100,000 m increments are replaced with alphabetic characters. Complete designators in MGRS are made by concatenating easting zone number, northing zone character, easting 100 km square designator, northing 100 km square designator, followed by easting and northing sub-100 km numeric values. If two digits follow the northing 100 km square designator, they represent values in increments of 10 km of easting. Four digits imply two easting digits, representing 10 and 1 km values, followed by two northing digits. Six digits include three easting digits followed by three northing digits to a precision of 100 m. Eight digits represent easting and northing digits to a precision of 10 m. Ten digits are used for precision of 1 m. Although not normal practice, additional digits can be used for increased precision without limit.

As an example, the position 16:45:00E/40:15:00 N in WGS-84 longitude/latitude can be represented in MGRS as 33TXE4883756974: Easting Zone 33, Northing Zone T, 100 km easting square X (6), 100 km northing square E (44), with least significant easting digits of 48837 m and least significant northing digits of 56974 m.

UTM is often used outside MGRS with only an easting zone number and easting and northing values given in numeric form to the desired precision without concatenation. The non-MGRS, UTM-only designator for 16:45:00E/40:15:00N would thus be Zone 33, Easting 648,837 m, Northing 4,456,974 m. These values indicate that the position is within zone 33, spanning 12 to 18° east longitude. Since the CM is 15° east longitude, the position is 148, 837 m east of the CM (the false easting of 500,000 m is subtracted from the easting to yield the distance from the CM on the TM projection plane). The position is 4,456,974 m from the equator on the projection plane. The same values of easting and northing can be found in each of the 60 UTM easting zones, in exactly the same place on the projection plane with respect to the 60 CM values and the equator.

Evaluation of the Current Metaponto Coordinate System

The geodetic datum selected for the current coordinate system is perfectly valid for GIS analysis. WGS-84 is a global, earth-centered datum that is generally accepted as an estimate for the size and shape of the Earth, as well as a definition for the origin and orientation of coordinate systems. Parameters exist for conversion from many local systems to WGS-84, and, while there are local maps referenced to other datums, the Italian national cartographic agency (IGM) produces maps referenced to WGS-84.

UTM is perfectly acceptable for defining point positions on the ground in a coordinate system. Conversion between UTM and other georeferencing systems in use such as longitude/latitude can be accomplished in both forward and inverse senses without loss of accuracy, as long as the methods are selected appropriately. However, the UTM projection plane is only moderately useful for accurate distance, area, and azimuth calculations in a GIS environment. Any Transverse Mercator projection is conformal, i.e., it can represent local shapes accurately because local and relative angular relationships are preserved. The projection is not equidistant, equal-angle, or equal-area.[1] TM projections have correct scale only along two lines of longitude equidistant from the CM if the scale factor at the CM is defined as less than one (<1.0). If the scale at the CM is set to 1.0, then scale is correct only along the CM. In the UTM version of the TM projection, the scale at the CM is 0.9996, which forces the lines of correct scale to fall along north-south lines approximately 180 km on either side of the CM. While this distributes scale errors nearly equally within the six-degree zones of the UTM projection, distances computed using simple Pythagorean methods (square root of the sum of the easting distance squared and the northing distance squared) are always incorrect except along paths where the scale factor averages to 1.0. Azimuthal relationships, while locally and relatively faithful to shapes, are not correct except along the CM. Azimuths computed using simple arctangent methods are subject to the local difference between grid north (the direction of UTM easting lines) and true north (the direction of longitudinal lines). Since distance and direction are everywhere incorrect (except as noted), area computations are erroneous everywhere as well.

In the UTM version of the Transverse Mercator projection, scale and azimuth errors can be computed at any point. Thus it is possible to estimate the errors in distance, azimuth, and area that will occur in any local region using simple Pythagorean methods for distance or arctangent methods for azimuths. Computing distance or azimuth errors, or correcting for them, over long distances on oblique paths is quite difficult (so difficult that accurate results are really possible only by conversion to longitude/latitude followed by computation of ellipsoidal geodesic path distances and azimuths).

The Metaponto study area can be defined as a region including the chora, sites, and related sites within a bounding rectangle with the following limits: 16.40–16.92° east longitude, and 40.1–40.7° north latitude. The UTM grid used within the Metaponto study area experiences scale errors ranging from 0.999772 to 0.99993 (*Fig. 4.6*).

Angle differences between true north and grid north (the convergence angle) vary between 0.9 and 1.2 degrees over the study area (*Fig. 4.7*).

Alternate Metaponto Coordinate System: The Pantanello Transverse Mercator (PTM)

It is relatively easy to design a coordinate system and implement it in ArcMap. A properly designed system can minimize distortions of distance, angle, shape, and area over the study area. The following coordinate system, designated Pantanello Transverse Mercator (PTM), is an attempt to solve distortion problems in GIS analysis. It has the following parameters: CM at 16.733333 degrees (16°:44′:00″) east longitude, scale at CM of 0.9999990, units in meters, latitude of origin at 38.80° degrees (38°:48′:00″), false easting of 40,000 m, and false northing of 0.0 m.

The TM projection is appropriate for a number of reasons. It is familiar to most researchers, used in Italy,

[1] Snyder 1987.

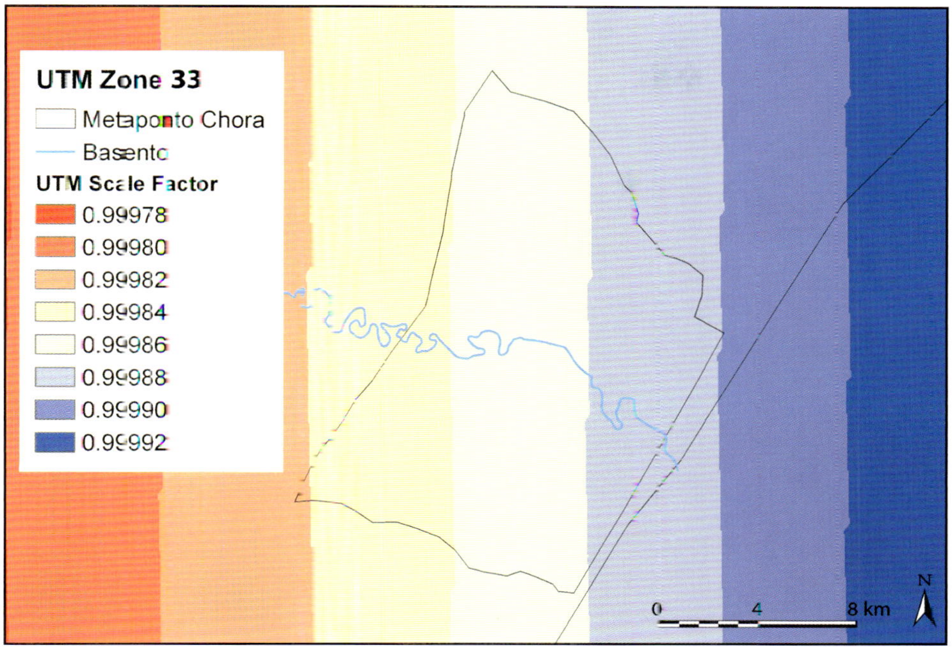

Figure 4.6 UTM scale factor.

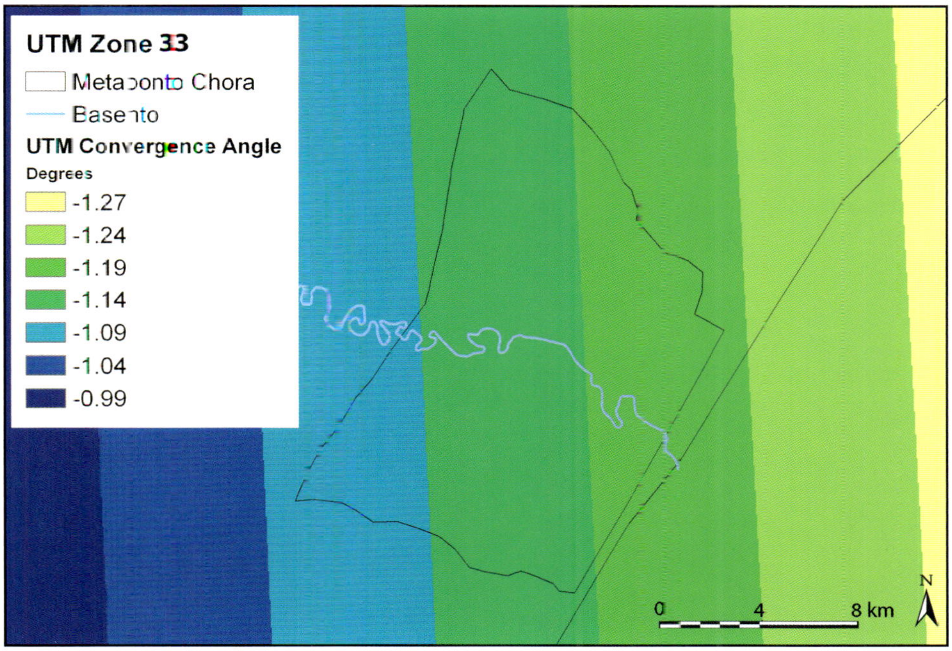

Figure 4.7 UTM convergence angle.

and based on UTM. It is often used for regions with longer latitudinal extent than longitudinal extent. It can be designed to minimize scale errors over the region of use and is conformal, thus maintaining local shapes.

The CM for PTM has been selected to pass near the center of the study area, at 16:44:00 east longitude. The scale of 1/1,000,000 (k = 0.999999) places the eastings of correct scale (k = 1.0) approximately 10 km east and west of the CM; the scale rises to 1.000004 ca. 20 km east and west of the CM. PTM distributes scale errors over the study area with distance errors of better than one part in one million (1 cm in 100 km). Fig. 4.8 illustrates the scale errors in PTM over the study area. (Note the relative im-

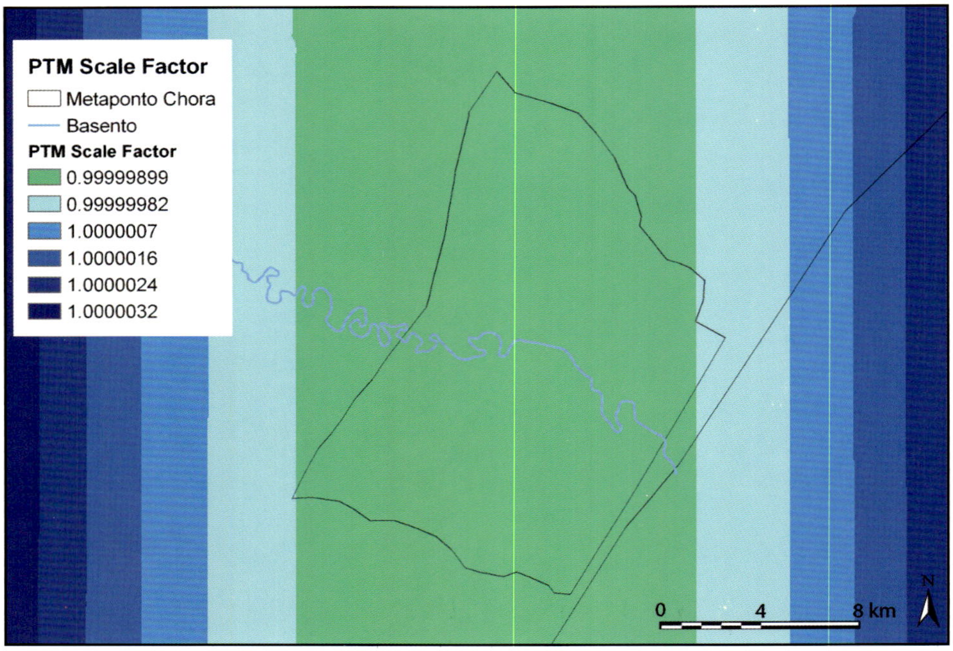

Figure 4.8 PTM scale error distribution.

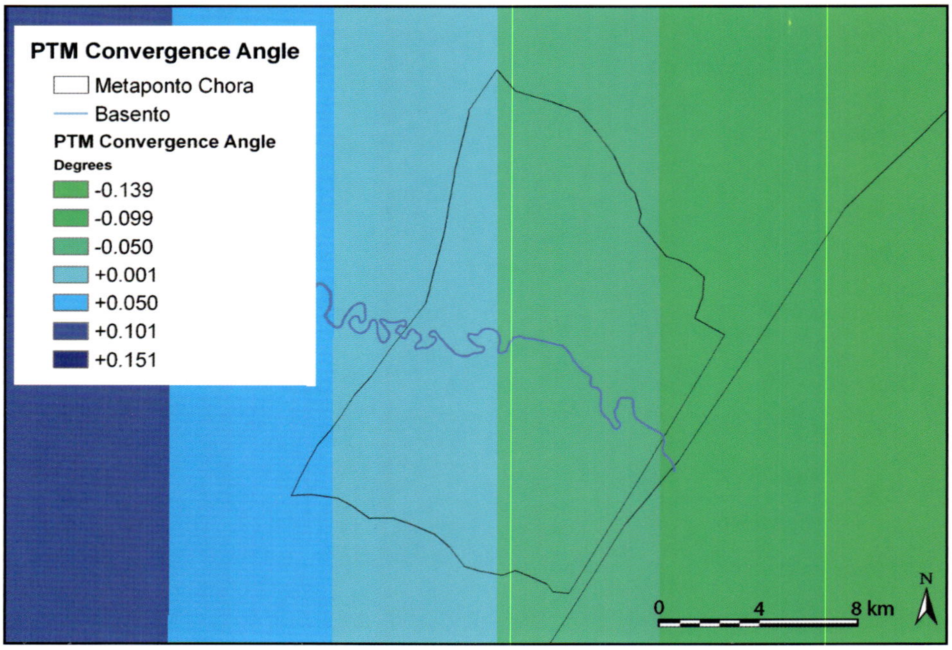

Figure 4.9 PTM convergence angle.

provement in the range of scale errors compared to those illustrated in Fig. 4.6.)

Angle distortion is minimized near the CM in a TM projection. In PTM, the convergence angle errors are less than ± 0.2 degrees within 20 km of the CM (*Fig. 4.9*).

With a false easting of 40,000 m and a latitude origin at 38:48:00N, PTM has easting and northing numeric values that are difficult to confuse with UTM coordinates. The UTM coordinates currently used have six-digit easting values in the range of 600 km within the study area. The false easting in PTM forces easting values within the chora to the range of 20–50 km; these are five-digit eastings, and they do not begin with the numeric character 6. North-

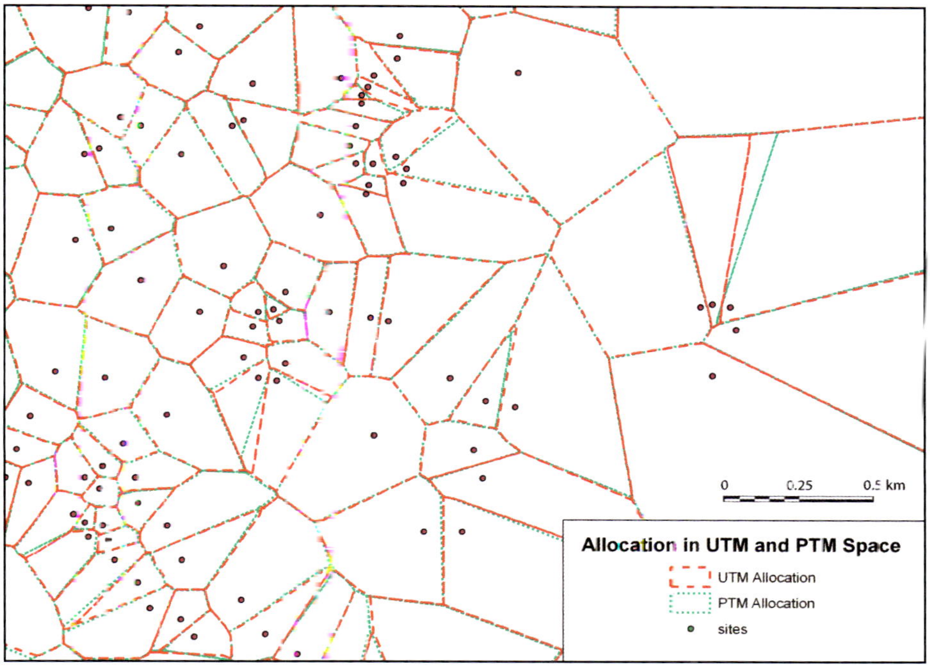

Figure 4.10 Allocation in UTM and PTM space.

ing values in UTM zone 33 within the chora are all seven-digit values in the range of 4400–4500 km, while in PTM the northing values in the chora range from 160–190 km. These are six-digit values, none beginning with 44 or 45. Thus PTM eastings and northings are easily differentiated from each other and from UTM values within the study area.

Advantages of PTM for GIS Analysis

The primary advantage of PTM is the reduction of UTM scale and angle errors. In all cases, point, line and polygon vertices remain the same with respect to the ground and to other layers in PTM or UTM. With changes in map projection, paths between points may change, and so polygon edges may fall more accurately in the PTM projection. It is in spatial analysis that the coordinate system of individual layers or the frame projection makes a difference and where PTM is advantageous.

Should analysis and mapping of the Metaponto data be combined with survey-quality data sets, the UTM scale and angle errors would be far too large. UTM scale errors of 1 m in 2500 m at the CM would be reduced to 1 m in 6666 m in PTM at the center of the study area, corresponding to second- or third-order land surveys. PTM scale reductions would result in a maximum error of 1 m in 100,000 m at the extent of current project mapping. In the chora, the scale errors would be better than one part in 500,000, far exceeding first-order land survey specifications.[2] Angular distortion would be minimized near the CM: in UTM errors exceed 1.2 degrees within the study area, but PTM has angular distortions of less than 0.2 degrees at the extent of the study area.

The implications of these scale and angle distortions for GIS analysis have been outlined above. Three examples of GIS spatial analysis may help demonstrate the improvement in PTM.

Allocation (sometimes called "proximity") is the process of creating a raster grid of cells containing a value corresponding to the nearest feature in a point layer to that grid cell. Allocation results in polygonal areas known as Thiessen or Voronoi polygons.[3] The point layer of the survey sites in UTM coordinates results in polygonal regions that differ from those created by the same layer converted to PTM. Since the allocation function operates within the coordinate space of the spatial database, the layer projection has no influence on the process in ArcMap. Fig. 4.10 shows the edge differences between two sets of proximity polygons, one in UTM, the other in PTM.

---

2 Kavanagh and Bird 2000.
3 O'Sullivan and Unwin 2003, 43

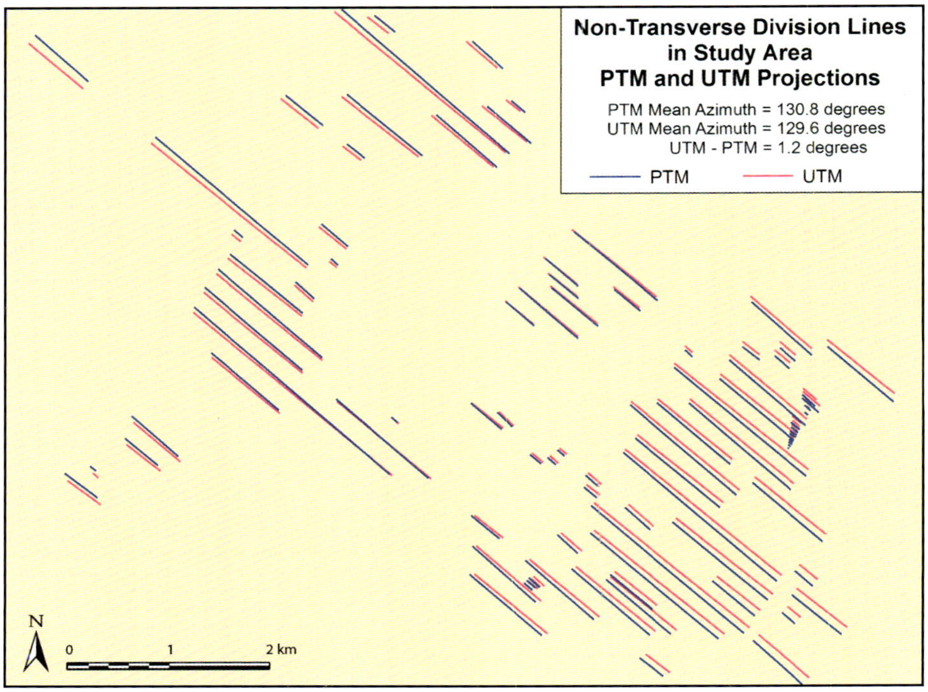

Figure 4.11 Direction differences for non-transverse division lines in the study area.

Note the occurrence of edge disparities between these two sets of polygons ranging from 10 to 100 m. Crucial here is that the scale and angle distortions in the UTM-based polygons have rendered them a less accurate representation of nearest neighbor position than polygons formed in PTM space.

Another example is the influence of distance and angle distortion on area and perimeter calculations. The chora extent layer in UTM results in an area computation of 194.807 km^2. When converted to PTM, it produces a more accurate result of 194.857 km^2. The perimeter length changes too, from a UTM estimate of 60.596 km to a more accurate PTM estimate of 60.604 km.

A third example illustrates the direction (angle) errors of the UTM grid within the study area. Fig. 4.11 shows the result of the difference between the computed mean direction for division lines (non-transverse) in the study area. The mean angular difference between the two sets of lines, one projected in UTM coordinates, the other in PTM is 1.2 degrees, the UTM grid convergence error in the study area.

Recommendations

A new GIS data set based on PTM could be introduced in place of the current UTM-based one with minimal effort. Conversion of point, line, and polygon layers from UTM to PTM is a task that can be accomplished in just a few hours. If analysis will never require accuracy better than 10 m in the horizontal plane or five degrees of azimuth, conversion may not be necessary. The computational risks in conversion are minimal. On the other hand, a new coordinate system would introduce a complexity that does not now exist. Confusion, while minimized by the specific parameters selected for the proposed PTM system, could result between datasets referenced to UTM and those referenced to PTM.

If project researchers decide that survey-quality distance and direction analysis will be required of the Metaponto data set, or that results will be combined with first- or second-order survey data, then this proposed coordinate system should be implemented before analysis is begun. Once new raster or vector layers are created in spatial analysis based on values derived from distance or directions, it is too late to make any meaningful conversion. The rule of thumb in GIS analysis is that results are never better than the spatial errors in the least accurate layer. Conversion to PTM

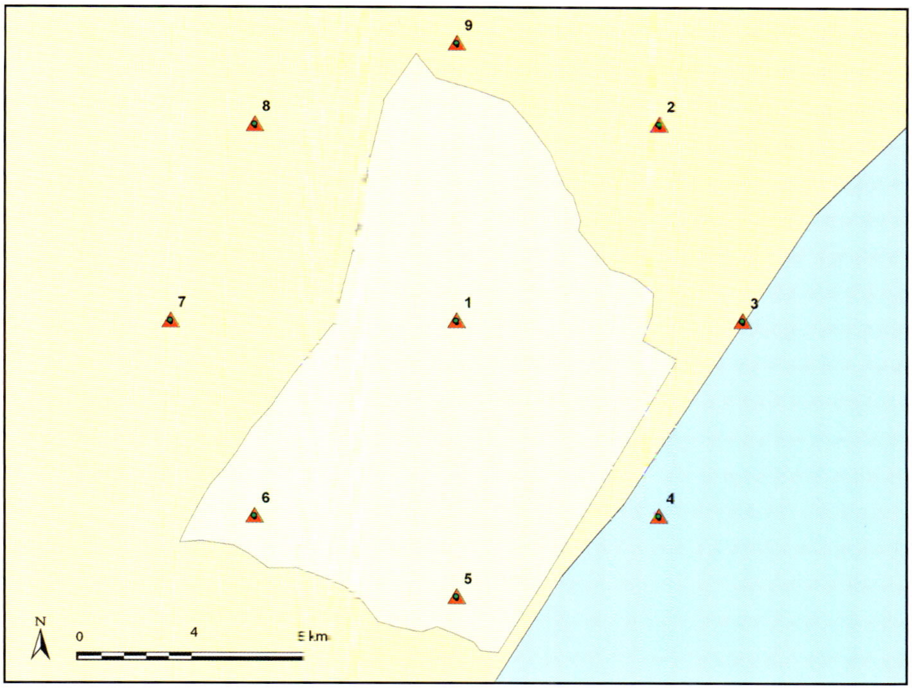

Figure 4.12 Nine test points within the study area.

would insure that future layers resulting from spatial analysis will be the most accurate possible.

GIS Platform Tests

Before implementing the proposed PTM coordinate system, we performed tests to ensure that conversions among longitude/latitude, UTM, and PTM could be accomplished within ArcMap with reasonable accuracy.

Description of the Test Data Set

In MatLab we computed the coordinates of nine points: a position near the center of the study area and eight points equally spaced in geodetic azimuth (45 degrees apart) around the center point. Each point was computed at a distance of 10,000 m from the center point on the surface of the WGS-84 ellipsoid using Vincenty's direct method for producing longitude and latitude at a distance and azimuth from a point (*Fig. 4.12*).

Coordinate Conversion and Results

The numeric values in decimal degrees were converted using methods specified in Snyder (1987) into UTM zone 33 and PTM coordinates. The nine points were then imported into ArcMap and converted to point feature layers using the Display XY Events tool with the coordinate system defined as WGS-84 with UTM and PTM. The Toolbox Project utility was used to convert the longitude/latitude layer into a UTM zone 33N/WGS-84 layer and a PTM layer.

Comparison showed that the PTM eastings and northings differ by less than 1 cm from the Snyder/MatLab coordinates, and the UTM eastings and northings differ by less than 1 dm (decimeter). The difference in the values is insignificant for any potential use of ICA's Metaponto data sets in GIS analysis. Of course, conversion to PTM, as helpful as it may be for accurate spatial analysis, is of little use if results cannot be exported as GIS layers in other formats. To insure that conversion using the Toolbox and the Projection wizard works equally well in both the forward and inverse directions, the point files in UTM and PTM were converted back to longitude/latitude in new layers. In addition, the PTM file was converted to UTM. In all cases the Toolbox conversions match the original longitude/latitude points and the UTM points to better than 1 cm.

Location Dating

We decided to date sites and plots solely by the fine-ware pottery. The first step in the process was to assign the dates provided by the artifact specialists to the sites and plots. All records for fine-ware artifacts were extracted from the master artifact database. The survey plots were converted to point features by computing a centroid X/Y value for each polygon, and these centroids were added to the database of site positions to create a new layer containing all locations. The location layer and fine-ware artifact database were then joined in ArcMap to give the artifacts geographic locations.

Artifact Date Range Processing

We then tested the differences between sites and site revisit polygons and decided that plot polygons and original site locations (not revisit locations) represent the best point locations for all artifacts.

We produced date histograms for each location based on the ensemble of dated artifacts found at each site. We computed histogram values for twelve date bins, each spanning fifty years, through the range from 625 BC to 25 BC. For example the 400 BC bin would contain the relative number of artifacts with date ranges falling within years more than or equal to 375 BC to less than 425 BC.

We first adjusted each artifact date by adding twenty years to each end of the artifact date span, to account for the uncertainty of artifact dating. We experimented with weighting of individual artifacts in an attempt to allow more weight to the center of an artifact date range without using a simple mid-point date to dominate the histogram for all artifacts at a location.

We computed individual artifact weights based on mid-point, rectangular, triangular, sinusoidal, and normal distribution curve shapes. After consultation with experts (see Dana 2004) we settled on the shape of a normal distribution curve that concentrated most of the weight near the mid-point of the assigned date range. We divided each artifact's extended date range by six and used the normal distribution curve cumulative probability function to assign 68% of the artifact weight in the middle third of the range, 27% to the third of the range beyond the middle third, and 5% to the outer third of the range.

The cumulative probability function we used is a numerical solution to the integral method:[4]

```
Public Function pnorm(x)
   Dim t As Double
   Dim dblQ As Double
          Dim pi As Double
   Dim fx As Double
          lngFeatureCount = 0
          pi=4 * Atn(1)' Calculate the value of pi.

   t=1# / (1# + Abs(x) * 0.2316419)
   fx=1 / Sqr(2 * pi) * Exp(-x ^ 2 / 2)
   dblQ = fx * (0.31938153 * t - 0.356563782 * t ^ 2 +
1.781477937 * t ^ 3 - 1.821255978 * t ^ 4 + 1.330274429 * t ^ 5)
   If x > 0 Then
          dblQ = 1# - dblQ
   End If
   pnorm = dblQ
End Function
```

The artifact weights are computed using a Visual Basic for Applications (VBA) script, ProcessDates.bas. The script allows the selection of any of the weighting mechanisms, and results in an artifact list with weights in each of the fifty-year date ranges for each artifact. Fig. 4.13 shows the five different weighting methods.

All the dated artifacts found at each location contribute to fifty-year date histogram bins for that location. Each bin is the result of summing the weights for each artifact of which the extended date range spans a date bin. The resulting sums are normalized to equivalent numbers of artifacts because each artifact weight is normalized to 1.0 over its date range through the cumulative probability function. Each location date bin represents the equivalent number of artifacts found at a location with a weighted date range within that fifty-year span. Fig. 4.14 shows the artifact weights and the resulting histogram bin sums for location 730.

Location Dating Analysis ("Best Date")

We asked several questions based on dated artifacts and their locations: When was a location occupied? When was the period of maximum occupation? Was the location occupied over a long period or a short period? Was the location occupied at different periods with intervening periods of lesser occupation?

To help answer these questions we took an approach that allowed us to weight evidence based on

[4] Abramowitz and Stegun 1968; Hewlett Packard 1974

an interpretation of location-date histograms. We used the notion of a Kernel Density Function and a resulting Kernel Density Evaluation (KDE) value. We developed an algorithm that revealed clusters of sequential date bins indicating periods of site occupation separated by periods of lesser occupation. We evaluated the size and shape of the clusters using a KDF that results in a high KDE value where and when there are a significantly large number of artifacts in a narrow date range.

A Visual Basic script (*GetBestDate.bas*) was used to process the weighted date histograms. For each location, the average number of artifacts in all date-bins was used as a cluster start and stop detection threshold. The threshold for cluster start- and end-detection was based on a hysteresis zone separating the level at which a new cluster begins (more than 90% of the mean of all weights for the location) and when a cluster ends (less than 80% of the weight mean).

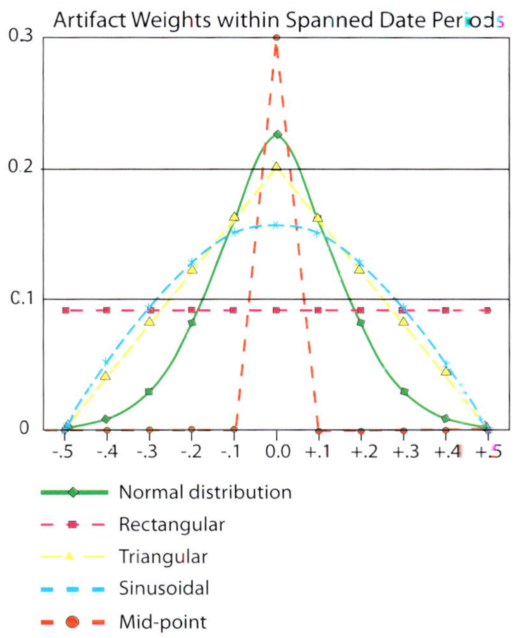

Figure 4.13 Artifact weight shapes.

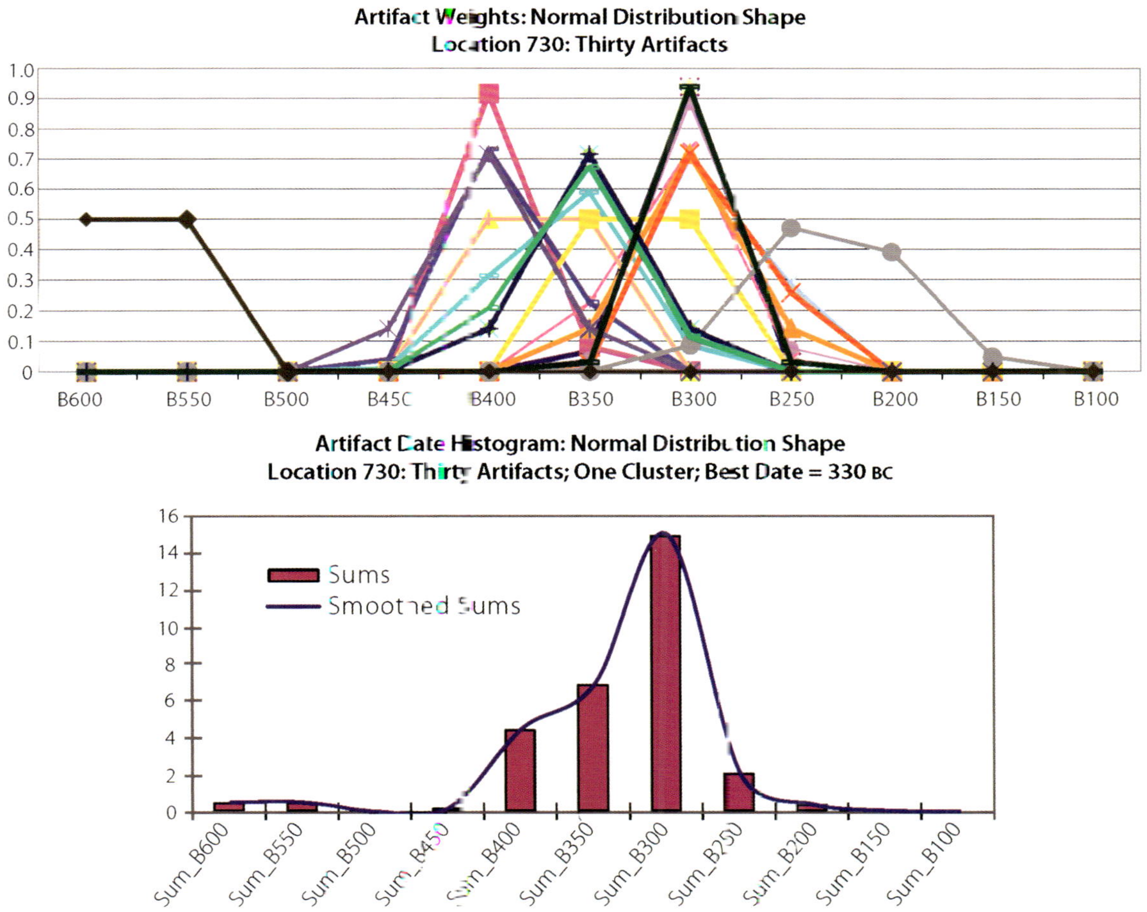

Figure 4.14 Normal distribution shape: artifacts and histogram for site 730.

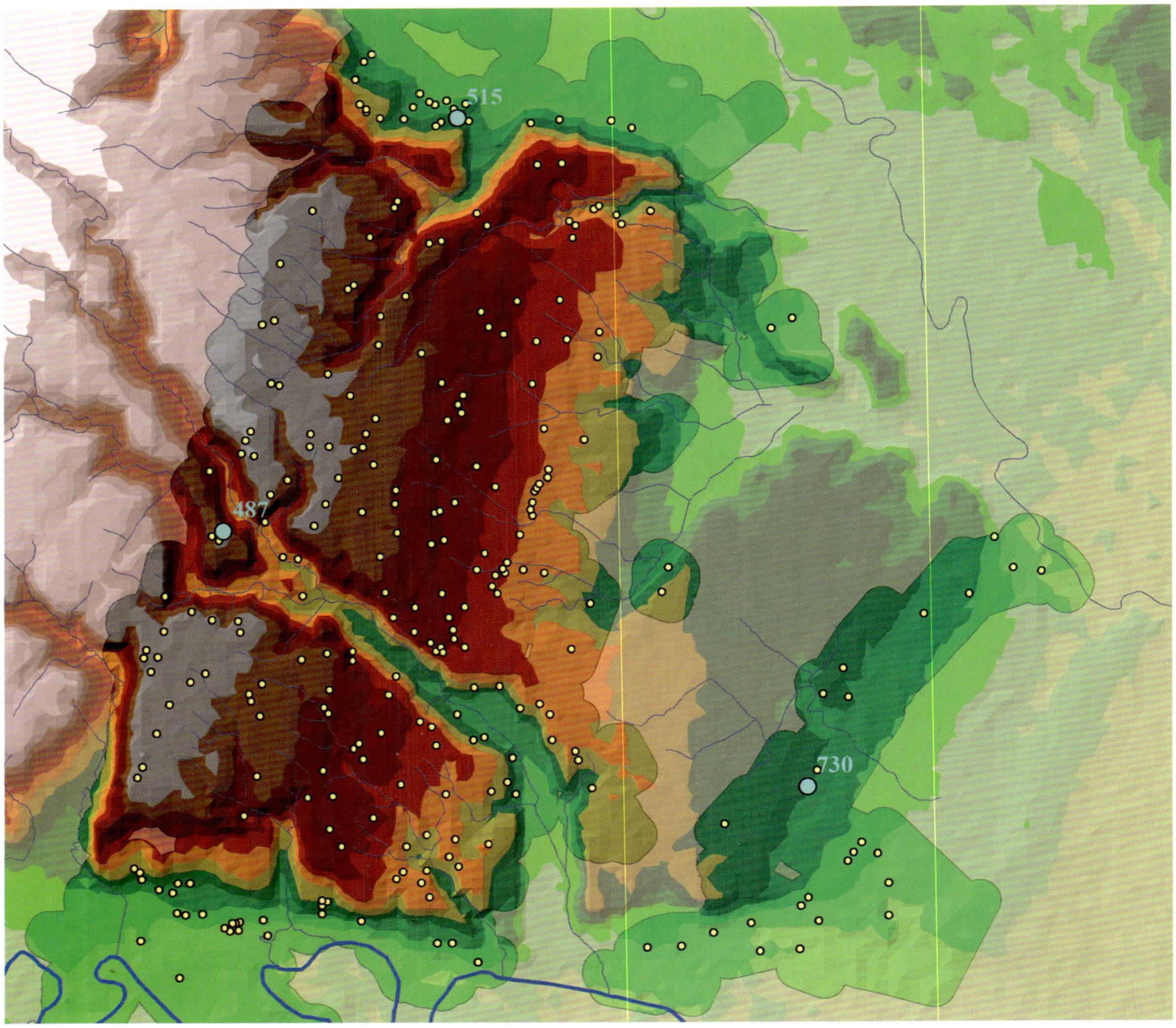

Figure 4.15 Case study area.

The number of bins and the number of artifacts in each cluster was computed. This process resulted in one to four possible artifact date clusters for each location. Two weights were computed for each cluster. A number-of-artifacts weight was computed as a value from 0.0 to 1.0 formed by the ratio of artifacts in each cluster to the number of artifacts found at each location. A deviation-weight was computed from the reciprocal of the number of date-bins in each cluster.

For each date cluster a cluster-weight (the KDE value) was computed by multiplying the number-of-artifacts weight by the deviation-weight. For a date where all the artifacts fell within a single bin, the resulting weight was 1.0. The date cluster with the highest KDE was selected as the "best date" cluster. The "best date" is the mean date for bins within the highest KDE cluster. This "best date" is an estimate for the date of maximum site occupation based on all dated artifacts found at the site.

Case Studies

To illustrate this process we have selected a set of three locations and the artifacts found at those locations (*Fig. 4.15*).

Forty-four artifacts classified as fine ware and dated were found in these three locations (487, 515, and 730). Table 4.1 lists the artifacts at these sites.

Fig. 4.16 shows the artifacts at site 515 weighted by the five different shapes. Note that the normal distribution curve weights artifacts with more smoothing than the mid-point technique, but less smoothing that the triangular, rectangular, or sine shaped weights.

LOC_ID	Form	Function	Part	Type	ILL	Beg. date	End date
487	sk	Drinking	Sh	PSk3b	cf. 622-01	550	480
487	sk	Drinking	R	C-S20/28	cf. 043-08	320	280
487	sk	Drinking	R	C-S20/28	cf. 521-54	320	280
487	CC	Food Serv	F	CC2	cf. 521-07	420	300
487	C	Drinking	RSh	IC5b	cf. 622-02	600	550
515	sk	Drinking	F	ionic	cf. 395-19	480	440
515	sk	Drinking	F	C-S25/S33	cf. 523-23	440	400
515	SA	Food Serv	B	SA3a	cf. 260-24	350	270
515	C	Drinking	RSh	IC4	cf. 397-47	580	540
515	C	Drinking	Sh	IC6	cf. 477-43	460	430
515	C	Drinking	F	CO6c	cf. 454-05	350	250
515	C	Drinking	F3	B6	515-03	200	100
515	C	Drinking	F3		cf. GW_372	200	100
515	B	Food Serv	R	B1	cf. 501-17	300	200
730	large ves	Drinking	R		730-17	320	300
730	sk	Drinking	F	C-S33/S49	cf. 521-11	410	375
730	sk	Drinking	F	A-S71	cf. 241-07	400	350
730	sk	Drinking	B	?	NO	400	300
730	sk	Drinking	R	C-S20/28	cf. 043-08	320	280
730	sk	Drinking	RF	C-S20/28	cf. 521-54	320	280
730	sk	Drinking	R	C-S20/28	cf. 521-54	320	280
730	sk	Drinking	R	C-S20/28	cf. 521-54	320	280
730	sk	Drinking	R	C-S20/28	cf. 578-06	320	280
730	sk	Drinking	F	C-S20/28	cf. 578-06	320	280
730	sk	Drinking	FH	C-S20/28	cf. 395-05	320	280
730	sk	Drinking	F	C-S65	cf. 454-08	320	280
730	SB	Food Serv	R	SB6	cf. 354-03	320	250
730	SA	Food Serv	B	SA3a	cf. 260-24	350	270
730	SA	Food Serv	B	SA3a	cf. 494-01	325	270
730	op?	?	R		NO	400	350
730	op	?	F		NO	430	350
730	op	?	B	rb		425	300
730	op	?	B	rb		400	300
730	op	?	F		NO	350	300
730	op	?	R		NO	350	250
730	op	?	B	rb		325	250
730	Kr?	Drinking	R		NO	450	350
730	Dish	Food Serv	F	D22	cf. 501-16	310	150
730	CC	Food Serv	R	CC1-rb	cf. 521-03	400	300
730	C	Drinking	FB	stl	cf. 428-01	410	300
730	sk	Drinking	R	C-S20/28	cf. 521-54	320	280
730	C	Drinking	F	IC4-5	cf. 397-53	600	550

Table 4.1 Case study site artifacts.

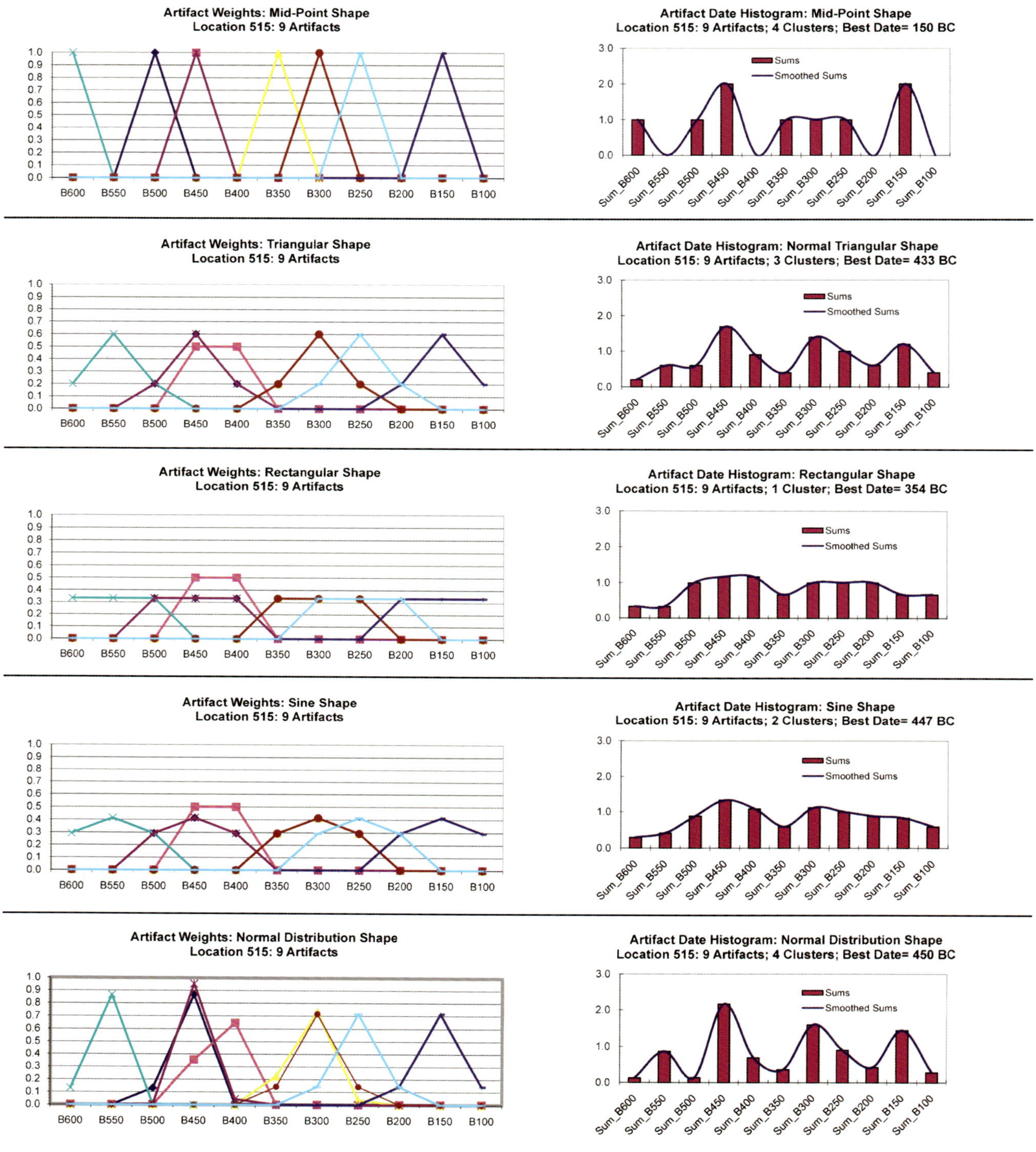

Figure 4.16 Location 515 artifacts and "best dates."

Code	Type
GFW	Glazed Fine Ware
FFW	Fig.d Fine Ware
TWC	Coarse Table Ware
COW	Cooking Ware
STW	Storage Wares
LMW	Loom Weights
TCF	Terracotta Figurines
TOT	Tomb Tiles

Table 4.2 Code definitions.

The normal distribution curve shape tends to be sensitive to changes in artifact weights over the range of dates, while not smoothing so much that rises and dips in "occupation" are ignored. The normal curve shape seems to allow the "best date" algorithm to pick out the date of most probable maximum occupation with considerable sensitivity.

Site Typing

In addition to site dating, we attempted to use statistical methods to identify clusters of correlated ratio patterns in all types of ceramics found at sites. We searched for ratio patterns that might represent meaningful signatures of location types such as farmhouses, tombs, or other kinds of sites. We produced a list of survey locations (sites and plots) with counts of artifacts distributed in eight categories (*Table 4.2*) considered to provide the most discrimination between different types of sites.

Correlations between one set of ratios (or artifact counts) and another were determined by computing a simple correlation coefficient from two sets of ratios or counts. The procedure works with ratios or counts in any combination because the correlation coefficient is identical for all cases. As long as the order of the categories is the same for both ratio (or count) sets, the procedure results in the same coefficient.

The correlation coefficient, r, is computed from two sets of ratios or counts consisting of pairs of X and Y values:

r = Sum XY / (Sqrt(Sum X^2) * Sqrt(Sum Y^2)[5]

The range of r is -1 to +1, with -1 values suggesting negative correlation, zero values indicating no correlation, and +1 values suggesting positive correlation. To evaluate r for different numbers of categories, we computed a t-distribution value from r and the degrees of freedom. Eight categories yield 6 degrees of freedom (df).

t = r * Sqrt(df) / Sqrt(1-r^2)[6]

For a 95% probability that X and Y are correlated in eight categories, the critical value for t (CTV) is 2.447. For a 99% significance level, the CTV is 3.707.[7]

The method is a variation and subset of unsupervised classification using the chain method.[8] It finds a mean set of category ratios from the best matches to the composite ratio as it passes through the database of ratios or counts of artifacts in each category. The method picks the same clusters and same mean ratios from data sets sorted in arbitrary orders.

Location Artifact Category Correlation

We computed the correlation coefficients between categories for all locations to see if any categories were so highly correlated with each other that their mutual inclusion was redundant (*Table 4.3*).

Method Details

The method was implemented with a MatLab script. The program flow is illustrated by the following pseudo-code:

Set critical t-distribution value for level of significance of correlation Loop
Pass One:
  Find mean pattern of all unused locations
Pass Two:
  Find location that is best match to that mean (best match is highest correlation coefficient).
  Mark that location as used.
  That location is the start of a group.
  The pattern for that location is the start for a group signature.
  Loop to populate a group.
    Locations are added to the group (and marked as used) if they are the next best match and their t-scores are over the critical t-value for the level of significance.

    The patterns for added locations are incorporated into the composite signature.

[5] J.C. Davis 1986, 40, No. 2.24.

[6] J.C. Davis 1986, 67, No. 2.35.
[7] J.C. Davis 1986, 62.
[8] Jensen 1986, 231–34.

	GFW	FFW	TWC	COW	STW	LMW	TCF	TOT
GFW	1.000	0.217	0.779	0.797	0.636	0.306	0.330	0.027
FFW	0.217	1.000	0.114	0.073	0.133	0.045	0.018	0.182
TWC	0.779	0.114	1.000	0.795	0.620	0.323	0.199	-0.010
COW	0.797	0.073	0.795	1.000	0.646	0.301	0.226	-0.018
STW	0.636	0.133	0.620	0.646	1.000	0.289	0.138	0.083
LMW	0.306	0.045	0.323	0.301	0.289	1.000	0.128	-0.018
TCF	0.330	0.018	0.199	0.226	0.138	0.128	1.000	-0.017
TOT	0.027	0.182	-0.010	-0.018	0.083	-0.018	-0.017	1.000

Table 4.3 Cross correlation between categories (see Table 4.2 for code defintions).

Pattern	GFW	FFW	TWC	COW	STW	LMW	TCF	TOT	Sum
P1	3	1	4	1	5	9	2	6	31.000
P2	6	2	9	5	1	4	1	3	31.000
P3	5	9	2	6	3	1	4	1	31.000
P4	1	4	1	3	6	2	9	5	31.000
P5	4	4	4	4	4	4	4	3	31.000
P6	1	1	2	3	4	5	6	9	31.000
P7	8	0	12	6	5	0	0	0	31.000
P8	14	4	9	2	2	0	0	0	31.000

Table 4.4 Patterns in random location artifact counts (see Table 4.2 for code defintions).

Pattern	p1	p2	p3	p4	p5	p6	p7	p8
P1	1.000	0.035	-0.835	-0.191	-0.312	0.489	-0.193	-0.267
P2	0.035	1.000	-0.191	-0.835	0.129	-0.418	0.799	0.663
P3	-0.835	-0.191	1.000	0.035	0.423	-0.645	-0.058	0.224
P4	-0.191	-0.835	0.035	1.000	-0.165	0.546	-0.588	-0.645
P5	-0.312	0.129	0.423	-0.165	1.000	-0.753	0.340	0.308
P6	0.489	-0.418	-0.645	0.546	-0.753	1.000	-0.565	-0.717
P7	-0.193	0.799	-0.058	-0.588	0.340	-0.565	1.000	0.743
P8	-0.267	0.663	0.224	-0.645	0.308	-0.717	0.743	1.000

Table 4.5 Cross correlations for patterns.

When no more unused locations fit the t-criteria the group is complete.
The composite signature is added to the list of signatures.
Pass Three:
Each location is matched to the signature that it best matches.
Pass Four
Locations in each group are used to form a composite signature

Five files are produced:
wareclusters.wk1 is a list of groups and the locations used in the formation of those groups' signatures.
waregroups.wk1 is a list of locations and the group they best match.
waresignatures.wk1 is a list of the group signatures.
groupclusters.wk1 is a list of the groups, the number in each group and the locations within each group.
groupsignatures.wk1 is a list of the final signatures for each group.

Tests

A sample file was produced as a check of the method's accuracy. The artificial test locations have a SUM value of =INT(RAND()*100)+11. The patterns use that SUM value and the following multipliers to compute artifact

	GFW	FFW	TWC	COW	STW	LMW	TCF	TOT
P1	0.097	0.032	0.129	0.032	0.161	0.290	0.065	0.194
P2	0.194	0.065	0.290	0.161	0.032	0.129	0.032	0.097
P3	0.161	0.290	0.065	0.194	0.097	0.032	0.129	0.032
P4	0.032	0.129	0.032	0.097	0.194	0.065	0.290	0.161
P5	0.129	0.129	0.129	0.129	0.129	0.129	0.129	0.097
P6	0.032	0.032	0.065	0.097	0.129	0.161	0.194	0.290
P7	0.258	0.000	0.387	0.194	0.161	0.000	0.000	0.000
P8	0.452	0.129	0.290	0.065	0.065	0.000	0.000	0.000

Table 4.6 Expected signatures from test patterns.

Group	Pattern	Total	Included	Excluded	Mis_Cat
G1	P7	100	100	0	0
G2	P8	100	100	0	0
G3	P6	100	100	0	0
G37	P2	100	100	0	0
G38	P4	100	100	0	0
G46	P3	100	100	0	0
G47	P1	100	100	0	0

Table 4.7 Detected clusters from algorithm on test data.

counts in each category =INT(SUM/8*(RAND()*M)) with the following multipliers (M):

Patterns P1-4 and P6 are variations on the digits 31415926. Pattern P5 is a 44444443 pattern representing almost equal values in each category. Pattern P7 is based on a hypothetical farmhouse signature composited from four excavated farmhouses (Sant'Angelo Grieco, San Biagio, Fattoria Stefan, and Fattoria Fabrizio). Pattern P8 is based on a hypothetical composite tomb/necropolis signature derived from the Pantanello necropolis burials (*Tables 4.4* and *4.5*).

Larger cross correlations (r>.50) exist for P4 and P6, P2 and P7, P2 and P8, and for P7 and P8. This suggests that the farmhouse (P7) and tomb (P8) signatures are similar and that P2 is similar to both the tomb/necropolis and the farmhouse. The expected ratio signatures from these groups are shown in Table 4.6.

The script was run with a CTV of 5.208 (99.9%) on the test data. The algorithm grouped the test location patterns into the clusters shown in Table 4.7.

The signatures for each group are shown in Table 4.8. Pattern P5 was distributed into 91 different groups with 91 unique signatures.

Signature Analysis of All Location Artifact Counts

The script was run using the list of 912 locations and their artifact counts. 43 signatures resulted (*Table 4.9*). Of these, 15 were associated with ten or more locations.

The detected signatures have correlation coefficients with respect to the composite farmhouse and tomb signatures; these are converted to t-values. The groups in bold in Table 4.10 show the detected signatures that correspond to farmhouse and tomb/necropolis signatures with a CTV of more than 5.207 for a 99.9% significance.

This suggests that signatures 1 and 4 match a farmhouse and that signature 3 matches a tomb/necropolis. Note that signature 1 is very close to matching both a farmhouse and a tomb/necropolis, reflecting the strong cross correlation between the two site-types.

Approximately one-third (332) of the locations were classified as farmhouses (in two categories) or tombs/necropoleis. The others were unclassified at this (99.9%) level of significance.

We ran similar cluster analysis tests using the Two-Step and K-Means processes in SPSS. Similar signatures resulted, but with differences in the assignment of locations to detected signatures.

Pattern	GFW	FFW	TWC	COW	STW	LMW	TCF	TOT
P1	0.097	0.032	0.129	0.032	0.161	0.290	0.065	0.194
G47	0.094	0.029	0.130	0.031	0.163	0.299	0.062	0.192
P2	0.194	0.065	0.290	0.161	0.032	0.129	0.032	0.097
G37	0.195	0.063	0.294	0.164	0.027	0.131	0.030	0.096
P3	0.161	0.290	0.065	0.194	0.097	0.032	0.129	0.032
G46	0.161	0.295	0.059	0.191	0.097	0.032	0.133	0.031
P4	0.032	0.129	0.032	0.097	0.194	0.065	0.290	0.161
G38	0.033	0.125	0.034	0.093	0.196	0.056	0.305	0.158
P5	0.129	0.129	0.129	0.129	0.129	0.129	0.129	0.097
P6	0.032	0.032	0.065	0.097	0.129	0.161	0.194	0.290
G3	0.031	0.028	0.065	0.095	0.130	0.160	0.195	0.296
P7	0.258	0.000	0.387	0.194	0.161	0.000	0.000	0.000
G1	0.241	0.014	0.369	0.180	0.152	0.015	0.015	0.016
P8	0.452	0.129	0.290	0.065	0.065	0.000	0.000	0.000
G2	0.440	0.123	0.284	0.056	0.057	0.015	0.013	0.012

Table 4.8 Expected and detected signatures from random artifact counts.

Group	GFW	FFW	TWC	COW	STW	LMW	TCF	TOT
1	0.377	0.005	0.469	0.092	0.047	0.004	0.004	0.001
2	0.117	0.002	0.765	0.064	0.047	0.004	0.002	0.000
3	0.572	0.008	0.265	0.086	0.052	0.004	0.010	0.003
4	0.249	0.000	0.309	0.162	0.278	0.000	0.003	0.000
5	0.318	0.000	0.338	0.295	0.041	0.004	0.003	0.001
6	0.134	0.005	0.495	0.075	0.277	0.012	0.002	0.000
7	0.578	0.000	0.047	0.037	0.336	0.000	0.002	0.000
9	0.380	0.008	0.339	0.017	0.247	0.007	0.000	0.001
10	0.113	0.000	0.456	0.226	0.190	0.014	0.000	0.000
13	0.028	0.008	0.456	0.479	0.030	0.000	0.000	0.000
14	0.092	0.000	0.021	0.004	0.875	0.008	0.000	0.000
15	0.454	0.000	0.130	0.391	0.025	0.000	0.000	0.000
17	0.935	0.010	0.017	0.014	0.021	0.000	0.000	0.003
21	0.107	0.005	0.591	0.288	0.007	0.002	0.000	0.000
29	0.047	0.000	0.423	0.004	0.526	0.000	0.000	0.000

Table 4.9 Signatures detected by algorithm from location data.

Error Analysis and Quality Control

The estimated spatial resolution for each of the layers used in the farm distribution analysis is presented in Table 4.11.

Based on these spatial resolutions, we can expect 32 m of meaningful spatial resolution for analysis that excludes the geomorphology layer and the DEM, 54 m excluding the DEM alone, and 116 m if all layers are used.

Date-bins have a quantization size of fifty years, so we expect date analysis to be meaningful at the 100-year level.

Division Line Analysis

There are two distinct sets of division lines; those north, and those south of the Basento. The division lines in each set are mostly aligned along one direction. This analysis does not consider the "transverse" division

Group	Count	r FH	r Tomb	t- FH	T - Tomb
1	228	0.919	0.898	5.708	5.000
2	290	0.835	0.548	3.715	1.607
3	91	0.703	0.951	2.423	7.503
4	12	0.924	0.678	5.932	2.258
5	32	0.899	0.780	5.029	3.055
6	29	0.884	0.526	4.634	1.517
7	14	0.436	0.685	1.186	2.301
9	12	0.845	0.851	3.870	3.974
10	17	0.931	0.495	6.230	1.396
13	16	0.710	0.266	2.469	0.675
14	15	0.145	-0.065	0.359	-0.159
15	10	0.588	0.678	1.779	2.258
17	42	0.354	0.795	0.928	3.205
21	23	0.860	0.505	4.135	1.433
29	12	0.613	0.225	1.902	0.566

Table 4.10 Detected signatures and correlation with farmhouse and tomb signatures.

Layer	Spatial resolution (m)	Source
Site positions	10	GPS or 1:10,000 maps
Survey plots	10	GPS or 1:10,0000 maps
Site elevations	2.5	5 meter elevation contours
Streams	20	Polylines from 1:10,000 maps
Division lines	20	Georegistered aerial photographs
Geomorphology	50	Digitized from aerial photographs
DEM	100	Radar DEM

Table 4.11 Spatial resolution of layers.

lines, only those aligned in the principle direction within each set. We focus here on the northern set of lines within the study area north of the Basento.

Proximity of Farms to Division Lines

We attempted to answer the question of whether or not farm sites were near to division lines.

We began with a subset of division lines, those within the study area aligned along the primary division line axis. We constructed a set of extended division lines, connecting line segments that were aligned. With a layer containing all sites identified as farms, we performed a spatial join, resulting in a table of perpendicular distances between each farm and the nearest division line. The resulting distribution is multi-modal, with peak at about 35 and 45 m and others at 75 and 95 m (*Fig. 4.17*).

A chi-square test for independence was performed that suggested that for 20-m histogram bins the observed distribution had a 12.9% probability of occurring by chance (*Fig. 4.18*).

Necropoleis and Division Lines

We attempted to answer the question of whether or not necropoleis were near to division lines.

We began with a subset of extended division lines, those within the study area aligned along the primary division line axis. With a layer containing all sites identified as necropoleis, we performed a spatial join, resulting in a table of perpendicular distances between each necropolis and the nearest division line. We then selected those 75 necropoleis within 120 m (about half the distance between parallel division lines) of a division line. The resulting distribution is bi-modal, with a peak at about 5 m and another at 60 m (*Fig. 4.19*).

Fifty necropoleis are within 60 m of a division line and twenty-five between 60 and 120 m from the nearest line. A chi-square test for independence was

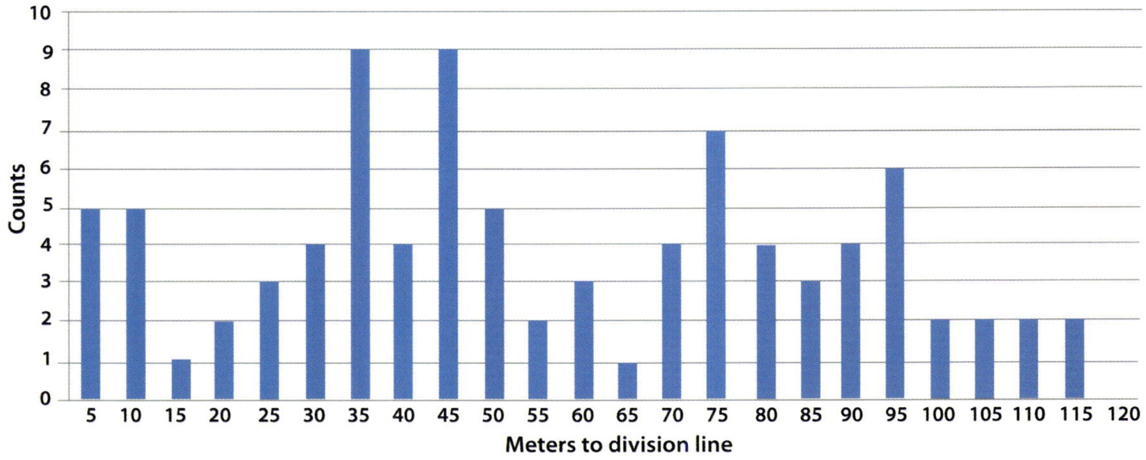

Figure 4.17 Distance from farms to division lines.

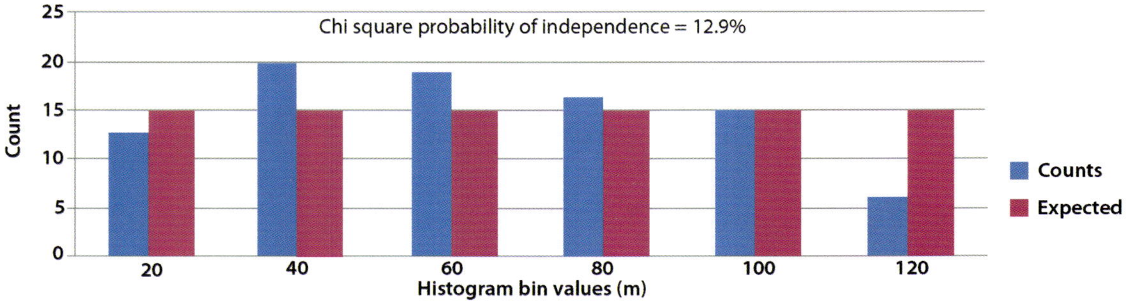

Figure 4.18 Significance of distance distribution (farms to division lines).

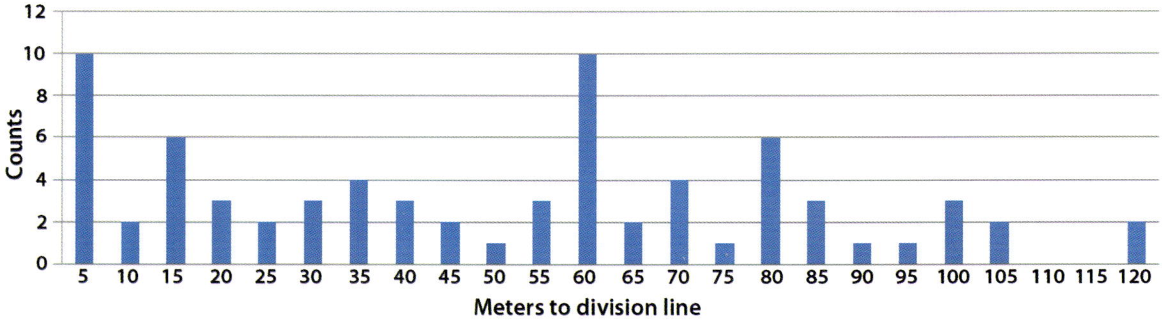

Figure 4.19 Distance from necropoleis to division lines.

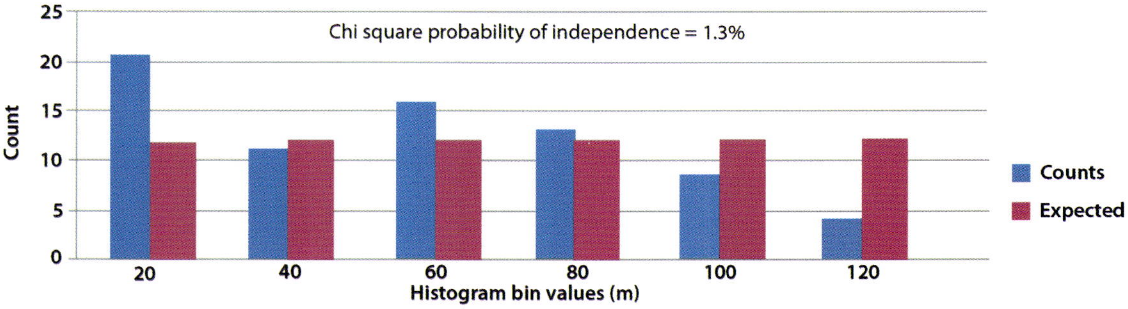

Figure 4.20 Significance of distance distribution (necropoleis to division lines).

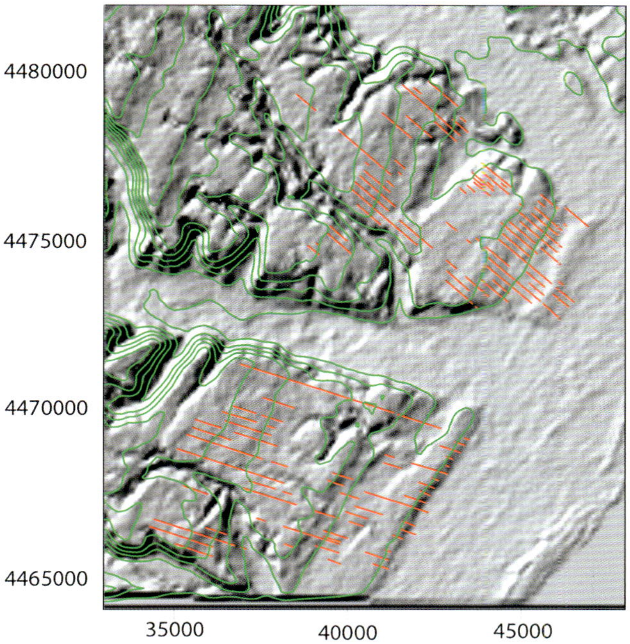

Figure 4.21 Division lines and elevation contours.

performed that suggested that for 20-m histogram bins the observed distribution had a 1.3% probability of occurring by chance (*Fig. 4.20*).

Division Line Azimuths and Slopes

Next we investigated the directions and slopes of the division lines with respect to the terrain surface. This analysis is highly speculative because we only have modern terrain surfaces to work with and our knowledge of those surfaces is derived from aerial photographs used to make topographic maps or digital elevation models derived from modern satellite imagery.

Division lines were extracted from the original data set resulting in South and North division line geodatabase feature classes projected in PTM coordinates. These contained only those division lines parallel to the primary axes for each region.

First we established the directions of the division lines with respect to true north. Azimuths are angles between 0 and 360° measured clockwise from true north. To avoid the convergence angle errors associated with some map projections we performed all this analysis in the Pantanello Transverse Mercator (PTM) coordinate system described above.

This analysis of division line slopes and azimuths and terrain is based on two digital elevation models: A NASA Advanced Spaceborne Thermal Emission and Reflection Radiometer (ASTER) DEM and a NASA Shuttle Radar Topography Mission (SRTM) DEM. The SRTM DEM in the 3 arc second finished format was downloaded from the USGS Seamless server in June of 2009. The ASTER Global Digital Elevation Model (GDEM) was downloaded from NASA's EOS data archive in July of 2009. Both original DEM tiles were projected into the PTM coordinate system. These two DEMs come from different sources and methods. The ASTER DEM has cell sizes of about 27 m and the SRTM cell size is about 80 m. Over the region containing division lines they are very similar in their elevation values. The SRTM mean value is about 5 m higher than the ASTER mean.

As digitized from the original aerial photographs, projected with the PTM system, and measured using the COGO functions of ArcGIS, the northern set of lines in the principal direction have a mean azimuth of 131.0° (or a back azimuth of 311.0°). The standard deviation of azimuth is 1.0°. For the southern set of lines the principle axis azimuth is 108.5° (back azimuth 288.5°) with a standard deviation of 1.1°. The two sets of lines then are very much parallel lines within each set, but the mean azimuths of the two sets differ by 22.5°. One of the enduring division line mysteries is this considerable difference in the primary direction of these two sets of parallel lines.

Fig. 4.21 shows the orientation of the north and south division lines. Note that the southern set of lines is close to perpendicular to the contour lines while the northern set point at an oblique angle to the contour lines. A line perpendicular to a contour line describes a direction that is the aspect of the terrain at the point of intersection.

In trying to make some sense of this we attempted to see what the relationship of the division line azimuths was to the aspects of the two terrain surfaces on which they lie. The aspect of a terrain surface is the direction that a downhill line faces. Of course for almost any real surface this idea is not so simple. Downhill paths often follow complex paths and so the aspect of a surface is also complex.

It is difficult to find the average of cyclical attributes such as direction. Simple averaging fails because simple averaging does not take into account the wrapping of azimuths between 360 and 0°. For example, a northwestern azimuth of 340° and a northeastern azimuth of 15° when averaged results in a numeric azimuth of 177.5°, almost due south rather than the actual mean direction of 357.5°. The arctangent (a four-quadrant atan2(x,y)) of the summed sines and summed cosines of angles results in a directional mean. This directional mean suffers from equal weights of each angle irrespective of slope. A steep hillside facing northeast (45°) and a slight sloping hillside facing northwest (315°) will have a directional mean of north (0°) using the summed sines and cosines approach.

Using the directional mean approach for the surface under the SRTM DEM within 200 m of division lines we get the SRTM terrain aspects shown in Table 4.12.

DL Group	Sum Sin	Sum Cos	Arctangent	Terrain Aspect
North	1282.68	-411.67	-72.206	107.8 degrees
South	1648.57	-829.85	-63.280	116.7 degrees

Table 4.12 SRTM directional means.

SRTM Directional Means

There is another approach for estimating average terrain aspects. We can make a generalized, smoothed surface from a complex one and then compute the terrain aspect of the smoothed elevation surface using the directional mean approach. To do this we took the ASTER DEM and the SRTM DEM and performed a neighborhood mean function with a neighborhood distance of 1000 m. The resulting grids were used to extract a generalized aspect for the terrain on which the division lines lie (*Table 4.13*).

DL Group	DEM	Terrain Aspect	Std. Deviation
North	ASTER	97.8 degrees	60 degrees
North	SRTM	100.9 degrees	55 degrees
South	ASTER	108.9 degrees	55 degrees
South	SRTM	104.8 degrees	62 degrees

Table 4.13 Smoothed ASTER and SRTM directional means.

Smoothed ASTER and SRTM Directional Means

It is clear from these computations that the aspect of the terraces on which the division lines are situated is complicated. While there is a general aspect of the terrace slopes, there is a large variation in terrain aspect along the division lines.

The southern set of division lines has a direction (108.5°) very close to the southern terrain aspect estimates

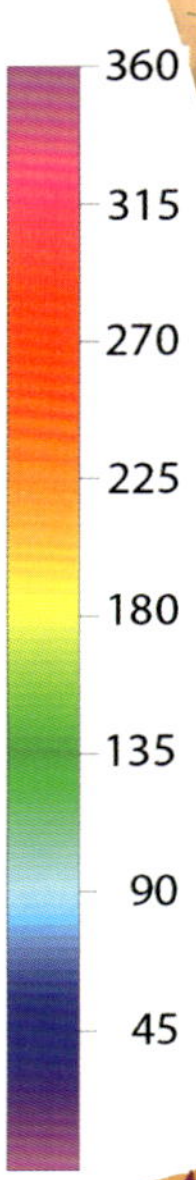

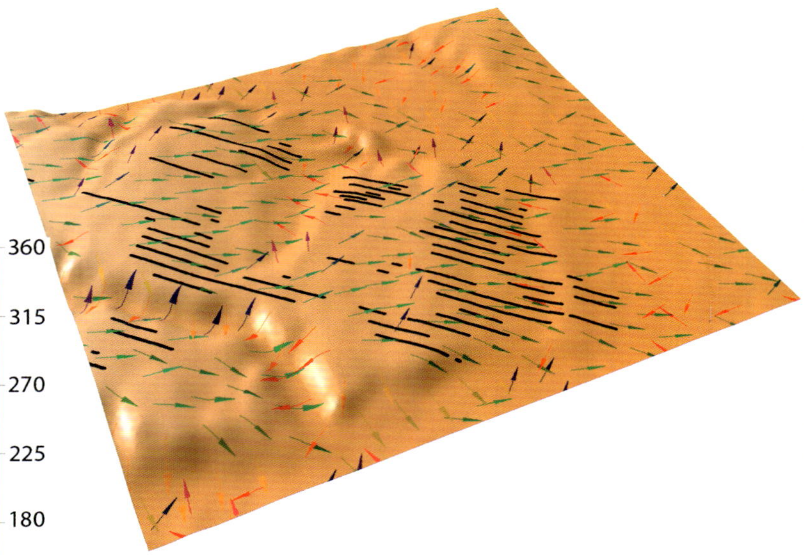

Figure 4.22 North division lines and terrain aspect (arrows point downhill).

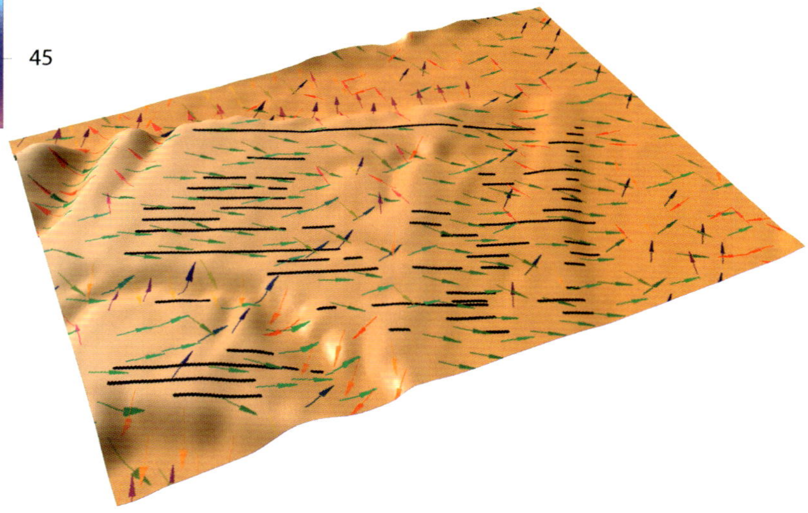

Figure 4.23 South division lines and terrain aspect (arrows point downhill).

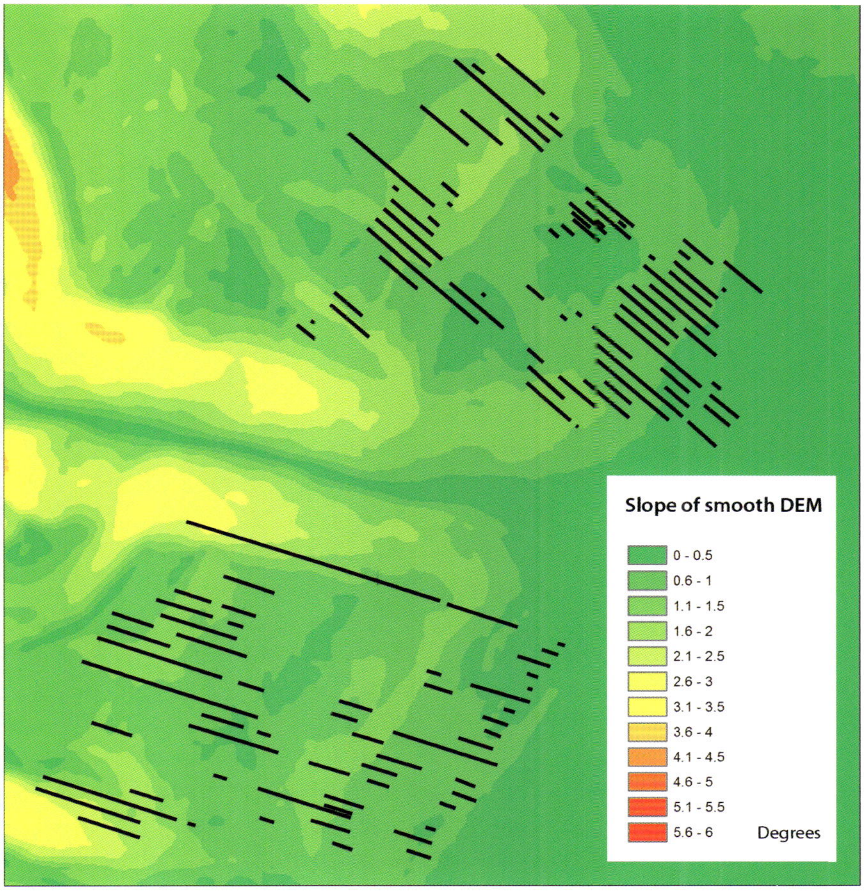

Figure 4.24 Slopes of the smoothed terrain.

DL Group	DEM	Slope Degrees	Slope Percent
North	ASTER	-0.8223	-1.44
North	SRTM	-0.9146	-1.59
South	ASTER	-0.9443	-1.64
South	SRTM	-0.9905	-1.72

Table 4.14 Terrain slopes within 200 m of division lines.

(104.8 to 116.7°), while the northern division lines have a direction (131.0°) that is oblique to the northern terrain aspect estimates (97.8 to 107.8°).

Figs. 4.22 and 4.23 show the complex relationship between terrain space and slope. The arrows show terrain aspect, their length indicating slope (longer = steeper).

It is clear that the northern division lines have a pronounced angle oblique to the terrain aspect. The southern lines are much closer to the terrain aspect angle.

Next we looked at the slopes of division lines with respect to the terrain slopes. Fig. 4.24 shows the angle of the smoothed terrain over the north and south regions in degrees.

In both regions where we find division lines the terrain slopes are gentle and generally less than 2.0° (3.5%).

For the ASTER and SRTM DEMs, the terrain slopes (measured downhill) within 200 m of the division lines are shown in Table 4.14.

Golden Software's Surfer platform was used to compute the slope of each division line by producing a terrain "slice" along each division line. The slices contain X, Y, distance, and elevation for each intersection between the DEM cells and the lines. Using the distance and elevations a linear regression function was used in Excel to

DL Group	DEM	Mean Slope Ratio	Slope STDEV	Slope Min	Slope Max	Slope %
North	ASTER	-0.01307	0.012	-0.0504	0.013	-1.3
North	SRTM	-0.01744	0.015	-0.0853	0.015	-1.7
South	ASTER	-0.02222	0.013	-0.0637	0.0058	-2.2
South	SRTM	-0.02153	0.014	-0.0823	0.0035	-2.2

Table 4.15 Division line slope statistics for northern division lines.

compute a percentage slope for the "best fit" line along the terrain profile for each division line. Slope statistics for the northern division lines are shown in Table 4.15.

Division Line as Berms

The division lines can serve as drainage features when there is excessive water on the terraces. The general downhill slope of the lines is such that excess water can be drained by the shallow sloping ditches. The same features might serve as berms as well when there is a shortage of water.

The northern and southern lines were analyzed through a direction derivative function in Golden's Surfer platform. To look at the fine grain slope of the division lines, the ASTER surface was used for both an elevation "slice" and a surface first derivative "slice."

Surfer was used to produce elevation and elevation derivative surfaces. For the northern lines a derivative of the ASTER surface was created with a direction of 319° (the "math" equivalent of the 131° azimuth mean). For the southern lines the angle used was 341.5° (for the 108.5° southern line azimuth mean).

Converted to ESRI floating point grids, these surfaces were used within ArcGIS to produce maps of the locations of first derivative positive peaks corresponding to berm highpoints that might trap water not flowing with a high enough volume to traverse the entire length of the division lines. Fig. 4.25 shows the locations of potential berms along the northern lines. Fig. 4.26 shows the southern berm locations.

There are many places in both regions where sufficient ponding could occur in times of low water.

Conclusions

This analysis has looked at division line proximity to farm sites and necropoleis, terrain slopes and aspects, and division line slopes and aspects. Drawing conclusions from modern terrain surfaces derived from aerial photographs or satellite imagery is dangerous because we have no way of knowing what the early Greek period surface was like.

We can conclude that the division lines have a gentle slope, perhaps by design. To achieve these slopes of less than 2.2% may have required the two different azimuths of the two sets of parallel lines be built on two different terrain aspects. The gentle slopes correspond to the critical slope angles suggested by several authors.

Prehistoric canals in many parts of the world are constructed to flow down gradients rarely greater than 2%. The prehistoric engineer went to great lengths to construct tall aqueducts against hillsides or across valleys in order to maintain a gentle gradient. If the slope is steepened, the whole range of hydraulic properties at a given discharge is altered to such an extent that local erosion and scouring may occur and/or the flow becomes supercritical, and there is spillage as well. This must have been recognized by prehistoric engineers, albeit empirically, in order for them to adapt their constructions to accommodate the problems introduced by the slope. Various solutions to problems of slope can be illustrated by reference to the author's work in both the Moche and Cusichaca valleys. These studies reveal an acute awareness by the prehistoric engineer of the velocity/discharge relationship and of the overriding importance of the nature of the material forming the bed and banks.[9]

The Agra Canal completed in 1874 also draws its supply of 1500 cubic feet per second from the Jumna, a weir half a mile long having been placed across the river at Okhla 10 miles below Delhi and it can irrigate 240,000 acres of land on the right bank of the Jumna between Delhi and Agra. This canal, 137 miles long, has a bottom width of 70 feet, a depth of water of 10 feet, and a gradient of 1 in 10,560.[10]

If the Greeks intended the division lines to serve as property boundaries they might have designed the equidistant parallel lines to serve that purpose. Drainage channels with stable sides would be ideal places to bury the dead, out of agricultural fields and away from homes.

[9] Farrington 1980, 295.
[10] Vernon-Harcourt 1896, 437.

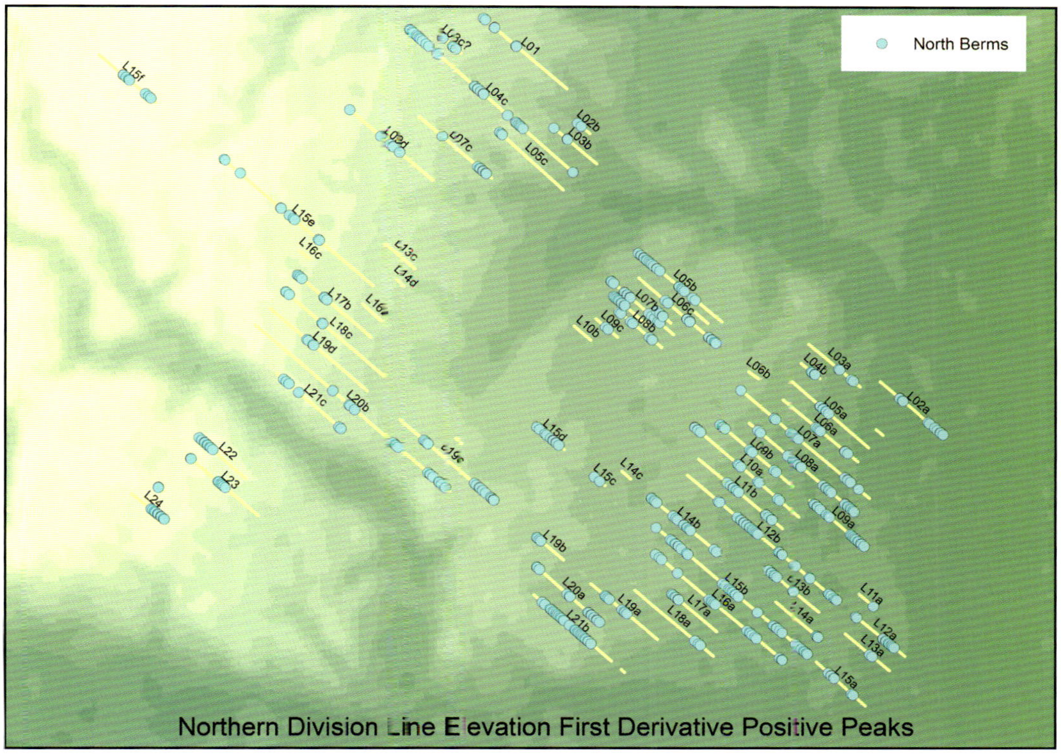

Figure 4.25 Locations of ponding berms along northern lines.

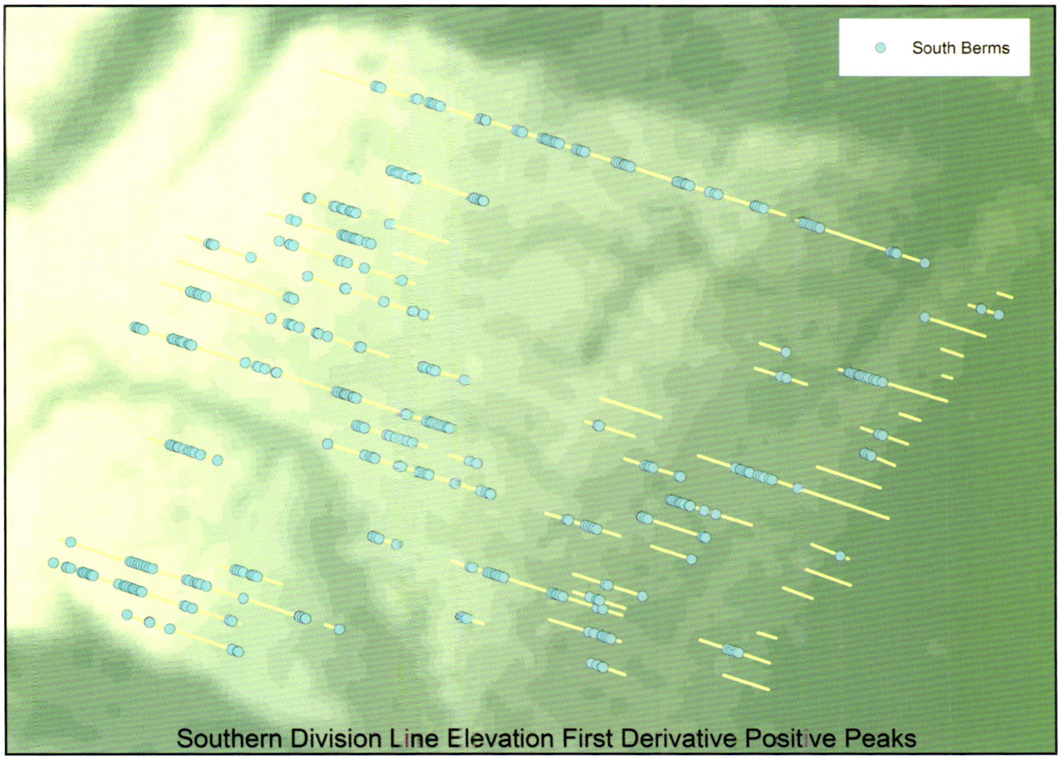

Figure 4.26 Locations of ponding berms along southern lines.

Best Date Range	Number of Sites	Mean ARW	MWD
250 (275–225)	15	4.2	178
300 (325–275)	92	4.4	184
350 (375–325)	56	3.9	170
400 (425–375)	23	3.6	162
450 (475–425)	21	3.5	158
500 (525–475)	5	3.9	170
550 (575–525)	42	3.2	148
600 (625–575)	2	3.3	152

Table 4.16 Maximum distance weight.

If they intended the lines to serve as drainage features in times of excess water they might have selected these critical slope angles that allow for runoff without channel scouring while keeping the channels clear of silt. If their intention was to retain water in times of drought, the ponding areas behind the berms along the lines might have been advantageous. Any conclusions must remain speculative without some way of reconstructing the terrain when the lines were constructed.

Farm Site Distributions

We have based our analysis of farm site distributions and farm importance on sites identified as farms by field investigators, rather than by the statistical methods explained above, because the on-site observations offer a more complete interpretive picture.

We employed the idea of Multiple Criteria Evaluation (MCE), based on a weighted linear combination of expert opinions, subjective judgments, quantitative measurements, and fuzzy estimation.

The basis of this analysis is a layer of farm sites that could be dated by fine-ware artifacts. The layer consists of sites and survey plot centroids described as definite farmhouses, probable farmhouses, and combination farmhouse/tomb sites. Attributes used include best date, date histograms, scatter area, density of tile fragments, and density of all other ceramic artifacts.

Farm Significance and Distribution for Each Location for All Date Periods

The estimate of farm size or importance is based on the size of the scatter, ceramic density, tile density, number of dated fine-ware artifacts, and the size of scatters in each best date period.

We calculated weights for these fields based on the following fuzzy sets for density (tile and ceramic):

Very Heavy = 9
Heavy = 7
Moderate = 5
Light = 3
No Data = 0
<Null> = 0

The approximate scatter size (length x width) for all locations was reclassified into values from 1 to 9. The distribution of scatter sizes from 0 to 5000 is almost linear; above 5000 the distribution makes a rapid change to a maximum of 42,000. The threshold for inclusion in category 9 was therefore set to 5000, with the values from 0 to 4999 distributed evenly between 1 and 8.

```
The Area Weight (ARW) =
IIF( [AREA_LW] >=5000, 9, IIF ( [AREA_LW] > 0, [AREA_
LW] * .0016 +1,0))

The Tile Density Weight (TDW) =
IIF( [TILE_DENS]="Very Heavy", 9,IIF( [TILE_DENS]
="Heavy", 7,IIF( [TILE_DENS] ="Moderate", 5,IIF( [TILE_
DENS] ="Light", 3,IIF( [TILE_DENS] ="None", 1,0)))))

The Ceramic Density Weight (CDW) =
IIF( [CERAM_DENS]="Very Heavy", 9,
IIF( [CERAM_DENS] ="Heavy", 7,
IIF( [CERAM_DENS] ="Moderate", 5,
IIF( [CERAM_DENS] ="Light", 3,
IIF( [CERAM_DENS] ="No Data", 0,0)))))]
```

Farm-Size Weight for Each Date Period

We computed a maximum weighted-distance (MWD) value for use in a cost-weighted allocation

Case	ARW	CDW	TDW	AW	CW	TW
0	0	0	0	X	X	X
1	0	0	NZ	X	X	1
2	0	NZ	0	X	1	X
3	NZ	0	0	1	X	X
4	NZ	0	NZ	.5	X	.5
5	NZ	NZ	0	.5	.5	X
6	NZ	NZ	NZ	.5	.25	.25

Table 4.17 Cases for IMCE categories. (NZ = non-zero, X = don't care).

Date Period	Number of Sites
600	23
550	69
500	18
450	79
400	89
350	105
300	143
250	56

Table 4.18 Locations estimated to be occupied in each period.

(see below). We selected the farmhouse locations with a best date within the fifty-year period of the date histogram. The MWD value (*Table 4.16*) is four times the square root of the mean area, or 4* SQT((Mean ARW-) / 0.016.

Date-Independent Multiple Criteria Evaluation for Each Location

A date-independent multiple criteria evaluation (IMCE) value from 0 to 10 was computed for each location based on the following weights (which sum to 1.0) for relative importance of the categories:

ARW= .50 Area
TDW= .25 Tile density
CDW= .25 Ceramic density

We computed the IMCE weight based on tile and ceramic density and scatter size/area. For the five sites where area was not recorded, leaving a value of 0, a value of 1943 was inserted instead, the average area of all farmhouses (except for the two highest values, 42,000 and 22,500). Weights are equally split between ceramic and tile density (50%) and area (50%). The intention of the method is to give area a weight of 1.0 where there are no density values, and to give artifact density a total weight of 0.5 and area a weight of 0.5 where there are both area and tile/ceramic density measurements. Table 4.17 shows the possible cases.

The IMCE value is a result of the linear combination of CDW, TDW, and ARW weighted when all are present by

TDW * .25 + CDW * .25 + ARW * .5 or when one or more values are not present by the following:

```
( IIF (  [CDW] + [TDW]  <1, [ARW], IIF(  [ARW] < 1   And
( [CDW] * [TDW]  <  1) , [CDW] + [TDW], IIF(  [ARW] >
0  And  ( [CDW] * [TDW]  >  0) , [ARW] * .5 + ( [CDW] +
[TDW] ) * .25, [ARW] * .5 + ( [CDW] + [TDW] ) * .5 ))))
```

This weight is high (10) for farmhouse locations with very heavy tile and ceramic density and large scatter sizes. The value is low (0) for farmhouse locations with light density and small scatter sizes.

Farm Site Locations for Each Date Period

We formed layers for eight of the twelve date periods based on the date histogram bins. Locations within a date bin were selected where the date bin sum was more than or equal to 0.95, just less than one artifact's total weight. Table 4.18 shows the date periods and the number of sites estimated to be occupied in that period.

Location Artifact Weight for Each Date Period

For each location a weight from 0 to 10 was computed for a specific date period based on the ratio of the number of dated artifacts found at that location to the number of artifacts dated within that specific date period.

LAW = Sum of dated artifacts at this location / Sum of all dated artifacts * 10
[SUM_Bnrn] / [SUM_NARTS] *10

Location Bin Artifact Weight for All Date Periods

For each location a weight from 0 to 10 was computed from the ratio of all artifacts within that date period to the number of artifacts at each location (*Table 4.19*).

BAW = Sum of dated artifacts at this location for this date period / Sum of all dated artifact histograms for this date period * 10

[SUM_Bnnn] / [n] *10

Date range	Histogram sums of dated artifacts
250	187
300	845
350	340
400	275
450	207
500	143
550	280
600	31

Table 4.19 Histogram sums of dated artifacts for each date range

Location Total Artifact Weight for Each Date Period

For each location a weight from 0 to 10 was computed from the ratio of dated artifacts to the number of all artifacts (dated or not), in the AdjustedCounts.dbf file, in the field ADJSUM.

TAW = Sum of dated artifacts for this date period / Sum of all artifacts found at this location

[SUM_Bnnn] / [ADJSUM] *10

Date-Dependent MCE Value for Each Location and Each Date Period

A date-dependent MCE value (DMCE) was computed for each site. This was based on these notions:

1. A location with many fine-ware artifacts is more significant than a location with fewer fine-ware artifacts.
2. For a particular fifty-year period a location is more significant when the number of fine-ware artifacts dated within that period is high with respect to the number of fine-ware artifacts dated to all periods found at that location.
3. The number of fine-ware artifacts dated to a particular period is further weighted by the number of fine-ware artifacts of that period at all locations.

Weight	Criteria	Weight
CDW	Ceramic Density	0.0625
TDW	Tile Density	0.0625
ARW	Area	0.1250
LAW	Ratio of all dated artifacts to each date bin	0.2500
BAW	Ratio of dated artifacts at this date to each date bin	0.2500
TAW	Ratio of all artifacts to each date bin	0.2500

Table 4.20 Weighting for final MCE value.

4. The number of fine-ware artifacts dated to a particular period is further modified by the relative number of all artifacts (whether fine-ware or not) found at that location.

The function for the date-dependent MCE value is:

DMCE = LAW * .33333 + BAW * .33333 + TAW *.33333

Final MCE Weight for Each Location

After a period of evaluation we finally settled on the appropriate weighting for the two components of the MCE process, the date-dependent weight (DMCE) and the date-independent weight (IMCE). To keep the date-independent weight based on tile and ceramic density and area from dominating the relative significance region size, we weighted the linear weighted combination so that DMCE accounted for 75% and IMCE for 25% of the final multiple criteria evaluation (FMCE) weight.

A final multiple criteria weight (FMCE) was computed for all locations:

FMCE = IMCE * 0.25 + DMCE * 0.75

The resulting value from 0 to 10 is computed from the weighting scheme shown in Table 4.20.

Farm Cost Value

Farm cost (COST) is a value ranging from 0 for a low-cost (large and important) farm site to 1.0 for a high-cost (small and less important) farm site.

COST = (10-FMCE) / 10

The COST value is the basis for a weight surface for each date period. The weight surface is used in a cost-weighted allocation, so that a large farmhouse with high density values is a "low-cost" source, claiming more grid cells than a small farmhouse site with low tile and ceramic density resulting in "high cost."

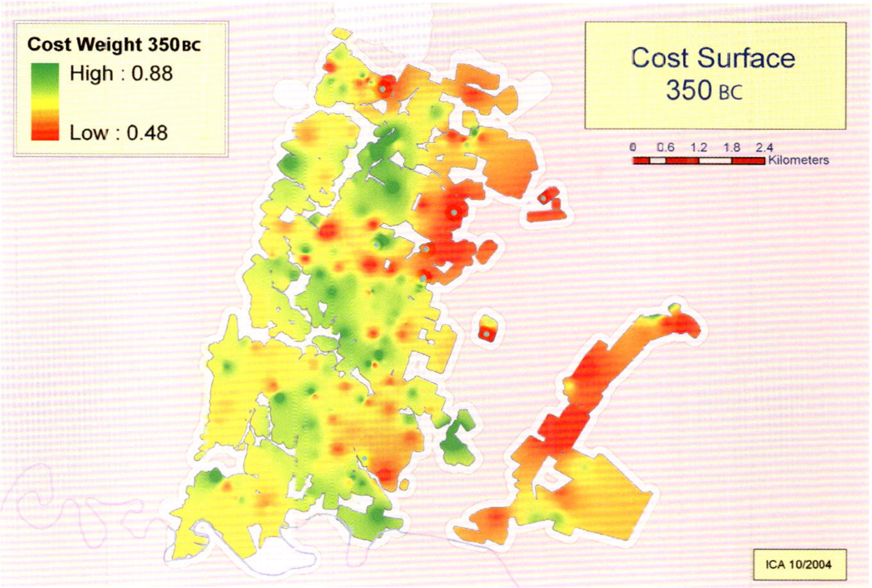

Figure 4.27 Cost-weighted surface: 350 BC.

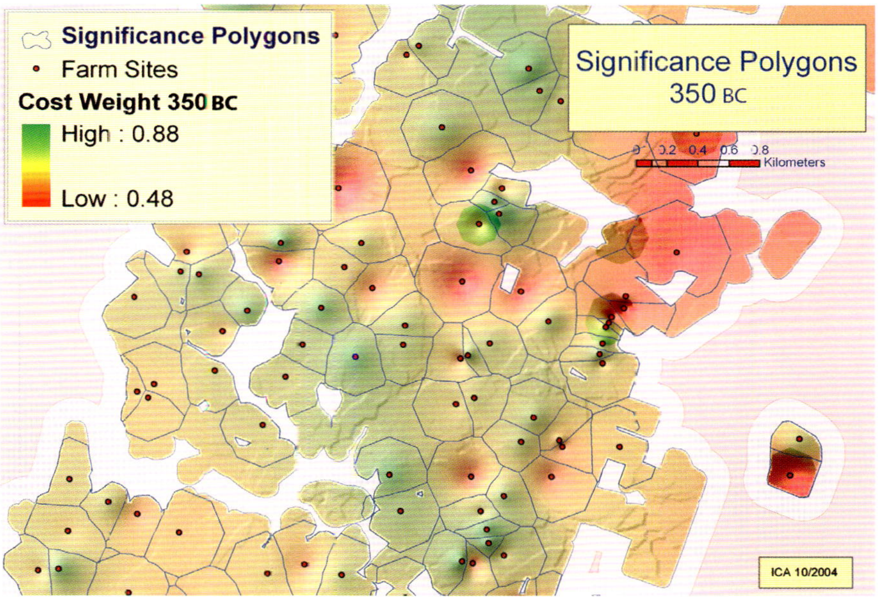

Figure 4.28 Cost-weighted allocation polygons: 350 BC.

Interpolating a Cost-Weighted Surface

Farm importance was interpolated from the COST value using the ArcGIS Spatial Analyst extension's Inverse Distance Weighting (IDW) function (parameters: power 2, cell size 10 m, 9 neighbors) within the 50 m study area buffer layer. This resulted in a cost-weight raster output file for each period. Fig. 4.27 shows the cost weighted surface for 350 BC. Here the high values are associated with high cost, low values with low cost. For the spatial allocation process described below we want more space allocated to low cost, more significant, sites (*Fig. 4.28*).

Cost-Weighted Allocation by Farm Importance

Each set of farmhouse locations and cost-weights was then fed into ArcGIS Spatial Analyst's Distance/Cost-Weighted Allocation function, producing allocation grids with lowest cost for large farmhouses and highest cost for small farmhouses. The farm site point layer is used as the source; the cost raster layer is used as the cost-weighted surface, and the maximum distance is the MWD computed above. The output grids were converted to polygon (vector) features, resulting in farm distribution and significance polygon layers. Fig. 4.28 shows the polygons resulting from the cost-weighted allocation process for 350 BC.

Mapping Farm Distribution and Significance

Using a gradient fill for the farm polygons, a solid fill for farm site buffers, and small point symbols for farm site centroids, each date period is mapped. Three sequential date periods are shown in Figs. 4.29–31.

A sequence of eight images was used to produce a QuickTime video showing sequential distributions for the periods from 600 to 250 BC (*Fig. 4.32*).

The video allowed researchers to examine change in settlement pattern over time by viewing the entire video or by manually shifting time using the QuickTime controls.

Conclusions

In this research GIS was used to assist in the understanding of history by providing a means for archaeologists to model spatial and temporal distributions through weighted criteria from a variety of archeological evidence. We set up a system with multiple layers and appropriate coordinate systems. We used statistical approaches and multiple criteria evaluation based on the advice of experts. Site types and settlement patterns were considered. Farm sites and the dated artifacts found at them were used to estimate change in space over time. The resulting polygons are not intended to represent the boundaries of farm sites; rather they are the approximate extents of regions of relative spatial significance. They are estimated influence regions, based on weighted evaluations of artifacts counts, artifact typology, artifact date ranges, site scatter size and density evaluations, and the weighted opinions of archaeologists.

The approach uses an ensemble of sites and artifacts, not in an attempt to characterize individual sites, but to estimate settlement pattern change over time. The resulting patterns of change within the Metaponto chora over time and space are intended to provide insights into history through further interpretation by historians, classicists, and archaeologists.

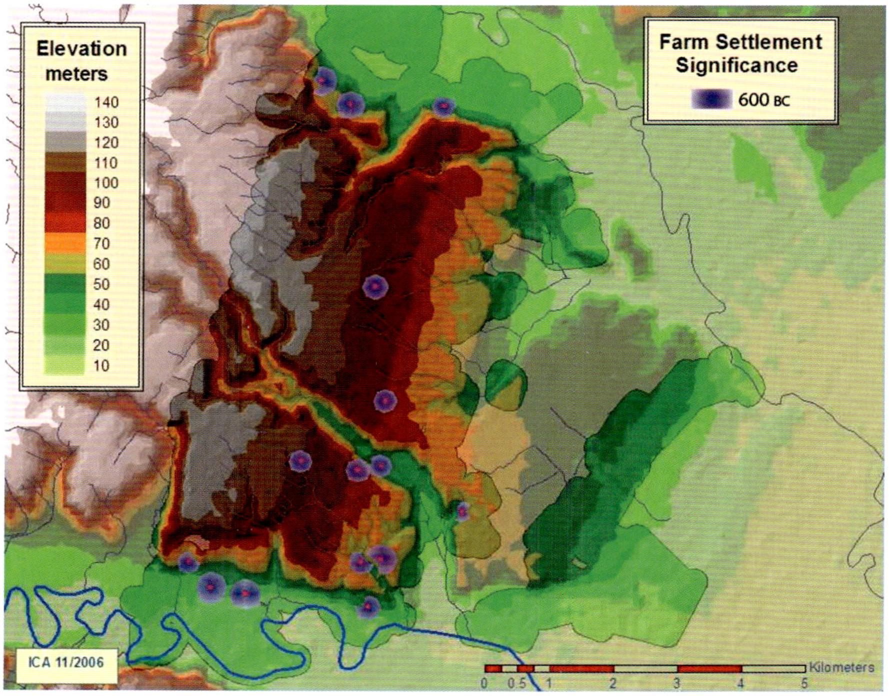

Figure 4.29 Settlement patterns from sinusoidal histograms: 600 BC.

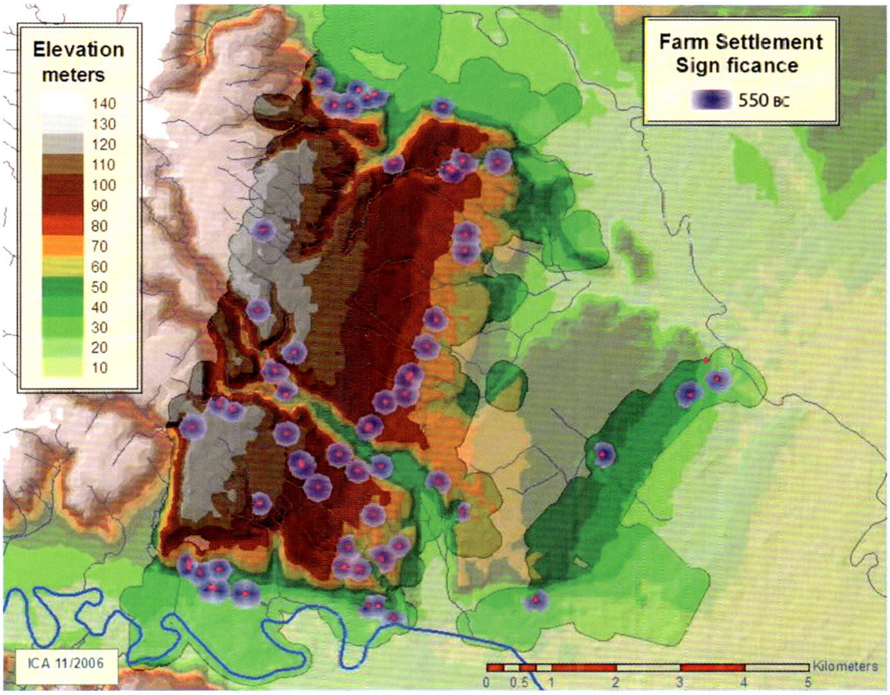

Figure 4.30 Settlement patterns from sinusoidal histograms: 550 BC.

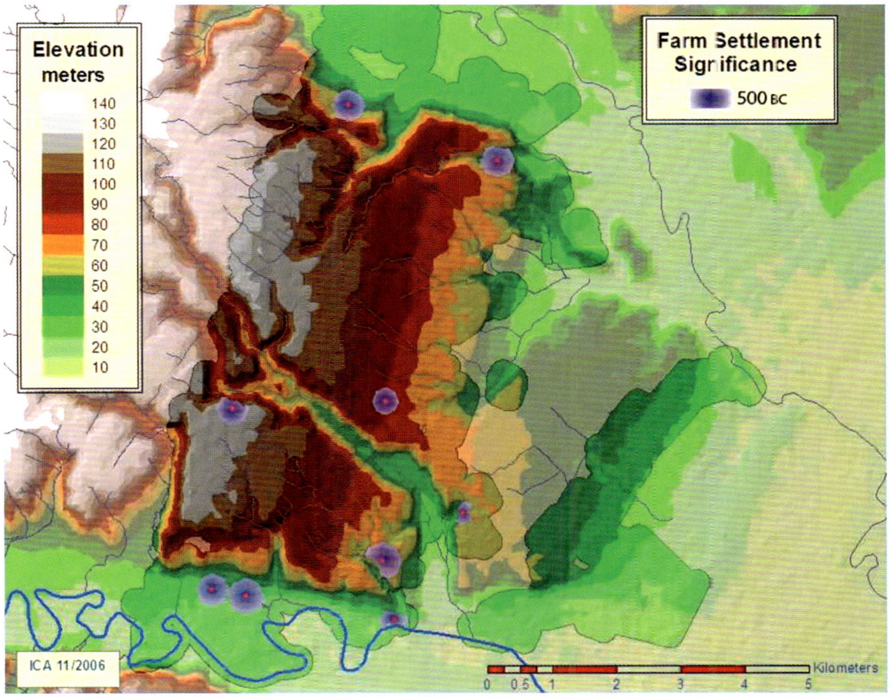

Figure 4.31 Settlement patterns from sinusoidal histograms: 500 BC.

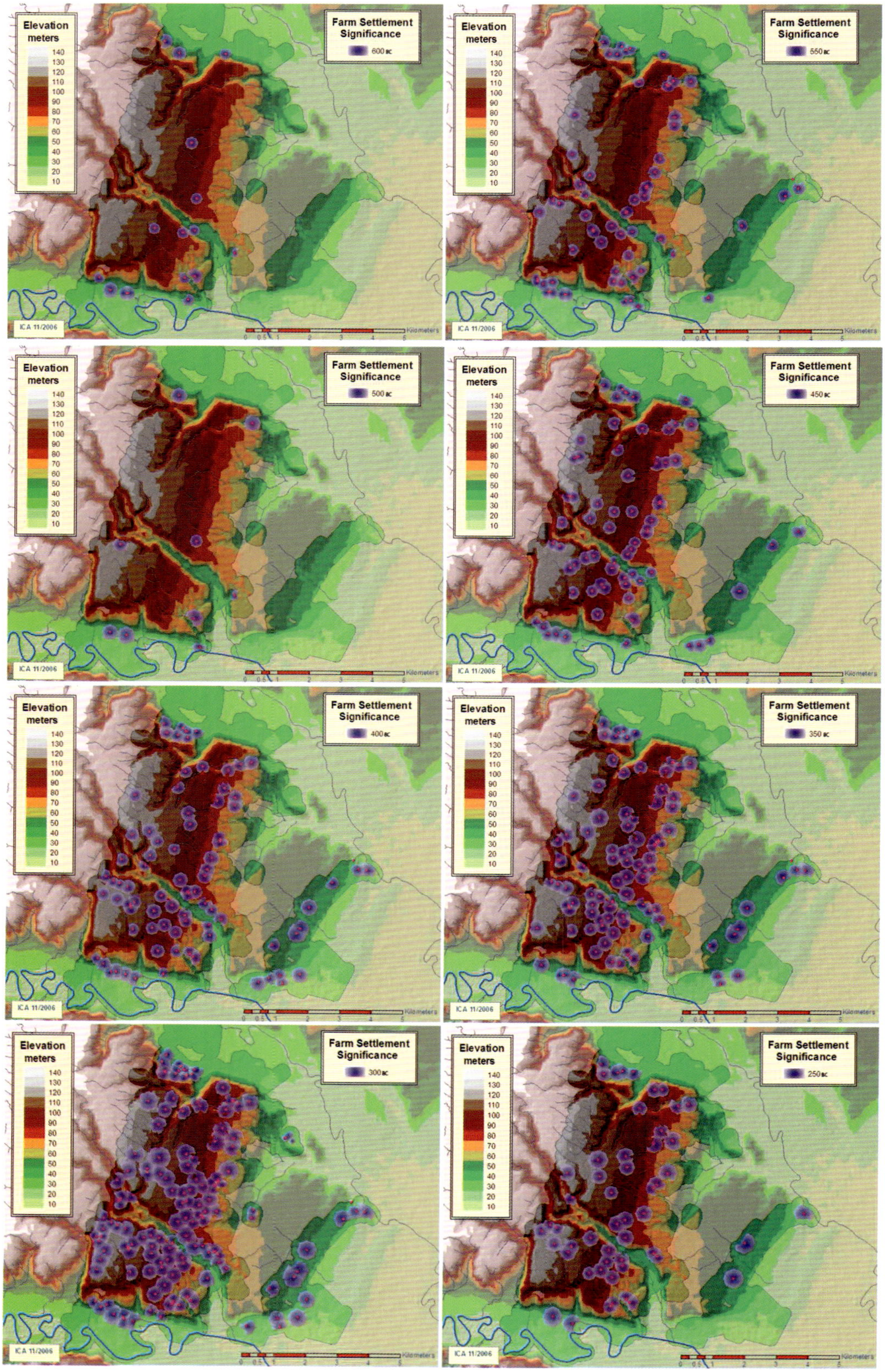

Figure 4.32 Video frames, 600 to 250 BC.

III. Survey Materials and Assemblages

5
The Survey Assemblages

Keith Swift

Pottery fragments in surface scatters constitute a primary source of evidence for sites identified through survey. This chapter examines the ceramic materials recovered by the ICA survey from Greek-period sites,[1] with the aim of identifying broad differences in the types and quantities of vessels found in the various assemblages attributed to necropoleis/tombs, farmhouses, and sanctuaries. Identification of site types was made in the field on the basis of criteria described in Chapter 3.

Due to the limited repertoire of the ceramic objects produced during the time periods under study, surface assemblages from different kinds of sites inherently share similarities in the range of ceramic types represented. Although the ranges of types are similar, aggregate quantification of the combined assemblages suggests that the relative proportions of the ceramic types vary between sites of different character. The ways in which they do so are outlined for each ceramic type in the first part of this chapter. The second part examines the ceramic assemblages by the type of site—necropoleis/tombs, farmhouses, and sanctuaries—and gives selected examples to illustrate the key features of each kind and to show the quantities, range, and variation in the surface assemblages from survey in the Metapontine chora.

Quantification of Sites and Assemblages

This is a synchronic analysis considering the colonial Greek period down to the Early Empire (625–25 BC), including all sites with material dating from the Archaic through Late Hellenistic periods. A total of 525 sites with ceramic assemblages dating to these periods were recorded by the survey.[2] Of these, 484 were attributed a function (site type) using the criteria outlined in Chapter 3. The total number of sites and total count of ceramics are shown in Table 5.1 along with their relative percentages.

The uneven distribution of the ceramics among the survey sites becomes clear when viewed in Figure 5.1. The farmhouses account for the majority of sites (271 sites, or over 51%), but an even greater majority of the ceramics (19,212 pieces, or almost 69%). In addition, 29 sites have been identified as farmhouses with tombs, 9 as farmhouses with kilns, and 1 site as a farmhouse associated with both a kiln and tomb(s). Together, these account for 86.5% of all ceramic forms from the survey.[3] By contrast, the sites identified as necropoleis and tombs (133 sites, just over 25%) account for less than 10% of all ceramics. Most of these assemblages come from individual or small groups of tombs (see below), which partly accounts for the lower proportion. Sanctuary assemblages tend to be proportionally larger: 4 sites comprise less than 1% of sites, yet their ceramics account for over 4% of the total.

Quantification of Ceramic Types by Site Function

For the purposes of aggregate quantification, four main ceramic groups have been used—fine wares, plain and banded wares, cooking and storage vessels, and other ceramics (*Table 5.2, Fig. 5.2*)[4]—the quantification of which will be discussed in turn. Each of these groups are further refined into the most common shapes, which in most cases allow for inter-period comparisons.[5] This is a synchronic analysis of the

[1] Fewer ceramic materials were recovered from the prehistoric and Post-Greek periods and did not lend themselves to the same type of analysis; these materials are discussed in Ch. 17 (prehistory) and Chs. 32–35 (Roman, Medieval, and Post-Medieval periods). The location and condition of all sites mentioned are described in the Gazetteer.

[2] From a total of 565 sites identified as having been active in the Greek period, ceramics were collected from 525. Of the remaining 40 sites, 33 yielded no artifacts, 4 yielded no ceramic materials, 1 had only a tile fragment of indeterminate Greco-Roman date, and 2 yielded only amphora body sherds.

[3] For the purposes of quantification and comparison with assemblages from sites identified as necropoleis/tombs and sanctuaries, these sites with dual or triple functions have not been included in the combined farmhouse assemblages. Ceramic production sites are particularly problematic from the point of view of quantification.

[4] The ceramic typologies and chronologies from the survey used in the dating of sites and for the GIS component are discussed in Chs. 6–14. The groupings used for quantification here apply only to the combined quantification of the survey assemblages undertaken in this chapter.

[5] Subdivisions by types and sub-types tend to be chronologically specific

Site attribution	Site count	% total sites	Ceramic count	% total ceramics
Necropoleis and tombs	133	25.3	2,569	9.2
Farmhouses	271	51.6	19,212	68.8
Farmhouses with tombs	29	5.5	2835	10.2
Farmhouses, kiln, and tombs	1	0.2	367	1.3
Farmhouses with kilns	9	1.7	881	3.2
Kilns	2	0.4	12	0
Greek agricultural structures	34	6.5	465	1.7
Sanctuaries	4	0.8	1,142	4.1
Other	1	0.2	2	0
Unidentified Greek feature	41	7.8	419	1.5
Total	**525**	**100**	**27,904**	**100**

Table 5.1 Total attributed sites from the survey, percentage of all sites, count, and percentage of ceramics.

Ceramic groups	Shapes	All sites		Necropoleis		Farmhouses		Sanctuaries	
		Count	%	Count	%	Count	%	Count	%
Fine wares	Cups	3,324	11.9	281	10.9	2,303	12	234	20.5
	Bowls	4,286	15.4	337	13.1	3,022	15.7	171	15
	Closed forms	1,638	5.9	429	16.7	822	4.3	54	4.7
	Other forms	461	1.7	90	3.5	259	1.3	17	1.5
All fine wares		9,709	34.8	1,137	44.3	6,406	33.3	476	41.6
Plain and banded wares	Bowls	4,305	15.4	379	14.8	3,019	15.7	152	13.3
	Mortaria	234	0.8	12	0.5	168	0.9	9	0.8
	Basins	240	0.9	16	0.6	176	0.9	12	1
	Louteria	28	0.1	0	0	19	0.1	1	0.1
	Closed forms	7,791	27.9	618	24.1	5,514	28.7	219	19.2
	Other forms	1,102	3.9	84	3.3	727	3.8	17	1.5
All plain and banded wares		13,700	49.1	1,109	43.2	9,623	50.1	410	35.9
Cooking and storage	Cooking ware	2,719	9.7	195	7.6	2,019	10.5	53	4.6
	Pithoi	333	1.2	17	0.7	256	1.3	1	0.1
Transport amphorae		563	2	69	2.7	393	2	6	0.6
Other ceramics	Unguentaria	136	0.5	14	0.5	100	0.5	4	0.3
	Lamps	82	0.3	3	0.1	63	0.3	2	0.2
	Loom weights	117	0.4	6	0.2	94	0.5	1	0.1
	Miniatures	231	0.8	9	0.4	71	0.4	133	11.6
	Terracottas	312	1.1	10	0.4	186	1	56	4.9
	Thymiateria	3	0	0	0	1	0	0	0
All ceramics		**27,905**	**100**	**2569**	**100**	**19,212**	**100**	**1142**	**100**

Table 5.2 Breakdown of ceramic types by sites identified as necropoleis/tombs, farmhouses, and sanctuaries. The total of "All sites" includes multi-function sites not listed.

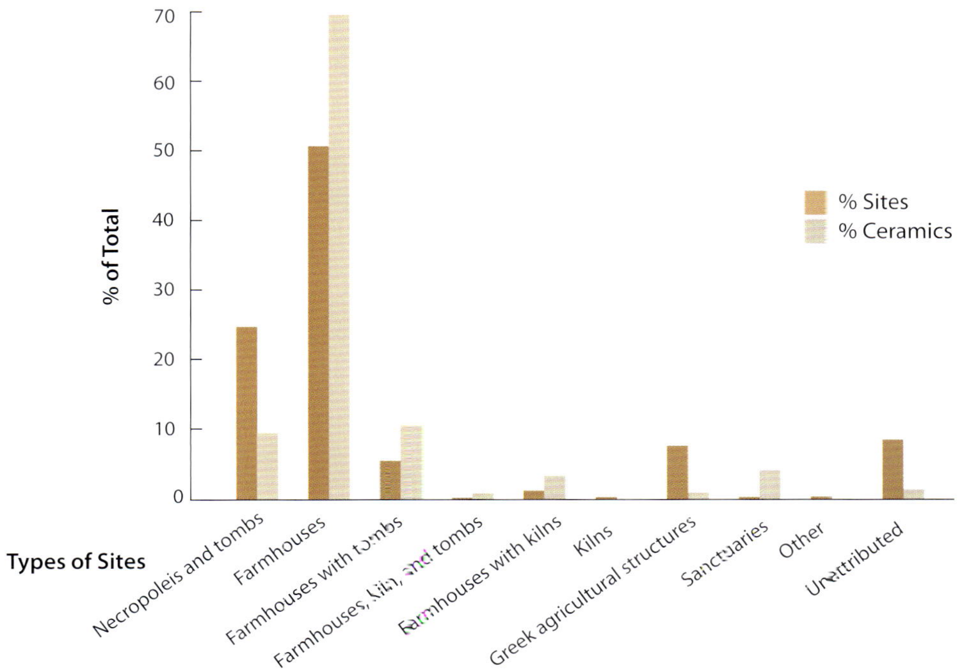

Figure 5.1 Breakdown of site types vs. ceramics by percentage.

colonial Greek period as a whole, including all sites with material dating from the Archaic through Late Hellenistic periods.

Fine wares

(Archaic fine wares, black gloss, and grey ware)

This group includes banded and semi-slipped fine wares of the Archaic period (Ch. 6), black-gloss ware (Ch. 6) and figured ware (Ch. 8) typical of the Classical and Hellenistic periods, as well as the grey ware typical of the Late Hellenistic period (Ch. 7).[6] Fine-ware cups comprise approximately 10% of identifiable forms from necropoleis/tombs and farmhouses, but twice that proportion (20%) of the assemblages from sanctuaries.[7] Fine-ware bowls are represented in the combined assemblages from all three types of site in similar proportions, roughly 13–16% of all identifiable forms. They are more frequent than fine-ware cups in the combined assemblages from necropoleis/tombs and farmhouses, but are outnumbered by fine-ware cups in the combined sanctuary assemblages.[8] Fine-ware closed forms predominate in the combined assemblages from necropoleis and tombs, where they account for just under 17% of all ceramics, outnumbering both fine-ware cups and bowls.[9] In both the combined farmhouse and sanctuary assemblages, these account for just less than 5% of all form-sherds and in both cases are outnumbered by both cups and bowls. Other fine-ware forms comprise a small proportion of the combined assemblages from all sites, and less than 1.5% from the farmhouses and sanctuaries, but they are better represented in necropoleis/tombs, where they comprise approximately 3.5% of all identifiable form-sherds.[10]

Plain and banded wares

The plain and banded wares are here grouped together since these are usually produced from similar fabrics, and in many cases it is difficult to distinguish between plain and banded or semi-slipped wares in fragmentary surface survey assemblages (Ch. 9). As is usual for

with restricted production spans and do not allow for comparisons between sites of varying dates. See Ch. 4 for the methods and approaches to the dating of sites from the survey.

[6] For the purposes of aggregate quantification, these groups do not include black-gloss lamps, miniatures, or thymiateria, which have been differentiated as specialized forms and grouped with their plain- and banded-ware counterparts (see below).

[7] Fine-ware cups include a wide range of relatively well-established types; see Chs. 6–8.

[8] Fine-ware bowls include all open shapes in fine wares, including the bowl, bolsal, concave-convex, dish, one-handler, and small bowls.

[9] Fine-ware closed forms consist mostly of liquid containers and include the table amphora, lekythos, oinochoe, and pelike.

[10] Other fine-ware forms include miscellaneous shapes, grouped together for the purposes of quantification, including the epichysis, feeder or lamp-filler, guttus, kothon, and lebes gamikos.

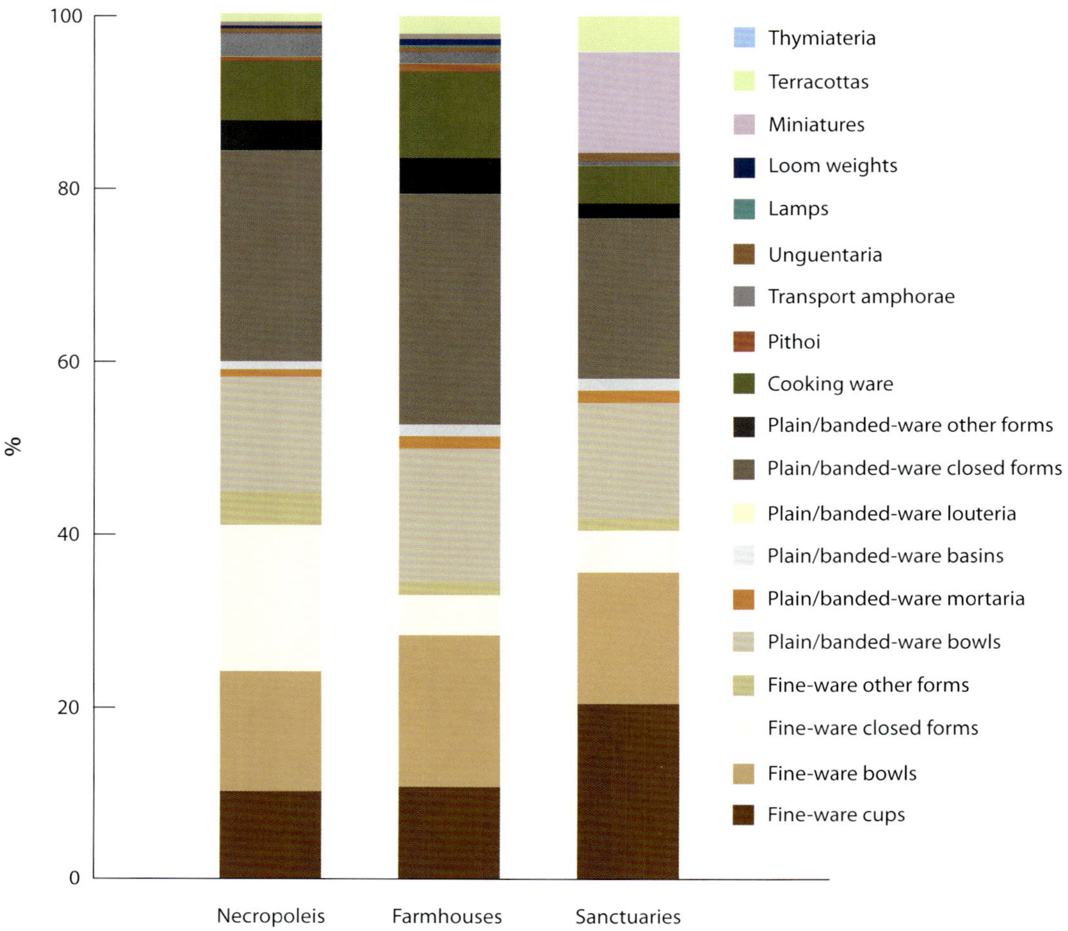

Figure 5.2 Percentages of shapes by site types.

plain wares, they are more frequent than fine wares in the combined assemblages from all sites (49.1% versus 34.8%) and make up nearly half (50.1%) of all forms in the combined farmhouse assemblages (*Table 5.2, Fig. 5.2*). In the combined necropoleis/tomb assemblages they occur in similar proportions to fine wares, at just less than 45%, while in the combined sanctuary assemblages they appear to be less frequent than fine wares, 35.9% and 41.6%, respectively. Plain- and banded-ware bowls comprise a similar proportion of forms in the combined assemblages, approximately 15%.[11] They are present in similar proportions to fine-ware bowls and other open shapes. While there is a certain degree of functional overlap in the small- and medium-sized forms, the repertoire of plain- and banded-ware bowls includes larger forms, particularly lekanai. Given this, it can be inferred that in the combined assemblages small fine-ware bowls make up a higher proportion of the combined assemblages than their plain- and banded-ware counterparts.

Mortaria have been differentiated for the purposes of quantification to assess whether these have a purely domestic function as food-preparation vessels, as is often supposed. They make up a similar proportion of the combined form-assemblages from all three types of site: 0.9% in farmhouses, 0.8% in sanctuaries, and 0.5% in necropoleis/tombs. The twelve examples from necropoleis/tombs and nine examples from sanctuaries suggest that this vessel type should not be interpreted solely as a food-preparation vessel associated with domestic contexts.

[11] Plain- and banded-ware bowls (see Ch. 9) include the entire range of open shapes, with the exception of basins and louteria, which have been quantified separately. Included in this category are the carinated lekane (CLne), flanged lekane (FLne), lekane (Lne), lekanis (Lnis), unidentified open forms (op), plate (Pl), plate/lekane (Pl/Lne), saltcellar (Sa), and small bowls (SB). Mortaria (Mor), louteria (Lou), and basins (Bas) have been differentiated for the purposes of quantification and are discussed separately below, though all three types of vessel have been included in the plain- and banded-ware totals.

Basins have been differentiated, since they are associated primarily with processing and preparation, and therefore the question arises as to whether these substantial vessels are, like pithoi, more frequent in farmhouse assemblages than in other types of site.[12] They are represented in the three main types of site, making up 1% or less of the combined form-assemblages.

Although similar in basic shape to basins, louteria have been differentiated from basins for the purposes of quantification, since they are often associated with, or interpreted as having, a ritual function and are thought to have been particularly common in sanctuary assemblages. While louteria are a typical component of sanctuary assemblages, it is not the case that they are exclusive to these contexts. The overwhelming majority of louteria found in the survey come from sites interpreted as farmhouses, with 19 examples, while a single example comes from one of the four sanctuary sites identified in the survey (*Table 5.2, Fig. 5.2*). The presence of louteria in domestic assemblages from farmhouses identified in survey is supported by evidence from excavations of farmhouses in the Metapontine chora, where they are, for example, present in the assemblages from the Fattoria Fabrizio farmhouse excavated by ICA.[13] There are no examples of louteria from sites interpreted as necropoleis/tombs, and this type of vessel was not present in the Pantanello necropolis. They form a small part of the total survey assemblages, around 0.1% of combined assemblages from all sites and from farmhouses and sanctuaries, where they occur in similarly small proportions.

Plain- and banded-ware closed forms are, as for the fine wares, mostly liquid containers, but they also encompass closed globular forms.[14] Closed forms make up the bulk of the plain and banded wares from all sites, being much more frequent than open shapes. They account for 28.7% of the assemblages at identified farmhouse sites, 24.1% of all ceramic forms in the combined necropoleis/tomb assemblages, and 19.2% of all ceramic forms from sites identified as sanctuaries. In all three types of site they are much more frequent than closed shapes in fine wares; this is especially the case for farmhouses, where they are over seven times as common, and sanctuaries, where they are five times as common.

Other plain- and banded-ware forms include a range of very different forms that are not readily grouped with the other basic shapes.[15] These miscellaneous forms, the majority being lids, account for less than 4% of the total form-assemblages from all sites, from farmhouses, and from necropoleis/tombs, and only 1.5% of forms from sites identified as sanctuaries.

Cooking wares

The relative proportions of cooking wares in ceramic assemblages are often taken as an indicator of domestic function (see Ch. 11).[16] The proportion of cooking wares in sites identified as farmhouses, about 10%, is twice that of cooking wares from sanctuaries (about 5%), and goes some way to supporting this. On the other hand, cooking wares make up just under 8% of all ceramics in assemblages from sites identified as necropoleis and tombs. It is therefore not possible to draw the conclusion that the frequency of cooking wares provides a reliable indicator of sites with a strictly domestic function, although generally a higher proportion of cooking wares might be expected in assemblages from such sites, as appears to be the case in the quantification of the combined survey assemblages. For necropoleis and tombs, cooking wares were well represented outside (and occasionally inside) the tombs in the Pantanello necropolis, and have been interpreted as evidence for celebration at the tomb.[17]

Pithoi

These large containers were used primarily for storage and agricultural processing, and as might be expected they are almost twice as frequent in farmhouses (1.3%) as in necropoleis/tombs (0.7%). They are very poorly represented in the four Archaic sanctuaries (0.1%).

[12] Three form-types have been included in this category: basins (Bas), deep basins (DBas), and lekane/basins (Lne/Bas).

[13] The complete publication of the site is forthcoming.

[14] They include the amphoriskos (A.skos), bottle/boccaletto (Bo), *dinos*/olla (Dn), globular urn (GU), hydria (H), jug/pelike/table amphora (J), jar (Ja), and unidentified closed forms (cl).

[15] These include the lid (Ld), which may have accompanied a range of shape-types, the pestle (Pst), and the pyxis (Py).

[16] The quantification used in this chapter is based on shapes/forms rather than fabric; therefore non-cooking-ware shapes in cooking-ware fabrics, though included in the cooking-ware typology, are quantified with the plain and banded wares rather than cooking wares, while the rare cooking-ware forms produced in fabrics typical of plain and banded wares are included in this quantification.

[17] Carter 1998d, 118–21.

Transport amphorae

Transport amphorae make up a small proportion of the total assemblages from all site types, though they are less frequent in sanctuaries (0.6%) than in sites identified as farmhouses (2%) and necropoleis/tombs (2.7%). Their slightly higher frequency in necropoleis and tombs relative to other types of sites is probably a consequence of their use as grave markers and grave goods (see Ch. 12).[18]

Other ceramics

Unguentaria include slipped black-gloss fine-ware and plain- and banded-ware forms that comprise a small proportion (<0.5%) of the total assemblages from necropoleis/tombs, farmhouses, and sanctuaries. These small containers served a variety of functions, or rather, their function as containers of low-volume commodities was suited to activities associated with all three types of site.

Lamps include black-gloss, banded, and plain examples, and these also form a small part of the total assemblages in necropoleis/tombs, farmhouses, and sanctuaries, where they make up <0.5% of all ceramic forms in the combined assemblages.

Loom weights appear in a range of different contexts beyond the domestic sphere (see Ch. 14). They make up a slightly higher proportion of the combined assemblages in farmhouses (0.5% of all ceramic forms) relative to necropoleis (0.2%) and sanctuaries (0.1%).

Miniature vessels, often interpreted as votive objects, are much more frequent in sanctuaries than in farmhouses and necropoleis.[19] In the combined sanctuary sites they account for over 10% of all ceramic forms, in contrast to necropoleis/tombs (0.4%) and farmhouses (0.4%), where they are nevertheless present. Miniatures are clearly not restricted to sanctuary deposits, since they are present also in necropoleis/tombs and farmhouses, but their frequency within the total assemblages is correlated to site function, with a higher proportion of the total ceramic assemblages typical of sanctuary sites. Terracottas account for a much higher proportion of the combined ceramic assemblages from sanctuaries (4.9%) than from either farmhouses (1%) or necropoleis/tombs (0.4%), where they nevertheless make up a small proportion of the site assemblages (see Ch. 13). Thymiateria are often thought to be associated with sanctuaries, but the evidence from the survey and from four excavated farmhouses in the chora suggests that these can also be found in domestic contexts, and are therefore not a diagnostic type indicating a sanctuary. Nevertheless, they are rare in the survey assemblages (three forms in total), with a single example from a site identified as a farmhouse (site 390) and none from necropoleis/tombs or sanctuaries. The other two thymiateria from the survey come from sites identified as farmhouses with tombs (sites 730 and 742).

Quantified Assemblages from Different Types of Site

The ceramic assemblages from different types of sites are examined here in more detail. While the relative proportions of different ceramic types from combined assemblages can be used to characterize the assemblages in general terms and elucidate the key features apparent in sites of different character, there is still significant variation among sites of the same type. Selected examples of site assemblages are therefore presented to give an idea of the nature of ceramic assemblages from necropoleis and tombs, from farmhouses and Greek agricultural structures, and from sanctuaries, and to provide examples illustrating key points about the identification of site types (see Ch. 3; also Ch. 18 on excavated sites and literary evidence). The site assemblages are discussed in the following order: (1) necropoleis and tombs, (2) farmhouses and Greek agricultural structures, and (3) sanctuaries.

Site assemblages from necropoleis/tombs

The non-ceramic criteria for the identification of necropoleis, individual tombs, and small groups of tombs were discussed in Chapter 3 (see also Ch. 18). A total of 133 sites in the survey area were identified as necropoleis and individual or small groups of tombs. As shown in Table 5.3 and Figure 5.3, there is a considerable range in the number of sherds from which forms could be identified. The majority of these sites (71) have fewer than 10 form-sherds; 12 of these had only a single form-sherd among the counts, and may derive from small groups of tombs. At the other extreme is a sole example of a site with more than 200 form-sherds, along with four sites having more than 100 sherds. Three very large necropoleis

[18] See Carter 1998d, 125–27, for amphorae as grave markers; see Morter and Leonard 1998 for a detailed study of the transport amphorae from the Pantanello necropolis.

[19] On miniatures in burials, see Carter 1998f, 196.

Assemblage size	No. sites	Total forms
<10	71	309
11 to 20	26	368
21 to 50	24	748
51 to 100	8	518
101 to 150	2	242
151 to 200	1	153
>200	1	231
Total	**133**	**2,569**

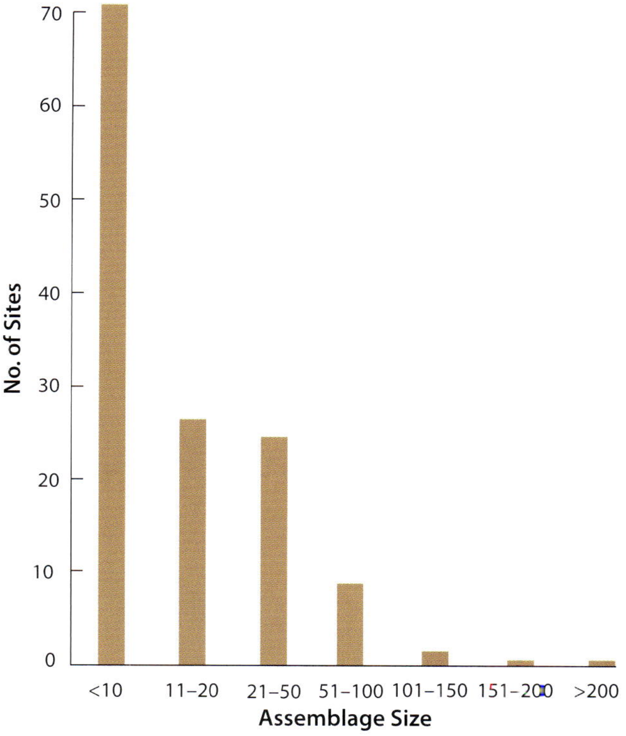

Table 5.3, Figure 5.3 Necropoleis, individual and small groups of tombs: assemblage sizes and number of sites.

that cover most of the period from the 5th to the 4th century have also been identified in the survey. For the 50 sites with between 11 and 50 sherds, it is difficult to determine the original size of the necropoleis because of variation in the numbers of ceramic vessels deposited in individual tombs.

This quantification does not include the 30 sites identified as farmhouses with associated tombs (*Table 5.1*) because it was not possible to differentiate the context of the ceramics. They have been excluded from the aggregated totals for necropoleis and tombs, as well as from the totals for farmhouses.

Quantification of the combined survey assemblages permits a broad characterization of the composition of ceramic assemblages from sites identified as necropoleis, individual tombs, and small groups of tombs. A number of key features are apparent relative to the combined assemblages from other types of site. Fine wares are more or less as frequent as plain and banded wares in the combined necropolis/tomb assemblages (44.3% and 43.2%). Fine-ware bowls make up a higher proportion than cups, 13.1% versus 10.9%. Closed forms make up a higher proportion of the fine-ware forms than cups and bowls (16.7% versus 10.9% and 13.1%, respectively), in contrast to the combined assemblages from farmhouses and sanctuaries, where closed forms in fine ware are less frequent than either cups or bowls. Other fine-ware forms occur in a higher proportion in the combined necropoleis and tomb assemblages than in those from sanctuaries and farmhouses (ca. 3.5% versus 1.5% and 1.3%). Transport amphorae are slightly better represented in the combined assemblages from necropoleis/tombs (2.7%) compared to farmhouses (2%) and sanctuaries (0.6%). Their higher proportions probably represent their use as grave markers and grave goods, a common feature of necropoleis in Magna Grecia (see above).

The ceramic assemblages from three necropolis/tomb sites are presented to illustrate some of these characteristics, as well as variations (*Table 5.4, Fig. 5.4*) and the issues inherent in identifying these types of sites strictly on the basis of ceramic forms. The sites encompass a range of sizes in the surface assemblages and range in date from the Archaic to Early Hellenistic periods. Site 70 is the largest assemblage (153 sherds) and dates to the Archaic and Early Classical periods; site 81, dating to the 4th century BC, is medium-sized (61 sherds); site 11, interpreted as a small necropolis (15 sherds) or small cluster of tombs, or perhaps even a relatively rich individual tomb, dates to the mid-5th century BC.

Site 70 has been identified as a large necropolis, dating primarily to the Archaic and Early Classical periods. In contrast to the combined assemblages from necropolis and tomb assemblages, fine-ware cups make up a greater proportion of the assemblage than bowls, while other fine-ware forms are less well-represented than they are in the combined assemblages as a whole. Two mortaria are also present in the assemblage from site 70. Since they come from the surface, it cannot be

Ceramic groups	All	%	Site 70	%	Site 81	%	Site 11	%
Fine-ware cups	281	10.9	28	18.3	11	17.7	1	6.7
Fine-ware bowls	337	13.1	20	13.1	11	17.7	1	6.7
Fine-ware closed forms	429	16.7	3	2	2	3.2	0	0
Other fine-ware forms	90	3.5	2	1.3	1	1.6	0	0
Plain/banded-ware bowls	379	14.8	19	12.4	9	14.5	1	6.7
Mortaria	12	0.5	2	1.3	0	0	0	0
Basins	16	0.6	0	0	0	0	0	0
Louteria	0	0	0	0	0	0	0	0
Plain/banded closed forms	618	24.1	56	36.6	16	25.8	6	40
Other plain/banded forms	84	3.3	9	5.9	1	1.6	1	6.7
Cooking ware	195	7.6	10	6.5	5	8.1	1	6.7
Pithoi	17	0.7	0	0	3	4.8	0	0
Transport amphorae	69	2.7	3	2	0	0	4	26.7
Unguentaria	14	0.5	0	0	2	3.2	0	0
Lamps	3	0.1	0	0	1	1.6	0	0
Loom weights	6	0.2	0	0	0	0	0	0
Miniatures	9	0.4	0	0	0	0	0	0
Terracottas	10	0.4	1	0.7	0	0	0	0
Thymiateria	0	0	0	0	0	0	0	0
All fine wares	1137	44.3	53	34.6	25	40.3	2	13.3
All plain wares	1109	43.2	86	56.2	26	41.9	8	53.3
All ceramics	**2569**	**100**	**153**	**100**	**62**	**100**	**15**	**100**

Table 5.4 Quantification of ceramic groups from the selected necropoleis assemblages, alongside quantification of the combined assemblages from all sites identified as necropoleis and individual or small groups of tombs.

demonstrated that they belong to a mortuary context, but they support the caution made that this type of vessel should not be used as an indicator of domestic function of sites or seen solely as a food-preparation vessel associated exclusively with domestic contexts. The presence of cooking wares, making up ca. 6.5% of the assemblage from site 70, may represent generalized material in the area of the necropolis but the best explanation is that cooking wares, including mortars, are the remains of meals or offerings beside the tomb (see above). The transport amphorae include an Archaic Type B toe (perhaps Calabrian),[20] consistent with the frequent use of amphorae as grave markers or grave goods in the Archaic period (see above).

The ceramic assemblage from site 81, a medium-sized necropolis assemblage dating to the 4th century BC,[21] is remarkable only because it does not appear to contain any ceramic types associated primarily with mortuary contexts. All of the types represented in the survey assemblage would not be out of place in a domestic assemblage, while the relative proportions of the different types vary less from those of farmhouses than do the combined necropolis/tomb assemblages. Site 81 therefore aptly illustrates the need for identification of sites on the basis of additional criteria.

Site 11 has been interpreted as a small necropolis or a relatively rich individual tomb dating to the mid-5th century BC,[22] with an assemblage consisting of 15 identifiable form-sherds (*Table 5.4*). As shown in Table 5.3, over 97 sites have been identified as small necropoleis or groups of tombs that have yielded fewer than twenty identifiable vessel shapes. In the case of

[20] For discussion of this type of amphora, see Ch. 12.

[21] See the Gazetteer.

[22] See the Gazetteer.

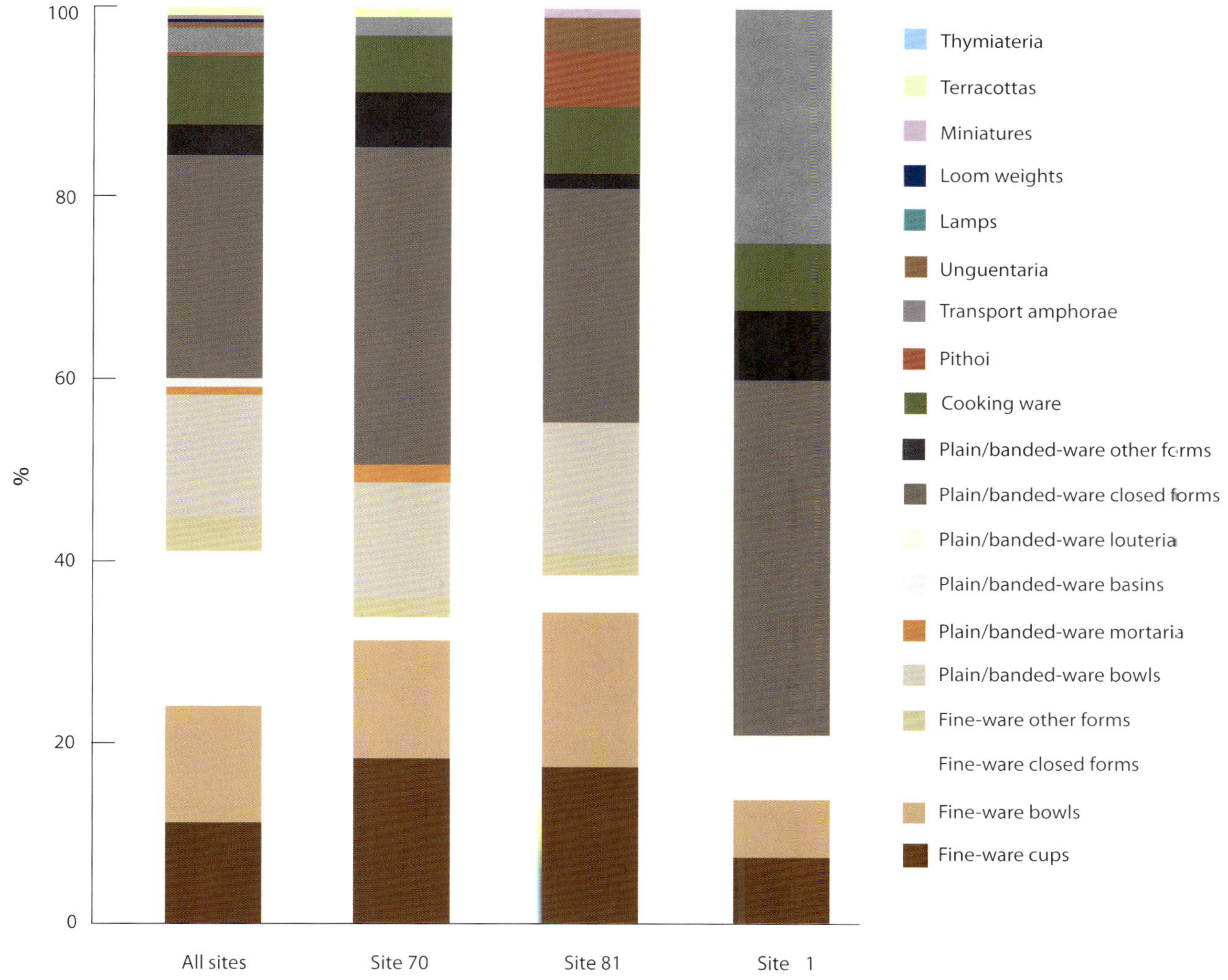

Figure 5.4 Quantification of ceramic groups from the selected necropoleis assemblages, alongside quantification of the combined assemblages from all sites identified as necropoleis and individual or small groups of tombs.

necropoleis, which are often more deeply buried than other structures, it is especially difficult to ascertain the original size of the site or to estimate the number of tombs on the basis of the size of the ceramic assemblages (see Ch. 3).

Site assemblages from farmhouses

Chapter 3 discusses the identification of Greek farmhouses, the range of such sites, and the nature of their ceramic assemblages (see also Ch. 18). As noted earlier, there are few vessel types or classes that are exclusive to, or through their presence or absence reliable indicators of, the function or type of a given site, and this is particularly true of sites identified as farmhouses. The entire range of vessel types is found in farmhouse assemblages, including those more typical of sanctuaries (miniatures, "votives," thymiateria, and louteria), and this is supported by assemblages from excavated farmhouses in the Metapontine chora.[23]

The nine sites presented show the range and variation in farmhouse assemblages and provide examples of high FMCE values for particular half-centuries ranging from 600 to 100 BC.[24] These assemblages are: site 285 (600 BC), site 336 (550 BC), site 182 (450 BC), site 549 (400 BC), site 449 (350 BC), site 542 (300 BC), site 83 (250 BC), site 745 (200 BC), and site 556 (100 BC). Quantification of their relative proportions of ceramic types is shown in Table 5.5 and Figure 5.5.

[23] Forthcoming studies include the farmhouses at Ponte Fabrizio, Sant'Angelo Vecchio, Sant'Angelo Grieco, and Fattoria Stefan.

[24] See Ch. 4 for discussion of FMCE.

Ceramic groups	All	%	285	%	336	%	182	%	549	%	449	%	542	%	83	%	745	%	556	%
Fine-ware cups	2303	12.0	29	15.4	18	10.9	13	6.2	6	9.5	11	13.3	5	8.1	5	16.1	2	3.5	6	7.7
Fine-ware bowls	3022	15.7	9	4.8	13	7.9	25	11.9	11	17.5	7	8.4	7	11.3	3	9.7	7	12.3	30	38.5
Fine-ware closed forms	822	4.3	5	2.7	7	4.2	5	2.4	1	1.6	7	8.4	1	1.6	1	3.2	6	10.5	13	16.7
Other fine-ware forms	259	1.3	4	2.1	3	1.8	0	0	2	3.2	3	3.6	2	3.2	0	0	2	3.5	3	3.8
Plain/banded-ware bowls	3019	15.7	32	17	24	14.5	62	29.5	7	11.1	19	22.9	17	27.4	3	9.7	10	17.5	9	11.5
Mortaria	168	0.9	2	1.1	0	0	7	3.3	1	1.6	0	0	2	3.2	0	0	0	0	0	0
Basins	176	0.9	1	0.5	0	0	4	1.9	1	1.6	0	0	2	3.2	0	0	0	0	0	0
Louteria	19	0.1	1	0.5	0	0	0	0	0	0	0	0	0	0	0	0	0	0	0	0
Plain/banded closed forms	5514	28.7	73	38.8	80	48.5	53	25.2	20	31.7	20	24.1	17	27.4	13	41.9	12	21.1	12	15.4
Other plain/ banded forms	727	3.8	5	2.7	4	2.4	4	1.9	6	9.5	3	3.6	0	0	3	9.7	0	0	1	1.3
Cooking ware	2019	10.5	18	9.6	10	6.1	28	13.3	6	9.5	9	10.8	4	6.5	2	6.5	10	17.5	3	3.8
Pithoi	256	1.3	4	2.1	0	0	5	2.4	1	1.6	2	2.4	4	6.5	0	0	1	1.8	1	1.3
Transport amphorae	393	2.0	5	2.7	4	2.4	3	1.4	1	1.6	1	1.2	1	1.6	1	3.2	3	5.3	0	0
Unguentaria	100	0.5	0	0	0	0	0	0	0	0	1	1.2	0	0	0	0	2	3.5	0	0
Lamps	63	0.3	0	0	0	0	0	0	0	0	0	0	0	0	0	0	1	1.8	0	0
Loom weights	94	0.5	0	0	2	1.2	0	0	0	0	0	0	0	0	0	0	0	0	0	0
Miniatures	71	0.4	0	0	0	0	1	0.5	0	0	0	0	0	0	0	0	0	0	0	0
Terracottas	186	1.0	0	0	0	0	0	0	0	0	0	0	0	0	0	0	1	1.8	0	0
Thymiateria	1	0	0	0	0	0	0	0	0	0	0	0	0	0	0	0	0	0	0	0
All fine wares	6406	33.3	47	25.0	41	24.8	43	20.5	20	31.7	28	33.7	15	24.2	9	29	17	29.8	52	66.7
All plain/ banded wares	9623	50.1	114	60.6	108	65.5	130	61.9	35	55.6	42	50.6	38	61.3	19	61	22	38.6	22	28.2
All ceramics	**19212**	**100**	**188**	**100**	**165**	**100**	**210**	**100**	**63**	**100**	**83**	**100**	**62**	**100**	**31**	**100**	**57**	**100**	**78**	**100**

Table 5.5 Quantification of ceramic groups from the selected farmhouse assemblages, alongside quantification of the combined assemblages from all sites identified as farmhouses.

Black-gloss fine wares are a constant feature of assemblages from farmhouses throughout the Metapontine chora. These wares had a wide distribution within the chora, being present at 476 out of 525 sites, just over 90% of sites dating to the Archaic, Classical, and Hellenistic periods that have associated ceramics (although there is danger of circularity here, since in most cases the survey assemblages are dated on the basis of the fine wares, and there are very few sites without any fine wares at all—49 out of 525). This tends to suggest that there was a ready supply of these wares, produced on a large scale certainly in the kerameikos of Metaponto[25] and probably also in the chora itself (see Ch. 6, on evidence for production). There are forms in black-gloss fine wares, particularly cups, which were not produced on a large scale in plain or banded wares, with the suggestion being that black-gloss fine wares were sufficiently plentiful to negate the need to make up a shortfall in cup-forms using unslipped or semi-slipped vessels.

Fine wares account for about a third of all ceramics from sites identified as farmhouses, with the functional plain and banded wares making up the largest class of ceramics, just over half of the aggregated assemblages. Cooking wares are consistently represented, just over 10%, and both account for a larger proportion of the ceramic assemblages in farmhouses than in either necropolis/tomb sites or sanctuaries. Pithoi (storage vessels) are better represented too; although not making up a large proportion of the assemblages

[25] See Ch. 6 on production.

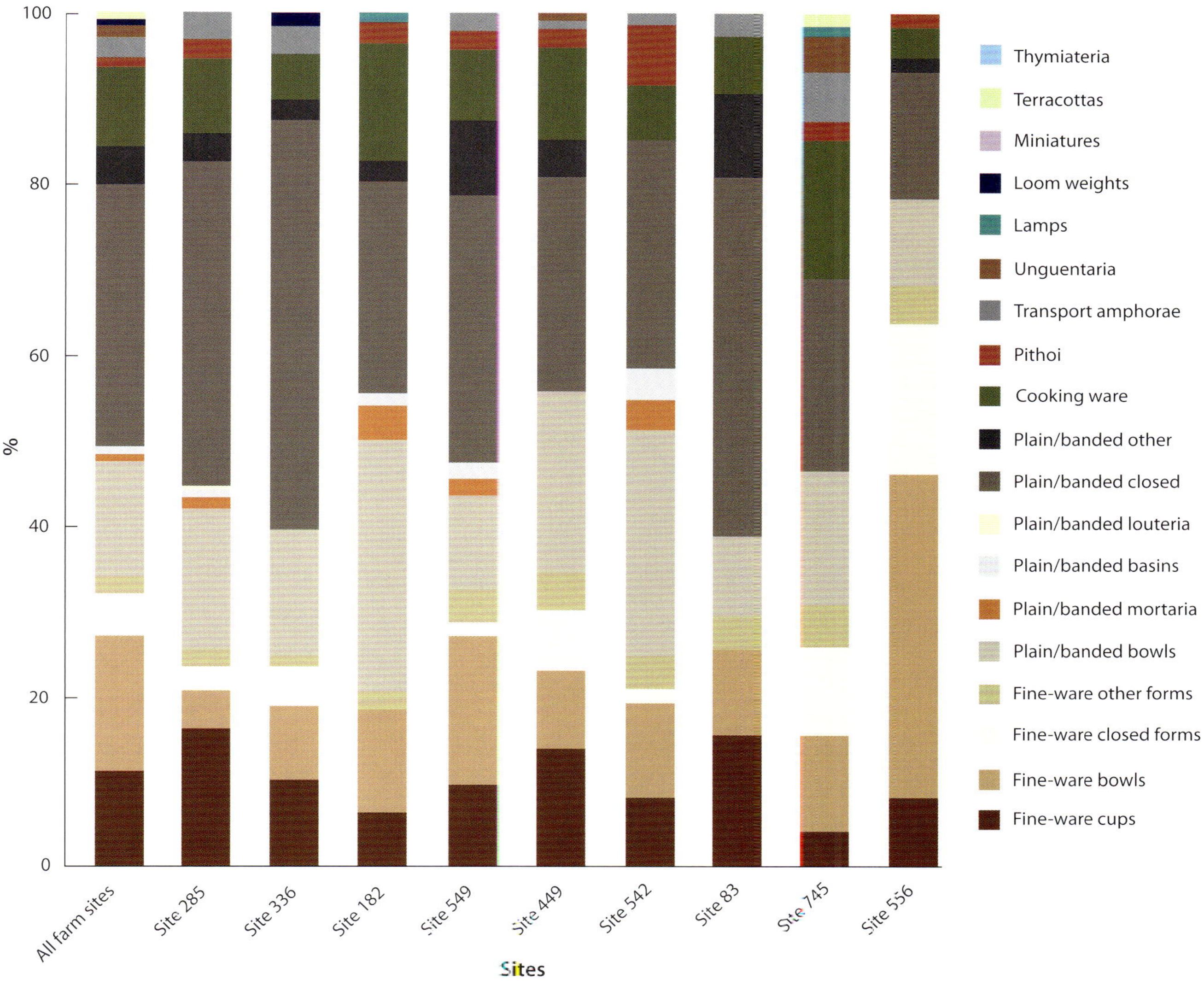

Figure 5.5 Quantification of ceramic groups from the selected farmhouse assemblages, alongside quantification of the combined assemblages from all sites identified as farmhouses.

from domestic contexts (1.3%), they are almost twice as frequent at sites identified as farmhouses as at necropoleis/tombs. Together, the higher frequency of these classes in sites identified as farmhouses is consistent with diverse assemblages from sites associated primarily with a domestic function.

Ceramic assemblages from Greek agricultural structures

The surface assemblages of sites identified as Greek agricultural structures tend to be relatively small, and as a whole the composition of the ceramic assemblages from these sites can generally be characterized as uneven compared to those from farmhouses. The wide variation in the composition of the assemblages likely reflects the broad range of activities associated with agrarian installations, but which may have been relatively restricted at any given site—different requirements for viticulture versus cereal processing, for example (see Ch. 18 for discussion of literary evidence and excavated farmhouses).

A key feature of assemblages from these sites is that fine wares are scarce or absent, in contrast to their ubiquity in ceramic assemblages from other types of sites, particularly farmhouses. Here fine wares account for just over 10% of the total assemblages, with plain and banded wares making up 74% of the combined assemblages (*Table 5.7, Fig.5.7*). Mortaria, unguentaria, miniatures, terracottas, and thymiateria are not

Ceramic groups	Count	%
Fine-ware cups	25	5.4
Fine-ware bowls	16	3.4
Fine-ware closed forms	3	0.6
Other fine-ware forms	6	1.3
Plain/banded-ware bowls	83	17.8
Mortaria	0	0
Basins	5	1.1
Louteria	0	0
Plain/banded closed forms	203	43.7
Other plain/banded forms	53	11.4
Cooking ware	53	11.4
Pithoi	7	1.5
Transport amphorae	8	1.7
Unguentaria	0	0
Lamps	1	0.2
Loom weights	2	0.4
Miniatures	0	0
Terracottas	0	0
Thymiateria	0	0
All fine wares	50	10.8
All plain/banded wares	344	74.0
All ceramics	**465**	**100**

Table 5.6 Quantification of assemblages from Greek agricultural structures.

Ceramic groups	All	%	397	%	477	%	334	%	270	%
Fine-ware cups	234	20.5	95	22.4	64	20.3	38	18.2	37	19.2
Fine-ware bowls	171	15	53	12.5	51	16.1	37	17.7	30	15.5
Fine-ware closed forms	54	4.7	21	4.9	11	3.5	8	3.8	14	7.3
Other fine-ware forms	17	1.5	5	1.2	6	1.9	0	0	6	3.1
Plain/banded-ware bowls	152	13.3	52	12.3	43	13.6	40	19.1	17	8.8
Mortaria	9	0.8	3	0.7	3	0.9	1	0.5	2	1
Basins	12	1	7	1.7	2	0.6	1	0.5	2	1
Louteria	1	0.1	1	0.2	0	0	0	0	0	0
Plain/banded closed forms	219	19.2	68	16	64	20.3	38	18.2	49	25.4
Other plain/banded forms	17	1.5	8	1.9	4	1.3	5	2.4	0	0
Cooking ware	53	4.6	11	2.6	14	4.4	13	6.2	15	7.8
Pithoi	1	0.1	1	0.2	0	0	0	0	0	0
Transport amphorae	6	0.6	2	0.5	1	0.3	0	0	3	1.6
Unguentaria	4	0.3	1	0.2	3	0.9	0	0	0	0
Lamps	2	0.2	1	0.2	0	0	0	0	1	0.5
Loom weights	1	0.1	0	0	1	0.3	0	0	0	0
Miniatures	133	11.6	85	20	10	3.2	24	11.4	14	7.3
Terracottas	56	4.9	10	2.4	39	12.3	4	1.9	3	1.6
Thymiateria	0	0	0	0	0	0	0	0	0	0
All fine wares	476	41.6	174	41	132	41.8	83	39.7	87	45.1
All plain/banded wares	410	35.9	139	32.8	116	36.7	85	40.7	70	36.3
All ceramics	**1142**	**100**	**424**	**100**	**316**	**100**	**209**	**100**	**193**	**100**

Table 5.7 Quantification of ceramic groups from sanctuary assemblages.

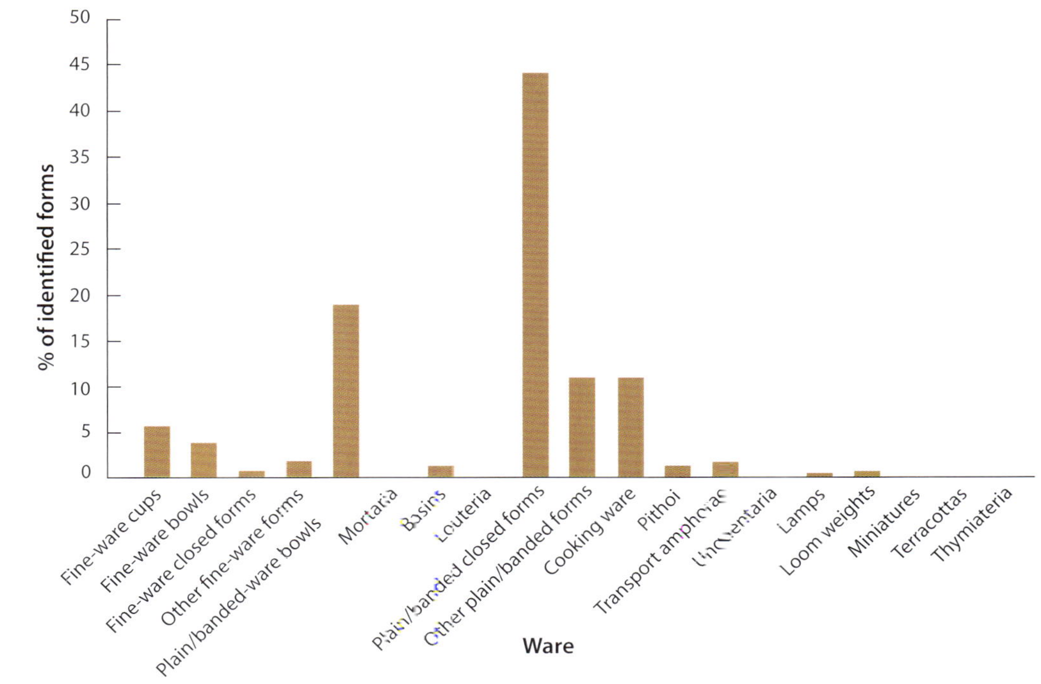

Figure 5.6 Quantification of assemblages from Greek agricultural structures.

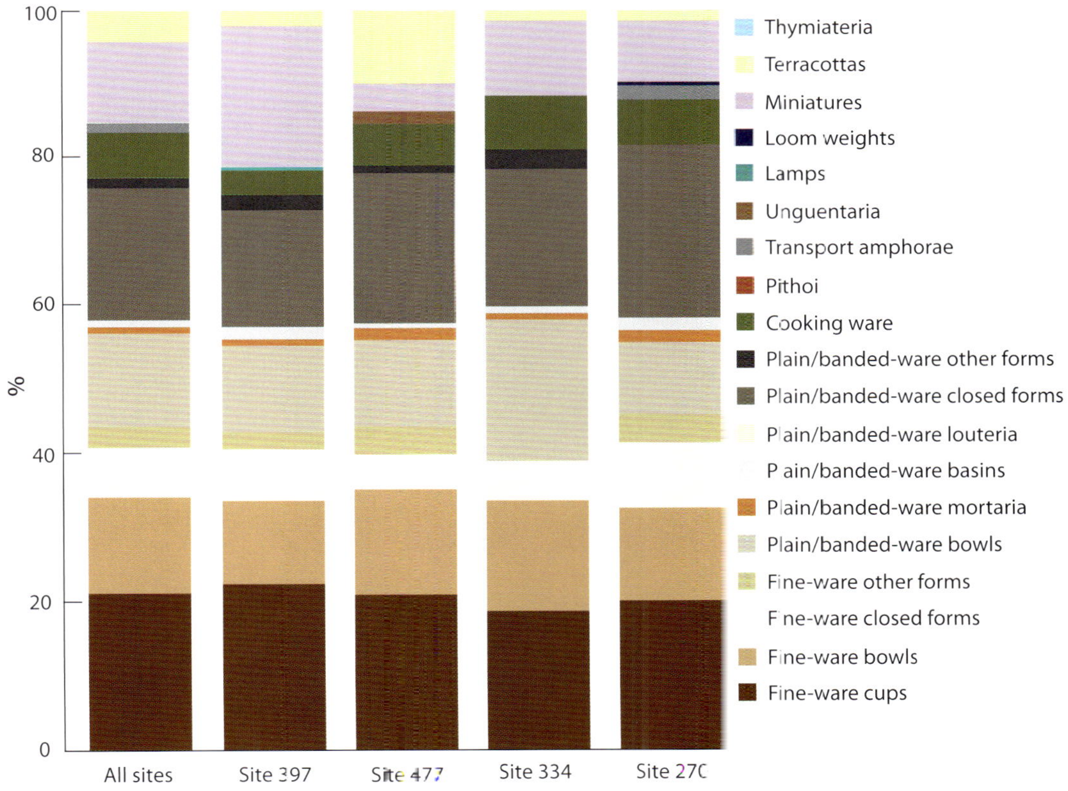

Figure 5.7 Quantification of ceramic groups from sanctuary assemblages.

present in assemblages from these sites. Site 219 has no fine ware and a relatively limited range of plain and banded wares. Site 25 is characterized by basins and jugs. The assemblage from site 252 includes pithoi and a jug, a lekane, and a (plain or banded?) small bowl/one-handler.

Site assemblages from sanctuaries

Four sites dated to the Archaic period were identified as sanctuaries in the survey,[26] all with relatively large surface assemblages, ranging from 193 to 1,142 form-sherds. All four of the Archaic sanctuaries are quantified in Table 5.7 and Figure 5.7. Quantification of the combined assemblages from the four sanctuary sites identified in the survey suggests some key features. Perhaps the most striking feature overall is that fine wares are slightly more frequent than plain and banded wares, 41.6% and 35.9%, respectively, in contrast to farmhouses, where plain and banded wares are more prevalent.

Fine-ware cups are much more frequent than bowls and closed forms in the fine-ware assemblages from all of the sites identified as (Archaic-period) sanctuaries than in the combined assemblages from necropoleis/tombs and farmhouses (*Table 5.3, Fig. 5.3*). This contrasts with the necropolis/tomb site assemblages, where closed forms are the most common fine-ware type, and with necropoleis/tombs and farmhouses, where fine-ware bowls outnumber fine-ware cups. As noted above, while louteria are often associated with sanctuary contexts, they make up a similarly small proportion of the total assemblages from both sanctuaries and farmhouses (ca. 0.1%). Although they can be associated with sanctuaries, these vessels should not, therefore, be used as markers for the identification of sanctuaries in survey. Transport amphorae make up a smaller proportion of the total assemblages in sanctuaries than in either necropoleis/tombs or farmhouses. Miniatures make up a much higher proportion of the total assemblages from sanctuaries than in necropoleis/tombs and farmhouses, comprising 11.6% of all ceramic forms by count. As noted in Chapter 3, miniatures were the only example of vessel types used to identify site types in the field (when found along with terracottas), and only when there was seen to be a relatively large number of them in the scatters. This was particularly true for site 397, where 20% of the survey assemblage was made up of miniatures. Terracottas also make up a higher proportion of ceramics in sanctuary assemblages than in other types of sites, where they are nevertheless also represented (see Ch. 13).

[26] One additional site identified as a sanctuary, site 266, dated to the 4th c. BC, is probably a ceramic production center with evidence for the production of transport amphorae, plain wares, and perhaps also terracottas. For discussion and presentation of the evidence for the production of transport amphorae at the site in the late 4th or early 3rd c. BC, see Ch. 12.

Conclusion

Sites were attributed different functions based on the survey methods outlined in Chapter 3, and then compared here to the quantification of their ceramic materials. It is in the relative proportions of different ceramic classes and form-types, rather than in the simple presence or absence of particular types of vessel, that differences are apparent in the survey assemblages.

6
Archaic and Black-Gloss Fine Ware

Elisa Lanza Catti, Francesca Silvestrelli, Keith Swift, Amelia Tubelli, and Eloisa Vittoria

This chapter presents analyses of Archaic-period fine wares and of Classical- and Hellenistic-period black-gloss fine wares. Together, they constitute the primary ceramic evidence for dating the assemblages recovered by the ICA survey of the Metapontine chora, and the introductory discussion of production in the chora applies to both.

The Archaic-period fine wares, however, are characterized by semi-slipped bodies and a repertoire of shapes that is, with few exceptions, discrete from those of the Classical period. This is a distinct tradition that continued through to the large-scale production of black-gloss fine wares, and as such merits separate discussion. The black-gloss pottery, produced on a large scale from the early 5th century BC, provides the primary means of recognizing Classical- and Early Hellenistic period assemblages. The chapter is therefore divided into the following sections:

I. Archaic Fine-Ware and Black-Gloss Production in the Chora
II. Archaic Fine Wares and Catalog
III. Black-Gloss Fine Ware and Catalog[1]

I. Evidence for Black-Gloss Fine-Ware Production in the Chora

F. Silvestrelli and K. Swift

Although the Metapontino is well-known as a pottery manufacturing center, evidence for production has hitherto been concentrated in the kerameikos (potters' quarter) of the asty and derives from excavations undertaken in the 1970s. These provided direct evidence for pottery production from the second half of the 6th to the first half of the 3rd century BC.[2] Excavated waste deposits yielded large quantities of black-gloss pottery, which, along with other classes of material (red-figure pottery, banded ware, plain wares and cooking wares), were produced in substantial quantities at the site. Beyond this, wasters and other indicators of ceramic production suggesting the presence of pottery workshops have also been found in other parts of the city, and much remains to be done on the subject of the topography of pottery production at Metaponto.[3]

The territory surrounding the city also appears to have been involved in pottery production: the recovery of wasters of Ionic cups of the 6th century BC has been reported from Cogno del Pero and Fattoria Musillo;[4] both Pantanello (Pizzica) and Sant'Angelo Vecchio have yielded evidence of Hellenistic-period grey-ware production (see Ch. 7). In contrast to the Metapontine kerameikos, however, the number of black-gloss production centers positively identified in the chora is small. The evidence is slight and dependent upon fortuitous preservation and recovery of key pieces in field survey—wasters, test-firing pieces, and kiln spacers—that may not be particularly prevalent even on excavated sites with large assemblages, as can be seen at the rural sanctuary and Early Hellenistic-period farmhouse at Pantanello (see below). The only information known prior to the Metaponto survey concerns the Torretta di Pisticci locality at the edge of the chora.[5] Here, on a spur overlooking the Cavone valley, a large rectangular kiln (preceded by two large *praefurnia* and probably connected with a nearby farmhouse) was discovered in 1993.[6] The site and the material are unpublished and only the presence of Ionic cups and black-gloss pottery is recorded. The kiln is said to have been active from the end of the 6th to the end of the 5th century BC. (FS)

A test-firing piece (**No. 7**, Catalog of Black-Gloss Fabrics) provides evidence for the production of black-gloss fine ware in the chora at site 105 in the Basento valley (*Fig. 6.1*), which, like Pantanello, seems to date to the late 4th or early 3rd century BC. The most likely sources of raw material in the vicinity of site 105 are fluvial clays deposited by the Basento River; as such,

[1] Special thanks to Margherita Nigro for her work on the Catalog of Black-Gloss Fine Ware. For the difference between black-gloss ware and black-glaze pottery, see n. 72.

[2] D'Andria 1980a; Silvestrelli forthcoming a.

[3] Silvestrelli 2004, 110 (with previous bibliography), and Fig. 1.

[4] Carter 2006a, 150, with bibliography.

[5] It is situated nearby the area where the tomb with 'Helen's egg' was discovered: A. Bottini 1988b; 1992, 70–72.

[6] A. Bottini 1994, 705–06, Pl. L.

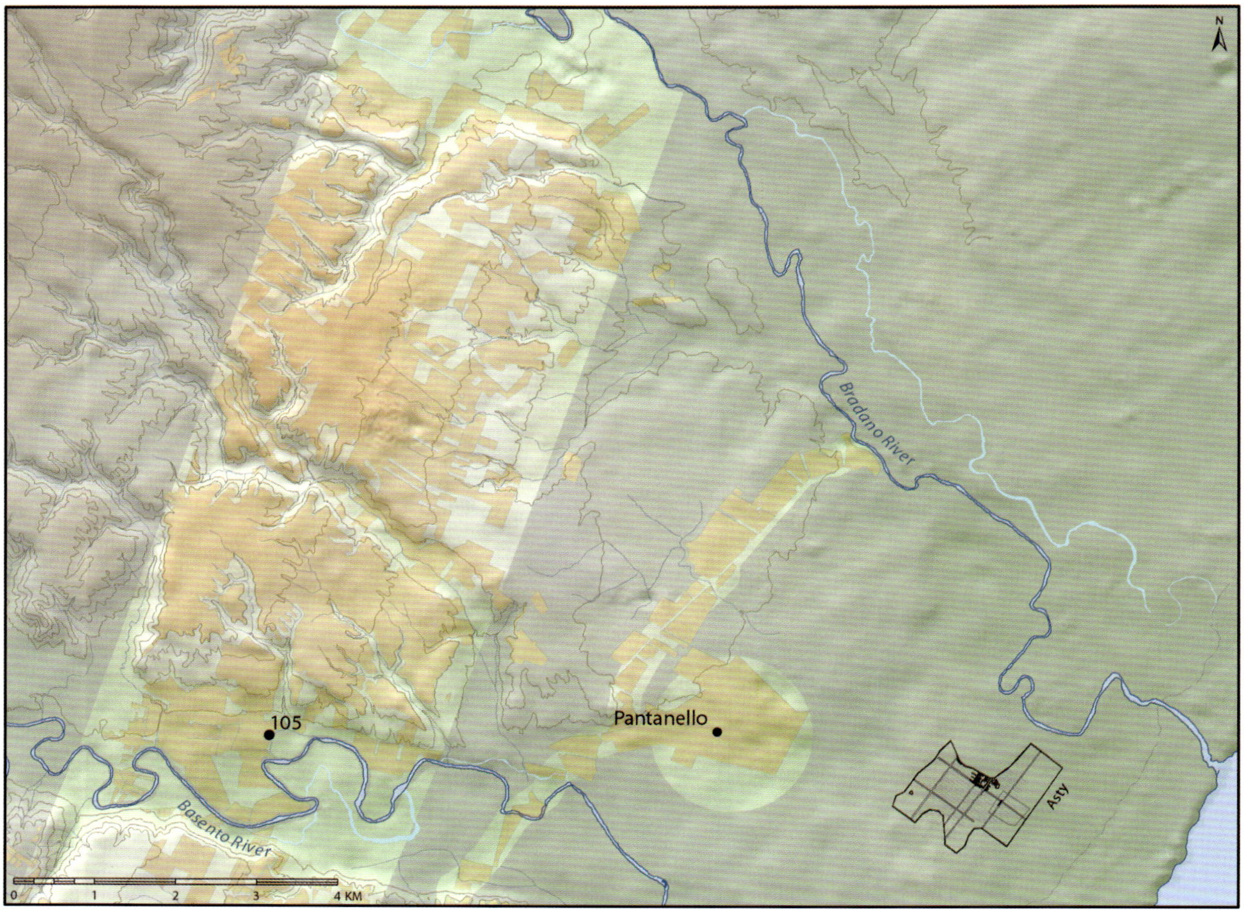

Figure 6.1 Location of site 105 and the site of Pantanello. (JT)

the black gloss from site 105 probably gives us the clearest idea of what fine-ware fabrics produced from these sediments look like, and these fall well within the usual range of Metapontine black-gloss fabrics. Given the similarities in composition of the river valley and lower valley/coastal plain sediments, these fabrics may be expected to appear similar to those produced in the kerameikos at Metaponto.

Black-gloss production sites in the Metapontine chora may also include Pantanello, where some recent evidence suggests fine-ware production in the vicinity, probably dating to the late 4th or early 3rd centuries BC. Three kiln spacers of late 4th century BC types were recovered in residual deposits from excavations in the area of the Late Hellenistic-period ceramic manufactory.[7] These can be identified and dated through parallels from the Metapontine kerameikos.[8] A series of misfired, though still functional, late C-type skyphos bases, again of late 4th or early 3rd century BC date, offer circumstantial support for black-gloss production at the site during this period. A series of 4th-century BC wasters of mold-made mortaria and a pyramidal loom weight waster provide further evidence for ceramic production at the site during the Late Classical or Early Hellenistic period. Combined, these finds support the inference of black-gloss production in the area of Pantanello.[9]

The black-gloss fabrics appear similar to those from site 105; Pantanello is situated at the inland margins of the coastal plain, and source sediments for the production of black gloss are therefore likely to be broadly similar to those in the vicinity of site 105 (fluvial clays) and also, given its geomorphological

[7] The kiln spacers were recovered during limited excavations undertaken during 2008. For a report on the large-scale excavations of the ceramic manufactory at Pantanello during 1974–76, see Carter 1983a.

[8] Cracolici 2004a.

[9] Preliminary results, including these finds, were presented at the 2009 Annual Meeting of the Archaeological Institute of America in Philadelphia.

location, the kerameikos of Metaponto.[10] This feature of the local geology, combined with similar production practices, particularly in the preparation of clays and slips, might account for the apparent similarities between Metapontine black-gloss fabrics produced in the three different, but sedimentologically similar, locations. While the fabrics are typically Metapontine, the evidence from site 105 and from Pantanello illustrates the difficulties inherent in differentiating between fine-ware fabrics produced from similar source sediments, but also indicates that, when found in the chora, black-gloss fabrics typical of those from the Metapontine kerameikos cannot be assumed to have originated at that production center, and many may indeed have been produced at rural workshops like those attested at site 105 and Pantanello. (KS)

II. Archaic Fine Ware: Introduction

A. Tubelli

The Archaic fine wares from the survey are the primary ceramic evidence for settlement of the Metapontine chora, from the first phases of Greek colonization of this part of southern Italy through to the production of black gloss proper in the early 5th century BC.[11] Key pieces, including all reference form-sherds used to co-identify similar pieces from the assemblages, are presented in the Catalog of Archaic Fine Ware.

The overwhelming majority of identifiable Archaic pieces from the survey appear to be local or regional in origin (see discussion of fabrics below), with extra-regional imports represented by a single fragment of a Corinthian kothon (**No. 7**) in a typical Corinthian fine-ware fabric. The influence of imported wares is variously apparent in the forms, styles, and production techniques of Metapontine pottery of the Archaic period (discussed for each class and form, below).

Small quantities of imitation proto-Corinthian pottery (**Nos. 1–2**), closely emulating Corinthian products (and inferring their presence in the region, though none were recovered in survey) and Colonial wares (**Nos. 3–6**) provide evidence for settlement in the 7th century BC. Metapontine pottery of the 6th century BC is characterized by the ubiquitous Ionic cups (**Nos. 8–26**). These are accompanied by the Archaic-period skyphoi with offset rims (**Nos. 27–40**), forms which continued to be produced into the Classical period; their successors in black gloss are discussed in Part III.[12] Together, these cup forms provide the best markers of the 6th century BC in the Metapontine chora, though it is not always possible to distinguish between them in fragmentary survey assemblages. Archaic line-decorated wares (**Nos. 41–45**), including Panionion-type cups (**Nos. 42–44**), are also represented in the survey, though these are not common.

Quantification

K. Swift, A. Tubelli, E. Vittoria[13]

Just under 500 Archaic fine-ware forms were identified in the survey assemblages (*Tables 6.1–2*). The identification of Archaic-period pottery in fragmentary surface survey assemblages is frequently problematic since the slips and decoration, which often serve to identify these wares, are relatively fragile and easily weathered compared to later black-gloss and banded wares. Differentiation of semi-slipped Archaic fine wares from later Classical-period banded wares usually depends also on identification of form types, particularly the ubiquitous Ionic cups and Archaic-period skyphoi with offset rims, for which sufficiently well-preserved fragments of the rim and shoulder, or bases, are required. As noted above, the quality and application of the slips may give some indication of Archaic-period material, but such equivocal features have not been used for the identification and dating of the survey assemblages (though they can provide pointers during recording of the survey assemblages); this study relies instead on the positive identification of Archaic-period forms—on sites where they are represented, these tend to be sufficiently clear as to leave no doubt about Archaic-period activity. Because it is based in most cases on the positive identification of forms, quantification of Archaic-period material represents a minimum in the survey corpus as a whole.

[10] For the geological and geomorphological context of the Metapontine chora, see Ch. 1 and Ch. 2.

[11] Translated by Alberto Prieto with editing by Keith Swift and Amelia Tubelli. Recording and quantification of all of the Archaic pottery from the survey assemblages was undertaken by E. Vittoria during 2002–03; re-examination of the Archaic fine wares from the survey was undertaken in 2009 by A. Tubelli, in light of study of the substantial excavated assemblages of Archaic fine wares from ICA excavations of the Archaic-period settlement at Incoronata in the chora of Metaponto, situated on the west side of the lower Basento River valley and outside the Bradano-Basento transect.

[12] See black-gloss forms "cup with offset rim" and "skyphos with offset rim" and Nos. 1–4 and No. 55 (Catalog of Black-Gloss Fine Ware).

[13] Initial identification, recording, and quantification of Archaic-period fine wares from the survey were undertaken by E. Vittoria in 2001–02. Final classification of the types was undertaken during a review of this material by A. Tubelli in 2009. Quantified data derived from both studies was collated by K. Swift.

Ware/form	Count	%
Imitation proto-Corinthian (**Nos. 1–2**)	2	0.4%
Colonial wares (**Nos. 3–6**)	8	1.6%
Corinthian fine ware (**No.7**)	1	0.2%
Ionic cups (**Nos. 8–26**)	399	81.3%
Skyphoi with offset rims (**Nos. 27–40**)	67	13.6%
Archaic line-decorated ware (**Nos. 41–45**)	14	2.9%
All Archaic-period forms	**491**	**100%**

Table 6.1 Quantification of Archaic-period pottery from the Metaponto survey with breakdown by cataloged types.

For the purposes of quantification, it is particularly important to note the overlap and degree of uncertainty in many cases between the Ionic cups (**Nos. 8–26**) and the skyphoi with offset rims (**Nos. 27–40**). It was often not possible to distinguish between the two shapes on the basis of rim and shoulder fragments, which, along with the bases, are the most readily identifiable parts. These can nevertheless be attributed a Late Archaic date regardless of which two shapes the fragments may have originally represented.

Ionic cups are by far the most frequent Archaic-period shape encountered in the survey, with almost 400 examples accounting for over 80% of all identified Archaic fine wares. Skyphoi with offset rims make up the majority of the remaining Archaic-period forms, with just fewer than 67 examples accounting for just under 14% of all Archaic fine-ware forms. Together, these 6th-century BC forms account for over 90% of all identified Archaic-period fine wares and together provide the marker par excellence for (Late) Archaic-period assemblages in the Metapontine chora.

Archaic Metapontine Fine-ware Fabrics

K. Swift

Archaic-period fine-ware fabrics comprise a relatively homogeneous group when compared to later Metapontine black gloss of the Classical and Hellenistic periods. Two typical examples from Ionic cups are presented here (**Nos. 1–2**, Catalog of Archaic Fine-Ware Fabrics), and similar fabrics predominate in the corpus of Archaic wares from the survey, and also among the substantial excavated assemblages from Incoronata and the Late Archaic sanctuary at Pantanello.[14]

Archaic-period fine-ware fabrics tend to have been fired to a range of pale yellowish colors when oxidized, and these generally show a more restricted range than is apparent in later black-gloss wares of the Classical and Hellenistic periods (see the discussion of fabrics in Part III). The surfaces are usually fired yellowish and tend to be paler than the body of the fabrics. This is perhaps a consequence of the use of water rich in calcium carbonates for preparation, forming, and smoothing of the vessel surfaces, with an increased concentration of calcium carbonate precipitated on or close to the surfaces as the vessel dried.[15]

The fabrics are usually consistent and relatively hard, though rarely approaching the hardness of the high-fired examples of black-gloss fabrics. The fracture tends to be clean, and harder-fired examples may have semi-conchoidal fractures, similar to later black-gloss fabrics, though more commonly in lower-fired examples there is the tendency for a slightly powdery fracture as a result of the abundant, extremely fine non-plastic grains in the fabrics. These give the fine-ware fabrics a finely granular texture when viewed microscopically, and consist of extremely fine crystalline grains that are sufficiently small to be considered as part of the matrix rather than inclusions, though larger grains, usually less than 0.1 mm, are occasionally present. Most are likely to be quartz, but feldspars may be represented too, mostly by slightly pinkish or yellowish crystalline grains. Examples in Metapontine fine-ware fabrics are extremely fine, but larger inclusions have been identified petrographically in coarser fabrics from the Metapontine chora.[16] The fabrics cannot be described as micaceous: fine flakes of biotite (gold mica) are rarely present and visible on the reserved surfaces. Although the fabrics may appear micaceous in hand specimen, with numerous extremely fine glistening inclusions, microscopically these can be seen to be crystalline quartz grains and feldspar lathes, rather than mica.

Few textural concentration features are apparent in Metapontine fine-ware fabrics of the Archaic period, and where present these tend to be fine and well-rounded. These are iron oxide-rich or calcareous clay concentrations which fire darker or lighter,

[14] The Archaic-period settlement at Incoronata and sanctuary at Pantanello are the subject of forthcoming studies.

[15] This feature is very common and more pronounced on mold-made terracottas from the Metapontine chora, where the undersides of the clay bodies rest for longer in a mold.

[16] See in particular the characterizations of Metapontine amphora fabrics, Ch. 12.

Ware	Catalog no.	Description	Chronology	Count
Imitation proto-Corinthian	1	Imitation proto-Corinthian cup rim	7th century BC	1
	2	Imitation proto-Corinthian base of unidentified closed form	7th century BC, second half	1
		All Imitation proto-Corinthian pottery		**2**
Colonial wares	3	Colonial-ware cup rim	7th century BC, second half	4
	4	Colonial-ware cup rim	7th century BC	2
	5	Colonial-ware base of unidentified open form	7th century BC	1
	6	Colonial-ware base of unidentified open form	7th century BC	1
		All Colonial ware		**8**
Corinthian fine ware	7	Corinthian kothon wall and handle fragment	Mid-6th century BC	1
		All Corinthian fine ware		**1**
Ionic cups	8	Ionic type B2 cup rim with handle terminus	580–500 BC	2
	9	Ionic type B2 cup rim	550–500 BC	7
	10	Ionic type B2 cup rim	550–500 BC	75
	11	Ionic type B2 cup rim	550–500 BC	1
	12	Ionic type B2 cup rim	550–500 BC	18
	13	Ionic type B2 cup rim	550–500 BC	51
	14	Ionic type B2 cup rim	550–500 BC	4
	15	Ionic type B2 cup rim	550–500 BC	7
	16	Ionic type B2 cup rim	550–500 BC	1
	17	Ionic type B2 cup rim	550–500 BC	30
	18	Ionic type B2 cup rim	550–500 BC	2
	19	Ionic type B2 cup rim	550–500 BC	2
	20	Small Ionic type B2 cup rim	550–500 BC	2
		Ionic type B2 cup rims (bulk, undifferentiated)		40
		All Ionic cup rims		**242**
	21	Ionic type B2 cup base	550–500 BC	72
	22	Ionic type B2 cup base	550–500 BC	6
	23	Ionic type B2 cup base	550–500 BC	9
	24	Ionic type B2 cup base	550–500 BC	3
	25	Ionic type B2 cup base	550–500 BC	1
	26	Ionic type B2 cup base	550–500 BC	1
		Ionic type B2 cup bases (bulk, undifferentiated)		8
		All Ionic cup bases		**100**
		Ionic type B2 cup handles (bulk)		6
		Ionic type B2 cup body sherds (bulk)		51
		All Ionic cups		**399**
Skyphoi with offset rims	27	Rim of skyphos with offset rim	550–500 BC	2
	28	Rim of skyphos with offset rim	550–500 BC	1
	29	Rim of skyphos with offset rim	550–500 BC	2
	30	Rim of skyphos with offset rim	550–500 BC	1
	31	Rim and handle terminus of skyphos with offset rim	550–500 BC	2
	32	Rim of skyphos with offset rim	550–500 BC	10
	33	Rim of skyphos with offset rim	550–500 BC	
	34	Rim of skyphos with offset rim	550–500 BC	
	35	Rim of skyphos with offset rim	550–500 BC	
	36	Rim of skyphos with offset rim	550–500 BC	
		Rims of skyphoi with offset rims (bulk)		
		All rims of skyphoi with offset rims		
	37	Base of skyphos with offset rim	550–480 BC	

Table 6.2 Quantification of Archaic fine wares by form, showing number of examples from the survey assemblages co-identified with the cataloged forms.

Ware	Catalog no.	Description	Chronology	Count
Skyphoi with offset rims cont.	38	Base of skyphos with offset rim	550–480 BC	1
	39	Base of skyphos with offset rim	550–480 BC	7
	40	Base of skyphos with offset rim	550–480 BC	1
		Bases of skyphoi with offset rims (bulk)		6
		All bases of skyphoi with offset rims		**18**
		Body sherds of skyphoi with offset rims (bulk)		9
		All skyphoi with offset rims		**67**
Archaic line-decorated ware	41	Archaic line-decorated cup rim	6th century BC	3
	42	Archaic line-decorated Panionion-type cup rim	550–500 BC	1
	43	Archaic line-decorated Panionion-type cup rim	550–500 BC	8
	44	Archaic line-decorated Panionion-type cup rim	550–500 BC	1
	45	Archaic line-decorated skyphos rim	550–500 BC	1
		All Archaic line-decorated ware		**14**
		All Archaic-period pottery		**491**

Table 6.2 cont. Quantification of Archaic fine wares by form, with number of examples co-identified with the cataloged forms.

respectively, than the main component of the clay body and which may either be present in the clay source or result from the mixing of different sediments during clay preparation; their scarcity in Metapontine fabrics therefore points to the use of relatively consistent source clays and/or the thorough application of clay preparation practices.

Fine, irregular, elongated expansion voids are common, and occasionally larger vughs (irregular voids) are apparent in some examples, particularly on bases and other areas where wall thickness varies widely.

The quality of slips on Archaic fine wares varies widely and poorer-quality slips predominate, usually matte and often fired light grey rather than the darker hues typical of Classical and Hellenistic black gloss. Partially oxidized slips, fired reddish-brown, are common, often accompanied by streaky or dilute slips. Although there are high-quality examples among the Archaic fine wares, comparable to those of later black gloss, particularly in terms of the well-sintered, semi-lustrous surfaces, these are a minority and often unevenly oxidized. These are perhaps indications of a lesser degree of control over the firing program, in terms of time/temperature combined with the control over the reducing and oxidizing phases of the kiln atmosphere, during the Archaic period compared to the Classical and Hellenistic periods.

Further, the intention during the Archaic period may not have been to produce slips of the same appearance, as was clearly the intention in the more-or-less entirely slipped vessels of later periods. The overall impression from these pieces is that obtaining a semi-lustrous slip may have been more important than producing a consistent dark hue so characteristically the aim of fine-ware production in the Classical and Hellenistic periods, since well-sintered slips often coincide with partially oxidized examples. This was perhaps more difficult to obtain on semi-slipped vessels rather than the mostly or wholly slipped vessels of the Classical and Hellenistic periods, and there may have been technical limits to the production of slips during this period, with a trade-off between luster and consistent, dark firing colors.

It appears that it was not until the early 5th century BC that these technical limitations were consistently surmounted, when the intention became to produce black gloss *sensu stricto*. Developments in production technologies, and particularly pyrotechnologies, seem to have accompanied the fundamental and broad-ranging changes in concepts of shapes and form repertoires which heralded the end of the Late Archaic tradition of ceramic production and gave way to the black-gloss traditions of the Early Classical period.

Archaic Fine-Ware Forms

A. Tubelli

Imitation proto-Corinthian pottery

A few fragments from the survey represent fairly faithful reproductions of proto-Corinthian models, the most interesting being a fragment of a small cup with slightly flaring lip forming a continuous profile with the bowl (**No. 1**); the profile accentuates the bowl, making it appear carinated. Decoration consists of reddish-brown horizontal lines on the lip and at the transition to the bowl, beneath which there is a reserved area. Given the poor state of preservation, it is unclear whether this area was completely reserved or decorated with metope motifs or vertical strokes painted in the handle zone, as common on proto-Corinthain pottery. A descendant of the Geometric-period Thapsos cups with a long production history through the end of the 7th century BC, this type of cup has a reserved band, bowl of varying depth, profile of varying curvature, and variable size.[17] Given its poor condition, this Metapontine example can be dated only very generally to the 7th century BC.

A fragment of the foot and lower body of a closed form, probably an olpe or oinochoe, is decorated with a crown of rays just above the seam between foot and body (**No. 2**). This fragment, too, is a local imitation of Corinthian models that can be placed, on the basis of the relatively short and tightly grouped rays, between the Late proto-Corinthian and Early Corinthian periods, with a likely date somewhere in the second half of the 7th century BC.[18]

Late proto-Corinthian and Early Corinthian pottery, both imported and locally produced, are attested at Metaponto at proprietà Andrisani, at Siris, at Sybaris on Timpone della Motta, and especially at the site of Incoronata greca in the Metapontine chora; in the latter case the numerous local imitations appear to cease at the beginning of the Late proto-Corinthian period.

Colonial wares

There are a few fragments of Colonial ware produced in the 7th century BC. Among these the only identifiable form in these wares from the survey is that of a cup with a short, out-turned rim and pronounced shoulder (**Nos. 3–4**), along with two fragments of unidentifiable open forms (**Nos. 5–6**). The exterior surfaces of the cups are particularly abraded and the decoration is consequently difficult to interpret, but the essential features are evident: the lip bears horizontal lines painted in brown gloss, the shoulders are reserved, and the gloss appears to resume on the bowl. On the interior the abrasion and flaking of the gloss is even more pronounced, but nonetheless the area immediately beneath the rim appears to be reserved. These fragments are difficult to classify typologically because of their deviations from the *a filetti* cups common elsewhere (for example, at Incoronata) and typical of colonial production and inspired by the proto-Corinthian tradition. These can be distinguished by a vertical or only slightly oblique lip, rounded shoulder, and completely glossed bowl of varying depth.[19]

A series of cups characterized by horizontal lines on the lip and glossed bowl with reserved band on the shoulder (occasionally decorated also with a wave motif) provide broad comparanda for the examples of Colonial-ware cups from the survey and date to the second half of the 7th century BC. Examples of this series have been discovered on the south side of the urban area of Metaponto in proprietà Andrisani, in the Pantanello necropolis, and in several contexts at Sybaris, such as the excavations at Stombi and Timpone della Motta and in the Berna-Malibù collection. Production of this type of cups is documented also at Croton and Kaulonia.[20] These cups are defined as *a filetti* cups, the proto-Corinthian types; it has been suggested, however, that these can be placed in East Greek rather than the Corinthian tradition.[21]

At the moment there is no organic and comprehensive study or typological overview that follows the chronological development of these cups, not only in terms of their decorative variation (stripes of varying thickness and proximity, reserved lip, reserved band

[17] *Perachora* II, 75–79, Fig. 3, Nos. 668, 691, 698. For the history of the type, see D'Agostino 1968, 96–97, types 12–13, Fig. 15, *Cuma*, 140 TA60, Pl. 3.25; Greco and Mermati 2007, 146–47, Fig. 3; *Mégara Hyblaea* 2, 143–44, Pl. 125.3–4; *Dea di Sibari* I.1 245–46, 248, Nos. 15–16.

[18] Payne 1931, 298–99; *Corinth* VII.2 138, 140, 145, Nos. 201, 220, 233, Pls. 47–48, 63; *Mégara Hyblaea* 2, 53, Pl. 35; *Dea di Sibari* I.1, 207–9 (*oinochoai a piede stretto*).

[19] Tocco Sciarelli 1981, 228, Pls. XLIII–XLV; A. Bottini 1982, 67–68; Berlingò 1986 (Madonnelle necropolis at Policoro), 119–25, Pls. 13–16; M. Tagliente 1986, 131, Pl. 29 (proprietà Cospito, Policoro); Tomay/Munzi/Gentile 1996, 213–19; *Dea di Sibari* I.2 85, 97–101; *Greci sul Basento* 144, 156–57, *Incoronata* 3, 70, 208; *Incoronata* 5, 57, Fig. 70; Macnamara 1983, 322–25, Nos. 149–57. Kiln wasters demonstrate the local production of these cups: see Sabbione 1982, 263–69; De Siena 1986b, 144–46, Pls. 42–45.

[20] De Siena 1986b, Pl. 45; Carter and Toxey 1998a, 697, CD 30-3; *Sibari* II, 157–60, 265–67; Tomay/Munzi/Gentile 1996, 216, No. 3.83; *Dea di Sibari* I.2, 92–93, 110–11 (cups with *a filetti* rim and reserved band on the shoulder, in particular group d2, in association with building Vc, dated to the second half of the 7th c. BC). See Sabbione 1982, 267–69; *Kaulonia* I, 44–45, Fig. 28.

[21] Many attested examples from southern Italy evidently derive from East Greek cups, especially in the area of Sybaris, both at the Stombi excavations and on Timpone Motta at Francavilla Marittima. The Sybarite examples have a particularly developed lip bearing *a filetti* decoration with a convex exterior profile clearly distinguished from the deep body, with its pronounced shoulder bearing a reserved band. See *Sibari* II, 264–66; *Dea di Sibari* I.2. In this case the reference types are Ionic cups of the last quarter of the 7th/early 6th c. BC. See *Gravisca* 4; Schlotzhauer 2000.

on the shoulder with or without wave motif), but also in terms of their morphological characteristics, which probably reflect a multitude of stylistic components.[22] The fragments found in the chora can therefore only be generically related to a colonial series and time span encompassing the second half of the 7th century BC.

The two fragments of disk feet with slightly concave base and mostly abraded gloss (**Nos. 5–6**) are not particularly diagnostic; they are attributable only very generally to Colonial ware or imitation proto-Corinthian cups or skyphoi.

Corinthian fine ware

Archaic-period Corinthian fine ware is represented in the survey by a single fragment of a kothon (**No. 7**), consisting of part of the wall and handle terminus with faint traces of brown slip. The fragment from the survey constitutes a rare example of imported ceramic products among the survey sample and the only imported Corinthian fine ware.

Precise parallels cannot be found because of the fragment's poor condition and almost completely abraded decoration, but the kothon with simple ribbon-handle was produced beginning in the Middle Corinthian and continued through the Late Corinthian II both in the "white style" and with simple linear decoration.[23] Kothons with ribbon-handle and linear decoration, produced at Corinth, are attested in southern Italy at the sanctuary on Timpone Motta at Francavilla Marittima, as well as at indigenous centers of the Agri River valley such as the necropolis at Santa Maria la Stella near Aliano, where the kothon was found in association with a Corinthian skyphos, Ionic B2 cups, Panionion-type cups, and one-handled cups dating to the second half of the 6th century BC.[24] Among the 6th-century materials in the territories of Sybaris and Metaponto there are numerous examples of kothons which imitate Corinthian models.[25]

[22] For a synthesis of the ceramic industries in the territories of Sybaris, Siris, and Metaponto in the 8th–7th c. BC, see Semeraro 1996, 273–75.

[23] For the form, see Payne 1931, 335, Fig. 183; Amyx 1988, 470.

[24] *Dea di Sibari* I.1, 264–65, No. 2, Fig. 16.2; Tocco Sciarelli 1981, 232, Pl. L. Examples of imitations with simple ribbon-handle were found at the necropolis of Macchiabate at Francavilla Marittima: see Zancani Montuoro 1980–1982, 118, Pl. 85.

[25] Tomay/Munzi/Gentile 1996, 219–20, 3.99 (with reference to other attestations of imitation Corinthian kothons in southern Italy); P. Bottini 1988, 121, Fig 24; Lo Porto 1973, Pl. XXV, No. 2.5; Lissi Caronna 1980, 148, Figs. 43–44; *Greci, Enotri e Lucani,* 144 No. 2.10.28, 159 No. 2.16.16.

Ionic cups

A majority of Archaic-period forms identified in the survey consists of so-called "Ionic" cups. The fragments are very small, so that in many cases the Ionic cups could not be clearly distinguished from the skyphoi with distinct rim and banded decoration. Almost 400 fragments of Vallet and Villard's type B2 cups can be positively identified, usually consisting of fragments of the lip and shoulder. No examples of the older types A1, A2, and B2, or the B3 type were positively identified.[26]

There is an overall homogeneity in the technical features of the fragments, particularly the reddish brown-to-black gloss that is always fairly dilute and opaque, occasionally distributed irregularly (especially on the exterior bands), or mostly abraded. The morphological and decorative features, however, are more variable. The Ionic cup forms from the survey have the characteristic step-like offset that distinguishes the rim from the shoulder; on most of the examples this offset is particularly accentuated with a tool-turned articulation, coinciding with a clear distinction between the lip and bowl even on the interior. The shoulder is pronounced and rounded, while the body-wall tapers as it approaches the foot (**Nos. 8–9, 11–15**). Such standardization of the form nevertheless leaves room for some variation: in some cases the shoulder is less pronounced, and the lip is higher (**No. 10**) or more inclined and tapering as it approaches the rim. Rarely, the shoulder is barely perceptible, and the bowl has unflared walls (**No. 19**, two examples)—features recalling the type B3 cup—but the lip is short, and the offset is still very marked, resulting in a typical, but somewhat less well-defined Ionic cup profile.

In the majority of cases the banded decorative scheme is that typical of Ionic type B2 cups. On the exterior, the lip and a strip on the shoulder at the level of the handle are reserved, while the handle, a band on the shoulder, and the body are glossed. The interior of the cup is completely slipped, with a thin reserved band immediately below the rim. On several of the foot fragments there is a reserved tondo (**Nos. 21** and **25**), and in one case the floor is slipped in the center producing a circle (**No. 24**).

Several Ionic cups from the survey have decorative schemes that diverge from those typical of type B2 cups as commonly standardized and circulated throughout the western Mediterranean. Though having the same

[26] Vallet and Villard 1955.

shape and similar rim and body proportions these cups have a completely slipped lip, reserved offset (though in some cases slipped), slipped body, and a reserved area at the level of the handle (**Nos. 16–17**). With only the rim and shoulder preserved, it is unclear whether there were other reserved strips on the lower part of the body.

There are also examples of completely slipped Ionic cups (**No. 18**). In these cases the gloss is reddish-brown, opaque, and thick, and the shape of the bowl is more compressed, with a particularly pronounced shoulder.

There are also examples of significantly smaller Ionic type B2 cups, fractional versions of the usual Ionic cups rather than miniatures, with a diameter of approximately 7 cm, a barely pronounced shoulder, and a low and compressed body (**No. 20**). The cup was dipped in a dense and opaque red-orange gloss; the bowl is partially reserved on the lower part with dribbles of gloss probably corresponding to the points where the vessel was held during immersion.

Several feet of B2 cups with deliberately perforated floors (**Nos. 23, 25**) are probably associated with ritual practice, a feature also attested in the Pantanello necropolis and known from other parts of southern Italy.[27]

The survey fragments have parallels in the cups from excavations in both the urban center and the territory of Metaponto dating from the second quarter to the end of the 6th century BC. The majority of the examples recovered in the Metapontino have, like the survey material, the offset at the transition between the lip and shoulder, the bowl of varying depth, the conical foot, and the classic decorative scheme of reserved lip and band at the level of the handle. The fragmentary nature of the examples from surface survey precludes reconstruction of the morphological development of the B2 cups based on complete shapes of the bodies and, consequently, the chronological sequence of the numerous variants.

The Ionic cup variant with slipped rim (**Nos. 16–17**) has a precise parallel in a cup from the Pantanello necropolis, which is attributed to the first half of the 6th century BC, following examples from Cozzo Presepe.[28] The fragments from Cozzo Presepe, however, do not have the typical shape of the B2 cups. The lip is particularly developed and inclined, the transition between the lip and bowl is not emphasized by a marked offset, and the bowl is deeper and completely glossed with only a reserved strip at the level of the handle. These features appear also on colonial cups produced at the end of the 7th and the beginning of the 6th century BC, which appear to combine motifs from proto-Corinthian ceramics with the decorative schemes of the earliest types of Ionic cups.[29] The morphological features of the examples from the Pantanello necropolis, on the other hand, are more similar to those of the cup found in the Favale votive deposit at Metaponto, which is dated to the second half of the 6th century BC; the latter differs from the more widespread type B2 only in the glossed lip.[30] A similar date can be proposed for the examples from surface survey (**Nos. 16–17**).

Variants that have relatively shallow, compressed bodies and are completely covered in poor-quality, opaque red-orange gloss are much less common (**Nos. 18, 20**). Smaller examples decorated via immersion are also attested in the Favale deposit from the urban area of Metaponto and also at the site of Incoronata greca, where they are probably associated with an Archaic rural sanctuary.

In the Greek West, the wide circulation of this form, including its related local versions and the proliferation of numerous variants, occupies a broad span from 580 to the end of the 6th century BC, although the colonial products appear to concentrate especially in the second half of that century.[31] The Ionic cups found in the survey probably date to the period 550–500 BC. Ionic type B2 cups are prevalent not only in the urban center and its territory but also at numerous settlements in the interior of Basilicata such as Pisticci, Montescaglioso, Matera, Pomarico Vecchio, and Garaguso.[32] Metapontine Ionic cups may have

[27] Carter 1998d, 121–22; Liseno 2004, 71; *Sibari* II, 263, 264, 330, 364, Nos. 160, 167, Figs. 280, 356, 386.

[28] Cozzo Presepe Phase IIIA: Carter and Hall 1998, 297–98.

[29] There is still no systematic and comprehensive study of this type of cup, which circulated widely in the areas centered on Siris, Metaponto, and Sybaris, and for which different definitions have been advanced, from imitations of types A1 and B1 Ionic cups to type B2 Ionic cups. See Macnamara 1983, 324–25; *Dea di Sibari* I.1, 19–20, Nos. E16–E20, Figs. 29a–32.

[30] Liseno 2004, Pl. XXXVIIc–d.

[31] Zancani Montuoro 1980–1982, 82–83. On the circulation of Ionic cups in the West, see Martelli Cristofani 1973, 6; Pierro 1984, 9–67; *Gravisca* 4, 137–235; Van Compernolle 1996 and 2000; *Cuma*, 43–51.

[32] Lo Porto 1973, Pl. V, Fig. 1, Nos. 4–5 (from Pisticci, middle of the 6th c. BC), Pl. VI, Fig. 1, Nos. 1 and 3, Pl. VIII, Fig. 1, No. 5 (end of the 6th c. BC), Pl. IX, Fig. 1, No. 3, Fig 3, No. 3; Pl. XVII, Fig. 3, No. 3, Pl. XXIII, Fig. 1, No. 1 (from Matera), Pl. LV, Fig. 2, No. 7 (end of the 6th/beginning of the 5th c. BC, banded B2 cup associated with a completely glossed cup); Pl. LXX, Fig. 1, No. 5 (end of the 6th/beginning of the 5th c. BC); Lo Porto 1988–1989, 409–10, Fig. 127, No. 3 (tomb 2 from Pisticci, middle of the 6th c. BC), Fig. 32, Nos. 18 and 20 (tomb 1 from

been distributed as far as Messapia, as indicated by several finds at mid-6th-century Oria and Manduria in the Salentine peninsula.[33]

As early as the 1970s, discoveries of kiln waste deposits with fragments of misfired Ionic cups not only at Italiote colonies such as Siris and Metaponto, but also at small settlements such as Termitito, demonstrated that the majority of the Greek and indigenous settlements produced Ionic cups, in particular the type B2, in local workshops.[34] Along with these other sites, production of Ionic cups was attested in the kerameikos of Metaponto by waste deposits with misfired Ionic type B2 cups.[35] Subsequent excavations in the urban center and in the necropoleis yielded numerous fragments of Ionic type B2 cups reflecting similar morphological and decorative variations apparent in the corpus from the survey, as well as by another decorative scheme consisting of overpainted red stripes on the interior or on elements of the form, such as the transition between lip and body, typically defined by a rounded shoulder rather than an articulated offset.[36]

Given their ubiquity on Late Archaic-period sites throughout the chora, the artisans' quarter of the urban center is unlikely to have been the only center of production of Ionic cups: other workshops were probably located also in the chora.[37] Few black-gloss production centers have been identified in the chora. As discussed in the first section of this chapter, the evidence is slight and dependent upon fortuitous preservation and recovery of key pieces—wasters, test-firing pieces, and kiln spacers—which may not be particularly prevalent even at known centers of production. The relative dearth of positively identified Archaic-period production centers in the chora is therefore likely to have much to do with archaeological factors, while the range and ubiquity of regionally-produced Ionic cups apparent in the survey and at other sites in the chora, such as the Archaic farmhouse of Fattoria Fabrizio excavated by ICA, circumstantially suggest production beyond the urban area.

Skyphoi with offset rims

Widespread in the Metapontino between the middle of the 6th and the first quarter of the 5th century BC, this cup shape is characterized by a deep skyphoid body and a rim distinguished from the body by an offset—a feature borrowed from the Ionic type B2 cups—and, especially in the earliest phases of production, semi-slipped decoration of slipped or reserved bands which, like the Ionic cups, was firmly in the Archaic tradition.[38] This complicates the attribution of small fragments to one shape or the other. For the purposes of this study, the fragments preserving the upper parts of a deep body, indicated by a more vertical wall descending from the shoulder, were classified as skyphoi with offset rims rather than Ionic cups. There is some variation in the nomenclature for this shape, variously called "skyphoid cups,"[39] "lip-kotylai,"[40] or "Ionic skyphoi."[41] One of the first major studies of a large corpus of these forms investigated the examples from the Pantanello necropolis in the chora of Metaponto, which were called simply "skyphoi."[42] For the material from the survey, the term "skyphos with offset rim" has been used, following the examples from Cozzo Presepe,[43] to identify the form on the basis of its morphological features, which remain constant during the course of its development, rather than the places where it was found or produced.

Examples from the survey include several variants already attested in the Metapontino. The variant with reserved lip and reserved band at the level of the handle is attested (**No. 28**), as is the completely slipped version with reserved band in the handle zone and, occasionally, one or more small reserved strips on the lower body. Among the examples with reserved lip is an example with three vertical strokes painted on the reserved band in the handle zone (**No. 27**). The latter variant is widely attested in Archaic votive

Montescaglioso, end of the 6th c. BC); Adamesteanu 1974a, 129–86; Barberis 1997, 57–59, Pl. 28 (for other attestations in the Metapontino and in Basilicata in general, see notes 35 and 36 below).

33 Semeraro 1997, 214–15, 396.

34 *Céramiques*, 312–16; Van Compernolle 1996, 299.

35 Adamesteanu 1980, 276–77; D'Andria 1980a, 370–446; *Céramiques*, 315; Maruggi 1996b, 250–51; Silvestrelli forthcoming a.

36 For the examples found in the urban center see Adamesteanu 1980, 276–77; Macnamara 1983, 324–27; Lo Porto 1981b, 292–304, 334–82, Figs. 5, 8, 110; Van Compernolle 1996, 299; Liseno 2004, 72–75, Pl. XXXVII.

37 Carter 2006a, 216–18. The author wishes to thank E. Lanza Catti for providing access to the Ionic cups found in the excavations at Fattoria Fabrizio, and K. Swift for providing helpful information about the Ionic cups from the Pantanello sanctuary.

38 For a synthesis of the development of the type and its circulation, see Semeraro 1997, 225–28; Elliott 1998, 677–78.

39 Liseno 2004, 70.

40 Semeraro 1997, 225.

41 Lo Porto 1981b, 304.

42 Elliott 1998, 677–78.

43 Macnamara 1983, 327.

deposits at Crucinia and Favale at Metaponto, where there are also examples bearing series of aligned dots, dotted rosettes, wavy segments, and painted myrtle boughs.[44] In the votive deposit at Timmari and in the votive deposit at Oria there is a variant with a torch between vertical strokes on the reserved strip in the handle zone; a date in the first quarter of the 5th century has been proposed for this variant.[45]

The most well-represented skyphos with offset rim variants among the survey materials have slipped rims and offsets, a reserved band at the level of the handles, and slipped bodies. Some are rather large, with diameters in the range 14–16 cm, and have a gentler transition from rims to bodies, rounded shoulders, and broader bodies (**No. 29**). On others, however, the rims have convex profiles or are particularly angled towards the exterior (**Nos. 30–32**).[46]

The most common variant in the Metapontine chora is smaller, with diameter between 11 and 13 cm, and it has the lip neatly distinguished from the bowl by a marked offset, a conical body, and the handle-attachments at the point of maximum expansion of the shoulder (**Nos. 33–36**). The slip covering the lip, offset, and body, excluding the strips of variable width in the handle-zone and on the lower body near the foot, varies from black to red-orange and is always rather opaque and diluted.

Several foot fragments recovered in the survey can also be ascribed to skyphoi with offset rim (**Nos. 37–40**). These are ring feet with flat or inwardly-inclined resting-surface and low, conical feet with rounded edge. Because of their small size they can only be attributed to the general shape and not to specific variants of skyphoi with offset rims.

Production of skyphoi with offset rims in the Archaic tradition lasted from the middle of the 6th to the beginning of the 5th century BC and appears to have been associated with that of the Ionic type B2 cups. The two shapes are often found in association, e.g., in the grave goods of the necropoleis at Pisticci and Pomarico Vecchio, in the Macchiabate necropolis at Francavilla Marittima near Sybaris, or at Laterza in Peucetia.[47] The type of skyphos with offset rim bearing a reserved strip in the handle-zone has been encountered in multiple contexts in the Metapontino: in the waste deposits and votive deposits of the urban center dating to the second half of the 6th century BC, at Cozzo Presepe, in the necropolis and Archaic-period sanctuary at Pantanello, and in the Archaic levels of several excavated farmhouses.[48] The form developed and continued to be produced in black gloss *sensu stricto*, completely slipped, in the 5th century BC (see section III and Catalog of Black-Gloss Fine Ware for *skyphos with offset rim*).

As with Ionic cups, Metaponto was surely the most important center of production of the skyphos with offset rim, amply attested by the kiln waste deposits in the urban kerameikos.[49] Numerous examples were produced in the course of the second half of the 6th century and the first quarter of the 5th century BC in both the asty and—presumably—the chora (see comments on the production of Ionic cups in the chora, above, which apply equally to Archaic-period skyphoi with offset rims). These same products were distributed widely in the area immediately south of Matera, in the indigenous settlements of the interior, and in the direction of Messapia, especially at Oria, where the examples recovered appear to have all of the technical features of the Metapontine series.[50]

Archaic line-decorated ware

Several fragments from the survey are characterized by horizontal glossed lines and bands created with wide brush-strokes on well-levigated fabrics, a decorative scheme which identifies them as locally-produced Archaic line-decorated wares (**Nos. 41–45**).[51] These materials have been distinguished from the larger classes of plain/banded wares since they

[44] Lo Porto 1981b, 304–5. A skyphos with vertical strokes painted on the reserved strip is attested also in tomb 4 at Pisticci: see Lo Porto 1973, 171, No. 9, Fig. XIX 3.1.

[45] Semeraro 1997, 228, No. 781.

[46] These last features recur also among the skyphoi of Phases IIIB and IV at Cozzo Presepe and in the Uomo Morto necropolis at Amendolara: Macnamara 1983, 328–29, Fig. 111; De La Genière 1971, 464–71.

[47] Lo Porto 1973, 157, Pls. V 1.2 and VIII 1.1; Lo Porto 1988–1989, 360–63, Fig. 66.5; Barberis 1997, 59–60, Pl. 29; Zancani Montuoro 1980–1982, 82–83, Fig. 30 B7.

[48] Liseno 2004, 274, Nos. 7–11, Pl. XXXVI; Silvestrelli forthcoming a; Macnamara 1983, 327–29, Nos. 170–72, Fig. 111; Carter and Hall 1998, 373, Fig. 1; Elliott 1998, 681 S57, Fig. 4.9; Adamesteanu 1974a, 82–84 (from Cogno del Pero). The author wishes to thank E. Lanza and K. Swift for the helpful information about the skyphoi with offset rim recovered in the Metapontine farmhouses excavated by ICA and in the Pantanello sanctuary.

[49] D'Andria 1980a, 373, Figs. 13 and 26.

[50] Semeraro 1997, 225–28.

[51] For a definition of the class, see *Cuma*, 88–89. On the various criteria used to define well-levigated line-decorated wares, see Bagnasco Gianni 1999, 106–07. Consideration of the problems inherent in the recognition of well-levigated, non-figured ware can be found in Cerchiai 2007.

belong to a morphological repertory that is typical of the Late Archaic phase and distinct from that of plain/banded ware.

This ceramic class has traditionally been defined as "Ionic" by virtue of its affinity with the decorative scheme of stripes on a reserved background characteristic of several series produced in Ionia. In reality, however, they were western Greek products that circulated widely, from Etruria to southern Italy. In Magna Grecia successive discoveries of materials and kilns at both Greek and indigenous sites have demonstrated the existence of an ever-larger number of local series.

The morphological repertoire of Archaic line-decorated wares in southern Italy can generally be characterized as relatively heterogeneous: the forms constitute the convergence of diverse stylistic influences, among which East Greek influences represent just one aspect. Alongside standardized and widely circulated forms there are also re-elaborations incorporating elements of more localized series.[52]

Cup

A rim fragment belonging to a locally-produced cup, whose form is inspired by that of the "Ionic" cups but lacks the characteristic distinction between lip and bowl, comes from site 285 in the chora (**No. 41**). The short, outward-turned lip, in fact, has a continuous profile with the body and a concave exterior profile, while the shoulder is rounded and pronounced, and the body is shallow. The type has affinities with the banded cups found in the fortifications of Kaulonia and in the urban center of Kyme, where the contexts point generally to the 6th century BC but currently cannot be dated more precisely.[53]

Panionion-type cup

The survey yielded several fragments of skyphoi having a lip with continuous profile, a rounded rim that is slightly flared towards the exterior, and a deep bowl. The interior is glossed with a reserved band immediately beneath the rim; on the exterior the banded decoration emphasizes the lip and the lower part of the bowl, leaving a wide reserved strip in the handle zone. The gloss is rather dense, opaque, and variable in color between brown and red (**Nos. 42–44**). This type of skyphos, which is widely attested in southern Italy, is classified on the basis of various definitions ranging from "two-handled cup" to "cup with deep bowl," from "skyphoid cup" to "skyphos with unflared rim" or "Ionic skyphos."[54] The shape is most immediately recognized as the Panionion type, however, and its morphological features in fact find precise parallels in those of the cups found at the Panionion in Ionia, in particular Hommel's type II dated to the end of the 7th century BC.[55] In reality, this type did not circulate widely in the East Greek world beyond the attestations at the pan-Ionian sanctuary. In the Greek West, on the other hand, it was widely produced during the second half of the 6th century BC, also in smaller local variants that had a lip with straight profile, ring foot, and several reserved strips on the bowl.[56]

In the Metapontino, the Panionion-type cup has been found in both the variant with flaring lip, closely matching the examples from the survey, and the variant with straight or slightly inward-projecting lip, attested in particular at Cozzo Presepe.[57] An example comparable to the mid-6th century *Becherschalen* from Palinuro was found in the votive deposit at Crucinia.[58] The form also reached many settlements further inland, such as the sites in the Agri valley and the area of Melfi, which were influenced not only by the Ionian Greek world but also by the Etruscan-Campanian area via the Vallo di Diano, where this form is particularly widespread.[59] Examples with slightly flaring lip and "reverse S" profile are attested in the Archaic levels of even Sybaris and Locri.[60]

Skyphos with indistinct lip

A single fragment can be ascribed to a skyphos with indistinct lip, forming a continuous curve with the body, slipped on the exterior save for a reserved band in the handle zone occupied by a painted vertical

[52] Bagnasco Gianni 2007, 94–99.

[53] *Kaulonia* I 49, 53, Fig. 31, No. 62; Munzi 2007, 126, Fig. 15.

[54] *Cuma*, 89–90, 179, Pl. 21.2–3; Munzi 2007, 126, Fig. 15, No. 25040.1020, 127; Lo Porto 1981b, 304; Macnamara 1983, 330.

[55] *Panionion und Melie*, 149–53.

[56] For a synthesis of its circulation in southern Italy, see Munzi 2007, 127, n. 78, and *Cuma*, 89–90, n. 225.

[57] From Cozzo Presepe Phases IIB and IV: Macnamara 1983, 329–30, Nos. 178–80, Fig. 111.

[58] Lo Porto 1981b, 304, No. 93.

[59] Adamesteanu et al. 1971, 52, Pl. XII (tomb 2); Giorgi and Martinelli 1988, 72, and Osanna 1988, 155, and Pl. 28.1 (Forentum, tomb 58 No. 5); A. Bottini 1981, 227, No. 62, Figs. 20–21 (tomb 8); De La Genière 1968, 294 and 298; Naumann and Neutsch 1960, 109.

[60] *Sibari* II 262, 265, 358, Fig. 275; Tocco Sciarelli 1981, 232; *Locri* II, 81–82, Fig. 25, Nos. 11–12 (second half/end of the 6th c. BC); Rubinich 1992, 91–92, 99, Pl. XLIV, No. 29 (second half/end of the 6th c. BC).

stroke (**No. 45**). This type of small skyphos with the handles attached to the lip immediately below the rim, ring foot, and flat floor circulated widely in Magna Grecia in the last quarter of the 6th century BC.[61] It usually has banded decoration with vertical tongues or a series of dots on the lip, but the decoration can also be as simple as partial immersion of the vessel in the gloss. The miniature version, too, is found frequently in necropoleis and votive deposits throughout southern Italy.[62]

The type borrows the morphology of the Late Corinthian skyphoi of the middle to third quarter of the 6th century BC,[63] although the shape is re-worked incorporating local influences and continues to be produced in some cases until the second quarter of the 5th century. The examples known from the Metapontino are few, but in the hinterland they are attested in 5th-century tombs, especially in the necropolis of Alianello.[64] The skyphoi produced at Locri, on the other hand, can be assigned to the second half to end of the 6th century BC.[65]

Summary

A. Tubelli

The earliest phase of Greek occupation of the chora is documented in the Metaponto survey by a few fragments found for the most part at site 285. Despite their poor physical condition, it is evident that these materials were locally made and represent re-elaborations of proto-Corinthian and East Greek models, produced probably throughout the second half of the 7th century BC. The fragment of an imitation proto-Corinthian cup found at site 530 dates generically to the 7th century. Because of the paucity of examples, only broad affinities can be identified in relation to the archaeological documentation from elsewhere in the chora—at the site that would later be occupied by the Pantanello necropolis and in proprietà Andrisani in the urban center—or in relation to the ceramic evidence recovered on the hill of Policoro, as well as on Timpone della Motta (building V), where the material dates to the second half of the 7th century BC.[66]

Evidence for Archaic settlement collected during the ICA survey of the chora consists for the most part of Ionic type B2 cups and skyphoi with offset rim and striped decoration, attested in Magna Grecia beginning in the second quarter of the 6th century BC, and produced locally in the Metapontino on a grand scale in the second half of the 6th century BC. The widespread nature of these products in the Late Archaic period is documented in both the asty and the chora, as well as in the necropoleis of the settlements of the interior through to the beginning of the 5th century BC. The few pieces of locally-produced line-decorated ware, and probably also the lone example of imported Corinthian fine ware, date to the third quarter/end of the 6th century BC, and are massively outnumbered by the Ionic cups and skyphoi with offset rims which constitute the primary marker of this period in the Metapontine chora.

61 Cuozzo and D'Andrea 1991, 83, Fig. 8, type 37A.

62 *Fratte*, 227–29, No. 4, Fig. 381 (tomb IV/1927, last quarter of the 6th c. BC), 255–57, No. 6, Fig. 432 (tomb 26/1963, dated by context 480–460 BC); Naumann and Neutsch 1960, No. 2, Pl. 36.1 and 36.5 (tomb VIII, tomb IX, tomb XXXIII, end of the 6th c. BC); Johannowsky 1983, 196–98 ("kotylai"); Menard 1990, 307–09, Nos. 352–54, Fig. 14.

63 *Corinth* XIII, 106, 204, Pl. 33 221.1–2 (tomb 221).

64 Adamesteanu et al. 1971, 53, Pl. XII (tomb 2); Rubinich 1992, 89, 96–97, Nos. 6–15, Pl. LIV.G.

65 *Locri* II, 76.

66 Carter 1998f, 167–71; Carter 2006a, 109–15; De Siena 1986b and 1996; M. Tagliente 1986; *Dea di Sibari* I.2.

Catalog of Archaic Fine-Ware Fabrics

Keith Swift

The two samples from the survey shown here are typical of Metapontine Archaic fine-ware fabrics, with their yellowish surfaces, yellowish-brown firing color, and finely granular texture with frequent, extremely fine carbonate inclusions powdered throughout. Frequent crystalline grains, probably quartz and feldspars, are sufficiently small to be considered constituent part of the matrix and give the fabrics their very finely granular texture and apparent hardness. Color was recorded with reference to the Munsell Soil Color Charts.

1. AFW 397-47 Fabric sample from Ionic cup rim

Color: Yellow surfaces, very pale brown to yellow 10YR 8/4-6, and section yellow to brownish yellow 10YR 6-7/6.

Texture: Finely granular texture, moderately hard with powdery fracture.

Inclusions: Matrix contains abundant extremely fine crystalline grains (quartz/feldspars), and fine carbonates are powdered throughout the fabric, with occasional larger inclusions (< 0.1 mm) apparent also on the surfaces.

Voids: Frequent elongated expansion voids orientated parallel to the vessel walls, most < 0.1 mm.

Slip: Thin, dilute, matte black slip.

A. Hand specimen: Scale 10 mm. Exterior showing surface texture and slip.

B. Microscopy: Rim section showing typical finely granular texture and fracture typical of Metapontine Archaic fine-ware fabrics, width of field ca. 20 mm.

C. Microscopy: Section showing extremely fine crystalline inclusions and frequent elongated expansion voids, width of field ca. 10 mm.

D. Microscopy: External surface showing dilute, streaky, matte band of slip on shoulder, width of field ca. 20 mm.

A

B

C

D

2. BG 477-56 Fabric sample from Ionic cup rim
Color: Fired yellow, reddish-yellow 7.5YR 7/6, with paler yellow surfaces, pink 7.5YR 7/4.
Texture: Very finely granular texture with powdery fracture.
Inclusions: Abundant very fine calcium carbonate inclusions, most < 0.1 mm.
Voids: Frequent fine rounded to elongated expansion voids, most < 0.2 mm.
Slip: Thin, streaky matte black slip.

A. Hand specimen: Scale 10 mm (1:1). Exterior showing yellowish surfaces and thin, streaky matte black slip.
B. Microscopy: Section of rim showing finely granular texture, width of field 20 mm.
C. Microscopy: Section of rim, width of field 10 mm.

A

B

C

Catalog of Archaic Fine-Wares

Amelia Tubelli

The catalog of Archaic fine wares presents the key pieces used to date the survey sites. These were used as reference pieces for the co-identification and quantification of further examples in the site assemblages. Description of the features of the reference sherds are given but may vary from the co-identified "bulk" survey material given in the quantification, particularly in terms of the features and extent of the slips, though these were sufficiently similar to permit co-identification of similar examples which were otherwise unremarkable. For Archaic-period pottery, dating is of necessity relatively broad compared to that of later black-gloss fine wares, but selected comparanda are given for each of the reference sherds to outline and support their chronology. Since many of the pieces can be dated only very generally, if relatively securely, the likely chronological range is given, along with selected comparanda where applicable. For the small number where none are cited, the fragments are not particularly diagnostic; these fall within a broad range of comparable specimens, and are dated by ware and/or basic shape.

Dimensions of the reference sherd are given: *rim diam.* and *base diam.* refer to rim and base diameters respectively, with *l*, *h* and *w* referring to the extant dimensions of the fragment. All dimensions are given in cm. The identifier code AFW is used in the data for Archaic fine wares.

Study, identification, and re-examination were undertaken by A. Tubelli in 2009 following original selection of the key reference pieces made by E. Vittoria during recording and quantification of the survey assemblages in 2001 and 2002. The catalog was translated by A. Prieto and edited by K. Swift.

The Archaic fine wares from the survey are presented by ware as follows:

Imitation proto-Corinthian pottery **(Nos. 1–2)**
Colonial wares **(Nos. 3–6)**
Corinthian fine ware **(No. 7)**
Ionic cups **(Nos. 8–26)**
Skyphoi with offset rims **(Nos. 27–40)**
Archaic line-decorated wares **(Nos. 41–45)**

Imitation Proto-Corinthian Pottery

1:2

1. AFW 530-20 Cup rim (single example)
Dimensions: rim diam. 17.5 h 2.6
Reference Sherd Description: Dilute, matte reddish-brown slip, four small strips painted on the lip and shoulder on the exterior; slipped interior with reserved rim. Lip forms a continuous profile with the body, slightly flaring at the rim; concave exterior profile; pronounced shoulder above a carinated body.
Selected Comparanda: D'Agostino 1968, 97 Nos. 12–13 (Pontecagnano, necropolis of contrada S. Antonio)
Dating: 7th c. BC.

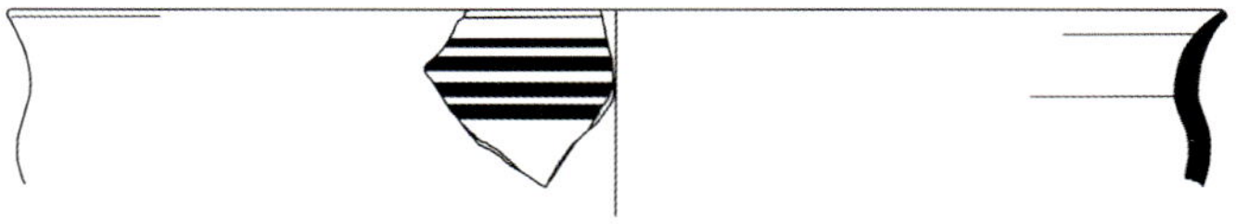

1:2

2. AFW 584-02 Base of unidentified closed form (single example)
Dimensions: base diam. 10.6 h 3.3
Reference Sherd Description: Ring foot with flat resting-surface and ovoid lower body. Dilute, matte reddish-brown slip. Slipped exterior profile of the foot with reserved resting-surface, reserved belly, decorated on the exterior with a crown of rays positioned slightly above the junction with the foot. The ray decoration is typical of closed forms in proto-Corinthian wares; the thick wall suggests either an oinochoe or perhaps an olpe. Reconstructed from two joining fragments.
Dating: Second half of 7th c. BC.

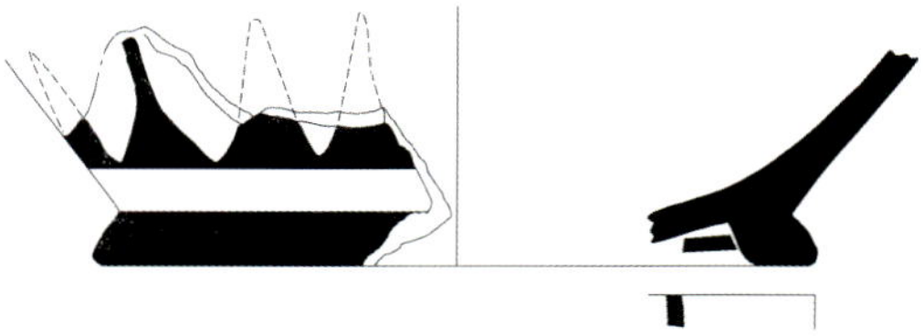

Colonial Wares

1:2

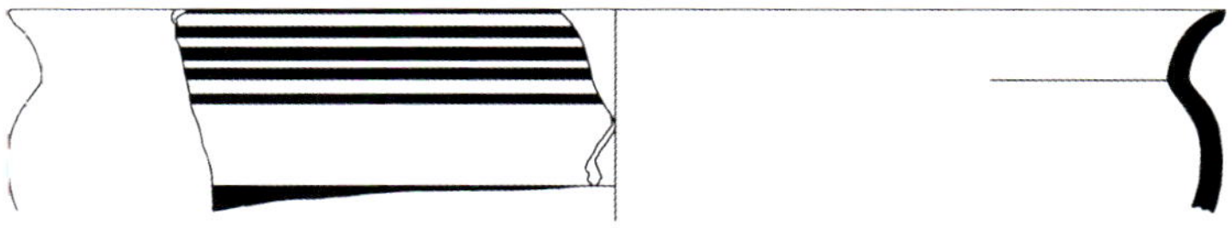

3. AFW 285-33 Cup rim (4 examples)
Dimensions: rim diam. 18.8 h 3.2 w 6.8
Reference Sherd Description: Misfired with dilute, matte reddish-brown slip, abraded. Slipped interior. Four or five thin bands of slip on the exterior rim and shoulder, reserved band at the level of the handle, below which are traces of slip. Short lip above pronounced rounded shoulder.
Selected Comparanda: *Dea di Sibari* 1.2, 110–11, No. C64, Fig. 76a–b (Sybaris, acropolis on the Timpone della Motta; second half of 7th c. BC); *Sibari* II 157–60, 265–70 (Sybaris, Parco del Cavallo, Stombi); *Kaulonia* I 44 Fig. 28 (Kaulonia, North Fortification).
Dating: Second half of 7th c. BC.

1:2

4. AFW 285-22 Cup rim (2 examples)
Dimensions: w 4.7 h 3.55
Reference Sherd Description: Dilute, matte reddish-brown slip. Broad band of slip on the exterior of the rim and shoulder, illegible decorative scheme on the body (reserved band at the level of the handle?); slipped interior with reserved inner rim. The fragment cannot be precisely identified, a situation complicated by the uncertainty of the decorative scheme.
Dating: 7th c. BC.

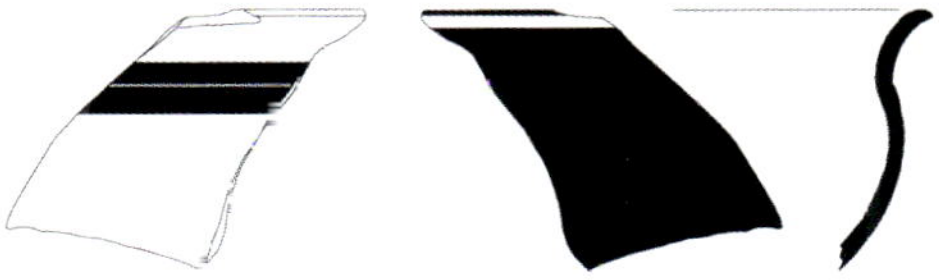

1:2

5. AFW 285-36 Base of unidentified open form (single example)
Dimensions: base diam. 5.5 h 3.9
Reference Sherd Description: Disk base with slightly concave underside and deep lower body. Dilute, matte reddish-brown slip. Slipped interior. Exterior of the body and foot slipped with reserved underside of the foot. Reconstructed from two joining fragments.
Dating: 7th c. BC.

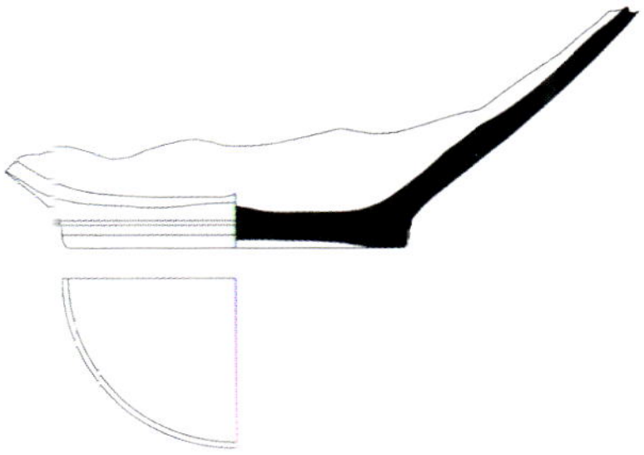

1:2

6. AFW 285-45 Base of unidentified open form (single example)
Dimensions: base diam. 5.8 h 2.0
Reference Sherd Description: Disk base with slightly concave underside. Dilute, matte reddish slip on interior, mostly worn.
Dating: 7th c. BC.

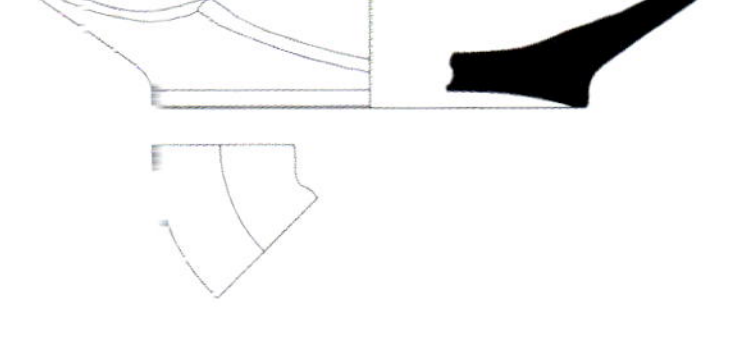

Corinthian Fine Ware

1:2

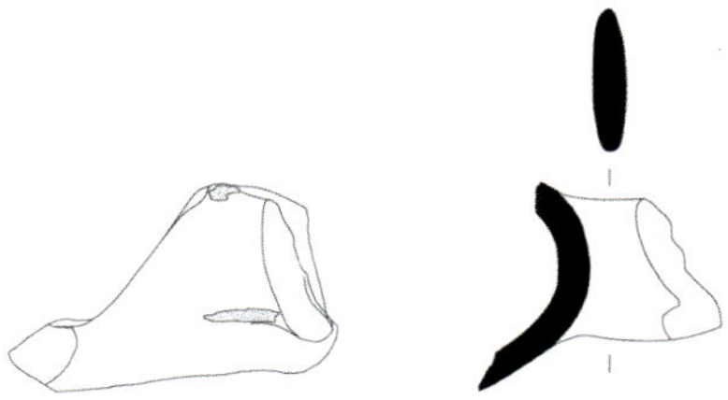

7. AFW 397-06 revisit 90 Kothon wall and handle terminus (single example)
Dimensions: h 4.65 w 3.05
Reference Sherd Description: Rounded shoulder with terminus of horizontal ribbon handle. Light yellowish fabric with dilute, matte reddish-brown slip, mostly worn with traces of slip on the handle.
Selected Comparanda: Payne 1931, 335, Fig. 183; Amyx 1988, 470; Tocco Sciarelli 1981, 232, Pl. L (Aliano, località S. Maria la Stella, T. 3); *Dea di Sibari* I.1, 264–65, No. 2, Fig. 16.2 (Sybaris, acropolis on the Timpone della Motta, Middle Corinthian)
Dating: Middle or Late Corinthian, middle to last third of 6th c. BC.

Ionic type B2 cups

1:1

1:2

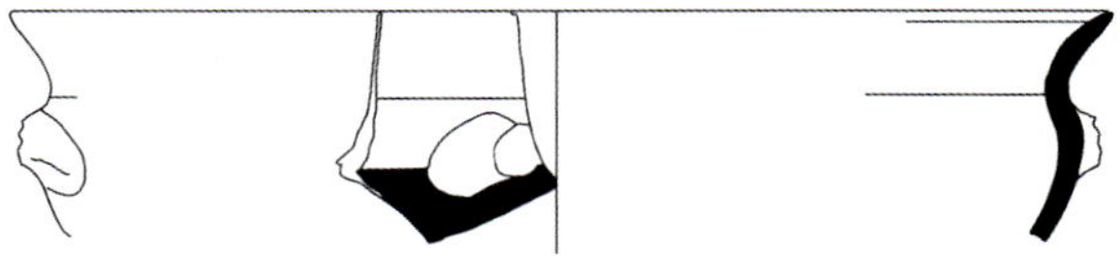

8. AFW 285-42 Rim with handle terminus (2 examples)
Dimensions: rim diam. 15.2 h 3.3
Reference Sherd Description: High, flaring lip with distinct articulation from slight shoulder. Thick, matte black slip. Reserved lip and shoulder on the exterior, slipped handle and shoulder; interior glossed with reserved inner rim.
Dating: 580–500 BC.

1:1

1:2

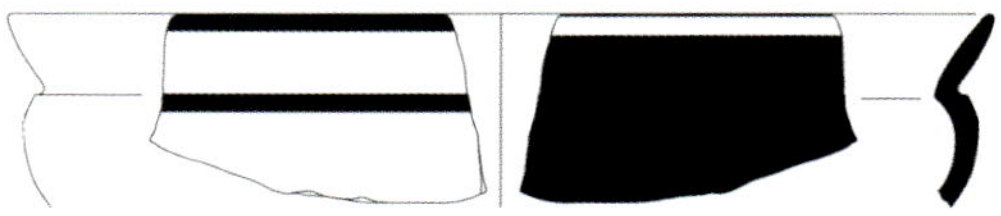

9. AFW 397-46 Rim (7 examples)
Dimensions: rim diam. 13.0 h 2.6
Reference Sherd Description: Outer edge of rim slightly incurved, marked offset with shoulder. Dilute, matte reddish-brown slip, interior bears reserved band on inner rim. Reserved exterior with thin band on the upper rim and on the shoulder immediately below articulation with rim.
Selected Comparanda: Adamesteanu 1980, 277, Fig. 290 (Metaponto, prop. Favale); Hänsel 1973, 467, No. 1 (Policoro); du Plat Taylor and Small 1992, 66, 285, No. 608 (Gravina, Botromagno hill).
Dating: 550–500 BC.

1:1

1:2

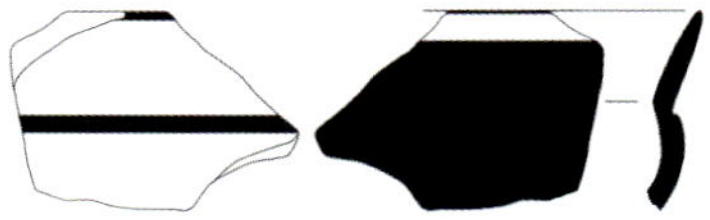

10. AFW 397-47 Rim (75 examples)
Dimensions: w 3.8 h 2.7
Reference Sherd Description: Straight, tapered lip, nearing vertical on the upper edge, marked offset between the rim and shoulder. Dilute, matte black slip, reserved band on the upper interior, reserved exterior with thin band on the outer lip and on the shoulder immediately below the pronounced articulation with rim.
Selected Comparanda: Adamesteanu 1980, 277, Fig. 290 (Metaponto, prop. Favale); Macnamara 1983, 326, No. 161, Fig. 110 (Cozzo Presepe, middle of the 6th c. BC); *Kaulonia* I, 48, No. 35 (Kaulonia, North Fortification).
Dating: 550–500 BC.

Corinthian fine ware, Ionic type B2 cups cont.

1:1

11. AFW 334-14 Rim (single example)
Dimensions: w 5.2 h 3.8
Reference Sherd Description: Marked offset between rim and shoulder. Dilute, matte black slip, reserved band on interior immediately beneath the rim, exterior reserved with slipped rim and thin band on the offset and shoulder.
Selected Comparanda: Adamesteanu 1980, 277, Fig. 290 (Metaponto, prop. Favale); Liseno 2004, 75, No. 14, Pl. XXXVII.b (Metaponto, votive deposit at Favale, second half of 6th c. BC).
Dating: 550–500 BC.

1:2

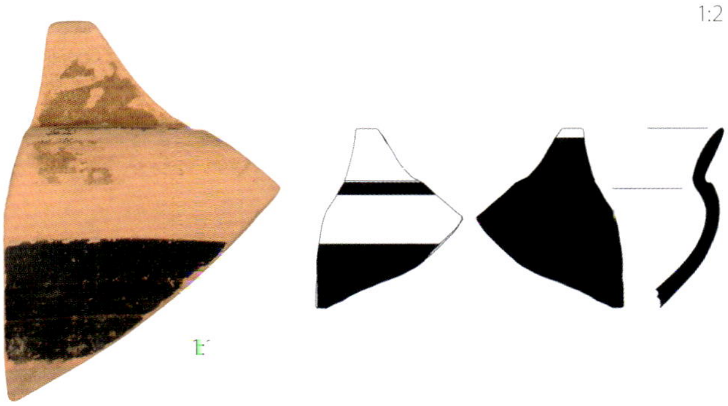

12. AFW 477-42 Rim (18 examples)
Dimensions: w 3.3 h 4.2
Reference Sherd Description: Offset between rim and shouder. Consistent, thick, semi-lustrous black slip, reserved band on the lip and on the body at the level of the handle.
Selected Comparanda: Lo Porto 1981b, 297–99, Figs. 5.1 and 8.2–5 (Metaponto, votive deposit at contrada Crucinia); Liseno 2004, 75, No. 16, Pl. XXXVII.e–f (Metaponto, votive deposit at Favale, second half of 6th c. BC).
Dating: 550–500 BC.

1:2

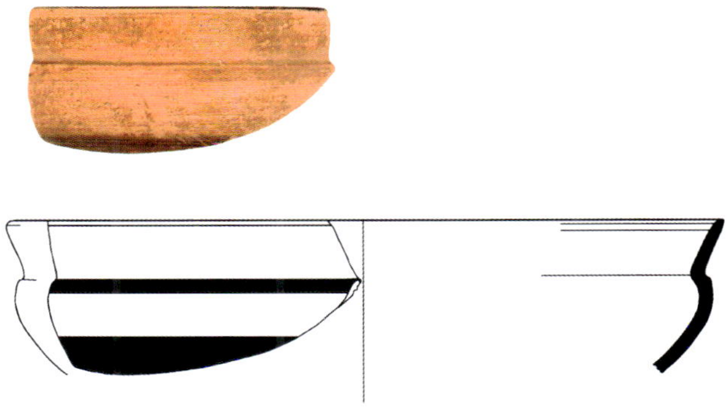

13. AFW 622-02 Rim (51 examples)
Dimensions: rim diam. 16.4 h 3.6
Reference Sherd Description: Tapered rim with marked offset at junction with shoulder, corresponding pronounced concavity on interior. Very dilute, matte reddish-brown slip, thin reserved band on the inner lip, thick reserved band on the body below the shoulder.
Selected Comparanda: Semeraro 1997, 150–51, Fig. 96, No. 298 (Oria, end 6th c. BC); Lo Porto 1981b, 297–99, Figs. 5.1 and 8.2–5 (Metaponto, votive deposit at contrada Crucinia)
Dating: 550–500 BC.

1:2

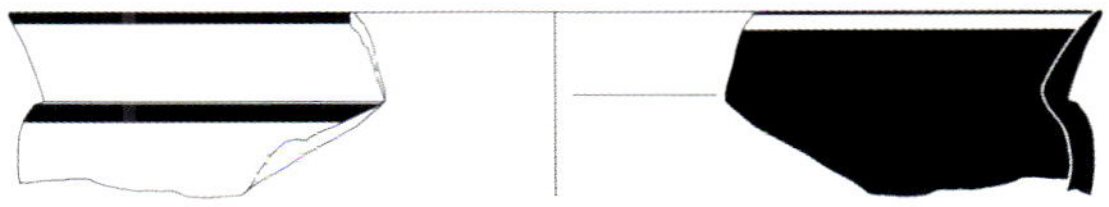

14. AFW 584-01 Rim (4 examples)
Dimensions: rim diam. 14.5 h 2.5
Reference Sherd Description: Out-turned, tapered rim, marked offset with shoulder. Dilute, matte reddish-brown slip, thin reserved band on the slipped interior, reserved exterior with thin band of slip on the lip and on the shoulder beneath the articulated junction with rim.
Selected Comparanda: Van Compernolle 2000, Fig. 2, No. 3 (Metaponto).
Dating: 550–500 BC.

Corinthian fine ware, Ionic type B2 cups cont.

1:1

1:2

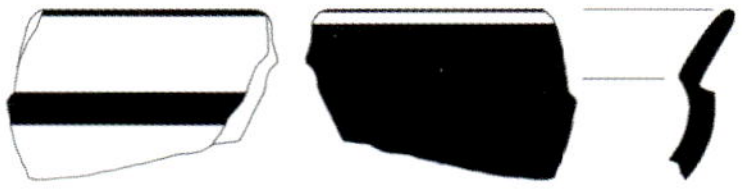

15. AFW 454-04 Rim (7 examples)
Dimensions: w 3.6 h 2.3
Reference Sherd Description: Relatively thick-walled with rounded, slightly out-turned lip. Pronounced offset between rim and shoulder demarcated by a relatively deep concave step. Dilute, matte black slip, reserved band on the inner lip, reserved exterior with thin band on the outer lip and thicker band on the upper shoulder and junction with rim.
Selected Comparanda: Adamesteanu 1980, 278, Fig. 291 (Metaponto, proprietà Favale); Lo Porto 1981b, 297–99, Figs. 5.1, 8.2–5 (Metaponto, votive deposit at contrada Crucinia).
Dating: 550–500 BC.

1:2

16. AFW 477-55 Rim (single example)
Dimensions: rim diam. 14.9 h 4.7 w 5.2
Reference Sherd Description: Slightly flared lip and offset with the wide, tapering bowl. Dilute, matte black slip, reserved band on inner rim, reserved band on the exterior on the shoulder at the level of the handle.
Selected Comparanda: *Necropoleis*, 697–98, T 266-8 (Pantanello, pottery lot outside T. 266); Liseno 2004, 75, No. 15, Pl. XXXVII.c–d (Metaponto, votive deposit at Favale, second half of 6th c. BC).
Dating: 550–500 BC.

1:2

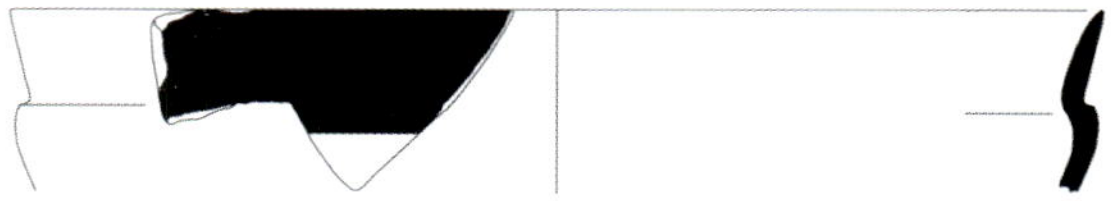

17. AFW 524-03 Rim (30 examples)
Dimensions: rim diam. 14.5 w 4.8 h 2.4
Reference Sherd Description: Tapered rim with marked offset with shoulder. Thick, consistent, matte black slip, completely slipped interior, exterior slipped to lower shoulder with reserved body.
Selected Comparanda: *Necropoleis*, 697–98, T 266-8 (Pantanello, pottery lot outside T. 266); Liseno 2004, 75, No. 15, Pl. XXXVII.c–d (Metaponto, votive deposit at Favale, second half of 6th c. BC).
Dating: 550–500 BC.

1:2

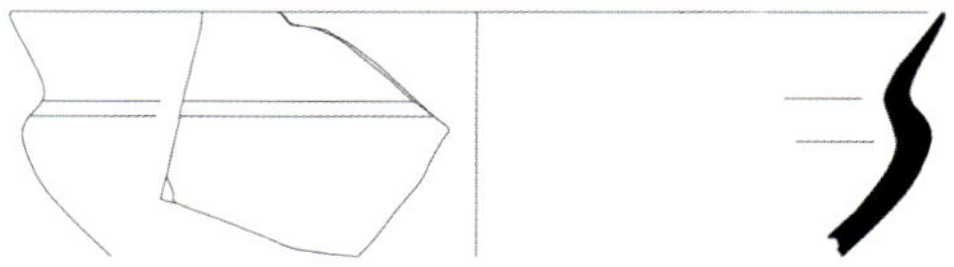

18. AFW 285-13 Rim (2 examples)
Dimensions: rim diam. 12.4 w 3.8 h 3.3
Reference Sherd Description: Very thin, tapered rim with elongated, curved junction with relatively tightly curved shoulder leading into a straight-walled lower body. Thick, matte reddish-brown slip, almost entirely abraded, with traces of slip on both the interior and exterior of the rim and body—the form was probably entirely slipped.
Dating: 550–500 BC.

Corinthian fine ware, Ionic type B2 cups cont.

1:1

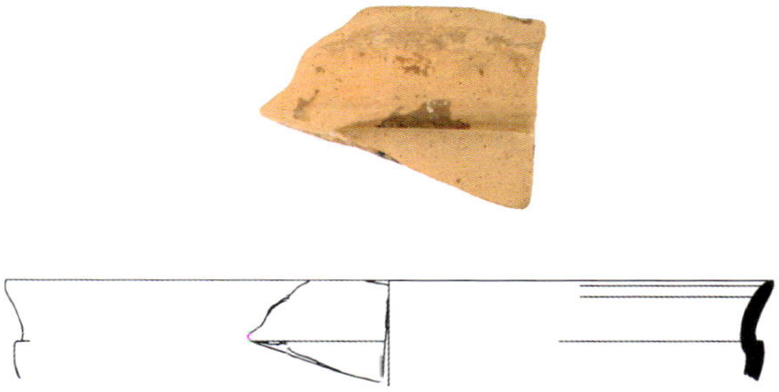

1:2

19. AFW 285-23 Rim (2 examples)
Dimensions: rim diam. 14.8 h 2
Reference Sherd Description: Flared rim distinguished from the shoulder by a pronounced grooved offset. Dilute, matte black slip, worn on the exterior with a thin reserved band below the lip.
Selected Comparanda: *Sibari* II 263–64, 279, No. 165, Figs. 279 and 375 (Sybaris, Parco del Cavallo, Stombi).
Dating: 550–500 BC.

1:2

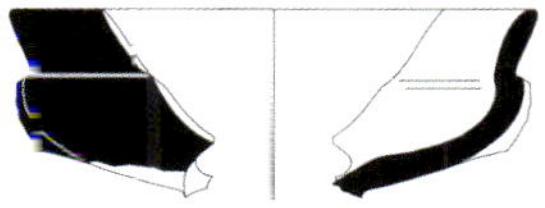

20. AFW 155-19 Rim with handle terminus (2 examples)
Dimensions: rim diam. 7.0 w 5.3 h 2.6
Reference Sherd Description: Relatively thick, rounded rim distinguished from the body by an offset above shallow body. Shoulder-mounted handle with circular section, oversized in relation to the small-scale body of the vessel. Dilute, matte red slip, entirely slipped interior, reserved lower body showing dribbles of gloss.
Selected Comparanda: Liseno 2004, 75, No. 17, Pl. XXXVII.g (Metaponto, votive deposit at Favale, second half of 6th c. BC).
Dating: 550–500 BC.

1:2

21. AFW 477-44 Base (72 examples)
Dimensions: base diam. 5.3 h 1.8
Reference Sherd Description: Conical foot with flat resting surface. Dilute, matte black slip, exterior slipped with worn slip on outer edge of resting surface, slipped underside with reserved tondo.
Selected Comparanda: Adamesteanu 1980, 279, Fig. 292 e–f (Metaponto, proprietà Favale); Liseno 2004, 71, Pl. XXXVII.h (Metaponto, votive deposit at Favale, second half of 6th c. BC).
Dating: 550–500 BC.

1:2

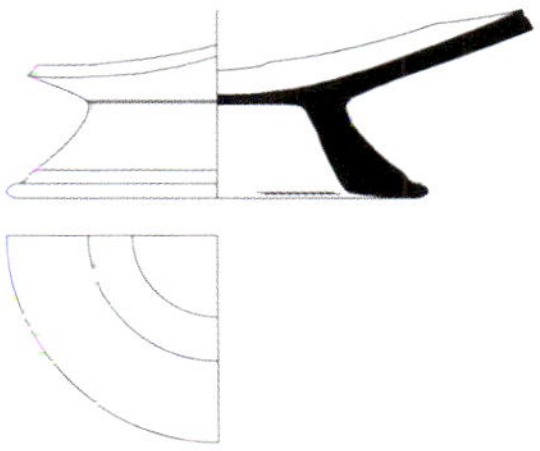

22. AFW 070-07 Base (6 examples)
Dimensions: base diam. 5.6 w 7.0 h 2.55
Reference Sherd Description: Conical foot with resting-surface and fluting at the base of the foot. Dilute, matte black slip, completely slipped exterior, reserved underside.
Selected Comparanda: Adamesteanu 1980, 279, Fig. 292.e–f (Metaponto, proprietà Favale); Liseno 2004, 71, Pl. XXXVII.h (Metaponto, votive deposit at Favale, second half of 6th c. BC).
Dating: 550–500 BC.

Corinthian fine ware, Ionic type B2 cups cont.

1:2

23. AFW 397-53 Base (9 examples)
Dimensions: base diam. 5.5 h 1.4
Reference Sherd Description: Conical foot with flat resting surface and concave exterior. Thick, consistent matte black slip, completely slipped exterior with reserved underside.
Selected Comparanda: Lo Porto 1981b, 297, Fig. 5.1 (Metaponto, votive deposit at contrada Crucinia); *Kaulonia* I, 48, Fig. 30, Nos. 48 and 52 (Kaulonia, North Fortification).
Dating: 550–500 BC.

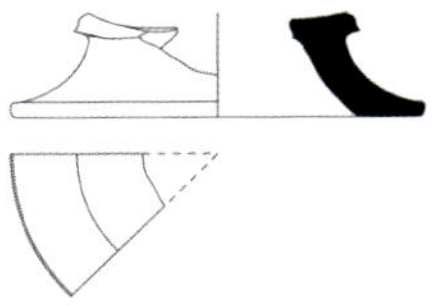

1:2

24. AFW 155-01 Base (3 examples)
Dimensions: base diam. 6.5 h 1.8
Reference Sherd Description: Wide, conical foot with resting-surface inclined towards the interior. Thick, consistent, matte black slip, completely slipped exterior, reserved underside. Slipped center of floor creating a circle.
Selected Comparanda: *Kaulonia* I, 46–47, Fig. 29, Nos. 26–27 (Kaulonia, North Fortification); for the decoration on the floor of the bowl see Liseno 2004, 71, Pl. XXXVII.h (Metaponto, votive deposit at Favale, second half of 6th c. BC).
Dating: 550–500 BC.

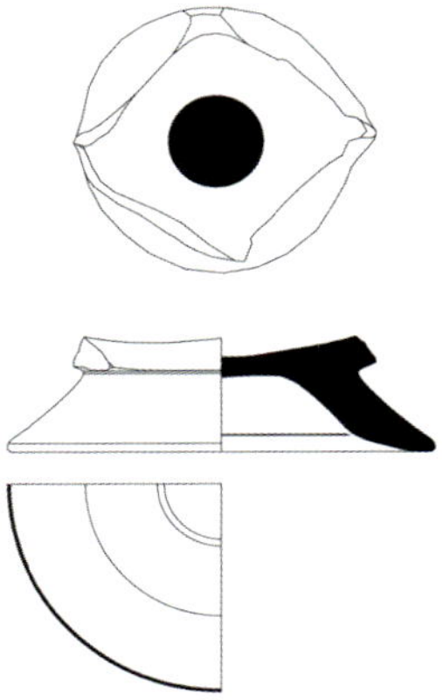

1:2

25. AFW 477-45 Base (single example)
Dimensions: base diam. 3.85 h 1.8
Reference Sherd Description: Small conical foot with convex exterior profile and flat resting-surface; floor of the bowl is perforated. Dilute, matte black slip, exterior slipped, floor slipped with reserved tondo.
Selected Comparanda: *Sibari* II 263–64, 330, 364, Nos. 160 and 167, Figs. 280, 356, and 386 (Sybaris, Parco del Cavallo, Stombi); Liseno 2004, 71, Pl. XXXVII.h (Metaponto, votive deposit at Favale, second half of 6th c. BC).
Dating: 550–500 BC.

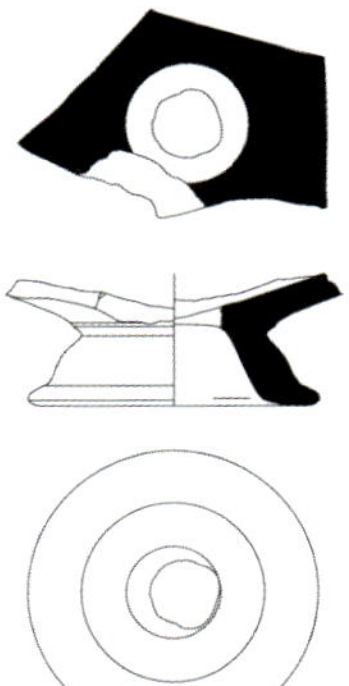

1:2

26. AFW 477-06 Base (single example)
Dimensions: base 4.0 h 1.5
Reference Sherd Description: Small conical foot with convex exterior profile and fluting on the rim; flat resting-surface. Dilute, matte reddish-brown slip, exterior slipped, floor slipped with reserved tondo.
Selected Comparanda: *Sibari* II 263, 330, 364, Nos. 160 and 167, Figs. 356 and 386 (Sybaris, Parco del Cavallo, Stombi).
Dating: 550–500 BC.

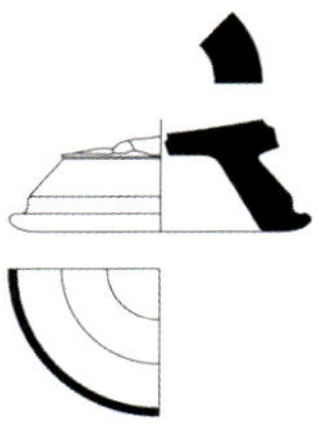

Skyphoi with Offset Rims

1:2

27. AFW 477-33 Shoulder (2 examples)
Dimensions: max. diam. 15.0 h 3.2 w 5.2
Reference Sherd Description: Dilute, matte brownish-black slip, reserved band on the exterior at the level of the handle decorated with three vertical brush strokes, small band of slip on the offset, slipped body and interior.
Selected Comparanda: Liseno 2004, 74, No. 6, Pl. XXXV.g–h (Metaponto, votive deposit at Favale); Lo Porto 1973, 171, Pl. XIX.3.1 (Pisticci, T 4); Lo Porto 1981b, 304, No. 102, Fig. 14.7 (Metaponto, votive deposit at contrada Crucinia).
Dating: 550–500 BC.

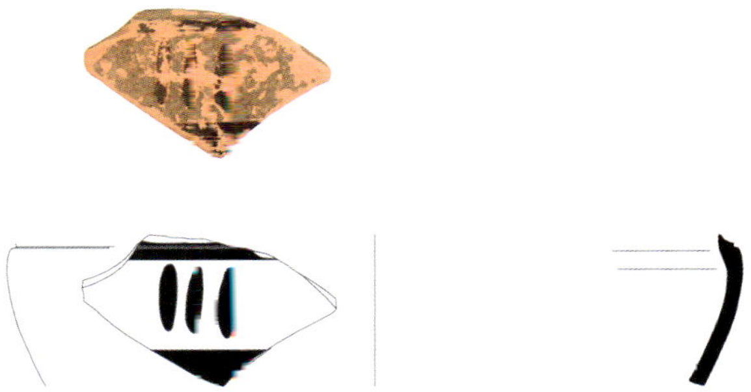

1:2

28. AFW 068-01 Rim and handle terminus (single example)
Dimensions: rim diam. 15.0 h 3.0 w 6.7
Reference Sherd Description: Relatively short, outturned rim; deep bowl with rounded shoulder and part of the handle terminus. Dilute, matte black slip, mostly worn; offset probably slipped, reserved band on the lip and on the body in the handle zone.
Selected Comparanda: Liseno 2004, 73, No. 5, Pl. XXXV.f (Metaponto, Favale votive deposit); Lo Porto 1981b, 304, No. 94, Fig. 14.1 (Metaponto, votive deposit at contrada Crucinia). For the form of the expanded bowl, see De La Genière 1971, Fig. 46 (Amendolara, Uomo Morto necropolis, T. 17 bis 1).
Dating: 550–500 BC.

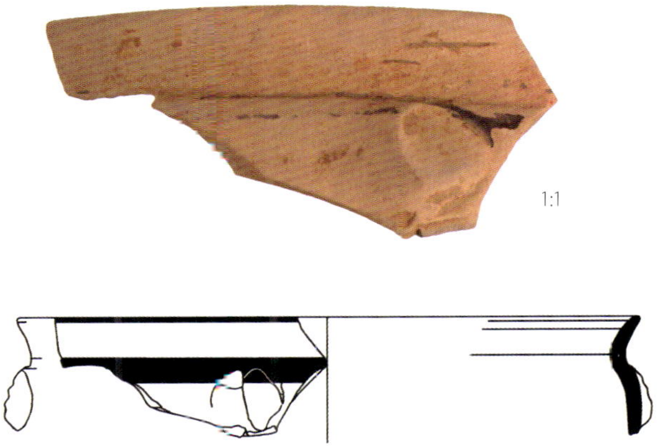

1:1

1:2

29. AFW 622-03 Rim (2 examples)
Dimensions: rim diam. 15.5 h 2.9 w 3.2
Reference Sherd Description: Distinct lip with flattened internal face above a deep body with slightly rounded shoulder. Dilute, matte black slip; rim and junction with shoulder slipped on the exterior, entirely slipped interior.
Selected Comparanda: Macnamara 1983, 329, No. 173 (Cozzo Presepe, 550–480 BC). For the form of the expanded bowl, see De La Genière 1971, Fig. 45 (Amendolara, Uomo Morto necropolis, T. 17 bis 1).
Dating: 550–500 BC.

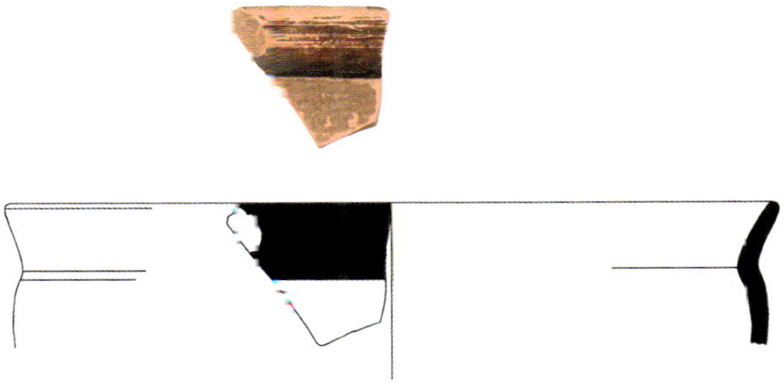

1:2

30. AFW 397-02 revisit 90 Rim (single example)
Dimensions: rim diam. 14.95 h 2.4 w 4.8
Reference Sherd Description: Near-vertical rim with pronounced convex profile distinguished from the near-vertical shoulder by a slight offset. Terminus of circular-sectioned handle. Dilute, matte reddish-brown slip; slipped rim to upper shoulder, slipped interior, almost completely worn.
Selected Comparanda: For the convex lip see De La Genière 1971, 464–65, Figs. 37 and 46 (Amendolara, Uomo Morto necropolis, T. 17 bis 1).
Dating: 550–500 BC.

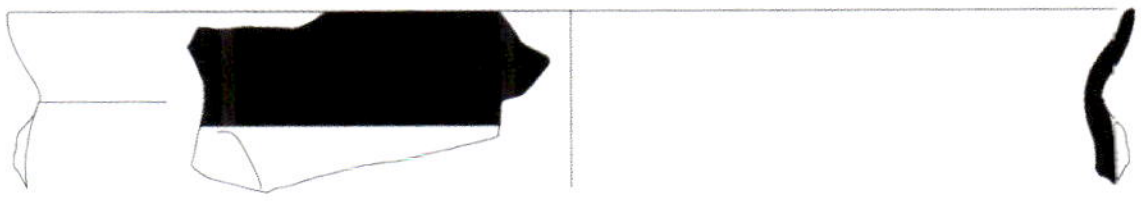

Skyphoi with offset rims cont.

1:2

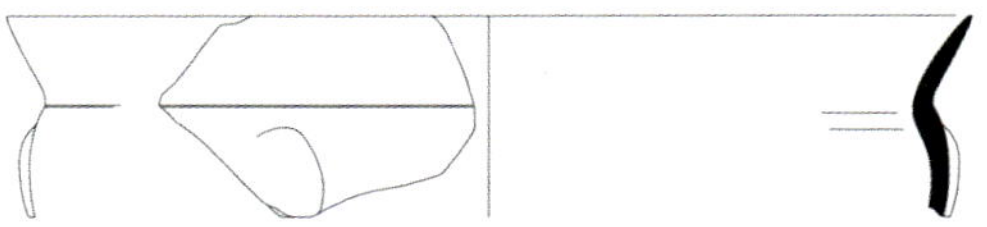

31. AFW 362-01 Rim with handle terminus (2 examples)
Dimensions: rim diam. 12.8 w 4.2 h 2.8
Reference Sherd Description: Thin, tapered rim with rounded junction between rim and shoulder forming a continuous curve rather than a sharp articulation. Traces of dilute, reddish-brown slip, mostly abraded, apparently reserved with traces of slip visible on the exterior at the junction of rim and shoulder.
Dating: 550–500 BC.

1:2

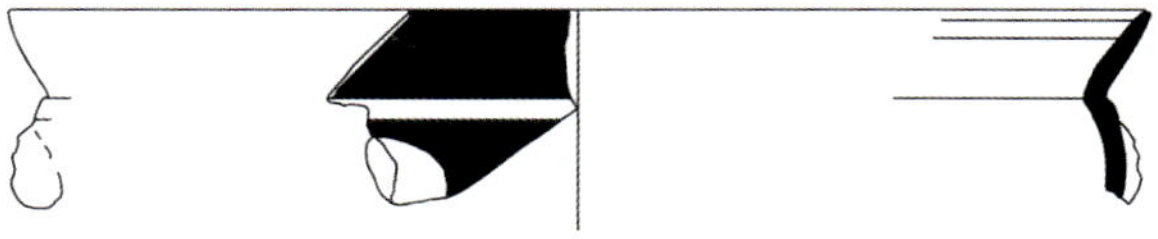

32. AFW 397-01 revisit 90 Rim (10 examples)
Dimensions: rim diam. 15.2 h 2.7 w 5.1
Reference Sherd Description: Straight, tapered rim strongly out-turned, distinguished from the bowl by a reserved offset; gently rounded shoulder with rod-shaped attachment of the handle. Dilute, opaque reddish-brown gloss on the exterior, traces of gloss on the lip and around the attachment of the handle; thin reserved band on the slipped interior.
Selected Comparanda: De La Genière 1971, 466, Figs. 44 and 46 (Amendolara, Uomo Morto necropolis, T. 20 bis 2); Macnamara 1983, 328–29, No. 170, Fig. 111 (Cozzo Presepe, 550–480 BC).
Dating: 550–500 BC.

1:2

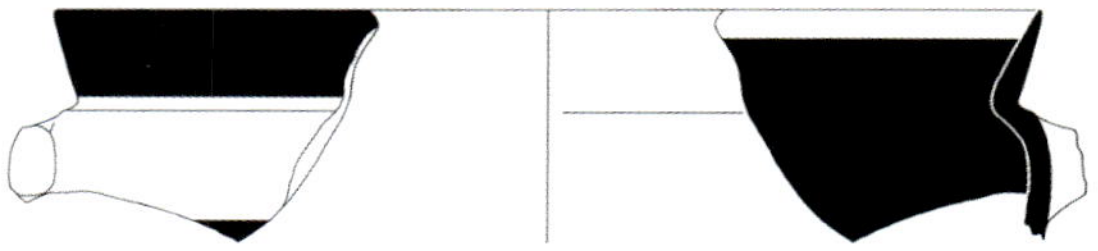

33. AFW 622-01 Rim with handle terminus (7 examples)
Dimensions: rim diam. 13.1 h 3.15
Reference Sherd Description: Straight, tapered rim distinguished from the bowl by a marked offset; pronounced shoulder with rod-shaped handle attachment. Dilute, matte black slip, reserved shoulder and junction with rim, slipped lower body; slipped interior with reserved band on the interior of the lip.
Selected Comparanda: Lo Porto 1988–1989, 362, Fig. 66, No. 5 (Montescaglioso, T. 2, end 6th–early 5th c. BC); Macnamara 1983, 328–29, No. 172 (Cozzo Presepe).
Dating: 550–500 BC.

1:2

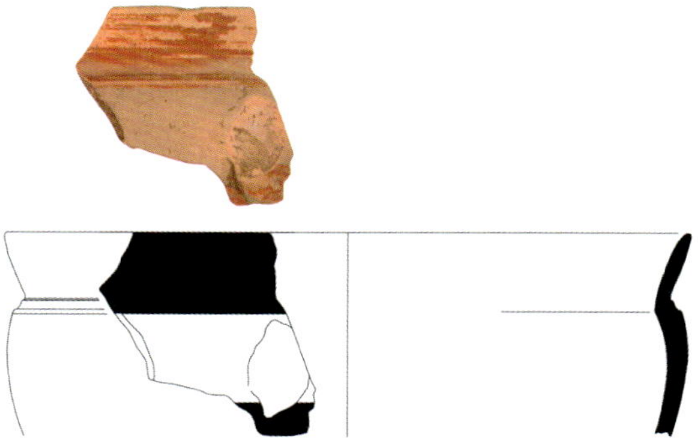

34. AFW 523-15 Rim (single example)
Dimensions: rim diam. 13.9 h 4.2 w 4.4
Reference Sherd Description: Rounded, thickened rim distinguished from the bowl by a marked offset articulated by two grooved steps above slightly rounded shoulder. Dilute, matte reddish slip, thick reserved band on the shoulder at the level of the handle, slipped interior.
Selected Comparanda: Macnamara 1983, 328–29, No. 172, Fig. 111 (Cozzo Presepe).
Dating: 550–500 BC.

1:2

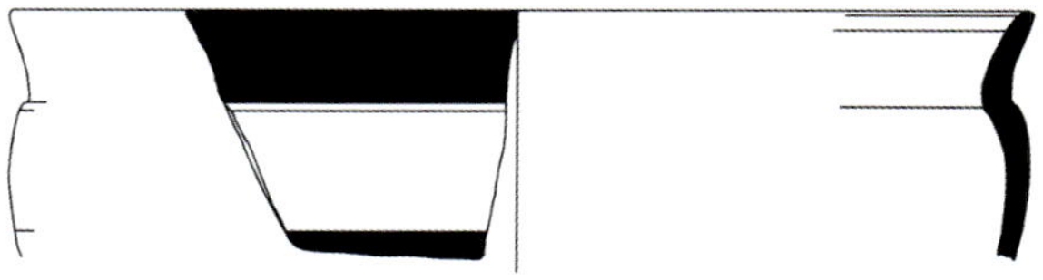

35. AFW 071-01 Rim (single example)
Dimensions: rim diam. 13.8 h 3.8 w 4.9
Reference Sherd Description: Tapered rim distinguished from the bowl by an offset; slightly rounded shoulder; conical bowl. Dilute, opaque yellowish-red slip, reserved band at the level of the shoulder; slipped interior with reserved band on the interior of the lip.
Selected Comparanda: Semeraro 1997, 227–28, No. 785, Fig. 197 (Oria); Liseno 2004, 74, No. 7, Pl. XXXVI.a–b (Metaponto, votive deposit at Favale, 575–480 BC).
Dating: 550–500 BC.

Skyphoi with offset rims cont.

1:2

36. AFW 477-56 Rim (single example)
Dimensions: rim diam. 15 h 3.7
Reference Sherd Description: Marked offset at the junction of rim and slightly rounded shoulder; conical bowl. Dilute, matte black slip with reserved strip at the level of the handle; slipped interior with reserved rim.
Selected Comparanda: Macnamara 1983, 328–29, No. 172, Fig. 111 (Cozzo Presepe); Lo Porto 1981b, 365, Fig. 86 (Metaponto, contrada Ricotta necropolis, T. 19, early 5th c. BC); Lo Porto 1988–1989, 362, Fig. 66, No. 5 (Montescaglioso, T. 2, end 6th–early 5th c. BC).
Dating: 550–500 BC.

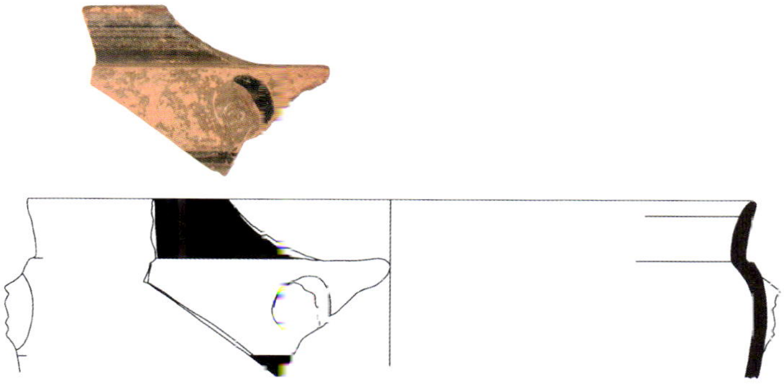

1:2

37. AFW 397-49 Base (3 examples)
Dimensions: base diam. 6.05 h 2.0
Reference Sherd Description: Ring foot with flat resting-surface, slightly concave exterior profile; conical bowl. Perhaps reserved, or slip completely abraded.
Selected Comparanda: Macnamara 1983, 329, No. 171 (Cozzo Presepe); Liseno 2004, 74, No. 9, Pl. XXXVI.d (Metaponto, votive deposit at Favale, 550–480 BC).
Dating: 550–480 BC.

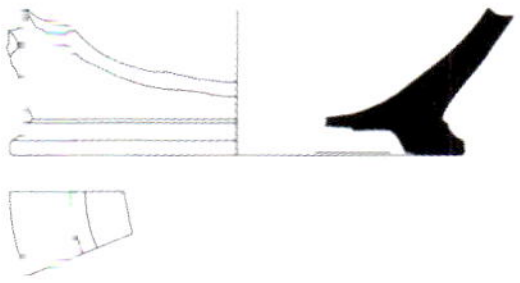

1:2

38. AFW 523-29 Base (single example)
Dimensions: base diam. 7.65 h 2.3
Reference Sherd Description: Ring foot with flat resting-surface; fluting at the attachment of the foot on both interior and exterior. Slip completely worn.
Selected Comparanda: D'Andria 1980a, 366, Fig. 13, No. 44a (Metaponto, kerameikos); Macnamara 1983, 329, No. 171 (Cozzo Presepe); Liseno 2004, 74, No. 9, Pl. XXXVI.d (Metaponto, votive deposit at Favale, 550–480 BC).
Dating: 550–480 BC.

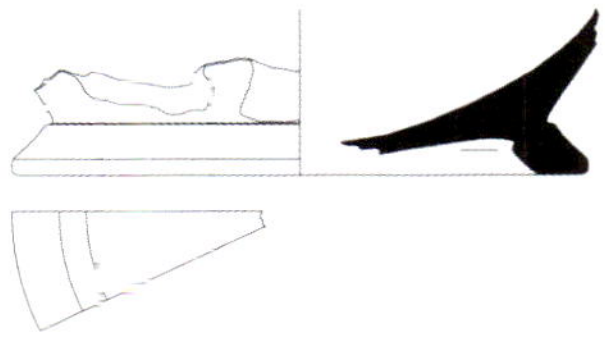

1:2

39. AFW 477-50 Base (7 examples)
Dimensions: base diam. 6.2 h 2.5
Reference Sherd Description: Ring foot with resting-surface inclined towards the interior. Dilute, matte reddish-brown slip; foot slipped on the exterior; body slipped on both the interior and the exterior.
Selected Comparanda: Macnamara 1983, 329, No. 172 (Cozzo Presepe).
Dating: 550–480 BC.

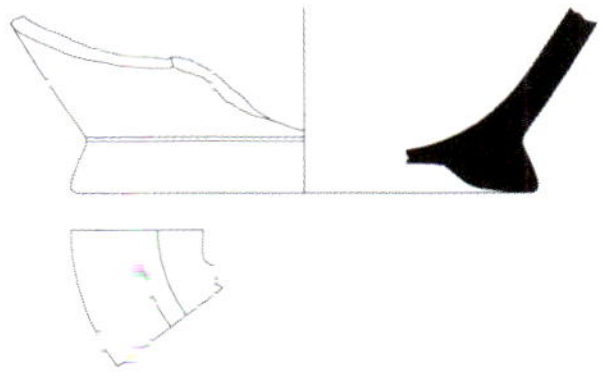

1:2

40. AFW 397-51 Base (single example)
Dimensions: base diam. 8.8 h 1.3
Reference Sherd Description: Low, conical foot with rounded exterior edge and flat resting-surface. Consistent, thick, matte black slip, partially worn; exterior of foot slipped.
Selected Comparanda: Liseno 2004, 74, No. 10, Pl. XXXVI.f (Metaponto, votive deposit at Favale).
Dating: 550–480 BC.

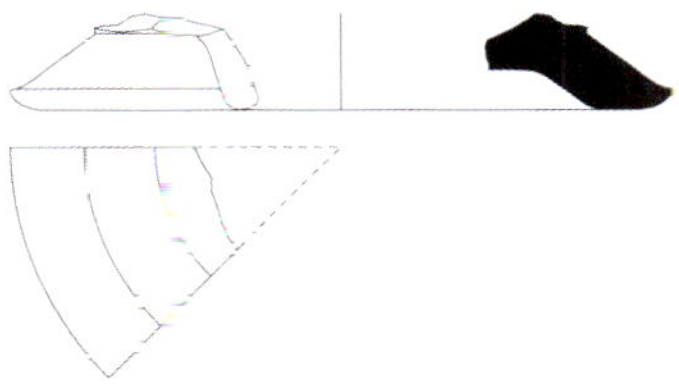

Archaic Line-Decorated Wares

1:2

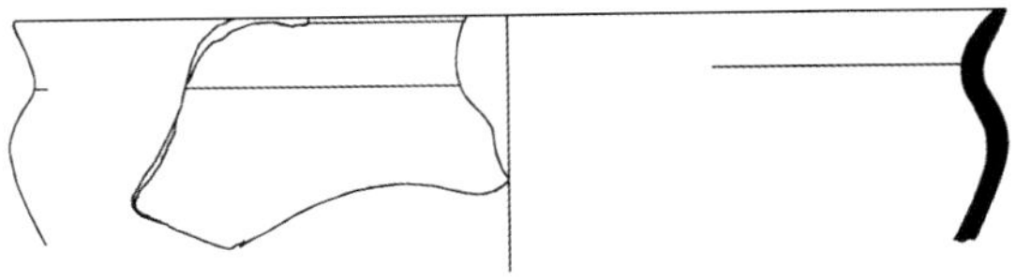

41. AFW 285-51 Cup rim (3 examples)
Dimensions: rim diam. 14 h 3.3 w 5.2
Reference Sherd Description: Flaring lip with concave profile and sharp break of angle forming the junction with the lower body. Dilute, matte reddish slip, mostly worn. Perhaps slipped on the exterior or with a reserved band at the shoulder; interior glossed.
Selected Comparanda: *Kaulonia* I, 49, 53, Fig. 31, No. 62 (Kaulonia, North Fortification, 6th c. BC); Munzi 2007, 126–27, Fig. 15, No. 25040.1021 (Kyme).
Dating: 6th c. BC.

1:2

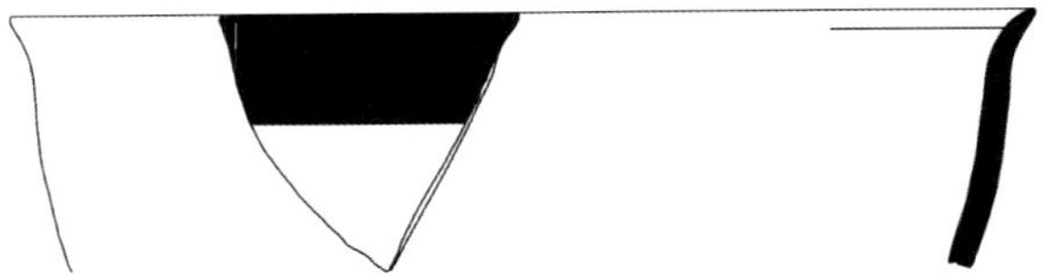

42. AFW 286-04 Panionion-type cup rim (single example)
Dimensions: rim diam. 15 h 3.9 w 4.4
Reference Sherd Description: Lip out-turned with profile that continues the line of the bowl; rounded rim; deep bowl. Thick, consistent, matte reddish-brown slip, thick reserved band on the mid-body, interior slipped with thin reserved band near the lip.
Selected Comparanda: Macnamara 1983, 329–30, No. 178, Fig. 111 (Cozzo Presepe); Lo Porto 1981b, 304, No. 93 (Metaponto, votive deposit at contrada Crucinia, middle of 6th c. BC).
Dating: 550–500 BC.

1:2

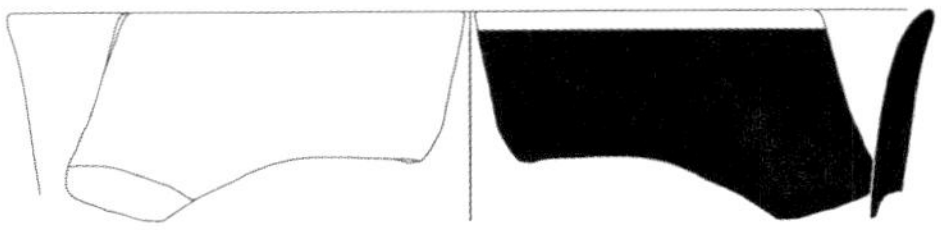

43. AFW 477-28 Panionion-type cup rim (8 examples)
Dimensions: rim diam. 12.3 h 2.85 w 5.3
Reference Sherd Description: Tapered lip with relatively straight wall. Thick, consistent reddish slip with reserved band on the interior of the lip.
Selected Comparanda: Macnamara 1983, 329–30, No. 178, Fig. 111 (Cozzo Presepe); Lo Porto 1981b, 304, No. 93 (Metaponto, votive deposit at contrada Crucinia, middle of 6th c. BC).
Dating: 550–500 BC.

1:2

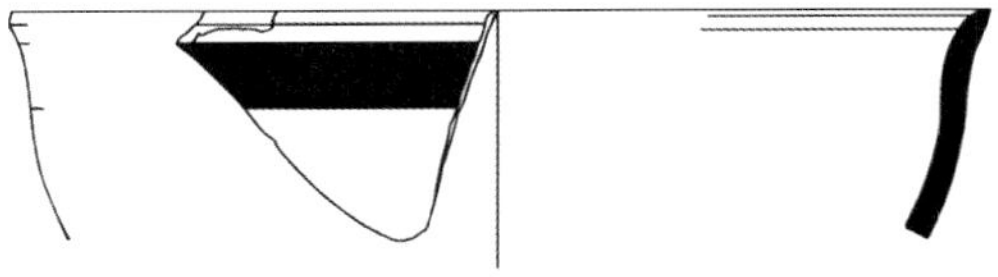

44. AFW 509-02 Panionion-type cup rim (single example)
Dimensions: rim diam. 13.0 h 3.1 w 4.2
Reference Sherd Description: Dilute, matte reddish-brown slip, abraded in many places; slipped band on the upper rim on the exterior, reserved body; interior slipped with small reserved strip beneath the rim.
Selected Comparanda: Macnamara 1983, 329–30, No. 178, Fig. 111 (Cozzo Presepe); Lo Porto 1981b, 304, No. 93 (Metaponto, votive deposit at contrada Crucinia, middle 6th c. BC).
Dating: 550–500 BC.

1:2

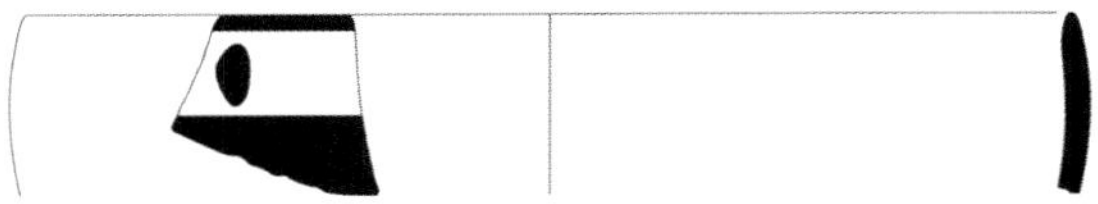

45. AFW 477-32 Skyphos rim (single example)
Dimensions: rim diam. 13.6 h 2.45 w 2.75
Reference Sherd Description: Indistinct, slightly inclined rim. Thick, consistent, matte black slip; slipped exterior with thick reserved band on the upper body at the line of the handle, bearing a vertical painted stroke.
Selected Comparanda: Cuozzo and D'Andrea 1991, 83, Fig. 8 (type 37A) (necropolis at Pontecagnano, last quarter of 6th–second quarter of 5th c. BC); Adamesteanu et al. 1971, 53, Pl. XII (Alianello, T. 2, 5th c. BC); *Locri* II, 76 (Centocamere), Rubinich 1992, 89 and 96–97, Nos. 6–15, Pl. LIV (Locri, Marasà Sud; second half–end of 6th c. BC).
Dating: 550–500 BC.

II. Black-Gloss Fine Ware
Introduction

K. Swift and E. Vittoria

Black-gloss fine wares constitute the primary source of ceramic evidence from the Metaponto survey for the Classical and Hellenistic periods, beginning with the large-scale production of these wares in the early 5th century BC and continuing into the Late Hellenistic period and the production of grey wares.[67]

The corpus of black gloss from the Bradano-Basento survey comprises approximately 8,750 sherds, of which 285 black-gloss form-sherds are presented in the catalog. These represent the de facto reference collection for the co-identification and quantification of similar examples from the bulk of the survey assemblages. During the primary recording of the survey assemblages, each form recorded and quantified for integration into the GIS was co-identified with a specific reference sherd—each illustrated in the catalog—to allow for subsequent refinements in classifications and typology necessitated by a continuously-developing field of study like that of southern Italian black-gloss fine ware.[68]

Quantification

K. Swift and E. Vittoria

The ICA survey recovered a black-gloss corpus of over 8,000 sherds. The primary aim of the quantification presented here is to provide estimates of the relative importance of different forms for dating as they are found in the Metapontine chora, i.e. the nature of the black-gloss evidence from the survey.[69] Quantification by shape is shown in Table 6.3; quantification of the pieces presented in the Catalog of Black-Gloss Fine Ware is shown in Table 6.4, which gives the number of pieces co-identified with examples from the catalog. Key features of the quantification are discussed by form in the following section.

Black-gloss form	Total	% all BG	% identifiable forms
Skyphoi with offset rims	22	0.25%	1.19%
Cups with offset rims	14	0.16%	0.76%
Cup-skyphoi	36	0.41%	1.94%
Kylikes	17	0.19%	0.92%
A-type skyphoi	230	2.63%	12.41%
C-type skyphoi	671	7.66%	36.21%
Stemless kylikes	23	0.26%	1.24%
Hellenistic cup-skyphoi	3	0.03%	0.16%
Bolsals	5	0.06%	0.27%
Stemmed cups	19	0.22%	1.03%
Ring-handled cups	112	1.28%	6.04%
Kantharoi	2	0.02%	0.11%
Cups/one-handlers	71	0.81%	3.83%
Cups	7	0.08%	0.38%
Dishes	296	3.38%	15.97%
Bowls	54	0.62%	2.91%
Small bowls	79	0.90%	4.26%
Concave-convexes	21	0.24%	1.13%
Salt cellars	46	0.53%	2.48%
Kraters	13	0.15%	0.70%
Deep bowls	1	0.01%	0.05%
Lids	7	0.08%	1.24%
Open forms (unidentified)	5,024	57.38%	n/a
Mugs	13	0.15%	0.70%
Oinochoai	2	0.02%	0.11%
Amphorae	11	0.13%	0.59%
Pelikai	25	0.29%	1.35%
Hydriai	3	0.03%	0.16%
Lebetes gamikoi	18	0.21%	0.97%
Gutti	2	0.02%	0.11%
Squat lekythoi	23	0.26%	1.24%
Net-decorated lekythoi	3	0.03%	0.16%
Pyxides	4	0.05%	0.22%
Closed forms (unidentified)	1,388	15.85%	n/a
Unidentified	490	5.60%	
All black-gloss forms	**8,755**	**100%**	**1,853**

Table 6.3 Quantification and relative frequencies of black-gloss shapes from the ICA survey.

[67] Unlike in many other black-gloss-producing regions of Italy and the Mediterranean, regional production of grey wares in Magna Grecia largely replaced that of Hellenistic black gloss in the 2nd c. BC, though initially continuing many of the popular forms in black gloss of the Hellenistic period. For grey ware, see Ch. 7.

[68] Recording and quantification of the black gloss from the survey assemblages, data used in the GIS analyses, was undertaken by E. Vittoria during 2002–03. Subsequent re-examination of the corpus of black-gloss forms, undertaken with the aim of presenting the key reference pieces in a more conventional catalog format, has involved some reclassifications and, in particular, some changes in nomenclature; in many cases this was due to further information and parallels having subsequently come to light following initial recording of the survey assemblages.

[69] Recording and quantification of the black-gloss assemblages from the survey was undertaken by E. Vittoria.

Form	Catalog no.	Description	Count
Skyphos with offset rims	4	Skyphos with offset rim	1
	5	Skyphos with offset rim, reserved band on exterior	5
	6	Skyphos with offset rim	8
	7	Base of skyphos with offset rim	8
		Skyphos rims	14
		Skyphos bases	8
		All skyphoi with offset rims	**22**
Cup with offset rims	8	Rim of cup with offset rim	14
		All cups with offset rims	**14**
Cup-skyphos	9	Cup-skyphos rim	1
	10	Cup-skyphos rim	1
	11	Cup-skyphos rim	1
	12	Cup-skyphos base	1
	13	Cup-skyphos base	1
	14	Cup-skyphos base	8
	15	Cup-skyphos base	10
	16	Cup-skyphos base	3
	17	Cup-skyphos base	3
		Unattributed/unidentifiable cup-skyphos rim	1
		Unattributed/unidentifiable cup-skyphos bases	5
		Unattributed/unidentifiable cup-skyphos handle	1
		All cup-skyphos rims	4
		All cup-skyphos bases	31
		All cup-skyphos handles	1
		All cup-skyphoi	**36**
Kylix	1	Attic kylix rim	1
	18	Kylix rim	7
	19	Kylix stem	1
	20	Kylix foot	1
	21	Kylix foot	5
		Unattributed/unidentifiable kylix handles	2
		All kylix rims	8
		All kylix bases	6
		All kylix handles	2
		All kylix body sherds	1
		All kylikes	**17**
A-type skyphos	23	A-type skyphos base	2
	24	A-type skyphos base	2
	25	A-type skyphos rim	29
	26	A-type skyphos base	14
	27	A-type skyphos base	14
	28	A-type skyphos base	8
	29	A-type skyphos base	1
	30	A-type skyphos rim	27
	31	A-type skyphos base	12
	32	A-type skyphos base	1
A-type skyphos	33	A-type skyphos base	10
	34	A-type skyphos base	18
	35	A-type skyphos rim	12
	36	A-type skyphos rim	9
	37	A-type skyphos rim and handle	3
	38	A-type skyphos rim and handle terminus	1
	39	A-type skyphos rim	26
	40	A-type skyphos base	16
	41	A-type skyphos base	5
	42	A-type skyphos base	2
	43	A-type skyphos base	1
	44	A-type skyphos base	6
	45	A-type skyphos base	6
		Unattributed/unidentifiable A-type skyphos bases	3
		Unattributed/unidentifiable A-type skyphos body sherds	2
		All A-type skyphos rims	107
		All A-type skyphos bases	121
		All A-type skyphos body sherds	2
		All A-type skyphoi	**230**
C-type skyphos	46	C-type skyphos base	3
	47	C-type skyphos rim	116
	48	C-type skyphos rim and upper body.	1
	49	C-type skyphos base	1
	50	C-type skyphos base	17
	51	C-type skyphos base	6
	52	C-type skyphos base	27
	53	C-type skyphos base	22
	54	C-type skyphos base	15
	55	C-type skyphos base	13
	56	C-type skyphos rim and upper body	8
	57	C-type skyphos base	1
	58	C-type skyphos base	22
	59	C-type skyphos base	5
	60	C-type skyphos base	5
	61	C-type skyphos base	7
	62	C-type skyphos base	15
	63	C-type skyphos base with perforated floor	1
	64	C-type skyphos rim and handle	144
	65	C-type skyphos rim	3
	66	C-type skyphos rim and upper body with handle terminus	1
	67	C-type skyphos rim	114
	68	C-type skyphos rim and body with handle terminus	4

Table 6.4 Quantification of all cataloged black-gloss forms including all catalog reference sherds with the number of co-identified fragments from the survey assemblages.

Form	Catalog no.	Description	Count
C-type skyphos	69	C-type skyphos rim	19
	70	C-type skyphos base	19
	71	C-type skyphos base	17
	72	C-type skyphos base	1
	73	C-type skyphos base	1
	74	C-type skyphos base	10
	75	C-type skyphos base	11
	76	C-type skyphos base	5
	77	C-type skyphos base	11
	78	C-type skyphos base	1
	79	C-type skyphos base	5
	80	C-type skyphos base	2
	81	C-type skyphos base	2
		Unattributed/unidentifiable C-type skyphos rims	12
		Unattributed/unidentifiable C-type skyphos bases	4
		All C-type skyphos rims	422
		All C-type skyphos bases	249
		All C-type skyphoi	**671**
Stemless kylix	2	Attic stemless kylix rim	1
	82	Stemless kylix base	1
	83	Stemless kylix base	2
	84	Stemless kylix base	4
	85	Stemless kylix floor	1
	86	Stemless kylix base	1
		Unattributed/unidentifiable stemless kylix rims	2
		Unattributed/unidentifiable stemless kylix base	1
		Unattributed/unidentifiable stemless kylix body sherds	10
		All stemless kylix rims	3
		All stemless kylix bases	9
		All stemless kylix body sherds	11
		All stemless kylikes	**23**
Hellenistic cup-skyphos	87	Cup-skyphos base	1
	88	Hellenistic cup-skyphos base	1
	89	Cup-skyphos base	1
		All Hellenistic cup-skyphoi	**3**
Bolsal	90	Bolsal base	5
		All bolsals	**5**
Stemmed cup	91	Stemmed cup rim with handle terminus and upper part of stem	10
	92	Stemmed cup rim	4
	93	Stemmed cup rim with handle terminus	2

Form	Catalog no.	Description	Count
Stemmed cup	94	Stemmed cup base	1
	95	Stemmed cup lower body and upper part of stem	1
	96	Stemmed cup base	1
		All stemmed cup rims	16
		All stemmed cup bases	3
		All stemmed cups	**19**
Ring-handled cup	97	Ring-handled cup rim with handle terminus	10
	98	Ring-handled cup rim	21
	99	Ring-handled cup rim with handle terminus	4
	100	Ring-handled cup rim	15
	101	Ring-handled cup base	1
	102	Ring-handled cup base	10
	103	Ring-handled cup base	9
	104	Ring-handled cup base	12
	105	Ring-handled cup base	17
	106	Ring-handled cup base	12
	107	Ring-handled cup base	1
		All ring-handled cup rims	50
		All ring-handled cup bases	62
		All ring-handled cups	**112**
Kantharos	108	Kantharos stem	1
	109	Kantharos wall with vertical ribbing	1
		All kantharoi	**2**
Cup/one-handler	110	Cup/one-handler rim and handle	1
	111	Cup/one-handler rim and handle	1
	112	Cup/one-handler rim	3
	113	Cup/one-handler rim	15
	114	Cup/one-handler rim and handle	2
	115	Cup/one-handler rim	10
	116	Cup/one-handler rim	3
	117	Cup/one-handler rim	7
	118	Cup/one-handler rim	1
	119	Cup/one-handler rim	8
	120	Cup/one-handler rim	1
	121	Cup/one-handler rim	2
	122	Cup/one-handler rim	17
		All cup/one-handlers	**71**
Cup	123	Cup rim	1
	124	Cup rim	4
	125	Cup base	2
		All cup rims	5
		All cup bases	2
		All cups	**7**

Table 6.4 cont.

Form	Catalog no.	Description	Count
Dishes	126	Dish rim	12
	127	Dish rim	1
	128	Dish rim	4
	129	Dish base	9
	130	Dish base	3
	131	Dish rim	2
	132	Dish rim	1
	133	Dish rim	1
	134	Dish rim	28
	135	Dish rim	38
	136	Dish rim	5
	137	Dish base	1
	138	Dish base	22
	139	Dish rim	1
	140	Dish base	3
	141	Dish rim	1
	142	Dish rim	2
	143	Dish rim	1
	144	Dish rim	1
	145	Dish	3
	146	Dish rim	1
	147	Dish rim	3
	148	Dish rim	5
	149	Dish rim	5
	150	Dish rim	3
	151	Dish rim	9
	152	Dish rim	2
	153	Dish rim	28
	154	Dish rim	2
	155	Dish rim	22
	156	Dish rim	1
	157	Dish rim	1
	158	Dish rim	1
	159	Dish rim	5
	160	Dish rim	1
	161	Dish rim	1
	162	Dish rim	1
	163	Dish rim	8
	164	Dish rim	3
	165	Dish rim	1
	166	Dish base	12
	167	Dish base	9
	168	Dish base	5
	169	Dish base	10

Forms	Catalog no.	Description	Count
	170	Dish base	1
	171	Dish floor with stamp	1
		Unattributed/unidentifiable dish rims	8
		Unattributed/unidentifiable dish bases	4
		Unattributed/unidentifiable dish body sherds	4
		All dish rims	212
		All dish bases	79
		All dish body sherds	5
		All dishes	**296**
Bowls	172	Bowl rim	1
	173	Bowl rim	11
	174	Bowl rim	5
	175	Bowl rim	2
	176	Bowl rim	5
	177	Bowl rim	1
	178	Bowl rim	18
	179	Bowl rim	1
	180	Bowl rim	1
	181	Bowl rim	1
	182	Bowl base	1
		Unattributed/unidentifiable bowl rims	6
		Unattributed/unidentifiable bowl bases	1
		All bowl rims	52
		All bowl bases	2
		All bowls	**54**
Small bowls	183	Small bowl rim	2
	184	Small bowl rim	32
	185	Small bowl rim	1
	186	Small bowl rim	1
	187	Small bowl rim	6
	188	Small bowl rim	1
	189	Small bowl rim	6
	190	Small bowl rim	14
	191	Small bowl rim	1
	192	Small bowl rim	4
	193	Small bowl rim	1
	194	Small bowl rim	1
		Unattributed/unidentifiable small bowl rims	9
		All small bowls	**79**

Table 6.4 cont.

Form	Catalog no.	Description	Count
Concave-convex	195	Concave-convex base	1
	196	Concave-convex rim	1
	197	Concave-convex rim	9
	198	Concave-convex rim	1
	199	Concave-convex rim	4
	200	Concave-convex rim	2
	201	Concave-convex rim	2
		Unattributed/unidentifiable concave-convex rim	1
		All concave-convex rims	20
		All concave-convex bases	1
		All concave-convexes	**21**
Salt cellars	202	Salt cellar	28
	203	Salt cellar rim	2
	204	Salt cellar	4
	205	Salt cellar rim	1
	206	Salt cellar base	3
	207	Salt cellar foot	1
	208	Salt cellar base to upper body	1
	209	Salt cellar rim	3
		Unattributed/unidentifiable salt cellar rim	1
		Unattributed/unidentifiable salt cellar base	1
		Unattributed/unidentifiable salt cellar body sherd	1
		All salt cellar rims	39
		All salt cellar bases	6
		All salt cellars	**46**
Krater	22	Archaic krater rim	2
	210	Krater rim and upper body	1
	211	Krater rim	1
	212	Krater rim	1
		Unattributable/unidentified krater rims	8
		All kraters	**13**
Deep bowl	213	Deep bowl rim	1
		All deep bowls	**1**
Lids	214	Lid handle	1
	215	Lid rim	1
		Unattributable/unidentifiable lid rims	2
		Unattributable/unidentifiable lid handles	2
		Unattributable/unidentifiable lid body sherd	1
		All lids	**7**

Forms	Catalog no.	Description	Count
Open forms (unidentified)	216	Open form ring base	1
	217	Open form ring base	2
	218	Open form ring base	31
	219	Open form ring base	2
	220	Open form ring base	4
	221	Open form ring base	3
	222	Open form ring base	1
	223	Open form ring base	1
	224	Open form ring base	1
	225	Open form ring base	3
	226	Open form ring base	2
	227	Open form ring base	3
	228	Open form ring base	2
	229	Open form pedestal base	5
	230	Open form ring base	1
	231	Open form ring base	1
	232	Open form ring base	1
	233	Open form ring base	1
	234	Open form base	1
	235	Open form ring base	1
	236	Open form ring base	3
	237	Open form disk base	1
	238	Open form disk base	2
		Unattributed/unidentified open form rims	16
		Unattributed/unidentified open form bases	63
		Unattributed/unidentified open form handles	83
		Unattributed/unidentifiable open form body sherds	4,789
		All open form rims	16
		All open form bases	136
		All open form handles	83
		All open form body sherds	4,789
		All open forms	**5,024**
Mug	239	Mug rim	1
	240	Mug rim	11
		Unattributed/unidentifiable mug rim	1
		All mug rims	**13**
Oinochoe	241	Oinochoe rim	1
	242	Oinochoe base	1
		Unattributed/unidentifiable oinochoe rim	1
		All oinochoai	**2**

Table 6.4 cont.

Forms	Catalog no.	Description	Count
Amphora	243	Amphora rim	1
	244	Amphora rim	1
	245	Amphora rim	2
	246	Amphora rim	7
		All amphorae	**11**
Pelike	247	Pelike rim	5
	248	Pelike rim	1
	249	Pelike rim	2
	250	Pelike rim	4
	251	Pelike (or perhaps a lekythos?) base	1
	252	Pelike base	1
	253	Pelike neck with handle terminus	3
	254	Pelike rim	2
	255	Pelike rim	2
	256	Pelike rim	1
		Unattributed/unidentifiable pelike bases	3
		All pelike rims	17
		All pelike bases	5
		All pelikai	**25**
Hydria	257	Hydria rim	2
		Hydria body sherd	1
		All hydriai	**3**
Lebes gamikos	258	Lebes gamikos rim	1
	259	Lebes gamikos rim, shoulder and handle terminus	4
	260	Lebes gamikos rim and shoulder	2
	261	Lebes gamikos rim	1
	262	Lebes gamikos body sherd	1
		Unattributed/unidentified lebes gamikos rims	2
		Unattributed/unidentified lebes gamikos bases	4
		Unattributed/unidentified lebes gamikos handles	2
		Unattributed/unidentified lebes gamikos body sherds	1
		All lebes gamikos rims	10
		All lebes gamikos bases	4
		All lebes gamikos handles	2
		All lebes gamikos body sherds	2
		All lebetes gamikoi	**18**
Guttus	263	Guttus rim	1
	264	Guttus fragment with molded decoration; bearded face (Herakles?)	1
		All gutti	**2**

Forms	Catalog no.	Description	Count
Squat lekythos	3	Attic squat-lekythos base	1
	265	Lekythos base	10
	266	Squat-lekythos body with vertical ribbing	2
	267	Squat-lekythos rim, upper neck and handle terminus	1
	268	Squat-lekythos base	2
	269	Squat-lekythos base	3
	270	Squat-lekythos base	1
		Unattributed lekythos bases	3
		All squat lekythoi	**23**
Net-decorated lekythos	271	Net-decorated lekythos, rim, neck, and upper body	1
	272	Net-decorated lekythos base	2
		All net-decorated lekythoi	**3**
Pyxis	273	Pyxis (or perhaps epichysis?) body sherd	2
	274	Pyxis base	1
		Unattributed pyxis body sherd	1
		All pyxides	**4**
Closed form (unidentified)	275	Closed form base	1
	277	Closed form rim	2
	276	Closed form rim	1
	278	Closed form rim	1
	279	Closed form base	1
	280	Closed form base	1
	281	Closed form base	4
	282	Closed form base	2
	283	Closed form base	1
	284	Closed form base and lower body	1
	285	Closed form disk base	1
		Closed form rims (not cataloged)	2
		Closed form body sherds	1,311
		Closed form bases	49
		Closed form handles	10
		All closed form rims	6
		All closed form bases	61
		All closed form handles	10
		All closed form body sherds	1,311
		All closed forms	**1,388**
Unidentified		All unidentified rims	29
		All unidentified handles	398
		All unidentified bases	0
		All unidentified body sherds	63
		All unidentified pieces	**490**

Table 6.4 cont.

The relative frequencies of forms in different periods is primarily a reflection of the number of sites represented in each period, and to a certain extent site function.[70] A straightforward quantification of the black-gloss forms in the survey assemblages without consideration of the frequency and size of survey sites and their individual chronologies is therefore of limited use, but it does allow for an assessment of the relative frequencies of forms in surface assemblages in the chora as a whole, and identification of the most frequent forms so crucial for understanding and interpreting ancient settlement in the chora.

Metapontine Black-Gloss Fabrics

K. Swift

Samples of Metapontine black-gloss fabric examples from the survey, selected to be representative of the common range and variation of fabrics and slips, are presented in the Catalog of Black-Gloss Fabrics at the end of this chapter. Although the examples of black-gloss fabrics shown in the catalog comprise a very small proportion of the number of fine-ware fabrics from the Metapontine chora examined microscopically, they were selected to encompass the chronological range of black gloss, taken from dateable reference sherds shown in the catalog, and include examples from forms dated to the 5th (Catalog of Black-Gloss Fabrics, **Nos. 1, 2,** and **5**), late 4th/early 3rd (**No. 3**), and 3rd centuries BC (**No. 4**).

The short summary here aims to describe the usual features of Metapontine black-gloss fabrics in hand specimen. Metapontine fine-ware fabrics are usually fired hard to very hard with clear fracture, usually semi-conchoidal, to conchoidal in the highest-fired and most consistent examples. Texturally, they are usually finely granular and consistent, with the majority of non-plastic inclusions sufficiently small to constitute part of the matrix.

Typical Metapontine black-gloss fabrics of the Classical to Hellenistic periods tend to appear pale reddish, less frequently yellowish like those of Archaic fine wares. Although fabrics similar to Archaic fine wares continued to be produced (**Nos. 1–2**) Classical and Hellenistic black-gloss fabrics show a wider range of firing colors, and the reddish variants tend to be typical during these periods (**Nos. 3–6**). Most black-gloss fabrics tend to be fired to colors falling within the range of Munsell light red (10R to 2.5YR 6/6-8) to reddish-yellow (5YR 6-7/6-8) when oxidized. Reduced fabrics tend to have similar color values with chroma varying towards grey, usually falling within the range of Munsell pale red to light reddish-brown (10R to 2.5YR 6/1-4). These reduced fabrics, or more usually reduced parts of otherwise oxidized fabrics, show a similar range of firing colors to those of the later Hellenistic-period grey wares.[71] Reduced cores or margins are relatively common (**Nos. 4–6**). Although this may be mistaken for irregular or misfiring, it may be an indication of a strictly controlled firing program with time/temperature phases necessary to produce the well-sintered, relatively high-quality, semi-lustrous slips typical of Metapontine black gloss; such a program may not always have allowed a sufficiently long oxidizing phase necessary to fully oxidize the cores of the thicker parts of the vessels, particularly the junctions of foot ring bases, but would have had no bearing on the external appearance of the fired vessels.

As with the Late Archaic fine-ware fabrics, there are no mineral inclusions that are diagnostic of Metapontine fabrics, with the majority appearing to consist of quartz and probably also feldspars. Most inclusions visible microscopically are extremely fine, most less than 0.2 mm and more usually less than 0.1 mm, sufficiently small to be considered part of the granular (non-plastic) components of the matrices, of which they make up a high proportion. These are predominantly crystalline, consisting of quartz and likely also feldspars (indicated by yellowish or pinkish crystalline grains), giving a finely granular, sugary consistency to the fabrics when viewed microscopically. Relatively rare, coarser examples are shown in **Nos. 4** and **6**. Along with consistent firing, the prevalence of these non-plastic components of the matrix probably contributes to the apparent hardness of the fabrics.

Very rarely, sparse lathes of biotite (gold mica) are present, but generally the fabrics are not micaceous; this is perhaps due to the use of sediments which have been through significant transport processes like those of the coastal plain and alluvial sediments of the river valleys.

Very fine calcium carbonates are often present in the fabrics or indicated by voids with white halos (Catalog of Black-Gloss Fabrics, **Nos. 4–5**). These may represent calcium carbonate consolidates within

[70] For the variations in relative proportions of different shape-types in assemblages from different types of site, see Ch. 5.

[71] For description of Hellenistic-period grey-ware fabrics from the survey, see Ch. 7.

the clay source, though the addition of calcareous material is known to improve the plasticity and therefore the workability of clays. Textural concentration features (consolidated clay varying slightly in firing color and/or texture to the main part of the matrix) are occasionally apparent in the typical range of Metapontine fabrics. These tend to be iron oxide-rich, rounded, and fired dark reddish-brown, and are probably consolidates from the source clay (example shown in **No. 5**).

Although the quality of black-gloss slips varies, these are usually of relatively high quality compared to black gloss produced in many other regions of the Mediterranean. The majority can be described as consistent and semi-lustrous, usually fired to a good, dark hue, though deep red when wholly or partially oxidized. These are usually well-sintered, and appear consistent in spite of being relatively thin. Lower quality examples are also represented, with thin, dilute slips, appearing matte since they lack the well-sintered surfaces which give the better-fired examples their semi-lustrous appearance.

The consistent and relatively restricted range of textures apparent in Metapontine black-gloss fabrics points to well-developed and applied clay preparation practices—thorough mixing and levigation with a good degree of control over the texture and granulometry of the clay used for black-gloss production. As the end-product of a technological process, the features and properties of fine-ware fabrics are also determined by pottery fabric technologies—clay exploitation and manipulation practices and pyrotechnology. The consistent application of these practices for fine-ware production often results in localized variation but also results in consistent differences between regional products. These technological factors, combined with the restrictions imposed by the geological and sedimentological context, result in a certain degree of homogeneity in the regional fine-ware fabrics, as is apparent in the textural properties of the range of Metapontine black-gloss fabrics.

Source sediments vary in mineralogical composition, determined by the solid geology and sedimentological processes. The Metapontine chora forms part of the lower reaches of the Bradano Trough, which extends from the edge of the Apulian limestone plateau in the east to the massifs of Calabria in the southwest (see Chs. 1 and 2). From the perspective of ceramic petrology, there is little to distinguish fabrics from within this zone, and this is particularly the case with the fine-grained fabrics of Archaic fine ware and Classical- and Hellenistic-period black gloss. Lacking sizable diagnostic rock fragment inclusions, the fine-ware fabrics from the region are difficult to distinguish on petrological grounds from those produced elsewhere in southern Italy. Where larger inclusions are present, these are usually crystalline, probably quartz or perhaps feldspars (**No. 4**); neither mineral is geographically diagnostic in the context of this part of southern Italy (see Ch. 12 on transport amphora fabrics). Consequently, identifications of fine-ware fabrics from the region (as is usually the case in other parts of the Mediterranean) are based on macro- and microscopic features of texture and color combined with the properties of the black "gloss" slip.

A broad development is apparent in the firing colors of Metapontine fine ware from the Late Archaic to Classical periods, with most Archaic fine wares tending to be fired to a relatively restricted range of colors, with a wider range characteristic of the Classical and Hellenistic periods. Reddish-fired fabrics are extremely common during this period, though textural and compositional features appear to remain similar.

The key question is whether this change represents the exploitation of different kinds of clay source, which might indicate changes in the locations of production in the Metapontine chora, or if it represents developments in production technologies, particularly in terms of the manipulation of clays and pyrotechnology. Considering the broader context of ceramic production in the Classical and Hellenistic periods, plain/banded wares continued to be produced predominantly in lighter yellowish-firing fabrics like those typical of Late Archaic fine wares—many of these Classical-period examples have similar fabrics and slips to those of the Archaic fine wares. Indeed, these might be regarded as continuing the Archaic tradition of pottery production in terms of semi-slipped, light-firing vessels, frequently with partially oxidized slips. These suggest a continuation throughout the chora of producing fine-grained, yellowish-firing fabrics similar to those of Late Archaic fine wares, and would therefore seem to imply that trends apparent with black-gloss fabrics are primarily technological—arising from developments in clay preparation and firing practices—rather than forming part of wholesale changes to the use of different clay sources.

These longer-term trends may therefore largely be the result of efforts to create a better-quality slip—black gloss *sensu stricto*—in the Classical period, with increasing emphasis on entirely slipped vessels. In this case, the fired color of the body was not a primary concern since it was not usually visible over large parts of the vessel, in contrast to the semi-slipped Archaic fine wares, where a yellowish-fired body may have been as important as a dark slip (Archaic fine wares frequently have partially oxidized slips, often streaky). This in itself points to pyrotechnological developments, but there also appear to have been developments in clay exploitation and preparation practices in the Classical period with the goal of improved gloss.

No chronological trends or developments are apparent in Metapontine fine-ware fabrics from the Early Classical production of black gloss to the final phases of production in the Hellenistic period. Microscopic examination of reference sherds used to date the survey sites suggested that the entire range of Metapontine fine-ware fabrics is represented from the earliest phase of black-gloss production.

Black gloss of the Hellenistic period developed into Late Hellenistic/Late Republican grey ware, and there are close similarities between the two wares in terms of the composition of the fabrics (see Ch. 7). Indeed, the production of grey wares largely involved the firing of fabrics similar to those of black gloss in a reducing kiln atmosphere without the oxidizing phase, with consequent differences in firing color but no real differences in texture or composition apparent in the range of Metapontine fabrics of this period. Broad differences are, however, apparent, in the higher frequency of poorer-quality, dilute matte slips apparent in grey wares than is evident in the earlier black-gloss fabrics, where these are typically relatively thick, consistent, and semi-lustrous. This probably reflects an increasing trend towards efficiency in production, with abandonment of oxidizing phases in the firing program along with a general lessening of the quality of the fired slips. This probably formed part of a broader trend towards standardization and efficiency of production first apparent in the standardization and simplification of base forms during Hellenistic-period black gloss production.

Imported Attic Black-Glaze Pottery

F. Silvestrelli, A. Tubelli

The survey has yielded three black-gloss fragments that can be attributed to Attic black-glaze production.[72] These include a type C cup (**No. 1**) produced in a hard, consistent and finely granular fabric which can be distinguished from those produced locally and regionally (see above for characterization of typical Metapontine black-gloss fabrics). The rim is slightly out-turned, with a marked concave profile forming a sharply defined carination with the body. Since the first phase of production of type C cups is characterized by vessels with a higher and only slightly concave lip, the example from the survey, with its shorter lip with pronounced concavity, can probably be attributed to the later forms produced around the beginning of the 5th century BC. Attic vessels that can be dated to the beginning phases of production, in the last quarter of the 6th century BC, have been found in tombs from the urban necropolis of contrada Crucinia; other fragments were discovered in the votive deposit of contrada Favale.[73] (AT)

The rim of **No. 2** can be probably interpreted as a stemless kylix with a plain rim.[74] The shape is attested in Sicily[75] and in Campania[76] from the middle of the 5th until the end of the century, possibly also with local imitations,[77] but few specimens are published from other areas of southern Italy.[78]

---

[72] The term 'glaze' rather than 'gloss' is used here for Attic examples from the survey, following the usual nomenclature applied in Greece and the eastern Mediterranean. The two terms refer to wares with broadly similar technical features of fabric and slip, produced using similar techniques.

[73] Lo Porto 1966, tomb 43, 215–16, Fig. 71; *Metaponto* I, 281, Fig. 296a; Liseno 2004, 83, 85, No. 13, Pl. XLIV.d, (end of 6th–beginning 5th c. BC).

[74] *Agora* XII, 102, Nos. 474–82, Fig. 5, Pl. 22.

75 *Meligunìs Lipàra* II, 211, Pl. XLI, 2c (Lipari, T45, 5th c. BC). Kustermann Graf 2002, 276, Pl. CXXII (Selinous, contrada Gaggera necropolis).

[76] For Fratte, see Donnarumma and Tomay 1990, 257, No. 3, Fig. 435a, 437,3 (T51/1963, second quarter of 5th c. BC); 258, No. 1, Fig. 519 (T15/1963, 460–450 BC); 263, No. 1, Fig. 450 (T146/1974, 450 BC); 267, No. 6, Fig. 457 (T58/1963, 440–430 BC); 272–73, No. 2, Fig. 464 (TXCVI/1929, 420–400 BC); 273, No. 3, Fig. 465 (T15/1972, end of 5th c. BC); 274, No. 4, Fig. 466 (T79/1963, end of 5th c. BC).

[77] Pontrandolfo and Rouveret 2002, 420 (Poseidonia, first half of 4th c. BC).

[78] From Messapia, see Mannino 2006, 107, No. 99 (Oria, T4, 460–450 BC); for Taranto, see 51, Nos. 30–32, Pl. XX (via Dante tomb, second half of 5th c. BC). For Ruvo del Monte, see A. Bottini 1981, 203, Fig. 74 (T28, 311, second–third quarter of 5th c. BC). A kylix of Apulian production from Cairano (Calvario necropolis, T46, end of 5th–beginning of 4th c. BC) decorated by a palmette in black figure could belong to the same type; see Palmentola 2006, 323, No. 5.

The base of the squat-lekythos **No. 3** can also be attributed to Attic production. The last two fragments are of particular interest due to the scarcity of Attic pottery at Metaponto after the second quarter of the 5th century, when a substantial decline in imports, which had begun in the preceding period, appears to have intensified.[79] This decrease has been explained in terms of the secondary role played by the colony in networks distributing Attic products, and especially as an effect of the increase in local production, which was able to reflect more closely the needs of the Metapontine consumers.[80] (FS)

Black Gloss of the Late Archaic and Classical Periods

F. Silvestrelli

The material from the survey sample that can be dated to the Late Archaic and Classical periods consists of 632 fragments. The morphological repertoire from the survey does not appear to be entirely representative of the variety of shapes produced in Metaponto during this time span; the fragmentary state of most of the specimens makes the establishment of chronological ranges a difficult task and many specimens can be dated only broadly. The chronology for the 5th and 4th century black gloss rests on finds from the Pantanello necropolis, while the published deposits from the kerameikos provide evidence for the second quarter of the 5th century BC[81] and for the first quarter of the 4th century BC.[82]

The range of shapes identified for the first half of the 5th century BC consists of skyphoi and cups with offset rim, cup-skyphoi, A-type skyphoi, kylikes, and a single example of a Laconian-type krater. This represents just a selection of the Metapontine morphological repertoire, which also encompassed cylinder lekythoi,[83] amphorae, and lebetes gamikoi.[84] Shapes are derived both from the local Archaic tradition, as exemplified by completely glossed cups and skyphoi with offset rim, and from Attic production, the influence of which is evident in cylinder lekythoi and type C cups.

During the second half of the 5th century BC the Metapontine morphological repertoire underwent a substantial transformation. New shapes, such as stemless kylikes, squat lekythoi, concave-convex bowls, bolsals, and dishes, derived chiefly from the Attic repertoire, were adopted by local workshops. In many cases the earliest examples show a substantial dependence on Attic prototypes. The process of acquisition can still be detected at the beginning of the 4th century BC, showing that the relationships and the morphological exchanges between the two productive areas were active for a long period. Nonetheless, evidence also suggests continuity in the morphological repertoire of the black-gloss pottery; shapes already in use in the previous period (such as the lebetes gamikoi or the skyphoi with offset rim) evolved, suggesting that the artisans still remained firmly rooted in the local traditions.

The identification of the morphological repertoire of the second and the third quarter of the 4th century BC is still an open problem.[85] The limited amount of tombs from Pantanello dated to this period show a decrease in the number of grave goods.[86] Yet black-gloss production continued in the kerameikos,[87] and new evidence from recently excavated nuclei of the urban necropolis of Metaponto suggests the possibility of future developments in the research.[88] More detailed study of squat lekythoi, pyxides, dishes, and especially pelikai, which are the best attested shape of the period, could be of great help in outlining the manufacture, which seems to continue following the lines established in the preceding period. In the last quarter of the 4th century BC, red-figure production was revitalized and new black-gloss shapes became part of the local repertoire, opening the way to the major transformation which would take place at the end of the century.[89]

Metapontine black-glazed vessels can be decorated with incised and stamped patterns. This type

[79] G. Giudice 2007, 338–39; 401–04. To the red-figure vases listed by G. Giudice 2007 (154, No. 275, 174, No. 336, 224, both from Casinello San Teodoro, in the Metapontine chora, No. 463, and Burn 1998, 612, T330.8, from the Pantanello necropolis) can also be added a recently found pelike attributed to the Meidias Painter; see De Siena 2007, 441–42, Pl. XIV.a–b (Metaponto, urban necropolis, proprietà Giasi, T428). It is important to emphasize that all the counts of Attic imports at Metaponto are based on published figured material and do not include black gloss.

[80] Mannino 1996, 364.

[81] For deposit 12, see D'Andria 1980b.

[82] For deposit 1, see D'Andria 1980a, 375–418.

[83] Elliott 1998, 664–66.

[84] For amphorae, see Elliott 1998, 648–49; these two shapes are represented in the survey sample with specimens of later chronology.

[85] Elliott 1998, 644.

[86] J. Hall 1998, 573.

[87] As attested by deposit 11; see Cracolici 2004a, 84.

[88] See, for instance, M. Tagliente 2006, 733, Pl. IV, 3.

[89] See the overview of the Hellenistic period by E. Lanza Catti, below.

of decoration appears in Attic black-glaze pottery in the late second quarter of the 5th century BC and was used chiefly for drinking cups and oil flasks.[90] It was perhaps already adopted at Metaponto after the middle of the century[91] but became especially common in the last quarter of the 5th century BC. The black gloss from the survey has yielded only few fragments bearing this type of decoration; nonetheless, finds from the Pantanello necropolis and from the urban kerameikos allow the general development to be sketched. In the second half of the 5th and in the first quarter of the 4th century BC stamped decoration, consisting mainly of chains of palmettes,[92] was applied to stemless kylikes, dishes, and, among closed shapes, to squat lekythoi, epichyseis, and lebetes gamikoi.[93] In the following period, this technique seems to have become simpler: while stemless kylikes and concave-convex bowls from deposit 1 in the kerameikos[94] are decorated by a chain of six or twelve palmettes connected by incised arches, and dishes have a chain of fourteen palmettes,[95] in later specimens of stemless kylikes a chain of four palmettes (**No. 85**) is more common. In the second half of the 4th century, stamped decoration is confined to stemless kylikes, which can have chains composed of four palmettes,[96] and to dishes and concave-convex bowls with a single rosette in relief (**No. 171**).[97]

Ribbed decoration is also seen in Metapontine production. This was applied to closed shapes such as shape 2 oinochoai, a decorative system that is also apparent in Attic pottery.

Vertical ribbing, which appears on Attic pottery at the beginning of the second quarter of the 5th century BC,[98] was adopted by Metapontine artisans by the mid-5th century BC.[99] It was first used for oinochoai, then for mugs, lebetes gamikoi, kantharoi, and squat lekythoi. Horizontal ribbing occurs only on cylindrical mugs, which are not found in the survey sample, but are well-known from the kerameikos and the Pantanello necropolis.[100] Ribs lost their plasticity in the second half of the 4th century BC but were still used on oinochoai, mugs, concave-convex bowls, and gutti.

Black Gloss of the Hellenistic Period

E. Lanza Catti

The evidence collected by the field survey provides a broad and interesting view of black-gloss pottery in Metapontino during the Hellenistic period, both in terms of quantity and evidence for morphological seriation and classification.[101] More than 1,200 fragments can be dated between the last quarter of the 4th and the mid-2nd century BC. Beginning in the late 4th century it appears that the morphological repertoire underwent considerable alterations: some Classical shapes disappeared, new vessel forms were introduced, and a new taste in decorative patterns developed

Open shapes were definitely more common than closed vessels, and the latter would become even rarer over the course of the Hellenistic period. Many containers without handles appeared, both large and small. These features are consistent with what is known from contemporary excavated contexts in the urban area of Metaponto and the chora. The evidence provided by the kerameikos shows that, between the end of the 4th and the early decades of the 3rd century BC, the number of skyphoi decreased, while small bowls, cups, and dishes became the most common shapes.[102] Hellenistic funerary contexts in the Metapontino record the prevalence of small containers (small bowl, salt cellar), drinking vessels (ring-handled cup, one-handler), service vessels (dish, hemispherical cup), and some closed shapes (pelike, unguentarium, and "baby-feeder").[103] In rural domestic contexts, the shapes are more commonly skyphoi, one-handlers, dishes, bowls, and salt cellars, while closed vessels appear to be rarer.[104]

[90] See *Agora* XII, 22–30.

[91] Elliott 1998, 684, SL1 (460–440 BC).

[92] Occasionally, other motifs such as lotus flowers, hearts, and spirals are also attested; see Elliott 1998, 646 675, PY1.

[93] D'Andria 1980a, 406, No. 193, Fig. 54. Elliott 1998 653, D11; 660–61; 675–76; 684; 688.

[94] D'Andria 1980a, 411, Nos. 216–17 Fig. 57. Silvestrelli 2004, 112, Fig. 5. Silvestrelli forthcoming a.

[95] Silvestrelli forthcoming a.

[96] Silvestrelli forthcoming a.

[97] D'Andria 1980a, 429, Nos. 324–25, Fig. 67. Silvestrelli forthcoming a; Elliott 1998, 690, SA2.

[98] *Agora* XII, 21–22.

[99] For example, the hydria in Elliott 1998, 655, H2.

[100] D'Andria 1980a, 411, No. 223, Fig. 57, and Elliott 1998, 667, M15.

[101] I wish to thank Professor Joseph C Carter for generously involving me in this stimulating research. My sincere thanks also go to my colleagues, F. Silvestrelli, E. Vittoria, and K. Swift, for the knowledge they shared with me and for the pleasant discussions, from which I have benefited deeply.

[102] Silvestrelli forthcoming a.

[103] Scarano 1992. Rocchietti 2002, 101–29. For the tombs excavated at Pizzica in 1999 see Chs. 28–29.

[104] Studies of the black gloss from farmhouses excavated by ICA will appear in future volumes in this series.

The chronology of 3rd- and 2nd-century-BC black gloss still presents some difficulties. The evidence from the Pantanello necropolis is restricted to the Early Hellenistic period: the last phase has traditionally been dated to the beginning of the 3rd century BC.[105] It has been suggested that the absolute chronology be down-dated to the second half of the 3rd century.[106] This would imply a down-dating of the latest graves in the chora (especially nucleus 6), as already suggested by Carter.[107] Such a chronological adjustment would help spread the late burials and their grave goods—which otherwise seem compressed—over a longer period.

The sector of the urban necropolis excavated in 1986 covers a chronological range from 350 to 250 BC,[108] and the evidence from the kerameikos does not include ceramic production after the mid-3rd century BC.[109] With the exception of the assemblages from Cozzo Presepe in the Metapontine chora,[110] the black gloss from the survey dating after 250 BC must be studied within the context of well-known sequences from outside the region. Two other important Greek poleis, Herakleia and Taranto, offer good comparisons.[111] Excavated sites from inland Basilicata, such as Pomarico Vecchio, Satriano, and Civita di Tricarico, also provide wide contemporary sequences of black-gloss ware.[112] In Apulia, the settlement of Botromagno (modern Gravina in Puglia) shows interesting links with the black gloss from the Metapontino.[113]

Although not derived from stratified contexts, the black gloss from the ICA survey is of great importance since it helps to define the picture of late 3rd–2nd century-BC black gloss in the Metapontine chora, providing a wide spectrum of materials for analysis. An overview of Hellenistic black gloss for study of the survey assemblages points out the need to refine some aspects of the traditional chronologies to further the continual process of readjustments. For example, the terminal date for the salt cellar (**Nos. 202–209**) can be down-dated slightly,[114] while a higher chronology for the appearance of new Hellenistic shapes such as the ring-handled cup (**Nos. 97–107**)[115] and the mastoid cup (**No. 125**) can be supported.[116]

In general, the last quarter of the 4th and the beginning of the 3rd century can be characterized as a transitional period, in which many later Classical shapes developed into their final stages before ceasing to be produced. During this time, standardization of production brought with it less careful attention to details of both shape and decoration. By the second half of the 4th century BC, a change in firing techniques is evident in the low number of testers discovered at the kerameikos of Metaponto.[117] This suggests that a lower number of tests was needed during the long and difficult process of firing. The drastic reduction in the number of types of kiln supports in the 3rd century BC at the kerameikos corresponds to the development of more standardized bases for stacking during firing.[118] Indeed, as shown by the evidence from the kerameikos, from the mid-4th century BC onward vessel bases were designed to suit the firing supports, an inverse trend compared to the Classical period, when around fifty different types of kiln supports were introduced to accommodate the already existing base features.[119]

Technical innovations combined with a process of standardization are evident in new features of vessel bases. From the beginning of the 3rd century BC, the undersides became convex, pointed, or nippled, with an abrupt transition to the inner face of the foot. Fundamental changes in the morphologies of some shapes are also apparent: for example, many open and closed shapes adopted a stemmed foot made separately and then added to the pot.[120] Since the bases of differ-

105 *Necropoleis*, 222–36.

106 This suggestion was made by M. Jackson (2004), who discusses the chronology of T 71, the tomb of a young woman. Among the grave goods were a Gnathia bottle (300 BC, which could be considered an heirloom piece), three other Gnathia vessels (ca. 250–240 BC), a black-gloss ring-handled cup (250–230 BC), a black-gloss unguentarium type Forti IIIa (320–225 BC, according to the new chronology), and two gold earrings (250–200 BC). The deceased had a worn bronze diobol (300–270 BC) in her mouth.

107 *Necropoleis*, 236.

108 Scarano 1992.

109 The latest workshops excavated to date were ascribed to the first half of the 3rd c.: Silvestrelli 2004, 107.

110 For black-gloss ware from Cozzo Presepe, see Morel 1970a and Prag 1983a.

111 For Herakleia, see L. Giardino 1990, Pianu 1990, Casagrande 2002. For Taranto, see: Dell'Aglio 1996b, Prazzoli 1996, Graepler 1997, *Museo Taranto* III.2, Hempel 2001.

112 For Pomarico Vecchio, see *Pomarico Vecchio* I, passim. For Satriano, see Bruscella and Virtuoso 2005. For Civita di Tricarico, see De Cazanove 2008.

113 Prag 1992b.

114 On the down-dating of the lower term of the traditional chronology for the salt cellar (ca 350–270 BC), see Prag 1992b, 91.

115 The high chronology was suggested by Dell'Aglio 1996b, 326, No. 322, as far as Taranto is concerned.

116 For the chronology of the mastoid cup see **No. 125**.

117 Cracolici 2001; Cracolici 2004a; Silvestrelli 2004, 115.

118 Cracolici 2001; Silvestrelli 2004, 115.

119 Cracolici 2001.

120 For instance, the molded foot of the stemmed cup (**No. 95**; Ch. 29, site 736, T 20-02), that of the Hellenistic skyphos, and that of the kantharos

ent shapes share the same features, it is often hard to identify a vessel shape by a foot fragment.[121] (For the implications of this for the survey assemblages, see the discussion of stemmed cup, ring-handled cup, cup-skyphos, dish, and bowl below.) Broad developments are also apparent elsewhere in southern Italy and other black gloss-producing regions of the Mediterranean, including Attica.

The progressive simplification of the production techniques, already begun at the end of the 4th century, is especially evident in the 3rd century BC and led to the use of the quicker dipping method as a glossing technique, which in turn ended in the semi-glazing technique. This new slipping method left the lower parts of the vessels completely reserved.[122] The application of miltos painted on reserved surfaces was abandoned in the course of the 3rd century BC. During the first half of the century ribbed decoration was rarer and less fine than in the previous century, and repeated horizontal grooves were often applied to the surfaces.[123]

Stamped decoration was rarely used in Metapontine black gloss of this period, differing both from the preceding phase of production and from contemporary black-gloss production in other parts of southern Italy, such as those identified in Calabria at Thourioi, Locri, and Rhegion.[124] Coroplastic decorations and appliques are almost absent from the survey black-gloss corpus, as is also the case in the Hellenistic necropoleis and at the kerameikos of Metaponto. Such ornamentation was restricted specifically to the guttus,[125] a single example of which is attested (**No. 264**).

Extra-regional influences on Metapontine production during this period are limited and tend to come from the eastern Mediterranean rather than Tyrrhenian Italy. This contrasts to production in other parts of southern Italy, where the influence of Campana A wares appears to be stronger. Some connections with the eastern Mediterranean can be traced from the presence of imported West Slope pottery attested at Metaponto, though in extremely small quantities, as well as in Apulia.[126] Examples of Greek influence can be seen in other typically Hellenistic-period shapes, such as the 3rd-century thin-walled bowl with inward curving rim[127] and the mastoid cup, for which a derivation from East Greek models has also been suggested.[128]

In the 2nd century BC the morphological repertoire of Metapontine black gloss appears drastically reduced compared to the preceding period. Shapes attested in the survey consist mostly of table vessels, in particular dishes, cups with horizontal grooves on upper exterior, bowls, ring-handled cups with offset lip, and small bowls. While the number of basic shapes decreases, new shapes appear, with many sub-types and morphological variants: for instance the mastoid cup, the ring-handled cup with conical offset upper body, and the dishes with curved rim.

The initial phase of grey-ware production took place shortly before the mid-2nd century BC. It is clearly connected to the black-gloss tradition, which it superseded during the Late Hellenistic period, and offers important clues in better understanding the final stages in the development of black gloss. In the first phase of grey-ware production (160/50–120/110 BC), many shapes relate to those of Hellenistic black gloss. This is evident in common grey-ware forms such as dishes with curved rims, unguentaria, ring-handled cups, bowls, and mastoid cups.[129] To judge from the survey evidence, the likely lower chronological terminus for the presence of black gloss in the Metapontine chora can be placed around 100 BC. The evidence provided by the farmhouses excavated by ICA in the chora, as well as the data from the Late Hellenistic tile factory at Pantanello, support such a lower limit.[130]

Black-Gloss Forms

E. Lanza Catti, F. Silvestrelli, K. Swift, and A. Tubelli

This section discusses the black-gloss forms from the survey, referring to and following the order in which they are presented in the Catalog of Black-Gloss Fine

(Ch. 8, Catalog of Figured Wares, **No. 99** [GN 247-01]) are very similar.

[121] Therefore many bases of open forms had to be considered together in the present catalog. *Vide infra*, Nos. 218–38.

[122] For instance, Attic-type skyphos **No. 42**, Corinthian-type skyphos **No. 75**, stemmed cup **No. 96**, cup/one-handler **No. 118**, and dish **No. 155**.

[123] For the horizontal grooving, see the cup **No. 124**.

[124] Preacco Ancona 1996, 347–50.

[125] For the Hellenistic necropolis, see Scarano 1992, Fig. 2, Nos. 1 (plain ware) and 2 (Gnathia ware, mid-3rd c. BC). For the kerameikos, see D'Andria 1980a, 360, Fig. 5, 24 (black-gloss krater with relief decoration depicting a vine branch), and 430, Fig. 65 (black-gloss gutti from deposit 3). Silvestrelli forthcoming a.

[126] Giannotta 1983.

[127] The thin-walled bowl with inward curving rim recalls the Attic echinus bowl in its deep variant (*Agora* XXIX, 1021; 250–200 BC?). See **No. 173**.

[128] The origin of the mastoid cup has been connected with West Slope cups, with glass vessels from Alexandria and the Syrian-Palestine area, with metal vessels, and with so-called Megarian bowls; Lippolis 1994b, 272.

[129] Yntema 2005, forms 1c (dish with curved rim), 41 (unguentarium), 36 (ring-handled cup), 26 (bowl), 31c (mastoid cup).

[130] On the Pantanello tile factory, see Carter 1977 and 2006a, 150–51.

Ware, which gives descriptions of the reference sherds discussed here along with selected comparanda and attested dates. The quantities and relative proportions of each of the forms are shown in Tables 6.2–6.3.

Skyphos with offset rim

Already a common shape in Late Archaic Metapontine fine-ware production (see Part II, and **Nos. 29–41**, Catalog of Archaic Fine Wares), in the first quarter of the 5th century BC the skyphos with offset rim began to be produced with completely glossed surfaces, though often with a thin reserved band at the join between rim and shoulder, and can be regarded as black gloss *sensu stricto*. Compared to the earlier Archaic-period forms the rim tends to be straighter and more vertical, with a flatter shoulder, and the body has a more conical form. The 22 examples from the Metaponto survey, accounting for 0.25% of all black gloss and 1.19% of all attributed form-sherds, are characterized by short rims and rounded shoulders; some have a reserved line on the join with the shoulder (**Nos. 4–5**), but most of the fragments appear to relate to the completely glossed variant (**No. 6**). The technical features of fabric and slip fall within the range typical of Metapontine black gloss,[131] rather than Archaic-period fine wares.[132]

This type of skyphos is attested in contexts of the first half of the 5th century BC in both the asty and in the chora.[133] Funerary assemblages from the Pantanello necropolis indicate that the skyphos with offset rim can be dated down to the second quarter of the 5th century BC.[134] Skyphoi with offset rims of likely Metapontine production were exported along the Bradano valley, in the indigenous settlements of the interior, such as Pisticci, Montescaglioso, Matera, Timmari, and also Peucetia and Messapia, where they have been found in association with cups with offset rims.[135] (AT)

Cup with offset rim

The survey yielded 14 fragments of rims and bodies of completely glossed cups with offset rims, making up 0.16% of all black gloss. In their profiles they show a derivation from the earlier Ionic type B2 cups of the Archaic period, with flared offset rim and shallow bodies relative to the skyphoi with offset rims (e.g., **No. 8**). These types of cups have been interpreted as a hybridization of Ionic type B2 cups and the Attic type C cups.[136] They are well attested in the Metapontine kerameikos in contexts of the first half of the 5th century BC and at sites located further into the interior of Basilicata in the necropoleis of Pisticci, Montescaglioso, and Matera.[137] The type is also attested in Tarentine tombs in association with Attic cylinder lekythoi of the second quarter of the 5th century BC, and at Oria, where examples can perhaps be attributed to Metapontine production, and in other Salentine settlements of Tarentine influence.[138] (AT)

Cup-skyphos

Produced in Metapontine fabrics, the forms of 36 examples of cup-skyphoi from the survey, 0.4% of all black gloss and just below 2% of all attributed form-sherds, can all be related to an early cup-skyphos type of the 6th and early 5th centuries BC known from the Athenian Agora.[139] Most of the identifiable fragments of this shape belong to feet and bodies, with only three examples of rims.

Two of them are out-turned and slightly thickened with a concave profile continuing to the body (**No. 9**), which is deep and rounded; in one case the lip seems to be more flared and concave (**No. 10**). This variant, characterized by a lip not marked off from the body, can be related to Attic examples of the first quarter of the 5th century BC, with local imitations attested in Peucetia and Messapia.[140] The third example has a flared, concave, and shorter rim with a distinct and sharply defined junction between the rim and the body (**No. 11**), a variant which can find parallels in the Attic cup-skyphoi dated 490–480 BC.[141]

[131] See discussion above on Metapontine black-gloss fabrics.

[132] See Part II on Archaic fine-ware fabrics.

[133] Lo Porto 1966, 210, Fig. 64.3; Lo Porto 1981b, Tb 18, 364–65, Fig. 85.2.

[134] Elliott 1998, type A1, 677–78, 680–81.

[135] Lo Porto 1973, 167, No. 8, 171, No. 12, Pl. XIX, 2, 225, No. 4, Pl. LXX, 3; Lo Porto 1988–1989, 330, 332–33, Figs. 32, 10, 35.1, 360–62, Fig. 66.3; A. Bottini 1993b, Pisticci-Matina Soprano, tomb 11, 145–46, No. 22; Adamesteanu et al. 1971, 42, T. 23, Pl. IX; Semeraro 1997, 225, 227–28, 400, Nos. 783–84, 786–89, Fig. 197; *Museo Taranto* II.1, 50–53, tomb 25, No. 19.8; tomb 14, No. 20.5; *Arte e artigianato*, 131–32, No. 94.24.

[136] Van Compernolle 2000, 92, Fig. 2,5.

[137] D'Andria 1980b; Silvestrelli forthcoming a; Lo Porto 1973, Pl. LV, 2,5; Lo Porto 1988–1989, 410–11, Fig. 128; A. Bottini 1993b, Pisticci-Matina Soprano, 145, tomb 11.

[138] Moreschini 1988, 922–23, Tb 333, Pl. CCXXIV, 1c (this cup has a reserved line on the join with the shoulder). Semeraro 1997, 216–19, 400, Fig. 191, Nos. 717–21.

[139] *Agora* XII, 109–10.

[140] *Agora* XII, 276, n. 569, Figs. 6, 22; *Bitonto*, 117, No. 15; Semeraro 1997, 202, Nos. 601–02.

[141] *Agora* XII, 276, n. 573, Pl. 25.

Some bases from the survey compare well to examples from the Metapontine farmhouses at Sant'Angelo Grieco and Fattoria Fabrizio excavated by ICA (**Nos. 12–14**). The ring feet with rounded external profile or with a thin resting surface, decorated on the underside with one or more concentric circles (**No. 15**), are paralleled by cup-skyphoi from the Pantanello necropolis, which are dated to the middle of the 5th century BC.[142]

Beyond the Metapontino, the cup-skyphos is attested at indigenous settlements in the area of Melfi and the upper Bradano valley, where both Attic and colonial (probably Metapontine) cup-skyphoi dating to the first half of the 5th century BC are attested.[143] They are found also in Peucetia, with examples from Ginosa and Laterza (dated 500–480 BC), Rutigliano (in association with type C *kylikes*), and Valenzano.[144] In the Messapian settlements at Cavallino and Oria, it is believed that examples of cup-skyphoi are of Tarentine manufacture.[145] (AT)

Kylix

The black-gloss type C cup with concave lip was one of the most popular black-gloss shapes in southern Italy from the end of the 6th to the first quarter of the 5th century BC, both at colonial and indigenous settlements where the shape occurs in Attic imports and local imitations. Production in the kerameikos of Metaponto is attested in the first half of the 5th century BC.[146]

An Attic example (**No. 1**) was found by the survey, while the remaining 16 fragments appear to be Metapontine on the basis of fabric and slip. The forms can be attributed to the first quarter of the 5th century BC on the basis of their shorter, concave lips similar to the Attic example, and thickened flared rims (**No. 16**).[147] A concave ring foot with reserved rim, with a continuous curve to the stem (**No. 21**), might perhaps also belong to a type B cup, which is more elegant and slender than the type C cup, and is a more common type in figured and black-gloss fine wares.

The popularity of this type of kylix in the Metapontine chora is apparent in the examples from the Pantanello necropolis, which contained forms dated from the end of the 6th century BC to the second quarter of the 5th century BC.[148] This type of cup was also popular in settlements of the upper Bradano valley, as is apparent from the necropoleis of the indigenous settlement at Irsina, where an Attic example was also found. It is also found closer to Metaponto in funerary assemblages dating to the late 6th and early 5th century BC from Pisticci and Montescaglioso, with both Attic examples and Metapontine imitations.[149] Regional imitations from indigenous settlements in Basilicata are consistent with those apparent in Metapontine production, as is the case with examples from Pomarico Vecchio, Oppido Lucano, Ruvo del Monte, Lavello, and from the necropoleis of Chiaromonte, where they are attested in funerary assemblages of the middle of the 5th century BC.[150] Further afield, Attic and regional imitations are common in Peucetia, attested for example by type C kylikes in tombs at Ginosa and Laterza.[151] (AT)

Laconian-type krater

A single Laconian-type krater from the survey (**No. 22**) is characterized by a projecting lip formed by a plane on the collar rim, set above a flared neck. The morphology of the rim is consistent with Laconian prototypes of the last quarter of the 6th century BC, which are completely glossed and without the thick lip typical of earlier examples, but with a rim with an inclined upper surface and a high neck flared towards the top. Since nothing is preserved of the rest of the body, the fragment can only be generically dated to the second half of the 6th century BC; completely glossed stirrup kraters of Laconian production seem to concentrate in the second half of the 6th century

[142] *Necropoleis*, 682, CS3.

[143] Lo Porto 1973, 188, 222, Pls. XXXVI 4,1, LXVI, 2,5–7; *Greci, Enotri e Lucani*, 162–64; *Forentum* I, 195, type 3; other examples from the Melfi area are cited in n. 63.

[144] *Museo Taranto* II.1, 177–78, Nos. 18.5, 19.9, 49.17; *Museo Taranto* II.2, 496, Pl. 40c, 91a; Ciancio 1985, 75–79, tomb 7, No. 143, tomb 8, 81–82, No. 172, Pl. XLII,3 (late first half of the 5th c. BC, colonial production).

[145] Lo Porto 1977, 737–45; Semeraro 1997, 202–04, 218–19, 265–67, 398.

[146] Wasters from deposit 12 of the kerameikos. D'Andria 1980b, 122, Pl. XI, No. 14; Silvestrelli forthcoming a.

[147] There are numerous comparanda for the characteristic rim form amongst examples from Messapia: Semeraro 1997, 113, No. 225, fig, 66; 148–49, No. 292, Fig. 94; 216–17, No. 711, Fig. 191.

[148] Elliott 1998, 657–58.

[149] Lo Porto 1973, 230, Pl. LXXIII,5, 159, Pl. VI,2,5, 169, Pl. XVIII,3, 188, Pl. XXXVI, 3,5,8 (in association with black-glazed skyphoi with offset rim). Lo Porto 1988–1989, 330–33.

[150] Barberis 1997, 60, Pl. 29, Nos. 15–16; Lissi Caronna 1980, 151–52, Figs. 44, 47; Lissi Caronna 1990–1991, 293–95, Figs. 118–19, 398, Figs. 199–201; A. Bottini 1981, 202–03, 244–45, 252–53, 267–68, Figs. 43, 54, 74; Giorgi 1985, 193–94; *Greci, Enotri e Lucani*, 162–68, No. 2.30.22.

[151] *Museo Taranto* II.1, 177, Nos. 18.11, 40.22 (from Ginosa), Nos. 47.10–11, 48.8, 49.14–15 (from Laterza).

BC; in the last quarter of the 6th and first quarter of the 5th century BC they take the place of black-figure and patterned kraters.[152]

The fabric and slip of the example from the survey appears to fall within the range of Metapontine products, and is probably a local imitation rather than a genuine Laconian import—in southern Italy this type of krater is not only imported but also imitated. In the Metapontino, where few specimens of this type occur, imitations seem to be better-attested than imports, as indicated by the fragment found by the survey and by a second specimen from an Archaic votive deposit at contrada Favale.[153] A likely Laconian import is attested in an excavated farmhouse in the Metapontine chora.[154] A krater of the Laconian type was found in a tomb of the second half of the 6th century BC from Pisticci.[155] Tombs excavated in Enotrian settlements located along the Agri and Sinni valleys, such as Alianello, Rivello, and Castelluccio, have yielded kraters modelled upon Laconian prototypes for which a Metapontine production has been proposed.[156]

Together with cups, kraters were the most popular and widespread Laconian exports: examples from the Italian peninsula and especially from southern Italy are largely imported. In the first half of the 6th century BC distribution appears to have been mainly directed towards Etruria, but in the second half of the century they are concentrated in Sicily and in southern Italy, especially in Apulia, with an apparent increase in frequency in the last quarter of the 6th century BC.[157] Laconian imports and some local imitations are attested in indigenous settlements of Peucetia and Messapia (e.g., tombs found on the acropolis of Monte Sannace), but also at Rutigliano, Bari, Cavallino, Oria, Otranto, Valesio, and Vaste.[158] (AT)

Skyphoi from the survey

The survey corpus for the 5th and 4th centuries BC is characterized by skyphoi, with 901 examples accounting for 48.7% of all identifiable black-gloss form-sherds and just over 10% of all black gloss. The relatively diagnostic bases of the A- and C-type skyphoi make most examples readily identifiable, and together they account for just under half (48.7%) of all attributed black-gloss forms. Gradual developments of this shape allow these to be dated with some confidence, since they are sufficiently ubiquitous in assemblages of the Classical and Early Hellenistic periods.[159] Together, the A- and C-type skyphoi, and particularly their relatively durable bases, which can be readily identified from fragments, constitute the markers par excellence of the Classical and Early Hellenistic periods in the Metapontine chora. (KS)

A-type skyphos

The 230 fragments of A-type skyphoi (**Nos. 23–45**), consisting of 107 rims and 121 bases, account for 12.4% of all identified black-gloss forms, though they are less than half as frequent as C-type skyphoi. (KS)

A-type skyphoi were introduced into the Attic morphological repertoire in the middle of the 6th century BC and soon adopted in south Italian workshops.[160] The shape is widely attested in the Metapontino and generally in southern Italy during the Classical period. Its popularity is also demonstrated by the adoption of the shape in the morphological repertoire of red-figure and Xenon Group wares. Local production of the shape is attested at least from the second quarter of the 5th century BC, and specimens have been found in deposit 12 of the kerameikos;[161] in the same period they appear in Metapontine funerary assemblages.[162] These skyphoi have a convex profile with the maximum diameter at the rim and a large torus foot; they depend on Attic examples of the first quarter of the 5th century. In the second half of the century the single curve from foot to rim was abandoned; the lower part of the body became straight, while the upper part was still convex. This type of skyphos finds comparisons with south Italian vessels attested from the middle of the 5th century BC.[163] In the first half of the 4th century the body progressively elongated, with an out-turning of the rim (see **No.**

[152] For the shape and its development, see Stibbe 1986 and 1989, 40–41.

[153] Liseno 2004, 85, No. 11, Pl. XLIV.a–b.

[154] I am grateful to E. Lanza Catti for this information.

[155] Lo Porto 1973, 169, Pl. XVII, 3,2.

[156] *Greci, Enotri e Lucani*, 91.

[157] Nafissi 1989, 68–72.

[158] *Monte Sannace*, 41, Pls. 200,1 and 221,2; *Museo Taranto* II.2, 470, Pl. 31b; M. Tagliente 1982, 24, Pls. 16,6 e 8, 17, 1–2; Semeraro 1983, 158–60, Pl. 84; Semeraro 1997, 57, 59, 73, 84, 91, 111, 170–71, 243, 245–46, 251, 281, 283–84, 290–91, 300, 304.

[159] Outlined for the Classical period by Silvestrelli, with the final phases of the morphological development of the skyphos forms in the Early Hellenistic period discussed by Lanza Catti.

[160] For the history and origin of this type see *Agora* XII, 84–86. On the adoption in southern Italy: Yntema 2001, 185–87.

[161] Silvestrelli forthcoming a.

[162] Elliott 1998, 679.

[163] Morel 1981a, 4382, Giorgi 1988b, 192–93, Type 1.2 and 201 n. 46.

30); handles tended to become more angular. A progressive reduction in the diameter of the foot can also be observed. In the second half of the 4th century BC skyphoi acquired a double curve; this profile is typical of the skyphoi in Morel's series 4373[164] which can be considered, as in Campania, a development of Morel 4382.[165] (FS)

The A-type skyphos seems to have been less popular in the Hellenistic period.[166] In the Early Hellenistic period the shape was still common in both black-gloss and Gnathia wares, but disappeared shortly thereafter. The latest examples from the necropolis of Taranto date to the last quarter of the 4th century BC[167] and the shape seems to go out of production around the turn of the century.[168] By contrast, in Messapia the A-type skyphos appears to have been particularly popular during the Early Hellenistic period, especially in funerary contexts.[169] The form was also still popular in funerary, sanctuary, and especially residential contexts in inland Lucania between the late 4th and early 3rd centuries BC,[170] while it became extremely rare in contemporary tombs of the Greek colonies Metaponto and Herakleia.[171] With the beginning of the 3rd century new shapes, such as the ring-handled cup, took the place of skyphoi in Metapontine graves. The A-type skyphos, however, was probably still produced in the Metapontine kerameikos in the early decades of the 3rd century BC.[172] It is likely that the shape was used mainly in residential and sanctuary contexts in the Metapontino during the first half of the 3rd century BC.[173] The evidence from the survey, consisting of 83 examples of Hellenistic A-type skyphoi from 67 different sites, along with a presence in excavated assemblages from farmhouses in the chora, shows that the A-type skyphos was still widely in use until the first quarter of the 3rd century BC. Of the 67 survey sites, only three have been interpreted as funerary contexts (sites 81, 294, and 415) and seven as Greek farmhouses with tombs. While the A-type skyphos appears to have become rarer in funerary contexts of this period and to have been replaced to some extent by other shapes, it continued to be used in domestic contexts.

The development of the shape is clear. During the second half of the 4th century BC the A-type skyphos continued to develop a taller and more slender profile; the rim was characterized by a pronounced outward turn (**Nos. 36–39**), very close to the attachment of the horseshoe handles (**No. 37**), while the torus ring foot became narrower than in the Classical period (**Nos. 40–43**). The underside of the base was still quite flat (**No. 42**) or slightly convex (**No. 43**). In the last stage of development the shape underwent a process that brought it closer to the C-type skyphos (below and **Nos. 65–68**), as can be seen from **Nos. 39, 44,** and **45**. In the first quarter of the 3rd century BC the profile became very sinuous, with a thin wall ending in an even narrower foot. The lower part of the body was usually reserved and the base became thicker, with a convex or pointed underside, characterized by an abrupt transition to the inner face of the foot. A similar development is apparent around the same time in Athens and in Tyrrhenian Italy,[174] where the shape instead assumed a more angular profile in its final phase in the first quarter of the 3rd century BC. (ELC)

164 Morel 1981a, 311, Pl. 131.

165 See Pontrandolfo 2000, 129 and Table 2. For the presence of this type of skyphos in Apulia and Lucania, see Labellarte 1995, 119 n. 2.1.25; *Pomarico Vecchio* I, 1, 76. Yntema 2001, 187, Subtype 43b.

166 Morel 1981a, species 4360, 4370, and 4380.

167 From phase B1 of the Tarentine necropolis, dated 325–300 BC; Lippolis 1994b, 248, Fig. 185.

168 Kiln waste deposits from via Leonida; Dell'Aglio 1996, 326 and 330, Cat. No. 316.

169 See, for instance, the funerary areas of Vaste (D'Andria 1990, 134. Delli Ponti 1996) and Rocavecchia (Giannotta 1996, 56–78).

170 Between the late 4th and early 3rd c. BC, A-type skyphoi are attested in funerary contexts in the Sannite center of Località Casalini Sottana, near Venosa (Cracolici 2004c, 55, Fig. 49) and at Pomarico Nuovo (Lattanzi 1978, 444, T 3/1976 and T 10/1976, Gnathia and black-gloss specimens). Evidence from a settlement is provided by Pomarico Vecchio (Surra and Preacco Ancona 1997, 76-77, Nos. 68–71). In the residential area of Civita di Tricarico the Hellenistic A-type skyphos surpasses the C-type skyphos; the last examples of A-type skyphoi date to the first third of the 3rd c. BC (De Cazanove 2008, 402). At Grumentum the A-type skyphos is attested in a votive deposit of the early 3rd c. BC (Alta Val d'Agri, 120, No. 4). Other examples are known from Roccagloriosa (Fracchia 1990a, 235–36, Fig. 179, Nos. 74–75) and Oppido Lucano (Lissi Caronna 1983, 297–328).

171 Elliott (1998, 677–79) records examples dated from 480 to 440 BC. Scarano (1992) does not mention any A-type skyphos from the 65 Hellenistic tombs excavated in 1986 at the crossing between the SS106 and SS175 roads. Pianu (1990) does not mention any A-type skyphos from the 278 tombs included in his research dedicated to the 4th–3rd c. BC necropolis of Herakleia. However, an example is known from the Pizzica-Pantanello necropolis (Morard 2002, T 5/1983).

172 A-type skyphoi are present in deposit 2 of the kerameikos, dated from end of the 4th to the early decades of the 3rd c. BC; Silvestrelli forthcoming a.

173 At Paestum a local production of A-type skyphoi is attested during the 3rd c. BC, but they are used only in residential contexts; in funerary areas the C-type skyphos is still present exclusively; Pontrandolfo 2000, 129.

174 For Athens, see *Agora* XXIX, Fig. 12, 151–54. For the development of the shape in Campania from the late 5th to early 3rd c., see Pontrandolfo 2000, 128–29.

C-type skyphos

C-type skyphoi are the best-represented black-gloss forms in the survey, more than twice as common as A-type skyphoi, with 671 fragments (of which 422 are rims and 249 are bases) making up over 7.7% of all black gloss and over a third (36.1%) of all attributed black-gloss forms from the survey. (KS)

In southern Italy, this shape, derived from the Attic morphological repertoire,[175] was very common. Metapontine production, confirmed by specimens found in the kerameikos, is attested at least from the second quarter of the 5th century BC[176] with specimens characterized by a convex profile with the maximum diameter located just below the slightly pointed rim, where also the thin, horizontal horseshoe handles are attached. The body can be completely glossed or can have a reserved area just above the foot, as in **No. 46**. In the second half of the 5th century BC, the greatest diameter drops lower (**No. 48**). From the end of the 5th the diameter is below the handle, and the rim turns in, giving the shape an ovoid form. This type, to which **Nos. 53** and **45** can possibly be connected, lasts into the first half of the 4th century BC.

In the Early Hellenistic period the C-type skyphos was still one of the most frequent shapes and was produced with many techniques, including red-figure, Gnathia, black-gloss, banded, and plain wares. More than 500 examples were collected by the survey. In the Metapontino the evidence comes mostly from excavated sanctuaries and residential contexts dated between the late 4th and early 3rd centuries BC. While a great number of Hellenistic C-type skyphoi were discovered during excavations of the Pantanello sanctuary and farmhouses in the chora, only a few contemporary examples were found in the Pantanello necropolis.[177] The shape was still in production during the last phase of the Metapontine kerameikos (end of the 4th to early decades of the 3rd century) and was still attested, though in reduced numbers, in dumps created during the 3rd century BC.[178]

The general development of the C-type skyphos is well attested in studies of the Tarentine and Campanian material.[179] Hellenistic examples are generally smaller than those of the Classical period, though with some exceptions (**No. 56**); the ring foot is narrower, thick and flaring (**Nos. 57–58**). At the end of the 4th century BC miltos was still used to decorate the reserved surfaces of the lower body and underside (**No. 57**). Miltos was abandoned when the foot later became narrower and thicker (**No. 77**). In the latest examples the base was almost converted into a thick disk (**No. 79**). The lower body became narrower and was reserved (**Nos. 76, 81**). The distinctive body profile became increasingly S-shaped, with a more graceful rim sloping outwards (**No. 68**). The handles narrowed towards the junction with the lip, creating a triangular arrangement when viewed from above (**No. 64**). The rim diameter was often very narrow (**No. 67**). A less common contemporary variant had an almost vertical lip (**Nos. 64–69**). The latest examples from the survey can be dated to the first half of the 3rd century BC (**Nos. 74–81**). (ELC)

Stemless kylix

Stemless kylikes (**Nos. 82–86**) are represented by 23 fragments, accounting for 1.2% of all identified black-gloss forms from the survey. Nine fragments can be attributed to the period of the second half of the 5th to the first half of the 4th century BC (**Nos. 71–73, 75–76**), one to the 4th century BC (**No. 74**), and one with a date range extending later, perhaps into the 3rd century BC (**No.77**). Thirteen further fragments are attributable only in general terms to this shape. (KS)

The stemless kylix with plain rim outside, an offset inside, and canted handles is borrowed from stemless kylikes of the *Delicate Class* produced in Attica from the second quarter of the 5th[180] down to the middle of the 4th century BC.[181] The shape attained great favor in southern Italy and was probably produced in many centers from the middle of the 5th down through the 4th century BC.[182] This type of kylix is well attested at Metaponto in the necropoleis of both the asty and the

[175] For the history of the shape in black gloss and red-figure, see *Agora* XII, 81–83, and Oakley 1988.

[176] D'Andria 1980b, 123, No. 21, Pl. VIII. Silvestrelli forthcoming a.

[177] Elliott 1998, 679. Scarano 1992.

[178] Silvestrelli forthcoming a.

[179] For the Tarentine necropolis, see Lippolis 1994, 248, Fig. 185. Graepler 1997, 93. Hoffmann 2002, Pls. 79–80. For the Tyrrhenian area, see Pontrandolfo 2000, 127.

[180] *Agora* XII, 102–05, Nos. 483–517, Pl. 22–23, Fig. 5.

[181] Campenon 1994, 73, Figs. 3, 9.

[182] Kylikes with a black-figure decoration were probably produced in Peucetia beginning in the middle of the 5th c. BC (Palmentola 2006). For the distribution of the black-gloss shape in Messapia: Yntema 2001, 183. For Daunia and Peucetia see Giorgi 1988b, 194, type 2.1 and *Museo Taranto* II.2, 495, where also vessels from Taranto are cited, to which can be added Hoffmann 2002, 43, Type 425/1, Pl. 84 (380–350 BC). Fragments of local production have been identified at Locri: Bitti 1989, 160, No. 137, Pl. XXIV (second quarter 4th c. BC).

chora[183] and at farmhouses.[184] The shape was produced locally beginning at least in the third quarter of the 5th and into the 4th century BC, as demonstrated by wasters found in the local kerameikos: the form is found in large quantities in all the deposits, along with examples decorated by the Xenon Group.[185] The development of the shape can be compared to that already reconstructed for vessels from Lavello.[186] The earliest specimens have a wide and shallow bowl with a substantial offset inside; the handles rise at the level or just above the rim, as in Attic examples of the central decades of the 5th century BC. The feet are large, with a certain variety of molding on external surfaces. In the first half of the 4th century BC, when this type of kylix is also adopted by red-figure painters,[187] there is a notable diminution of the rim diameter, while the molded feet seem to be more standardized. The handles assume an angular profile that will be accentuated in stemless kylikes of the second half of the 4th century BC, when the bowl becomes more angular and has a lighter internal offset; the molded feet have a smaller diameter and are connected to the body by a higher groove. The survey examples consist mostly of fragmentary feet; while **No. 82** could be dated to the last quarter of the 5th century BC, **No. 83** can be compared with kylikes found in deposit 1 (see catalog below) **No. 84** with the oblique foot and shrunken foot diameter is representative of the shape in the second half of the 4th century BC. Only a few pieces preserve the stamped decoration these kylikes often bore (**Nos. 82, 85–86**). (FS)

Classical and Hellenistic cup-skyphos

The molded ring foot **No. 87** can possibly be attributed to a cup-skyphos derived from Attic light-wall cup-skyphoi.[188] The shape is common in Sicily and at Locri, with both imported and locally produced vessels[189] while in Magna Grecia it is less popular.[190] On the whole, the chronology seems to be consistent with that already established for Attic prototypes; firmly dated funerary assemblages containing this type of cup-skyphoi mostly belong to the first half of the 4th century BC.[191]

The shape, rare at Metaponto, comprises around 1.5% of all attributed black-gloss forms from the survey; nonetheless, a cup-skyphos has been identified in the Pantanello necropolis,[192] and fragments found in deposits of the kerameikos support the hypothesis of a limited, local production of the shape.[193] (FS)

The modern term "Hellenistic cup-skyphos" is used here, but it refers to a completely different shape than the Classical one, to which it is not related. Hellenistic cup-skyphoi are barely identifiable from rim and base fragments: the 4th-century-BC rims can readily be confused with those of skyphoi, with stemmed cups, or with Hellenistic bolsals, unless the more curved lower wall is extant.[194] Three bases (**Nos. 87–89**) identified as cup-skyphoi are characterized by a large molded ring foot joining with the body (reserved on **Nos. 88–89**, painted in **No. 87**). The underside is reserved, while the interior surface of the ring foot is painted. The shape is typical of the Gnathia morphological repertoire, but is also attested, albeit rarely, in

[183] Chora: Lo Porto 1988-1989, 320–22, No. 2, Fig. 20,2; 22 (Casinello-San Teodoro, Tomb 3, second half of the 5th c. BC), and Elliott 1998, 657–58 (Pantanello necropolis). Urban necropolis: Sestieri 1940, 74–75, Fig. 20 (Metaponto, Casa Ricotta, tomb 11). Scarano 1992, 21, Fig. 11.

[184] Fattoria Fabrizio, SantAngelo Grieco and Fattoria Stefano (C. Lanza and E. Vittoria, pers. comm. 2009).

[185] Cracolici 2004a, 54, Fig. 15. Silvestrelli forthcoming a.

[186] Giorgi 1988b, 193–94, Type 1.2.

[187] Denoyelle 2002c, Pl. 26a–b; Pl. 27a (Creusa Painter); and Pl. 27b–c (Dolon Painter). Fragments of this shape have also been found in the kerameikos: Silvestrelli 2008, 281, Fig. 4.

[188] *Agora* XII, 110–11, Nos. 580–611, Fig. 6. Pl. 26, 54–55. Their production spans from the third quarter of the 5th c. BC to the middle of the 4th c. BC.

[189] For South Italian imitation: Morel 1981a, 219, 4213, Pl. 120. Sicily: *Meligunìs-Lipára* II, 224, Pl. d,1 (T22bis, 13, Pl. d,4, Pl. LXX, 5b, first quarter 4th c. BC) and 4 (T198, 65, Pl. d, 1 e, Pl. LXIV, 2, second quarter 4th c. BC, possibly later); *Meligunìs-Lipára* V, 396, Pl. XV, T2636; 390, Pl. XV, T2584; 393–94, Pl. XV, T2620 (Lipari, contrada Portinenti necropolis, middle of the 4th c. BC); 397, Pl. XXVI (T2560, contrada Portinenti necropolis, second quarter 3rd c. BC). Pl. XXX (sporadic finds, 1993-95 excavations) De Miro 1963, 130, AG. 1366–67 bis, Figs. 43c, 48a (Agrigento). Rizza 1955, 311, No. 2, Figs. 21, 13 (Lentini, Valle San Mauro necropolis, T13), 335, No. 3, Fig. 47 (Lentini, Valle San Mauro necropolis, T264). Adamesteanu 1958b, 317, Nos. 1–2, Fig. 22 (Manfria). Kustermann Graf 2002, 43, Pls. CXXV–CXXVI (Selinous, contrada Gaggera necropolis, Attic examples). Terravecchia di Grammichele: Elia 2006, 167, Fig. 7, FI167–FI171 (400–360 BC). Locri: Buzzi Auxilia 1989, 120–25, Nos. 77–82, Pl. XXII (Locri, Centocamere, 420–350 BC).

[190] Specimens of Attic import (*Museo Taranto* I.3, 356, No. 125.30) or colonial production (*Museo Taranto* I.3, 392, No. 165.7) are known from Taranto; a limited distribution can be recognized for Peucetia, where vases of this shape have been found at Ceglie (Labellarte 1988, 332–33, No. 726, Fig. 463, Ipogeo via Giuseppe Martino). See also Pontrandolfo and Rouveret 1992, 314, No. 6 and 420.

[191] A possible evolution of the shape, which could continue in the second half of the century is represented by the cup-skyphos from Lavello (Giorgi 1988b, 195, Type 2, Pl. 30) and by a vessel from Ordona (*Ordona* VII.2, 671, No. 3, Fig. 376).

[192] Elliott 1998, 658, SK2.

[193] Silvestrelli forthcoming a.

[194] Hellenistic bolsals are also known as "two-handled cups." No examples have been identified in the survey assemblages.

red-figure production and in plain black versions.[195] Cup-skyphoi are common among the drinking vessels in Apulian and Lucanian funerary contexts.[196] (ELC)

Bolsal

Dating to the first half of the 4th century BC, five bolsal bases are represented in the survey assemblages, four co-identified with the base shown in the black-gloss catalog (**No. 90**).[197] Together they account for 0.27% of all identified black-gloss forms. (KS)

Derived from the Attic bolsal,[198] two-handled cups were probably adopted by Metapontine workshops already in the last quarter of the 5th century BC;[199] production continued into the 4th century BC. These cups have a plain, continuous rim, horizontal bell-shaped handles attached just below the rim, a groove in the lower part of the bowl, and (differing from Attic prototypes) a ring foot separated from the body by a throat. The external surface of the foot is usually decorated with concentric circles (**No. 90**). The body has a convex profile that becomes progressively straight with a vertical rim;[200] during the 4th century BC the bowl becomes shallower.[201] The shape occasionally bears a red-figure decoration, especially owls.[202] Bolsals occur with a certain frequency in Metapontine tombs until the second half of the 4th century BC.[203]

Specimens of Attic import[204] and especially of south Italian production are attested in funerary assemblages of Magna Grecia beginning in the last decades of the 5th century BC. They seem to be more popular in colonial centers than in the indigenous world.[205] (FS)

Stemmed cup

The vessel traditionally known as the 'stemmed cup' (Morel 4122, 4125, and 4131) is a typical Hellenistic-period form. Nineteen fragments (16 rims and 3 bases) together account for 1.0% of all identifiable black-gloss forms. Three examples may date to the later 4th century BC (**Nos. 94–96**), with ten examples dating to the first half of the 3rd century BC (co-identified with **No. 91**), and four to the 3rd to early 2nd century BC (co-identified with **No. 92**). (KS)

Typical of Gnathia ware, the shape is rare in black gloss.[206] Attested in southern Italy between the last decades of the 4th and the early 3rd century BC, the shape is characterized by a thin-walled, bowl-like body, two horizontal handles, and a stemmed base with molded foot (**No. 91**). The exterior of the body is usually plain (**Nos. 91–93**) but can sometimes be ribbed.[207] The stem is always reserved (**Nos. 91, 94–95**). An example with a completely reserved lower exterior and a lower, less well-articulated foot (**No. 96**) is likely to be slightly later. Although Morel associates this shape with the more common southern Italian two-handled cup

195 An example of red-figure cup-skyphos was recovered in T 12 at Montescaglioso (Lo Porto 1973, Pl. XLII,1.1, 330–320 BC). A Gnathia cup-skyphos is attested in T 13 of the same site (ibid., Pl. XLIV,1.1). A black-gloss specimen is present in T 1/2000 of Altamura, last decades of the 4th c. BC (Venturo 2001, Pl. III).

196 Graepler 1997, 93, type 422/1 (phase B1, 325–300 BC). Hoffmann 2002, 42, type 422/1 and Pl. 81 (phase A of the necropolis of Taranto, 375–325 BC). Pianu 1990, Pl. XII, I (Herakleia). Canosa 2007, Pl. XXXIX, 70–71 (Timmari).

197 See also Ch. 29, site 737 T 18-8.

198 Attic production starts at the end of the third quarter of the 5th c. BC and cups of this type are attested until the middle of the 4th c. BC: *Agora* XII, 107–8, Nos. 532–61, Pl. 24,58, Figs. 6, 22.

199 Silvestrelli forthcoming a.

200 Compare Morel 1981a, F4122, Pl. 117. Silvestrelli forthcoming a.

201 Silvestrelli forthcoming a. The evolution of the Metapontine *bolsal* is consistent with that outlined for Taranto, where it is attested until the 2nd c. BC: Maruggi 1982, 60–61; D'Amicis 1984, 55 and Pl. XV; *Museo Taranto* II.1, 185; *Museo Taranto* III.1, 246, Fig. 184.

202 *LCS* Suppl. 3, 2/338b (Schwerin Group). Owl bolsals: Burn 1998, 622, T 330-13 (Pantanello, tomb 330, Spanner Group, 410–270 BC). Possibly of Metapontine production are also two vessels from Ceglie, Ipogeo G. Martino (Labellarte 1988, 325, Nos. 695–96, Figs. 442–43, Spanner Group). An owl bolsal possibly produced at Herakleia is illustrated at Pianu 1990, 57, No. 40, Pl. XXV,4 (tomb 16, second half 4th c. BC).

203 See Elliott 1998, 650 and, in this publication, Ch. 29, site 737 T 18-8. Nava 2004, 961–62, Pl. XXVII,1 (Metaponto, Torre di Mare, tomb 76). Scarano 1992, Fig. 12, presents two-handled cups dated to the second half of the 4th c. BC. No later specimens are known.

204 *Museo Taranto* II.2, 499, Fig. 41,c (Rutigliano, Tomb 29/1978, *ripostiglio*, end of the 5th c. BC). To Attic production are also ascribed the bolsals from Taranto, Via Dante Tomb: Patera 1986, 52–53, Nos. 34–38 (end of the 5th c. BC) and Ginosa, contrada Lama: *Museo Taranto* II.1, 58, No. 27.9 (tomb 21, bolsal dated to the end of the 5th–beginning of the 4th c. BC).

205 On the success encountered by bolsals in southern Italy, where they were adopted by many productions which imitated Attic prototypes and continued the manufacture well beyond the occurrence of the shape in Attic workshops, see Patera 1986, 52–54, Pl. XX,34. For Taranto see n. 200, Ginosa: *Museo Taranto* II.1, 104, No. 43.46 (first half of the 4th c. BC). Laterza: *Museo Taranto* II.1, Nos. 52.73–76. Herakleia: Degrassi 1965, 20, Nos. 16–17, Fig. 48 (tomb of the Policoro Painter, end of the 5th c. BC) and Pianu 1990, 223–24 (4th–beginning of the 3rd c. BC). Kaulonia: Gagliardi 2001, 282, Nos. 11–13, Pl. 267 (Kaulonia, Punta Stilo, Attic, last quarter of the 5th c. BC); Gagliardi 2001, 300 (Kaulonia, Punta Stilo sanctuary, local production); Gagliardi 2004, 513, Nos. 64–65, Fig. 186 (Monasterace, contrada Lupa, kerameikos). Locri: *Locri* II, 109–20 (last quarter of the 5th–third quarter of the 4th c. BC). Poseidonia: Pontrandolfo and Rouveret 1992, 419–20. Pomarico: Osanna 2008, 925, Pl. XVII,2. Bolsals from the Lavello necropolis, attributed to Apulian production, have a similar development as the Metapontine specimens and occur between the last quarter of the 5th c. and 325 BC: Giorgi 1988b, 195, Type 3.1.

206 Forti 1965, 73. Webster 1968, 2. Green 1976, Pl. 5, and 1989, Pl. 41,5.

207 For ribbed examples see Pianu 1990, Pl. XXVIII,1 (from Herakleia) and Lippolis 1994b, 248, Fig. 185 (from Taranto, 300–275 BC).

with horizontal handles, derived from the Classical bolsal,[208] the stemmed cup differs from the late version of the bolsal in its high molded foot and deeper body. The shape is similar to the 4th century BC version of the cup-skyphos (Morel 4242), but with straight horizontal handles and with a higher and narrower foot.[209] It also recalls the latest evolution of C-type skyphoi, usually stemmed and with a ribbed body (Morel 4312), though it differs in the shallower body.[210]

The stemmed cup has been associated with "ring-handled kantharoi", a typical later Gnathia shape, usually almost completely ribbed with overpainted decoration on the lip.[211] The different names used to refer to this Gnathia shape suggest that there is no universal agreement on its origins.[212] The stemmed cup and the ring-handled kantharos appear to differ only in the types of handles and the depth of the body. It is hard to distinguish between the two forms on the basis of rim, body, or foot fragments.[213]

On closer examination it appears clear that the stemmed cup and ring-handled kantharos come from different origins. Even though they share some morphological features, as is usual in Hellenistic shapes in general, these similarities are the result of convergent development rather than common origins. While the stemmed cup seems connected with the later evolution of the bolsal (with the typical Hellenistic introduction of a separated stem/foot),[214] the ring-handled kantharos comes from a different tradition, probably developing as a different form of skyphos.[215]

The stemmed cup is found throughout Apulia[216] and along the Ionian coast, from Taranto[217] to Metaponto[218] and Herakleia.[219] Close to the Metapontino, it is also found at indigenous sites such as Pomarico,[220] Irsina,[221] and Calciano.[222] Stemmed cups were produced in the kerameikos of Metaponto, as attested by black-gloss and overpainted examples.[223] (ELC)

Ring-handled cup

Typical of the Hellenistic period, 112 fragments (50 rims and 62 bases) of ring-handled cups (**Nos. 97–107**) make up 1.3% of all black gloss and 6.1% of attributed black-gloss forms. Eleven rims and eleven bases can be dated to the first half of the 3rd century BC (co-identified with **Nos. 97** and **102**). A single base (**No. 101**) may be as early as the second half of the 4th century BC, with the remaining 90 fragments attributable to the period from the 3rd to the first half of the 2nd century BC (**Nos. 98–100, 103–107**). (KS)

A typical western Mediterranean shape featured in black gloss, Gnathia, black-gloss "Hard-Fired Red" ware,[224] grey ware, and banded ware, the ring-

[208] Morel 4120 series (Morel 1981a, 290).

[209] Morel 4242 series (Morel 1981a, 297).

[210] An unribbed stemmed cup is included in the plate illustrating the different morphological features of the "skyphos" and "cup-skyphos" in the necropolis of Taranto during its various phases (Lippolis 1994b, 248, Fig. 185, phase C, 275–225 BC). The shape is considered a type of skyphos also by Pianu (1990, Pl. V, type 1 skyphos) in his study of the necropolis of Herakleia. A cautious compromise solution is adopted by Labellarte (1995, 120) in her terminology: "tazza skyphoide" (skyphoid cup).

[211] Morel 3111 and 3112a. In the *Necropoleis* volume such a kantharos is considered a different type of stemmed cup: Elliott 1998, 651, No. SC3 (T 194-11). For a Gnathia example, see *supra* Cat. No. 101

[212] In the Anglo-American tradition the shape is known as 'ring-handled kantharos" (Green 1977, 554, Fig. 5a). The ring-handled kantharos was labelled 'F-type' by J.R. Green (pers. comm. 2005). See Lanza Catti (2008, 267, Pls. I–II) for different names and shapes of Gnathia vessels. Yntema (2005, 74) refers to this shape as "kantharos with stemmed foot" or "ring-handled kantharos". Morel (1981a, 247) includes his series 3111 and 3112 among the "coupes" and not among the "canthares et vases apparentés" as his genus 3700. The Italian literature usually refers to this shape as "skyphos" probably following Forti's definition (Forti 1965, 74, Figs. 19 and 22).

[213] To compare the morphological similarities among the different shapes—all labelled "skyphos"—see Delli Ponti 1996, 118, Figs. 32 (cup-skyphos), 33 (ring-handled kantharos) and 34 (stemmed cup), from Vaste.

[214] A good specimen testifying to this passage is the Early Hellenistic two handled-cup from a tomb in Taranto (Lippolis 1994, 258, Fig. 192).

[215] Since the first quarter of the 3rd c. BC, the rounded body shape (Forti 1965, 74, Fig. 22) developed with a narrower lower body (Forti 1965, 74, Fig. 19). The shape was particularly successful in Tarentine workshops producing Gnathia pottery of the Alexandria Group, which was exported to northern Africa (Green 2001, 61, 70-1).

[216] The overpainted stemmed cup is quite common in Peucetia, with examples from Bitonto (*Bitonto*, 70, Fig. 75), Conversano (L'Abbate 1979, 48, No. 6; 78, Fig. 22.3; Chieco Bianchi Martini 1964, 105, Figs. 4.8 and 108, Fig. 4.11), Rutigliano (*Museo Taranto* II.2, 156, T 54.31), and Santo Mola (Donvito 1991, 115, Fig. 46). In Messapia, many overpainted stemmed cups were brought to light at Vaste (Delli Ponti 1996, 118, Fig. 34; 132, Fig. 72; 158, Fig. 134). The shape is attested in the area of Taranto as well (Fedele/Alessio/Del Monaco 1992, Fig. 105, from Grottaglie, and Fig. 139, from Manduria).

[217] Graepler 1997, 93, type 422/2 (ribbed and overpainted).

[218] Lo Porto 1981b, 362, Fig. 83, No. 2 (T 12/1957 Casa Ricotta, late 4th c. BC). Elliott 1998, 652, specimens Nos. SC1 (T 14-6) and SC2 (T 94-3), dated to the 3rd c. BC, from the Pantanello necropolis. See Ch. 19, site 736, T 20-C2 (black-gloss specimen from the Pizzica necropolis).

[219] Pianu 1990, Pl. XXVIII,1 (ribbed and overpainted); Pl. XXX,4 (overpainted); Pl. XLVI,3 (overpainted); Pl. XLVIII,3 (overpainted); Pl. LII,3 (overpainted); Pl. LVIII,165.1 (black gloss); Pl. LXII,1 (black gloss); Pl. LXIV,167.1 (black gloss).

[220] Macchioro 1981, Pl. VI, 1 (T 3, last quarter 4th c. BC, black gloss).

[221] Califano Ascenzi 1998, 102–3, Fig. 28 (overpainted).

[222] Lo Porto 1988–1989, 417–18, Fig. 135, No. 2 (black-gloss specimen from T 1/1955, 330–320 BC).

[223] Silvestrelli forthcoming a.

[224] The Hard-Fired Red pottery (HFR) is a particular sub-class of the late black-gloss ware, first identified by Yntema at Valesio (Yntema 1990b). The name reflects the red-orange color of the fabric, which is

handled cup does not seem to have derived from Attic or eastern Mediterranean traditions.[225] The black-gloss version is characterized by two vertical ring handles, an out-turned lip above a deep, thin-walled body, usually widest below the rim, and a conical ring foot. The lower body and foot are usually reserved.

Independent trajectories of development are apparent in each class and region. The ring-handled cup was a widespread form in southern Apulia and around the Gulf of Taranto in the first half of the 3rd century BC. The shape, known from settlement, sanctuary, and funerary contexts, appears to have been particularly popular in Messapia[226] and Taranto,[227] as well as in the Metapontino,[228] while it does not appear to have been popular in Peucetia.[229] An analogous shape was manufactured in Taranto in Hard-Fired Red (HFR) ware from the last quarter of the 3rd century BC.[230] In the late 2nd century BC a similar shape appeared in grey ware, this furnished with a central spout.[231]

The chronology of the ring-handled cup is at present controversial. Black-gloss examples seem to appear in Taranto and in Messapia at the end of the 4th century.[232] Unfortunately the Messapian graves include multiple depositions, which do not provide truly closed contexts, though most of the Hellenistic tombs containing ring-handled cups can be dated to a short chronological range between the end of the 4th and the early 3rd century BC.[233] Some graves containing a single deposition are more instructive.[234] It has been proposed that the typically Apulian black-gloss cup originated from the Gnathia ring-handled kantharos of the Alexandria Group around 280/70 BC.[235] This would imply that the shape needs to be down-dated, though this is far from certain.

Relationships with the C-type skyphos variant with ring-handles, not represented in the survey, require further investigation beyond the scope of this work. The morphological origins of this new variant C-type skyphos—which could alternatively be considered as a different type of kantharos—need to be defined for multiple regions during the Early Hellenistic period.[236] Any chronological reconstruction

particularly hard and covered by a high-quality black gloss. The class is attested in Taranto and in Messapia between ca. 225 and 125 BC. The typical morphological repertoire includes plates, dishes, salt cellars, cups, mastoid cups, two-handled cups, and lamps.

225 The shape is also known in modern literature as "tazza biansata," "tazza skyphoide," "coppa skyphoide," "skyphos," "kantharos," and "cup." The Attic "bowl shaped" or "goblet kantharos"—rarely attested in the third quarter of the 4th c. BC (*Agora* XII, Fig. 7, No. 722)—does not seem to be the predecessor of the western Mediterranean shape. A vessel similar to the later version of the ring-handled cup was manufactured in 2nd-c. BC Campanian-A, but with strap handles and higher upper handle attachment (Morel 1981a, 256, series 3221). The two shapes seem to derive from different origins, however, even though they share some features, as typical in the Hellenistic koiné. A similar shape, with large strap handles, is attested at Locri (*Locri* II, 223, No. 252, Pl. XXIX, end of the 4th–first half of the3rd c. BC).

226 Bernardini 1961, Pl. 66, 14–21 (various examples, Lecce Museum). For Manduria, see: Forti 1972, 23, Fig. 16. Carovigno: Forti 1965, 73–74, Fig. 22, Pl. XIV; Marinazzo 1978, 211, Fig. 2 (T 2, 310–275 BC). Francavilla Fontana: Marinazzo 1979, 37, Nos. 24–26, Fig. 11, Pl. VIII (T 1). Rocavecchia: Delli Ponti 1981, 62–78, Figs. 17 (T 1a/1970), 25 (T 4/1970), 30 (T 8/1970), 36 (T 17/1970). Giannotta 1995, 56, Fig. 24, Cat. No. 25 (T 11/1934, end 4th–early 3rd c. BC). Giannotta 1996, 47–49, No. 18 (T 19/1934, 325–300 BC). For Vaste: Campagna 1995, 228–29, No. 52, Fig. 8,3 (from the settlement, end 4th c. BC). Delli Ponti 1996, 195, No. 242 (T 18/1968, end 4th–early 3rd c. BC); 206, No. 278 (plain ware, T 3/1970, 275–250 BC?); 209, No. 288 (T 4/1970, end 4th c. BC). Giannotta 1997a, 167. For the settlement of Valesio: Boersma and Yntema 1989, 144, Figs. 15–17 (second half 3rd c. BC). Yntema 2001, 191–93, form K44A.

227 Lippolis 1994b, 249, Fig. 186 (from the necropolis). Prazzoli 1996, 334, types 11 and 11A (from the asty).

228 In the 3rd c. BC the vertical handled-cup was the preferred drinking vessel in the Pantanello necropolis, where it occurred in combination with the amphora (*Necropoleis*, 228, and Elliott 1998, 651). For other Metapontine funerary contexts see Galli 1926–28, 63, Fig. 1; Sestieri 1940, 75; Lo Porto 1981b, 377–78, Nos. 2–5, Fig. 104, 1–4 (early decades of the 3rd c. BC); Scarano 1992, 19 and Fig. 5,1–7 (end of the 4th c.–250 BC); Rocchietti 2003, Cat. No. 128 (first half 3rd c. BC); Osanna 2008, Pls. XXIIb (T 626) and XXIII (T 558). For Cozzo Presepe, see Morel 1970a, 102, Fig. 23.4 (320–270 BC). For Pomarico, see *Pomarico Vecchio* I, 93, Pl. 43, No. 156 (end of the 4th–first half of the 3rd c. BC). For Matera, see Lo Porto 1973, Pl. LXX,3,7, early 3rd c. BC (from a building, perhaps a workshop, at località Serra Pollara).

229 A few black-gloss and Gnathia specimens are attested at Ginosa (*Museo Taranto* II.1, 102, No. 43.32, late 4th–early 3rd c. BC; No. 43.64, 4th c. BC), Santo Mola (Donvito 1991, 76, No. 5 and 104, Fig. 38,8) and Acquaviva delle Fonti (*Archeoviva*, 89, T 25/1978, 320–300 BC).

230 Hempel 1996a, 338, form No. 14.

231 Hempel 1996a, 338, form No. 16. These classes and shapes are attested also at Valesio: Yntema 1990b, Pl. 82, forms 2 and 4 (HFR). Yntema 2005, 75, form 36 (Grey Ware).

232 Dell'Aglio 1996b, 330, No. 322; Prazzoli 1996, 334, type 11A (from the asty). Hempel 2001, 112 and 239, form 321 (attested in the necropolis since phase B, 325–275 BC).

233 Delli Ponti 1996. Giannotta 1996.

234 See, for instance, the tomb from Mesagne in which a ring-handled cup is associated with a Messapian *trozzella*, a black-gloss lekythos, and a black-gloss unguentarium (Forti type 2), ascribed to the end of the 4th c. BC (Nitti 1983, 85, No. 170).

235 Yntema 2005, 74, form 36 (Morel 3134), variant with knot handle.

236 See, e.g., D'Amicis 1995, 82, No. 1.3.30. This type of skyphos is attested also at Vaste (Giannotta 1997, 172, Fig. 6). This shape (Morel 3112b 1—drawing with modified foot, different from the cited original source: Forti 1965, 74, Fig. 19) is rarer than the stemmed ring-handled kantharos (Morel 3111a1 and 3112a1) and was especially common in Messapia. It was the typical shape manufactured by the Alexandria Group workshop in Taranto (Green 2001, 61, 70-1). A Gnathia ring-handled vessel—which clearly stands between a skyphos (for its foot) and a kantharos (for its handles)—is known from Daunia as well, attributed to the Knudsen Group (Lanza Catti 2008, 158, Fig. 137, 320–310 BC) and, later on, by the Late Canosan Group (Corrente 2003, 68, Fig. 5.4, from Canosa, Ipogeo dei Serpenti Piumati, advanced or late 3rd c. BC). Such a link between the ring-handled skyphos and the ring-handled cup

must be undertaken with caution, since it is risky to crystallize such processes into rigid schemes, especially in a period such as the Hellenistic, when different trends and fashions circulated quickly and forms exerted a reciprocal influence on each other.

The evidence from the urban necropolis of Metaponto supports the higher chronology for the ring-handled cup.[237] In the kerameikos the black-gloss ring-handled cup is documented in deposits of the late 4th and early decades of the 3rd century and later second half of the 3rd century BC.[238] Evidence from the necropoleis in southern Basilicata suggests that the ring-handled cup first accompanied the skyphos, subsequently replacing it in the early 3rd century BC funerary contexts.[239] In contrast, both shapes are attested in Metapontine settlement and sanctuary contexts of the same period.[240]

The development of the form can be traced in the assemblages from the necropolis of Taranto, where the ring-handled cup shows a morphological evolution from a continuous to a keeled profile, with a larger and shallower body.[241] An analogous development can be seen at Valesio, where in the later 3rd century the fairly broad shape assumed a more or less conical upper body.[242]

In the absence of preserved handle attachments it is difficult to recognize the shape from fragments since the body has the same profile as the stemmed cup, the Hellenistic cup-skyphos, and some types of one-handlers; nevertheless 62 base and 51 rim fragments from the survey have been confidently identified as belonging to ring-handled cups. The nearly-vertical wall with slight outward turn at the lip evident on some examples (e.g. **No. 99**) could easily be confused with a skyphos if the upper handle attachment were not present. The body can be shallower (**No. 97**) or deeper (**No. 98**) and rests on a ring foot with pointed underside (**Nos. 101–107**).[243] A slight offset of the lip appears in the latest specimens (**No. 100**). (ELC)

Kantharos

Although a relatively distinctive shape in terms of morphological elements, kantharoi are rare in the survey assemblages. Two fragments, a stem and a ribbed body sherd (**Nos. 108–109**), both dateable to the 4th century BC, together account for just 0.1% of all attributed black-gloss forms. (KS)

Among the drinking vessels the kantharos was particularly popular in the Hellenistic period, in Greece and in southern Italy, especially in Messapia, where different shapes and types developed.[244] In contrast, while attested in Gnathia ware, the shape is very rare in black gloss in the Metapontino, with only two examples from the Metaponto survey, both relat-

would follow Yntema's proposed shape history, without down-dating the origin of the ring-handled cup. Antonietta Dell'Aglio has instead argued that the ring-handled cup is a later variant of the typical Tarentine *tazza biansata* with vertical—instead of horizontal—handles, developed by the end of the 4th c. BC. Dell'Aglio 1996b, 326 and 330, Cat. No. 322, from the workshop deposit of Via Leonida: "Le tazze biansate vengono integrate e progressivamente sostituite dalla tazza ad anse verticali." For these two shapes, a common origin with the Late Classical bolsal is quite likely, especially in Taranto. See, e.g., the type 424/3-bolsal attested in Taranto in phase A (375–325 BC): Hoffmann 2002, Pl. 82. However the *tazza biansata* and the ring-handled cup follow a different development. While the *tazza biansata* becomes more shallow after the last quarter of the 4th c. BC (Hoffmann 2002, Pl. 82, type 424/4), the ring-handled cup keeps its deep bowl with a gently curved profile, becoming slender and more akin to a skyphos in the 3rd c. BC. In the 2nd c. BC, when the *tazza biansata* in its turn begins to have a rounded profile, the ring-handled cup has already developed a slight offset lip. Compare the profile of the two shapes in phase E1 (200–150 BC) of the necropolis of Taranto (Hempel 2001, 239, forms 321, 240, and 424).

[237] Three cups with ring handles are associated with a bolsal of the last quarter of the 4th c. BC in T 35/1986 (Scarano 1992, 19, Fig. 5). Two cups with ring handles are associated with Gnathia vases which can be dated to the first decades of the 3rd c. in T 558/2007 (Osanna 2008, Pl. XXIII).

[238] Deposit 2 (end of the 4th–early decades of the 3rd c. BC) and the abandonment layers of the stenopos (advanced second half of the 3rd c. BC); Silvestrelli forthcoming a.

[239] For the Pantanello necropolis, see Elliott 1998, 651. For Herakleia see Pianu 1990, 228, type 2 skyphos. At Taranto instead both the ring-handled cup and the skyphos are attested in phase C (275–225 BC), while only the second shape disappears in phase D (225–200 BC): cfr. Lippolis 1994b, 248–49.

[240] Both the ring-handled cup and the C-type skyphos are attested in deposit 2 of the kerameikos: Silvestrelli forthcoming a. Also the great number of Hellenistic skyphoi discovered from the Pantanello sanctuary confirms their continued production after the introduction of the new shape (Elliott 1998, 651). Black-gloss and overpainted ring-handled cups associated with black-gloss skyphoi were discovered in the votive deposit of Cozzo Presepe (Morel 1970a, Fig. 23, Nos. 4, 5, 8; 320–270 BC). The overpainted ring-handled cup is also attested in the residential area of that site (*Cozzo Presepe* 347, Fig. 122, No. 287). A great number of banded examples were brought to light in the excavation of the farmhouse known as Fattoria Stefan (A. Cavallo, pers. comm. 2009)

[241] Lippolis 1994b, 249–50, Fig. 186.

[242] Boersma and Yntema 198[illegible]; Yntema 2001, 193–95, form K4-3B. From this final step probably originated the mid-2nd c. BC spouted grey-ware form (Yntema 2005, 74, form 36A). The later variant 36B, attested in the second half of the 2nd c. BC, provided with a spout, abandoned the carination again and adopted a molded foot.

[243] For a banded example similar to **No. 98** (BG 113-04) from the same site, see Ch. 9, **No. 1** (PWBW 113-09), Catalog of Plain/Banded Wares.

[244] In Athens, the main drinking vessel of late 4th and early 3rd c. BC was the kantharos with its various shapes (kantharos, cup-kantharos, calyx-cup, bowl kantharos, and kantharos with special handles (*Agora* XXIX, 83)). In Messapia many types of black-gloss kantharoi are attested: Morel 1981a, series 3111, 3112, 3532, 3713, 3714, 3734, 4646.

ing to a different type of kantharos. The first type, attested by the kantharos stem (**No. 108**) can be associated with the type with high overhanging handles (similar to Morel's type 3714a), probably derived from metal vessels. This form was typical of northern Apulian red-figure pottery and was manufactured also in other classes, such as Gnathia and washed pottery (*scialbata* and *dorata*). These kantharoi were also common in central Apulia[245] and are attested at Taranto[246] and in Messapia.[247] Also of interest is a depiction of this shape on a stone feature from the Hellenistic buildings discovered in contrada Sansone of Metaponto in 1955.[248]

The second type of kantharos is attested by a ribbed body sherd (**No. 109**) characterized by a concave neck and convex body.[249] An example of this shape in Gnathia ware was discovered in the Metapontine chora in a tomb located near the Tavole Palatine.[250] The survey fragment shows a particular treatment of the decorative pattern, consisting of individual stamped arches above vertical lines, instead of the traditional ribs with continuous lines connected to the upper arches. (ELC)

Cup/one-handler

Since cups and one-handlers share similar morphological elements (rim, bowl, base), fragmentary rims cannot be assigned with confidence to either shape, and both are therefore incorporated into the category "cup/one-handler." Although having a relatively long production span from the second half of the 5th well into the 3rd century BC, cups and one-handlers do not initially appear to be particularly well-represented in the survey assemblages, with 71 examples accounting for just less than 4% of all attributed black-gloss forms and less than 1% of all black gloss. These forms, however, are identified solely on the basis of their rims (or rims and handles), in contrast to most other cup/bowl forms from the Classical and Early Hellenistic periods, and may therefore be considered to be better represented than a simple count of total fragments might otherwise suggest. (*KS*)

One-handlers are known in two varieties, the first with a convex body, the second with a deeper body and a flat, straight rim. One-handlers with convex body, direct rim turning inward, and base-ring with a reserved outer surface treated with miltos or completely glazed and horizontal handle attached just below the rim are exemplified by **No. 110**. This type of cup, also present in the Attic morphological repertoire,[251] appears in southern Italy in contexts dated to the second quarter of the 5th century BC[252] with vessels that reveal a possible derivation from banded one-handlers.[253] The form occurs in funerary contexts within the middle of the 4th century BC, but never achieved a wide popularity.[254] Continuity of production into the first half of the 2nd century BC has nonetheless been suggested for vessels from the Salento and Taranto.[255] They are attested,

245 Bitonto T 4/1981, red-figure kantharos, 330–310 BC (*Bitonto*, 154, No. 157). Ceglie Peuceta T 1/1908 vico Loseto (Cassano 1989, Fig. 16). The high-stemmed kantharos was common in the *argentata* and *dorata* ware (De Palma 1989, 82–83, Pl. V, Nos. 100 and 106; Montanaro 2007, 276, Fig. 131, No. 21.27 kantharos *scialbato* from Ruvo di Puglia, tomb in via delle Monache, second half–end of the 4th c. BC).

246 An example is attested in the chamber tomb No. 1 of Viale Virgilio (*Vecchi scavi*, 65, 78 No. 6.52). See also Hoffmann 2002, Pl. 74, form 323, type 323/3.

247 Oria, T 1/1994, 330–early 3rd c. BC (*Oria*, Fig. 64).

248 Lo Porto 1966, 156, Fig. 16, No. 2.

249 Morel 1981a, species 3530 and 4640. See also Hoffmann 2002, Pl. 74, type 323/2.

250 Lo Porto 1966, 225, Fig. 85, No. 2 (T 2/1943, early 3rd c. BC, rural cemetery of contrada San Salvatore).

251 *Agora* XII, 124–27, Nos. 764–70, Pl. 30 and Fig. 8 (examples from just after the middle of the 5th c. BC. to the last quarter of the 4th c. BC). *Agora* XXIX, 861–65, Fig. 58, Pl. 71 (last quarter 4th–first quarter 3rd c. BC).

252 *Museo Policoro*, 84, Pl. 28 (Latronico, Colle dei Greci, second quarter of the 5th c. BC).

253 Andreassi/Scattarella/De Lucia 1988, 280–81, 635–37 (480–470 BC, with a flat base); Miroslav Marin 1982, 78, Pl. IV,5 (Ceglie, FXI,5, second quarter of the 5th c. BC). Fornaro 1988, 198, Fig. 221.1 (Bari, Tomb II, first half of the 5th BC). For the connections with banded-ware one-handlers see Yntema 2001, 163–64.

254 Ciancio 1997, 218, No. 272; 221, No. 282; 223, No. 287 (Botromagno, tomb 3/1994, second half of the 5th c. BC). *Museo Taranto* II.2, 503–4, Pl. 42,c (Rutigliano, Type 12.2). Cassano 1982a, 68, Pl. III, A I 7, and Cassano 1988, 355, No. 763, Fig. 510 (Ceglie Peuceta, second half of the 5th c. BC). Cassano 1982a, 170, Pl. XXVI,6 (Ceglie Peuceta, second half of the 5th c. BC). Miroslav Marin 1982, 15–16, 112, Pl. XI (Ceglie Peuceta, F XIV); 98, Pl. VIII (F XIII 1–2); Cassano 1982a, Pl. XXIV,2 (Ceglie Peuceta, FXXII, first quarter of the 4th c. BC). Scarfì 1961, 252, Fig. 89, nn. 8–9 (Monte Sannace, tomb 5, middle 4th c. BC). Tinè Bertocchi 1985, 108–9 No. 8 (Ascoli Satriano, tomb 41, end 5th–beginning 4th c. BC). Rossi and van der Wielen-van Ommeren 1983, Pl. I,3,11 a VI.1 (Canosa, Toppicelli tomb 1, 1979, end of the 5th–beginning of the 4th c. BC); Giorgi 1988b, 197, Type 4.1 and 204, n. 80. A. Bottini 1987a, 13, No. 7, Pl. 1 (Banzi, T 164, second deposition, end of the 5th c. BC). Oppido Lucano: Lissi Caronna 1972, 496, Figs. 6, 4 and 11,4 (T 1, third quarter of the 5th c. BC); Lissi Caronna 1980, 171, Figs. 74,2 and 77,2 (T 46, 5th c. BC); Lissi Caronna 1980, 125, Figs. 8,2 and 10,2 (T 20, second half of the 5th–beginning of the 4th c. BC); Lissi Caronna 1972, 523, Figs. 44,3 and 46,3 (T 14, first half of the 4th c. BC); Lissi Caronna 1983, 234, No. 4. Lissi Caronna 1990–1991, 406–08, Figs. 205–5, No. 93. For one-handlers of this variant dated to the end of 4th c. BC see De Cazanove 2008, 409, Nos. 276–77, Fig. 296. Lo Porto 1973, 218 Pl. LXIII, 3,3 (Matera, T 8, dated to the end of the 6th c. BC but probably of later chronology). For an isolated presence in Sannio: Bonifacio 2004, 239, Fig. 9 (Carife, T 89, first quarter of the 4th c. BC) and 240, Fig. 40 (Carife, T 90, first quarter of the 4th c. BC).

255 Yntema 2001, 163–64. *Museo Taranto* III.1, 270, and Fig. 201 (Phase

albeit infrequently, in Metapontine necropoles both in the asty and in the chora within the first half of the 4th century BC[256] and occur also at farmhouses.[257] Local Metapontine production is confirmed by wasters from deposits of the kerameikos dated to the second half of the 5th and the first half of the 4th century BC.[258]

The second variety of one-handler cups has a flat rim, bell-shaped handles, and a ring foot. These banded cups were produced in Attica between 520 and 470 BC and then substituted by black-gloss versions beginning around 480.[259] South Italian imitations, collected by Morel in Especie 6210, came into use during the second half of the 5th century BC; in the 4th century BC they were very popular and are attested in many contexts.[260] The shape is part of the Metapontine morphological repertoire from at least the last quarter of the 5th century BC,[261] and, with transformations, continued into the Hellenistic period. The earliest examples, not represented among the survey material, have a large and shallow bowl, a flat rim, and bell-shaped handles attached just below the rim.[262] With the first decades of the 4th century the bowl is smaller and the external wall has a slightly concave profile.[263] (FS)

This bowl-like vessel could probably have been used both for solids and liquids, and this versatility can perhaps account for its success. One-handlers are known in two varities, the first with a convex body, the second with a deeper body and a flat, straight rim. The one-handler was common in the 4th century BC in Greece and in Magna Grecia, appearing in both black gloss and banded ware.[264] In contrast to the bolsal, the shape is not known in red-figure, overpainted, or Gnathia pottery. The one-handler was particularly popular in Lucania and in central and northern Apulia during the 4th century BC. In Metapontino the one-handlers with flattened rim became particularly popular in funerary contexts of the second half of the 4th century BC.[265] In Taranto the form is scarce in funerary contexts, while in the asty it is documented mostly by examples in banded ware.[266] Black-gloss one-handlers are rare in the necropolis of Herakleia.[267] While the latest examples in the Attic sequence date to the early 3rd century BC, in southern Italy the shape seems to go out of production one century later. The last examples from the necropolis of Taranto indeed belong to the first half of the 2nd century BC.[268]

Early Hellenistic one-handlers have flat rims, which can be horizontal (**Nos. 112–114**) or slightly sloping towards the inside (**No. 115**). The lip can otherwise be slightly downward-sloping towards the outside (**Nos. 117, 119–120**). From the beginning of the 3rd century the rim slopes more inwards (**No. 116**). A gentle concave curve appears on the upper wall in a different type, characterized by an S-shaped profile (**Nos. 121–122**). The handles show the usual development from the Classical rounded shape (**No. 111**) to the triangular pattern (**No. 114**). (ELC)

Cup

The cup is a typical Hellenistic drinking vessel, widely diffused throughout the eastern and western Greek world and generally in Italy during the 3rd century BC.[269] The shape was manufactured in metal, glass, and different types of wheel-made and mold-made pottery, with possible interconnections in form between the different productions.

Four fragments of hemispherical cups are exemplified by **No. 124**. Analogous shapes are attested

D: 225–175 BC).

[256] Contrada San Salvatore, T 7: Lo Porto 1966, 230, Fig. 53,3. Pantanello, Group i (460–440 BC): Elliott 1998, 619; Metaponto, Saldone T 7 (third quarter 5th c. BC): *Necropoleis*, 431; T 37 (410–370 BC): *Necropoleis*, 437.

[257] Fattoria Stefan has yielded 9 fragments (E. Vittoria, pers. comm. 2009).

[258] D'Andria 1980a, 411, No. 215, Fig. 56; Silvestrelli forthcoming a.

[259] *Agora* XII, 290, Nos. 758–60, Pl. 31.

[260] For a list of occurences see Giorgi 1988b, 198 and 204 n. 32. One-handlers with flat rim are known from Ginosa: *Museo Taranto* I.1, 187 (end 5th–beginning 4th c. BC, T 30, 66, No. 30.26, in association with a bell krater by the Tarporley Painter, Xenon Group vases and Shape II oinochoe), and the second half of the 4th c. BC (Nos. 31.2, 15–17 and 32.19). From Rutigliano are also one-handled examples of the second half of the 4th c. BC (T. 102). At Conversano (Tomb 2/79 7; *Arte e artigianato*, 413 and 415–17: end 5th–first quarter of the 4th c. BC) one-handled and two-handled cups with flat rim are associated in the same context. For the development of the shape see *Ceglie Peuceta* I, 78, ssg. No. 5, Pl. XI, and 137, n. 433. See also Giorgi 1988b, 197–98.

[261] D'Andria 1980b.

[262] Silvestrelli forthcoming a.

[263] Silvestrelli forthcoming a.

[264] *Agora* XII, 124–27, Fig. 8 Nos. 726–69. Pls. 30–31. Morel 1981a, series 6231. *Ceglie Peuceta* I, 78–79 and 137, No. 433; Pl. XI, No. 15. Giorgi 1988b, 197–98. Bruscella and Virtuoso 2005, 262.

[265] Sestieri 1940, Fig 25. Osanna 2008, Pl. XXIII (Metaponto, urban necropolis).

[266] Dell'Aglio 1996b, 324

[267] Only three examples dated to the second half of the 4th c. BC are recorded by Pianu (1990, 224).

[268] Hempel 2001, 238, form 221 (necropolis of Taranto, phase E1, 200–150 BC).

[269] *Agora* XXIX, Figs 20–22.

in Taranto,[270] in the interior of Basilicata,[271] along the Bradano River valley,[272] in Campania,[273] and in Calabria.[274] Gnathia examples are known as well.[275] The form, which recalls the 2nd-century Attic cups decorated with a net-pattern, is worked with a low ridge and scraped grooves below the rim on the exterior wall.[276] An analogous glass-and-gold specimen is known from the Tomba degli Ori at Canosa.[277] A similar shape in Campana A wares points to the diffuse Hellenistic *koiné* for this shape.[278]

Another type of cup, characterized by straight thick walls, conical profile, and two small grooves on upper exterior, is attested by a single example (**No. 123**). The shape recalls mastoid cups, but with a thicker wall. Analogous shapes were recovered at Civita di Tricarico in Gnathia pottery.[279]

Specimen **No. 125** exemplifies a type of depressed base, with shallow groove, which might be part of a hemispherical bowl,[280] but could also belong to the mastoid cup.[281] Both shapes, which were also executed in glass and in metalwork, are attested in Athens from the advanced 3rd to the third quarter of the 2nd century BC. These vessels were manufactured in southern Italy as well, also in the overpainted technique.[282] An example of an overpainted parabolic cup with a similar base was discovered in 3rd-century BC deposits of the Metapontine kerameikos.[283] There is some uncertainty about the chronology of these forms: ranges span from traditional high dating (second quarter of the 3rd century BC)[284] to a recent down-dating, formulated on the basis of Tarentine funerary assemblages, as well as the forms' presumed origin from the West Slope cups (end of the 3rd century BC).[285] A Gnathia cup from the Pantanello necropolis may provide important evidence for the absolute dating of this shape, which already appears to have been in use in the Metapontino in the second half of the 3rd century BC.[286] (ELC)

Dish

After skyphoi, dishes are the second best attested shape in the survey assemblages, with 296 examples, including 212 rims and 79 bases, making up over 15% of all attributed black-gloss forms and 3.4% of all black gloss. Forms dating to the 5th century BC are relatively few compared to later forms, with 16 examples that may date to the 5th century BC versus 263 which have attested dates of 400 BC or later. Of the latter, 250 belong to the mid-4th century BC or later, with 55 of these with date ranges falling into the 3rd century BC. (KS)

Dishes entered the Metapontine repertoire in the last quarter of the 5th century BC. The type exemplified by **No. 126**, derived from Attic models,[287] is characterized by a convex body and low base-ring with a convex outer profile; the inner surface often carries an impressed decoration. Produced at Metaponto from the last quarter of the 5th century BC,[288] it occurs in tombs of the Pantanello necropolis of the late 5th to the first half of the 4th century BC[289] and is also attested in domestic contexts.[290] The shape also seems to have gained a certain popularity in other sites of southern Italy.[291]

270 Hempel 2001, 165, T D/E1-1, No. 5, and 256, Pl. 25 (necropolis of Taranto, phase E1, 200–150 BC). Prazzoli 1996, 334, type 4A, from the deposit at Via Nitti, in the asty.

271 *Torre di Satriano* I, 275–76, Pl. XXXIX, Nos. 253–54 (3rd c. BC). De Cazanove 2008, 594, Fig. 305, Nos. 507–8 and 511 (from Civita di Tricarico, 3rd c. BC).

272 Prag 1983a, Fig. 130, No. 340 (from Cozzo Presepe). Prag 1992b, 97, Fig. 32, No. 776 (from Silvium, 2nd c. BC).

273 Fracchia 1990a, 242, Fig. 180, Nos. 127–28, from Roccagloriosa, 275–100 BC.

274 Preacco Ancona 1996, 349, No. 16, from Oppido Mamertina.

275 De Cazanove 2008, Nos. 142–43 and 148–49 (from Civita di Tricarico).

276 *Agora* XXIX, Fig. 20, Nos. 320 and 322, 200–160 BC.

277 Corrente 1992, 344, Fig. No. 45 (late 3rd–first decades 2nd c. BC).

278 See, for instance, Morel 1981a, 2574a 1.

279 De Cazanove 2008, 386–87, Nos. 144 and 147 (3rd c. BC).

280 *Agora* XXIX, Fig. 20, No. 321.

281 *Agora* XXIX, Fig. 20, No. 331; Morel 2153.

282 Overpainted examples are attested at Taranto (Hempel 2001, 51, Figs. 9 and 10), Metaponto (Elliott 1998, 630, T71-1) and Botromagno/Silvium (*Gravina* II, Nos. 495 and 499).

283 Abandonment phase of the kerameikos, advanced 3rd c. BC; Silvestrelli forthcoming a.

284 Green 1976, 13.

285 Lippolis 1994b, 272 and Fig. 203. Hempel 2001, type 625/3. Yntema 2001, form K46. For a punctual discussion on this matter, see Silvestrelli forthcoming a.

286 *Necropoleis*, 265, T 71, although the suggested chronology of the tomb (300–270 BC) has been recently down-dated (Jackson 2004).

287 *Agora* XII, 130, Nos. 809–15, Fig. 8, Pl. 32 (*bowl, deep wall and convex-concave profile*, beginning to the last quarter of the 5th c. BC). The shape, never popular in Attica, was occasionally exported to southern Italy: see *Agora* XII, 130, n. 12 and D'Henry 1981, 166, No. 8, Fig. 28, 40.3 (430 BC).

288 Silvestrelli forthcoming a.

289 Elliott 1998, 653, Group I (410–380/370 BC); Scarano 1992, 21, Fig. 11,3.

290 Specimens have been identified at the farmhouses of Ponte Fabrizio and Fattoria Stefan in the Metapontine chora (E. Lanza Catti and E. Vittoria, pers. comm. 2009).

291 South Italian and Sicilian imitations are collected in Morel 1981a, 2963, Pl. 81. Specimens from Peucetia, where the shape is frequently

Funerary assemblages at Pantanello from the first half of the 4th century BC[292] show the introduction of dishes similar in shape to **No. 128**, which have a body with a continuous profile and an almost flat or slightly downward-sloping rim.[293] Similar dishes have been found in association with vases attributed to the Dolon Painter.[294] Dishes in Metapontine contexts occur in increasing quantity and morphological variety throughout the 4th century BC and especially in the Hellenistic period; more than 200 Hellenistic-period examples come from the Metaponto survey, confirming the popularity of the form in this area. (FS)

In the Hellenistic period in southern Italy the shape was manufactured not only in black gloss, but also in red-figure, Gnathia, plain/banded ware, Hard-Fired Red (HFR), and grey ware. At the beginning of the Hellenistic period (325–300 BC) the dish appears to be better attested in domestic rather than funerary contexts.[295] This differs from Peucetia, where black-gloss, Apulian red-gloss and brown-gloss dishes are usually present in late 4th century graves. The black-gloss dish is rarer than its banded version in Messapian funerary assemblages.[296] At Taranto dishes are common in domestic areas, but appear only occasionally in funerary contexts.[297]

In contrast, different types and sizes of dish are present in the Hellenistic necropoleis of Metaponto.[298] Here examples are also found outside tombs, raising the possibility that they functioned as grave markers or, more likely, were used during post-burial funerary rites.[299] In the Metapontine kerameikos the form was largely manufactured in the last phase of production (end of the 4th–first decades 3rd c. BC).[300]

Hellenistic dishes are clearly distinguishable from their Classical ancestors by their narrower bases and straight walls, but a considerable variation can be seen in shapes, types, sizes, proportions and details, with many different contemporaneous types developing. Since one of the key identifiers is the ratio between rim diameter and total height, it is difficult to build a typology on the basis of rim or base fragments like those from the Metaponto survey. Differences between a "plate," a "dish," and a "bowl" are based on the relative width and depth of the body.[301] In the study of the better preserved materials from the Pantanello necropolis, the Hellenistic dishes were subdivided into two main types. A chronological development can be traced from the first type, which has a straight body and flaring ring base (Group v, dated 325–300 BC), to the second type, which has a similar body but also an out-turned lip and narrower foot in relation to the rim diameter (Group vi, dated 310–275 BC).[302]

attested, are listed by Labellarte (1988, 333–34, No. 719 and 1995, 129–130). More examples have been published from Gravina (T 3/1994: Ciancio 1997, 218, No. 271, 219, No. 274, and 220, No. 277: second half of the 5th c. BC), Monte Sannace (T 11, *Arte e artigianato*, 414–15, in association with a bell-krater by the Amykos Painter), from Rutigliano (*Museo Taranto* II.2, 505, Pl. 42h, where they occur from the second half of the 5th to the third quarter of the 4th c. BC), from Montescaglioso (T 2 of Vigna Pesca della Zingara: Lo Porto 1988–1989, 334 and 338 n. 15, Fig. 36,20). The shape occurs at Taranto in the second (phase A1: Hoffmann 2002, 48, type 622.1, Pl. 99) and third quarter of the 4th c. BC (*Gli ori*, 377, No. 26); the possible continuation of shape is suggested by specimens dated within the first half of the 3rd c. BC from Pomarico Vecchio (Surra and Preacco Ancona 1997, 74, Nos. 53–54, Pl. 33). Local productions have also been recognized at Locri (Bitti 1989, 148–49, No. 111, Pl. XXIII, first half 4th c. BC), Kaulonia (Gagliardi 2001, 293–95, No. 15, Fig. 275b), and Poseidonia (Pontrandolfo and Rouveret 1992, 423–24).

292 Elliott 1998, 653, D16 (410-380-370).

293 Elliott 1998, 653 (D16).

294 Matera, tomb 4 Rione Gradoni Pianella: Lo Porto 1973, 209, Pl. LVII,2; for the bell krater: *LCS Suppl.* 3, 3/D38. See also Tomb 2/1994 from Botromagno: Ciancio 1997, 32. Ciancio 2005, 47–48

295 For instance, in the necropolis of Herakleia only a few specimens are recorded, dated between the late 4th and the early 3rd c. BC: Pianu 1990, Pl. LIII, 3 (overpainted); Pl. LXXIV, 236.1 (black gloss). Pl. LXXV, 278.1 (plain ware with traces of gloss on the interior). The first two examples were used to cover the cinerary urn. For evidence from settlements see: Prag 1992b, 286–89 (ill.) (from Botromagno), Yntema 2001 (from Valesio) and De Cazanove 2008, 586–89 (from Civita di Tricarico).

296 See, for instance, D'Andria 1990, 123 and 144 (from tombs at Vaste, banded ware).

297 This in spite of the greater number of tomb assemblages with dishes from the Tarantine necropolis dating to the final phases (from the 3rd to 1st c. BC); Lippolis 1994b, 271, Fig. 204; 272.

298 Lo Porto 1981b, 370, Fig. 92.2 and 376, Fig. 101, No. 1 (late 4th–early 3rd c. BC), 374, Fig. 97, No. 1 (advanced second half 4th c. BC). Scarano (1992, 19) subdivides the dishes in five types, four pertinent to the black-gloss class and the fifth including mid-3rd c. BC banded- and plain-ware specimens.

299 See, for instance, the case of the black-gloss dish on top of T 32/1957 from the urban necropolis of Casa Ricotta (Lo Porto 1981b, 378, Fig. 104, No. 32, 3rd c. BC). Scarano (1992, 17) argues that black-gloss small dishes, small containers, and unguentaria were used to perform post-burial rites. Rocchietti (2002, 87–88, n. 24) has suggested that there is evidence of a cultural or ritual practice, stressing the association between the small patera and the small bowl, which is attested only outside the graves, but not among the grave goods. The author mentions the case of a dish that was intentionally destroyed upon the covering of T 9 in the necropolis of contrada Ricotta (Lo Porto 1981b, 355, mid-3rd c. BC).

300 D'Andria 1980a, 410, Fig. 56, No. 221 (deposit 1). Silvestrelli forthcoming a.

301 According to Elliott's classification, a 'plate' has a height of less than one-fifth its maximum diameter. The height of a 'dish' is more than one-fifth, but less than one-third its maximum diameter. Finally the 'bowl' is a deeper vase (Elliott 1998, 652).

302 Elliott 1998, 653.

Some early examples in the Hellenistic sequence from the Metaponto survey still show a slightly convex profile: the lip can be thickened with a horizontal rounded rim (four specimens exemplified by **Nos. 131–33**) or, more frequently, tapering with inwardly sloping rim (more than 60 specimens exemplified by **Nos. 134–35**). With the beginning of the 3rd century BC the lip starts to protrude from the wall (**No. 136**); the base has a flat or slightly convex underside and begins to show a steeper angle with the ring foot, which tapers slightly towards the resting surface (**Nos. 137–138**).

A less common type of dish has a slightly concave underside, with grooves on the ring foot and raised rings on underside (**No. 140**). This shape is also documented by two other base fragments from the survey and is present in the Metapontine kerameikos.[303] To judge by its fabric, however, the dish from site 43 (**No. 140**) was very likely imported from outside the region. A particular type of rim (**No. 139**) might have been inspired by Attic prototypes.[304]

During the first decades of the 3rd century BC dishes became shallower and flatter, with a straight profile. The tapering lip, almost triangular in section, became longer and more extended, creating a stronger angle with both the interior and exterior of the bowl. The rim is quite flat and the ring foot is high; the floor is thicker and the underside can be nippled (**No. 145**; rims **Nos. 146–50**; bases **Nos. 166–67**, dated 310–275 BC). This type is close to Morel 1514 and is typical of central and southern Italy around 300 BC.[305] A different type of dish, common in the 3rd century BC, has a similar internal junction between lip and body, but the lip is extended further and curved downwards (**Nos. 151–52**). A flattened version of this shape was also found (**Nos. 154–55**). A more angular profile, with a longer and more downwards curved lip, is typical of another group (**Nos. 156–58**). The latest specimens date to the mid-2nd century BC, in which period semi-glazing technique was common and the dish developed a shallow and broad shape, with a distinctive long and strongly downturned lip (**No. 165**). Sometimes the interior may be decorated by an impressed motif such as the rosette in **No. 171**. (ELC)

[303] D'Andria 1980a, Fig. 56, No. 221 (deposit 1, mid-4th c. BC).

[304] Compare Morel 1981a, series 2222.

[305] Morel 1981a, 118.

Bowl

Here the term "bowl" refers to the relatively large Hellenistic vessel of the 4th to 2nd centuries BC with offset lip, and the thin-walled bowl that is deeper than a dish and usually stands on a ring foot. A different and earlier form, with disk foot and straight convex body, dating to the late 5th–early 4th centuries BC, is represented in the Pantanello necropolis where it is called "bowl," but this is an unrelated shape.[306]

Ranging in date from roughly 400 BC down to the late 2nd century BC, the 54 identifiable bowl fragments (52 rims and 2 bases) represent 2.9% of all attributed black-gloss forms and 0.6% of all black gloss from the survey. A single base (**No. 182**) may belong to the mid-3rd century BC; 46 other fragments can be attributed to the mid-4th century BC or later, of which 19 post-date approximately 325 BC, 7 post-date 275 BC, 19 date from the last quarter of the 3rd to the third quarter of the 2nd century BC, and two examples may belong to the last quarter of the 2nd century BC. Seven fragments are attributable only to the general shape. (KS)

The number of fragments and different types of bowls represented in the survey assemblage demonstrate the popularity of this Hellenistic shape. Among these, a single example (**No. 172**, dated to the second half of the 4th c. BC), recalls the larger ribbed concave-convex form (**No. 196**), though with the lower part of the body being convex rather than concave. Similar examples are attested in northern Apulia in the second half of the 4th century BC.[307]

With the beginning of the 3rd century BC a thin-walled bowl with convex wall and inward curving rim appears, recalling the Attic echinus bowl in its deep variant.[308] The most common type has a groove on outer edge of the lip and an almost hemispherical bowl (**No. 173**, similar to Morel 2782 but with deeper bowl). This type, present at ten sites identified by the survey, is attested until the mid-2nd century BC in southern Italy. The type was also produced in Gnathia ware and, with further evolution, in grey ware.[309] Black-gloss parallels dated to the first half of the 3rd century BC were discovered in domestic and sanctuary contexts at Metaponto,[310] Gravina/

[306] Elliott 1998, 688, examples BO1 and BO2.

[307] De Cazanove 2008, 439 and 595, Fig. 306, Nos. 528–30.

[308] *Agora* XXIX, 1021 (250–200 BC?).

[309] Prag 1992b, 96–97.

[310] Postrioti 1996, Fig. 4d (votive deposit of Temple E, 4th c. BC, local

Botromagno,[311] San Brancato,[312] and Cozzo Presepe.[313] Another type (**No. 174**, Morel 2783), attested at five different survey sites, appeared at the end of the 4th century BC and was widespread in central and southern Italy during the first half of the 3rd century BC.[314]

The remaining bowls are characterized by a pendant projecting rim. A variant (**No. 175**, 3rd c. BC) has a slightly concave profile and rim grooved on the upper surface; this variant is not common and is attested only by two examples from the survey. It can be compared to 3rd-century BC Attic deep bowls with projecting rims,[315] and other examples in southern Italy are known from Botromagno (ancient Silvium).[316]

The most common bowl form during the Hellenistic period had a convex profile with a pendant lip set off from the wall by a groove (Morel's species 2530). The down-turned lip could be near vertical (**Nos. 178–80**, early 3rd–mid-2nd c. BC) or project outwards (**Nos. 176–177**, third-quarter 3rd c. BC). In the Metaponto kerameikos the bowl with down-turned lip is attested by a fragment from deposit 2 (end of the 4th–early decades of the 3rd c. BC), which is similar to our specimen **No. 178** in the shape of its rim and size.[317] The bowl with down-turned lip is well attested in domestic and sanctuary contexts of the Metapontino and also appears to have been common in central Italy during the 2nd century BC.[318]

One rim fragment (**No. 181**), also accompanied by a banded example from the survey (FMBW 535-09), is an example of a particular kind of bowl. It bears a flange just below the grooved rim which has two perforations. To judge from the comparanda, the flange was not continuous, but repeated several times at regular intervals. It might have supported a lid. Analogous shapes are known from Apulia and Calabria,[319] where this type of bowl was also produced in banded ware.[320] The extant part of **No. 181** is completely slipped and is characterized by good-quality black gloss rather than the dilute slips typical of banded ware.

All of these bowl forms in the survey were identified by rim fragments, which are the best diagnostic means for these shapes. Only in a few cases can they be identified from base fragments, since the usual ring foot is not sufficiently distinctive (e.g., **No. 182**); the latter feature is connected with shape changes and standardization in the Hellenistic period and presents a methodological problem, especially in fragment assemblages collected during survey campaigns. (ELC)

Small bowl

A total of 79 examples of small bowls were identified in the survey corpus, accounting for 0.9% of all black gloss and under 5% of all attributed forms. As with the cups and cup/one-handlers, these forms were only identified on the basis of their rims. (KS)

The number of small bowls that can be soundly dated to the Classical period is limited. Two fragments (**Nos. 183–84**) can be linked to a type characterized by an incurving thickened rim and a low body associated with a ring foot. This form probably derived from Attic prototypes[321] of the 5th century BC, which were also imitated in other areas of southern Italy.[322] Metapontine production is confirmed by wasters found in the kerameikos. The two examples from the survey suggest a possible development of the shape during the 4th century BC, from a vessel with a slightly incurving rim and a wide and low ring

or regional production).

311 Prag 1992b, 290 (ill.), Fig. 32, No. 770 (300–260 BC).

312 Quilici and Quilici Gigli 2002, 170, Fig. 20, No. 1.

313 Morel 1970a, Fig. 24, No. 11 (from a votive deposit at Cozzo Presepe).

314 Bruscella and Virtuoso 2005, 276–77 (from the Lucanian sanctuary of Torre di Satriano, first half 3rd c. BC). Casagrande 2002, 139, Cat. No. 76 (end 4th–early 3rd c. BC).

315 *Agora* XXIX, Fig. 65, No. 1091 (275–250 BC), No. 1095 (ca 200 BC?).

316 Prag 1992b, 103, Fig. 33, No. 809 (context dated to the late 2nd–1st c. BC). Small et al. 1994, 208, Fig. 7, No. 16 (from a deposit in the settlement, grey ware, 1st c. BC).

317 Silvestrelli forthcoming a.

318 Morel 1981a, series 2534 (Rome, 2nd c. BC). *Gravina*, 86, Nos 764–68. Morel 1970a, 104, Fig. 24.4 (from the votive deposit at Cozzo Presepe, 320–270 BC). Fabbricotti 1979, Fig. 16, No. 28 (from a perhaps sacred area at Ruoti, Potenza). An analogous shape is known in central Italy in the production of the *petites estampilles*: Brecciaroli Taborelli 2005, 84, Pl. 9, d.

319 *Forentum* I, Pl. 30 "coppa a labbro interrotto No. 2." *Gravina*, 296, Fig. 38, No. 915 (320–250 BC). Gagliardi 2001, 292, "ciotola" types A and B.

320 A banded example is attested at Cozzo Lategana (Castoldi 2008, Fig. 6, 5th–4th c. BC). For an example from Otranto, see Semeraro 1983, Pl. 86, No. 142.

321 *Agora* XII, 132–33, Nos. 843–89, especially Nos. 850 (430–420 BC) and 861 (450–425 BC), Fig. 9 and Pl. 33.

322 Morel 1981a, 2714, Pl. 67. A south Italian (Metapontine?) example has been found at Nocera, tomb 21: D'Henry 1981, 165, No. 7, Fig. 28,7 (around 430 BC, in association with a pseudo-Panathenaic amphora by the Amykos Painter). Melfese: A. Russo 1988, 217, type 5.1, nn. 59–60. They are also attested in Peucetia: see examples listed in *Museo Taranto* II.2, 505, Type 2, Fig. 43b; to these add *Bitonto*, 118, No. 22 (T 4/1982, end 5th–beginning 4th c. BC) and Cassano 1988, 355, No. 764, Fig. 511 (Ceglie, second half 5th c. BC). For Locri and Kaulonia, where similar small bowls occur from the end of the 5th c. BC, see Oliveto Ferrero 1989, 95, No. 44, Pl. XX, and Gagliardi 2001, 298–99, type C. Poseidonia: Pontrandolfo and Rouveret 1992, 422.

foot (as in the small bowl from deposit 1,[323] to which **No. 183** can be compared) to specimens with a more concave profile and higher cylindrical foot (as the one in deposit 3,[324] almost identical to **No. 184**). (FS)

The small bowl was a popular shape in southern Italy during the Hellenistic period; the shape is widespread in Magna Grecia and Sicily, with a range of different types and variants. It appears to have been a popular form in Athens as well; the Attic shape, known as an 'echinus' bowl, also gained popularity in the Hellenistic period, particularly in the 3rd and 2nd centuries.[325] Parallel development of the shape can be traced between Attica and southern Italy.

The basic features of the small bowl are a completely slipped surface, an inward-curving rim, and a shallow body with convex profile that rests on a convergent ring base. The shape developed from the Classical version, but adopted a thinner wall and an inward-curving rim.

In the Hellenistic period, small bowls are represented in a range of functional contexts in Apulian and Lucanian residential areas,[326] but also occur among funerary goods and at sanctuaries.[327] In the necropolis of Taranto these vessels are typical of the phase dating from approximately 225 to 150 BC.[328] In the asty of Taranto the small bowl is attested from a deposit of the 3rd to 1st centuries BC.[329] In the necropolis of Herakleia small bowls are present between roughly 330 and 270 BC.[330] At the Pantanello necropolis the shape is barely attested, probably because this funerary area went out of use at the beginning of the 3rd century BC,[331] but it is known from the urban necropolis, which dates slightly later into the second quarter of the 3rd century BC.[332] In the kerameikos of Metaponto the shape is evinced by different variants in deposits dated from the first half of the 4th century to the late 3rd century BC.[333]

Differences in the thickness of the wall and the ratio between rim diameter and height are solid criteria to indicate different morphological variants in this shape; a number of different types are attested from the survey and a broad chronological trend is apparent. In general, the profile develops from a steeper and thicker wall (**No. 186**) to a more convex and thinner wall (**Nos. 187–188, 190–191**).

At the end of the 4th century BC new types appeared. Example **No. 192** is characterized by a straight wall and a sharply inward-curving lip with rounded rim; it is attested by four specimens from the survey, and is also found on sites in the interior of Basilicata.[334] The form became very successful in the Campana A ware during the 2nd century BC (Morel 2737c1 and 2744c1–Lamboglia 34b).

Small bowl **No. 193** is the only example of this morphological variant, which features a shallow body. A completely different type of small bowl (**No. 194**) is characterized by a horizontal lip projecting outwards with a molded rim, and a deep and nearly hemispherical body that rests on a convergent ring base. Analogous forms are known from Apulia and Calabria, between the late 4th and 3rd century.[335] (ELC)

Concave-convex

Relatively distinctive in fragmentary assemblages like those from survey, concave-convexes are represented by 21 examples (20 rims and a single base fragment, **No. 195**), together accounting for just over 1.1% of all attributed black-gloss forms and 0.2% of all black gloss. Four examples (**Nos. 199-203**) may be earlier and belong to the middle of the 5th century BC, though the majority (16 examples) of the concave-convexes date from the late 5th to the early 3rd centuries BC, of which two examples seem to date from around the middle of the 4th to the early 3rd centuries BC (**Nos. 200–201**). (KS)

The concave-convex bowl, derived from the At-

323 Silvestrelli forthcoming a.

324 D'Andria 1980b, Fig. 67, No. 323, and Silvestrelli forthcoming a.

325 *Agora* XXIX, Fig. 64, Nos. 1026–33.

326 Surra and Preacco Ancona 1997, Pl. 33, Nos. 42–51 (from Pomarico Vecchio).

327 See, for instance, the small bowls from the Lucanian sanctuary of Torre di Satriano (Bruscella and Virtuoso 2005, 296, Pl. XLII, Nos. 281–88, mid-4th–early-3rd c. BC).

328 Phases D and E1 (225–150 BC); Lippolis 1994b, 268, Fig. 202; 270–2. Hempel 2001, 245, forms 624.4–5.

329 Deposit at via Nitti (3rd–1st c. BC); Prazzoli 1996, 334, types 7 and 8.

330 Pianu 1990, Pl. I, types 2 and 4 "ciotole" (Morel 2710).

331 Elliott 1998, 682. Only five Hellenistic small bowls are attested: SB1, SB2, SB5, SB6, SB7.

332 Lo Porto 1966, 215, Fig. 70, No. 2 (T 42/1955, 3rd c. BC). Scarano 1992, 22.

333 Deposits 1 (first half 4th c. BC) and 2 (end 4th–early decades 3rd c. BC) and abandonment levels (advanced 3rd c. BC. Cfr. D'Andria 1980a, 410, Fig. 56, No. 215. Silvestrelli forthcoming a.

334 See, for instance, De Cazanove 2008, 596, No. 572 (from the settlement of Civita di Tricarico).

335 Morel 1981a, 133, F 1735a1 (from Sybaris, 250–250 BC). Preacco Ancona 1992, 133, Pl. XLIX, No. 99 (from Locri, 360–330 BC). Prag 1992b, 81, Fig. 29, No. 685 (from Botromagno, late 4th–3rd c. BC). Giorgi 1988b, 197, Pl. 32, No. 9 (350–325 BC).

tic repertoire,[336] is a popular shape in southern Italy both in indigenous and Greeks sites.[337] The wide distribution and the morphological variety[338] suggest a plurality of production centers, probably not limited to the colonial world. These bowls were manufactured by Metapontine workshops beginning in the last quarter of the 5th century BC and produced until the Early Hellenistic period.[339] Two series are attested, which share a similar profile but have different dimensions. The most popular series is small in size, with a rim diameter of around 10 cm; the medium-large series is exemplified by **No. 195**, with a rim diameter between 16 and 20 cm and sometimes featuring an impressed decoration on the inner surface[340] (large concave-convex bowls, never very popular, are also known from Locri,[341] Civita di Tricarico, Ordona, and Canosa).[342]

The earliest examples from the survey (**No. 197**), which can be compared to Morel 2433,[343] have a large foot and a groove separating the thick convex rim from the concave body;[344] these derive from Attic prototypes of the later 5th century BC. In the first half of the 4th century, the diameter of the foot decreases, the body becomes higher, and the external surface of the rim can be characterized by a ridge (as in **No. 198**) or by a reed (as in **No. 199**, Morel 2441[345]). Examples of concave-convex bowls have been found at farmhouses[346] and in votive deposits,[347] but are rarely attested in Metapontine funerary assemblages.

The concave-convex appears to have gone largely out of fashion by the Early Hellenistic period. The survey evidence indeed records a decrease in the number of examples, a phenomenon that seems to parallel the contemporary situations at Metaponto and Taranto.[348]

One variant of the concave-convex appears to have persisted into the Hellenistic period: the shallow ribbed version, attested from the beginning of the 4th century BC, is still present in southern Italy until the early 3rd century BC. This type is very shallow, with a ribbed convex upper wall ending in a slight groove where it joins with the concave foot (e.g. **No. 200**). It is derived from the Classical Attic tradition.[349] The shape is attested by two examples from the survey and is well documented in Apulian sites in the second half of the 4th century BC.[350] This type is still found in Metapontine funerary contexts of the 3rd century BC[351] and is present in deposits of the kerameikos dated to the third quarter of the 4th century BC.[352] A larger version (e.g. **No. 201**, Morel 2433) recalls the Early Hellenistic bowl (e.g. **No. 172**), but differs from it in the distinctively concave lower body. Analogous large concave-convex bowls are attested in 4th century BC deposits in the kerameikos of Metaponto.[353]

The latest concave-convex specimens developed a narrower body and a less pronounced groove (e.g. **No. 196**); this form very likely led to the creation of the new typically western shape, commonly known as a 'salt cellar' (Morel 2424). (ELC)

[336] *Agora* XII, 130–31, Nos. 817–22, Fig. 8, Pl. 32: bowl, shallow wall and convex-concave profile (second quarter 5th–first half 4th c. BC).

[337] For concave-convex bowls found in Daunia and Peucetia, see Giorgi 1988b, 196, Labellarte 1995, 125, and *Museo Taranto* II.2, 506–7 (second quarter 5th–4th c. BC). For Messapia, where specimens are recorded both from tombs and from sanctuaries, see D'Andria 1990, 114, Nos. 147–48 (Vaste Fondo Melliche, Tomb 558, last quarter 4th c. BC) and 302–4, Nos. 255–67 (Oria, Monte Papalucio sanctuary); Mannino 2006, 120, Nos. 174–75, Fig. 129 (Attic, from Oria, 430 BC). See also Yntema 2001, 157–60, with further bibliography. Herakleia: Casagrande 2002, 137, No. 69 (Type 5a), Fig. 38 (375–325 BC). Local productions have been recognized at Locri (Olivero Ferrero 1989, 91–92, Pl. XX and Bitti 1989, 151, No. 117, Pl. XXIV) and Kaulonia (Gagliardi 2004, 511, No. 59, Fig. 185, end 5th–beginning 4th BC). Ample bibliography on find spots is given in Bruscella and Virtuoso 2005, 279–80 (Morel 1981a, F 2422 and 2433), 280–81 (Morel 1981a, F 2424), and 281–82 (Morel 1981a, F 2433). For two examples from the necropolis of Carife, in Samnium, see Bonifacio 2004, 238, Fig. 8 (Tomb 89) and 240, Fig. 46 (Tomb 90).

[338] The different types are gathered by Morel 1981a, 169–72, Pl. 50 in Especie 2420, 2430, and 2440.

[339] D'Andria 1980a, 411, Nos. 216–19, Figs. 56–57 (diam. 18 cm). Silvestrelli forthcoming a.

[340] D'Andria 1980a, 411, No. 217, Fig. 57 (deposit 1).

[341] Bitti 1989, 150–51, Nos. 114–16, Pl. XXIV (second-quarter 4th c. BC).

[342] De Cazanove 2008, 439, with bibliography.

[343] Morel 1981a, 170–71, Pl. 50.

[344] D'Andria 1980b; Silvestrelli forthcoming a.

[345] Morel 1981a, 172, Pl. 50.

[346] Specimens have been identified in a farm excavated by the Superintendent of Basilicata in 1979 at Sant'Angelo Vecchio (Lattanzi 1980, 406–7; E. Vittoria, pers. comm. 2009), at Sant'Angelo Grieco (E. Lanza, pers. comm. 2009), and Fattoria Stefan (E. Vittoria, pers. comm. 2009).

[347] Liseno 2004, 84, No. 6, Pl. XLIII.g–h (Metaponto, Favale votive deposit, end 5th c. BC) (FS).

[348] Hoffmann 2002, P. 100, types 624/5 (350–325 BC) and 624/6 (325–275 BC).

[349] *Agora* XII, 130–31, No. 819, Pl. 32 (shallow wall and convex-concave profile, end 5th c. BC).

[350] Prag 1992b, 95, Fig. 32, No. 758 (from the settlement of Botromagno, 335–300 BC). Montanaro 2007, 271, Fig. 122, 21.20 (from a tomb at Ruvo di Puglia, 350–300 BC). *Museo Taranto* II.2, 506, Pl. 43g, type 14.6a (from the necropolis of Rutigliano, end 5th–4th c. BC).

[351] See for instance T 444, discovered along the SS106 road in proprietà Vitelli (tomb goods on display at the National Archaeological Museum of Metaponto).

[352] Deposit 3 of the kerameikos (350–325 BC). Silvestrelli forthcoming a.

[353] D'Andria 1980a, 411, Nos. 216–19 and 412, Fig. 57, No. 217 (4th c. BC).

Salt cellar

Small and relatively durable, 46 salt cellars were identified in the survey corpus, including 39 rims and 6 bases, making up 2.5% of all black-gloss forms. Two examples representing the earlier version of these forms probable date to the 4th century BC (**No. 207**), while the majority (41) of dated examples belong to the late 4th and early 3rd centuries BC. (KS)

The salt cellar (series Morel 2424) was popular in the repertoire of typically Hellenistic small containers, particularly in Apulia and Lucania. The shape is characterized by a concave-convex profile, with a hollow foot usually equal in height to the convex upper body. The vessel is generally completely slipped by the typical Hellenistic-period dipping technique, but its underside may also be reserved with painted rings (**No. 203**). The form derives from the Late Classical version of the concave-convex, which developed a higher and narrower foot in the last quarter of the 4th century (see above and **No. 201**). While the presence of the ribless concave-convex progressively decreases in the second half of the 4th century BC and eventually seems to disappear at the end of the century, the salt cellar is widely spread in southern Italy between the late 4th and the first decades of the 3rd century BC. The shape is attested mostly in eastern Lucania, but also in central and northern Apulia and in northern Campania.[354] In contrast, it appears to be scarce in Taranto.[355] A practical small container for table condiments and votive offerings, in the Metapontino and interior Lucania the shape is well known from funerary as well as domestic and sanctuary contexts.[356] The salt cellar is still attested in the last phase of the activity of the kerameikos (end of the 4th–first decades of the 3rd c. BC);[357] it is also documented at Herakleia and in its hinterland.[358] The 41 examples collected in the survey are a good indicator of the success of the shape in the Metapontine region.

The most common type from the survey, with 28 examples, is exemplified by **No. 202**. Slight differences can be detected among the more common examples. The underside can be flat (**No. 202**) or pointed (**No. 206**). The rim usually curves slightly inwards (**Nos. 204–205**), but can also be direct (**No. 203**). The outer surface of the foot is smooth (**Nos. 202, 206**), infrequently decorated by a horizontal groove (**Nos. 204, 208**). The general development of the form sees a progressive growth of the foot in height, while the rim becomes correspondingly narrower. **No. 209** could be assigned to an earlier stage than **No. 202**, but it may also represent a coeval variation.

In traditional chronology, the lower terminus of the shape (roughly 350–270 BC) should probably be reconsidered. It has been suggested that an abrupt disappearance of the shape around 270 BC is incompatible with the subsequent emergence of the 2nd-century "small bowls with ribbon band rim."[359] In fact, the salt cellar is also documented in other classes after 270 BC: a plain ware version is attested in the first half of the 3rd century BC in central Apulia,[360] and the form appears again in Hard-Fired-Red (HFR) production of the late 3rd–first half of the 2nd century BC.[361] It is therefore likely that during the 3rd century BC the black-gloss version was still produced and that its final phase of development led to the production of the form in Apulian Hard-Fired-Red ware (HFR).[362] (ELC)

Krater

Krater rims are relatively durable and distinctive, but not well represented in the survey assemblages. Only 13

[354] Morel 1981a, 169. In Campania, for instance, the salt cellar is documented at Roccagloriosa (Fracchia 1990a, 239, No. 98).

[355] Hoffmann 2002, Pl. 100, type 624/6 (only two specimens, dated between 325 and 275 BC).

[356] For the necropoleis of Metaponto, see: Lo Porto 1966, 206–7, Fig. 60, No. 2 (T 2) and 210, Fig. 65, No. 2 (T 31). Elliott 1998, 690. Rocchietti 2003, Pl. XLVIII, Nos. 19–22. Osanna 2008, Pl. XXIII, T 558/2007 (first quarter 3rd c. BC). For the Pantanello sanctuary, see Wade 2002, 68, Fig. 3. For Cozzo Presepe: Morel 1970a, 99, Fig. 25 a–d. Prag 1983a, 357–58, Nos. 326–28. Pomarico Vecchio: Surra and Preacco Ancona 1997, 40 (ill.), Pl. 32, Nos. 24–28. Civita di Tricarico: De Cazanove 2008, 596, Fig. 307, Nos. 544–67. Serra di Vaglio: G. Greco 1991, 50–51, Fig. 115; 72, Fig. 145–50. Satriano: *Satriano*, 46, Pl. X, 1–4. For the sanctuary of Torre di Satriano: Bruscella and Virtuoso 2005, Pl. XLI, 270–79. Oppido Lucano: Lissi Caronna 1980, Figs. 61, 79, 94, 116, 123, 144. For the sanctuary of Chiaromonte: Barra Bagnasco and Russo Tagliente 1996, 187. Ruoti: Fabbricotti 1979, 359, Fig. 16, No. 56. In Peucetia the shape is documented at Botromagno (Prag 1992b, 290, Fig. 32, Nos. 739–55), Ginosa (*Museo Taranto* II.1, 104, Fig. 43.49), Monte Sannace (Scarfì 1961, 204, Fig. 34), Ceglie (*Ceglie Peuceta* I, 112, Pl. XI, No. 16). In Messapia at Otranto (Semeraro 1983, Pl. 91); Valesio (Yntema 1990b, 173, HFR group).

[357] Silvestrelli forthcoming a.

[358] Pianu 1990, 222, type 3 "ciotola" (340–280 BC). For the interior Val d'Agri, see for instance the evidence from Sant'Arcangelo (*Greci, Enotri e Lucani*, 240).

[359] Prag 1992b, 91.

[360] Depalo 1984, 179, No. 7.

[361] Yntema 1990b, 173 and Pl. 82, form 5.

[362] The specimen from T 31/1953 of the urban necropolis can likely be down-dated to the early 3rd c. BC, since the form seems more developed when compared to the earlier type represented by the salt cellar from T 20/1952 (late 4th c. BC). Compare Fig. 65 with Fig. 60, in Lo Porto 1966, 206–10.

examples account for 0.7% of all attributed black-gloss forms. The four examples of Archaic-period black-gloss krater rims make up almost a quarter of all kraters. (KS)

In the Late Classical and Hellenistic periods, the krater is a shape usually produced in Gnathia ware; black-gloss examples are definitely rare. In the fragmentary survey assemblages, where no traces of overpainted decoration are preserved, it is impossible to determine whether fragments come from black-gloss or Gnathia ware.

Although a relatively rare shape in Greece during the Hellenistic period, the krater is attested in Athens with different types, manufactured in black-gloss or West Slope ware.[363] In the same period in southern Italy, the krater developed different types. This vessel was often placed among the grave goods by native Apulians, while used as a tomb marker in some Greek colonies, such as Taranto and Metaponto.[364] The plain black-gloss krater is well attested in Messapian funerary contexts between the end of the 4th and the early 3rd century BC, with different shapes: bell-krater (with different types), skyphoid-krater, and bowl-krater (Italian *zuppiera*).[365] A rim fragment that probably belonged to a bowl-krater was recovered in the field survey (**No. 210**). This shape, which was particularly successful in Messapia, was manufactured mostly in Gnathia ware but also in plain black gloss. Relief ovoli were typical Hellenistic decoration for rims of both open and closed shapes (e.g. **No. 211**), a feature also attested in red-figure and Gnathia ware.[366] Even though the rim diameter of our specimen is quite small,[367] the shape has been tentatively interpreted as a small krater, bearing some similarities to the 4th century BC series of black-gloss calyx-kraters with overpainted decoration of Attic tradition;[368] the ovoli could be either in relief or painted in black on the reserved surface of the rim.[369] The distinctive rim diameter of about 30 cm suggests that another fragment should be identified as part of a krater (**No. 212**). (ELC)

Deep bowl

The deep bowl (Morel Especie 4731) is a form typically associated with Gnathia ware. Like the krater, it functioned as a mixing bowl. The shape is characterized by an almost vertical wall with two horizontal lugs (usually one or both pierced) below the rim. The deep bowl was popular in Messapia during the Hellenistic period.[370] A single fragment (**No. 213**) from the survey can be tentatively assigned to this form, even though the reconstructed rim diameter—to judge from the small and worn sherd—is quite small (13 cm).[371] The only preserved lug is not pierced. (ELC)

Lid

The survey black-gloss assemblage includes an interesting example of a Hellenistic vessel (**No. 215**), which likely served the twofold function of deep plate and lid. The shape, defined as a "lid-plate" (French *assiette-couvercle*) was typical of grey-ware and Roman productions, but some black specimens are also known. The lid-plate is attested in the native settlements of Botromagno and Civita di Tricarico, where it has a standard diameter of around 15 cm, like the specimen from the survey.[372] Following Prag, the form could derive from the lid of the 4th-century figured lekanai.[373] (ELC)

Mug

Mugs are not particularly frequent in the survey assemblages, with 13 rims making up 0.7% of all identified black-gloss forms. A single mug rim (**No. 243**) can be dated between 500 and 400 BC while the majority (11 examples, co-identified with **No. 244**) can be attributed to 400–300 BC. (KS)

The fragments from the survey most likely belong to a mug shape, a semi-open form with offset rim, a flat bottom separated from the body by a groove, and a band handle. The form derived from the so-called Pheidias mugs of Attic production.[374] This shape, also

[363] Calyx krater, bell krater, bolster, and lug-handled krater; *Agora* XXIX, 135–36.

[364] Toxey 1998, 692.

[365] Black-gloss kraters are attested, for instance, at Rocavecchia (Giannotta 1996, RF type bell-kraters, 47, Fig. 14, Cat. No. 17 and 54, Fig. 20, Cat. No. 28) and Rudiae (Bernardini 1961, bell-krater Pl. X,1 and bowl-krater XI,2).

[366] Stamped ovolo decoration is also seen in ceramic productions of Asia Minor in the 2nd and 1st c. BC (*Agora* XXIX, Pl. 135, Nos. 1707–9).

[367] An analogous problem is indicated by Prag (1992b, 110, No. 849) as far as a rim fragment from Botromagno is concerned.

[368] Koepcke 1964.

[369] See, for instance, Forti 1965, Pl. IV.

[370] Bernardini 1961, Pl. 50, Nos. 1 and 3.

[371] An overpainted specimen with small diameter (18 cm) is known from Valesio (Bernardini 1961, Pl. XVI, No. 3).

[372] Prag 1992b, 88–89; De Cazanove 2008, 425–26.

[373] Prag 1992c, 141.

[374] *Agora* XII, 72, Nos. 201–14.

part of the Xenon Group repertoire,[375] was adopted by local Metapontine workshops at least from the second half of the 5th century,[376] but the best known examples come from deposit 1, where plain and ribbed mugs have been found in large quantities.[377] The production continues until the last decades of the 4th century BC with a progressive elongation of the body and a diminution of rim and foot diameters.[378] The two fragments in the catalog could belong to the first (**No. 239**) and to the second half of the 4th century BC (**No. 240**). The shape is uncommon in Metapontine tombs, where it occurs just in the plain version,[379] but it appears at the farmhouses.[380] Plain and ribbed mugs are instead common in tombs of Daunia and Peucetia, where the existence of local productions has been hypothesized.[381] (FS)

Closed forms

Although encompassing a broad range of shapes, identifiable closed forms are less frequent in the black-gloss corpus than open forms. The total of 90 fragments includes 2 oinochoai, 11 amphorae, 25 pelikai, 3 hydriai, 18 lebetes gamikoi, 2 gutti, 22 squat lekythoi, 3 net-decorated lekythoi and 4 pyxides, which together account for just over 1% of all black gloss and just under 5% of all attributed form-sherds.

Fragments of closed vessels for which the shape cannot be determined overwhelmingly consist of body sherds, with 1,311 examples, accompanied by 6 rim fragments, 62 bases and 10 handles. Together these account for 15.9% of all black gloss from the survey. (KS)

Oinochoe

No. 241 can be possibly attributed to a shape 2 oinochoe; the form, derived from Attic prototypes,[382] was produced at Metaponto by the 5th century BC[383] in black gloss and, to a lesser extent, in red-figure.[384] Specimins by the Xenon Group are also known.[385] The shape is uncommon in Metapontine funerary assemblages, but as suggested by fragments found in farmhouses of the chora, it occurs in domestic contexts.[386] This type of oinochoe seems to be especially popular in Apulia.[387] (FS)

Amphora

Black-glaze amphorae are part of the Metapontine funerary assemblages, with evidence mainly concentrated in the first half of the 5th century BC;[388] in the central decades of the 5th they seem to be substituted by pelikai (see below). Their possible function as wine containers may also explain their occurrence in farmhouses excavated in the chora.[389]

The fragments from the survey sample are too worn to be fully interpreted. One rim with concave outer side (**No. 244**) can possibly be compared to the Pantanello group iii, which is dated to the central decades of the 5th century BC,[390] while another fragmentary rim (**No. 243**) belies its derivation from Attic neck-amphorae of the first half of the 5th century BC. Local production in black glaze with the same type of rim but a different type of base is confirmed by complete examples found in Metapontine tombs.[391]

With its larger diameter, **No. 246** can possibly be identified as the rim of a monumental pseudo-Panathenaic amphora. The absence of any trace of decoration on the outer surface suggests the fragments

[375] E.G.D. Robinson 1996, 450.

[376] Silvestrelli forthcoming a.

[377] D'Andria 1980b, 407–10, Nos. 200–10, Figs. 54–55.

[378] D'Andria 1980b, No. 310b, Fig. 65, of the Xenon Group, which also adopted the shape.

[379] Elliott 1998, 667, M5.

[380] Sant'Angelo Grieco, Fattoria Stefan, and Fattoria Fabrizio (E. Vittoria and E. Lanza Catti, pers. comm. 2009).

[381] Labellarte 1995, 108, No. 2.1.12.

[382] D'Andria 1980a, 406, Nos. 196–98, Figs. 54 and 56; Silvestrelli forthcoming a.

[383] For the Xenon Group, where the shape is attested already in the last quarter of the 5th c. BC, see E.G.D. Robinson 1996, 447.

[384] *LCS Suppl.* 3, 3/A37 (Anabates Painter).

[385] Green 1972, 6. For the origin of the Metapontine version of the shape, also characterized in black-gloss specimens by the presence of a ribbed decoration on the shoulder, see E.G.D. Robinson 1990, 259–62.

[386] Black-gloss specimens have been found at Sant'Angelo Grieco and Fattoria Stefan, which has yielded also a Gnathia example (E. Lanza Catti, E. Vittoria, pers. comm. 2009).

[387] Labellarte 1995, 103; in addition to the examples listed by Labellarte add *Museo Taranto* II.2, 473, Type 3, Pl. 33a (Rutigliano) and Hoffmann 2002, 35, Type 115/1, Pl. 56 (Taranto, phase A1, 375–350 BC). An oinochoe of possible Metapontine manufacture was found in association with a pseudo-Panathenaic amphora by the Amykos Painter at Nocera, T 21: D'Henry 1981, 166, No. 10, Fig. 29.

[388] Elliott 1998, 648–49; see also Lo Porto 1966, 209–10, Fig. 64 (urban necropolis, T 29, Attic, dated to the middle of the 6th c. BC) and 231, No. 1, Fig. 93.1 (rural necropolis of contrada San Salvatore, T 7, dated to the middle of the 6th c. BC). Vessels without secure context from the necropolis of Metaponto: Sestieri 1940, 115, Fig. 54. The amphora in the Chini Collection at Bassano del Grappa has been attributed to Metapontine production: Labellarte 1995, 100, No. 2.1.1 (second half 6th c. BC).

[389] Amphorae have been recognized in the assemblages of Fattoria Stefan and Sant'Angelo Grieco (E. Lanza and E. Vittoria, pers. comm. 2009).

[390] Elliott 1998, 648.

[391] See Ch. 19, site 737, T 22 (second half 5th c. BC). For a similar amphora from the urban necropolis, found in association with a lebes gamikos by the Pisticci Painter: Osanna 2008, 928, Pl. XX,2.

were part of a black-glazed vessel of a type frequently recovered in deposits outside tombs.[392] (FS)

Pelike

The survey yielded 25 recognizable pelikai fragments, making it the most numerous closed shape, although it accounts for only 1.4% of all attributable black-gloss forms.[393] Pelikai are normally figured and less popular in black-glaze versions,[394] but Metaponto seems to be an exception: beginning in the central decades of the 5th and for the entire 4th century BC black-gloss pelikai occur in funerary assemblages both in the asty and in the chora; the shape is also attested at farms.[395] The types in use from the second half of the 5th are known from tombs found at Pantanello and Saldone.[396] They have a flaring[397] (or, more typically, out-turned) rim angled on the outer surface, flat handles that are sometimes concave in the middle,[398] and a low base-ring. The maximum diameter can be just above the foot[399] or, more commonly, at the mid-point of the body.[400] The inside of the neck is glazed as it is also the outside, with the exception of a reserved area in the lower part of the body at the junction with the foot, whose outer surface is partially glazed. During the 4th century BC the neck becomes higher and the pot assumes a slender appearance.[401] This type, to which we can compare **Nos. 247–248**, is especially popular during the first half of the 4th century BC[402] and seems to be adopted only in black gloss. A second type[403] (represented in the survey sample by **No. 250** and possibly by bases **Nos. 251–252**) is characterized by a high base and an out-turned rim, and also occurs in red-figure pelikai of local manufacture. Some examples bear a patterned decoration that can be easily connected to local red-figure workshops,[404] but most of the specimens are completely glazed.[405] As for the plain pelike, they seem to be especially common within the first half of the 4th century BC; in the second half of the century the most popular type seems to be the pelike with molded rim.[406] (FS)

The pelike is a typical western shape, adopted in red-figure, Gnathia, and black-gloss productions. The plain black-gloss version is quite unusual during the Hellenistic period in southern Italy in general, with the Gnathia-ware version being more common. Still, the black-gloss form was well represented in the Hellenistic urban necropolis of Metaponto.[407] Only five specimens are attested from the survey by small rim fragments and it is not possible to determine which types are represented. For example, the large

[392] Carter 1998d, 124–25. For figured pseudo-Panatheanaic amphorae see Ch. 8.

[393] On use as a vessel for oil (and possibly also for wine or water), see Hoffmann 2002, 105–8, particularly 106–7, with previous bibliography. See also Lynch 2001, which reports the presence of black-glazed pelikai in deposits of Athenian wells, which suggests their use as vessels to fetch water. For the possible use of Metapontine pelikai as subtitutes for table amphorae see Elliott 1998, 670. A special connection between pelikai and female tombs has been suggested for Taranto (Taranto: Hoffmann 2002, 107–8); this association is not exclusive (Hoffmann 2002, 106 n. 584) and does not seem to be confirmed for Metapontine tombs: J. Hall 1998, 585.

[394] For the history of the shape in Attic production and evidence of black-glazed pelikai from the Agora, mainly dated to the first half of the 5th c. BC, see *Agorà* XII, 48–51, Nos. 13–33, Fig. 2, Pls. 1–2. For Attic black-glazed examples in the 4th c. BC and their connections with red-figure pelikai, see Papanastasiou 2004, 12–23.

[395] Sant'Angelo Grieco (E. Lanza, pers. comm. 2009) and Sant'Angelo Vecchio, House on Hill (E. Vittoria, pers. comm. 2009).

[396] Pantanello: Elliott 1998, 670–74. Saldone: *Necropoleis*, 431 (T 5, 420–370 BC); 434 (T 17, first half 4th c. BC); 435 (T 19, last quarter 5th or first half 4th c. BC); 436 (T 23, second half 5th c. BC); 437 (T 27, last quarter 5th–first quarter 4th c. BC). See also Ch. 19, site 737, T 23. Other tombs with pelikai from Metaponto are published by Lo Porto 1966, 226, (T 3, not illustrated) and 227, No. 1, Fig. 86,1 (Metaponto, San Salvatore, T 4, dated to the first half of the 5th c. BC but possibly of later date), and De Siena 2007, 439, Pl. XI,3 (rural necropolis at proprietà Bonaventura, T 295). The shape in black glaze is rarely attested outside Metaponto; examples are known from Pisticci (Lo Porto 1973, 161, No. 24, Pl. VIII,3,2), from Taranto (Via Dante Tomb, *Museo Taranto* I.3, 358, 125.36–37, end 5th–beginning 4th c. BC), and from Armento (Serra Mauta, T 10: *Greci, Enotri, Lucani*, 3.32.10, end 5th–beginning 4th c. BC).

[397] Elliot 1998, P24 (second half 5th c. BC).

[398] Elliot 1998, P18 (second half 5th c.BC).

[399] Elliot 1998, P24, *Necropoleis* (from Saldone, T 23, last quarter 5th c. BC).

[400] See Ch. 19, site 737, T 23. *Necropoleis*, Saldone T 19 (last quarter 5th c. BC).

[401] Elliot 1998, P15, P 29, P30, and P32. *Necropoleis*, Saldone T 25 (end 5th–first quarter 4th c. BC), T 27, T 17.

[402] Elliott 1998, 671 and 674, P30 is nonetheless dated to the second half of the 4th c. BC

[403] Elliott 1998, 670–71 (elaborate pelike).

[404] Elliott 1998, 670, P27 and P12, group i (second half 5th c. BC).

[405] Elliott 1998, 670–71, group ii and iii (end 5th–first half 4th c. BC). See also Lo Porto 1966, 217, No. 1, Fig. 74,1 (urban necropolis, T 45, dated to the 3rd c. BC). This type of pelike is also rarely found outside Metaponto. Examples are known from Montescaglioso (Lo Porto 1988–1989, 339–40, No. 2, Fig. 44, (T 2, first half 4th c. BC), Policoro (Degrassi 1965, 19, No. 14, Fig. 46), from Pomarico (Piana di Pacilio, T 8, first half 4th c. BC: Lattanzi 1976, 862, Pl. CXLI,1) and from Ginosa (T 27, middle of the 4th c. BC: *Museo Taranto* II.1, 70, No. 31.14). A miniature example of possible local production was found at Vaste (Fondo Melliche, T 597: D'Andria 1990, 141, No. 215, first half 4th c. BC).

[406] Elliott 1998, 671 group iv.

[407] Scarano 1992, 2[illegible]. Three black-gloss pelikai, characterized by the typical Lucanian foot (high cylindrical ring base), were discovered in the chora in 1935 (Sestieri 1940, 110, Fig. 50). Another example is from T 1/2002 in the Basento valley (Nava 2003 a, Pl XXXII, 2, 350–300 BC).

flaring mouth with downturned thickened lip (**Nos. 254–255**) may belong to a number of different types of Early Hellenistic pelikai, which are distinguished by different bases (cylindrical or bell-shaped), bodies (globular or pear-shaped), rims (more or less flaring and down-turned), and necks (with or without the horizontal molding). In the necropolis of Taranto pelikai with this rim form are attested in the second half of the 4th century BC.[408] Also produced in red-figure and Gnathia wares, this shape is known in southern Basilicata in the last three decades of the 4th century BC.[409] A black-gloss specimen from Metaponto with analogous rim is present in the urban necropolis.[410] A type known from the Gnathia-ware version[411] and attested during the last decades of the 4th century BC[412] is present in the survey (**No. 256**) and attributable by the horizontal ridge below the rim. (ELC)

Hydria

No. 257 can be associated with the rim of a black-glazed hydria, dateable to the second half of the 5th or the first part of the 4th century BC, as the large straight neck and the profile of the rim suggest. Hydriai are usually produced in red-figure, but black-glazed examples are also known.[413] The diameter of this fragment (5 cm) suggests a vessel of reduced dimensions, such as the two red-figure hydriai, found in a recently excavated Metapontine tomb, that can be connected to the Pisticci Painter workshop.[414] (FS)

Lebes gamikos

The survey has yielded 18 highly fragmentary examples of lebetes gamikoi, a shape common in Metapontine funerary assemblages from the Late Archaic period to the second quarter of the 4th century BC.[415] Local production continues until the second half of the 4th century BC.[416] The shape is also present in the local red-figure morphological repertoire,[417] and attested at farmhouses in the chora.[418] While **No. 258** can be possibly dated to the last quarter of the 5th or to the first quarter of the 4th century BC, the concave shoulder of **No. 261** is typical of vessels of the second half of the 4th century BC. **Nos. 259–60** are unusual for their carinated shoulder, which is not attested in vessels of Metapontine manufacture but can instead be found in examples from Taranto. (FS)

In the Hellenistic period the evidence for black-gloss lebetes gamikoi in the Metapontino is scarce, even though the shape is still well attested in Gnathia ware.[419] A ribbed wall fragment (**No. 262**) was discovered at site 401, where the remains of a small necropolis were identified on the edge of a plateau overlooking the Basento valley. (ELC)

Guttus

Inspired by Attic models, the production of gutti started in southern Italy at the beginning of the 4th century BC, mostly in Campania, Lucania, and Apulia, with no clear differences in forms between these regions.[420] The main *floruit* of the guttus appears to have been between about 340 and 300 BC. While the guttus was a common shape in the necropoleis of the interior of Basilicata[421] and in Apulia,[422] no gutti were found either in Pantanello or the urban necropolis of Metaponto.[423] This is significant, since the shape is

[408] Hoffmann 2002, Pl. 66, type 311/4, and Pl. 68, types 311/18 and 311/19, phases A2 (350–325 BC) and B1 (325–300 BC).

[409] Pianu 1990, 228.

[410] Lo Porto 1966, 217, Fig. 74, No. 1 (T 45/1956, 3rd c. BC).

[411] Sammis 1968, 517, Fig. 13c.

[412] Pianu 1990, Pl. XV,3.

[413] Elliott 1998, 655–56 with examples dated to the second and third quarter of the 5th c. BC.

[414] Metapontino, lavori del consorzio di bonifica, sito, tomb 55: Nava 2003a, 667, Pl. XXXIII,1.

[415] Elliott 1998, 639–63.

[416] Fragments of lebetes gamikoi have been found in deposit 3 of the kerameikos.

[417] Pisticci Painter: Burn 1998, 622–23, T354-2. For the Dolon and Creusa Painters workshop see De Siena 2007, 443, Pl. XV,a and Silvestrelli 2008, 282, Fig. 2.

[418] Fattoria Fabrizio (E. Lanza Catti, pers. comm. 2009).

[419] In the Pantanello necropolis the last black-gloss lebetes gamikoi date to about 370 BC, while the overpainted specimens continue until the first quarter of the 3rd c. BC (Elliott 1998, 632–35 and 661). Gnathia lebetes gamikoi are attested in the first half of the 3rd c. BC in the urban necropolis of Metaponto (Lo Porto 1966, 214, Fig. 69, Nos. 1–2).

[420] Morel 1981a, Nos. 8111–73; Gilotta 1985, 108.

[421] Examples are known from the necropolis of S. Arcangelo di S. Brancato (S. Bianco 1994, 120, Fig. 7, early third quarter 4th c. BC), from Irsina (Lo Porto 1973, Pl. LXXVII, Nos. 5 and 6) and from Montescaglioso (Lo Porto 1988–1989, 379, Fig. 88, No. 1, T 14/1957, 325–300 BC).

[422] Morel 1981a, 422.

[423] Scarano (1992, 20) lists four different types of 'gutti' from the necropolis, dated to the first half of the 3rd c. BC, which are indeed 'baby-feeders.' However Rocchietti (2003, 44–45 and 116) lists six gutti, which were once at the Museum of Reggio Calabria and are said to come from the necropolis of Metaponto together with other finds discovered in the 1920s and 1930s. Some specimens could have been brought to light in the chora (Rocchietti 2003, 18; Pesce 1936, 447; Califano Ascenti 1998, 99). A miniature black-gloss guttus was discovered in T 440/2006 of the urban section of the necropolis along road SS106, in proprietà Vitelli, dated to the 3rd c. BC (now on display at the National Museum of Metaponto).

well attested in the kerameikos during the second half of the 4th century BC. Some examples are also known from the urban sanctuary[424] and from excavated farmhouses in the Metapontine chora.[425] The shape is therefore present in the Metapontino, but in contrast to other parts of southern Italy, it was apparently not used in funerary contexts.[426] Although produced in the kerameikos, production may have been mostly geared towards the extra-urban markets in Lucania and Apulia, since the form is not as well represented in the urban area of Metaponto as in areas beyond it. Metapontine production appears stylistically original, since many subjects depicted on the gutti from the kerameikos differ from the Apulian tradition.[427]

The more common guttus attested in Metapontino is the type with a ribbed body that rests on a ring foot, and which was provided with ring handle and elongated spout (Morel series 8141). The top of the body could hold either a filter (e.g. **No. 263**, likely covered by a lid) or a small relief (e.g. **No. 264**). The plastic 'medaillon' is a typical feature of western production. The central round surface was used to depict single heads or figured scenes. The molded medaillon from the survey likely depicts Herakles' head crowned by his leonté (lion skin) or perhaps the head of a Silenos. A similar specimen at the Gotha Museum is said to come from Basilicata.[428] The lidded type with shallow body (**No. 263**) shows a stronger Attic influence; the shape is attested in Apulia and in Basilicata in the second half of the 4th century BC.[429]

The precise function of the guttus is still unclear. It could have been an unguent container as well as a lamp-feeder, with its spout particularly suited for pouring dense and viscous liquids. (ELC)

Squat lekythos

Black-glazed squat lekythoi were produced in Metapontine workshops starting at least in the last decades of the 5th century BC.[430] The variety of forms attests the great popularity of this shape, which is well exemplified by the lekythoi found at the Pantanello necropolis;[431] lekythoi are also attested in domestic contexts.[432] Specimens from the survey sample dated to the Classical period belong to two different types. Base **No. 265** bears similarities to black-gloss and figured lekythoi that are usually dated between the second quarter and the last quarter of the 4th century BC;[433] similar vessels have been found in the kerameikos[434] and in a tomb from Metaponto in association with a pelike by the Creusa Painter.[435] These attest that lekythoi with a pronounced squat body were also produced in the late 5th and first decades of the 4th century BC. One body fragment with ribbed decoration (**No. 266**) may be compared to the squat lekythoi attested at Pantanello in tombs of the second half of the 5th and first quarter of the 4th century BC,[436] as well as to others from sites in Lucania and Peucetia.[437]

During the Hellenistic period, plain black-gloss squat lekythoi seem less popular than their counterparts with overpainted Gnathia-style or black net-pattern decoration (see 'net-decorated' lekythoi below and **No. 271**). Only seven fragments of squat lekythoi were identified in the corpus of black gloss from the survey. The vase was still typically used in Apulian and Lucanian funerary contexts of the last quarter of the 4th century,[438] while its presence was rarer in residential contexts.[439]

No. 267 has a completely-painted bell-shaped mouth, common in Gnathia ware and net-decorated lekythoi, but also found on black-gloss examples in

[424] *Metaponto* I, 103, Fig. 97, a–b, gutti of the second half of the 4th c. BC.

[425] Three different gutti were recovered at the 4th-c. BC farmhouse at Ponte Fabrizio excavated by ICA: a medallion (depicting a gorgoneion), a molded spout in the shape of a lion head, and a plain spout A-rim with three parallel moldings belonging to a guttus was found at Sant'Angelo Grieco and a fragmentary specimen was also discovered at Fattoria Stefan.

[426] Many 4th-c. BC gutti were discovered among the kiln wasters: D'Andria 1980a, 430, Fig. 66, Nos. 311–17 (deposit 3); 433, Fig. 70, Nos. 370–71 (deposit 5); 439, Fig. 71, No. 382 (deposit 6); 444, Fig. 75, No. 404 (deposit 8).

[427] Deposit 3 of the kerameikos (350–325 BC): Silvestrelli forthcoming a.

[428] *CVA Gotha* 2, Pl. 91, 6–7, 4th–3rd c. BC.

[429] *Museo Taranto* II.2, 517, Pl. 46, b–d (from Rutigliano). P. Bottini 1998, 79–82 (from the necropolis of Serra Città). P. Bottini 1988, 169, Fig. 8 (from Armento). Rocchietti 2003, Pl. LXI, No. 34 (from Metaponto, second half 4th–early 3rd c. BC).

[430] Silvestrelli forthcoming a.

[431] Elliott 1998, 685; J. Hall 1998, 566, table 12.3.

[432] Fattoria Fabrizio and Fattoria Stefan, with vessels dated between the second half of the 5th and the first quarter of the 4th c. BC (E. Lanza and E. Vittoria, pers. comm. 2009).

[433] Elia and Meirano 1997, 84, No. 120, Pl. 40, with list of references. See also Hoffmann 2002, 31 Type 111/120–122, Pl. 47 (Taranto, Phases A1–A2, 375–325 BC).

[434] Deposits 19 and 22, Silvestrelli forthcoming a.

[435] See catalog below.

[436] Elliott 1998; Morel 1981a, 5411, 360, Pl. 167.

[437] Examples listed in *Museo Taranto* II.2, 485–86. (FS)

[438] Scarano 1992, 18–9 and Fig. 3 (T 17/1986).

[439] Some examples are known, for instance, from the settlements of Botromagno (Prag 1992b, 297 (ill.), Fig. 39, No. 931), Pomarico Vecchio (Elia and Meirano 1997, 83, Nos. 117–19, Pl. 40) and Civita di Tricarico (De Cazanove 2008, 377).

the second half of the 4th century BC. The typical squat lekythos base consists of a high, molded conical foot. Two lekythoi are represented by base **No. 268** which is particularly notable for its large diameter. The fragment may belong to a lekythos with slender body, such as those attested in the necropolis of Taranto during the second half of the 4th century BC and in the Apulo-Lucanian region during the same period.[440] The narrower type of base, exemplified by **No. 269**, is represented by three specimens and is well known in Basilicata as well as in Taranto.[441] A type with ribbed body (**No. 270**) was also recovered by the survey. (ELC)

Net-decorated lekythos

Net-decorated lekythoi were produced in Greece in the first half of the 4th century BC and subsequently in southern Italy starting in the the mid-4th century BC or slightly earlier. These vessels were widely used in Magna Grecia and Sicily, particularly in funerary and votive contexts. Each region seems to have developed its own variations. For example, Campanian and Sicilian net-decorated lekythoi are closer to the Attic prototypes, while shapes and types which developed in Apulia and eastern Lucania tend to show more originality.[442] Net-decorated lekythoi are quite common in eastern Lucania[443] and in the Metapontino between the second half of the 4th and the first half of the 3rd century, with an apparent peak around 330–300 BC.[444] The manufacture of this class of lekythoi at Metaponto is attested in the kerameikos around the mid-4th century BC.[445] The single net-decorated lekythos rim from the survey (**No. 271**) is a typical Apulo-Lucanian lekythos, belonging to type IVb of Hurschmann's classification, the main arrangement dedicated to this specific class.[446] This form is characterized by an out-turned lip with interior projection and convex rim, a long concave neck with angular junction with the shoulder, and a slender body mounted on a low and thick ring foot. The foot usually has a profile analogous to **No. 272**. (ELC)

Pyxis

Two fragments (**Nos. 273–274**) are characterized by the cylindrical body typical of both epichyseis and pyxides; in absence of the upper part of the vessel the identification of the shape is a difficult task. Epichyseis were locally produced in black-gloss, red-figure, and Gnathia wares,[447] but this shape does not seem to be popular. It is possible that these two fragments can be identified as pyxides, which are found in Metapontine tombs between the last quarter of the 5th and the 4th century BC.[448] (FS)

440 *Necropoleis*, 668, NL4 (net-decorated lekythos, 350–300 BC). *Pomarico Vecchio* I, Pl. 108, No. 63 (Gnathia lekythos, 340–320 BC). Lippolis 1994b, 253, Fig. 189 (Taranto, phases A2 and B1, 350–300 BC).

441 *Pomarico Vecchio* I, Pl. 40, No. 118 (340–320 BC). Pianu 1990, Pl. XXIV, No. 32.2 (340–320 BC) and Pl. XXXVII, No. 75.1 (340–320 BC). Lippolis 1994b, 253, Fig. 189 (from Taranto, phase B1, 300–275 BC).

442 Lanza 2004, 34–48.

443 Pianu 1990, Pl. XXIX, No. 4 and XXXV, No. 48.1 (340–310 BC).

444 Sestieri 1940, 109, Fig. 48 (from the chora). Lo Porto 1966, 206, Fig. 60, No. 1 (late 4th c. BC). D'Andria 1980a, 428, No. 305 (from the kerameikos, late 4th c. BC). Lo Porto 1988–1989, Fig. 89, No. 2 (320 BC). *Pomarico Vecchio* I, Pl. 49, No. 1 (340–320 BC). Elliott 1998, 668 (from the Pantanello necropolis). Scarano 1992, 18 and Fig. 3 (from the urban necropolis).

445 Deposit 3 of the kerameikos; D'Andria 1980a, 428, Fig. 65, Nos. 302–5 (350–325 BC).

446 Hurschmann 1995, 683.

447 D'Andria 1980a, 406, No. 198, Fig. 54; 425, No. 294, Fig. 64, and Silvestrelli forthcoming a.

448 Elliott 1998, 675–76. For the shape and its possible evolution, see Ch. 19, site 736, T 19-04.

Catalog of Black-Gloss Fabrics

Keith Swift

These examples of Metapontine black-gloss fabrics show the range and variation in the commonly occurring fine-ware fabrics in the Metapontine chora. They were selected to cover the chronological range of black-gloss fabrics—Early Classical to Hellenistic period—with all but one from identifiable and dateable forms. Color was recorded with reference to the Munsell Soil Color Charts.

1. BG 131-01 Metapontine black-gloss fabric sample from a kylix rim (Catalog of Black-Gloss Fine Ware, **No. 18**, 470–440 BC).

Description: This piece dates to the early phase of production of Metapontine black gloss *sensu stricto*. It is very similar to those typical of Archaic fine ware, with a similar firing color and consistency, particularly in terms of the abundant, extremely fine carbonates throughout the fabric and in the frequency of the fine irregular voids. The slip is misfired, unevenly oxidized, though it is of relatively high quality in terms of consistency, appearing semi-lustrous where evenly fired, and could certainly be described as good quality in relation to the usual slips of Archaic-period fine wares. It is tempting to interpret this fabric and slip as representing the early, and not always perfected, production of Metapontine black gloss during the Early Classical period.

Color: Yellow 10YR 7/6 on surfaces and much of matrix, with darker, not-fully-oxidized core grey 10YR 7/1 in places.

Texture: The fabric has a finely-granular consistency similar to the typical fine-ware fabrics of the Late Archaic period, with a somewhat powdery appearance and texture, though it is in fact fired relatively hard.

Inclusions: Matrix contains abundant extremely fine crystalline grains (quartz/feldspars) giving it the typical finely granular consistency of Metapontine fabrics. Abundant fine carbonates are powdered throughout the fabric, with occasional larger inclusions < 0.1 mm.

Voids: Frequent fine, elongated voids orientated parallel to the vessel walls, most < 0.1 mm.

Slip: Semi-lustrous slip, patchy and partially oxidized in places to dark reddish brown 2.5YR 3/4. Where not misfired or of uneven consistency, the slip reaches the usual relatively high quality of Metapontine black gloss. Numerous, extremely fine crystalline lathes are apparent on the surfaces of the slip; this is not a common feature of most Metapontine black-gloss slips, and indicates some variation in the preparation of the slip for this vessel, using a deflocculated source clay rich in very fine quartz/feldspar grains.

A

B

C

D

E

A. Hand Specimen: Scale 10 mm. Exterior with misfired slip.
B. Microscopy: Rim section, width of field ca. 15 mm.
C. Microscopy: Rim section, width of field ca. 5 mm.
D. Microscopy: Exterior surface showing oxidized slip, width of field ca. 20 mm.
E. Microscopy: Exterior surface showing consistent, semi-lustrous black slip, flaked in patches showing surface of fabric, width of field ca. 20 mm.

2. BG 477-49 Metapontine black-gloss fabric sample from an A-type skyphos rim (Catalog of Black-Gloss Fine Ware, **No. 25**, dated 460–440 BC).

Description: Fabric similar to **No. 1** in firing color and consistency, comparing well to Archaic fine ware fabrics associated with Ionic cups.

Color: Fired yellowish, varying throughout the fabric from reddish yellow to pink to light brown 7.5YR 6-7/4-6 from surface to core, and generally similar in color to the majority of Archaic fabrics.

Texture: Finely granular, powdery consistency similar to 131-01, with a clean, though not conchoidal, fracture.

Inclusions: Extremely fine crystalline grains, sufficiently small to be considered as a component of the matrix, giving it a finely granular texture. Carbonate inclusions or calcareous clay components are similarly fine, powdered throughout the fabric, occasionally larger, < 0.1 mm. An ovoid iron-rich concentration is also visible in the fabric, < 0.2 mm. These concentrates may have been present in the clay source or may point to the admixture of clays.

Voids: Frequent fine voids, AFWAF < 0.2 mm, most < 0.1 mm, with occasional larger elongated voids throughout.

Slip: Matte black slip, thin and streaky on the interior, almost completely worn on the exterior. Flecks of crystalline grains (quartz/feldspars) are apparent on the surfaces of the fabric where the slip has flaked.

A. Hand Specimen: Scale 10 mm. Interior.

B. Microscopy: Section of the rim showing clean fracture, width of field ca. 20 mm.

C. Microscopy: Section of rim showing iron-oxide rich inclusion, width of field ca. 10 mm.

D. Microscopy: External surface showing matte black slip, flaked, and extremely fine crystalline grains apparent on the exposed surfaces of the matrix, width of field ca. 10 mm.

A

B

C

D

3. BG 413-04 Metapontine black-gloss fabric sample from a small bowl (Catalog of Black-Gloss Fine Ware, **No. 187**, dated 330–275 BC).

Description: A good example of typical Metapontine black-gloss fabric of the late 4th and early 3rd c. BC, fired reddish, with similar texture to, though harder than, the yellowish fired fabrics **Nos. 1** and **2**.

Color: Light reddish variant, reddish yellow 5YR 6/6 to light reddish brown 5YR 6/4.

Texture: Finely granular, hard fabric with a clean, semi-conchoidal fracture.

Inclusions: Extremely fine crystalline grains, sufficiently small to be considered as a component of the matrix, giving it a finely granular texture. Rare rounded carbonates < 0.2 mm and occasional crystalline grains (quartz/feldspars) < 0.1 mm.

Voids: Frequent vughs and elongated voids throughout the fabric.

Slip: Consistent, semi-lustrous black slip.

A. Hand Specimen: Scale 10 mm. Exterior with semi-lustrous black slip.

B. Microscopy: Section showing typical finely granular texture of Metapontine black-gloss fabrics, width of field ca. 20 mm.

C. Microscopy: Section showing rounded carbonate inclusion and frequent vughs, width of field ca. 10 mm.

D. Microscopy: Exterior surface showing semi-lustrous slip and exposed surface of the matrix where flaked, width of field ca. 20 mm.

A

B

C

D

4. BG 501-09 Metapontine black-gloss fabric sample from a dish rim (Catalog of Black-Gloss Fine Ware, **No. 160**, dated to the 3rd c. BC).

Description: Granular Metapontine black-gloss fabric, representing a slightly coarser variant, fired reddish.

Color: Reddish variant, red 2.5YR 5/8, partially reduced to reddish brown 2.5YR 5/4.

Texture: Finely granular, hard fabric with a clean, semiconchoidal fracture.

Inclusions: As with other examples of Metapontine fabrics, extremely fine crystalline grains are abundant in this fabric. Abundant very fine carbonates are powdered throughout the matrix, with occasional larger rounded carbonate reaction halos and crystalline grains, probably quartz, both < 0.2 mm. This variant is a noticeably coarser variant of Metapontine black-gloss fabrics, though the matrix is of a similar consistency.

Voids: Frequent elongated voids, most < 0.1 mm throughout the fabric, with occasional larger irregular voids, < 0.3 mm.

Slip: Dilute, thin matte grey slip.

A

B

C

A. *Hand Specimen*: Scale 10 mm. Interior with dilute matte black slip.

B. *Microscopy*: Section showing granular texture and range of firing colors, width of field ca. 15 mm.

C. *Microscopy*: Section showing frequent extremely fine carbonate inclusions, with larger rounded carbonate voids with reaction halos and sparse large crystalline grain (quartz), < 0.2 mm, and abundant voids, width of field ca. 5 mm.

5. BG 477-47 Metapontine black-gloss fabric sample from a large A-type skyphos base (BG 477-47, dated ca. 460–440 BC).

Color: Unevenly oxidized, varying from reddish yellow to pink to light brown, 7.5YR 6-7/4-6.

Texture: Finer variant of Metapontine black-gloss fabrics, hard with a semi-conchoidal fracture.

Inclusions: Matrix appears to have fewer of the extremely fine crystalline grains and lacks the abundant very fine calcium carbonate inclusions. Sparse crystalline (quartz/feldspar) inclusions, < 0.1 mm. Rare, iron-rich, rounded textural concentration features, < 0.1 mm.

Voids: Frequent irregular voids with white reaction rims from decomposed calcium carbonate inclusions, most < 0.1 mm.

Slip: Thin, matte dark grey slip.

A. Hand Specimen: Scale 10 mm. Interior.
B. Microscopy: Section of foot ring showing range of firing colors and microscopic texture, width of field ca. 15 mm.
C. Microscopy: Section of foot ring showing carbonate inclusions and voids, width of field ca. 10 mm.

A

B

C

6. BG 153-00 Metapontine black-gloss fabric

Description: This fabric is representative of the Metapontine black-gloss variant fabric fired to a deep reddish color. The fabric bears some similarities to the fabrics of Metapontine amphora wasters from site 153, though these contain much coarser inclusions (see discussion in Ch. 12 and Catalog of Amphorae **Nos. 34** and **57**). Although production of black gloss at the site cannot be demonstrated this fabric predominates in the corpus of black-gloss sherds from site 153 and may have been produced in the area—it compares well to the black-gloss fabrics of test-firing pieces from site 105 (Catalog of Black-Gloss Fabrics, **No. 7**), also located in the Basento valley.

Color: Red 10R 5/6 to light red 2.5YR 6.6 to pale red 10R6/4 from external surface to core.

Texture: Hard, very finely granular with clean fracture.

Inclusions: Abundant very fine carbonate inclusions, < 0.05 mm and frequent crystalline inclusions, < 0.1mm, occasional larger examples, probably quartz, < 0.2 mm.

Voids: Frequent fine elongated voids, < 0.2 mm in length, orientated parallel to the vessel walls, probably resulting from air trapped in the fabric during throwing and not apparently related to the carbonate inclusions.

Slip: Consistent, relatively thick semi-matte black slip.

A. Hand Specimen: Scale 10 mm. Black-gloss slip on the interior.

B. Microscopy: Section showing texture in hand specimen and range of firing colors common to reddish-fired Metapontine black-gloss fabrics, width of field 25 mm.

C. Microscopy: Section showing frequent elongated expansion voids and rare larger crystalline inclusions, width of field 10 mm.

D. Microscopy: Section showing abundant, extremely fine carbonate inclusions, width of field 5 mm.

A

B

C

D

7. BG 105-001 Test-firing piece from site 105.
Description: Perforated test-firing piece with vertical ribbing from site 105 in the Basento valley. The fabric is similar to Metapontine black-gloss fabrics fired to a reddish range of colors (e.g. **No. 6**, above), and may have been produced from locally available fluvial sediments. The fabric falls within the range of the fabrics of other black-gloss sherds from site 105, which have similar textures and inclusions.
Color: Red 10R 5/6 to light red 2.5YR 6.6 to pale red 10R6/4 with greyish core.
Texture: Hard, very finely granular with clean fracture.
Inclusions: Frequent crystalline inclusions, < 0.1 mm, probably quartz and feldspars, occasional larger examples, probably quartz, < 0.2 mm.
Voids: Frequent fine elongated voids, < 0.1 mm in length, orientated parallel to the vessel walls.
Slip: Thin, matte dark grey slip, oxidized red over much of the surface, poorly adhered to the matrix, abraded or flaked.

A. Hand Specimen: Scale 10 mm. Exterior of test-firing piece showing slip, vertical ribbing, and test-firing perforation.
B. Hand specimen: Section showing texture of the fabric in hand specimen, scale 10 mm.
C. Microscopy: Section showing microscopic texture and very fine crystalline grains, width of field 25 mm.
D. Microscopy: Section showing frequent, extremely fine crystalline inclusions, width of field 5 mm.

A

B

C

D

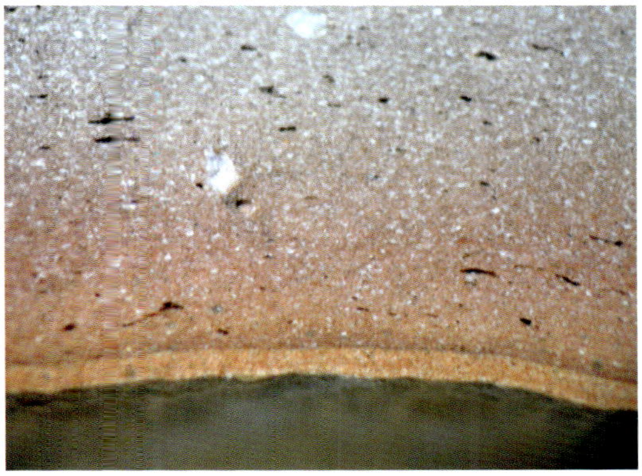

Catalog of Black-Gloss Fine Ware

Elisa Lanza Catti, Margherita Nigro, Keith Swift, and Amelia Tubelli

Attic Black-Gloss Imports

1:1

1:2

1. BG 178-01 Attic kylix rim (single example)
Dimensions: rim diam. 17.7 h 2.05 w 3.8
Reference Sherd Description: Very good quality, consistent lustrous black slip, typical finely-granular Attic fabric, reddish orange.
Selected Comparanda: Elliott 1998, 657, K6 (Metaponto, Pantanello necropolis, 470–440 BC); *Agora* XII, 91–92, No. 413, Fig. 4, Pl. 19 (480–450 BC); Lo Porto 1973, 230, Pl. LXXIII, No. 5 (Irsina, 470–460 BC, Attic); Lo Porto 1966, 216, No. 2, Fig. 71.2 (Metaponto, urban necropolis, T 43, 520–510 BC, Attic); Lo Porto 1988–1989, 328–34, Fig. 35.2 (Montescaglioso, T 1, second quarter of the 5th c. BC, Attic) and 403, Fig. 117.1 (Laterza, T 1, second quarter of the 5th c. BC, Attic); *Museo Taranto* II.2, 492, Pl. 39.base diam., 7.2 (first half of the 5th c. BC).
Attested Dates: 500–450 BC.

1:2

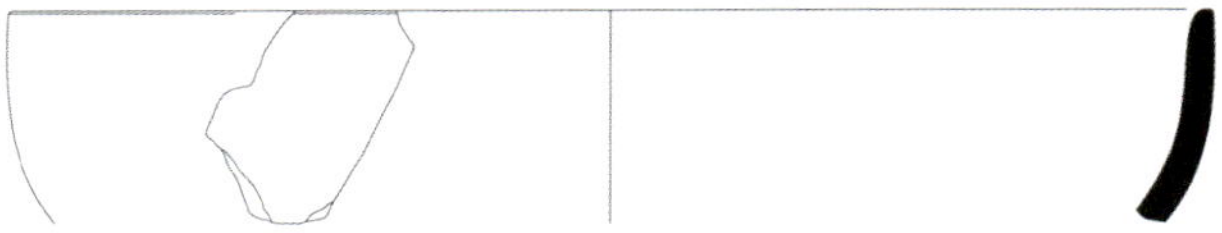

2. BG 557-01 Attic stemless kylix rim (single example)
Dimensions: rim diam. 16.0 h 2.9 w 2.75
Reference Sherd Description: Consistent, lustrous black slip, typical finely-granular Attic fabric.
Selected Comparanda: *Agora* XII, 102, No. 476, Fig. 5, Pl. 49 (stemless kylix: large, plain rim, 440–430 BC). Patera 1986, 51, Nos. 30–32, Pl. XX (Taranto, via Dante Tomb, second half 5th c. BC); Donnarumma and Tomay 1990, 267, No. 6, Fig. 457 (Fratte, T58/1963, 440–430 BC).
Attested Dates: 430–400 BC.

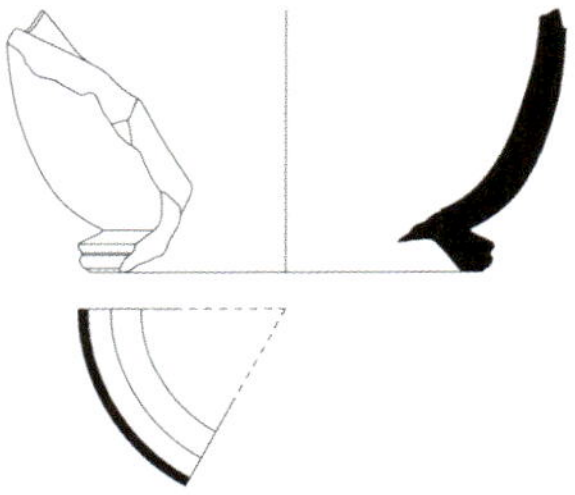

1:2

3. BG 791-01 Attic squat lekythos base (single example)
Dimensions: base diam. 5.5 h 3.6 w 5.8
Reference Sherd Description: Consistent, semi-lustrous black slip, reserved underside and unslipped interior.
Selected Comparanda: *CVA Leiden* 3, 67–68, Pl. 161,1 (Attic, ca. 425 BC); D'Andria 1990, 117, No. 158 (Vaste, Fondo Melliche, funerary deposit 563, Attic, end of the 5th c. BC); *Museo Taranto* I.3, 309, 93.3 (Taranto, tomb from via Messapia, 7-2-1947, Attic, 430–420 BC); for a similar lekythos of local production see Elliott 1998, 684, SL27 (Metaponto, Pantanello necropolis, 400–370 BC).
Attested Dates: 430–400 BC.

Skyphoi with Offset Rims

1:2

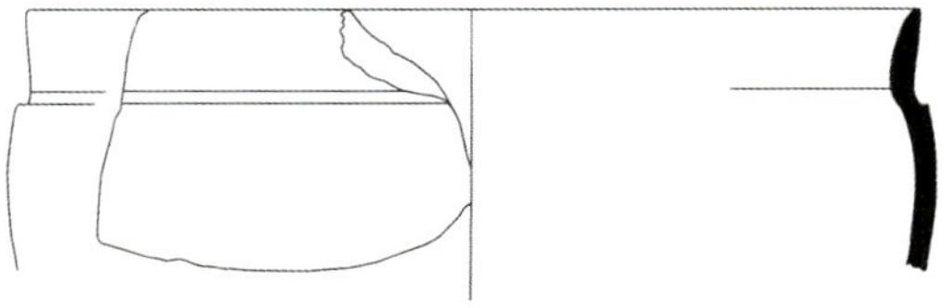

4. BG 477-54 Skyphos with offset rim (single example)
Dimensions: rim diam. 12.8 h 3.1 w 3.7
Reference Sherd Description: Consistent, semi-lustrous black slip, worn in places, reserved lower lip on the exterior.
Selected Comparanda: Semeraro 1997, 227–28, No. 788, Fig. 197 (Oria, Monte Papalucio sanctuary, ascribed to Metapontine production).
Attested Dates: 500–450 BC.

Skyphoi /cups with offset rims cont.

1:2

5. BG 335-01 Skyphos with offset rim, reserved band on exterior (5 examples)
Dimensions: rim diam. 13.0 h 3.2 w 4.2
Reference Sherd Description: Upper body of skyphos with offset rim, reserved band on exterior at the junction of the rim and shoulder.
Selected Comparanda: Adamesteanu et al. 1971, 42, T 23, Pl. IX; Lo Porto 1973, Pl. XIX, No. 2 (Pisticci, T 4/1937, end 6th–early 5th c. BC) and Pl. LXX, No. 1.3 (Matera, "Rifeccia" necropolis, around mid-6th c. BC).
Attested Dates: 500–450 BC.

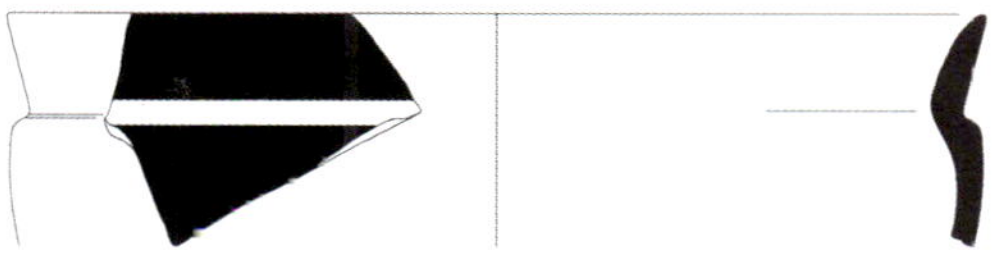

1:2

6. BG 155-12 Skyphos with offset rim (8 examples)
Dimensions: rim diam. 11.3 h 2.9 w 3.3
Reference Sherd Description: Upper body and rim of skyphos with offset rim, entirely slipped with pronounced articulation of junction between rim and shoulder.
Selected Comparanda: Lo Porto 1981b, 364, Fig. 85.2 (Metaponto, T 18, end 6th c. BC); Lo Porto 1988–1989, 332, No. 17, Figs. 33.10 and 35.1 (Montescaglioso, T 1, end of 6th–beginning 5th c. BC); *Arte e artigianato*, 131–32, No. 94.25 (Ginosa, via Vittorio Emanuele II, tomb excavated in January 1934, end 6th–first quarter 5th c. BC); Semeraro 1997, No. 786, Fig. 197 (Oria, Monte Papalucio sanctuary, ascribed to Metapontine production).
Attested Dates: 500–450 BC.

1:1

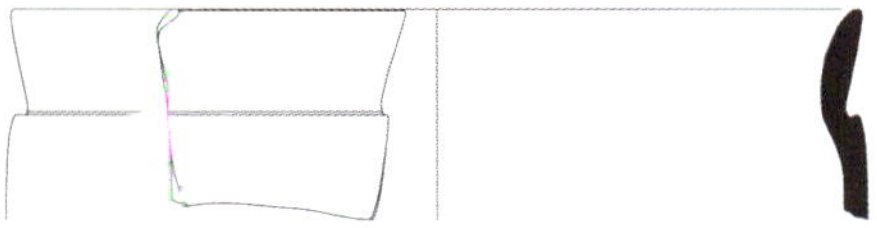

1:2

7. BG 397-42 Base of skyphos with offset rim (3 examples)
Dimensions: base diam. 6.3 h 1.6
Reference Sherd Description: Reserved band on the exterior of the footring.
Selected Comparanda: Elliott 1998, 678, S23 (Metaponto, Pantanello necropolis, 460–440 BC); Macnamara 1983, 329, Nos. 171–72 (Cozzo Presepe, 550–480 BC); Lo Porto 1973, Pl. VIII, No. 1.1 (Pisticci, dated to the 6th c. BC).
Attested Dates: 460–440 BC.

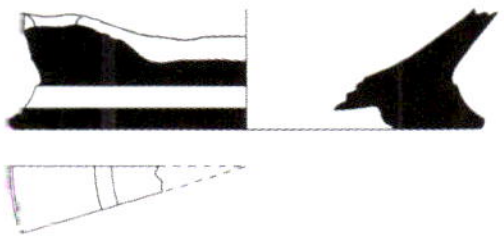

Cups with Offset Rims

1:1

8. BG 477-43 Rim of cup with offset rim (14 examples)
Dimensions: h 2.4 w 2.9
Reference Sherd Description: Consistent, semi-lustrous black slip. Completely slipped.
Selected Comparanda: Lo Porto 1988–1989, 410–11, Fig. 128 (Pisticci, T 3, end 6th c. BC); Van Compernolle 2000, 92, Figs. 2, 5; Semeraro 1997, 216–19, Fig. 191, Nos. 717–21 (Oria, Monte Papalucio sanctuary, first quarter of the 5th c. BC, probably of Metapontine production).
Attested Dates: 500–450 BC.

1:2

Cup-Skyphoi (Late Archaic/Early Classical)

1:2

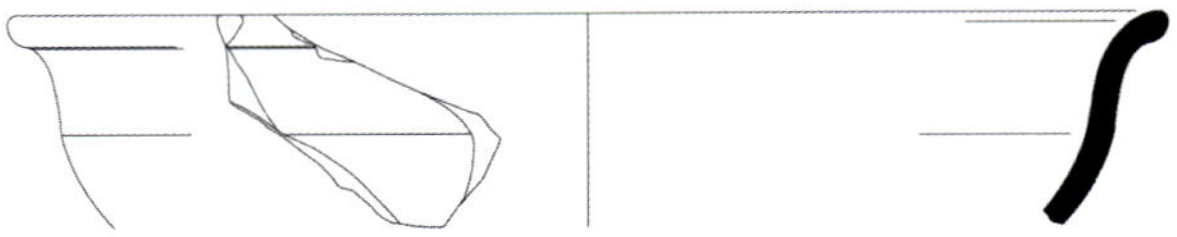

9. BG 414-06 Cup-skyphos rim (single example)
Dimensions: rim diam. 15.4 h 2.9 w 3.7
Reference Sherd Description: Consistent, semi-lustrous black slip.
Selected Comparanda: Lo Porto 1973, 222, Pl. LXVI, Nos. 2.5–7 (Matera, T 6, 500–480 BC); *Bitonto*, 117, No. 15 (Bitonto, T 4/1982, vase dated to the second half of the 5th c. BC); Semeraro 1997, 202, Nos. 601–2 (Oria, Monte Papalucio sanctuary, 500–480 BC, Attic); *Agora* XII, 109–10, No. 569, Figs. 6, 22 (500–480 BC).
Attested Dates: 500–475 BC.

1:1

1:2

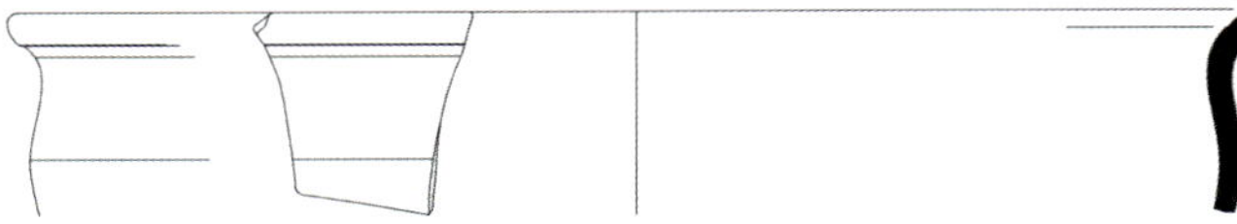

10. BG 288-02 Cup-skyphos rim (single example)
Dimensions: rim diam. 17.2 h 2.85 w 3.0
Reference Sherd Description: Matte black slip, almost totally worn.
Selected Comparanda: See **No. 9**.
Attested Dates: 500–450 BC.

1:2

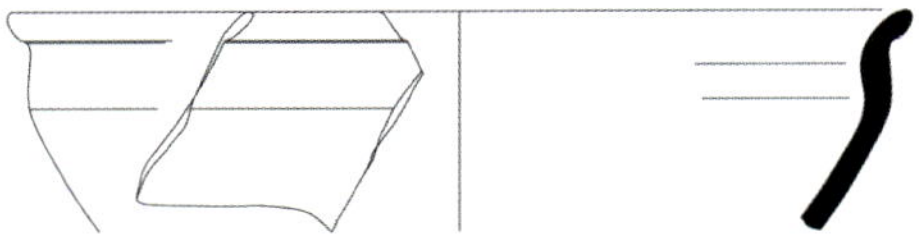

11. BG 482-01 Cup-skyphos rim (single example)
Dimensions: rim diam. 12.0 h 3.0 w 3.8
Reference Sherd Description: Consistent, semi-lustrous black slip.
Selected Comparanda: Lo Porto 1973, 222, Pl. LXVI, Nos. 2.5–7 (Matera, T 6, 500–480 BC); Riccardi 2008, 61, No. 23 (second half of the 4th c. BC).
Attested Dates: 500–475 BC.

1:2

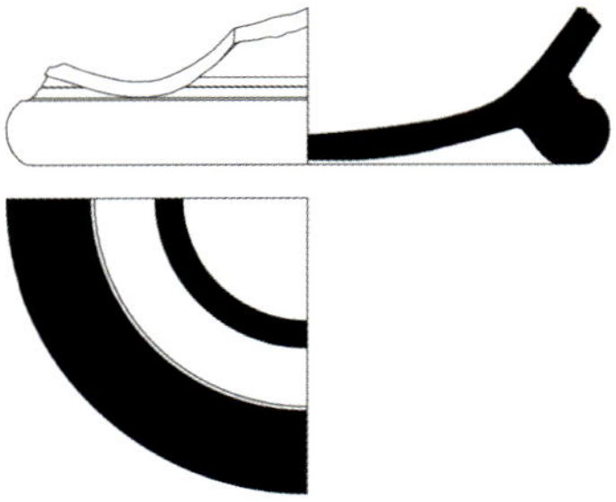

12. BG 319-02 Cup-skyphos base (single example)
Dimensions: base diam. 8.9 h 2.4 w 8.9
Reference Sherd Description: Reserved underside of floor with thin band of slip inside footring. Consistent, semi-lustrous black slip, worn on the inside of the foot.
Selected Comparanda: Elliott 1998, 682, CS3 (Metaponto, Pantanello necropolis, 460–440 BC); Lo Porto 1973, 221–22, Pl. LXVI, Nos. 2.5–7 (Matera, 500–480 BC).
Attested Dates: 500–450 BC.

1:2

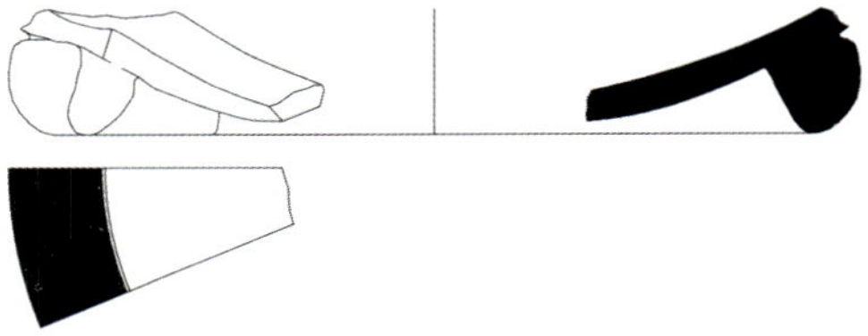

13. BG 524-01 Cup-skyphos base (single example)
Dimensions: base diam. 12.6 h 1.9
Reference Sherd Description: Consistent, semi-lustrous black slip, reserved floor inside foot ring.
Selected Comparanda: Sant'Angelo Grieco, SG233.8; Lo Porto 1973, 222, Pl. LXVI, Nos. 2.5–7 (Matera, T 6, 500–480 BC); Semeraro 1997, 203, No. 617 (Oria, Monte Papalucio sanctuary, Attic, first quarter of the 5th c. BC); *Agora* XII, 276, No. 572, Fig. 6, Pl. 25 (500 BC).
Attested Dates: 460–440 BC.

Cup-skyphoi (Late Archaic/Early Classical) cont.

1:2

14. BG 477-07 Cup-skyphos base (8 examples)
Dimensions: base diam. 10.5 h 3.35 w 10.3
Reference Sherd Description: Reserved exterior of foot, slipped underside of foot ring and reserved interior with thin band of slip.
Selected Comparanda: Lo Porto 1973, 222, Pl. LXVI, Nos. 2.5–7 (Matera, T 6, 500–480 BC).
Attested Dates: 480–440 BC.

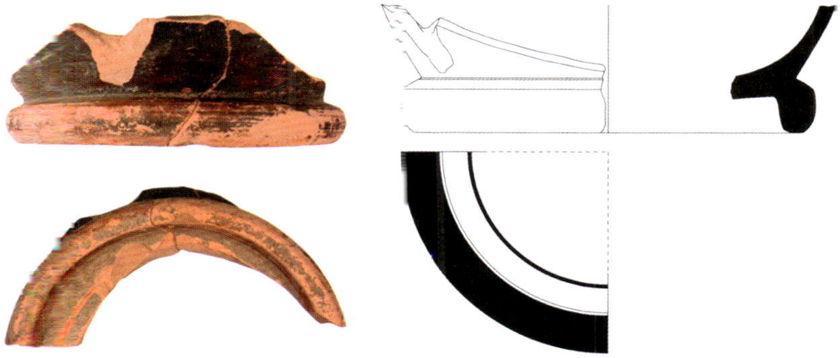

1:2

15. BG 477-08 Cup-skyphos base (10 examples)
Dimensions: base diam. 8.5 h 1.9 w 4.9
Reference Sherd Description: Consistent, matte black slip, reserved underside of floor inside foot ring.
Selected Comparanda: Fattoria Fabrizio, FF80-124PL-I; Sant'Angelo Grieco, SG205.11; Lo Porto 1973, 222, Pl. LXVI, Nos. 2.5–7 (Matera, T 6, 500–480 BC); *Museo Taranto* II.2, 498, Pl. 40e, 9.2 (Rutigliano, second half of the 5th c. BC); Semeraro 1997, 22, No. 756, Fig. 194 (Oria, Monte Papalucio sanctuary).
Attested Dates: 480–440 BC.

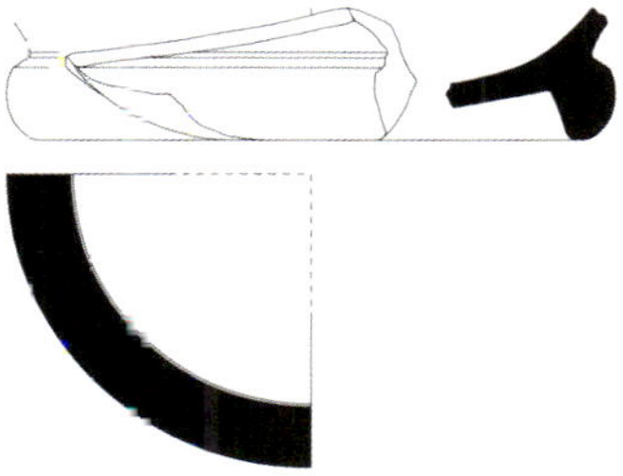

1:2

16. BG 397-48 Cup-skyphos base (3 examples)
Dimensions: base diam. 10.95 h 2.1
Reference Sherd Description: Reserved inner face of foot ring, slipped underside. Dilute, matte brown slip, partially worn.
Selected Comparanda: Sant'Angelo Grieco, SG117.3; *Agora* XII, 109–10, 572, Fig. 6, Pl. 25 (500 BC); Lo Porto 1973, 188, Pl. XXXVI, Nos. 4.1 (Montescaglioso, Attic example with reserved exterior of foot, early decades of the 5th c. BC) and 222, Pl. LXVI, Nos. 2.5–7 (Matera, T 6, Attic examples, 500–480 BC). Bears similarities to Govi 1999, 74, No. 48, Pl. VII (Attic, 500–475 BC).
Attested Dates: 500–450 BC.

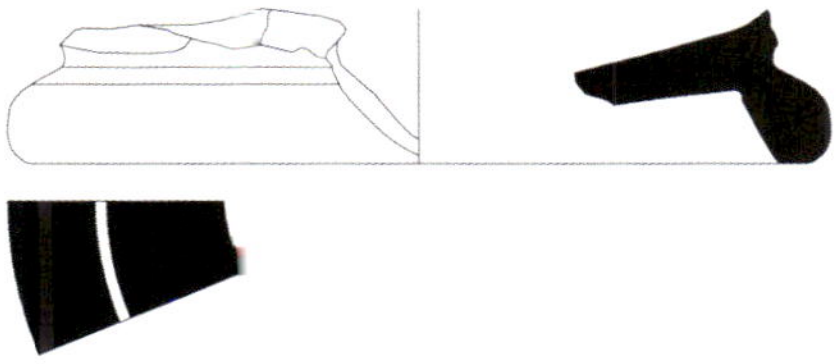

1:2

17. BG 477-48 Cup-skyphos base (3 examples)
Dimensions: base diam. 9.1 h 1.9
Reference Sherd Description: Consistent, matte black slip, reserved lower exterior and underside with two thin bands of slip extant inside foot ring.
Selected Comparanda: Bears similarities to Govi 1999, 75, No. 51, Pl. VII (500–475 BC).
Attested Dates: 460–440 BC.

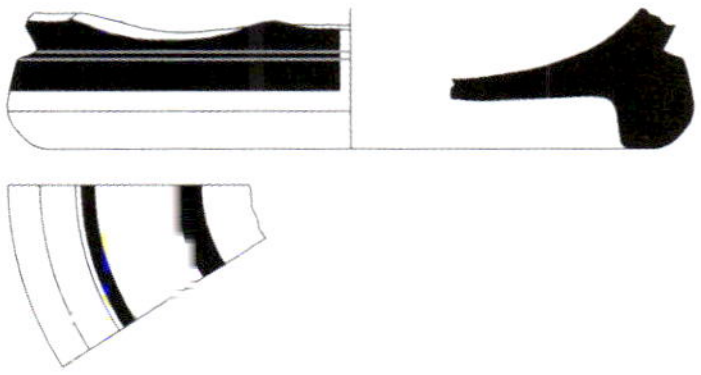

Kylikes

1:2

18. BG 131-01 Kylix rim (7 examples)
Dimensions: rim diam. 16.5 h 2.4 w 2.95
Reference Sherd Description: Dilute, semi-lustrous black slip.
Selected Comparanda: Elliott 1998, 657, K6 (Metaponto, Pantanello necropolis, 470–440 BC); Fattoria Fabrizio, FF80-55PL-B (500–450 BC); Lo Porto 1988–1989, 328–34, Fig. 35.2 (Attic, from Montescaglioso, T 1, second quarter of 5th c. BC); Riccardi 2008, 57, No. 1 (not Attic, second half of 5th c. BC); A. Bottini 1993b, 121, Nos. 9–10 (Satriano, first half of 5th c. BC).
Attested Dates: 470–440 BC.

1:1 1:2

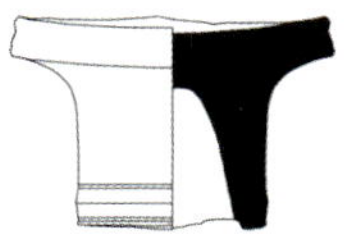

19. BG 342-31 Kylix stem (single example)
Dimensions: h 2.9 w 4.6
Reference Sherd Description: Terminus of hollow stem and central part of the floor.
Selected Comparanda: Elliott 1998, 657, K3 (Metaponto, Pantanello necropolis, 470–440 BC); Semeraro 1983, Pl. 84, No. 72 (Otranto); *Agora* XII, No. 413 (late 6th to first quarter of 5th c. BC).
Attested Dates: 470–440 BC.

1:2

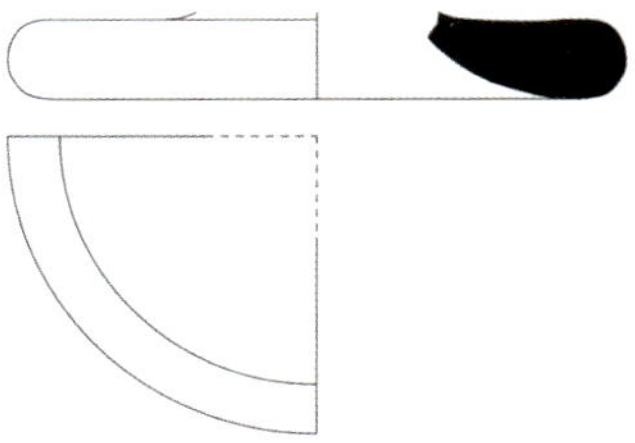

20. BG 328-01 Kylix foot (single example)
Dimensions: base diam. 8.2 h 1.2 w 7.0
Reference Sherd Description: Slip almost completey abraded but traces indicate that foot was not completely reserved.
Selected Comparanda: Elliott 1998, 657, K6 (Metaponto, Pantanello necropolis, 470–440 BC); Lo Porto 1966, 216, No. 2, Fig. 71.2 (Metaponto, urban necropolis, Attic, 520–510 BC); Surra and Preacco Ancona 1997, No. 16, Pl. 29 (end of 6th–early 5th c. BC).
Attested Dates: 470–440 BC.

1:2

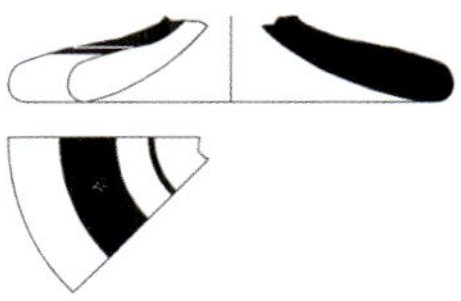

21. BG 285-12 Kylix foot (5 examples)
Dimensions: base diam. 5.9 h 1.2
Reference Sherd Description: Consistent, semi-lustrous black slip, reserved exterior of foot with concentric bands of slip on the underside.
Selected Comparanda: Elliott 1998, 657, K6 (Metaponto, Pantanello necropolis, 470–440 BC); *Agora* XII, 92, No. 432, Fig. 4 (500–480 BC).
Attested Dates: 470–440 BC.

Krater (Late Archaic/Early Classical)

1:2

22. BG 635-14 Krater rim (single example)
Dimensions: rim diam. 28.7 h 3.3 w 7.6
Reference Sherd Description: Streaky, semi-lustrous slip, partially oxidized red.
Selected Comparanda: Stibbe 1989, F58, Fig. 73; Boitani 1990, 53, No. 74, Fig. 117; Semeraro 1997, 283–84, No. 1049 (Valesio, Laconian, second half of 6th c. BC).
Attested Dates: 550–500 BC.

A-Type Skyphoi

1:2

23. BG 082-05 A-type skyphos base (2 examples)
Dimensions: base diam. 9.0 h 2.9 w 7.4
Reference Sherd Description: Reserved lower exterior of foot and underside.
Selected Comparanda: Liseno 2004, 87, No. 32, Pl. XLVIII.e (Metaponto, Deposito Favale, end of the 6th–first quarter of the 5th c. BC).
Attested Dates: 500–450 BC.

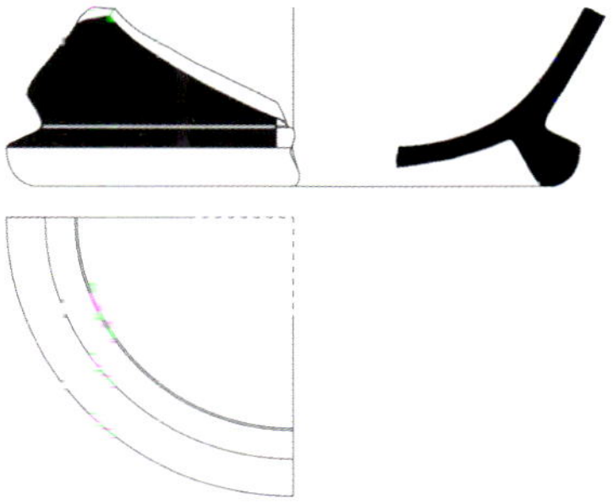

1:2

24. BG 477-47 A-type skyphos base (2 examples)
Dimensions: base diam. 5 h 1.65
Reference Sherd Description: Dilute, matte black slip almost completely worn on the interior. Reserved lower foot with concentric bands of slip on the underside.
Selected Comparanda: Elliott 1998, S42 (Metaponto, Pantanello necropolis, 460–440 BC); Sant'Angelo Grieco, SG139.1; for a similar decoration: Lo Porto 1981b, 308, Fig. 18.4 (Metaponto, votive deposit at contrada Crucinia, context dated between 600 and 450 BC); *Agora* XII, 84–86, No. 338, Pl. 16 (500–480 BC).
Attested Dates: 460–440 BC.

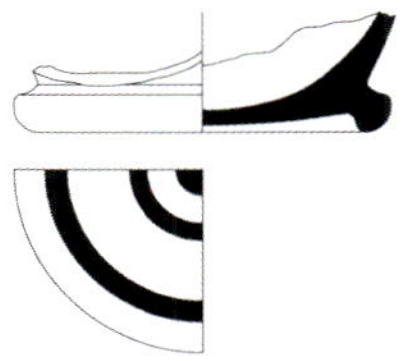

1:2

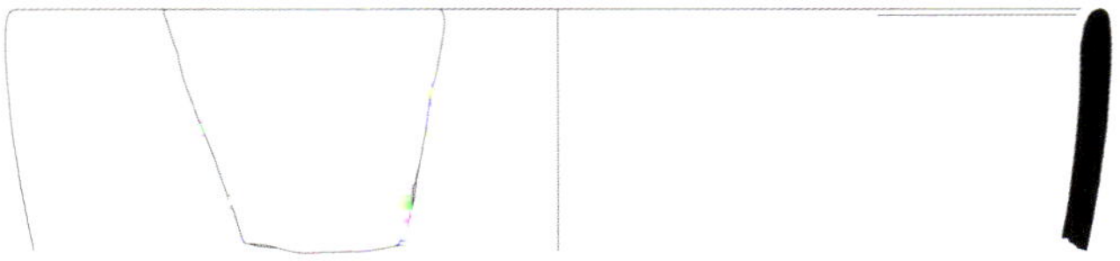

25. BG 477-49 A-type skyphos rim (29 examples)
Dimensions: rim diam. 16.7 h 3.8 w 4.2
Reference Sherd Description: Thin, matte black slip almost completed faded on the exterior.
Selected Comparanda: Elliott 1998, 681, S61 (Metaponto, Pantanello necropolis, 460–440 BC); Fattoria Fabrizio FF80-5PL-U (460–400 BC); Sant'Angelo Grieco SG319.1; Lo Porto 1973, 221, Pl. LXVI, No. 21 (Matera, T 6, 500–480 BC); *Bitonto*, 117, Nos. 16–17 (Bitonto, T 4/1982, second half of the 5th c. BC); Olivero Ferrero 1989, 97, Pl. XX, No. 47 (Locri, 470–460 BC); Gagliardi 2001, 302–3, Fl. 279f (Kaulonia, mid to third quarter of 5th ce. BC); *Agora* XII, 84–86, No. 342, Fig. 4, Pl. 16 (470–460 BC).
Attested Dates: 460–440 BC.

1:2

26. BG 397-04 revisit A-type skyphos base (14 examples)
Dimensions: base diam. 11.9 h 1.65
Reference Sherd Description: Thin black slip, almost completely worn. Reserved underside of floor inside foot ring.
Selected Comparanda: Elliott 1998, 681, S61 (Metaponto, Pantanello necropolis, 460–440 BC); Olivero Ferrero 1989, 96–97, Pl. XX, No. 47 (Locri, 470–460 BC); Semeraro 1983, 167, Pl. 85, No. 86 (Attic, from Otranto, 470–460 BC) *Agora* XII, 84–6, No. 342, Fig. 4, Pl. 16 (470–460 BC).
Attested Dates: 460–440 BC

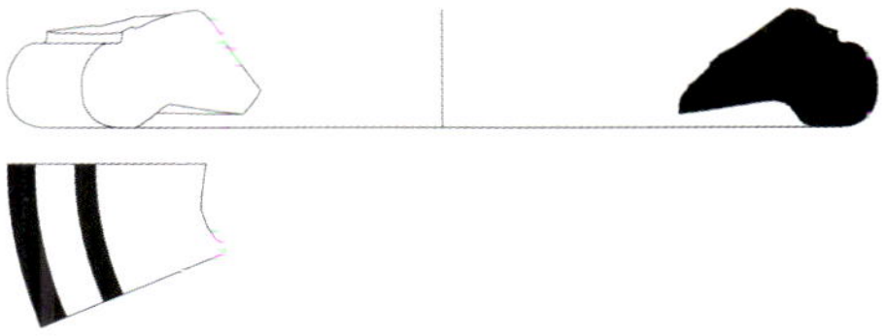

A-type skyphoi cont.

1:2

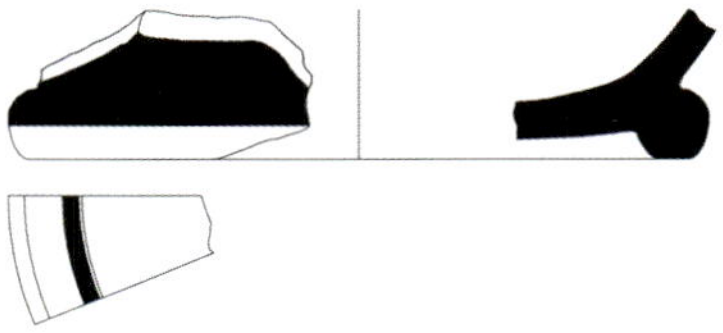

27. BG 578-03 A-type skyphos base (14 examples)
Dimensions: base diam. 9.3 h 2.05
Reference Sherd Description: Consistent, matte black slip. Reserved lower foot and underside, with slipped band on the internal face of the foot ring.
Selected Comparanda: *Necropoleis*, 431 SS 6-2 (Metaponto, Saldone, T 6, 460–440 BC); Sant'Angelo Grieco, SG133.6; Gagliardi 2001, 303, No. 46, Fig. 279g (variant A3, second and third quarters of 5th c. BC, from Kaulonia, Punta Stilo sanctuary). Bears similarities with Attic examples of the second quarter of 5th c. BC: *Agora* XII, 84–85, No. 342, Fig. 4, Pl. 16 (470–460 BC); Semeraro 1983, 167, Pl. 85, No. 86 (Otranto, 470–460 BC).
Attested Dates: 460–440 BC.

1:2

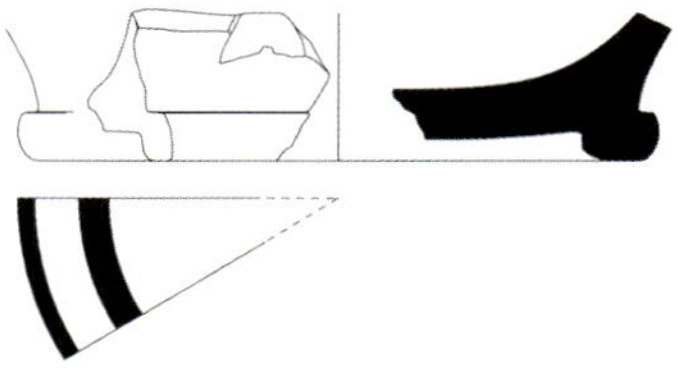

28. BG 241-07 A-type skyphos base (7 examples)
Dimensions: base diam. 8.5 h 2.1 w 3.2
Reference Sherd Description: Dilute, semi-lustrous black slip. Reserved underside of floor inside foot ring.
Selected Comparanda: *Bari*, 287–88, Nos. 648–51, Fig. 383, 1-2-5-11 (Bari, via Trento, mid-5th c. BC); Prag 1992b, 116, No. 871, Fig. 37 (Botromagno, ca. 450 BC following Morel); *Greci, Enotri e Lucani*, 166, No. 2.30.18 (Chiaromonte–S. Pasquale, T 227, last quarter of 5th c. BC); Donnarumma and Tomay 1990, 262, No. 12, Fig. 447 (Fratte, T 17, third quarter of 5th c. BC); Gagliardi 2001, 302, No. 43, Fig. 279d (variant A1, second and third quarter of 5th c. BC, from Kaulonia, Punta Stilo sanctuary).
Attested Dates: 450–400 BC.

---

29. BG 465-12 A-type skyphos base (single example)
Dimensions: base diam. 9.9 h 2.15
Reference Sherd Description: Consistent, matte reddish (oxidized) slip. Reserved underside of floor inside foot ring.
Selected Comparanda: *Museo Taranto* II.2, 488, 490, Pl. 38a, 6.2a (Rutigliano, beginning of 5th c. BC); *Agora* XII, 4–86, No. 344, Pl. 16 (440–425 BC); Semeraro 1983, 167, Pl. 85, No. 88 (Otranto, 470–460 BC).
Attested Dates: 470–350 BC.

1:2

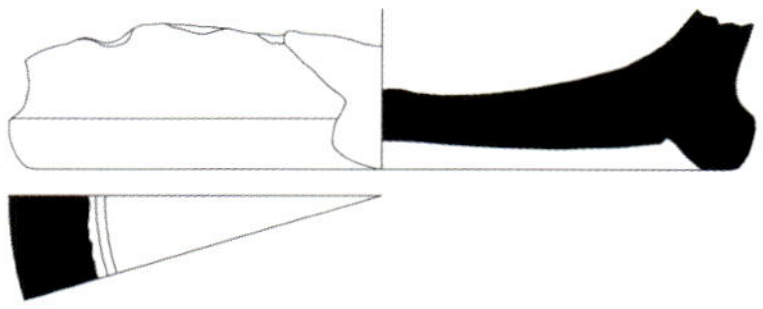

1:2

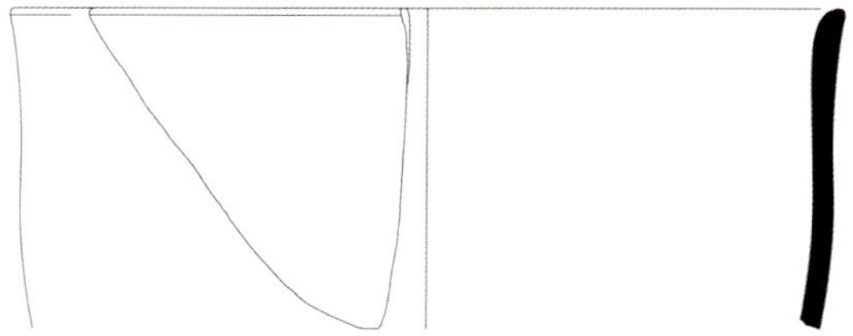

30. BG 352-08 A-type skyphos rim (27 examples)
Dimensions: rim diam. 13.8 h 5.4 w 5.3
Reference Sherd Description: Good quality, consistent, semi-lustrous black slip.
Selected Comparanda: Buzzi Auxilia 1989, 132, Pl. XXII, No. 88 (Locri, mid-4th c. BC); Semeraro 1983, 180, Pl. 89, No. 164 (Otranto, 350–300 BC); De Juliis 1990, 74, No. 33, Figs. 285–86 (Canosa, Ipogeo dei Vimini, Cella B, right deposition, second quarter of the 4th c. BC); *Nobili e guerrieri*, 53, Figs. 4.c and 49 (Palazzo San Gervasio, T 13/2002, end 4th–early 3rd c. BC following Morel's chronology). Montanaro 2007, 601, Fig. 529, 147.15 (Ruvo, early 4th c. BC); Delli Ponti 1996, 173, No. 172 (Vaste, T 13/1968, last quarter of the 4th c. BC); Giannotta 1997, 167, No. 22 (Vaste, T 3, late 5th c. BC); Morel 1981a, type 4373, 311, Pl. 131.
Attested Dates: 400–340 BC.

A-type skyphoi cont.

1:2

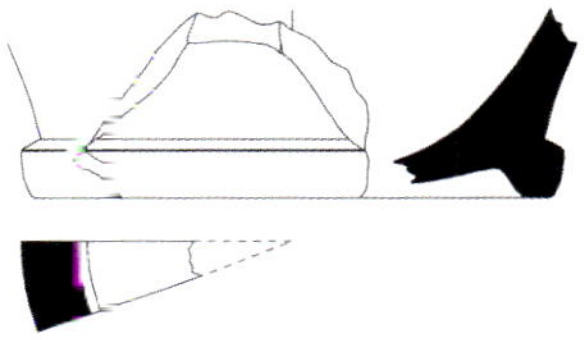

31. BG 043-06 A-type skyphos base (12 examples)
Dimensions: base diam. 7.2 h 2.6 w 3.9
Reference Sherd Description: Consistent, semi-lustrous black slip, almost completely worn on the interior. Reserved underside inside foot.
Selected Comparanda: De Juliis 1990, 74, No. 33, Figs. 285–86 (Canosa, Ipogeo dei Vimini, Cella B, second quarter of 4th c. BC); *Agora* XII, 84–86, No. 348, Pl. 16 (400 BC); Morel 1981a, 4382a1, 313, Pl. 132 (Pontecagnano, T 569, third quarter of 4th c. BC).
Attested Dates: 410–325 BC.

1:2

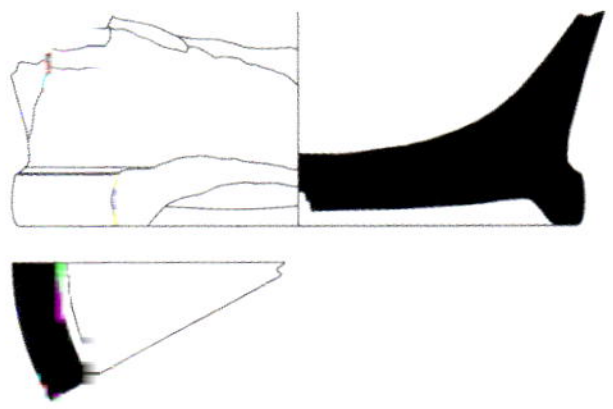

32. BG 413-03 A-type skyphos base (single example)
Dimensions: base diam. 7.6 h 2.9
Reference Sherd Description: Consistent, matte black slip. Reserved underside of floor inside foot ring.
Selected Comparanda: Lo Porto 1973, 210, Pl. LVII, No. 2.3 (Matera, T 4, beginning of 4th c. BC); De Juliis 1990, 74, No. 34, Figs. 286–87 (Canosa, Ipogeo dei Vimini, Cella B, right deposition, second quarter of 4th c. BC); Gagliardi 2001, 281, No. 5, Fig. 266e (Attic, first quarter of 4th c. BC, from Kaulonia, Punta Stilo sanctuary); *Agora* XII, 84–86, No. 349, Fig. 4, Pl. 16 (400–375 BC).
Attested Dates: 400–350 BC.

1:2

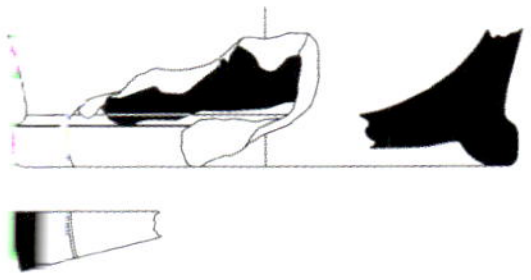

33. BG 516-03 A-type skyphos base (10 examples)
Dimensions: base diam. 6.7 h 1.9 w 3.4
Reference Sherd Description: Consistent, matte black slip, worn on the interior. Reserved inner foot and underside.
Selected Comparanda: Lo Porto 1973, 210, Pl. LVII, No. 2.3 (Montescaglioso, T 9, first quarter of 4th c. BC); J.W. Hayes 1984, 196, No. 198 (400–350 BC); *Agora* XII, 260, No. 349, Fig. 4, Pl. 16 (400–375 BC); Morel 1981a, 4382a1, 313, Pl. 132 (Pontecagnano, T 569, 350–325 BC).
Attested Dates: 400–350 BC.

1:2

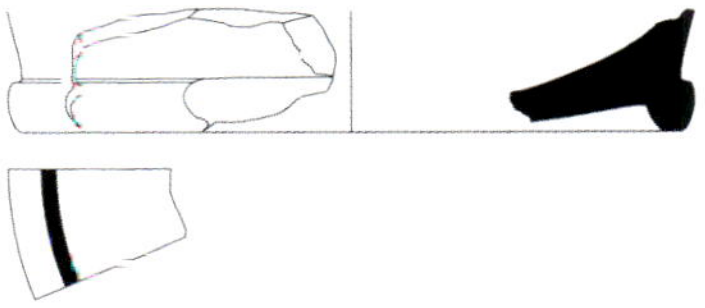

34. BG 566-02 A-type skyphos base (18 examples)
Dimensions: base diam. 9.1 h 1.7 w 3.6
Reference Sherd Description: Reserved lower exterior of foot and underside, except for band of slip on the internal face of the foot.
Selected Comparanda: Lo Porto 1988–1989, 372, No. 1, Fig. 81.1 (Montescaglioso, T 9, first quarter of 4th c. BC); A. Bottini 1993b, 193, No. 7 (Montescaglioso, Steppinia, first half of 4th c. BC); *Agora* XII, Fig. 4, No. 349 (4th c. BC); J.W. Hayes 1984, 196, No. 198 (first half of 4th c. BC).
Attested Dates: 400–350 BC.

1:2

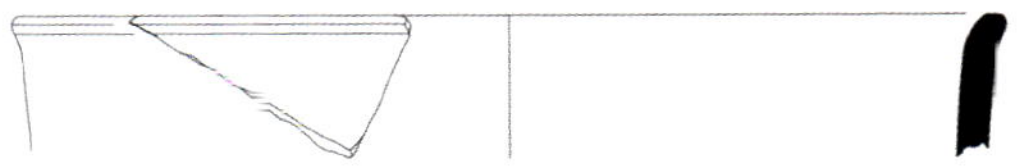

35. BG 043-07 A-type skyphos rim (12 examples)
Dimensions: rim diam. 13.2 h 1.95 w 3.8
Reference Sherd Description: Consistent, semi-lustrous black slip, almost completely worn on the interior.
Selected Comparanda: Prag 1983a, 352, No. 293 (Cozzo Presepe, ca. 300 BC); Prag 1992b, 118, No. 885, Fig. 36 (Botromagno, 340–290 BC); Semeraro 1983, 180, Pl. 89, No. 164 (Otranto, 350–300 BC); Giannotta 1996, 44, Fig. 9, No. 9 (Rocavecchia, T 12/1934, end of 5th to early 4th c. BC); Morel 1981a, series 4373, 311, Pl. 131 (exemplified by vessels of the second half of 4th c. BC).
Attested Dates: 375–300 BC.

A-type skyphoi cont.

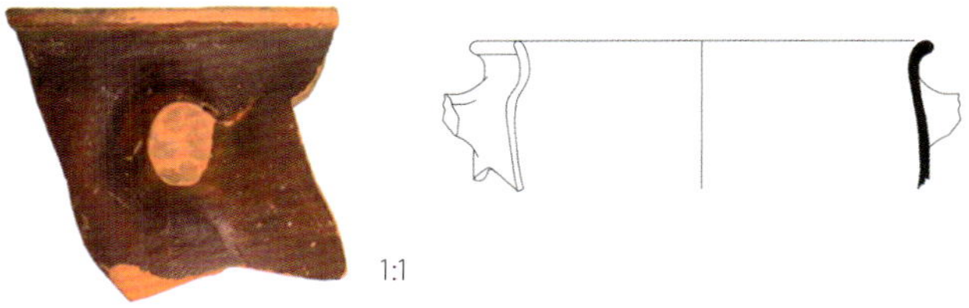

36. BG 128-26 A-type skyphos rim (9 examples)
Dimensions: rim diam. 7.9 h 2.6
Reference Sherd Description: Consistent, matte, partially oxidized slip.
Selected Comparanda: Prag 1983a, 352, No. 298 (Cozzo Presepe, second half of 4th c. BC); Prag 1992b, 116, No. 872, Fig. 36 (Botromagno, 360–300 BC, following Morel's chronology); De Cazanove 2008, 583, Fig. 294, No. 211 (Civita di Tricarico, 350–300 BC).
Attested Dates: 350–300 BC.

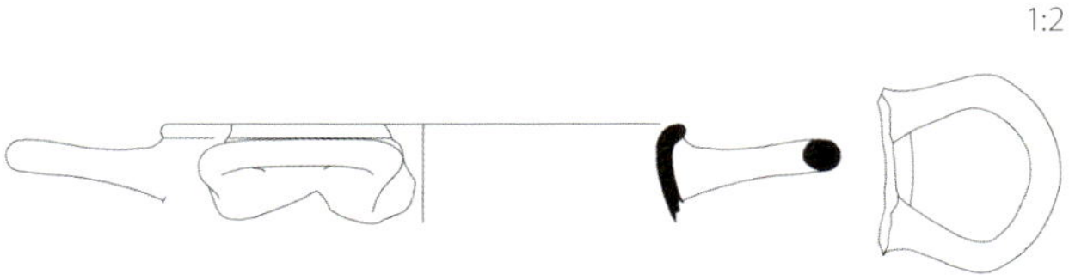

37. BG 258-11 A-type skyphos rim and handle (3 examples)
Dimensions: rim diam. 8.1 h 1.6 w 3.7
Reference Sherd Description: Rim preserved only above the single extant handle.
Selected Comparanda: Prag 1992b, 120, No. 891, Fig. 36 (Botromagno, 335–300 BC); Morel 1981a, 4373a1, 311, Pl. 131 (Ugento, 335–300 BC). For Attic skyphoi with a similar profile, *Agora* XXIX, 208, No. 1396, Fig. 85, Pl. 107 (300–275 BC).
Attested Dates: 335–300 BC.

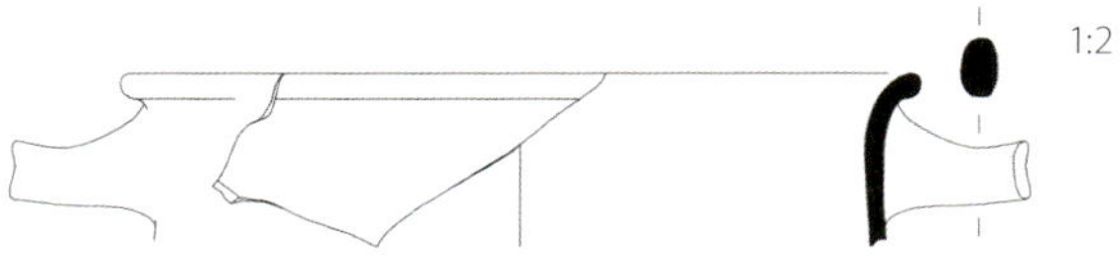

38. BG 497-23 A-type skyphos rim and handle terminus (single example)
Dimensions: rim diam. 10.7 h 2.4 w 5.2
Reference Sherd Description: Dilute, matte slip, black on the interior, partially oxidized brownish on the exterior.
Selected Comparanda: Prag 1992b, 117, No. 878, Fig. 36 (Botromagno, 325–300 BC); De Cazanove 2008, 583, Fig. 294, No. 233 (Civita di Tricarico, 350–300 BC); Morel 1981a, 4373c1, 311, Pl. 131 (said to be from Ruvo, end of the 4th–early 3rd c. BC).
Attested Dates: 325–300 BC.

39. BG 501-14 A-type skyphos rim (26 examples)
Dimensions: rim diam. 8.7 h 2.8
Reference Sherd Description: Consistent, matte black slip.
Selected Comparanda: Sant'Angelo Grieco, SG274.5; Prag 1983a, 352, No. 296 (Cozzo Presepe, 375–325 BC); Prag 1992b, 120, No. 891, Fig. 36 (Botromagno, 335–300 BC, following Morel's chronology); *Museo Taranto* II.2, 491, Pl. 38d, 6.2d (Rutigliano, second half of the 4th c. BC); Montanaro 2007, 419, Fig. 321, 79.1 (Ruvo, late 5th c. BC); *Bitonto*, 124, No. 44 (Bitonto, T 3/1981, second half of the 4th c. BC); Giannotta 1996, 49, Fig. 16, No. 19 (Rocavecchia, T 19/1934, last quarter of the 4th c. BC); Morel 1981a, 4373a2, 311, Pl. 131 (Ordona, TLIII, 335–300 BC).
Attested Dates: 335–275 BC.

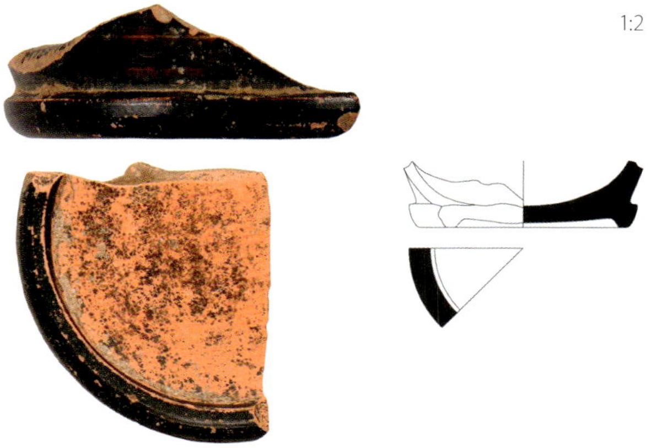

40. BG 528-13 A-type skyphos base (16 examples)
Dimensions: base diam. 5.5 h 1.7
Reference Sherd Description: Consistent, semi-lustrous black slip, unevenly fired to reddish brown on the exterior. Reserved underside of floor inside foot ring.
Selected Comparanda: Morel 1981a, type 4373b, 311, Pl. 131 (last quarter of the 4th c. BC); Sant'Angelo Grieco, SG233.8; Prag 1992b, 122, No. 901, Fig. 37 (Botromagno, end of the 4th to first quarter of the 3rd c. BC, following Morel's chronology); *Museo Taranto* II.2, 491, Pl. 38c, 6.2c (second half of the 4th c. BC).
Attested Dates: 340–300 BC.

A-type skyphoi cont.

1:2

41. BG 353-06 A-type skyphos base (5 examples)
Dimensions: base diam. 6.3 h 5 w 8.7
Reference Sherd Description: Dilute, semi-lustrous black slip. Reserved band on lower body and underside of floor inside foot ring.
Selected Comparanda: Preacco 1989, 235, Pl. XXX, XXXV, No. 268 (Locri, 325 to beginning of 3rd c. BC); *Museo Taranto* II.2, 491, Pl. 38c, type 6.2b (Rutigliano, second half of 4th c. BC); *Gli ori*, 393, No. 4 (Taranto, T 5 via Argentina, 29.V.1956, last quarter of 4th c. BC); Morel 1981a, 4373c1, 311, Pl. 131 (said to be from Ruvo, end 4th or beginning of 3rd c. BC).
Attested Dates: 320–290 BC.

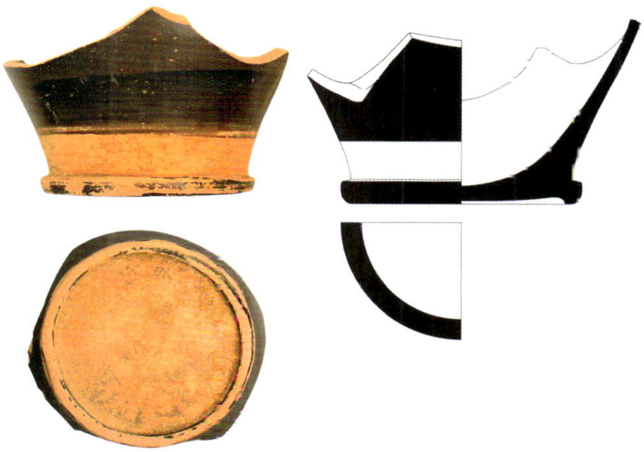

1:2

42. BG 528-20 A-type skyphos base (2 examples)
Dimensions: base diam. 5.6 h 3.1
Reference Sherd Description: Dilute, matte black slip, worn on exterior, reserved underside and (?) lower body.
Selected Comparanda: Semeraro 1983, Pl. 89, No. 156 (Otranto, Hellenistic); Morel 1981a, type 4373a, 311, Pl. 131 (third quarter of the 4th c. BC). Very similar is also the Attic specimen in *Agora* XXIX, 258, No. 154, Fig. 12, Pl. 14 (300–275 BC).
Attested Dates: 335–300 BC.

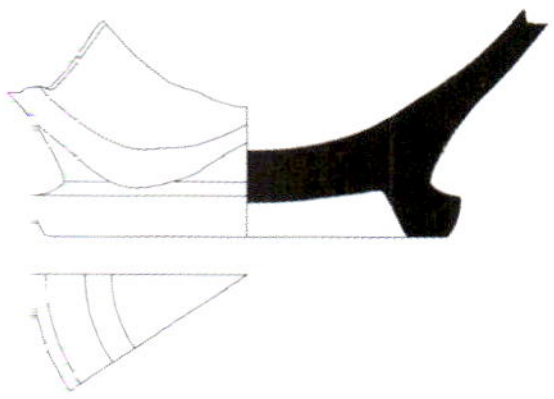

1:2

43. BG 566-03 A-type skyphos base (1 example)
Dimensions: base diam. 4.4 h 3.3
Reference Sherd Description: Reserved lower exterior of foot and underside, with inner rings.
Selected Comparanda: Prag 1983a, 352, No. 300 (Cozzo Presepe, ca. 300 BC); Cracolici 2004a, 10, Fig. 4c (Palazzo San Gervasio, T 13/2002, end of the 4th–early 3rd c. BC); *Museo Taranto* II.2, 491, Pl. 38c, type 6.2c (Rutigliano, 350–300 BC); Prag 1992b, 121–22, No. 898, Fig. 37 (Botromagno, 340–290 BC); Preacco Ancona 1992, 132, No. 89, Pl. XLIX (Locri, Casa dei Leoni, end of the 4th–first quarter of the 3rd c. BC); Morel 1981a, 4373a1, 311, Pl. 131 (Ugento, 333–300 BC).
Attested Dates: 330–280 BC.

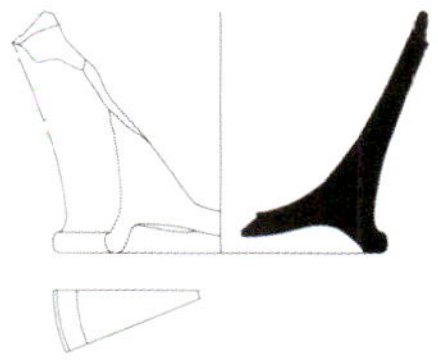

1:2

44. BG 395-17 A-type skyphos base (5 examples)
Dimensions: base diam. 4.2 h 2.9
Reference Sherd Description: Good quality, consistent, semi-lustrous black slip. Reserved band on lower body and underside of floor inside foot ring.
Selected Comparanda: Prag 1983a, 352, No. 300 (Cozzo Presepe, ca. 300 BC); Lo Porto 1973, 194–95, Pl. XLIII, No. 5.3 (Montescaglioso, T 12, 330–320 BC); De Cazanove 2008, 583, Fig. 294, No. 228 (Civita di Tricarico, second half of 4th c. BC); Giannotta 1996, 75–77, Figs. 37–38, No. 77 (Rocavecchia, T 27/1934, last quarter of 4th c. BC); Attic examples can share a similar profile: *Agora* XXIX, 94, No. 152, Fig. 12, Pl. 14 (300–275 BC).
Attested Dates: 330–300 BC

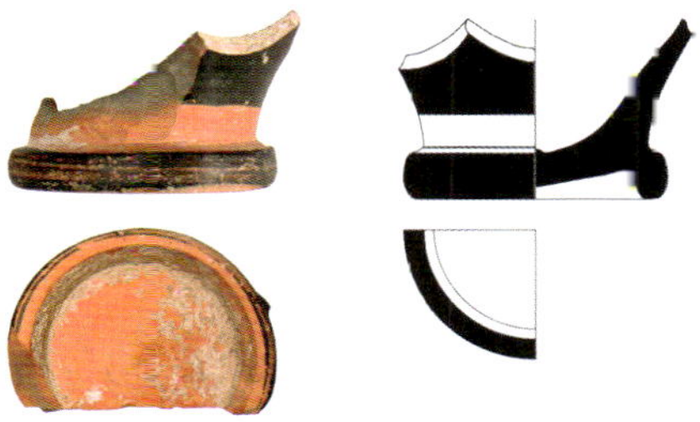

A-type skyphoi cont.

1:2

45. BG 173-13 A-type skyphos base (6 examples)
Dimensions: base diam. 4.4 h 8.1 w 4.6
Reference Sherd Description: Consistent, matte reddish (oxidized) slip. Reserved lower body and underside of floor inside foot ring.
Selected Comparanda: Prag 1983a, 352, No. 300 (Cozzo Presepe, ca. 300 BC); G. Greco 1991, 77, Fig. 180 (Serra di Vaglio, Casa dei Pithoi, 350–300 BC); Prag 1992b, 121–22, No. 898, Fig. 37 (Botromagno, 340–290 BC); Prag 1992b, 120, No. 892, Fig. 37 (Botromagno, 325–300 BC); *Poseidonia-Paestum* III, 132, No. 482, Fig. 93 (Poseidonia, second half of 4th c. BC); Morel 1981a, series 4343, 308, Pl. 129 (end of 4th c. BC).
Attested Dates: 310–280 BC.

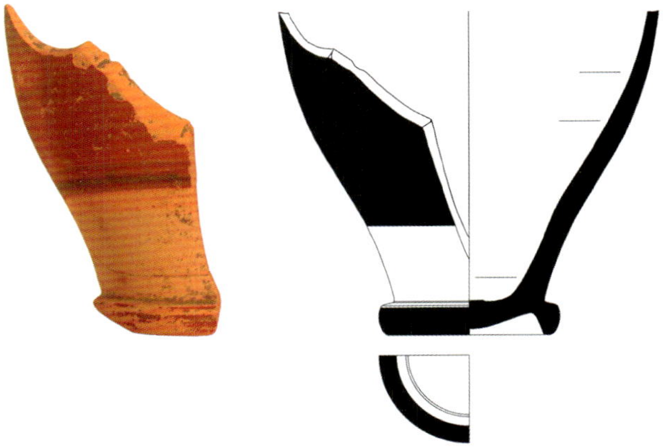

C-Type Skyphoi

1:2

46. BG 395-19 C-type skyphos base (3 examples)
Dimensions: base diam. 8.1 h 2.8 w 7.6
Reference Sherd Description: Dilute, semi-lustrous black slip, reserved lower body and underside of floor, slipped foot ring.
Selected Comparanda: Elliott 1998, 677, S56 (Metaponto, Pantanello necropolis, 480–440 BC).
Attested Dates: 480–440 BC.

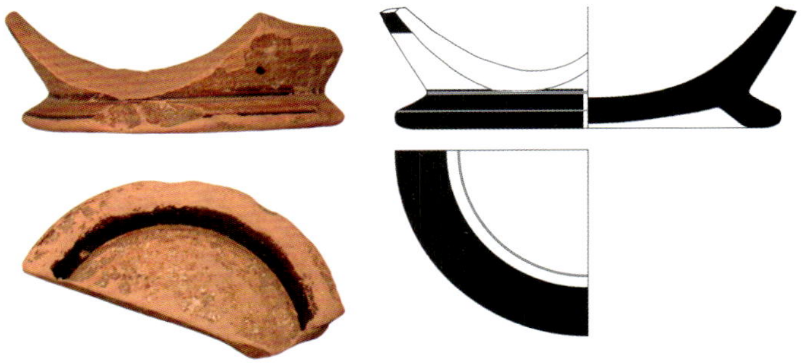

1:2

47. BG 578-04 C-type skyphos rim (116 examples)
Dimensions: rim diam. 10.6 h 2.8
Reference Sherd Description: Dilute, matte black slip.
Selected Comparanda: Elliott 1998, 679, S32 (Metaponto, Pantanello necropolis, 440–400 BC); Fattoria Fabrizio, FF80-94P+130PL+165PL (400–375 BC); Lo Porto 1973, 176, Pl. XXVI, No. 1.7 (Pisticci, T 13, ca. 380 BC); Prag 1992b, 116, No. 870, Fig. 36 (Botromagno, 450–375 BC); Morel 1981a, 4314a1, 306, Pl. 127 (said to be from Ruvo, second half of the 5th c. BC); *Agora* XII, No. 359, 86, Fig. 4 (480–450 BC).
Attested Dates: 440–400 BC.

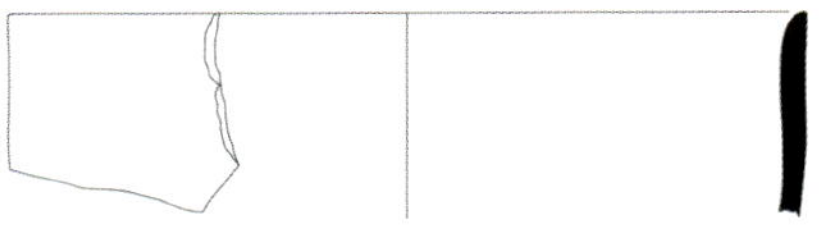

1:2

48. BG 503-01 C-type skyphos rim and upper body (single example)
Dimensions: rim diam. 10.0 h 4.1 w 3.1
Reference Sherd Description: Dilute, matte black slip.
Selected Comparanda: Sant'Angelo Grieco, SG1.81-1PL; Lo Porto 1973, Pl. VII, No. 4.2 (Pisticci, first quarter of 4th c. BC); Lo Porto 1988–1989, 337, No. 6, Fig. 40.3 (Montescaglioso, T 2, first quarter of 4th c. BC); *Museo Taranto* II.2, 487–488, Pl. 37, 6.1c (Rutigliano, 4th c. BC); Gagliardi 2001, 305, Nos. 54–55, Fig. 280 i–h (type B, Variant B3, second half of 4th c. BC, from Kaulonia, Punta Stilo sanctuary); *Agora* XII, No. 318 (late 5th c. BC).
Attested Dates: 400–380 BC.

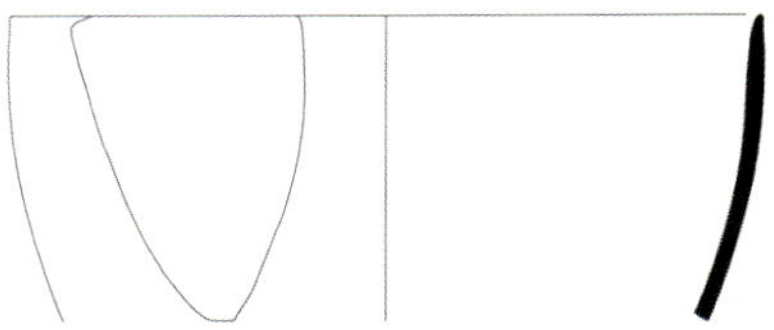

C-type skyphoi cont.

1:1 1:2

49. BG 013–03 revisit C–type skyphos base (single example)
Dimensions: base diam. 6.0 h 2.2 w 4.5
Reference Sherd Description: Dilute, semi–lustrous slip, completely oxidized on the interior, reserved underside of floor inside foot ring.
Selected Comparanda: Elliott 1998, 677, S3 (Metaponto, Pantanello necropolis, 440–400 BC); *Agora* XII, 81–83 and 258, No. 322, Fig. 4, Pl. 15 (ca. 400 BC).
Attested Dates: 440–400 BC.

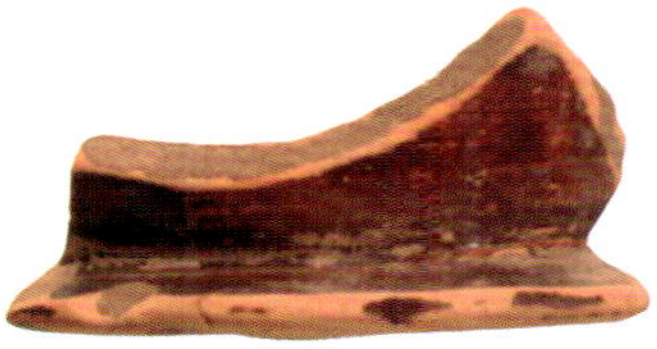

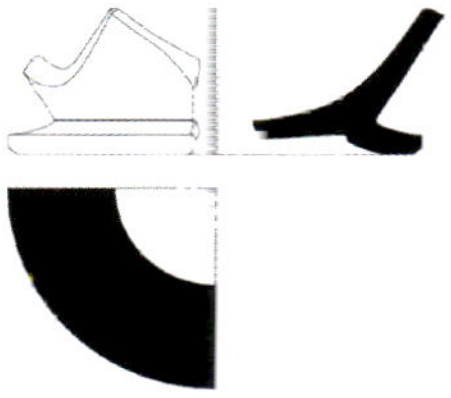

---

50. BG 043-05 C-type skyphos base (17 examples)
Dimensions: base diam. 5.4 h 2.0
Reference Sherd Description: Consistent, matte reddish brown (partially oxidized) slip, reserved underside of floor inside foot ring.
Selected Comparanda: Elliott 1998, 678, S25 (Metaponto, Pantanello necropolis, 440–400 BC); *Museo Taranto* I.3, 356, No. 125.20 (T Agosto 1934, Taranto via Dante, last quarter of 5th c. BC); *Agora* XII, 81–83 and 258, Fig. 4, Pl. 15, No. 322 (ca. 400 BC).
Attested Dates: 440–400 BC.

1:2

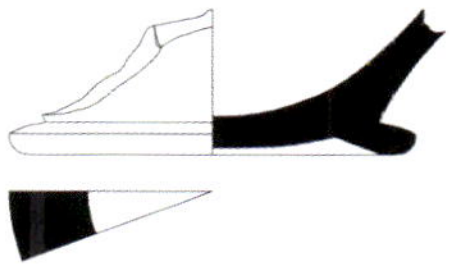

---

51. BG 285-37 C-type skyphos base (6 examples)
Dimensions: base diam. 7.1 h 1.0
Reference Sherd Description: Consistent, matte black slip, thin reserved band on underside of foot ring and reserved underside of floor inside foot ring.
Selected Comparanda: Elliott 1998, 678, S25 (Metaponto, Pantanello necropolis, 440–400 BC); Fattoria Fabrizio, FF80-175PL-A (430–400 BC).
Attested Dates: 440–400 BC.

1:2

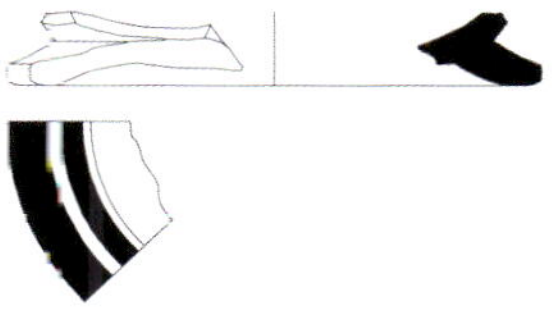

---

52. BG 523-23 C-type skyphos base (27 examples)
Dimensions: base diam. 6.0 h 1.45
Reference Sherd Description: Consistent, semi-lustrous black slip, entirely slipped.
Selected Comparanda: Elliott 1998, 678, S25 (Metaponto, Pantanello necropolis, 440–400 BC).
Attested Dates: 440–400 BC.

1:2

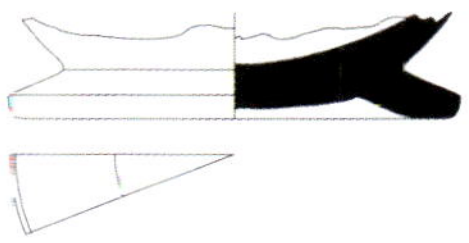

---

53. BG 529-22 C-type skyphos base (22 examples)
Dimensions: base diam. 4.05 h 2.7
Reference Sherd Description: Dilute, matte black slip, reserved underside of floor inside foot ring.
Selected Comparanda: Elliott 1998, 681, S32 (Metaponto, Pantanello necropolis, 440–400 BC); *Agora* XII, 81–83 and 258, No. 318, Fig. 4, Pl. 15 (450–430 BC).
Attested Dates: 440–400 BC.

1:2

---

54. BG 509-01 C-type skyphos base (15 examples)
Dimensions: base diam. 5.6 h 1.5
Reference Sherd Description: Consistent, faded, matte black slip, reserved underside of floor inside foot ring.
Selected Comparanda: Elliott 1998, 677, S33 (440–375 BC); Fattoria Fabrizio, FF80-99 PL-B (450–425 BC); Lo Porto 1973, 160, Pl. VII, No. 4.2 (Pisticci, early 4th c. BC, with reserved underside).
Attested Dates: 420–380 BC.

1:2

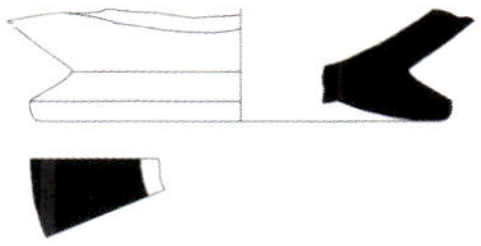

C-type skyphoi cont.

1:2

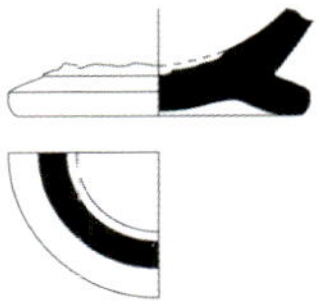

55. BG 521-114 C-type skyphos base (13 examples)
Dimensions: base diam. 4.0 h 1.5
Reference Sherd Description: Consistent, semi-lustrous black slip, reserved underside with band of slip on the underside of floor adjacent to junction with foot ring.
Selected Comparanda: Elliott 1998, 681, S33 (440–375 BC); Lo Porto 1973, 177, Pl. XXVI, No. 1.7 (Pisticci, T 13, 380 BC); Lo Porto 1988–1989, 337, No. 3, Figs. 36.6 and 40.3 (Montescaglioso T 2, beginning of the 4th c. BC).
Attested Dates: 410–375 BC.

1:2

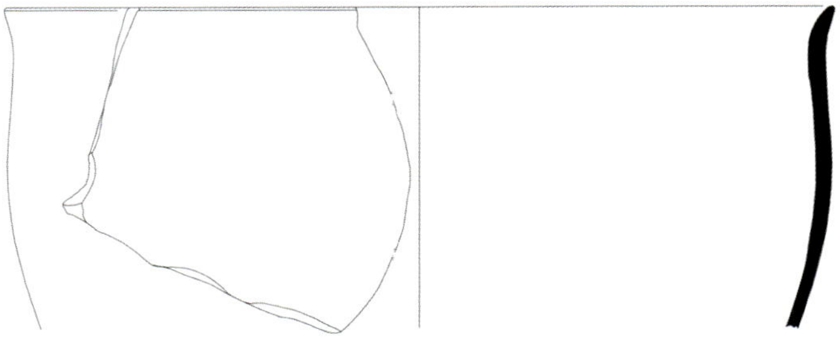

56. BG 488-96 C-type skyphos rim and upper body (8 examples)
Dimensions: rim diam. 17.5 h 7.0 w 7.3
Reference Sherd Description: Consistent, matte black slip.
Selected Comparanda: Prag 1992b, 117, No. 875, Fig. 36 (Botromagno, 350–?300 BC); Semeraro 1983, Pl. 88, No. 157 (Otranto, 330–300 BC).
Attested Dates: 330–300 BC.

1:2

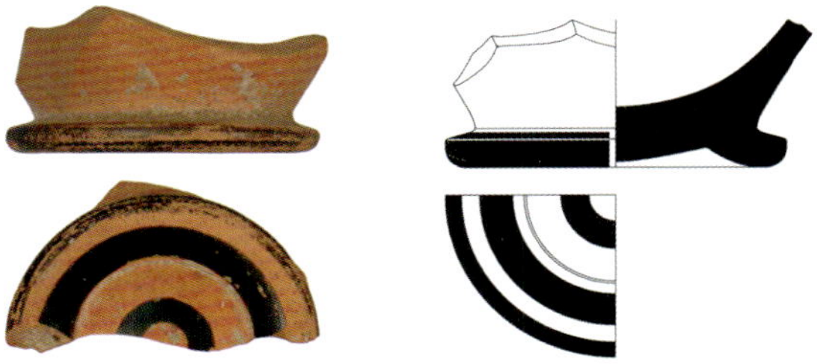

57. BG 269-33 C-type skyphos base (single example)
Dimensions: base diam. 6.0 h 2.7 w 6.1
Reference Sherd Description: Dilute, semi-lustrous reddish-brown slip on body, consistent black slip over base, with concentric bands of slip on underside of foot ring and floor and reserved lower body.
Selected Comparanda: Prag 1983a, 352, No. 305 (Cozzo Presepe, late 4th to first quarter of 3rd c. BC); *Museo Taranto* II.2, 486–87, Pls. 37.b, 6.1b (Rutigliano, 4th c. BC); Prag 1992b, 116, No. 869, Fig. 37 (Botromagno, late 4th to early 3rd c. BC); Pianu 1990, 144–45, Pl. V, No. 160.1 (330–300 BC).
Attested Dates: 350–300 BC.

1:2

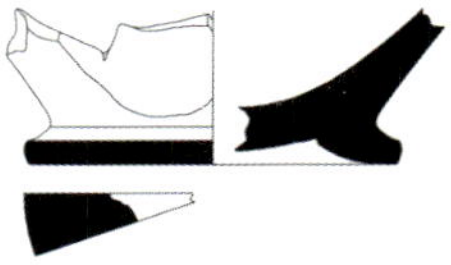

58. BG 413-05 C-type skyphos base (22 examples)
Dimensions: base diam. 5.0 h 2.1
Reference Sherd Description: Consistent, semi-lustrous black slip, reserved lower body and underside of floor inside foot ring.
Selected Comparanda: Prag 1983a, 352, No. 305 (Cozzo Presepe, late 4th to first quarter of 3rd c. BC); *Museo Taranto* II.2, 486–87, Pl. 37.base diam., 6.1b (Rutigliano, 4th c. BC).
Attested Dates: 330–300 BC.

1:2

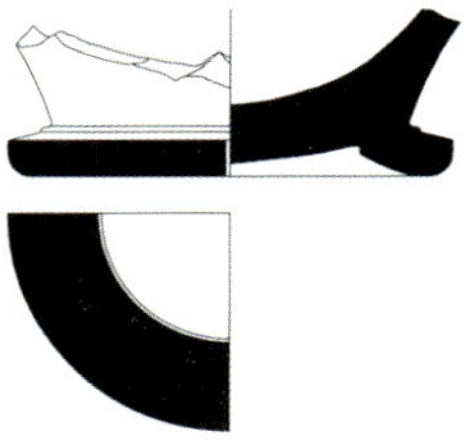

59. BG 521-131 C-type skyphos base (5 examples)
Dimensions: base diam. 5.9 h 2.225
Reference Sherd Description: Consistent, semi-lustrous black slip, reserved lower body and underside of floor inside foot ring.
Selected Comparanda: Fattoria Fabrizio, FF80-145PL-A (330–300 BC); Sant'Angelo Grieco, SG205.3; Prag 1992b, 115–16, No. 869, Fig. 37 (Botromagno, 325–300 BC); *Museo Taranto* II.2, 486–87, Pl. 37b, 6.1b (Rutigliano, 4th c. BC).
Attested Dates: 340–300 BC.

C-type skyphoi cont.

1:2

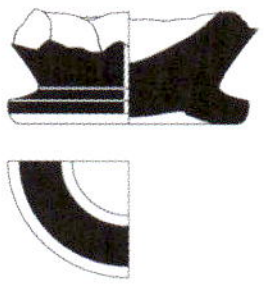

60. BG 521-130 C-type skyphos base (5 examples)
Dimensions: base diam. 3.2 h 1.6
Reference Sherd Description: Consistent, semi-lustrous black slip, reserved underside of floor inside and the lower exterior and outer edge of foot ring.
Selected Comparanda: Pianu 1990, 115–16, Pl. LI, No. 113.1 (Herakleia, Gnathia example, 340–300 BC); De Cazanove 2008, 584, Fig. 295, No. 247 (Civita di Tricarico, second half of the 4th c. BC).
Attested Dates: 350–300 BC.

1:2

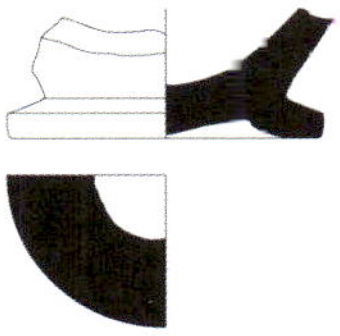

61. BG 460-02 C-type skyphos base (7 examples)
Dimensions: base diam. 4.2 h 1.8
Reference Sherd Description: Consistent, semi-lustrous black slip, reserved underside of floor inside foot ring.
Selected Comparanda: Sant'Angelo Grieco, SG204.1 (330–300 BC); De Cazanove 2008, 584, Fig. 295, No. 249 (Civita di Tricarico, 350–300 BC); Bruscella and Virtuoso 2005, 291, Pl. XXXVII, No. 210 (Torre di Satriano, sanctuary, 330–300 BC).
Attested Dates: 330–300 BC.

1:2

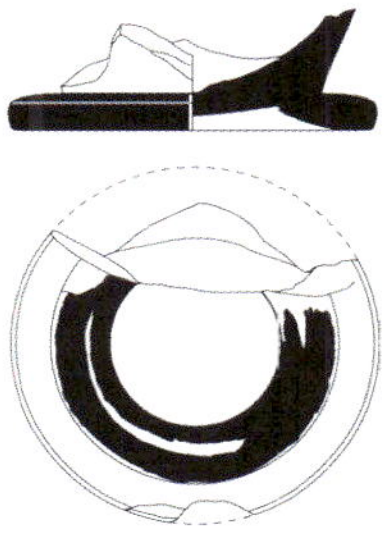

62. BG 516-02 C-type skyphos base (15 examples)
Dimensions: base diam. 4.9 h 1.7
Reference Sherd Description: Dilute, unevenly-fired, matte reddish brown slip on body, oxidized red on the underside of the base. Reserved lower body and underside of floor inside foot.
Selected Comparanda: Prag 1983a, 352, No. 305 (Cozzo Presepe, ca. 300 BC); Prag 1992b, 115, No. 868, Fig. 37 (Botromagno, 325–300 BC); Surra and Preacco Ancona 1997, Pl. 36, No. 82 (Pomarico Vecchio, 330–300 BC); Morel 1981a, series 4311, 305, Pl. 126 (second half of 4th c. BC).
Attested Dates: 340–300 BC.

1:1

63. BG 341-26 C-type skyphos base with perforated floor (single example)
Dimensions: base diam. 5.2 h 1.9
Reference Sherd Description: Dilute, matte black slip.
Selected Comparanda: Prag 1983a, 352, No. 305 (Cozzo Presepe, ca. 300 BC); Prag 1992b, 115, No. 868, Fig. 37 (Botromagno, 325–300 BC); Surra and Preacco Ancona 1997, Pl. 36, No. 82 (Pomarico Vecchio, 330–300 BC); Pianu 1990, 144–45, Pl. V, No. 160.1 (Herakleia, 330–300 BC); *Poseidonia-Paestum* II, 146, No. 333 (Poseidonia, Plateia AN, mid-4th c. BC); Morel 1981a, 4311b2, 305, Pl. 126 (said to be from Ruvo, end of the 4th c. BC).
Attested Dates: 340–300 BC.

C-type skyphoi cont.

1:2

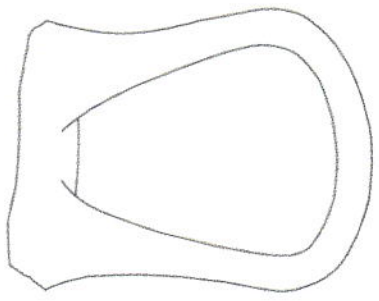

64. BG 521-54 C-type skyphos rim and handle (144 examples)
Dimensions: rim diam. 13.8 h 2.05 w 4.0
Reference Sherd Description: Consistent, semi-lustrous black slip.
Selected Comparanda: Elliott 1998, 679, S28 (Metaponto, Pantanello necropolis, 320–280 BC); Sant'Angelo Grieco, SG249.3; Prag 1983a, 352, No. 295 (Cozzo Presepe, 320–270 BC, following Morel 1970a); De Cazanove 2008, 584, Fig. 295, No. 266 (Civita di Tricarico, 300–275 BC).
Attested Dates: 320–280 BC.

1:2

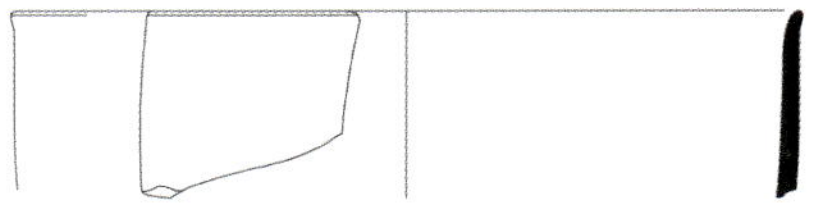

65. BG 395-05 C-type skyphos rim (3 examples)
Dimensions: rim diam. 10.5 h 2.55 w 2.9
Reference Sherd Description: Consistent, semi-lustrous grey slip.
Selected Comparanda: Prag 1983a, 354, No. 308a (Cozzo Presepe, late 4th to first quarter of 3rd c. BC); Prag 1992b, 118, No. 888, Fig. 36 (Botromagno, 350 to early 3rd c. BC); De Cazanove 2008, 584, Fig. 295, No. 264 (Civita di Tricarico).
Attested Dates: 320–280 BC.

1:2

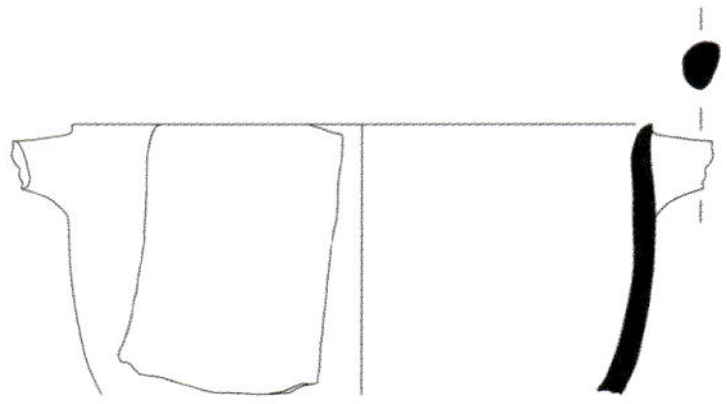

66. BG 578-06 C-type skyphos rim and upper body with handle terminus (single example)
Dimensions: rim diam. 7.7 h 3.7 w 3.0
Reference Sherd Description: Consistent, matte black slip.
Selected Comparanda: Elliott 1998, 679, S28 (Metaponto, Pantanello necropolis, 320–280 BC); Prag 1983a, 352, No. 295 (Cozzo Presepe, 320–270 BC, following Morel); Surra and Preacco Ancona 1997, 70, Pl. 36, No. 80 (Pomarico Vecchio, 330–300 BC); Bruscella and Virtuoso 2005, 270, Pl. XXXVII, No. 206 (Torre di Satriano, last three decades of the 4th c. BC, following Morel); Prag 1992b, 121, No. 897, Fig. 36 (Botromagno, 335–300 BC).
Attested Dates: 320–280 BC.

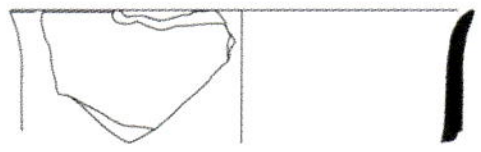

1:2

67. BG 043-08 C-type skyphos rim (114 examples)
Dimensions: rim diam. 6.2 h 1.8 w 2.6
Reference Sherd Description: Good quality, consistent, semi-lustrous black slip.
Selected Comparanda: Elliott 1998, 679, S28, S20 (Metaponto, Pantanello necropolis, 320–280 BC); Prag 1983a, 352, No. 295 (Cozzo Presepe, 320–270 BC); Prag 1992b, 118, No. 888, Fig. 36 (Botromagno, 350 to early 3rd c. BC).
Attested Dates: 320–280 BC.

1:2

68. BG 113-17 C-type skyphos rim and body with handle terminus (4 examples)
Dimensions: rim diam. 7.5 h 4.9 w 5.2
Reference Sherd Description: Consistent, semi-lustrous black slip.
Selected Comparanda: Elliott 1998, 679, S28 (Metaponto, Pantanello necropolis, 320–280 BC); Pianu 1990, 144–45, Pl. V, No. 160.1 (Herakleia, 330–300 BC); De Cazanove 2008, 584, Fig. 295, No. 265 (Civita di Tricarico, 300–275 BC).
Attested Dates: 320–280 BC.

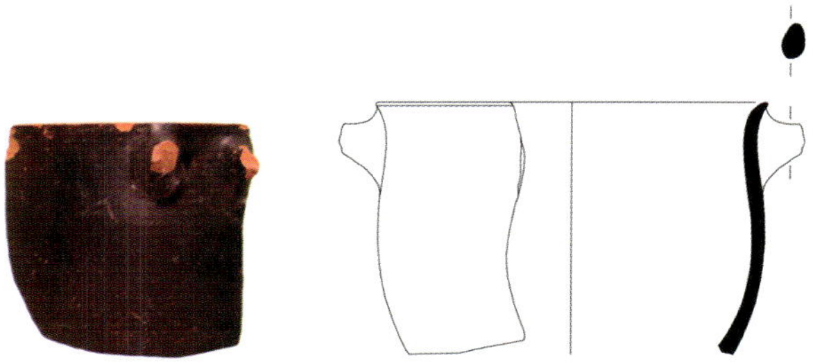

C-type skyphoi cont.

1:2

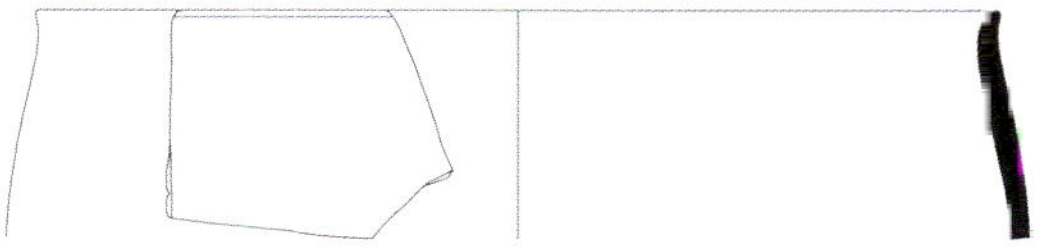

69. BG 454-08 C-type skyphos rim (19 examples)
Dimensions: rim diam. 12.8 h 3.1 w 3.8
Reference Sherd Description: Dilute, matte brownish slip, partly faded.
Selected Comparanda: Fattoria Fabrizio, FF80-10PL-C2 (330–300 BC); Sant'Angelo Grieco, SG50.13; Prag 1983a, 352, No. 292 (Cozzo Presepe, late 4th to first quarter of the 3rd c. BC); Prag 1992b, 121, No. 893, Fig. 36 (Botromagno, 335–300 BC).
Attested Dates: 320–280 BC.

1:2

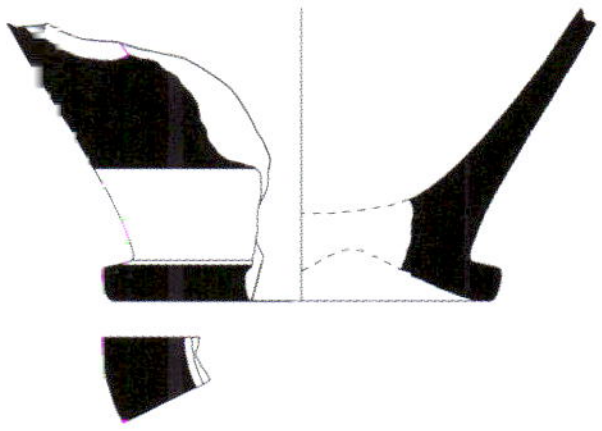

70. BG 503-09 C-type skyphos base (19 examples)
Dimensions: base diam. 5.3 h 4.0
Reference Sherd Description: Consistent, matte black slip, reserved lower body and extant underside of floor inside foot ring.
Selected Comparanda: Prag 1992b, 115–16, No. 869, Fig. 37 (Botromagno, 325–300 BC); Lo Porto 1973, Pl. LIV, No. 1.1 (Miglionico, late 4th c. BC) and LVII, No. 3 (Matera, T 6, 350 BC); Bruscella and Virtuoso 2005, 270, Pl. XXXVII, No. 211 (Torre di Satriano, last three decades of 4th c. BC, following Morel); *Museo Taranto* II.2, 488, Pl. 37d, 6.1d (Rutigliano, second half of the 4th c. BC).
Attested Dates: 340–300 BC.

1:2

71. BG 529-12 C-type skyphos base (17 examples)
Dimensions: base diam. 4.45 h 3.3
Reference Sherd Description: Dilute, matte black slip, reserved lower body and band on underside of foot ring and central underside of floor.
Selected Comparanda: Lo Porto 1973, 205, Pl. LIV, No. 1.1 (Miglionico, 320–310 BC); Bruscella and Virtuoso 2005, 270, Pl. XXXVII, No. 210 (Torre di Satriano, last three decades of the 4th c. BC, following Morel); Prag 1992b, 115, No. 868, Fig. 37 (Botromagno, 325–300 BC); *Museo Taranto* II.2, 488, Pl. 37d, 6.1d (Rutigliano, second half of the 4th c. BC); Morel 1981a, series 4311, 305, Pl. 126 (350–300 BC).
Attested Dates: 340–300 BC.

1:2

72. BG 498-17 C-type skyphos base (single example)
Dimensions: base diam. 4.5 h 2.7
Reference Sherd Description: Consistent, semi-lustrous black slip.
Selected Comparanda: Prag 1983a, 352, No. 301 (Cozzo Presepe, ca. 300 BC); Prag 1992b, 115–16, No. 868, Fig. 37 (Botromagno, 325–300 BC); *Poseidonia-Paestum* II, 146, No. 332 (Poseidonia, Plateia AN, mid-4th c. BC).
Attested Dates: 340–300 BC.

C-type skyphoi cont.

1:2

73. BG 352-04 C-type skyphos base (single example)
Dimensions: base diam. 5 h 3.4
Reference Sherd Description: Consistent, semi-lustrous black slip.
Selected Comparanda: Prag 1983a, 352, No. 301 (Cozzo Presepe, ca. 300 BC); Pianu 1990, Pl. V, No. 160.1 (Herakleia, 330–300 BC); Lo Porto 1973, Pl. LIV, No. 1.1 (Miglionico, 320–310 BC); *Sibari* IV, 249, No. 196, Fig. 229 (Parco del Cavallo, second half of 4th c. BC).
Attested Dates: 340–300 BC.

1:2

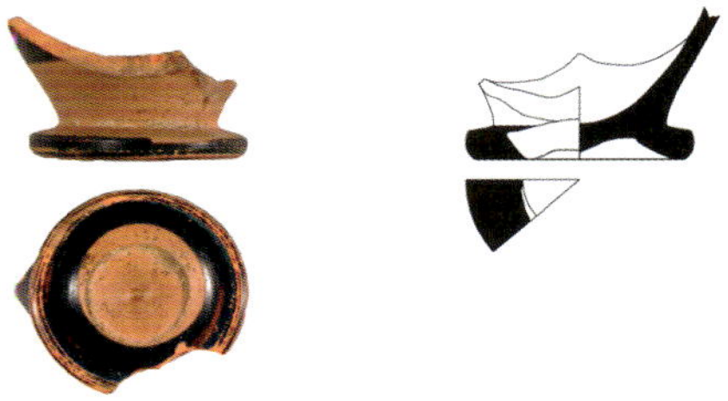

74. BG 528-17 C-type skyphos base (10 examples)
Dimensions: base diam. 3.5 h 2.4
Reference Sherd Description: Consistent, semi-lustrous black slip, reserved lower body and underside of floor inside foot.
Selected Comparanda: Prag 1983a, 352, No. 302 (Cozzo Presepe); Lo Porto 1988–1989, 377, 380, Fig. 88.8 (Montescaglioso, T 14, 320 BC); Prag 1992b, 118, No. 868, Fig. 37 (Botromagno, 325–300 BC); *Museo Taranto* II.2, 488, Pl. 37d, 6.1d (Rutigliano, second half of 4th c. BC); Bruscella and Virtuoso 2005, 270, Pl. XXXVII, No. 210 (Torre di Satriano, last three decades of 4th c. BC, following Morel); Morel 1981a, series 4311, 305, Pl. 126 (350–300 BC).
Attested Dates: 320–280 BC.

1:2

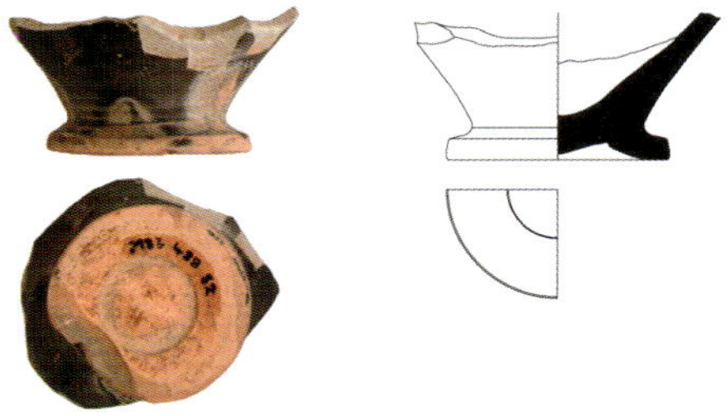

75. BG 488-82 C-type skyphos base (11 examples)
Dimensions: base diam. 3.6 h 2.45
Reference Sherd Description: Dilute, semi-lustrous reddish brown slip, reserved underside.
Selected Comparanda: Prag 1992b, 115, No. 868, Fig. 37 (Botromagno, 325–300 BC); *Museo Taranto* II.2, 488, Pl. 37d, 6.1d (Rutigliano, second half of 4th c. BC).
Attested Dates: 320–280 BC.

1:2

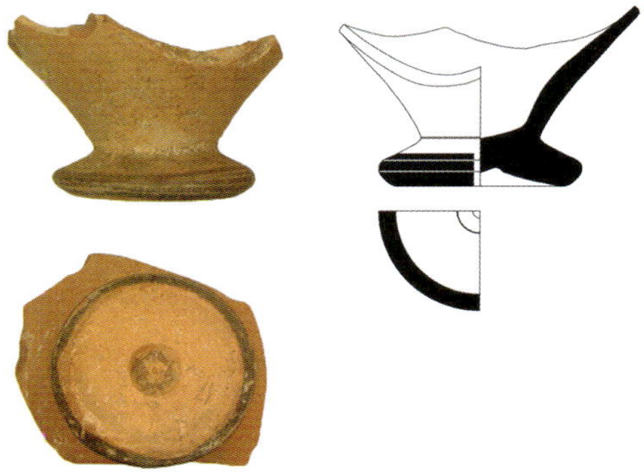

76. BG 266-17 C-type skyphos base (5 examples)
Dimensions: base diam. 4.05 h 3.65
Reference Sherd Description: Consistent, matte reddish brown slip, greyish misfired fabric, reserved lower body and interior of foot ring and floor.
Selected Comparanda: Prag 1983a, 352, No. 306a (Cozzo Presepe, late 4th to first quarter of 3rd c. BC); Prag 1992b, 117, No. 880, Fig. 37 (Botromagno, 325–300 BC, following Morel); De Cazanove 2008, 584, Fig. 295, No. 247 (Civita di Tricarico, second half of 4th c. BC); Scarfì 1961, 211, Fig. 54, No. 2 (Monte Sannace, T17, early 3rd c. BC).
Attested Dates: 325–280 BC.

C-type skyphoi cont.

77. BG 566-01 C-type skyphos base (11 examples)
Dimensions: base diam. 4.0 h 3.1
Reference Sherd Description: Dilute, matte black slip, reserved lower body, internal underside of foot and floor.
Selected Comparanda: Elliott 1998, 678, S20 (Metaponto, Pantanello necropolis, 300 BC); Prag 1992b, 117, No. 879, Fig. 37 (Botromagno, 325–300 BC); Lissi Caronna 1990–91, 206, Fig. 26, No. 16 (Oppido Lucano, house C, context dated to the second half 4th–early 3rd c. BC).
Attested Dates: 320–280 BC.

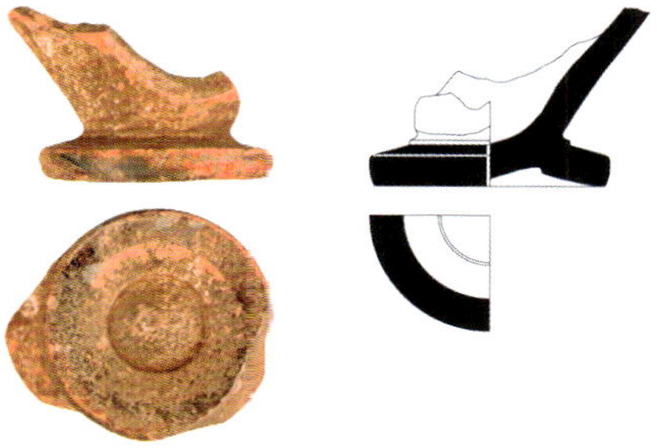

1:2

78. BG 128-23 C-type skyphos base (single example)
Dimensions: base diam. 3.3 h 2.3
Reference Sherd Description: Dilute, matte reddish brown slip almost completely faded, reserved lower body and underside.
Selected Comparanda: Elliott 1998, 679, S28 (Metaponto, Pantanello necropolis, 320–280 BC); Frag 1983a, 352, No. 302 (Cozzo Presepe); Lissi Caronna 1990–91, 223, Fig. 46, second layer, No. 3, and 227, No. 3 (Oppido Lucano, house C, context dated to second half 4th–early 3rd c. BC).
Attested Dates: 320–280 BC.

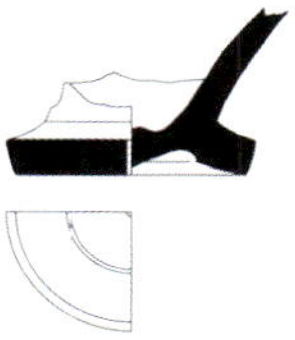

1:2

79. BG 497-26 C-type skyphos base (5 examples)
Dimensions: base diam. 4.5 h 1.15
Reference Sherd Description: Consistent, semi-lustrous black slip, reserved extant lower body and underside, with band of slip only on the external face of the foot.
Selected Comparanda: Elliott 1998, 678, S20 (Metaponto, Pantanello necropolis, 320–280 BC); Prag 1983a, 352, No. 306a (Cozzo Presepe, late 4th to first quarter of 3rd c. BC); Prag 1992b, 117, No. 879, Fig. 36 (Botromagno, 325–300 BC); Morel 1981a, 4353a1, 309, Pl. 129 (Norba, ca. 300 BC).
Attested Dates: 320–280 BC.

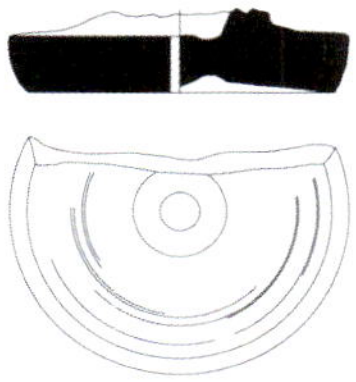

1:2

80. BG 183-14 C-type skyphos base (2 examples)
Dimensions: base diam. 4.4 h 3.2
Reference Sherd Description: Dilute, matte reddish brown slip, reserved extant exterior and underside.
Selected Comparanda: Elliott 1998, 679, S20 (Metaponto, Pantanello necropolis, 320–280 BC); Lissi Caronna 1990–91, 223, Fig. 46, ambitus, No. 2 (Oppido Lucano, house C, context dated to the second half of 4th–early 3rd c. BC).
Attested Dates: 320–280 BC.

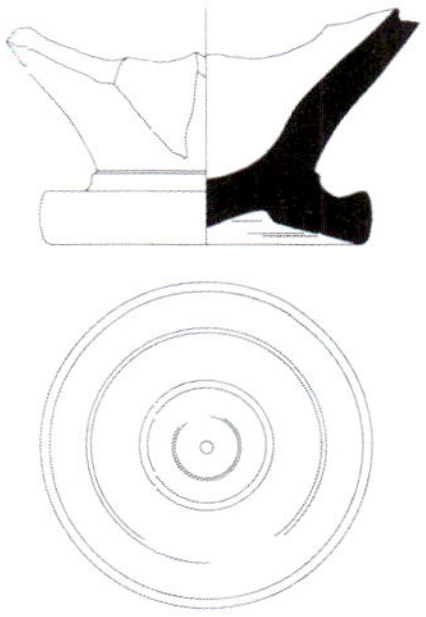

1:2

81. BG 113-12 C-type skyphos base (2 examples)
Dimensions: base diam. 3.6 h 4.7
Reference Sherd Description: Consistent, semi-lustrous black slip.
Selected Comparanda: Lo Porto 1966, 214–15, Fig. 69, No. 9 (Metaponto, urban necropolis, 3rd c. BC); Pianu 1990, Pl. V, No. 103.1 (Herakleia, 340–300 BC); Lo Porto 1988–1989, 379, Fig. 88, No. 8 (Montescaglioso, T 2/1957, 320 BC); Prag 1992b, 117, No. 879, Fig. 37 (Botromagno, 325–300 BC).
Attested Dates: 320–280 BC.

1:2

Stemless Kylikes

1:2

82. BG 250-09 Stemless kylix base (single example)
Dimensions: base diam. 10.8 h 2.0 w 4.0
Reference Sherd Description: Dilute, matte black slip, reserved lower foot ring and underside except for thin band of slip on the internal facet of foot ring. Part of a palmette stamp extant on the interior of the floor.
Selected Comparanda: *Museo Taranto* II.2, 495, Pl. 39f, 7.6 (Rutigliano, second half of 5th c. BC); Morel 1981, 4221a1, 295, Pl. 120 (Capua, 4th c. BC).
Attested Dates: 450–400 BC.

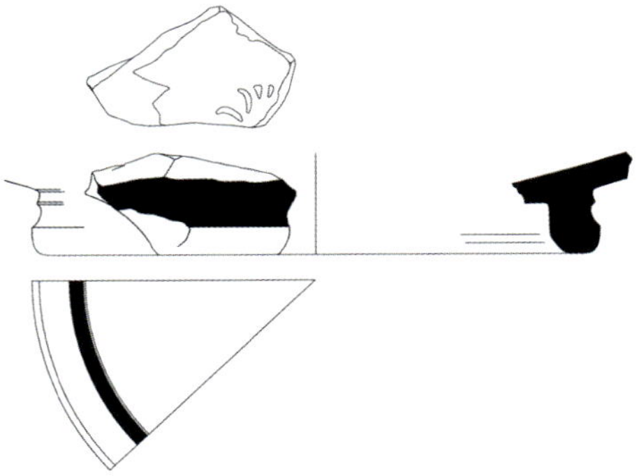

1:2

83. BG 283-01 Stemless kylix base (2 examples)
Dimensions: base diam. 10.0 h 2.2 w 6.4
Reference Sherd Description: Consistent, semi-lustous black slip, reserved underside of floor inside foot ring. On the inner surface, very small portion of the impressed decoration and *disco di impilamento*.
Selected Comparanda: Elliott 1998, 658, SK5 (Metaponto, Pantanello necropolis, 430–370 BC); Fattoria Fabrizio, FF80-81PL-A (430–350 BC).
Attested Dates: 430–390 BC.

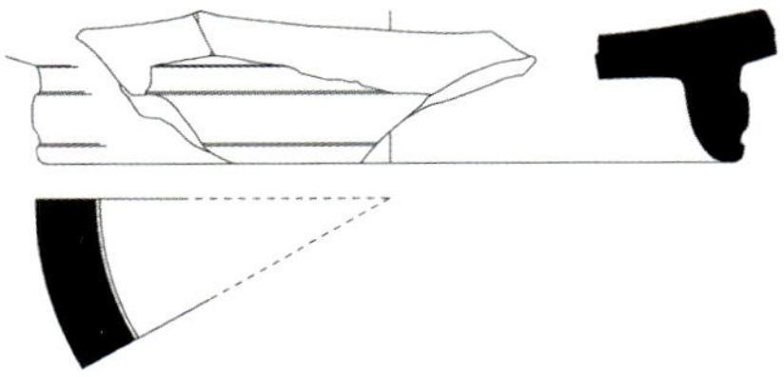

1:2

84. BG 521-122 Stemless kylix base (4 examples)
Dimensions: base diam. 7.65 h 2.05
Reference Sherd Description: Dilute, matte black slip, reserved underside with traces of a thin band of slip on the internal facet of the foot ring.
Selected Comparanda: Elliott 1998, 658, SK5 (Metaponto, Pantanello necropolis, 430–370 BC); Prag 1992b, 113, No. 858, Fig. 35 (Botromagno, 400–350 BC, following Morel); *Museo Taranto* II.2, 494–95, Pl. 39e, 7.5 (Rutigliano, second half of 5th to second half of 4th c. BC); Bitti 1989, 146, Pl. XXIII, No. 107 (Locri, 350–325 BC).
Attested Dates: 430–350 BC.

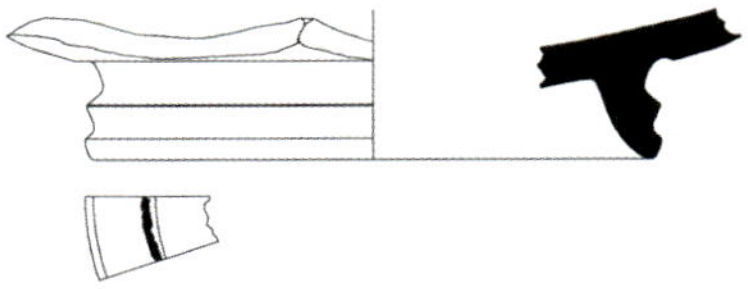

1:2

85. BG 272-06 Stemless kylix floor (single example)
Dimensions: h l 4.4 w 5.4
Reference Sherd Description: Dilute, matte reddish brown slip, four palmette stamps extant on the floor.
Selected Comparanda: For palmette decoration: De Juliis 1990, 102, No. 57, Fig. 456 (Canosa, Ipogeo dei Vimini, Cella B, left deposition, second quarter of 4th c. BC); Labellarte 1988, 335, No. 723, Fig. 470 (Ceglie, Ipogeo di via Giuseppe Martino, mid-4th c. BC).
Attested Dates: 380–330 BC.

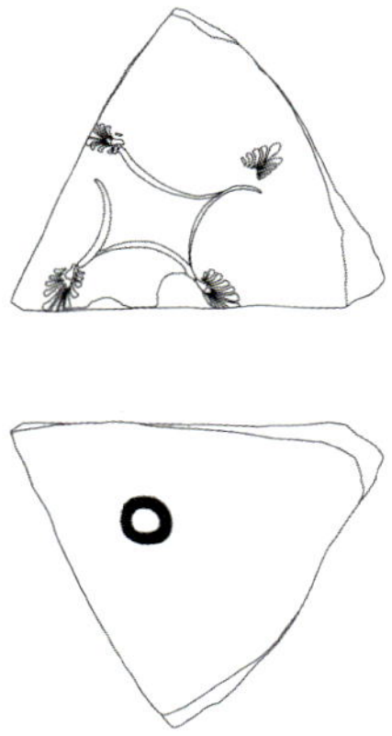

Stemless kylikes cont.

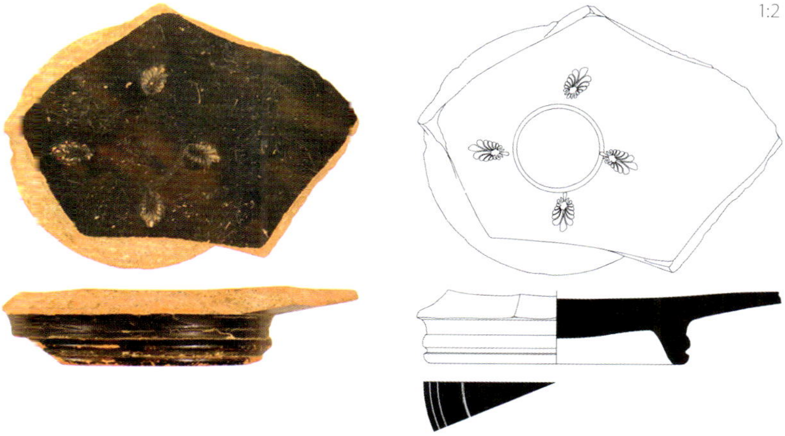

1:2

86. BG 488-02 Stemless kylix base (single example)
Dimensions: base diam 7.1 h 2.0 w 9.6
Reference Sherd Description: Consistent, semi-lustrous black slip, four palmette stamps on the interior of the floor.
Selected Comparanda: Elliott 1998, 658, SK5 (Metaponto, Pantanello necropolis, 430–370 BC); *Museo Taranto* I.2, 494–95, Pl. 39e, 7.5 (Rutigliano, second half of 5th to second half of 4th c. BC).
Attested Dates: 430–350 BC.

Cup-Skyphos

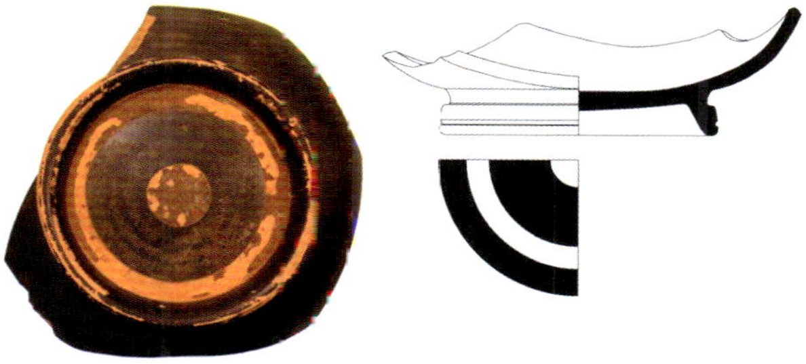

1:2

87. BG 176-02 Cup-skyphos base (single example)
Dimensions: base diam. 5.45 h 2.6 w 8.3
Reference Sherd Description: Dilute, matte black slip with reserved band and center of underside.
Selected Comparanda: Bearing some similarities to feet associated with Attic cup-skyphoi, *light wall* of end of 5th c. BC: *Agora* XII, 110–11 and 278, No. 593, Fig. 6, Pl. 26 (410 BC). See Patera 1986, 61, No. 58 (Attic, end of 5th c. BC, from Taranto, tomb of Via Dante). Also Pontrandolfo and Rouveret 1992, 314, No. 6 (Poseidonia, T 88a, beginning of second quarter of 4th c. BC). Morel 1981a, 4213 Pl. 120.
Attested Dates: 400–350 BC.

1:2

88. BG 272-05 Cup-skyphos base (single example)
Dimensions: base diam. 6.45 h 2.6
Reference Sherd Description: Consistent, semi-lustrous black slip, reserved band on the upper concavity on exterior of foot ring, slipped internal facet of foot with reserved underside of floor.
Selected Comparanda: D'Andria 1980a, 428, No. 310, Fig. 65 (Metaponto, kerameikos, deposit 3, third quarter of 4th c. BC); Pianu 1990, 31, No. 3, Pl. XII, 1 (Herakleia, T 32, 350–310 BC for the vase, while tomb is dated to third quarter of 4th c. BC); *Museo Taranto* I.3, 412, No. 194.2 (T 9. V. 1950, August 1934, third quarter of 4th c. BC); Canosa 2007, 89–90, Pl. XXXIX, Nos. 70–71, Fig. 10 (Timmari, T 33/1984, ca. 340 BC).
Attested Dates: 350–300 BC.

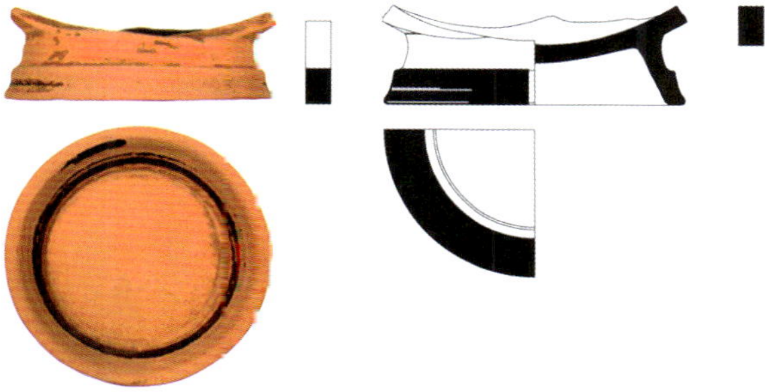

1:2

89. BG 291-01 Cup-skyphos base (single example)
Dimensions: base diam. 5.95 h 2.0 w 6.1
Reference Sherd Description: Consistent, matte black slip, reserved lower body and upper foot ring, slipped lower exterior and interior foot ring, reserved underside.
Selected Comparanda: Graepler 1997, 94, type 423 (Taranto, 375–325 BC); *Bitonto*, 156, No. 164 (Bitonto, T 4/1981, Gnathia example, 330–320 BC); Morel 1981a, 4242d1, 297, Pl. 121 (Aléria, ca. 370–275 BC).
Attested Dates: 300–280 BC.

Bolsals

1:2

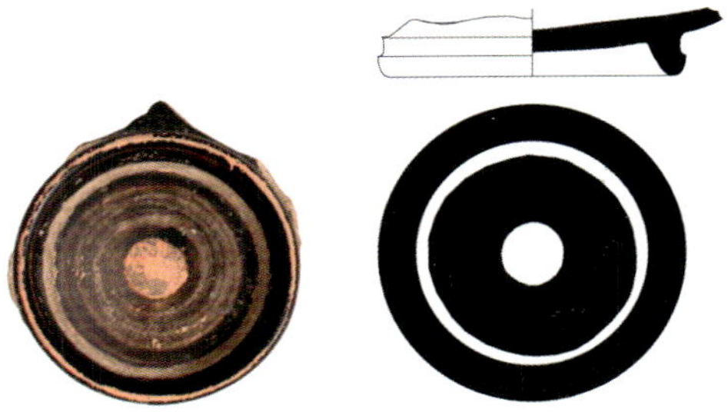

90. BG 401-22 Bolsal base (5 examples)
Dimensions: base diam. 5.3 h 1.2
Reference Sherd Description: Dilute, semi-lustrous black slip with reserved band and center of floor inside foot ring.
Selected Comparanda: Elliott 1998, 650, B1 (Metaponto, Pantanello necropolis, 420–400 BC).
Attested Dates: 400–350 BC.

Stemmed Cups

1:2

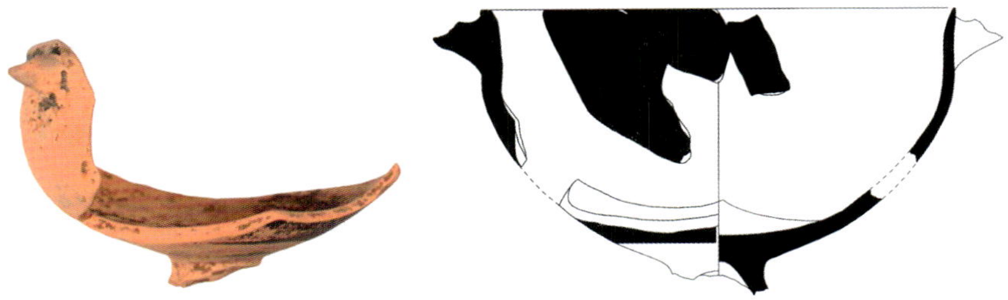

91. BG 402-01 Stemmed cup rim with handle terminus and upper part of stem (10 examples)
Dimensions: rim diam. 10.6 h 6.6
Reference Sherd Description: Consistent, matte black slip almost totally faded, stem appears to be reserved.
Selected Comparanda: Elliott 1998, 651, SC2 (Metaponto, Pantanello necropolis, 3rd c. BC); Elia and Meirano 1997, 93 (ill.), Pl. 43, No. 159 (Pomarico Vecchio, end of 4th to first half of 3rd c. BC); Pianu 1990, 116, Pl. V, No. 114.1 (Herakleia, 330–280); Morel 1981a, 4131a1, 292, Pl. 118 (unknown provenance, ca. 300 BC).
Attested Dates: 300–250 BC.

1:2

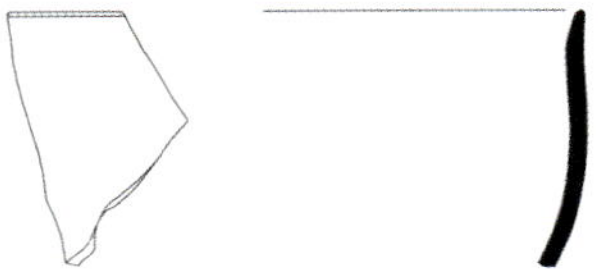

92. BG 413-07 Stemmed cup rim (4 examples)
Dimensions: h 3.55 w 2.4
Reference Sherd Description: Consistent, semi-lustrous black slip.
Selected Comparanda: Pianu 1990, 116, Pl. V, No. 114.1 (Herakleia, 330–280 BC); De Cazanove 2008, 579, Fig. 290, No. 132 (Civita di Tricarico, Gnathia example, 300–225 BC); bearing some similarities also to Delli Ponti 1996, 195–96, No. 242 (Vaste, T 18/1968, ring-handled cup, late 4th to early 3rd c. BC).
Attested Dates: 300–150 BC.

1:2

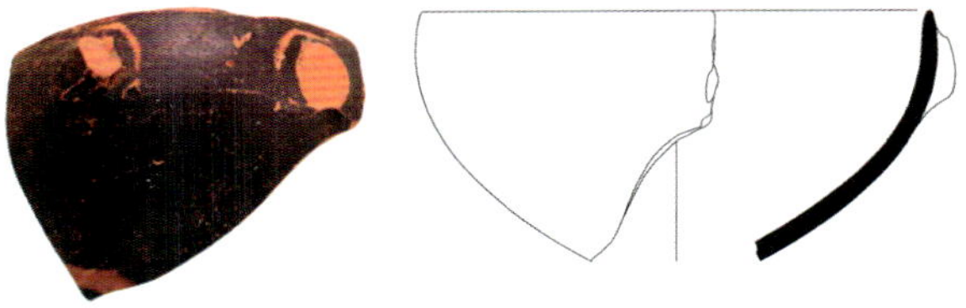

93. BG 113-02 Stemmed cup rim with handle terminus (2 examples)
Dimensions: rim diam. 8.0 h 4.1 w 6.2
Reference Sherd Description: Consistent, semi-lustrous black slip.
Selected Comparanda: See Ch. 19, site 736, T 20-02.
Attested Dates: 350–275 BC.

1:2

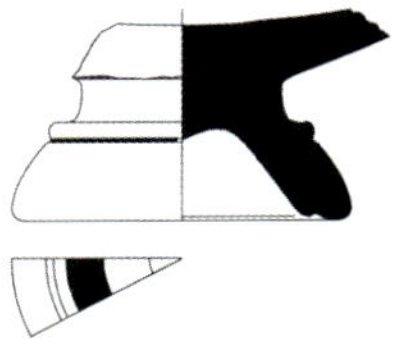

94. BG 224-01 Stemmed cup base (single example)
Dimensions: base diam. 4.5 h 2.85 w 5.0
Reference Sherd Description: Consistent, semi-lustrous black slip, almost completely worn. Reserved stem with band of slip on interior of stem.
Selected Comparanda: Lo Porto 1981b, Fig. 3, No. 2 (from a tomb at Metaponto, end of 4th c. BC); Labellarte 1995, 120, No. 2.1.27 (black-gloss stemmed cup from the Chini Collection, late 4th c. BC); D'Amicis 1995, 87, No. 1.3.36 (Gnathia stemmed cup from the Chini Collection, late 4th c. BC).
Attested Dates: 325–300 BC.

Stemmed cups cont.

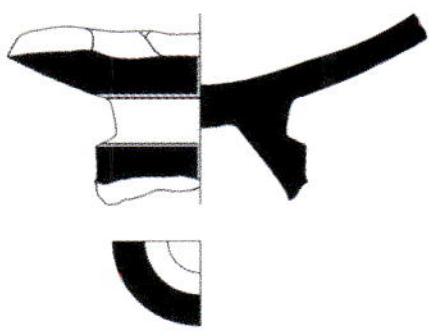

1:2

95. BG 152-04 Stemmed cup lower body and upper part of stem (single example)
Dimensions: base diam. 2.8 h 2.6 w 4.4
Reference Sherd Description: Dilute, semi-lustrous black slip, reddish brown on the interior, reserved external mid-section and interior of stem.
Selected Comparanda: Elliott 1998, 651, SC2 (Metaponto, Pantanello necropolis, first half of 3rd c. BC); Lo Porto 1988–1989, 417, Fig. 135, No. 2 (Calciano, T 1, 330–320 BC).
Attested Dates: 320–280 BC.

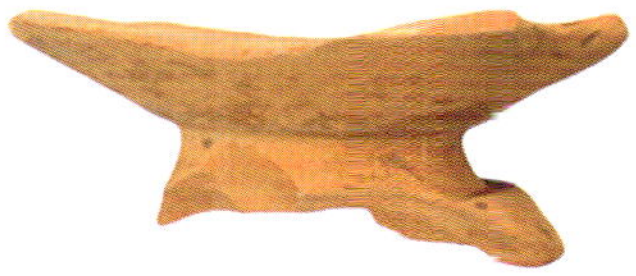

1:1

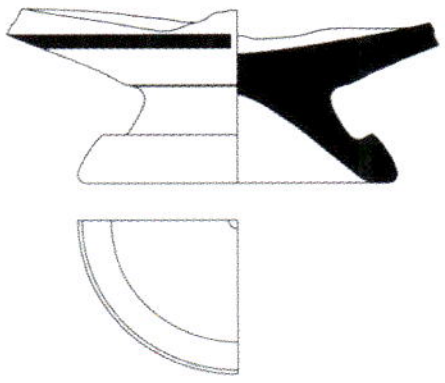

1:2

96. BG 043-21 revisit Stemmed cup base (single example)
Dimensions: base diam. 4.25 h 2.4 w 5.7
Reference Sherd Description: Consistent, semi-lustrous black slip almost completely worn, reserved stem and underside.
Selected Comparanda: Bearing similarities to Pianu 1990, 68–69, Pl. XXXV, No. 55.2 (Herakleia, late 4th c. BC); Riccardi 1989b, 73, Pl. 282, No. 27 (Monte Sannace, Gnathia, 3rd c. BC).
Attested Dates: 320–275 BC.

Ring-Handled Cups

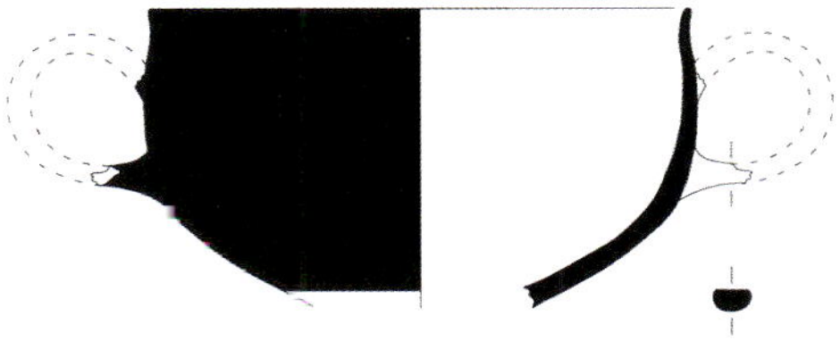

1:2

97. BG 401-18 Ring-handled cup rim with handle terminus (11 examples)
Dimensions: rim diam. 8.05 h 4.65
Reference Sherd Description: Consistent, semi-lustrous black slip, reserved extant lower body.
Selected Comparanda: Elliott 1998, 651, C17 (Metaponto, Pantanello necropolis, T 193-5, beginning of 3rd c. BC); Scarano 1992, Fig. 5.7 (Metaponto, T 22/1986, early 3rd c. BC); Lo Porto 1981b, 377–78, No. 2, Fig. 104.1 (Metaponto, contrada Ricotta, T32, 300–250 BC); Morel 1970a, 100, Fig. 23.4 (Cozzo Presepe, sacred deposit, 3rd c. BC); Pianu 1990, 145, Pls. V, LXI No. 2, No. 162.1 (Herakleia, 300–250 BC); Campagna 1995, 229, Fig. 8, No. 3 (Vaste, end of 4th–first half of 3rd c. BC).
Attested Dates: 300–250 BC.

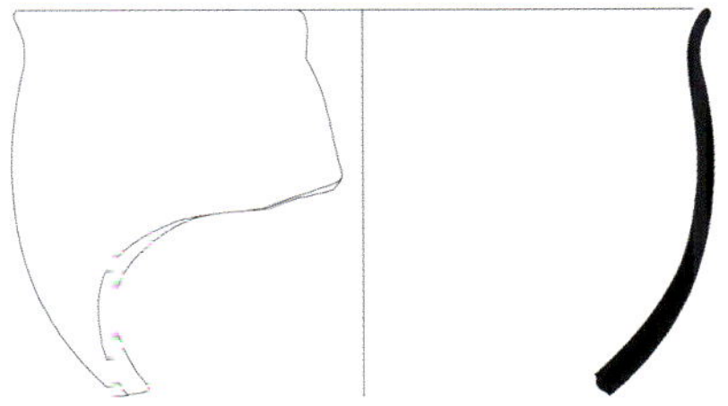

1:2

98. BG 113-04 Ring-handled cup rim (21 examples)
Dimensions: rim diam. 10.3 h 5.85 w 8.0
Reference Sherd Description: Traces of dilute, matte reddish-brown slip.
Selected Comparanda: Elliott 1998, 651, C17 (Metaponto, Pantanello necropolis, T 193-5, beginning of 3rd c. BC); Lo Porto 1981b, 377-78, Fig. 104.2 (Metaponto, T 32, 300–250 BC); Scarano 1992, Fig. 5.4 (Metaponto, T 35/1986, end of 4th–mid-3rd c. BC); *Arte e Artigianato*, 330, No. 323 (Taranto, deposit of via Leonida, 3rd c. BC).
Attested Dates: 300–200 BC.

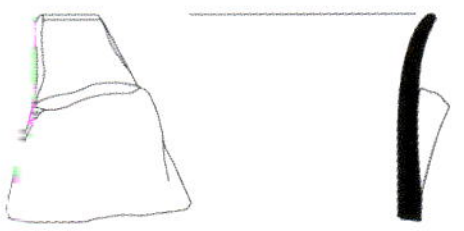

1:2

99. BG 352-01 Ring-handled cup rim with handle terminus (4 examples)
Dimensions: h 2.8 w 2.45
Reference Sherd Description: High quality, consistent, semi-lustrous black slip.
Selected Comparanda: Lo Porto 1981b, 377–78, Fig. 104.3 (Metaponto, T 32, 300–250 BC); Scarano 1992, Fig. 5.3 (Metaponto, T 29/1986, end of 4th–mid-3rd c. BC); Graepler 1997, 91, type 321/1 (Taranto, necropolis, phase C, 275-225 BC); Giannotta 1996, 49, Fig. 16, No. 18 (Rocavecchia, T 19/1934, 3rd c. BC); Marinazzo 1979, 211, Fig. 2 (Francavilla Fontana, T 2, 310–275 BC).
Attested Dates: 270–150 BC.

Ring-handled cups cont.

1:2

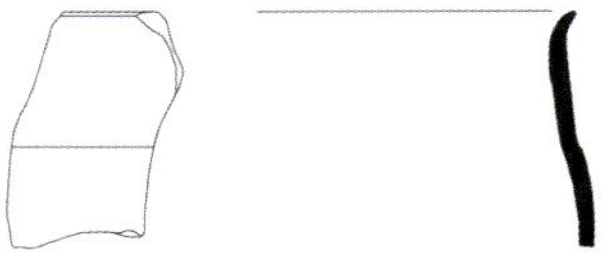

100. BG 667-03 Ring-handled cup rim (15 examples)
Dimensions: h 3.2 w 2.3
Reference Sherd Description: Consistent, semi-lustrous black slip.
Selected Comparanda: Hempel 2001, 239, type 321-2 (Taranto, 225–150 BC); *Museo Taranto* III.2, 93, Fig. 7.21 (Taranto, T 7, late 3rd to second half of 2nd c. BC); Yntema 2005, 75, Pl. 36b, form 36A (Pizzica Pantanello, Grey Ware example, 150–125 BC).
Attested Dates: 225–150 BC.

1:2

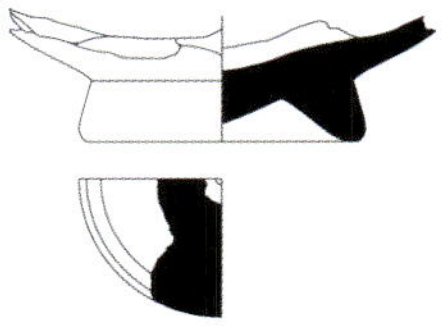

101. BG 528-10 Ring-handled cup base (single example)
Dimensions: base diam. 3.8 h 1.8
Reference Sherd Description: Consistent, semi-lustrous black slip, reddish-brown and misfired on the exterior, irregularly dipped on the underside.
Selected Comparanda: Elliott 1998, 651, C7 (Metaponto, Pantanello necropolis, T 80-6, early 3rd c. BC); Scarano 1992, Fig. 5.3 (Metaponto, end of 4th–mid-3rd c. BC).
Attested Dates: 350–250 BC.

1:1

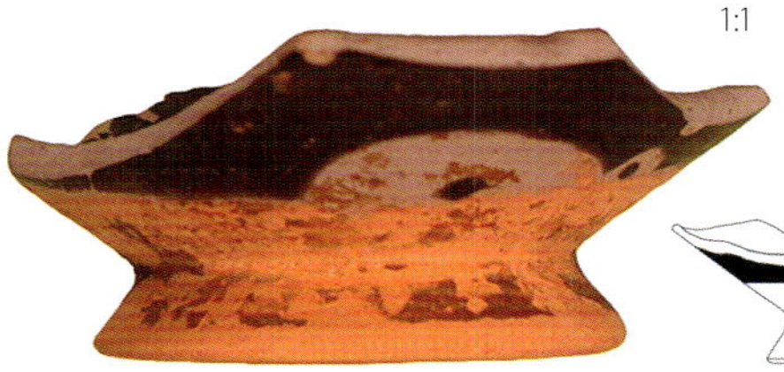

1:2

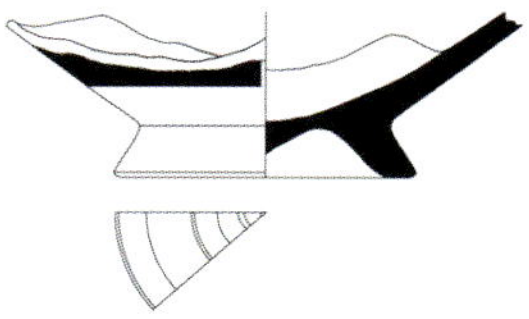

102. BG 128-22 Ring-handled cup base (10 examples)
Dimensions: base diam. 4.0 h 2.25
Reference Sherd Description: Consistent, matte black slip, irregularly dipped on the lower wall with unslipped base.
Selected Comparanda: Scarano 1992, Fig. 5.5 (Metaponto, end of 4th–mid-3rd c. BC).
Attested Dates: 300–250 BC.

1:2

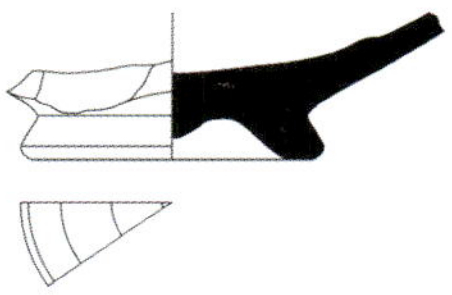

103. BG 460-03 Ring-handled cup base (9 examples)
Dimensions: base diam. 3.4 h 1.7
Reference Sherd Description: Consistent, matte black slip on interior and lower wall, unslipped base.
Selected Comparanda: Elliott 1998, 651, C7 (Metaponto, Pantanello necropolis, T 80-6, early 3rd c. BC); Scarano 1992, Fig. 5.3 (Metaponto, end of 4th–mid-3rd c. BC); L. Giardino 1990, 98, Pl. II.2 (Herakleia, 225–200 BC).
Attested Dates: 300–190 BC.

1:2

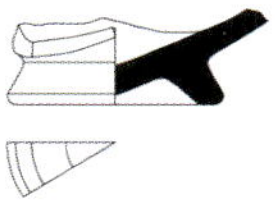

104. BG 260-27 Ring-handled cup base (12 examples)
Dimensions: base diam. 2.85 h 1.3
Reference Sherd Description: Consistent, matte black slip on the interior, unslipped base.
Selected Comparanda: Elliott 1998, 651, C3 (Metaponto, Pantanello necropolis, T 71-15, early 3rd c. BC); Scarano 1992, Fig. 5.6 (Metaponto, end of 4th–mid-3rd c. BC); Marinazzo 1979, 211, Fig. 2 (Francavilla Fontana, 310–275 BC).
Attested Dates: 300–150 BC.

1:2

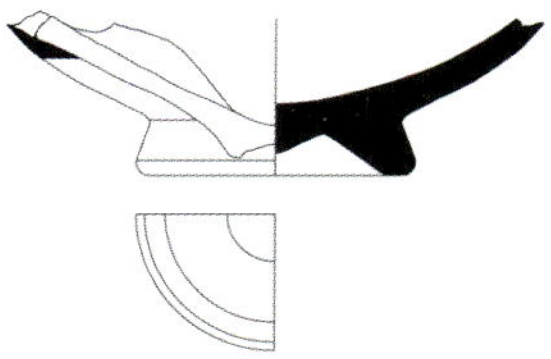

105. BG 498-16 Ring-handled cup base (17 examples)
Dimensions: base diam. 3.7 h 2.1
Reference Sherd Description: Dilute, matte reddish-brown slip on interior and lower wall, unslipped base.
Selected Comparanda: Elliott 1998, 651, C7 (Metaponto, Pantanello necropolis, T 80-6, early 3rd c. BC); Lo Porto 1981b, 368, Fig. 89.2 (Metaponto, T 23, mid-3rd c. BC); *Arte e Artigianato*, 331, Fig. 322 (Taranto, via Leonida, end of the 4th–3rd c. BC); Mastronuzzi 1995, 217, Fig. 43, No. 6 (Nardò, 4th–3rd c. BC).
Attested Dates: 300–150 BC.

Ring-handled cups cont.

106. BG 042-02 Ring-handled cup base (12 examples)
Dimensions: base diam. 5.4 h 4.15
Reference Sherd Description: Thin, matte slip, oxidized to brown, with unslipped lower exterior of wall and base.
Selected Comparanda: Elliott 1998, 651, C3 (Metaponto, Pantanello necropolis, early 3rd c. BC); Scarano 1992, Fig. 5.2 (Metaponto, early 3rd c. BC); *Arte e Artigianato*, 331, Fig. 323 (Taranto, via Leonida, 3rd c. BC); Giannotta 1996, 49, Fig. 16, No. 18 (Rocavecchia, T 19/1934, 3rd c. BC).
Attested Dates: 300–150 BC.

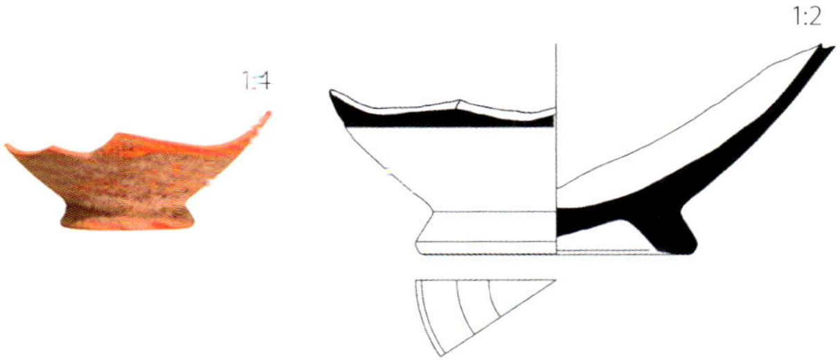

107. BG 395-09 Ring-handled cup base (single example)
Dimensions: base diam. 5.3 h 2.8
Reference Sherd Description: Consistent, semi-lustrous black slip on interior, unslipped extant exterior and base.
Selected Comparanda: Elliott 1998, 651, C17 (Metaponto, Pantanello necropolis, T193-5, beginning of 3rd c. BC); Scarano 1992, Fig. 5.4 (Metaponto, end of 4th–mid-3rd c. BC).
Attested Dates: 300–190 BC.

Kantharoi

108. BG 390-08 Kantharos stem (single example)
Dimensions: base diam. 4.6 h 5.7
Reference Sherd Description: Dilute, matte black slip, partly faded, with reserved interior of stem.
Selected Comparanda: Corrente and Labellarte 1992, 464 No. 128 (Canosa, T 2, Vico San Martino, red-figure kantharos, 330–300 BC); Morel 1981a, 3714a1, 283, Pl. 112 (Rudiae, overpainted example, second half of 4th c. BC).
Attested Dates: 350–300 BC.

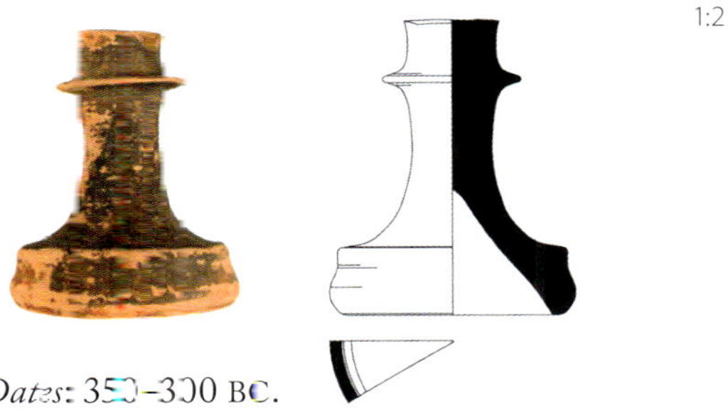

109. BG 081-02 revisit Kantharos wall with vertical ribbing (single example)
Dimensions: d 11 h 3.4 w 3.2
Reference Sherd Description: Dilute, matte grey slip over reduced greyish-fired fabric.
Selected Comparanda: De Juliis 1990, 48, Nos. 43–44, Figs. 144–48 (Canosa, Ipogeo dei Vimini, Cella A, first quarter 4th c. BC); Lo Porto 1966, 225, Fig. 85, No. 2 (Metaponto, rural cemetery of contrada San Salvatore, T 2/1943, early 3rd c. BC).
Attested Dates: 370–300 BC.

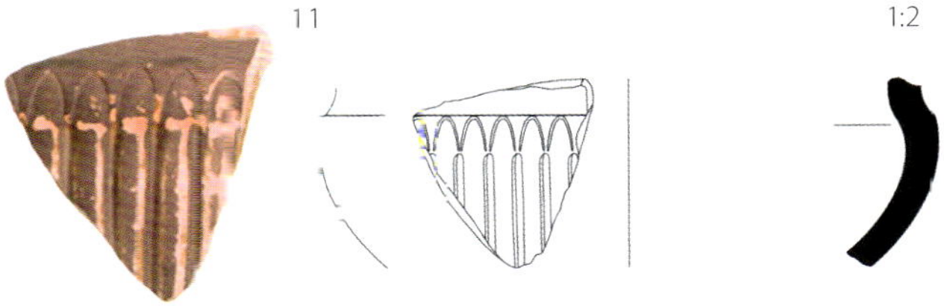

Cup/One-Handlers

110. BG 637-09 Cup/one-handler rim and handle (single example)
Dimensions: rim diam. 8.0 h 1.3 w 4.5
Reference Sherd Description: Consistent, semi-lustrous black slip.
Selected Comparanda: Elliott 1998, 669, O4 (Metaponto, Pantanello necropolis, 410–370 BC); Lo Porto 1988–1989, 341–52, Fig. 56.5 (Montescaglioso, T 5, mid-4th c. BC); Ciancio 1997, 221, No. 282 (Botromagno, T 2/1994, second half of the 5th c. BC); Scarfì 1961, 252, Fig. 89, Nos. 8–9 (Monte Sannace, T 5, mid-4th c. BC); Ricchetti 1992b, 253 No. 50 (Canosa, Ipogeo Varrese, 350–300 BC); Morel 1981, 6241a1, 395, Pl. 195 (Sulmona, end of the 4th to early 3rd c. BC).
Attested Dates: 410–300 BC.

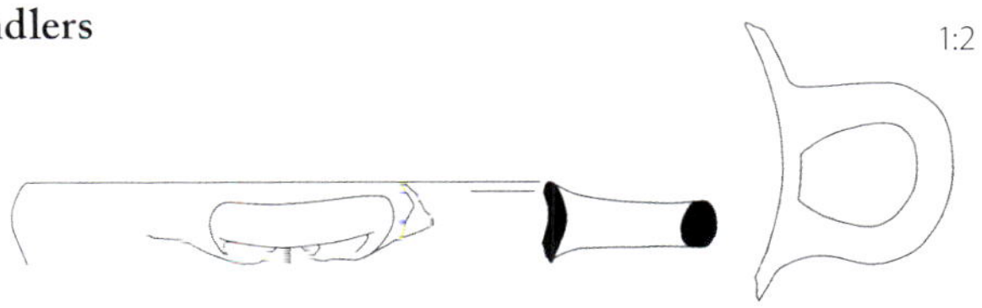

Cup/one-handlers, cont.

1:2

111. BG 488-06 Cup/one-handler rim (single example)
Dimensions: rim diam. 9.9 h 1.7 w 5.8
Reference Sherd Description: Consistent, semi-lustrous black slip.
Selected Comparanda: Elliott 1998, 669, O4 (Metaponto, Pantanello necropolis, 410–370 BC).
Attested Dates: 410–370 BC.

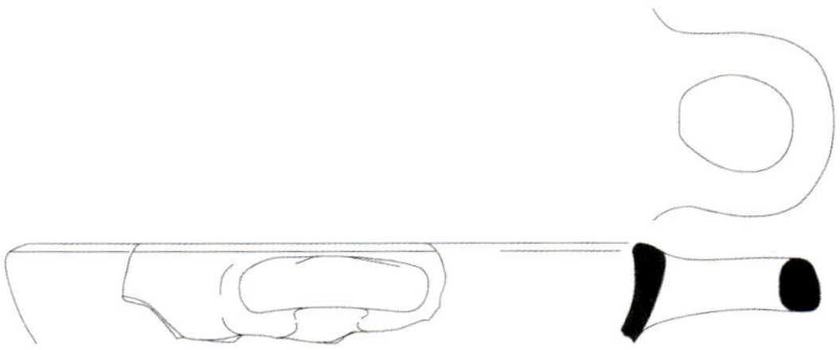

1:2

112. BG 488-97 Cup/one-handler rim (3 examples)
Dimensions: rim diam. 16.1 h 3.0 w 3.8
Reference Sherd Description: Dilute, matte black slip, partially worn.
Selected Comparanda: Surra and Preacco Ancona 1997, 79, Pl. 37, Nos. 94–95 (Pomarico Vecchio, abandonment deposit); Cracolici 2004c, 52, Fig. 46, Cat. No. 8 (Palazzo San Gervasio, località Casalini Sottana, T4/2002, mid-4th c. BC).
Attested Dates: 350–270 BC.

1:2

113. BG 578-05 Cup/one-handler rim (15 examples)
Dimensions: rim diam. 8.2 h 3.2 w 3.6
Reference Sherd Description: Consistent, matte black slip, partially worn.
Selected Comparanda: Fattoria Fabrizio, FF80-253PL-A (350–300 BC); Prag 1983a, 354, No. 316 (Cozzo Presepe, 350–200 BC); *Bitonto*, 124, No. 42 (Bitonto, T 3/1981, second half of 4th c. BC); Prag 1992b, 104, No. 815, Fig. 34 (Botromagno, late 4th to first quarter of 3rd c. BC); Montanaro 2007, 783–84, Figs. 738–39, 179.6–7 (Ruvo, 330–300 BC) 834, and 837, Figs. 798–99, 805, 260.3–4, 9; Morel 1981a, series 2672, 203–4, Pl. 65 (early 3rd c. BC).
Attested Dates: 350–300 BC.

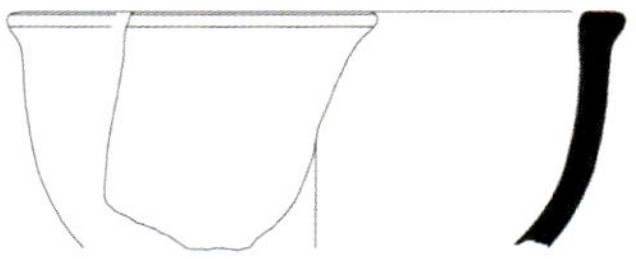

1:2

114. BG 395-22 Cup/one-handler rim and handle (2 examples)
Dimensions: rim diam. 9.5 h 3.2 w 6.8
Reference Sherd Description: Consistent, semi-lustrous black slip.
Selected Comparanda: Elliott 1998, 669, O3 (Metaponto, Pantanello necropolis, 325–275 BC); Lo Porto 1981b, 362, No. 3, Fig. 83.3 (Metaponto, Casa Ricotta, T 12, late 4th c. BC); Prag 1983a, 354, No. 312 (Cozzo Presepe); Prag 1992b, 100, No. 790, Fig. 33 (Botromagno, 325–250 BC).
Attested Dates: 325–275 BC.

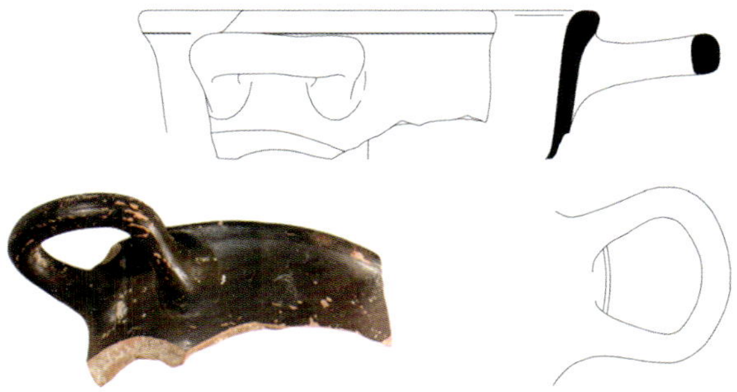

1:2

115. BG 122-02 Cup/one-handler rim (10 examples)
Dimensions: rim diam. 10.4 h 4.4
Reference Sherd Description: Dilute, matte black slip, completely faded on the exterior.
Selected Comparanda: Hoffmann 2002, 37, type 222/1, Pl. 60 (Taranto, 350–325 BC); Prag 1992b, 104, No. 817, Fig. 34 (Botromagno, 310–275 BC); Ricchetti 1992b, 253, No. 49 (Canosa, Ipogeo Varrese, 4th c. BC); Morel 1981a, 2672a2, 204, Pl. 65 (Trebula Mutuesca, around the beginning of 3rd c. BC).
Attested Dates: 350–300 BC.

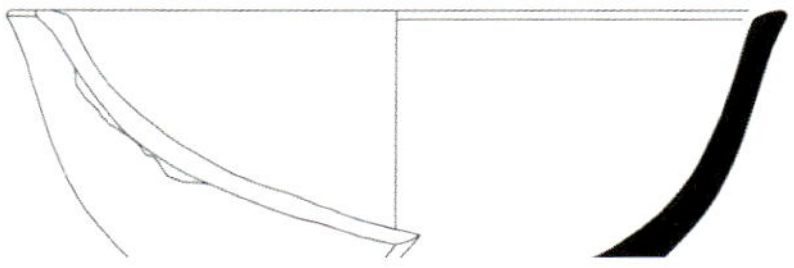

Cup/one-handlers cont.

1:2

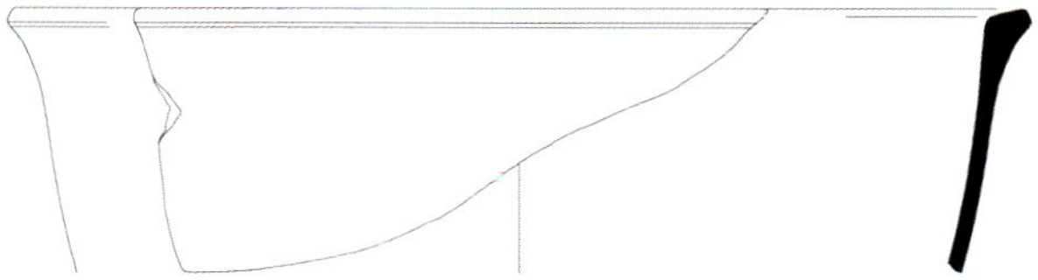

116. BG 113-16 Cup/one-handler rim (3 examples)
Dimensions: rim diam. 15.5 h 4.5 w 9.6
Reference Sherd Description: Consistent, semi-lustrous black slip.
Selected Comparanda: Prag 1992b, 102, No. 803, Fig. 33 (Botromagno, broadly dated 3rd–2nd c. BC); De Cazanove 2008, 535, Fig. 296, Nos. 285 and 592, Fig. 303, No. 474 (Civita di Tricarico, first half of 3rd c. BC); Morel 1981a, series 5231, 395, Pl. 194 (4th c. BC).
Attested Dates: 300–250 BC.

1:2

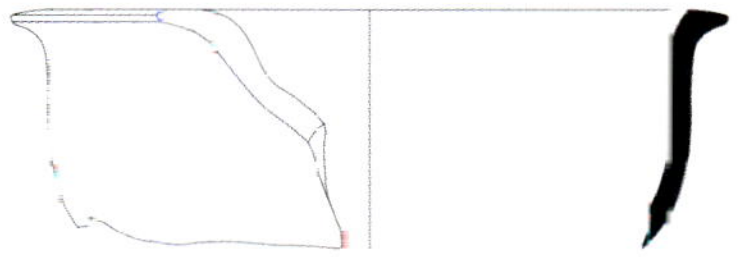

117. BG 451-07 Cup/one-handler rim (7 examples)
Dimensions: rim diam. 9.5 h 3.3
Reference Sherd Description: Consistent, matte black slip.
Selected Comparanda: Lo Porto 1981b, 361–62, No. 3, Fig. 83.3 (Metaponto, contrada Ricotta, T 12, late 4th c. BC); Fattoria Fabrizio, FF 80-184 PL-A (350–300 BC); Prag 1983a, 354, No. 312 (Cozzo Presepe); Riccardi 2008, 82, No. 102–4 (Bitonto, T 7/2003, last quarter of 4th c. BC); Prag 1992b, 100, No. 792, Fig. 33 (Botromagno, 325–250 BC); F. Rossi 1989, 165, No. 4a, Pl. 307,5 (Monte Sannace); Morel 1981a, 1551a1, 123, Pl. 23 (unknown provenance, probably southern Italian fabric, first half of 3rd c. BC).
Attested Dates: 330–280 BC.

1:2

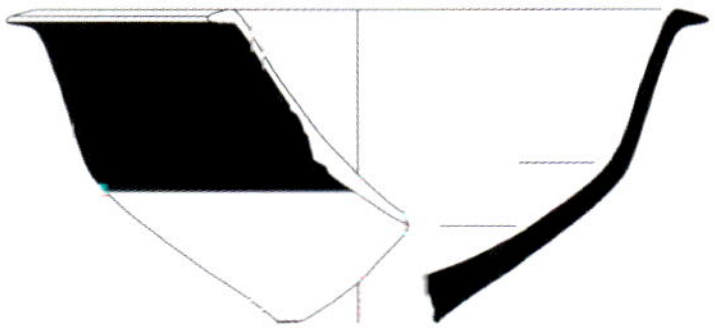

118. BG 266-06 Cup/one-handler rim (single example)
Dimensions: rim diam. 9.3 h 4.4 w 5.8
Reference Sherd Description: Consistent, semi-lustrous black slip, irregularly dipped with unslipped lower exterior.
Selected Comparanda: Prag 1983a, 354, No. 313 (Cozzo Presepe, 350–250 BC); Prag 1992b, 99–100, No. 788, Fig. 33 (Botromagno, cup, 325–250 BC).
Attested Dates: 350–275 BC.

1:2

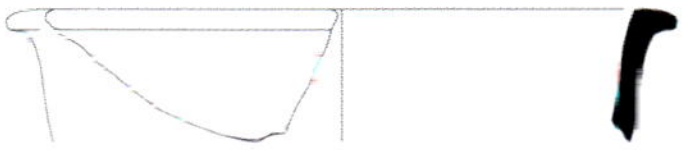

119. BG 667-02 Cup/one-handler rim (8 examples)
Dimensions: rim diam. 8.9 h 1.8 w 3.9
Reference Sherd Description: Dilute, matte reddish-brown slip, partly worn.
Selected Comparanda: Prag 1983a, 354, No. 312 (Cozzo Presepe); Prag 1992b, 100, No. 792, Fig. 33 (Botromagno, 325–250 BC); Riccardi 2008, 82, No. 102–4 (T 7/2003, last quarter of the 4th c. BC).
Attested Dates: 325–250 BC.

1:2

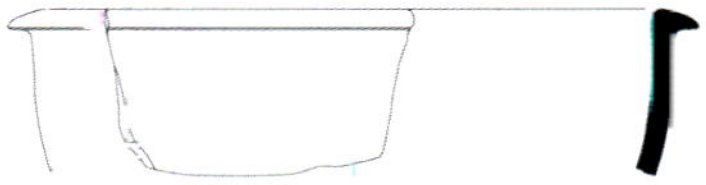

120. BG 343-03 Cup/one-handler rim (single example)
Dimensions: rim diam. 9.2 h 2.3 w 4.1
Reference Sherd Description: Reddish brown slip almost completely worn.
Selected Comparanda: Bears some similarities to Prag 1992b, 100, No. 792, Fig. 33 (Botromagno, 325–250 BC).
Attested Dates: 325–250 BC.

1:2

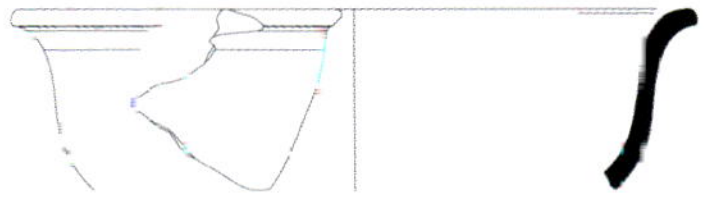

121. BG 376-02 Cup/one-handler rim (2 examples)
Dimensions: rim diam. 9.1 h 2.5 w 2.8
Reference Sherd Description: Compact, semi-lustrous black slip.
Selected Comparanda: Prag 1992b, 111, No. 850, Fig. 35 (Botromagno, 350–250 BC); Ricchetti 1992b, 253, No. 45 (Canosa, Ipogeo Varrese, 4th c. BC).
Attested Dates: 350–200 BC.

Cup/one-handlers cont.

1:2

122. BG 454-06 Cup/one-handler rim (17 examples)
Dimensions: rim diam. 9.6 h 2.95
Reference Sherd Description: Consistent, matte black slip
Selected Comparanda: Prag 1983a, 354, Nos. 313–14 (Cozzo Presepe, 350–250 BC); Riccardi 2008, 73, No. 69 (Bitonto, T 3/2002, second half of 4th c. BC); *Museo Taranto* II.2, 504, Pl. 42e, 12.3b (Rutigliano, second half of 4th c. BC); Scarfì 1961, 203, Fig. 34, Nos. 20–21 (Monte Sannace, T 3, second half of 4th c. BC); Ricchetti 1992b, 253, No. 45 (Canosa, Ipogeo Varrese, 4th c. BC).
Attested Dates: 350–250 BC.

Cups

1:2

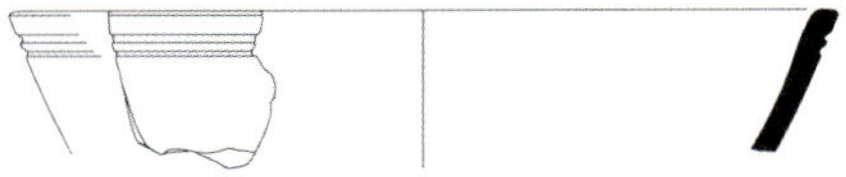

123. BG 183-04 Cup rim (single example)
Dimensions: rim diam. 11.0 h 2.2 w 2.2
Reference Sherd Description: Traces of reddish brown slip, almost completely worn.
Selected Comparanda: De Cazanove 2008, 580, Fig. 291, No. 147 (Civita di Tricarico, Gnathia cup with painted handles, 3rd c. BC); Prag 1992b, 290, No. 779, Fig. 32 (Botromagno, 2nd c. BC); Morel 1981a, 2575a1, 186, Pl. 57 (Volterra, 3rd to 2nd c. BC) and 2573a1, 186, Pl. 57 (Sutri, first half of the 2nd c. BC).
Attested Dates: 300–100 BC.

1:2

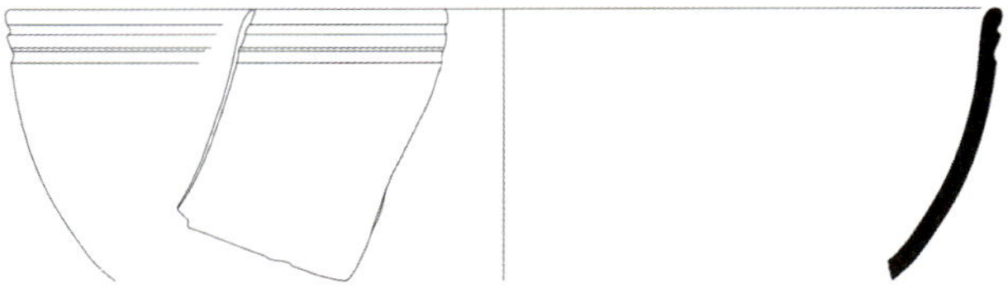

124. BG 250-07 Cup rim (4 examples)
Dimensions: rim diam. 16.0 h 4.5 w 4.4
Reference Sherd Description: Consistent, semi-lustrous black slip.
Selected Comparanda: Bruscella and Virtuoso 2005, 275, Pl. XXXIX, No. 253 (Torre di Satriano, 3rd c. BC); Prag 1992b, 97, No. 776, Fig. 32 (Botromagno, 2nd c. BC); Preacco Ancona 1992, 131, Pl. XLVIII, No. 77 (Locri, Casa dei Leoni, 2nd half of 4th c. BC); Morel 1981a, 2572a1, 186, Pl. 57 (Adria, 2nd c. BC).
Attested Dates: 200–100 BC.

1:2

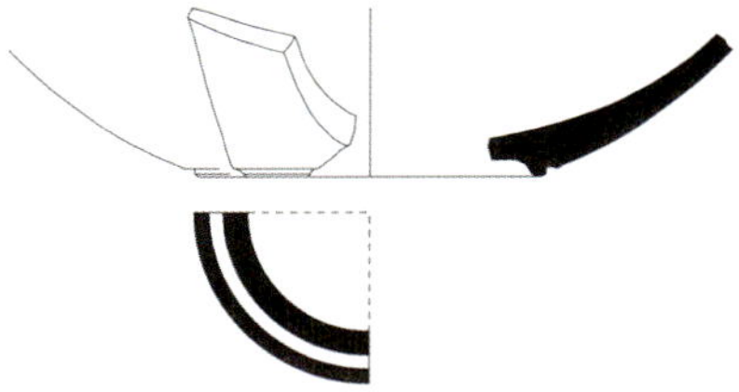

125. BG 515-03 Cup base (2 examples)
Dimensions: base diam. 5.0 (?) h 2.5 w 2.4
Reference Sherd Description: Consistent, matte reddish-brown slip, worn on resting surface of foot.
Selected Comparanda: Burgers 1991, form 18, PZ81. NW.735.1 (Metaponto, Pantanello, grey ware, 150–100 BC); Elliott 1977, 472, No. 63; *Gli ori*, 465, No. 17 (Taranto, necropolis, first quarter of 2nd c. BC, mastos); Morel 1981a, 2153a1, 142, Pl. 32 (Tripoli, first quarter of 2nd c. BC); *Agora* XXIX, Fig. 20, No. 321 (175–150 BC).
Attested Dates: 200–100 BC.

Dishes

1:2

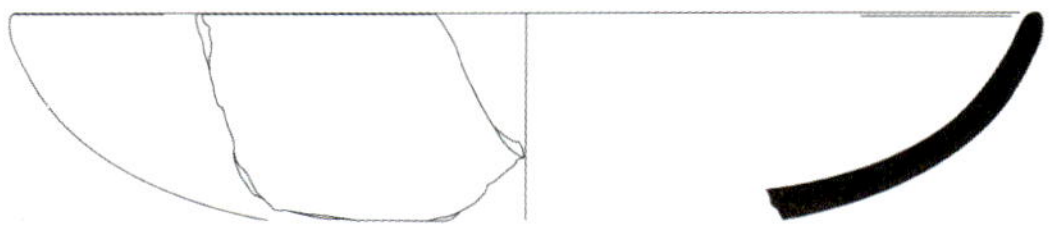

126. BG 188-01 Dish rim (12 examples)
Dimensions: rim diam. 13.7 h 2.8 w 4.35
Reference Sherd Description: Consistent, semi-lustrous black slip over reduced greyish fabric.
Selected Comparanda: Elliott 1998, 653, D11 (Metaponto, Pantanello necropolis, 410-380/70 BC); Ciancio 1997, 220, No. 277 (Botromagno, T 3, second half 5th c. BC). *Museo Taranto* II.2, 509, Pl. 44a, 15.1 (late 5th to 4th c. BC).
Attested Dates: 420–350 BC

Dishes cont.

1:2

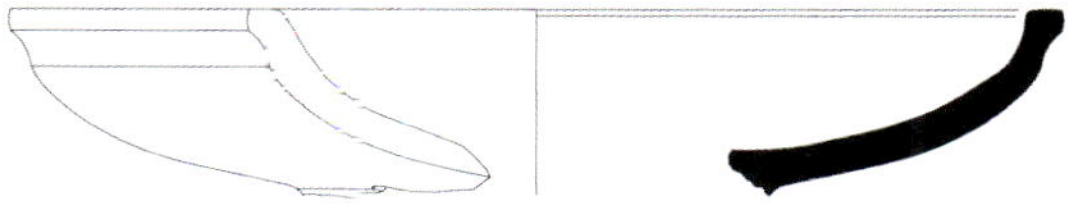

127. BG 372-04 Dish rim (single example)
Dimensions: rim diam. 14.0 h 2.5
Reference Sherd Description: Consistent, semi-lustrous slip, black on the exterior and fired reddish-brown on the interior.
Selected Comparanda: Elliott 1998, 653, D10 (Metaponto, Pantanello necropolis, 410-380/70 BC); Lo Porto 1973, 209–10, Pl. LVII, No. 2.1 (Matera, T 4, early 4th c. BC).
Attested Dates: 410–370 BC.

1:2

128. BG 329-04 Dish rim (4 examples)
Dimensions: rim diam. 15.3 h 1.95 w 3.4
Reference Sherd Description: Consistent, semi-lustrous black slip.
Selected Comparanda: Elliott 1998, 653, D16 (Metaponto, Pantanello necropolis, 410-380/70 BC); Lo Porto 1981b, 375, No. 1, Figs. 69.1, 101.1 (Metaponto, contrada Ricotta, T 30, late 4th to early 3rd c. BC); Fattoria Fabrizio, FF80-233PL-A; Morel 1981a, 2963d1, 240, Pl. 81 (Ordona, late 4th c. BC).
Attested Dates: 410–300 BC.

129. BG 428-01 Dish base (9 examples)
Dimensions: base diam. 6.0 h 2.0 w 4.3
Reference Sherd Description: Consistent, matte black slip, unslipped underside.
Selected Comparanda: Lo Porto 1981b, 369–70, No. 2, Figs. 69.2, 92.2 (Metaponto, contrada Ricotta, T 26, late 4th to first decades of the 3rd c. BC); Lo Porto 1988–1989, 341–52, Fig. 54.3 (Montescaglioso, T 5, mid-4th c. BC).
Attested Dates: 410–300 BC.

1:2

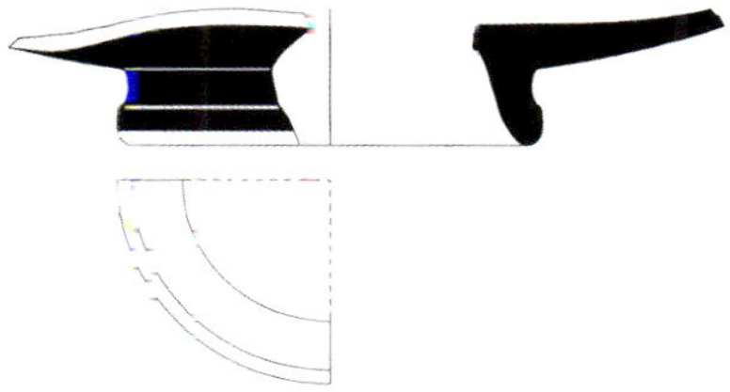

130. BG 320-21 Dish base (3 examples)
Dimensions: base diam. 7.6 h 1.7 w 4.0
Reference Sherd Description: Consistent, semi-lustrous black slip, a good part of which is worn.
Selected Comparanda: Lo Porto 1981b, 370, Figs. 69.2 and 92.2 (Metaponto, contrada Ricotta, T 26, late 4th to early 3rd c. BC).
Attested Dates: 350–280 BC.

1:2

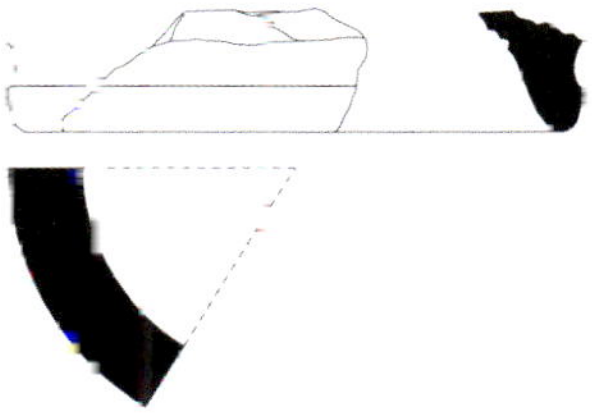

131. BG 542-01 Dish rim (2 examples)
Dimensions: rim diam. 14.2 h 1.3 w 3.4
Reference Sherd Description: Consistent, semi-lustrous black slip.
Selected Comparanda: Elliott 1998, 653, D18 (Metaponto, Pantanello necropolis, 325–275 BC); Prag 1992b, 75, No. 650, Fig. 28 (Botromagno, 300–265 BC, following Morel); *Nobili e guerrieri*, 54, Figs. 6.5 and 50 (Palazzo San Gervasio, late 4th c. BC); Morel 1981a, 2237a2, 151, Pl. 39 (Botromagno, TS15, late 4th c. BC).
Attested Dates: 325–275 BC.

1:2

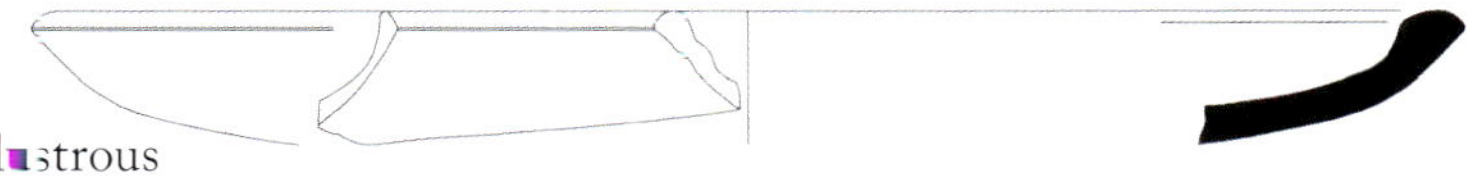

132. BG 365-06 Dish rim (single example)
Dimensions: rim diam. 19.0 h 1.8 w 5.6
Reference Sherd Description: Consistent, semi-lustrous black slip.
Selected Comparanda: Elliott 1998, 653, D10 (Metaponto, Pantanello necropolis, 410-380/70 BC); Prag 1992b, 87, No. 720, Fig. 31 (Botromagno, late 4th c. BC); Cracolici 2004a, 12, Fig. 6.5, Cat. 39 (Palazzo S. Gervasio, T 15/2002, end of 4th c. BC); Morel 1981a, 2237a2, 151, Pl. 38 (Botromagno, TS15 late 4th c. BC).
Attested Dates: 325–300 BC.

Dishes cont.

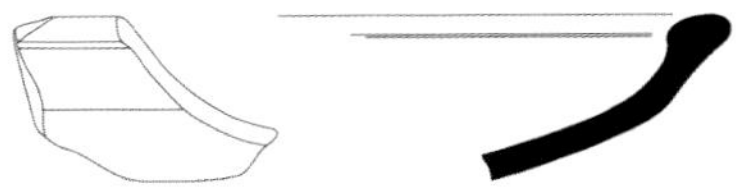

1:2

133. BG 048-02 Dish rim (single example)
Dimensions: h 2.25 w 3.5
Reference Sherd Description: Dilute, semi-lustrous reddish-brown slip.
Selected Comparanda: Prag 1992b, 75, No. 651, Fig. 28 (Botromagno, 300–265 BC).
Attested Dates: 300–250 BC.

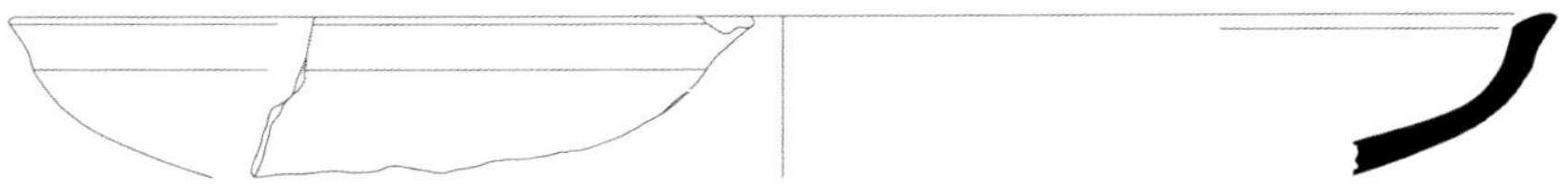

1:2

134. BG 476-02 Dish rim (28 examples)
Dimensions: rim diam. 22.2 h 2.4 w 7.2
Reference Sherd Description: Consistent, matte black slip.
Selected Comparanda: Elliott 1998, 653, D18 (Metaponto, Pantanello necropolis, 325–275 BC); Lo Porto 1982, 369–70, No. 2, Fig. 69.2, 92.2 (Metaponto, contrada Ricotta, T 26, late 4th to early 3rd c. BC); Surra and Preacco 1997, 69, Pl. 32, No. 23 (Pomarico Vecchio, mid to late 4th c. BC); Prag 1992b, 87–88, No. 721, Fig. 31 (Botromagno, late 4th c. BC); Montanaro 2007, 272, Figs. 125–26, 21.23 (Ruvo, second half of 4th c. BC).
Attested Dates: 325–275 BC.

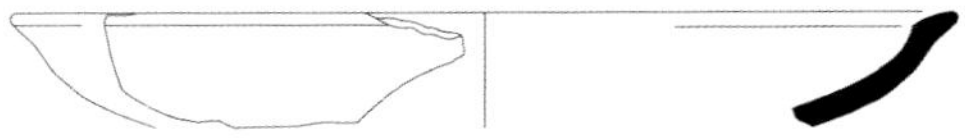

1:2

135. BG 413-06 Dish rim (38 examples)
Dimensions: rim diam. 12.55 h 1.6 w 4.8
Reference Sherd Description: Consistent, matte black slip.
Selected Comparanda: Elliott 1998, 653, D18 (Metaponto, Pantanello necropolis, 325–275 BC); Lo Porto 1982, 369–70, Figs. 69.2, 92.2 (Metaponto, contrada Ricotta, T26, late 4th to early 3rd c. BC); Prag 1992b, 87–88, No. 721, Fig. 31 (Botromagno, late 4th c. BC); Montanaro 2007, 272, Figs. 125–126, 21.23 (Ruvo, second half of the 4th c. BC).
Attested Dates: 325–275 BC.

1:2

136. BG 454-05 Dish rim (5 examples)
Dimensions: rim diam. 18.9 h 2.1 w 4.7
Reference Sherd Description: Consistent, matte black slip, partly worn on the lip
Selected Comparanda: Prag 1992b, 75, No. 648, Fig. 28 (Botromagno, 300–270 BC); Montanaro 2007, 272, Figs. 125–26, 21.23 (Ruvo, second half of 4th c. BC); Pianu 1990, 118–9, No. 1, Pl. LIII, 3 (Herakleia, T120, over-painted example, 320–270 BC).
Attested Dates: 300–270 BC.

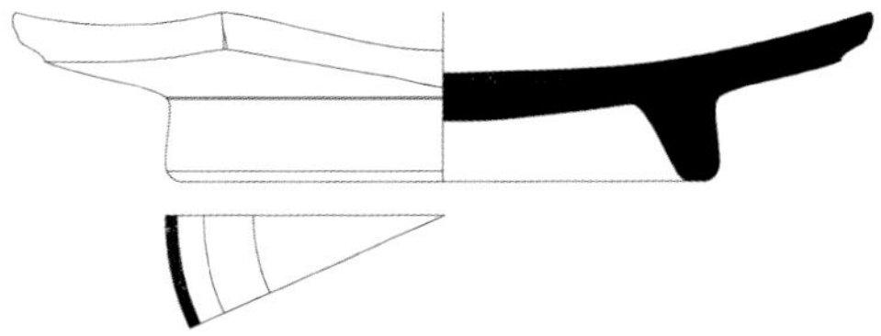

1:2

137. BG 355-01 revisit Dish base (single example)
Dimensions: base diam. 8.0 h 2.5 w 7.7
Reference Sherd Description: Traces of black slip, almost completely abraded, unslipped underside.
Selected Comparanda: Elliott 1998, 653, D2 (Metaponto, Pantanello necropolis, 310–275 BC).
Attested Dates: 310–275 BC.

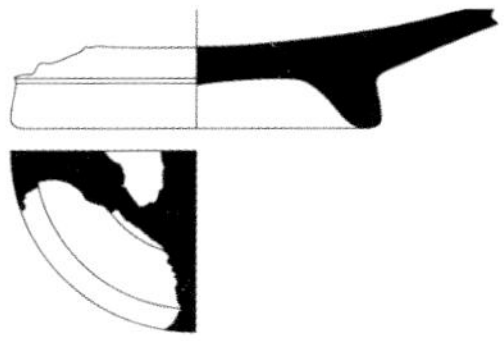

1:2

138. BG 501-16 Dish base (22 examples)
Dimensions: base diam. 4.9 h 1.7
Reference Sherd Description: Dilute, matte reddish brown slip, irregularly dipped on underside.
Selected Comparanda: Elliott 1998, 653, D2 (Metaponto, Pantanello necropolis, 310–275 BC); Fattoria Fabrizio, FF80-139 PL-B (end of the 4th–early 3rd c. BC); Prag 1983a, 360, No. 346 (Cozzo Presepe, late 4th c. BC).
Attested Dates: 310–150 BC.

Dishes cont.

1:2

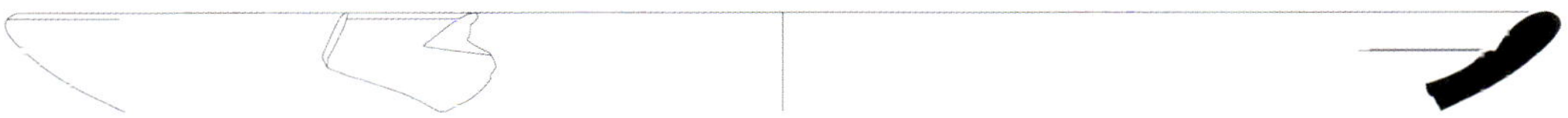

139. BG 213-01 Dish rim (single example)
Dimensions: rim diam. 27.0 h 1.8 w 3.1
Reference Sherd Description: Dilute, semi-lustrous black slip, worn on the interior.
Selected Comparanda: Morel 1981a, 2222a1, 148, Pl. 34 (Cordoba, Attic, 300 BC ± 50). Some similarities to Campagna 1995, 241, Fig. 14, No. 4 (Vaste, end of 4th–early 3rd c. BC).
Attested Dates: 350–250 BC.

---

140. BG 043-18 Dish base (3 examples)
Dimensions: base diam. 15.0 h 2.5 w 9.45
Reference Sherd Description: Dilute, semi-lustrous black slip on the interior, unslipped exterior and underside. Not local fabric.
Selected Comparanda: A similar molding of the outer surface of the base is in D'Andria 1980a, 411, Fig. 56, No. 221 (Metaponto, kerameikos, deposit 1, 4th c. BC); see also Prag 1983a, 358, No. 330 (Cozzo Presepe, late 4th to first half of 3rd c. BC); *Pomarico Vecchio* I, 95, Pl. 44, No. 163 (mid-to-late 4th c. BC).
Attested Dates: 350–300 BC.

1:4

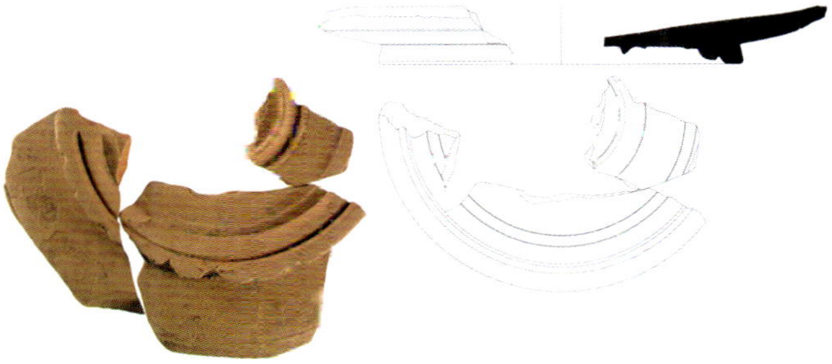

---

141. BG 460-04 Dish rim (single example)
Dimensions: rim diam. 15.7 h 1.8 w 4.5
Reference Sherd Description: Consistent, matte black slip.
Selected Comparanda: Bitti 1989, Pl. XXIII, No. 113 (Locri, 4th c. BC), No 149, Pl. XXIII, No. 113 (Locri, 325–300 BC).
Attested Dates: 400–300 BC.

1:2

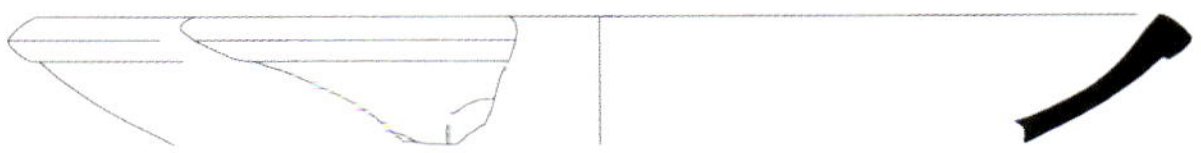

---

142. BG 365-02 Dish rim (2 examples)
Dimensions: rim diam. 10.0 h 1.6 w 3.0
Reference Sherd Description: Consistent, semi-lustrous black slip.
Selected Comparanda: Bitti 1989, 144–45, Pl. XXIII, No. 103 (Locri, second half of 4th c. BC); Morel 1981a, 1514h1, 118, Pl. 20 (unknown provenance, 310 ± 40 BC ?).
Attested Dates: 310–290 BC.

1:2

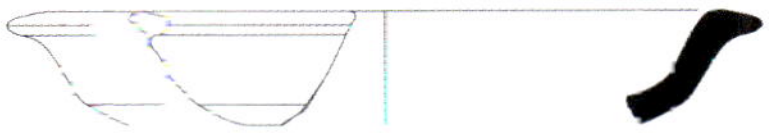

---

143. BG 341-04 Dish rim (single example)
Dimensions: rim diam. 14.3 h 1.6 w 3.1
Reference Sherd Description: Consistent, semi-lustrous black slip, worn on the interior.
Selected Comparanda: Morel 1970a, 104, Fig. 24.5 (Cozzo Presepe, 320–270 BC).
Attested Dates: 320–270 BC.

1:2

1:2

144. BG 248-03 Dish rim (single example)
Dimensions: rim diam. 19.0 h 2.1 w 3.5
Reference Sherd Description: Dilute, semi-lustrous reddish brown slip, fired red on the upper face of the rim.
Selected Comparanda: Prag 1992b, 79, No. 673, Fig. 29 (Botromagno, 250–200 BC); Morel 1981a, 1534h2, 121, Pl. 22 (Rome?, second half of the 3rd c. BC).
Attested Dates: 300–200 BC.

Dishes cont.

1:2

145. BG 113-270 Dish (3 examples)
Dimensions: rim diam. 14.6 base diam. 5.0 h 3.2 w 9.7
Reference Sherd Description: Dilute, semi-lustrous black slip, unslipped interior of foot ring.
Selected Comparanda: Elliott 1998, 653, D2 (Metaponto, Pantanello necropolis, 310–275 BC); Sant'Angelo Grieco, SG3.81-6PL; Prag 1992b, 77, Nos. 657 and 654, Fig. 29 (Botromagno, ca. 300 BC); Surra and Preacco Ancona 1997, 67–68, Pl. 31, No. 9 (Pomarico Vecchio, late 4th c. BC); Morel 1970a, 99, Fig. 24.6 (Cozzo Presepe, 320–270 BC); Morel 1981a, 1514c1, 118, Pl. 20 (Cales, ca. 300 BC).
Attested Dates: 310–275 BC.

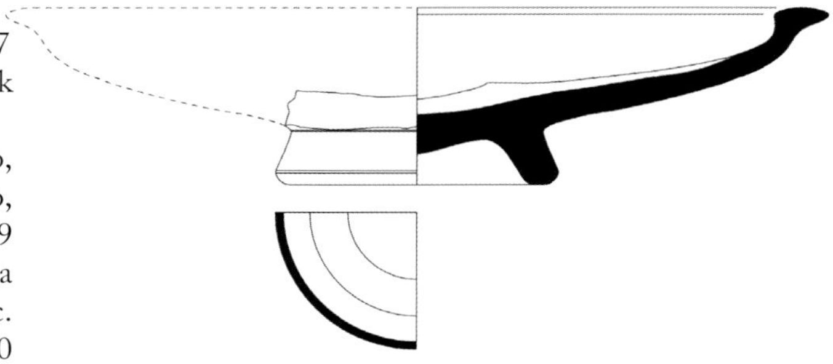

1:2

146. BG 469-02 Dish rim (single example)
Dimensions: h 2.45 w 2.0
Reference Sherd Description: Consistent, semi-lustrous black slip.
Selected Comparanda: Prag 1983a, 358, No. 335 (Cozzo Presepe, first half of the 3rd c. BC); Prag 1992b, 76, No. 653, Fig. 28 (Botromagno, 300–265 BC).
Attested Dates: 300–250 BC.

1:2

147. BG 260-17 Dish rim (3 examples)
Dimensions: rim diam. 16.25 h 2.05
Reference Sherd Description: Dilute, matte black slip, almost totally worn, particularly on the exterior.
Selected Comparanda: Prag 1992b, 75, No. 648, Fig. 28 (Botromagno, Period VI, late 4th to early 3rd c. BC; 300–270 BC in Morel's chronology); G. Greco 1991, 73, Fig. 158 (Serra di Vaglio, Casa dei Pithoi, last quarter of 4th to first quarter of 3rd c. BC).
Attested Dates: 300–280 BC.

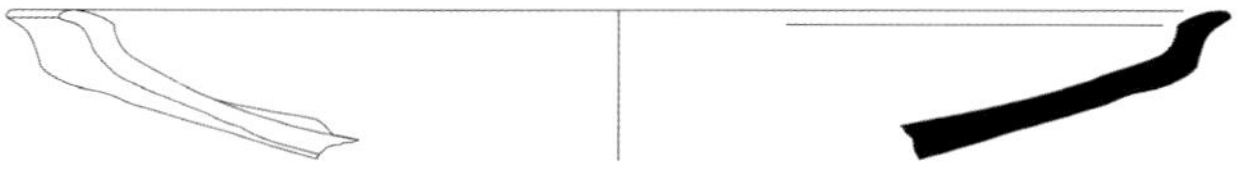

1:2

148. BG 296-07 Dish rim (5 examples)
Dimensions: rim diam. 12.8 h 1.55 w 4.4
Reference Sherd Description: Dilute, matte black slip, almost totally worn.
Selected Comparanda: Prag 1983a, 358, No. 334 (Cozzo Presepe, 225 BC); Preacco 1989, 205, Pl. XXVIII, No. 226 (Locri, late 4th to early 3rd c. BC); Delli Ponti 1996, 154, No. 122 (Vaste, T 10/1968, end of 4th c. BC); Semeraro 1983, 184, No. 202, Pl. 91 (Otranto, second half of 4th c. BC); Morel 1981a, 1514b1, 118, Pl. 20 (Ugento, late 4th c. BC).
Attested Dates: 310–225 BC.

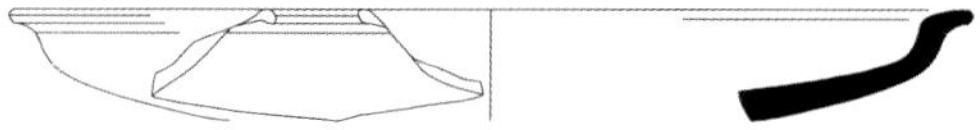

1:2

149. BG 043-09 Dish rim (5 examples)
Dimensions: rim diam. 20.6 h 1.05 w 2.9
Reference Sherd Description: Consistent, lustrous black slip.
Selected Comparanda: Prag 1992b, 82, No. 690, Fig. 29 (Botromagno, late 4th to first quarter of 3rd c. BC); Prag 1983a, 358, No. 335 (Cozzo Presepe, first half of 3rd c. BC).
Attested Dates: 300–250 BC.

1:2

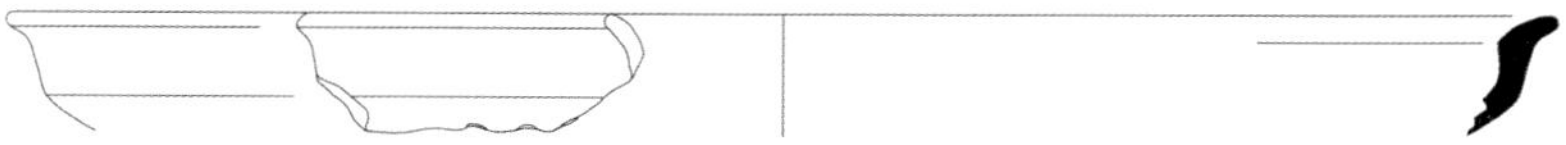

150. BG 476-03 Dish rim (3 examples)
Dimensions: rim diam. 20.6 h 1.6 w 4.6
Reference Sherd Description: Consistent, semi-lustrous black slip.
Selected Comparanda: Elliott 1998, 653, D2 (Metaponto, Pantanello necropolis, 310–275 BC); Lo Porto 1981b, 353, No. 1, Fig. 69.3 (Metaponto, contrada Ricotta, T 6, late 4th to first quarter of 3rd c. BC); Morel 1981a, 1514b1, 118, Pl. 20 (Ugento, late 4th c. BC).
Attested Dates: 310–275 BC.

Dishes cont.

1:2

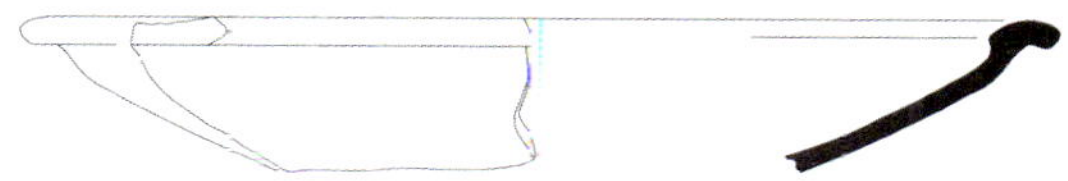

151. BG 128-02 Dish rim (9 examples)
Dimensions: rim diam. 13.8 h 2.1
Reference Sherd Description: Dilute, matte slip, black on the interior, fired reddish-brown on the exterior.
Selected Comparanda: Prag 1983a, 358, No. 335 (Cozzo Presepe, first half of the 3rd c. BC); *Gravisca* 9, 119, No. 277, Pl. 29 (first quarter of the 3rd c. BC).
Attested Dates: 300–250 BC.

1:2

152. BG 506-42 Dish rim (2 examples)
Dimensions: rim diam. 14.3 h 1.4 w 2.9
Reference Sherd Description: Consistent, matte black slip.
Selected Comparanda: Prag 1983a, 358, No. 337 (Cozzo Presepe, 150–100 BC); Prag 1992b, 79, No. 673, Fig. 29 (Botromagno, 250–200 BC); Morel 1981a, 1534f1, 121, Pl. 21 (Rome, around the second half of 3rd c. BC).
Attested Dates: 300–200 BC.

1:2

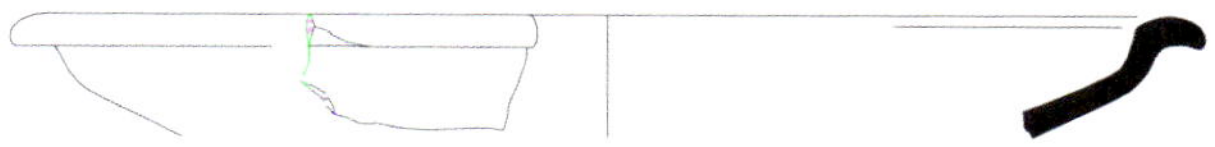

153. BG 566-04 Dish rim (28 examples)
Dimensions: rim diam. 15.9 h 1.7 w 3.1
Reference Sherd Description: Consistent, semi-lustrous black slip.
Selected Comparanda: Elliott 1998, 653, D2 (310–275 BC); Prag 1983a, 358, No. 333 (Cozzo Presepe, beginning of 3rd c. BC); Prag 1992b, 74, No. 540, Fig. 74 (Botromagno, late 4th to first quarter of 3rd c. BC); Preacco 1989, 205, Pl. XXVIII, No. 226 (Locri, late 4th to early 3rd c. BC); Morel 1981a, 1322b1, 105, Pl. 13 (Lipari, T196, 335–300 BC).
Attested Dates: 310–270 BC.

1:2

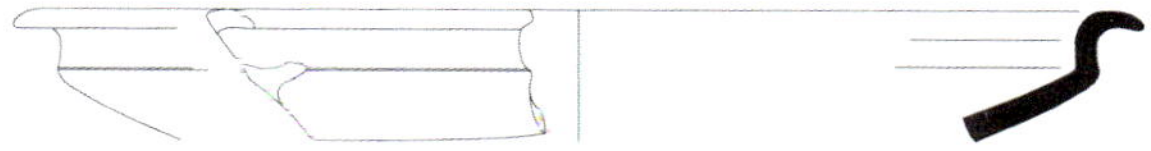

154. BG 309-03 Dish rim (2 examples)
Dimensions: rim diam. 15.0 h 1.8 w 4.5
Reference Sherd Description: Consistent, matte black slip.
Selected Comparanda: Prag 1983a, 358, No. 332 (Cozzo Presepe, first quarter of 3rd c. BC); Bruscella and Virtuoso 2005, 283, Pl. XLII, No. 295 (Torre di Satriano, late 4th to first half of 3rd c. BC); Prag 1992b, 78, No. 667, Fig. 29 (Botromagno, late 4th to first quarter of 3rd c. BC); Morel 1981a, 1523c1, 119, Pl. 20 (Lacco Ameno, 280 BC ± 30).
Attested Dates: 310–250 BC

1:2

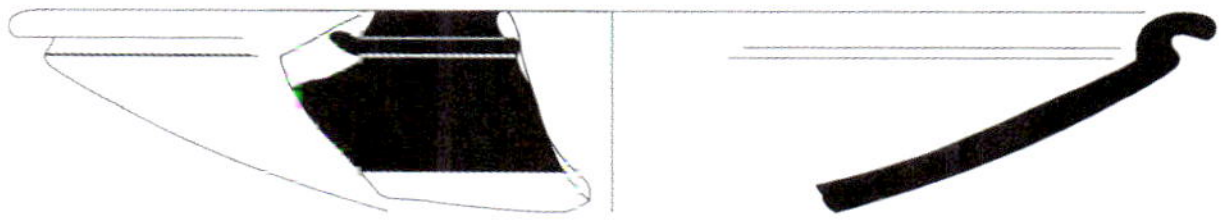

155. BG 043-17 Dish rim (22 examples)
Dimensions: rim diam. 16.0 h 2.75 w 4.1
Reference Sherd Description: Dilute, matte reddish-brown slip, almost completely worn on the exterior but showing unslipped extant lower body.
Selected Comparanda: Sant'Angelo Grieco, SG205.5; Prag 1983a, 358, No. 332 (Cozzo Presepe, first quarter of 3rd c. BC); Lissi Caronna 1990–91, 206, Fig. 26, No. 2 (Oppido Lucano, context dated to second half of the 4th–early 3rd c. BC); Prag 1992b, 78, No. 666, Fig. 29 (Botromagno, late 4th to early 3rd c. BC).
Attested Dates: 300–250 BC.

1:2

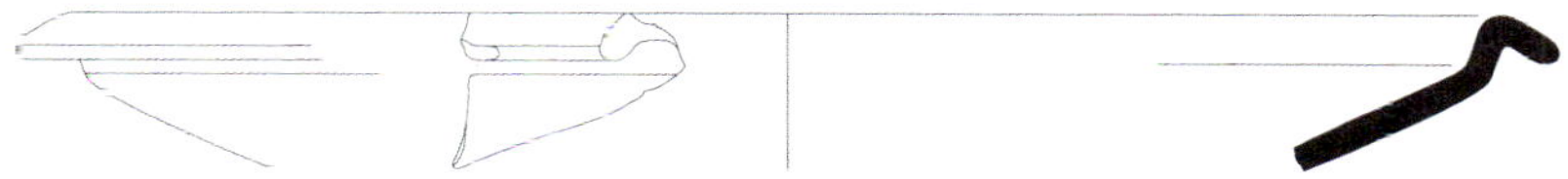

156. BG 341-07 Dish rim (single example)
Dimensions: rim diam. ca 21.0 h 2.2 w 3.2
Reference Sherd Description: Dilute, matte black slip, almost completely worn.
Selected Comparanda: Prag 1983b, 359, No. 332, Fig. 129 (Cozzo Presepe, 3rd c. BC or later); Prag 1992b, 78, No. 667, Fig. 29 (Botromagno, 350–300 BC) and 78, No. 669 (Botromagno, late 4th c. BC); Morel 1981a, 1314e1, 104, Pl. 12 (Teano, around 300 BC).
Attested Dates: 310–280 BC.

Dishes cont.

1:2

157. BG 043-10 Dish rim (single example)
Dimensions: rim diam. 12.9 h 0.9 w 2.2
Reference Sherd Description: High quality, consistent, matte black slip.
Selected Comparanda: Prag 1983a, 358, No. 337 (Cozzo Presepe, 3rd c. BC); Prag 1992b, 78, No. 666, Fig. 29 (Botromagno, late 4th to early 3rd c. BC).
Attested Dates: 310–280 BC.

1:2

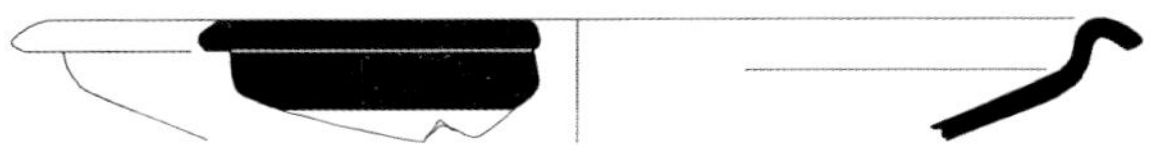

158. BG 128-25 Dish rim (single example)
Dimensions: rim diam. 15.0 h 1.7 w 4
Reference Sherd Description: Consistent, matte black slip extant on the interior, with unslipped lower exterior.
Selected Comparanda: Prag 1983a, 358, No. 332 (Cozzo Presepe, first half of 3rd c. BC); Prag 1992b, 78, No. 666, Fig. 29 (Botromagno, late 4th to early 3rd c. BC); Preacco 1989, 197, Pl. XXVIII, No. 213 (Locri, late 4th to late 3rd c. BC); Morel 1981a, 1314e1, 104, Pl. 12 (Teano, around 300 BC).
Attested Dates: 310–250 BC.

1:2

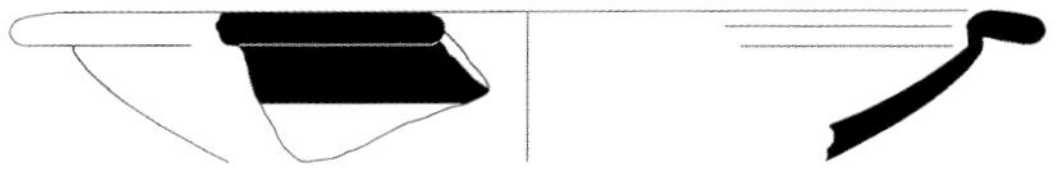

159. BG 501-15 Dish rim (5 examples)
Dimensions: rim diam. 13.7 h 2.1 w 3.6
Reference Sherd Description: Consistent, semi-lustrous black slip, unslipped lower wall, irregularly dipped.
Selected Comparanda: Prag 1992b, 81, No. 687, Fig. 29 (Botromagno, 350–300 BC); Morel 1981a, 1742c1, 134, Pl. 29 (Lipari, late 4th c. BC).
Attested Dates: 320–300 BC.

1:2

160. BG 501-09 Dish rim (single example)
Dimensions: rim diam. 16.0 h 1.45 w 3.85
Reference Sherd Description: Consistent, matte black slip.
Selected Comparanda: Prag 1983a, 358, No. 334 (Cozzo Presepe, last quarter of 3rd c. BC); Prag 1992b, 76, No. 655, Fig. 29 (Botromagno, ca. 300 BC); Morel 1981a, 1514d1, 118, Pl. 20 (unknown provenance, ca. 300 BC).
Attested Dates: 300–200 BC.

1:2

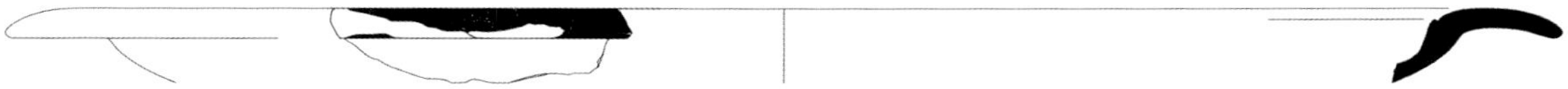

161. BG 416-09 Dish rim (single example)
Dimensions: rim diam. 30? h 1.5 w 5.8
Reference Sherd Description: Dilute, semi-lustrous reddish brown slip, worn on the exterior.
Selected Comparanda: Prag 1992b, 75, No. 638, Fig. 28 (Botromagno, 3rd to 2nd c. BC); Preacco 1989, 199, Pl. XXVIII, No. 217 (Locri, second half of 3rd c. BC); Preacco Ancona 1992, 132–33, Pl. XLIX, No. 96 (Locri, Casa dei Leoni, first half of 3rd c. BC); Lo Porto 1981b, 350–51, No. 4, Fig. 64.2 (Metaponto, necropolis of Casa Ricotta, T 5, grey-ware example, mid-3rd c. BC); Morel 1981a, 1314d1, 104, Pl. 12 (Akrai, 3rd c. BC).
Attested Dates: 300–200 BC.

1:2

162. BG 502-05 Dish rim (single example)
Dimensions: rim diam. 25? h 1.1 w 3.9
Reference Sherd Description: Consistent, matte black slip, unslipped exterior of wall.
Selected Comparanda: Bruscella and Virtuoso 2005, 283, Pl. XLII, No. 289 (Torre di Satriano, first half of 2nd c. BC); Morel 1981a, 1315e1, 104, Pl. 13 (Vienne, 150 ± 30 BC).
Attested Dates: 200–120 BC.

Dishes cont.

1:2

163. BG 260-16 Dish rim (8 examples)
Dimensions: rim diam. 13.1 h 1.5 w 4.6
Reference Sherd Description: Consistent, well-preserved, semi-lustrous black slip.
Selected Comparanda: Prag 1983a, 358, No. 336 (Cozzo Presepe, 3rd c. BC); Morel 1970a, 99, Fig. 24.5 (Cozzo Presepe, 320–270 BC); Preacco 1989, 197, Pl. XXVIII, No. 213 (Locri, late 4th to first half of 3rd c. BC); Prag 1992b, 78, No. 666, Fig. 29 (Botromagno, Period VI, early 3rd c. BC); Morel 1981a, 1311a1, 102, Pl. 11 (La Escuera, 150 ± 30 BC).
Attested Dates: 310–120 BC.

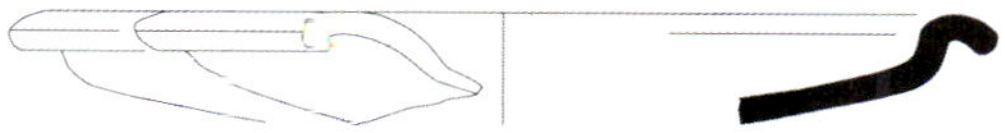

1:2

164. BG 278-03 Dish rim (3 examples)
Dimensions: rim diam. 16.0 h 1.05 w 5.25
Reference Sherd Description: Dilute, matte black slip almost completely worn on the exterior.
Selected Comparanda: Prag 1992b, 72, No. 631, Fig. 28 (Botromagno, first half of 2nd c. BC); Morel 1981a, 1536a1, 122, Pl. 22 (Arezzo, second half of 3rd c. BC?).
Attested Dates: 250–150 BC.

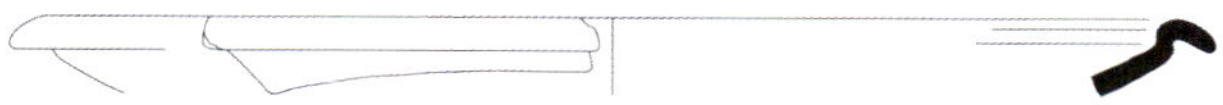

1:2

165. BG 278-01 Dish rim (single example)
Dimensions: rim diam. 23? h 0.9 w 4.7
Reference Sherd Description: Dilute, matte black slip almost completely worn on the interior.
Selected Comparanda: F. Rossi 1989, 164, No. 2, Pl. 306.2 (Monte Sannace, 3rd–2nd c. BC, Morel 1312 = Lamboglia 36); Morel 1981a, 1312b1, 104, Pl. 13 (Ensérune, Campana A, first half of 2nd c. BC).
Attested Dates: 250–150 BC.

166. BG 497-25 Dish base (12 examples)
Dimensions: base diam. 4.5 h 1.5
Reference Sherd Description: Consistent, semi-lustrous black slip, unslipped interior of foot ring.
Selected Comparanda: Elliott 1998, 653, D18 (Metaponto, Pantanello necropolis, 325–300 BC); Prag 1992b, 76, No. 655, Fig. 29 (Botromagno, 310–275 BC, following Morel).
Attested Dates: 310–275 BC.

1:2

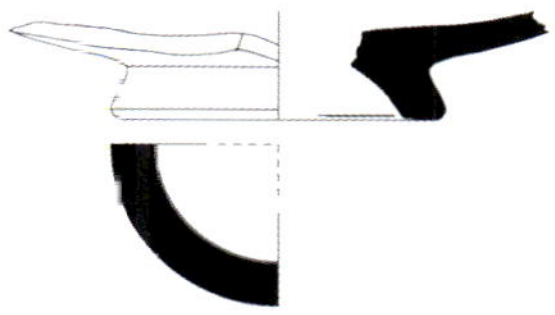

167. BG 260-25 Dish base (9 examples)
Dimensions: base diam. 4.6 h 1.8
Reference Sherd Description: Consistent, semi-lustrous black slip on interior only.
Selected Comparanda: Elliott 1998, 653, D2 (Metaponto, Pantanello necropolis, 310–275 BC).
Attested Dates: 310–275 BC.

1:2

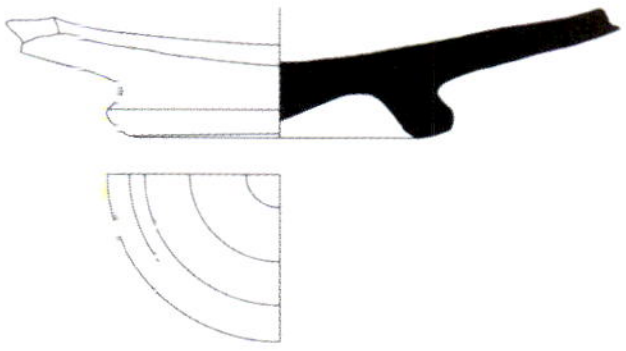

168. BG 260-26 Dish base (5 examples)
Dimensions: base diam. 3.95 h 1.65
Reference Sherd Description: Consistent, semi-lustrous black slip, entirely slipped.
Selected Comparanda: Elliott 1998, 653, D18 (Metaponto, Pantanello necropolis, 325–275 BC).
Attested Dates: 310–275 BC.

1:2

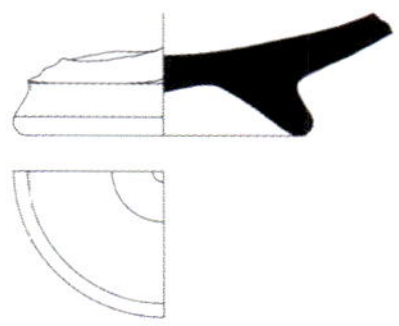

Dishes cont.

169. BG 183-15 Dish base (10 examples)
Dimensions: base diam. 4.0 h 1.65
Reference Sherd Description: Consistent, semi-lustrous black slip, irregularly dipped, unslipped underside with unfinished surfaces and throw removal marks.
Selected Comparanda: Elliott 1998, 653, D18 (Metaponto, Pantanello necropolis, 325–275 BC).
Attested Dates: 325–275 BC.

1:2

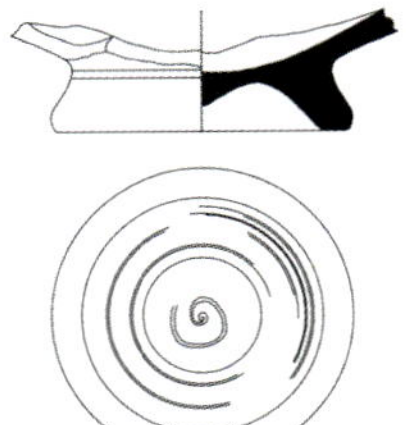

170. BG 417-01 Dish base (single example)
Dimensions: base diam. 3.0 h 1.4
Reference Sherd Description: Consistent, semi-lustrous black slip, unslipped base with wheel-throw removal marks.
Selected Comparanda: Prag 1992b, 90, No. 735, Fig. 31 (Botromagno, 3rd to 1st c. BC); Lissi Caronna 1990–91, 201, 203, Fig. 23, humus, No. 4 (Oppido Lucano, house C, context dated to second half of 4th–early 3rd c. BC).

1:2

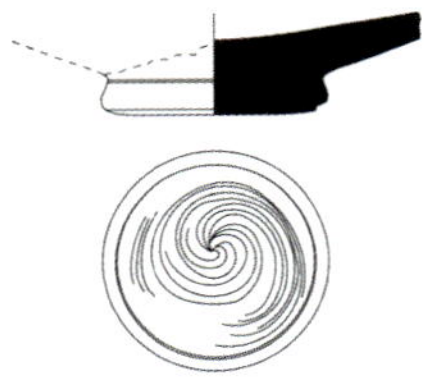

Attested Dates: 300–150 BC.

171. BG 312-08 Dish floor with stamp (single example)
Dimensions: l 4.1 w 4.2
Reference Sherd Description: Consistent, semi-lustrous black slip on the interior only. Circular stamp, half extant, probably originally with 12 radials.
Selected Comparanda: D'Andria 1980a, 429, Nos. 324–25, Fig. 67 (Metaponto, kerameikos, deposit 3, second half of 4th c. BC); *Gravina* III.2, 117, Fig. 48, No. 245 (early 3rd c. BC). Analogous specimens were found during the excavations of Fattoria S. Angelo Grieco (SG204.2) and Fattoria Stefan (No. 78ST172).
Attested Dates: 350–290 BC.

1:1 1:2

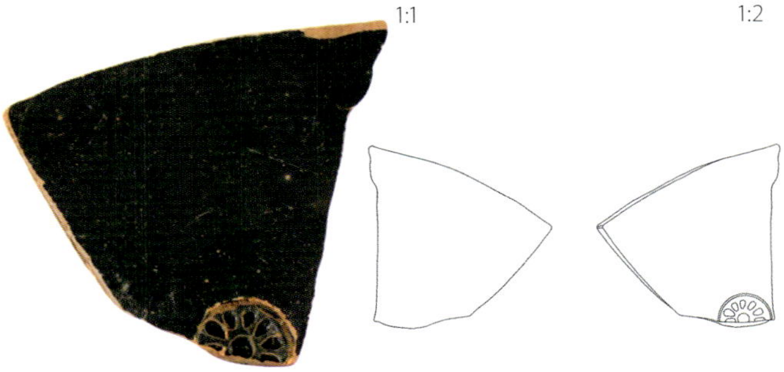

Bowls

172. BG 038-123 Bowl rim (single example)
Dimensions: rim diam. 19.0 h 3.4 w 7.8
Reference Sherd Description: Consistent, semi-lustrous black slip.
Selected Comparanda: *Ordona* VII.2, 433, Fig. 235, No. 5 (T 105, end of second quarter of 4th c. BC); *Forentum* I, Pl. 32, No. 4.1 (4th c. BC). It bears some similarities to Morel 1981a, 2544d1, 182, Pl. 55 (Ensérune, late 4th or early 3rd c. BC, with deeper body).
Attested Dates: 320–280 BC. 360–300 BC.

1:2

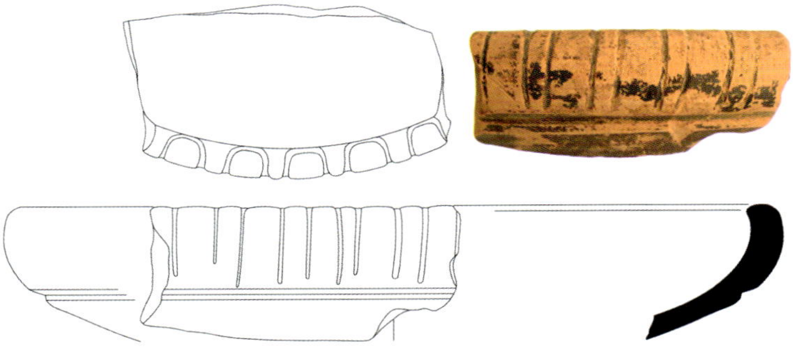

173. BG 258-04 Bowl rim (11 examples)
Dimensions: rim diam. 12.9 h 1.85 w 2.95
Reference Sherd Description: Dilute, matte black slip.
Selected Comparanda: Prag 1983a, 360, No. 340 (Cozzo Presepe, 3rd to 2nd c. BC); Prag 1992b, 97, No. 770, Fig. 32 (Botromagno, 300–260 BC); Fattoria Fabrizio, FF80-248PL-A (300–260 BC).
Attested Dates: 300–200 BC.

1:2

Bowls cont.

1:2

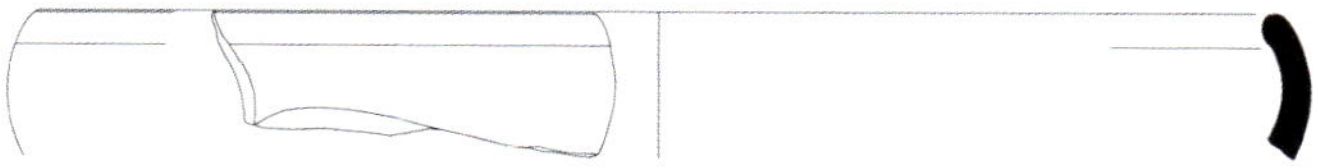

174. BG 167-06 Bowl rim (5 examples)
Dimensions: rim diam. 16.0 h 2.0 w 5.4
Reference Sherd Description: Dilute, matte reddish-brown slip, partly worn.
Selected Comparanda: Bruscella and Virtuoso 20[illegible], 277, Pl. XL, No. 258 (Torre di Satriano, first quarter of 3rd c. BC); Casagrande 2002, 139 and 391 (ill.), Fig. 40, No. 76 (Herakleia, agora, end of 4th–early 3rd c. BC); Morel 1981a, 2783h1, 233, Pl. 72 (Aléria, T 2, 285 ± 20 BC).
Attested Dates: 300–150 BC.

1:2

175. BG 501-17 Bowl rim (2 examples)
Dimensions: rim diam. 16.4 h 1.4 w 4.9
Reference Sherd Description: Dilute, matte reddish-brown slip.
Selected Comparanda: Prag 1992b, 103, No. 809, Fig. 33 (Botromagno, 3rd c. BC); Preacco 1989, Pl. XXIX, No. 230 (Locri, late 4th to late 3rd c. BC); bears some similarities to Lissi Caronna 1990–91, 292, Fig. 1[illegible], No. 4 (Oppido Lucano, banded example, from a context dated to second half of 4th–early 3rd c. BC); for the detail of the molded rim see Morel 1981a, series 1634, 128, Pl. 25 (Attic examples, ca 300 BC).
Attested Dates: 300–200 BC.

1:2

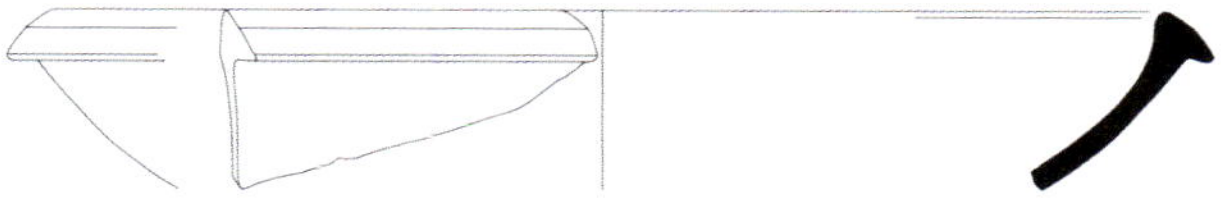

176. BG 354-08 revisit Bowl rim (5 examples)
Dimensions: rim diam. 16.0 h 2.45 w 5.0
Reference Sherd Description: Dilute, matte black slip, almost completely worn on the interior.
Selected Comparanda: De Cazanove 2008, 590, Fig. 301, No. 440 (Civita di Tricarico, last quarter of 3rd c. BC).
Attested Dates: 250–225 BC.

1:2

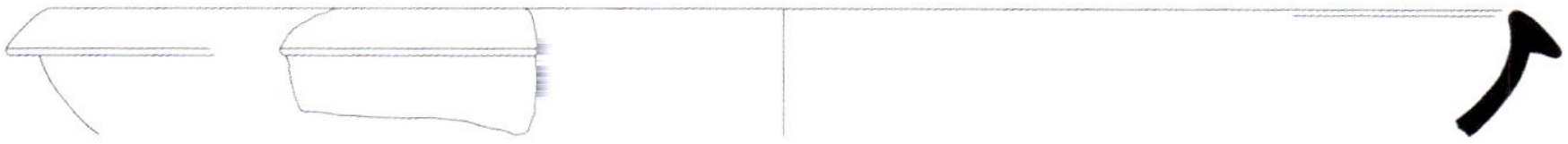

177. BG 354-07 revisit Bowl rim (single example)
Dimensions: rim diam. 23.0 ? h 1.95 w 3.8
Reference Sherd Description: Dilute, matte black slip.
Selected Comparanda: De Cazanove 2008, 590, Fig. 301, No. 441 (Civita di Tricarico, last quarter of 3rd c. BC); Lissi Caronna 1990–91, 252, Fig. 79, No. 10 (Oppido Lucano, house D, context dated to second half of 4th–early 3rd c. BC); Morel 1981a, series 2538, 181, Pl. 54 (3rd c. BC).
Attested Dates: 250–225 BC.

1:2

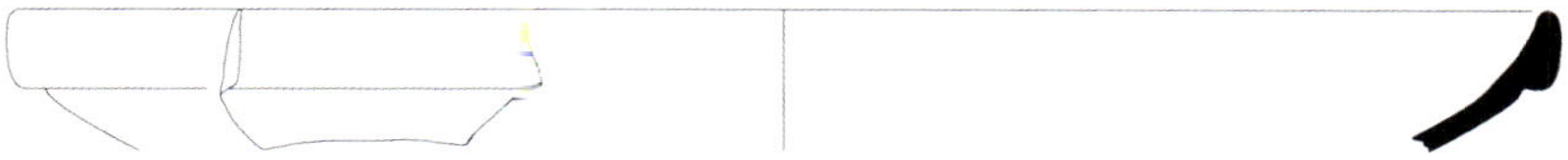

178. BG 128-24 Bowl rim (18 examples)
Dimensions: rim diam. 21.6 h 2.0
Reference Sherd Description: Dilute, matte black slip, almost completely worn, on reduced greyish-fired fabric.
Selected Comparanda: Prag 1992b, 96, No. 767, Fig. 32 (Botromagno, 200–125 BC); De Cazanove 2008, 590, Fig. 301, No. 430 (Civita di Tricarico); Morel 1981a, series 2538, 181, Pl. 54 (area under Etruscan influence, 3rd c. BC). Bears some similarities to Silvestrelli forthcoming a, Fig. 15, No. 5 (Metaponto, kerameikos, deposit 2, end of 4th–early decades of 3rd c. BC, with deeper bowl).
Attested Dates: 3rd c. BC.

1:2

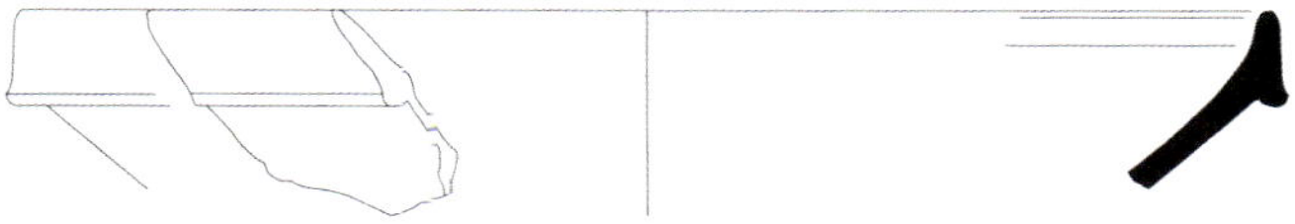

179. BG 292-04 Bowl rim (single example)
Dimensions: rim diam. 17.2 h 2.8 w 4.15
Reference Sherd Description: Consistent, semi-lustrous black slip.
Selected Comparanda: Prag 1983a, 358, No. 3[illegible] (Cozzo Presepe, ca 300 BC); De Cazanove 2008, 590, Fig. 301, No. 431 (Civita di Tricarico); Prag 1992b, 96, Nos. 767–68, Fig. 32 (Botromagno, 2nd c. BC). Bears some similarities to Morel 1981a, series 2534 (2nd c. BC).
Attested Dates: 2nd c. BC.

Bowls cont.

1:2

180. BG 042-03 Bowl rim (single example)
Dimensions: rim diam. 18.0 h 3.0 w 8.1
Reference Sherd Description: Consistent, matte black slip.
Selected Comparanda: De Cazanove 2008, 590, No. 429 (Civita di Tricarico); Prag 1992b, 290 (ill.), Fig. 32, No. 766 (Botromagno, 200–125 BC).
Attested Dates: 200–150 BC.

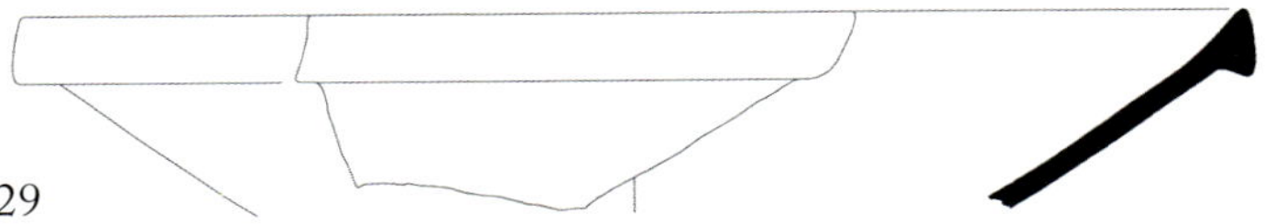

1:2

181. BG 523-06 Bowl rim (single example)
Dimensions: rim diam. 19.5 h 3.1 w 5.9
Reference Sherd Description: Consistent, matte black slip, almost completely worn.
Selected Comparanda: Prag 1992b, 124, No. 915, Fig. 38 (Botromagno, 320–250 BC); Gagliardi 2001, 292, Fig. 274,b. (Kaulonia, sanctuary of Punta Stilo, dated to second third of 5th c. BC on the basis of the associated materials).
Attested Dates: 400–250 BC.

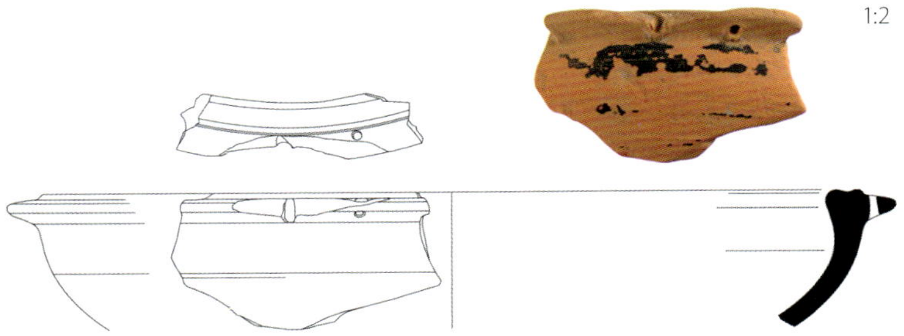

1:2

182. BG 354-04 Bowl base (single example)
Dimensions: base diam. 9.8 h 3.15 w 6.6
Reference Sherd Description: Dilute, matte black slip, almost completely worn on the interior, unslipped lower exterior and underside.
Selected Comparanda: Morel 1981a, 2538f1, 181, Pl. 54 (Volterra, around second half of 3rd c. BC).
Attested Dates: 250–225 BC.

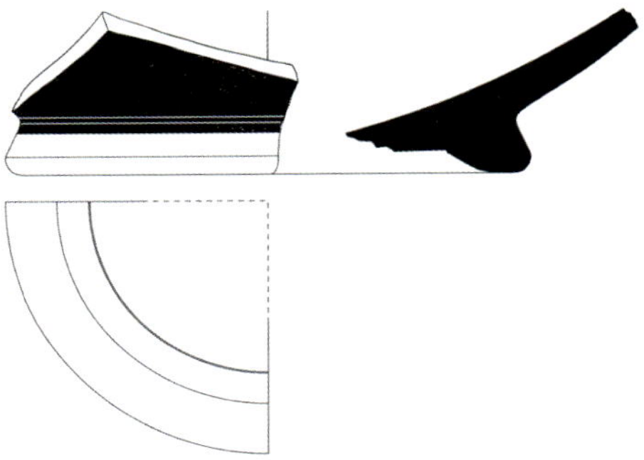

Small Bowls

1:2

183. BG 352-16 Small bowl rim (2 examples)
Dimensions: rim diam. 6.6 h 1.7 w 6.4
Reference Sherd Description: Dilute, matte reddish brown slip, partly worn.
Selected Comparanda: Bitti 1989, 156, Pl. XXIV, No. 129 (Locri, early 4th c. BC).
Attested Dates: 400–350 BC.

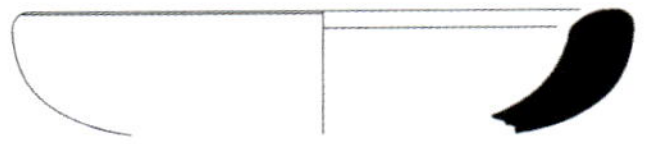

1:2

184. BG 488-65 Small bowl rim (32 examples)
Dimensions: rim diam. 6.55 h 5.25 w 2.7
Reference Sherd Description: Consistent, semi-lustrous black slip.
Selected Comparanda: D'Andria 1980a, 429, Fig. 67, No. 323 (Metaponto, kerameikos, deposit 3, 350–300 BC); Prag 1983a, 356, No. 325 (Cozzo Presepe); Lo Porto 1988–1989, 350, Nos. 22–23, Figs. 46,14–15 and 56,3 (Montescaglioso, T 5/1909, 350 BC); Surra and Preacco Ancona 1997, 72–73, Pl. 33, No. 42 (Pomarico Vecchio, ca. 340 BC); Montanaro 2007, 835, Fig. 801, 260.5 (Ruvo, contrada Colaianna, T 259, 330–300 BC). *Laos* II, 79, Fig. 12, No. 74 (Marcellina, chamber tomb, second or third quarter of 4th c. BC); Bitti 1989, 156, Pl. XXIV, No. 129 (Locri, early 4th c. BC); Morel 1981a, 2715a1, 209, Pl. 67 (Siracusa, 320 ± 30 BC).
Attested Dates: 350–300 BC.

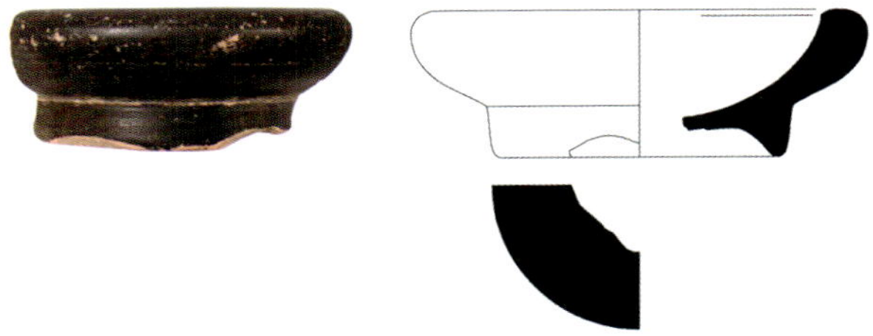

Small bowls cont.

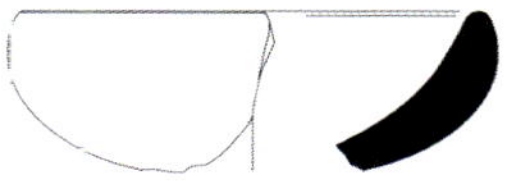

1:2

185. BG 187-01 Small bowl rim (single example)
Dimensions: rim diam. 6.0 h 2.2 w 4.0
Reference Sherd Description: Consistent, semi-lustrous black slip, almost completely worn.
Selected Comparanda: Bitti 1989, 157, Pl. XXIV, No. 132 (Locri, 4th c. BC); Morel 1981a, 2781a1, 223, Pl. 72 (El Cigarralejo, T227, Attic example, 400–375 BC or perhaps later).
Attested Dates: 400–360 BC.

1:1

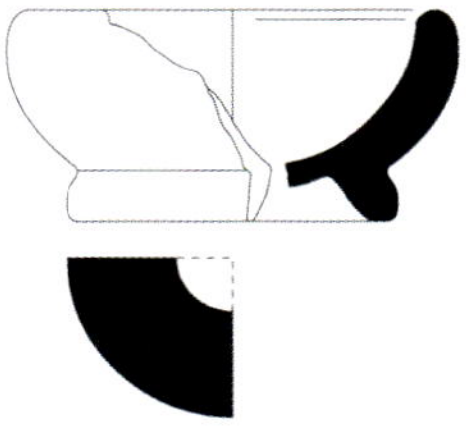

1:2

186. BG 111-262 Small bowl rim (single example)
Dimensions: rim diam. 6.0 h 2.9 w 4.8
Reference Sherd Description: Dilute, matte black slip, almost completely worn on the interior.
Selected Comparanda: Surra and Preacco Ancona 1997, 73, Pl. 33, No. 50 (Pomarico Vecchio, first quarter of 3rd c. BC); Morel 1981a, 2715a1, 209, Pl. 67 (Siracusa, 300 ± 30 BC).
Attested Dates: 330–275 BC.

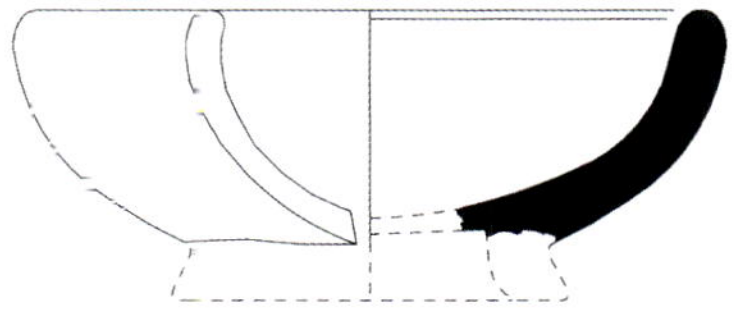

1:2

187. BG 413-04 Small bowl rim (6 examples)
Dimensions: rim diam. 6.7 h 2.25
Reference Sherd Description: Consistent, semi-lustrous black slip
Selected Comparanda: Prag 1983a, 356, No. 323 (Cozzo Presepe, 350–300 BC); Bitti 1989, Pl. XXIV, No. 135 (Locri, 4th c. BC); *Roccagloriosa* I, Fig. 180, No. 104 (second half of 4th to first quarter of 3rd c. BC); Morel 1981a, 2715a1, 209, Pl. 67 (Siracusa, 300 ± 30 BC).
Attested Dates: 330–275 BC.

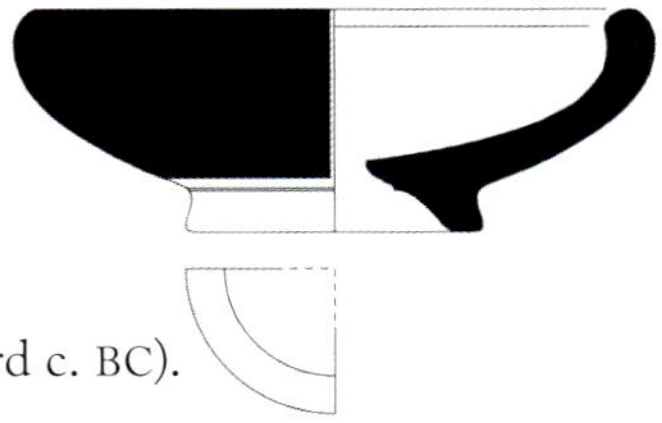

1:2

188. BG 354-03 Small bowl rim (single example)
Dimensions: rim diam. 8.5 base diam. 3.95 h 3.05 w 7.6
Reference Sherd Description: Dilute, matte black slip, worn on the exterior, with unslipped lower exterior and underside.
Selected Comparanda: Elliott 1998, 682, SB6 (Metaponto, Pantanello necropolis, 310–275 BC); Prag 1983a, 356, No. 324 (Cozzo Presepe, 320–270 BC); G. Greco 1991, 72, Fig. 151 (Serra di Vaglio, Casa dei Pithoi, last quarter of 4th to first quarter of 3rd c. BC).
Attested Dates: 320–270 BC.

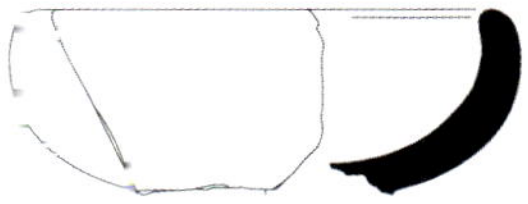

1:2

189. BG 320-06 Small bowl rim (6 examples)
Dimensions: rim diam. 5.7 h 2.5 w 3.7
Reference Sherd Description: Consistent, matte black slip, almost completely worn.
Selected Comparanda: Elliott 1998, 682, SB5 (Metaponto, Pantanello necropolis, 310–275 BC); Prag 1983a, 356, No. 323 (Cozzo Presepe, 350–300 BC).
Attested Dates: 340–275 BC.

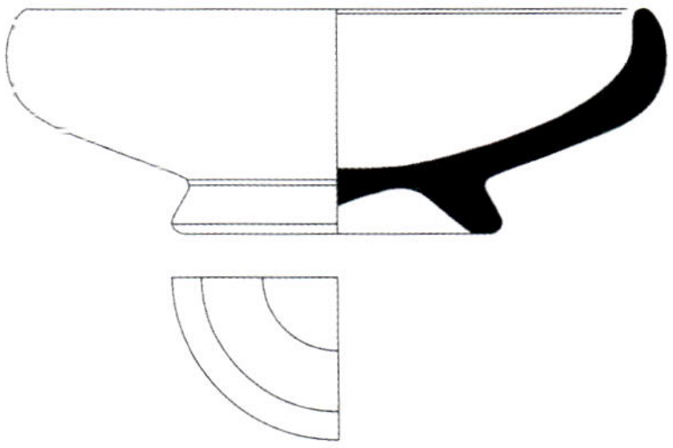

1:2

190. BG 312-06 Small bowl rim (14 examples)
Dimensions: rim diam. 7.4 base diam. 3.7 h 2.8 w 4.0
Reference Sherd Description: Slip completely worn, greyish reduced fabric, misfired.
Selected Comparanda: Elliott 1998, 682, SB6 (Metaponto, Pantanello necropolis, 310–275 BC); Prazzoli 1996, 333–34, type 7 (Taranto, via Nitti well, late 4th to 2nd c. BC); Morel 1981a, 2714a1, 209, Pl. 67 (Lacco Ameno, ca 270 ± 50 BC).
Attested Dates: 310–250 BC.

Small bowls cont.

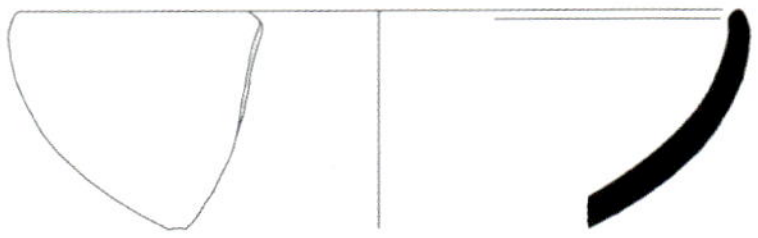

1:2

191. BG 521-53 Small bowl rim (single example)
Dimensions: rim diam. 9.3 h 2.3
Reference Sherd Description: Dilute, matte reddish-brown slip.
Selected Comparanda: Elliott 1998, 682, SB1 (Metaponto, Pantanello necropolis, 310–275 BC); Prag 1983a, 356, No. 322 (Cozzo Presepe, first quarter of 4th c. BC); Prag 1992b, 107, No. 832, Fig. 34 (Botromagno, late 3rd to 2nd c. BC); *Gravisca* 9, Pl. 20, No. 182 (late 4th to first half of 3rd c. BC); Morel 1981a, 2788g1, 226, Pl. 74 (Kouass, 3rd c. BC).
Attested Dates: 310–200 BC.

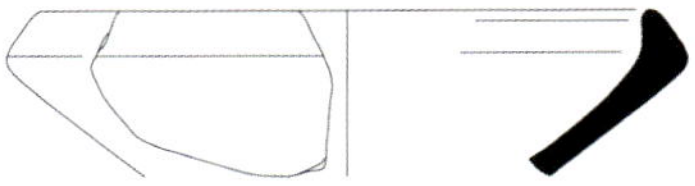

1:2

192. BG 260-14 Small bowl rim (4 examples)
Dimensions: rim diam. 7.8 h 2.25 w 3.2
Reference Sherd Description: Consistent, matte red slip, almost completely worn on the exterior.
Selected Comparanda: De Cazanove 2008, 596, No. 572 (Civita di Tricarico, 280 BC–early 2nd c. BC); Bacci and Tigano 2002, 34, Fig. 13, No. 41 (first half of 3rd c. BC); Morel 1981a, 2737c1, 212, Pl. 68 (Lacco Ameno, 160 ± 40 BC, Campana A); *Corinth* VII.3, 35, No. 95, Pl. 3 and 44 (300–275 BC).
Attested Dates: 300–175 BC.

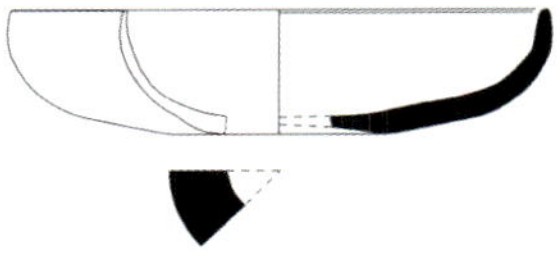

1:2

193. BG 397-44 Small bowl rim (single example)
Dimensions: rim diam. 7.2 base diam. 2.8 h 1.7
Reference Sherd Description: Dilute, matte black slip, partly worn.
Selected Comparanda: Morel 1981a, species 2140, 139, Pl. 31 (early 3rd c. BC); Preacco 1989, 208, Pl. XXIX, No. 231 (Locri, late 4th to early 3rd c. BC).
Attested Dates: 310–250 BC.

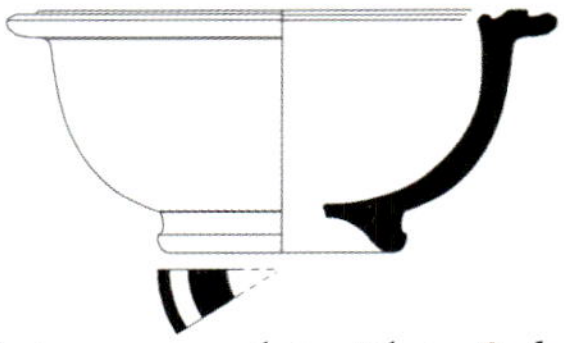

1:2

194. BG 154-19 Small bowl rim (single example)
Dimensions: rim diam. 7.3 base diam. 3.3 h 3.3 w 5.0
Reference Sherd Description: Consistent, semi-lustrous black slip, with worn slip on resting surface of base.
Selected Comparanda: *Greci, Enotri e Lucani*, 254, No. 3.37.11 (Sant'Arcangelo-San Brancato, T 371, 375–325 BC); Giorgi 1988b, 197, Pl. 32, No. 9 (Forentum, third quarter of 4th c. BC); *Gli ori*, 377, Nos. 23–24 (Taranto, T 2 via Bellini, 15.IX.1956, third quarter of 4th c. BC); Prag 1992b, 81, No. 685, Fig. 29 (Botromagno, late 4th to 3rd c. BC); *Kaulonia* I, Fig. 39, No. 177; Morel 1981a, 1734a1, 133, Pl. 29 (unknown provenance, 4th c. BC, with flat rim without moldings).
Attested Dates: 360–300 BC.

Concave-Convexes

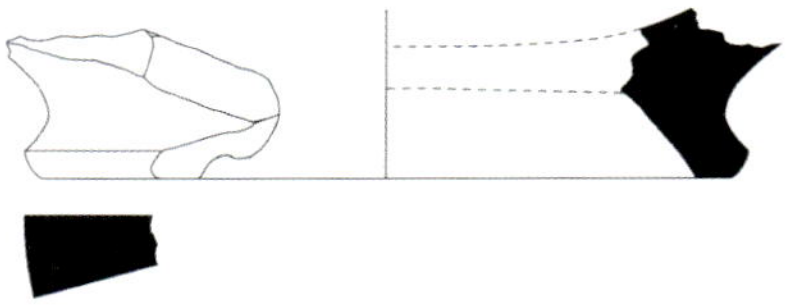

1:2

195. BG 638-09 Concave-convex base (single example)
Dimensions: base diam. 9.6 h 2.3
Reference Sherd Description: Consistent, semi-lustrous black slip.
Selected Comparanda: D'Andria 1980a, 411, Fig. 57, Nos. 217–18 (Metaponto, kerameikos, deposit 1, first quarter 4th c. BC); Morel 1981a, 2431, 170, Pl. 50 (exemplified by a vessel from Ordona, second half of 4th c. BC).
Attested Dates: 400–350 BC.

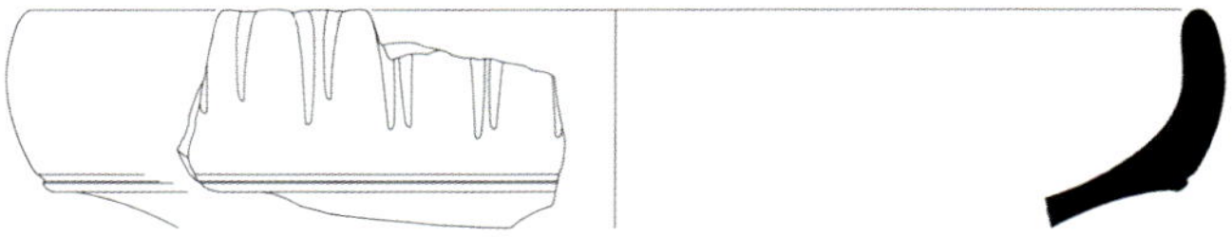

1:2

196. BG 169-10 Concave-convex rim (single example)
Dimensions: rim diam. 15.8 h 3.1 w 5.4
Reference Sherd Description: Dilute, matte black slip.
Selected Comparanda: D'Andria 1980a, 411, No. 217, Fig. 57 (Metaponto, kerameikos, deposit 1, first quarter of 4th c. BC); De Cazanove 2008, 595, Fig. 306, No. 529 (350–300 BC); Morel 1981a, 2433c1, 170, Pl. 50 (Ordona, late 4th c. BC).
Attested Dates: 350–300 BC.

Concave-convexes cont.

1:2

197. BG 521-07 Concave-convex rim (9 examples)
Dimensions: rim diam. 10.5 base diam. 7.75 h 2.7
Reference Sherd Description: Consistent, semi-lustrous black slip, reserved underside and worn slip on resting surface.
Selected Comparanda: Lo Porto 1973, Pl. LXXIII, No. 4 (Irsina, contrada Cappuccini, T 6/1950, said to be Attic, 470–460 BC); Ciancio 1997, 206, No. 225 (Gravina, T 1/1967, end of 5th–first half of 4th c. BC); Bitti 1989, 150, Pl. XXIV, No. 114 (Locri, 4th c. BC); Labellarte 1995, 125, No. 2.1.35 (Chini Collection, first half of 4th c. BC); Morel 1981a, series 2433, 170, Pl. 50 (second half of 4th c. BC).
Attested Dates: 400–350 BC.

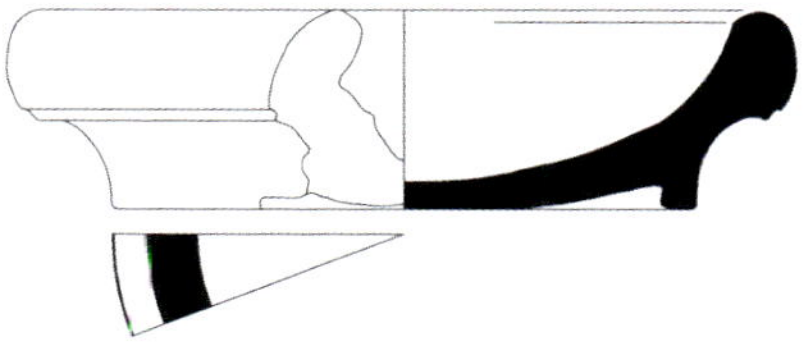

1:2

198. BG 266-15 Concave-convex rim (single example)
Dimensions: rim diam. 7.0 h 2.8 w 8.7
Reference Sherd Description: Dilute, semi-lustrous reddish-brown slip, fired red on the rim and interior of foot.
Selected Comparanda: D'Andria 1980a, 411, Nos. 216–18, Figs. 56–57 (Metaponto, kerameikos, deposit L, 400–350 BC); Sant'Angelo Grieco, SG4.81-21PL; Bitti 1989, 150, Pl. XXIV, No. 114 (Locri, 380–350 BC); De Juliis 1990, 102, No. 55, Figs. 451–52 (Canosa, Ipogeo dei Vimini, Cella B, right deposition, second quarter of 4th c. BC); Morel 1981a, 2433b1, 170, Pl. 50 (Ordona, late 4th c. BC).
Attested Dates: 400–350 BC.

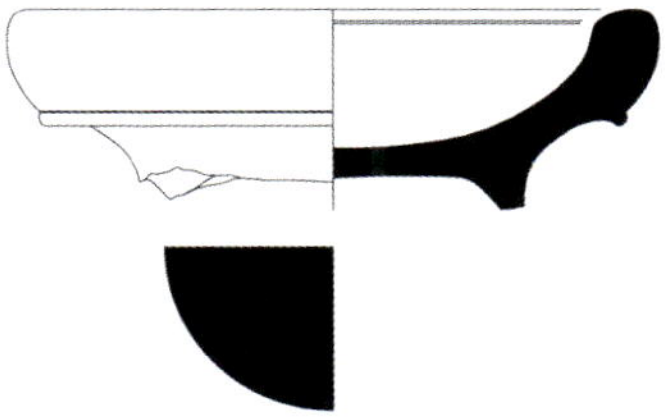

1:2

199. BG 521-03 Concave-convex rim (2 examples)
Dimensions: rim diam. 12.35 h 1.8 w 3.4
Reference Sherd Description: Consistent, semi-lustrous black slip.
Selected Comparanda: Montanaro 2007, 281, Fig. 137 (Ruvo, via Moro, T 33, first half of 4th c. BC); F. Rossi 1992, 485, No. 19 (Canosa, via Lavello, T 3); Giorgi 1988b, Pl. 31, No. 3.1a (Forentum, 425–375 BC); Morel 1981a, 2433c1, 170, Pl. 50 (Ordona, late 4th c. BC).
Attested Dates: 425–300 BC.

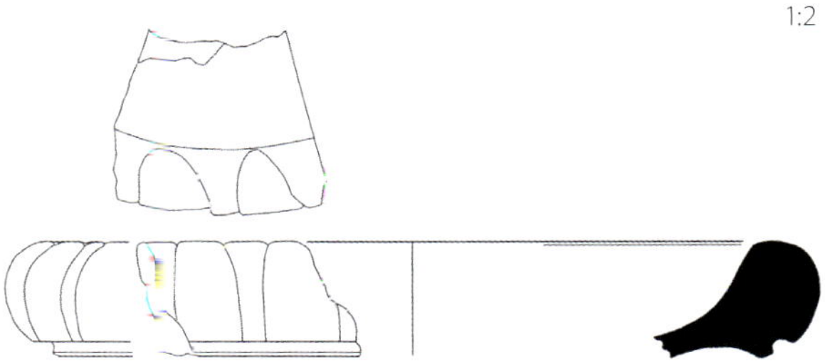

1:1 1:2

200. BG 446-01 Concave-convex rim (2 examples)
Dimensions: rim diam. 8.0 h 2.0 w 2.5
Reference Sherd Description: Consistent, matte black slip.
Selected Comparanda: *Greci, Enotri e Lucani*, 258, No. 3.39.28 (Sant'Arcangelo-San Brancato, T 33b, 375–325 BC); Montanaro 2007, 271, Fig. 122, 21.20 (Ruvo, via delle Monache, second half of 4th c. BC); *Forentum* II, Pl. CXXXIX, No. 3.2a, T 598-4 (first half of 4th c. BC); Morel 1981a, 2421d1, 168, Pl. 49 (Aquinum, second half of 4th c. BC).
Attested Dates: 350–300 BC.

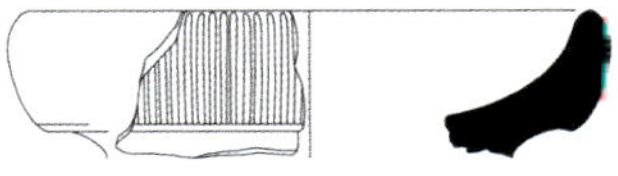

Concave-convexes cont.

1:2

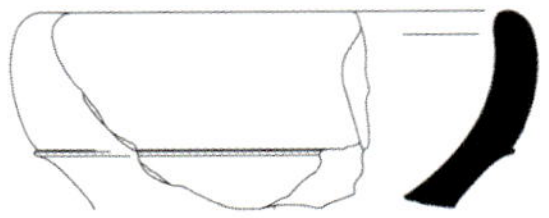

201. BG 258-01 Concave-convex rim (2 examples)
Dimensions: rim diam. 5.8 h 2.7 w 4.2
Reference Sherd Description: Dilute, matte black slip, almost completely worn on the exterior.
Selected Comparanda: Bitti 1989, 150, Pl. XXIV, No. 114 (Locri, 4th c. BC); Preacco Ancona 1992, 128, No. 50, Pl. XLVI (Locri, Casa dei Leoni, end 4th–beginning of 3rd c. BC); Surra and Preacco Ancona 1997, 71, Pl. 32, No. 34 (Pomarico Vecchio, late 4th c. BC); *Bitonto*, 157, Nos. 169–70 (Bitonto, T 4/1981, second half of 4th c. BC); Riccardi 2008, 73, No. 70 (T 3/2002, second half of 4th c. BC); Morel 1981a, 2435b1, 171, Pl. 50 (Pompei, ca. 300 BC).
Attested Dates: 350–300 BC.

Salt Cellars

1:2

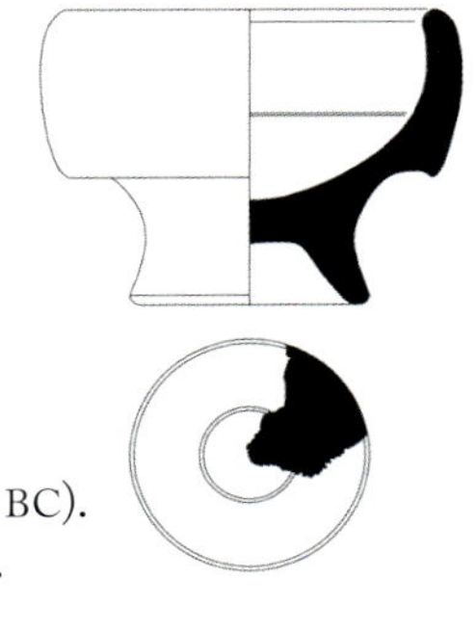

202. BG 494-01 Salt cellar (28 examples)
Dimensions: rim diam. 4.75 base diam. 3.3 h 4.2
Reference Sherd Description: Consistent, matte black slip, partially dipped on underside of foot.
Selected Comparanda: Elliott 1998, 690, SA3 (Metaponto, Pantanello necropolis, 325–300 BC); Lo Porto 1966, 211, No. 2, Fig. 65.2 (Metaponto, urban necropolis, T 31, late 4th c. BC); Prag 1983a, 358, No. 326 (Cozzo Presepe, 350–300 BC); Pianu 1990, 183, Pl. I, No. 245.1 (Herakleia, 340–280 BC); Morel 1981a, 2424a1, 169, Pl. 49 (Cozzo Presepe, 320–270 BC).
Attested Dates: 325–270 BC.

1:2

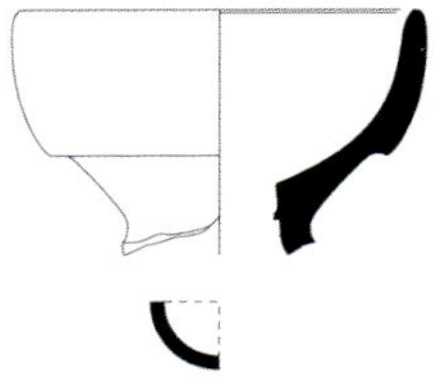

203. BG 320-26 Salt cellar rim (2 examples)
Dimensions: rim diam. 4.9 h 3.4 w 3.6
Reference Sherd Description: Salt cellar rim to upper foot. Consistent, matte black slip.
Selected Comparanda: Prag 1992b, 91, No. 742, Fig. 32 (Botromagno, 320–270 BC); *Monte Sannace*, Pl. 277, No. 4 (phase III, 4th–3rd c. BC); Morel 1981a, 2441b1, 172, Pl. 50 (Spina, second half of 4th c. BC).
Attested Dates: 320–270 BC.

1:2

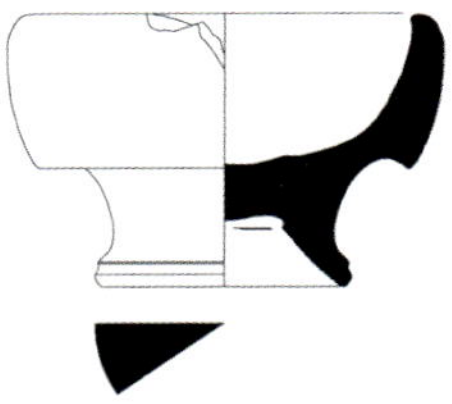

204. BG 258-03 Salt cellar (4 examples)
Dimensions: rim diam. 4.9 base diam. 3.4 h 3.7
Reference Sherd Description: Complete salt cellar profile. Dilute, semi-lustrous red (oxidized) slip, worn on the upper exterior of the rim.
Selected Comparanda: Prag 1983a, 358, No. 326 (Cozzo Presepe, late 4th c. BC); Prag 1992b, 91, No. 739, Fig. 32 (Botromagno, 320–270 BC); Morel 1981a, 2424c1, 169, Pl. 49 (Capua, ca. 300 BC).
Attested Dates: 320–270 BC.

1:2

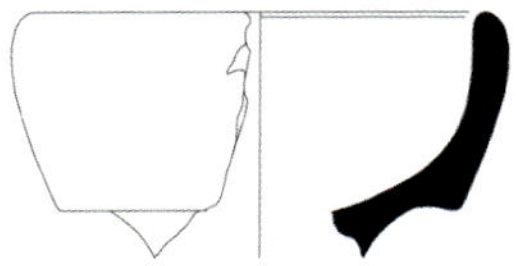

205. BG 260-24 Salt cellar rim (single example)
Dimensions: rim diam. 5.65 h 3.3
Reference Sherd Description: Consistent, matte reddish-brown slip.
Selected Comparanda: Wade 2002, 68, Fig. 3.2 (Metaponto, Pantanello sanctuary); Elliott 1998, 690, SA3 (Metaponto, Pantanello necropolis, 325–300 BC); Sant'Angelo Grieco, SG302.2; Prag 1983a, 358, No. 328 (Cozzo Presepe, 350–300 BC); Pianu 1990, 83, Pl. I, No. 70.2 and 185, Pl. LXXIV, No. 248.1 (Herakleia, 340–280 BC); G. Greco 1991, 72, Fig. 150 (Serra di Vaglio, Casa dei Pithoi, 325–250 BC).
Attested Dates: 350–270 BC.

Salt cellar, cont.

1:2

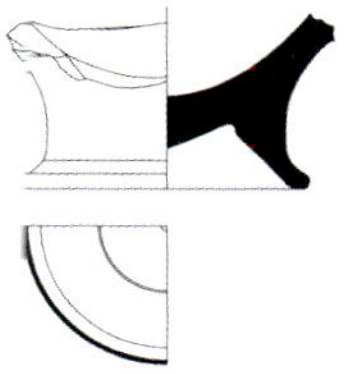

206. BG 465-08 Salt cellar base (3 examples)
Dimensions: base diam. 3.8 h 2.5 w 4.6
Reference Sherd Description: Consistent, semi-lustrous black slip, unslipped underside.
Selected Comparanda: Wade 2002, 68, Fig. 3.3 (Metaponto, Pantanello sanctuary); Morel 1981a, 2424a1, 169, Pl. 49 (Cozzo Presepe, 320–270 BC); Prag 1983a, 358, No. 327 (Cozzo Presepe, late 4th to first quarter of 3rd c. BC); Prag 1992b, 92, No. 744, Fig. 32 (Botromagno, 320–270); Lo Porto 1988–1989, No. 14, Fig. 88.7 (Montescaglioso, T 14, ca. 320 BC); Surra and Preacco Ancona 1997, Pl. 32, No. 26 (Pomarico Vecchio, late 4th to first quarter of 3rd c. BC).
Attested Dates: 320–270 BC.

1:2

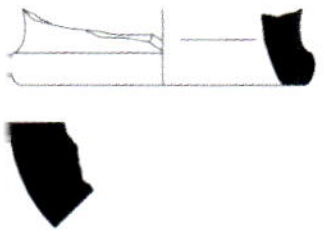

207. BG 068-03 Salt cellar foot (single example)
Dimensions: base diam. 4.1 h 1.1
Reference Sherd Description: Consistent, matte black slip, partially worn.
Selected Comparanda: Prag 1983a, 358, No. 327 (Cozzo Presepe, late 4th to first quarter of 3rd c. BC); Lo Porto 1988–1989, 380, No. 14, Fig. 88.7 (Montescaglioso, T 14, ca. 320 BC); Surra and Preacco Ancona 1997, 69–70, Pl. 32, No. 25 (Pomarico Vecchio, late 4th to first quarter of 3rd c. BC); Morel 1981a, 2424a1, 169, Pl. 49 (Cozzo Presepe, 320–270 BC).
Attested Dates: 320–290 BC.

1:1

1:2

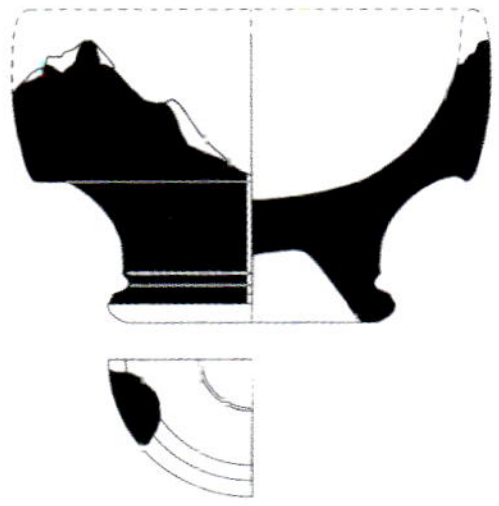

208. BG 342-30 Salt cellar base to upper body (single example)
Dimensions: base diam. 3.8 h 3.9 w 4.8
Reference Sherd Description: Dilute, matte black slip, unslipped lower foot and underside with irregular dipping evident on the underside of the foot.
Selected Comparanda: Wade 2002, 68, Fig. 3.1 (Metaponto, Pantanello sanctuary); Prag 1992b, 91, No. 743, Fig. 32 (Botromagno, 320–270 BC); Morel 1981a, 2424a1, 169, Pl. 49 (Cozzo Presepe, 320–270 BC).
Attested Dates: 320–270 BC.

1:2

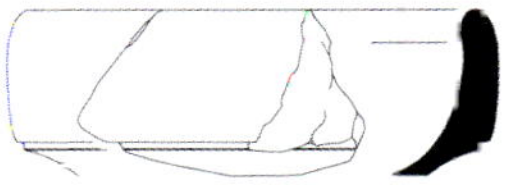

209. BG 296-02 Salt cellar rim (3 examples)
Dimensions: rim diam. 5.5 h 2.25 w 3.75
Reference Sherd Description: Consistent, matte black slip.
Selected Comparanda: Prag 1983a, 358, No. 326 (Cozzo Presepe, 350–275 BC); Prag 1992b, 92, No. 745, Fig. 32 (Botromagno, 320–270 BC); G. Greco 1991, 7, Fig. 148 (Serra di Vaglio, Casa dei Pithoi, last quarter of 4th to first half of 3rd c. BC).
Attested Dates: 350–270 BC.

Kraters

1:2

210. BG 042-01 Krater rim and upper body (single example)
Dimensions: rim diam. 26.75 h 6.9 w 10.5
Reference Sherd Description: High quality, consistent lustrous black slip.
Selected Comparanda: Bernardini 1961, Pl. XI, No. 1 (Rocavecchia); Morel 1981a, 4742a1, 328, Pl. 145 (unknown provenance, end 4th–first quarter of 3rd c. BC).
Attested Dates: 400–300 BC.

Kraters cont.

1:2

211. BG 464-01 Krater rim (single example)
Dimensions: rim diam. 16.5 h 1.0 w 2.9
Reference Sherd Description: Slipped rim, with reserved wall under rim. Consistent, semi-lustrous black slip.
Selected Comparanda: Bears some similarities to De Cazanove 2008, 579, Fig. 290, No. 106 (Civita di Tricarico, perhaps Gnathia example, second half of 4th c. BC); *CVA Naples* 3, IV.E.03, Pl. 42.2; *CVA Schwerin, Staatliches Museum* 1, 35, Pl. 58.1. Bears also some similarities to a pelike rim of the same size in Delli Ponti 1996, 208, No. 285 (Vaste, T4/1970, mid-4th c. BC).
Attested Dates: 350–300 BC.

1:2

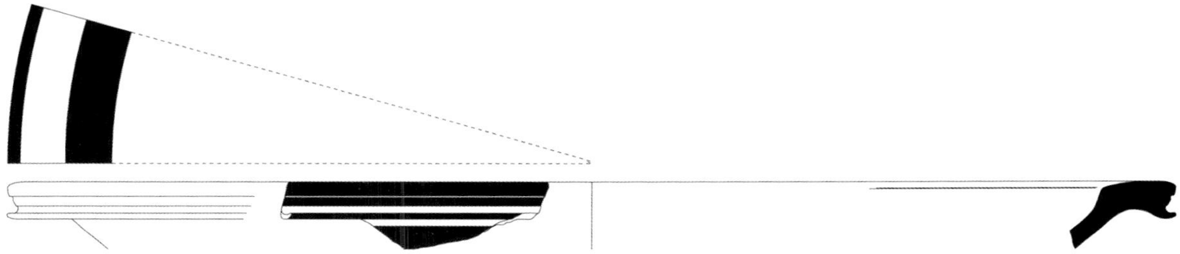

212. BG 260-12 Krater rim (single example)
Dimensions: rim diam. 31.0 h 1.85 w 7.1
Reference Sherd Description: Consistent, semi-lustrous black slip, with reserved band on the upper face of the rim, thin reserved band on outer face and underside of projecting rim.
Selected Comparanda: See Gnathia krater type C1, from site 105-01.
Attested Dates: 300–250 BC.

Deep Bowl

1:2

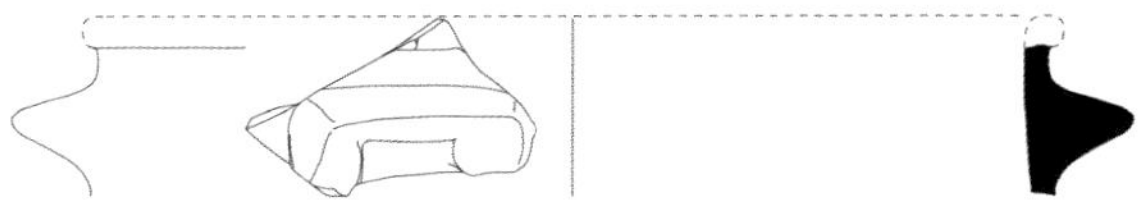

213. BG 043-207 Deep bowl rim (single example)
Dimensions: rim diam. 13.0 h 2.5 w 3.8
Reference Sherd Description: Consistent, semi-lustrous black slip.
Selected Comparanda: Liseno 2004, Pl. XLV.1 (Metaponto, late 4th c. BC); Morel 1970a, Fig. 24, No. 10 (Cozzo Presepe, end of 4th–first third of 3rd c. BC); Elia and Meirano 1997, 91, Pl. 42, Nos. 146 and 97, Pl. 44, No. 52 (Pomarico Vecchio, last decades of 4th–first half of 3rd c. BC); Giannotta 1996, 88, Fig. 44 (Rocavecchia, T 28/1934, Gnathia example, 330–320 BC); Bernardini 1961, Pl. 60, No. 1 (Rudiae); Morel 1981a, 4731b2, 328, Pl. 144 (Pomarico Vecchio, necropolis, last decades of 4th–first half of 3rd c. BC).
Attested Dates: 310–250 BC.

Lids

1:2

214. BG 296-09 Lid handle (single example)
Dimensions: h 2.1 w 2.4
Reference Sherd Description: Consistent, semi-lustrous black slip on exterior, unslipped interior.
Selected Comparanda: For a similar lid of a lebes gamikos see Elliott 1998, 660, LG8 (Metaponto, Pantanello necropolis, 450–370 BC).
Attested Dates: 450–200 BC.

1:2

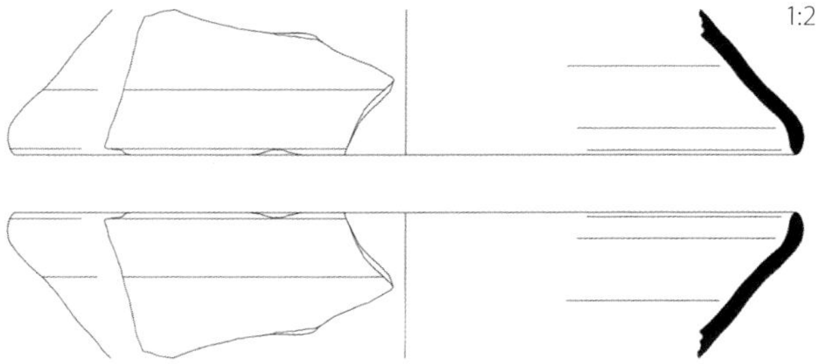

215. BG 476-04 Lid rim (single example)
Dimensions: rim diam. 15.3 h 2.9 w 5.7
Reference Sherd Description: Consistent, semi-lustrous black slip.
Selected Comparanda: Bears some similarities to: De Cazanove 2008, 589, Fig. 300, No. 414 (Civita di Tricarico, mid-3rd c. BC).
Attested Dates: 3rd c. BC

Open Forms

1:2

216. BG 266-16 Open form ring base (single example)
Dimensions: base diam. 10.0 h 1.8 w 3.6
Reference Sherd Description: Dilute, matte black slip, almost completely worn, unslipped lower foot and underside, with slipped internal face of foot ring.
Attested Dates: 500–400 BC.

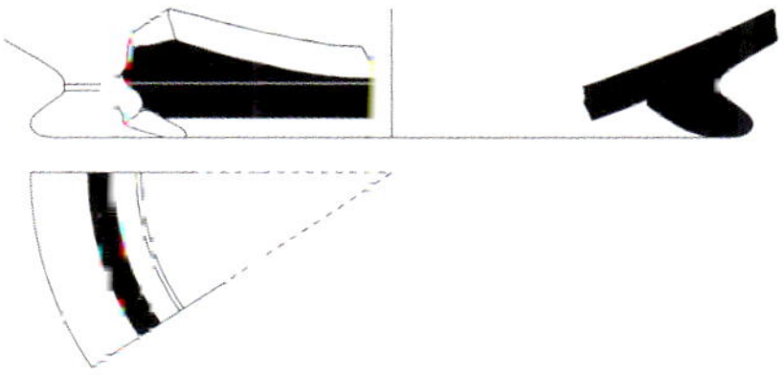

1:1

217. BG 326-27 Open form ring base (2 examples)
Dimensions: base diam. 6.3 h 1.9 w 6.2
Reference Sherd Description: Dilute, matte black slip, unslipped exterior and underside, with perforation?
Attested Dates: Classic–Hellenistic period.

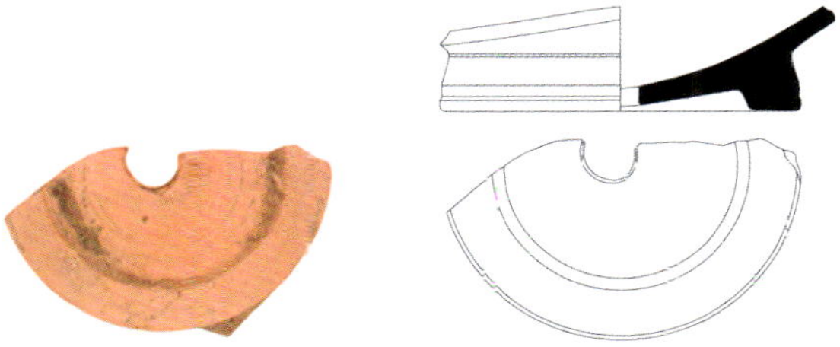

1:2

218. BG 343-08 Open form ring base (31 examples)
Dimensions: base diam. 5.9 h 1.6
Reference Sherd Description: Consistent, semi-lustrous black slip, reserved lower wall and interior of foot ring.
Selected Comparanda: Prag 1983a, 360, Fig. 130, No. 344 (Cozzo Presepe, bowl, phase VI, 4th–3rd c. BC).
Attested Dates: 350–300 BC.

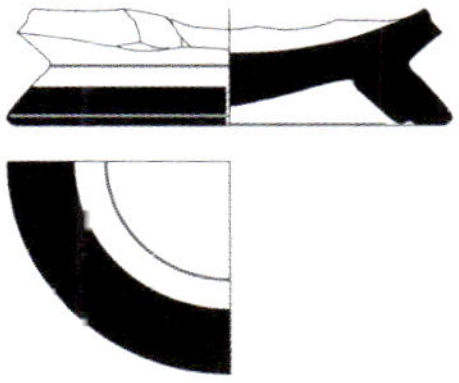

1:2

219. BG 343-07 Open form ring base (2 examples)
Dimensions: base diam. 4.8 h 1.9
Reference Sherd Description: Consistent, semi-lustrous black slip, unslipped exterior and underside.
Selected Comparanda: Similar in general form to ring-handled cup bases: Pianu 1990, Pl. V, No. 162.1 (Herakleia, 300–250 BC) and Scarano 1992, Fig. 5.5 (Metaponto, T 16/1986, end of 4th–mid-3rd c. BC); similar also to one-handler bases: Bruscella and Virtuoso 2005, Pl. XL, No. 264 (Torre di Satriano, 350–300 BC), to bolsal bases: Hoffmann 2002, 43, Pl. 83, No. 424/9 (Taranto, 350–275 BC), and to pyxis bases: Hoffmann 2002, 45, Pl. 88, No. 427/17 (Taranto, 300–275 BC).
Attested Dates: 300–250 BC.

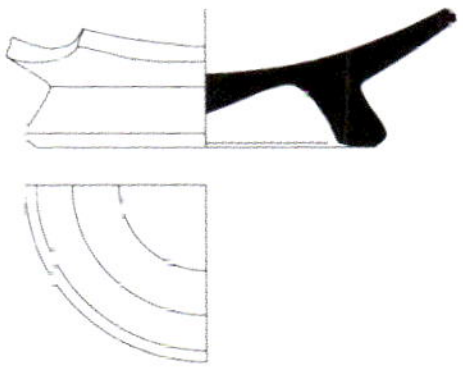

1:2

220. BG 497-24 Open form ring base (4 examples)
Dimensions: base diam. 4.7 h 2.2
Reference Sherd Description: Consistent, semi-lustrous black slip, reserved lower body and interior of foot ring.
Selected Comparanda: Similar in general form to cup-skyphos bases: Elliott 1998, 651, C3 (Metaponto, Pantanello necropolis, first quarter of 3rd c. BC); Pianu 1990, Pl. V, No. 169.2 (Herakleia, 300–250 BC); similar also to two-handled cup base: Scarano 1992, Fig. 5.6 (Metaponto, T 22/1986, early 3rd c. BC). See also Morel 1981a, 3221a1, 256, Pl. 91 (shipwreck of Grand Congloué, Campana A, ca. 190 BC) for a later form.
Attested Dates: 300–190 BC.

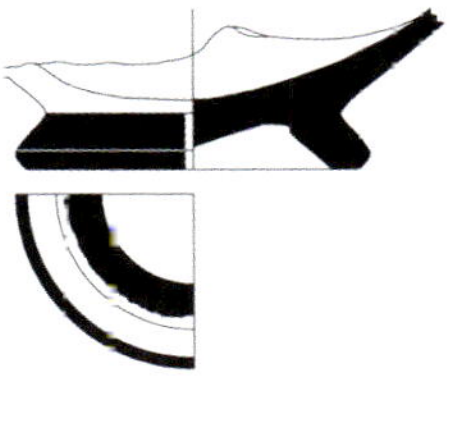

Open forms cont.

1:2

221. BG 465-10 Open form ring base (3 examples)
Dimensions: base diam. 8.7 h 3.15 w 7.4
Reference Sherd Description: Dilute, matte reddish-brown slip on base, black on lower body. Reserved underside with thin band of slip at junction of foot ring and floor.
Selected Comparanda: Scarano 1992, Fig. 5.1 (ring-handled cup from Metaponto, T 18/1986, end of 4th–mid-3rd c. BC).
Attested Dates: 325–275 BC.

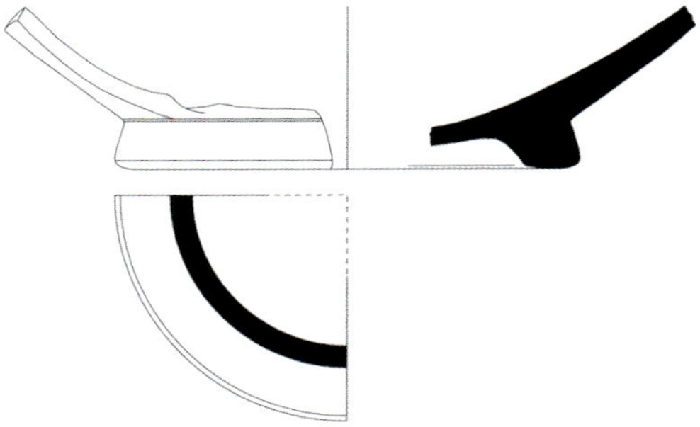

1:2

222. BG 498-15 Open form ring base (single example)
Dimensions: base diam. 6.2 h 2.35
Reference Sherd Description: Dilute, matte black slip, reserved lower body and underside.
Attested Dates: Classical period?

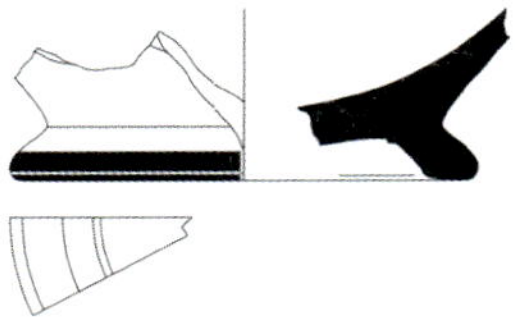

1:2

223. BG 516-01 Open form ring base (single example)
Dimensions: base diam. 5.7 h 2.5
Reference Sherd Description: Consistent, matte black slip, almost completely worn on the exterior, with reserved underside.
Selected Comparanda: Bears some similarities to Giannotta 1996, 68, Fig. 32, No. 59 (Rocavecchia, T 25/1934, Gnathia kantharos, second half of 3rd c. BC); *Agora* XXIX, 260, No. 170, Fig. 13 (Hellenistic kantharos, 310–290 BC or possibly earlier).
Attested Dates: 325–200 BC.

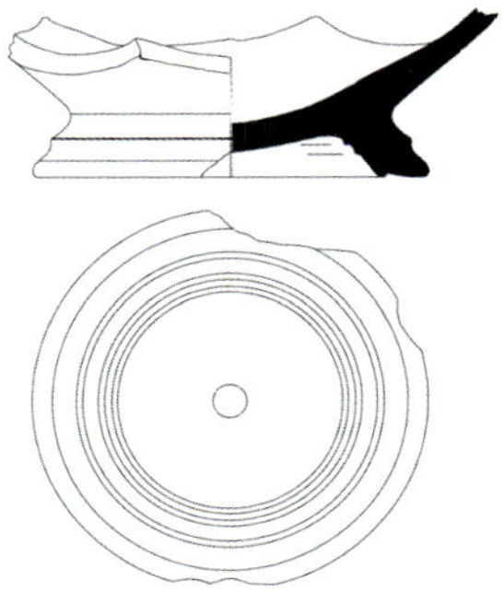

1:2

224. BG 221-01 Open form ring base (single example)
Dimensions: base diam. 9.4 h 1.9 w 7.0
Reference Sherd Description: Consistent, semi-lustrous black slip, almost completely worn. Lower exterior and underside reserved, with slipped foot ring.
Selected Comparanda: Bears some similarities to a lebes gamikos base in Elliott 1998, 633, T 193-10 (Metaponto, Pantanello necropolis, ca. 300 BC), a lebes-pyxis base in Hoffmann 2002, Pl. 98, No. 614/1 (Taranto, 350–325 BC) and a pelike base in Pianu 1990, 84, Pl. IV, No. 71.1 (Herakleia, 340–300 BC).
Attested Dates: 340–300 BC.

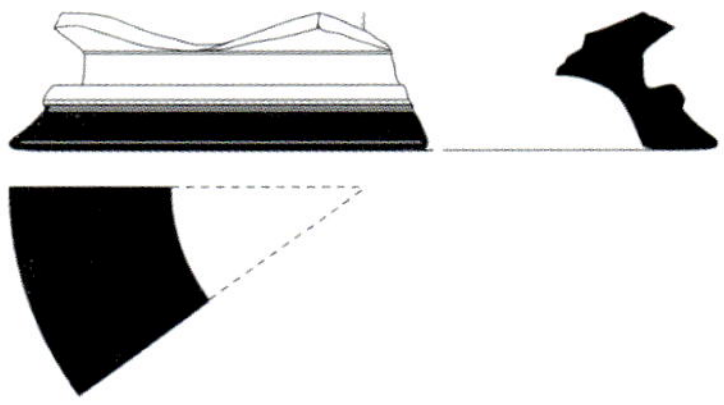

1:2

225. BG 320-27 Open form ring base (3 examples)
Dimensions: base diam. 4.5 h 1.5 w 4.4
Reference Sherd Description: Consistent, semi-lustrous black slip, with reserved internal face of foot ring and center of underside.
Selected Comparanda: Bears some similarities to *Pomarico Vecchio* I, 94, Pl. 43, No. 157 (fragment ascribed to a ring-handled cup, Morel species 3210, last quarter of 4th–beginning of 3rd c. BC); Preacco 1989, 221, Pl. XXIX, No. 249 (Locri, end of 4th–first half of 3rd c. BC); Delli Ponti 1996, 158, No. 132 (Vaste, T 12/1968, Gnathia cup-skyphos, end of 4th c. BC).
Attested Dates: 350–300 BC

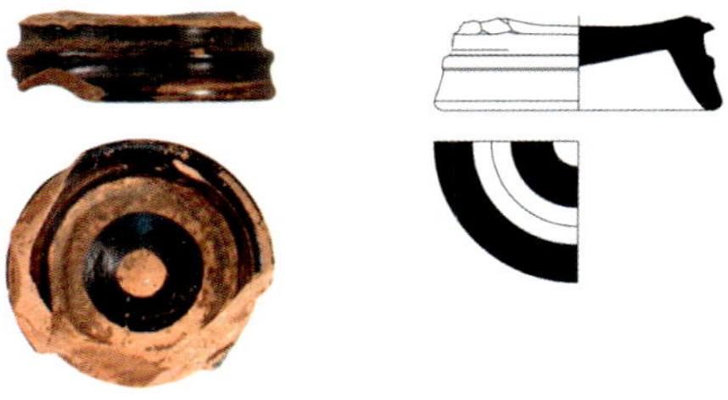

Open forms cont.

1:2

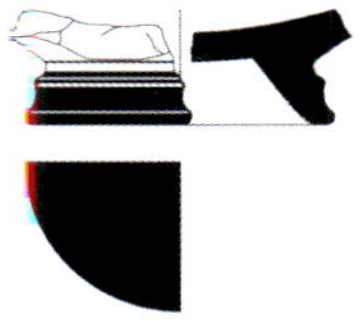

226. BG 269-14 Open form ring base (2 examples)
Dimensions: base diam. 4.1 h 1.6 w 4.0
Reference Sherd Description: Consistent, semi-lustrous black slip, reserved lower body with slipped foot ring, underside and interior.
Selected Comparanda: Prag 1983a, 354, No. 318 (Cozzo Presepe, late 4th c. BC); Prag 1992b, 113, No. 859, Fig. 35 (Botromagno, 335–300 BC); Pianu 1990, 117, Pl. V, No. 116.1 (Gnathia cup-skyphos from Herakleia, 330–300 BC).
Attested Dates: 350–300 BC.

1:2

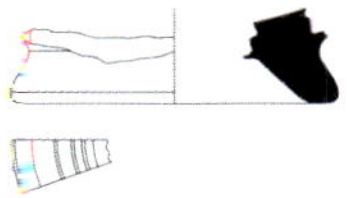

227. BG 516-04 Open form ring base (3 examples)
Dimensions: base diam. 4.35 h 1.3
Reference Sherd Description: Consistent, matte black slip, almost completely worn, unslipped underside.
Selected Comparanda: Bears some similarities to Morel 1981a, 3221c1, 256, Pl. 91 (Naples, Campana A, second quarter of 2nd c. BC); Prag 1992b, 81, No. 687, Fig. 29 (Botromagno, 350–300 BC); Fracchia 1990a, 256, No. 182, Fig. 182 (second half 4th–early 3rd c. BC); Pianu 1990, 147–8, Pl. V, No. 166.1 (stemmed cup from Herakleia, 320–280 BC); Hoffmann 2002, Pl. 83, No. 424/11 (bolsal from Taranto, 325–275 BC).
Attested Dates: 300–150 BC.

1:2

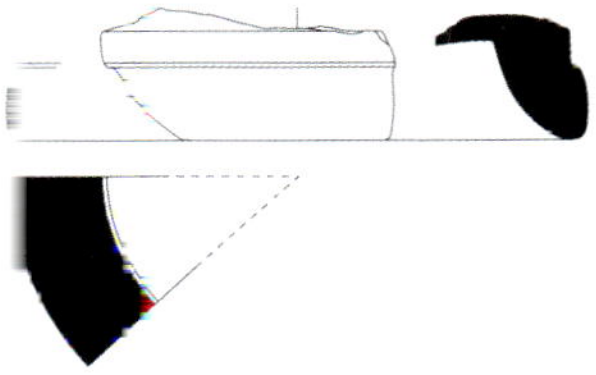

228. BG 306-01 Open form ring base (2 examples)
Dimensions: base diam. 7.7 h 1.8 w 4.3
Reference Sherd Description: Consistent, semi-lustrous black slip, with reserved underside inside foot ring.
Selected Comparanda: Bears some similarities to: D'Andria 1980a, 429, No. 325, Fig. 67 (dish from Metaponto, kerameikos, deposit 3, second half of 4th c. BC); *Pomarico Vecchio* I, 93, Pl. 43, No. 154 (dish, second half–end of 4th c. BC); Hoffmann 2002, Pl. 84, No. 425/3 (stemless kylix from Taranto, 325–300 BC).
Attested Dates: 350–300 BC.

1:2

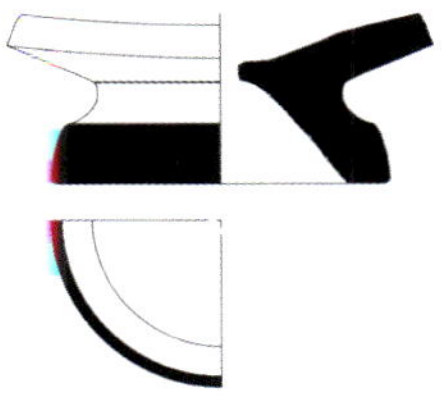

229. BG 266-13 Open form pedestal base (5 examples)
Dimensions: base diam. 4.5 h 2.3 w 3.6
Reference Sherd Description: Consistent, semi-lustrous black slip, reserved exterior, slipped lower foot ring and resting surface, reserved underside.
Selected Comparanda: Bearing some similarities to the lekanis bases: Elliott 1998, 689, LE1 (Metaponto, Pantanello necropolis, 325–300 BC); Pianu 1990, 48, No. 3, Pl. XVIII, 4 (Herakleia, T28, plain example, 350–325 BC) and Hoffmann 2002, Pl. 87, No. 427/13 (Taranto, 350–325 BC). Also similarities to the lebes gamikos bases: Pianu 1990, 89, No. 1, Pl. XXXIX, 4 (Herakleia, T 80, Gnathia example, 310–250 BC) and to the cup bases: Surra and Preacco Ancona 1997, 75, Pl. 34, No. 66 (Pomarico Vecchio, end of 4th–3rd c. BC).
Attested Dates: 325–300 BC.

1:2

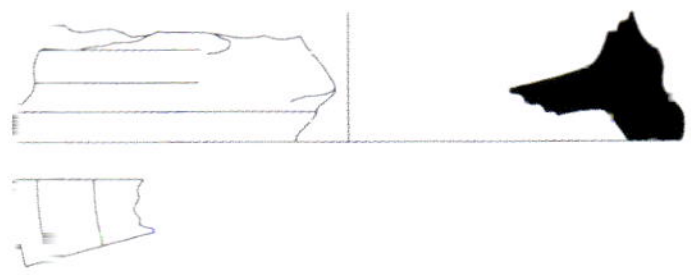

230. BG 638-08 Open form ring base (single example)
Dimensions: base diam. 8.9 h 1.8
Reference Sherd Description: Consistent, semi-lustrous black slip, reserved underside, traces of irregular dipped slip on lower exterior of foot ring.
Selected Comparanda: Bears some similarities to Morel 1981a, Pl. 113, 3721a1 (overpainted kantharos from Aléria, T 53, 320–280 BC).
Attested Dates: Hellenistic?

Open forms cont.

1:2

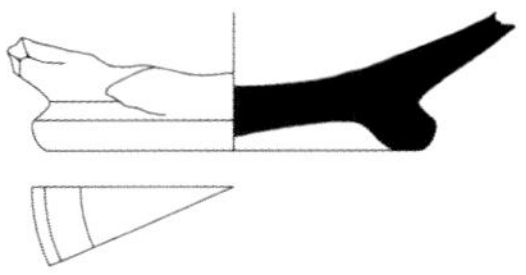

231. BG 638-07 Open form ring base (single example)
Dimensions: base diam. 5.35 h 1.9
Reference Sherd Description: Consistent, semi-lustrous black slip, unslipped exterior and underside, slipped interior.
Selected Comparanda: Bearing some similarities to: Pianu 1990, 114, Pl. LI, No. 111.2 (pyxis from Herakleia, T 111, end of 4th–early 3rd c. BC); Fabbricotti 1979, 357, No. 90 (base of Hellenistic cup or patera from Ruoti); Bruscella and Virtuoso 2005, Pl. XLIV, No. 313 (cup from Torre di Satriano, end of 4th–second half of 3rd c. BC); Surra and Preacco Ancona 1997, 82, Pl. 39, No. 113 (Pomarico Vecchio, open form base, last decades of 4th c. BC); Lissi Caronna 1990–91, 206, Fig. 26, No. 18 (Oppido Lucano, house C, context dated to second half of 4th–early 3rd c. BC); Preacco 1989, 199, Pl. XXVIII, No. 218 (Locri, dish or cup base, 3rd c. BC).
Attested Dates: 325–200 BC

1:2

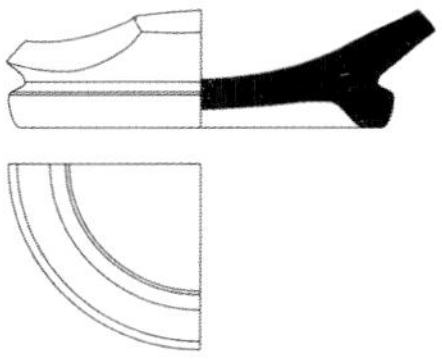

232. BG 348-02 Open form ring base (single example)
Dimensions: base diam. 5.1 h 1.6 w 5.7
Reference Sherd Description: Dilute, semi-lustrous black slip, reserved exterior and underside.
Attested Dates: Hellenistic?

1:2

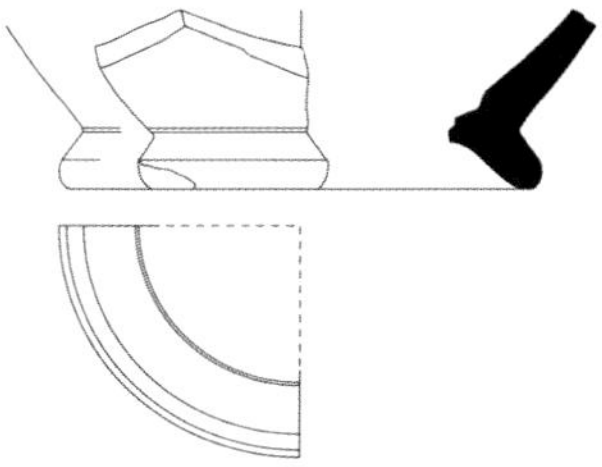

233. BG 016-28 Open form ring base (single example)
Dimensions: base diam. 6.4 h 2.45 w 3.1
Reference Sherd Description: Consistent, matte black slip, completely worn on the exterior, unslipped underside.
Selected Comparanda: Bears some similarities to the cup form Elliott 1998, 651, C8 (Metaponto, Pantanello necropolis, early 3rd c. BC); Delli Ponti 1996, 195, No. 242 (Vaste, 18/1968, ring-handled cup, end of 4th–early 3rd c. BC); Semeraro 1983, Pl. 89, No. 171 (Hellenistic skyphos from Otranto); Surra and Preacco Ancona 1997, 81–82, Pl. 39, No. 114 (Pomarico Vecchio, open form base, advanced 4th–3rd c. BC); Morel 1981a, series 3221, 257–58, Pl. 91 (Campana A ring-handled cup, first half of 2nd c. BC).
Attested Dates: 300–150 BC.

1:2

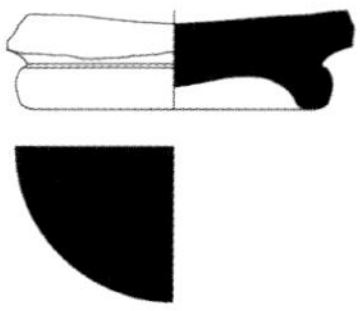

234. BG 506-62 Open form base (single example)
Dimensions: base diam. 4.2 h 1.4
Reference Sherd Description: Consistent, semi-lustrous black slip.
Selected Comparanda: Casagrande 2002, 136, No. 66, Fig. 38 at p. 389, "ciotola" type 2 (agora of Herakleia, 350–300 BC); Campagna 1995, 219, Fig. 5, No. 5 (Vaste, second half of 4th–early 3rd c. BC). Bears also some similarities to Giannotta 1996, 74, Fig. 36, No. 73 (Rocavecchia, T26/1934, cup Morel species 2780, 3rd–2nd c. BC).
Attested Dates: 350–300 BC.

1:2

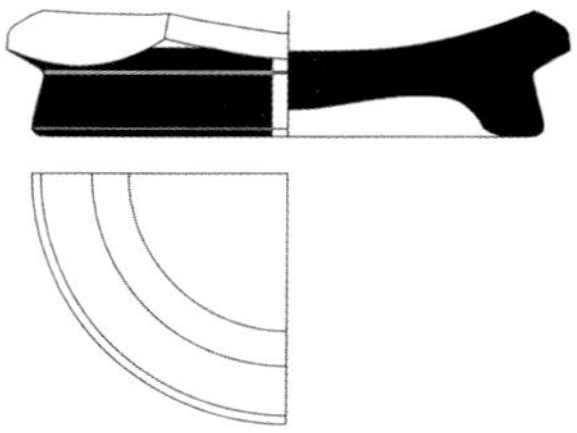

235. BG 318-01 Open form ring base (single example)
Dimensions: base diam. 6.8 h 1.7 w 7.2
Reference Sherd Description: Consistent, semi-lustrous black slip.
Selected Comparanda: Surra and Preacco Ancona 1997, 72, Pl. 33, No. 42 (Pomarico Vecchio, small bowl, around 340 or shortly later); Morel 1981a, 2714f1, 209, Pl. 67 (Segesta, small bowl, 310 ± 30).
Attested Dates: 350–300 BC.

Open forms cont.

1:2

236. BG 395-09 revisit Open form ring base (3 examples)
Dimensions: base diam. 5.3 h 2.8
Reference Sherd Description: Consistent, semi-lustrous black slip, unslipped exterior and underside.
Selected Comparanda: Bears some similarities to the cup base Morel 1981a, Pl. 85, 2987c1 (Volterra, 210 ± 20 BC ?).
Attested Dates: 300–200 BC

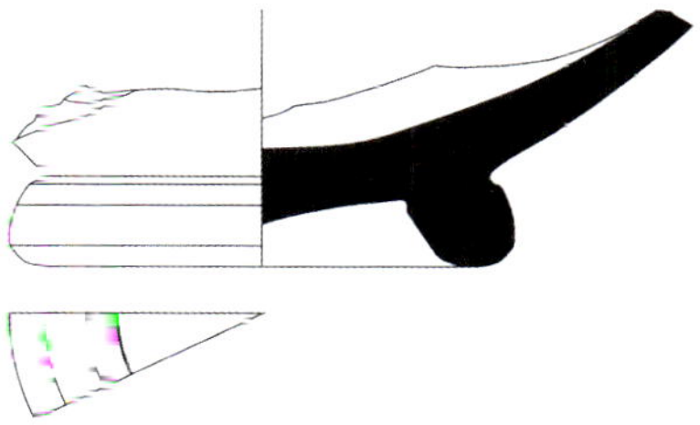

1:2

237. BG 454-01 Open form disk base (single example)
Dimensions: base diam. 4.8 h 1.9
Reference Sherd Description: Dilute, matte reddish-brown slip, unslipped lower exterior and underside.
Selected Comparanda: Bears some similarities to a stamnos base: Hoffmann 2002, Pl. 77, No. 413/1 (Taranto, 375–350 BC); a mug base (Morel 1981a, 5331a1, from Ordona, second half of 4th c. BC).
Attested Dates: 375–300 BC.

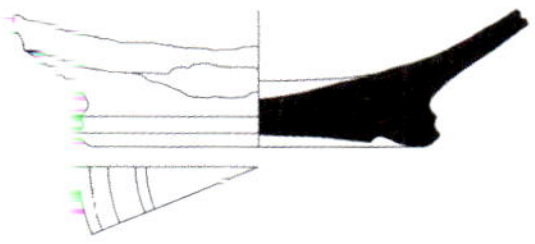

1:2

238. BG 633-03 Open form disk base (2 examples)
Dimensions: base diam. 4.4 h 1.9 w 6.1
Reference Sherd Description: Consistent, semi-lustrous black slip, unslipped exterior and underside.
Selected Comparanda: Bearing some similarities to a disk base in Prag 1992b, 90, No. 735, Fig. 31 (Botromagno, 3rd to 1st c. BC); *Gli ori*, 460, No. 26 (Taranto, 22.X.1934, via C. Battisti, 3rd c. BC); Carter 1983a, 477, Fig. 52, No. 115 (bowl from Pizzica Pantanello); Semeraro 1983, Pl. 91, No. 201 (cup from Otranto, first half of 2nd c. BC); Morel 1981a, 2767b1, 220, Pl. 71 (cup from Vado di Camaiore, 3rd c. BC); Hoffmann 2002, Pl. 77, No. 413/2 (stamnos from Taranto, 375–350 BC).
Attested Dates: 300–100 BC.

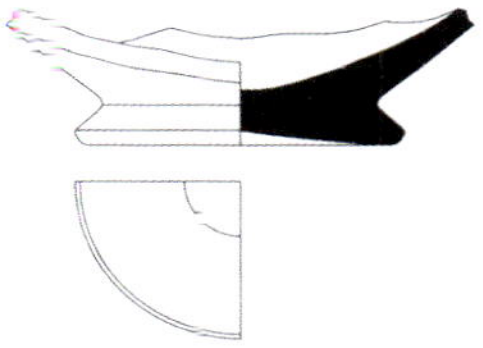

Mugs

1:2

239. BG 128-10 Mug rim (single example)
Dimensions: rim diam. 7.0 h 1.1 w 2.6
Reference Sherd Description: Consistent, semi-lustrous black slip.
Selected Comparanda: D'Andria 1980a, 408, No. 204, Fig. 55 (Metaponto, kerameikos, deposit 1).
Attested Dates: 400–350 BC.

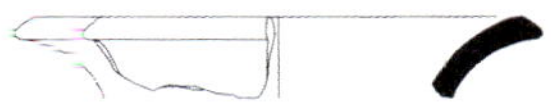

1:2

240. BG 164-01 Mug rim (11 examples)
Dimensions: rim diam. 4.5 h 1.7 w 3.5
Reference Sherd Description: Consistent, semi-lustrous black slip.
Selected Comparanda: Elliott 1998, 667, M5 (Metaponto, Pantanello necropolis, 410–380 BC); A. Bottini 1993b, 185–86, No. 14 (Metaponto, proprietà D'Onofrio, urban necropolis, T 15, last quarter of 4th c. BC); D'Andria 1980a, 408, Fig. 55, Nos. 206–7 (Metaponto, kerameikos, 4th c. BC); Riccardi 2008, 73, No. 67 (Bitonto, T 3/2002, second half of 4th c. BC); *Museo Taranto* II.2, 479, Pl. 34.g, 2–g (Rutigliano, second half 4th c. BC); Morel 1981a, 5335b2, 354, Pl. 164 (Botromagno, TS17, late 4th c. BC).
Attested Dates: 350–300 BC.

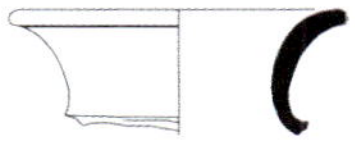

Oinochoai

1:2

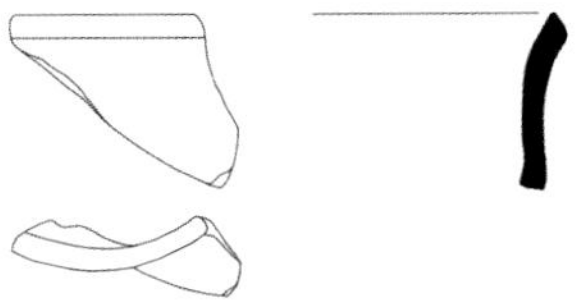

241. BG 397-45 Oinochoe rim (single example)
Dimensions: w 3.1 h 2.4
Reference Sherd Description: Consistent, semi-lustrous black slip.
Selected Comparanda: Bitti 1989, Pl. XXVI, No. 175 (Locri, 4th c. BC); Morel 1981a, 5623c1, 377, Pl. 181 (Ugento, around 300 BC).
Attested Dates: 400–300 BC.

1:2

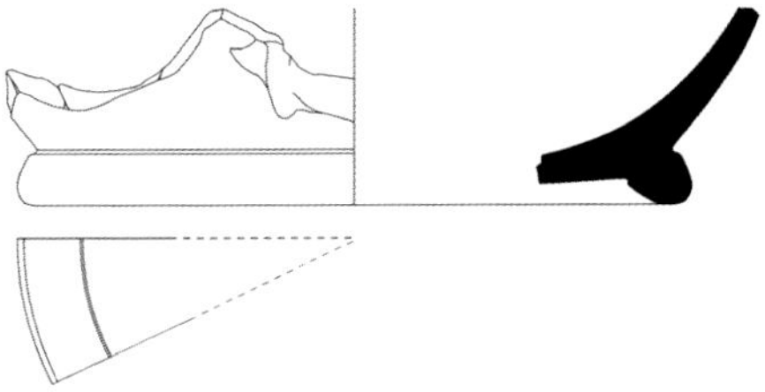

242. BG 488-08 Oinochoe base (single example)
Dimensions: base diam. 10.0 h 3.0 w 9.5
Reference Sherd Description: Consistent, semi-lustrous black slip, unslipped underside and interior.
Selected Comparanda: Pianu 1990, Pl. 50, No. 101.1 (Herakleia, T 101, 330–300 BC); Hoffmann 2002, 34, Pl. 53, type 113/6 (Taranto, 375–300 BC).
Attested Dates: 400–300 BC.

Amphorae

1:2

243. BG 529-01 revisit Amphora rim (single example)
Dimensions: rim diam. 11.5 h 1.0 w 3.15
Reference Sherd Description: Consistent, semi-lustrous black slip.
Selected Comparanda: See amphora from site 737, T 22-01. Osanna 2008, Pl. XX (Metaponto, proprietà Mutinati, T 597, in association with a lebes gamikos by the Pisticci Painter). For similar type of rim on a hydria see Elliott 1998, 655, H2 (Metaponto, Pantanello necropolis, 450–430 BC).
Attested Dates: 480–430 BC

1:2

244. BG 488-91 Amphora rim (single example)
Dimensions: w 2.6 h 2.3
Reference Sherd Description: Consistent, semi-lustrous black slip.
Selected Comparanda: Elliott 1998, 649, A4 (Metaponto, Pantanello necropolis, 460–440 BC). See also Sant'Angelo Vecchio, T 2: *Necropoleis* 444 and Saldone, T 8, *Necropoleis* 433 (460–440 BC), both of larger dimensions.
Attested Dates: 460–440 BC.

1:2

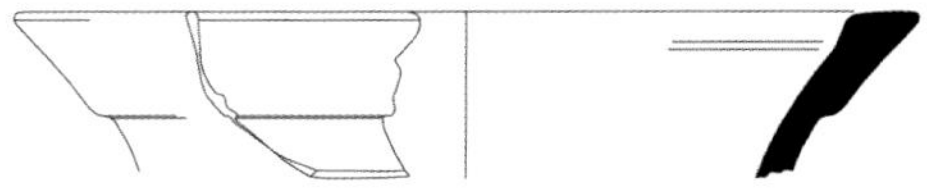

245. BG 266-04 Amphora rim (2 examples)
Dimensions: rim diam. 12.0 h 2.3 w 3.1
Reference Sherd Description: Consistent, semi-lustrous black slip.
Selected Comparanda: Bears similarities to Elliot 1998, 648, A1 (Metaponto, Pantanello necropolis, T27-2, 2nd or 3rd c. BC).
Attested Dates: 450–400 BC.

1:2

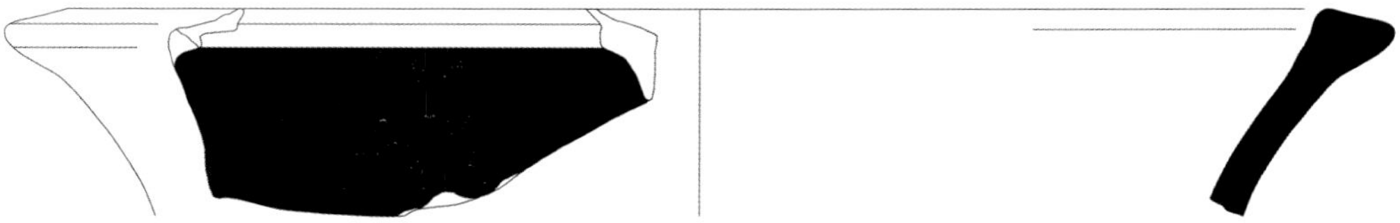

246. BG 497-22 Amphora rim (7 examples)
Dimensions: rim diam. 26.5 h 4.0 w 9.3
Reference Sherd Description: Consistent, semi-lustrous black slip, reserved upper face of rim.
Selected Comparanda: See Ch. 19, site 736, trench 6, Tr6-01 and Ch. 11, No. 9.
Attested Dates: 350–300 BC.

Pelikai

1:2

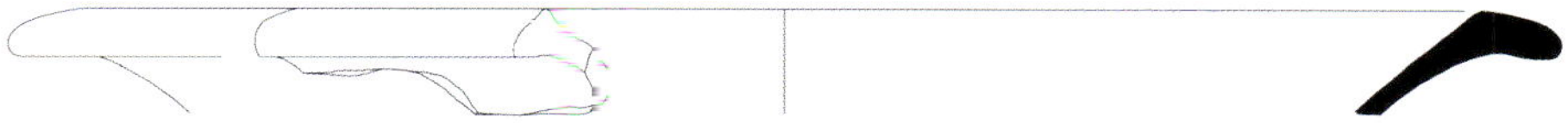

247. BG 258-10 Pelike rim (3 examples)
Dimensions: rim diam. 23.9 h 1.65 w 5.45
Reference Sherd Description: Consistent, semi-lustrous black slip.
Selected Comparanda: Elliott 1998, 674, P15, 673 (Metaponto, Pantanello necropolis, 400–360 BC); *Museo Taranto* I.3, 358, No. 125.36–37 (last quarter of 5th–first half of 4th c. BC).
Attested Dates: 450–350 BC.

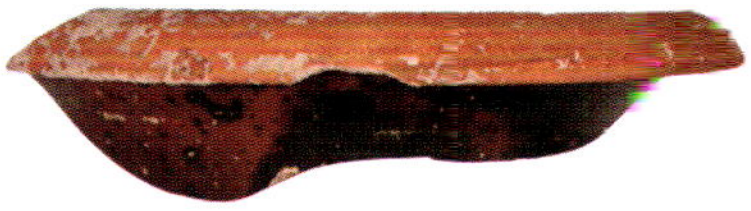

1:2

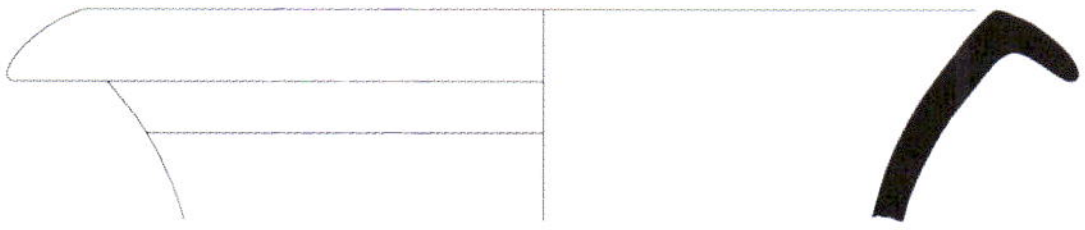

248. BG 272-07 Pelike rim (single example)
Dimensions: rim diam. 14.2 h 2.9
Reference Sherd Description: Dilute, semi-lustrous reddish-brown slip.
Selected Comparanda: Elliott 1998, 674, P32 (Metaponto, Pantanello necropolis, 400–360 BC); *Necropoleis*, 437, SS 27-1 (Metaponto, Saldone, T 27, last quarter of 5th–first quarter of 4th c. BC).
Attested Dates: 400–350 BC.

1:2

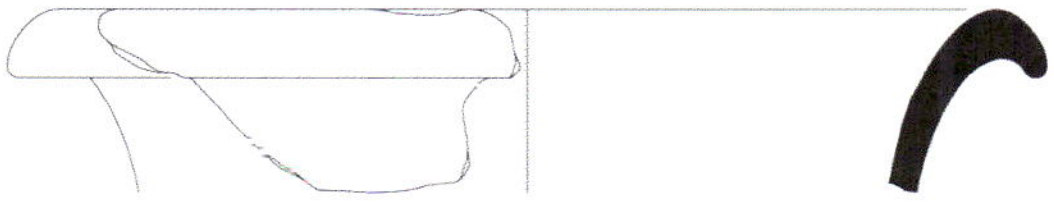

249. BG 395-18 Pelike rim (2 examples)
Dimensions: rim diam. 13.8 h 2.55 w 5.6
Reference Sherd Description: Consistent, semi-lustrous black slip.
Selected Comparanda: Bearing similarities with Elliott 1998, 673, P24 (Metaponto, Pantanello necropolis, 440–400 BC).
Attested Dates: 400–380 BC.

1:2

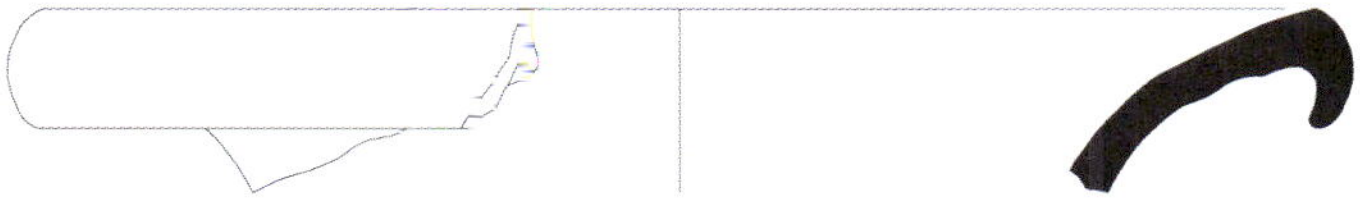

250. BG 401-19 Pelike rim (4 examples)
Dimensions: rim diam. 17.9 h 2.5
Reference Sherd Description: Consistent, semi-lustrous black slip.
Selected Comparanda: Elliott 1998, 674, P37 (Metaponto, Pantanello necropolis, 420–370 BC); this type of rim is also common in red-figure pelikai of the first half of the 4th c. BC; see, as an example, *CVA Philadelphia*, Pl. 5, Fig. 1.5 (Hoppin Painter, end of first quarter of 4th c. BC); *CVA Sydney, Nicholson Museum*, Fig. 3.2 (Monash Group) and Fig. 3.3 (Thyrsos Painter). See also Hoffmann 2002, 38, type 311/2, Pl. 61 (first quarter of 4th c. BC) and type 311/8, Pl. 63 (Iris Painter, second quarter of 4th c. BC).
Attested Dates: 400–350 BC.

1:2

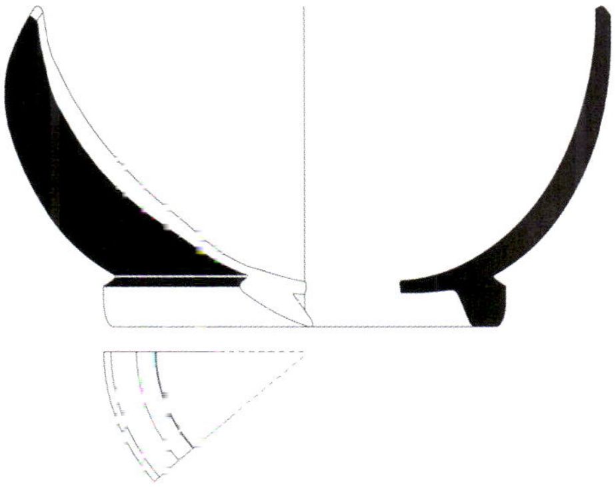

251. BG 763-02 Pelike base (single example)
Dimensions: base diam. 8.0 h 6.5 w 6.0
Reference Sherd Description: Consistent, semi-lustrous black slip, reserved underside and exterior of foot ring, unslipped interior.
Selected Comparanda: Elliott 1998, 672, P10 (Metaponto, Pantanello necropolis, 370–350 BC); this type of foot is also common in red-figure pelikai: *CVA Sydney, Nicholson Museum*, Fig. 3.2 (Monash Group).
Attested Dates: 400–350 BC.

Pelikai cont.

1:2

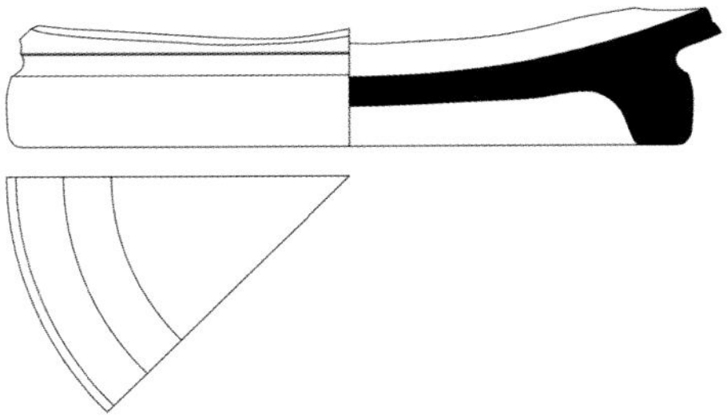

252. BG 375-01 Pelike base (single example)
Dimensions: base diam. 11.4 h 2.4
Reference Sherd Description: Unslipped interior, reserved exterior and underside.
Selected Comparanda: Elliott 1998, 672, P10 (Metaponto, Pantanello necropolis, 370–350 BC).
Attested Dates: 370–350 BC.

1:2

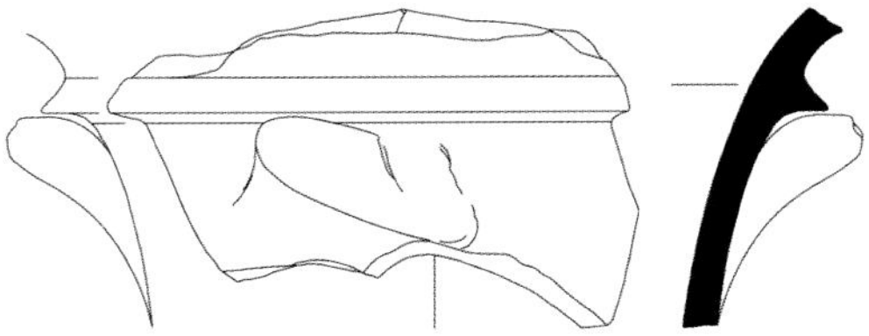

253. BG 250-06 Pelike neck with handle terminus (3 examples)
Dimensions: d 11.9 h 5.0 w 8.0
Reference Sherd Description: Consistent, semi-lustrous black slip.
Selected Comparanda: Elliott 1998, 672, P9 (Pantanello necropolis, 350–330 BC); Preacco Ancona 1997, Pl. 109, No. 62 (Gnathia, second half of 4th c. BC).
Attested Dates: 350–320 BC.

1:2

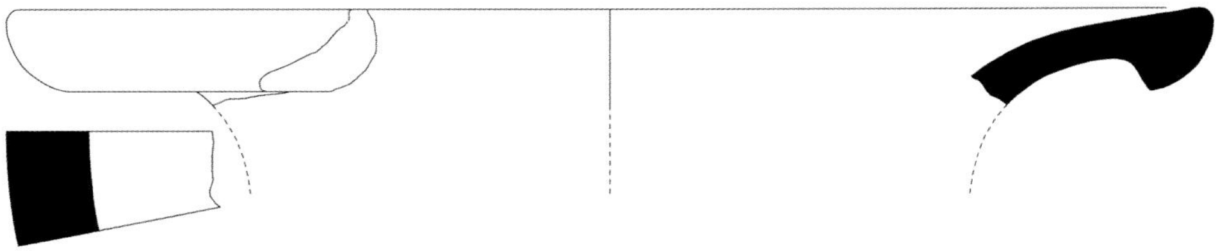

254. BG 471-13 Pelike rim (2 examples)
Dimensions: rim diam. 25.0 w 2. 1
Reference Sherd Description: Consistent, semi-lustrous black slip, reserved underside of rim.
Selected Comparanda: Pianu 1990, 127, Pl. IV, No. 63.1 (Herakleia, T 63, 350–320 BC)
Attested Dates: 330–300 BC.

1:2

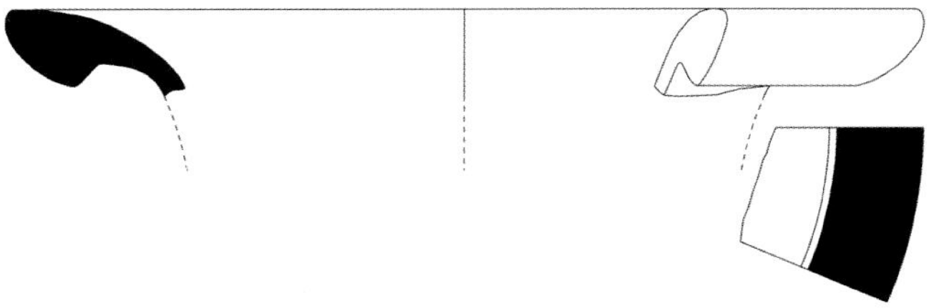

255. BG 497-21 Pelike rim (2 examples)
Dimensions: rim diam. 17.1 h 1.7
Reference Sherd Description: Consistent, semi-lustrous black slip, reserved underside of rim.
Selected Comparanda: Sestieri 1940, 111, Fig. 50 (Metaponto, 330–300 BC); Pianu 1990, 84, Pl. IV, No. 71.1 (Herakleia, T 71, overpainted example, 340–300 BC).
Attested Dates: 340–300 BC.

1:2

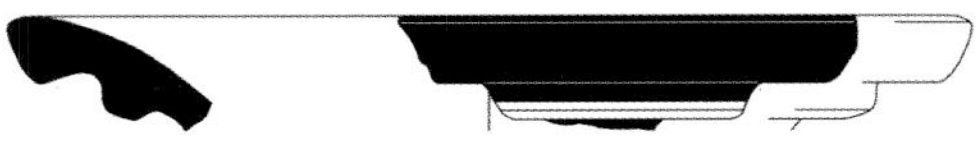

256. BG 415-01 Pelike rim (single example)
Dimensions: rim diam. 12.8 h 1.6 w 6.2
Reference Sherd Description: Consistent, semi-lustrous black slip, reserved band on underside of rim.
Selected Comparanda: Pianu 1990, 83, Pl. XXXVII, No. 69.2 (Herakleia, T 69, 340–300 BC).
Attested Dates: 340–300 BC.

Hydriai

1:2

257. BG 266-03 Hydria rim (2 examples)
Dimensions: rim diam. 5.0 h 1.5 w 2.7
Reference Sherd Description: Consistent, semi-lustrous black slip, irregularly slipped on upper interior.

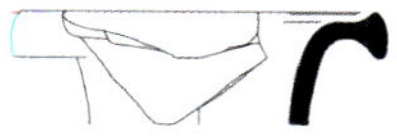

Attested Dates: 450–350 BC.

Lebetes Gamikoi

1:2

258. BG 415-10 Lebes gamikos rim (single example)
Dimensions: rim diam. 3.5 h 0.5
Reference Sherd Description: Consistent, semi-lustrous black slip, unslipped interior beyond rim.
Selected Comparanda: Elliott 1998, 662, LG30 (Metaponto, Pantanello necropolis, T 209-3, second half of 5th c. BC) and 663, LG45 (T 330-9).

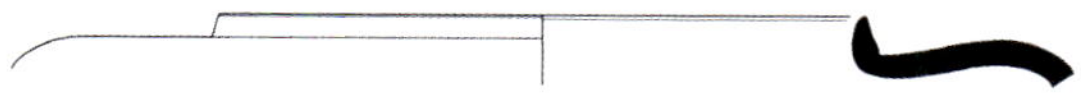

Attested Dates: 430–380 BC.

259. BG 320-28 Lebes gamikos rim, shoulder, and handle terminus (4 examples)
Dimensions: rim diam. 5.6 h 2.7 w 2.8
Reference Sherd Description: Consistent, semi-lustrous black slip.
Attested Dates: Classical period

1:2

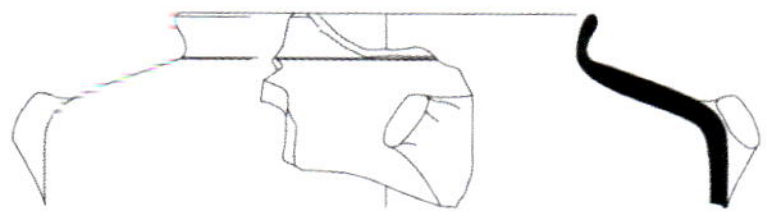

260. BG 353-07 Lebes gamikos rim and shoulder (2 examples)
Dimensions: rim diam. 7.6 h 2.7 w 6.4
Reference Sherd Description: Dilute, matte reddish brown slip, unslipped interior below the upper rim.
Attested Dates: 350–300 BC.

1:2

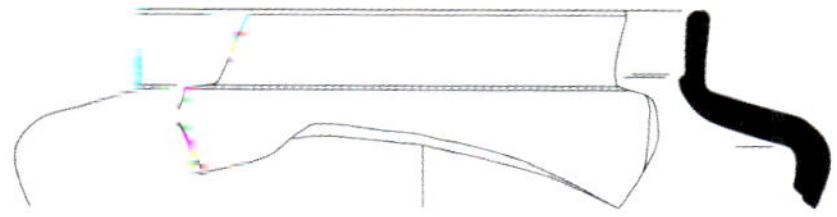

261. BG 742-07 Lebes gamikos rim (single example)
Dimensions: rim diam. 5.2 h 1.1 w 4.5
Reference Sherd Description: Consistent, semi-lustrous black slip, unslipped lower interior of rim.
Selected Comparanda: Elliott 1998, 660, LG1 (Metaponto, Pantanello necropolis, T 7-3, 360–330 BC); Sant'Angelo Grieco SG204 – SG243 (330–300 BC); Carando 1997, 294, Pl. 109, No. 65 (Pomarico Vecchio, T 14, third quarter of 4th c. BC).

1:2

Attested Dates: 350–300 BC.

262. BG 401-17 Lebes gamikos body sherd (single example)
Dimensions: w 8.4 h 4.4
Reference Sherd Description: Consistent, semi-lustrous black slip, unslipped extant lower body below ribbing, slipped interior.
Selected Comparanda: Lo Porto 1966, 214, Fig. 69 Nos. 1–2 (Metaponto, Gnathia examples from the urban necropolis); Lo Porto 1973, Pl. XLIV, No. 3.1 (Montescaglioso, Gnathia example, first quarter of 3rd c. BC); Pianu 1990, 99–100, No. 1, Pl. XLV,4 (Herakleia, T 94, Gnathia example, 300–250 BC). Some similarities to an oinochoe in Elia and Meirano 1997, Pl. 41, No. 129 (Pomarico Vecchio, 3rd c. BC).

1:2

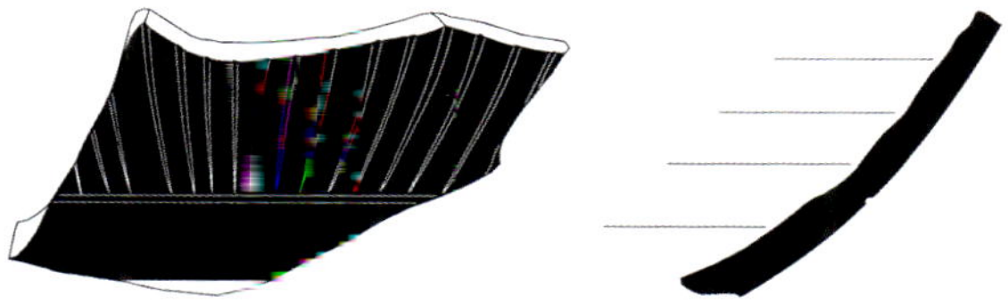

Attested Dates: 340–250 BC.

Gutti

1:2

263. BG 128-294 Guttus rim (single example)
Dimensions: rim diam. 3.6 h 2.3
Reference Sherd Description: Consistent, matte black slip, reserved interior below lower rim.
Selected Comparanda: Morel 1981a, series 8162, 24, Pl. 210 (4th c. BC); J.W. Hayes 1984, 193, No. 75 (400–370 BC).

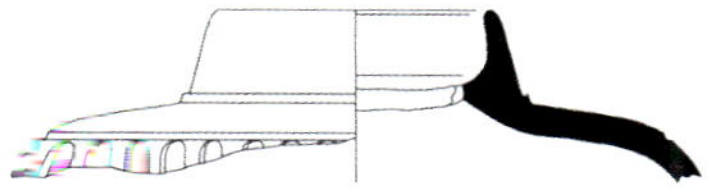

Attested Dates: 400–300 BC.

Gutti cont.

1:1

264. BG 528-06 Guttus fragment with molded decoration; bearded face (Herakles?) (single example)
Dimensions: w 3.0 l 4.8 th 1.0
Reference Sherd Description: Relief medallion depicting a bearded face (Herakles? Silenos?).
Selected Comparanda: Morel 1970a, Fig. 27 (Cozzo Presepe, sacred deposit, late 4th–early 3rd c. BC); Lo Porto 1991, Pl. XVII.1.1 (Timmari, Camposanto, T 12/1911, around 340 BC); Labellarte 1995, 137, No. 2.1.59 (Chini collection, 350–300 BC); *CVA Gotha* 2, Pl. 91, Nos. 6–7 (said to be from Basilicata, between 4th and 3rd c. BC).
Attested Dates: 350–280 BC.

Squat Lekythoi

1:2

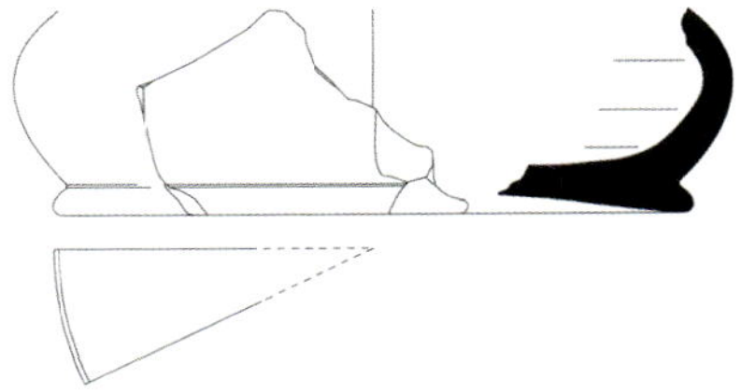

265. BG 288-12 Lekythos base (10 examples)
Dimensions: base diam. 9.1 h 3.0 w 4.7
Reference Sherd Description: Consistent, semi-lustrous black slip, unslipped interior and underside.
Selected Comparanda: Nava 2002, 742, Pl. LXIX (Pantanello, proprietà Sellitti, in association with a pelike by the Creusa Painter, first half of 4th c. BC); *CVA Zürich* 1, Pl. 53,18 (Apulian black gloss, first half of 4th c. BC). Bears similarities to *Museo Taranto* I.3, 356, No. 125.24 (Taranto, tomb from via Dante, last quarter of 5th c. BC).
Attested Dates: 440–400 BC.

1:2

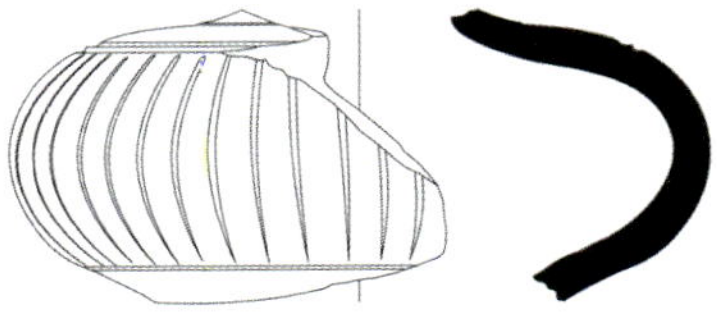

266. BG 341-29 Squat lekythos body with vertical ribbing (2 examples)
Dimensions: d 9.3 h 4.0 w 9.0
Reference Sherd Description: Consistent, semi-lustrous black slip, unslipped interior.
Selected Comparanda: Elliott 1998, 684, SL20 (Metaponto, Pantanello necropolis, 440–390 BC); Lo Porto 1981b, 372, No. 1, Fig. 94.1 (Metaponto, contrada Ricotta, T 27); *Museo Taranto* II.2, 485, Pl. 36.e, 5.3a (Rutigliano, late 5th to early 4th c. BC); Hoffmann 2002, 34, type 111/14, Pl. 34 (Taranto, 350–325 BC).
Attested Dates: 440–300 BC.

1:1

1:2

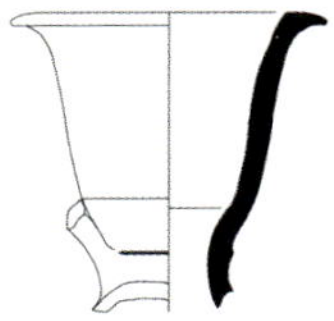

267. BG 454-03 Squat lekythos rim, upper neck and handle terminus (single example)
Dimensions: rim diam. 4.2 h 4.0
Reference Sherd Description: Consistent, matte black slip, unslipped interior from neck.
Selected Comparanda: Elliott 1998, 668, NL2 (Metaponto, Pantanello necropolis, net-decorated lekythos, 350–325 BC); Hoffmann 2002, Pl. 47, type 111/122 (Taranto, necropolis, 350–325 BC).
Attested Dates: 350–300 BC.

Squat lekythoi cont.

1:2

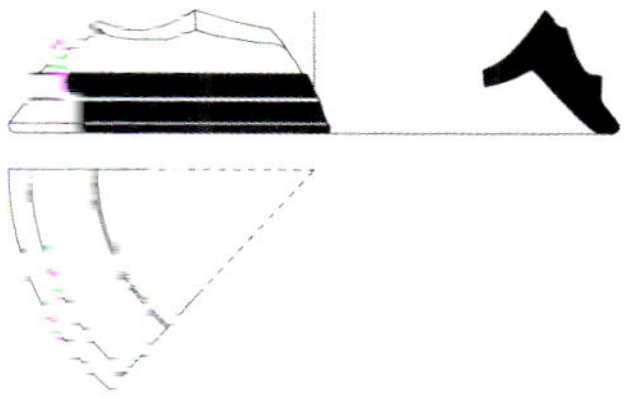

268. BG 343-09 Squat lekythos base (2 examples)
Dimensions: base diam. 8.1 h 1.7 w 3.7
Reference Sherd Description: Consistent, semi-lustrous black slip, band of slip on lower exterior of foot ring, reserved underside and unslipped interior.
Selected Comparanda: Elliott 1998, 668, NL4 (Metaponto, Pantanello necropolis, net-decorated lekythos, 350–300 BC); Preacco Ancona 1997, Pl. 108, No. 63 (Gnathia lekythos, 340–320 BC); Hoffmann 2002, Pl. 45, type 111/106 (Taranto, necropolis, 350–300 BC).
Attested Dates: 350–300 BC.

1:2

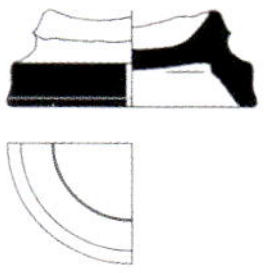

269. BG 258-09 Squat lekythos base (3 examples)
Dimensions: base diam. 3.3 h 1.4
Reference Sherd Description: Consistent, semi-lustrous black slip, reserved upper pedestal and underside, unslipped interior.
Selected Comparanda: Elia and Meirano 1997, Pl. 40, No. 118 (Pomarico Vecchio, 340–320 BC). Bears some similarities to Pianu 1990, 51, Pl. XXIV, No. 32.2 (Herakleia, T 32, net-decorated lekythos, 340–320 BC) and 86, Pl. XXXVII, No. 75.1 (T 75, 340–320 BC).
Attested Dates: 340–320 BC.

1:2

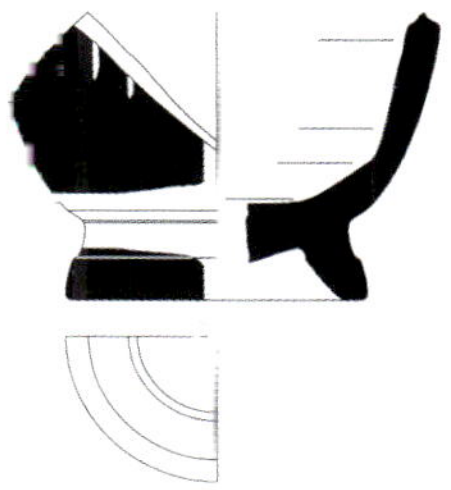

270. BG 103-03 revisit Squat lekythos base (single example)
Dimensions: base diam. 4.0 h 2.1
Reference Sherd Description: Consistent, semi-lustrous black slip, unslipped underside and interior.
Selected Comparanda: Pianu 1990, 63–64, Pl. XXXV, No. 48.1 (Herakleia, 340–320 BC); Morel 1981a, 5416g1, 361, Pl. 168 (Capua or surroundings, around second half of 4th c. BC).
Attested Dates: 340–300 BC.

Net-Decorated Lekythoi

1:1

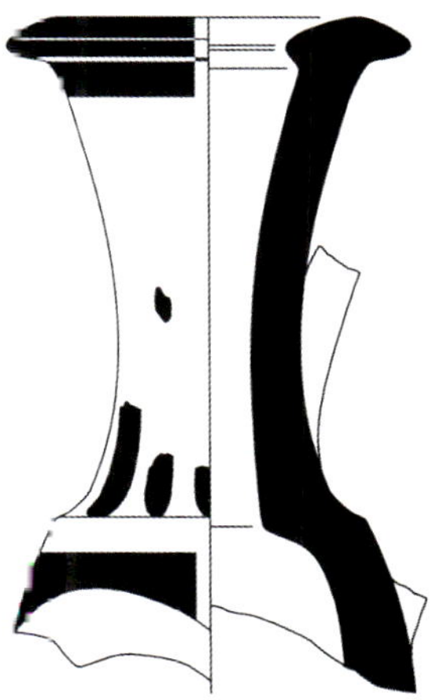

271. BG Plot 99-91-01 Net-decorated lekythos, rim, neck and upper body (single example)
Dimensions: rim diam. 3.4 h 5.8
Reference Sherd Description: Consistent, semi-lustrous black slip with reserved neck to shoulder with traces of slipped decoration, unslipped interior.
Selected Comparanda: Elliott 1998, 668, NL5 (Metaponto, Pantanello necropolis, 350–325 BC); Lo Porto 1966, 206, No. 1, Fig. 60.1 (Metaponto, urban necropolis, T 20, late 4th c. BC); Lo Porto 1988–1989, 380, No. 7, Fig. 89.2 (Montescaglioso, T 14, 320 BC); *Pomarico Vecchio I*, Pl. 49, No. 1 (340–320 BC); Pianu 1990, 70, Pl. XXIX,4 (Herakleia, T 54, 340–310 BC) and 63–4, Pl. XXXV, No. 57.2 (T 57, 340–310 BC); *Gli ori*, 415, No. 16 (Taranto, T 23 via Umbria, 22.II.1958, second half of 4th c. BC); Dell Ponti 1996, 112, No. 16 (Vaste, T 5/1968, second half of 4th c. BC).
Attested Dates: 350–300 BC.

Net-decorated lekythoi cont.

1:2

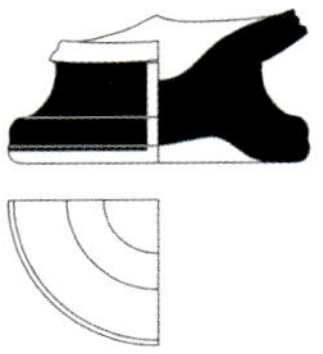

272. BG 319-01 Net-decorated lekythos base (2 examples)
Dimensions: base diam. 4.0 h 4.0 w 4.8
Reference Sherd Description: Consistent, semi-lustrous black slip, irregularly slipped lower exterior, unslipped underside and interior.
Selected Comparanda: Pianu 1990, 63, Pl. II, No. 46.1 (Herakleia, T 46, 350–325 BC); S. Bianco 1994, 123, Fig. 10, No. r.1 (Sant'Arcangelo-San Brancato, T95, end of third quarter of 4th c. BC); Hurschmann 1995, 681, Fig. 15 (type IVa, 350–300 BC).
Attested Dates: 350–325 BC.

Pyxides

1:2

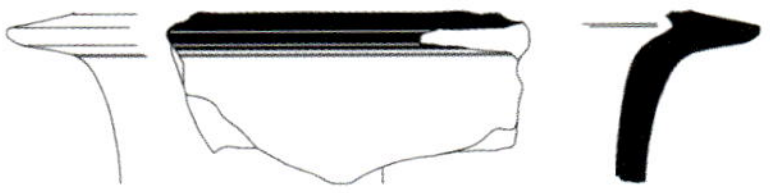

273. BG 326-28 Pyxis (or perhaps epichysis?) body sherd. (2 examples)
Dimensions: base diam. 10.0 h 2.4 w 4.8
Reference Sherd Description: Consistent, semi-lustrous black slip, reserved body below rim.
Selected Comparanda: Elliott 1998, 676, PY5 (Metaponto, Pantanello necropolis, 360–340 BC); Lo Porto 1981b, 372, No. 2, Fig. 94.2 (Metaponto, contrada Ricotta, late 4th to first quarter of 3rd c. BC); Corrente and Labellarte 1992, 474, No. 36 (Canosa, Vico San Martino, T 2, Cella C, late 4th to first quarter of 3rd c. BC).
Attested Dates: 360–300 BC.

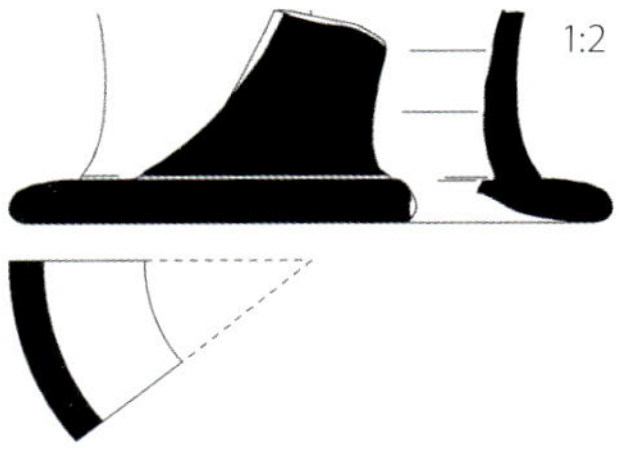

1:2

274. BG 043-27 Pyxis base (single example)
Dimensions: base diam. 8.0 h 2.9 w 5.4
Reference Sherd Description: Consistent, semi-lustrous black slip, reserved underside and unslipped interior.
Selected Comparanda: Elliott 1998, 676, PY12 (Metaponto, Pantanello necropolis, 400–380 BC).
Attested Dates: 440–340 BC.

Closed Forms

1:2

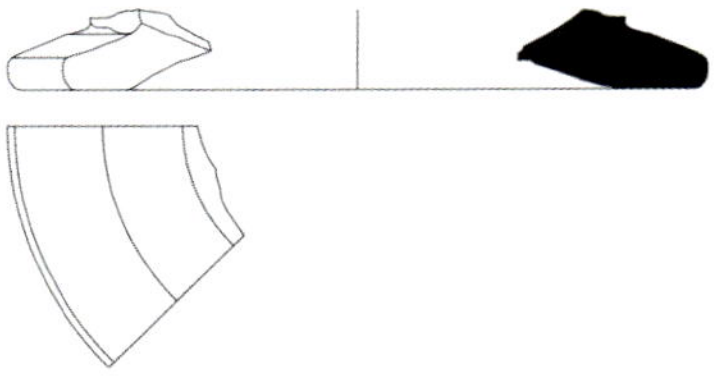

275. BG 397-52 Closed form base (single example)
Dimensions: base diam. 9.3 h 1.1
Reference Sherd Description: Dilute, matte black slip, reserved underside.
Selected Comparanda: Bears similarities to feet associated with oinochoai (Lo Porto 1988–1989, Figs. 32,8 and 34,1 (Montescaglioso, T 1, second deposition, end of 6th–beginning of 5th c. BC), pelikai (see, in this volume, Ch. 19, site 737, T 23-01).
Attested Dates: 500–400 BC.

1:2

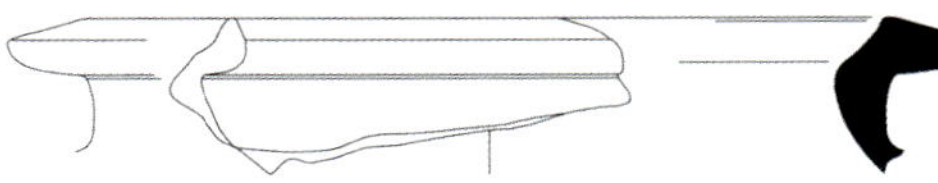

276. BG 105-03 revisit Closed form rim (single example)
Dimensions: rim diam. 12.8 h 2.1 w 6.1
Reference Sherd Description: Consistent, semi-lustrous black slip.
Selected Comparanda: Bears some similarities to Morel 1981a, 5112a1, 336, Pl. 150 (Volterra, olpe, 4th to 3rd c. BC).
Attested Dates: 400–200 BC.

1:2

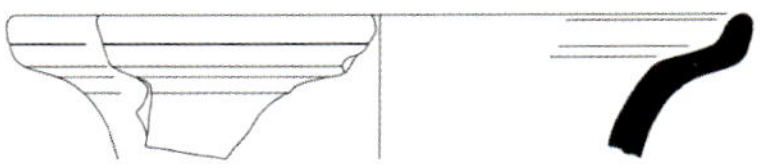

277. BG 128-11 Closed form rim (2 examples)
Dimensions: rim diam. 9.9 h 1.95 w 3.7
Reference Sherd Description: Consistent, semi-lustrous black slip.
Selected Comparanda: Bears some similarities to a type of olpe attested in Daunia: Morel 1981a, 5211, 339, Pl. 155 (examples from Ordona of second half of 4th c. BC).
Attested Dates: 350–250 BC.

Closed forms cont.

1:2

278. BG 070-09 Closed form rim (single example)
Dimensions: rim diam. 9.7 h 2.1 w 3.85
Reference Sherd Description: Consistent, semi-lustrous black slip.
Attested Dates: Classical-Hellenistic period?

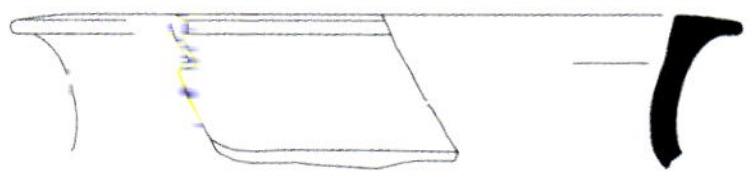

1:2

279. BG 397-43 Closed form base (single example)
Dimensions: base diam. 7.8 h 2.35
Reference Sherd Description: Consistent, semi-lustrous black slip, unslipped underside and interior.
Selected Comparanda: Bears some light similarities to a red-figure oinochoe base from Ortelle (Giannotta 1994, 99, Fig. 11, No. 11, 325–300 BC).
Attested Dates: Hellenistic?

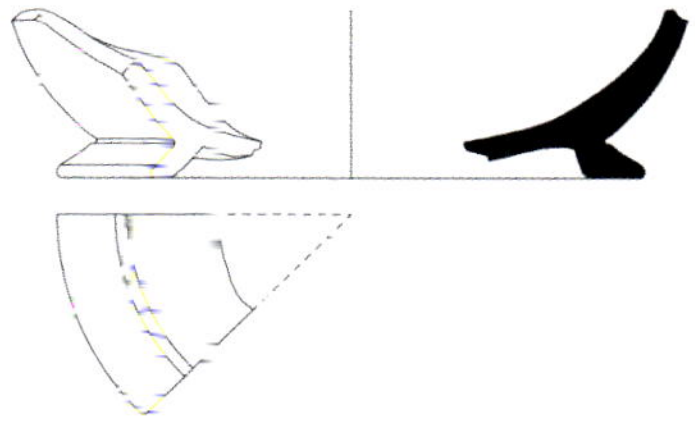

1:2

280. BG 487-03 Closed form base (single example)
Dimensions: base diam. 4.8 h 2.2 w 4.8
Reference Sherd Description: Consistent, semi-lustrous black slip, reserved band on lower body, reserved interior of foot ring, unslipped interior.
Selected Comparanda: Bears some similarity to a lekythos base: *Greci, Enotri e Lucani*, 267, No 3.41.4 (Sant'Arcangelo-San Brancato, T 371, 350–300 BC).
Attested Dates: 350–300 BC.

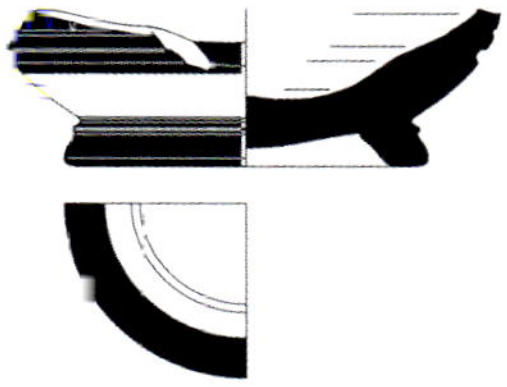

1:2

281. BG 521-125 Closed form base (4 examples)
Dimensions: base diam. 5.5 h 1.7
Reference Sherd Description: Consistent, semi-lustrous black slip, reserved band on upper base, irregular slip on underside, unslipped interior.
Selected Comparanda: Bears some similarities to: Elia and Meirano 1997, 84–85, Pl. 40, No. 123 (Pomarico Vecchio, bottle, with shorter stem, first quarter of 3rd c. BC); Pianu 1990, 86, Pl. XXXVII, No. 75.1 (Herakleia, lekythos, 340–320 BC); Cassano 1992a, 264, No. 8 (Canosa, Ipogeo Varrese, red-figure lekythos, 325–300 BC); J.W. Hayes 1984, 192, No. 245 (net-decorated lekythos, late 4th c. BC).
Attested Dates: 340–300 BC.

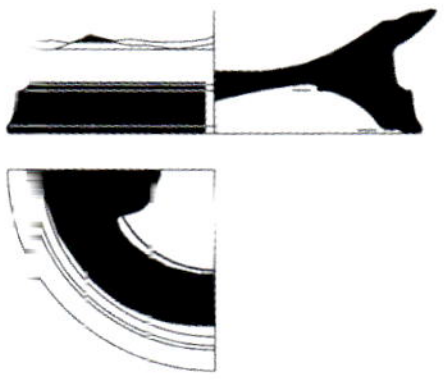

1:2

282. BG 258-08 Closed form base (2 examples)
Dimensions: base diam. 4.7 h 2.0
Reference Sherd Description: Consistent, semi-lustrous black slip, band of slip on lower exterior of foot ring, unslipped interior.
Selected Comparanda: Bears some similarities to: Elia and Meirano 1997, Pl. 44, No. 170 (Pomarico Vecchio, kyathos or juglet, Morel genus 5300, 300–250 BC); Pianu 1990, Pl. XXVI, No. 3 (Herakleia, net-decorated lekythos, 350–325 BC); Delli Ponti 1996, 122, No. 43 (Vaste, T 6/1968, Gnathia bottle, end of 4th c. BC); J.W. Hayes 1984, 192, No. 245 (net-decorated lekythos, late 4th c. BC).
Attested Dates: 350–300 BC.

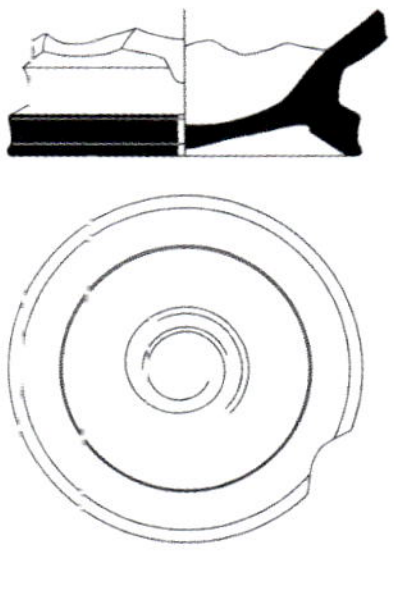

Closed forms cont.

1:2

283. BG 352-15 Closed form base (pelike?) (single example)
Dimensions: base diam. 8.0 h 1.7 w 6.9
Reference Sherd Description: Consistent, semi-lustrous black slip, band of slip on lower exterior base.
Selected Comparanda: The fragment shows resemblances to bases associated with pelikai. See *Necropoleis*, 435, SS19-1 (Metaponto, Saldone T 19, second half of 5th c. BC); Patera 1986, 56, No. 43 (Taranto, tomb of via Dante, 5th c. BC).
Attested Dates: Classical period.

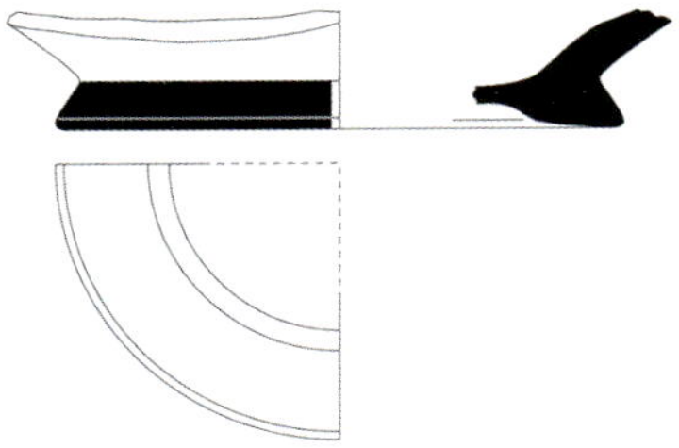

1:2

284. BG 465-14 Closed form base and lower body (single example)
Dimensions: base diam. 4.7 h 3.5
Reference Sherd Description: Consistent, semi-lustrous black slip, reserved band on lower body, reserved underside and slipped interior.
Selected Comparanda: Bears some similarities to Elia and Meirano 1997, 87–88, Pl. 41, No. 133 (Pomarico Vecchio, jug, Morel species 5150, 4th c. BC); Lissi Caronna 1990–91, 252, Fig. 79, No. 20 (Oppido Lucano, house D, brown gloss lekythos? from a context dated to second half of 4th–early 3rd c. BC); De Cazanove 2008, 578, Pl. 289, No. 76 (Civita di Tricarico, oinochoe); Olivero Ferrero 1989, 104, Pl. XXI, No. 62 ("brocchetta" Morel genus 5200, from Locri, end of 5th c. BC); Morel 1981a, 2591a1, 189, Pl. 58 (Agrigento, pyxis, second half of 4th c. BC).
Attested Dates: 4th c. BC.

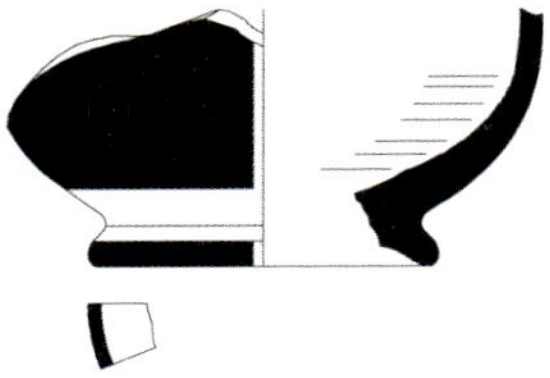

1:1

1:2

285. BG 128-20 Closed form disk base (single example)
Dimensions: base diam. 7.0 h 1.95 w 3.4
Reference Sherd Description: Consistent, semi-lustrous black slip, reserved lower exterior and outer edge of extant underside.
Selected Comparanda: Bearing some similarities to: Pianu 1990, 86, Pl. XXXVII, No. 75.1 (Herakleia, lekythos, 340–320 BC) and 109, Pl. L, No. 105.1 (bottle, 350–320 BC); *Greci* 1996, No. 3.40.14 (bowl); Lissi Caronna 1990–91, 203, Fig. 23, No. 10 (Oppido Lucano, base of pyxis?, context dated to second half of 4th–early 3rd c. BC); Myrto 1995, 269, Fig. 11, No. 3 (lekythoss from Dürres, context dated to 4th and 3rd c. BC).
Attested Dates: 350-320 BC.

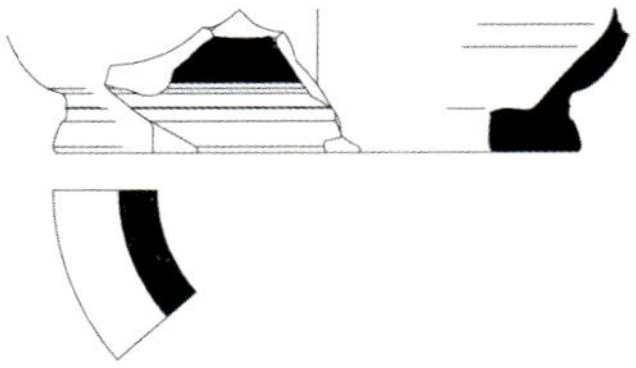

7
Grey Ware

Eloisa Vittoria

The ICA survey between Bradano and Basento recovered more than 200 fragments of grey ware.[1] Before turning to the materials themselves, the discussion will first take a general look at grey ware in southern Italy, specifically the issues of defining grey ware, its chronology, and its morphological relationship to other ceramics. Analysis of the survey grey ware follows, with a catalog.

Grey Ware in Southern Italy

Grey ware is limited chronologically to the 2nd and 1st century BC.[2] According to Lamboglia,[3] it coincides with the circulation period of *ceramica campana* C (Campana C), with which it was long identified.[4] By the 1970s, however, grey ware and Campana C were recognized as distinct wares.[5] Carter, for example, defined grey ware as a local variant of Campana C,[6] while Giardino emphasized the adoption of the term *pasta grigia*, which was used by Pelagatti to describe materials from Akrai,[7] thus superseding Lamboglia's definition of Campana C, since in truth, the only common elements in the two wares are the color of the clay, and occasionally, the gloss.[8] They are therefore two distinct ware classes: Campana C is but one of the many types of grey ware produced.[9] It is possible, however, to find points of contact between them as is the case between grey ware and other products.[10]

Today grey ware is studied independently as material evidence of the Late Republican period in the settlements of southern Italy and Sicily, and so much so that it is defined as a link to the black-gloss

[1] I would like to express my gratitude to Professor J.C. Carter for giving me a chance to study this material within the greater survey project, as well as to his research group, who were always open-minded and amenable to clarification and profitable discussion. I wish furthermore to thank M.L. Nava and A. De Siena (Soprintendenza per i Beni Archeologici per la Basilicata) for granting me access to the museum storerooms, and V. Cracolici for information and advice that helped me carry out the study. Finally, I would like to take this opportunity to remember my professor, M. Trojani, whose words of encouragement over the years helped me improve both my work and life.

[2] L. Giardino 1980a, 286; 1982, 168; Small 1977c, 124; F. Rossi 1989, 173; *Otranto* II, 21; Mazzei and Pietropaolo 1995, 249; Hempel 2001, 113. Several authors date the appearance of grey ware to the end of the 3rd c. BC: Prag (1992c, 134); Small (1992b, 27); Hempel (1996a, 337; 1996b, 343).

[3] Lamboglia 1952, 205. See also Small 1977c, 124; L. Giardino 1978b, 130; 1980a, 252; Stone 1982, 99; Morel 1981a, 47; Hempel 1996a, 337.

[4] "Argilla grigia ('buccheroide') per effetto di cottura, con vernice nera od olivastra," obtained "in conseguenza del sistema di cottura a riduzione," i.e., in the absence of oxygen (Lamboglia 1952, 140, 156). Marabini Moevs 1973, 213; Morel 1980, 105; 1981a, 520; Cuomo di Caprio 1988, 67; 1992, 89; Morel and Picon 1994, 44; *Agora* XXIX, 232.

[5] For the three types indicated by Lamboglia (Campana A, B, and C), Fabbricotti (1970, 57) used the term *Campana* to refer to only the color of the clay, and not the place of manufacture.

[6] Carter 1983a, 413, 488; *Territory of Metaponto*, 38.

[7] Pelagatti 1970, 470.

[8] L. Giardino 1978b, 124 n. 22, 129; 1980a, 231. See also the section "Final Considerations," below.

[9] According to Morel (1980, 104): "toutes les céramiques à pâte grise ne doivent pas être considérées comme de la Campanienne C" and "la plupart des céramiques à pâte grise ne s'inspirent nullement de la Campanienne C" (105); see also Morel 1981a, 520. Pavolini (1981, 160) remarks, "Bisogna guardarsi dal vedere una Campana C in qualsiasi prodotto contraddistinto da argilla grigia." The same considerations apply to other grey-ware productions of the 2nd/1st c. BC widespread in the western Mediterranean, in particular type D from Albintimilium, the African grey wares (type C from Hippo, Moroccan type D, types E and F from Tipasa), the Roman type D, the Sardinian type E and other grey wares from Sardinia, and the so-called *gris ibicenco* from Ampurias (Emporion) and Ibiza. For all of these types, see Morel 1981a, 45–52, 517–18, and 627–41, with relevant bibliography. To these can be added type III from Cosa, assimilated by D.M. Taylor (1957, 164–73) to Campana C, the products of southern France (Stone 1982, 99), and those of the western Mediterranean (see *Agora* XXIX, 232 n. 57). Other indications come from Cyprus and Chersonesos in Crimea (*Meligunìs-Lipára* II, 249 n. 4). Small (1977c, 124) refers to "the Mediterranean world in the Late Hellenistic period," and does Prag (1992c, 134), who speaks of "the Hellenistic Mediterranean." For Campana C ware there are secure attestations at, e.g., Benghazi-Berenice (Yntema 1995a, 389), Thamusida in Morocco (Callu et al. 1965) and Pollentia (Morel 1981a, 165–66, types 2323i1 and 2352c1), as well as on Malta (D'Andria 1972, 92–95).

[10] See the section on "Relationships with Other Ceramic Classes," below. Yntema (1993, 37, and *Valesio* I, 138) recognizes the substantial differences between Campana C and Apulian grey-gloss ware caused by the type of clay and gloss, the morphology, and the decorative syntax, but he also emphasizes that it would be an exaggeration to suggest that the two wares have nothing in common, despite admitting the independence of Apulian grey gloss. In this context he mentions the example of Oria, where the first forms similar to those in Campana C appear only from the second quarter of the 1st c. BC onwards. For Morgantina, however, Stone (1982, 100, 127) maintains that "gray ware represents early Campana C" because the excavated contexts date to the 2nd c. BC, while the production of Campana C is attested beginning in the last quarter of the 2nd c. BC.

pottery of the Hellenistic and Republican periods.[11] In fact, the largest southern Italian centers in which grey ware was produced—or even simply consumed on a large scale—had their roots in Greek culture; they were struck by the "whirlwind" of Romanization in a later moment of history.[12] As this study will demonstrate in detail, the morphological repertory of grey ware is permeated as much by influences from Greek products as from typically Roman ones. In both cases the Hellenistic and Roman elements combine without excluding each other.[13] Because of this, one should speak of grey ware's pertinence to the Late Hellenistic/Republican period, although there is still a preference for designating this ceramic class as Late Republican in order to distinguish it from the real Hellenistic wares foreign to the Roman world.[14]

Chronology

Recent studies have better defined—and in several cases narrowed—the time span during which grey ware represented one of the principal classes of objects for the table. Yntema defines the period of the circulation of grey ware, placing the start of production in the middle of the 2nd century BC or shortly thereafter.[15] In particular, the data obtained from the multiyear research project conducted at Valesio (Brindisi) by the Free University of Amsterdam have produced a chronology of 160/150 BC–90/80 BC for the largest collection of grey-gloss ware, while the more recent phase has been generally ascribed to the beginning of the Augustan period based on the link with the "relatively well-dated Campana C wares."[16] At Gravina, on the other hand, the floruit of grey ware appears to have begun circa 140/130 BC, slipping to circa 70/60 BC for the latest date, provided by the closed context of trench F202.[17] This hypothesis is based also on analysis of the materials recovered there, which appear to be slightly more recent than those of Valesio.[18]

At Cosa, Taylor maintains that fabric type III was assimilated to Campana C and was imported by the middle of the 2nd century BC or perhaps earlier;[19] thus she arrived at a chronology close to that proposed by Yntema—all the more so because fragments not securely belonging to type III were recovered from the oldest deposit (A, 273–180/160? BC).[20] In the case of Herakleia, the chronology does not currently appear as definable, because it is based almost exclusively on grave goods that are dateable from the third quarter of the 2nd century BC onwards, yet which often belong to graves that were reused at length between the 3rd and 1st century BC.[21] At Taranto the most recent work, particularly that of Hempel on the necropoleis, has expanded the chronologies of the introduction and use of grey ware that were accepted until a decade ago. By combining the typologies of the associated materials in the funerary contexts, Hempel has arrived at dates independent of those proposed for other sites. Thus phases D and E of the necropoleis of Taranto have been better defined, with a consequent down-dating of the new ceramic class to the beginning of the 2nd century BC; its decline probably began in the course of the third quarter of the 1st century BC (phase E3).[22]

---

[11] L. Giardino 1976b, 20; F. Rossi 1989, 173. For a synthesis of the grey-ware materials recovered in southern Italy, see the catalog. Currently the most detailed studies on this topic are those dedicated to the materials from Metaponto (L. Giardino 1980a, esp. 249 n. 9, with relevant bibliography, and 281; Burgers 1991), from Taranto (Hempel 2001), from Gravina (Prag 1992c; Small et al. 1994, 207–11), from Monte Irsi (Small 1977c, 124–27), from Valesio (*Valesio* I, 213–35), and from Leuca (L. Giardino 1978b, 128–40).

[12] The Italian centers had their roots in Greek culture insofar as they were Greek colonies, such as Metaponto, Taranto, Sybaris, and Herakleia, or strongly Hellenized centers such as Gravina, Valesio, Monte Sannace, Oria, and Leuca, to name just the centers richest in grey ware that are currently known.

[13] There was a gradual transformation and assimilation of Roman culture by Greek tradition: see Hempel 2001, 85, 90 n. 7, 115.

[14] For Campana C, too, Morel (1980, 104) believes in a direct relationship between the Roman conquest of Syracuse and the wide diffusion of the ware. See also Hempel 2001, 110.

[15] *Valesio* I, 139; Yntema 1993, 37; 1990b, 168–69. The success of grey ware ended with the rise of the Italic sigillata wares.

[16] *Valesio* I, 213.

[17] A worn denarius of 85 BC was recovered from this trench. Moreover, the ceramic finds provided a date of 75±15 BC (Small et al. 1994, 203).

[18] *Valesio* I, 214. Prag (1992c, 134) assigns the chronology of grey ware at Gravina to the period between the end of the 3rd and middle of the 1st c. BC, principally on the basis of the stratigraphy, but also of the comparanda; the material presenting a higher chronology is that more closely linked to the black-gloss tradition, like the plates with recurving rim (298, Fig. 40, No. 980), the lid-plates (300, Fig. 42, No. 994), the bowls with grooved walls (299, Fig. 41, No. 985), and the krater fragments (304, Fig. 46, Nos. 1065–66).

[19] D.M. Taylor 1957, 164 n. 23, 191.

[20] More precisely, the bowl A24 (D.M. Taylor 1957, 86), the pitcher A42 (D.M. Taylor 1957, 90), and the fragment A49 (D.M. Taylor 1957, 91).

[21] See, for example, tomb Umbria 141. In addition, L. Giardino herself (1990, 74, 77, 81, 87) emphasizes "la mancata definizione della cronologia iniziale della ceramica a pasta grigia." F. Rossi (1989, 174) maintains that the Herakleian context dates to the full 2nd and beginning of the 1st c. BC.

[22] Hempel 2001, 93, 100, 114. For preceding studies, see Hempel and Mattioli 1994, 359, and Hempel 1996a, 339. Likewise, the grey ware from Monte Sannace seems to date to the first half of the 1st c. BC as it is not found with TSI (F. Rossi 1989, 175). At Oppido Mamertina (contrada Mella), the material appears to concentrate between the second half of the 2nd and beginning of the 1st c. BC (Preacco Ancona 1999b, 292).

The first attempt at organizing a typology of the grey ware from Metaponto was a systematic analysis of the materials recovered in the excavation of the castrum area, for which Giardino suggested a chronology between the first decades of the 2nd century BC and the end of the Republican period.[23] Though at the same time Carter described grey ware as "ubiquitous, but difficult to date," he was nevertheless aware of the precious information that could be derived from the excavation of the stratified waste deposit associated with a Late Republican kiln at the site of Pantanello.[24] The kiln was linked to a houselike structure, forming a true industrial ceramic facility. The location was favored by the presence of a spring and clay pit, which was later filled by the kiln waste deposit.[25]

Except for general references whenever the topic has demanded such,[26] the systematic study of the Metapontine grey ware has been sporadic, despite the considerable quantity of it recovered in subsequent excavations. A special case is the grey ware from the waste deposit at Pantanello, which was the subject of a still unpublished preliminary study by G.-J. Burgers.[27] This assemblage is highly interesting because of its array of practically intact or only slightly incomplete forms. The dates provided by the stratigraphic sequence and comparanda seem to assign the Metapontine grey ware produced in the chora to the period between the middle of the 2nd and the 1st century BC, with the greatest abundance produced before the middle of the 1st century.

On the basis of what has been stated so far, it is not possible to accept in general the middle of the 2nd century BC as the moment in which the production of grey ware began;[28] this date, however, can be considered the moment in which the "new" ware achieved a primary role in many southern Italian centers,[29] so much as to become the trace artifact of the Late Republican period for these areas.[30] The study of the technical characteristics of the materials from various sites has demonstrated that, beyond the apparent similarities in morphology and the color of the clays and glosses, grey-ware production was distributed across numerous production centers, as has been demonstrated for Campana C as well.[31] However, it should be emphasized that the existence of workshops producing grey ware has been archaeologically confirmed only in the territories of Metaponto and Oria.[32]

[23] L. Giardino 1980a, esp. 286. See also S. Bianco 1993, 207.

[24] Carter 1980a, 17

[25] Carter 1983a, 415. Cf. n. 10 above.

[26] See, e.g., several burials in the urban necropolis at Crucinia (De Siena 1993, 126–34), a general treatment in Hempel 1996a, and the excavation of an agrarian structure in contrada Casa Teresa (Lissi Caronna 1998–99).

[27] Burgers 1991.

[28] Furthermore, not all researchers agree on the establishment of a precise chronological limit to the period in which grey ware circulated: see, e.g., Russo Tagliente 1995, 88; A. Bottini 1983–84, 218.

[29] Cf. n. 10 above.

[30] Hempel 1996a, 337; 2001, 113; Yntema 1990b, 149; *Valesio* I, 139; L.

Giardino 1990, 74. Hayes (in Small et al. 1994, 207–8) further restricts the field, considering the Metapontine grey ware specifically the "type-fossil" of the Late Republican period in Basilicata and much of Puglia. On this hypothesis, see also Hempel 1996b, 343; 2001, 113, 124 n. 172. Carter (1983a, 446) defines grey ware as "a key to more precise dating" for the agrarian structure found at Pantanello. On the other hand, Morel (1965, 233) identifies a clear superiority even for Campana C in Sicily in the 2nd and 1st c. BC. Clearly, the situation is not the same everywhere: at Cosa, for example, D.M. Taylor (1957, 191) affirms that type III, assimilated to Campana C, never achieved the popularity of type I (Campana A) or type II (Campana B). However, the importance of grey ware is attested also by the occasional adoption of its absence in a stratigraphic sequence as an *argumentum ex silentio* indicating that the use of an area ceased at the end of the 3rd c. BC—see, e.g., L. Giardino 1978a, 416—where a surface of the *plateiai* in the castrum of Metaponto is dated to the end of the 3rd c. because it lies beneath an occupation layer with grey ware—and Costamagna 1983, 118–19, which describes "un sicuro limite cronologico inferiore della colmata dato dalla assoluta mancanza di ceramica a pasta grigia."

[31] Lamboglia 1952, 161; D.M. Taylor 1957, 71; *Meligunìs-Lipára* II, 249 n. 4; Pelagatti 1970, 471; *Sibari* II, 544; *Sibari* III, 436; Fallico 1971, 602; Prag et al. 1974, 155; Pavolini 1981, 160; Small 1977c, 124; L. Giardino 1976a, 553; 1978b, 130; 1980a, 249, 281; 1982, 168 n. 76; Stone 1982, 98; A. Bottini 1983–84, 218; F. Rossi 1989, 173; Prag 1992c, 134; Small 1992a, 17; Yntema 1993, 37; *Valesio* I, 214; Hempel 1996a, 337; 1996b, 343; 2001, 113; *Chiori*, 487; Morel 1981b, 83; 1989, 118; Morel and Picon 1994, 45. On the other hand, it is recognized that within the same form there can be details that give the piece uniqueness and suggest local production (see the description of the typological structure, below).

[32] The identification of Metaponto as a center of grey-ware production has now been established. Although this could be supposed on the basis of the homogeneity of the clays and glosses that has characterized the materials recovered over the course of the years (L. Giardino 1978a, 424; 1980a, 281), as well as the articulated morphology, for which it is often impossible to identify a precise correspondence with Lamboglia's Campana C (L. Giardino 1980a, 282, 284), it can be more reasonably confirmed following the discovery of kilns used to produce grey ware in both the city and the chora: D'Andria 1976, 541 n. 9; Carter 1977, 852; 1979, 62; 1983a, 447; *Territory of Metaponto*, 38; L. Giardino 1980a, 281 n. 59; Giannotta 1980, 62, 91; Edlund 1986, 120; Hempel 1996b, 343. Marletta (1987, 221) mentions Gravina di Puglia also as a production center and suggests production at Grumentum because of the quantity and uniqueness of certain forms. Chemical-mineralogical analyses performed on samples of grey ware from Gravina have indicated that the latter was equipped with a production facility on the hill of Botromagno or nearby (Ciancio/Dell'Anna/Laviano 1994, 273–74; see also *Otranto* II, 21.) For Oria, see Maruggi 1996c, 70; also *Valesio* I, 214. Volpe (1990, 76–77) suggests production at Canosa between the end of the 2nd and first half of the 1st c. BC because of the presence of kiln waste deposits with grey ware. Yntema (1995b, 268) instead suggests Brindisi and/or Taranto as the production centers of Apulian grey gloss in the 1st c. BC. Hempel (2001, 121) emphasizes that Taranto produced the ceramics needed to satisfy the city's demand in loco, including grey ware.

Relationships with other ceramic classes

The morphological relationships of grey-ware vessels with vessels in other wares—black gloss, HFR ("hard-fired red"), Campana A/B/C, the sigillatas—are frequently evident,[33] yet it is impossible to speak of a true imitation of known models, both ceramic and metal. One should rather speak of an assimilation and reworking of these models via the introduction of variants and details that make forms unique, and distinguish grey-ware production as a class.[34] The supposed inheritance of certain forms in grey ware from black gloss may have induced Yntema to consider Apulian grey gloss the final phase of Apulian black gloss, or even two separate productions of the same Apulian black-gloss tradition.[35] Yntema also hints at different circulation patterns for the two classes because grey ware is found consistently only in Basilicata, south-central Puglia, northern Calabria, and Sicily.[36] On the other hand, the ICA survey registers a quantitatively limited production of grey ware compared to black gloss: grey ware comprises only 2% of all collected black-gloss materials pertaining to the entire chronological span in consideration (Archaic to Republican), and 4.5% of the Hellenistic to Republican materials.[37]

The proposed borrowing of morphological characteristics from other productions was perhaps motivated by a desire to "ennoble" this tableware of common use and excellent quality, which followed fashion without losing sight of practicality.[38] There was likely a strong desire (present by the end of the 4th c. BC at least) to imitate the sheen of metalwares; the use of grooves and indentations, too, is perhaps to be attributed to the influence of these products.[39]

The Survey Materials

Glosses and fabrics

The survey of the Bradano-Basento region recovered 203 fragments of grey ware, mostly small pieces. Of this total, 179 pieces had visible traces of grey gloss with the color of anthracite (Munsell N4/ and N5/), generally diluted and matte, occasionally with olive tinges, more or less light (Munsell 5Y 4/1, 5/1, 6/1), very rarely with blue-black (Munsell 5B 4/1) or nearly black (Munsell 7.5YR N3/) tinges.[40] The larger pieces display the usual technique for applying the gloss—immersion—which often leaves telltale trickles.[41] Several distinctions in the gloss were recognized by visual inspection. Besides having a variable consistency associated with greater or lesser homogeneity, these can range from more or less shiny to matte and were classified as follows:[42]

a1: homogeneous and matte
a2: diluted and matte
a3: very diluted and matte
b1: homogeneous and semi-shiny
b2: diluted and semi-shiny
c: homogeneous and shiny

Table 7.1 and Figure 7.1 demonstrate the greater frequency of the diluted and matte gloss, which has

[33] See the catalog below for detailed explanation. In general, see Lamboglia 1952, 156; L. Giardino 1978b, 127–29; 1980a, 252 and 282; Yntema 1993, 37; Hempel 1996a, 337; 1996b, 343; 2001, 113–14; D.M. Taylor 1957, 166; *Sibari* III, 258, 436; Carter 1983a, 488–89; Prag 1992c, 134; S. Bianco 1993, 207; Marletta 1997, 221.

[34] Or, as Morel (1981a, 515) explains it, of *imitation-influence*, not *imitation-contrefaçon*. See also Hempel 1996a, 337, 339; Hempel elsewhere (2001, 114–15) speaks of the "adaptability" of grey ware, which borrowed and reworked the forms of other wares, and of its independence from specific paradigms. Morel (1981a, 22 n. 37) clearly defines the concept of "ceramic class" as based on uniform technical characteristics (physical, chemical, tactile, or visual) that are not always completely identical.

[35] Yntema 1993, 37; *Valesio* I, 139. On the other hand, even Small (1977c, 124) and Prag (1992c, 134) emphasize the sharing of certain characteristics between "grey glaze" and "black semi-glaze."

[36] *Valesio* I, 213–14.

[37] At many sites the presence of grey ware is only sporadic, numerically insignificant in comparison with the quantities of black-gloss or coarse ceramics. It is significant from a cultural-historical point of view, however, because it allows secure identification of a chronological span (2nd–1st c. BC) that could otherwise be identified only with difficulty on the basis of black-gloss fragments alone (all the more true in a study such as this one, which is obviously not supported by stratigraphic sequences). From this perspective, the study of the forms is a valid aid: once extrapolated from contexts dated by materials with well-defined chronology and subsequently compared to similar morphologies, the forms are the key to defining and narrowing the dates of the material collected during a field survey, as in the current case.

[38] L. Giardino 1976b, 20; 1980a, 285–86; 1982, 168.

[39] Prag et al. 1974, 155; Small 1977c, 124; Prag 1992c, 134. The same hypothesis has been proposed for Campana C ware: see, e.g., Morel 1981a, 526; Stone 1982, 104.

[40] The determination of the colors of the fabrics and glosses is based on visual inspection with the naked eye and a microscope, assisted by the Munsell Soil Color Charts (1990). Given the highly fragmentary nature of the specimens, the colors of the fabrics were read from the section/core. Moreover, the description of a fragment's gloss refers to the color on both the interior and exterior surfaces, unless otherwise specified.

[41] In this case the gloss covers roughly three-quarters of the body, a detail that has led some researchers to compare "grey glaze" with "black semi-glaze" (see n. 33 above). See, e.g., the photograph of type GW-Cl1, where the trickles of gloss are clearly visible.

[42] The gloss consists of very well purified and refined clay, which has identical chemical and mineralogical characteristics to the clay mass from which the vase was formed, but is enriched with compounds containing iron and carbon (Ciancio/Dell'Anna/Laviano 1994, 266, 274, with relevant bibliography).

Gloss color	Gloss qualities						
	a1 homogeneous matte	a2 diluted matte	a3 very diluted matte	b1 homogeneous semi-shiny	b2 diluted semi-shiny	c homogeneous shiny	Total
Greyish-brown (2.5Y 5/2)	0	1	0	0	0	0	1
Black (7.5YR N2/)	0	0	0	1	0	0	1
Very dark grey (7.5YR N3/)	3	0	0	0	0	1	4
Brown (7.5YR 5/4)	1	0	0	0	0	0	1
Dark grey (N4/)	29	51	3	3	1	1	88
Grey (N5/)	1	29	4	0	0	0	34
Grey (N6/)	0	2	0	0	0	0	2
Dark blue-grey (5B 4/1)	3	0	0	0	1	0	4
Dark grey (5Y 4/1)	2	13	1	0	1	0	17
Grey (5Y 5/1)	2	19	1	[illegible]	0	0	23
Grey (5Y 6/1)	0	3	1	0	0	0	4
Total	**41**	**118**	**10**	**5**	**3**	**2**	**179**

Table 7.1 Quantification of qualities and colors among grey-ware fragments with traces of gloss.

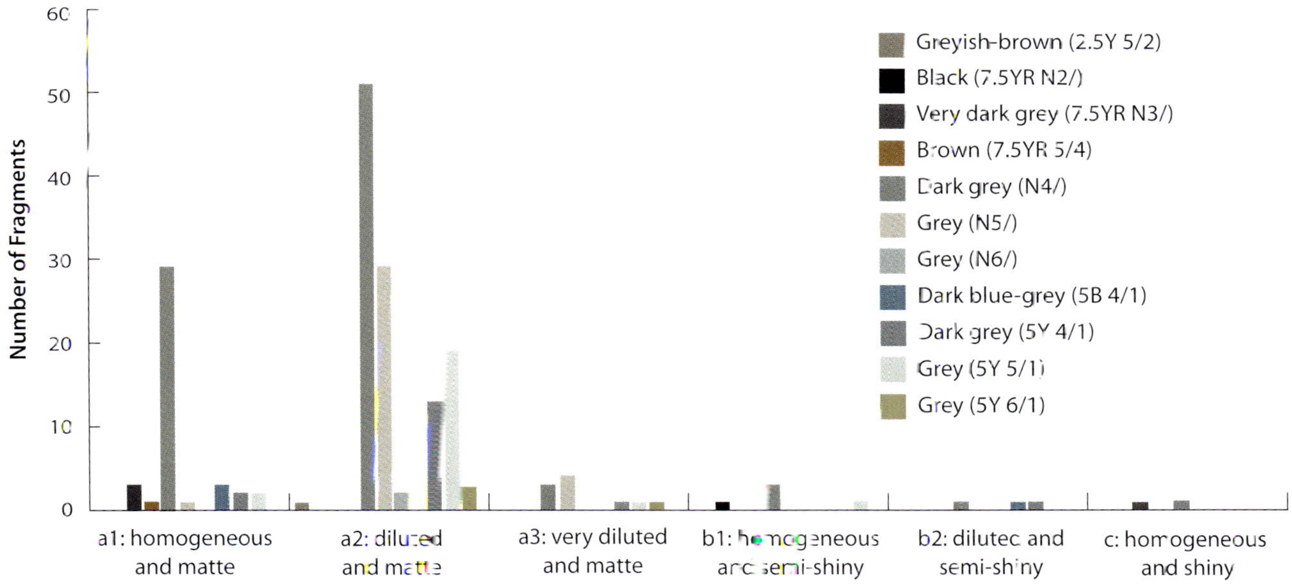

Figure 7.1 Frequency of gloss qualities and colors among grey-ware fragments.

both light and dark tones; the generally more homogeneous shiny or semi-shiny gloss is unmistakably an exception, appearing in a very dark grey tone in nine of the ten fragments among b1, b2, and c. In general, the homogeneous gloss is predominantly dark in color, while the diluted gloss can be either grey or dark

Fabric Color	A1 pure, hard, compact	A2 pure, soft, compact	B1 hard, small inclusions	B1a hard, large inclusions	B2 soft, small inclusions	Total
7.5YR N5/ (grey)	9	12	1	0	0	22
7.5YR N6/ (grey)	30	47	2	0	7	86
7.5YR N7/ (light grey)	27	46	1	1	6	81
7.5YR N5/, core N6/	1	0	0	0	0	1
7.5YR N6/, core N7/	2	4	0	0	1	7
7.5YR N7/, core N6/	0	4	0	0	1	5
10YR 6/1 (grey)	0	1	0	0	0	1
Total	**69**	**114**	**4**	**1**	**15**	**203**

Table 7.2 Quantification of qualities and colors of fabrics of grey-ware fragments.

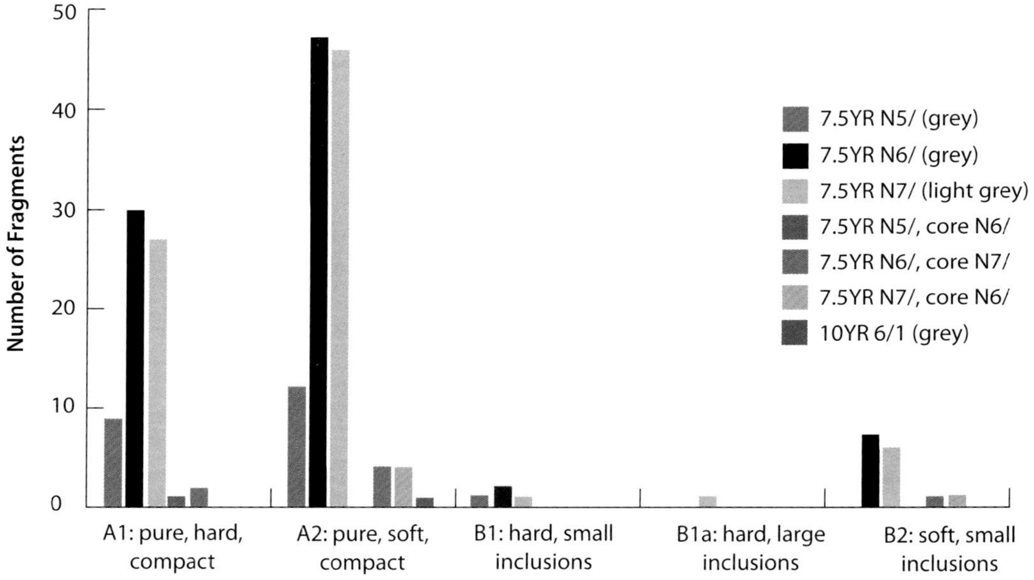

Figure 7.2 Frequency of qualities and colors of fabrics.

grey without preference, thus confirming the clear preponderance of the anthracite grey tone in more or less dark gradations (see above).

The color of the fabric itself varies only slightly from grey to light grey (Munsell 7.5YR N5/, N6/, N7/); only in rare cases is a slightly different core attested, although it always remains grey in color. Within the two principal variants—the presence or absence of inclusions—a distinction was made between soft and hard fabrics, classified as follows:

A1: well purified, hard, and compact
A2: well purified, soft, and compact
B1: hard, with (very) small micaceous inclusions
B1a: hard, with large micaceous inclusions
B2: soft, with (very) small micaceous inclusions

In general, the fabric is well purified (fragments with small inclusions do not exceed 10%) and soft (roughly 64% of the collection), although there are numerous examples of a somewhat harder fabric (*Table 7.2* and *Fig. 7.2*).

Form	Type	Description	No. frags.	% Total by form	% Open vs. closed
Dish	GW-D1	Dish with short vertical rim	1	3.94	Open form 67%
	GW-D2a	Dish with vertical rim	3		
	GW-D2b	Dish with vertical rim & slightly sagging wall	1		
	GW-D3	Dish with offset vertical rim	1		
	GW-D4	Dish with horizontal rim	1		
	GW-D5	Dish with offset horizontal rim	1		
Cup	GW-C1	Cup with flaring wall and faint rim	2	8.37	
	GW-C2a	Cup with offset rim	1		
	GW-C2b	Cup with offset grooved rim	2		
	GW-C3	Hemispherical cup	3		
	GW-C4	Cup with vertical rim	1		
	GW-C5	Cup with offset vertical rim	1		
	GW-C6	Cup with offset vertical, grooved rim	1		
	GW-C1-C3	Disk foot, hemispherical body	5		
	Unidentifiable	Rim	1		
Cup or bowl	GW-C1/B1	Slightly outward-turned, indistinct rim	1	0.49	
Bowl	GW-B1	Bowl with concave/convex body	2	1.97	
	GW-B2	Bowl with grooved rim	2		
Mug	GW-Bo1	Mug	3	1.48	
Open form	GW-Op1	Flat resting surface & oblique wall	2	50.74	
	GW-Op2	Flat resting surface & horizontal wall	1		
	GW-Op3	Slightly oblique resting surface	1		
	GW-dec. 1a	Drop rouletting	1		
	GW-dec. 1b	Delta rouletting	1		
	GW-dec. 2	Two impressed concentric circles	1		
	Unidentifiable	Feet	2		
	Unidentifiable	Bodies	84		
Open (?) form	GW-Op?1	Slightly oblique resting surface & horizontal wall	1		
	GW-Op?2	Oblique resting surface and wall	1		
	Unidentifiable	Bodies	8		
Closed form	GW-Cl1	Flat resting surface & hemispherical wall	2	30.05	Closed form 30%
	Unidentifiable	Foot	1		
	Unidentifiable	Bodies	43		
Closed (?) form	GW-Cl?1	Nearly flat resting surface	4		
	Unidentifiable	Rim and foot	2		
	Unidentifiable	Bodies	9		
Open (?)/closed (?)	GW-Op?/Cl?1	Slightly oblique resting surface	3	1.48	
Unidentifiable		2 molded handles & 1 spout	3	1.48	
Total			**203**	**100%**	

Table 7.3 Quantification of grey-ware fragments by form and type.

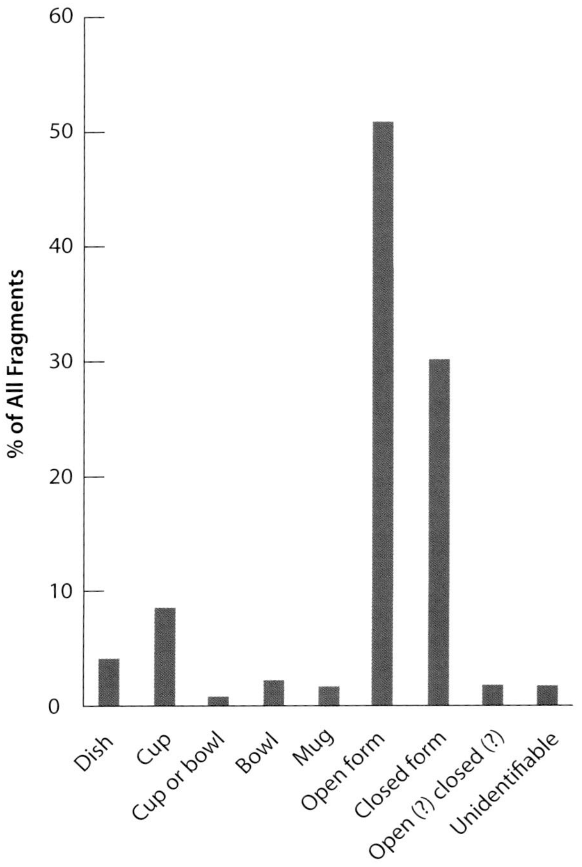

Figure 7.3 Frequency of forms by percentage of total fragments.

The fragments examined seem to suggest a reasonable homogeneity of Metapontine production. In the Bradano-Basento region, in fact, there is no evidence for the existence of areas within which one type of fabric or gloss was particularly concentrated. The slight variations in color and consistency in the fabrics and glosses—perhaps to be attributed to different firing temperatures—create the impression of an essentially uniform production.

Typology

The following classification is based on a morphological analysis of the vessels, in which four primary shape types were distinguished along with secondary types: dish, cup, bowl, mug (*Table 7.3, Fig. 7.3*).[43] The dish and cup are characterized by a considerable series of subtypes, which create a well-developed picture of this production and, in certain cases, suggest local reworking of known types, if not genuine artistic inspiration (e.g., dish GW-D3 and cup GW-C5).[44]

This typology demonstrates the marked predominance of open forms (dishes, cups, bowls), while closed forms are limited to fragments from parts of vessels that cannot always be identified with certainty. Despite the small quantity of material examined, this result seems to confirm the hypothesis of a sharp contraction in fine pottery for holding or pouring liquids, in favor of plain/banded wares (see, for example, the text accompanying mug GW-Bo1).[45]

The cup in particular, whether hemispherical or with offset or vertical rim, is present in 17 examples (counting C1-3 and the unidentified rim), corresponding to 32.1% of the total number of diagnostic fragments (53); one should keep in mind, however, that all of the feet comparable to type GW-C1–3—and therefore securely belonging to cups—are also included in this count. For the dish, which is also one of the more frequent forms (8 examples, equivalent to 15.1% of the diagnostic group), the count cannot include the bases and thus considers only the rims (in many cases, in fact, it is impossible to attribute a foot securely to a dish). In general, the proportion of open forms is 62%, versus 23% for closed forms, to which should be added the forms of uncertain attribution (4.9% possibly open, 7.4% possibly closed).

A more difficult task is inserting into a typology the fragments of vessels that preserve only portions of the base. It is not always clear whether such fragments are to be ascribed to open or closed forms, partly because of the condition of the gloss, which occasionally facilitates classification but is often not clearly recognizable. Beyond the subdivision between open and closed forms based on the shape of the wall—where preserved—and the presence or absence of gloss on the interior surface, the subtypes were defined by treating as a distinctive element the resting surface of the foot, i.e., whether flat or oblique. In general, considering also the fragments that cannot be inserted into the typology because of insufficient size, the feet are always rather low and squat and belong to fairly thick walls.[46]

[43] It is useful to recall the definition of *type* supplied by Morel (1981a, 22): "un ensemble de vases ayant en commun un certain nombre de caractéristiques formelles . . . selon le même modèle idéal."

[44] Each type is preceded by GW to distinguish the grey-ware forms from the forms of the black-gloss typology (see Ch. 6).

[45] Closed vessels in black gloss appear to become rare products after the first third of the 3rd c. BC, as happens to drinking vessels after the second third of the 2nd c. BC (Morel 1981a, 503–4). A preference for open forms is found at Taranto as well, at least by the early 2nd c. BC (Hempel 2001, 98, 113).

[46] The description of an uncertain fragment proceeds from the resting

There are many fragments belonging to various parts of vessels that cannot be attributed to specific forms. These were analyzed simply to complete the picture of the grey ware collected during the survey and are therefore included in the calculation and analysis of the distribution of open and closed forms; they also fill out the fairly uniform picture of the fabrics. They have not been included in the catalog, but their quantities by type are reported below.

Only three fragments preserve traces of decoration: two portions of body walls with roulette decoration (GW-dec.1a, drop rouletting; GW-dec.1b, delta rouletting), presumably belonging to dishes, and a ring foot (GW-dec.2) with two impressed concentric circles on the interior surface.

Final Considerations

Despite the fragmentary nature of the recognized typologies, comparative study of grey ware confirms its permeability with respect to the morphologies derived from other ceramic classes, including Campana C, with which grey ware is contemporary. Nonetheless, it must be distinguished by fabric. Morel points out the typical feature of Campana C produced at Syracuse and in surrounding areas: grainy, light grey fabric, often with a light brown or pink core.[47] As demonstrated in Table 7.2 and in the catalog below, the fabric of the materials recovered by the survey is in no way similar: it is never grainy, and the color of the core never varies from grey.

From the morphological perspective, too, the grey ware from the Metaponto survey bears only sporadic resemblance to Campana C.[48] Slightly more than 25% of the Campana C forms identified by Lamboglia appear in the survey material (see in particular the dishes with vertical rim, similar to Lamboglia's C7; the cup with offset vertical rim, similar to Lamboglia form C19b; and the bowl with grooved rim, similar to Lamboglia form C1). Likewise Giardino notes that only two of the Metapontine forms from the castrum show "corrispondenza con quelle considerate da Lamboglia come tipiche della 'campana C.'"[49] Analysis of the materials recovered from the kiln waste deposit at Pantanello also tends toward the same conclusion: only four forms (3, 4/5, 15, and 27) out of a total of 51 bear resemblance to Lamboglia's Campana C.[50]

Grey ware's independence from Campana C is further attested by the absence of the grooving on the resting surface that Lamboglia considered typical of the majority of the forms he examined.[51] It is certainly not a distinctive element of grey ware: even if it appears on several examples of Metapontine production, it is entirely absent from the fragments recovered by the survey.

The relationship with black-gloss prototypes, not only of the 2nd century BC but also of the mature Hellenistic period (in some cases common to Campana C as well), is evident above all in the oldest forms, in particular the dishes with vertical rim, the cups with offset rim, the hemispherical cups, and the bowls with concave/convex body.

On the other hand, the dish with offset vertical rim, which represents the latest phase of the grey ware from the survey, is clearly related to the terra sigillata productions. The same analogy seems to apply in the case of the cup with offset vertical rim, which (in the variant with a groove near the rim—GW-C6) also has its closest parallels in the sigillata ware. The type GW-C5, which does not have a groove, could instead be a local interpretation of black-gloss and grey-ware models with slightly different characteristics (less pronounced rim, more accentuated carination of the body).

The mug, with its characteristic concave upper rim surface, for the moment seems to be exclusive to the Metapontino. It was produced late and harks not far back to several families of thin-walled wares.

surface towards the wall (where present).

[47] Morel 1981a, 47; 1965, 16.

[48] The Syracusan production, on the contrary, can be retraced to a typological repertory that belongs to the area in which it was made. See Morel 1981a, 521.

[49] L. Giardino 1980a, 281–82.

[50] Burgers 1991; Hempel 2001, 113.

[51] Lamboglia 1952, 156.

Catalog of Grey Ware

The references cited cannot be considered exhaustive: the intention was to create as wide a view of the subject as possible, linking not only grey-ware productions, but also contemporary productions of other wares, as well as possible prototypes (especially in black gloss). This approach has led to a more precise understanding of the relationships between grey ware and chronologically related productions, which the catalog can help to elucidate.

The dates have been proposed on the basis of comparison with materials from the same class, preferably originating in stratified or closed contexts. Priority was obviously given to materials recovered in the areas subject to Metapontine influence. At this point it should be noted that an important point of reference for this study is Burgers's unpublished preliminary report (1991) on the grey ware from the kiln deposit at Pantanello. The author hopes that this material will see full publication soon (only some of the excavated material has been published to date, see Carter 1983a, 470–89) because it expands the typology delineated by L. Giardino (1980a) and also because the limited amount of grey ware collected in the survey fails to do justice to Metaponto's important position as a center of grey-ware production.

Reference is often made to the material recovered in five deposits at Cosa (D.M. Taylor 1957). It should be noted that the relevant dates indicated here reflect Morel's revisions (Morel 1965, 21–22; 1981a, 57–58 and Nos. 195–99). The date proposed by D.M. Taylor (1957) of deposit A (225–150 BC) should be raised by virtue of the presence of the so-called *piatti di Genucilia*, a *poculum*, and the bowls *à petites estampilles* (273–180/160? BC), while deposit B (dated to 170/160–140 BC by D.M. Taylor) begins after the closing of deposit A (180/160?–140 BC). A stamped Rhodian amphora handle in deposit C (167–140 BC by D.M. Taylor) should move its beginning to at least 190/180 BC. The ceramics of deposit D (dated by D.M. Taylor to 130/120–70/60 BC) appear to be older than the ceramics of deposit E, and therefore the beginning of deposit D should be raised slightly (150/130–70/60 BC). The date of deposit E (110/100–30 BC) is based on comparison with the others and on the presence of Arretine ware but should perhaps be moved upwards by a decade (to 120/110–40/30 BC) because of the presence of Rhodian amphora handles and Lamboglia's form A33.

All dimensions are in cm. Maximum height is indicated by "h," maximum width indicated by "w," maximum thickness indicated by "th," and maximum diameter indicated by "diam." Color is described according to the Munsell Soil Color Charts. An asterisk (*) after the site number indicates materials from the 1980 ICA pilot survey (sites numbered 1–44), as well as at the intensive survey sites and plots located south of the Basento River through the 2001 season (most of the sites numbered 562 and higher). Additional information about the sites can be found in the Gazetteer.

For fragments that could not be assigned to a specific form, the "Date" field has been omitted; in these cases a span between 2nd–1st c. BC is implied.

Dish with Vertical Rim (GW-D1 and GW-D2)

This is a dish with a vertical or slightly oblique rim, mated to a body that is generally oblique and shallow. Two types of dishes with vertical rim were identified among the survey materials. Type GW-D1, with its very low rim and oblique body, recalls Prag's proposed classification as "lid-plates," according to which these dishes probably continue the tradition of the red-figure lekanis lids of the 4th c. BC, used as either container or cover as needed (Prag 1992c, 141). On lekanis lids see *Agora* XXIX, 197, and also the so-called reversible lids, often associated with pyxides and therefore linked to the feminine world; the earliest examples appear to belong to the 5th c. BC: see *Agora* XXIX, 192 n. 15.

Type GW-D2, instead, with its high and more or less vertical rim, is represented by two subtypes: GW-D2a, with nearly straight body line, and GW-D2b, with a slightly curved upper body.

Among the various grey-ware forms, this type of dish is probably the one most frequently produced also in Arretine and other Roman wares (Prag 1992c, 141; L. Giardino 1980a, 254). Its relationship with preceding and following forms is already clear in Lamboglia's classification: form B5 (from Rome, Ventimiglia, and Ensérune) has an obvious connection with Campana A (form A5) and represents the moment of passage to form B7 (from San Miguel de Sorba), which is closer to the typology in this study by virtue of the more oblique body; the author also emphasizes its affinities with the subsequent production in Arretine ware (Lamboglia 1952, 146, 148, 167). At Cosa it is produced in the local type IV fabric (D.M. Taylor 1957, 72, Pl. 26 B9, from a deposit dated 180–140 BC), which in several cases imitates forms of the *ceramica campana* type II (= Campana B). D.M. Taylor (1957, 156), moreover, emphasizes the practical functionality and ease of execution of this type of dish, which is imported into Cosa in type II in the first

Dish with vertical rim (GW-D1 and GW-D2) cont.

half of the 2nd c. BC and remains in use throughout the 1st c. BC. The success of this dish is further confirmed by production in red gloss, as, for example, at Gravina (Cotton 1992a, Fig. 49, Nos. 1119 and 1120; Small et al. 1994, Fig. 9, Nos. 67 and 69).

According to Rotroff, Attic production of large-diameter (20–31 cm) versions of these dishes was influenced by western productions, especially from Campania and Etruria, and limited to the period 110–86 BC (*Agora* XXIX, 155).

Morel's dishes 2250–80, typologically related to those considered here, facilitate reconstruction of the evolution of this form. From a low, generally vertical rim with a soft rim/body transition (*espèce* 2250, dated ca. 200–50 BC, with a few older examples from the 3rd c. BC; according to J.W. Hayes [1985, 14], in fact, the dish with short vertical rim, type GW-D1 here, is an old form that probably precedes the forms with more pronounced rim) there was a change to a well-defined carination (*espèce* 2260, dated ca. 150–50 BC), and then to one that is even more precise and angular (*espèces* 2270 and 2280, dated between 160 BC and the Tiberian period). The dish with oblique rim from the House of Livia in Rome, for example, figures in the analysis of the morphological evolution and in the series of mutual influences among different wares (Morel 1965, 157 and 205, Pl. 26, No. 380, Campana A, dated 65±20 BC; see also Morel 1981a, 2282a1). It is clear that the presence of one does not exclude the presence of the other. In fact, at Taranto the dish with flat body (form 622/4b, phase E2, dated to the second half of the 2nd c. BC) appears to be slightly older than the dish with more oblique walls (form 622/4a, phase E3, dated to the first half of the 1st c. BC; see Hempel 2001, 142). The final result is a dish with vertical rim, well-defined carination, and a nearly horizontal body typical of sigillata ware (see, e.g., Carettoni 1959, 316, Fig. 23 f, from Halaesa; *Agora* V, Pl. 62 G34; Goudineau 1968, 279–80, type 1; Deschenes 1977, Fig. 34, No. 280, from Monte Irsi; Mazzeo Saracino 1985, Pl. 55, Nos. 11–13; Pucci 1985, Pl. 115, Nos. 8–11); Vanderhoeven 1988, 205, Fig. 38; Marchi/Sabbatini/Salvatore 1990, Pl. 23, No. 2, from Venosa; Melissano 1990, Pl. 113, No. 109, from Cutrofiano; *Conspectus*, 52–53, form 1; Yntema 1993, 100, Fig. 37, No. 214, from Oria).

Parallels in the contemporary black-gloss production can be found at Valesio, too (*Valesio* I, 158, form K05, dated end of the 2nd–1st c. BC).

On the basis of materials recovered from Gravina and comparison with the Cosan material (D.M. Taylor 1957, Pls. 34 D5cIII and 41 E5cI, dated mid-2nd–mid-1st c. BC), Prag (1992c, 141) proposes for these grey-ware dishes a chronology in the 2nd c. BC, while L. Giardino (1980a, 253–54) dates them generally between the beginning of the 1st c. BC and the Augustan period, dividing them into three groups: *a*, with low and thick rim; *b*, with higher and more oblique rim; *c*, with vertical rim (the dishes from the Metaponto survey mostly belong to group *c* because of their nearly vertical rims). The date, however, should be raised to the second half of the 2nd c. BC on the basis of the materials from the Pantanello kiln deposit (see the catalog). Yntema, too, inclines towards a chronology for these "carinated dishes" in the 1st c. BC—more precisely, in the years between 90/70 and the Augustan period. He maintains that this type of dish (form L05) develops from form L04 (dish with very low rim and deeper body in relation to form L05) in roughly the second or third decade of the 1st c. BC and remains in use until the Augustan period, being typical of the late phase of Apulian grey gloss (*Valesio* I, 219).

Type GW-D1 is comparable to L. Giardino's type *a* (1980a, 253, Pl. 81, No. 1) in the low rim, although the body of the survey piece is deeper. Precise parallels among the material recovered in the kiln deposit at Pantanello (Burgers's form 4/5) have not yet been found; nevertheless, it is possible to accept the hypothesis of a double function for this form, to judge from the very low rim, allowing it to be identified as a lid-dish, a form not foreign to classifications of ancient ceramics (see introduction to this form, above). The interpretation as dish is furthermore favored by the presence of gloss on the interior surface of the fragment.

There are three fragments of type GW-D2a, one of type GW-D1, and one of type GW-D2b. The dish with vertical rim is therefore the most widespread form (alongside the cup with offset rim) both inside and outside the survey zone, as demonstrated by the substantial number of attestations that can be found. Beyond Metaponto (see also L. Giardino 1980a, 253, and Burgers's form 4/5), the dish with vertical rim is one of the most common forms also at Herakleia (L. Giardino 1990, Pl. 24, type A), Taranto (Hempel 2001, 114), and Gravina (Prag 1992c, 141–43). The same is true for the Campana C production at Morgantina (Stone 1982, 101) and Syracuse (Pelagatti 1970, 470).

The diameter of this form is generally very large and can vary from 18–37 cm, although there are smaller examples (14 cm) and larger examples (40–55 cm). The dishes presented here have diameters in the intermediate range.

Dish with vertical rim (GW-D1 and GW-D2) cont.

1:2

GW-D1

1. GW 502-09 Dish rim/body

Dimensions: diam. 31.0 w 4.5 h 2.0 th 0.5/0.55

Condition: Roughly 7% of the rim and part of the body are preserved. Gloss is mostly detached.

Description: Low, indistinct rim with elliptical section. At ca. 0.8 cm the rim curves into the straight body wall with a gentle carination. The fragment was presumably completely covered in gloss. Grey fabric (7.5YR N6/), well purified, compact, and soft. Dark grey gloss (N4/), matte, diluted, and thin.

Comparisons: See L. Giardino's type *a* (1980a, 253, Pl. 81, No. 1). A dish similar to this was retrieved from the Crucinia necropolis (De Siena 1993, 129, No. 176). In Morel's classification the closest type to the survey's GW-D1 is 2252b1 in Campana A, from the wreck at Punta Scaletta, dated 140/130 BC. In Cotton 1979 (Fig. 27, No. 16), there is another similar dish, in black gloss and with a slightly lower rim, called "platter or lid," recovered in the levels of periods I, II, I/II, and III (roughly end of the 2nd c. BC–AD 160) of the Late Republican villa at Posto. Two dishes with low rim from Monte Irsi, too, are in black gloss (Small 1977c, Fig. 24, No. 102, also present in a grey-ware variant, and Fig. 24, No. 103, especially from levels before 100 BC), as well as one from Paphos (J.W. Hayes 1985, Pl. 1, No. 5, form 2A, ca. 150–100 BC).

Date: Middle of the 2nd–beginning of the 1st c. BC, because of contemporaneity with (but a shorter span than) type GW-D2 with a higher vertical rim (second half of the 2nd–1st c. BC), which presumably replaced it.

1:2

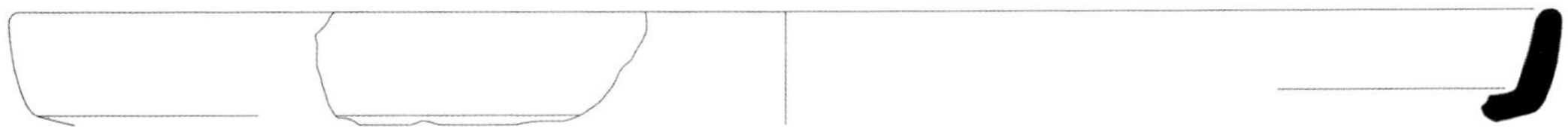

GW-D2a

2. GW 231-02 Dish rim

Dimensions: diam. 27.4 w 5.85 h 2.05 body th 0.5

Condition: Roughly 7% of the rim is preserved, along with the attachment to the body wall. The gloss is partially detached; on the interior surface it is present in scales.

Description: High, nearly vertical rim with elliptical section, attached to the body with a fairly well-defined carination; the body wall is slightly oblique. The fragment is completely covered in gloss. Light grey fabric (7.5YR N7/), well purified, compact, and hard. Dark grey gloss (N4/), matte, slightly diluted, and worn.

Comparisons: Some of the materials recovered in the castrum of Metaponto are similar (D'Andria 1976, Pl. 36, No. 31413, generally dated to the 2nd–1st c. BC; L. Giardino 1980a, Pl. 81, No. 5). A dish from tomb 324 at Crucinia is identical (De Siena 1993, 129, No. 178), as is another dish from the Pantanello kiln deposit (Burgers's form 4-PZ81.W.20.1, dated from the 1st c. BC until the Augustan period, but with several examples also in the second half of the 2nd c. BC); there is another similar fragment from the same area (Carter 1983a, 473, Fig. 49, No. 150, 150–50 BC, compared to both Lamboglia's C7 and an example from deposit E at Cosa in Campana B [D.M. Taylor 1957, Pl. 41 E5bI], dated 120/110–40/30 BC).

Date: Comparison with materials from the Metapontino suggests that production of this form began in the second half of the 2nd c. BC but achieved maximum circulation only in the 1st c. BC (at least until the Augustan period).

3. GW 105-01 Dish rim (see **No. 2**)

Dimensions: diam. ca. 23.0 w 3.35 h 2.15 rim th 0.65

Condition: Less than 10% of the rim is preserved, along with a trace of the attachment to the body. The gloss is worn in places.

Description: Slightly oblique rim with elliptical section, attached to the body with a gentle curve. The fragment is completely covered in gloss. Grey fabric (7.5YR N6/), well purified, compact, and soft. Dark grey gloss (N4/), matte, diluted, and thin.

4. GW 105-07 Dish rim (see **No. 2**)

Dimensions: diam. ca. 22.0 w 4.6 h 1.65 rim th 0.6

Condition: Less than 10% of the rim is preserved, along with the attachment to the body. Gloss is lost in places.

Description: Rim with elliptical section and slightly oblique form. The attachment to the body creates a gently curved carination on both the interior and exterior. The fragment is completely covered in gloss. Grey fabric (7.5YR N6/), well purified, compact, and soft. Grey gloss (N5/), matte and diluted.

Dish with vertical rim (GW-D1 and GW-D2) cont.

1:2

GW-D2b

5. GW 042-10 Dish rim/body

Dimensions: h 2.2 w 2.75 body th 0.5

Condition: A small part of the rim and body is preserved. The gloss is fairly well-preserved on the interior partially worn on the exterior.

Description: Nearly vertical rim with subrectangular section, receding at the edge. At ca. 0.5 cm from the rim the body wall tends to curve inwards slightly. The fragment is completely covered in gloss. Light grey fabric (7.5YR N7/), well purified, compact, and soft. Dark grey gloss (N4/), matte, diluted, and present in small scales on the interior body surface; dark olive-grey gloss (5Y 4/1), matte and very diluted on the exterior surface.

Comparisons: This variant is known at Metaponto (L. Giardino 1980a, Pl. 81, No. 6, with a subtle curvature of the body, dated to the 1st c. BC; Burgers 1991, form 4-SG35.15, from the farmhouse at Sant'Angelo Nuovo, dated to the 1st c. BC, with more oblique rim; also from the Pantanello kiln deposit is form 4-PZ81.W.327P, which Burgers dates to the second half of the 2nd c. BC on the basis of comparison to a dish from a Metapontine tomb).

Date: See type GW-D2a.

Dish with Offset Vertical Rim (GW-D3)

This particular form of dish does not yet have precise parallels in the Metapontino or elsewhere at least as far as grey-ware production is concerned. The thickened rim, the groove on the upper surface, and the concave wall of this dish recall the unpainted and banded plates well known even in the Metapontino beginning at the end of the 4th c. (see the plates in Ch. 9; this type of plate is found at Monte Irsi also in "grey glaze": see Small 1977c, 125 and Fig. 26, Nos. 126–27), but the closest examples are found among the sigillatas. It is therefore possible that this dish was a reworking of traditional forms accompanied by the assimilation of new models. Such an evolution perhaps occurred via the wide dish with thickened and variously modeled rim (see Burgers's form 14, dated to the second half of the 2nd c. BC). The width of this dish appears to vary from 16 to 40 cm and therefore is not far from that of the dish with vertical rim (D1 and D2). Only one fragment of this type was recovered in the survey.

1:2

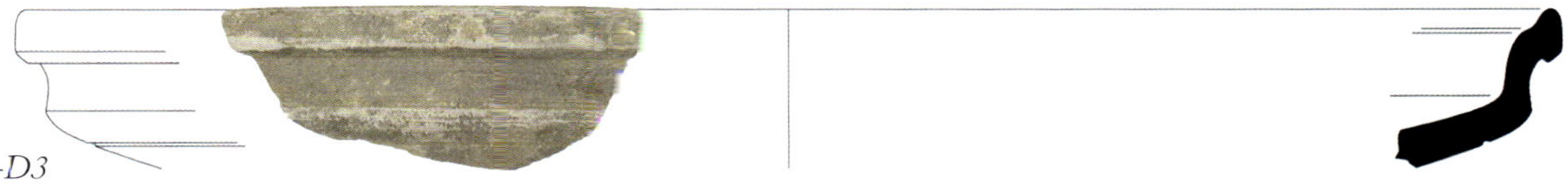

GW-D3

6. GW 372-03 rev93 Dish rim/body

Dimensions: diam. 30.95 w 8.2 h 3.3 body th 0.5/0.75

Condition: Roughly 10% of the rim is preserved, slightly chipped. The gloss is partially worn.

Description: Vertical, folded rim with elliptical section, distinguished from the body on the interior by a small depression. At 2.1 cm below the rim there is a well-defined carination in the body wall, which has a slightly concave profile on the upper surface, while the wall thickens and becomes oblique. A concentric groove 0.15 cm wide lies 0.9 cm below the carination on the exterior. The fragment is completely covered in gloss. Light grey fabric (7.5YR N7/), well purified, compact, and hard. Dark grey gloss (N4/), matte and homogeneous on the interior; dark olive-grey gloss (5Y 4/1), matte and diluted on exterior.

Comparisons: The rim shape recalls a cup from the Pantanello kiln deposit (Carter 1983a, 472, Fig. 48, No. 65, 150–50 BC; Burgers's form 27-PZ76.345, cf. dishes in Campana C from the area of Syracuse [Gentili 1956, 157, Fig. 8, Nos. 1–4, from a Hellenistic tomb in contrada San Giuliano]). The closest grey-ware shape is Morel's dish 1163a1 in the Cagliari Museum, perhaps of Sardinian origin (2nd c. BC). This type of rim is well known in terra sigillata (*Sibari* III, 330, No. 77bis and 422, Fig. 454, No. 18459) cf. Goudineau's form 17, dated after AD 12 (Goudineau 1968, 291); the body of the Sybarite fragment is curved, rather than carinated like the survey piece. See also De Boe and Vanderhoeven 1979, 109, Fig. 25, No. 8, from Ordona; Vanderhoeven 1988, 204, Fig. 37, No. 109; Mazzeo Saracino 1985, Pl. 57, No. 7, dated 20 BC–AD 15; Pucci 1985, Pl. 118, No. 1, form VIII variant 2, with rounded body, beginning 15 BC; *Conspectus*, 72–73 form 12.1), and in similar productions (Stone 1982, Pl. 27, No. 325, from Morgantina, a Sicilian imitation of TSI).

Date: Lacking more precise parallels, Burgers's form 27 in a level dated to the first half of the 1st c. BC leaves open the possibility of a date in the 1st c. BC, reinforced also by an obvious morphological affinity with dishes in sigillata, although the only suitable example, that from Cagliari, could move the date towards the 2nd c. BC.

Dish with Horizontal Rim (GW-D4)

This type of dish with thick horizontal rim and shallow body does not have exact parallels in other grey wares. It is close to a type with a more or less thickened rim in the black-gloss repertory of the first half of the 3rd c. BC, but probably did not achieve wide circulation. Function as both plate and lid is possible. The small size (ca. 12–20 cm) relative to other dishes (GW-D1–3 and GW-D5) also suggests a different function. One fragment was recovered.

1:2

GW-D4

7. GW 049-01 rev90 Dish rim

Dimensions: diam. 18.0 w 4.05 h 1.1 body th 0.4

Condition: Less than 10% of the rim and a small portion of the body are preserved. There are no traces of gloss.

Description: Flat rim with rectangular section, clearly distinguished from body on interior; on exterior rim/wall transition is gently curved. Grey fabric (10YR 6/1), well purified, compact, and soft.

Comparisons: Similar example, with deeper body, from Metaponto (L. Giardino 1980a, Pl. 82, No. 16, called *patera a bordo espanso*), as well as from Sant'Angelo Nuovo, with less pronounced rim (Burgers's form 7-SG35.11). Comparanda found among black-gloss dishes from Gravina (Prag 1992b, Fig. 28, No. 652), with less pronounced rim (300–265 BC); at Gravina in red gloss (Cotton 1992a, Fig. 49, No. 1112, belonging to phases of the 4th–3rd c. BC and end of the 2nd–1st c. BC); from Valesio (*Valesio* I, 148 form K03b, end of the 4th–beginning of the 3rd c. BC, with thinner rim); and similar from Etruscan area (Morel type 1315f1, from Roselle and dating probably to 2nd c. BC). J.W. Hayes 1972, 54, Fig. 9, No. 5, form 33, in ARSW with short, flat rim (diam. ca. 22 cm) although the rim of the ARSW example is turned slightly outwards (ca. AD 200–250).

Date: 2nd–1st c. BC.

Dish with Offset Horizontal Rim (GW-D5)

The dish with offset horizontal rim is characterized by a deep groove on the upper surface. Almost never produced in series, it has variants (L. Giardino 1980a, 256). It is well documented in the grey ware from Metaponto, and a similar dish in grey ware, with oblique rim, comes from Herakleia.

In several cases there is impressed roulette decoration on the rim with small parallel strokes. L. Giardino distinguishes two types. Type *a* with wider rim is more strongly linked to the Campana tradition (L. Giardino 1980a, 256): A, B, and C, cf. Lamboglia's forms A6 (1952, 168), from Ventimiglia, Entremont, and Saint-Remy, described as typical of the 2nd and 1st c. BC, and B6 (1952, 147), from Ventimiglia and the museums of Syracuse and Alba (uncertain provenience), dated ca. 200–50 BC; and C6 (1952, 158), from Syracuse and Tindari, dated 2nd–1st c. BC. In type *b* the rim is more contracted and thickened. Closest parallels are in Arretine and Eastern Sigillata A wares (L. Giardino 1980a, 256–57). The strongest link is to forms of the northern Etruscan area and northern Italy in general, e.g., Goudineau's type 12 in Arretine (Goudineau 1968, 288). The grooved horizontal rim is typical of Hayes's form 6 in ARSW (J.W. Hayes 1972, 28, No. 16, dated ca. AD 150–200). At Cosa the form is common in Campana type II (Campana B), less so in type I (Campana A) between the end of the 2nd–1st c. BC (D.M. Taylor 1957, 144, 154). The form appears to be absent at Gravina. The diameter generally varies from 19–36 cm, but there are smaller (ca. 15–16 cm) and larger (ca. 44 cm) examples. One fragment was recovered.

1:3

GW-D5

8. GW 105-06 rev90 Dish rim

Dimensions: diam. 34.95 w 4.5 h 1.3 body th 0.55

Condition: Roughly 6% of rim is preserved. Gloss is partially detached.

Description: Horizontal molded rim with subrectangular section, separated from body wall by groove ending in sharp angle at attachment point to body. Gloss covers entire fragment. Grey fabric (7.5YR N6/), well purified, compact, and hard. Very dark grey gloss (7.5 YR N3/), matte, homogeneous on exterior, shiny, homogeneous on interior, tending to form scales.

Comparisons: Among variants from the Pantanello kiln deposit, the closest parallel is PZ81.W.134.3 (Burgers's form 3, first half of 1st c. BC). From same deposit come two other dishes (Carter 1983a, 473, Fig. 49, Nos. 58 and 94, dated 150–50 BC), one (**No. 94**) with similar rim, but turned further outwards. Dishes with *orlo modanato orizzontale* from the castrum (L. Giardino 1980a, Pl. 81, second quarter of the 2nd–1st c. BC) are slightly different, but **No. 10**, with impressed decoration, is closest comparison; another has impressed decoration on rim (D'Andria 1976, Pl. 36, No. 3133, 2nd–1st c. BC).

Date: Middle of the 2nd–middle of the 1st c. BC, contemporary with D1–D2, which seems to linger slightly longer.

Cup with Offset Rim (GW-C1 and GW-C2)

The deep cup with oblique walls and offset pronounced rim was widely circulated and produced in several ceramic classes (black gloss, overpainted ware, grey ware, red gloss, relief ware, thin-walled ware, plain ware). Several types include: cups without feet; cups with variously modeled supports or with pronounced ring feet; simple cups; cups with grooves on the interior and exterior of the body.

Morel's *espèces* 2110–30 and 2150 (and type 1212a1) shows the cup's presence from northern Italy to Sicily, Spain, and Africa, and its circulation between the 3rd and mid-2nd c. BC. Lamboglia also places its appearance at the beginning of the 3rd c. BC as it is not found among Attic ceramics (Lamboglia 1952, 180, form A31, A31b from Ensérune; 181, form A33, from Ampurias, Ventimiglia, and Ensérune, dated 3rd–1st c. BC; cf. 149 form B9 [= Morel 2111a1], at the Tarragona Museum). The first examples, usually decorated with grooves and ridges, are well known in Athens and the eastern Mediterranean from the end of the 4th c. BC and persist through the beginning of the 1st c. BC (Prag 1992b, 82; Yntema 1990b, 181–83; 1993, 122, No. 320; *Valesio* I, 229; Hempel 2001, 120; Mazzei and Pietropaolo 1995, 246; *Corinth* VII.3, 46–47, Pl. 47, No. 190 "hemispherical krater," and 90–92, Pls. 17, 40, 55, Nos. 532–49 "conical bowl"; *Agora* XXIX, 107–10, Fig. 20 "hemispherical cup and parabolic cup"; Small 1977c, 125, No. 129).

The morphological structure of the cups with offset rim—thin walls, grooves, ridges, metallic gloss—suggests metal prototypes, upon which followed independent development of the ceramic production (e.g., *Corinth* XII, 69, Fig. 1, No. 531; Tassinari 1993, 153, type 1131) and in glass (J.W. Hayes 1985, 22; Prag 1992b, 82–84; Prag 1992c, 139; Yntema 1990b, 181; Lippolis 1994b, 272; *Agora* XXIX, 109 n. 107; Hempel 2001, 112; cf. H.A. Thompson 1934, 427, Fig. 113 E159; J.W. Hayes 1975, 166, Fig. 1; *Corinth* XII, 94, Fig. 6; *Otranto* I, 77, No. 7:11; *Gli ori*, 449–451, Nos. 42–50, from a tomb at Canosa dated end of the 3rd–beginning of the 2nd c. BC, cf. Lippolis and Mazzei 1984, 223, and Corrente 1992, 344).

Yntema emphasizes the direct descent of this form, which he calls form K45, a *mastos*, and includes the hemispherical cups) from similar forms in black gloss dated between the end of the 3rd and the mid-2nd c. BC. The term *mastos* (see Kanowski 1984, 105) originally defined cups lacking a resting surface and having a conical termination reminiscent of a breast (cf. No. 550 in *Corinth* VII.3). Yntema suggests that its success as a drinking vessel was due to the decline of the ring-handled kantharos (form L36), making it a chronological marker for contexts of the 2nd c. BC. He maintains that the absence of this form from L. Giardino's study of the Metapontine material must indicate the disappearance of the "mastoid" cup roughly at the end of the 2nd/beginning of the 1st c. BC, the period of grey ware's great success in the Metapontino (*Valesio* I, 195–98, form K45; 229, 232, form L36; cf. L. Giardino 1980a, 286). Yntema (1990b, 183 n. 62) describes it as the most common grey-ware form in contexts dated between the middle and third quarter of the 2nd c. BC. New stratigraphic data from the kiln deposit at Pantanello indicates the second half of the 2nd c. BC as the production span for this drinking vessel, confirming the hypothesis proposed by Yntema (see Burgers's forms 18 "mastos or hemispherical bowl" and 20 "rounded bowl").

At Taranto circulation of this cup in black gloss and HFR seems to belong exclusively to phases D/E1 and E1 (end of the 3rd–first half of the 2nd c. BC), while the grey-ware version is found in phases E1/E2 and E2 (middle–end of the 2nd c. BC). A single example is attested in phase E2 (second half of the 2nd c. BC) and one in E3 (first half of the 1st c. BC): see Hempel 2001, 143, E2-1:1, E3-44:17. D'Amicis, referring to tomb CXXX, assigns the "mastoid" cup in black gloss to the first quarter of the 2nd c. BC (*Gli ori*, 465 and 467, No. 17). Yntema (1990b, 176) and Hempel (2001, 176, E1-24:4) describe this cup as pertaining to the HFR group, while Graepler (*Gli ori*, 494, No. 74) dates the example in grey ware, with a groove on the exterior immediately below the rim, to the third quarter of the 2nd c. BC on the basis of the associated grave goods from tomb CXXXIX; see also Prazzoli 1996, 334 type 1, in black gloss. At Herakleia hemispherical cups in black gloss are found in tombs dated to the third quarter of the 2nd c. BC, but no examples in grey ware have been found (L. Giardino 1990, Pl. 28, Umbria 13–14).

In the sites above listed the cup with offset rim was in use at the same time (second half of the 2nd c. BC), while at Gravina this form appears to have been introduced earlier, at the end of the 3rd/beginning of the 2nd c. BC, perhaps as a replacement for the exhausted black-gloss production, which did not go beyond the first quarter of the 2nd c. BC (Prag 1992c, 139–40, Fig. 41, No. 989; see a hemispherical cup in grey ware, No. 988, which the stratigraphic context dates between the end of the 3rd and end of the 1st c. BC).

Associations with precious objects in glass and metal suggest this shape was not exclusive to table service, but perhaps also used in the female toilette to mix powdered cosmetics (Lippolis 1994b, 272; Mazzei and Pietropaolo 1995, 246).

Among the fragments there are two principal types: type GW-C1 with a rim that is turned outwards only slightly and distinguished from the body wall by a slight thinning; type GW-C2 with a rim that is clearly turned outward. Type GW-C2 has two subtypes. Despite the fragmentary nature of the assemblage, this "mastoid" cup can be differentiated from the hemispherical variety

Cup with offset rim (GW-C1 and GW-C2) cont.

(types GW-C3 and GW-C1-3), which has less rigidity in the body wall, especially near the foot; still, the boundary between the two forms is not always clear-cut (Yntema classifies both as mastoi). There are two fragments of type GW-C1, one of type GW-C2a, and two of type GW-C2b. The diameter of the rim varies from 11–25 cm. As mentioned above, this is the most common form among the grey-ware fragments from the survey, alongside the dish with vertical rim (see the introduction to types GW-D1 and GW-D2). This finding is confirmed by the Pantanello kiln deposit (Burgers's forms 18 and 20). The same situation occurs at Taranto (Hempel 2001, 114), Valesio (*Valesio* I, 229), and Gravina (Prag 1992c, 139-40).

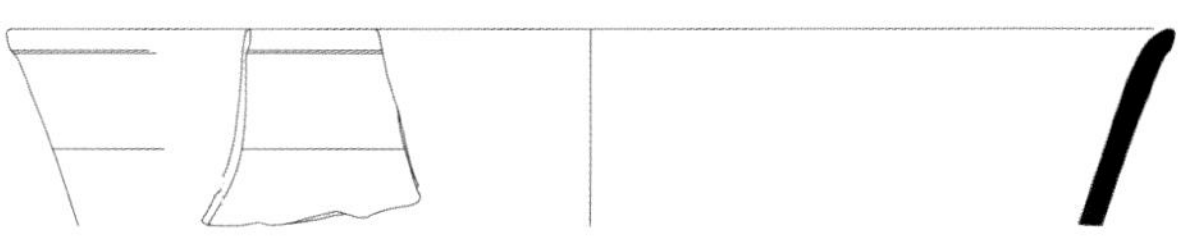

GW-C1
9. GW 476-05 Cup rim
Dimensions: diam. 15.5 w 2.9 h 2.75 body th 0.3
Condition: Less than 5% of rim is preserved, chipped, along with traces of the body wall. Gloss is partially detached.
Description: Slightly outward-turned, indistinct rim with subtriangular section; the body wall is deep and straight. The fragment is completely covered in gloss. Grey fabric (7.5YR N6/), well purified, compact, and soft. Dark grey gloss (N4/), matte and diluted.
Comparisons: From the Pantanello kiln deposit the closest example is PZ81.NW.752.12 (Burgers's form 18, second half of the 2nd c. BC). A similar cup from tomb 5 in contrada Casa Ricotta, with 100 fragments of "greyish clay," is dated second half of the 3rd/perhaps part of the 2nd c. BC (Lo Porto 1981a, Fig. 64, No. 1); cf. fragment from the area of the castrum (D'Andria 1976, Pl. 36, No. 31636); also Difesa S. Biagio (Roubis 1996, Fig. 9, No. 12).
Date: Second half of the 2nd c. BC.

---

10. GW 047-01 Cup rim (see **No. 9**)
Dimensions: w 2.7 h 2.45 th 0.25
Condition: A small portion of the rim is preserved, insufficient to determine diameter. Gloss almost completely lost.
Description: Thin, oblique rim with triangular section.Very small traces of gloss suggest that the fragment was completely covered in gloss. Grey fabric (7.5YR N6/), well purified, compact, and soft. Grey gloss (N5/), matte.

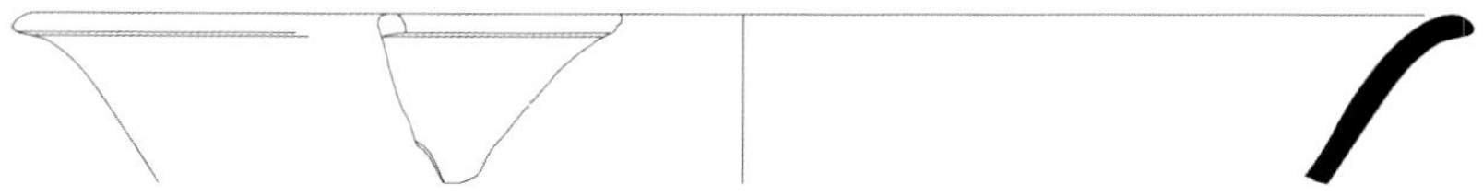

GW-C2a
11. GW 105-05 rev90 Cup rim
Dimensions: diam. 21.0 w 3.5 h 2.5 body th 0.35/0.4
Condition: Roughly 6% of the rim is preserved. The gloss is worn in places.
Description: Thin, outward-turned rim with subtriangular section, indistinct from the deep body, which has a straight wall as preserved. The fragment is completely covered in gloss. Grey fabric (7.5YR N6/), well purified, compact, and soft. Dark grey gloss (N4/), matte and diluted.
Comparisons: From the Pantanello kiln deposit, the cup closest to this piece, albeit with thicker walls, is PZ81. NW.971.1 (Burgers's form 18, second half of the 2nd c. BC). Similar examples from Valesio (*Valesio* I, 228, No. 405, form L31) were found in a context of the 1st c. BC, although the form is generally dateable between the middle and end of the 2nd c. BC. Also found at Monte Irsi (Small 1977c, Fig. 26, No. 128) without the thin rim, from phase A1, dated before 100 BC, as well as at Posto (Cotton 1979, Fig. 25, No. 14, in black gloss, called "medium sized bowl") from the period I levels of the Late Republican villa dated end of the 2nd c.–30 BC. At Sybaris it is present in black gloss with indentations on the interior of the body (*Sibari* II, 483, No. 613, 536, Fig. 590, No. 20823, compared to Lamboglia's A33); cf. Morel's cup 1212a1 from Lilybaeum, dated perhaps to the beginning of the 3rd c. BC.
Date: Second half of the 2nd c. BC.

Cup with offset rim (GW-C1 and GW-C2) cont. 1:2

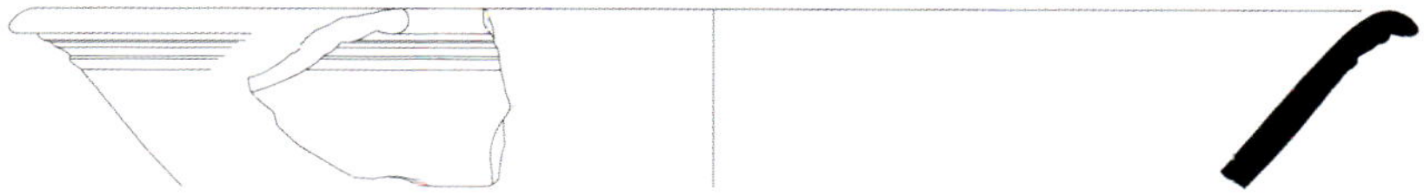

GW-C2b

12. GW 556-03 Cup rim

Dimensions: diam. 21.5 w 4.0 h 2.8 body th 0.45/0.5

Condition: A small portion of rim is preserved, chipped, as well as part of the body wall. Gloss is partially detached.

Description: Outward-turned, indistinct rim with triangular section, differentiated on exterior by four parallel grooves (0.15, 0.15, 0.2 , and 0.1 cm). Body wall is oblique and straight. Fragment presumably completely covered in gloss. Grey fabric (7.5YR N6/), well purified, compact, and soft. Dark grey gloss (N4/), matte, diluted, and thin.

Comparisons: Cups with oblique rim from the Fantanello kiln deposit generally lack grooving beneath the rim: the closest example is PZ81.NW.791.5 (Burgers's form 18, second half of the 2nd c. BC); see also cup with practically everted rim and grooves from Valesio (*Valesio* I, 228, No. 401, form L31, dated to the middle–end of the 2nd c. BC); another, with a single groove from field survey at Oria (Yntema 1993, 123, Fig. 46, No. 320) is dated middle–end of the 2nd c. BC); in grey fabric from excavation in Via Margherita in Taranto (Costamagna 1983, Pl. 49, No. 17).

Date: See **No. 11**.

13. GW 556-04 Cup rim (see **No. 12**)

Dimensions: w 1.85 h 1.75 th 0.35/0.4

Condition: A small portion of the rim and body wall is preserved. The gloss is worn on the interior.

Description: Indistinct, slightly outward-turned rim with subrectangular section, emphasized on the exterior by two shallow parallel grooves (0.1 and 0.15 cm). The rim is slightly less pronounced than **No. 12**. The body wall appears to be straight. The fragment is completely covered in gloss. Grey fabric (7.5YR N6/), well purified, compact, and hard. Grey gloss (N5/), matte, diluted, and thin.

Hemispherical Cup (GW-C3)

This study distinguishes the two cups—the mastoid and the hemispherical—despite their close relationship and common origins (see introduction to the cup with offset rim). It has also noted how several authors group them under the same entry. Prag, for example, speaks generally of "bowls with grooved walls," Yntema of *mastoi*, and Hempel of *hemisphärischer Becher*. Burgers, however, calls the mastoid type a "hemispherical bowl" and the other type a "rounded bowl," while, on the other hand, Marabini Moevs speaks of a "conical bowl" and "hemispherical bowl" (Prag 1992c, 139–40, Nos. 985–91; *Valesio* I, 229; Hempel 2001, 143; Burgers's forms 18 and 20; Marabini Moevs 1973, forms XXXVI–XXXVII).

The characteristic element of this form is the nearly constant presence of one or two grooves immediately beneath the rim, usually indistinct and mated to a body with vertical walls forming a soft, wide curve near the base, giving the vessel its unique hemispherical shape. In general the foot is barely marked, with the creation of a hollow at the resting surface and a groove near it (the "false ring foot": see L. Giardino 1978b, 122 B1).

Study of the ceramic forms in circulation during the Late Republican period shows an affinity between the hemispherical cup and several cups that are more shallow but have the same form of rim. This latter type will be examined in more detail in the section on the bowl with grooved rim (type GW-B2), which, given its circulation between the end of the 2nd and first half of the 1st c. BC, seems almost to replace the hemispherical cup as a drinking vessel. (The earliest bowls—5th c.—were instead attempts to create a new type of dish, therefore designed to contain solid foods rather than beverages: see *Agora* XII, 128.)

The survey collected three rim fragments and five base fragments of hemispherical bowls, confirming the abundant presence of this type of deep cup (which, alongside the dish with vertical rim, is the best-attested form in the material recovered: see the introduction to the cup with offset rim). Occasionally, however, the poor state of preservation of the foot fragments casts doubt on the secure attribution to one form or another—i.e., mastoid or hemispherical—a situation rendered all the more complicated by the weak morphological boundary between them.

The diameter of the rim varies from 9.5–13.5 cm, although some are as large as 18 cm. The fragments from the survey fall in the middle of this range of values.

Hemispherical cup (GW-C3) cont.

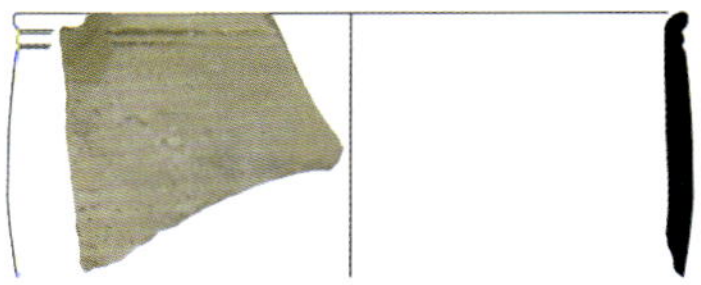
1:2

GW-C3
14. GW 502-11 Cup rim
Dimensions: diam. 8.95 w 3.75 h 3.6 body th 0.35
Condition: Roughly 8% of rim and part of body wall are preserved. Gloss is preserved only in exterior grooves and in a thin band immediately beneath rim on interior.
Description: Indistinct, slightly outward-turned rim with triangular section, emphasized at 0.25 cm by two narrow parallel grooves, each 0.5 cm and separated by 0.1 cm. The line of the body wall perhaps hemispherical, the curvature very soft. The fragment was likely covered in gloss. Grey fabric (7.5YR N6/), well purified, compact, and hard. Grey gloss (N5/), matte, diluted, and thin.
Comparisons: The hemispherical cup is well documented in the Pantanello kiln deposit, where it is present in several variants defined mostly by the line of the rim and the number of grooves beneath it; the closest comparison is with PZ81.NW.963.3 (second half of the 2nd c. BC). Other examples come from the area of the castrum: D'Andria 1976, Pl. 36 A, No. 31957 (also with ribbed body: No. 31975). At Metaponto it is present in an unpainted variety (Scarano 1992, Fig. 12 S.T.29 R4, with shell-shaped feet).
Date: Second half of the 2nd c. BC, with a possible extension into the first half of the same century.

15. GW 197-01 Cup rim (see **No. 14**)
Dimensions: diam. ca. 12.0 w 2.6 h 2.85 body th 0.35
Condition: Roughly 5% of rim is preserved. Gloss is perhaps detached on interior, lost in part on exterior.
Description: Indistinct rim with subtriangular section and deep body. At 0.65 cm beneath the rim there is a narrow groove (0.15 cm). Absence of gloss on interior creates doubt if vessel was completely covered in gloss. Light grey fabric (7.5YR N7/), well purified, compact, and soft. Dark grey gloss (N4/), matte, diluted, and thin.

1:1

16. GW 339-01 Cup rim
Dimensions: diam. ca. 12.0 w 4.35 h 2.55 body th 0.4
Condition: Roughly 15% of the rim is preserved, chipped. The gloss is lost in part, part preserved in scales.
Description: Rim with circular section emphasized by two wide, parallel grooves, each 0.2 cm wide and separated by 0.1 cm. The body wall is straight in the preserved section. The vessel was presumably completely covered in gloss, although on the exterior the gloss is preserved only in the grooves. Grey fabric (7.5YR N6/), well purified, hard, and slightly cracked. Dark grey gloss (N4/), matte, slightly diluted, and tending to flake on the exterior; olive-grey gloss (5Y 4/1), matte and diluted on the interior.

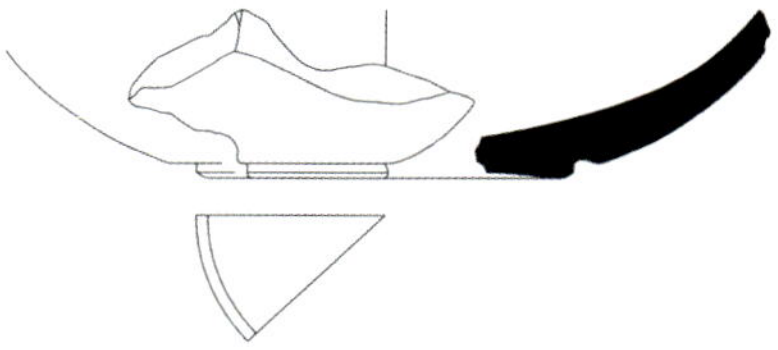
1:2

GW-C1–C3
17. GW 372-04 rev93 Cup foot
Dimensions: diam. 5.0 w 4.65 h 2.3 body th 0.5/0.75
Condition: Roughly 17% is preserved, along with lower part of body wall. Gloss present in traces, especially above the foot.
Description: Very low disk foot clearly distinguished from the hemispherical body wall via a wide and deep groove (0.45 x 0.15 cm). Fragment presumably covered in gloss. Grey fabric (7.5YR N5/), well purified, compact, and hard. Dark grey gloss (N4/), matte, diluted, and thin.
Comparisons: Likely a hemispherical cup, but there are several parallels with offset rim. In the Pantanello kiln deposit nearly all of the rounded bowls have this type of foot: the closest one is PZ81.NW.855.21; several mastoi are comparable, e.g. PZ81.NW.735.1 (Burgers's forms 20 and 18, second half of the 2nd c. BC), and a foot (Carter 1983a, 472, Fig. 48, No. 63, "conical cup," 150–50 BC). From the castrum: D'Andria 1976, Pl. 36A, No. 31957.
Date: See type GW-C3 above.

18. GW 317-01 Cup foot (see **No. 17**)
Dimensions: diam. 5.0 w 4.3 h 2.5 body th 0.45/0.8
Condition: A small portion of the foot and body wall are preserved. There are traces of gloss only near the foot.
Description: Low, concave disk foot distinguished from hemispherical body via a groove 0.35 cm wide and 0.1 cm deep. Presumably completely covered in gloss, which is almost completely detached. Traces of exposure to fire. Grey fabric (7.5YR N6/), well purified, compact, and hard, with a light grey core (7.5 YR N7/). Dark grey gloss (N4/), homogeneous and slightly shiny.

Hemispherical cup (GW-C3) cont.

19. GW 502-02 Cup foot (see **No. 17**)
Dimensions: diam. 4.7 w 4.1 h 1.0 body th 0.75
Condition: Roughly 20% preserved, with traces of body wall. Traces of gloss on the exterior; interior very worn.
Description: Very low, slightly concave (0.5 cm) disk foot distinguished from presumably hemispherical body by small groove 0.3 cm wide and 0.05 cm deep. A trickle of gloss reaches foot. Grey fabric (7.5YR N6/), well purified, compact, and hard. Grey gloss (N5/), matte, diluted, and worn.

20. GW 539-01 Cup foot (see **No. 17**)
Dimensions: diam. 4.8 w 4.65 h 1.5 body th 0.8/0.9
Condition: Roughly 25% of foot is preserved, chipped, along with base of the body wall. Gloss is partially lost.
Description: Very low, slightly concave disk foot clearly distinguished from the presumably hemispherical body by a groove 0.35 cm wide and 0.15 cm deep. The fragment is completely covered in gloss. Light grey fabric (7.5YR N7/), well purified, compact, and hard. Dark grey gloss (N4/), matte, diluted, and thin.

21. GW 556-02 Cup foot (see **No. 17**)
Dimensions: diam. ca. 4.8 w 3.15 h 1.0 body th 0.7
Condition: Roughly 20% preserved. Gloss is preserved.
Description: Slightly concave disk foot clearly distinguished from the body, which is carinated at the base. The fragment is completely covered in gloss. On the underside there are traces of manufacture on a wheel. Grey fabric (7.5YR N5/), well purified, compact, and soft. Dark blue-grey gloss (5B 4/1), matte, homogeneous, and compact.

Cup with Vertical Rim (GW-C4)

This cup is as deep as the preceding ones, but its vertical rim is clearly distinguished from the body via a carination located a short distance from the rim. The closest parallels come from black-gloss materials, especially from shallow cups (bowls). A link with productions in north-central Italy in particular can be identified: very similar to the survey piece is a black-gloss cup from Vado di Camaiore, dated to the middle of the 3rd c. BC (Morel type 2722a1). The form is known in southern Italy as well (e.g., Morel's bowl 2731a1, from Teano) and there are attestations of it in Greece (H.A. Thompson 1934, 436, Fig. 117 D9). It is fairly common in black gloss in the 3rd c. BC, with a tendency for the carination to be accentuated clearly in the two following centuries.

The form in grey ware is found only in the hinterland of Metaponto and is perhaps slightly older than the mastoid and hemispherical cups, although confirmation of this hypothesis would require more data from stratified contexts. The unique form of the rim could suggest a link with the dishes/lids, but in this case the depth of the form appears to preclude a double function. The diameter generally varies between 15 and 20 cm, but the diameter of the survey piece—the only one of its kind recovered—is smaller.

GW-C4
22. GW 502-04 Cup rim

1:2

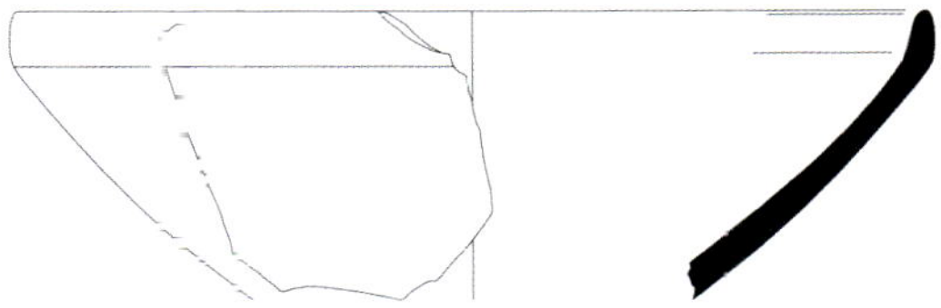

Dimensions: diam. 13.5 w 4.85 h 4.35 body th 0.45/0.5
Condition: Roughly 6% of the rim is preserved, along with part of the body wall; the fragment is heavily worn.
Description: Rim with elliptical section; at 0.6 cm below rim there is a soft carination, from which body wall opens in a nearly straight line. Gloss has disappeared almost completely, being partially recognizable only on exterior. Grey fabric (7.5YR N6/), soft, with very small inclusions. Olive-grey gloss (5Y 6/1), matte, diluted, and very thin.
Comparisons: It is not yet possible to identify precise parallels for this piece in grey ware. The closest is a form from Gravina, although the latter has a less pronounced carination (Prag 1992c, Fig. 45, No. 1061, dated to the end of the 3rd–beginning of the 2nd c. BC, but recovered from period VIIIa levels dated end of the 2nd–1st c. BC). At Gravina it is found also in unpainted ware, but it is shallower and has a slightly inward-turned rim (Small et al. 1994, Fig. 10, No. 94, from trench F202).
Date: Lacking further parallels, first half of the 2nd c. BC.

Cup with Offset Vertical Rim (GW-C5 and GW-C6)

This is a medium-sized cup with a strongly outward-turned and nearly vertical rim. Its relationship to black-gloss forms of the Greek tradition is clear, as demonstrated by a bowl found at Athens, and as Lamboglia suggests for his form C19 (*Agora* XII, No. 808, dated 310 BC; Lamboglia 1952, 160, 177. See introduction to type GW-B1, above).

Two types appear in the survey material: type GW-C5, with simple rim, is associated with the black-gloss tradition; type GW-C6, with a groove near the rim, is typical of later production in both grey ware and sigillata.

The type GW-C5 does not have precise parallels at the moment: some comparisons can be found at Gravina (Prag 1992c, Fig. 44, No. 1045, dated to the end of the 2nd c. BC rather than the 1st c. BC), at Oria (Yntema 1993, 99, Fig. 36, No. 200, and 128, Fig. 48, No. 344, dated to the 1st c. BC), and at Cosa (D.M. Taylor 1957, 168, Pl. 28, B36b, from a deposit dated 180–140 BC; the type was imported at Cosa towards the middle of the 2nd c. BC and is common in the deposits dated to the end of the 2nd–1st c. BC).

For type GW-C6 a probable link to the small cup with offset rim and concave wall should not be ignored; examples are Morel's *espèces* 1220–40 without the groove on the rim, and 1250 with the groove—these were in many cases local grey-ware products—and in the examples from Gravina without the groove on the rim (Prag 1992c, Fig. 40, No. 953, in grey ware, a reworking of a similar black-gloss form, dated 2nd–1st c. BC; cf. 286, Fig. 28, No. 623). There are also affinities with fish plates, a form that circulated between the end of the 5th and the 2nd c. BC (see Morel 1981a, 83; *Agora* XII, 147; *Agora* XXIX, 146–49; *Corinth* VII.3 40-41). This is especially evident in the vertical rim and the groove on the upper surface (D.M. Taylor 1957, 178; Carter 1983a, 474, No. 65. Compare e.g., D.M. Taylor 1957, Pl. 21 A2; Morel 1965, 51, form 81, Pl. 4, No. 65). The absence of precise parallels suggests, as for type GW-D3, a local interpretation of well-known models in grey ware and black gloss.

Comparable examples to type GW-C5 have a diameter ranging from 20 to 30 cm, although there are occasionally cups of smaller size like the survey piece (ca. 7.5–13 cm). Type GW-C6 has a diameter between ca. 10.5 and 15 cm. A single fragment of each type was found.

GW-C5

23. GW 042-12 Cup rim/body

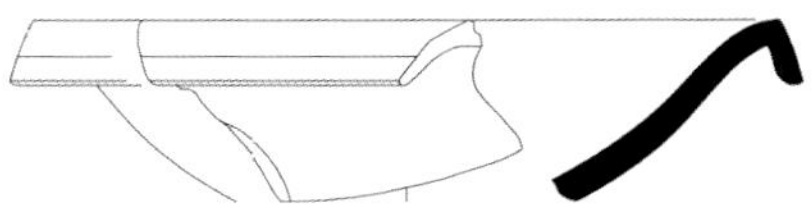

1:2

Dimensions: diam. 10.5 h 2.5 w 5.1 body th 0.35/0.4

Condition: Roughly 10% of the rim is preserved, chipped, along with part of the body wall. The gloss is preserved.

Description: Folded rim with triangular section; body wall is gently curved, nearly hemispherical, with irregularities in exterior finishing. Fragment completely covered in gloss. Grey fabric (7.5YR N6/), well purified, compact, and hard. Dark grey gloss (N4/), matte, uniform on interior, irregular on rim, and very diluted on exterior.

Comparisons: Precise parallels for this piece could not be found, but the form can be attributed to L. Giardino's *coppe a parete espansa e orlo estroflesso* (L. Giardino 1980a, 267–72, Pl. 83, No. 43, type *c*, with less pronounced rim, dated to the second half of the 2nd c. BC), as well as a cup from the Pantanello kiln deposit (Burgers's form 13-PZ81. NW.769.1 dated to the first half of the 1st c. BC), and another from Sant'Angelo Nuovo (Burgers SG90/1.5).

Date: 150–50 BC.

GW-C6

24. GW 372-03 Cup rim

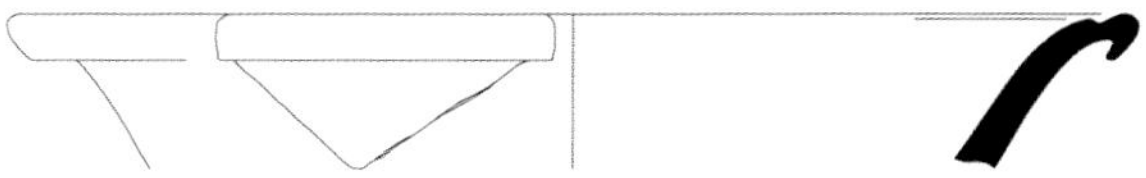

1:2

Dimensions: diam. 15.0 w 4.5 h 2.15 body th 0.45/0.5

Condition: Roughly 10% of the rim (chipped) and body wall is preserved. The gloss is worn in places.

Description: Oblique, folded rim with elliptical section, clearly distinguished from body via a groove on upper surface; body wall is straight and slightly concave. Fragment is completely covered in gloss. Grey fabric (7.5YR N6/), soft, with very small inclusions. Very dark grey gloss (7.5YR N3/), matte and homogeneous.

Comparisons: A cup from Sant'Angelo Nuovo is similar, but its rim is more oblique (Burgers's form 27-SG304.4). The type is generally comparable to L. Giardino's *coppa con orlo rovescio*, described as typical of the 1st c. BC and similar to Goudineau's type 12 dish in Arretine (L. Giardino 1980a, 266–67, especially Pl. 82, No. 31), but the thin rim does not correspond to Goudineau type 12 (Goudineau 1968, 288, ca. 20 BC), to which type GW-D5 comes close. The closest parallel among Goudineau's materials is type 13 (Goudineau 1968, 289, ca. 20 BC).

Date: 1st c. BC.

Bowl with Concave/Convex Body (GW-B1)

The shallow bowl with concave/convex body features a more or less distinct rim with a generally subtriangular section. The rim attaches to oblique walls that close the body with a soft carination at about two-thirds the height.

The dimensions and structure of this bowl suggest a strict morphological and chronological link to the bowl with offset rim (see Burgers's form 12, dated to the second half of the 2nd c. BC). Its model is in black gloss: Lamboglia's form A28a is an example, originating perhaps in the 4th c. BC (Lamboglia 1952, 172, 177–78), an evolution of Attic form A22. The bowl with offset rim is well known at Athens beginning in the 5th c. BC (*Agora* XII, 128–30 and Fig. 8, Nos. 777–808) with a form that has a long life in the Hellenistic period (Nos. 802–8; *Agora* XXIX, 156–60 and Figs. 59–61; cf. H.A. Thompson 1934, 436, Fig. 117 A9, A71, D5). The closest parallel is the Hellenistic type which began production in the 3rd c. BC (*Agora* XXIX, 158 n. 44. Dimensions vary between 8 and 27 cm. These objects could be serving plates or dishes to hold condiments or fruit, but were probably not drinking vessels; cf. H.A. Thompson 1934, 436, Fig. 117, E33).

Morel's classification facilitates identification of the variants of this shallow bowl, illustrated in *genre* 2600 *vases à courbe et contre-courbe*. The form is found chiefly in the area of Etruscan influence, in northern Italy, and in the western Mediterranean, with just a few examples in southern Italy, all dating between the 4th and 2nd c. BC, although several types last into the 1st c. BC. Morel's *genre* 2600 has two comparable cups, both dated to the 3rd–2nd c. BC: type 2614c1 from Aléria and type 2645a1 from Ornavasso. The form is found at Minturnae (*Minturnae*, Pl. 3, No. 18, middle of the 3rd c. BC; see Morel 2643b1); at Cosa it is produced in type IV, a local ware, and described as the most popular form of bowl from 170/60 to 40/30 BC (D.M. Taylor 1957, 179). The same form occurs in ARSW, but with a much more rigid body wall (J.W. Hayes 1972, 28, Fig. 3, form 5, No. 26, perhaps middle of the 2nd c. AD).

In grey ware the form belongs to the late second half of the 2nd c. BC. It corresponds to Burgers's forms 12A–B and 13: he describes an evolution from a bowl with nearly indistinct rim to one with a hanging rim (L. Giardino 1980a, 267–72; Burgers 1991, second half of the 2nd–first half of the 1st c. BC; cf. type GW-C5 above). In general all parallels lack the groove beneath the rim. Two examples come from Gravina: the first with a more accentuated rim is dated end of the 2nd–1st c. BC (Prag 1992c, Fig. 45, No. 1054, beginning of the 1st c. BC based on the moderate carination); the second is dated 75±15 BC (Small et al. 1994, Fig. 7, No. 10). Other examples come from Leuca (L. Giardino 1978b, Pl. 68, B30) and Sybaris (*Sibari* IV, 376, No. 265, and 408, Fig. 375, No. 11609), to mention a few.

The diameter of the rim varies from 13 to 30 cm. Only one fragment of this type was recovered in the survey.

1:2

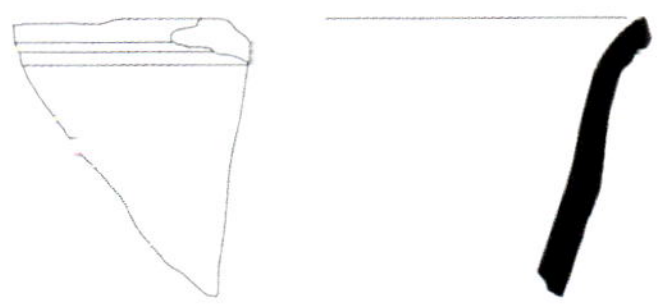

GW-B1

25. GW 502-10 Bowl rim

Dimensions: w 3.15 h 3.75 body th 0.4/0.45

Condition: Roughly 7% of rim is preserved with part of body wall. Gloss is detached on interior and worn on exterior.

Description: Indistinct, outward-turned rim with square section; the body wall is oblique and doubly curved. Exterior and rim covered in gloss, while gloss on interior is hypothetical. Grey gloss (7.5YR N6/), soft, with very small micaceous inclusions. Dark grey gloss (N4/), matte, diluted, and thin.

Comparisons: The bowl closest to this fragment is PZ81.W.377.2, though lacking the groove beneath the rim (Burgers's form 12B, second half of the 2nd c. BC). It recalls a cup in L. Giardino's classification, although the rim is indistinct; the groove beneath it suggests that it belongs to type *b* (L. Giardino 1980a, Pl. 83, No. 38), compared to Goudineau's type 14 (Goudineau 1968, 289–90, dated 20–15 BC, with outward-turned rim and carinated body).

Date: Second half of 2nd–beginning of the 1st c. BC.

Bowl with Grooved Rim

The rim of this shallow bowl is emphasized by one or two parallel grooves. The form is well known in grey ware and corresponds to L. Giardino's *coppa con orlo liscio scanalato* (L. Giardino 1980a, 261–62). Lamboglia believes the form B1 to be derived from the Attic and Campanian kylix, deprived of its handles, and characteristic of the 1st-c. BC levels (form C1, by contrast, has a less oblique foot and thicker walls); from the B1 was then derived the deeper B8 (Lamboglia 1952, 143–44, 148, and 157). This point recalls form L33 from Valesio, belonging to a form that Yntema describes as enigmatic and perhaps linked to the hemispherical cups (see type GW-C3 above). Yntema proposes a date in the 1st c. BC because of its absence from the main contexts at Valesio and its presence at Metaponto (see L. Giardino above). Cf. also *Valesio* I, 230, No. 410, from a 1st-c. BC context.

Among the materials from the unpublished kiln deposit at Pantanello, the "bowl with grooved rim" originates essentially in level I, dated to the first half of the 1st c. BC (Burgers's form 15), confirming the date proposed by Yntema but inverting the hypothesis of Lamboglia, according to whom the deeper cup derived from the cup with lower walls (see above). As support for this conclusion one can consider Morel's *espèces* 2320–50, dated mid-2nd–1st c. BC, produced most of all in Campana B and C and circulated in Africa, Spain, south-central Italy, Sicily, and areas under Etruscan influence; these types generally appear slightly later than the deeper *espèces* 2550–70, dated 3rd c.–mid-1st c. BC, present in all of Italy (including the islands), Spain, and Africa, and mostly locally produced. At Taranto, too, this bowl belongs to the more recent phases of the necropolis and could be a unique form derived from models in Campana C (Hempel 2001, 115 and 142, form 622/6, phase E3/F, second half of the 1st c. BC).

Well known in grey ware in southern Italy and Sicily, the form, with variants, is successful in other wares such as TSI and ARSW. The rim is found in light sigillata (Fallico 1971, 607, Fig. 30 A142, from Syracuse, second half of the 2nd c. AD). The rim emphasized by a pair of grooves, generally found on shallow cups, is found also in ARSW (J.W. Hayes 1972, Fig. 4, form 9, type B, second half of the 2nd c. AD).

It was possible to measure the rim diameter of one of the two fragments. It falls in the middle of measurements for this form, which vary from 11 to 20 cm.

GW-B2

26. GW 105-253 rev82 Bowl rim

Dimensions: w 1.95 h 2.1 body th 0.45

Condition: A small portion of the rim is preserved, insufficient to measure the diameter. Gloss is worn in places.

Description: Oblique rim with subcircular section, not distinguished from body, emphasized by two small parallel grooves. Fragment is completely covered in gloss. Grey fabric (7.5YR N6/), well purified, compact, and soft. Olive-grey gloss (5Y 5/1), matte and diluted.

Comparisons: This form is well documented at Pantanello (Burgers's form 15, first half of the 1st c. BC), the castrum (D'Andria 1976, Pl. 36A, No. 31373; L. Giardino 1980a, Pl. 82, No. 20, dated 1st c. BC by an example from Rome), and contrada Casa Teresa (Lissi Caronna 1998–99, 194, Fig. 16, No. 1, with a slightly thinner rim and more oblique wall, 1st c. BC.)

Date: First half of the 1st c. BC.

1:2

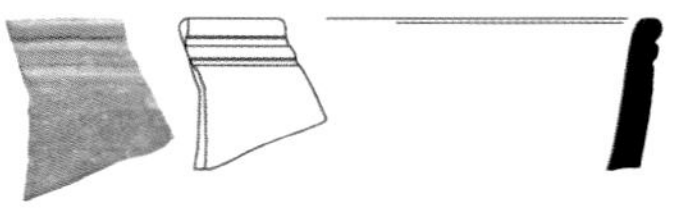

27. GW Plot 1999.177-01 Bowl rim (see **No. 26**)

Dimensions: diam. ca. 16.0 w 2.45 h 3.0 body th 0.45

Condition: Less than 5% of the rim and a trace of the body wall are preserved, both chipped. Gloss is mostly lost.

Description: Indistinct rim with square section and nearly flat upper surface; workmanship is irregular; on exterior rim is emphasized by two grooves 0.2 cm wide, one at 0.4 cm below the rim, the second at 1.1 cm. Body wall is vertical until ca. 2.5 cm below rim, where it begins to curve. The fragment is completely covered in gloss. Light grey fabric (7.5YR N7/), well purified, compact, and soft. Grey gloss (N5/), matte, diluted, and thin.

Mugs

1:2

This mug belongs to a type with upturned rim and a concave upper surface, attached to a slightly concave neck closed by a nearly globular body. It is generally equipped with two handles, although no trace of either could be found in the survey pieces. The form recalls the cup-skyphos in black gloss with strongly outward-turned rim and neck distinguished from the body, which circulated in the 3rd and 2nd c. BC; see also *Valesio* I, 194, form K44b. A rim with concave upper surface is fairly frequent in the forms suitable for pouring, such as the jug, but appears to be absent until

Mugs cont.

the 4th c. BC; in this case the vessel has a narrower rim compared to the body and is equipped with a single handle. See, for example, Morel's *espèce* 5210, which presents products typical of the Etruscan with more elaborate rims, and series 5211 and 5215 from Ordona and Minturnae respectively, dated second half of the 4th–middle of the 3rd c. BC. Marabini Moevs (1973, 77) also associates her form XVII in Cosan thin-walled ware ("ovoid, two handled pitcher") with the late-4th–3rd c. pelike.

Rotroff defines the "biconical jug" as a rare form with influences from metalworking, especially in bronze. The evolution of the type is based on the point of greatest expansion in the body, which tends to become increasingly distant from the foot with time. The upturned rim is described as typical of the Late Hellenistic period, as well as of thin-walled ware in Italy in the 2nd c. BC (*Agora* XXIX, 129–30. For the thin-walled ware see, e.g., Marabini Moevs 1973, Pl. 4, and Pl. 10, No. 113, dated second quarter of the 2nd c. BC–first half of the 1st c. AD; Small et al. 1994, Fig. 10, No. 82, from trench F202 at Gravina; A. Ricci 1985, 253, and Pl. 80, No. 11, suggesting central Italian and Iberian production for similar forms). On the Late Hellenistic chronology for this type of rim with concave upper surface, see the jug in Hempel 2001, 136, and Pl. 3, form 115/3 Kanne, attributed to phases E2–E3, second half of the 2nd/middle of the 1st c. BC. At Monte Irsi there is a "small bowl with projecting rim" in black gloss that appears in profile to be very similar to the survey piece (Small 1977c, Fig. 28, No. 160). From Paphos comes a mug in Eastern Sigillata B with a rim similar to the survey piece, but the carination of the body is accentuated (J.W. Hayes 1985, Pl. 15, No. 14, form 79; on p. 69 it is described as an uncommon form, dated perhaps to the end of the 1st/beginning of the 2nd c. AD).

The boundary between jug (or pitcher) and mug is not definite, considering forms 21 and 34/36 from Pantanello; the same can be said of two different jugs from the waste deposit at Minturnae, a one-handle example with rim smaller than the body and a two-handled example with very wide rim (see Burgers's boccaletto and "jug"; *Minturnae*, Pl. 2, No. 4 and Pl. 3, No. 11, middle of the 3rd c. BC; Morel 5215a1 and 3682a1). Cotton (1979, 153) identifies the jug as a container equipped with not only handles but also a spout. Since the survey pieces lack both, it is better to speak of a mug. The grey-ware mug is a frequent form in the Metapontino, if not actually exclusive to it. It appears in several variants, as shown by the materials from Pantanello and Sant'Angelo Nuovo, but also the castrum and contrada Casa Teresa (see Burgers's form 21; Carter 1983a, 475, Fig. 50, Nos. 54 and 154; L. Giardino 1980a, 273–75, emphasizing its link with the olpai of the second half of the 2nd c. BC; Lissi Caronna 1998–99, 194, Fig. 16, No. 3).

The diameter falls generally between 8 and 16 cm (at least for the rims belonging to mugs, since rims of jugs can be even 6 cm). The survey recovered two fragments securely attributable to this form, and a third, very poorly preserved fragment of uncertain attribution; the latter, from site 105, could belong to a mug of this type and retains the attachment of a handle, but its far worse state of preservation does not allow for a more precise identification.

GW-Bo1

28. GW 502-08 Mug rim

Dimensions: diam. 11.5 w 3.8 h 3.1 body th 0.35/0.4

Condition: A small portion of the rim (chipped) and body is preserved. Gloss is detached and worn in places.

Description: Vertical rim with triangular section and concave upper surface, attached to a neck with concave profile, which is in turn attached to a presumably globular body. The exterior is completely covered in gloss, except for the reserved top of the rim (ca. 0.4 cm); there is a small reserved zone on the interior at the rim/neck interface (0.25 cm). Light grey fabric (7.5YR N7/), well purified, compact, and soft, with grey core (7.5YR N6/). Dark olive-grey gloss (5Y 4/1), diluted and barely shiny on the exterior; dark grey gloss (N4/), dense and worn on the interior, present in scales near the rim.

Comparisons: The piece is very similar to one of the mugs recovered from the Pantanello deposit (see Burgers's form 21-PZ81.W.217.1 from level I, dated to the second half of the 1st c. BC; and Carter 1983a, 475, Fig. 50, Nos. 54 and 154, "kantharoi"), and also to a mug from the castrum (L. Giardino 1980a, Pl. 84, No. 50, identified as *boccale* of type *b*) and an example from the agrarian structure in Casa Teresa (Lissi Caronna 1998–99, 194, Fig. 16, No. 3, decorated with false ribbing).

1:2

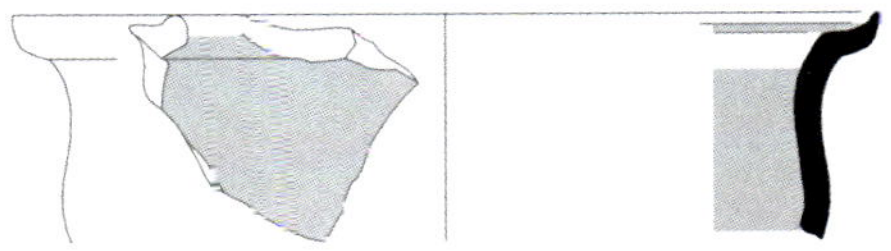

Date: Precise parallels are rare because olpai and jugs predominate. The best date is the stratified context from the Pantanello kiln deposit, second half of the 1st c. BC.

Mugs cont.

29. GW 372-03 rev01 Mug rim (see **No. 28**)
Dimensions: diam. ca. 10.0 w 3.0 h 1.55 body th 0.3
Condition: Roughly 10% of rim is preserved with a trace of body wall. Gloss is present in traces only on rim exterior.
Description: Slightly oblique rim with triangular section, concave on interior and distinguished from body, which is straight and oblique. Light grey fabric (7.5YR N7/), well purified, compact, and soft, with grey core (7.5YR N6/). Dark olive-grey gloss (5Y 4/1), matte, diluted, and thin.

30. GW 105-07 rev90 Mug (?) neck
Dimensions: diam. ca. 10.0 w 3.75 h 2.4 th 0.35/0.4
Condition: A small portion of neck is preserved with traces of the handle attachment. Gloss is fairly well preserved.
Description: Concave neck with traces of the convex body wall. The fragment is completely covered in gloss. Grey fabric (7.5YR N5/), misfired and hard, with very small inclusions at the attachment to the handle. Dark grey gloss (N4/), matte and homogeneous.

Ambiguous Fragments

GW-C1/B1

1:2

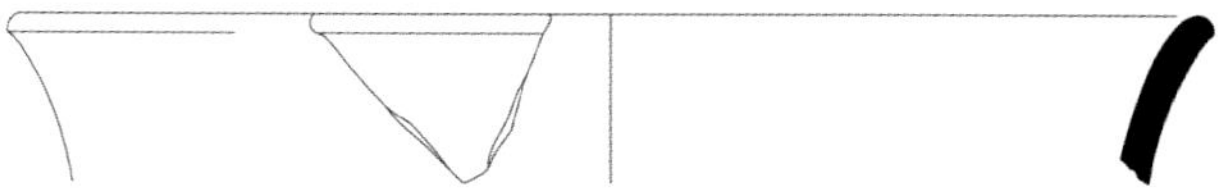

31. GW 726-02 Cup/bowl rim
Dimensions: diam. 16.0 w 3.2 h 2.3 body th 0.5
Condition: Roughly 7% of the rim is preserved, along with traces of the body wall. The gloss is partially lost.
Description: Slightly outward-turned, indistinct rim with elliptical section. Body wall is oblique and straight. Fragment was presumably completely covered in gloss, which is poorly preserved. Light grey fabric (7.5YR N7/), well purified, compact, and hard. Dark grey gloss (N4/), matte, diluted, and thin.
Comparisons: The preserved rim can be attributed to either type GW-C1 (cup with offset rim) or type GW-B1 (bowl with concave/convex body), which are contemporary productions (see introductions for each regarding details and origins of the forms). Compare Burgers's forms 12A, *Bowl with Convex/Concave Wall*, and 20, *Rounded Bowl* (PZ81.W.339.1 and PZ81.W.419.1) from the Pantanello kiln deposit, dated to the second half of the 2nd c. BC; there is a similar fragment from the castrum (L. Giardino 1980a, 267–72, Pl. 83, No. 36) belonging to type *a* of the *coppe a parete espansa e ad orlo estroflesso*, dated to the second half of the 2nd c. BC; see also Vaste (Campagna 1995, Fig. 5, No. 9, from cistern 121, and Fig. 23 Nos. 8–9, from pit 291). From the acropolis of Monte Sannace come similar cups with slightly enlarged and oblique rims (F. Rossi 1989, Pl. 311, No. 3); see also Leuca (L. Giardino 1978b, Pl. 68 B26). At Monte Irsi there is the same type of slightly outward-turned, indistinct rim on both a cup and a bowl (Small 1977c, Fig. 26, Nos. 128 and 133, from levels dated roughly to the second half of the 2nd–1st c. BC). A cup from Cosa is also similar, dated 180–140 BC, and belonging to a form very popular at Cosa between 170/160 and 40/30 BC (D.M. Taylor 1957, 179, Pl. 28 B37d in fabric type IV, a local form).
Date: Second half of the 2nd c. BC.

GW-B1?

32. GW 198-01 Bowl (?) rim
Dimensions: diam. ca. 16.0 w 3.3 h 1.05 rim th 0.5
Condition: Roughly 5% preserved. Gloss lost in places.
Description: Outward-turned, indistinct rim with subrectangular section; the very small preserved portion of the body wall suggests that it was straight and oblique near the top. The fragment is completely covered in gloss. Light grey fabric (7.5YR N7/), well purified, compact, and soft. Grey gloss (N5/), matte, diluted, and thin.
Comparisons: Burgers's form 12A (PZ81.W.359.4) from the Pantanello kiln deposit, dated to the second half of the 2nd c. BC, akin to type GW-C1/B1, but it has a more pronounced and strongly outward-turned rim. Also comparable to L. Giardino's *coppa a parete espansa e ad orlo estroflesso* (L. Giardino 1980a, 267–72, and especially Pl. 83, No. 33, dated to the second half of the 2nd c. BC). See also a fragment from Leuca that recalls type GW-C1/B1, although this seems associated with a more oblique body wall (L. Giardino 1978b, Pl. 68 B26). The same can be said of a bowl from Gravina that, however, recalls type GW-B1 (Small et al. 1994, Fig. 7, No. 10, from trench F202, dated 75±15 BC); a "bowl with S-shaped profile" from Gravina is also similar (Prag 1992c, Fig. 43, No. 1035, dated 175–50 BC). A bowl from Cosa has a similar rim (D.M. Taylor 1957, Pl. 31 C28b, produced in type IV, a local ware; Morel 2614a1, roughly second quarter of the 2nd c. BC). The form may belong to Lamboglia's form A28 (Lamboglia 1952, 177).
Date: The proposed parallels suggest a date between the second half of the 2nd and the middle of the 1st c. BC.

Body of Open Shape

Among the indeterminate grey-ware sherds from the survey, five fragments could be identified as belonging to open forms, either bowls or dishes. In all five cases this determination is based on fracture in apparently close proximity to the carination of the vessel, thus indicating the probable direction of the body wall.

33. GW 502-12 Dish or bowl (?) body
Dimensions: w 4.15 h 3.75 th 0.35/0.6
Description: Fragment of wall from a dish or bowl, porous and worn, straight, and preserving traces of attachment to rim. There are traces of gloss on both interior and exterior. Light grey fabric (7.5YR N7/), well purified, compact, and hard. Dark grey gloss (N4/), matte and diluted.
Comparisons: The preserved portion is minimal and could be attributed to several vessel forms with straight body wall, such as the dish with vertical rim (type GW-D2) or the "dish with broad horizontal offset rim" (not present in this typology; see Burgers's form 2, second half of the 2nd c. BC), and even the "dish with variously articulated rim" (Burgers's form 14, also second half of the 2nd c. BC). On the other hand, the same type of carination is found on several bowls, such as Morel's series 2654 or Lamboglia's type B16, both dated to the 1st c. BC.
Date: 2nd c. (second half?)–1st c. BC.

---

34. GW 042-09 Dish (?) body
Dimensions: w 8.5 h 7.4 th 0.5/0.8
Condition: Fragment of wall preserved. Gloss is worn.
Description: Straight wall of a dish (?) bearing signs of manufacture on a wheel; on the interior the attachment to the rim is emphasized by a slight depression. The fragment is completely covered in gloss. Light grey fabric (7.5YR N7/), well purified, compact, and hard. Olive-grey gloss (5Y 5/1), matte and diluted.
Comparisons: The comments for **No. 33** apply here, with the difference that the width of the point in which the body bends upward (corresponding to the exterior carination, ca. 24–26 cm) is more suggestive of a dish than a bowl.
Date: 2nd c. (second half?)–1st c. BC.

---

35. GW 042-13 Bowl (?) body
Dimensions: w 5.5 h 5.25 th 0.5/0.7
Condition: Fragment of wall preserved. Gloss is lost in places.
Description: This is perhaps a fragment of a bowl, which has a thick body wall tending to become thinner at the point where the wall curves softly. The interior of the fragment is completely covered in gloss, while the exterior is only partially covered. Grey fabric (7.5YR N5/), well purified, compact, and hard. Dark olive-grey gloss (5Y 4/1), matte and diluted on the interior; dark grey gloss (N4/), compact, shiny, and present only in traces on the exterior.
Comparisons: The direction of the body wall suggests Burgers's forms 12 and 13 from the Pantanello deposit or L. Giardino's *coppa a parete espansa e ad orlo estroflesso* (L. Giardino 1980a, 267–72), comparable to type GW-B1.
Date: Second half of the 2nd–beginning of the 1st c. BC.

---

36. GW 042-14 Open-shape body
Dimensions: w 5.9 h 3.5 th 0.35/0.6
Condition: Fragment of wall preserved. Gloss mostly lost.
Description: This is perhaps a fragment of a bowl, which has a softly curving body wall tending to become thinner towards the rim. The fragment was perhaps completely covered in gloss on the interior, and only partially covered on the exterior. Light grey fabric (7.5YR N7/), well purified and hard, cracked in places. Grey gloss (N5/), matte, very diluted, and thin.
Comparisons: See **No. 35**.
Date: Second half of the 2nd–beginning of the 1st c. BC.

---

37. GW 062-01 Open-shape rim
Dimensions: w 4.1 h 2.5 th 0.4/0.55
Condition: A portion of rim preserved, chipped. Traces of gloss on the interior only at the attachment of the body wall; on the exterior the gloss is partially lost.
Description: Rim of an oblique open shape for which a profile cannot be described because of the presence of the fracture at the extremity. The rim is distinct from the body, of which there are no traces. The fragment was presumably completely covered in gloss. Grey fabric (7.5YR N6/), well purified, compact, and soft. Dark grey gloss (N4/), matte and homogeneous on the interior; olive-grey gloss (5Y 5/1), matte and very diluted on the exterior.
Comparisons: The poor state of preservation does not permit detailed description of the fragment; the only reasonable suggestion is a bowl of modest size (the diameter at the carination is ca. 8.0 cm) similar to Burgers's form 29 from the Pantanello deposit (from the same deposit see also Carter 1983a, 475, Fig. 50, Nos. 82–83, 150–50 BC) and a dish with high rim like Morel's series 2277 (Sardinian grey ware dated to the 1st c. BC), and akin to Lamboglia's forms B16 and C16 (Lamboglia 1952, 151 and 159).
Date: Second half of the 2nd–1st c. BC.

Decorated Body Fragments of Unknown Shape (GW-dec)

Motifs include rouletting with bands of light impressions arranged in multiple rows, and impressed concentric circles that occasionally frame the rouletting. This scheme is found also in black gloss and sigillatas. The poor condition of the fragments prevents any link between type of decoration to a specific vessel form, as well as precise dating. Decoration, however, seems more common from the second half of the 2nd c. through the end of the 1st c. BC, but more study of whole vessels from stratified contexts is needed.

GW-dec.1a

38. GW 428-04 Dish (?) body

1:2

Dimensions: w 4.9 h 2.8 th 1.0

Condition: Chipped. Gloss preserved inside decoration.

Description: Straight body wall suggests a dish. Roulette decoration consists of droplike impressions, each measuring 0.25 x 0.1 cm and separated by 0.15 cm, arranged in two concentric rows and surrounded by an impressed circle 0.1 cm wide. Gloss covered interior, no traces left on exterior. Light grey fabric (7.5YR N7/), well purified, compact, and hard. Dark grey gloss (N4/), matte and homogeneous.

Comparisons: See similar decoration on open-shape fragment from castrum (L. Giardino 1980a, Pl. 86, No. 64; cf. also D'Andria 1976, Pl. 36, No. 31413); without surrounding circle on interior of bowl from Pantanello (Burgers's form 12A, PZ81.W.413.1, second half of 2nd c. BC); dish from Monte Irsi (Small 1977c, Pl. 33, No. 124); black-gloss dish from Bolsena (Morel type 1542a1, end of 3rd–first half of 2nd c. BC); dish with "drops" arranged irregularly from Gravina (Prag 1992c, Fig. 40, No. 984, end of 2nd–beginning of 1st c. BC); two concentric circles on dish from Sybaris (*Sibari* II, 116, No. 15, and 183, Fig. 201, No. 8377, in grey fabric); fragment in Campana C from Monte Iato (*Studia Ietina* IV, Pl. 34, No. 981, second half of 2nd–first half of 1st c. BC).

Date: Second half of the 2nd–end of the 1st c. BC.

GW-dec.1b

39. GW 042-08 Dish (?) body

1:2

Dimensions: w 7.5 h 8.0 th 0.8/1.05

Condition: Gloss mostly lost on interior, worn on exterior.

Description: The impressed roulette decoration, likely restricted to the middle of a dish, consists of a circle 0.15 cm wide around two rows of subtriangular impressions, each 0.25 x 0.2 cm, with vertices turned toward the center. Original was completely covered in gloss. Grey fabric (7.5YR N6/), well purified and soft, with light grey core (7.5YR N7/). Grey gloss (N5/), matte and slightly diluted.

Comparisons: Triangular impressions are found on an open-form fragment from the castrum, although the vertices do not appear turned towards the center (L. Giardino 1980a, Pl. 86, No. 66). See similar decoration on floor of dish from Pantanello (Carter 1983a, 473, Fig. 49, No. 59, 150–50 BC); dish in grey ware from Herakleia (L. Giardino 1990, Pl. 24, Umbria 96, second half of 2nd c. BC), dish from Sybaris (*Sibari* IV, 241, No. 161, and 295, Fig. 285, No. 10286), and example in Campana B from Ventimiglia (Lamboglia 1952, 154, No. 1, 2nd c. BC).

Date: Second half of the 2nd–end of the 1st c. BC.

GW-dec.2 and Op?2

40. GW 416-08 Dish (?) foot

1:2

Dimensions: diam. 10.0 w 4.8 h 1.7 foot th 1.0 body th 0.8

Condition: Roughly 16% of the foot is preserved, along with a small portion of the body floor. Gloss partially detached.

Description: Ring foot with subtriangular section and vertically oriented exterior face (ca. 0.5 cm), becoming oblique until attachment to body, of which a part of floor is preserved. Floor has two concentric impressed circles, each 0.15 cm wide and separated by 0.3 cm. Interior completely covered in gloss, foot is reserved. On underside are traces of wheel manufacture. Grey fabric (7.5YR N6/), soft, with rare and very small inclusions. Dark grey gloss (N4/), matte, slightly diluted, and thin.

Comparisons: Two similar circles on stamped dish from castrum (L. Giardino 1980a, Pl. 86, No. 70); Pantanello (Carter 1983a, 480, and 483, No. 156, in TSI); Grumentum (L. Giardino 1980b, Pl. 7 D8); Taranto (Alessio 1988, 340, 30.11b); see also a cup from Sybaris (*Sibari* IV, 242, No. 164, and 299, Fig. 288, No. 11459). On another dish from Pantanello circles are separated (Burgers's form 4/5-PZ81.NW.843.1, second half of 2nd–1st c. BC).

Date: Second half of the 2nd–end of the 1st c. BC.

Feet of Open, Closed, and Unknown Shapes (GW-Op and GW-Cl)

1:2

GW-Op1

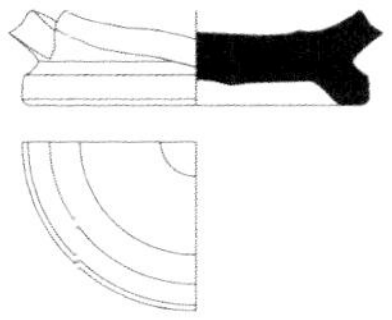

41. GW 042-15 Open-shape foot
Dimensions: diam. 4.6 h 1.32 w 4.1 body th 0.35/0.6
Condition: Roughly 30% of the foot and a minimal portion of the body are preserved. The gloss is worn.
Description: Ring foot with slightly rounded resting surface; interior of body irregular and covered in gloss; a trickle of gloss on foot, rest of fragment reserved. Traces of wheel manufacture on interior and exterior. Grey fabric (7.5YR N6/), well purified, compact, and soft. Dark grey gloss (N4/), matte, in part lost, present in small scales on interior.
Comparisons: Shape of the preserved portion suggests a bowl, but cf. "dish with vertical rim" from Gravina, corresponding to type GW-D2, with identical foot (Prag 1992c, Fig. 42, No. 994); also similar is foot of a cup or dish from Pantanello (Carter 1983a, 475, Fig. 50, No. 83, 150–50 BC).
Date: Generally 2nd–1st c. BC.

42. GW 502-14 Open-shape foot (see **No. 41**)
Dimensions: diam. 5.0 w 4.45 h 1.1 body th 0.4/0.75
Condition: Small portion of foot is preserved with central part of body. Gloss present in traces on interior of body.
Description: Ring foot with trapezoidal section, slightly oblique resting surface, and irregular exterior face; at center the body has a conical section; on exterior (underside) of floor are traces of wheel manufacture. Interior was covered in gloss, exterior reserved. Light grey fabric (7.5YR N7/), well purified, compact, and soft. Dark grey gloss (N4/), matte, diluted, and thin.

GW-Op2

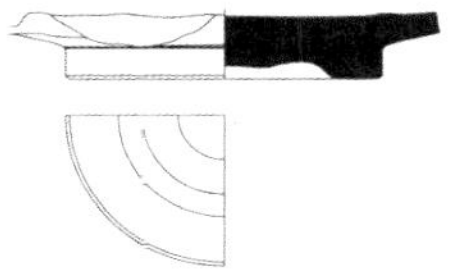

1:2

43. GW Plot 2000.087-01 Open-shape foot
Dimensions: diam. 4.2 w 5.35 h 0.9 body th 0.3/0.4
Condition: Two fragments. Entire foot and a small portion of the body preserved. Gloss in part detached.
Description: Low ring foot with nearly vertical exterior surface and flat resting surface. Foot was completely covered in gloss, except for interior (underside) and resting surface. Execution of interior is not regular, traces of wheel manufacture on exterior. Grey fabric (7.5YR N6/), hard, with occasional very small inclusions. Dark olive-grey gloss (5Y 4/1), matte, diluted, and thin.
Comparisons: A similar foot, belonging to a small cup, comes from Otranto (Semeraro 1983, Pl. 95, No. 309).
Date: Generally 2nd–1st c. BC.

GW-Op3

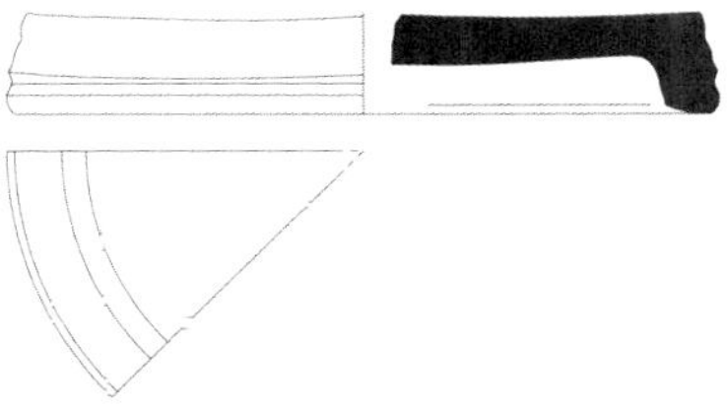

1:2

44. GW 001-1 Pl. Open-shape foot
Dimensions: diam. 9.6 w 9.3 h 1.4 body th 0.6/0.65
Condition: Less than half of floor and foot is preserved. Gloss in scales in several places but generally well preserved.
Description: Ring foot with receding resting surface and irregular exterior profile; interior floor is flat. Fragment completely covered in gloss on interior, with a trickle of gloss on foot; exterior (underside) of floor and foot reserved. Traces of wheel manufacture on all parts. Grey fabric (7.5YR N5/), well purified, compact, and hard, with clean fractures. Dark grey gloss (N4/), matte and uniform.
Comparisons: A similar foot is found on a dish from the Crucinia necropolis (De Siena 1993, 129, No. 177).
Date: Generally 2nd–1st c. BC.

GW-Op?1 (= Op3)

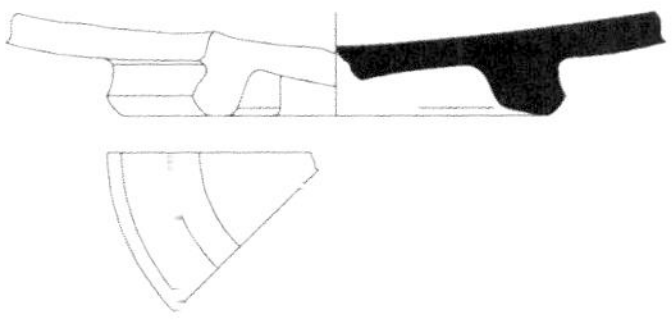

1:2

45. GW 502-06 Open(?)-shape foot
Dimensions: diam. 6.1 w 5.25 h 1.4 body th 0.55/0.6
Condition: Roughly 25% of foot, chipped, and part of body are preserved. There are no traces of gloss.
Description: Ring foot with slightly oblique resting surface, which has a trapezoidal section (interior and exterior surfaces are parallel). Body wall straight, floor exterior (underside) is conical. Grey fabric (7.5YR N6/), well purified, compact, and soft, with clean fractures.
Comparisons: Despite absence of gloss, wall line suggests an open shape: cf. Carter 1983a, 473, Fig. 49, No. 59 "plate"; Burgers's form 1-PZ81.W.35.2 "dish with broad convex lip-rim"; Small 1977c, Fig. 26, No. 118 "plate with up-turned rim."
Date: Generally 2nd–1st c. BC.

Feet of open, closed, and unknown shapes (GW-Op and GW-Cl) cont.

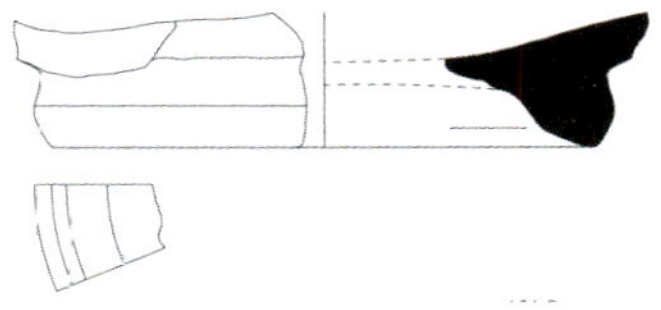

1:2

GW-Op?2

46. GW 502-07 Open (?)-shape foot

Dimensions: diam. 7.7 w 4.1 h 1.9 body th 0.6

Condition: Roughly 15% of the foot and a very small portion of the body are preserved. There is no trace of gloss.

Description: Ring foot with triangular section and oblique resting surface; interior and exterior surfaces are parallel. Body wall is straight. Grey fabric (7.5YR N6/), compact and soft, with very small micaceous inclusions.

Comparisons: The parallels recall open forms, despite the absence of gloss—cf. a dish from Pantanello (Burgers's form 14-PZ81.NW.795.3); also similar, but with a flat resting surface, is the foot of a *coppa a parete espansa e ad orlo estroflesso* from Metaponto (L. Giardino 1980a, Pl. 83, No. 37).

Date: Generally 2nd–1st c. BC.

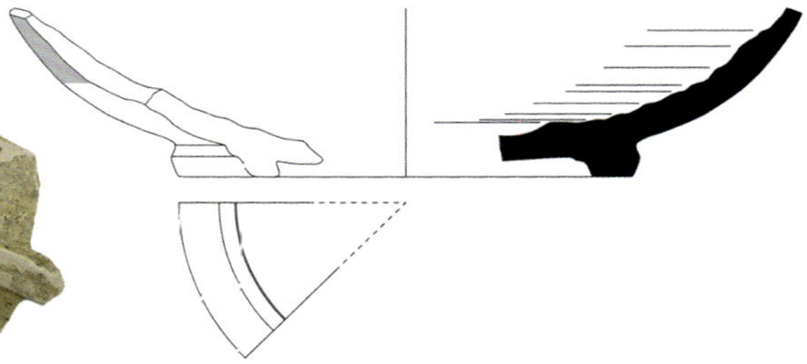

1:2

GW-Cl1

47. GW 095-01 Closed-shape foot/body

Dimensions: diam. 8.6 h 3.3 w 6.0 body th 0.35/0.6

Condition: Roughly 20% of the foot and a part of the body are preserved. The gloss is partially detached.

Description: Ring foot with slightly oblique exterior face and carination immediately before attachment to hemispherical body wall. Interior has ridges created by wheel manufacture. Interior floor concave; resting surface of foot flat. Piece could belong to a closed form, such as a jug. Gloss present only on exterior, with two large trickles reaching foot. Grey fabric (7.5YR N6/), soft, with occasional small inclusions. Dark blue-grey gloss (5B 4/1), matte, homogeneous, and thick.

Comparisons: Form 35 "jug" from Pantanello has similar foot and irregular workmanship on the interior (Burgers's PZ81.NW.573P). Cf. also pitcher from Cosa, although not directly belonging to type III (identified by D.M. Taylor as Campana C: D.M. Taylor 1957, 90, and Pl. 25 A42, from the deposit dated 273–180/160? BC).

Date: Generally 2nd–1st c. BC.

---

48. GW 556-01 Closed-shape foot

Dimensions: diam. 8.5 w 5.8 h 1.5 body th 0.4/0.45

Condition: Roughly 20% of foot is preserved, chipped, along with traces of body wall. No visible traces of gloss.

Description: Low ring foot with subrectangular section, oblique resting surface, and rounded exterior face. Preserved section of the body is oblique. Interior, exterior, and foot show traces of wheel manufacture. Light grey fabric (7.5YR N7/), well purified, compact, and soft.

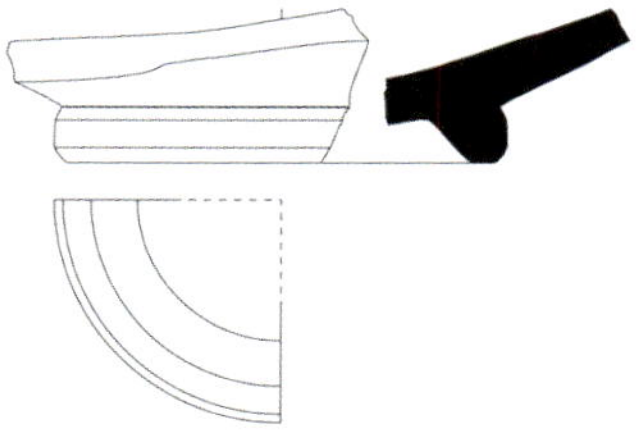

1:2

GW-Cl?1

49. GW 502-01 Closed (?)-shape foot

Dimensions: diam. 6.0 w 4.65 h 2.1 body th 0.5/0.7

Condition: Roughly 20% of foot and part of body are preserved. Visible trickles of gloss only on exterior.

Description: Ring foot with subtrapezoidal section, nearly flat resting surface, oblique interior surface, and nearly vertical and faceted exterior surface; the base of the body wall is straight. The upper part of the exterior was presumably covered in gloss, given the presence of two small trickles. There are no traces of gloss on the interior, but the absence of gloss cannot be proved given the poor state of preservation. There are apparent traces of exposure to flame. Grey fabric (7.5YR N6/), soft, with small micaceous inclusions and light grey core (7.5YR N7/). Dark grey gloss (N4/), thin and worn.

Comparisons: Notwithstanding the absence of gloss on the interior of the body, the closest parallels are open forms: a "broad-rimmed plate" from Gravina with a similar foot (Prag 1992c, Fig. 40, No. 974) and a dish from Herakleia that is similar to the latter piece (L. Giardino 1990, Pl. 25, Umbria 91 D). A bowl from Valesio also has the same characteristics (*Valesio* I, 220, No. 388, form L16).

Date: Generally 2nd–1st c. BC.

Feet of open, closed, and unknown shapes (GW-Op and GW-Cl) cont.

GW-Cl?1
50. GW 042-17 Closed (?)-shape foot (see **No. 49**)
Dimensions: diam. 6.0 h 1.75 w 3.8 body th 0.8/1.0
Condition: Roughly 10% of foot preserved, with a small portion of body wall, both worn. Gloss is almost completely lost.
Description: Small ring foot to which attaches a straight wall 0.9 cm thick. Traces of gloss at the attachment of the foot and wall. This was presumably a closed form, but the poor state of preservation prevents more precise identification. Light grey fabric (7.5YR N7/), well purified, compact, and fairly soft. Dark grey gloss (N4/), matte.
Date: Generally 2nd–1st c. BC.

---

51. GW 230-01 Dish (?) foot (see **No. 49**)
Dimensions: diam. ca. 6.0 w 3.95 h 1.35 body th 0.5
Condition: Roughly 20% preserved. Gloss worn in part.
Description: Ring foot with subtrapezoidal section, flat resting surface, and slightly oblique exterior surface. The foot is reserved, while the interior of the body is covered in gloss (although only a minimal amount of the gloss is preserved). The fragment could belong to a dish. Light grey fabric (7.5YR N7/), well purified, compact, and hard. Dark grey gloss (N4/), matte and diluted.

---

52. GW 231-01 Dish foot (see **No. 49**)
Dimensions: diam. 6.6 w 6.1 h 2.25 body th 0.6/1.0
Condition: Roughly 10% of the foot is preserved, along with a portion of the body. The gloss is worn.
Description: Ring foot with subrectangular section, slightly rounded resting surface, and slightly oblique exterior face; the foot is clearly distinguished from the straight, oblique body. The interior of the body is covered in gloss, while on the exterior the gloss is present only in trickles, some of which reach the foot. Grey fabric (7.5YR N6/), well purified, compact, and soft. Dark grey gloss (N4/), matte, diluted, and thin, tending to form scales.

GW-Op?/Cl?1
53. GW 339-03 Foot, unknown shape
Dimensions: diam. 8.9 h 1.9 w 4.5 body th 0.75/1.1
Condition: Roughly 20% of foot is preserved, along with traces of the wall. There are traces of gloss only at the attachment of the foot to the body and on the foot itself.
Description: Ring foot clearly distinguished from body, with oblique resting surface; interior and exterior surfaces are inclined and parallel. Exterior of foot covered in gloss, while interior is reserved. Interior of the body, which was apparently not covered in gloss, is poorly preserved. Grey fabric (7.5YR N6/), compact and hard, with rare and very small inclusions. Dark grey gloss (N4/), homogeneous.
Comparisons: A bowl from Pantanello has a similar foot (Burgers's form 12A-PZ81.W.359.4), as does a dish from the Crucinia necropolis (De Siena 1993, 134, No. 215).
Date: Generally 2nd–1st c. BC.

1:2

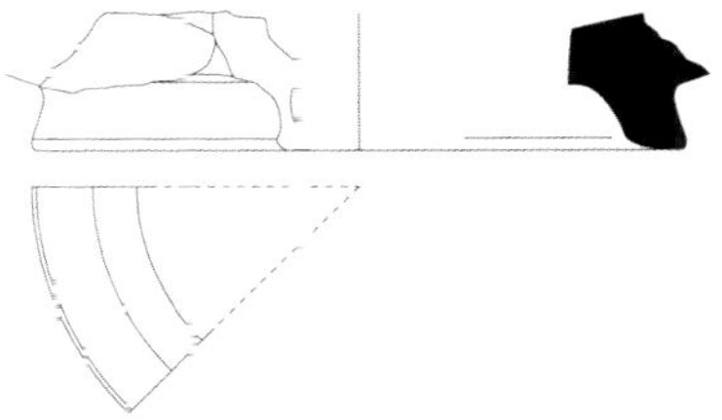

---

54. GW 317-02 Open-shape foot (see **No. 53**)
Dimensions: diam. 7.8 w 5.35 h 1.8 body th 0.7/0.85
Condition: Roughly 15% of foot preserved, along with part of the body. The gloss is almost completely detached.
Description: Ring foot with parallelepipedal section, slightly oblique resting surface, straight and nearly vertical exterior surface, oblique body distinguished from foot. Gloss traces on interior, none on exterior. Traces of exposure to flame. Light grey fabric (7.5YR N7), soft, with very small micaceous inclusions concentrated on body interior. Dark grey gloss (N4/), matte.

Feet of open, closed, and unknown shapes (GW-Op and GW-Cl) cont.

55. GW 502-15 Closed-shape foot (see **No. 53**)
Dimensions: w 2.3 h 1.6 foot th 0.9 body th 0.7
Condition: A small portion of foot is preserved (insufficient for diam.), with attachment of body. No traces of gloss.
Description: Ring foot with parallelepipedal section and oblique resting surface; body is straight and nearly horizontal. Grey fabric (7.5YR N6/), well purified, compact, and hard.

---

56. GW 042-16 Closed-shape foot
Dimensions: diam. 4.5 h 1.45 w 4.95 body th 0.65/0.8
Condition: Roughly half of foot is preserved in poor state: almost completely chipped. No traces of gloss.
Description: Very poor manufacture, with excess clay on the resting surface and the interior (underside) of the ring foot. The preserved part of the body is massive (ca. 0.8 cm at the attachment to the foot). Light grey fabric (7.5YR N7/), well purified, compact, and soft.
Date: Generally 2nd–1st c. BC.

---

57. GW 197a Closed (?)-shape foot (?)
Dimensions: w 3.35 h 3.35 th 0.75/0.8
Description: Foot (?) fragment, closed (?) form. Gloss on exterior. Grey fabric (7.5YR N6/), well purified, compact, and hard. Dark grey gloss (N4/), matte and homogeneous.
Date: Generally 2nd–1st c. BC.

---

58. GW 317-03 Open-shape foot
Dimensions: diam. 8.0 w 4.15 h 1.6 body th 0.6/0.8
Condition: Roughly 10% of the foot, very worn, and a small portion of the body are preserved. Only traces of the gloss are visible. The fragment is in very poor condition.
Description: Low ring foot with irregular exterior surface and very oblique interior surface; the foot is distinguished from the body, which is also oblique and straight. The vessel was probably completely covered in gloss, except for the interior (underside) of the foot. Light grey fabric (7.5YR N7), soft, with rare and very small inclusions. Dark grey gloss (N4/), matte and homogeneous. There are traces of exposure to flame.
Comparisons: The same foot with very oblique interior surface is found on a cup from the castrum (L. Giardino 1980a, Pl. 82, No. 21), a dish from the Pantanello deposit (Burgers's form 1-PZ81.W.35.2, second half of the 2nd c. BC), and a dish from Gravina (Prag 1992c, Fig. 42, No. 996, end of the 2nd–beginning of the 1st c. BC).
Date: Generally 2nd–1st c. BC.

---

59. GW 428-05 Open-shape foot
Dimensions: diam. ca. 4.2 h 1.45 w 4.95 foot th 0.45 body th 0.8 cm at the attachment to the foot.
Condition: The entire foot is preserved, but only in the portion near the attachment to the body wall; the resting surface is missing. The gloss on the interior is slightly worn, and on the exterior it is present in a single trickle.
Description: Oblique and fairly thin ring foot. The exterior (underside) of the floor is conical, with traces of manufacture on a wheel; the attachment to the body is soft. The interior was presumably completely covered in gloss, while the lower part of the exterior was probably reserved; the interior (underside) of the foot is also reserved. Light grey fabric (7.5YR N7/), well purified, compact, and hard. Dark grey gloss (N4/), slightly diluted and barely shiny.
Date: Generally 2nd–1st c. BC.

Other Unknown Shapes

60. GW 042-11 Spout of unidentifiable form
Dimensions: w 1.3/2.4 h 5.2 th 0.3/0.45
Condition: Only the spout is preserved.
Description: Conical spout, chipped. No traces of gloss. Grey fabric (7.5YR N6/), compact and soft; well purified in section, but small inclusions on surface evident to the touch.
Comparisons: A "jug with strainer at the base of the neck" in grey ware from Pantanello has a similar spout on the shoulder (Burgers 1991, Pls. 17B and 60A); cf. also a fragment from the castrum (L. Giardino 1980a, Pl. 84, No. 56, belonging to an amphora or jug), a fragment from Gravina (Prag 1992c, Fig. 46, No. 1070, perhaps from a jug, dated 2nd–1st c. BC), and a fragment from Oria (Yntema 1993, 136, No. 379, second half of the 2nd c. BC). A fragmentary guttus from Monte Sannace has same conical spout (F. Rossi 1989, Pl. 313, No. 3), also a fragment of a globular vase from Sybaris (*Sibari* I, 59, Fig. 53, No. 22, compared to *Olynthus* XIII, 264–68, Pls. 178–79, Nos. 478–87 "infant feeders," perhaps "lamp fillers" or "feeding cups," 5th–first half of the 4th c. BC), a fragment from Monte Irsi (Small 1977c, Pl. 33, No. 138), and a fragment in Campana A from Sybaris (*Sibari* IV, 472, No. 108, Figs. 442 and 478).

---

61. GW 095-02 Molded handle of unknown shape
Dimensions: w 1.6/1.9 h 3.9 th 0.7/0.9
Condition: Chipped in several places. Gloss partially lost.
Description: Irregularly ribbed vertical handle; ribs measure 0.4, 0.7, and 0.8 cm. Grey fabric (7.5YR N6/), well purified compact, and soft. Grey gloss (N5/), matte and diluted.

---

62. GW 558 b Molded handle of unknown shape
Dimensions: w 1.35 h 1.65 th 0.45
Description: Small handle with three ribs measuring 0.45, 0.35, and 0.55 cm wide. Gloss is partially worn. Grey fabric (7.5YR N6/), well purified, compact, and soft. Dark grey gloss (N4/), matte and thin, tending to form scales.

---

63. GW 502-03 Lid (?) shoulder (?)
Dimensions: diam. 14.0 w 4.95 h 1.8 th 0.3/0.6
Condition: Shoulder (?), perhaps from a lid. The rim is not preserved. Gloss is present in traces on the exterior.
Description: Convex wall is distinguished from straight, oblique rim via a slight carination. Gloss probably covered exterior, traces on interior. Light grey fabric (7.5YR N7/), well purified, compact, and soft, with grey core (7.5YR N6/). Dark grey gloss (N4/), matte, very diluted, and thin

---

64. GW 556-09 Closed (?)-shape vessel rim
Dimensions: diam. 4.0 w 1.35 h 1.75 th 0.5/0.6
Condition: Roughly 10% of the rim is preserved, worn. There are no visible traces of gloss.
Description: Flat rim with rectangular section, the exterior profile is ribbed. Grey fabric (7.5YR N6/), well purified, compact, and soft.
Comparisons: A lagynos from Taranto has a very similar rim (*Gli ori*, 486, No. 36, from a tomb dated to the third quarter of the 2nd c. BC).

8
Figured Ware

Francesca Silvestrelli

The corpus of figured pottery from the survey is relatively small, consisting of 280 fragments belonging to 238 vessels (*Fig. 8.1*). Most of the fragments are Lucanian and Apulian red-figure,[1] along with a few fragments of Attic black- and red-figure, and a small amount of Gnathia-style pottery. In most cases the sherds are very poorly preserved, and the often worn or abraded surfaces and limited dimensions greatly restrict the range of information that can be obtained from study. Nonetheless, this sample may provide some clues as to the distribution patterns and functional range of figured pottery in the ancient countryside. These aspects have never been explored in detail because of the scarcity of systematic archaeological research in the Greek chorai of southern Italy. It was therefore difficult to compare evidence from different contexts in order to detect patterns in the selection of shapes and subjects.

Metaponto, however, is an exception in several respects. First, the discovery of red-figure workshops—dated to the end of the 5th/beginning of the 4th century BC, and the third quarter of the 4th century BC—has provided valuable information about local production.[2] Second, extensive excavations of sites in the chora, including the Pantanello necropolis,[3] other necropoleis, both urban and rural,[4] and sanctuaries[5] have already given us detailed knowledge of the figured wares, to which can now be added the evidence from a large area of the chora.

Attic Black- and Red-Figure Pottery

Attic pottery is represented scarcely in the sample from the survey, with figured vessels whose poor preservation precludes secure attribution.[6] Three Attic black-figure sherds (**Nos. 1–3**) can be dated to the Late Archaic period. Two fragments belong to shouldered lekythoi, a widespread shape in Metapontine funerary assemblages of this period and one of the few shapes imported in the area.[7] Attic red-figure pottery is attested by a single fragment of a calyx-krater (**No. 4**), probably from the second quarter of the 5th century BC.

Lucanian and Apulian Red-Figure Pottery

The red-figure sherds cover almost the entire chronological range of southern Italian production, albeit with different frequencies. Just over a third of the corpus (38.8%) cannot be dated. Among the dateable vessels, the frequency of Lucanian vases within the first quarter of the 4th is low (8.1%). Most of the fragments (58 pieces, making up 35.2%) are Late Apulian and can be dated to the second half of the 4th century BC.

There are very few vessels well enough preserved to be attributed to a precise workshop or painter, and even to the Lucanian or Apulian group. Among the Lucanian fragments, a bell-krater by the Pisticci Painter (**No. 41**) has been recognized, as well as a pseudo-Panathenaic amphora probably by the Dolon Painter (**No. 6**). The fragmentary pelike found at site 737 (SF-01, see Ch. 29) can be attributed to the Creusa-Dolon workshop. Other fragments—a pseudo-Panathenaic amphora (**No. 5**), a skyphos (**No. 12**), bell-kraters from T 2 at site 736 (T2-01, T2-03, and T2-04, see Ch. 29) and **Nos. 42–43**, a volute-krater (**No. 44**), a calyx-krater (**No. 58**), and two kraters (**Nos. 64–65**)—can be ascribed more generally to the first phases of Lucanian production.

Vases in Apulian style from the second quarter of

[1] The terms 'Lucanian' and 'Apulian' refer to the red-figure styles rather than to the geographical area of production. For the possible Metapontine production of most of the fragments illustrated in the present chapter, see the discussion of Lucanian and Apulian red-figure pottery below.

[2] D'Andria 1980a; Silvestrelli 2005, and Silvestrelli forthcoming a.

[3] Burn 1998.

[4] M. Tagliente 2006; De Siena 2007, 435–44.

[5] Adamesteanu 1980, 59, Figs. 47e–g; 105, Figs. 97c, e and Fig. 99a; 130, Fig. 136; 131, Fig. 138; 219, Fig. 225a; 130, Fig. 135. Pantanello: Jircik 1983.

[6] On Attic pottery from Metaponto asty, see Lo Porto 1981b, 300–301 (contrada Crucinia votive deposit; Burn 1998, 594–96; Rocchietti 2002, 130–33 (urban necropoleis); Liseno 2004, 79–83 (Favale votive deposit). For a general view, see F. Giudice 1997 and G. Giudice 2007, 333–42, 401–04.

[7] F. Giudice 1997, 409, Pl. VIII.3. Burn 1998, 595.

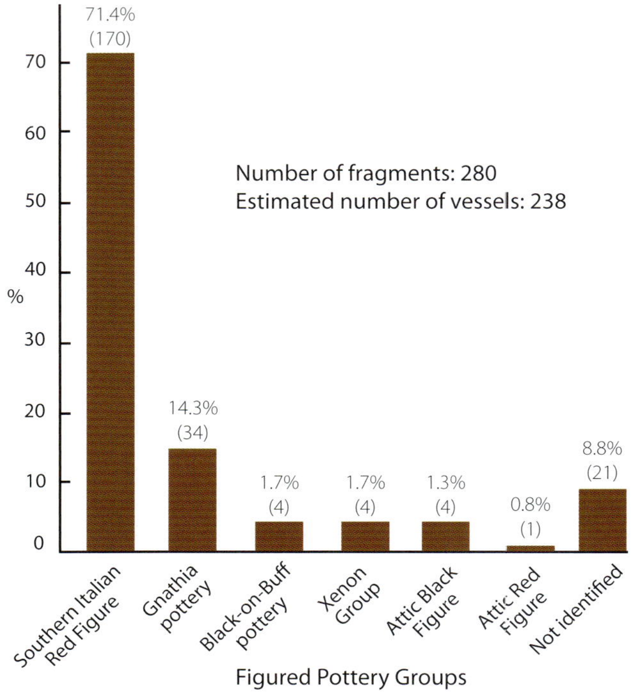

Figure 8.1 Frequencies of figured pottery from the Bradano-Basento survey. Percentages are calculated on the estimated number of vessels.

the 4th century BC are attested by a fragment of a pelike found in T 18 at site 737 (T18-10, Ch. 29) that depicts a woman in front of an altar. The drapery recalls followers of the Tarporley Painter. This earlier example seems to be isolated; most of the material in Apulian style can, in fact, be dated to the second half of the 4th century BC.

The volute-krater fragments found at site 401 (**Nos. 46–50**) and a krater (**No. 73**) are as interesting as they are tantalizing. The high quality of the drawing and potting, along with comparanda, and morphological considerations suggest attribution in both cases to a painter influenced by the Darius Painter and his school, and the same environment can perhaps be suggested for other volute-krater fragments such as **No. 52** and **No. 54**. These workshops of red-figure production seem to be responsible for most of the recognizable fragments found in the chora.[8] The decorative patterns of bell-kraters **Nos. 35-37** are comparable to those usually attributed to the Chevron Group.[9] The chevron pattern, not common in Apulian red-figure pottery,[10] is also attested at the urban necropolis of Metaponto.[11] Nonetheless, the poor preservation of the main decoration, limited to a part of a *kekryphalos* (hair-net) does not permit a secure attribution to the group. Other fragments can also be generally ascribed to artisans traditionally connected to the Circle of the Darius and Underworld Painters, as is the case for two squat lekythoi (**Nos. 27–28**) and an epichysis (**No. 29**).

The impossibility of a sound attribution for so many fragments is disappointing. The fabric is very homogeneous in appearance and does not help to identify the production centers,[12] making it difficult to decide if the fragments are imports or, as is more likely, local products. It has long been known, in fact, that the activity of Metapontine workshops lasted into the second half of the 4th century BC, and the deposits from the urban kerameikos have provided a very vivid picture of a group of artisans producing, besides black-gloss and other classes of pottery, high-quality red-figure vessels influenced by the Apulian tradition, and particularly by the Darius and Underworld Painters.[13] The fragmentary kraters from Pantanello are ascribed to the Circle of the Darius and Underworld Painters,[14] and the volute-krater from T 5/83 at Pantanello, whose Metapontine production has been rightly asserted, can be attributed to the Underworld Painter.[15] But the specific characteristics of Metapontine production in the second half of the 4th century BC have yet to be defined clearly with a sufficient level of detail, and a complete revision of the red-figure pottery attested at the Metaponto asty and in its chora in this period is needed.

Subjects

Most of the sherds preserve just decorative patterns, or they belong to the reverse of the vessels and show draped youths, e.g. the fragment from site 737 (Tr7-01, see Ch. 29) and **Nos. 64–65**. The meager dimensions of the few fragments that belong to the main

8 The same pattern has been suggested for other sites of the Bradano watershed, such as Pomarico: Macchioro 1984, 62–63 and Elia 1997, 120.

9 On the problems of classification of the Chevron Group, see now Lippolis 2006, 442–43.

10 Schauenburg 1995, 64, No. 5.

11 Lo Porto 1981b, 382, No. 4 (Crucinia, Area B, sporadic find).

12 On the problems encountered in distinguishing the clays of Taranto and Metaponto, even on the basis of chemical composition, see Grave et al. 1996–1997 and E.G.D. Robinson forthcoming.

13 See D'Andria 1980a, 422–40; Silvestrelli forthcoming b. On the distribution pattern of this part of the production, see Elia 1997, 120–22, and No. 133.

14 Burn 1998, 625–29.

15 Morard 2002, 49–50.

decorative field preclude a sound interpretation of the subjects, but some suggestions may be ventured.

The bell-krater fragment by the Pisticci Painter (**No. 41**), collected from a site where a necropolis with at least ten tombs has been recognized, was probably part of a pursuit scene. This subject, frequently chosen for vases deposited in rural burials at Metaponto,[16] probably had an eschatological meaning.[17] If the staff visible on the fragment can truly be interpreted as a thyrsos, the pursuit scene would have had a Dionysiac significance.[18] An open-air symposiastic scene could be depicted on another krater (**No. 67**). The Apulian bell-krater sherd (**No. 74**) shows a man standing in front of a burning thymiaterion; the presence of the second figure and the general organization of the scene are typical of cult scenes (see the catalog below), often connected to the Dionysiac sphere.

Some of the fragments show part of a building. The identification of the type of structure on two fragmentary kraters (**Nos. 68** and **70**) is uncertain, but a funerary naiskos might have been represented on two Late Apulian hydriai (**Nos. 23–24**). The probable funerary function of one of the vases (**No. 23**) is made clear also by the absence of the bottom. Both vessels were found in sites where tombs have been identified, and they were probably part of an assemblage of graves dating to the second half of the 4th century BC. Naiskos scenes do not seem to occur frequently at Metaponto nor in its chora, but at least one example is known from a fragment of a pseudo-Panathenaic amphora, a sporadic find from Area B of the contrada Ricotta necropolis.[19]

Other fragments, mainly belonging to Late Apulian kraters, suggest the presence of more complex iconographic subjects. A sphinx, probably perched on a column capital, is depicted on a krater from a site identified as a farmhouse (**No. 73**). The earliest evidence for this type of capital in southern Italian figured pottery is a calyx-krater, decorated by an artist near to the Painter of the Birth of Dionysos, on which Hades' palace is represented.[20] Similar capitals are part of the decoration (also Hades' palace) on the Middle Apulian volute-krater Karlsruhe B4,[21] and they decorate the fountain house in a representation of the ambush of Troilus on a volute-krater in St. Petersburg,[22] as well as the naiskos on a fragment by the Darius Painter probably representing Achilles and Phoenix.[23]

Mythological scenes could be represented on the two fragments depicting horses (**Nos. 71–72**), as well as on a krater fragment with a male figure standing on a chariot (**No. 75**).[24] All three fragments come from sites where tombs have been detected. The lack of mythological scenes in the choice of the vases found in Metapontine necropoleis has often been noted:[25] female genre scenes, pursuits, nuptials scenes, and Dionysiac representations seem to constitute the favorite subjects of the vases found inside the tombs.[26] In a few cases, however, there are larger vessels, dating to the second half of the 4th century BC, with complex figurative compositions of mythological subjects. These were probably grave-markers,[27] as they were found outside of the burials.[28] The sample from the survey could suggest that vases depicting mythological scenes were not so rare, at least as far as the chora is concerned. A similar phenomenon has already been observed at Taras.[29]

Shapes

Of the 595 sites identified in the survey, 256 were occupied during the period of red-figure pottery production. Fragments of vessels of this class have been recognized at 34% of the sites, and their homogeneous spatial distribution in the survey study area (*Fig. 8.2*) seems to suggest a diffuse consumption of red-figure pottery in the chora.

Determination of the range of functions supplied by red-figure pottery in a rural landscape, if

[16] Carter and Hall 1998, 433–34 and Burn 1998, 637, SS 12-1 (Saldone, T 12); Carter and Hall 1998, 446–47 and Burn 1998, 640, SAV 4-3 (Sant'Angelo Vecchio, T 4). See also the bell-krater from tomb 415 in proprietà La Torre: De Siena 2007, 440, Pl. XIIIa. Amykos Painter: Carter and Hall 1998, 418 and Burn 1998, 618–19, T 175-2 (Pantanello, T 175).

[17] Denoyelle 2002a, 105.

[18] For the subject in Lucanian vases, see Söldner 2007, 38.

[19] Lo Porto 1981b, 383, No. 7, Fig. 114.

[20] Taranto inv. 54959: *RVAp* I 2/26.

[21] *RVAp* I 1[illegible]/81, Pl. 160.1.

[22] *RVAp* I 8/103 (associated with Iliupersis Painter).

[23] *RVAp* II 18/77, Pl. 181.2.

[24] See, for examples, representations of Triptolemos: *RVAp* I 15/60, Pl. 144.2 (associated with Iliupersis Painter).

[25] Burn 1998, 597; Pouzadoux 2002, 121.

[26] With noteworthy exceptions, such as the pelike depicting the apotheosis of Herakles from T 306 at Pantanello (Burn 1998, 621–22) or the recently discovered vases from a tomb in proprietà Andrisani (De Siena 2007, 442–44, Pl. XIVc and XVa).

[27] Morard 2002, 22–24.

[28] See the vessels cited in n. 14.

[29] Lippolis 1995, 357.

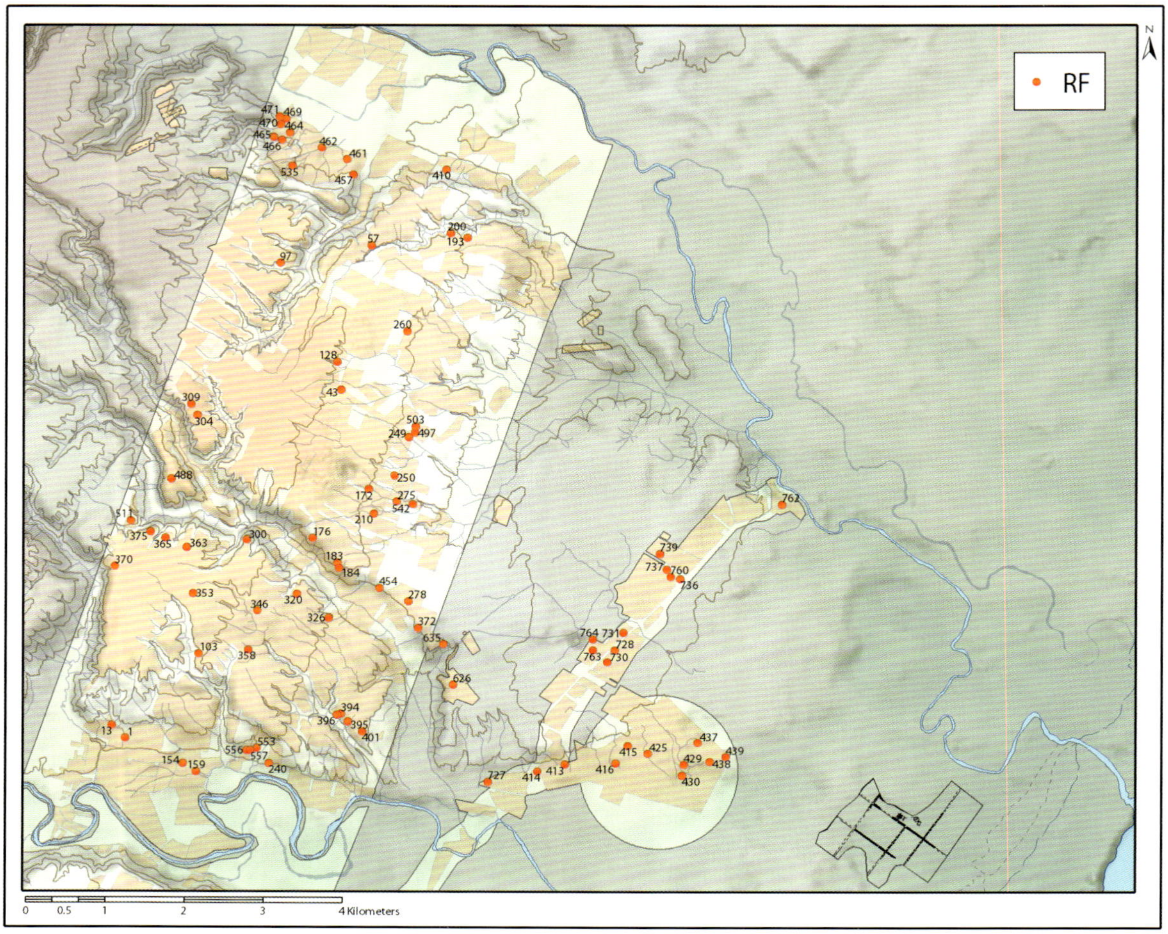

Figure 8.2 Sites with red-figure pottery (450–300 BC). (JT)

based mainly on material derived from field survey, as in this case, is especially challenging. The analysis of sites and shapes can provide only a limited basis upon which to formulate interpretative hypotheses that will have to be verified with further study of evidence from excavated context. The number of vases identified rarely exceeds two vessels per site, and of the eight sites with more than five red-figure vases, seven provide evidence for the presence of graves.[30] Furthermore, the range of shapes identified in the territory (*Fig. 8.3*) is for the most part consistent with the selections encountered in Metapontine funerary assemblages. Pelikai, hydriai, lekythoi, and skyphoi are deposited in Metapontine graves in all of the periods considered,[31] and most of the vessels have been found at sites where the presence of tombs has been hypothesized (*Fig. 8.4*).

A similar interpretation can be advanced for the pseudo-Panathenaic amphora, a shape that probably played an important role in Metapontine funerary rites and customs. These amphorae have been found (albeit rarely) inside tombs,[32] and numerous fragments were recovered in the ceramic deposits and pottery lots of the Pantanello necropolis.[33] Some of the vessels of the second half of the 4th century BC are monumental in size and have been interpreted as grave-markers.

Kraters comprise 44.7% of the vases (*Fig. 8.3*). Over half of these, accounting for 27.6% of identified red-figure vases, can be attributed to Lucanian and Apulian bell-kraters of average size. To these can be added the fragments belonging to at least nine volute-kraters (see Selective Catalog), eight calyx-kraters,

[30] Sites 401, 415, 439, 471, 497, 635, and 728 (see Gazetteer).

[31] J. Hall 1998, 567–73.

[32] Examples are known from tombs excavated along the SS106 "Jonica" coastal road: see, e.g. T 376 proprietà Latorre (last quarter 5th c. BC) and T 713 proprietà Tarulli (first half of the 4th c. BC). The author is grateful to A. De Siena and L. Cossalter for the information.

[33] Carter 1998d, 125; Elliott 1998, 648, 692 (black-gloss amphorae).

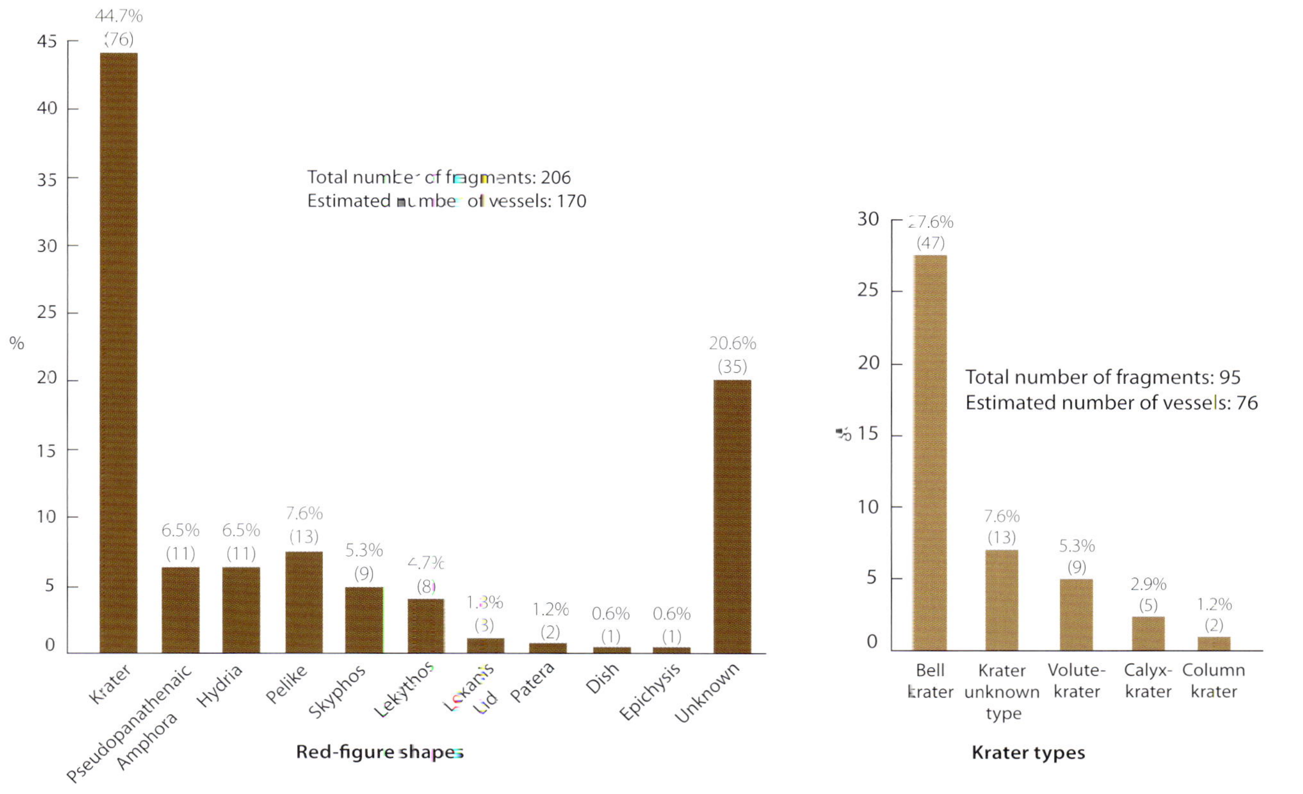

Figure 8.3 Frequencies of southern Italian red-figure shapes.

and one column-krater. Red-figure kraters have been found both at survey sites where tombs were recognized and at sites where tombs were not detected (*Fig. 8.4*).[34] In the first instance, the presence of this shape is consistent with what we know from other excavated sites. It has long been known that kraters were not, as a rule, deposited inside Greek tombs,[35] but krater fragments are frequently recovered in layers covering tombs and in the deposits of many Greek cemeteries,[36] where these vessels could be employed as grave markers or, as the latest evidence appears to indicate,[37] in rites performed at the tomb at the moment of burial and subsequent visits.

The evidence from Metaponto and its territory, until recently very meager,[38] has now been augmented substantially by the corpus of kraters from the survey. Nevertheless, the quality of the new evidence is not sufficient to identify the essential function of kraters in Metapontine necropoleis, where the large volute-kraters in particular are usually interpreted as grave-markers.[39] However, since many of them come from sites where tombs have not been detected, the possibility of different spheres of consumption must always be borne in mind.[40]

34 Beyond the examples cited in the catalog that follows, there are also the fragments of a bell-krater from site 736 T 2, while a second bell-krater fragment was found sporadically around the same site: see Ch. 29 for the foot from the fill of T 18, site 737: 18.02.

35 Elia 2003, 146, with bibliography.

36 See most recently Elia 2003, 146–47 for a detailed discussion of the problem and a list of the evidence. Two exceptions are known from Metapontion: Galli 1926–28, 63–64, Figs. 4–5 (bell-krater from a tomb from contrada Ricotta) and De Siena 2007, 440, Pl. XIIIa (bell-krater by the Pisticci Painter from T 415 in proprietà La Torre).

37 Elia 2003, 153.

38 Urban necropolis: Sestieri 1940, 79–80 (Apulian kraters from contrada Ricotta). Lo Porto 1981b, 383, No. 4, Fig. 113.8 (fragmentary Apulian bell-krater rim of the second half of the 4th c. BC from contrada Ricotta, Area B). Lo Porto 1981b, 376, Nos. 2–3, Fig. 102.2 (mascaroon of a volute-krater of the second half of the 4th c. BC from the fill of T 31 in Crucinia, Area A). A. Bottini 1993a, 767 (amphorae and kraters used as offerings or as *semata* at the necropolis of Crucinia). Pantanello: Carter 1983a, 459, Nos. 39 and 41, Fig. 35 (two Late Apulian mascaroons from volute-kraters). Carter 1998d, 123–25.

39 Carter 1998d, 116–17 and 125. The author does not exclude, however, the possibility of their use in libation rites. The monumental volute-krater from T 5/1983 at Pantanello has been interpreted as a grave-marker (Morard 2002, 22–24). For the urban necropolis, see Rocchietti 2002, 85–86, who leaves the question open because of the limited quantity of available data.

40 Elia 1997, 117–20; Lippolis and Mazzei 2005, 15–16; Söldner 2007, 231–32. See also, for Civita di Tricarico, Pouzadoux 2009, 352–53. At Metaponto few fragments of figured pottery from sanctuaries have been

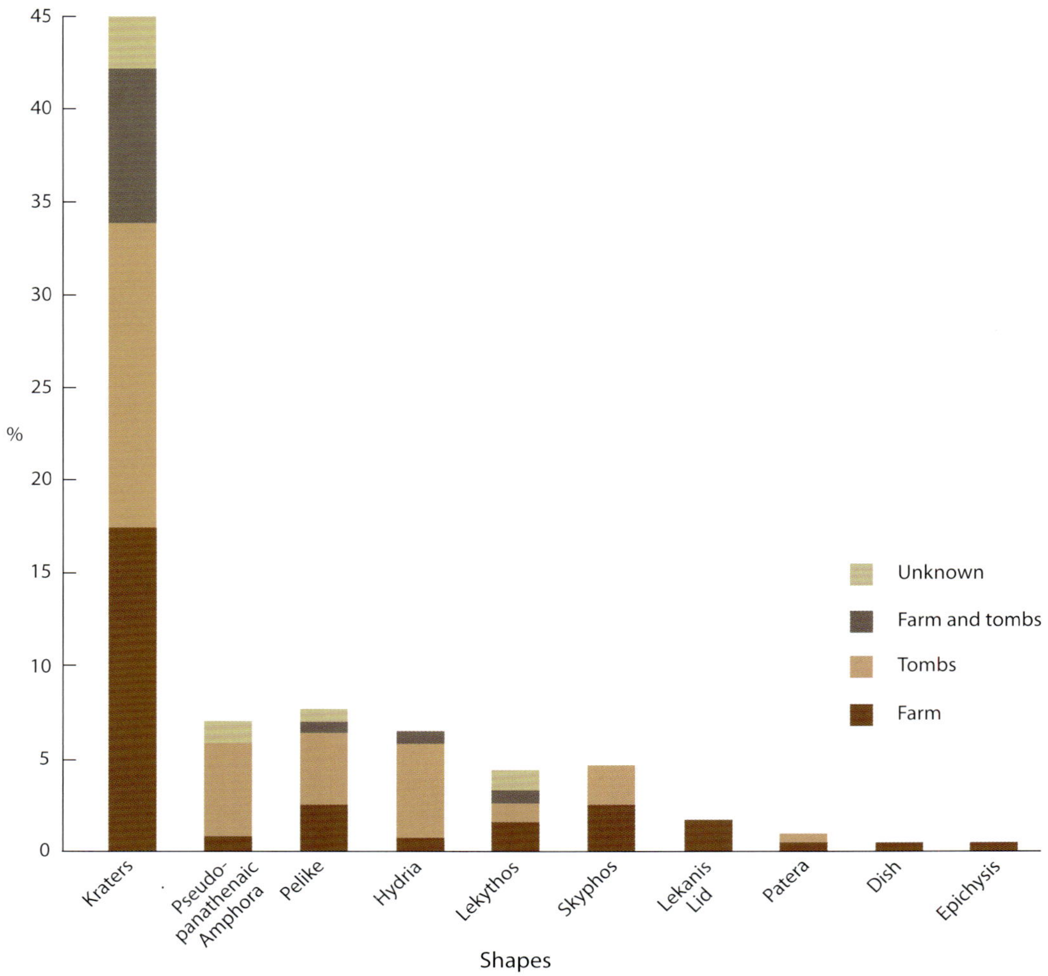

Figure 8.4 Southern Italian red figure. Shapes by site.

Xenon Group

This class of vessels may have begun after the middle of the 5th century BC in the asty of Metaponto, where it was produced at least until the third quarter of the 4th century, as confirmed by kiln wasters found in deposits 3, 5, and 6 of the kerameikos.[41] In the 4th century BC, Xenon Group vases were imitated also in other areas.[42] These vessels are mainly found at indigenous sites in eastern Basilicata and north-central Puglia. Nonetheless, recent research has brought to light evidence from the Metapontine urban and rural necropoleis: two oinochoai (forms 2 and 8) were found in unpublished graves from proprietà Giasi;[43] a stemless kylix of the Red Swan Group (second half of the 4th c. BC) was part of the funerary assemblage of T 38 in proprietà Venezia,[44] while a second kylix of the same type was found in T 145 of the rural necropolis at Pantanello.[45] The fragment of the Red Swan Group from the Pantanello sanctuary area ("deposit over the claypit"[46]) also belongs to an open shape. To the previously cited evidence can now also be added the sample from the field survey, consisting of two fragments of closed shapes (**No. 76**) and a fragment of a kantharoid vase (**No. 77**) found at a site where the

published; evidence from the domestic sphere is lacking. Nonetheless, study of the pottery found in the farmhouses (in progress, to appear in forthcoming volumes of the *Chora of Metaponto* series) reveals the presence, even if limited, of red-figure fragments; a consistent number of sherds from kraters have been identified. For a preliminary note, see Silvestrelli forthcoming c.

[41] Silvestrelli forthcoming b.

[42] For the Metapontine origin of the class and its distribution, see E.G.D. Robinson 1996 and De Juliis 2006, 532.

[43] T 286 and T 467. The author is grateful to L. Cossalter and A. De Siena for the information. For this area, see M. Tagliente 2006, 732, Fig. 5.

[44] M. Tagliente 2006, 732–33, Pl. IV.1.

[45] Elliott 1998, 647, Fig. 19.

[46] Carter 1983a, 470, No. 56, Fig. 35.

presence of tombs has been suggested.[47] This shape, certainly produced at the urban center of Metaponto[48] probably following indigenous models,[49] was distributed especially in Peucetia and, to a lesser extent, in Daunia.[50]

The presence of this type of pottery in the asty of Metaponto fills a gap in the Xenon Group distribution, supporting evidence from a Greek area,[51] but it does not appear to invalidate the hypothesis that these vases were made mainly for indigenous consumption. The percentage of Metapontine tombs in which Xenon Group vases have been found is in fact extremely small when compared to the total number of excavated graves. Their occurrence could also be explained by the occasional presence of indigenous burials in Metapontine necropoleis.[52]

Gnathia Pottery

The details regarding characteristics and distribution of Gnathia pottery at the Metaponto asty and in its chora still need further clarification. Workshops producing Gnathia pottery were active in the Achaian colony in the third quarter of the 4th century BC, and a certain number of wasters belonging to skyphoi, cup-skyphoi, Type B and C bell-kraters, calyx-kraters, and hydriai with decorative patterns influenced by contemporary Apulian workshops (Konnakis Group and Group of the Rose Painter) were found in deposit 3 of the local kerameikos, which is dated to the late third quarter of the 4th century BC.[53] J.R. Green has recognized a 3rd-century BC production from a workshop connected to the Alexandria Group.[54] Nonetheless, the available evidence does not allow the reconstruction of a sequence or the identification of a specific style in the choice of shapes and decorative patterns.

The number of vessels decorated in Gnathia style from the survey is very small, and most of them were probably manufactured locally. The sherds are small and worn, and their contribution to the comprehension of the Gnathia pottery distribution is limited (*Fig. 8.1*). The fragments can be attributed to Middle (Type B and Type C bell-kraters and Type C skyphoi of the last third of the 4th c. BC) and especially Late Gnathia vessels of the 3rd century BC (skyphoi, hydriai, bombylioi, and lebetes gamikoi). On the whole, the range of identified shapes is wide and consistent with the vessels deposited in Metapontine tombs. The only exceptions are the bell-kraters (e.g., **Nos. 90–91** and **93–95**), found at sites interpreted as farmhouses, but also at those where tombs have been detected; they could have shared the same function as red-figure bell-kraters, that is, as part of the domestic assemblage or for rites outside the tombs.

Black-on-Buff Ware

An unusual situation is presented by four sherds, probably from skyphoi (**No. 102** with two sherds and **Nos. 103–104**), with a more or less pronounced, outward-turned rim, convex body and black-gloss decoration. They can be related to a peculiar type of fine ware—mainly cups and skyphoi—possibly of Metapontine production and known from very few examples found at Cozzo Presepe, Pomarico, and Gravina.[55] They are characterized by a well-levigated, reddish-yellow clay, a thin, grayish, brush-painted gloss on the interior, and a reserved exterior surface. Just below the rim there is a black-gloss decoration, typically ivy branches, but also rows of dots or stylized branches.[56] The decorative patterns appear to derive from the red-figure and Gnathia styles, and both the patterns and shapes fit comfortably in the second half of the 4th century BC. The evidence from Pomarico confirms that these vessels continued to be produced into the early 3rd century BC.[57]

[47] See Gazetteer, site 521.

[48] Silvestrelli 2008, 281–82, with further bibliography.

[49] The evolution of the form is traced in E.G.D. Robinson 1990, to which can now be added also Colivicchi 2006.

[50] E.G.D. Robinson 1996, 448; De Juliis 2002, 135.

[51] For the occasional (and uncertain) presence of Xenon Group vases in Taras, see De Juliis 2002, 97 and 188, No. 4.

[52] This view is supported by E.G.D. Robinson (2004, 202, No. 60), who cites other indigenous graves. A similar explanation has been proposed for the group buried in the tombs of proprietà Venezia (M. Tagliente 2006, 733).

[53] D'Andria 1980a, 427. Silvestrelli forthcoming a, Fig. 14.

[54] Green 2001, 64.

[55] Cozzo Presepe: *Cozzo Presepe* 346–48, Nos. 273–82, Fig. 122. Pomarico: *Pomarico Vecchio* I 1.128, Nos. 6–9, 2.57, Pl. 49. Botromagno: *Gravina* I 45, No. 449, 385, Pl. VII C. Other sherds found in the excavation of the Pantanello sanctuary and of Fattoria Stefan will be published in a forthcoming volume of the *Chora of Metaponto* series.

[56] *Pomarico Vecchio* I 1.128, Nos. 8–9, 2.57, Pl. 49.

[57] *Ibid.*

Selective Catalog of Figured Pottery

All dimensions are in cm. Maximum height is indicated by "h," maximum width indicated by "w," maximum thickness by "th." An asterisk (*) after the site number indicates materials fromm the 1980 ICA pilot survey (sites numbered 1–44), as well as at the intensive survey sites and plots located south of the Basento River through the 2001 season (most of the sites numbered 562 and higher). Additional information about the sites can be found in the Gazetteer. The nature of the survey sample has led the author to organize the catalog of the material by shapes.

Attic Black-figure Pottery

1:2

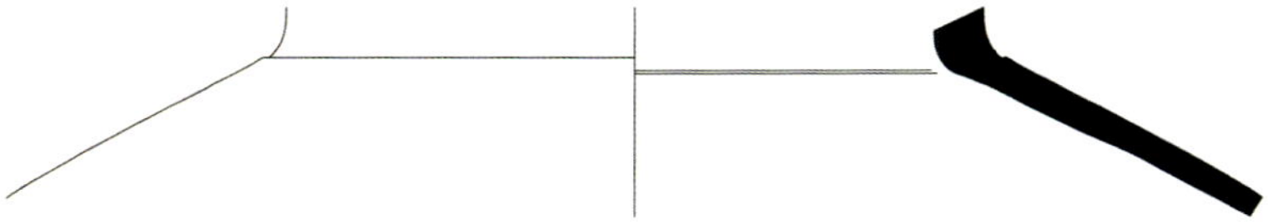

1. BF 050-01 Neck and part of shoulder of closed shape
Dimensions: neck diam. 11.1 h 3.4 w 6.7
Description: Clay red (5YR 6/6) to reddish-yellow on the surface (7.5YR 7/6). Upper body of a hoplite looking left with his right arm raised, probably holding a spear. He wears a crested Corinthian helmet. The scene probably represents a duel between two hoplites.
Comparisons: See the shoulder of the hydria Naples 81104: *CVA* Napoli, Museo Nazionale 1, III.H.E.18, Pls. 41.1–4 (510–490 BC).
Date: 510–480 BC.

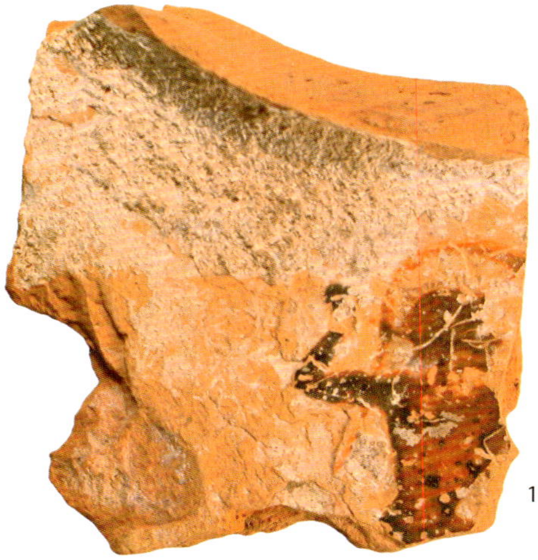

1:1

1:2

2. BF 265-01 Attic black-figure lekythos
Dimensions: body diam. 4.39 h 3.65 base diam. 2.08
Description: Clay reddish-yellow (5YR 7/6). Worn exterior surface. Thick and lustrous black gloss on the exterior. Reserved line on the lower part of the body.
Comparisons: Burn 1998, 602–603 T 137-1 (Class of Athens 581, 500 BC). Burow 2000, 284, Nos. 472–73, Pl. 87 (500–490 BC). Semeraro 1997, 147, Nos. 288–89 (480–470 BC).
Date: 500–480 BC, as suggested by L. Burn.

1:2

3. BF 797-01 and **BF 797-02** Shoulder and body fragments of an Attic black-figure shouldered lekythos
Dimensions:
BF 797-01 h 3.0 w 4.5
BF 797-02 h 1.6 w 2.7
Description: Clay reddish-yellow (5YR 7/6). Short rays at the base of the neck; the shoulder was probably decorated with palmettes.
Comparisons: For a similar decorative pattern see Burn 1998, 604, T 196-9 (500–490 BC); 605, T 237-4 (490–480 BC).
Date: 500–480 BC.

Attic Red-figure Pottery

1:3

4. RF Plot 2001.062-01 Rim of a calyx-krater
Dimensions: rim diam. ca. 43.5 h 3.63 w 4.26
Description: Clay red (5YR 6/6). Lustrous black gloss inside and out, partially worn. On the lip, inclined palmettes.
Comparisons: For the shape, characterized by a marked vertical rim, see the calyx-krater by the Altamura Painter ARV2 591.25, *CVA* Baltimore, Walters Art Gallery 1.16–17, Fig. 4.4, Pls. 20.1–4 (470–450 BC). For the decoration, see the calyx-krater in the manner of the Berlin Painter, Cincinnati Art Museum XXXX0.2136: ARV2 215.10; for the vases of the same shape by the Altamura Painter see Frank 1990, 194, No. 79, Pls. 16.1–2, 470–460 BC; Lo Porto 1973, 231, No. 3, Pls. LXXIV-LXXV (from Irsina); and the bell-krater (with lugs) by the Mykonos Painter, ARV2 515.9bis (Neils 1997, 237, Fig. 12). A similar pattern, but with very slender leaves, was still in use at the end of the 5th c. BC. See the volute-krater by the Kadmos Painter, ARV2 1184.1 and Sichtermann 1966, 20, No. K10.
Date: First half of the 5th c. BC, probably second quarter.

Southern Italian Red-figure Pottery

Pseudo-Panathenaic amphorae

Pseudo-Panathenaic amphorae are one of the best-represented shapes in the survey corpus. Some of the fragments (**Nos. 5** and **6**) can be dated to the first phases of Lucanian red-figure production. Except for a fragment that could belong to the middle of the 4th c. BC (**No. 7**), the other sherds probably date to the second half of the 4th c. BC, when amphorae of two different sizes can be distinguished. The rim (**No. 9**) and the foot (**No. 8**) might belong to monumental Late Apulian amphorae, while **Nos. 10–11** belong to smaller amphorae. The fabric is consistent with the local clay, and the shape fits in the morphological repertoire of Metapontine workshops of the second half of the 4th c. BC. See Metapontion, kerameikos deposit **No. 3** in Silvestrelli forthcoming a, Fig. 13.14.

1:2

5. RF 401-12 Fragment of lower part of body
Dimensions: h 5.2 w 7.2
Description: Clay pink (5YR 7/3). Miltos on reserved surfaces. Above the foot (not preserved), vertical rays.
Comparisons: Rays appear on Lucanian red-figure amphorae through the first quarter of the 4th c. BC. See the amphorae Napoli 2416, *LCS* 48, No. 246, by the Amykos Painter, and Munich 3262, *LCS* 124, No. 590, Pls. 59.4–5, by the Brooklyn-Budapest Painter. The large, short conjunction between the body and the foot recalls vases from the beginning of Lucanian production, e.g., a black-figure panathenaic amphora by the Pisticci Painter from T 1 at Casinello, Pisticci (Denoyelle 1997, 403, Figs. 11–12 and *Archeologia di un sapere*, 401 III.168).
Date: Second half of the 5th c. BC.

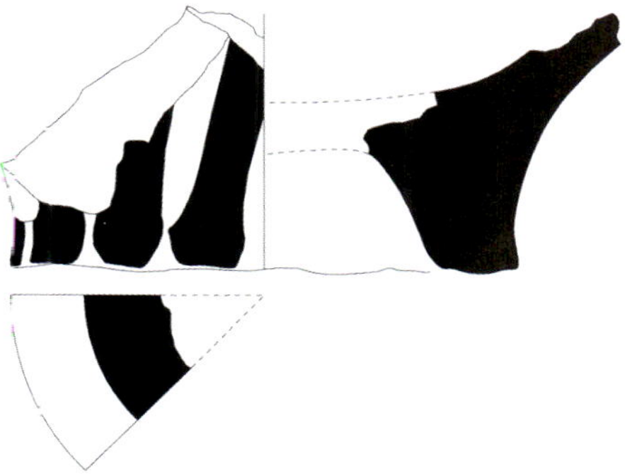

Pseudo-Panathenaic amphora cont.

1:1

6. RF 471-04 Body fragment
Dimensions: h 3.8 w 5.0 th 0.6
Description: Clay pinkish gray (7.5YR 7/2). Miltos on the reserved surfaces. Black gloss on the exterior. Torso of a young man facing right. The left arm is raised, perhaps leaning on a diphros (in the case of a seated figure) or on a pilaster (in the case of a standing figure). A cloak passes behind his shoulders.
Comparisons: The same anatomy and pose are found in the Dionysos of the bell-krater in Sydney, *LCS Suppl.* III D44 (Dolon Painter).
Date: Beginning of the 4th c. BC.

1:2

7. RF 176-01 Body fragment
Dimensions: body diam. 13.5 h 8.8 w 6.3
Description: Clay pink (7.5YR 7/4). Rays.
Comparisons: For the shape and decoration see Middle Apulian pseudo-Panathenaic amphorae such as Ruvo, Museo Jatta 423, *RVAp* I 15/41 and Sichtermann 1966, 49–50, No. K 71, Pls. 114–115, Group of Ruvo 423.
Date: Middle to third quarter of the 4th c. BC.

1:2

8. RF 471-07 Foot
Dimensions: base diam 20.5 h 5.65 w 8.15
Description: Clay reddish yellow (7.5YR 7/6) to pinkish gray (core 7.5YR 7/2), powdery and depurated. Dull brownish gloss on the exterior. Misfired fragment.
Date: Second half of the 4th c. BC.

1:2

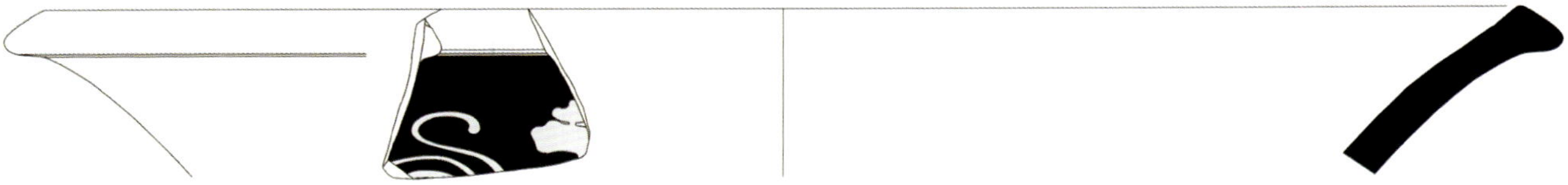

9. RF 250-02 Rim
Dimensions: rim diam 30–34 h 3.74 w 4.55
Description: Clay pink (5YR 7/4). Lustrous brownish to reddish gloss on the interior and exterior. Faint vine branch. Accurate calculation of the diameter is impossible because of the small preserved portion; monumental examples can have a diameter of 30 cm, like the example from the survey.
Comparisons: For the profile, see the pseudo-Panathenaic amphora in Gnathia style at Mannheim, Reiss-Engelhorn Museum Cg 460, *CVA* Mannheim 2, Pl. 47, Fig. 18.1, workshop of the Baltimore Painter (325–300 BC).
Date: Second half of the 4th c. BC.

Pseudo-Panathenaic amphora cont.

1:2

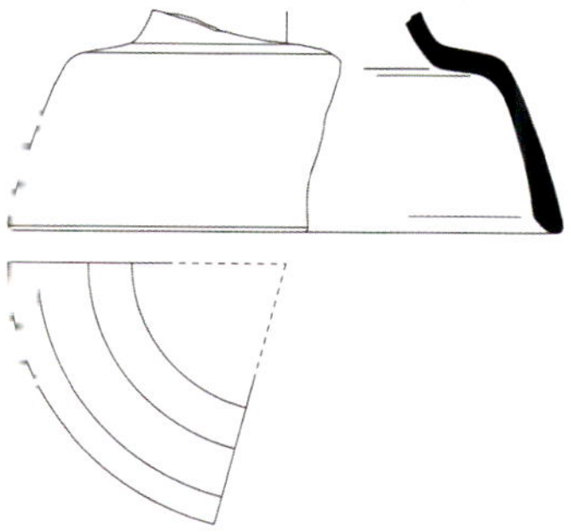

10. RF 184-01 Foot
Dimensions: base diam 8.5 h 3.5 w 5.7
Description: Clay pink (5YR 7/4). Thick brownish gloss on the exterior.
Date: Second half of the 4th c. BC.

1:2

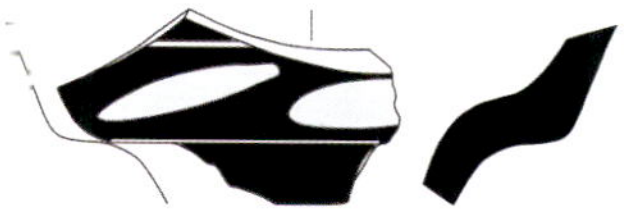

11. RF 415-07 Rim
Dimensions: rim diam. 8.1 h 2.66 w 4.55
Description: Clay reddish yellow (5YR 7/6). Interior partly reserved, partly painted in a black to reddish, lustrous gloss. Thick, brownish gloss on the exterior. On the rim, overpainted laurel branch to left, now faint.
Date: Second half of the 4th c. BC.

Type C skyphos

1:1

12. RF 462-08 Foot
Dimensions: h 3.4 w 5.3 th 0.5
Description: Clay pink (5YR 8/4). Reddish to black gloss on the interior; on the bottom upper side, stacking ring (Ø 2.7). On the lower part of the body, rays.
Comparisons: The decorative pattern is borrowed from Attic vessels, which adopted it in 520–510 BC and in the last quarter of the 5th c. (Oakley 1997, 170); this type of decoration, not frequently found in southern Italian figured pottery, is mainly attested at the end of the 5th c. BC (see the Type C skyphos by the Via Dante Group, LCS 66, No. 325, Pl. 31.1. For Apulian examples of the second half of the 4th c. BC, see the Alabastra Group, *RVAp* II 21/27–28, Figs. 233.3–4). The adoption by Metapontine workshops, to which the fragment can be attributed, is confirmed by fragments found in deposits No. 16 and No. 1 of the local kerameikos.
Date: Second half of the 5th c. BC.

1:2

1:1

13. RF 503-10 Rim
Dimensions: rim diam. 11.6 h 3.35 w 5.7
Description: Clay reddish yellow (5YR 7/6). Thick black gloss on the interior and exterior. On the rim, egg-and-dot pattern; below, reserved line. On the body, volute, part of the decorative pattern of the handle area.
Date: Probably second or third quarter of the 4th c. BC.
Comparisons: For the shape, see the fragment from deposit No. 3 of the kerameikos (Silvestrelli forthcoming a, Fig. 11.5). Similar shape and decorative pattern are found on skyphoi from the agora of Herakleia (Casagrande 2002, 121, No. 14, Fig. 51, 375–350 BC) and a skyphos from Pomarico (Elia 1997, 113, No. 12, Pl. 46, Fig. 48, Group of the Dresden Amphora, second quarter of the 4th c. BC. For the problems associated with the chronology of this group, which could be postponed to the third quarter of the c., see Elia 1997, 113, nn. 31 and 33).

Type C skyphos cont.

1:2

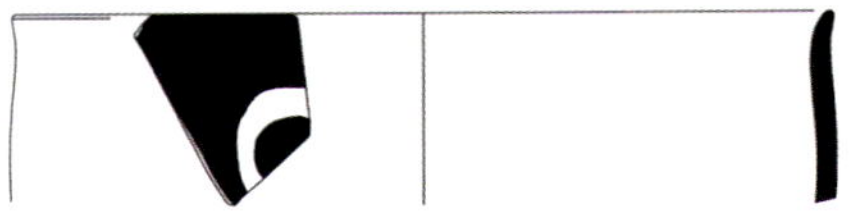

14. RF 410-03 Rim
Dimensions: rim diam. 10.9 h 2.63 w 2.29
Description: Clay pink (7.5YR 8/3). Thick, dark gray to green gloss on the interior and exterior.
Comparisons: For shape, see fragment from deposit No. 3 of the kerameikos (Silvestrelli forthcoming a, Fig. 13.17).
Date: Second half of the 4th c. BC.

Dish

1:1

15. RF 103-13 rev90 Body fragment
Dimensions: h 2.5 w 2.6 th 0.4
Description: Clay pink (5YR 8/6). Lustrous, thick black gloss on the interior and exterior. Incised branch, probably originally completed with added decoration now completely vanished. A reserved line encircles an unidentified red-figure decoration.
Comparisons: It is not possible to identify the typology, subject, and production center. The dish is a shape produced in Metapontine red-figure workshops, as attested by a dish found in an unpublished kerameikos deposit excavated in 2005.
Date: Second half of the 4th c. BC.

Patera

Paterae imitating bronze vessels were introduced in the red-figure morphological repertoire at the beginning of the second quarter of the 4th c. BC and quickly attained considerable popularity. This shape seems to have been produced at Metaponto, too: a patera of the middle of the 4th c. BC was in fact found in deposit No. 6 of the kerameikos (D'Andria 1980a, 440, No. 381, Fig. 71). The sporadic fragment from the kerameikos depicting two seated women attributed to the workshop of the Darius Painter (third quarter of the 4th c. BC) also belongs to a patera (D'Andria 1980a, 358, No. 14, Fig. 4). The shape does not occur, as far as is known, in Metapontine necropoleis.

1:2

16. RF 057-01 rev90 Foot and body
Dimensions: base diam. 8.2 int h 2.6 w 6.9
Description: Clay pink. Draped figure (?).
Date: 375–330 BC.

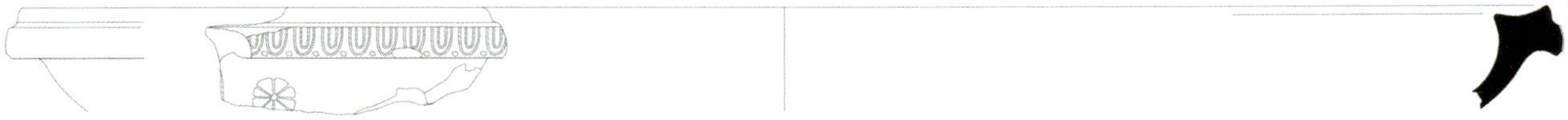

1:3

1:2

17. RF 068-186P (89) Rim and part of the body
Dimensions: rim diam. 49.5 h 3.35 w 9.6
Description: Clay pink (7.5YR 7/4). Thick, lustrous black gloss on the interior and exterior. Interior of the rim reserved. On the edge, egg-and-dot pattern, almost completely abraded. On the body, a rosette, probably part of the handle-area decoration.
Comparisons: For a similar rim see the patera by the Oakleaf Painter Malibu 71 AE 236 (*CVA* Malibu 3, Pl. 163, Fig. 3, 330 BC).
Date: Second or third quarter of the 4th c. BC.

Lekanis lids

1:2

18. RF 043-137P Rim of a lekanis lid
Dimensions: rim diam. 14.7 h 2.9 w 7.7
Description: Clay pink (7.5YR 7/4). Thick, lustrous glaze on the interior and exterior. Crested wave pattern on rim. On the upper surface, part of a hand holding a phiale with ribbing added in white. On the right, feet of a figure seated to left. The two figures are separated by a volute from which sprouts a rhomboidal flower; at the base, reserved dots (small stones?).
Date: Second half 4th c. BC
Comparisons: A similar composition appears on the lekanis lid found outside tomb 33/1978 at Rutigliano (*Museo Taranto* II.2, 105, No. 33.41, Fig. p. 101, second half of the 4th c. BC) although the survey vessel lacks the palmette dividing the two figures.

1:2

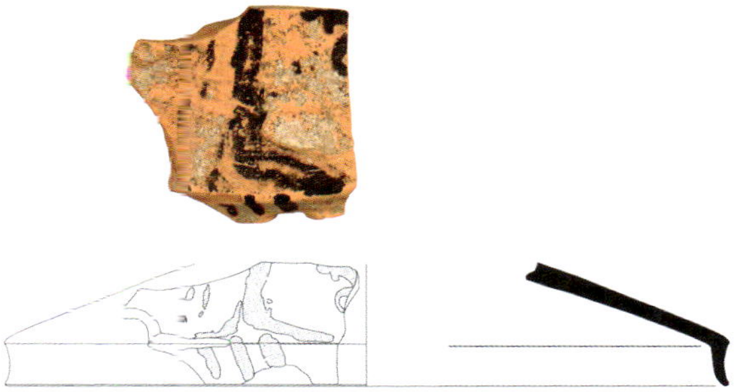

19. RF 454-02 Rim of a lekanis lid
Dimensions: rim diam. 14.6 h 2.6 w 4.75
Description: Clay reddish yellow (5YR 7/6). Thick, opaque black gloss on the exterior. Vertical rays on the edge of the rim. On the body, female head facing left.
Date: Last quarter of the 4th c. BC.
Comparisons: For the shape see the lekanis lid from Torre di Satriano (*Torre di Satriano* I, 312, No. 393, Pl. 52, Ivy Leaf Painter, end of the 4th c. BC).

Hydria

1:2

20. RF 401-01 rev90 Rim
Dimensions: h 1.6 w 3.2
Description: Clay pink (7.5YR 7/4). Thick, lustrous black gloss. Misfired. Egg-and-dot decoration on the edge.
Date: Last quarter of the 5th–beginning of the 4th c. BC.
Comparisons: For the profile see the hydria *RVAp* 1/81 and *CVA* Bonn 3, Pl. 17.1–5, Fig. 16, by the Sisyphus Painter.

1:2

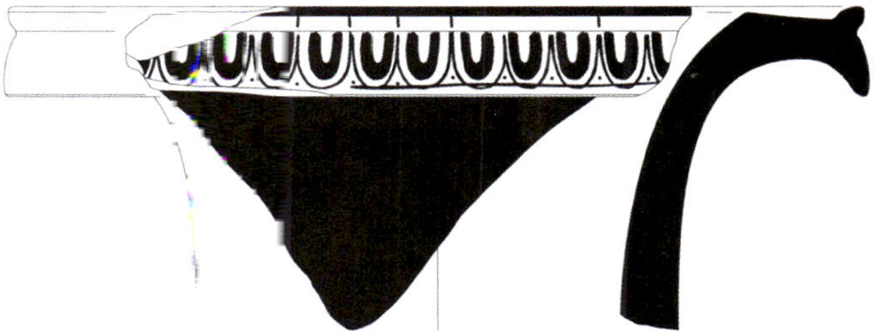

21. RF 764-02 Rim and neck
Dimensions: rim diam. 17.3 h 6.6 w 11.3
Description: Clay reddish yellow (5YR 7/6). Thick, lustrous black gloss on the interior and exterior. Egg-and-dot decoration on the edge.
Date: First half of the 4th c. BC.

1:2

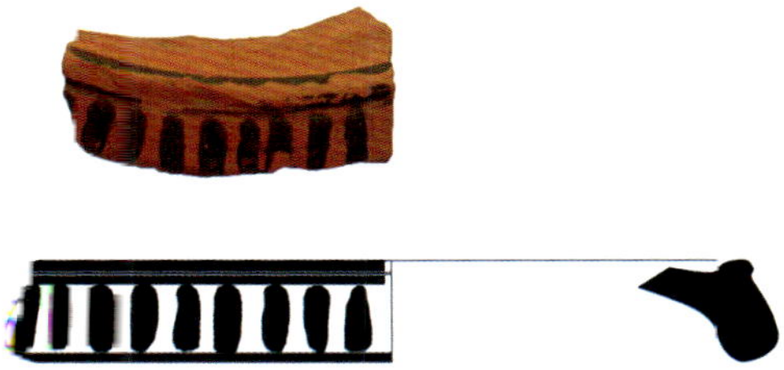

22. RF 557-01 rev90 Rim
Dimensions: rim diam. 12.6 h 1.7 cam (?) w 5.1
Description: Clay pink (7.5YR 7/4). Thick, lustrous black to brownish gloss on the exterior. On the edge, vertical lines.
Date: 4th c. BC.

Hydria cont. 1:2

23. RF 497-12 through **497-20** Nine non-joining fragments (wall and foot) of a single red-figure hydria

Dimensions:
foot RF 497-12 h 7.2 diam. 11.1
body sherd RF 497-13 h 9.0 w 5.9
body sherd RF 497-14 h 6.7 w 8.0
body sherd RF 497-15 h 8.5 w 6.5
body sherd RF 497-16 h 8.5 w 8.8
body sherd RF 497-17 h 5.0 w 6.4
body sherd RF 497-18 h 2.6 w 3.6
body sherd RF 497-19 h 1.8 w 4.5
body sherd RF 497-20 h 6.0 w 10.2

Description: Clay pink (5YR 7/4), hard and depurated. Thin, lustrous black gloss on the exterior. Miltos on the reserved surfaces. On the handle area, palmettes (RF 497, 16–17, and RF 497–20). Around the horizontal handles, stylized tongues (RF 497 14–15). The figured field is bordered by a meander (RF 497-18). Three fragments (RF 497-13, RF 497-15, and RF 497-19) belong to the figured scene. On RF 497-15, to the right of the handle, an Aeolic capital and part of the fluted column in added white, probably part of a funerary monument. On RF 497-13, a taenia and a draped figure with a stripe in added red, probably one of the men visiting the tomb.

Date: Second half of the 4th c. BC.

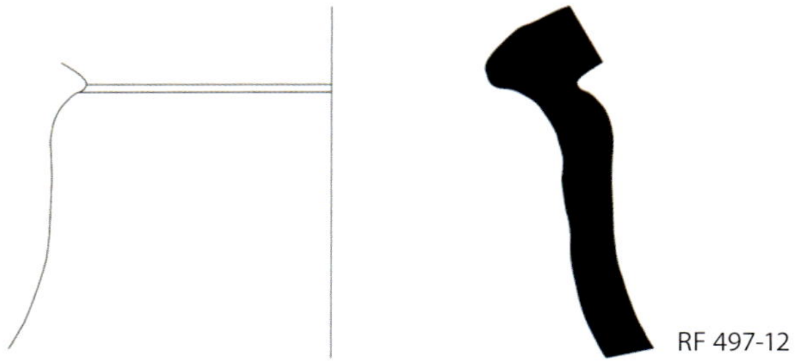

24. RF 184-03 Shoulder

Dimensions: h 3.7 w 4.1

Description: Clay pink (core 5YR 7/4) to reddish yellow (surface 5YR 7/6). Thick, dull gray gloss on the exterior. Column and lintel of a naiskos. Faint, unidentified overpainted decoration on the sides.

Date: 375–320 BC.

1:2

25. RF 471-08 Shoulder

Dimensions: h 4.2 w 3.3 th 0.6

Description: Clay pink (7.5YR 7/3). Black gloss on the exterior. Torso of a seated male facing right, with a cloak on the shoulders.

Date: 4th c. BC.

1:1

Hydria cont.

1:2

26. RF 471-03 Body
Dimensions: h 8.5 w 8.8 th 0.6
Description: Clay pink (5YR 7/3). Thin, lustrous black gloss on the exterior. Miltos on the reserved surfaces. Left hand, arm, and part of the dress of a seated woman. On the handle area, anthemion composed of scrolls and part of the palmette.
Date: 330–300 BC.

Squat lekythos

1:1

27. RF 320-36 Body
Dimensions: h 3.7 body diam. 4.2 th 0.3
Description: Clay reddish yellow (5YR 7/6), hard and depurated. Thin, lustrous black gloss on the exterior. Palmette flanked by scrolls.
Comparisons: For the shape see Hurschmann 1995, 677–79, Type III (Tonnenförmige Netzlekythoi, 330–300 BC). A very similar decorative pattern is found in D'Andria 1980a, 427, No. 299, Fig. 65 (from deposit No. 3 of the Metapontine kerameikos).
Date: Second half of the 4th c. BC.

1:1

28. RF 553-01 Neck and part of the body
Dimensions: body diam. 4.7 h 2.9 w 3.35
Description: Clay pink (5YR 8/4). Thin, lustrous black gloss on the exterior. At the base of the neck, vertical rays; in the handle area, volute; in the field, rear part of a sakkos or kekryphalos on a woman's head facing left.
Comparisons: For the shape and decoration see **No. 27.**
Date: Second half of the 4th c. BC.

Epichysis

This shape has a long tradition at Metaponto, where it is produced beginning in the second half of the 5th c. BC first in black gloss (see Silvestrelli, forthcoming a, Fig. 7 for the examples from deposit 22; for the epichyseis from deposit 1 of the kerameikos see D'Andria 1980a, 406, No. 198, Fig. 54; and Silvestrelli forthcoming b, Fig. 10) and then also in red-figure (for a red-figure epichysis of the Creusa-Dolon workshop, see Silvestrelli 2005, 120, Fig. 8 a–b; fragments of the second half of the 4th c. BC are also known from kerameikos deposit 3, D'Andria 1980a, 425, No. 294, Fig. 64). They are also part of the morphological repertoire of later Lucanian painters (Manchester IV E 31, *LCS* 884, related in style to the Painter of Naples 1959; *LCS* 1069–1077 and 117, Primato Painter). The shape is also known in Apulian red-figure starting in the second quarter of the 4th c. BC; its number increases considerably in the second half of the century (see McPhee and Pemberton, 2004, 57) when it also becomes part of the Gnathia repertoire (Lanza 2005, 46, with previous bibliography).

The distribution of the red-figure epichysis is quite difficult to detect: most of the vessels cited in the bibliography are of unknown provenience. As far as is known, they do not seem to occur either in Metapontine necropoleis nor have they been found in Tarantine graves, while they do appear at indigenous sites in the Agri valley (*Lagonegrese*, Pl. XV.6, from Serra Città; *RVAp* 11/251 and 16/40, both from Roccanova; *Greci, Enotri e Lucani*, 257, No. 3.39.4 and S. Bianco 1994, 120, 330–320 BC, from Sant'Arcangelo, contrada S. Brancato) and beyond (*Roccagloriosa* I, No. 219); at Torre di Satriano (*Torre di Satriano* I 311); in the area between Bradano and

Epichysis cont.

Basento (Lo Porto 1973, 192, No. 9, Pl. XLII.5 from Montescaglioso and *Pomarico Vecchio* I T 9, 289, No. 41, Pls. 107 and 127), in Messapia (Forti 1965, Pl. XV, from Manduria), and Daunia (*RVAp* 21/384 from Canosa).

1:1

29. RF 542-04 Shoulder fragment
Dimensions: h 2.1 w 4.5 th 0.5
Description: Clay pink (7.5YR 8/4). Thick, depurated, and lustrous black gloss outside. Enclosed palmette.
Comparisons: See the palmettes of the epichysis by the Zaandam Group at Adolphseck, Schloss Fasanerie 175 (*RVAp* 11/31 and *CVA* Adolphseck, Pls. 75.11–12, second quarter of the 4th c. BC), and especially those decorating some epichyseis attributed to the Liverpool Group (Edinburgh, National Museum of Scotland 1872. 23.24: *RVAp* 21/384 and *CVA* Edinburgh, Pls. 32.5–6; Corinth, Sanctuary of Demeter and Kore: *RVAp* 1 21/372b and McPhee and Pemberton 2004, 56–57, Pls. 7.3–4).
Date: The flatness of the shoulder and the type of palmette recall epichyseis of the second and especially the third quarter of the 4th c. BC.

Pelike

1:2

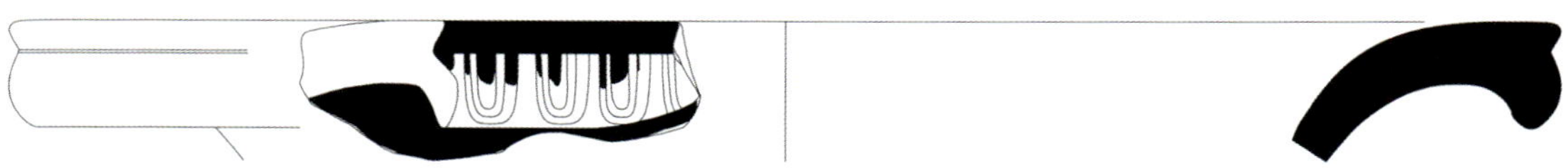

30. RF 154-22 Rim
Dimensions: rim diam. 29.0 h 2.72 w ca 5.07
Description: Clay pink (5YR 8/4). Lustrous, brownish to red gloss on the interior and exterior. On the edge, egg motif.
Date: End of the 5th–beginning of the 4th c. BC.

31. RF 300-01 Fragment of the body
Dimensions: h 4.0 w 2.2 th 0.4
Description: Clay reddish yellow (5YR 7/6). Thick, lustrous black gloss on the exterior; gray contour line. Miltos on the reserved surfaces. Meander pattern and part of a scroll.
Comparisons: For similar scroll see the Lucanian hydria Naples 81950 (H3247, *LCS* 45.22) by the Amykos Painter and the volute on the Apulian bell-krater from tomb 45/1978 at Rutigliano (Ciancio 2006, 417, Figs. 32–33, 380–370 BC) attributed to the workshop of the Tarporley Painter.
Date: Last quarter of the 5th–first quarter of the 4th c. BC.

1:1

32. RF 465-11 Neck
Dimensions: h 4.9 w 5.15
Description: Clay pink (5YR 8/4). Black gloss on the interior; brownish to black gloss on the exterior. Rosettes with faint dots in added white. Below, double reserved line and pendants.
Comparisons: For the decorative pattern see, e.g., the pelike Copenhagen 2252: Fischer-Hansen et al. 1992, 138–139, No. 107 (Apulian, 350–340 BC); *RVAp* 18/15, Pl. 172 (Laodamia Painter).
Date: 340–320 BC.

1:2

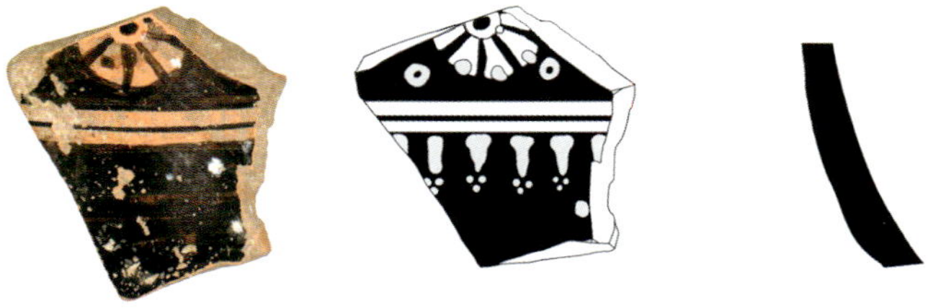

Bell-krater

1:3

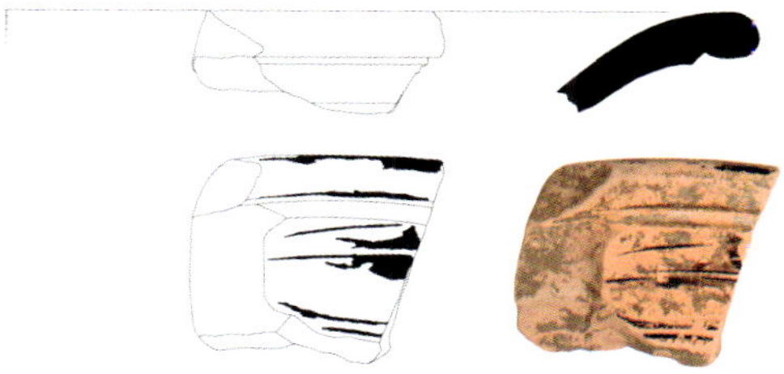

33. RF 401-13 Rim fragment
Dimensions: rim diam. ca. 37.0 h 2.55
Description: Clay pink (5YR 7/4). Black gloss on the interior with reserved band. Exterior decoration abraded. Laurel wreath to right.
Comparisons: Similar bell-kraters are Dresden Dr 377 (*RVAp* 162, No. 221 and *CVA* Dresden, Pl. 21, Fig. 5.2, late work or workshop of the Graz Painter, 370–360 BC) and Würzburg, Martin-von-Wagner-Museum HA 169 (*CVA* Deutschland 71, Würzburg, Martin-von-Wagner-Museum, Pl. 2, Fig. 1.2, workshop of the Hoppin Painter, 380–370 BC).
Date: First half of the 4th c. BC.

1:3

34. RF 471-12 Rim
Dimensions: h 2.65 w 11.7
Description: Clay pink (5YR 8/4). Brownish to red black gloss on the exterior, grey gloss on the interior. Laurel wreath to right. Misfired fragment.
Comparisons: *Torre di Satriano* I, 312, No. 389, Pl. LI (Ivy Leaf Painter, end of the 4th c. BC).
Date: Second half of the 4th c. BC.

1:3

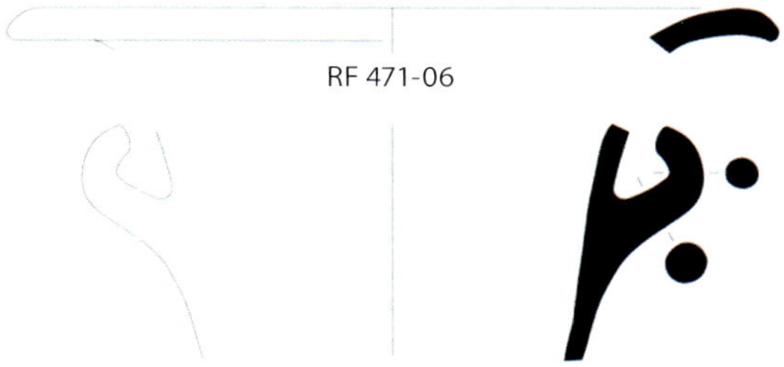

RF 471-06

RF 471-02

RF 471-02 (1:2)

RF 471-06 (1:1)

35. RF 471-02 Handle and part of the body
Dimensions: body diam. 12.8 h 6.57 w 6.51
Description: Clay reddish yellow (5YR 7/6). Thin, brownish gloss on the interior; thick, dull, brownish gloss on the exterior. Dots around the handles. Palmette on the handle area. Probably part of same vessel as **Nos. 36–37**.
Comparisons: For the decoration, see the bell-krater by the Chevron Group in a German private collection, Schauenburg 2005, 21, Fig. 40 c.
Date: Last quarter of the 4th c. BC.

36. RF 471-06 Body
Dimensions: h 4.0 w 4.2 th 0.5
Description: For clay and gloss see **No. 35**. Female head in profile facing left wearing a kekryphalos decorated with cross-shaped motifs, a radiate wreath, and earrings, both overpainted in yellowish white. Probably part of same vessel as **Nos. 35** and **37**.
Comparisons: For kekryphaloi decorated in a similar manner, see the lekanis lid by the Monopoli Group: *RVAp* 22/685, Pl. 262.
Date: Last third of the 4th c. BC (suggested by L. Burn).

37. RF 471-06 rev01 Rim
Dimensions: rim diam. 20.9 h 1.26 w 6.3
Description: For clay and gloss see **No. 35**. On the rim, band of chevrons. Could be part of same vessel as **Nos. 35–36**.
Comparions: The same decoration appears on a bell-krater from Pomarico, Elia 1997, 112, Pl. 45.6.
Date: Last third of the 4th c. BC.

Bell-krater cont.

1:3

38. RF 401-16 Rim
Dimensions: rim diam. 36.6 h 3.7 w 8.7
Description: Clay reddish yellow (5YR 7/6). Brownish to red gloss on the interior and exterior. Laurel branch to right.
Comparisons: Cahn Collection inv. 254 A-B (Cambitoglou and Chamay 1997, 88–90, No. 32, drawing at p. 327, connected to the Eumenides Group, first quarter of the 4th c. BC); Canosa, Ipogeo dei Vimini, cella B, right deposition (De Juliis 1990, 77, No. 42, Fig. 309, Dijon Painter, 370 BC); Agrigento R 179 (*RVAp* 4/119 and *CVA* Agrigento, Museo Archeologico Regionale, Pl. 2, Fig. 2, Long Overfalls Painter, Calvet Sub-Group, 380–370 BC).
Date: First quarter of the 4th c. BC.

39. RF 401-02 rev90 Rim
Dimensions: h 2.8 w 2.65
Description: Clay pink (5YR 8/4). Thin, lustrous black gloss on interior and exterior. Laurel branch and eggs.
Comparisons: A similar decorative pattern is found mainly on Apulian vases of the first half of the 4th c. (see bell-krater Taranto, Collezione Rotundo Vinc. No. 112/6: *RVAp* I 5/211a and *CVA* Italia 70, Museo Nazionale di Taranto, Collezione Rotundo, IV.D Pl. 35, by the Lecce Painter, 360 BC; and bell-krater Dresden DR 376: *RVAp* 5/279, Pls. 43.1–2, by the Painter of Sydney 64, 360 BC).
Date: First half of the 4th c. BC.

1:2

40. RF 401-15 Rim
Dimensions: rim diam. 46.5 h 2.5 w 7.4
Description: Clay reddish yellow (5YR 7/6). Black to brownish gloss on the interior and exterior. Reserved band on the interior. Miltos on the reserved surfaces.
Comparisons: Dresden Dr. 516 by the Wolfenbüttel Painter: *RVAp* 356, No. 187 and *CVA* Dresden, Pl. 22, Fig. 6.2 (350–340 BC).
Date: Third quarter of the 4th c. BC.

1:3

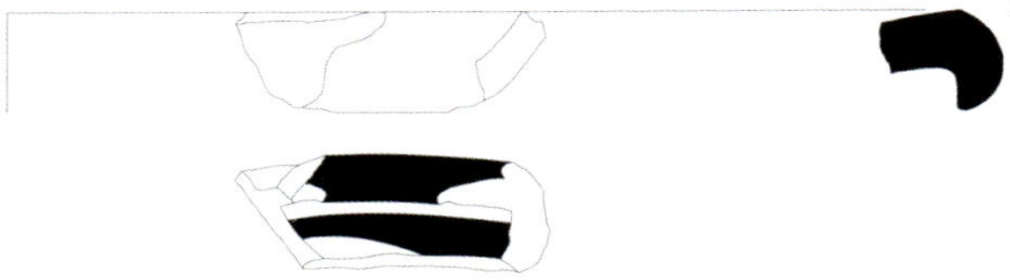

41. RF 401-08 Body
Dimensions: h 4.0 w 7.4 th 0.4
Description: Clay pink (5YR 8/4). Thick, lustrous black gloss on the interior and exterior. Miltos on the reserved surfaces. Draped figure running to right with folds of the mantle hanging over the outstretched left forearm; in front of the right hand, part of a staff.
Comparisons: Figures with similar drapery are usually found in pursuit-scenes by the Pisticci Painter (e.g. bell-krater *LCS* 16, No. 12, Pls. 2.5–6 and calyx-krater *LCS* 17, No. 19, Pls. 3.5–6). The staff held in the left hand could be a thyrsos (as in the bell-krater *LCS Suppl.* 2 1/69a, Pl. 29.1 or in the just-published bell-krater from tomb 415 of proprietà La Torre: De Siena 2007, 440, Pl. XIIIa). Less likely is the interpretation of the figure as a frontal variant of the Type B1 draped youth as, e.g., the central figure in the bell-krater in Schauenburg 2000b, 25, Fig. 77.
Date: Pisticci Painter, 450–420 BC (also suggested by L. Burn).

1:1

Bell-krater cont.

1:1

42. RF 413-02 Body
Dimensions: h 3.6 w 4.5 th 0.5/0.6
Description: Clay reddish yellow (7.5YR 8/6). Thick, dull, brownish gloss on the interior and exterior, with a reserved line on the interior. Miltos on the reserved surfaces. Column with Doric capital and abacus.
Comparisons: A similar column appears on the bell-krater by the Creusa Painter: *LCS Suppl.* 3 3/C 79.
Date: First half of the 4th c. BC.

Lug-handled bell-krater

This small rim fragment, characterized by a strong verticality and a continuous profile, belongs probably to a type of krater (the so-called lug-handled krater) derived from the Attic tradition of the 5th c. BC (for the adoption of this peculiar type of bell-krater in southern Italian productions, see ESIV 15). Used in Lucanian red-figure by the Pisticci Painter (e.g., *LCS* 14, No. 1, Pls. 1–2 from Pisticci; *LCS* 17, No. 15, Pl. 3.3; *LCS* 19, No. 33, Pl. 4.3–4) and in Apulian red-figure by the Painter of the Berlin Dancing Girl (*RVAp* 1/5), it was soon abandoned, and no examples of the 4th c. BC are known.

1:3

43. RF 240-01 Rim
Dimensions: rim diam. 38.5 h 3.8 w 5.5
Description: Clay pink (5YR 7/4). Thick and lustrous, brownish to black gloss inside and out. Laurel wreath to right with midrib.
Comparisons: For a similar rim see the Attic red-figure bell-krater from Vaste by Polignotos (Mannino 2006, 187–89 No. 277b, Fig. 226, 440 BC) and the vessel from Vaste (Mannino 2006, 212, No. 313, Fig. 266, third quarter of the 5th c. BC).
Date: Second half of the 5th c. BC.

Volute-krater

1:3

44. RF 275-01 Handle in the shape of a volute
Dimensions: h 13 w 7.2
Description: Clay light gray (core 5YR 7/1) to reddish yellow (5YR 7/6). Diluted brownish gloss on the interior and exterior. Miltos on the reserved surfaces. Lower part formed by hand. On the sides, ivy.
Comparisons: This type of handle and its decoration, derived from Attic prototypes, are used on Lucanian and Apulian volute-kraters from the last quarter of the 5th c. BC (when the shape is introduced into the morphological repertoire of both productions: *RVAp* 186) and the first half of the 4th c.; volute-handles are also occasionally employed on Late Apulian kraters (as in *RVAp Suppl.* 2 18/17a, by the Darius Painter), but with different decorative patterns. For similar ivy leaves see the volute-krater of Malibu *RVAp Suppl.* 1 2/90a (Sisyphus Painter) for which has been recently proposed an attribution to the Painter A of the Karneia krater, perhaps of Metapontine production (Denoyelle 2002b) or the vessel by the Creusa Painter (*LCS Suppl.* 3 3/C24). Possible Metapontine production, as also suggested by L. Burn.
Date:Last quarter of the 5th–beginning of the 4th c. BC.

Volute-krater cont.

1:2

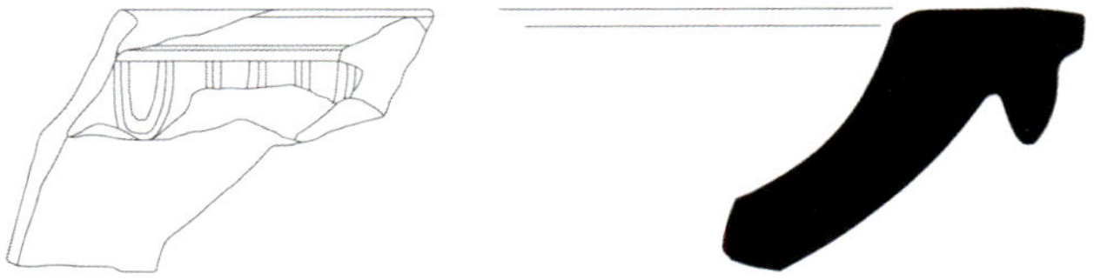

45. RF Plot 1999.118-02 Rim
Dimensions: h 4.1 w 6.5
Description: Clay light gray (nucleus 5YR 7/1) to reddish yellow (5YR 7/8). Gloss almost completely abraded. Surface very worn. On the edge of the rim, eggs; below the rim, wave pattern.
Comparisons: The wave pattern in this position makes its first appearance with the Iliupersis Painter (cfr. *RVAp* 8/7, Pl. 61.1), becoming very popular in Late Apulian volute-kraters.
Date: Second half of the 4th c. BC.

1:2

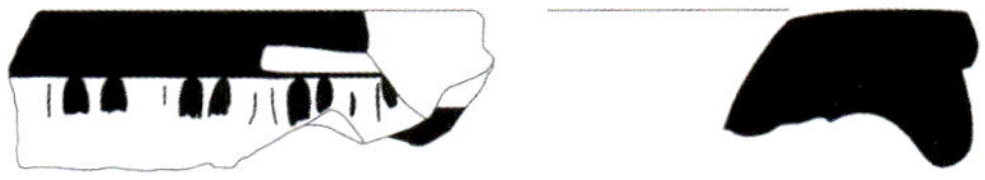

46. RF 401-14 Rim
Dimensions: h 2.2 w 6.45
Description: Clay pink (5YR 8/4). Thick, lustrous black gloss on the interior. On the edge, eggs with added white in the center, partially lost. Probably from the same vessel as **No. 47** (401-09).
Comparisons: For a very similar profile see the volute-krater by the Underworld Painter from Metaponto, Pizzica, tomb 5/83: Morard 2002, 25–26, Pl. 4.1 (Underworld Painter, 330–320 BC).
Date: Third quarter of the 4th c. BC.

1:2

1:4

47. RF Plot 2001.046-08 and **RF 401-09** Two joining neck fragments
Dimensions: h 10 w 6.6
Description: Clay reddish yellow (5YR 7/6). Brownish to black gloss on the interior; lower part of the neck reserved. Black gloss on the exterior. Reserved acanthus leaf and tendrils in added color, now faint. The decoration was completed by flowers, one of them partially visible on the right side of the fragment. Probably from the same vessel as **No. 46**.
Comparisons: The high neck and decorative pattern are typical of Late Apulian volute-kraters, but the portion of decoration preserved is too small to attempt an attribution.
Date: Late Apulian (suggested by L. Burn), third quarter of the 4th c. BC.

1:1

48. RF 401-29 Fragment of the upper part of a mascaroon
Dimensions: h 2.6 w 4.0 th 1.3
Description: Clay pink (5YR 7/4) to reddish yellow (surface 5YR 7/6). Part of the hair of a head, mold-made. Probably from the same vessel as **No. 46**.
Date: Late Apulian, third quarter of the 4th c. BC.

1:2

49. RF 401-24 Lower part of a mascaroon handle
Dimensions: h 5.2 w 10.0 th 0.9
Description: Clay pink (5YR 7/4). Thick, lustrous black gloss on the exterior. On the hidden surface, gloss almost completely abraded. Probably from the same vessel as **No. 46**.
Date: Late Apulian, third quarter of the 4th c. BC.

Volute-krater cont.

1:2

50. RF 401-01 Body sherd
Dimensions: h 5.0 w 6.0 th 0.7
Description: Clay pink (5YR 7/4), well-depurated and hard. Thick, lustrous black gloss on the exterior. Miltos on the reserved surfaces. Meander pattern. Probably from the same vessel as **No. 46**.
Date: Third quarter of the 4th c. BC.

51. RF 401-02 and **RF 401-03** Two non-adjoining body fragments, probably part of the handle area
Dimensions:
RF 401-02: h 4.1 w 4.2 th 0.6–0.8
RF 401-03: h 3.3 w 6.6 th 0.6–0.8
Description: Clay reddish yellow (5YR 7/6). Thick black gloss on the exterior. Upper part of an enclosed palmette-branch flanked by a half-branch on the right and a flower on the left.
Comparisons: For the decoration see the pseudo-Panathenaic amphora by the Underworld Painter: Schauenburg 2000a, 255, Fig. 3.
Date: Third quarter of the 4th c. BC.

1:2

RF 401-03

RF 401-02

52. RF 464-04 Neck fragment
Dimensions: h 6.5 w 5.0 th 0.6
Description: Clay light reddish brown (5YR 6/3). Faint black gloss on upper part; the remaining part of the surface is reserved. Thin, lustrous black gloss on the exterior. Reserved acanthus leaf, tendrils, and rosette, with faint tendrils originally in added white.
Date: Second half of the 4th c. BC.

1:2

53. RF 728B-03 Neck fragment
Dimensions: h 4.0 w 5.5 th 0.8
Description: Clay reddish yellow (5YR 7/6), powdery. Gloss not preserved; reserved decoration abraded, except for the relief-line delimiting the figures. Leg of a man leaning on a staff (?).
Date: End of the 5th–first half of the 4th c. BC.

1:2

54. RF 728B-01 Fragment of the neck
Dimensions: h 6.3 w 3.5 th 1.0
Condition: Red-figure parts abraded.
Description: Clay pink (5YR 7/4) to pinkish gray (7.5YR 7/2). Thick, dull, very dark gray gloss on the interior; lustrous black to brownish gloss on the exterior. Reserved acanthus leaf and scroll.
Date: Probably third quarter of the 4th c. BC.

1:2

Volute-krater cont.

1:2

55. RF 365-22 Handle/volute
Dimensions: h 4.37 w 8.35
Description: Clay pink (5YR 8/4) with gray core. The volute is hand-made with visible fingerprints. The surface decoration is an overpainted (now faint) palmette with long petals.
Comparison: Overpainted decoration on volute-kraters is not uncommon: see, for instance, the example by the Lycurgus Painter in the Banca Intesa Collection No. 111 (*RVAp* 16/4, Pl. 60.3 and *Collezione Banca Intesa* 306–310, No. 110, stars in added white, 360–350 BC).
Date: Second or third quarter of the 4th c. BC.

Calyx-krater

1:3

56. RF 425-01 Rim
Dimensions: rim diam. 54.8 h 2.3 w 11.7
Description: Clay pink (5YR 7/4), hard and well-depurated. Thin black gloss on the interior; grayish to black gloss on the exterior. Misfired fragment.
Date: Second half of the 5th–beginning of the 4th c. BC.

1:2

57. RF 272-01 and **RF 272-02** Two body sherds probably belonging to two sides of a single vessel
Dimensions:
RF 272-01 h 2.9 w 2.7 th 0.7
RF 272-02 h 3.5 w 3.9 th 0.7
Description: Clay pink (5YR 7/3). Fairly thick, lustrous black gloss on the interior. Enclosed palmette and meander pattern.
Comparisons: For the shape see **No. 58**.
Date: First quarter of the 4th c. BC.

1:3

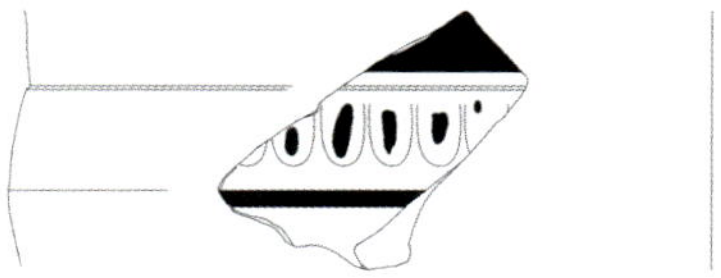

58. RF 734-01 Body
Dimensions: body diam. 28.0 h 5.22 w 6.12
Description: Egg-and-dot decoration, almost entirely abraded. The decorative pattern could have been completed by a meander or an anthemion (not preserved).
Comparisons: The shape, characterized by a vertical body-wall and a barely accentuated curvature, recalls calyx-kraters of the Lucanian and Apulian productions from the beginning of the 4th c. See, e.g., the fragment from the Cahn Collection No. 234: *RVAp* I 7/27 and Cambitoglou and Chamay 1997, No. 33, drawing at p. 327 (Black Fury Group, 390–380 BC). A similar decorative pattern was adopted on calyx-kraters by the Painter of Taranto 102547 (*LCS* 74, No. 372, Pl. 35.1) and by the Dolon Painter (Paris, Cabinet des Médailles 422, *LCS Suppl.* 3, 3/D19 and Cambridge, Fitzwilliam Museum GR70.1970, *LCS Suppl.* 3, 3/D20). For the egg decoration placed in the same position, see also the calyx-kraters in the Cahn Collection: inv. 1340, Cambitoglou and Chamay 1997, 96–97, No. 35, drawing at p. 327; and inv. 255, Cambitoglou and Chamay 1997, 106–107, No. 40, drawing at p. 328 (both attributed to the Painter of the Black Fury).
Date: Beginning of the 4th c. BC.

Calyx-krater cont.

1:2

59. RF 395-01 Lip
Dimensions: h 3.96 w 5.21
Description: Clay reddish yellow (7.5YR 8/6 exterior) to pink (7.5YR 8/6 core). Relatively thick, dark gray gloss on the interior, black, thick, and lustrous gloss on the exterior. Miltos on reserved surfaces. Laurel wreath with berries to left. Perhaps same vessel as No. 60.
Date: First half of the 4th c. BC.

1:2

60. RF 395-20 Foot
Dimensions: h 4.7 w 9.5 base diam. 13.3
Description: Clay pink (5YR 8/4). Thick, lustrous black gloss on the exterior. Miltos on reserved surfaces. Perhaps same vessel as No. 59.
Comparisons: A similar foot is associated with calyx-kraters of the Dolon Painter and is comparable to a fragment from deposit No. 1 of the kerameikos of Metaponto (workshop of the Dolon and Creusa Painters, inv. 29306). Similar, too, is the foot of the calyx-krater by the Painter of Athens 1714: Schauenburg 2000b, 34–35, Pls. XVII-XVIII.125–28 (second quarter of the 4th c. BC).
Date: First half of the 4th c. BC.

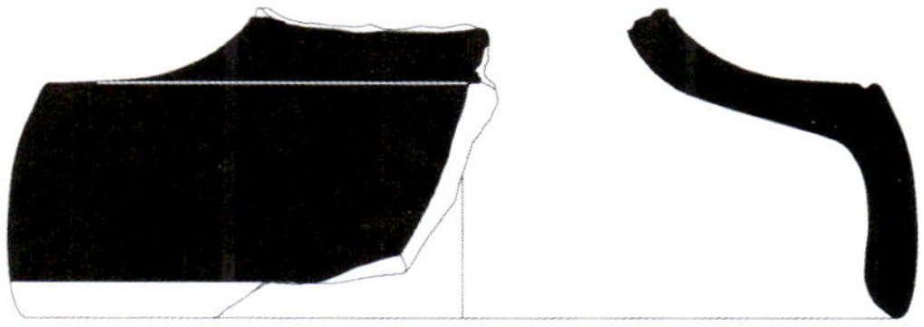

1:2

61. RF 375-02 Lip
Dimensions: h 3.58 w 5.13
Description: Clay pink (7.5YR 8/4). Thick, lustrous black gloss on the interior and exterior. Laurel wreath to left with berries.
Comparisons: A similar laurel wreath was used on Early Apulian vases (see Kymninon Group: *RVAp* 7/111, Pl. 59.5) and especially on Middle (Painter of Athens 1714: *RVAp* 8/153, Pl. 67.3–4, Bari, Lagioia Collection; Lykurgus Painter: *RVAp* 16/1, Pl. 145.1–2; Painter of Boston 76.65: *RVAp* 16/35, Pl. 153.3–4) and Late Apulian calyx-kraters (vase from Irsina, tomb in contrada Le Croci, *RVAp* 18/65 and Lo Porto 2001, 94–97, No. 13, Pl. 2, Darius Painter, 330 BC).
Date: A date in the second or third quarter of the 4th c. BC is confirmed also by the marked inclination of the lip.

1:2

62. RF 471-09 Foot
Dimensions: base diam. 15.6 h 3.3 w 7.4
Description: Clay pink (5YR 7/3). Relatively thick, lustrous black gloss on the exterior. Miltos on reserved surfaces.
Comparisons: Hoffmann 2002, Type 430/1, Pl. 92 (Painter of Wien 1072, *RVAp* 14/161). The calyx-krater is attributed to the Tarantine Phase A2 (350–325 BC). See also the foot of the calyx-krater from Irsina tomb in contrada Le Croci, *RVAp* 18/65 and Lo Porto 2001, 94–97, No. 13, Pl. 2 (Darius Painter, 330 BC).
Date: 4th c. BC

Column-krater

1:2

63. RF 415-03 rev93 Rim
Dimensions: h 1.95 w 3.71
Description: Clay reddish yellow (5YR 7/6). Black gloss almost entirely abraded. On the edge of the rim is a double row of black dots divided by a horizontal line, a decorative pattern used in Apulian column-kraters beginning with the Berlin Dancing Girl (*RVAp Suppl.* 2 1/12c; and Cambitoglou and Harari 1997, Pl. 3, No. 1 B) and in Lucanian vessels with the Amykos Painter; it becomes popular starting with the Tarporley Painter (first quarter of the 4th c. BC) and is used until the end of production.
Comparisons: Precise comparisons have not been found. See, in general, the rim profiles of column-kraters such as Dresden 518 (*RVAp* 14/240 and *CVA* Dresden 1, Pl. 18 and Fig. 4.2, by the Barletta Painter, 340–330 BC) or Würzburg, Martin-von-Wagner-Museum H 4518 (*RVAp* 14/80, and *CVA* Deutschland 71, Martin-von-Wagner-Museum, Pl. 6, Fig. 3.1, Painter of Geneva MF 290, 350–340 BC).
Date: Second half of the 4th c. BC.

Krater (uncertain types)

1:1

64. RF 200-01 Body
Dimensions: h 4.1 w 6.0 th 0.5
Description: Clay reddish yellow (5YR 7/6). Thick, lustrous black gloss on the interior; dull and brownish gloss on the exterior. Head and shoulders of a draped youth facing right. Metapontine production.
Date: Lucanian, second half (probably last quarter) of the 5th c. BC.

1:1

65. RF 395-02 rev01 Body
Dimensions: h 4.5 w 2.9 th 0.6
Description: Clay pink (5YR 7/4), hard and depurated. Thin and dull black gloss on the interior. Miltos on the reserved surfaces. Upper part of the body of a Type B draped youth facing right.
Date: Lucanian, second half of the 5th c. BC.

1:1

66. RF 395-08 rev90 Body fragment
Dimensions: h 2.2 w 3.7 th 0.6
Description: Clay pink (5YR 8/4). Thick, lustrous black gloss on the interior and exterior. Hand holding a branch and part of the leg of a youth.
Comparisons: A similar branch is sometimes held by young men or women in Lucanian vases of the first generation: see, e.g., the hydria by the Amykos Painter *LCS Suppl.* 2 1/231b, Pl. 30.3.
Date: 420–400 BC (?).

Krater (uncertain types) cont.

67. RF 728-01 rev00 Body fragment

Dimensions: h 3.0 w 3.2

Description: Clay light brown (5YR 6/4) with darker nucleus, depurated and hard. Thick, lustrous black gloss on the interior; brownish to black gloss on the exterior. Ivy branch bearing fruit.

Comparisons: For similar branches, which usually appear in Dionysiac scenes to suggest an open-air location, see, e.g., *RVAp* 1/54, Pl. 5.2 (Sisyphus Painter), the calyx-krater Ruvo, Jatta Museum 1495, *RVAp* 3/40, and Sichtermann 1966, 42, No. K, 52, Pl. 85 (Tarporley Painter), and the calyx-krater New York L63: *RVAp* 8/152, Pl. 67 (Painter of Athens 1714). An ivy-stem (but with the stem and flowers in added yellow on white) running horizontally along the upper section of the vase as a sort of pergola appears on the bell-krater with Dionysos by the Hoppin Painter in the Cahn Collection, inv. No. 235: Cambitoglou and Chamay 1997, 124–28, No. 48 (380 BC).

Date: First half of the 4th c. BC.

1:1

68. RF 210-02 Body fragment

Dimensions: h 3.4 w 4.3 th 1.0

Description: Clay pink (7.5YR 8/3). Diluted, lustrous black gloss on the interior; dull gray gloss on the exterior. Miltos on the reserved surfaces. Column and architrave. On the right, faint vegetal decoration.

Date: 4th c. BC.

1:1

69. RF 326-33 Body

Dimensions: h 5.4 w 3.1 th 0.6

Description: Clay pink (5YR 8/6). Thick, lustrous black gloss on the interior and exterior. Head of a woman with a sakkos or in a kekryphalos (?), with a crown, necklace, and earrings in added white.

Comparisons: Ganymede Painter: *RVAp* 25/24–25, Pls. 297.2–3, or Berkeley Group: *RVAp* 8/61, Pls. 282.2–4 (suggested by L. Burn).

Date: Second half of the 4th c. BC.

1:1

70. RF 415-04 Body

Dimensions: h 4.0 w 4.6 th 0.8

Description: Clay pink (7.5YR 8/4). Fairly thick, lustrous black gloss on the interior and exterior. Miltos on the reserved surfaces. Upper part of the tympanon of a building decorated with a palmette acroterion. The reduced dimensions of the building could suggest, more than a naiskos, a fountain-house or a small temple.

Comparisons: See similar architectural form on the pseudo-Panathenaic amphora from Policoro by the Brooklyn-Budapest Painter (Pianu 1990, 23–24, No. 1, Pls. VI.1–2) and on the hydria *RVAp* 1 18/11a, Pl. 9.1 (in added white).

Date: 4th c. BC, probably first half.

1:1

Krater (uncertain types) cont.

1:1

71. RF Plot2001.046-09 (= site 401) Body
Dimensions: h 3.0 w 2.5 th 0.6
Description: Clay pink (5YR 7/4). Thick, lustrous black gloss on the interior. Miltos on the reserved surfaces. Hind legs of a galloping horse.
Date: Second half of the 4th c. BC.

1:2

72. RF 457-01 Body fragment
Dimensions: h 5.6 w 3.6 th 1.0
Description: Clay reddish yellow (5YR 7/6). Fairly thick, lustrous black gloss on the interior and exterior. Neck, mane, and reins of a horse facing left.
Date: Second half of the 4th c. BC.

1:1

73. RF 097-232 Body fragment
Dimensions: h 4.9 w 4.7 th 0.6
Description: Clay reddish yellow (5YR 7/6). Thick, lustrous black gloss on the interior and exterior. Sphinx with wings spread. To her right, part of a pilaster.
Comparisons: See the fragmentary krater by the Darius Painter representing Achilles and Phoinix: *RVAp* 18/77, Pl. 181.2.
Date: Third quarter of the 4th c. BC.

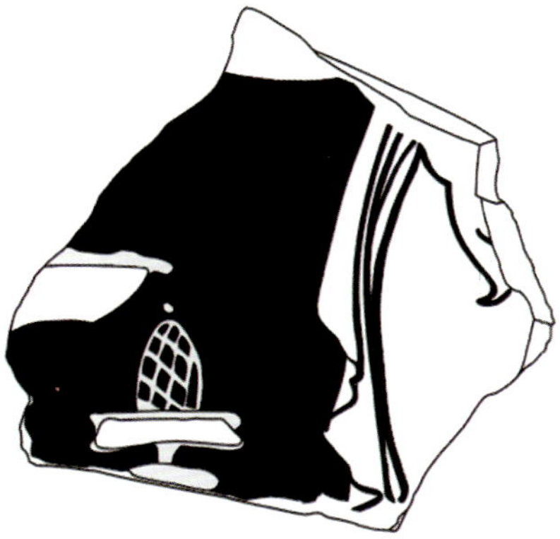

1:1

74. RF 726-01 Body
Dimensions: h 4.8 w 4.9 th 0.7
Description: Clay reddish yellow (5YR 7/8), depurated and hard. Thick, lustrous black gloss on the interior; lustrous, thin brownish gloss on the exterior. Lower torso of a young male with a mantle flowing along the body. In front of him, a patera with food in added white (now faded), held by another figure (not preserved). Between them (as suggested by L. Burn), a Type E1 thymiaterion (Zaccagnino 1998: 72–73, Pl. 2); upper part reserved, lid and lower part in added color (now faded).
Comparisons: The fragment is probably part of a cult scene; for the iconography, often related to Dionysos, see *RVAp* 15/61, Pls. 144.3–4 (Group of Vatican W 4). For a similar thymiaterion see the loutrophoros Basel S 21 (*RVAp* 18/16, Pl. 171.4, Laodamia Painter, middle of the 4th c. BC).
Date: Second half of the 4th c. BC.

Krater (uncertain types) cont.

1:1

75. RF 728A-01 Body fragment
Dimensions: h 6.7 w 6.3 th 1.0
Description: Clay pink (5YR 8/4). Thick, lustrous black gloss on the interior; brownish gloss on the exterior. On the right, a male figure with a naked torso wearing a mantle around the chest and standing on a chariot (?) in added red. He extends his right arm towards a second figure, of whom just the two hands are preserved.
Date: Second half of the 4th c. BC.

Xenon Group

1:2

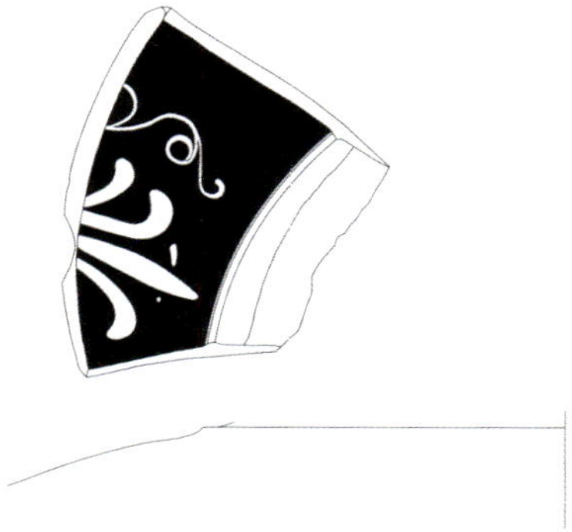

1:1

76. RF 266-45 Closed shape
Dimensions: body diam. 12.5 h 2.09 w 6.29
Description: Clay pinkish-gray (5YR 7/2) to reddish-yellow (7.5YR 8/6). Thick, lustrous black gloss on exterior surface. On the shoulder, palmette and volute added in red.
Date: 4th c. BC.

1:2

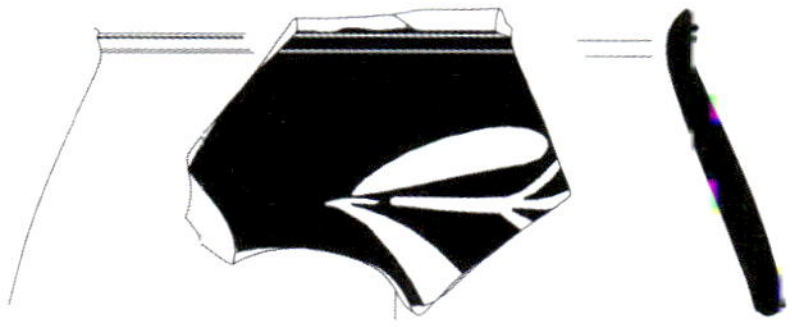

1:1

77. RF 521-02 Body fragment of a kantharoid vase
Dimensions: body diam. 10.3 h 4.2 w 5.1
Description: Clay pink. Thick, black to brownish gloss inside and outside. On the shoulder, ghostly laurel branch added in pale yellow.
Comparisons: See the kantharoid vase from Canosa, Ipogeo dei Vimini, cella B, right deposition: De Juliis 1990, 75, No. 38, Fig. 296 (second quarter of the 4th c. BC).
Date: The ovoid body recalls examples of the second quarter of the 4th c. BC.

Gnathia Pottery

Skyphos

78. GN 272-04 Rim
Dimensions: rim diam. 10.0 h 1.65 w 2.6
Description: Clay pink (5YR 8/4), hard and depurated. Thick, lustrous black gloss inside and outside. Incised eggs and, below, fillet.
Comparisons: Shape and decorative pattern are attested in skyphoi from deposit No. 3 of the Metapontine kerameikos: Silvestrelli forthcoming a, Fig. 12.5.
Date: Second half of the 4th c. BC, probably third quarter.

1:2

1:1

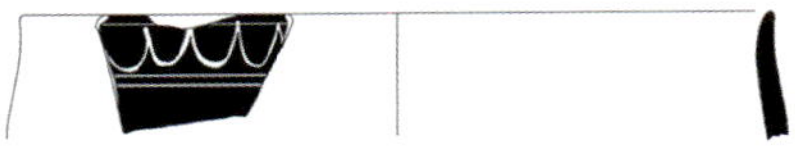

79. GN 394-09 Rim
Dimensions: rim diam. 10 h 2.46 w 1.29
Description: Clay pink (5YR 8/4). Reddish gloss inside, black outside. Incised eggs limited above and below by a double incised line; below, painted band with dots, now faded, limited by an incised double line.
Comparisons: For shape and decoration see **No. 78**.
Date: Second half of the 4th BC, probably third quarter.

1:1

1:2

80. GN 521-06 Rim
Dimensions: rim diam. 8.8 h 2.37 w 2
Description: Clay pink (5YR 8/4). Brownish gloss inside and outside. Incised eggs delimited by a double line.
Comparisons: For shape and decoration see **No. 78**.
Date: Second half of the 4th BC, probably third quarter.

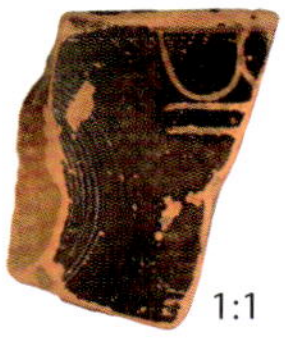

1:1

1:2

81. GN 395-21 Rim
Dimensions: rim 9.3 h 3.3 w 4.3
Description: Clay pink (5YR 8/4). Thick and lustrous black gloss inside and outside. Waves between horizontal lines. Below, line of dots from which hangs a stylized branch.
Comparisons: Sidewinder Group (suggested by E. Lanza). *CVA* Mainz 2, Pl. 22.2 (Middle Gnathia, 340–330 BC).
Date: Second half of the 4th c. BC.

1:2

82. GN 342-29 Rim
Dimensions: rim diam. 9.1 h 2.5 w 2.0
Description: Clay light red (2.5YR 6/8), hard, and well-depurated. Thick and lustrous black gloss inside, thick, dull, dark brown gloss outside. Stylized branch in added white.
Date: End of the 4th c. BC.

1:1

1:2

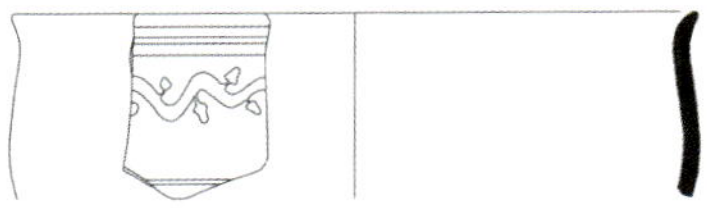

83. GN 269-21 Rim
Dimensions: rim diam. 10.9 h 2.57 w 1.33
Description: Clay from pink (5YR 7/4) to white (5YR 8/1), powdery, well-depurated. Thick, dull gray gloss inside and outside. Stylized branch in added faint color.
Comparisons: See skyphos from Botromagno, inv. P1577: *Gravina* II, 57, 388, No. D 522, Pl. X.
Date: 3rd c. BC.

1:2

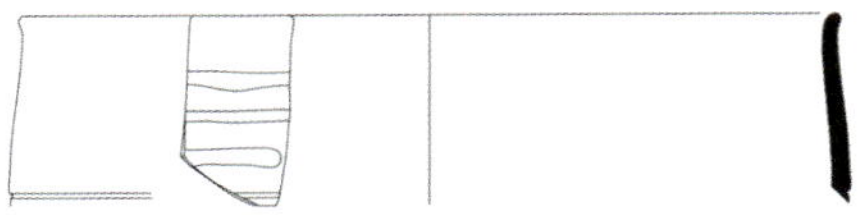

Gnathia skyphos cont.

1:2

1:1

84. GN 128-19 Rim
Dimensions: rim diam. 9.5 h 1.75 w 2.1
Description: Clay light brown (7.5YR 6/4), hard and well-depurated. Thick, dull black gloss inside and outside. Overpainted in yellow is an ivy branch with heart-shaped leaves and berries (New Ivy).
Date: end of the 4th–beginning of the 3rd c. BC.
Comparisons: For the decoration see the cup-skyphos from Rutigliano, T 54: *Museo Taranto* II.2, 160, No. 54.31; and 547, Pl. 50d (end of the 4th–beginning of the 3rd c. BC). See skyphos from Botromagno, inv. P1588: *Gravina* II, 55, 388, No. C510, Pl. X (3rd–2nd c. BC).

Type A Skyphos

1:2

1:1

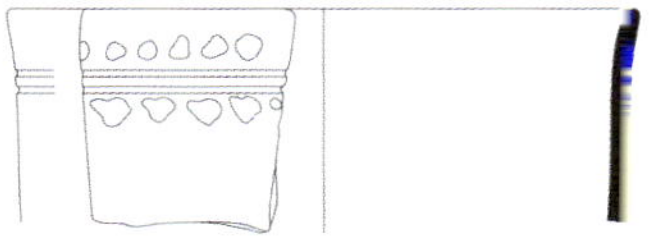

85. GN 341-28 Rim
Dimensions: rim diam. 8.4 h 3.1 w 2.85
Description: Clay light red (2.5YR 6/6), with grayish stains, hard and depurated. Lustrous and quite thick black gloss inside and outside. Faint and very stylized ivy leaves flanking both sides of an incised linear double line.
Comparisons: For the decorative pattern see a cup-skyphos from Torre di Satriano: *Torre di Satriano* I, 320, No. 425, Fig. on p. 324 (second half of the 4th–beginning of the 3rd c. BC) and the sherd from the Sanctuary of Dionysos in the agora of Herakleia: *Eraclea lucana* 89, Fig. 37B.
Date: End of the 4th–beginning of the 3rd c. BC.

Hydria

1:3

1:2

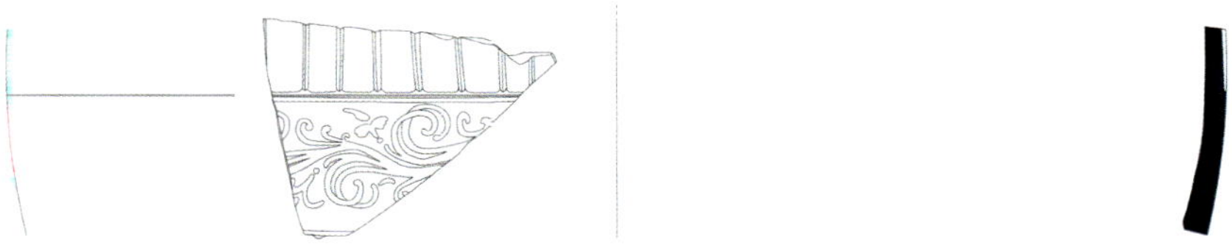

86. GN 105-252PA Body
Dimensions: body diam. 32.9 h 6.08 w 7.88
Description: Clay pink (5YR 7/6). Brownish to black gloss on exterior. Ribbed body interrupted by belly-band decorated with a faint overpainted scrolling floral design.
Comparisons: For the decoration see *Necropoleis* CD 267 (first half of the 3rd c. BC) and hydria *CVA* Philadelphia 1, Pl. 31 (inv. L-64-195) attributed to the Alexandria Group by R. Green (first quarter of the 3rd c. BC). For the shape, see the hydria Copenhagen 2260: Fischer-Hansen et al. 1992, 152–53, No. 117.
Date: First half of the 3rd c. BC.

1:3

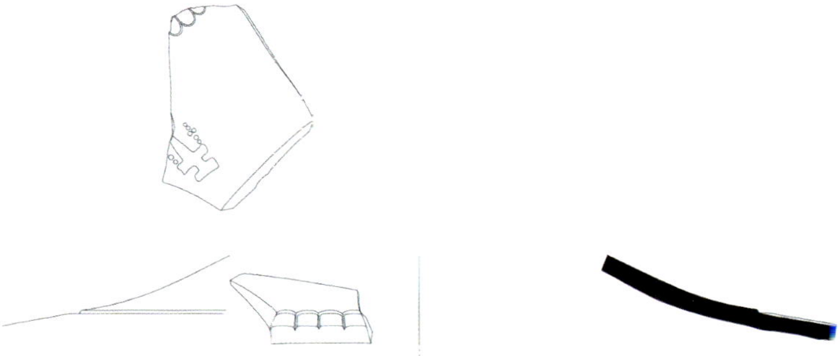

87. GN 105-05 Two shoulder fragments
Dimensions: body diam. 28.85 h 3.0 w 3.35
Description: Clay pink (5YR 7/6). Brownish to red gloss on the exterior. Shoulder and beginning of the body. On the shoulder, faint decoration in added white, perhaps a thymiaterion. The ribs end in an arch decorated with a line on the body in added cream.
Date: First half of the 3rd c. BC.

Lebes gamikos

1:2

88. GN 375-06 Body and shoulder of ribbed lebes gamikos
Dimensions: body diam. 12 h 1.6 w 4.27
Description: Clay pink (5YR 7/3). Dull black gloss on the interior; gray gloss on the exterior. Ribbing on the body, faint overpainted meander on the shoulder.
Comparisons: Burn 1998, 633, T 192-8 (Metapontine production, 300 BC) and 634 T 200-4 (300–275 BC).
Date: First half of the 3rd c. BC.

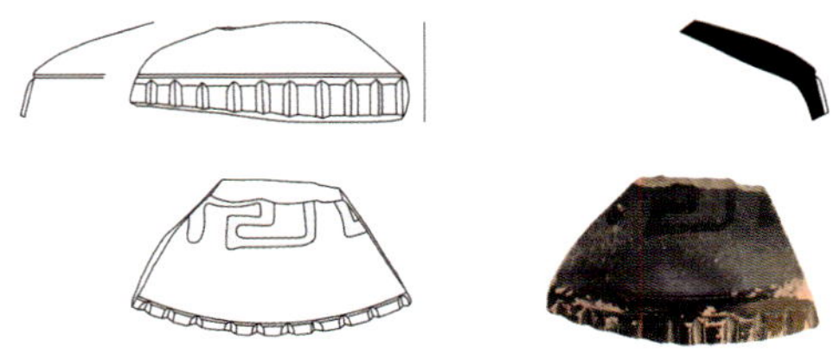

Pelike

1:2

89. GN 038-124P Neck
Dimensions: h 4.16 w 4.96
Description: Clay reddish yellow (7.5YR 7/6), hard and depurated. Dark gray gloss outside. Incised eggs and a faint zig-zag decoration delimited above and below by two double-incised lines.
Date: Second half of 4th c. BC, probably last quarter.

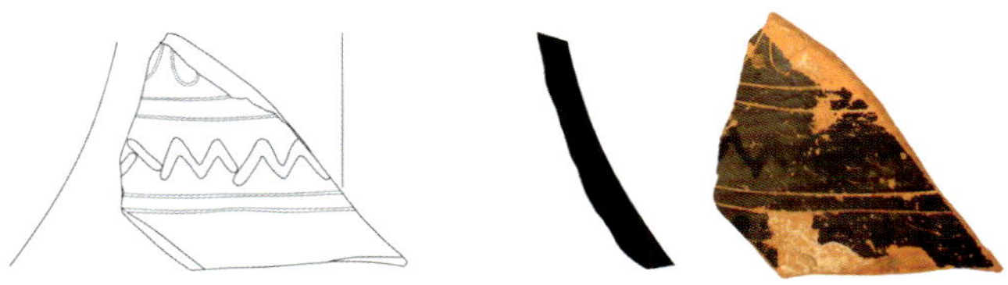

Bell-krater

No. 90 belongs to a bell-krater of the second half of the 4th c. BC. The shape is unusual in Gnathia pottery, which preferred Type C kraters (see items below), but not completely unknown (see bell-krater in the National Archaeological Museum, Bari: Forti 1965, Pl. XXVI c). Also rare, for its position on the edge of the rim, is the overpainted egg-and-dot decoration, which is found occasionally on Type C bell-kraters as in the vessel in Boston, Museum of Fine Arts: Forti 1965, Pl. XX a.

1:3

1:2

90. GN 464-02 Rim
Dimensions: rim diam. 35 h 1.85 w 6.5
Description: Clay reddish yellow (7.5 YR 7/6). Thin, dull brownish gloss on the interior and exterior.
Comparisons: For a similar rim see the rim fragment attributed to the Primato Painter from the sanctuary of Torre di Satriano: *Dea lucana*, 99, Fig. 65; and 100, No. 185 (third quarter of 4th c. BC), and the red-figure bell-krater by the Chevron Group *RVAp* II 22/17 and Hoffmann 2002, 317.1, 45, Type 429/2, Pl. 91 (last quarter of 4th c. BC).
Date: Last quarter of the 4th c. BC.

Bell-krater (Type B)

This fragment can be attributed to a Type B bell-krater, probably derived from Corinthian vessels and known beginning from the initial phase of the Tarantine Gnathia production (370–360/330–320 BC). See Green 1979, 81 and 87, n. 3; 1986, 137–38; also McPhee 1997, 134–37, with bibliography and a list of all known examples. For the attribution to Taras, see McPhee 1997, 136. Wasters from the Metapontine kerameikos (deposit 5, Silvestrelli 2004, 110, Fig. 3.3; deposit 3, Silvestrelli forthcoming a, Fig. 12.3; other fragments come from deposit 3) confirm that the shape was produced also by Metapontine workshops, to which this fragment can probably be ascribed. The shape is not common and seems to be attested in Messapia (Rudiae, McPhee 1997, 134, No. AP1, Pls. 43 and 135; 135, No. AP17) and Peucetia (Ruvo di Puglia, McPhee 1997, 135, No. AP16).

Gnathia bell-krater (Type B) cont.

1:3

91. GN 727-01 Rim and protrusion flanking the horizontal handle (lost)
Dimensions: rim diam. 28.3 h 4.25 w 4.84
Description: Clay pink, well-depurated and hard. Brownish gloss inside and outside. Sprays of ivy with incised stems; the overpainted decoration is completely lost.
Comparisons: For the shape and decoration see the krater Malibu 82.AE.15 (Green 1986, 137, Fig. 31 a–b, Konnakis Painter) and the fragment from Metaponto, kerameikos deposit 3 (D'Andria 1980a, 364, No. 35, Fig. 13; and Silvestrelli forthcoming a, Fig. 12.3).
Date: Second half 4th c. BC.

Bell-krater (Type C)

The survey corpus yielded a certain number of vessels that can be ascribed to the standard type of Gnathia bell-krater (Type C: Morel 1981a, 4618). Red-figure examples of this shape are known, even if in a restricted number: e.g., the vase in the National Archaeological Museum, Bari (Phlyax Vases, 28, No. 19), dated to the middle of the 4th c. BC. This type of krater is part of the morphological repertoire of Metapontine workshops involved in the production of overpainted pottery in Gnathia style, and fragments of it have been found in the deposits of the kerameikos dating to the third quarter of the 4th c. BC (Deposit inside Kiln B: D'Andria 1980a, 364, No. 35, Fig. 13. Deposit No. 3: Silvestrelli forthcoming b). Rim fragments **Nos. 92–93** were probably locally manufactured.

1:3

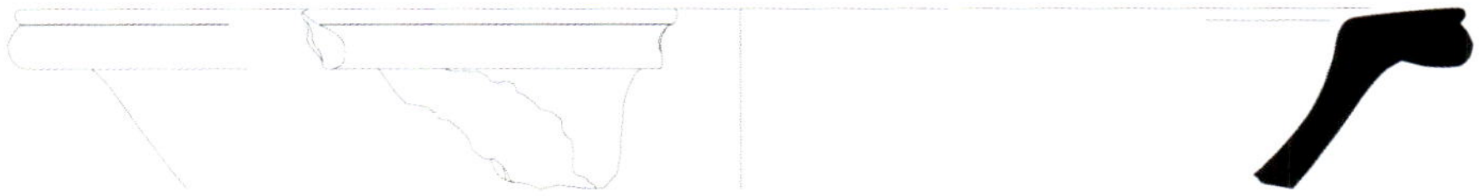

92. GN 326-26 Rim
Dimensions: rim diam. 35 h 4.5 w 9
Description: Clay pink (5YR 8/6). Thick, lustrous black gloss inside and outside. Incised ivy branch with ghostly leaves in added color.
Comparisons: Metaponto, kerameikos deposit No. 3: Silvestrelli forthcoming b. See also the Gnathia bell-krater Bochum S 1044: *CVA* Deutschland 82, Bochum, Kunstsammlung der Ruhr-Universität (3), Pl. 58, Fig. 8.2 (middle of the 4th c. BC).
Date: Middle to third quarter of the 4th c. BC.

93. GN 250-01 Rim

1:3

Dimensions: rim diam. 41.5 h 2.15 w 8.44
Description: Clay pink (5YR 8/6).
Comparisons: See **No. 92.**
Date: Middle to third quarter of 4th c. BC.

1:3

94. GN 439-01 Rim
Dimensions: rim diam. 51 (?) h 3.7 w 9.1
Description: Clay pink (5YR 7/4). Black gloss inside and outside.
Date: Second half of the 4th c. BC.

Bell-krater (Type C1)

1:2

95. GN 105-01 Rim
Dimensions: h 3.9 w 5.15
Description: Clay pink (5YR 7/4). Black gloss inside and outside.
Date: Late Gnathia, end of the 4th c. BC.

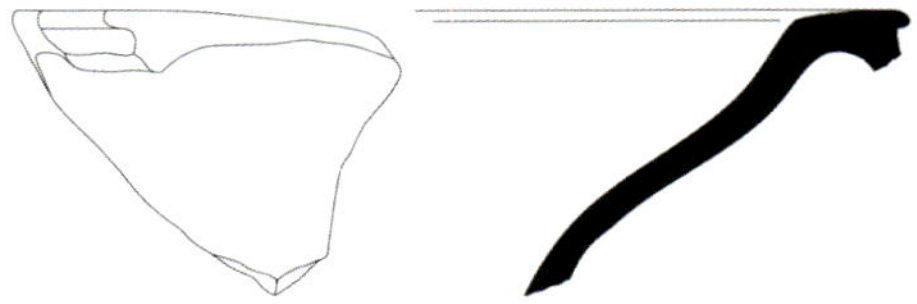

Lekanis

1:2

96. GN 460-14 Rim
Dimensions: rim diam. ca. 12.4 h 1.85 w 2.4
Description: Clay pink (5YR 7/4). Black gloss inside and outside. Tendrils in added white now faded.
Date: Second half of the 4th c. BC.

1:1

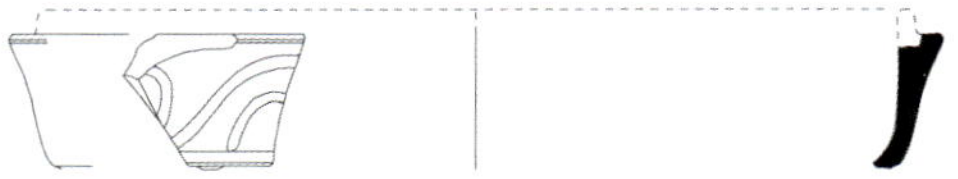

Bombylios

On the evolution of the form, see D'Amicis 1995, 60. This shape is frequently found in urban and rural tombs of Metaponto both in red-figure (e.g. Lo Porto 1966, 218, Nos. 2–3 T 49, Figs. 77.2–3, 340–330 BC; *Necropoleis*, T 58-2) and in Gnathia pottery (Galli 1926–1928, 63, No. 1; Burn 1998, 631, Nos. T 71-6, T71-11, and T71-13, attributed to the Alexandria Group and dated to the last quarter of the 4th c. BC). The shape is well known also from Tarentine graves (Graepler 1997, Type 613/15, first half of the 3rd c. BC) and attested in indigenous sites of Lucania (Torre di Satriano, sanctuary: *Torre di Satriano* I, 318, Nos. 402–409, Pl. LIII), Apulia (Canosa: *Principi imperatori vescovi*, 474, No. 44. Messapia: Delli Ponti 1996, 121, No. 43 (Vaste, Fondo Aia, T 6/1968, end of the 4th–beginning of the 3rd c. BC); 138, No. 88 (Vaste, Fondo Aia, T 7/1968, last quarter of the 4th c. BC); 164, No. 147 (Vaste, Fondo Aia, T 11/1968, last quarter of the 4th c. BC); 209, No. 286 (Vaste, Fondo Aia, T 4/1970, second quarter of the 4th c. BC); and Campania (*Fratte*, 127, Figs. 246a, 247c. Pontrandolfo and Rouveret 1992, 345–46, Nos. 1–2).

1:2

97. GN 635-07 Complete bombylios
Dimensions: rim diam. 4.9 base diam. 3.9 h 13.65
Description: Grey to reddish gloss outside. On the neck, overpainted rays now faded. Ribbing on the body. Foot and lower part of the body reserved. Misfired vessel.
Comparisons: Very similar from the morphological point of view is the bombylios from Torre di Satriano: *Torre di Satriano* I, 405, Pl. LIII. For shape and decoration see the bombylios in the Lagioia Collection: *Collezione Lagioia* 292, No. 232 (320–275 BC).
Date: Last quarter 4th–first quarter 3rd c. BC

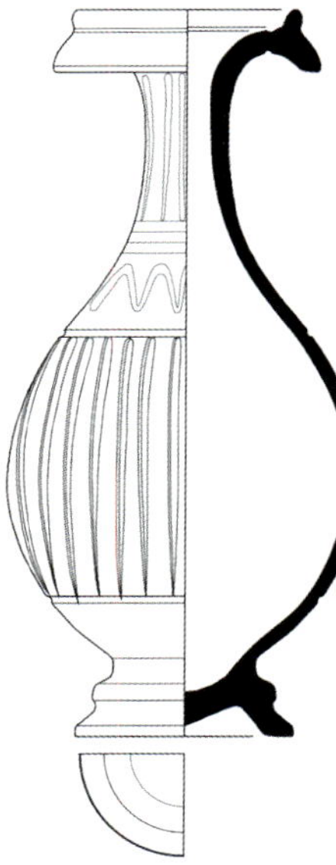

Kantharos

1:2

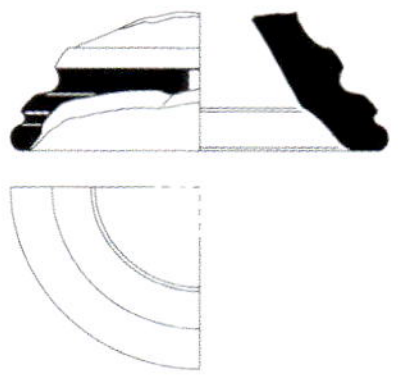

98. GN 266-14 Base
Dimensions: base diam. 5.0 h 1.9 w 4.3
Description: Light red clay (2.5YR 6/6). Brown gloss on the lower grove, black gloss on the upper groove. Faded dots on the lower groove. Local or regional production.
Comparisons: Type B kantharos from Oria (Franco 1989, 258, Fig. 19). Kantharos from Botromagno, late 4th c. BC (*Gravina* II, 283, Fig. 25, No. 526). Stemmed cups from Rutigliano (*Museo Taranto* II.2, 546, Pl. 50.d, T 54/1978, 335–315 BC) and Vaste (Delli Ponti 1996, 132, No. 72, end 4th–early 3rd BC). (EL)
Date: Late 4th–early 3rd c. BC

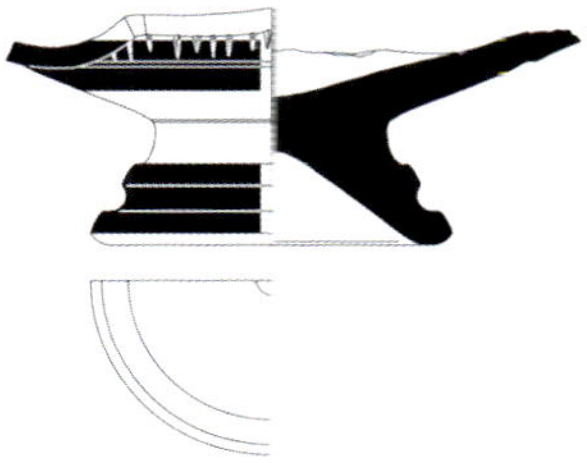

1:2

99. GN 247-01 Base and body
Dimensions: base diam. 4.8 h 3.25
Description: Reddish-orange clay (2.5YR 5/8) with grey-brown core (5YR 5/1). Black gloss inside and black-brown outside. Stemmed foot, ribbed body. Reserved lower body, stem and underside. The original presence of painted dots is indicated by black gloss preserved only where it was covered by the overpainted color. Local or regional production.
Comparisons: Foot fragment from Cozzo Presepe, 300–275 BC (*Cozzo Presepe*, 343, Fig. 123, No. 289). Ring-handled kantharos from a pit at Botromagno, 250–200 BC (*Gravina*, 388, Pl. X, No. B 504). Kantharos from Egnazia, 300–250 BC (Donvito 1988, 112, Fig. 83). (EL)
Date: First half of the 3rd c. BC.

1:1

1:2

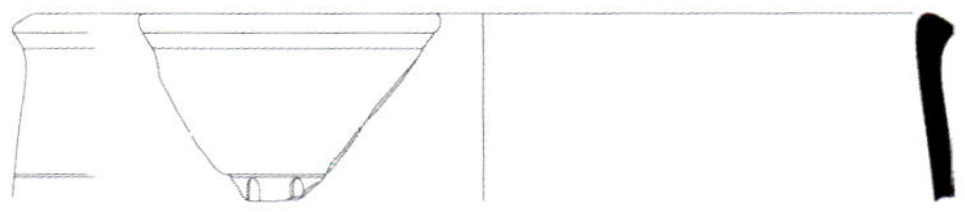

100. GN 128-21 Rim
Dimensions: rim diam. 12.5 h 2.6 w 4.0
Description: Red-orange clay (2.5YR 5/8). Black gloss inside and outside. Thickened lip, ribbed body. Faded spray with elongated leaves. Probably local or regional production.
Comparisons: Small 'deep bowl' from Cozzo Presepe, 4th–3rd c. BC (*Cozzo Presepe*, 360, No. 342). Ring-handled kantharoi from Rocavecchia, T 25/1934, second half 4th–3rd c. BC (Giannotta 1996, 67, Fig. 31) and Valesio (Yntema 2001, 134, No. 201, and, for the decoration, No. 203, late 3rd–early 2nd c. BC). Example from Roccagloriosa, 400–350 BC (Fracchia 1990a, 256, No. 188). Rim fragment from Botromagno (*Gravina* II, 57, No. 522). (EL)
Date: 3rd c. BC

1:1

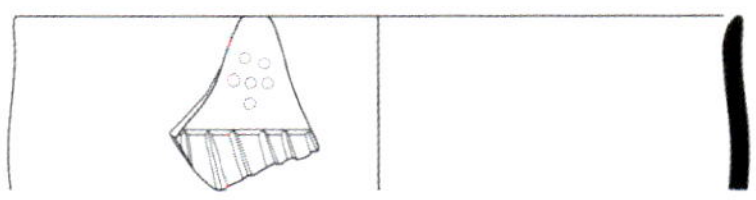

1:2

101. GN 173-12 Kantharos or skyphos rim
Dimensions: rim diam. 9.5 h 2.3
Description: Reddish yellow clay (7.5YR 7/6). Brown gloss inside, black gloss outside. Slight outward turn of the rim at the lip, ribbed body. Faded dotted rosette on unribbed lip. Perhaps not local production.
Comparisons: Ring-handled kantharoi from Manduria, 300–275 BC (Fedele/Alessio/Del Monaco 1992, 287, Fig. 131) and from Egnazia, 300–250 BC (Donvito 1988, 112, Fig. 83). Skyphoi from Botromagno, 320–280 BC (*Gravina in Puglia*, 79) and Gravina, T 200/1989, 310–280 BC (Ciancio 2005, 24). Rim fragment from Civita di Tricarico (De Cazanove 2008, Fig. 290, No. 125). (EL)
Date: Late 4th–first half of the 3rd c. BC.

Black-on-Buff Ware

1:1

Skyphos

102. BoB 082-206P Two joining fragments
Dimensions: h 3.0 w 4.2
Description: Clay reddish-yellow (5YR 7/6). Black gloss on interior. Ivy branch bearing fruit.
Comparisons: For the decoration see *Cozzo Presepe*, 346, No. 275, Fig. 122.
Date: Second half of the 4th c. BC.

1:1

1:2

103. BoB 312-03 Rim
Dimensions: rim diam. 9.8 h 2.45 w 2.35
Description: Clay powdery and reddish-yellow (5YR 7/6), without inclusions. Thin gray glaze inside. Below the rim, ivy branch with heart-shaped leaves and flowers.
Comparisons: For the decoration, see *Cozzo Presepe*, 346, No. 275, Fig. 122 (cup, Phase III), of Metapontine production.
Date: Second half of the 4th c. BC.

1:1

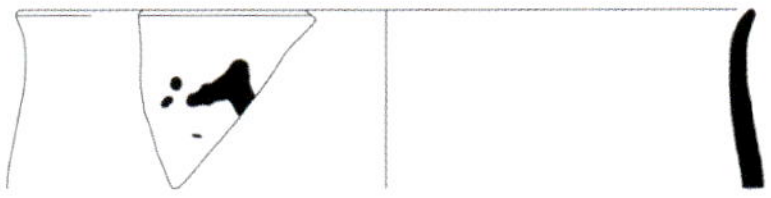

1:2

1:1

104. BoB 312-04 Rim
Dimensions: rim diam. 10.5 h 4.3 w 4.9
Description: Clay powdery and reddish-yellow (5YR 7/6), without inclusions. Thin gray gloss on interior. On exterior, below the rim, ivy branch with flowers.
Comparisons: For the decoration, see *Cozzo Presepe*, 346, No. 273, Fig. 122 ("stemless cup").
Date: Second half 4th c. BC.

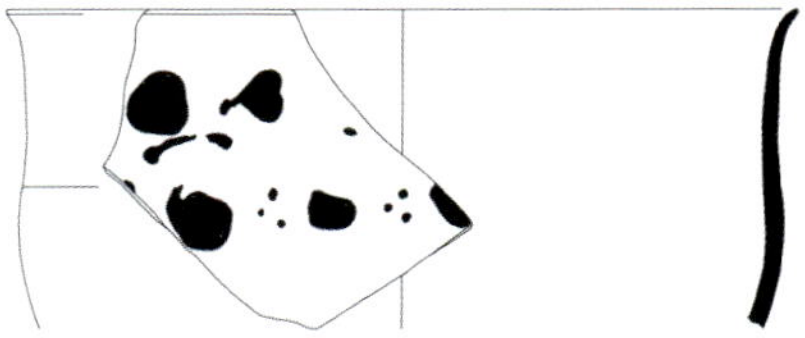

9
Plain and Banded Ware

Eloisa Vittoria

The ICA survey recovered 15,045 fragments of plain/banded wares ("table wares"). This chapter presents the principal forms and a typology for the most common ones.[1] The study additionally considers whether forms in this ware were used side by side with their counterparts in other classes (e.g, black gloss), and if banded- and plain-ware forms were used primarily in domestic contexts, or if they also served significant roles in funerary and religious contexts.[2]

Definition of Plain/Banded Wares

Plain/banded wares are generally classified as "utilitarian pottery," a category of ceramics intended for daily use. Although less prestigious than black-gloss ware, they are likewise characterized by well-levigated clay and local production.[3] To create the broader ceramic category of utilitarian ware,[4] these two classes are combined with those made of truly coarse fabric and then divided into distinctive functional groups: table wares (mostly small- and medium-sized vessels for eating and drinking), storage wares (mostly medium- and large-sized containers for preserving foodstuffs), and kitchen wares (mostly medium-sized vessels for cleaning, preparing, and cooking foodstuffs).[5]

Such a definition, however, has not gained universal acceptance among specialists. Cooking ware is often treated as a class in itself because of the unique physical-chemical properties that allow it to be exposed directly to fire.[6] Thus the generic term *ceramica da acqua* ("water ware") was born, which—despite the name—does not indicate a ware for holding liquids, but rather a ware produced in a well-levigated fabric[7] that was not intended for cooking directly over fire.[8] This definition allows one to extend the class to include what might be termed "domestic ware" (i.e., for daily use), thus broadening the range of shapes and fabrics that can be considered in this study to include vessels associated with the preparation of foodstuffs (such as the mortar for grinding, mincing, kneading, and mixing), and large containers for storing dry goods and liquids (such as the pithos).[9] Also in this class are some specialized shapes, such as the thymiaterion, a vessel used for burning incense in religious ceremonies, which could take place in domestic contexts. The louterion was also used (but not exclusively) in sacred

[1] The major reference is Morel's classification scheme and his precise distinction between form and type (Morel 1981a, 22). For a detailed discussion, see Ch. 7, nn. 34 and 43. C. D'Annibale first examined these wares in 1980, and I thank him for the draft of his typology.

[2] Vegas 1973, 7–8. Sparkes and Talcott (*Agora* XII, 33) are aware that the understanding of "domestic pottery" as specifically non-funerary is too restrictive. Archaeological evidence shows that the same pottery found in occupation contexts can be found also in burials. Cf. also Saunders and du Plat Taylor 1992, II.13; *Valesio* I, 63 and 237.

[3] Deodato and Bianco 1997, 1.175; Marchi 1991 171; Blasi 2001, 65; *Montemurro* 148; Cuozzo 2003, 54. This does not mean, however, that production was exclusively local; see, e.g., Bragantini/Setari/De Filippis 1991, 55, who challenge the *opinio communis* that utilitarian pottery generally had little value and was therefore not normally a part of commercial exchange. Cf. also Carter and Toxey 1998a, 695; Marchegiani 1997, 1.161; Gasperetti 1996, 20. The absence of decorative elements is responsible for the scant attention paid to this category of ceramics, which could instead shed new light on the daily life of ancient populations (cf. Panella 1996, 9–10). The superficiality of this initial approach has also severely hampered the process of comparison, seeing as how the materials, if at all published, are typically documented only by photographs and rarely have illustrations. Consequently it is not always possible to find comparanda among identical material from other sites, as the small but numerous and important variations of form within a given type are difficult to discern in the published documentation.

[4] Panella 1996, 10; Bats 1996, 481.

[5] *Montemurro,* 157–58; Deodato and Bianco 1997, 1.175; Panella 1996, 10; Marchi 1991, 171; Conti 1989a, 257; Vegas 1973, 7–8. Munzi (1999, 92), however, defines utilitarian pottery as a class distinguished by function and physical characteristics, with cooking ware and table ware as sub-classes. Cf. also Rinaldi 2005, 222–24. A domestic space (kitchen or pantry) is depicted on an Attic red-figure skyphos (cf. Tempesta 2002, 1160, Fig. 38). The term "preparation" includes various phases of the transformation and subsequent presentation of foodstuffs via mechanical and biochemical processes, as well as cooking, e.g., mixing, kneading, milling/grinding, filtering, maceration, drying, smoking, fermenting, direct or indirect cooking, arrangement (cf. Bats 1988, 20–21).

[6] For the cooking ware recovered in the Metaponto survey, see Ch. 11.

[7] That is, a fine-textured clay created by purifying with water to remove coarser particles and organic impurities.

[8] Cuomo di Caprio (1985, 65) emphasizes that a ware intended for exposure to fire cannot be classified among plain wares precisely because of the variations in color caused by the physical-chemical properties of the fabrics.

[9] Cf. *Agora* XII, 33, where the "domestic pottery" encompasses vessels for storage, pouring, mixing, grinding, cooking, and heating. True cooking ware and transport amphorae, characterized by a fabric rich in inclusions and very similar to that of a pithos and (in certain cases) a mortar, are treated separately (see Chs. 11–12).

rites; its fabric, however, along with its morphology and employment in daily activities make it a part of this group.

The decorative bands that give this ware its name suggest that it was an intermediate product between the more refined, decorated, and gloss-covered wares and the daily-use plain ware;[10] analysis of the materials, however, indicates that this was not the case, as the same forms are present in both classes. Because of its long period of usage, this ceramic class provides little information about chronology. In contrast to black-gloss ware, only rarely did a form, once adopted for daily use, undergo variation of enough significance to allow its evolution to be followed over the course of a century.[11] Whether it is described as "utilitarian ware," "kitchen ware," or "ceramica da acqua," the name asserts that this type of pottery was inspired not by fashion, but rather because it generally remained unchanged for a long time.[12] Still, this material can shed light on the daily life and habits of a culture.[13]

Fabrics and Glosses

Macroscopic examination of the fragments reveals three fundamental fabric groups:

Fabric 1. Well-levigated, compact, orange-pink in color, typical of the plain/banded service forms;

Fabric 2. Moderately levigated, often dusty, pale yellow in color, characteristic of certain types of mortars, deep basins, and louteria;

Fabric 3. Rich in medium-sized inclusions, orange-pink in color, usually encountered in the pithos and some mortar types.

The glosses are generally diluted and opaque, with a dark brown color that verges on black in some cases (black gloss is often found on the rims of some jug types and sometimes on the interior of the carinated lekane types, see below). The decoration consists essentially of one or more horizontal lines (maximum width 2 cm) that emphasize the principal parts of the vessel (rim, shoulder, handle, base, and point of maximum body expansion). Complete coating of the interior is rare, and found primarily on the small bowl. Sites 173 and 501, both interpreted as farmhouses, produced a handle each with the so-called *fiaccola di Demetra* ("torch of Demeter"),[14] but in general decorative motifs consist simply of two intersecting lines.

Functional Typology and Quantification

The objects have been divided by function into the general categories below. Some vessels—particularly those for food service and storage—likely had multiple uses, and these will be discussed in the analysis of specific forms. Even though these vessels were intended to be functional, there were still modifications—some more apparent than others—that arose from the imagination of the potter to create a rich variety of sub-types. Lamps have not been included here (see Ch. 10).

Food service

These forms were designed to serve foodstuffs at the table, and include communal dishes (medium-large plate/lekane, lekane, carinated lekane, the smaller varieties of lekane/basin, and krater), individual dishes (plate, small-medium plate/lekane, and small bowl/one-handler), and vessels for beverages (cup, cup-skyphos, and skyphos). This definition is sometimes expanded in the discussion of the individual forms: the small bowl/one-handler, for example, is a small container not restricted exclusively to the table; it could also be used in sacred rites and to contain cosmetics; it is also often found in burials.

Short-term storage

These vessels were used to store solid foodstuffs and liquids in the pantry for the short and medium term, as well as to hold or mix wine and water (jug/pelike/table amphora, along with the few examples of hydria). The jug is a good example of the functional interchangeability of utilitarian forms: beyond its probable use in the pantry for short-term storage, it was used at the table. The same applies to the lekane/basin that could

[10] Marchegiani 1997, 1.161 and n. 4. Some see banded ware as reminiscent of the Geometric period and associated with the age of colonial foundations; Yntema speaks of an "Ionian tradition" (*Valesio* I, 63). Cf. also Russo Tagliente 1995, 89; Giorgi 1988a, 172; Ciancio 1985, 94. Casagrande (2002, 154) emphasizes aesthetic motive behind the banded decoration.

[11] On the other hand, plain/banded-wares can also borrow forms from the classes with better chronological control (cf. Deodato and Bianco 1997, 1.179), especially black-gloss cups, dishes, and bowls (see Ch. 6). Regarding the concept of functionality, Bats (1996, 481) sees a utilitarian character in every type of ceramic; cf. also Conti 1989e, 345.

[12] Casagrande 2002, 155; *Valesio* I, 64, 237; Panella 1996, 9–10; Marchi 1991, 171; *Agora* XII 32; Martinelli 1988, 145; Rossignani 1973, 410.

[13] Deodato and Bianco 1997, 1.175; Conti 1989a, and n. 7. This situation is all the more true considering that utilitarian pottery is the kind most frequently encountered in excavation (cf. Gasperetti 1996, 20).

[14] Cf. Adamesteanu 1980, 103, Fig. 97d (Metaponto); Otto 2005, 18, Fig. 26 (Herakleia, on a Hydria dated 330–270 BC); Lissi Caronna 1990–1991, 295, Fig. 120, No. 6 (Oppido Lucano).

be employed in the pantry, while the flanged lekane could be brought to the table as a communal dish. Other forms include the chytra, jar, and *dinos*.

Food preparation

The item considered an essential part of the kitchen equipment is the mortar, which, along with the pestle, was employed for grinding and mincing. It should also be noted that the basin and lekane could also be used for kneading and mixing. The lekane could have a pouring spout just like a mortar.

Long-term storage

These were medium and large containers, such as the deep basin and pithos, designed to hold solid and liquid foodstuffs for long periods. Voluminous forms like the basin could also fulfill functions in the household beyond food preparation/storage, such as bathing and washing.

Personal objects

These are containers generally intended to accommodate small objects associated with personal care: the globular urn and lebes gamikos (generally with lids); vessels for oils and unguents, lekythos and unguentarium.

Ritual Objects

This group is perforce flexible, because the materials classified as louterion types could also be used as work surfaces in the kitchen for kneading, or as simple tubs for daily washing. Also included in this category are the miniature vessels and the thymiaterion.

Other Utility Objects

This category includes objects that were utilized in other activities in daily life and work: pestles, bellows, kiln supports, small tubs, and terracotta well liners.

Unknown Forms

This group includes fragments attributable generically to open or closed vessels. As shown in Table 9.1 and Figure 9.1, over half of the fragments can only be identified as closed or open forms, and almost a third of these are uncertain. The highest percentage of identifiable fragments belongs to food service objects, which account for approximately 20% of the total assemblage. Short-term storage vessels rank second at 16.55%. Food preparation and long-term storage vessels only account for 1.20% and 2.93% respectively, while personal, ritual, and other utility objects account for less than one percent each.

Analysis by Functional Groups, Forms, and Relationships to Other Classes

The following analysis examines each functional group's most common forms and its relationships to other classes. Also noted is whether some of the forms are concentrated at sites variously identified as farms, sanctuaries, and tombs. For in-depth discussion of distribution by site types, see Chapter 5 on assemblages.

Food service

Table 9.2 and Figure 9.2 show the breakdown of forms within the category of food service.

Cup (C). There are fragments belonging to open forms intended specifically for drinking, and are less common than the small bowl/one-handler that could be adapted for the same purpose (see below). These vessels were apparently used side by side with the finer black-gloss versions (see Ch. 6), as contemporary examples of both black-gloss and plain/banded-ware cups were found together at 26 sites, which range in date from the Archaic to Hellenistic periods.[15] The state of preservation in this group is generally poor.

Small Bowl/One-handler (SB/O). The only morphological difference between the small bowl and the one-handler is that the latter has a horizontal handle attached near the rim; the handle, however, is frequently lost, and so the two vessels are considered together here. Both have the same echinus-shaped bowl, which is produced in both plain/banded ware. The form not only had a long chronological span, but also a broad geographic circulation in Greece and Magna Grecia. Probably of East Greek origin,[16] the small bowl was produced beginning in the 7th century BC, but was most widely distributed between the 6th and the turn

[15] Sites with both black-gloss and plain/banded-ware cups: 43, 103, 109, 113, 167, 248, 258, 266, 269, 320, 341, 352, 354, 358, 365, 372, 375, 376, 394, 395, 454, 477, 488, 498, 506, 542. Sites with only banded or plain versions: 115, 125, 147, 276, 371, 443, 532. Compare a contrary situation at Locri, Conti 1989d, 314.

[16] Carter and Toxey 1998a, 702; Corrente and Labellarte 1992, 436, No. 8; *Kaulonia* I, 53; Giorgi 1988a, 180; Patera 1986, 59; Ciancio and Radina 1983, 39; Bailo Modesti 1980, 82; Pancrazzi 1979, 209, No. 1; E. Macnamara in du Plat Taylor 1983, 334; Naumann and Neutsch 1960, 112. According to Saunders and du Plat Taylor (1992, II.14) the immediate success of this form in banded ware was probably due to its similarity to a form and decorative style pre-existing in the indigenous tradition. Cf. also *Valesio* I, 63, and De Juliis 1990, 114.

Function	Form	No. of frags.	Subtotal	% of total frags.
Food service	Cup and skyphos	94	3,012	20.02%
	Small bowl	346		
	Plate	123		
	Plate/Lekane	348		
	Lekane	1,704		
	Lekane, carinated	213		
	Lekane, flanged	12		
	Lekane/Basin	168		
	Krater	4		
Short-term storage	Chytra	10	2,487	16.53%
	Jar	89		
	Dinos	184		
	Hydria	69		
	Jug /Pelike / Table amphora	2,135		
Food preparation	Mortar	176	335	2.23%
	Pestle	4		
	Basin	155		
Long-term storage	Deep basin	74	286	1.90%
	Pithos	212		
Personal	Globular urn	5	145	0.96%
	Lebes gamikos	26		
	Lekythos	22		
	Lid	13		
	Unguentarium	79		
Ritual	Louterion	28	132	0.88%
	Miniature	102		
	Thymiaterion	2		
Other utility	Kiln support	7	21	0.14%
	Bellows nozzle	1		
	Tub	12		
	Well liner	1		
Unknown	Open form	1,166	8,627	57.34%
	Open (?) form	248		
	Closed form	4,675		
	Closed (?) form	161		
	Open (?) / Closed (?)	2,377		
Total		**15,045**	**15,045**	**100.00%**

Table 9.1 Distribution of plain- and banded-ware fragments by functional groups and forms.

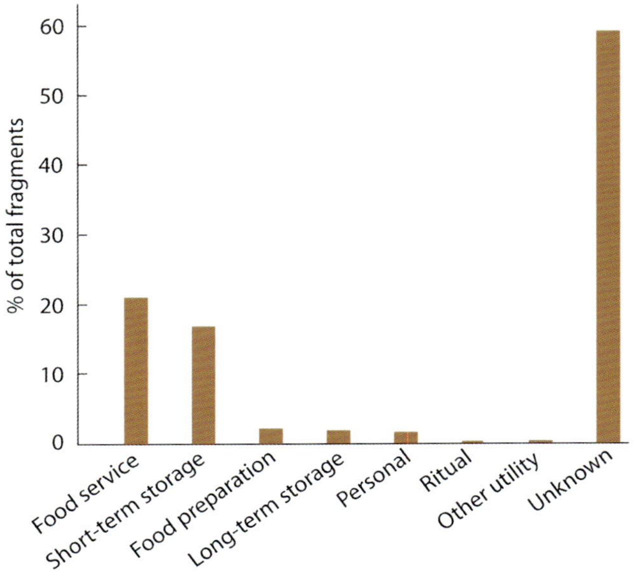

Figure 9.1 Percentages of plain- and banded-ware fragments by functional groups.

of the 4th–3rd centuries, with a significant presence through the 2nd century BC.[17] Its modest dimensions would favor its use as a container for sauces, spices, and small quantities of solid foodstuffs, while the horizontal handle renders it suited for drinking;[18] these same morphological characteristics have suggested use as a ladle,[19] or a feeder cup for children.[20] The banded decoration was an imitation of the contemporary Ionic cup (see Ch. 6),[21] with which the small bowl/one-handler is often found.[22] Almost 50% of the fragments have traces of banded decoration.

Among the 336 fragments of rims, bases, and handles recovered, 318 belong to type SB/O2. The other types are less frequent: 2 banded examples for type SB/O1; 1 plain for type SB/O3; and 2 plain for type SB/O4. The total number of recovered bases is 102, as well as 7 base fragments with completely reconstructable profile. Although the shape of type SB/O1 cannot be reconstructed in its entirety, a flat base like that of type SB/O2 is a reasonable hypothesis.

[17] See Valesio (*Valesio* I, 66, Form C11, No. 79, from a context of the end of the 3rd–first half of the 2nd c. BC), Taranto (Hempel and Mattioli 1994, 363, Fig. 309, T3, first quarter of the 2nd c. BC, phase D), Brindisi (Cocchiaro and Andreassi 1988, 112, No. 118, from a tomb dated to the second half of the 2nd c. BC).

[18] *Agora* XII, 124, where the authors also speculate about an appreciation of its versatility and practicality among travelers and soldiers.

[19] E. Macnamara in du Plat Taylor 1983, 336.

[20] Prag 1983a, 354–55.

[21] E. Macnamara (du Plat Taylor 1983, 336–37) notes that the oldest types were decorated by immersion, a simple technique in use between the end of the 7th and the 6th c. BC.

[22] Agresti 1988b, 346, No. 750; Giorgi 1988a, 180; Bailo Modesti 1980, 82. In Classical Athens banded decoration was surpassed by black gloss: cf. *Agora* XII, 125.

Form	No. of fragments	% of total food service frags.
Cup	94	3.12%
Small bowl/One-handler	346	11.50%
Plate	123	4.08%
Plate/Lekane	348	11.55%
Carinated lekane	213	7.07%
Lekane	1,704	56.57%
Lekane/Basin	168	5.58%
Flanged lekane	12	0.40%
Krater	4	0.13%
Total	**3,012**	**100.00%**

Table 9.2 Food service forms by count and percentage.

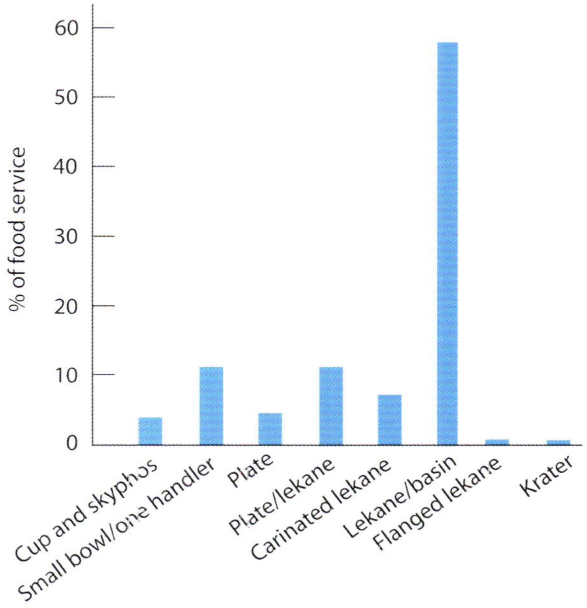

Figure 9.2 Food service forms by percentage.

The small bowl/one-handler is often found both inside and outside of tombs, suggesting that it was probably one of the objects employed in the ceremonies honoring the deceased. This multi-functional form was also recovered from domestic contexts, in sacred areas (sanctuaries, votive deposits), and in situations that presume the performance of propitiatory rites and libations in honor of a divinity. A fragment from site 235 has a perforated base, an element that recalls ritual practice connected to burial.[23] The plain version is also common. Its employment in funerary contexts is confirmed at 28 sites interpreted as tombs or farmhouses with associated tombs.[24] The form is documented at all five of the sites identified as sanctuaries.[25]

The presence of black-gloss one-handled cups does not preclude that of their plain/banded-ware counterparts. Of a total of 51 sites that yielded black-gloss types, 29 of these (mostly identified as farmhouses) had both black-gloss types (see Ch. 6) and plain- or banded-ware versions.[26]

Plate (Pl). The form initially connected the body wall and the foot via a rounded carination (type Pl2), achieving a sharp-edged profile in the later types (types Pl1 and Pl3). The oldest type, the Archaic Pl2, has a thin rim with elliptical section, while on types Pl1 and Pl3 the rim has a square section with flat and vertical exterior face. The evolution of the rim therefore reflects that of the general profile, which goes from curvilinear to sharp-edged with a flat base. The rim also tends to develop a slight downward slope in the lower exterior edge (**No. 13**, PWBW 113-272). Plates recovered inside T76, T102, and T110 at Chiaromonte-Sotto la Croce (among others), dated to the mid-6th century and demonstrating the same type of body wall with concave profile and bichrome banded decoration,[27] suggest that the footless plate types of the Metapontine chora go back to an indigenous tradition. In this case, however, as for the small bowl/one-handler, the existence of a similar footless form in the Greek world has also been documented.[28]

The most significant characteristic of these plates is an incised groove that runs along the edge of the outward-turned rim, very often associated with a sawtooth projection on the interior at the junction of the body wall and base (cf. **No. 8**, PWBW 260-19). Many examples from the Archaic period or the 5th century have a smooth interior surface. In other cases, the transition from the base to the rim is marked instead by a circular groove parallel to the base (**No. 7**, PWBW 578-02). **No. 14** (PWBW 477-57), a type

[23] See Carter 1998d, 121–23.

[24] It should be noted that these constitute only a modest proportion (ca. 14%) of the total number of sites identified as tombs and farmhouses with associated tombs.

[25] Sites 266, 270, 334, 397, and 477.

[26] Sites with both black-gloss cups and banded or plain versions: 43, 57, 113, 173, 183, 197, 200, 215, 260, 266, 269, 320, 326, 341, 343, 352, 353, 358, 365, 376, 395, 454, 465, 488, 506, 511, 515, 521, and 637.

[27] *Greci, Enotri e Lucani* 137, Nos. 2.9.27–28 (diam. 26.8 and 19.5 cm), 144, No. 2.10.32 (diam. 23.7 cm), 159, No. 2.16.17 (diam. 21.7 cm).

[28] *Corinth* VII.2, Pl. 112, An2 and An17 (diam. 19.0 cm).

Pl2, has a hole near the rim that, as for the lekane (see below), probably allowed it to be suspended on a wall or closed with a lid.

The plate is usually undecorated: only about 20% of the fragments show traces of gloss. The decoration, when present, consists of a wide band that ends near the groove on the rim; in some cases only the exterior face of the rim is decorated. When the interior of the vessel is decorated, the band ceases near the sawtooth projection, but there are also examples with interiors completely glossed (e.g., **No. 11**, PWBW 260-18).

The plate is one of the least common open forms among the survey materials, but is found at sites identified as farmhouses, tombs, and sanctuaries. A majority of plate fragments, both plain/banded, come from farmhouses.[29] The presence of this form in the rural sanctuaries (sites 266, 270, 334, 397, and 477) and in several funerary contexts (sites 81, 98, 405, 471, 513, and 626) attests the close link between these objects and ritual practice. Their appearance together at sites identified as farmhouses (sites 312, 344, 488, 506, 516, and 523) with other items associated with cult activity, such as terracotta figurines and plaques and miniature vessels, provides further proof of the existence of ritual activities within private domestic contexts.

Plate/Lekane (Pl/Lne). The plate/lekane was a form used to contain foodstuffs and serve them at the table. The smallest forms may have been used for individual service. Among the 348 fragments belonging to this type, including dishes and bowls, 78 are banded (and possibly 3 more on which the black gloss is not well preserved).

The various types and sub-types of the plate/lekane can be distinguished by their rims. Four types have been identified, ranging from those with a simple thickened rim to those with a wide, outward-turned rim. Type Pl/Lne4 is the most prevalent, with 211 examples, and it also has the most sub-types. On the basis of comparanda from various parts of Greece and Magna Grecia, a ring foot and, more rarely, a concave disk base are the most likely forms of resting surface (cf. type Pl/Lne4b3); this, however, appears to be a feature only of the lekane (see below).

Several forms clearly derive from black-gloss types, but these generally have a shallower bowl. The typology also contains a few fragments that recall the equivalent black-gloss versions and have a smaller diameter than the average for the plate/lekane.

The form is found primarily at sites identified as farmhouses, but examples also come from sites identified as tombs; in both contexts the form may also appear alongside the lekane (see below).

Carinated Lekane (CLne). Like the plate/lekane, the carinated lekane probably served to contain and serve foodstuffs. It was equipped with horizontal handles of elliptical section attached beneath the rim and could be accompanied by a filleted lid. Like the flanged lekane (see below), the carinated lekane could also be useful for other functions, perhaps as a storage container for small objects associated with personal care or games, or for performance of rituals. Its popularity in plain/banded wares should probably be attributed to its role as a practical alternative to the flanged lekane, which was instead more common in black-gloss and figured wares.

This form also had a likely origin in the Ionic or Attic tradition.[30] The name is derived from the characteristic carination of the body wall, which becomes straight near the bottom. The earliest examples, however, demonstrate a nearly hemispherical body wall, with a very subtle carination (type CLne1).

The vessel often bears banded decoration that puts particular emphasis on the rim and the higher part of the interior body wall: of the 213 fragments recovered, 117 are banded (including a doubtful fragment from site 485). The form is found at sites identified as farms, tombs, and sanctuaries.

Lekane (Lne). The lekane is a wide, open form that could be employed in the preparation of foodstuffs, particularly when kneading or mixing was required, but its primary function was like that of transporting food from the kitchen to the table. The clay is always well-levigated, and the walls can be concave or straight.

The form is divided into 13 types, several of which are further classified into sub-types. Among the types that could be identified precisely, type Lne11 is the most widespread in the study area (141 fragments), followed by type Lne9 (similar but with a less oblique rim, 131 fragments), then by types Lne12 and Lne13 (both having a nearly horizontal upper face, 103 and

[29] Three sites stand out for their higher yields of plate fragments: two sanctuaries (sites 397 and 477, 6 and 14 fragments, respectively) and one farm (site 506, 4 fragments). Other sites that yielded more than three fragments are farmhouses (sites 170, 404, and 523) and the sanctuaries (sites 270 and 334).

[30] Giorgi 1988a, 179.

113 fragments, respectively). It should be noted, however, that 75 rims have been attributed generically to type Lne12/13 because of their poor state of preservation and the concomitant difficulty in identifying their types; consequently, the most frequently attested type is the one with a nearly horizontal upper face and sharply drooping rim-edge. Types Lne2 and Lne4 also contribute to the preponderance of the lekane with an almost horizontal rim.

The form is generally plain. Traces of banded decoration were observed in only 23 cases, and on another two fragments the presence of decoration is doubtful.[31] In several cases there are small circular holes on the upper face of the rim, probably intended for hanging the vessel when not in use or closing it with a lid, which itself could be used as serving plate.[32]

The wide geographical and chronological diffusion of this form makes it difficult to date precisely. As is demonstrated by the comparanda the lekane is found in many parts of Greece and Magna Grecia, with numerous variations of the rim. The handles are horizontal with circular section. In the earlier types the handles are attached to the body wall immediately below the rim, while in the later types they are attached to the upper face of the rim.[33]

The type is very common in the Metapontino: of the 595 sites, 335 (56.3%) contain lekane fragments.

Lekane/Basin (Lne/Bas). The lekane/basin is an intermediate form between the plate/lekane, the lekane, and the basin. In two variants (Lne/Bas1 and Lne/Bas3) the lekane/basin is distinguished by the greater thickness of its body wall and rim; in another (Lne/Bas2), the diameter is generally smaller, and the body wall thinner. Several of the sub-types (e.g., Lne/Bas2) were probably used for storing foodstuffs, mostly for short periods, rather than for serving at the table.

Of the fragments identified as belonging to the lekane/basin, all were rims with the exception of one concave disk base (from site 222), which is comparable to type Mor12c. Of these, 100 fragments belong to type Lne/Bas1b, and only 32 to type Lne/Bas1a. Another 16 fragments can be ascribed to type Lne/Bas1, while only 8 fragments of type Lne/Bas2 and 4 fragments of type Lne/Bas3 were identified. The remaining 8 fragments can be attributed generically to the lekane/basin.

Decorative bands were observed on type Lne/Bas1, morphologically the closest type to the plate/lekane, but this is a rare occurrence. Banding is seen, in fact, on only six examples.[34]

Flanged Lekane (FLne). The flanged lekane has a rim designed to support a lid, but the latter object has never been found together with the form among the survey materials. The flanged lekane is closely linked with the feminine world, where it was used to contain cosmetics and small objects, but it also functioned at the table and as a generic container for various utensils.[35] It is the most common counterpart to the black-gloss and figured lekanai, but among the plain/banded wares it was largely replaced by the carinated lekane, which could also accommodate a lid. Traces of banded decoration appear in only three cases (from sites 43, 169, 521).

The body wall began as hemispherical, gradually becoming more compressed by the second quarter of the 5th c. BC. In the first phase (6th c. BC), moreover, the notch on which the lid rested was barely rendered; the form then split into two essential types: one created the resting surface for the lid from the flare of the body wall, while the other employed a thickened notch (5th–4th c. BC). The horizontal strap-handles were attached beneath the rim; in the earliest examples they were thick and bell-shaped, in the later examples thinner and horseshoe-shaped, becoming triangular by the 4th century BC.[36]

The form was recovered primarily from sites identified as farms, but fragments were also found in tombs.[37]

---

[31] From site 201, a rim belonging perhaps to a type Lne9a, and from site 479, a possible lekane base.

[32] Munzi 1999, 95.

[33] *Agora* XII, 211–14: basin usually equipped with horizontal handles. Fragments of this form were often used as ostraka in Athens. By the middle of the 5th c. BC the rim tends to become wider, and the handles are attached immediately below it; in the Hellenistic period the handles are attached directly to the rim, thereby reducing the chances of breakage and creating a better grip.

[34] From sites 258, 285, 320, 404, and 497, with two fragments coming from the last site, plus another two—from sites 154 and 501—possible but poorly preserved.

[35] *Agora* XII, 164 and 197: the form was used for both foodstuffs and spices, as well as for toys/games, spools of thread, and cosmetics. It is likely, furthermore, that the versions in black gloss or Attic red-figure were preferred for personal use, while the banded or plain versions were employed mostly at the table (although this distinction should not be applied rigidly).

[36] Cf. *Agora* XII 164–67 and Pl. 40.

[37] Farmhouse sites 43, 49, 68, 121, 169, 193, 282, and 465; farmhouse and tomb sites 182 and 521; tomb sites 430 and 443. To these can be added the sites that contain black gloss examples: sites 460 and 535.

Krater. Four fragments (three rims and one handle) that could belong to kraters, from sites 215, 238, 395, and 635. Little more can be said due to the small size of the sample and degree of preservation.

Short-term storage

Vessels used for the short-term food storage include the chytra and jar, containers that in some cases are morphologically similar to cooking-ware forms. Other forms—*dinos*, hydria, and jug—are the typical containers for liquids. As seen in Table 9.3 and Figure 9.3, the overwhelming majority of finds—over 2000—came from jugs. The fragmentary state of the recovered materials dictated the agglomeration under the category of "jug" not only of the one-handled pitchers (jug, olpe, oinochoe) but also the two-handled containers for the table (pelike and table amphora/amphoriskos).

Form	No. of frags.	% of short-term storage
Chytra	10	0.40%
Jar	89	3.58%
Dinos	184	7.40%
Hydria	69	2.77%
Jug/Pelike/Table Amph.	2,135	85.85%
Total	**2,487**	**100.00%**

Table 9.3 Short-term storage forms by count and percentage.

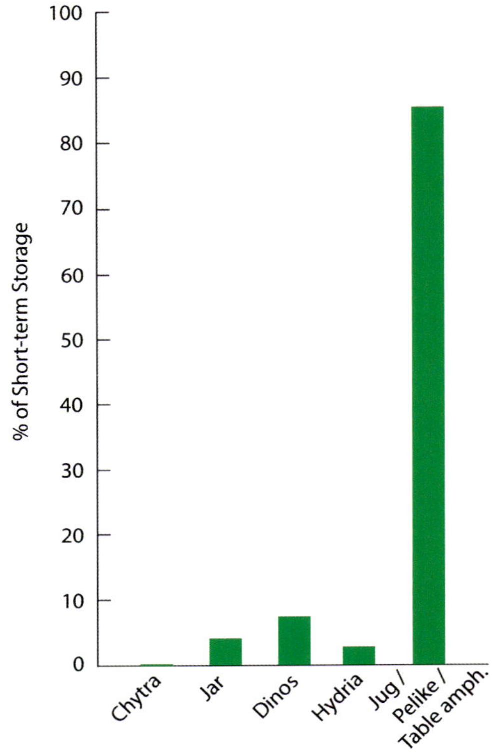

Figure 9.3 Short-term storage forms by percentage.

Chytra (Chy). The chytra is common in fabrics for cooking foods (see Ch. 11), but in some cases it was produced in levigated clay. It has an outward-turned rim, cylindrical neck, and globular body, and it also served to preserve foodstuffs for the short term. Nine chytrai fragments were recovered from nine different sites, to which can be added a fragment of type Chy2 from Plot 2001-119.

The presence of handles, otherwise not encountered among the survey materials, suggests that the chytra was utilized preferentially for the storage of liquids, although use on the table was also possible.[38] The chytra was apparently not an alternative to the jug, since every site containing a chytra fragment also had evidence for jugs.[39]

Jar (Ja). For the purposes of this study, the jar includes relatively deep containers with a flaring or outward-turned rim that is small in proportion to the body. The jar was likely employed for the short-term storage of solid or liquid foodstuffs. In many cases, as for the chytra, the form seems to recall cookware types. While the jug (see below) could also be employed in the pantry, sometimes accompanied by a small lid, the jar—especially the small versions—could have a role at the table, as is the case for the chytra (see above).

Three types of jars can be distinguished by the rim: types Ja1-2 have a high, concave rim, types Ja3-5 have an outward-turned rim, and type Ja6 has a barely perceptible rim. The poor preservation of the recovered fragments makes the identification of type Ja7 more tentative. The lone example of the type that preserves traces of the handle attachment also suggests a situla (hypothetically reconstructed in Plate 9.8), but the other fragments attributed to the same type could belong merely to a jug, in part by virtue of the dimensions of the rim, which in some cases appears to be too small for a situla.

[38] See discussion of the jar below. At two sites (416 and 511) the two forms, chytra and jar, are associated, a situation encountered also in Plot 2000-119.

[39] Sites 210, 306, and 423 contain, respectively, a single jug rim, two jug body-fragments, and a base perhaps belonging to a jug.

The most common groups are types Ja1–2, with a high rim and concave internal profile (50 fragments), followed by the group with more or less oblique, horizontal rim and straight, concave neck, represented by types Ja3 (18 fragments), Ja4 (6 fragments), and Ja5 (3 fragments). Type Ja6, with a thickened rim, is attested by a single example from site 358, while type Ja7 was identified at sites 482 and 488.[40]

Types Ja1 and Ja2 have a long history which begins in the Archaic period with the so-called banded stamnoid kraters and continues at least through the end of the 1st century BC, repeated also in reduced format in the kantharoid cups of the 4th century BC.[41]

Banded decoration is less common, appearing on only 22 fragments. Banding appears in particular on types with outward-turned rim. An example of type Ja1 (site 84) bears traces of a black band on the rim.

Both the types with high rim and the type with outward-turned rim recall cooking ware types (see Ch. 11). Nineteen sites yielded both plain-ware jars and their cooking-ware counterparts.

Dinos (Dn). The *dinos*, a jar with flattened rim, is a perfect example of an ancient utilitarian vessel that maintained its shape with little variation over time.[42] Its use, however, was not limited to the pantry: an Attic red-figure stamnos by Smikros dated to 510 BC[43] depicts a *dinos* employed as a container for mixing water and wine, a role that typically belonged to the krater.

Among the 184 fragments identified as belonging to dinoi, seven types were classified according to the degree of convexity of the rim and the shape of the body wall (straight or convex). Most common is type Dn6 (74 fragments) with a wide convex, outward-turned rim; also well attested is type Dn5 (52 fragments) with a generally inward-turning rim and straight body wall. Six are rims of type Dn1, eight are rims of type Dn2, and three are rims of type Dn3; there is also one rim each of types Dn4 and Dn7. There are three additional rims that cannot be ascribed to specific types, as well as a body fragment that appears to belong to type Dn3 or Dn4. Sixteen bases of closed forms, similar in size and form to more complete *dinos* fragments, are all disk concave in type, with the exception of the base from site 118, which, though poorly preserved, could be concave. The remaining 19 fragments include 1 handle, 11 body walls (one of which could belong to the *dinos* 352-27 but no joins were found), and 7 small rims ascribed generically to dinoi.

One-hundred and five sites contain at least one fragment of a *dinos*, most of them identified as farms. Among the five sites identified as sanctuaries (including site 266, both a farmhouse and sanctuary), three types of dinoi (Dn1, Dn2b, and Dn4) were recovered only at site 397, while at the other sites dinoi were either completely absent (site 270) or attested by a single example (site 266: type Dn3; sites 334 and 477: type Dn6a). Twelve sites identified as tombs yielded *dinos* fragments.[44]

Although less common than the plain-ware versions, banded dinoi are well attested (77 fragments). types Dn3, Dn4, and Dn7 occur exclusively in banded versions, and type Dn2 is encountered more frequently with banding (six fragments out of eight); type Dn1 is banded in only two out of six cases. Among the two most frequently found types of dinoi, there is a preference for the plain-ware version in the type with straight body wall (Dn5), of which only 25% of the fragments are banded; the type with globular body (Dn6) is banded in 54.0% of the cases.

Hydria (H). The hydria is traditionally used to carry water. The container has three handles: two horizontal handles for holding it when full, and one vertical handle for carrying it when empty. Although the hydria is one of the most common forms in every ceramic class,[45] only 69 fragments were recovered by the survey, perhaps because it had been largely replaced by the one- and two-handled jug types (see below).[46] The larger types of this form may have been equipped with lids and employed in the pantry to store liquids.[47] The hydria is found throughout the chora, at

[40] Besides the 80 fragments classified as belonging to jars, there are also six rims of doubtful interpretation, which are broadly comparable to type Ja7. Another three fragments, very small and poorly preserved, can be hypothetically attributed to jars.

[41] See Yntema 1990a, 342. D'Andria 1978a, 51. Cf. also Morel 1981a, 3431 and 3740.

[42] Cf. Marchi 1991, 171, Nos. 1–3.

[43] Franzoni 2002, 1235, Fig. 3

[44] Sites 11, 114, 132, 285, 239, 401, 438, 443, 482, 495, 520, and 524. In all cases the type attested is the one with wide, outward-turned rim and straight or globular body wall (types Dn5 and Dn6). Type Dn1 is found only at site 285.

[45] Cf. *Agora* XII, 53.

[46] It should be noted that the fragmentary state of the survey materials impeded the attribution of several rim types to the hydria form; these proved easier to incorporate into the jug typology. Among the fragments, some could belong to a jug (in particular the fragments of type H3).

[47] See, e.g., type H1, on which the shape of the rim appears ideally suited to accommodate a lid. Cf. also O. Bianco 1997a, 1.183.

tombs, sanctuaries, and farmhouses, but primarily at the latter. Despite the small number of recovered fragments, however, hydriai are well represented at sites identified as sanctuaries,[48] confirming their use in ritual practices as well.

The hydria is divided into three principal groups: type H1 has a rim with concave upper face, type H2 has an outward-turned rim, and type H3 has an oblique rim. Of the 69 fragments recovered, 43 are type H1, which appears to be the oldest.

The banded fragments total 34. Types H1a and H1d were apparently preferred in plain form (the only banded examples were found at sites 521 and 524). Type H1b is banded in only one case (from site 155). The fragments of type H1c can be either banded or plain, just like the few fragments of type H2. It is noteworthy that the only fragment of type H2a is banded. On type H3 banding is predominant, albeit barely.

Jug (J). As noted above, because of the poor preservation of the fragments, the category of "jug" includes one-handled pitchers (jug, olpe, oinochoe) and two-handled containers (pelike and table amphora/amphoriskos). These plain and banded mixing vessels, intended for the table were widely diffused beginning in the 3rd century BC at the expense of the black-gloss types. The one-handled forms had a secondary use as a dipper; the larger versions, as well as the two-handled forms, could have been employed in the pantry.

The survey recovered 2,135 jug fragments, of which 11 small fragments are difficult to attribute securely. The total number of fragments includes 220 fragments of body walls displaying a sharp edge at the transition between shoulder and neck, and therefore likely belonging to the various types of jug (e.g., **No. 145**, PWBW 113-21, attributed to type J16a).[49]

The jugs are classified into 24 types on the basis of rim shape. The most common group has a thickened rim with sub-rectangular section (335 examples, types J7, J8, J12, J14, and J20). Next is the group bearing a thickened rim with parallelepiped section (285 examples, types J15, J18, and J19), followed by the group bearing a thickened rim with sub-triangular section (174 examples, types J16, J17, and J22), and then the group with outward-turned rim (144 examples, types J1 and J2). The type with indistinct rim, generally identifiable as an olpe, comes next (93 examples, type J5), and last is the group with outward-turned triangular rim (73 examples, types J3 and J4). Type J23, bearing a thickened rim with sub-triangular section and a ridge near the neck, is attested by 21 examples. Precise parallels for several infrequently represented types proved difficult to find.[50]

Of the fragments analyzed, 582 are banded, and 18 preserve traces of probable decoration. In only 16 cases is the decoration visible on the base, in 28 cases on the body wall, and in another 5 cases on the handle. On all of the remaining fragments the decoration appears only on the rim, especially the upper face and the exterior, with a hint also on the interior. Division of the types into two macro-groups—one including the outward-turned rims (types J1–J5), the other including the thickened rims (types J6–J8 and J12–J23)—reveals a difference in the presence of decoration: in the first group the majority of the pitchers are plain (276 of 310 fragments, or 89.0%), while in the second group the rims are mostly banded (491 of 833 fragments, or 58.9%).

Food Preparation

Mortars were used exclusively for food preparation, whereas the basin was probably used for both the preparation and storage of foodstuffs. The counts and percentages for food preparation vessels are shown in Table 9.4 and Figure 9.4.

Mortar (Mor) and Pestle (Pst). Mortars, with their characteristic fabric rich in inclusions, relatively thick walls, and shallow body, were employed in the grinding,

[48] Site 270: five fragments, all banded ware. Site 397: five fragments, three in banded ware. Site 477: two banded fragments and one plain fragment. Only site 266, interpreted as a farmhouse-sanctuary, yielded a single fragment of a hydria (type H1c, plain).

[49] There are 61 rims that probably belonged to a jug/pelike/table amphora but cannot be certainly attributed to a precise form, and to these can be added two probable olpe rims (see type J5). There are, furthermore, 29 body walls (of which 11 are banded) and 10 handles (of which 5 are banded, and 1 from site 501 that bears the so-called "*fiaccola di Demetra*") that could belong to jug types, as well as 20 fragments (5 rims, 11 necks, 4 bodies) that have been attributed to amphoriskos types (see Ch. 11 on the cooking ware) by virtue of the poorly levigated fabrics full of small inclusions.

[50] Type J9, bearing parallel grooves on the exterior face, is attested in only two cases (sites 82 and 488). A single fully-preserved example facilitated the identification of type J10 (from site 635); likewise unique are the fragments of type J11 with indentation on the interior (site 394), which at the moment are comparable only with materials from Herakleia (Policoro), and the fragments of type J24 with concave profile on both the interior and exterior (site 505). 5 rims belong to type J6 with thickened elliptical rim (sites 66, 230, 498, Plot 2000-119), while 5 other fragments are attributable to type J13 (sites 43, 201, 323, 395, 471), and 8 fragments are ascribed to type J21 (sites 182, 238, 322, 334, 502, 523, Plot 2000-119), both of which have concave rims.

mashing, and even kneading[51] of foodstuffs with the help of pestles. These processes were often facilitated by the presence of large inclusions on the floor of the interior. Some, like several types of lekane, were equipped with a spout for pouring the prepared product into another container for additional processing or storage.[52] These vessels were likely the oldest of the kitchen equipment, preceding the addition of cooking vessels.[53]

The use of the mortar as an essential component for food processing is confirmed by its widespread presence at sites identified as farmhouses. It also appears at three sites interpreted as sanctuaries (sites 334, 397, and 477) and at site 266 (which has the double attribution of both farmhouse and sanctuary) suggesting that it played a part in rituals.

Foodstuffs were generally ground and crushed with the help of a pestle or a grinding stone.[54] Fragments of pestles were found at four sites identified as farmhouses (sites 327, 358, 485, and 491). Site 358 alone produced both a pestle and a mortar fragment (type Mor5a).

The survey recovered 176 mortar fragments. The rim fragments form a series of 11 types, to which can be added another two types identified by the bases. The rim can be indistinct in the amygdaloid variant (type Mor1), hemispherical (types Mor5, Mor7, Mor9, and Mor10), or rectangular (type Mor6); it can also be outward-turned with elliptical section (types Mor2, Mor8, and Mor11c), sub-triangular (types Mor3 and Mor11a), or rectangular (types Mor4, Mor11b, and Mor11d). The best-represented type is Mor5 (58 fragments),[55] while the oldest type (Mor1) is modestly represented by 19 examples. Types Mor4 and Mor11 with outward-turned rims are attested by 12 fragments. The remaining types are encountered only sporadically.

Type Mor5 is also known as "Corinthian" because it was probably created at Corinth, where it appears at the end of the 5th century BC. This type was imported

Form	No. of frags.	% of total food prepartion frags.
Mortar	176	52.53%
Pestle	4	1.19%
Basin	155	46.28%
Total	**335**	**100.00%**

Table 9.4 Food preparation forms by count and percentage.

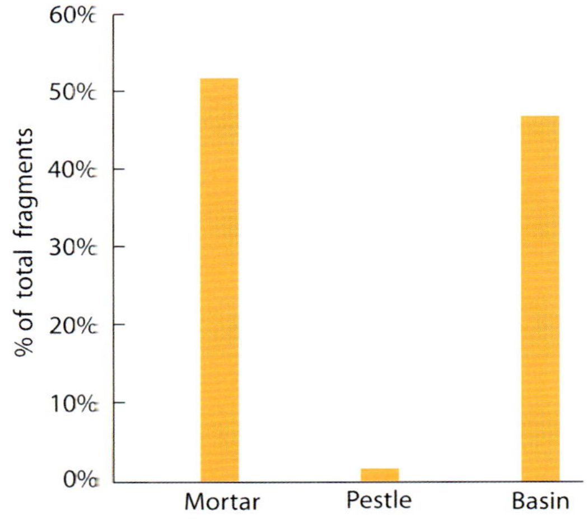

Figure 9.4 Food preparation forms by percentage.

at first, then later produced *in loco*.[56] The lower face of the base is generally flat or barely concave, but in certain cases there is a false ring foot delineated by a somewhat deep groove with triangular section that recalls the groove running beneath the rim.[57]

Another characteristic feature of this type, found on several examples, is a spool-shaped handle with incised vertical lines that is purely decorative (e.g., **No. 171**, PWBW 309-02, of type Mor5a).[58] Type Mor5 is also the only type on which a spout is attested.[59] The clay of this type of mortar is yellowish and sandy, similar to that used to make the louterion type Lou2 (which, among other shared features, is

[51] *Corinth* VII.3, 110. Mortars could also be used for processing milk or making cheese, as one sees in Switzerland today. The rough interior surface probably served this purpose by accelerating the coagulation of the milk, with the excess water eliminated via the spout.

[52] Vegas 1973, 28. Cf. *Agora* XII, 221 n. 1. These shallow objects were used daily and are to be distinguished from the larger mortar types in wood or stone, which were used for processing large quantities of foodstuffs and therefore were not part of the daily kitchen routine. For the spout cf. *Agora* XII, 223.

[53] *Agora* XII, 221, in particular n. 2.

[54] This implement was also employed in the kneading and refining of clay before it could be worked.

[55] If we consider also its variants, types Mor7 (7 fragments), Mor9 (5 fragments), and Mor10 (1 fragment), the total is 71 examples.

[56] Conti (1989c, 295) further specifies that the form does not change significantly between the end of the 5th and the 3rd c. BC. Cf. also *Agora* XII, 222–23: at Athens, too, the 5th-c. "Corinthian" type is an integral part of a farmhouse's ceramic complement.

[57] Cf. type Mor12a from site 309-06 and type Mor13b from site 319-03; on type Mor5b from site 395 there are two grooves on the resting surface.

[58] The same decoration appears on louterion type Lou2 (**No. 279**, PWBW 497-01). Four fragments with spool-handle were recovered in the survey (from sites 226, 309, 352, and 542), to which might be added one fragment (from site 521).

[59] Four such fragments were recovered (two from site 258, one from site 333, and one from site 495), and to these can be added the fragment from site 405 which has been provisionally assigned to the mortar form.

often decorated with the same spool-shaped handle). This fabric is distinct from the one found in almost all other classes of utilitarian pottery, which is generally orange-pink and compact, and could indicate a different place of origin. The examples repaired with lead clamps suggest that these were imported objects, restored and kept in use because of their durable and relatively lightweight sandy fabric.[60] From the chronological perspective, type Mor5 appears to be later than type Lou2, perhaps by as much as a century. On the other hand, the deep basin types DBas1–2 were also produced in a yellowish and sandy clay with a purely decorative handle (in the shape of a horseshoe). The comparanda suggest a date in the 4th century BC, therefore even later than either type Mor5 or type Lou2. There are, furthermore, several cases of types DBas1–2 with lead clamp repairs (e.g., site 99).

Basin (Bas). The basin is a thick-walled vessel employed primarily for the preparation and short-term conservation of solid foodstuffs. The body wall generally did not have to be very deep, and this characteristic suggests that, in addition to mixing and kneading, it could have been used to dry out legumes and cereals. The variety of types identified in this study probably reflects the range of the basin's uses. If types Bas1–2, distinguished by a shallow body, could also perform the functions of the louterion,[61] types Bas6–9, which have a decidedly deeper body, could likely accommodate water stored for domestic usage, such as for the washing of small quantities of foodstuffs and clothing.

The basin is represented by 155 fragments, 109 of which are rims.[62] The mainly domestic nature of the basin's function is demonstrated by its frequent presence at farm sites (116 fragments). Some fragments were also recovered at sites identified as tombs (sites 364, 495, and 635).

The basin rim is generally outward-turned with a section that is rectangular (types Bas1, Bas2, and Bas9), trapezoidal (types Bas3, Bas6, and Bas7a), subrectangular (types Bas5 and Bas7b), or sub-circular (type Bas8). Only type Bas4 has an indistinct rim. No traces of painted decoration were found, but in several cases shallow parallel grooves were observed on the rim (type Bas7a from site 215; type Bas9 from site 103) or on the body wall (type Bas6 from sites 128 and 312). Type Bas1b (No. 196, PWBW 489-08) has a tongue motif.

Type Bas1b has a thicker body wall than type Bas1a. Seven fragments of type Bas1a and eight of type Bas1b were recovered in the survey, as well as one fragment hypothetically attributed to a basin (site 353), for a total of 16 examples. Comparison with several examples from the Athenian Agora that share the same rim with rectangular section (in particular, see type Bas1a) suggests for these types a function as a louterion, therefore a use not strictly domestic, but also religious (see below). Type Bas7, with its 42 recovered examples (of which 36 belonging to type Bas7b), is the most frequently represented basin in this study. The lead-clamp repair of a type Bas7a at site 376 is noteworthy.

Long-term storage

Very deep containers were used for longterm storage. The deep basin was not employed exclusively for food preservation, since this form could be utilized also as a washtub (see below). The pithos was the standard container for large quantities of goods. The majority of the fragments collected came from pithoi (*Table 9.5, Fig. 9.5*).

Deep Basin (DBas). The generally thick, nearly vertical walls of the deep basin suggest use not only as a container for long-term storage in the pantry, but also as a tub for daily washing in the kitchen or bathing.[63] Its dimensions make it an intermediate form between the basin and the pithos. The decorations on several examples (e.g., DBas6, **No. 219**, from site 309) suggest that these objects were displayed as furnishings.[64]

The 74 deep basin fragments fall into two broad typological categories. The first, and larger, group is distinguished by a flat or rounded, indistinct rim (types DBas1 and DBas2, well known throughout the Metapontino); the second group, sometimes bearing decoration, is differentiated by a rim with variable profile and a thinner body wall (types DBas3–6), for which precise parallels are not readily available.

[60] See *Agora* XII, 222.

[61] In particular sites 334 and 397, identified as sanctuaries, which yielded fragments of types Bas1a and Bas1b, respectively.

[62] The total also includes six fragments from plots (1999-082, 1999-160, 2001-028, 2001-043, 2001-111, and 2001-127), as well as one fragment from site 24 (Basento-Cavone watershed) that, though beyond the scope of this volume, is useful for the definition of type Bas6.

[63] See the 5th/4th-c. BC Attic red-figure cup on which a youth is surprised in the act of bathing inside a large basin (Tempesta 2002, 1156, Fig. 33). An Attic red-figure pelike dated 480–450 BC depicts the use of the deep basin, for washing clothes (Tempesta 2002, 1213, Fig. 107).

[64] Deodato and Bianco 1997, 1.177; Conti 1989c, 310.

Forty-three fragments were identified as type DBas1.[65] The exterior face is sometimes decorated with two or more parallel grooves.[66] Several examples have horseshoe-shaped handles or outward projecting handles with rectangular section.[67] In this type and the following one (type DBas2) there are several examples with the yellowish and sandy fabric encountered in mortar type Mor5 and louterion type Lou2. The repair of PWBW 81-239P (site 99) with a lead clamp is noteworthy. Type DBas2 is attested in nine cases. This type, too, can be decorated with impressed grooves.[68] The remaining types are represented by at least one or more examples.

The deep basin's domestic function is confirmed by its presence at 42 sites identified as farmhouses, including five sites interpreted as farmhouses + tomb; eight tombs and one farmhouse + sanctuary (site 266) also had deep basin fragments.

Pithos (Pi). The pithos is the largest of the containers designed for the long-term storage of foodstuffs. It was generally made with poorly levigated clay that was rich in small-to-medium inclusions; this treatment increased the fabric strength and enabled the shape's considerable dimensions and large capacity. The vessel could be placed in the pantry of a domestic structure, or be partially or completely buried up to the rim to reinforce its sides and provide a more stable temperature for the stored goods.[69] Metal-clamp repairs found on several examples show the importance of prolonging the life of the vessel, and demonstrate the central role of the pithos in the domestic economy and the great effort involved in their production. The pithos was certainly furnished with a lid (see type Pi9), which was all the more necessary when the vessel was buried up to the rim.

Form	No. of frags.	% of total long-term storage vessel fragments
Deep Basin	74	25.87%
Pithos	212	74.13%
Total	**286**	**100.00%**

Table 9.5 Long-term storage forms by count and percentage.

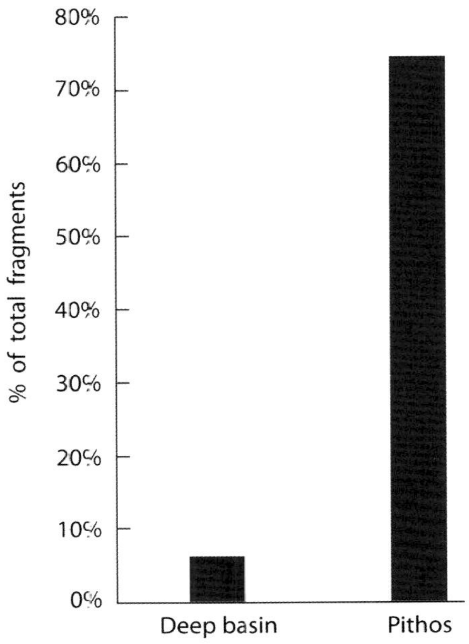

Figure 9.5 Long-term storage forms by percentage.

The great majority of pithos fragments were found at farm sites. Pithos fragments were found at five tomb sites (81, 276, 391, 638, and 776), but the only secure attributions among these are at sites 276 (one body wall) and 391 (one base, type Pi7b). The lone fragment from a sanctuary (site 397) belongs to either a pithos or a basin/deep basin.

The typological classification of the pithos is based on the shape of the rim, which has a variable width and section. The section is sub-rectangular or trapezoidal (types Pi1–3), elliptical (type Pi4), square (type Pi5), or rectangular (type Pi6). Furthermore, two types of base were identified: type Pi7 is a ring, while Pi8 is a disk. In several cases there is a decorative motif of concentric grooves on the rim.[70] A similar decorative pattern was found on the resting surface

[65] Two are poorly preserved and only attributed hypothetically to the type (from sites 131 and 302). Another five fragments could belong to this type but are too small to be identified securely (from sites 250 and 266).

[66] See in particular fragment FS82.225-01, on which appear three closely-spaced, parallel grooves at ca. 6.4 cm below the rim, each of which is ca. 0.5 cm wide; on fragment FS94.667-01 the grooves, located ca. 3.8 cm below the rim, are narrower (ca. 0.2 cm) and more widely spaced (ca. 1.0 cm).

[67] For the horseshoe type, cf. site 365; for the outward-projecting type, cf. FS82.225-01.

[68] On PWBW 060-05 the grooves (0.3 cm wide) lie 5.0 cm below the rim and are spaced 1.7 cm apart; on PWBW 309-04 the grooves are located 9.2 cm below the rim and are separated by 0.9 cm; the first groove is 0.4 cm wide, the second 0.6 cm; on PWBW 106-01, finally, below two grooves similar to those encountered on the other fragments there is an impressed undulating line framed on the lower side by another groove.

[69] Peluso (1992, 257 n. 4) describes other circumstances of employment for a pithos, particularly funerary contexts and the storage of clothing. Cf. also Vegas 1973, 117–18.

[70] See type Pi1d (**No. 228**, PWBW 097-233P and **No. 229**, PWBW 197-13) or type Pi3d (**No. 234**, PWBW 167-12).

(see type Pi8a, PWBW 12-04rev90) and shoulder of the lid type Pi9a.

Personal objects

Among the forms associated with personal care, the unguentarium was the most common by far, accounting for more than half of the fragments recovered (*Table 9.6*, *Fig. 9.6*).

Globular Urn (GU). The globular urn or pyxis probably originated in the Cycladic world and was produced in the indigenous Italic one. Adamesteanu noted its presence in the Metapontine ceramic typology and its absence in the Tarantine.[71]

The rim of the globular urn often has a circular perforation (see **No. 254**, PWBW 270-01rev90) to allow the vessel to be closed with a lid, which in general is similarly perforated. Banded decoration is present on the rim and the upper face of the shoulder; the body can also be enhanced by concentric circles, but this was not noted among the survey materials. The field survey recovered five globular urn fragments at three sites identified as farmhouses (two examples from site 323, one each from sites 394 and 488) and at a site identified as a sanctuary (one fragment from site 270). The form is divided into three sub-types.

Lebes Gamikos (LG). The name of this container—also known as a stamnos—refers to its use in wedding ceremonies.[72] The form is generally characterized by a low, vertical rim attached to a horizontal, convex shoulder, to which in turn are attached two vertically-oriented horizontal handles; the body is ovoid.

Comparative study shows that the lebes gamikos was derived from Greek and East Greek prototypes; there was a massive circulation of the form in Magna Grecia and Sicily that became particularly intense starting at the end of the 6th century BC.[73]

Recovery of the form from domestic contexts, as well as sanctuaries, reflects the object's use in daily life, especially in its simplified banded version. The smallest version was probably employed simply as a container for personal care items, while the larger versions may have been used in the pantry. Most of the 26 fragments of plain- or banded-ware lebetes recovered during the survey come from sites identified as farmhouses; a few others came from farmhouse + tombs or sanctuaries.[74]

Among the survey materials only one type was identified, with two sub-types. Fifteen fragments have banded decoration, and to these should be added the lone fragment from site 320 that could also be banded.

The survey also recovered five fragments of a type of lebes very common in black gloss, which could not be rendered graphically for this study because of the small size and highly degraded state of the fragments.[75] The shape is similar to type LG1, but the shoulder is entirely horizontal.[76]

Lekythos (L). The lekythos, a one-handled container with narrow neck and flaring rim, held oils and ointments, for both daily and funerary use. In mainland Greece and Magna Grecia the lekythos had a long history until the advent of the unguentarium (see below).[77]

There are 22 fragments of lekythoi, several of which cannot be securely attributed to the form because of their poor state of preservation. The majority of the sites that have yielded at least one fragment of a plain lekythos are interpreted as farmhouses.

Among the lekythos fragments a single example of type L1 with oblique outward-turned rim was identified (**No. 258**, PWBW 043-09, banded); other types included five examples with conical rim, of which three belong to type L3 (from sites 173 and 288, banded, and site 301, plain), and two of type L4 (from sites 250 and 417, banded). Type L2, the most commonly encountered, has an almost horizontal rim; seven examples of it were recovered: five securely attributed fragments (from sites 48, 197, 258, 383, and 521) and two hypothetically attributed fragments (from sites 291 and 341). Thirteen of the recovered fragments are banded.

The prestige black-gloss version is rarely present alongside the plain-ware version: at site 288 a type L3 banded-ware rim is accompanied by a squat-lekythos base; at site 341 a plain-ware rim, probably from a lekythos, is associated with a ribbed black-gloss body wall that is attributed hypothetically to a lekythos.

[71] Adamesteanu 1980, 59 and n. 166; Lo Porto 1959–1960, 189–90.

[72] Lippolis 1996, 262.

[73] Pizzo 1998–1999, 348; Patera 1986, 36.

[74] Plots (1999-168 and 2001-119); farmhouse + tombs (sites 182 and 509); tombs (sites 11, 48, 364, and 392); sanctuaries (sites 270 and 334).

[75] Site 48 (1990 revisit), banded; site 68 (1990 revisit), plain; site 286, banded; site 320, probably banded (only shoulder preserved); site 336, banded.

[76] See, e.g., Elliott 1998, 663, LG44 (Pantanello, 460–430 BC), and 700, T125-6 (325–300 BC).

[77] Lippolis 1994, 252–54; Bats 1988, 55.

Form	No. of frags.	% of total personal object frags.
Globular urn	5	3.45%
Lebes gamikos	26	17.93%
Lekythos	22	15.18%
Lid	13	8.97%
Unguentarium	79	54.48%
Total	**145**	**100.00%**

Table 9.6 Personal object forms by count and percentage.

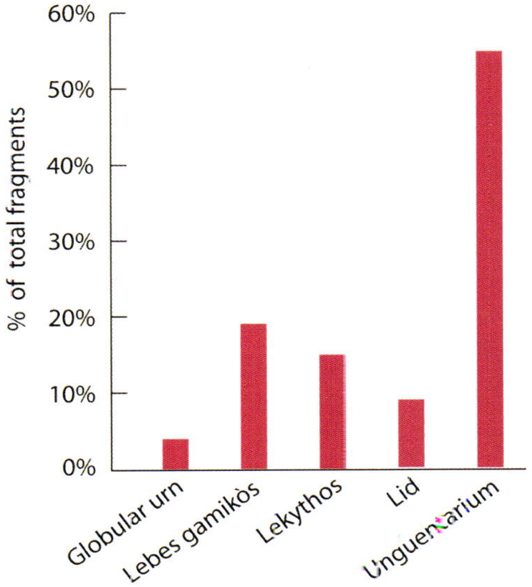

Figure 9.6 Personal object forms by percentage.

Unguentarium (U). The unguentarium is a small, generally narrow and elongated container, so-called because it was intended to contain unguents, oils, perfumes, and perhaps even powdered cosmetics.[78] It appears in the 4th century BC in funerary, religious, and domestic contexts, gradually replacing the lekythos.[79] In some cases, traces of organic remains such as woven straw, which probably served to hold and support these small objects, were found attached to the exterior.[80]

In the survey assemblage the unguentarium is represented by 79 fragments from 50 sites and 3 plots, the majority of them found at sites identified as farmhouses (38), with others coming from farmhouses + tombs (7) or simply tombs (5).

The form evolved from the type with globular body (U1), becoming elongated and ovoid (types U2 and U3, both of which correspond to Forti's type III), and eventually achieving the tapered style (types U4–6) corresponding to Forti's type IV. The best-represented types among the survey materials are types U2 and U3. The essential distinction between these two types is in the foot, which is a ring on the former and a conical disk on the latter. Types U1 and U2 are both represented by seven examples, type U3 by 24, type U4 by three, and types U5 and U6 by nine. The remaining 20 fragments cannot be identified securely by type because of their small size.

The most common decoration is black gloss (36 examples), followed by plain fabric (28 examples), and banding (8). The remaining 7 fragments could not be identified by decoration because of their poor state of preservation. Study of the artifacts in their find-contexts does not indicate a preference for the black-gloss version over the plain- or banded-ware versions in tombs: only at site 635 are all three of the unguentaria black gloss, while at sites 81 and 439 the plain-ware versions are also present. Type U1 is encountered equally in plain/banded-ware and rarely in black gloss, while type U2 is generally banded. Since the banding is applied on the shoulder, the rim, and sometimes the foot, it is practically impossible to detect unless the fragment in question is sufficiently large. Type U3, on the other hand, is generally black gloss: only four plain-ware fragments of this type were identified, and another fragment is hypothetically identified as banded ware. The fragments of types U4–6 appear to be mostly plain ware, but they are also frequently black gloss (never banded ware, however). The applied decoration on the shoulder of an unguentarium from site 57 is similar to the decoration on a banded-ware unguentarium from a tomb in the Pantanello necropolis.[81]

Lid (Ld). Several fragments belong to lids, for which a typology could not be formed because of their small size. Seven knobs were recovered at sites identified as farmhouses,[82] and five rim fragments (of which three are banded) were found at both farmhouses and tombs.[83] To these can be added the plain body-wall fragment from site 345, which may have belonged to a lid.

[78] *Agora* XXXIII, 137 n. 23.

[79] Cf. Camilli (1999) and Giglio (2005, 334–37), who carefully analyze the studies on the unguentarium and prefer instead the term *balsamario* based on the observation that unguents are strictly medicinal, while balsams/balms are both medicinal and cosmetic. Cf also Anderson-Stojanović 1987, 105–06.

[80] Lippolis 1996, 256; Anderson-Stojanović 1987, 114.

[81] See Elliott 1998, 686–87, U6, dated 325–310 BC based on context.

[82] Sites 154, 303, and 476 (PW); site 293 (PW); site 414, Plots 1999-086 and 1999-111 (BW).

[83] Sites 354, 416, and 459; sites 464 and 495.

Ritual objects

The objects used in ritual practices include the louterion, miniatures, and the thymiaterion. The great majority of finds came from miniatures (*Table 9.7, Fig. 9.7*), which are small containers of purely symbolic value. The thymiaterion is minimally represented in the survey assemblage.

Louterion (Lou). The louterion, a large basin with a shallow body and thick walls, was employed for purification and ablution at sacred rites; the form could also be used in private for personal care.[84] Several terracottas depict the louterion as a large cutting surface, while the deeper type is shown as a collecting basin for water. Impressed decoration on the interior body wall of louterion types intended for domestic use, moreover, could have facilitated kneading or mixing, while the ornamentation on the interior of louterion types intended for religious ceremonies held a purely symbolic value.[85] The louterion stood on a support, which could form an integral unit with the basin, or be separate.

Use of the louterion in domestic contexts appears to be confirmed by the materials from the survey. Of a total of 22 sites[86] with at least one fragment of this form, 15 have been interpreted as farmhouses; the proportion rises remarkably when also considering the 5 sites identified as farmhouses + tomb. On the other hand, despite the louterion's role in the ceremonial sphere, site 397 was the only sanctuary to produce an example (type Lou2a).

The 28 louterion fragments from the survey are mostly rims (only three fragments are hypothetically attributed to bases), divided into four distinct types. type Lou2 is the best represented (19 examples), while the others are rarer: type Lou1 occurs only four times, and types Lou3 and Lou4 are attested by a single fragment each. The type Lou2 has grooves on the exterior face that are generally four in number, but other examples were found with 2, 5, 6, or as many as 14 parallel lines. Several examples are decorated with spool-shaped handles that served only a decorative purpose and certainly did not have a practical function as grips (see types Lou2a–b). A typical feature of this type is the fabric, which is yellowish and sandy like that encountered in mortar type Mor5 and deep basin type DBas1 (see above).

Miniature (Min). Miniature ceramic vessels, so defined by their extremely small size, had a purely symbolic value associated with both public and private religious practices. For this reason they are often found in votive deposits, sacred areas, burials, and even private dwellings. The feet of these small objects typically bear the concentric circles characteristic of wheel-made ceramics (see types Min2 and Min5).

The analysis of the miniatures is based on 102 fragments from 22 sites.[87] There is a remarkable concentration at the five sites interpreted as sanctuaries, confirming the strong link with the ceremonial sphere. At the sanctuaries there is an average of 15 miniatures per site, with the highest single concentration at site 397, which contains 54 examples.[88] Association with the private sphere, however, is attested by the recovery of at least one miniature object at 15 farmhouses.[89]

The miniatures are divided into six types, each of which corresponds to a form: type Min1 is a krateriskos, type Min2 a lebes, type Min3 a phiale mesomphalos, type Min4 a dish, type Min5 a cup, and type Min6 a small mug. Types Min1–Min4, common to the Late Archaic/Early Classical period, are accompanied by later versions (types Min5 and Min6) produced exclusively in the miniature format.[90]

The krateriskos type Min1 is the most frequently encountered form among the miniatures, with 46 examples (37 from site 397). There are 13 examples of the lebes type Min2, although most consist of just the base. The phiale type Min3 is the rarest form among

[84] Cf. also Tempesta 2002, 1157, Fig. 34, and 1158, Fig. 35; Fabbri and Osanna 2005, 231; *Montemurro* 104 and 108; Maturo 1997, 1.251; Peluso 1992, 259.

[85] *Agora* XII, 218 and 220; Pizzo 1995a, 94.

[86] To these must be added site 679 in the Basento-Cavone watershed, which, though beyond the scope of this volume, contains a fragment useful for the typological definition and illustration of type Lou2c (from site 679-01).

[87] To these can be added a fragment from site 32 in the 1980 pilot season, which contains a fragment useful for the present typology, Plot 1999-024, and Plot 2001-129.

[88] From site 266 comes a single Miniature, three from site 477, six from site 270, and 11 from site 334.

[89] Sites 57, 209, 241, 307, 312, 344, 369, 449, 462, 516, 521, and 633. From sites 18 and 340 come two miniature objects, while site 362 has three.

[90] There are also types that reproduce in reduced scale black-gloss or utilitarian forms: see, for example, plate type Pl1 from site 348 which has a smaller than average diameter of 10.0 cm; dish PWBW 506-03, which is also smaller than the average (diameter of 10.0 cm), repeats the black-gloss form. The one-handler PWBW 523-10 has a diameter of 4.4 cm, while the two small dish types from site 532 have diameters of 7.0 and 8.3 cm and also repeat black-gloss forms (see Ch. 6).

Form	No. of frags.	% of total frags. of ritual objects
Louterion	28	21.21%
Miniature	102	77.27%
Thymiaterion	2	1.52%
Total	**132**	**100.00%**

Table 9.7 Ritual object forms by count and percentage.

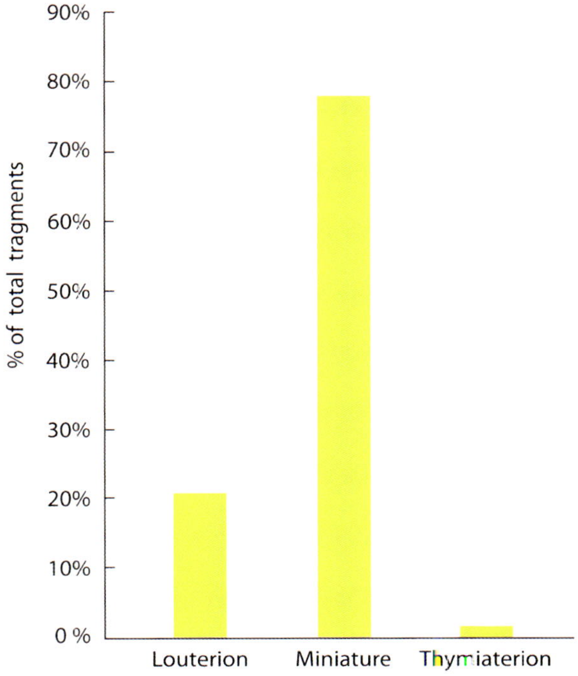

Figure 9.7 Ritual object forms by percentage.

the Metapontine miniatures, found only at site 334 (a sanctuary). The only example of type Min4 comes from site 312. The cup type Min5 yielded ten examples (from sites 18, 241, 270, 334, 397, 449), and to these can be added two fragments of uncertain attribution (from sites 18 and 57, both farmhouses). Type Min6 is represented by two fragments: one from site 32J-037 is glossed, while the example from Plot 2001-129 is plain. Given the poor condition of the only two recovered examples (from sites 270 and 307), the miniature oinochoe could not be illustrated satisfactorily for publication. The rim is trilobate with elliptical section.[91] These vessels are generally plain: among the materials recovered in the survey only **No. 287** (type Min2, PWBW 334-08) preserves traces of gloss on both the interior and exterior; **No. 289** (type Min4, PWBW 312-07) is completely glossed; one body fragment (type Min1a, PWBW 397-86) is banded; an oinochoe from site 270 has a band around the rim; and a small cup (PWBW 80-32(J)) is completely glossed.

Thymiaterion (Thy). The thymiaterion was a ritual object,[92] but also one with uses in the domestic environment as well as at symposia. To these can be added the divinatory art, which was practiced by observing the direction of the flow of the smoke.[93] The form varies in shape and dimensions[94] and is occasionally equipped with a perforated lid that permits a wider and slower diffusion of the incense and avoids accidental dispersion of the embers. It generally consists of a tall stem and a shallow body that served to contain the incense and other substances to be burned; only in rare cases was a true small cup placed on top of a thymiaterion. Another type has a low stem or no base.[95]

The thymiaterion is frequently found in sanctuaries, but also inside burials[96] and in domestic contexts (dwellings, deposits, wells). The two examples of the thymiaterion belong to two types. Both fragments recovered by the survey came from sites identified as farmhouses (sites 730 and 742), confirming the presence of spaces intended for common rituals inside the domestic environment.[97] At site 730, in particular, the suggestion of private cults attested by the recovered thymiaterion (type Thy2) is reinforced by the presence of a louterion (type Lou2a).[98] In this case, too, the clay is well-levigated, and the decorative motif is diluted bands.

[91] Adamesteanu 1980, 223, No. 231d (Metaponto); Lo Porto 1981b, 311, Fig. 221 (Metaponto, second half of the 6th c. BC).

[92] The intense odor produced by the incense burned in a thymiaterion during sacrifices was probably also intended to mask the equally strong odor produced by the burning meat: cf. Zaccagnino 1998, 51 n. 106.

[93] Zaccagnino 1998, 37.

[94] In general, the types pertaining to public rites are larger than those used in the private sphere: cf. Zaccagnino 1998, 67.

[95] Cf. Di Lieto (2001, 57), who refers to the origin of the name and the use of incense in the Greek world, as well as the types encountered in funerary and ceramic iconography. See also Zaccagnino 1998 and Di Lieto 2005b.

[96] Cf. Di Lieto 2001, 57; Lo Porto 1991, 159; Ricchetti 1992c, 258, Nos. 7–8; Cassano 1992a, 282, No. 75, and 300, No. 149 (Canosa, *ipogeo* Varrese); Cassano 1992c, 500, No. 4 (*ipogeo* 3).

[97] On domestic cults and the divinities venerated, see *Montemurro* 163 and 174–180 and Barra Bagnasco and Russo Tagliente 1996, 185. On an Attic red-figure kylix of the 5th c. BC a woman is depicted in the act of filling a thymiaterion with granules of incense (Tempesta 2002, 1181, Fig. 67). Cf. also Perrone 2003, 363–64.

[98] Di Lieto (2005a, 153) notes that, beyond being found in association with strictly ritual objects, the thymiaterion can also be found with vessels that have a secondary ritual use/function.

Other utility

Kiln Support (KS). Kiln-spacers are terracotta objects used to separate ceramic objects inside the kiln during firing.[99] The survey yielded seven fragments of kiln supports from the following sites: 64 (type KS2); 320 (type KS1); 372, 417 (type KS3), 529, 651 (type KS2); 741. Of these, the fragment from site 372 is glossed, probably as a firing test.[100]

Bellows Nozzle. The fragment of a bellows nozzle from site 43 is an unusual find.

Tub. 12 fragments of small tubs were found at nine sites (152, 217, 249, 326, 376, 395, 452 492, and 635). Three of these fragments are bean-shaped while the rest are of the elliptical type with straight rectangular rim.

Although precise parallels for the bean-shaped examples are not available (with the exception of the whole example from a necropolis at Pizzica[101]), a use associated with domestic activities can be proposed. The tub with straight rim, however, was associated with both the funerary and the domestic spheres: in the former it was used to hold the remains of infants and serve as a sarcophagus (multiple examples fitted together); in the latter it served as a bathtub for infants. This type is well known in southern Italy[102] and Sicily.[103]

Well Liner. The last fragment worthy of note is the terracotta well liner from site 100. Four well liners of similar shape were found near Temple B in Metaponto.[104]

Unknown forms

The survey recovered bases belonging to open forms that cannot be matched to a specific form. Two principal types have been identified by the shape of the body wall, which is horizontal on type Op1 and oblique on type Op2. The type Op1 base (one fragment) is attributed to the plate/lekane; type Op2 (23 fragments) is associated with drinking vessels; another 26 fragments were too small to be identified.

Ten fragments have banded decoration on the floor of the interior and occasionally on the exterior face of the foot. **No. 299** (PWBW 529-05) is unusual in that the exterior is completely glossed and the interior has a single concentric band on the floor.

Several bases of open forms do not fit within the typology but can be distinguished as disk concave (155 fragments, as well as 79 fragments of presumably open forms) or simply concave (3 fragments, of which 2 are perhaps open), disk flat (4 fragments, as well as 3 fragments that could be open forms) or flat (12 fragments, to which can be added 7 bases belonging probably to open containers), and ring (55 fragments, of which 21 are perhaps open). Banded decoration was observed on 11 fragments of the disk concave base type, on one fragment of the disk flat base type, on two fragments of the flat base type, and on nine fragments of the ring type.

The bases of closed forms that could be classified belong mostly to jug (540 examples) and *dinos* types (see above). The most common type is the disk (types Cl1 and Cl2), represented by 387 fragments. Rarer are types Cl3 (two fragments), Cl4 (nine fragments), and Cl5 (five fragments).

Beyond these types, there are other fragmentary bases of closed forms that are too small to be classified. The concave base is attested by 37 fragments, the disk concave by 11 fragments, and to these can be added the fragments probably belonging to closed forms (70). The disk flat base is represented by 12 fragments, to which can be added the two fragments of possible closed forms. The flat base has 56 examples, as well as another seven fragments of possible closed vessels. The ring base is present in 46 examples, with another five fragments belonging presumably to closed shapes.

Banding is rare in these cases, since the decoration is generally applied to the rim and shoulder of the vessel. Nevertheless, banded decoration was observed on nine fragments of type Cl2 and on three fragments of the disk concave base type belonging to forms that were probably closed. The disk flat base and flat base variants each yielded one banded example, while three examples of banded ring foot were encountered.

[99] For a detailed study of kiln supports in the Metapontine kerameikos, see Cracolici 2004 a.

[100] See Cracolici 2004a, 55–56.

[101] See Vittoria 2000, 351, Pl. 68.

[102] Rocchietti 2002, 62; Maruggi 1994, 77–85. See Carter 1998c, 103, for a bathtub.

[103] Bonanno 1998, 331, Nos. 1–2, 388, No. 1.

[104] For other comparanda, see Martorano 2001, 22–23 and nn. 40–41, Pls. XXI–XXII (Reggio Calabria, with section drawing of a well in the kerameikos of Athens). For well liners found in Metaponto, see De Siena 1980, 88 and 110, Pl. II, as well as Ch. 18 for use of well liners as tomb coverings.

Catalog of Plain/Banded Ware

This catalog describes the objects in the best state of preservation. The comparanda give precedence to materials recovered in the Metapontino and nearby areas. All dimensions are in cm. Maximum rim diameter is indicated by "r," maximum height by "h," maximum width by "w." An asterisk (*) after the site number indicates materials from the 1980 ICA pilot survey (sites numbered 1–44), as well as at the intensive survey sites and plots located south of the Basento River through the 2001 season (most of the sites numbered 562 and higher). Additional information about the sites can be found in the Gazetteer. For lamps, see Chapter 10. The following abbreviations are used for classes: BW (banded ware), PW (plain ware), BG (black gloss), and Ck (cooking ware).

Cup (C)

The rim diameter of the cup generally falls between 8.0 and 12.5 cm. Two examples from site 269 are smaller than the norm (6.0 and 6.2 cm), while the cup from site 326 has an unusually large rim diameter of 16 cm. The base diameter varies, ranging from 3.0 to 5.4 cm, but Archaic-Classical cup bases can be as large as 8.0 cm.

1:2

1. PWBW 113-09
Description: Indistinct, slightly outward-turned rim with elliptical section; body wall initially inward-projecting, then generally ovoid.
Dimensions: r 12.5 h 4.9 w 4.0
Comparisons: Carter and Toxey 1998, 69: T125-13 (Pantanello, one-handed cup, 325–300 BC); Marchegiani 1997, Pl. 58, No. 21 (Pomarico, BW *bicchiere*, diam. 9.3 cm, end of 4th c.–second half of 3rd c. BC); *Museo Taranto* II.1 107, No. 43.64 (Ginosa, kantharos from a burial context, 4th c. BC, diam. 12.3 cm) and 109, No. 43.71 (one-handler, 4th c. BC, diam. 10.1 cm); Saunders and du Plat Taylor 1992, Fig. 10, Nos. 220 (Gravina, cup type 2, period VIIIa-b, end of 2nd–1st c. BC) and 222 (cup type 3, period VIIIa, end of 2nd–1st c. BC); Cotton 1992b, Fig. 69, No. 1348 (Gravina, cup, period VI-VIIIb, 4th–1st c. BC). Comparanda suggest a date from middle of the 4th c. to at least the 1st c. BC.

Small Bowl/One-handler (SB/O)

The dimensions of the small bowls that are definitely one-handled (47 examples or 68.1%, of which 32 are banded, including two miniatures) are between 6.0 and 10.0 cm, with a preponderance of rims that approach 10.0 cm (sites 166, 285, 286, 326, 349, 358, 395, 465); to these should be added two small bowls with rims measuring 12.0 cm (sites 410 and 477) and another two, which are miniatures (PWBW 523-10, diam. 4.4 cm and site 358, diam. 4.8 cm). The bases all belong to SB/O2 (except for PWBW 365-06, which is SB/O3) and they have a diameter that varies between 3.0 cm (2 examples) and 4.0 cm (7 examples). Only one fragment (FS81.120PLb from site 36) has a larger diameter of 4.5 cm, and two bases are smaller, one from site 260 (2.4 cm) and the other from site 482 (1.7 cm). A fragment from site 235 that has a diameter of 2.8 cm also has a perforated base. A base of SB/O3 with a diameter of 3.2 cm also falls within this range.

1:2

SB/O1
2. PWBW 375-13
Description: Inward-projecting rim with elliptical section; hemispherical body wall.
Dimensions: r 10.0 h 4.0 w 5.6
Comparisons: G. Russo 1996, 329–31, No. 317 (Taranto, end of the 5th–beginning of 4th c. BC); *Sibari* IV, 68, Fig. 49, No. 47, and 144, Fig. 138, No. 3017 (Stombi, beginning of the 5th c. BC).

Small bowl/one-handler (SB/O) cont.

1:2

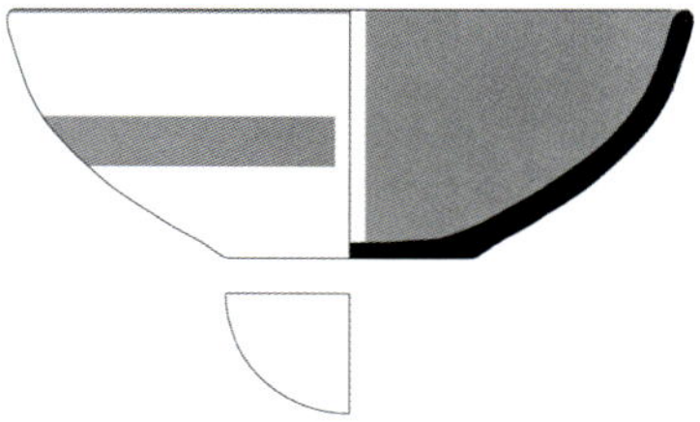

SB/O2

3. PWBW 482-02

Dimensions: r 9.4 base 1.7 h 3.5 w 6.3

1:2

4. PWBW 730-01

Dimensions: r 6.3 base 2.6 h 2.6 w 5.5

Description: Indistinct rim with elliptical section; flat or slightly concave base; hemispherical body wall.

Comparisons: Carter and Toxey 1998, 701–704 (Pantanello, 450–300 BC); Lo Porto 1988–1989, 330, Fig. 32, No. 2 (Metaponto, PW one-handler, tomb of the third quarter of 6th c. BC), and 372, Fig. 81, No. 3 (BW one-handler, tomb, 380–370 BC); Lo Porto 1981a, 305, Fig. 15, No. 1 (Metaponto, BW one-handler, votive deposit, second half of 6th c. BC, described as "local imitation of Ionian style"), 308, Fig. 19, No. 1 (PW one-handler, end of 6th–beginning of the 5th c. BC), 359, Fig. 77, No. 2 (BG small bowl, tomb dated to beginning of the 3rd c. BC, diam. 6.8 cm), and 363, Fig. 84, No. 1 (PW one-handler, tomb of the second half of 4th c. BC); D'Andria 1980b, Pl. XI, No. 29 (Metaponto, BW small bowl, *scarico* No. 12, 5th c. BC); Carter 1983a, 455, Fig. 30, No. 44 (Metaponto, tomb); Adamesteanu 1980, 285, Fig. 298a,e,c (Metaponto, BW one-handler, votive deposit, 4th c. BC); D'Andria 1980a, 412, Fig. 57, No. 229 (Metaponto, BW one-handler, *scarico* No. 1, 4th c. BC).

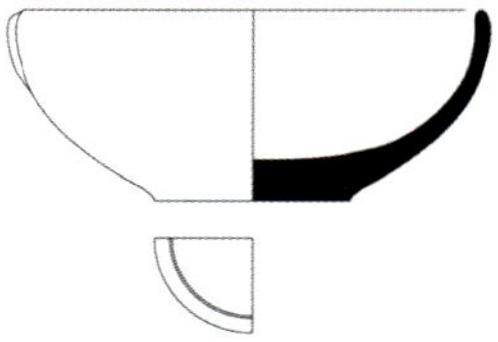

SB/O3

1:2

5. PWBW 365-06

Dimensions: r 6.65 base 3.2 h 2.65

Description: Indistinct rim with elliptical section, slightly thickened on the exterior; hemispherical body wall; disk foot with oblique exterior face and flat base.

Comparisons: Valesio I, 240, Form M01, No. 420 (PW one-handler, diam. 12.5 cm, context of the first half of the 2nd c. BC). The thickened rim makes this cup very similar to the BG examples recovered in the Metaponto survey (see Ch. 6). The comparanda suggest 4th c. BC as the period of this type's widest circulation.

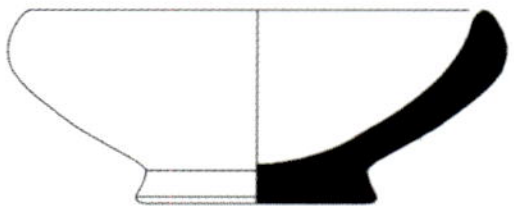

SB/O4

1:2

6. PWBW 395-40

Dimensions: r 9.5 h 2.3 w 4.6

Description: Outward-turned rim with triangular section; nearly hemispherical, slightly carinated body wall; horizontal handle attached at the point of greatest extent of the body wall.

Comparisons: Rinaldi 2005 232, 239, Pl. XXVIII, No. 126 (Torre di Satriano, diam. 10 cm). The type is well known in BG (see Ch. 6), distinguished by a slightly deeper body wall and dated from mid-4th to the mid-3rd c. BC.

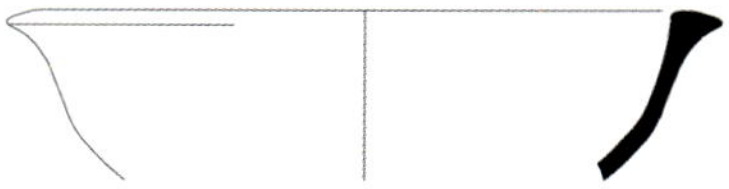

Plate (Pl)

1:2

The rim diameter of Pl1 and Pl3 generally falls between 16.0 and 20.0 cm, reaching a maximum of 20.8 cm (No. 13, PWBW 113-272), but there are also slightly smaller examples (13.7–15.9 cm), as well as others so small as to appear to be miniature types (10.0–13.0 cm). Pl2 appears to have been larger: the few fragments preserved (nine from site 477 and one from site 170) have diameters between 22.0 and 23.3 cm.

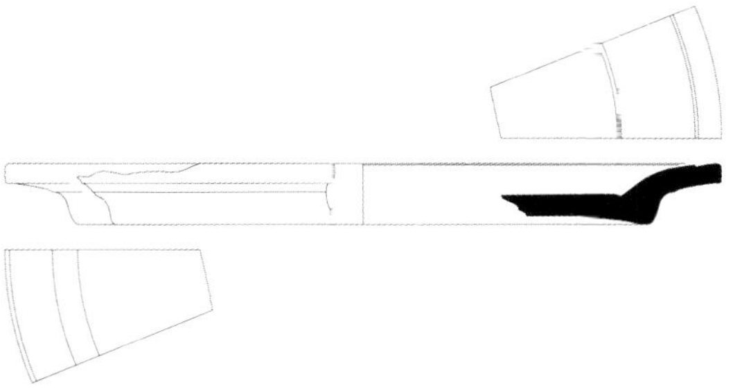

Pl1

7. PWBW 578-02

Dimensions: r 13.7 h 1.2 w 4.9

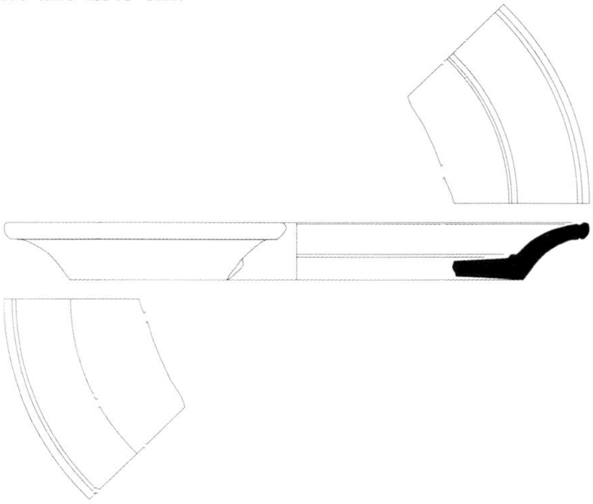

8. PWBW 260-19

Dimensions: r 14.8 base 11.5 h 1.5

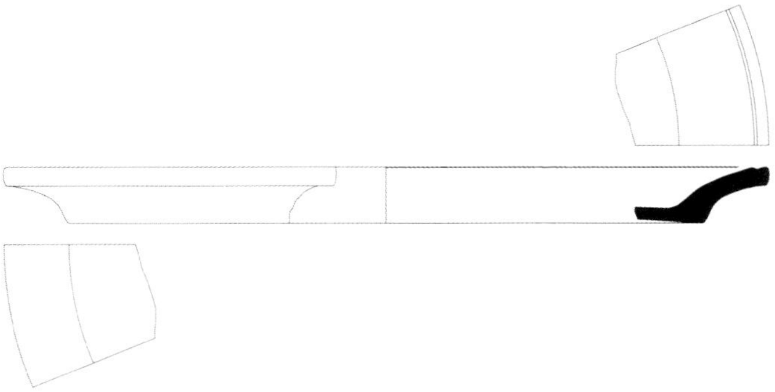

9. PWBW 122-01

Dimensions: r 17.5 base 14.5 h 1.35

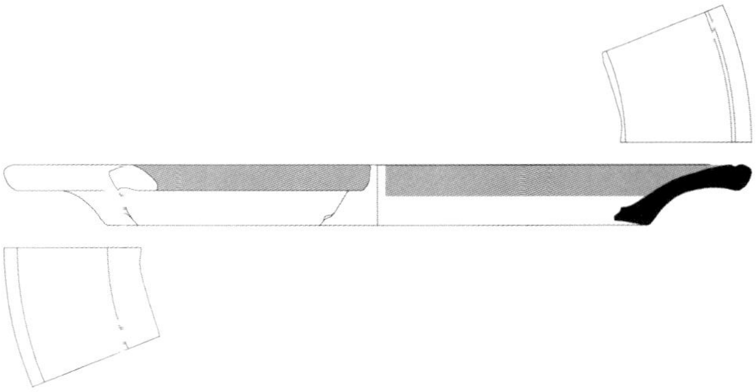

10. PWBW 445-01

Dimensions: r 16.3 base 11.8 h 1.4 w 5.7

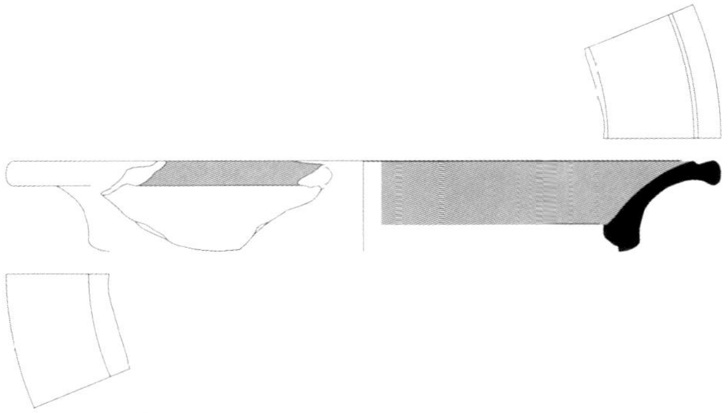

11. PWBW 260-18

Dimensions: r 15.9 base 12.3 h 2.07 w 5.15

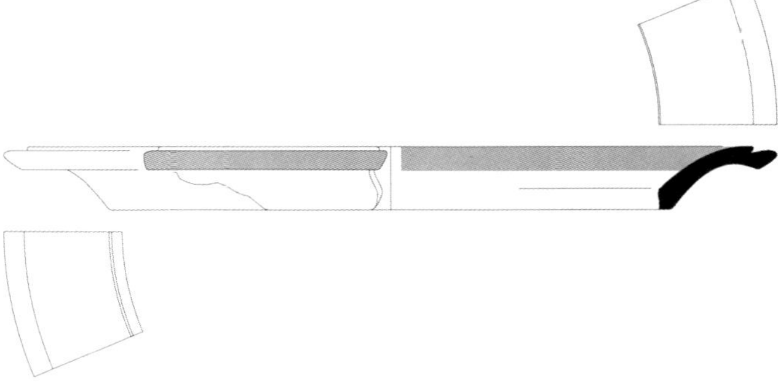

12. PWBW 513-01

Dimensions: r 17.5 h 1.45 w 5.5

Plate (Pl) cont.

1:2

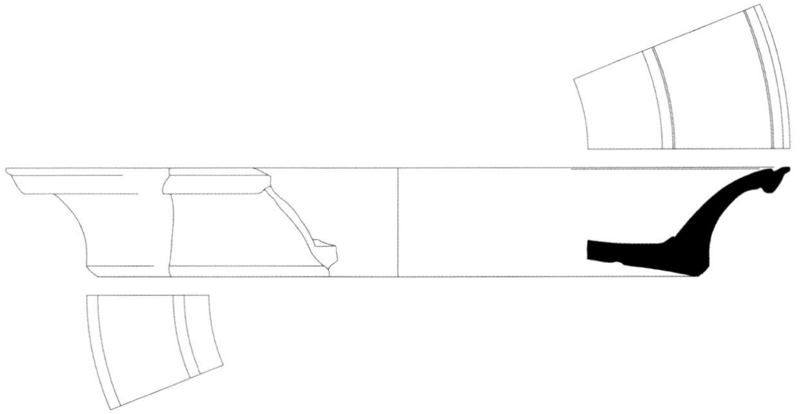

13. PWBW 113-272

Dimensions: r 20.8 b 16.5 h 3.0 w 4.6

Description: Outward-turned rim emphasized at the top by a groove, attached to a body wall with concave profile; flat or slightly convex floor connected to the rim via a sharp edge.

Comparisons: Adamesteanu 1980, 286, Fig. 299b (ca. 300 BC); Prag 1983a, 358, Fig. 129, No. 329 (Cozzo Presepe, 320–270 BC); Deodato and Bianco 1997, Pl. 75, Nos. 122–123 (Pomarico, PW *piatto*, diam. 17.2 cm and 22.4 cm, 4th–3rd c. BC); *Museo Taranto* III.2, 197, Fig. 38, No. 8 (double burial, last decades of 1st c. BC, diam. 18.0 cm); Prazzoli 1996, 334 (Taranto, type 16); G. Greco 1991, 51, Fig. 116 (Serra di Vaglio); Pianu 1990, Pl. LXXV, No. 278.1 (end of 4th–beginning of 3rd c. BC); L. Giardino 1990, 114, Pl. XVIII (Herakleia, last quarter of 3rd c. BC).

1:2

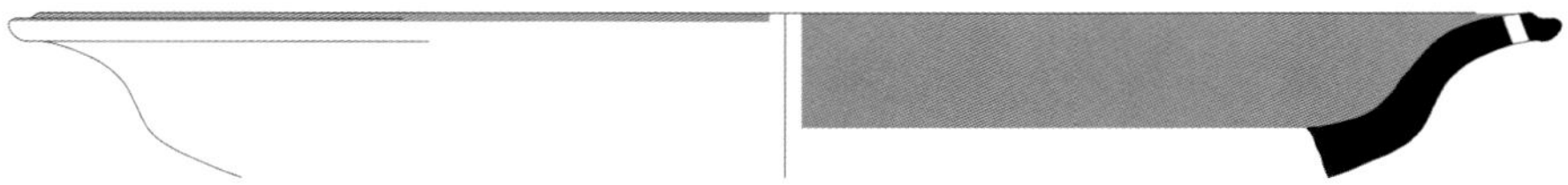

Pl2

14. PWBW 477-57

Dimensions: r 23.3 h 2.6 w 8.0

Description: Outward-turned rim emphasized at the top by a groove, attached to a body wall with concave profile; curvilinear transition from the rim to the slightly convex base. There is a hole near the rim that probably allowed it to be suspended on a wall or closed with a lid.

Comparisons: None found.

1:2

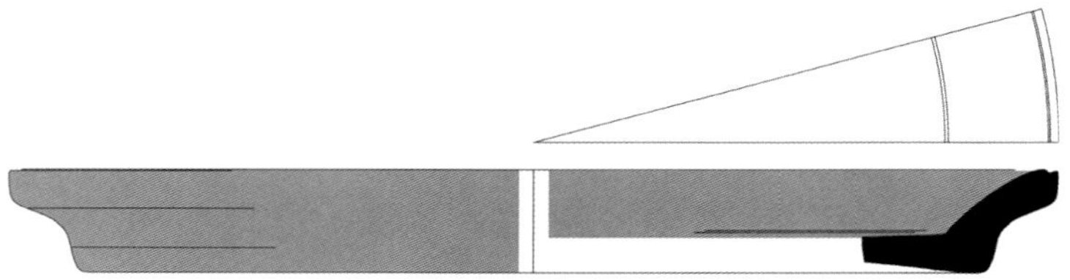

Pl3

15. PWBW 358-06 rev90

Dimensions: r 19.0 h 1.9 w 5.3

Description: Same as Pl1, but with a less accentuated rim and thicker walls.

Comparisons: Hempel 2001, Pl. 11, type 622/7 (Taranto, phase F, second half of 1st c. BC).

Plate/Lekane (Pl/Lne)

Of the total 232 rims examined, only 94 can be measured; of these, 62 fragments have a diameter that falls within 14.0–30.0 cm. The rest measure on average between 31.0–35.7 cm, although there are examples with diameters of 43.5 cm and even 50.0 cm. Another ten fragments of this type have a diameter that exceeds the average range (38.0–46.0 cm) with relatively shallow body wall. Three miniatures related to BG forms have diameters of 7.0–10.0 cm. Of the 73 base fragments examined, 47 were measurable; the majority of the fragments have a diameter between 8.0 and 14.0 cm (65.9%), but on the smaller forms the diameter is 4.0–7.0 cm.

1:2

Pl/Lne1

16. PWBW 521-162

Dimensions: r 21.7 h 2.1 w 8.2

Description: Nearly indistinct rim with sub-triangular section and slightly convex upper face (Pl/Lne1a) or indistinct rim with groove on the upper face and perforated handle (Pl/Lne1b); hemispherical body wall.

Comparisons: Pl/Lne1a: Marchegiani 1997, Pl. 57, No. 10 (Pomarico, BW *ciotola*, diam. 11.7 cm, second half of 3rd c. BC; cf. Morel 1981a, 2612a1, BG version from Delos, Campana A, mid-2nd c. BC); Saunders and du Plat Taylor 1992, Fig. 6, No. 128 (Gravina, period VI, 4th–3rd c. BC; II.19: this is the most common type of dish at Gravina, dated end of the 5th–4th c. BC). Cf. Morel 1981a 2237, BG version from southern Italy, second half of 4th–beginning of 3rd c. BC); also BG versions (Ch. 6). Pl/Lne1b: Riccardi 2006, 353, Pl. 13c (*coppa* da Rutigliano, tomb 39, end of 6th c. BC). Gagliardi 2001, 292, No. 274b (BG from Punta Stilo, *ciotola*, middle third of the 5th c. BC, 14.0 cm).

1:2

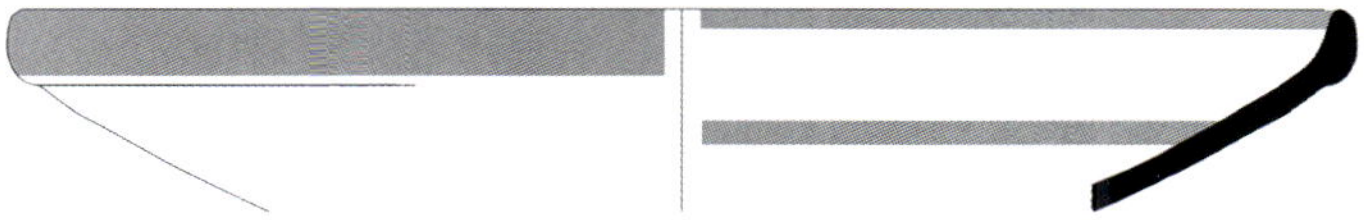

Pl/Lne2

17. PWBW 528-12

Dimensions: r 17.9 h 2.8 w 3.9

1:3

18. PWBW 128-29

Dimensions: r 35.7 h 3.6 w 5.0

Description: Thickened rim with elliptical section (Pl/Lne2a) or hanging rim with sub-triangular section (Pl/Lne2b); hemispherical body wall.

Comparisons: BG type dated 2nd c. BC (see Ch. 6).

1:2

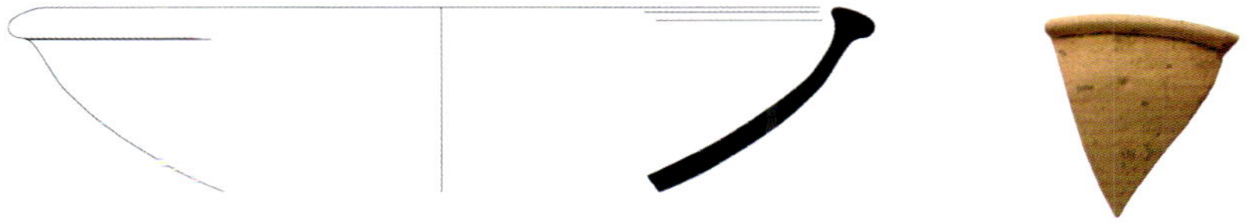

Pl/Lne3

19. PWBW 365-05

Dimensions: r 14.8 h 3.2 w 4.6

Description: Outward-turned, inward-projecting rim with elliptical section; hemispherical body wall.

Comparisons: Deodato and Bianco 1997, Pl. 72, No. 107 (Pomarico, PW *ciotola*, diam. 35 cm); Russo Tagliente 1995, 89, Fig. 90, No. 217 (Armento; derived from cups widely distributed in 5th c. BC and well-known in indigenous centers of Basilicata through third quarter of 4th c. BC); Andreassi 1988c, 275, No. 621 (Bari, "coppa biansata a decorazione lineare," tomb of the mid-5th c. BC, diam. 19.5 cm); Lissi Caronna 1983, 234, Fig. 19, No. 7 (Oppido Lucano, T57 dated end of 5th–beginning of 4th c. BC, diam. 17.8 cm); G. Greco 1980, 404, Pl. XVI (BG version from Serra di Vaglio, end of 4th–beginning of 3rd c. BC); Tigano 2000, 143 S/97 (Zancle-Messina, second half of 6th c. BC, diam. 19.6 cm).

Plate/Lekane (Pl/Lne) cont.

1:2

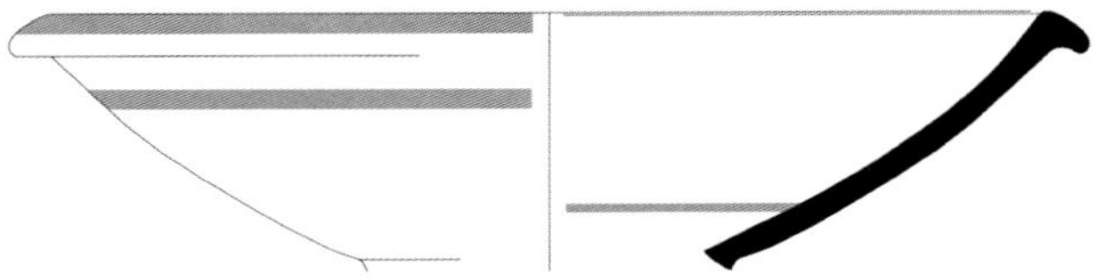

Pl/Lne4a

20. PWBW 309-07

Dimensions: r 16.1 h 4.0 w 7.8

Description: Outward-turned, hanging rim with triangular section and a hemispherical body wall.

Comparisons: Marchegiani 1997, Pl. 56, No. 7 (Pomarico, BW *piatto profondo*, diam. 22 cm, second half of 3rd c. BC); Lissi Caronna 1983, 249, Fig. 37, Nos. 15–16, and 317, Fig. 108, No. 2 (Oppido Lucano, rim diam. 13.0 cm and 16.0 cm, from T61 dated to second half of 5th c. BC; diam. 20.0 cm, from a kiln context); du Plat Taylor 1983, 340, Fig. 117, No. 220 (Cozzo Presepe, Site E, Phase III); Small 1977b, Fig. 22, No. 74 (Monte Irsi, BW "bowl," phase BV-AVI, 400 BC, from topsoil); the type is also known in BG: cf. Prag 1992b, Fig. 28, No. 636 (Gravina, ca. 300 BC); Morel 1981a, 1131b1 (Volterra, 200 BC?).

1:2

Pl/Lne4b1

21. PWBW 258-20

Dimensions: r 25.2 h 6.5 w 6.4

1:2

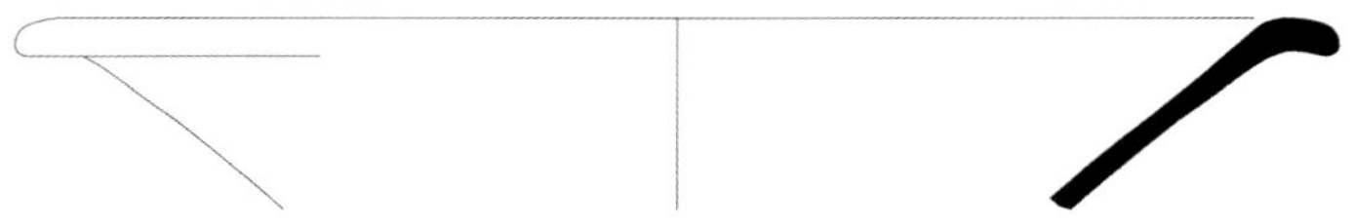

22. PWBW 358-05 rev90

Dimensions: r 18.0 h 2.7 w 6.8

Description: Horizontal, outward-turned rim with elliptical section; the body wall is deep and straight.

Comparisons: Marchegiani 1997, Pl. 56, No. 1 (Pomarico, BW *piatto profondo*, occasionally with perforated rim possibly for hanging when not in use, diam. 20.1 cm, last quarter of 4th c.–first half of 3rd c. BC); Morel 1981a, 1315c1 (BG, Etruria, 200±50 BC).

1:2

Pl/Lne4b2

23. PWBW 358-34 rev90

Dimensions: r 18.8 h 1.5 w 4.3

Description: A horizontal, outward-turned rim with elliptical section that is slightly indented; the body wall is deep and straight.

Comparisons: Dyson 1983, Fig. 259, No. 20 (Buccino, BG, diam. 17 cm). Cf. Morel 1981a, 1315a1 (BG, Etruria, 200±50 BC). Also BG 113-270, dated end of 4th–beginning of 3rd BC.

Plate/Lekane (Pl/Lne) cont.

1:2

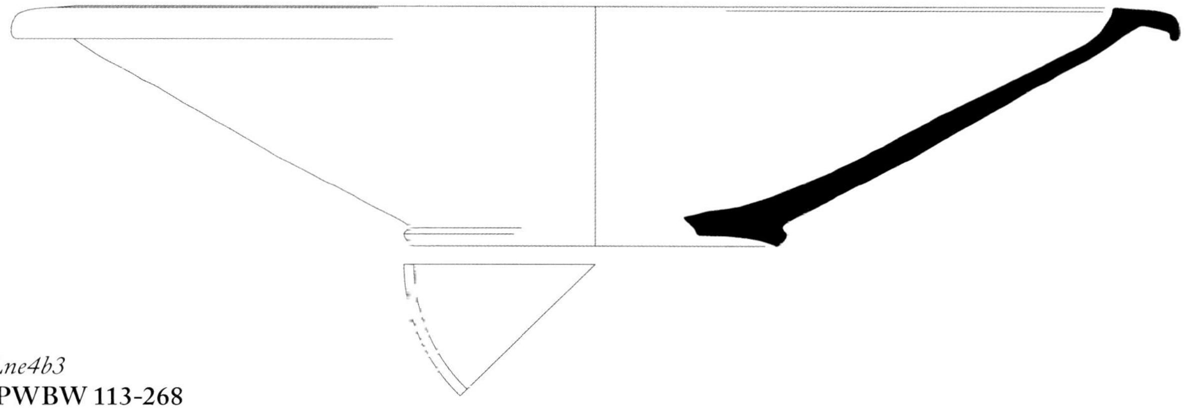

Pl/Lne4b3

24. PWBW 113-268

Dimensions: r 31.0 base 10.1 h 6.6 w 16.8

Description: Horizontal, outward-turned rim with elliptical section that is barely hanging; the body wall is deep and straight.

Comparisons: Prag 1992c, Fig. 40, No. 982 (Gravina, GW, shallower body, 2nd c. BC).

1:2

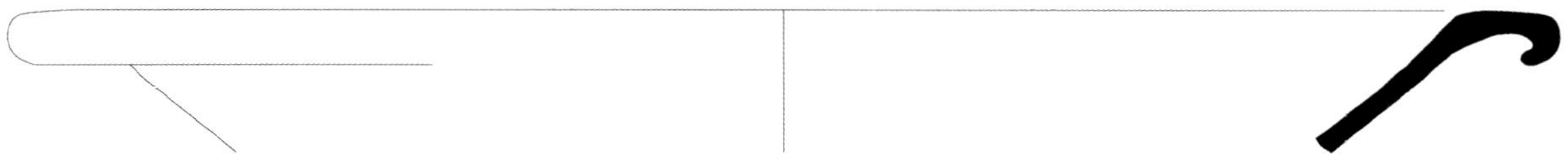

Pl/Lne4b4

25. PWBW 167-11

Dimensions: r 31.3 h 3.0 w 5.4

Description: Horizontal, outward-turned rim with elliptical section that is hooked; the body wall is deep and straight.

Comparisons: See Pl/Lne 4b3.

1:3

Pl/Lne4c

26. PWBW 521-187

Dimensions: r 35.5 h 2.7 w 8.2

Description: Wide horizontal, outward-turned rim with elliptical section and a hemispherical body wall.

Comparisons: Cotton 1983, 367, Fig. 136, No. 378 (Cozzo Presepe, Site B, phase VI, Site E, phase IVa, 3rd c. BC); du Plat Taylor 1983, 340, Fig. 117, No. 219 (Cozzo Presepe, Site E, Phase III); *Agora* XII, Fig. 15, No. 1806 (Lekane, 430–420 BC?); Deodato and Bianco 1997, Pl. 70, No. 95 (Pomarico, PW *bacile*, diam. 23.6 cm; 1.178: the diameter varies from 22 to 52 cm), and 2.82, Pl. 74, No. 118 (PW *ciotola*, diam. 23.6 cm; 1.180: the diameter varies from 17 to 24 cm); Marchegiani 1997, Pl. 61, No. 39 (Pomarico, BW *bacile*, diam. 28.2 cm, 4th–3rd c. BC); Cotton 1992b, Fig. 57, No. 1207 (Gravina, period VIIIb, unstratified; cf. *Gravina* III.2, 128, Fig. 52, No. 289); Costamagna 1983, Pl. XLVIII B5 (Taranto, HFR, diam. 40 cm, end of 3rd–first half of the 2nd c. BC); Jones/Sackett/Graham 1962, 94, Fig. 8, No. 47 (Dema, BW, diam. 40 cm, last quarter of 5th c. BC); Vittoria 2000, Pl. 66, No. 6.

1:2

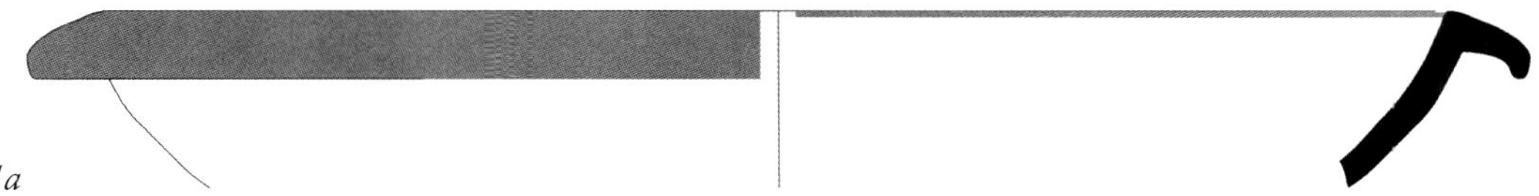

Pl/Lne4d1a

27. PWBW 363-09

Dimensions: r 27.2 h 3.8 w 8.1

Description: Outward-turned rim inclined towards the exterior with elliptical section.

Comparisons: See Pl/Lne4d1b.

Plate/Lekane (Pl/Lne) cont.

1:2

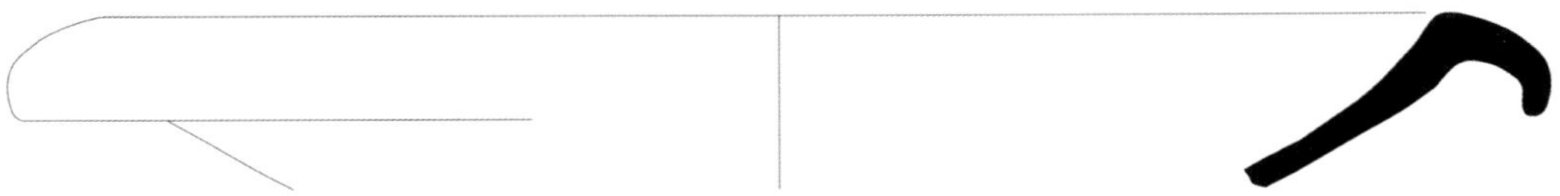

Pl/Lne4d1b

28. PWBW 048-03 rev90

Dimensions: r 27.3 h 3.2 w 11.3

Description: On this variant the hanging termination is more accentuated.

Comparisons: Lissi Caronna 1983, 234, Fig. 19, No. 10 (Oppido Lucano, T57 dated end of 5th–beginning of 4th c. BC, diam. 26.0 cm).

1:2

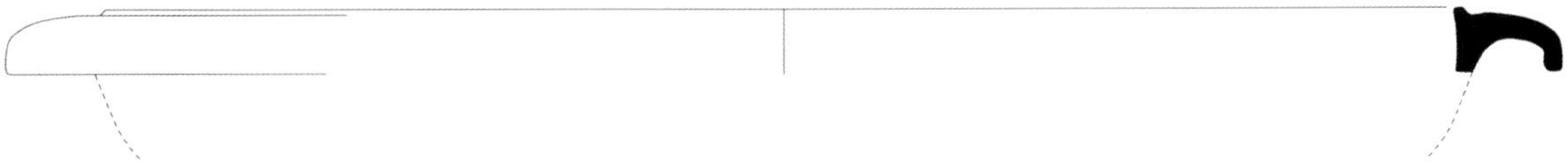

Pl/Lne4d1c

29. PWBW 326-37

Dimensions: r 30.0 h 1.3 w 5.9

Description: On this variant there is a groove near the body wall, which is generally hemispherical.

Comparisons: Cotton 1992b, Fig. 56, No. 1199 (Gravina, period VIIIa–b, end of 2nd–1st c. BC); Saunders and du Plat Taylor 1992, Fig. 6, No. 151 (Gravina, period VI, 4th–3rd c. BC; II.20: type 4, 4th c. BC); Preacco 1989, Pl. XXVIII, Nos. 221–22 (Locri, BG, 3rd c. BC); *Kaulonia* I, 61, Fig. 40, No. 167 (BG); *Sibari* IV, 502, No. 225, and 524, Fig. 485, No. 9160 (Gnathia); Preacco Ancona 1999a, 274, Fig. 289, No. 435 (Oppido Mamertina, BG plate, second half of 3rd c. BC; 273: the shape of the rim recalls Western Greek examples, particularly from eastern Sicily). Cf. Morel 1981a, 1333, in particular 1333e1 (Syracuse, BG, second half of 3rd c. BC).

1:3

Pl/Lne4d2

30. PWBW 043-132 rev82

Dimensions: r 41.5 h 2.4 w 7.5

Description: This variant has a notably thickened rim and body wall, almost like a true lekane.

Comparisons: Cotton 1992b, Fig. 60, No. 1241 (Gravina, period VIIIb, unstratified).

Carinated Lekane (CLne)

The diameter varies from 20.0 to 44.0 cm, with examples that reach as much as 65.0 cm (cf. CLne5, site 268-01). Of the 92 measurable fragments, 60 have a diameter between 20.0 and 33.0 cm (another two very small fragments probably measure 26.0 and 30.0 cm); 22 larger fragments measure 34.0–44.0 cm (there are two doubtful fragments that measure approximately 40.0 and 44.0 cm). Five other fragments measure between 14.0 cm and 16.0 cm.

1:3

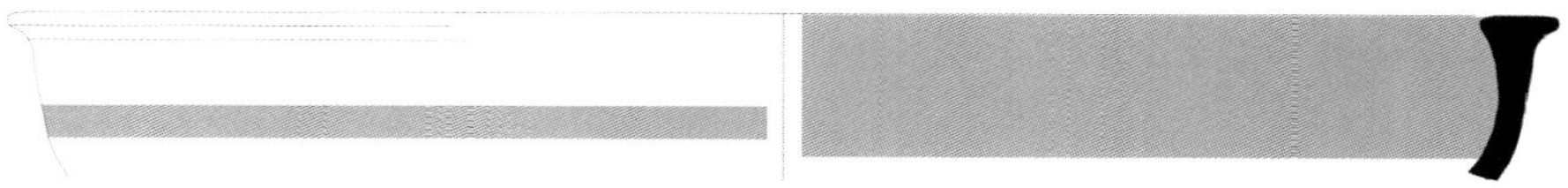

CLne1

31. PWBW 155-24

Dimensions: r 35.0 h 3.9 w 6.2

Description: Horizontal, outward-turned rim with elliptical section, slightly inward-projecting; nearly hemispherical body wall.

Comparisons: du Plat Taylor 1983, 338, Fig. 116, No. 207a (Cozzo Presepe, Site A, Phase IV, mid-6th c.–480 BC); cf. Lo Porto 1973, Pl. XLV, Fig. 2, No. 5; Saunders and du Plat Taylor 1992, Fig. 8, Nos. 186–187 (Gravina, bowl type 2, period VI, 4th–3rd c. BC; II.22: this is the most common type of bowl at Gravina; at Cozzo Presepe it appears beginning in mid-6th c. BC and continues through 4th c. BC); G. Greco 1991, 78, Fig. 193, (foot-cup from Serra di Vaglio, first half of 5th c. BC); Lissi Caronna 1990–1991, 234, Fig. 58, No. 7 (Oppido Lucano, *casa* C, *coppa in ceramica indigena*, diam. 18.0 cm); Lissi Caronna 1983, 342, Fig. 139, No. 6 (Oppido Lucano, *casa* B, *piatto in ceramica indigena*, diam. 20.5 cm); *Sibari* III, 157, Fig. 168, No. 22279 (diam. 12 cm, first half of 6th c. BC), and 160, Fig. 170, No. 44201 (BW *bacile*, diam. 30.6 cm).

1:3

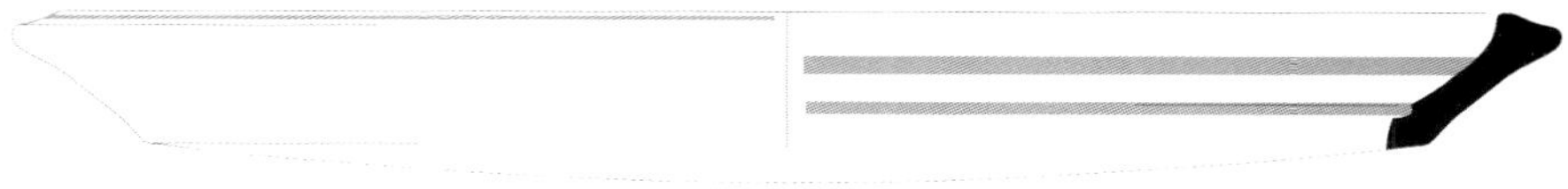

CLne2a–b

32. PWBW 365-24

Dimensions: r 34.8 h 3.2 w 5.3

1:2

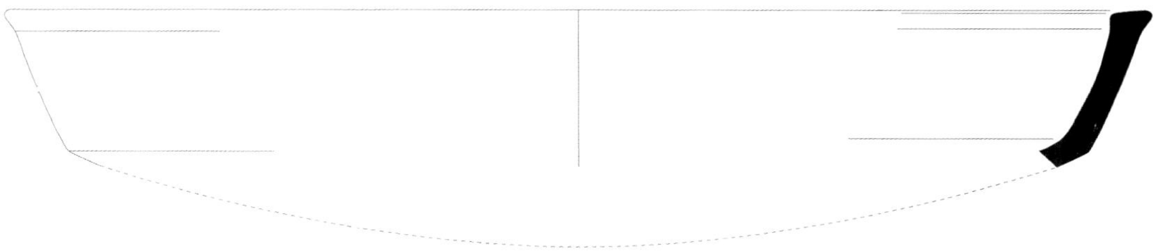

33. PWBW 014-04 rev90

Dimensions: r 28.0 h 3.9 w 6.9

Description: Nearly indistinct rim with sub-triangular section, oblique in CLne2a and nearly flat in CLne2b.

Comparisons: Deodato and Bianco 1997, Pl. 73, No. 115 (Pomarico, PW *ciotola*, diam. 21.2 cm); Cotton 1992b, Fig. 58, No. 1225 (Gravina, period VIIIa–b, end of 2nd–1st c. BC); *Kaulonia* I, 67, Fig. 45, No. 279; Lissi Caronna 1983, 252, Fig. 40, No. 5 (Oppido Lucano, T63, robbed, diam. 14.8 cm, without carination); Brizzi 1999b, 315, Fig. 315, No. 617 (Oppido Mamertina, Form IV, type 22, diam. 33 cm). At Pomarico another, similar type has a deep body wall (cf. Deodato and Bianco 1997, Pl. 71, No. 96, diam. 29.2 cm, belonging to the local tradition, irregular workmanship). The same type is known in BG and with a strongly carinated body wall in southern Italy between the second half of the 2nd c. and the 1st c. BC: cf. Morel 1981a, 2654.

Carinated Lekane (CLne) cont.

1:3

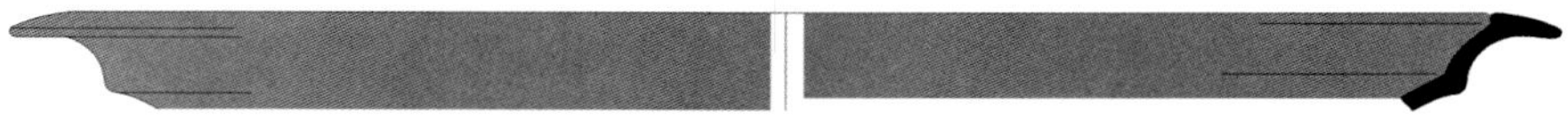

CLne3a–b

34. PWBW 081-07

Dimensions: r 34.7 h 2.3 w 5.7

1:2

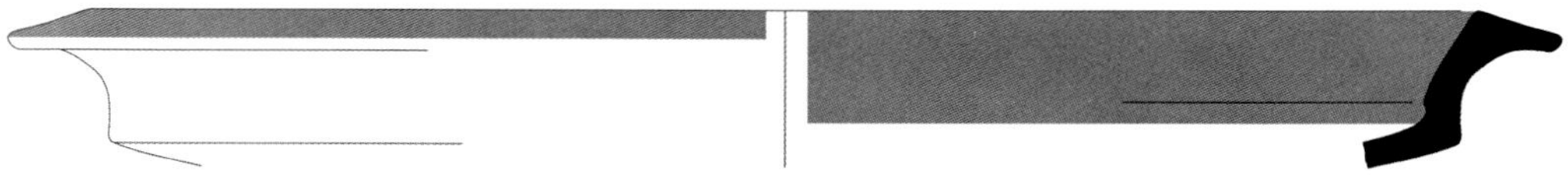

35. PWBW 509-06

Dimensions: r 26.8 h 2.8 w 8.4

Description: Oblique, outward-turned rim; upper body wall with concave profile; carination emphasized on the interior by a groove; CLne3a has thinner walls.

Comparisons: CLne3a: Giorgi 1988a, Pl. 31, C.B. 2.2 (tomb dated last quarter of 5th c. BC). CLne3b: Carter and Toxey 1998, 709, T343-1 (Pantanello, 420–360 BC); D'Andria 1980a, 410, Fig. 56, No. 228 (Metaponto, *scarico* 1); Marchegiani 1997, Pl. 61, No. 41 (Pomarico, BW *bacile*, diam. 39.2 cm, second half of the 3rd c. BC; 1.168: similar *bacili*, also in PW are widespread throughout southern Italy in 4th–3rd-c. contexts); Saunders and du Plat Taylor 1992, Fig. 8, No. 183 (Gravina, bowl type 1D, period VI, 4th–3rd c. BC); Conti 1989c, Pl. XLI, No. 361 (Locri, diam. 31.0 cm, 4th–3rd c. BC).

1:3

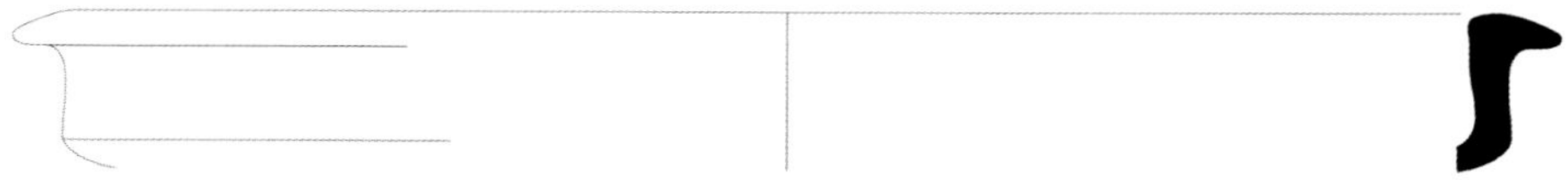

CLne4

36. PWBW 523-58

Dimensions: r 36.0 h 3.8 w 8.4

Description: Horizontal, slightly convex, outward-turned rim with sub-triangular section; the body wall is initially vertical, then decidedly oblique.

Comparisons: Sibari II, 333, Fig. 359, No. 23416 (BW *bacile*), and 357, Fig. 380, No. 25533 (BW *baciletto*); *Minturnae,* Pl. X, type 3 (Black-on-buff shallow bowl, from deposit 1 dated 295–245 BC); Semeraro 1983, Pl. 86, No. 133 (Otranto, BW, from *cantiere* 2, 6th–3rd c. BC); Tigano 2000, 142 S/92 (Zancle-Messina, second half of 6th c. BC, diam. 27 cm), and 143 S/100 (diam. 21 cm); *Kaulonia* I, 67, Fig. 45, No. 274; Naumann and Neutsch 1960, Pl. 13.1.4.11 (Palinuro). For examples with a high foot, cf. Lo Porto 1992, 90, No. 8 (Canosa, diam. 10.3 cm, tomb of middle–second half of 6th c. BC).

1:2

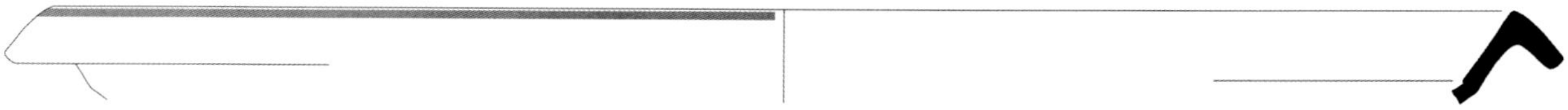

CLne5

37. PWBW 268-01

Dimensions: r 65.0 h 4.0 w 9.3

Description: Wide, hanging, outward-turned rim with rectangular section; the body wall is initially oblique, followed by a barely perceptible carination.

Comparisons: See CLne6b, on which the rim is less accentuated than the drooping rim of the CLne5.

Carinated Lekane (CLne) cont.

1:3

CLne6a

38. PWBW 178-01

Dimensions: r 40.0 h 5.7 w 10.3

Description: Outward-turned rim with horizontal rectangular section; the body wall is initially vertical, then decidedly oblique.

Comparisons: Carter and Toxey 1998, 709, T15-4 (Pantanello, with a more pronounced rim, 420–360 BC; 709: this is one of the most common forms in the chora); Cotton 1983, 367, Fig. 136, No. 383 (Cozzo Presepe, Site B, phase V, 4th–3rd c. BC); du Plat Taylor 1983, 338, Fig. 116, No. 209 (Cozzo Presepe, Site A, topsoil); Lissi Caronna 1983, 234, Fig. 19, No. 3 (Oppido Lucano, T56, *piatto in ceramica indigena*, diam. 12.0 cm); *Gravina* III.2, 122, Fig. 49, No. 261 (BW "stemmed dish"); Small 1977b, Fig. 22, No. 71 (Monte Irsi, BW "bowl," phase AVI, topsoil; 115: the form is present in 5th c. BC, but the closest comparanda belong to 4th c.); Bailo Modesti 1980, Pl. 95, Nos. 65–66 (Cairano, *tazza biansata*, 5th-c. BC tomb, diam. 15.2–16.4 cm); Dyson 1976, Fig. 31 PD27 (Cosa, from deposit 5 dated to 1st c. BC, corresponding to Taylor's deposit E dated 110/100–40/30 BC, diam. 22 cm); *Meligunìs-Lipára* II, Pl. LVIII, No. 11 (T190, 470–400 BC).

1:3

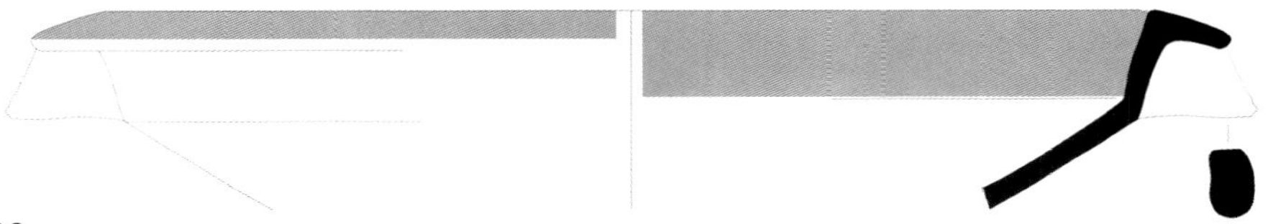

CLne6b

39. PWBW 062-02

Dimensions: r 33.2 h 5.7 w 24.9

Description: Outward-turned rim with wider rectangular section; the body wall is initially vertical, then decidedly oblique.

Comparisons: *Sibari* I, 38, Fig. 34, No. 112; *Meligunìs-Lipára* XI.I, Pl. CLXXVIII, No. 1c (BW *bacile di tipo ionico*, 5th-c. BC tomb).

1:2

CLne6c

40. PWBW 353-16

Dimensions: r 34.0 h 3.3. w 3.6

Description: Outward-turned rim with rectangular section, slightly inward-projecting; the body wall is initially vertical, then decidedly oblique.

Comparisons: Saunders and du Plat Taylor 1992, Fig. 7, No. 172 (Gravina; II.21: bowl type 1, common in 5th and beginning of 4th c. BC and probably continuing even into the 3rd c.; this example comes from period VIII levels, dated end of the 2nd–1st c. BC); Giorgi 1988a, Pl. 31, C.B. 2.1 (tomb of the second half of 5th c. BC).

1:3

CLne7

41. PWBW 395-37

Dimensions: r 38.8 h 4.5 w 12.7

Description: Outward-turned rim, oriented towards the exterior, with rectangular section; initially vertical, carinated body wall.

Comparisons: Carter and Toxey 1998, 709 T7-2 (Pantanello, 360–325 BC); du Plat Taylor 1983, 338, Fig. 116, No. 208 (Cozzo Presepe, Site A, Phase IIIb, 575–550 BC).

Lekane (Lne)

Of the 317 measurable rims, 283 (as well as 14 doubtful fragments) have a diameter between 30.0 and 58.0 cm; only 20 fragments fall in the range of 17.0–28.5 cm. A small fragment of uncertain attribution might have been used as a lid (Site FS82.169.19, diam. 8.0 cm). The foot is generally a concave disk, less commonly a ring. Given the absence of complete forms, plate/lekane **No. 24 (PWBW 113-268)** was selected as the model because its concave disk foot is identical to the foot commonly encountered on the lekane. Diameters could be calculated for 200 fragments: the ranges are 8.0–10.0 cm for the 23 smallest ones and 10.2–22.0 cm for the 177 remaining fragments.

Lne1

42. PWBW 353-15

Dimensions: r 48.7 h 2.9 w 7.3

Description: Horizontal, outward-turned rim with subtriangular section, vertical exterior face and flat upper face, decorated at the tips with two parallel grooves; deep body wall.

Comparisons: Kaulonia I, 62, Fig. 41, No. 205, with grooves (*plats profonds à marli horizontal*; 63: common in Classical and Hellenistic levels, PW, diam. 30–40 cm without rim; the examples with contracted rim perhaps derive from Late Archaic prototypes, with decoration of parallel lines on the rim).

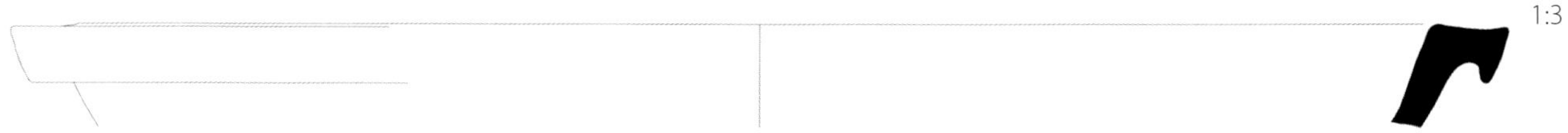

Lne2a–b

43. PWBW 059-01

Dimensions: r 48.0 h 3.4 w 5.7

44. PWBW 182-08

Dimensions: r 41.4 h 5.2 w 10. 4

Description: Horizontal, outward-turned rim with nearly vertical exterior face and upper face inclined towards the exterior, occasionally decorated near the edges with two parallel grooves (Lne2b); deep body wall.

Comparisons: L. Giardino 1996a, 41 (Herakleia); Cotton 1992b, Fig. 59, No. 1230 (Gravina, period VI, 4th–3rd c. BC); Conti 1989c, Pl. XL, No. 358 (Locri, diam. 32 cm, 4th–3rd c. BC); *Kaulonia* I, 62, Fig. 41, No. 208 (*plats profonds à marli horizontal*).

Lne3

45. PWBW 449-10

Dimensions: r 45.7 h 4.1 w 6.3

Description: Narrow, outward-turned rim inclined towards the exterior, with raised margins and central groove; deep body wall.

Comparisons: See Lne4a, which has a similar rim, albeit wider and with hanging tip.

Lekane (Lne) cont.

1:4

Lne4a

46. PWBW 359-01

Dimensions: r 58.0 h 5.9 w 8.3

Description: Horizontal, outward-turned rim with vertical exterior face and slightly concave upper face with central groove; deep body wall.

Comparisons: Marchegiani 1997, Pl. 61, No. 40 (Pomarico, BW *bacile*, diam. 35.6 cm, second half of 3rd c. BC; 1.168: similar *bacili* are widespread throughout southern Italy in 4th–3rd c. BC contexts, also in PW); Vittoria 2000, Pl. 66, No. 5.

1:3

Lne4b

47. PWBW 341-44

Dimensions: r 39.4 3.5 w 7.2

Description: Horizontal, outward-turned rim with vertical exterior face and flat upper face; deep body wall.

Comparisons: *Museo Taranto* III.2 214, Fig. 50, No. 12 (tomb of first half of 1st c. AD, diam. 33.0 cm); Cotton 1992b, Fig. 59, No. 1231 (Gravina, period VIIIa, end of 2nd–1st c. BC), and II.315, Fig. 57, No. 1206 (similar, period VIIIb, unstratified); *Gli ori*, 512, No. 19 (*bacino*, tomb of first half of 1st c. BC); Conti 1989c, Pl. XL, No. 357 (Locri, diam. 40.2 cm, 4th–3rd c. BC); *Sibari* II, 180, Fig. 198, No. 10942.

1:3

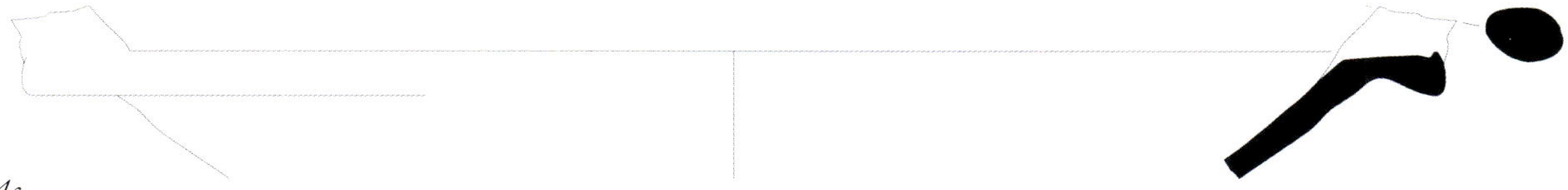

Lne4c

48. PWBW 043-123 rev82

Dimensions: r 43.5 h 4.0 w 10.2

Description: Horizontal, outward-turned rim with vertical exterior face and flat upper face and raised edge at the tip; deep body wall.

Comparisons: *Kaulonia* I, 62, Fig. 41, No. 202 (*plats profonds à marli horizontal*); Hänsel 1973, 451, Fig. 31, No. 3 (Herakleia, from a layer dateable to last third of 5th c. BC).

1:3

Lne5

49. PWBW 488-14

Dimensions: r 43.3 h 4.3 w 12.6

Description: Horizontal, outward-turned rim with vertical exterior face and flat upper face decorated with parallel grooves; deep body wall.

Comparisons: Cotton 1983, 367, Fig. 136, No. 382 (Cozzo Presepe, Site B, phase II, last quarter of 4th c. BC); *Sibari* II, 333, Fig. 359, No. 24103, 354, Fig. 378, No. 27065, and 360, Fig. 384, No. 25831 (diam. 38 cm); *Valesio* I, 264, Form M55, Nos. 464–465 (diam. 48 and 47 cm, context of end of the 3rd–beginning of 2nd/1st c. BC); Cotton 1992b, Fig. 56, Nos. 1196–1197 (Gravina, period VIIIa, end of 2nd–1st c. BC), and Fig. 59, No. 1232 (period VIIIb, unstratified); Fornaro 1988, 207, No. 487, (Bari, deposit of ca. 300 BC, diam. 26.5 cm); Pancrazzi 1979, Fig. 64, No. 1 (Cavallino); D'Andria 1978a, Pl. 27, No. 35 (Leuca, 5th–4th c. BC); Mastronuzzi 1995, 187, Fig. 4, No. 5, and 188, Fig. 5, No. 4 (Nardò, PW); *Agora* XII, Fig. 15, No. 1818 ("rim roughly grooved on top," 375–330 BC). Cf. also Lissi Caronna 1980, 209, Fig. 120, No. 1 (Oppido Lucano, diam. 19.4 cm).

Lekane (Lne) cont.

1:3

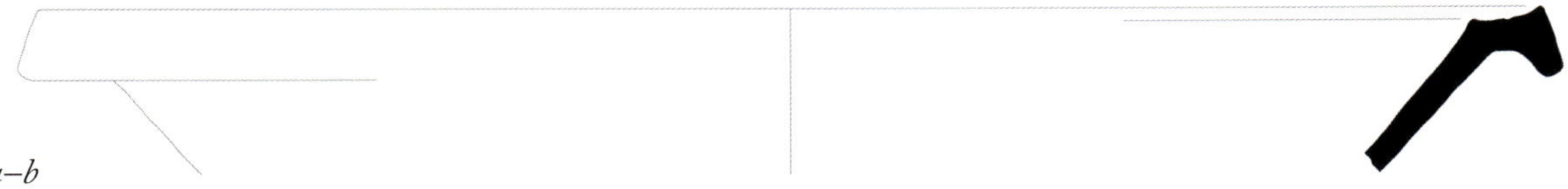

Lne6a–b

50. PWBW 398-01

Dimensions: r 45.0 h 5.0 w 6.8

1:3

51. PWBW 014-02 rev90

Dimensions: r 40.0 h 9.3 w 14.7

Description: Horizontal, outward-turned rim with oblique exterior face and upper face inclined towards the interior, generally with central groove; Lne6a has a sub-rectangular section, while Lne6b has a sub-triangular section; deep body wall.

Comparisons: Carter 1983a, 478, Fig. 53, No. 163 (Metaponto, diam. 24 cm); Hänsel 1973, 451, Fig. 31, No. 2 (Herakleia, from a stratum dateable to last third of 5th c. BC).

1:4

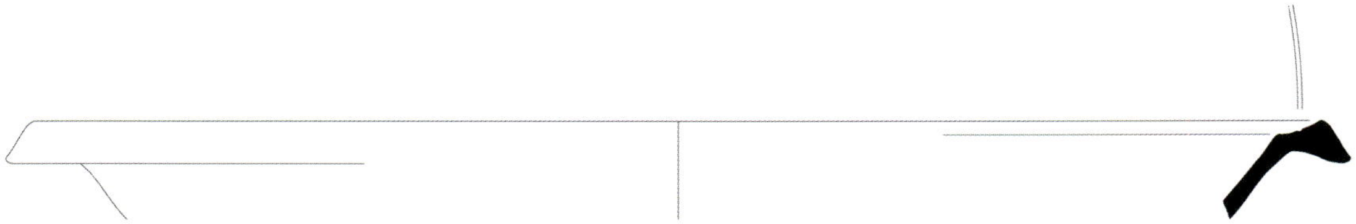

Lne7

52. PWBW 332-01

Dimensions: r 57.7 h 4.4 w 19.5

Description: Horizontal, outward-turned rim with oblique exterior face, slightly concave upper face inclined towards the interior and decorated with a groove at the center, and hanging tip with triangular section; deep body wall.

Comparisons: See Lne6b, which shares some features of the rim.

1:3

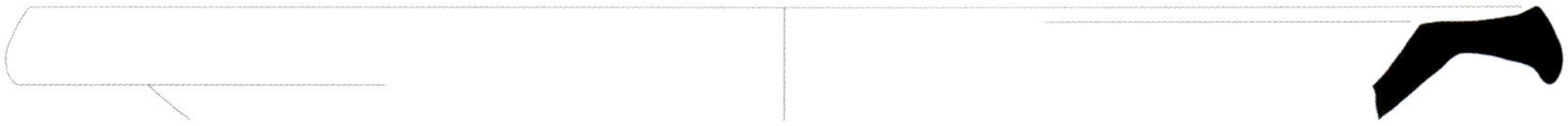

Lne8a

53. PWBW 258-19

Dimensions: r 42.5 h 3.2 w 7.3

Description: Horizontal, outward-turned rim with convex exterior face, concave upper and slightly hanging tip with elliptical section; deep body wall.

Comparisons: D'Andria 1980b, Pl. XIII, No. 34 (Metaponto, *scarico* 12, end of 5th–4th c. BC); Carter 1983a, 478, Fig. 53, No. 90 (Metaponto, diam. 38 cm); Cotton 1992b, Fig. 60, No. 1244 (Gravina, period VIIIa–b, end of 2nd–1st c. BC); *Sibari* III, 249, Fig. 255, No. 2477 (diam. 40 cm); *Kaulonia* I, 62, Fig. 41, No. 214 (*plats profonds à marli horizontal*).

Lekane (Lne) cont.

1:3

Lne8b

54. PWBW 337-01

Dimensions: r 45.3 h 4.3 w 13.0

Description: Horizontal, outward-turned rim with convex exterior face, concave upper face, strongly inclined towards the interior, and slightly hanging tip with elliptical section; deep body wall.

Comparisons: Deodato and Bianco 1997, Pl. 70, No. 92 (Pomarico, diam. 52 cm; 1.178: the diameter varies between 22 and 52 cm); Conti 1989c, Pl. XL, Nos. 354–55 (Locri, diam. 42.2 cm and 43.6 cm, end of 4th–3rd c. BC); *Kaulonia* I, 62, Fig. 41, No. 212 (*plats profonds à marli horizontal*); *Laos* I, Pl. 20, No. 23 (end of 4th–beginning of 3rd c. BC).

1:3

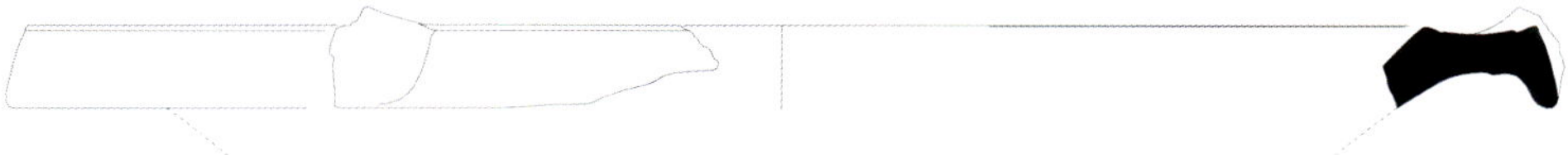

Lne9a

55. PWBW 365-03

Dimensions: r 37.3 h 2.1 w 9.3

Description: Horizontal, outward-turned rim with oblique exterior face, slightly concave upper face and decorated near the edges with two parallel grooves, and hanging tip with elliptical section; deep body wall.

Comparisons: Kaulonia I, 62, Fig. 41, No. 217 (*plats profonds à marli horizontal*).

1:3

Lne9b

56. PWBW 523-96

Dimensions: r 42.7 h 4.6 w 10.6

Description: Horizontal, outward-turned rim with oblique exterior face, slightly concave upper face inclined towards the interior; decorated near the edges with two parallel grooves, and hanging tip with elliptical section; deep body wall.

Comparions: Carter 1983a, 478, Fig. 53, No. 88 (Metaponto, diam. 42 cm); Deodato and Bianco 1997, Pl. 69, No. 89 (Pomarico, diam. 42.2 cm; 1.178: the diameter varies between 22 and 52 cm).

1:3

Lne10

57. PWBW 365-01

Dimensions: r 48.0 h 5.5 w 10.6

Description: Oblique, outward-turned rim with oblique exterior face, concave upper face, and hanging tip with elliptical section; deep body wall.

Comparisons: Vittoria 2000, Pl. 66, No. 2.

1:3

Lne11

58. PWBW 365-02

Dimensions: r 49.0 h 4.1 w 10.3

Lekane (Lne) cont.

1:4

Lne11 cont.

59. PWBW 394-02 rev90

Dimensions: r 56.0 h 4.8 w 16.0

1:3

60. PWBW 014-03 rev90

Dimensions: r 44.0 h 4.0 w 9.2

Description: Oblique, outward-turned rim with oblique exterior face, concave upper face with decoration of two parallel grooves near the edges, and hanging tip with elliptical section; deep body wall.

Comparisons: Toxey and Carter 1998, 722, T353 (Pantanello, 460–440 BC; similar vessels are very common in the Metapontino, particularly at the Pantanello sanctuary).

1:3

Lne12

61. PWBW 309-01

Dimensions: r 44.0 h 4.3 w 11.1

Description: Horizontal, outward-turned rim with vertical exterior face, slightly concave upper face that projects inwards, and hanging tip with elliptical section; deep body wall.

Comparisons: Deodato and Bianco 1997, Pl. 70, No. 93 (Pomarico, diam. 48.6 cm; 1.178: the diameter varies between 22 and 52 cm); Cotton 1992b, Fig. 57, No. 1205 (Gravina, period VIIIb, unstratified); Vittoria 2000, Pl. 66, No. 1.

1:3

Lne13a

62. PWBW 266-48

Dimensions: r 40.3 h 2.7 w 8.4

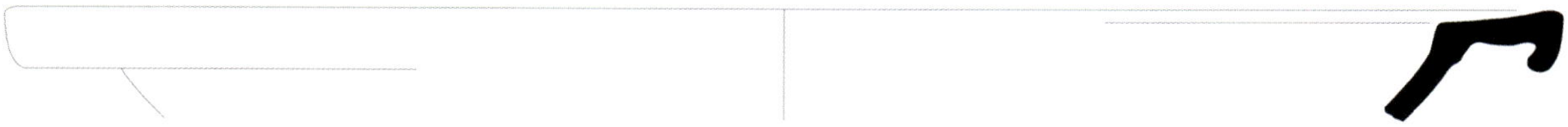

1:3

Lne13b

63. PWBW 506-89

Dimensions: r 47.9 h 3.5 w 9.5

Description: Horizontal, outward-turned rim with vertical exterior face, slightly concave upper face inclined towards the interior, and drooping tip with elliptical section; deep body wall. On Lne13b tip of rim is hook-shaped.

Comparisons: Deodato and Bianco 1997, Pl. 69, No. 90, and Pl. 70, No. 91 (Pomarico, diam. 42.2 and 48 cm; 1.178: diameter varies from 22–52 cm); Cotton 1983, 367, Fig. 136, No. 379 (Cozzo Presepe, site B, phase V, site E, phase IVb, 4th–3rd c. BC); Prag 1983a, 360, Fig. 131, No. 352 (Cozzo Presepe, site E, phase III, BW lekane in grey-faced ware); Vittoria 2000, Pl. 66, No. 3.

Lekane/Basin (Lne/Bas)

While the plate/lekane is characterized by a generally modest diameter and shallow body wall, the lekane/basin has a relatively wide rim diameter of 26.0–43.9 cm and a deep body wall. An example with larger rim belongs to Lne/Bas1b1 (site 390, diam. 48.0 cm). Two fragments of Lne/Bas1a1 with diameters of 18.0 cm (site 509) and 22.0 cm (site 398) were also identified, as well as one fragment of Lne/Bas1a2 with diameter of 20.0 cm (site 411).

1:3

Lne/Bas1a1

64. PWBW 254-04

Dimensions: r 35.0 h 4.6 w 7.3

Description: Outward-turned, slightly oblique rim with elliptical section; hemispherical body wall. This variant has a thinner rim than Lne/Bas1a2.

Comparisons: Lissi Caronna 1983, 317, Fig. 108, No. 9 (Oppido Lucano, diam. 23.0 cm).

1:3

Lne/Bas1a2

65. PWBW 524-06

Dimensions: r 37.5 h 3.9 w 9.0

Description: Outward-turned, slightly oblique rim with elliptical section; hemispherical body wall. This variant has a thicker rim than Lne/Bas1a1.

Comparisons: Lissi Caronna 1983, 328, Fig. 123, No. 3 (Oppido Lucano, *casa* B, diam. 24 cm); Lissi Caronna 1990–1991, 292, Fig. 117, No. 5 (Oppido Lucano, diam. 27.5 cm).

1:3

Lne/Bas1b1

66. PWBW 182-19

Dimensions: r 32.5 h 4.1 w 11.1

Description: Outward-turned, slightly oblique rim with elliptical section; hemispherical body wall; rim is hanging.

Comparisons: Deodato and Bianco 1997, Pl. 67, No. 75 (Pomarico, diam. 36.2 cm, 4th c. BC; 1.177: at Pomarico this is a type of light mortar with diameter between 30 and 36 cm); Hempel and Mattioli 1994, 363, Fig. 309 (Taranto, T3, phase D, first quarter of 2nd c. BC); Saunders and du Plat Taylor 1992, Fig. 7, No. 170 (Gravina; II.21: bowl type 1, common in 5th–beginning of 4th c. BC and probably continuing into 3rd c.; this example comes from a period VIII level, end of 2nd–1st c. BC); Cotton 1983, 367, Fig. 136, No. 376 (Cozzo Presepe, Site A, phase IV-V, mid-6th–4th c. BC); *Valesio* I, 260, Form M54, No. 458 (diam. 27 cm, context of end of 3rd–first half of 2nd c. BC); *Corinth* VII.3, Pl. 21, No. 616 ("krater," diam. 42 cm, deep, ca. 300 BC); *Agora* XII, Fig. 15, No. 1820 (Lekane, 320–290 BC); *Sibari* II, 190, Fig. 206, No. 8491 (*bacile* in micaceous yellow-green clay); Vittoria 2000, Pl. 66, No. 7.

Lne/Bas1b2

1:3

67. PWBW 390-10

Dimensions: r 34.2 h 2.6 w 10.6

Description: Outward-turned, slightly oblique rim with elliptical section; hemispherical body wall; rim is hooked.

Comparisons: Cotton 1983, 367, Fig. 136, No. 378 (Cozzo Presepe, Site B, phase VI, Site E, phase IVa, 3rd c. BC); du Plat Taylor 1983, 340, Fig. 117, No. 219 (Cozzo Presepe, Site E, Phase III); Deodato and Bianco 1997, Pl. 70, No. 95 (Pomarico, PW *bacile*, diam. 23.6 cm; 1.178: the diameter varies between 22 and 52 cm), and 2.82, Pl. 74, No. 118 (PW *ciotola*, diam. 23.6 cm; 1.180: diameter varies between 17 and 24 cm); Marchegiani 1997, 2.69, Pl. 61, No. 39 (Pomarico, BW *bacile*, diam. 28.2 cm, 4th–3rd c. BC); Cotton 1992b, Fig. 57, No. 1207 (Gravina, period VIIIb, unstratified; cf. *Gravina* III.2 128 Fig. 52 No. 289); Costamagna 1983, Pl. XLVIII, B5 (Taranto, HFR, diam. 40 cm, end of 3rd–first half of 2nd c. BC); *Agora* XII, Fig. 15, No. 1806 (lekane, 430–420 BC?); Jones/Sackett/Graham 1962, 94, Fig. 8, No. 47 (Dema, BW, diam. 40 cm, last quarter of 5th c. BC); Vittoria 2000 Pl. 66 No. 6.

Lekane/Basin (Lne/Bas) cont.

1:3

Lne/Bas2
68. PWBW 182-11
Dimensions: r 27.9 h 5.0 w 10.1

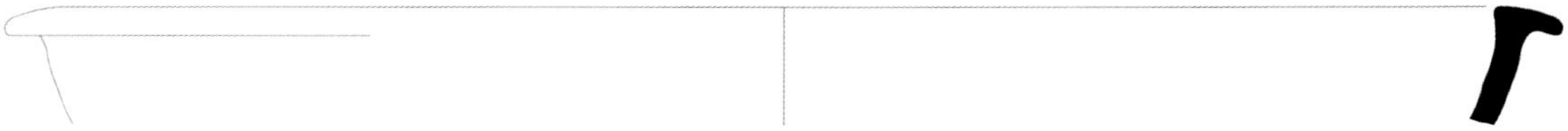

1:3

69. PWBW 182-12
Dimensions: r 43.9 h 3.4 w 9.4

1:3

70. PWBW 014-01 rev90
Dimensions: r 28.5 h 7.6 w 7.8
Description: Horizontal, outward-turned rim with elliptical section and slightly convex upper face; deep, nearly straight body wall.
Comparisons: Cotton 1983, 367, Fig. 136, No. 377 (Cozzo Presepe, Site A, Topsoil); Marchegiani 1997, Pl. 56, No. 3, and 2.67, Pl. 59, No. 31 (Pomarico, BW *piatto profondo*, second half of 3rd c. BC, diam. 19.1 and 17.5 cm); Cotton 1992b, Fig. 57, Nos. 1208 (Gravina; oblique rim and deep, straight body wall; period VI, 4th–3rd c. BC, and period VIIIa, end of 2nd–1st c. BC), and 1214 (period VII-VIIIa, 3rd–1st c. BC); Saunders and du Plat Taylor 1992, Fig. 8, No. 179 (Gravina, 4th c. BC; cf. also *Gravina* III.2, 122, Fig. 49, No. 257); *Valesio* I, 70, Form C12, No. 83 ("small bowl," diam. 15 cm, context and suggested dating at end of 4th–beginning of 3rd c. BC), and 72, Form C21c, No. 92 ("bowl," diam. 29.5 cm, end of 6th–5th c. BC); Pancrazzi 1979, Fig. 62, No. 2 (*vaso troncoconico*); D'Andria 1978a, Pl. 30, No. 154 (Leuca, PW *coppa*); *Corinth* VII.3, Pl. 21, No. 616 (krater, ca. 300 BC).

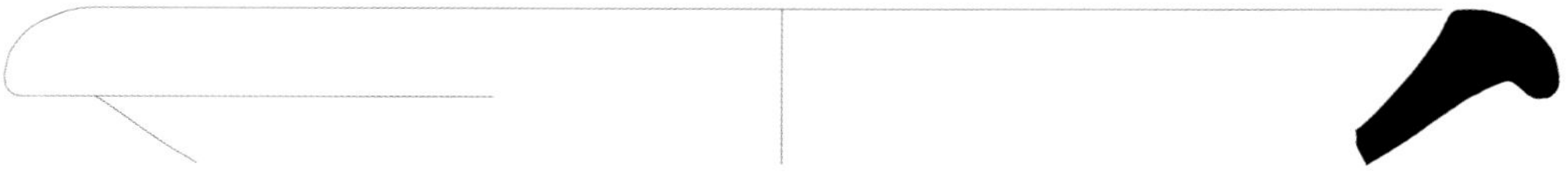

1:3

Lne/Bas3
71. PWBW 182-13
Dimensions: r 39.6 h 4.1 w 12.3

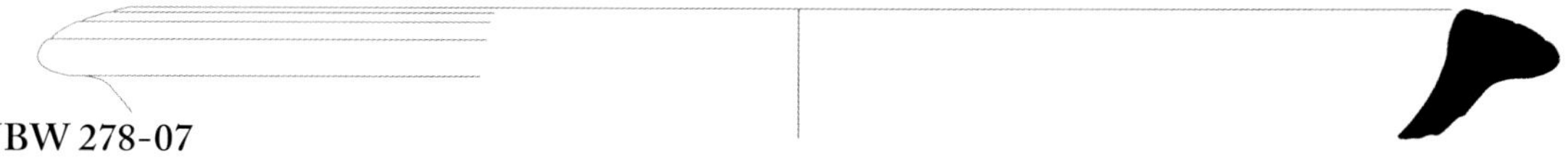

1:3

72. PWBW 278-07
Dimensions: r 41.8 h3.6 w 8.1
Description: Oblique, outward-turned rim, convex in Lne/Bas3a and flat in Lne/Bas3b, with elliptical section; nearly straight body wall.
Comparisons: Carter 1983a, 453, Fig. 28, No. 26 (Pantanello, "mortar," diam. 60 cm, kiln area); Deodato and Bianco 1997, Pl. 67, No. 74 (Pomarico, "mortar," diam. 26.1 cm, 4th–2nd c. BC); Marchegiani 1997, Pl. 56, No. 4 (Pomarico, BW *piatto profondo*, diam. 23.2 cm, second half of 3rd c. BC); *Kaulonia* I, 64, Fig. 42, No. 240; Semeraro 1983, Pl. 92, No. 236 (Otranto, PW lekane, shallower than Metapontine equivalent); *Agora* XII, Fig. 16, No. 1884 (Attic mortar, 565–550 BC; cf. Mor3 below); Jones/Graham/Sackett 1973, 385, Fig. 9, No. 76 (Vari, BW lekane, diam. 46.5 cm, 4th c. BC); cf. Morel 1981a, 1634c1 (BG, shallow, Attic production, 265–261 BC).

Flanged Lekane (FLne)

The examples recovered (12 fragments, of which two are of dubious attribution, all of which lack evidence of the handle attachment) generally have a rim diameter between 19.9 and 23.2 cm, while the fragment from site 68 has a diameter of 37 cm.

FLne1

73. PWBW 521-173

Dimensions: r 19.9 h 4.2 w 4.3

Description: Forked rim with resting groove for lid and elliptical section; convex body wall.

Comparisons: Marchegiani 1997, Pl. 60, No. 35 (Pomarico, BW *ciotola*, diam. 33 cm); Saunders and du Plat Taylor 1992, Fig. 10, No. 218 (Gravina, lekanis period VIIIb, unstratified; II.25: probably imitation of Attic lekanis types); *Kaulonia* I 68 Fig. 46, No. 285; *Agora* XII Fig. 11, No. 1213 (non-carinated body wall, 575 BC) and Fig. 11, No. 1220 (ca. 425 BC); *Corinth* VII.3, Pl. 18, No. 557 (3rd c. BC); Jones/Sackett/Graham (1962) 95, Fig. 9, No. 53 (Dema, BW, diam. 24 cm, last quarter of 5th c. BC).

1:3

74. PWBW 043-192

Dimensions: r 23.2 h 6.3 w 6.4

Description: Forked rim with resting groove for lid and elliptical section; convex body wall.

Comparisons: See No. 73.

1:3

FLne2

75. PWBW 169-16

Dimensions: r 22.1 h 2.5 w 7.0

Description: High, outward-turned rim with flat upper face; on the interior there is an oblique resting groove for the lid, with elliptical section. There is only one example of this type (site 169).

Comparisons: Marchegiani 1997, Pl. 60, No 34 (Pomarico, BW *vaso con coperchio*, diam. 15–25 cm, second half of the 3rd c. BC; 1.167: the forms were borrowed from Ck—cf. O. Bianco 1997b, Pl. 80, No. 169, and *Kaulonia* I, 84, Fig. 58, No. 410).

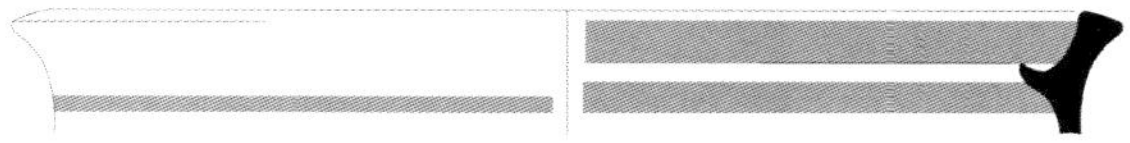

1:3

Chytra (Chy)

Because the two types are differentiated only by small rim details, the comparanda and bibliography are presented together. The size of the rim generally varies between 18.0 and 27.7 cm. The smallest example (12.0 cm, from Plot 2000–119) approaches in size the equivalent cooking-ware types, in which the diameter generally falls between 10.0–15.0 cm, although cooking-ware chytrai with smaller or larger rims (6.0–8.0 cm and ca. 18 cm) are also attested.

Chy1

76. PWBW 511-27

Dimensions: r 27.7 h 3.5 w 8.5

Description: Outward-turned rim with triangular section and slightly oblique interior face, attached to a concave neck distinguished from the globular body.

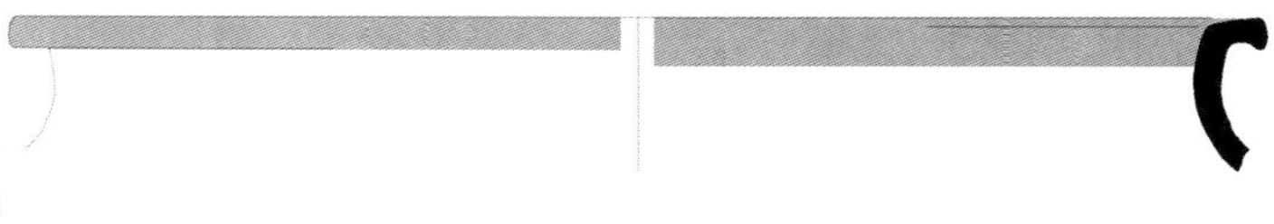

1:3

Chy2

77. PWBW 210-03

Dimensions: r 22.8 h 3.6 w 8.0

Description: Outward- and downward-turned rim with triangular section, attached to an initially straight neck.

Comparisons: Colangelo 2005, 252, 258, Pl. XXXIII, No. 157 (Torre di Satriano, *olla*, diam. 20.2 cm). For Etruscan BG examples with handles placed on top, dated 4th–3rd c. BC, see Morel 1981a, 5112. For the typical form in Ck, see O. Bianco 1997b, Pl. 78, No. 160 (Pomarico, *olla*, diam. 11.6 cm, 3rd c. BC); Cotton 1992c, Fig. 79, No. 1462 (Gravina, jar, period VI, 4th–3rd c. BC); Rinaldi 2005, 225, 234, Pl. XXIII, No. 61 (Torre di Satriano, 4th–3rd c. BC). Cf. also *Museo Taranto* II.1, 110, Nos. 43.76–78 (from a 5th–4th-c. BC tomb, diam. 9.0–9.7 cm).

1:3

Jar (Ja)

On Ja1-2 the rim size varies between 14.0 and 20.0 cm, although there are smaller examples, while larger forms are very rare. The only example with a wider rim comes from site 46 and measures 28.0 cm. Reduced dimensions are characteristic of Ja2 in particular, and range from 8.0–12 cm. On Ja3-5, although jars with diameters similar to those of Ja1-2 are known, the dimensions are generally larger (19.5–28.0 cm), except for one fragment measuring 13.0 cm. Ja6, attested here, measures 17.7 cm, while the two fragments belonging to Ja7 have very different diameters (11.8 cm and 25.8 cm). The long history of this type is confirmed by the comparanda cited below. In some cases it is also possible to identify similar forms in cooking ware.

1:3

Ja1a
78. PWBW 343-12
Dimensions: r 26.4 h 3.6 w 5.6

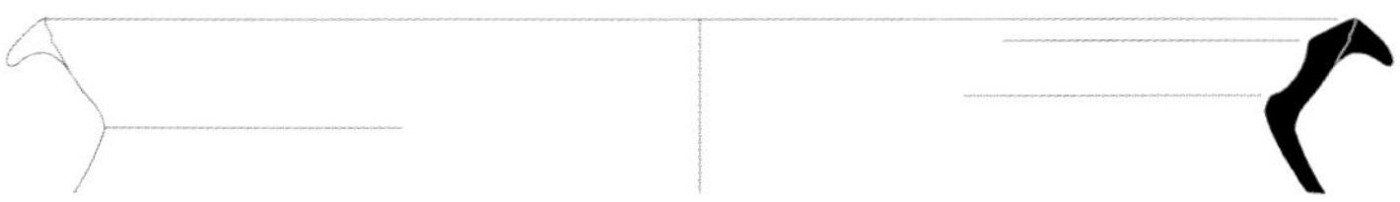

1:3

Ja1b
79. PWBW 169-15
Dimensions: r 22.6 h 3.1 w 5.1

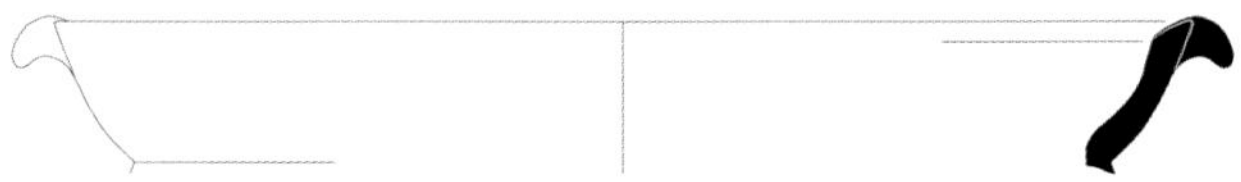

Description: High rim with triangular section and oblique upper face and concave interior, occasionally emphasized on the exterior by a groove and neatly distinguished from the globular body. The rim is often accompanied by a small, attached, outward-turned wing with elliptical section.

Comparisons: Vegas 1973, 113, Fig. 41, No. 7 (type 48, two vertical saddle-handles, AD 60–70). Corinth VII.3, Pl. 27, Nos. 651 (Ck chytra, deposit 112, 146 BC), and 656 (Ck stew pot with horizontal handles, deposit 46, 146 BC).

Ja2
80. PWBW 525-33
Dimensions: r 23.3 h 3.3 w 4.4

1:3

Description: High rim with triangular section and flat upper face and concave interior, occasionally emphasized on the exterior by a groove and neatly distinguished from the globular body.

Comparisons: Postrioti 1996, Fig. 8e (Metaponto, BW jug, diam. 12.2 cm, 3rd c. BC; cf. Morel 1981a, 5231a1, BG/Gnathia jug, end of 4th–beginning of 3rd c. BC); similar, but with horizontal handles, in Cotton 1992b, Fig. 64, No. 1296 (Gravina, "jar," period VIIIa-b, end of 2nd–1st c. BC); Semeraro 1983, Pl. 92, No. 239 (BW *cratere ollare*); *Kaulonia* I, 71, Fig. 48, No. 333; Pancrazzi 1979, Fig. 62, No. 11 (BW, diam. ca. 23 cm). *Corinth* VII.3, Pl. 27, Nos. 651 (Ck chytra, deposit 112, 146 BC), and 656 (Ck stew pot with horizontal handles, deposit 46, 146 BC).

Ja3a
81. PWBW 358-04 rev90
Dimensions: r 19.5 h 3.9 w 5.5

1:3

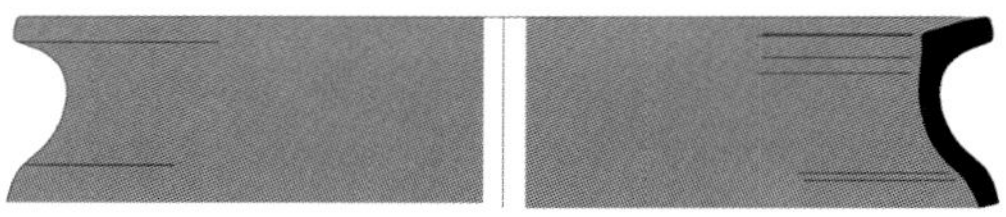

Description: Outward-turned rim with rectangular section, flat and oblique upper face, concave neck, and globular body.

Comparisons: Cotton 1983, 374, Fig. 145, No. 449 (Cozzo Presepe); Marchegiani 1997, Pl. 59, No. 32 (Pomarico, BW *vaso con coperchio*, diam. 15–25 cm, thick rim, second half of 3rd c. BC; 1.167: the type is a reworking of a Ck form); Colangelo 2005, 252, 257, Pl. XXXII, No. 155 (Torre di Satriano, *olla*, diam. 20 cm). Cf. O. Bianco 1997b, Pl. 79, No. 162; *Agora* XII, Fig. 17, No. 1659 ("jug," 500–480 BC); L. Giardino 1996a, 40 (Herakleia).

Jar (Ja) cont.

1:3

Ja3b

82. PWBW 043-58 rev82

Dimensions: r 20.9 h 2.4 w 5.1

Description: Outward-turned rim with rectangular section, slightly concave upper face, exterior face occasionally bearing a groove, concave neck, and globular body.

Comparisons: Cotton 1983, 369, Fig. 138, No. 401 (Cozzo Presepe, Site B, phase V, 4th–3rd c. BC; cf. *Corinth* VII.3, Pl. 21, No. 613, column krater, Deposit 115, ca. 300 BC); Marchegiani 1997, Pl. 59, No. 30 (Pomarico, BW *olla*, diam. 18 cm; 1.167: the type is well known in Ck); Cotton 1992b, Fig. 58, No. 1218 (Gravina, "basin or large bowl," period VIIIa, end of the 2nd–1st c. BC), and II.323, Fig. 65, No. 1300 (Gravina, "jar," period VIIIb, unstratified); *Valesio* I, 90, Form C44, No. 128 (BW "krater bowl/*zuppiera*," end of the 3rd–beginning of the 2nd c. BC), and 180 Form K32 No. 305 (BG, second half of the 3rd–beginning of the 2nd c. BC); Preacco Ancona 1999a, 286, Fig. 299, No. 509 (Oppido Mamertina, BG, vertical body wall, 3rd c. BC); De Juliis 1990, 87, Figs. 343–348 (Canosa, *olle biansate*, diam. 14.3–15.9 cm); Pancrazzi 1979, Fig. 29, cmr XIX/12 (Cavallino, *olla*, diam. 22 cm); *Corinth* VII.3, Pl. 33, No. 706 (bowl, Deposit 94, 150 BC).

1:3

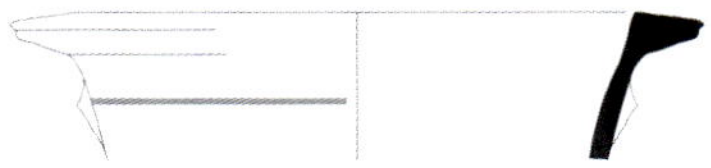

Ja4

83. PWBW 070-10

Dimensions: r 13.8 h 3.0 w 3.7

Description: Outward-turned rim with sub-rectangular section, flat upper face, exterior face with groove, and nearly straight neck. The reduced dimensions of several recovered fragments leave room for doubt as to the attribution to a jar, rather than to a jug (in this regard see J17d); only at site 416, in fact, is the rim diameter 25.0 cm, while the rim diameter of the piece belonging to site 70 is 13.8 cm.

Comparisons: O. Bianco 1997a, Pl. 77, No. 141 (Pomarico, PW jug, diam. 11 cm, 320–second half of the 3rd c. BC; 1.183: the rim diameter varies between 10 and 15 cm); Saunders and du Plat Taylor 1992, Fig. 11, No. 245 (Gravina, small jug type 3, period VI and perhaps VII, 4th–beginning of the 2nd c. BC); Naumann and Neutsch 1960, Fig. 11, No. 4 I 3 (Palinuro, *Kratere*).

1:3

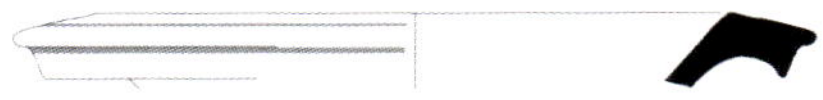

Ja5

84. PWBW 523-12

Dimensions: r 16.0 h 1.5 h 5.7

Description: Similar to Ja4 in the shape of the neck, which is oblique, but in this case the rim is inclined towards the exterior and hangs slightly. In the three examples recovered the diameter ranges from 16.0 to 20.0 cm (sites 305, 336, and 523).

Comparisons: See Ja3 and Ja4.

1:3

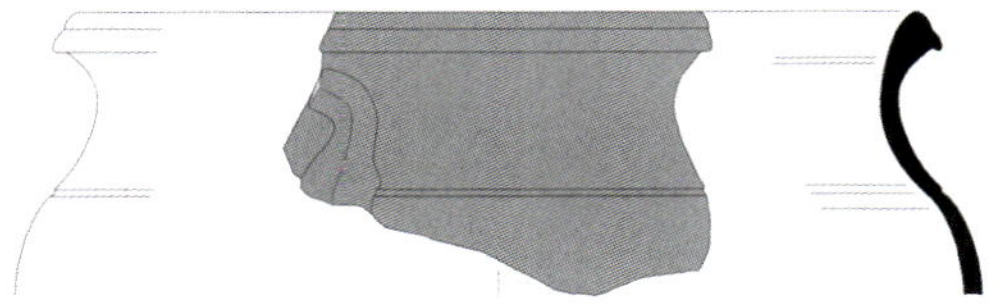

Ja6

85. PWBW 358-02 rev90

Dimensions: r 17.7 h 5.8 w 8.6

Description: Outward-turned rim with triangular section and groove on the exterior face, attached to a concave neck and globular body; horizontal, vertically-oriented handle attached at the point of maximum expansion of the body, against which it abuts.

Comparisons: Cotton 1992b, Fig. 63, No. 1274 (Gravina, "jug" with rim similar to the current example, period VIIIa-b, end of the 2nd–1st c. BC).

Jar (Ja) cont.

1:4

Ja7

86. PWBW 482-04

Dimensions: r 25.8 h 4.6 w 12.3

Description: Indistinct, slightly thickened rim with flat and oblique upper face; flared neck; vertical (?) handle attached at the rim. For details see the introduction to this section.

Comparisons: None found, at present.

Dinos (Dn)

Dn1 is 16.5–18.0 cm wide but was also produced in larger sizes (26.0 cm). Dn2 has a wider rim (27.6–34.7 cm) but could also be smaller (24.0 cm). The fragments of Dn3 preserving the diameter indicate great variation in size, ranging from 24.0 to 47.7 cm. The only fragment of Dn4 measures 21.3 cm, and the only fragment of Dn7 measures 11.4 cm. The best-represented types suggest an average size between 22.0 and 29.6 cm for Dn5 and between 20.0 and 31.6 cm for Dn6; in both cases there are also significantly larger examples (31.0–39.0 cm for Dn5 and 33.5–46.0 cm for Dn6) and, rarely, smaller examples (16.0–19.0 cm for Dn5 and 17.85 for Dn6). The diameter of the bases attributed to dinoi varies between 18.0 and 28.0 cm.

1:3

Dn1

87. PWBW 509-05

Dimensions: r 17.7 h 5.6 w 5.1

1:3

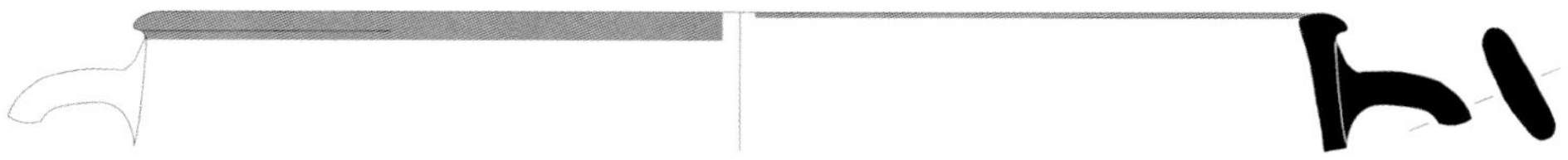

88. PWBW 285-32

Dimensions: r 29.1 h 3.5 w 11.0

Description: Narrow horizontal, outward-turned rim with elliptical section; globular body; vertical handle with elliptical section attached immediately below the rim.

Comparisons: Pizzo 1995b, 99, Fig. 97 (Incoronata, diam. 40 cm; 98: probable evocation of metallic forms); Pancrazzi 1979, Fig. 64, No. 3 (*vaso cilindrico*); Cotton 1983, 367, Fig. 136, No. 389 (Cozzo Presepe, Site A, phase VI, bead rim bowl); *Sibari* II, 333, Fig. 359, No. 27106 (BW, with wider rim); Naumann 1958, 44, Fig. 11, No. 7 (Palinuro).

1:3

Dn2a

89. PWBW 169-17

Dimensions: r 31.4 h 4.4 w 4.1

Description: Thickened, slightly inward-projecting rim with elliptical section and convex upper face; globular body.

Comparisons: Deodato and Bianco 1997, Pl. 68, No. 82 (Pomarico, *olla* with more pronounced rim, diam. 14.7 cm; 1.178: the type is distributed almost exclusively in burial contexts of the 5th–3rd c. BC); Marchi 1991, 172, Pl. XVI, No. 4 (Venosa, end of 1st c. BC–1st c. AD; but the form is also known at Oppido Lucano by the 5th c. BC).

Dinos (Dn) cont.

1:3

Dn2b
90. PWBW 183-18
Dimensions: r 34.7 h 4.8 w 6.9
Description: Thickened, slightly inward-projecting rim with elliptical section and convex upper face; globular body. This variant is distinguished by a thinning of the rim on the interior and of the body walls.
Comparisons: Russo Tagliente 1995, 89, Fig. 90, No. 212 (Armento); Mazzei 1985, Pl. LV, and 128, No. 13 (San Severo, tomb dated end of the 4th–beginning of the 3rd c. BC, diam. 9.5 cm); Naumann 1958, 44, Fig. 11, No. 6 (Palinuro); Keith and Fracchia 1990a, 272, Fig. 187, No. 287 (Roccagloriosa, 3rd c. BC, diam. 17.0 cm).

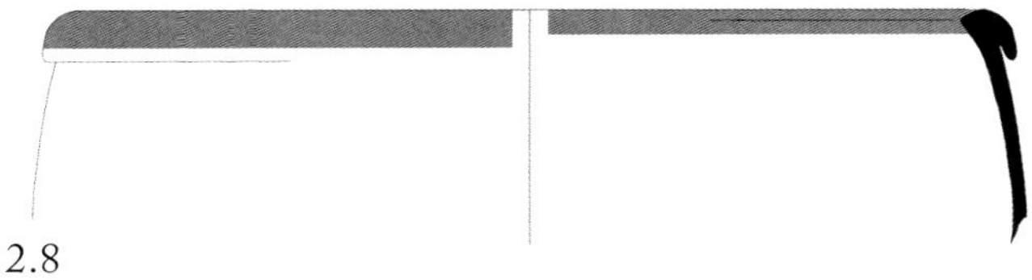

1:3

Dn3
91. PWBW 269-56
Dimensions: r 24.0 h 6.0 w 2.8
Description: Convex rim with elliptical section; globular body. Can be considered an evolution of Dn2 (in particular sub-type Dn2b), albeit with the lower part of the rim separated from the body and not merely thickened.
Comparisons: see Dn2b.

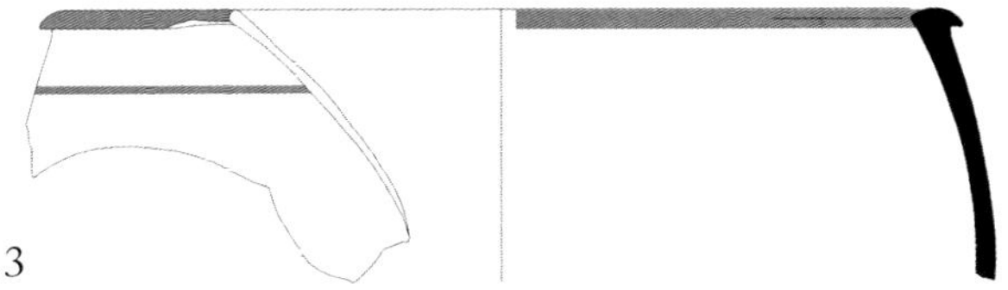

1:3

Dn4
92. PWBW 397-07 rev90
Dimensions: r 21.3 h 6.5 w 11.3
Description: Narrow horizontal, outward-turned rim with elliptical section; globular body. Similar to Dn1, except that the body of the former has a more globular profile in the area immediately below the rim.
Comparisons: see Dn1.

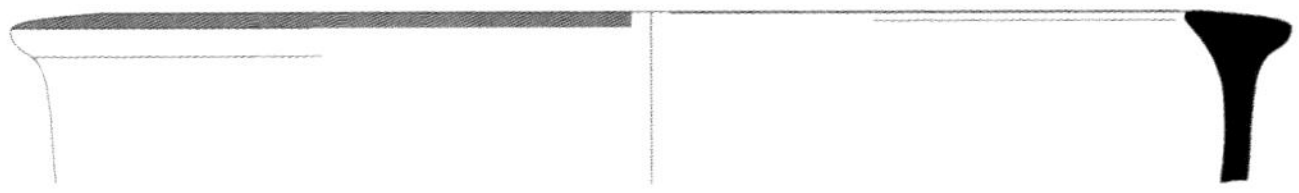

1:3

Dn5a
93. PWBW 521-15
Dimensions: r 25.5 h 3.5 w 7.3
Description: Outward-turned, inward-projecting rim, with generally elliptical section; straight body wall in the upper section; the handle, when present, is horizontal and vertically oriented, attached beneath the rim and abutting the body wall. Dn5a has a thickened rim with sub-triangular section and flat upper face. The vertical wall in particular recalls the so-called *sombrero de copa* (see detailed bibliography in Brizzi 1999b, 318, No. 631, 2nd c. BC).
Comparisons: Cotton 1992b, Fig. 67, No. 1320 (Gravina, "storage jar," 4th c. BC, period VIIIb, unstratified). The morphology is present in Middle Corinthian pottery: cf. *Sibari* I, 86, Fig. 76, No. 204, and 93; Lissi Caronna 1990–1991, 208, Fig. 27, No. 27 (Oppido Lucano, *casa* C, *scodellone biansato*, diam. 32.8 cm).

1:3

Dn5b
94. PWBW 521-180
Dimensions: r 29.0 h 4.2 w 9.0
Description: See Dn5a.
Comparisons: Greci, Enotri e Lucani, 257, No. 3.38.5 (PW, T124 from Sant'Arcangelo-San Brancato, end of 4th c. BC, diam. cm 10.1).

Dinos (Dn) cont.

1:3

Dn5c

95. PWBW 326-35

Dimensions: r 38.0 h 4.8 w 6.1

Description: See Dn5a. On Dn5c the rim projects inwards less and has a groove on the exterior face.

Comparisons: Greci, Enotri e Lucani, 257, No.3.38.5 (PW, T124 from Sant'Arcangelo-San Brancato, end of 4th BC, diam. 10.1 cm); Semanaro 1983, Pl. 89, No. 145 (Otranto, BW basin, 6th–5th c. BC, without grooves, from cantiere 3); Carter 1983a, 467, Fig. 43, No. 22 (Pantanello, similar but with vertical body wall).

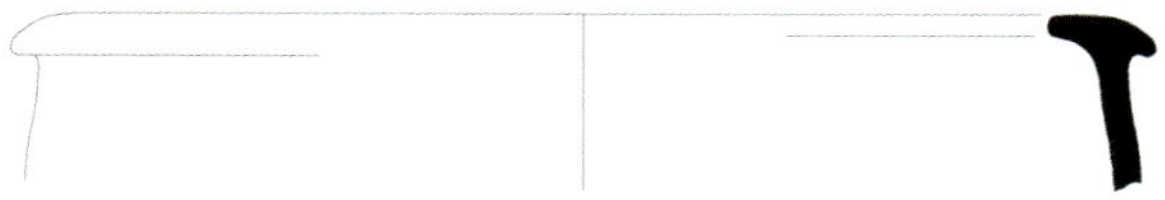

1:3

Dn5d

96. PWBW 518-04

Dimensions: r 22.8 h 3.6 w 8.5

Description: See Dn5a. The rim has a convex upper face and can project inwards strongly while projecting outwards slightly.

Comparisons: Deodato and Bianco 1997, Pl. 71, No. 98 (Pomarico, diam. 32.8 cm, belonging to the local tradition, of irregular workmanship); L. Giardino 1996a, 41 (Herakleia).

1:3

Dn5e

97. PWBW 326-36

Dimensions: r 28.0 h 4.2 w 6.4

Description: See Dn5a. The rim has a convex upper face and may project inwards slighty while projecting outwards strongly.

Comparisons: See Dn5d.

1:3

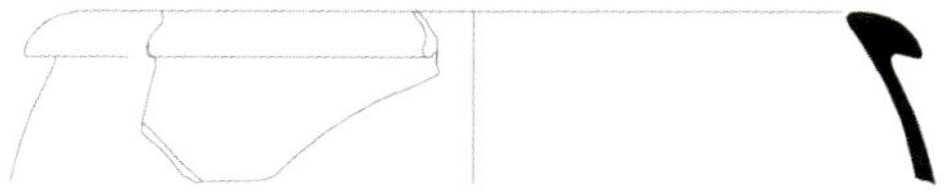

Dn6a

98. PWBW 068-02

Dimensions: r 17.85 h 3.6 w 5.9

Description: Wide, outward-turned, more or less hanging rim with elliptical section and convex, slightly inward-projecting upper face; globular body. On Dn6a the rim is smaller, while on Dn6c it is horizontal.

Comparisons: Valesio I, 94, Form C71, No. 145 ("jar," diam. 34.5 cm, context of end of 4th–beginning of 3rd c. BC, but perhaps produced in 4th–3rd c. BC; the diameter varies between 25 and 50 cm); *Kaulonia* I, 68, Fig. 46, No. 299; Vittoria 2000, Pl. 65, No. 1.

1:3

Dn6b

99. PWBW 488-13

Dimensions: r 40.5 h 5.8 w 11.8

Description: See Dn6a.

Comparisons: Naumann 1958, 40, Fig. 5, No. 2 (Palinuro, int. diam. 22–25 cm).

Dinos (Dn) cont.

1:3

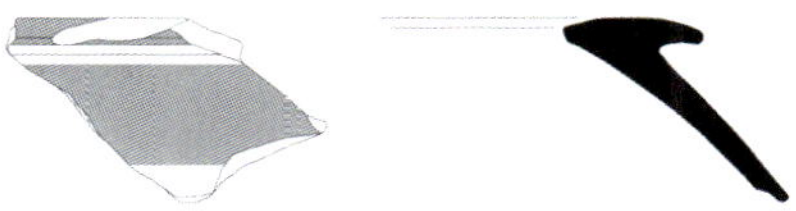

Dn6c
100. PWBW 445-02
Dimensions: h 3.8 w 6.4
Description: See Dn6a.
Comparisons: Carter 1983a, 455, Fig. 30, No. 73 (Pantanello, deposit in the clay pit, ext. diam. 20 cm); Lo Porto 1988–1989, 411, Fig. 129, No. 1 (tomb of beginning of 5th c. BC); Russo Tagliente 1995, 89, Fig. 90, No. 211 (Armento); Saunders and du Plat Taylor 1992, Fig. 14, No. 282 (Gravina; II.30: storage jar type 1, typical of periods VI-VII, 4th–beginning of 2nd c. BC); Naumann 1958, 40, Fig. 5, No. 3 (Palinuro, int. diam. 22–25 cm).

1:3

Dn7
101. PWBW 269-57
Dimensions: r 11.4 h3.1 w 10.4
Description: Thickened rim with irregular section, thickened also with respect to the globular body wall. The type of *dinos* with thickened rim is fairly common, while the type with molded rim, like the current example, is less common.
Comparisons: Brizzi 1999b, 318, Fig. 316, No. 635 (thinner rim); Lissi Caronna 1983, 256, Fig. 46, No. 8 (Oppido Lucano T64 dated end of 6th c. BC, diam. 13.7 cm, with smooth and almond-shaped rim and flat bottom), and 234, Fig. 19, No. 12 (T57 dated end of 5th–beginning of 4th c. BC, diam. 9.2 cm); also in De Juliis 1973, 331, Fig. 63, No. 1, and 334, Fig. 68, No. 1 (Ordona, TXXVIII and TXXIX both dated first quarter of 4th c. BC, diam. 15.0–16.0 cm). The *dinos* types found in burials are similar to the current example but have a smooth rim. The same applies at Cancellara (Fabbricotti 1976, 331, Fig. 4, No. 16, from T1, dated end of 5th–beginning of 4th c. BC).

Hydria (H)

The rim diameter of this type of container, measured on 41 fragments, is generally in the range of 16.0–25.6 cm (30 examples, equivalent to 73.2%). The larger hydriai belong mostly to H1 (which reaches a diameter of 36.0 cm at site 406). Hydriai with smaller rims (11.0–14.0 cm) were also identified, particularly in H3.

1:3

H1a
102. PWBW 509-04
Dimensions: r 25.6 h 5.9 w 6.2
Description: Rim with generally concave upper face; flaring neck. H1a appears to be the earliest version (round mouth oinochoe), with a rim characterized by a hemispherical section.
Comparisons: Carter and Toxey 1998, 707, CD289-25 (Pantanello, elliptical rim, 625–500 BC); Elliott 1998, 656, H3 (Pantanello, BG, 460–440 BC); Hänsel 1973, 452, Fig. 32, No. 7 (Herakleia, from a stratum dated to last third of 5th c. BC); *Sibari* II, 201, Fig. 216, No. 10903 (*anfora*, diam. 18 cm), and 333, Fig. 359, No. 25091 (BW *anfora*); Naumann and Neutsch 1960, Fig. 14, No. 2 I 4 (Palinuro).

1:3

H1b
103. PWBW 397-90 rev90
Dimensions: r 21.5 h 3.1 w 9.1
Description: See H1a. On H1b–c the rim is thickened and has an elliptical section.
Comparisons: du Plat Taylor 1983, 342, Fig. 119, No. 240 (Cozzo Presepe, Site A, Phase IV, mid-6th c.–480 BC); *Sibari* I, 101, Fig. 86, No. 159 (6th c. BC); *Sibari* II, 180, Fig. 198, No. 10765 (*hydria o anfora*, diam. 16 cm); Semeraro 1983, Pl. 89, No. 185 (Otranto, Gnathia oinochoe, from *cantiere* 2).

Hydria (H) cont.

1:3

H1c

104. PWBW 397-01 rev90

Dimensions: r 20.4 h 5.7 w 11.8

Description: See H1a and H1b.

Comparisons: du Plat Taylor 1983, 342, Fig. 119, Nos. 234, and 237 (Cozzo Presepe, Site A, Phase IIIb, 575–550 BC), and Fig. 119, No. 242 (Phase V); Saunders and du Plat Taylor 1992, Fig. 20, No. 363 (Gravina, large vessel type 3, period VI, 4th–3rd c. BC); *Kaulonia* I, 70, Fig. 47, No. 308; *Sibari* II, 333, Fig. 359, No. 25086 (BW *anfora*), and 343, Fig. 368, No. 24277 (BW *anfora*; cf. 305, No. 368: probable derivation from the so-called Ionic types); *Sibari* III, 160, Fig. 170, No. 43773 (PW hydria, diam. 15.3 cm); Semeraro 1983, Pl. 92, No. 238 (Otranto, Hellenistic BW hydria).

1:3

H1d

105. PWBW 358-03 rev90

Dimensions: r 11.0 h 3.8 w 9.2

Description: See H1a. On H1d the interior concavity is barely perceptible.

Comparisons: Semeraro 1983, Pl. 92, No. 241 (Otranto, Hellenistic PW *olla*?); *Corinth* VII.3, Pl. 12, No. 287 (two-handled "decanter II," 425 BC, diam. 9.2 cm).

1:3

H2a

106. PWBW 182-09

Dimensions: r 22.1 h 2.7 w 5.4

Description: Horizontal, outward-turned rim with elliptical section and nearly flat upper face; flaring neck.

Comparisons: *Meligunìs-Lipára* XI.II, Fig. 69b (BW pelike, 5th-c. BC tomb); *Kaulonia* I, 70, Fig. 47, No. 303; Bailo Modesti 1980, Pl. 56, No. 2-44 (Cairano, diam. 8.9 cm).

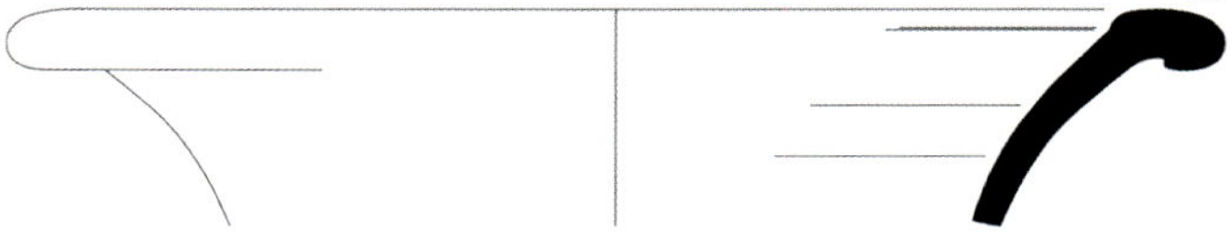

1:3

H2b

107. PWBW 397-02 rev90

Dimensions: r 17.4 h 3.2 w 10.6

Description: See H2a. On H2b the rim is nearly hooked.

Comparisons: Angelelli 1998–1999, 32, Fig. 15, No. 112 (Amelia, diam. 17.4 cm, 1st c. BC).

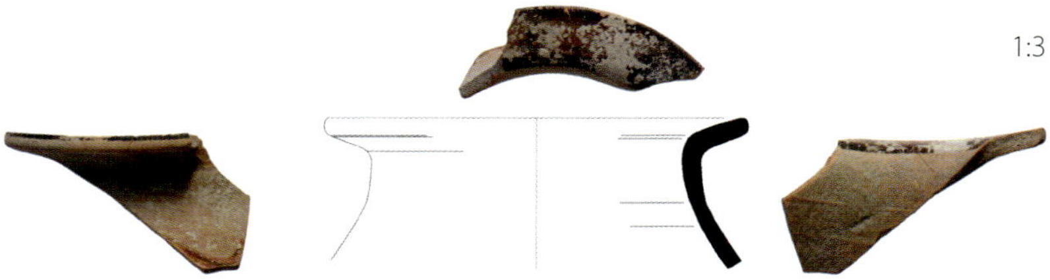

1:3

H3a

108. PWBW 465-01 rev90

Dimensions: r 13.2 h 4.8 w 8.4

Description: Outward-turned rim with rectangular section and flat upper face that becomes oblique towards the interior.

Comparisons: Marchegiani 1997, Pl. 63, No. 49 (Pomarico, BW *anfora da tavola*, diam. 13 cm, 3rd c. BC; 1.170: the profile recalls the banded hydria types of the 6th c. BC from Cozzo Presepe, and the diameter varies between 10–11 and 15–16 cm; cf. Cotton 1983, 369, Fig. 139, No. 407); Cotton 1992b, Fig. 63, No. 1273 (Gravina, "jug," period VIIIa-b, end of the 2nd–1st c. BC); Saunders and du Plat Taylor 1992, Fig. 11, No. 238 (Gravina, II.27: small jug type 2, period VI and perhaps VII, 4th–beginning of the 2nd c. BC: it is the most common type of small jug in this ceramic class at Gravina); Colangelo 2001, 75, Fig. 53, No. 97 (Torre di Satriano, BW jug, diam. 12.8 cm, end of the 5th c. BC); Keith and Fracchia 1990b, 274, Fig. 188, No. 310 (Roccagloriosa, diam. 10 cm); Bailo Modesti 1980, Pl. 56, No. 14-8 (Cairano, diam. 11 cm).

Hydria (H) cont.

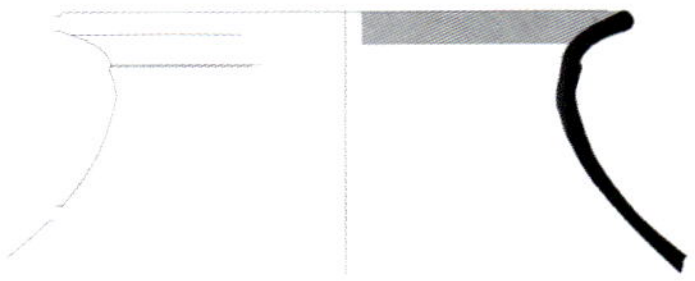

1:3

H3b

109. PWBW 477-58

Dimensions: r 11.5 h 5.4 w 7.2

Description: See H3a. On H3b the rim can be distinguished from the concave neck by a ridge.

Comparisons: du Plat Taylor 1983, 341, Fig. 118, No. 225 (BW "mug," Site A, Phase IV, 6th–beginning of the 5th c. BC; cf. Lo Porto 1973: Pl. VI, Fig. 1, No. 5, dated to second quarter of 6th c. BC).

Jug (J)

The average rim diameter of the jug ranges from 12.0 to 16.9 cm, but detailed analysis of the various types reveals the existence of several larger and smaller variants. The most attested group (J7, J8, J12, J14, and J20) falls within the average range (12.0–17.0 cm), although it appears also in a smaller version (9.0–11.6 cm) and even in a nearly miniature version (diam. 7.0 cm, from site 44). The second most frequent group (J15, J18, and J19), too, has a rim 13.0–16.8 cm wide; sometimes associated with it are pitchers with smaller or larger mouths (10.0–12.0 cm and 18.0–18.5 cm, respectively). J16, J17, and J22 bearing a thickened rim with sub-triangular section fall squarely within the norm (12.0–16.9 cm), and for this group too there are versions both smaller and larger (6.0–10.0 cm and 17.4–18.0 cm, respectively). J1–J2 can achieve slightly larger sizes (14.0–18.2 cm, up to 19.8–22.0 cm), and even this group has a smaller version (7.0–13.8 cm). The olpe (J5) is distinguished by its medium size tending towards small (9.0–16.3 cm) and becoming almost miniature (5.0–8.0 cm): in only one case the diameter reached 22.0 cm (from site 43), but the state of preservation of the fragment is poor. J3–J4 are similarly small (10.0–16.2 cm) and are attested in versions both smaller and larger (8.0–9.6 cm and 18.0 cm, respectively) only in the variant J4. J23 falls within the average (12.0–16.9 cm), although it was produced also with a smaller diameter (7.0–9.0 cm), and in only one case it reaches 18.3 cm (site 300). The group with concave rim (J13 and J21) generally has a large diameter (16.0–20.6 cm), which in only one case (from site 334) contracts to 12.0 cm. J6 is not far from the average diameter (11.5–16.0 cm). Among the poorly represented types there are diameters both small (J9: 7.0–10.0 cm; J10: 6.1 cm) and large (J11: 22.0 cm; J24: 19.0 cm).

Among the fragments of feet or bases that can be attributed generically to a jug there are 540 fragments belonging to C11–C15 (see below) divided as follows:

C11: 1 fragment from site 529
C12: 518 fragments
C13: 1 fragment from site 266
C14: 8 fragments
C15: 12 fragments

To these can be added:

15 feet, concave base type, all plain and characterized by two distinct diameters: on the larger versions the diameter is 8.0–12.0 cm, on the smaller ones 3.0–4.3 cm;
5 feet, disk concave base type (1 banded example from site 346 probably belongs to an olpe); diameter 8.0–12.0 cm;
39 feet, flat base type, plain; diameter of 8.0–13.8 cm (1 fragment from site 497 with diameter of 18.0 cm);
4 feet, disk flat base type, plain; diameter 9.0–12.0 cm;
30 ring feet (2 banded and 6 amphoriskos types); 8.8–13.0 cm:

The banded base from site 535 cannot be identified with certainty as a ring foot and has a greater diameter (16.0 cm) than the others.

1:2

J1a

110. PWBW 762-01

Dimensions: r 20.7 h 5.2 w 10.8

Description: Outward-turned rim with sub-rectangular, sub-triangular, or elliptical section attached to a concave neck. J1a has a flat upper face slightly inclined towards the interior, and the rim is distinguished from the neck by means of a ridge or groove.

Comparisons: D'Andria 1980b, Pl. XII, No. 24 (Metaponto, *scarico* 12, BW amphora); Marchegiani 1997, Pl. 62, No. 47 (Pomarico, BW *anfora da tavola*, diam. 16 cm, second half of 3rd c. BC); Russo Tagliente 1995, 87, Fig. 89, No. 207 (Armento); *Gravina* III.2, 128, Fig. 52, No. 296 (PW jug).

Jug (J) cont.

1:2

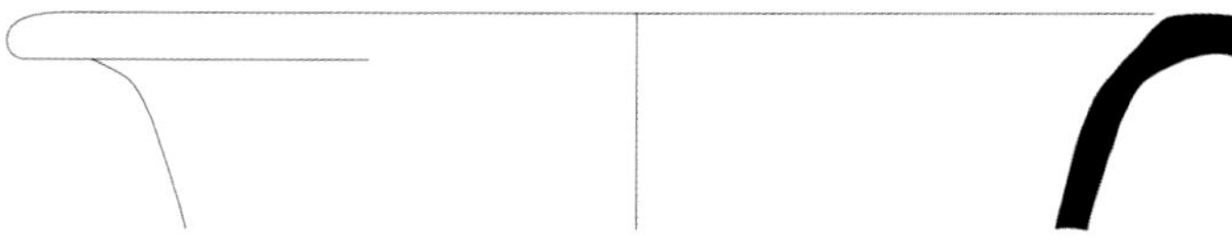

J1b

111. PWBW 397-87

Dimensions: r 18.2 h 3.2 w 5.2

1:2

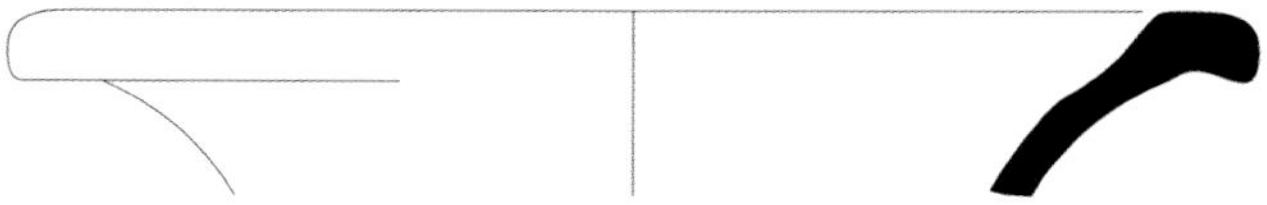

112. PWBW 530-07

Dimensions: r 16.6 h 2.5 w 9.3

Description: See J1a. J1b has a rim with a slightly hanging tip, sub-triangular section, and flat upper face.

Comparisons: Valesio I, 252, Form M23, No. 441 (PW "elegant jug," diam. 17.2 cm, from a context of the second/third quarter of the 2nd c. BC); similar, but with oblique upper face, is an example from Pomarico (O. Bianco 1997a, Pl. 77, No. 143, PW jug, diam. 19 cm, second half of the 3rd c. BC; 1.183: the rim diameter varies between 19 and 25 cm, and the large size suggests a use as a container for conserving liquids, rather than for pouring at the table; 1.184: at Cozzo Presepe and Cavallino the type is attested from the 6th c. BC to Roman times).

1:2

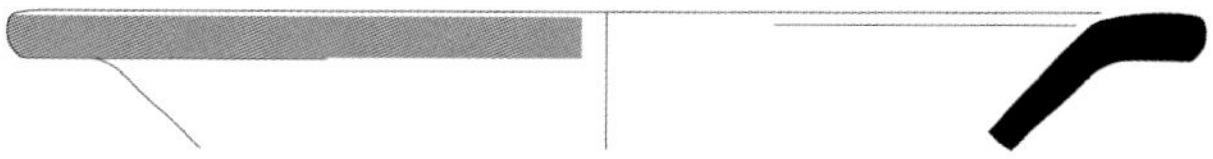

J1c

113. PWBW 515-08

Dimensions: r 15.9 h 1.9 w 7.3

Description: See J1a. J1c is similar to J1b but has a rim that hangs less.

Comparisons: Vittoria 2000, Pl. 50, No. 1 (T66 dated end of the 4th c. BC).

1:2

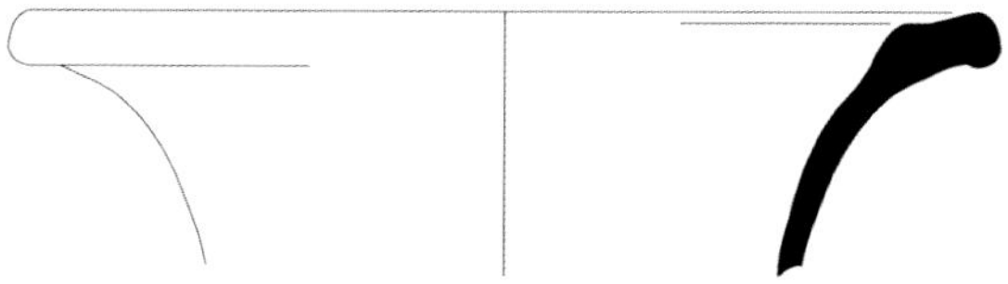

J1d1

114. PWBW 359-02

Dimensions: r 14.2 h 3.9 w 7.2

Description: See J1a and J1d2

Comparisons: See J1d2

1:2

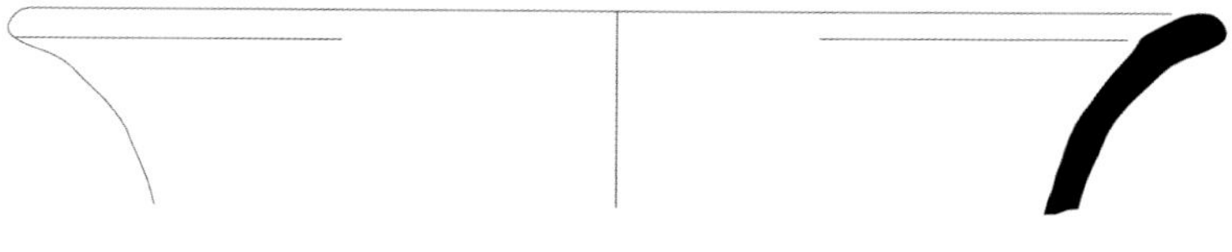

J1d2

115. PWBW 181-03

Dimensions: r 19.8 h 3.2 w 5.2

Description: See J1a. J1d has a rim with elliptical section and upper face inclined towards the interior; the vertical handle is generally attached to the rim; on J1d1 the upper face of the rim is slightly concave.

Comparisons: Saunders and du Plat Taylor 1992, Fig. 11, No. 234 (Gravina, II.27: small jug type 2, period VI and perhaps VII, 4th–beginning of 2nd c. BC; this is the most common type of small jug in this ceramic class at Gravina); Lissi Caronna 1983, 225, Fig. 9, No. 2 (Oppido Lucano, BG olpe from T53 dated end of 5th c. BC, diam. 7.3 cm), and 310, Fig. 99, No. 3 (from a plundered tomb, diam. 10.0 cm); *Valesio* I, 90, Form C52c, No. 133 ("hydria," diam. 19 cm, context of second half of 3rd c. BC but produced by end of 4th c. BC).

Jug (J) cont.

1:2

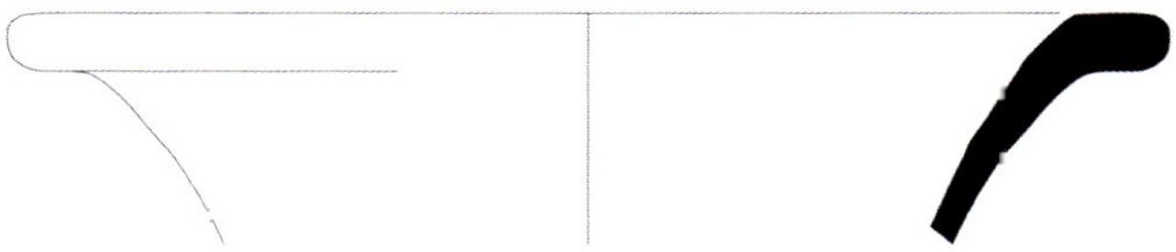

J1e

116. PWBW 178-03

Dimensions: r 16.8 h 3.5 w 5.8

Description: See J1a. J1e has a rim with sub-rectangular section and flat upper face.

Comparisons: Sibari II, 108, Fig. 89, No. 12591 (*hydria locale*); Valesio I, 90, Form C52c, No. 134 ("hydria," diam. 16 cm, from a context of end of 3rd–beginning of 2nd c. BC but produced by end of 4th c.); D'Andria 1978a, Pl. 30, No. 176 (Leuca, BW hydria).

1:2

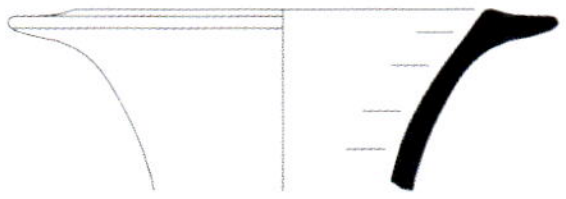

J1f

117. PWBW 084-01 rev90

Dimensions: r 7.3 h 2.5 w 3.1

Description: See J1a. J1f has an outward-turned rim with elliptical section and slightly concave upper face.

Comparisons: O. Bianco 1997a, Pl. 76, No. 136 (Pomarico, diam. 18.6 cm; in this case, too, on 1.183 it is emphasized that, given the dimensions, the vessel was perhaps used for storing liquids rather than pouring them at the table, and the level surface on the interior served perhaps to support a lid); Saunders and du Plat Taylor 1992, Fig. 11, No. 244 (Gravina, small jug type 3, period VI and perhaps VII, 4th–beginning of 2nd c. BC); *Kaulonia* I, 70, Fig. 47, No. 313; Hänsel 1973, 451, Fig. 31, No. 14 (Herakleia, from a stratum dateable to last third of 5th c.).

1:2

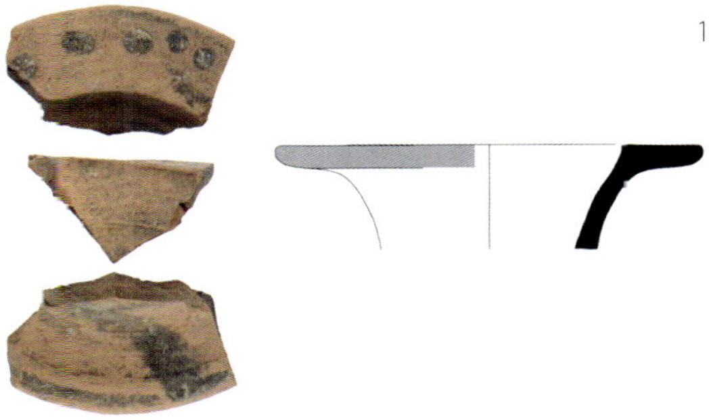

J1g

118. PWBW 488-63

Dimensions: r 8.4 h 2.1 w 4.3

Description: See J1a. J1g is similar to J1f but has a flat upper face.

Comparisons: Carter and Toxey 1998, 706, CD289-35 (Pantanello, "amphora," end of the 7th–beginning of the 6th c. BC), and 712 T70-1 ("pelike," rim with molded profile, 525–390 BC); O. Bianco 1997a, Pl. 77, No. 149 (Pomarico, PW *anforetta o anfora da tavola*, diam. 13.8 cm).

1:2

J2

119. PWBW 503-12

Dimensions: r 18.1 h 2.7 w 7.2

Description: Thickened, outward-turned, and hanging rim with elliptical section; straight and oblique neck walls.

Comparisons: Kaulonia I, 70, Fig. 47, No. 303; *Meligunìs-Lipára* XI.II, Fig. 69b (BW pelike, 5th-c. BC tomb); Bailo Modesti 1980, Pl. 56, No. 2-44 (Cairano, diam. 8.9 cm).

1:2

J3

120. PWBW 442-01

Dimensions: r 16.2 h 5.3 w 6.4

Description: Hanging, outward-turned rim with sub-triangular section, practically indistinguishable from the concave neck.

Comparisons: O. Bianco 1997a, Pl. 76 No. 135 (Pomarico, PW jug, diam. 11.3 cm, 320–second half of the 3rd c. BC; 1.183: the rim diameter varies between 10 and 15 cm); Dyson 1983, fig. 144, no. 13 (Buccino, carinated jug); *Agora* XII, fig. 14, no. 1613 (460-440 BC, "jug"; 204: the type is well-known at the end of the 5th c. BC). Cf. the BG example with overhead handle, dated end of the 4th–3rd c. BC: Morel 1981a, 5114b1.

Jug (J) cont.

1:4

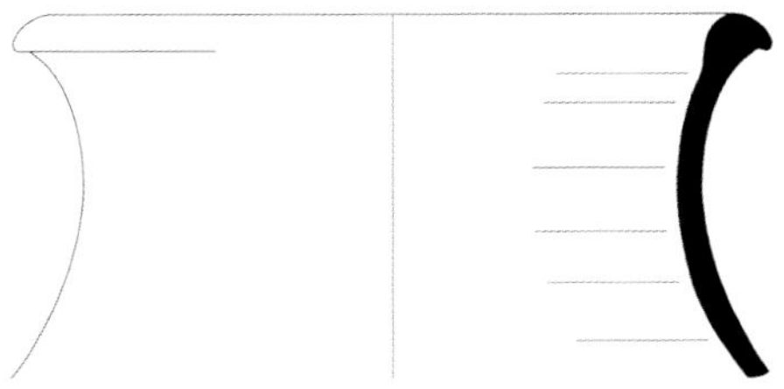

J4a

121. PWBW 358-01 rev90

Dimensions: r 11.5 h 5.7 w 7.3

Description: Outward-turned rim with sub-triangular section and convex upper part; concave neck.

Comparisons: O. Bianco 1997a, Pl. 76, No. 140 (Pomarico, PW jug, diam. 13.3 cm, 320–second half of the 3rd c. BC; 1.183: the rim diameter varies between 10 and 15 cm).

1:2

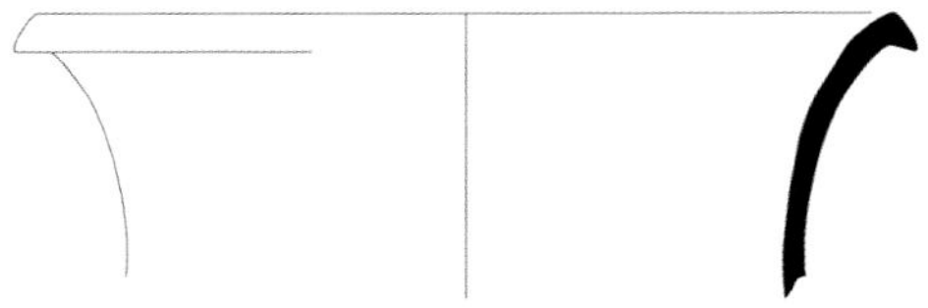

J4b

122. PWBW 043-71 rev82

Dimensions: r 12.2 h 4.0 w 4.0

Description: See J4a. On J4b the rim is hanging.

Comparisons: Cotton 1983, 369, Fig. 139, No. 409 (Cozzo Presepe, Site A, phase VI); Cotton 1992b, Fig. 63, No. 1272 (Gravina, "jug," 4th c. BC, periods VI–VIIIa, 4th–1st c. BC); Dyson 1976, Fig. 17, 16IV71 (Cosa, diam. 11 cm, from deposit 3 dated to middle of the 2nd c. BC, corresponding to Taylor's deposit C, 190–140 BC); *Sibari* III, 152, Fig. 164, No. 8123 (BF jug, end of 6th–first quarter of 5th c. BC); *Agora* XII, Fig. 3, No. 169 (425–400 BC, Corinthian "mushroom jug"; 67: at Corinth this type is known by beginning of 5th c. BC; it is also common in Ck but especially No. 260 (BW olpe, ca. 500 BC; 78: in several cases the word *demosios* painted on the shoulder suggests use as a unit of measure); Vittoria 2000, Pl. 47, No. 3 (Ck T69, first third of 4th c. BC).

1:2

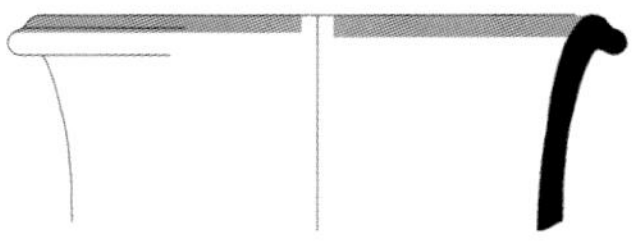

J4c

123. PWBW 395-44 rev90

Dimensions: r 8.2 h 2.9 w 3.2

Description: See J4a. On J4c the neck has nearly vertical walls, and there is a shallow groove on the upper face of the rim.

Comparisons: See J4a–b.

1:2

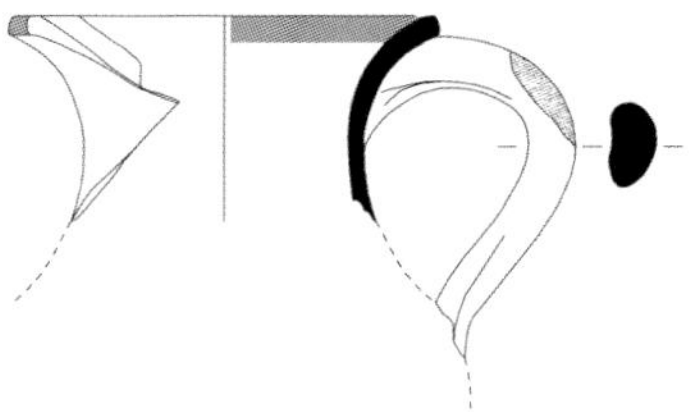

J5a

124. PWBW 395-64

Dimensions: r 5.7 h 4.7 w 2.8

Description: Indistinct, outward-turned rim with elliptical section; concave neck; vertical handle attached immediately beneath the rim.

Comparisons: Carter and Toxey 1998, 701, T193-9 (Pantanello, BW, 310–280 BC); De Siena 1980, 114, Pl. VI, No. 31 (Metaponto, *strato* 2b, second half of 6th–first quarter of 5th c. BC); Pianu 1990, Pl. XIX, No. 3, and Pl. XXIII, No. 30.3 (Herakleia, BW, tomb of 340–320 BC), as well as Pl. LIX, No. 4 (PW, tomb dated to end of 4th c. BC); Cotton 1992b, Fig. 62, No. 1262 (Gravina, period VIIIa-b, end of 2nd–1st c. BC); Saunders and du Plat Taylor 1992, Fig. 11, No. 223 (Gravina, 4th c. BC); Hempel 2001, Pl. 3, Form 1-2 (Taranto, phase D-E1, end of 3rd–first half of 2nd c. BC); *Gli ori,* 387, No. 12 (BW, overhead handle, tomb of second half of 4th c. BC), 407, No. 6 (PW, overhead handle, tomb of last quarter of 4th c. BC), and 474, No. 18 (BG, tomb of first half of 2nd c. BC); *Valesio* I, 246, Form M22, No. 433 (PW "plump jug," diam. 18 cm, context of end of 3rd–beginning of 2nd c. BC).

1:2

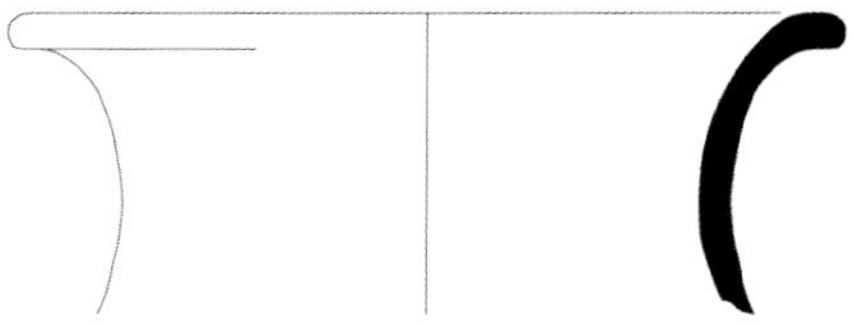

J5b

125. PWBW 416-12

Dimensions: r 11.1 h 4.1 w 6.5

Jug (J) cont.

126. PWBW 326-38

Dimensions: r 14.3 h 2.6 w 6.7

Description: See J5a. J5b is distinguished by a more pronounced rim.

Comparisons: Carter 1983a, 478, Fig. 53, No. 134 (Pantanello, "jug or pitcher"); O. Bianco 1997a, Pl. 76, No. 132 (Pomarico, PW jug, diam. 8.8 cm, second half of 3rd c. BC; 1.183: the rim diameter varies between 6 and 10 cm); Carando 1997, Pl. 107, No. 55 (Pomarico, *boccaletto*, T10d dated to middle of the 4th c. BC); Cotton 1992b, Fig. 63, No. 1268 (Gravina, "jug," end of 4th–mid-3rd c. BC, periods VII–VIIIb, 3rd–1st c. BC); *Valesio* I, 246,

1:2

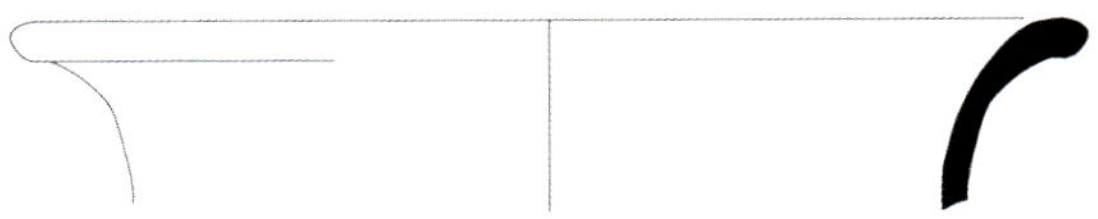

Form M22, No. 434 (PW "plump jug," diam. 14.5 cm, context of first half of 2nd c. BC); Cassano 1992c, 492, No. 3 (Canosa, PW jug, diam. 9.3 cm, end of 4th c. BC); Keith and Fracchia 1990b, 274, Fig. 188, No. 307 (Roccagloriosa, diam. 7.5 cm); *Gli ori*, 486, No. 42 (tomb of third quarter of 2nd c. BC); Lissi Caronna 1983, 288, Fig. 77, No. 1 (Oppido Lucano, casa A, diam. 12.0 cm); Vittoria 2000, Pl. 47, No. 2.

Several fragments, which could not be reproduced graphically, have an overhead handle attached to the interior of the rim (sites 84 and 305), a type known in Magna Grecia from at least the second half of the 6th c. BC.

For the comparanda see: O. Bianco 1997a, Pl. 76, No. 133 (Pomarico, PW jug, diam. 18.6 cm, 320/270–second half of the 3rd c. BC; 1.183: the rim diameter varies between 10 and 16 cm; in Greece, Sicily, and Magna Grecia the type is attested from the 6th through the 4th/3rd c. BC); Marchegiani 1997, Pl. 63, No. 52 (Pomarico, BW "olpe," diam. 6.3 cm, end of the 6th c. BC; 1.170: the use of the olpe is attested throughout the Mediterranean basin in the 6th–4th c. BC); Manzo 1989, Pl. XLVI, No. 403 (Locri, diam. 4.6 cm, usually 3–5 cm, height 9–11 cm; 332: the type is widespread in Magna Grecia from the second half of the 6th c. BC but appears at Locri in the 5th and lasts through the end of the 3rd and perhaps the beginning of the 2nd c. BC). For the probable 6th-c. Attic antecedents cf. *Agora* XII, Fig. 3, Nos. 144–45 (525–500 BC), 240–52 (olpe types, 575–550 BC), and 65.

J6

127. PWBW 230-02

Dimensions: r 11.5 h 2.5 w 2.6

Description: Thickened rim with elliptical section, distinguished from the flaring neck by a ridge.

Comparisons: Carter and Toxey 1998, 708, T350-11 (Pantanello, "jug," 460–440 BC).

1:2

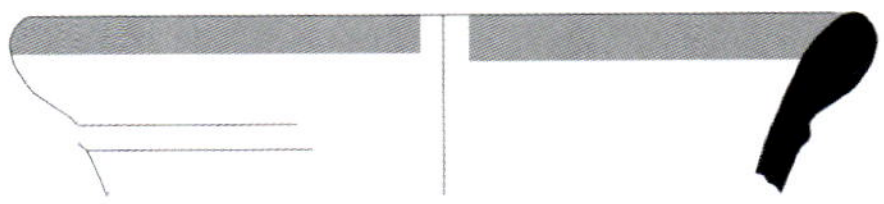

J7a1

128. PWBW 516-06

Dimensions: r 14.7 h 3.0 w 8.8

Description: Thickened rim with convex exterior profile, concave interior profile, and sub-square section; concave neck. On J7a1 the upper face is slightly inclined towards the interior, and the transition from the rim to the neck is

1:2

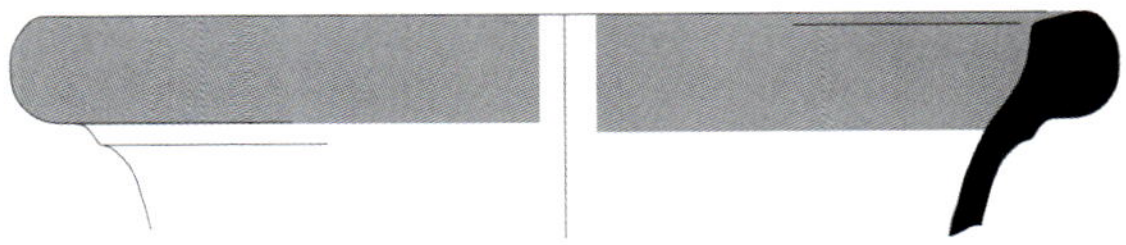

emphasized by a ridge.

Comparisons: See J7a2.

J7a2

129. PWBW 305-01

Dimensions: r 14.5 h 2.7 w 3.1

Description: See J7a1. J7a2 has a flat upper face.

Comparisons: Carter and Toxey 1998, 711, T324-2 (Pantanello, "pelike," 460–440 BC; there are similar examples from the kerameikos of Metapontion), and 713, T261-3 ("pelike," 460–440 BC); Saunders and du Plat Taylor

1:2

1992, Fig. 20, No. 359 (Gravina, large vessel type 2, period VI, 4th–3rd c. BC).

Jug (J) cont.

1:2

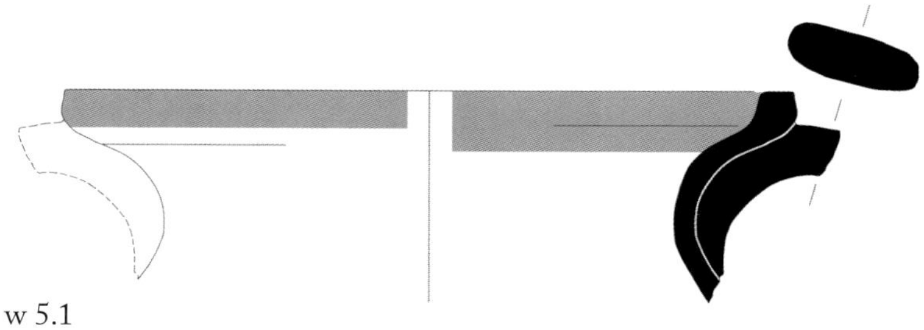

J7a3

130. PWBW 173-20

Dimensions: r 16.0 h 4.7 w 5.1

Description: See J7a1. J7a3 has a nearly indistinct rim with flat upper face and a vertical handle with elliptical section attached at the neck.

Comparisons: Carter and Toxey 1998, 706, T149-1 (Pantanello, hydria, 450–430 BC); Lo Porto 1966, 210, Fig. 64, No. 1 (Metaponto, *anfora*, mid-6th c. tomb, local production).

J7b

1:2

131. PWBW 045-02

Dimensions: r 15.0 h 4.5 w 7.4

Description: See J7a1. J7b is similar to J7a but has the upper face inclined towards the interior and a nearly straight neck.

Comparisons: See J7a2.

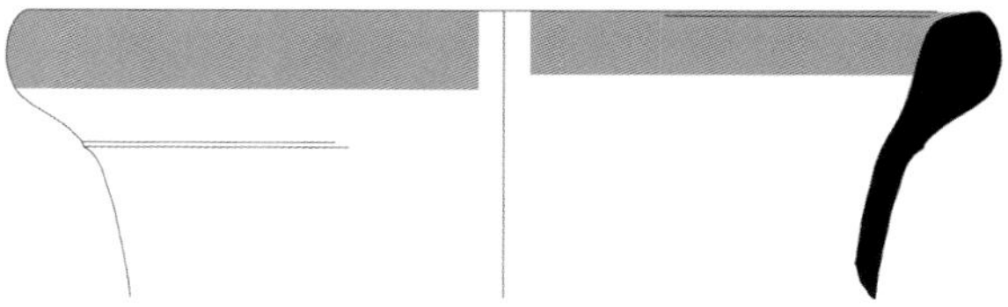

J8a

1:2

132. PWBW 521-165

Dimensions: r 15.0 h 3.0 w 7.2

Description: Thickened rim with convex exterior profile, nearly flat interior profile, and upper face inclined towards the interior; slightly concave neck.

Comparisons: see J7a2, which has the most similar rim.

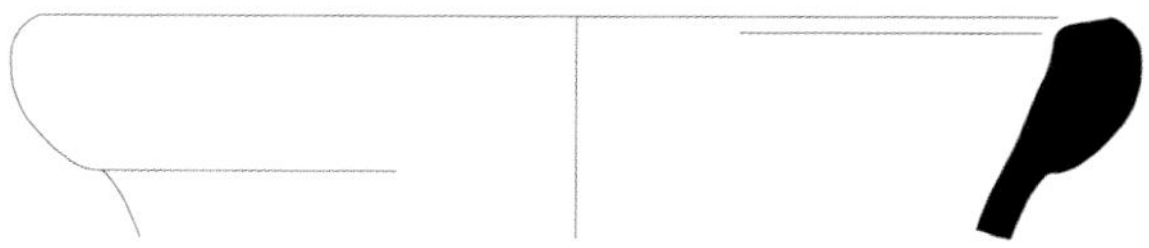

1:2

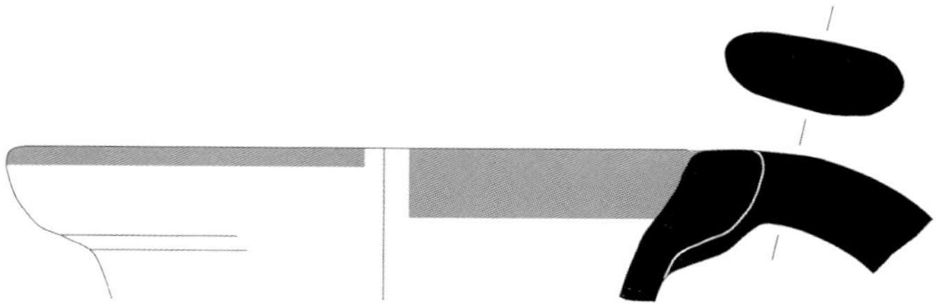

J8b

133. PWBW 521-88

Dimensions: r 14.4 h 3.0 w 4.9

Description: J8b is similar to J8a but is more compressed.

Comparisons: see J7a2, which has the most similar rim.

J9

1:2

134. PWBW 082-07

Dimensions: r 7.8 h 1.7 w 5.2

Description: Thickened rim with upper face inclined towards the exterior; exterior decoration of parallel grooves.

Comparisons: Marchegiani 1997, Pl. 63, No. 51b (Pomarico, BW *anfora da tavola*, thin rim, diam. 8.8 cm); Dyson 1983, Fig. 314, No. 3 (Buccino, PW jug, thin rim).

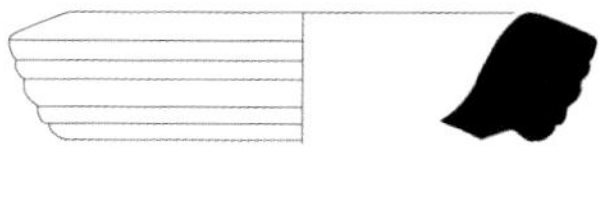

Jug (J) cont.

1:2

J10

135. PWBW 635f4-05

Dimensions: r 6.1 base 6.0 h 7.9

Description: Outward-turned rim with flat upper face inclined towards the interior and concave profile neatly distinguished from the globular body disk foot with concave base; vertical strap handles attached beneath the rim and at the belly.

Comparisons: Hempel 2001, Pl. 7, Form 311 (Taranto, phase E3, first half of 1st c. BC); Hempel and Mattioli 1994, 366, Fig. 314 (Taranto, T1); Mastronuzzi 1995, 204, Fig. 28, No. 33 (Nardò, GW "pelike," middle of the 2nd c. BC; cf. Dell'Aglio 1991b, 87 No. 73, Taranto, GW *anfora*, tomb, second half of 2nd c. BC, diam. 10.8 cm).

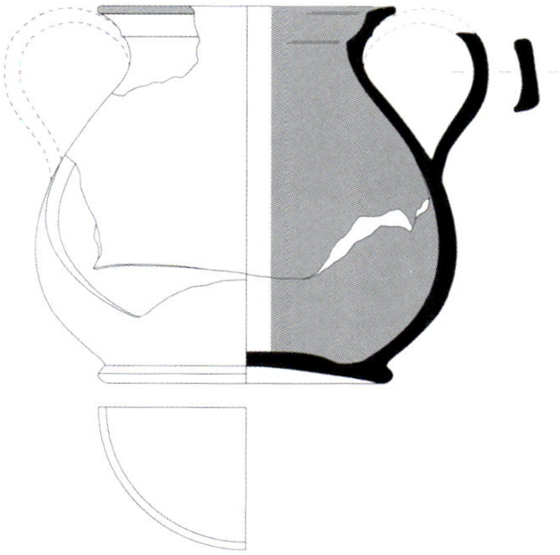

1:2

J11

136. PWBW 394-12

Dimensions: r 22.0 h 3.9 w 5.5

Description: Vertical rim that is concave on the interior and decorated on the exterior by a groove, attached to a concave neck; a sharp edge on the interior is probably a lid-supporting ledge.

Comparisons: Hänsel 1973, 452, Fig. 32, No. 8 (Herakleia, from a stratum dateable to last third of 5th c. BC).

1:2

J12

137. PWBW 549-07

Dimensions: r 13.0 h 2.5 w 4.8

Description: Slightly thickened rim with sub-rectangular section, convex exterior profile, and flat upper face.

Comparisons: See J14a.

1:2

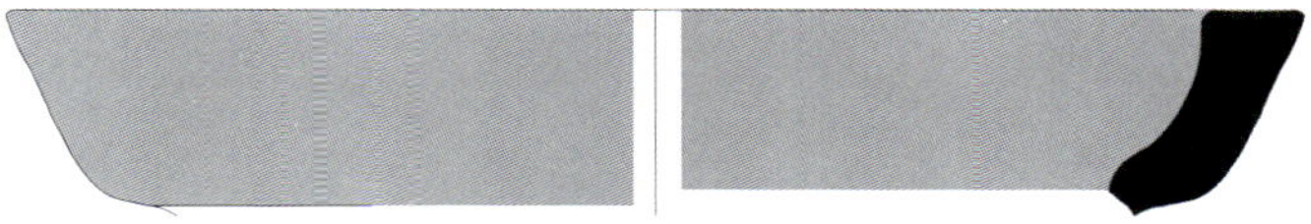

J13

138. PWBW 395-39

Dimensions: r 17.5 h 2.8 w 6.0

Description: Thickened rim with sub-rectangular section, flat upper face, and concave interior, neatly distinguished from the neck.

Comparisons: The type is probably close to J21b (see below); for the shape of the rim, see also lekythos L4.

Jug (J) cont.

1:2

J14a

139. PWBW 394-01 rev90

Dimensions: r 14.7 h 6.1 w 14.3

Description: Thickened rim with sub-rectangular section and flat upper face, neatly distinguished from the nearly straight neck wall; vertical handle with elliptical section attached to the rim.

Comparisons: Postrioti 1996, Fig. 8d (Metaponto, BW jug, diam. 13.3 cm, end of 4th–beginning of 3rd c. BC); O. Bianco 1997a, Pl. 78, No. 153 (Pomarico, PW *anforetta*, diam. 17 cm, second half of 3rd c. BC); Saunders and du Plat Taylor 1992, Fig. 20, No. 358 (Gravina, large vessel type 2, period VI, 4th–3rd c. BC); Hempel 2001, Pl. 8, Form 312 (Taranto, phase F, third quarter of 1st c. BC); *Gli ori,* 511, No. 17 (Taranto, *anfora*, tomb of first half of 1st c. BC); *Sibari* III, 157, Fig. 168, No. 25898 (BF amphora, diam. 18 cm), and 250, Fig. 256, No. 3504 (diam. 14 cm); *Corinth* VII.3, Pl. 12, No. 301 (two-handled "decanter III," first quarter of 3rd c. BC, diam. 9 cm).

J14b

1:2

140. PWBW 043-200 rev82

Dimensions: r 16.0 h 5.3 w 11.4

Description: See J14a. J14b has a groove on the lower part of the rim.

Comparisons: See J14a, and in particular Postrioti 1996, Fig. 9c (Metaponto, BW jug, diam. 16.1 cm, end of the 4th–beginning of 3rd c. BC).

J15

1:2

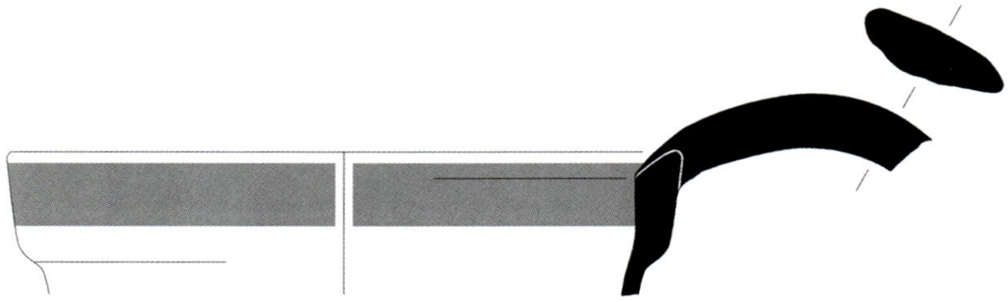

141. PWBW 258-16

Dimensions: r 14.0 h 3.0 w 6.5

Jug (J) cont.

1:2

142. PWBW 525-04
Dimensions: r 18.5 h 2.7 w 6.5

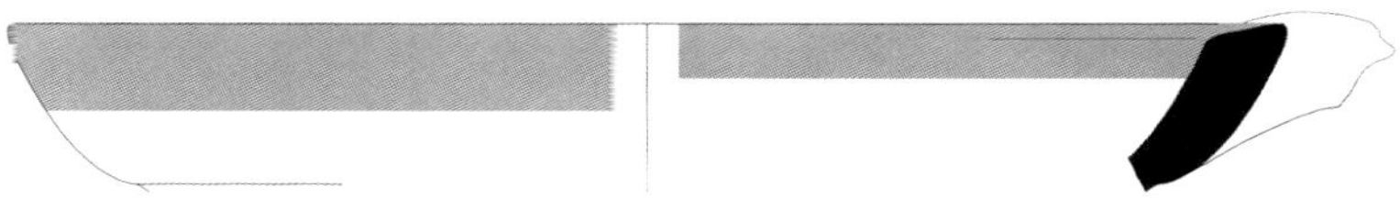

143. PWBW 735-08
Dimensions: r 15.5 h 2.4 w 2.6

1:2

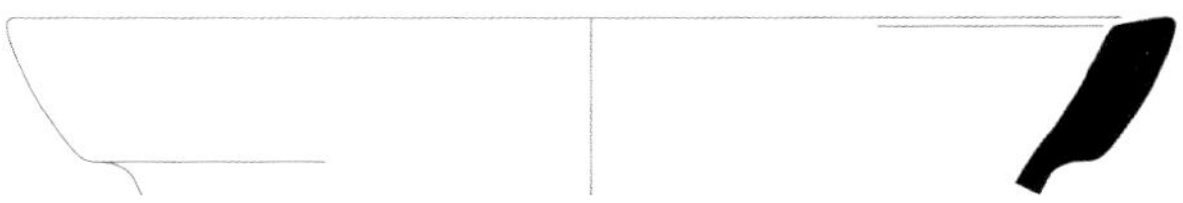

144. PWBW 258-15
Dimensions: r 15.7 h 5.6 w 5.3

1:2

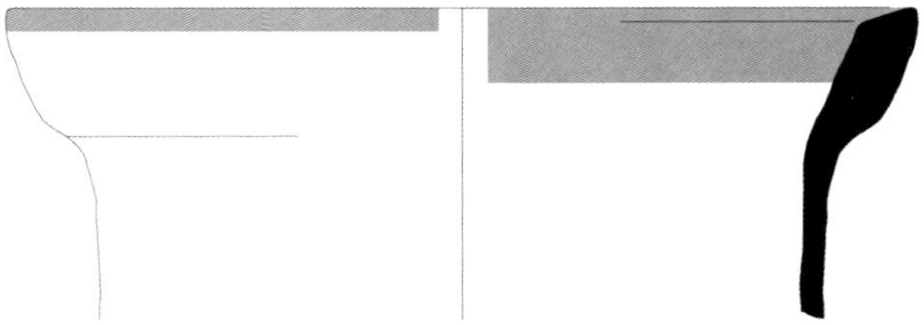

Description: Thickened rim with parallelepiped section, barely convex exterior profile, and flat upper face inclined towards the interior, neatly distinguished from the straight neck wall; vertical handle with elliptical section attached to the rim. The example 525-04 has an impressed rosette on the rim, at the point of attachment of the handle.

Comparisons: Postrioti 1996, Fig. 9b (Metaponto, BW jug, diam. 14.1 cm, end of 4th c. BC); D'Andria 1980a, 431, Fig. 67, No. 328 (Metaponto, *scarico* 3, third quarter of 4th c. BC, diam. 14.7 cm); Lo Porto 1966, 230, Fig. 93, No. 1 (Metaponto, *anfora*, mid-6th c. BC tomb); Marchegiani 1997, Pl. 62, No. 44 (Pomarico, BW, diam. 16.6 cm, end of the 6th c. BC; 1.169: the diameter is generally between 15 and 18 cm, and the type is often plain); O. Bianco 1997a, Pl. 78, No. 152 (Pomarico, PW *anforetta*, diam. 16.4 cm, second half of the 3rd c. BC); *Sibari* III, 250, Fig. 256, No. 4475 (diam. 14 cm).

J16a
145. PWBW 113-21
Dimensions: r 12.8 h 12.9 w 9.3

1:2

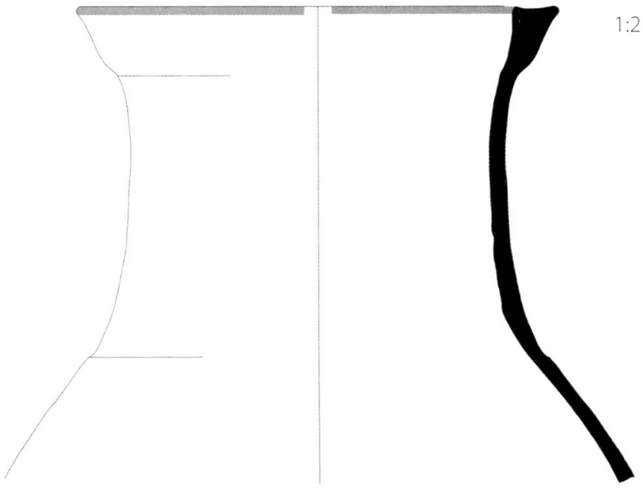

Jug (J) cont.

1:2

146. PWBW 365-04
Dimensions: r 12.6 h 7.6 w 8.2
Description: Thickened rim with sub-triangular section and concave upper face, distinguished from the neck, which is in turn distinguished from the shoulder via a slight carination.
Comparisons: Cotton 1983, 370, Fig. 140, No. 412 (Cozzo Presepe, PW table amphora); O. Bianco 1997a, Pl. 77, No. 150 (Pomarico, PW *anforetta*, diam. 12 cm, 3rd c. BC); *Sibari* I, 101, Fig. 86, No. 104; similar also in *Sibari* III, 259, Fig. 262, No. 11583 (amphora).

1:2

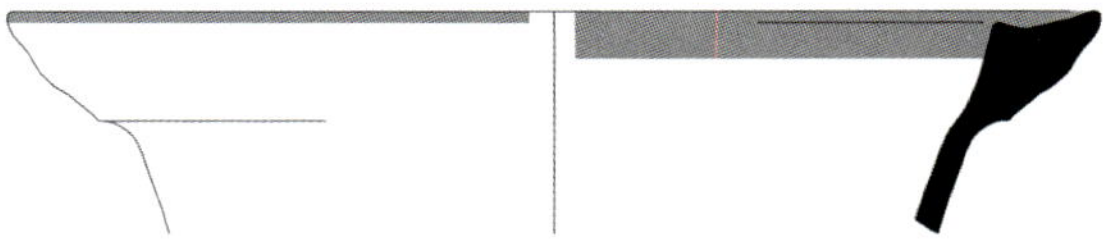

J16b
147. PWBW 352-25
Dimensions: r 14.5 h 3.0 w 3.5
Description: See J16a. J16b has a decidedly concave upper face.
Comparisons: See J16a.

1:2

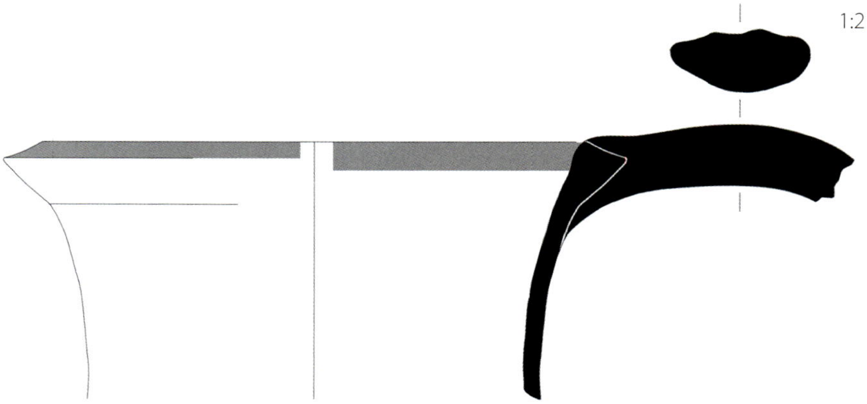

J16c
148. PWBW 043-63 rev82
Dimensions: r 16.5 h 7.0 w 8.4
Description: See J16a. On J16c the rim with sub-triangular section is nearly indistinct, attached to a neck with straight walls; the vertical, ribbed handle is attached at the rim.
Comparisons: None found, at present.

1:2

J17a
149. PWBW 365-25
Dimensions: r 17.0 h 2.9 w 6.9
Description: Thickened rim with sub-triangular section. J17a has a flat upper face with the rim neatly distinguished from the neck and a vertical handle with elliptical section attached at the rim.
Comparisons: For similar rim, see J16a; type present also in BG, dated approximately to middle of the 5th c. BC.

Jug (J) cont.

1:2

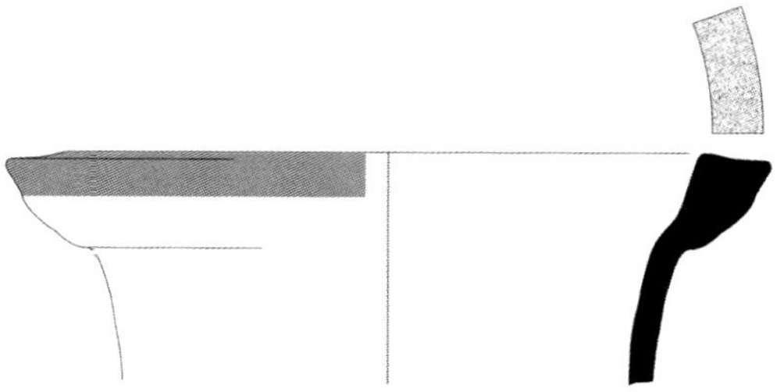

J17b
150. PWBW 394-10
Dimensions: r 13.0 h 4.1 w 7.5
Description: See J17a. J17b has a rim with slightly concave upper face, barely inclined towards the exterior.
Comparisons: See J17a.

J17c
151. PWBW 365-26

1:2

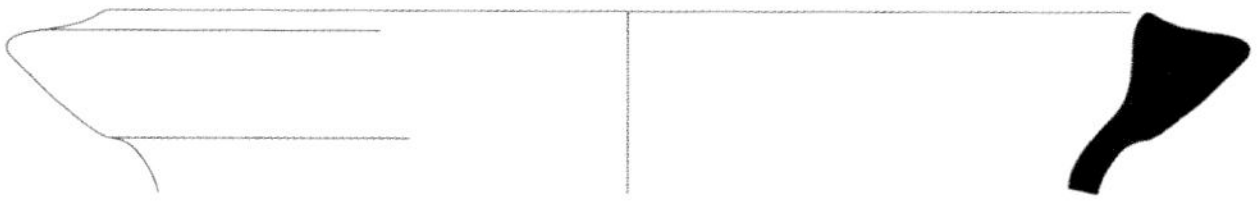

Dimensions: r 16.5 h 2.5 w 8.5
Description: See J17a. J17c is like J17b, but the face of its rim is decidedly inclined towards the exterior.
Comparisons: See J17a.

J17d
152. PWBW 762-16

1:2

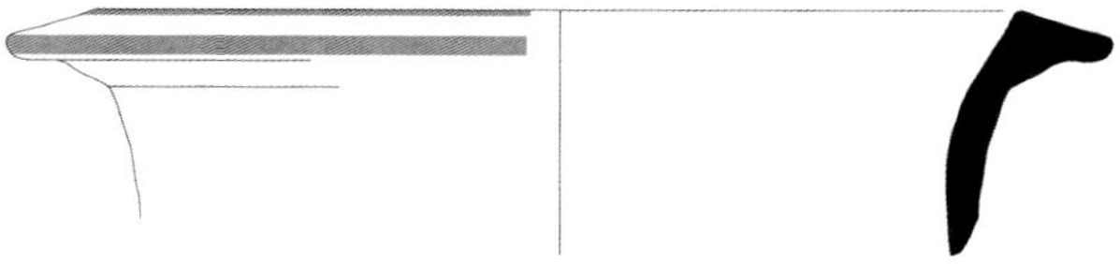

Dimensions: r 15.7 h 3.6 w 5.4
Description: See J17a. J17d has an outward-turned rim with sub-triangular section and flat upper face inclined towards the exterior.
Comparisons: See J1.

J18
153. PWBW 358-33 rev90

1:2

Dimensions: r 15.0 4.7 w 10.5
Description: Similar to J15 but with slightly concave upper face.
Comparisons: du Plat Taylor 1983, 345, Fig. 121, No. 265 (Cozzo Presepe, Site A, topsoil); cf. also J15.

J19a
154. PWBW 365-27

1:2

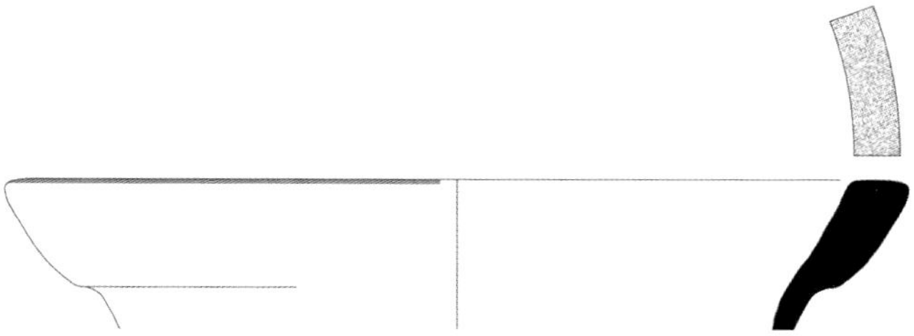

Dimensions: r 16.8 h 3.0 w 5.2
Description: Similar to J15 but with flat upper face.
Comparisons: Saunders and du Plat Taylor 1992, Fig. 20, No. 360 (Gravina, large vessel type 2, period VI, 4th–3rd c. BC; cf. also *Gravina* III.2 124, Fig. 50, No. 265).

J19b
155. PWBW 312-14

1:2

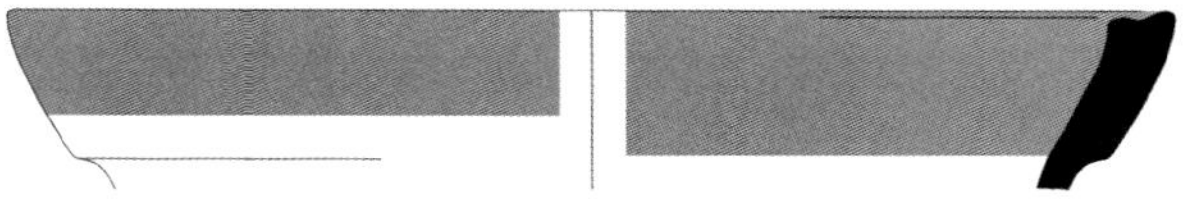

Dimensions: r 15.5 h 2.4 w 3.4
Description: Similar to J15 but with concave upper face.
Comparisons: See J19a.

Jug (J) cont.

1:2

J20

156. PWBW 529-42

Dimensions: r 15.0 h 3.5 w 4.9

Description: Slightly thickened rim with sub-rectangular section, straight exterior profile, and flat upper face, nearly indistinguishable from the neck.

Comparisons: See J14a.

1:2

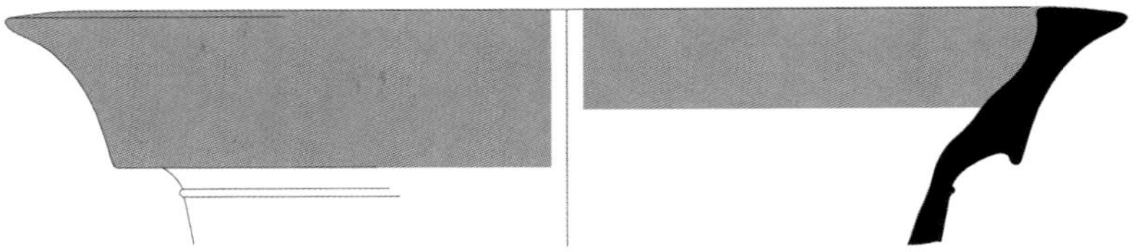

J21a

157. PWBW 182-20

Dimensions: r 20.6 h 4.5 w 4.8

Description: High, outward-turned rim with concave profile on both the exterior and the interior and flat upper face, neatly distinguished from the neck.

Comparisons: D'Andria 1990, 267, No. 103 (Oria, *hydria messapica a figure nere* with four-pointed torch on the handle, diam. 19.1 cm, second half of 6th c. BC).

1:2

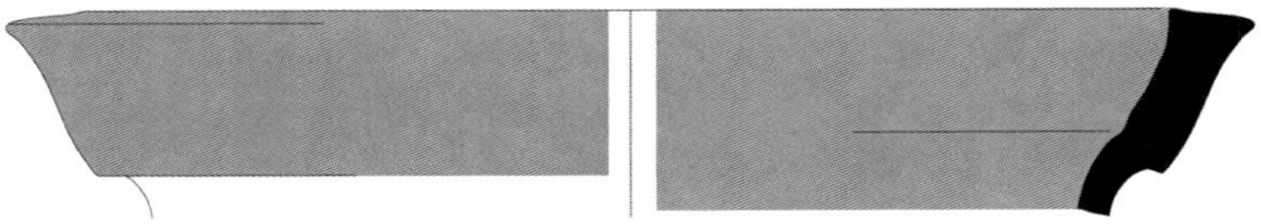

J21b

158. PWBW 334-22

Dimensions: r 19.5 h 3.4 w 7.9

Description: See J21a. On J21b both the exterior and interior faces are only slightly concave.

Comparisons: Elliott 1998, 649, A7 (Pantanello, black gloss, third quarter of 6th c. BC); *Sibari* II, 109, Fig. 90, No. 12638 (end of 7th–first half of 6th c. BC); Donnarumma and Tomay 1999, Pl. XXIV, Fig. 1 (San Brancato, BW hydria, tomb, 490–470 BC).

1:2

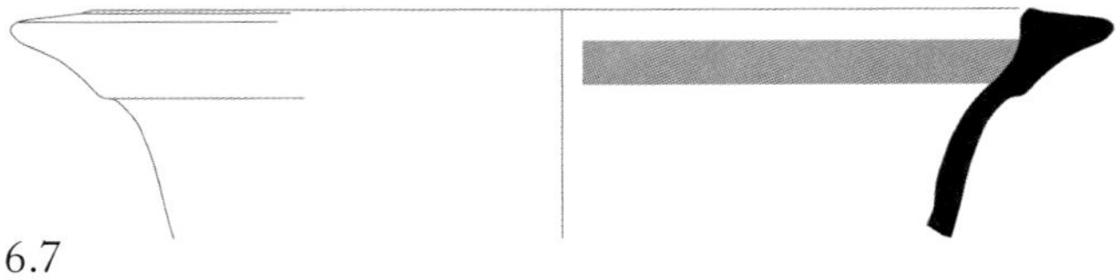

J22a

159. PWBW 038-125PLb

Dimensions: r 16.9 h 3.6 w 6.7

Description: Rim with sub-triangular section and concave faces (both exterior and interior), neatly distinguished from the straight neck wall.

Comparisons: Cf. BG fragment BG 529-01rev90 (Elliott 1998, 655, Pantanello, H2, 450–430 BC).

1:2

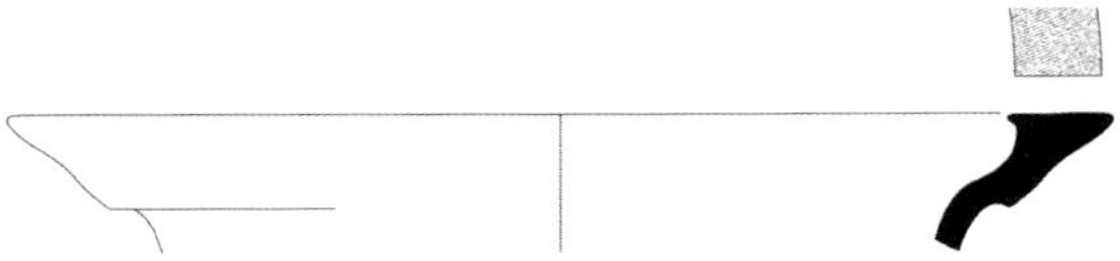

J22b

160. PWBW 183-19

Dimensions: r 14.8 h 1.9 w 5.8

Description: See J22a. On J22b the concavity of the interior face is notably accentuated.

Comparisons: See J22a.

Jug (J) cont.

1:2

J23a
161. PWBW 300c-03
Dimensions: r 18.3 h 3.0 w 5.1
Description: Thickened rim with sub-triangular section, concave upper face, and ridge near the straight neck walls.
Comparisons: Cotton 1983, 370, Fig. 140, No. 415 (Cozzo Presepe, unstratified); du Plat Taylor 1983, 345, Fig. 121, No. 261 (Cozzo Presepe, Site A, topsoil); Marchegiani 1997, Pl. 63, No. 48 (Pomarico, BW *anfora da tavola*, diam. 8.8 cm, second half of 3rd c. BC).

1:2

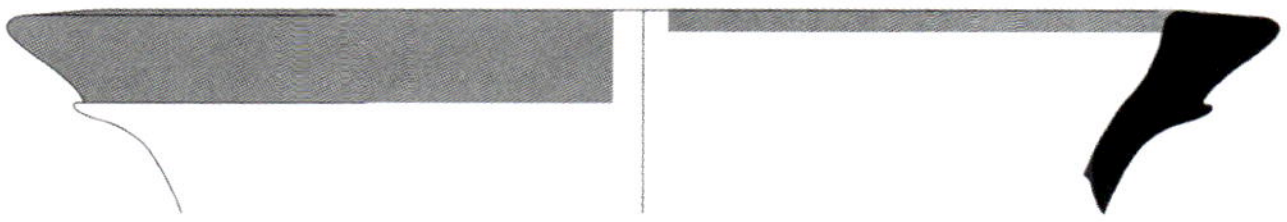

J23b
162. PWBW 221-02
Dimensions: r 16.9 h 2.8 w 5.7
Description: See J23a. On J23b the upper face is flat.
Comparisons: Elliott 1998, 649, A4 (Pantanello, BG, 470/460–440 BC); Postrioti 1996, Fig. 9d (Metaponto, BW jug, diam. 19.3 cm, end of 4th–beginning of 3rd c. BC); Marchegiani 1997, Pl. 62, No. 45 (Pomarico, BW *anfora da tavola*, diam. 12.6 cm, last quarter of 4th–beginning of 3rd c. BC); Costamagna 1983, Pl. XLIX, No. 6 (Taranto, BG, diam. 11.6 cm; cf. Morel 1981a, 5344a1, BG jug, Lipari, first half of 3rd c. BC); *Gravina* III.2, 117, Fig. 48, No. 251 (BG "olpe," overhead handle, second half of 5th c. BC); Dyson 1983, Fig. 144, No. 3 (Buccino, PW jug, diam. ca. 7.5 cm); Dyson 1976, Fig. 6, CF76 (Cosa, "pitcher," diam. 16 cm, from deposit dated 275–150 BC, corresponding to Taylor's deposit A, 273–180/160? BC); *Sibari* III, 263, Fig. 265, No. 2671 (BG jug or amphoriskos, diam. 9 cm).

1:2

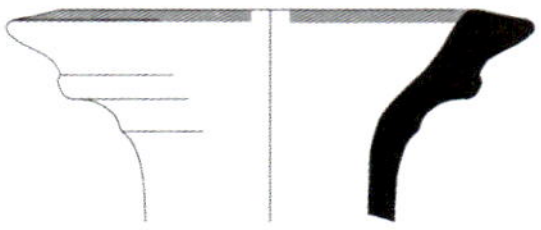

J23c
163. PWBW 353-14
Dimensions: r 7.0 h 2.9 w 4.7
Description: See J23a. J23c is like J23b but has an upper face decidedly inclined towards the exterior.
Comparisons: du Plat Taylor 1983, 345, Fig. 121, No. 262 (Cozzo Presepe, Site B, Phase VI, mid-3rd c. BC); Cotton 1992b, Fig. 62, No. 1279 (Gravina, "jug," period VIIIb, unstratified).

1:2

J23d
164. PWBW 449-11
Dimensions: r 13.0 h 2.9 w 6.8
Description: See J23a. J23d has a thickened rim with irregular section, flat upper face inclined towards the interior, and a groove on the exterior face.
Comparisons: None found, at present.

1:2

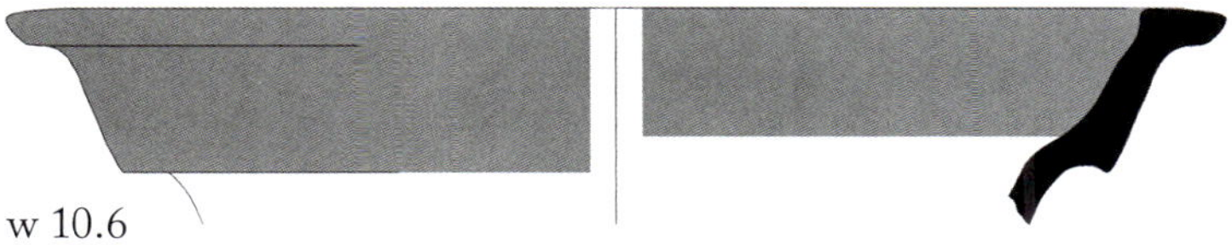

J24
165. PWBW 505-03
Dimensions: r 19.0 h 3.4 w 10.6
Description: Similar to J21, but the rim is flat and outward-turned with sub-rectangular section, concave on the exterior, and neatly distinguished from the neck.
Comparisons: Cotton 1992b, Fig. 62, No. 1275 (Gravina, "jug," period VI, 4th–3rd c. BC).

Mortar (Mor)

Type Mor1 has an average diameter in the range 26.0–34.0 cm (except for one example from site 239 measuring 16.0 cm). Mor2 falls within the same range (26.0–30 cm), and even in part Mor11, which attains a diameter of up to 42.0 cm. The type with indistinct hemispherical rim (Mor5), on the other hand, is larger (34.0–44.0 cm), although smaller examples have also been identified (32.0 cm at sites 352 and 516) as well as larger ones (59.0 cm at site 12); the variants, however, have smaller diameters (Mor7: 31.5–40.0 cm; Mor9: 30.0–35.6 cm; Mor10: 32.7 cm), just as Mor8 is smaller (28.0–40.0 cm). The few fragments of the other types have diameters of 36.6 cm (Mor3), 37.0–39.8 cm (Mor4), and 32.5 cm (Mor6). The bases have two types: on Mor12 the foot is a ring or a false ring (13 examples were recovered that have a diameter between 12.0 and 19.0 cm), while on Mor13 the foot is a disk (nine fragments are attributed to this type, measuring 16.0–25.3 cm).

1:4

Mor1

166. PWBW 073-01

Dimensions: r 30.1 h 7.2 w 11.0

Description: Thickened, amygdaloid rim; hemispherical body wall. As the earliest type of mortar encountered among the survey materials, Mor1 possibly evolved into Mor2 and Mor3.

Comparisons: Cotton 1983, 367, Fig. 136, Nos. 386–87 (Cozzo Presepe, Site A, phase IIIb, 575–550 BC); Rinaldi 2005, 229–30, 238, Pl. XXVII, No. 103 (Torre di Satriano); Blasi 2001, 64, Fig. 49, No. 65 (Torre di Satriano, middle of the 4th c. BC, diam. 22.6 cm); Cotton 1992d, Fig. 83, No. 1507 (Gravina, 5th–4th c. BC); Cuozzo and D'Andrea 1991, Fig. 9, No. 53A (Pontecagnano, Orientalizing period–first quarter of the 5th c. BC); *Kaulonia* I, 64, Fig. 42, Nos. 225–28 (from Archaic/5th-c. BC levels); *Sibari* III, 251, Fig. 257, No. 4246 (diam. 27 cm), and 255, Fig. 259, No. 5967 (diam. 45 cm); *Sibari* II, 336, Fig. 362, No. 23075 (*bacile*); Jones/Graham/Sackett 1973, 387, Fig. 10, No. 98 (Vari, mortar, diam. 23 cm, 4th c. BC).

1:4

Mor2a

167. PWBW 360-01

Dimensions: r 28.4 h 3.1 w 10.3

Description: Oblique, outward-turned rim with elliptical section; hemispherical body wall. Mor2a has a thickened rim. The type probably evolved from Mor1.

Comparisons: See Mor2b; the rim also recalls Lne/Bas3 (see above).

1:3

Mor2b

168. PWBW 074-02

Dimensions: r 39.5 h 3.6 w 8.6

Description: See Mor2a. On Mor2b the tip is tapered.

Comparisons: Deodato and Bianco 1997, Pl. 73, No. 110 (Pomarico, PW *ciotola* with wider rim, diam. 21.2 cm); *Kaulonia* I, 64, Fig. 42, No. 231; *Gravina* III.2 97, Fig. 41, No. XIX.2 ("mortar," 4th c. BC); *Valesio* I, 260, Form M54, No. 449 (diam. 30.5 cm, context of end of 5th–beginning of 4th c. BC); Jones/Graham/Sackett 1973, 387, Fig. 10, No. 97 (Vari, mortar, diam. 28 cm, 4th c. BC). G. Greco 1991, 74, Figs. 165–66 (Serra di Vaglio, 4th c. BC).

Mortar (Mor) cont.

1:4

Mor3

169. PWBW 064-01

Dimensions: r 36.6 h 5.5 w 9.2

Description: Outward-turned rim with sub-triangular section; hemispherical body wall. Like Mor2, this type also probably evolved from Mor1.

Comparisons: Cotton 1983, 367, Fig. 136, No. 384 (Cozzo Presepe, end of 6th c. BC–Hellenistic period, Site D); Saunders and du Plat Taylor 1992, Fig. 7, No. 173 (Gravina; II.21: bowl type 1, common in the 5th and early 4th c. BC, probably continuing also into 3rd c.; from period VIII levels, end of 2nd–1st c. BC); Keith and Fracchia 1990a, 272, Fig. 187, No. 278 (Roccagloriosa, PW "bowl," diam. 22–24 cm); Rouveret 1978, Pl. 53, No. A64 (Leuca, BW, Ionic-type ceramic); Threipland and Torelli 1970, 102, Fig. 17, B.4 (Casale Pian Roseto, Coarse Creamware, end of 5th–4th c. BC); Lissi Caronna 1983, 239, Fig. 24, No. 11 (Oppido Lucano, "pelvis," diam. 24.6 cm, from T57 dated end of 5th–beginning of 4th c. BC). Cf. also *Agora* XII, Fig. 16, No. 1884 (Attic mortar, 565–550 BC; this example is similar to Lne/Bas3b, also in part because of the concentric grooves on the rim).

Mor4

170. PWBW 432-01

1:4

Dimensions: r 39.8 base 22.7 h 12.4 w 19.7

Description: Horizontal, outward-turned rim with rectangular section, slightly oblique upper face, and nearly vertical exterior face; hemispherical body wall; disk foot; horizontal, vertically-oriented handle with elliptical section, attached at the rim. Although the thickness of the body walls indicates identification of this form as a mortar (related to Mor11b), the relative depth of the body suggests a possible second function as a basin (see especially Bas7).

Comparisons: For the shape of the rim, see Bas1b and especially Bas7; among the Metapontine mortar types see, as noted above, Mor11 (in particular Mor11b).

Mor5a

171. PWBW 309-02

1:4

Dimensions: r 44.0 h 6.2 w 17.3

Description: Indistinct rim with hemispherical section, emphasized on the lower side by a groove (occasionally there are two grooves below the rim: cf. site 395, Mor5a); nearly straight body wall; slightly concave disk base with sub-rectangular section.

Comparisons: Carter 1983a, 453, Fig. 28, No. 20 (Pantanello, kiln fill, diam. 28 cm); Deodato and Bianco 1997, Pl. 66, No. 68 (Pomarico, diam. 38.6 cm, Corinthian type, end of 4th–second half of 3rd c. BC; 1.176: the examples from Pomarico have a variable rim diameter between 20 and 40 cm); Cotton 1992d, Fig. 83, No. 1510 (Gravina, 5th–4th c. BC, period VIIIa, end of 2nd–1st c. BC); *Gravina* III.2, 129, Fig. 53, No. 290; *Sibari* III, 155, Fig. 166, No. 13848 (diam. 37.2 cm, with spout), and 413, Fig. 447, No. 12769 (diam. 38 cm); Conti 1989c, Pl. XXXIX, No. 345 (Locri, diam. 36.8 cm, end of 5th–3rd c. BC; 295: rim diameter varies between 36 and 39 cm, and height is 6–7 cm); *Kaulonia* I, 65, Fig. 43, Nos. 246 (type b) and 248 (type a), from 4th/3rd-c. BC levels; *Agora* XII, Fig. 16, No. 1912 (Corinthian, 450–425 BC; 370: comparison with a similar mortar from a mid-4th c. BC context as evidence of these objects' long life).

Mortar (Mor) cont.

1:4

Mor5b

172. PWBW 489-01

Dimensions: r 36.0 base 24.7 h 8.1 length 14.7 w 18.5

Description: See Mor5a. Mor5b is distinguished by a concave body wall and a base with sub-triangular section.

Comparisons: See Mor5a.

1:4

Mor5c

173. PWBW 012-16P

Dimensions: r 59.0 base 48.0 h 6.1 w 17.6

Description: See Mor5a. Mor5c has a more compressed rim with nearly elliptical section.

Comparisons: See Mor5a.

Mor6

1:4

174. PWBW 481-04

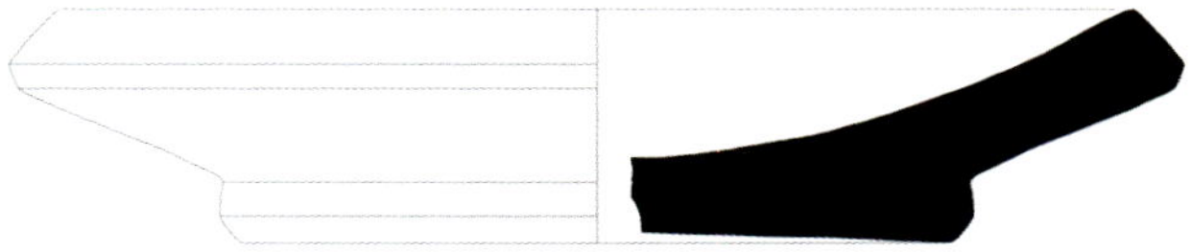

Dimensions: r 32.5 base 20.8 h 6.7 w 12.0

Description: Indistinct rim with flat, oblique upper face and rectangular section; straight body wall; carinated disk foot with slightly concave lower face. The thickness of the walls (2.0–2.5 cm) makes it particularly durable.

Comparisons: None found, at present.

1:4

Mor7a

175. PWBW 182-16

Dimensions: r 39.0 h 4.9 w 14.4

Description: Indistinct rim with nearly hemispherical section, emphasized on the underside by two grooves; hemispherical body wall. The type is certainly a smaller version of Mor5 (see above for the comparanda) but with deeper body, thinner walls, less pronounced rim, and smaller diameter.

Comparisons: See Mor5.

Mor7b

1:4

176. PWBW 182-17

Dimensions: r 31.5 h 4.4 w 9.2

Description: See Mor7a. On Mor7b the groove on the body wall is just barely perceptible.

Comparisons: See Mor5.

Mortar (Mor) cont.

1:4

Mor8

177. PWBW 635-01

Dimensions: r 28.0 base 17.0 h 5.3 w 23.5

Description: Outward-turned rim with elliptical section, emphasized on the underside by a groove; hemispherical body wall; disk foot with nearly vertical exterior face; concave resting surface bearing parallel grooves. The type can be considered an intermediate stage in the evolution of Mor5 towards Mor11. Among the materials recovered, the example from site 248 has the purely aesthetically-motivated spool-shaped handle already noted for Mor5.

Comparisons: Deodato and Bianco 1997, Pl. 66, No. 69 (Pomarico, diam. 28.3 cm, Corinthian type, end of 4th–second half of 3rd c. BC); *Kaulonia* I, 65, Fig. 43, No. 247 (from 4th/3rd-c. BC levels); Andreassi 1988b, 247–248, Nos. 524–525 (Bari, int. diam. 23–26.5 cm; a common object in 4th-c. Peucetian tombs); *Agora* XII, Fig. 16, No. 1891 (Attic Ck, 520–480 BC; 222: Athenian mortars, made on the wheel, appear not to have developed in later periods, having been supplanted by imported Corinthian mortars like Mor11c); *Corinth* VII.3, Pl. 22, No. 622 (Deposit 13, 430–420 BC).

1:4

Mor9a

178. PWBW 274-01

Dimensions: r 34.5 h 5.5 w 12.8

Description: Indistinct rim with hemispherical section, emphasized on the underside by two grooves (in several cases the grooves are completely absent: cf. site 395, Mor9a); hemispherical body wall. The shape of the rim makes this type similar to Mor5, although in this case the rim diameter and the body wall thickness are reduced, and the body is deeper.

Comparisons: *Kaulonia* I, 65, Fig. 43, No. 250 (from 4th/3rd-c. BC levels).

Mor9b

179. PWBW 370-08

Dimensions: r 35.6 h 5.0 w 6.6

Description: See Mor9a. This variant has a convex upper face, and it is emphasized on the underside by three grooves.

Comparisons: See Mor9a.

1:4

1:4

Mor10

180. PWBW 045-01

Dimensions: r 32.7 h 4.7 w 13.2

Description: Indistinct rim with compressed hemispherical section, emphasized on the underside by a groove; hemispherical body wall. This type is a variant of Mor5 (see above for the comparanda), but in this case, too (as for Mor7 and Mor9), the dimensions are reduced, and the body is slightly deeper.

Comparisons: See Mor5.

Mortar (Mor) cont.

1:4

Mor11a

181. PWBW 666-01*

Dimensions: r 33.6 h 3.7 w 10.6

Description: Outward-turned rim with sub-triangular section; generally hemispherical and shallow body wall.

Comparisons: Deodato and Bianco 1997, Pl. 66, No. 71 (Pomarico, diam. 35.2 cm, 4th–3rd c. BC); Keith and Fracchia 1990a, 272, Fig. 187, No. 289 (Roccagloriosa, diam. 26 cm); *Kaulonia* I, 64, Fig. 42, No. 236.

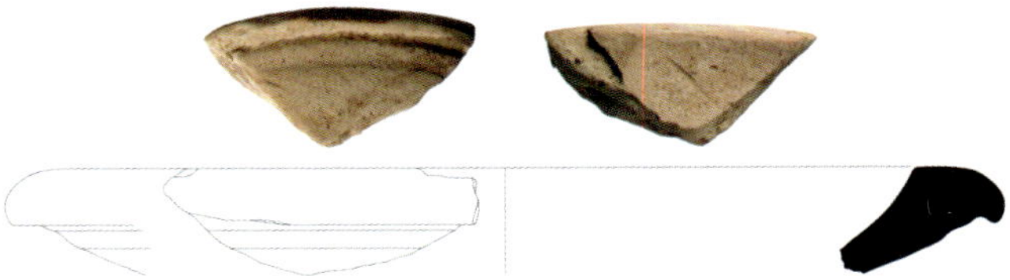

Mor11b

1:4

182. PWBW 173-01

Dimensions: r 36.0 h 5.7 length 12.55 w 20.1

Description: Outward-turned rim with sub-rectangular section; generally hemispherical and shallow body wall. Mor11b demonstrates a possible solution for the base as a false ring foot distinguished from the body wall, with sub-triangular section.

Comparisons: Kaulonia I, 64, Fig. 42, No. 232; *Corinth* VII.3, Pl. 22, No. 622 ("mortar I," 430–420 BC, diam. 26.4 cm).

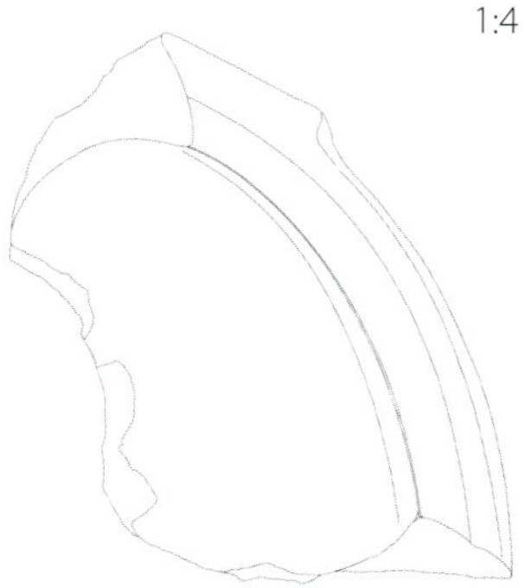

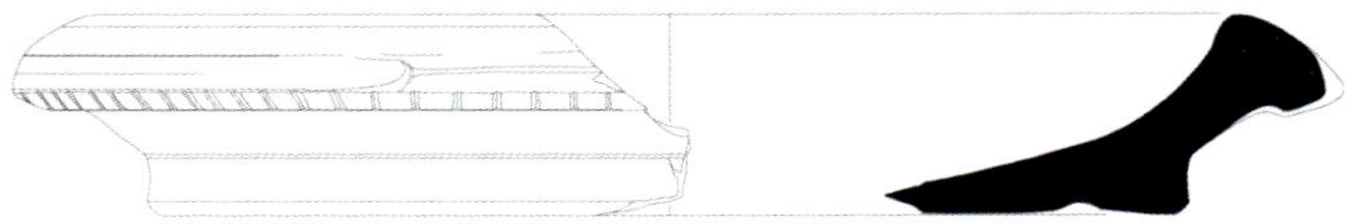

Mor11c

1:4

183. PWBW 695-01*

Dimensions: r 40.2 base 30.0 h 6.3 w 23.5

Description: Outward-turned rim with elliptical section; generally hemispherical and shallow body wall. Mor11c demonstrates a possible solution for the base as a false ring foot distinguished from the body wall, with sub-rectangular section.

Comparisons: Deodato and Bianco 1997, Pl. 66, No. 72 (Pomarico, diam. 26.6 cm, 4th–2nd c. BC); Cotton 1992d, Fig. 83, No. 1511 (Gravina, 5th–4th c. BC, period VI, 4th–3rd c. BC); Lo Porto 1991, Pl. XVIII, Fig. 1, No. 3 (Timmari, tomb dated 340 BC, diam. 24 cm); Small 1977a, 31, Fig. 6, No. 2 (Monte Irsi, tomb, 400 BC?); *Agora* XII, Fig. 16, No. 1898 (Corinthian, Sandy Class, second half of 4th c. BC; cf. 369: a 5th-c. example from the Agora, No. 1896, has the same type of clay; in the 4th c. the same form was also produced in Attica with a hard red clay); Vegas 1973, 30, Fig. 9, No. 6 (type 7, with grooves on the upper face, second half of 2nd c. BC); the example from T8 at Pisticci, dated to first decades of 5th c. BC, could be related to this type (Lo Porto 1973, Pl. XXV, No. 1.1).

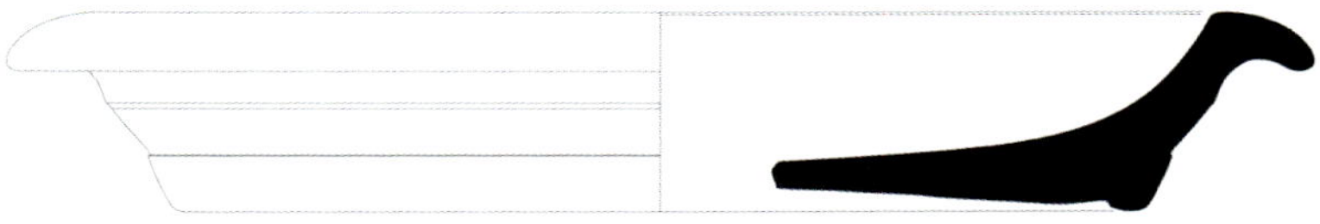

Mortar (Mor) cont.

1:4

Mor11d

184. PWBW 360-02

Dimensions: r 42.0 h 5.9 w 14.4

Description: Outward-turned rim with rectangular section. Mor11d has straight walls.

Comparisons: For the initially straight body wall and the rim which nearly forms a right angle with the wall, cf. *Agora* XII, Fig. 15, No. 1818 ("lekane," 375–330 BC, diam. 36.2 cm).

1:4

Mor12a

185. PWBW 309-06

Dimensions: base 19.0 h 3.4 w 13.4

Description: False ring foot with deep groove on the resting surface, trapezoidal section, and nearly vertical exterior face; oblique body wall.

Comparisons: None found, at present.

1:4

Mor12b1

186. PWBW 247-02

Dimensions: base 14.2 h 3.3 w 7.5

Description: Ring foot with square section, nearly vertical exterior face, and convex interior face; oblique body wall.

Comparisons: None found, at present.

1:4

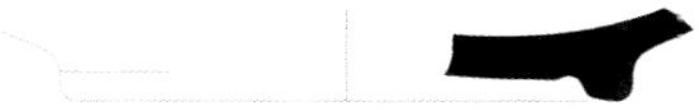

Mor12b2

187. PWBW 322-08

Dimensions: base 15.2 h 2.5 w 7.0

Description: Ring foot with square section, nearly vertical exterior face, and groove between interior face and underside; oblique body wall.

Comparisons: None found, at present.

1:4

Mor12c

188. PWBW 344-02

Dimensions: base 17.4 h 4.8 w 10.9

Description: Disk foot with triangular section and concave lower face; initially straight body wall.

Comparisons: Jones/Sackett/Graham 1962, 98, Fig. 11, No. 80 (Dema, diam. 12.8 cm, last quarter of 5th c. BC).

1:4

Mor12d

189. PWBW 334-23

Dimensions: base 14.4 h 4.0 w 9.4

Description: Ring foot with triangular section and oblique exterior face and resting surface; oblique body wall.

Comparisons: None found, at present.

Mortar (Mor) cont.

1:4

Mor13a
190. PWBW 169-20
Dimensions: base 25.3 h 3.8 w 10.6
Description: Disk foot with elliptical section barely distinguished from the oblique body wall by means of a shallow groove; flat resting surface.
Comparisons: None found, at present.

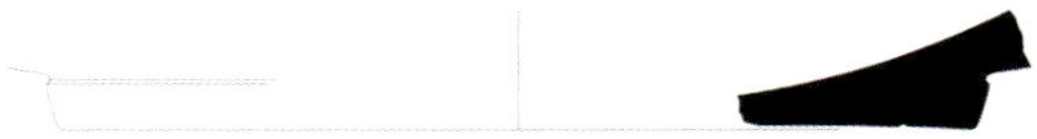

1:4

Mor13b
191. PWBW 319-03
Dimensions: base 25.0 h 3.3 w 7.1
Description: Disk foot with quadrangular section barely distinguished from the oblique body wall by means of a shallow groove; flat resting surface.
Comparisons: None found, at present.

1:4

Mor13c
192. PWBW 043-147 rev82
Dimensions: base 18.4 h 3.0 w 9.7
Description: Disk foot with triangular section barely distinguished from the oblique body wall by means of a shallow groove; flat resting surface.
Comparisons: None found, at present.

Pestle (Pst)

1:4

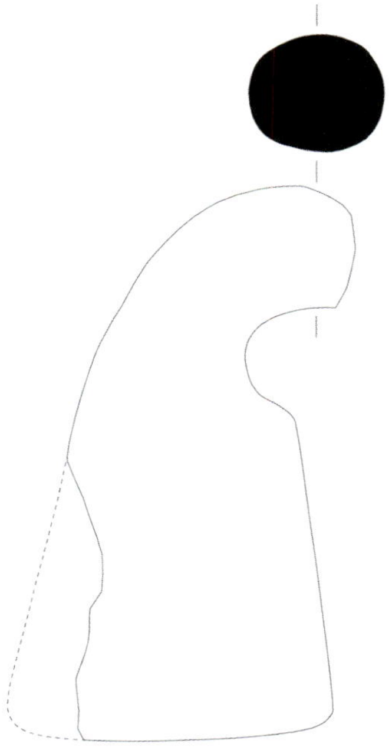

Pst1
193. PWBW 491-03
Dimensions: h 12.9 w 7.4
Description: Conical body; grip with arched, circular section.
Comparisons: *Arte e artigianato,* 70–71, Nos. 37–40 (Pestle types and their molds, from Taranto, 5th–4th c. BC, associated with the working of clay); Adamesteanu 1980, 60, No. 46c.

Basin (Bas)

The diameter of Bas1a lies generally between 50.0 and 59.5 cm, although a fragment recovered from site 164 belongs to a rim of ca. 40 cm diameter; the rim diameter of Bas1b, on the other hand, varies significantly, between 33.6 cm (from site 245) and 59.3 cm (PWBW 154-21), with a recorded maximum of 83 cm (**No. 196**, PWBW 489-08). The only example recovered of Bas2 (PWBW 182-18) has a diameter of 46.5 cm. The suggested employment as a louterion, mentioned in the introductory remarks, is reinforced by the similarity with Lou3 (see below).

Of the 18 examples of Bas3 recovered in the survey, ten are of Bas3b, three of Bas3a and Bas3d, two of Bas3c; the few fragments which preserve the diameter indicate an average between 30.0 and 45.2 cm, but there are examples that reach 56.2 cm (PWBW 344-01), 71.5 cm (PWBW 397-89), and 92.0 cm (PWBW 183-16). Bas4, of which six examples were recovered, has a rim diameter varying between 36.7 and 50 cm, but there is also a smaller variant (24.0 cm, from site 154). For the rim see also Bas1 above, namely the suggestion that this Basin could have been used also as a Louterion, a suggestion borne out by the comparanda. The only example recovered of Bas5 (PWBW 193-27) has a diameter of 49.0 cm.

Of the ten examples of Bas6 recovered in the survey only two preserve the rim diameter, which is generally smaller than in the previous types, apparently falling between 28.0 and 34.0 cm, while the two fragments preserving the base indicate dimensions of 15.3–18.0 cm. The three measurable fragments of Bas7a indicate a diameter varying between 41.6 and 46.0 cm; Bas7b is on average 40.0–51.5 cm wide, but a smaller type (26.0–30.0 cm) is also attested, which in one case (site 397) accompanies the larger form. Only three fragments belong to Bas8 (from sites 82, 231, and 364), of which only one (from site 82) gives an indication of the rim diameter: 47.9 cm.

In the preserved fragments of Bas9 the shape of the body wall, which appears to be somewhat deep, indicates a similarity to a deeper, albeit certainly smaller, basin: e.g., DBas4. Of the six fragments recovered, the example from site 103 (1990 revisit) has parallel grooves on the upper face; reference measurements are provided by fragments from site 358 (34.7 cm) and from Plot 2001-043 (38.0 cm).

1:4

Bas1a

194. PWBW 549-06

Dimensions: r 59.5 h 5.1 w 8.2

Description: Horizontal, outward-turned, and nearly indistinct rim with flat upper face, vertical exterior face, and rectangular section; straight body wall.

Comparisons: Pizzo 1995a, 95, Fig. 96 (Incoronata, type with support); *Sibari* II, 342, Fig. 367, No. 25713 (*bacile*, diam. 40 cm), and 360, Fig. 384, No. 27371 (*bacino*, diam. 46 cm); D'Andria 1978a, Pl. 27, No. 31 (Leuca, 5th–4th c. BC); *Agora* XII, Fig. 16, No. 1865, No. 1882 ("louterion," 425–400 BC).

1:4

Bas1b

195. PWBW 154-21

Dimensions: r 59.3 h 8.8 w 10.8

Description: Similar to Bas1a, but thicker body wall.

Comparisons: See Bas1a

196. PWBW 489-08

1:4

Dimensions: r 83.0 h 7.2 w 8.7

Description: Horizontal, outward-turned rim with flat upper face, vertical exterior face, and square section; concave body wall. Decorated with lightly impressed tongue pattern.

Comparisons: *Agora* XII, Fig. 16, No. 1865 ("louterion" with the same type of decoration as PWBW 489-08, Corinthian or near, 420–390 BC).

1:4

Bas2

197. PWBW 182-18

Dimensions: r 46.5 h 3.1 10.2

Description: Horizontal, outward-turned rim with flat upper face bearing two parallel grooves, vertical upper face, and rectangular section; shallow, presumably hemispherical body wall.

Comparisons: Cotton 1992b, Fig. 56, No. 1195 (Gravina, period VIIIa, end of 2nd–1st c. BC).

Basin (Bas) cont.

1:4

Bas3a

198. PWBW 382-01

Dimensions: r 45.2 h 5.4 w 12. 6

Description: Horizontal, outward-turned rim with trapezoidal section and oblique upper face; deep body. The rim of Bas3a is nearly indistinct.

Comparisons: Similar, but with grooves on the rim and the exterior, in Deodato and Bianco 1997, Pl. 68, No. 81 (Pomarico, diam. 35 cm; 1.177: these vessels are considered intermediate containers among the large basins and pithoi because of their shape, dimensions, and functions, and their diameter varies between 35 and 46 cm); Cotton 1992b, Fig. 61, Nos. 1248–1249 (Gravina, period VIIIb, unstratified).

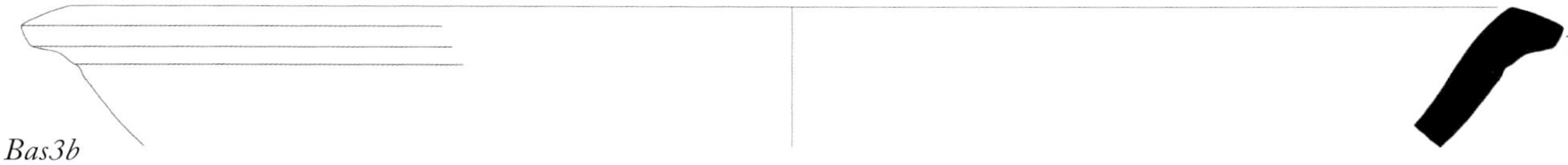

1:4

Bas3b

199. PWBW 397-89

Dimensions: r 71.5 h 6.8 w 16.5

Description: See Bas3a.

Comparisons: D'Andria 1980b Pl. XIII, No. 35 (Metaponto, *scarico* 12); *Agora* XII, Fig. 19, No. 1847 ("pre-Persian tub").

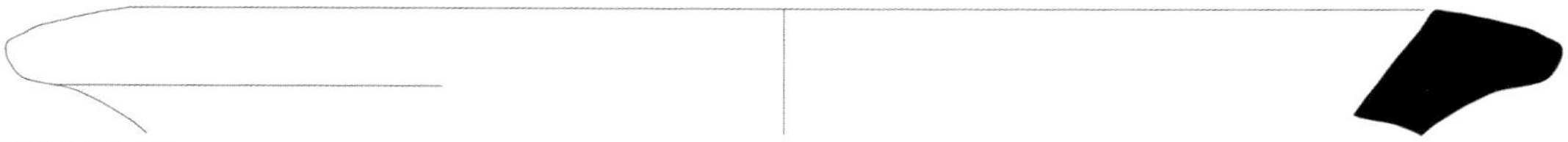

1:4

Bas3c

200. PWBW 344-01

Dimensions: r 56.2 h 4.7 w 21.7

Description: See Bas3a. Bas3c is similar to Bas3b but has a thicker rim.

Comparisons: *Sibari* III, 413, Fig. 44,7 No. 10083 (diam. 32 cm, flat base, shallow), and 422, Fig. 454, Nos. 16868–16869 (diam. 42,5/30 cm, misfired, deep); *Gravisca* 12.I, Pl. 13, No. 134 (L1, flat floor, end of 7th c.–beginning of 6th c. BC, diam. 53 cm).

1:6

Bas3d

201. PWBW 183-16

Dimensions: r 92.0 h 10.0 w 15.1

Description: See Bas3a. The characteristic feature of Bas3d is the rim with flat upper face.

Comparisons: see Bas3b.

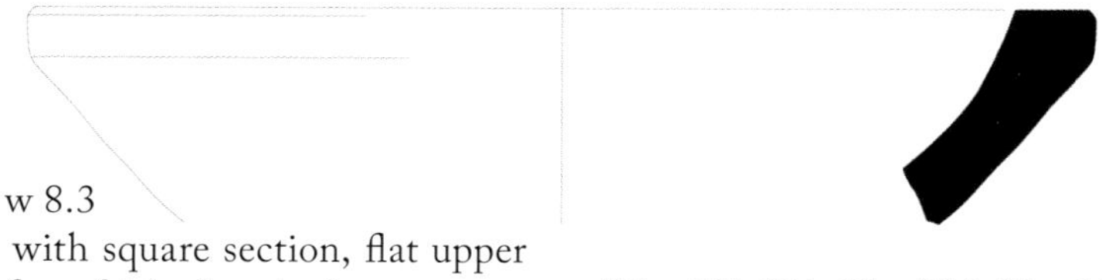

1:4

Bas4

202. PWBW 046-03

Dimensions: r 36.7 h 7.6 w 8.3

Description: Indistinct rim with square section, flat upper face, and vertical exterior face; fairly deep body.

Comparisons: Pizzo 1995a, 95, Fig. 95 (Incoronata, type with support); *Sibari* III, 403, Fig. 441, No. 4498 (marble); *Sibari* II, 201, Fig. 216, No. 9187 (*grosso bacile o mortaio a tre piedi*), and 342, Fig. 367, No. 27265 (*base di louterion*, with bead decoration).

1:4

Bas5

203. PWBW 193-27

Dimensions: r 49.0 h 3.7 w 10.8

Description: Horizontal, outward-turned rim with subrectangular section, flat upper face, and vertical lateral face; straight body wall.

Comparisons: None found, at present.

Basin (Bas) cont.

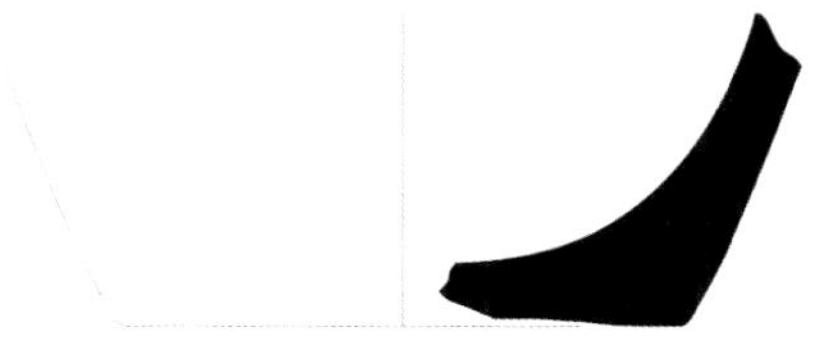

1:4

Bas6

204. PWBW 024-08

Dimensions: base 15.3 h 8.6 w 14.0

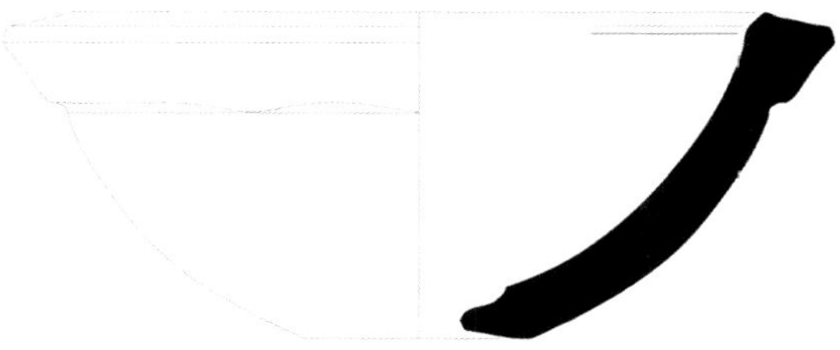

1:4

205. PWBW 342-01

Dimensions: r 28.0 base 7.6 h 11.4 w 13.0

Description: Thickened rim with variable section and concave upper face; hemispherical body wall; slightly concave floor.

Comparisons: Cotton 1992b, Fig. 61, No. 1246 (Gravina, lower rim, period VIIIa-b, end of 2nd c.–1st c. BC).

1:4

Bas7a

206. PWBW 349-03

Dimensions: r 41.6 h 4.6 w 7.5

Description: Oblique, outward-turned rim with trapezoidal section; deep body.

Comparisons: Cotton 1983, 372, Fig. 142, No. 424 (Cozzo Presepe, unstratified); *Sibari* II, 360, Fig. 384, No. 25460 (*bacile*, diam. 50 cm); *Kaulonia* I, 64 and 66, Figs. 42 and 44, Nos. 241 and 261; Dyson 1983, Fig. 222, No. 152 (Buccino); D'Andria 1978a, Pl. 27, No. 32 (Leuca, 5th–4th c. BC); *Agora* XII, Fig. 19, No. 1849 ("tub," 4th c. BC; 366: found buried with the rim at the occupation level).

1:4

Bas7b

207. PWBW 312-13

Dimensions: r 51.5 h 9.4 w 13.2

Description: Oblique, outward-turned rim with subrectangular section; deep body.

Comparisons: See Bas7a

1:4

Bas8

208. PWBW 082-06

Dimensions: r 47.9 h 6.1 w 10.0

Description: Horizontal, outward-turned rim with subcircular section and two parallel grooves on the upper face; hemispherical bowl.

Comparisons: Cotton 1983, 367, Fig. 136, No. 381 (Cozzo Presepe, Site A, phase IV, mid-6th c.–480 BC).

1:4

Bas9

209. PWBW 358-32

Dimensions: r 34.7 h 3.3 w 7.4

Description: Horizontal, outward-turned rim with rectangular section, flat upper face, and nearly vertical exterior face; initially straight body wall.

Comparisons: None found, at present.

Deep Basin (Dbas)

Only five fragments of DBas1 were large enough to preserve the diameter, which falls in the range of 43.0–61.0 cm. DBas2 is attested in nine cases. Three body fragments from sites 2, 250 (with horseshoe-shaped handle), and 394 (with parallel grooves) could belong to either DBas1 or DBas2. The rim dimensions can be reconstructed from only two fragments: sites 60 (diam. 49.0 cm) and 309 (diam. 65.0 cm). DBas3 (four fragments) has a diameter of 46.0–48.2 cm. DBas4 (site 354) has a diameter of 64.0 cm, DBas5 (site 773) 44.5 cm, and DBas6 (site 309) 34.0 cm.

1:4

DBas1
210. PWBW 225-01
Dimensions: r 61.0 h 11.2 w 20.2

211. PWBW 667-01*
Dimensions: h 10.9 w 8.6
Description: Thickened, slightly inward-projecting rim with flat upper face; deep body wall, occasionally decorated with two parallel grooves on the exterior face.
Comparisons: Cotton 1983, 372, Fig. 142, No. 422 (Cozzo Presepe, Site B, phase I, second half of 4th c. BC); Cotton 1992f, Fig. 87, No. 1562 (Gravina, "decorated pithos"); *Meligunìs-Lipára* XI.I, Pl. CLXXIX, No. 1a (with horseshoe-shaped handles, 4th-c. BC tomb).

1:4

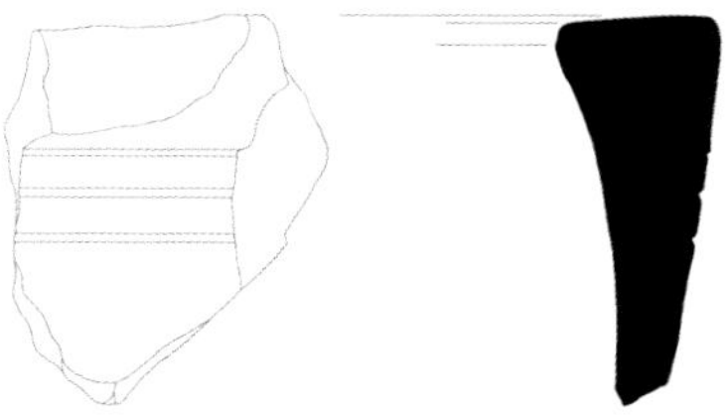

1:4

DBas2a
212. PWBW 060-05
Dimensions: r 49.0 h 12.8 w 15.5

Deep Basin (Dbas) cont.

213. PWBW 106-01

Dimensions: h 11.5 w 14.6

Description: Indistinct, rounded rim; deep body wall with slightly concave exterior face, occasionally decorated with two parallel grooves.

Comparisons: Pianu 1990, Pl. XIII, No. 2, and Pl. XXXIX, No. 3 (Herakleia, *conca*, tomb of third quarter of 4th c. BC).

1:4

1:4

DBas2b

214. PWBW 309-04

Dimensions: r 65.0 h 15.0 w 16.7

Description: See DBas2a. The variant DBas2b has slightly flaring body walls.

Comparisons: See DBas2a.

1:4

DBas3a

215. PWBW 038-122P

Dimensions: r 48.2 h 7.7 w 9.0

Description: Indistinct, outward-turned rim; the section is sub-triangular with concave exterior face and outward-projecting cyma; deep body.

Comparisons: Saunders and du Plat Taylor 1992, Fig. 19, No. 349 (Gravina, pithos type 2); Bailo Modesti 1980, Pl. 40, No. 12-2 (Cairano, *dolium*; Jones/Graham/Sackett 1973, 392, Fig. 12, No. 142 (Vari, PW kalathos, diam. 39 cm, 4th c. BC).

1:4

DBas3b

216. PWBW 309-03

Dimensions: r 46.0 h 7.6 w 12.9

Description: Indistinct, outward-turned rim; the section is circular with exterior ridge for emphasis; deep body.

Comparisons: Conti 1989c, Pl. XLIII, No. 370 (Locri, diam. 36 cm, 4th–3rd c. BC); *Kaulonia* I, 66, Fig. 44, No. 263; *Agora* XII, Fig. 15, No. 1786 (lekane, but with outward-turned and slightly inward-projecting rim, 520–480 BC); Angelelli 1998–1999, 47, Fig. 24, No. 193, (Amelia, *dolium*).

Deep Basin (Dbas) cont.

1:4

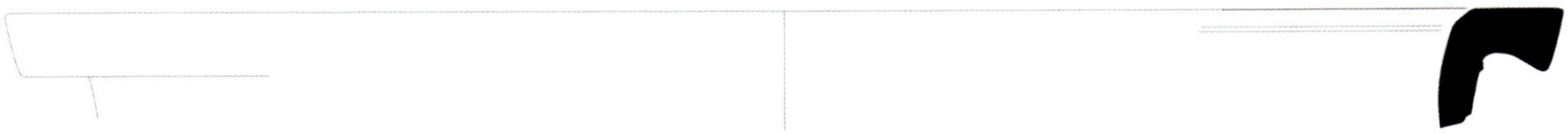

DBas4

217. PWBW 354-01 rev90

Dimensions: r 64.0 h 5.1 w 10.5

Description: Horizontal, outward-turned rim with flat upper face, nearly vertical exterior face, and sub-rectangular section; initially vertical body wall; exterior ridge.

Comparisons: See Bas7.

1:4

DBas5

218. PWBW 773-01

Dimensions: r 44.5 h 4.3 w 5.3

Description: Narrow, outward-turned rim with trapezoidal section and two parallel grooves on the upper face; deep body.

Comparisons: The narrow rim recalls Bas7.

1:4

DBas6

219. PWBW 309-05

Dimensions: r 34.0 h 8.9 w 12.8

Description: Thickened, inward-turned, and inward-projecting rim with square section, emphasized on the exterior by moldings and incised diagonal slashes; deep body.

Comparisons: The features of this fragment make it unique, at present.

Both the basin (Bas) and the deep basin (DBas) can have a ring, disk, or flat foot, classified as follows:

1:4

DBas/Bas7

220. PWBW 154-22

Dimensions: base 23.8 h 7.3 w 11.9

Description: High ring foot with sub-rectangular section.

1:4

DBas/Bas8a

221. PWBW 354-24

Dimensions: base 28.0 h 8.2 w 14.5

Description: Barely perceptible ring foot with sub-triangular section.

Deep Basin (Dbas) cont.

1:4

DBas/Bas8b
222. PWBW 521-133
Dimensions: base 31.5 h 4.2 w 8.9
Description: Disk foot with circular profile.

1:4

DBas/Bas8c
223. PWBW 245-01
Dimensions: base 21.5 h 7.5 w 14.6
Description: Flat or slightly concave base.
Comparisons: Cavagnera and Pizzo 1995, 108, Fig. 113 (Incoronata, diam. 13 cm), and 107, Fig. 110 (Incoronata, Corinthian pithos, diam. 13 cm).

Pithos (Pi)

The rim diameter generally falls in the range 45.0–75.4 cm, dimensions confirmed also by the only two lid fragments preserving the diameter (site 296, diam. 62.0 cm; site 462, diam. 48.0 cm). Nevertheless Metapontine pithoi could be smaller (cm 30.0–42.0) and, rarely, larger (the only example, a Pi2 measures 89.0 cm). Consequently, the resting surface, too, could be both smaller (14.0–24.3 cm) and larger (26.0–30.7 cm).

1:4

Pi1a
224. PWBW 057-04
Dimensions: r 47.5 h 8.4 w 17.8
Description: Horizontal, outward-turned rim with sub-rectangular section; the interior of the body wall is slightly or decidedly oblique; the exterior face is rounded; the short neck is neatly distinguished from the body.
Comparisons: *Meligunis-Lipára* II, Pl. LXXIX, No. 2 (T127, first two-thirds of 4th c. BC).

1:4

Pi1b
225. PWBW 542-05
Dimensions: r 61.0 h 13.0 w 37.9

Pithos (Pi) cont. 1:4

226. PWBW 394-13
Description: Horizontal, outward-turned rim with subrectangular section; the interior of the body wall is slightly or decidedly oblique; the exterior face is vertical; the short neck is neatly distinguished from the body.
Comparisons: Cotton 1992f, Fig. 89, No. 1579 (Gravina, period VIIIb, unstratified, but the type is common beginning in period VII, 3rd–2nd c. BC); *Sibari* III, 257, Fig. 261, No. 6917/25 (diam. 60 cm); Brizzi 1999c, 329, Fig. 319, No. 688 (Oppido Mamertina, diam. 24 cm); D'Andria 1990, 155, No. 250 (Vaste, diam. 33.3 cm, 5th c. BC?); Dyson 1983, Fig. 85, No. 3 (Buccino); Pancrazzi 1979, Fig. 66, No. 2 (Cavallino, int. diam. 27.6 cm); *Meligunìs-Lipára* XI.II, 468a (tomb, 5th c. BC); *Meligunìs-Lipára* II, Pl. XCI, No. 2 (T218, first two thirds of 4th c. BC).

1:5

Pi1c
227. PWBW 501-21
Dimensions: r 74.5 h 7.2 21.6
Description: Horizontal, outward-turned rim with subrectangular section; the interior of the body wall is vertical; the exterior face is vertical.
Comparisons: See Pi1b.

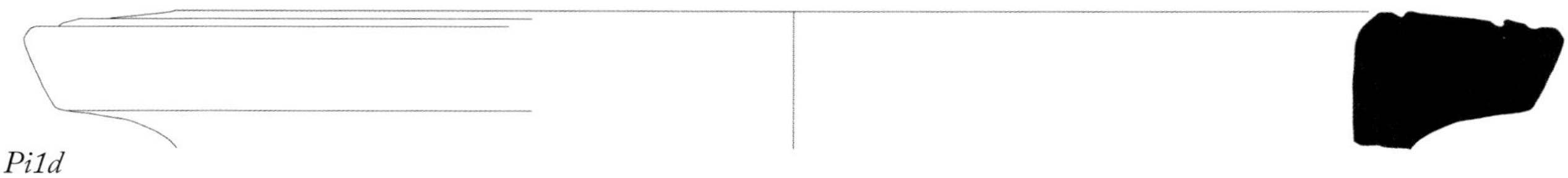

1:4

Pi1d
228. PWBW 097-233P
Dimensions: r 66.0 h 6.1 w 9.1

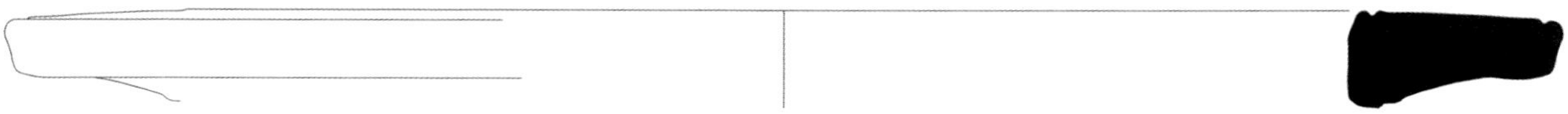

1:4

229. PWBW 197-13
Dimensions: r 64.0 h 4.1 w 11.2
Description: Horizontal, outward-turned rim with subrectangular section; the interior of the body wall is vertical; the exterior face is vertical. Pi1d has two concentric grooves on the upper face of the rim.
Comparisons: See Pi1b.

1:6

Pi2
230. PWBW 395-34
Dimensions: r 89.0 h 5.2 w 10.1
Description: Outward-turned rim with trapezoidal section and upper face inclined towards the exterior; the interior face, too, is slightly oblique.
Comparisons: See Pi1.

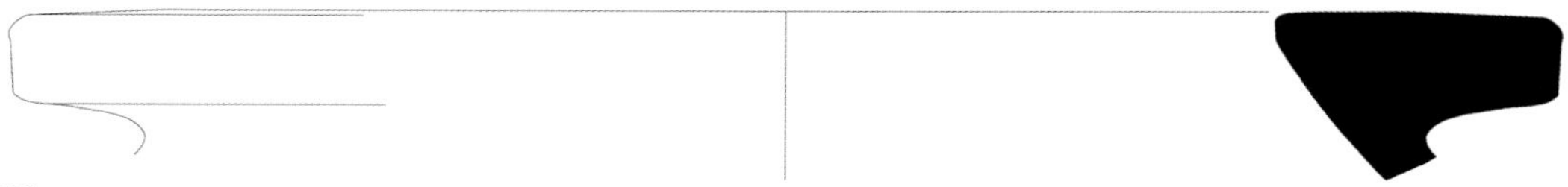

1:5

Pi3a
231. PWBW 384-01
Dimensions: r 69.0 h 7.6 w 25.4
Description: Horizontal, outward-turned rim with subrectangular section and oblique, inward-projecting interior wall. On Pi3a the rim is wide.
Comparisons: See Pi1b.

Pithos (Pi) cont.

1:4

Pi3b

232. PWBW 037-01

Dimensions: r 40.0 h 4.6 w 12.5

Description: Horizontal, outward-turned rim with subrectangular section and oblique, inward-projecting interior wall. The width is reduced compared with Pi3a.

Comparisons: Maturo 1997, Pl. 100, No. 19 (Pomarico, diam. 32 cm). Cotton 1992f, Fig. 88, No. 1576 (Gravina, period VIIIa-b, end of 2nd–1st c. BC, but the type is common beginning in period VII, 3rd–2nd c. BC); Dyson 1983, Fig. 222, No. 155 (Buccino).

1:5

Pi3c

233. PWBW 375-12

Dimensions: r 75.4 h 5.6 w 10.5

Description: Horizontal, outward-turned rim with subrectangular section and oblique, inward-projecting interior wall. The width is reduced compared with Pi3a, and the rim is distinguished on the exterior of the body by a groove.

Comparisons: Cotton 1983, 372, Fig. 142, No. 431 (Cozzo Presepe, Site B, phase V, 4th–3rd c. BC); Cotton 1992f, Fig. 88, No. 1577 (Gravina, period VIIIa-b, end of 2nd–1st c. BC); similar, but with more nearly vertical wall, in F. Rossi 1989, Pl. 351 (*fase* III, second half of 4th c. BC–1st c. AD): the same type of rim recurs on a jar from Pomarico (cf. Deodato and Bianco 1997, Pl. 68, No. 83, diam. 13.2 cm). Cf. Jones/Graham/Sackett 1973, 390, Fig. 12, No. 133 (Vari, diam. 65 cm, 4th c. BC); Jones/Sackett/Graham 1962, 98, Fig. 11 H (Dema, last quarter of 5th c. BC).

1:4

Pi3d

234. PWBW 167-12

Dimensions: r 65.0 h 7.8 w 15.5

Description: Horizontal, outward-turned rim with subrectangular section and oblique, inward-projecting interior wall. Pi3d is distinguished by a rim with oblique interior face.

Comparisons: See Pi3b.

1:4

Pi4a

235. PWBW 182-15

Dimensions: r 48.0 h 7.7 w 15.7

Description: Horizontal, outward-turned rim with elliptical section and oblique, inward-projecting interior wall.

Comparisons: Brizzi 1999c, 331, Fig. 320, No. 692 (Oppido Mamertina, diam. 25 cm).

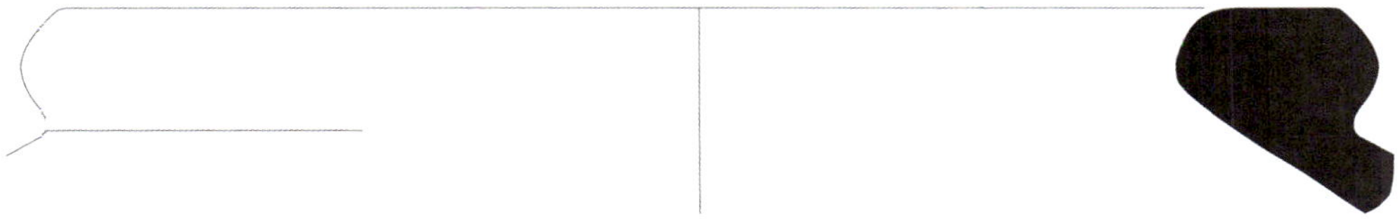

1:4

Pi4b

236. PWBW 559-01

Dimensions: r 51.3 h 7.9 w 16.0

Description: See Pi4a; Pi4b has a narrow rim.

Comparisons: See Pi4a.

Pithos (Pi) cont.

1:5

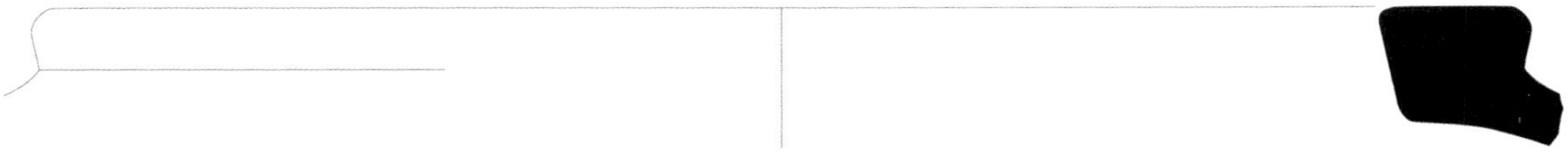

Pi5a

237. PWBW 377-01

Dimensions: r 73.9 h 7.2 w 10.2

Description: Vertical rim with square section attached to an initially nearly horizontal wall.

Comparisons: Sibari II, 360, Fig. 384, No. 25820 (diam. 25 cm).

1:4

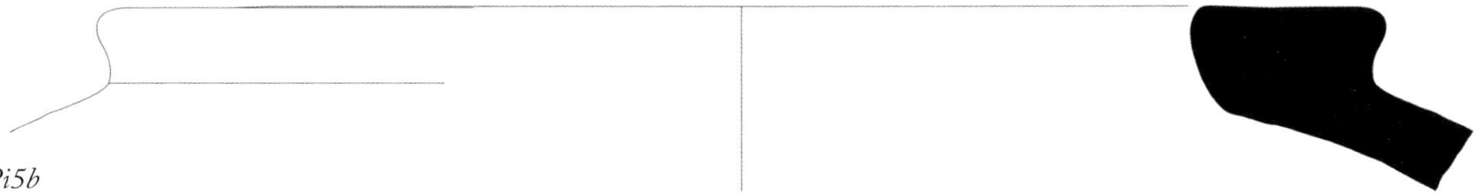

Pi5b

238. PWBW 394-15 rev90

Dimensions: r 57.6 h 8.6 w 15.5

Description: See Pi5a. On Pi5b the rim has a rectangular section.

Comparisons: See Pi5a.

1:4

Pi5c

239. PWBW 489-06

Dimensions: r 45.4 h 9.2 w 16.4

Description: See Pi5a. On Pi5c the body wall is initially more oblique.

Comparisons: See Pi5a.

1:4

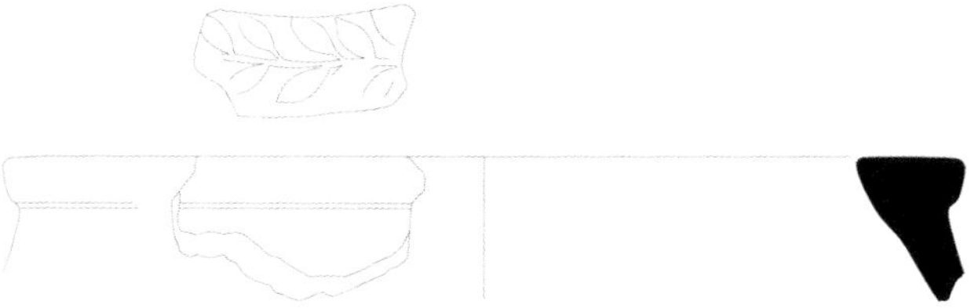

Pi6

240. PWBW 197-10

Dimensions: r 32.5 h 5.0 w 8.6

Description: Flat, slightly inward-projecting rim with rectangular section; initially straight body wall; the upper face is decorated with an olive branch in light relief.

Comparisons: See Pi5a.

1:4

Pi7a

241. PWBW 354-02 rev90

Dimensions: base 20.8 h 7.3

Description: Disk foot with concave or ring-shaped resting surface and carinated exterior profile.

Comparisons: Brizzi 1999c, 331, Fig. 320, No. 693 (Oppido Mamertina, diam. 14 cm).

Pithos (Pi) cont.

1:4

Pi7b
242. PWBW 104-248P
Dimensions: base 24.6 h 11.5 w 35.8
Description: See Pi7a.
Comparisons: Brizzi 1999c, 331, Fig. 320, No. 656 (Oppido Mamertina, diam. 11 cm).

1:4

Pi8a
243. PWBW 012-04 rev90
Dimensions: base 30.7 h 5.2 w 14.0
Description: Disk foot with flat resting surface and carinated exterior face. Pi8a has grooves on the resting surface of the foot and on the wall.
Comparisons: Cavagnera and Pizzo 1995, 108, Fig. 112 (Incoronata, levigated, diam. 16.5 cm).

Pi8b1
244. PWBW 481-03
Dimensions: base 24.3 h 5.9 w 17.4

1:4

245. PWBW 164-07
Dimensions: base 23.7 h 5.1 w 5.7
Description: See Pi8a.
Comparisons: See Pi8a.

1:4

Pi8b2
246. PWBW 490-11
Dimensions: base 20.7 h 4.7 w 11.5
Description: See Pi8a. On Pi8b2 the exterior face projects inwards.
Comparisons: See Pi8a.

1:4

Pithos (Pi) cont.

1:4

Pi8c1
247. PWBW 211-05
Dimensions: base 30.7 h 6.9 w 14.0
Description: See Pi8a. On Pi8c1 the foot is barely perceptible and convex and is distinguished from the body by a groove.

Comparisons: See Pi8a.

1:4

Pi8c2
248. PWBW 515-07
Dimensions: base 22.3 h 6.9
Description: See Pi8a. On Pi8c2 the foot is rounded and is distinguished from the body by a groove.

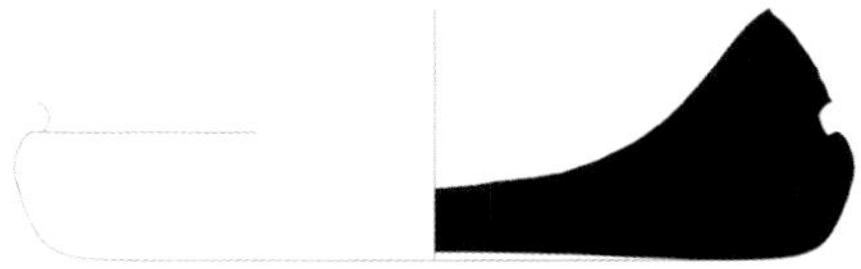

Comparisons: See Pi8a.

1:4

Pi8c3
249. PWBW 481-02
Dimensions: base 24.0 h 5.3 w 24.7
Description: See Pi8a. On Pi8c3 the foot is barely perceptible and flat and is distinguished from the body by a groove.

Comparisons: See Pi8a.

1:4

Pi8d
250. PWBW 286-05
Dimensions: base 18.0 h 3.0 w 9.0
Description: See Pi8a. On Pi8d the foot is a concave disk.
Comparisons: See Pi8a.

1:4

Pi9a
251. PWBW 296-15
Dimensions: r 62.0 h 2.7 w 10.8
Description: Lid for pithos with rectangular section, decorated with concentric grooves on the upper face.
Comparisons: Cotton 1983, 373, Fig. 143, No. 434 (Cozzo Presepe, unstratified); *Sibari* II, 336, Fig. 362, No. 26927 (*coperchio? o strumento di tornio*); Lissi Caronna 1983, 281, Fig. 70, No. 6 (Oppido Lucano, from a plundered tomb, reddish gloss, *piatto o disco?*, diam. 38.0 cm).

1:2

Pi9b
252. PWBW 296-16
Dimensions: r 5.25 h 4.2 w 7.7
Description: See Pi9a. Pi9b is a molded lid/knob.
Comparisons: See Pi9a.

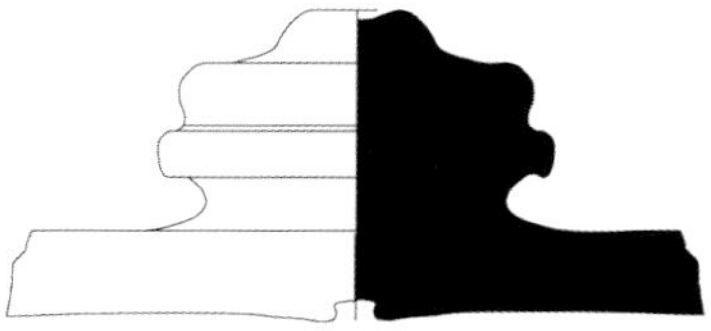

Globular Urn (GU)

1:2

Types GU1a–b have a diameter in the range 9.6–12.0 cm. The only measurable example of GU1c has a diameter of 21.6 cm.

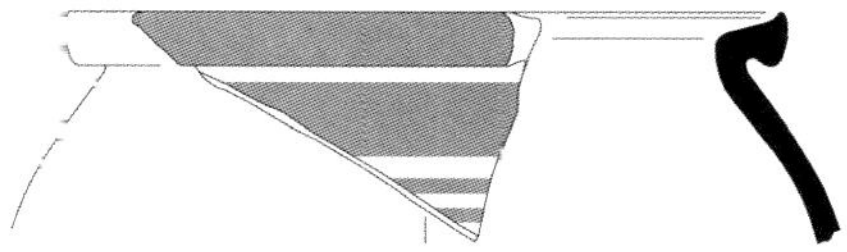

GU1a

253. PWBW 488-88

Dimensions: r 9.6 h 3.2 w 5.4

Description: Horizontal, outward-turned rim with sub-rectangular section; the tong-shaped extremity is variably accentuated (more in GU1a); narrow, concave neck; globular body.

Comparisons: Carter and Toxey, 1998, 699, CD 105-6 (Pantanello, ca. 550 BC); Adamesteanu 1980, 61, Fig. 47a (Metaponto, mid-6th c. BC); Maruggi 1996b, 257 (Tarantine tomb).

1:2

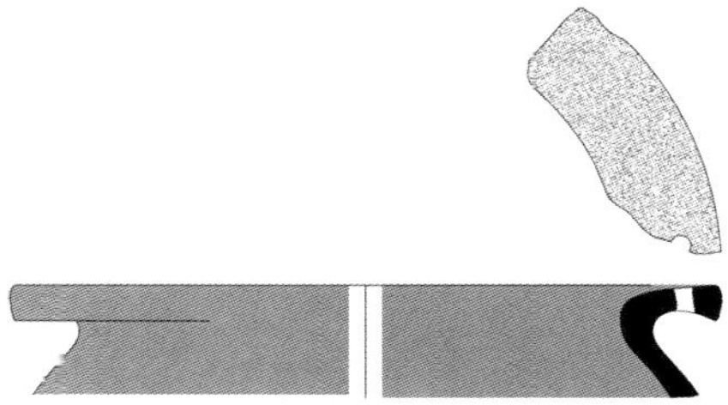

GU1b

254. PWBW 270-01 rev90

Dimensions: r 11.0 h 1.8 w 4.6

Description: See GU1a. The tong-shaped extremity is variably accentuated (less in GU1b). The rim has a circular perforation.

Comparisons: See GU1a.

1:2

GU1c

255. PWBW 394-11

Dimensions: r 21.6 h 2.2 w 7.2

Description: See GU1a. On GU1c the rim has a small groove on the upper face.

Comparisons: See GU1a.

Lebes Gamikos (LG)

1:2

The lebes gamikos generally varies in size from 10.0 to 16.0 cm but in several cases can reach 19.0 cm (see the examples from sites 392 and 506).

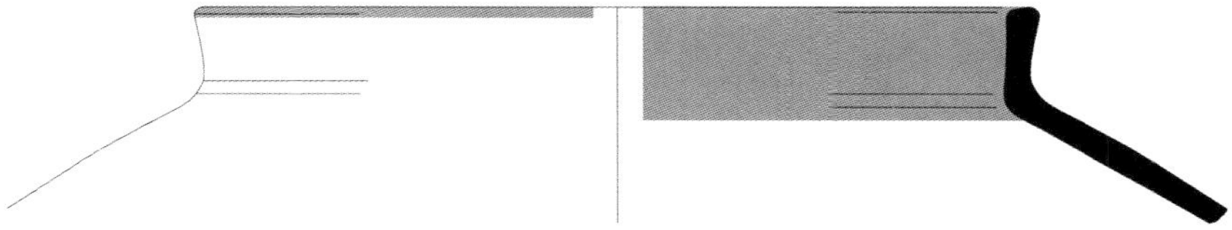

LG1a

256. PWBW 103-01

Dimensions: r 15.4 h 4.0 w 6.4

Description: High, vertical rim with sub-rectangular section and flat upper face, neatly distinguished from the oblique shoulder.

Comparisons: Carter and Toxey 1998, 700, T306-10/14 (Pantanello, with convex shoulder, 400–370 BC); D'Andria 1980a, 431, Fig. 67, No. 327 (Metaponto, *scarico* 3, last quarter of 4th c. BC, diam. 12.8 cm); Lo Porto 1973, Pl. LXV, Fig. 3, No. 1 (tomb dated 550–540 BC), and Pl. LXXII, Fig. 5, No. 3 (tomb dated to beginning of 5th c. BC); *Museo Taranto* I.3 221, No. 41.3 (diam. 5.9 cm, tomb dated 520–500 BC), and 352, No. 125.3 (diam. 7.5 cm, tomb of end of the 6th–beginning of 5th c. BC); Vittoria 2000, Pl. 22, No. 4 (T. 85); Patera 1986, Pl. XII, No. 1 (Taranto, tomb of end of 6th–beginning of 5th c. BC); *Agora* XII, 195, No. 1529 ("storage-bin," last quarter of 5th c. BC). For the examples with barely perceptible shoulder see: Marchegiani 1997, Pl. 62, No. 42 (Pomarico, BW *stamnos*, diam. 8.6 cm); Pianu 1990, Pl. LXXIV, No. 236.4 (Herakleia, PW *cinerario*, end of 4th c. BC).

Lebes Gamikos (LG) cont.

1:2

LG1b

257. PWBW 182-14

Dimensions: r 14.8 h 4.0 w 6.0

Description: See LG1a.

Comparisons: See LG1a.

Lekythos (L)

1:2

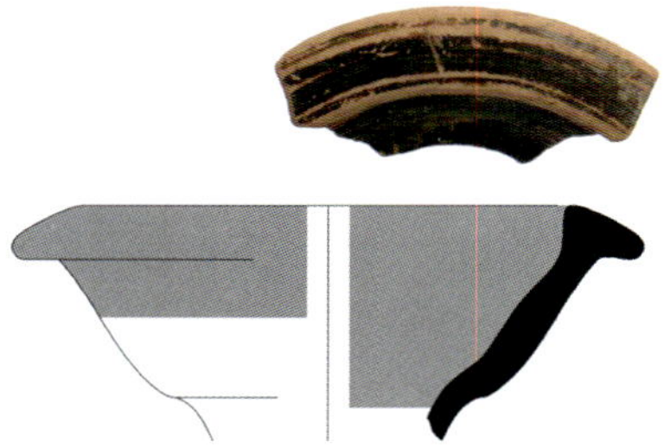

Type L1 has a diameter of 9.7 cm, L3 8.0 cm, and L4 6.5 cm, while the rim of L2 varies between 5.0 and 8.0 cm.

L1

258. PWBW 043-09 rev82

Dimensions: r 9.7 h 3.7 w 5.8

Description: High, oblique, outward-turned, and slightly inward-projecting rim with elliptical section and straight exterior profile, distinguished from the neck.

Comparisons: Carter 1998g, MINA1 (Pantanello, BG, mid-5th c. BC); Pizzo 1998–1999, 255, Fig. 39, No. 3, and 346, Fig. 103b (Vassallaggi, BG *amphoriskos*, tomb, last quarter of 5th c. BC, diam. 2.9 cm); Morel 1981a, 3621b1 (cf. *Agora* XII, Fig. 11, No. 1150, BG *amphoriskos*, 430–420 BC, with stamped decoration; p 156: production lasted through beginning of 4th c. BC).

1:2

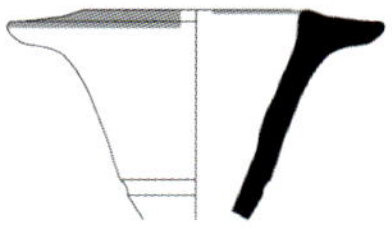

L2

259. PWBW 258-18

Dimensions: r 5.0 h 2.9 w 4.7

Description: High, outward-turned, nearly horizontal rim with elliptical section and straight exterior profile, distinguished from the neck.

Comparisons: BG 454-03, dated second half of 4th c. BC.

1:2

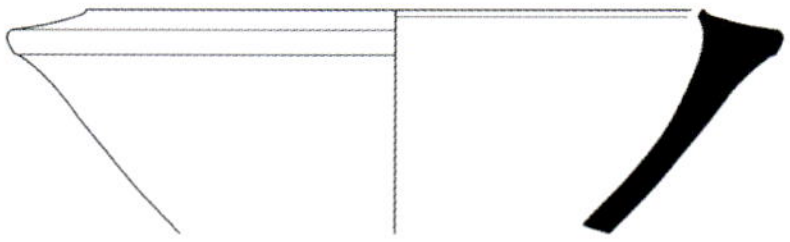

L3

260. PWBW 288-01

Dimensions: r 10.3 h 3.1 w 4.95

Description: High rim with conical section, upper face turned slightly towards the exterior, and concave interior face.

Comparisons: None found, at present.

1:2

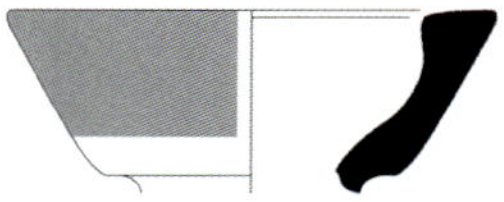

L4

261. PWBW 417-02

Dimensions: r 6.5 h 2.5 w 3.1

Description: High rim with conical section, straight exterior face, flat upper face, and concave interior face distinguished from the neck.

Comparisons: Carter and Toxey 1998, 705, T 208-5 (460–440 BC); *Cozzo Presepe,* 345, No. 265 (ca. 300); Lo Porto 1973, Pl. LXX, No. 1.2 (from Matera, Archaic); *Gravina* III.2, Fig. 50, No. 265; Marchegiani 1997, 171, Pl. 63, No. 53 (Pomarico, end of 4th c. BC).

1:2

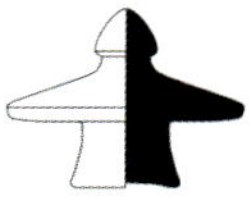

Ld1a

262. PWBW 341-45 (= LA 341-45, see Ch. 10)

Dimensions: r 3.15 h 2.4

1:2

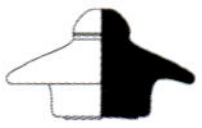

Ld1b

263. PWBW 638-01 (= LA 638-01, see Ch. 10)

Dimensions: r 2.5 h 1.5

Unguentarium (U)

The fragments large enough for measurement indicate a base diameter of 2.3–3.2 cm, with the lone exception of the fragment from site 216, which has a foot measuring 4.4 cm. The rims are slightly smaller, with a diameter of 1.8–2.0 cm, but there is a larger example (3.4 cm) from site 167.

U1

264. PWBW 358-11 rev90

Dimensions: base 2.5 h 3.5 w 4.4

Description: Globular body; concave disk foot.

Comparisons: Elliott 1998, 687, U51 (Pantanello, 325–300 BC); Prag 1983a, 362, Fig. 132, No. 356 (Cozzo Presepe, end of 4th–beginning of 3rd c. BC); Pianu 1990, Pl. XLIV, No. 4 (Herakleia, 330–300 BC); Forti 1962, type III (330–275 BC); *Agora* XXXIII, Fig. 60, No. 400 (325–260 BC).

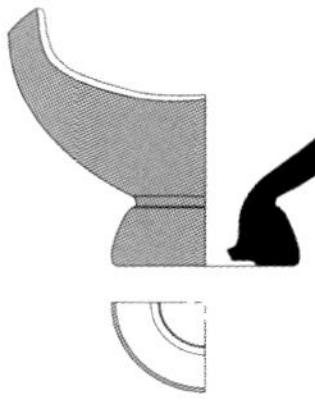

1:2

265. PWBW 173-11

Dimensions: base 2.8 h 2.95 w 4.0

Description: Globular body; concave disk foot.

Comparisons: See U1a.

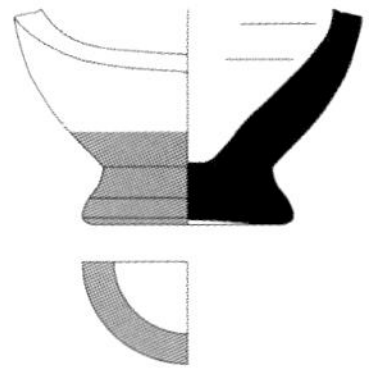

1:2

U2

266. PWBW 081-01 rev90

Dimensions: base 2.6 h 4.1 w 4.6

Description: Ovoid body; ring foot.

Comparisons: Elliott 1998, 687, U33 (Pantanello, 325–300 BC); Pianu 1990, Pl. XIV, No. 1 (Herakleia, BW, 330–310 BC); Lo Porto 1988–1989, 418, Fig. 135.4 (tomb 1, Calciano, 330–320 BC); Forti 1962, type III.

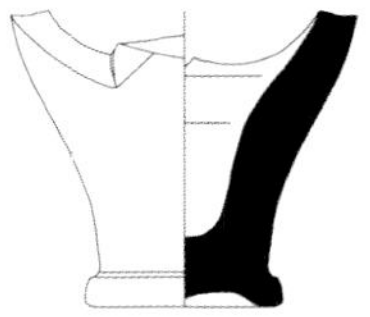

1:2

U3a

267. PWBW 113-265P

Dimensions: base 2.2 h 4.8 w 6.8

Description: Ovoid body; conical disk foot. On U3a the body is more compressed than U3b

Comparisons: Elliott 1998, 687, U44 (Pantanello, 300–275 BC); Pianu 1990, Pl. LXIX, No. 208.1, and Pl. LXIV, No. 181.1 (Herakleia, 320–280 BC); Hempel 2001, Pl. 11, Form 611/4 (Taranto, phase D, end of 3rd–beginning of 2nd c. BC); Forti 1962, type III (330–275 BC).

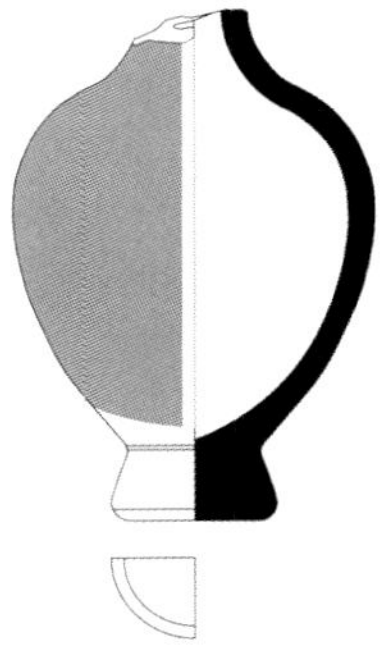

1:2

U3b

268. PWBW 167-04

Dimensions: r 1.8 base 2.0 h 4.2 w 8.6

Description: Outward-turned rim with triangular section attached to a high cylindrical neck; ovoid body; conical disk foot.

Comparisons: Elliott 1998, 687, U13 (Metaponto, 325–280 BC), and U40 (ca. 300 BC); Pianu 1990, Pl. XLII, No. 94.10, and Pl. LVII, No. 128.2 (Herakleia, 320–280 BC); Hempel 2001, Pl. 11, Form 611/1 (Taranto, phase D, end of 3rd–beginning of 2nd c. BC); Forti 1962 type III.

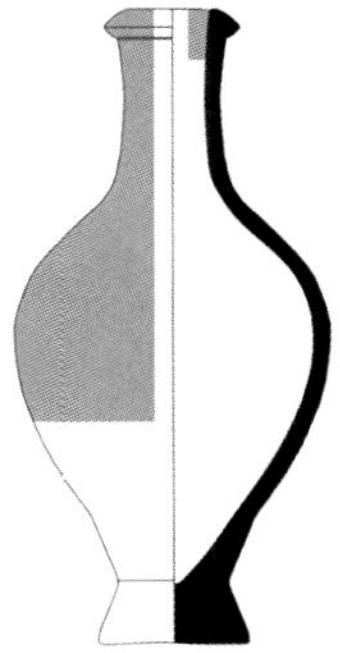

1:2

Unguentarium (U) cont.

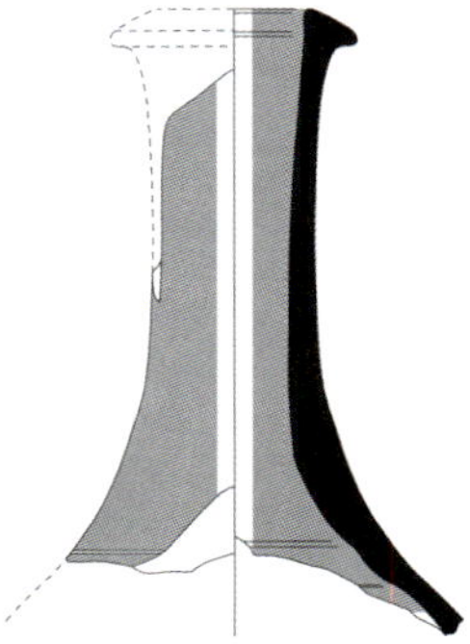
1:2

269. PWBW 167-05
Dimensions: r 3.4 h 8.9 w 5.5
Description: See **No. 268**.

1:2

270.**PWBW 358-10 rev90**
Description: See **No. 268.**

1:2

271. PWBW 635 f4-06
Description: See **No. 268.**

U4

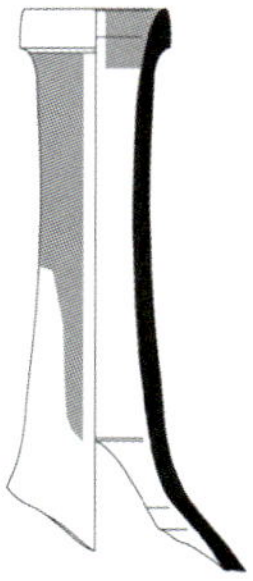
1:2

272. PWBW 272-03
Dimensions: r 2.0 h 7.6 w 3.0
Description: Thickened rim with triangular section attached to a high cylindrical neck.
Comparisons: Lo Porto 1966, Fig. 46, No. 4 (Metaponto, 3rd–2nd c. BC); Lo Porto 1981a, Fig. 64, No. 3 (Metaponto, third quarter of 3rd c. BC); L. Giardino 1990, 122 A (Herakleia, last quarter of 3rd c. BC); Hempel 2001, Pl. 11, Form 611/13 (Taranto, phase E3, first half of 1st c. BC); Prag and Broughton 1992, Fig. 47, No. 1087 (Gravina, end of 4th–3rd c. BC); *Sibari* IV, 326, Fig. 320, No. 120, and 403, Fig. 371, No. 1115 (*Prolungamento Strada*; cf. Forti 1962, type V, end of 3rd–2nd c. BC); Forti 1962, type IV.

Unguentarium (U) cont.

1:2

U5a

273. PWBW 047-01 rev90

Dimensions: base 2.4 h 5.25

Description: High, thin ovoid body; conical disk foot.

Comparisons: Elliott 1998, 687, U13 (Pantanello, 325–280 BC); Prag 1983a, 362, Fig. 132, No. 359 (Cozzo Presepe); Pianu 1990, Pl. XLII, No. 94.10 (Herakleia, 320–280 BC); *Sibari* IV, 468, Fig. 443, No. 54, and 507, Fig. 476, No. 1162 (Casa Bianca, 4th–2nd c. BC); Forti 1962, type III.

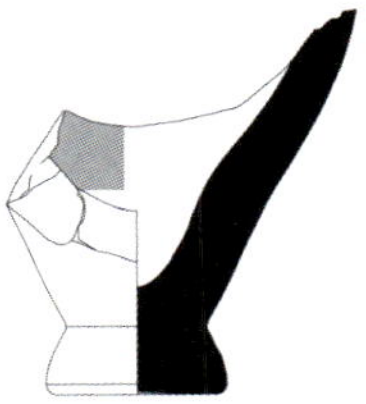

1:2

U5b

274. PWBW 278-04

Dimensions: base 2.3 h 5.2 w 4.1

Description: See U5a. On U5b the body is slimmer.

Comparisons: Elliott 1998, 687, U41 (Pantanello, 320–300 BC); L. Giardino 1990, Pl. XXVI, Umbra 1 (Herakleia, 200–175 BC); Prag and Broughton 1992, Fig. 47, No. 1094 (Gravina, last quarter of the 3rd–2nd c. BC); *Sibari* IV, 330, Fig. 323, No. 201, and 404, Fig. 372, No. 5446 (*Prolungamento Strada*, end of the 3rd c. BC); Forti 1962, type IV(?).

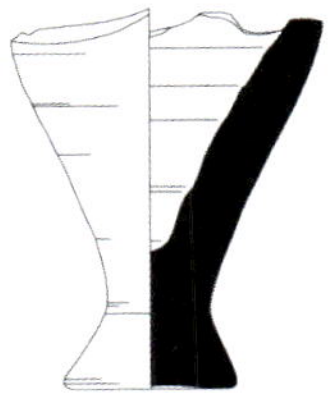

1:2

U6

275. PWBW 327-05

Dimensions: base 3.25 h 3.5

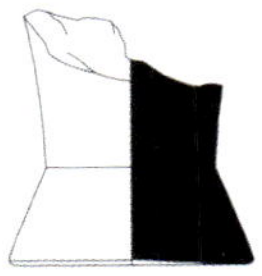

1:2

276. PWBW 728-06 rev00

Dimensions: base 2.7 h 3.9

Description: Cylindrical stem on conical disk foot.

Comparisons: Adamesteanu 1980, 104, Fig. 98g (Metaponto, end of the 4th–beginning of the 3rd c. BC); Lo Porto 1966, Fig. 66, No. 3 (Metaponto); Lo Porto 1981a, Fig. 64, No. 5 (Metaponto, last quarter of the 3rd c. BC); L. Giardino 1990, Pl. XXVI A (Herakleia, last quarter of the 3rd c. BC), and C-9A (225–200 BC); Hempel 2001, Pl. 11, Form 611/5 (Taranto, phase E1, first half of the 2nd c. BC); *Gli ori*, 466 (first quarter of the 2nd c. BC); Prag and Broughton 1992, Fig. 47, Nos. 1083 (Gravina, end of the 4th–end of the 3rd c. BC), and 1096 (end of the 4th–end of the 3rd c. BC); Giglio 2005, 338, 340, Pl. LVIII, No. 464 (Torre di Satriano, Forti type IV); Morel 1981a, 7111c1 (2nd c. BC), and 7111b1 (ca. 190 BC); Giannotta 1995, 59, Fig. 29, No. 35 (Rocavecchia, 3rd c. BC); Forti 1962, type IV.

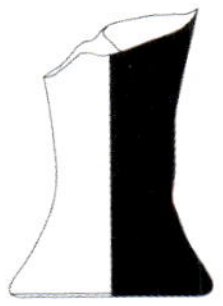

Louterion (Lou)

1:6

The size of the louterion, derived from the seven best-preserved examples, ranges from 50.0 to 61.7 cm.

Lou1

277. PWBW 191-03

Dimensions: r 89.0 h 10.0 w 14.3

Description: Flat, indistinct rim with sub-triangular section; hemispherical body wall. Impressed palmettes in light relief on the upper face; similar palmettes on fragment PWBW 183-17 (not illustrated) alternate and are connected to the base via small arches.

Comparisons: Sibari III, 415, Fig. 449, No. 15105 (marble; cf. *Corinth* XII, Fig. 19, No. 817, 1st c. BC); Hänsel 1973, 469, Fig. 44, No. 13 (Herakleia, from a ditch filled in the mid-7th c. BC); *Gravisca* 12.I, Pl. 10, No. 100 (F2, 6th–4th c. BC, diam. 20–50 cm); Threipland and Torelli 1970, 102, Fig. 17, B.2 (Casale Pian Roseto, Coarse Creamware, end of the 5th–4th c. BC). For a similar but undecorated example from the Athenian Acropolis cf. Pimpl 1997, Pl. 9, No. 7; cf. also *Agora* XII, Fig. 16, No. 1866 (from the Stoa, second half of 5th c. BC, diam. 93.0 cm).

1:4

Lou2a

278. PWBW 014-05 rev90

Dimensions: r 61.0 h 5.0 w 22.6

Description: Horizontal, outward-turned rim with parallel grooves on the vertical face; straight body wall; spool-shaped handle. On Lou2a the rim has a square section.

Comparisons: Cotton 1983, 372, Fig. 142, No. 425 (Cozzo Presepe, unstratified); Maturo 1997, Pl. 100, No. 21 (Pomarico, diam. 40 cm); *Agora* XII, Fig. 16, No. 1876 (grooved also on the top of the rim, 525–500 BC); *Sibari* IV, 71, Fig. 54, No. 53, and 150, Fig. 142, No. 9563 (Stombi), as well as 139, Fig. 133, No. 416, and 144, Fig. 138, No. 547 (Stombi).

1:4

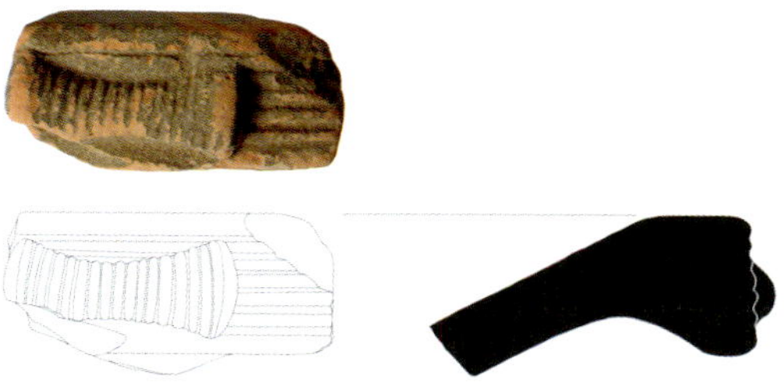

Lou2b

279. PWBW 497-01

Dimensions: h 5.6 w 11.2

Description: See Lou2a. On Lou2b the rim hangs and has a trapezoidal section.

Comparisons: See Lou2a.

Louterion (Lou) cont.

1:4

Lou2c

280. PWBW 679-01*

Dimensions: h 5.1 w 8.8

Description: See Lou2a. On Lou2c the rim is similar to the preceding but thinner.

Comparisons: Gualtieri 1990, 306, No. 572 (Roccagloriosa); *Gravina* III.2, 129, Fig. 53, No. 292.

1:2

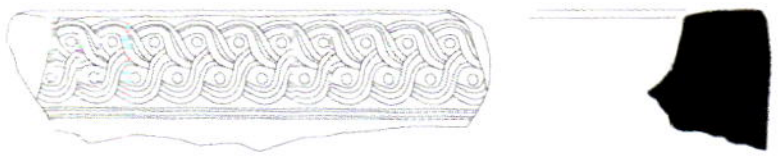

1:4

Lou3

281. PWBW 432-02

Dimensions: h 3.9 w 11.0

Description: Indistinct vertical rim with square section; probably concave body wall. Double-braid decoration on rim exterior.

Comparisons: G. Greco 1991, 62, Fig. 127 (Serra di Vaglio, from which site come cover-plates for the projecting beam-ends of the roof that have a very similar double-braid decoration: cf. 31 Fig. 82); *Sibari* II, 194, Fig. 209, No. 8779; same rim shape, but with parallel grooves on the exterior face, in *Corinth* VII.2, Pl. 110, An268.

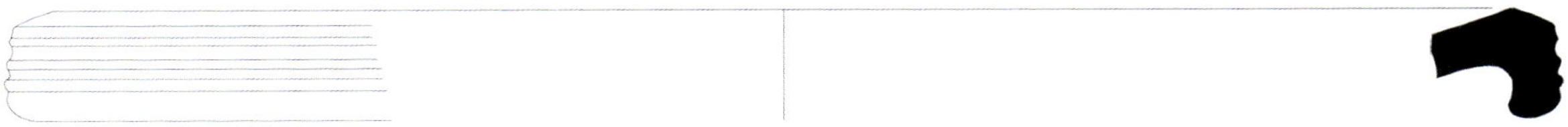

1:4

Lou4

282. PWBW 395-35

Dimensions: r 61.7 h 4.5 w 12.6

Description: Hanging, hooked, outward-turned rim with elliptical section; vertical face decorated with parallel grooves; straight body wall.

Comparisons: Agora XII, Fig. 16, No. 1861 (diam. 65.0 cm; 219: context dates to end of 4th c. BC, but type, described as "Corinthian" because of its fabric, is common at Athens in 5th and 4th c. BC).

Miniature

For Type Min1 the rim diameter varies between 3.7 and 6.3 cm, while the base diameter falls in the range 1.85–3.3 cm. **No. 285** (PWBW 397-07) is the only example preserving the full profile and is 2.9 cm high. Most examples of the lebes Min2 consist of just the base: the foot size varies between 2.2 and 2.8 cm. For the rim, the only source of information is the whole example **No. 287** (PWBW 334-08) with a diameter of 2.4 cm and a height of 3.55 cm.

The phiale Min3 comes from site 334 and is 6.7 cm at the rim; this is the only example of that miniature. The only example of Min4 comes from site 312 and has a rim 8.8 cm in diameter and 1.35 cm high. Min5 varies in size between 1.5 and 2.2 cm at the rim and 2.7 and 3.5 at the base, with a height in the range 2.75–3.2 cm. The small mug Min6 has a base 1.6 cm in diameter with a height of 1.5 cm.

1:2

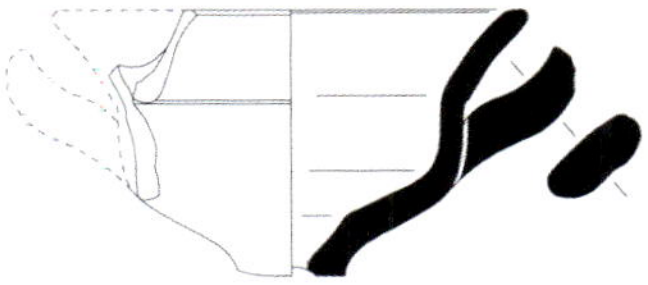

Min1a

283. PWBW 369-05

Dimensions: r 6.3 h 3.6

Description: Krateriskos with high outward-turned rim with elliptical section, distinguished from the almost carinated body; handles with elliptical section, attached at the point of maximum expansion of the body and oriented obliquely.

Comparisons: Cotton 1983, 368, Fig. 137, No. 397 (Cozzo Presepe, end of 4th–3rd c. BC); Lo Porto 1981a, Fig. 22, No. 1 (Metaponto, second half of 6th–second quarter of 5th c. BC); similar example from Norba: Perrone 2003, 363, Fig. 8, No. 51 (Temple of Juno Lacina, from the votive deposit). An unpublished and almost entirely reconstructable example comes from Fattoria Fabrizio in the Metapontino (currently under study).

Miniature (Min) cont.

1:2

Min1b

284. PWBW 397-06

Dimensions: r 4.7 h 2.1 w 3.3

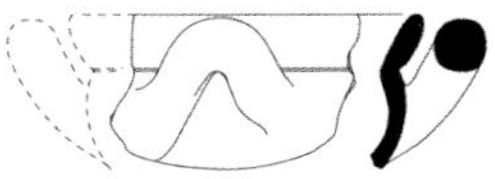

1:2

285. PWBW 397-07

Dimensions: r 3.7 base 2.0 h 2.9 w 2.7

Description: Krateriskos with low, outward-turned rim with elliptical section, distinguished from the almost hemispherical body; handles with circular section, attached at the point of maximum expansion of the body and oriented obliquely; concave or flat disk base.

Comparisons: Toxey and Carter 1998, 729, T29-6 (Pantanello, 460–440 BC); Lo Porto 1981a, Fig. 23, No. 11 (Metaponto, mid-6th–second quarter of 5th c. BC); Casagrande 2002, 199, and Fig. 66, No. 326 (Herakleia, 375–270 BC, by stratigraphy); *Sibari* IV, 456, Fig. 437, No. 298, and 507, Fig. 476, No. 5383 (Casa Bianca, diam. 6 cm, horizontal handles).

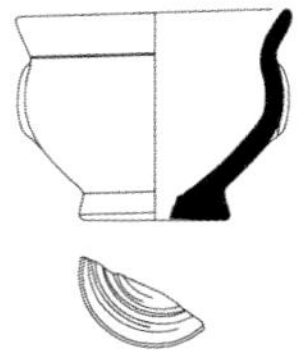

1:2

Min1c

286. PWBW 516-05

Dimensions: base 1.85 h 2.1

Description: Krateriskos with almost hemispherical body; concave disk base.

Comparisons: Toxey and Carter 1998, 729, T29-6 (Pantanello, 460–440 BC); Lo Porto 1981a, Fig. 23, No. 11 (Metaponto, mid-6th–second quarter of 5th c. BC); similar to Min1, but with vertical handles, are the Krateriskos types from Capo Colonna at Croton: Verbicaro 2006, 90, Fig. 102 (dated to 6th c. BC, from a waste deposit).

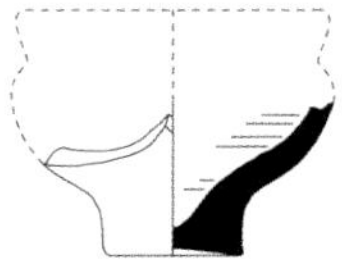

1:2

Min2

287. PWBW 334-08

Dimensions: r 2.4 base 2.2 h 3.55

Description: Lebes with narrow rim that has a vertical, elliptical section; convex shoulder; pyriform body; disk foot; horizontal, vertical handles attached to the shoulder.

Comparisons: Carter 1998g, MINLG6 (Pantanello, second half of 5th c. BC); Lo Porto 1981a, 313, Fig. 23, No. 4 (Metaponto, end of 6th–beginning of 5th c. BC); Fracchia 1990c, 228, Fig. 2, No. 17 (Roccagloriosa, BG lebes, rim diam. 1.3 cm, base diam. 2.0 cm); *Sibari* IV, 136, Fig. 131, No. 419, and 158, Fig. 147, No. 14819 (Stombi, diam. 4 cm, horizontal handles).

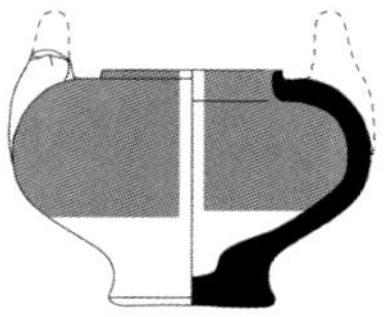

1:2

Min3

288. PWBW 334-10

Dimensions: r 6.7 h 1.75

Description: Dish/phiale mesomphalos with indistinct rim that has a sub-circular section; concave base.

Comparisons: Carter 2006, 166, Fig. 436-A (Pantanello sanctuary, 6th c. BC); Adamesteanu 1980, 102, Fig. 94 (Metaponto, 5th–4th c. BC); Lo Porto 1981a, Fig. 23, No. 5 (Metaponto, mid-6th c. BC); Di Noia 2005, 344, 347, Pl. LX, No. 491 (Torre di Satriano, diam. 8.4 cm).

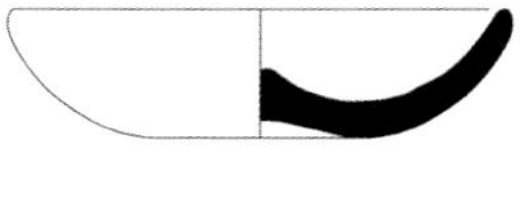

Miniature (Min) cont.

Min4
289. PWBW 312-07
Dimensions: r 8.8 h 1.35 w 4.9
Description: Dish with outward-turned rim which has an elliptical section; flat base.

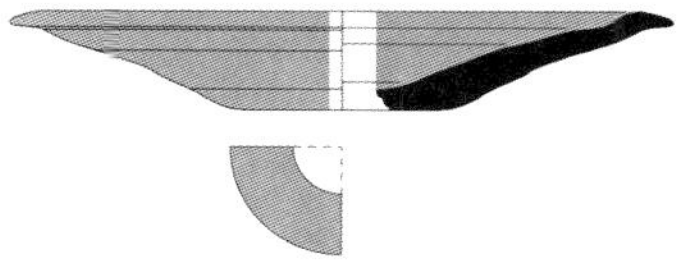
1:2

Comparison: Similar to Roccagloriosa: Fracchia 1990c, 229, Fig. 3, No. 27 (PW saucer, rim diam. 4.3–5.5 cm).

Min5
290. PWBW 334-09
Dimensions: r 3.0 base 1.7 h 2.75
Description: Cup with indistinct, outward-turned rim that has an elliptical section; carinated body wall; high disk foot with conical section.

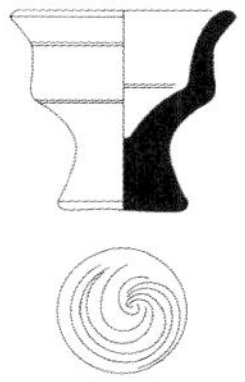
1:2

Comparisons: Toxey and Carter 1998, 730, T213-5, and 213-6 (Pantanello, 420–370 BC); Adamesteanu 1980, 103, Fig. 96d, and 220, Fig. 228 (Metaponto); Casagrande 2002, 199, Nos. 327 and 417 (Herakleia, end of 4th c. BC); *Sibari* IV, 224, Fig. 221, No. 91, and 283, Fig. 278, No. 2358 (Parco del Cavallo, first half of 4th c. BC); Di Noia 2005, 343, 347, Pl. LX, Nos. 484–485 (Torre di Satriano, Krateriskos types with vertical handles, diam. 4.1 and 4.0 cm, end of the 6th–end of 4th c. BC); Grasso 2005, 549, Pl. III a.f (Pompeii, *coppetta*, Hellenistic).

Min6
291. PWBW 032J-03*
Dimensions: base 1.6 body 2.2 h 1.5
Description: Small cup with pyriform profile; concave base.

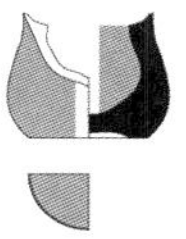
1:2

Comparisons: Lo Porto 1966, Fig. 3, No. 4 (Metaponto, 3rd–2nd c. BC); Pianu 1990, Pl. XXIV, No. 32.6 (Herakleia, PW, end of 4th c. BC); *Gli ori,* 443, No. 9 (end of 4th–3rd c. BC); Fracchia 1990c, 228, Fig. 2, No. 12 (Roccagloriosa, BG kalathos, base diam. 1.2 cm).

Thymiaterion (Thy)

Thy1
292. PWBW 742-11
Dimensions: stem 3.25 h 3.85
Description: Cylindrical stem with four small oblique perforations; banded decoration.

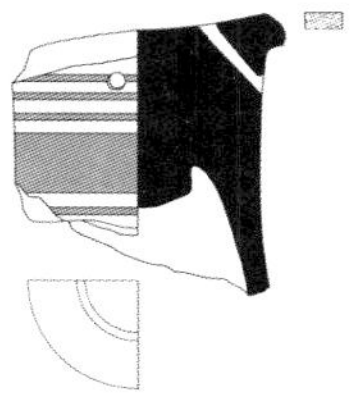
1:2

Comparisons: Lo Porto 1973, Pl. XVII, Fig. 2, No. 3 (6th c. BC), and Pl. XXXVI, Fig. 3, No. 2 (510 BC); Lo Porto 1988–1989, Fig. 75, No. 1 (470–430 BC). As in the Metapontine example, the thymiaterion from T102 at Chiaromonte-Sotto la Croce, dated to mid-6th c. BC, has two perforations visible near the rim: cf. *Greci, Enotri e Lucani* 160 No. 2.16.26.

Thy2
293. PWBW 730-29
Dimensions: stem 3.1 h 6.15 w 3.6
Description: Cylindrical stem with concave section in the lower part, full section in the higher part, and outward-projecting ring; banded decoration.

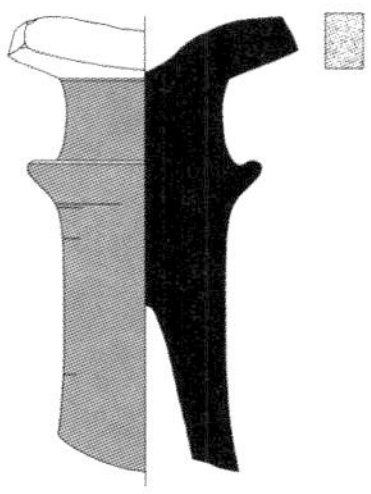
1:2

Comparisons: Di Lieto 2005b, 372, 385, Pl. LXX, No. 584 (Torre di Satriano, second half of 4th–first half of 3rd c. BC); Di Lieto 2001, 61, No. 52 (Torre di Satriano, end of the 4th c. BC); Fracchia 1990b, 256, Fig. 182, No. 224 (Roccagloriosa); Perrone 2003, 365, Fig. 13 (Norba, Temple of Juno Lucina, from the votive deposit); a lamp with stand from S. Maria del Cedro (Cosenza) has the same type of outward-projecting ring on the stem, which in the latter case is the element that connects the two pieces: cf. Guzzo and Greco 1978, 457, Fig. 44, No. 45, with comparanda dated end of 4th–beginning of 3rd c. BC.

Kiln Support (KS)

Three types of kiln supports are represented: the "leaf type," that with cuneiform section, and the ring type.

1:2

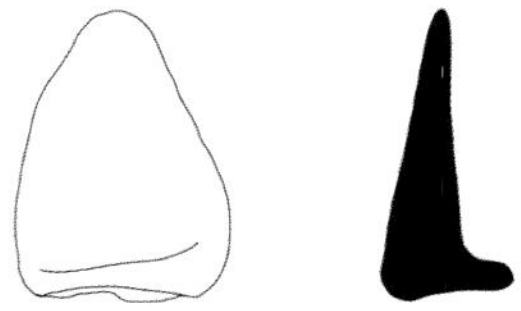

KS1

294. PWBW 320-37

Dimensions: h 4.0 w 2.8

Description: The so-called "leaf type," with sub-triangular profile, triangular section, and raised edge.

Comparisons: Cracolici 2004, 49, type VI-A, and 98, Fig. 53, No. 338 (Pisticci, lance-shaped wedge, 6th–3rd c. BC), cf. also 21, Fig. 2, MC 703 (Athens), and 23, Fig. 3 (Taranto); D'Andria 1980a, 415, Fig. 58, No. 263 (Metaponto); Dell'Aglio 1996, 68–69, No. 31 (Taranto, leaf type, 6th–5th c. BC).

1:2

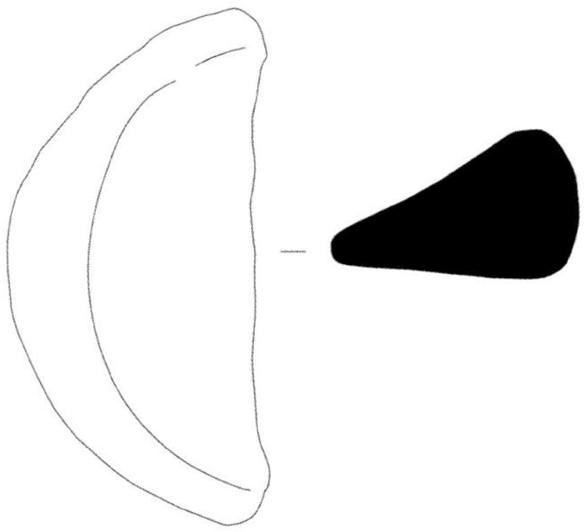

KS2

295. PWBW 651-01*

Dimensions: h 3.0 l 10.35 w 4.8

Description: Semi-circular profile with cuneiform section.

Comparisons: Cracolici 2004, 49, type VI-B1, and 98, Fig. 53, No. 337 (Pisticci, wedge with circular base, 6th c. BC), cf. also 23, Fig. 3 (Taranto); Dell'Aglio 1996, 68–69, No. 24 (Taranto, half-moon, 5th–4th c. BC).

1:2

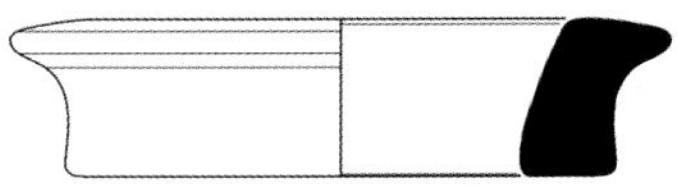

KS3

296. PWBW 417-03

Dimensions: r 8.75 h 2.2 w 2.75

Description: Ring with sub-rectangular section and outward-projecting upper edge.

Comparisons: Cracolici 2004, 39, type III-C3, and 78, Fig. 33, No. 179 (Metaponto, *scarico* 4, end of 5th–first half of 4th c. BC). D'Andria 1980a, 416, No. 59.253, 417, No. 60.333 (kerameikos, *scarichi* 1 and 3).

Bellows Fragment

Bellows Fragment

297. PWBW 043-221

Dimensions: diam. 3.5

Description: Fragment of cylindrical body with fragment of handle with elliptical section.

Comparisons: See Castoldi 1995, 110–11, and Fig. 127 (*tuyère o sostegno fittile*, with extensive bibliography).

Unknown Open Forms (Op)

The only example of Op1 has a diameter of 9.1 cm, while on Op2 the size ranges from 3.5 to 6.0 cm, although examples reaching 7.0 cm (from site 128), 8.0 cm (from site 298), and 9.0 cm (from site 182) are known.

Op1

298. PWBW 466-01 rev90

Dimensions: base 9.1 h 1.6 w 3.6

Description: Ring foot with elliptical section, distinguished from the body by a shallow groove. The body is initially horizontal and then curves obliquely; a sharp edge on the interior emphasizes the point of curvature.

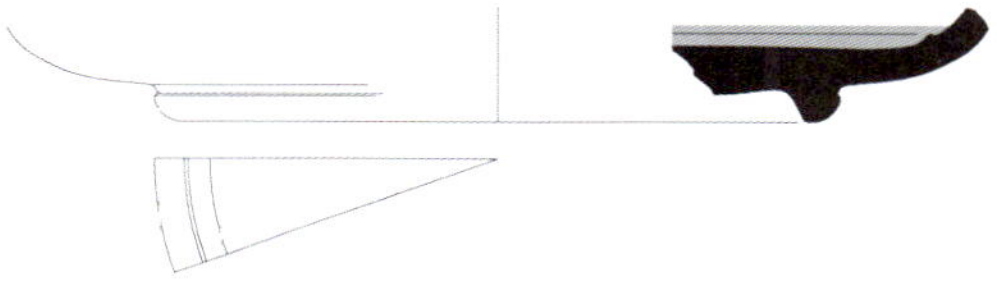

Op2a

299. PWBW 529-05

Dimensions: base 5.95 h 2.9

Description: Ring foot with sub-triangular section; hemispherical body wall, typical of cups.

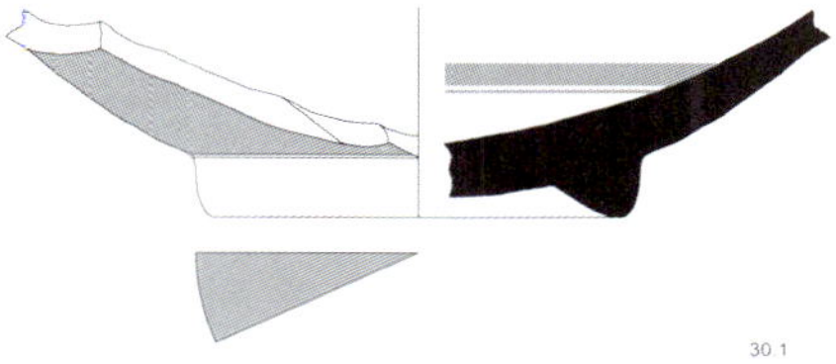

Op2b

300. PWBW 083-210PLa

Dimensions: base 4.9 h 3.6 w 9.8

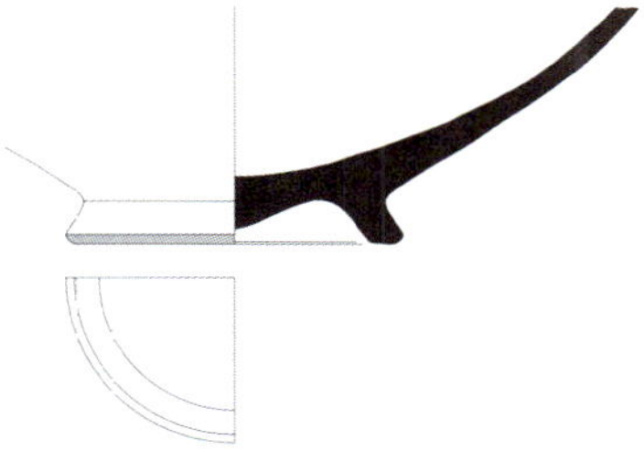

301. PWBW 395-38

Dimensions: base 5.0 h 2.7 w 6.6

Description: Ring foot with sub-rectangular section; hemispherical body wall, typical of cups.

Comparisons: Elliott 1998, 651, C3, C7, and C8 (Pantanello, beginning of 3rd c. BC); Lo Porto 1981a, Fig. 104, No. 3 (Metaponto, first half of 3rd c. BC); Morel 1981a, 3220 (first half of 2nd c. BC), and 3221e1 (second quarter of 2nd c. BC); Pianu 1990, Pl. V, Nos. 162.1, and 169.2 (Herakleia, 300–250 BC). Cf also BG ring-handled cups BG 042-02 and BG 128-22.

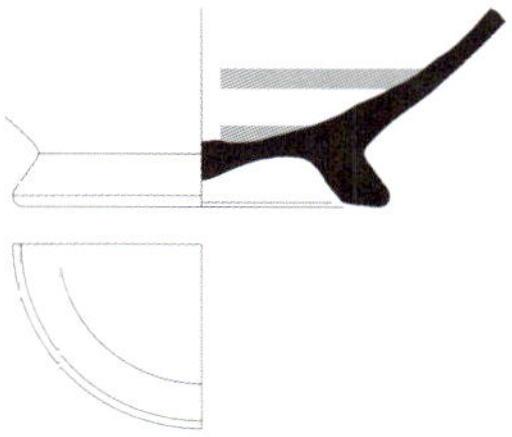

Unknown Closed Forms (Cl)

Cl1

302. PWBW 529-45

Dimensions: base 13.7 h 8.2 w 12.6

Description: Disk foot with elliptical section, concave base, and carinated exterior face; ovoid body.

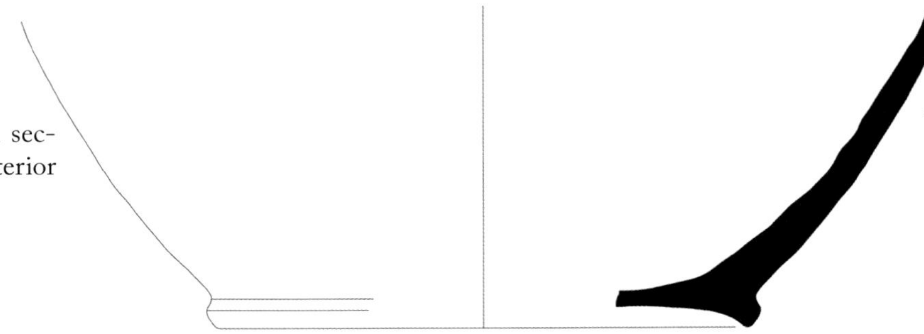

1:2

Cl2

1:2

303. PWBW 352-26

Dimensions: base 10.8 h 3.5 w 14.3

Description: Disk foot with elliptical section and concave base, distinguished from the ovoid body. The examples from sites 16, 173, 213, 221, 266, 296, and 341 have a groove on the lower face.

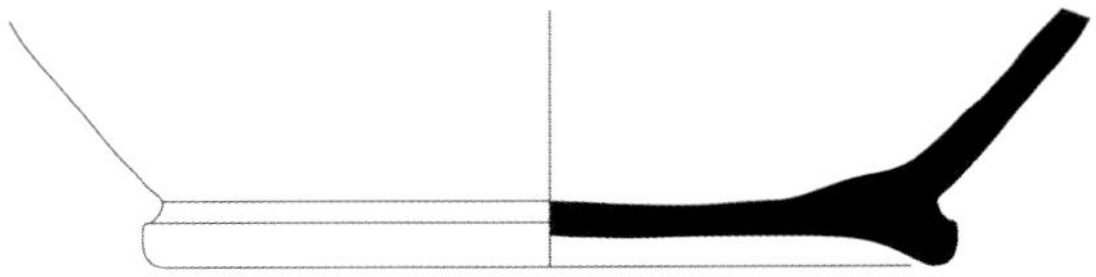

1:2

Cl3

304. PWBW 266-50

Dimensions: base 16.0 h 4.3 w 10.9

Description: False ring foot with elliptical section, defined by a deep groove with triangular section; initially straight body wall. Three examples were identified, from sites 4, 38, and 266, all plain and with diameter between 10.0 and 16.0 cm.

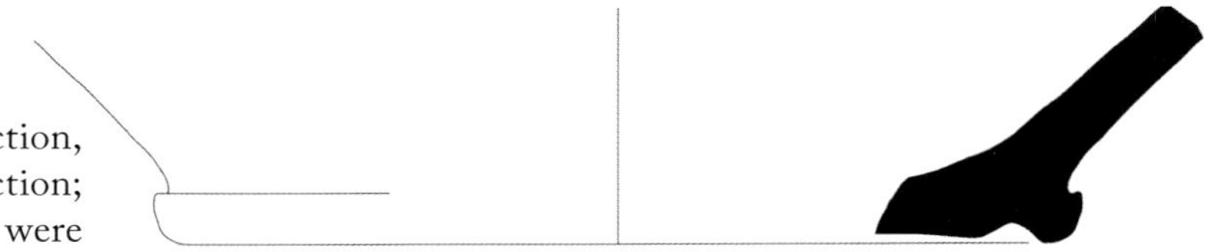

1:2

Cl4

305. PWBW 401-30

Dimensions: base 13.0 h 5.5

Description: Disk foot with triangular section and flat base; ovoid body.

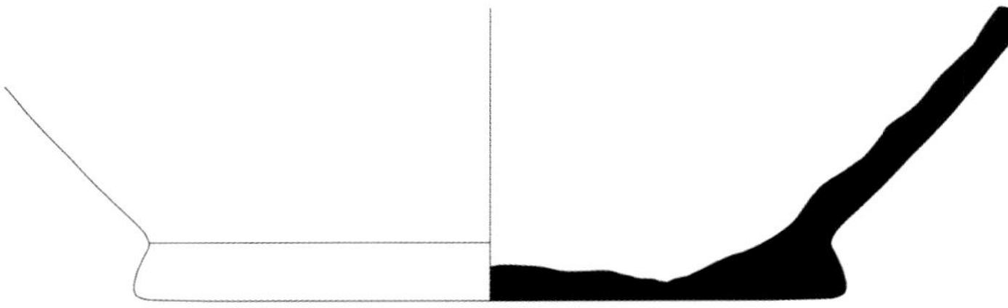

Cl5

306. PWBW 482-03

Dimensions: base 4.1 h 4.4 w 3.5

Description: Flat base; ovoid body. This type generally belongs to an olpe similar to J5a.

Comparisons: Museo Taranto II.1, 128, No. 51.17 (Laterza, 4th c. BC); Lissi Caronna 1980, 125, Fig. 8, Nos. 5–6 (Oppido Lucano, from a domestic structure).

1:2

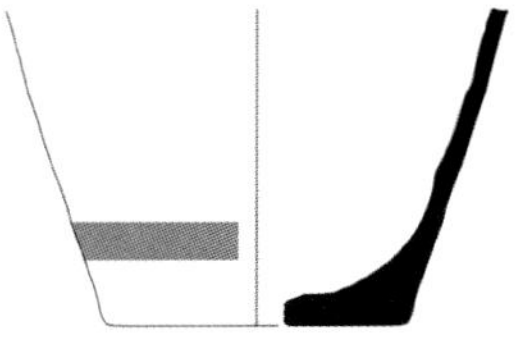

10
Lamps

Emanuela Conoci with Eloisa Vittoria

The ICA survey recovered 54 lamp fragments, all wheel-made and most dating to the Hellenistic period, particularly the 3rd century BC (*Fig. 10.1*).[1] Association with specific types has been possible for only 44 fragments (81.5%), due primarily to the varying states of preservation, which range from nearly complete examples with gloss that is thick and semi-lustrous (**No. 3**) to extremely small fragments almost completely lacking gloss and characterized by a powdery surface.

The data reflects the conditions in the Metapontino, which has been heavily subjected to post-depositional disturbances such as irrigation and deep plowing.[2] Such conditions have resulted in the substantial deterioration of the buried materials and their dispersion on the surface, which is not random but rather mirrors the nature and the internal articulation of the site.[3] The study of the lamp fragments therefore fits within the broader research framework of the survey and the site types that have been identified (see Ch. 5).

Fabrics

The clay fabrics share a similar consistency, usually hard or compact, pinkish to reddish in color, with levigation and smoothing dictated by the functional need for impermeability (*Fig. 10.2*).[4] Some fabrics were not always well-levigated and have small voids and occasionally black or reddish inclusions. Frequent firing defects resulted in a grey fabric color and oxidation blotches on the gloss, which is usually thick and semi-lustrous. Technical comparisons with typical Metapontine fabrics in other classes, particularly fine wares, suggest a local or regional production of the lamps from the survey.[5] Detailed analysis of the black-gloss, grey-gloss, and plain lamps has allowed the subdivision of fabrics into seven groups on the basis of color and degree of levigation.

Quantification and Distribution Among Sites

The 54 lamp sherds were found at 42 of the 595 total sites identified by the survey (*Fig. 10.3*). Most sites had only one fragment, with a maximum of four found together within the same context (*Fig. 10.4*), and each fragment corresponds to an individual lamp. Sites interpreted as Greek farmhouses yielded 93% of the sherds, as opposed to 7% found where tombs were also attested (sites 497, 638, and 742, with one sherd each). These findings, consistent with the scarcity of lamps from funerary contexts at Metaponto and the chora,[6] suggest that the examples found by the survey were used primarily in domestic contexts. The use of oil lamps for this purpose was common in the Greek colonies,[7] where imported versions were first circulated and later followed by locally made products.[8] Nearly all of the nozzle fragments show traces of blackening attributable to use.

It is noteworthy that the number of lamps retrieved is not directly proportional to the extent of the area surveyed: that is, the largest area did not yield the largest number of individual examples. This may well be coincidental, but might also be due to the fact that illumination in ancient houses must have been poor. Main activities were accomplished during the day;[9] moreover, it is possible that other forms of illumination aside from lamps were used, such as torches.[10]

---

[1] My special thanks to Professor J. C. Carter, Francesca Silvestrelli, Keith Swift, Eloisa Vittoria, Cesare Raho, Alessandro Quercia, Antonio R. De Blasi, and Giampiero Colaianni.

[2] Carter 2008, 73.

[3] Cambi and Terrenato 1994, 167–68.

[4] Theseider Dupré 1992, 188.

[5] Lamps, for example, are attested among the materials from the kerameikos of Metaponto (D'Andria 1980a; see below).

[6] Metaponto, contrada Ricotta, only one lamp in tomb 24 and fragments of "bronze tweezers for lamps" in tombs 11 and 23 (Sestieri 1940, 75, 78–79); Metaponto, contrada Ricotta, zone A, 2 lamps among the grave offerings of tomb 2, 1 in tomb 6 and 1 in tomb 9 (Lo Porto 1981b, 337, 339, 353, 355, 357); Pantanello, 1 lamp in tomb 75-1 (Elliott 1998, 689).

[7] Pavolini 1982, 292; *Locri* III, 14.

[8] Petitti 1997, 147.

[9] *Locri* III, 14.

[10] On the use of traditional systems of artificial illumination, see Pavolini 1982, 297, 301; on the presence of woods in the chora, see Carter 2008, 71.

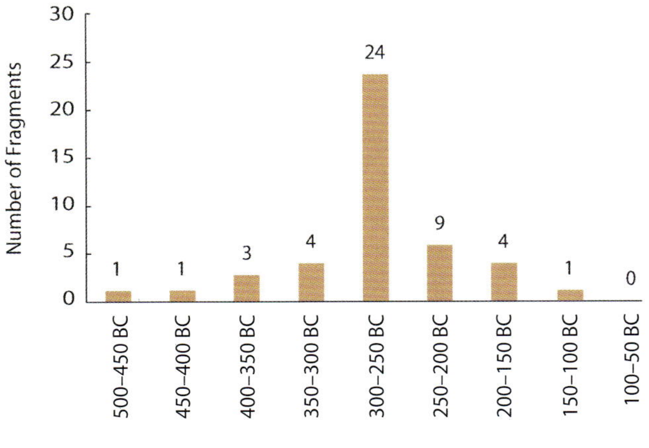

Figure 10.1 Distribution of fragments by time period.

Although there has been no analysis of residues from the lamp fragments, it is possible that the fuel used in the Metapontino was olive oil. [11]

Typology

The great majority of the lamp fragments are black gloss, with only three plain-ware fragments and one grey-ware fragment. These can be divided into four types: globular, circular, cylindrical, and Apulian (*Table 10.1*). The globular is the most common, with 31 examples (28 black gloss and 3 plain ware); two plain lids were also recovered. The sole nozzle in grey ware could not be classified with certainty.

Globular type

E. Conoci, E. Vittoria

There are 31 fragments that can be identified as belonging to globular lamps. Of the 28 individual examples in black gloss, only one is well preserved, lacking just its handle (**No. 3**). The most common variant has a vertical ring handle and a reservoir with rounded or oblate globular walls, sometimes with a rounded carination in the central part of the reservoir or near the base. The filling hole, which is not very wide, is often recessed and surrounded by a thickened rim; the terminus of the nozzle is anchor-shaped, more pronounced in some cases than others.[12] The examples from the survey show the common features of the type: a more-or-less narrow filling hole with a rounded rim marked by a slight groove, which could be interpreted as a stop for a small lid. The characteristic vertical ring handle is preserved in two cases. The disk-shaped bases are mostly concave with a marked bulge (omphalos); the only exceptions are two small and almost flat bases (**No. 9**), which can be associated with the globular type on the basis of comparanda.

Comparative materials also suggest that seven nozzles can be attributed to this type. These are morphologically distinguished by a greater splaying at the tips and include rounded-off (**Nos. 13** and **15**), pointed (**No. 14**), and markedly anchor-shaped nozzles (**Nos. 3** and **12**).

Three unslipped lamp fragments were recovered by the survey from three separate sites (site 172, rim and handle; site 312, foot; site 506, nozzle). Because of the complete absence of information regarding the vessel itself, little can be said about the nozzle fragment from site 506 (**No. 16**), although it is the only one that could be illustrated. From the comparanda one can hypothesize that it had a globular reservoir. The fragment with a ring handle from site 172 also belongs to the globular type, as does the fragment of a base from site 312, which can be compared to a lamp fragment from site 260 (**No. 5**).

The fabrics mostly belong to Groups 1–3 (*Fig. 10.2*) with slight differences in consistency and color. For the majority of the glazed examples, the gloss covers the entire surface of the lamp; exceptions include one example with gloss only on the exterior (**No. 9**), two bases with the outer surface reserved (**Nos. 7** and 8), and one worn base with gloss only on the inside.

The type is dated between 325–250 BC.[13] It occurs at Metaponto[14] and is also attested in the necropolis of Taranto during phase C (275–225 BC).[15] The survey examples, which can be dated between the 4th and 3rd centuries BC, fit this general chronological framework. Possible later examples, dating to the 2nd century BC (**Nos. 3, 12, 14,** and **15**), have also been documented on the basis of comparisons.

[11] Various kinds of fuels have been documented since prehistoric times: vegetal, animal, and mineral oils, animal grease, and beeswax (Pavolini 2000, 168; Bortolin 2008, 16; Quercia 2008, 213). Archaeobotanical data from the chora of Metaponto between the end of the 4th and the beginning of the 3rd c. BC show increased evidence for *Olea europaea* in the pollen record (Carter 2008, 67). This suggests a large availability of oil, part of which was probably used for illumination (*Locri* III, 14, n. 52; Pavolini 1982, 304). This may help to explain the peculiar concentration of lamp fragments from the survey in the same period.

[12] Masiello 1992, 62.

[13] Masiello 1992, 63–64, Pl. XXXII, 4.

[14] Hellenistic structures at contrada Sansone (Lo Porto 1966, Pl. III, 7–8, plain; 2–3, 5, black gloss); contrada Ricotta, tomb 2 (Lo Porto 1981b, Fig. 46, 4–5, black gloss); kerameikos, deposit 6 (D'Andria 1980a, 440, No. 383, Fig. 71, plain).

[15] Masiello 1994, 347.

Group 1. (Cat. **No. 21**) Width of field: 8.5 mm. Very hard, pink (5YR 7/4) matrix with semi-conchoidal fracture; well-levigated with sparse rounded white and dark extremely fine sand-sized inclusions; sparse fine voids.

Group 2. (Cat. **No. 13**) Width of field: 10 mm. Very hard, reddish-yellow (5YR 7/6) matrix with semi-conchoidal fracture; levigated with common rounded white and grey fine sand-sized inclusions; common rounded voids.

Group 3. (Sample 103-04, not in catalog) Width of field: 8 mm. Very hard, pink (5YR 7/4) matrix with semi-conchoidal fracture; levigated with sparse rounded white extremely fine sand-sized inclusions; frequent elongated voids.

Group 4. (Cat. **No. 14**) Width of field: 6.5 mm. Hard, reddish-yellow (5YR 7/6) matrix with semi-conchoidal fracture; abundant elongated white, greyish crystalline and red fine sand-sized inclusions.

Group 5. (Cat. **No. 26**) Width of field: 8.5 mm. Hard, grey (Grey 2 8/1) matrix with semi-conchoidal fracture, levigated with sparse rounded white (quartz/feldspar) fine sand-sized inclusions; common elongated voids.

Group 6. (Cat. **No. 27**, complete vessel, surface) Width of field: 8 mm. Powdery, pink (5YR 8/4); levigated with common red and quartz/feldspar fine sand-sized inclusions.

Group 7. (Cat. **No. 24**) Width of field: 9mm. Hard, pink (5YR 7/4) matrix with semi-conchoidal fracture; very abundant rounded white fine sand-sized inclusions; sparse elongated voids.

Figure 10.2 Lamp fabrics.

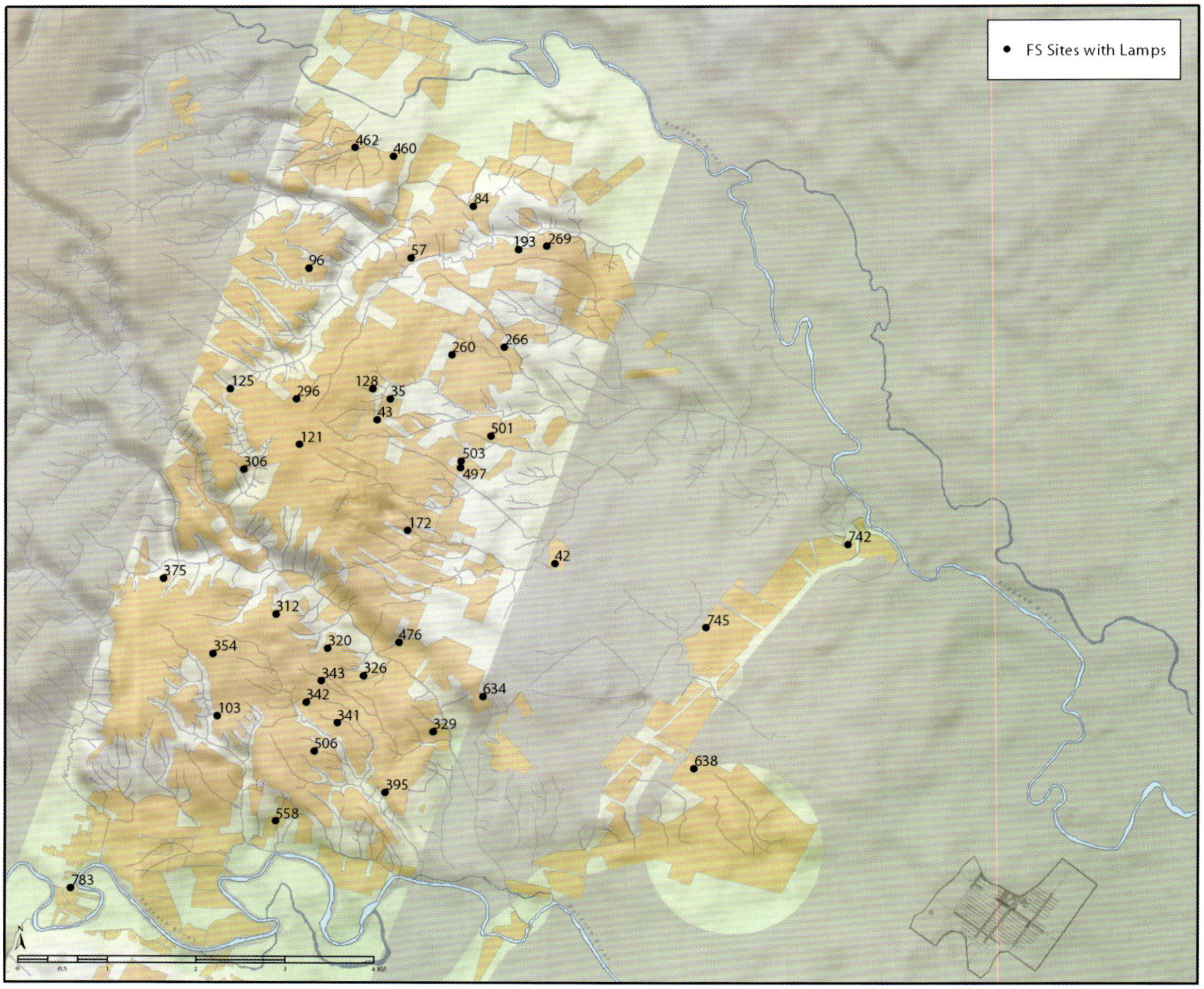

Figure 10.3 Sites with lamp fragments. (JT)

Circular type

In contrast to the overwhelming majority of globular lamps, the survey recovered only a few fragments belonging to other types. Lamps with a circular reservoir, similar to Classical Greek production documented among the finds of the Athenian Agora until the first quarter of the 3rd century BC,[16] were produced in southern Italy until the end of the 4th century BC.[17] These have a broad disk, an open, low-walled reservoir with a convex profile, a flat or slightly concave base, and a horizontal handle opposite the nozzle. This type is attested by five fragments.

No. 17 represents the most widely circulated shape, with a rounded, inward-sloping rim that is comparable to Howland's type 30B,[18] dated to the mid-4th to 3rd centuries BC. Based on comparisons, **No. 18** can also be considered as a circular type. This fragment has an almost vertical rim that is flat on its upper edge, and an open, low-walled reservoir to which a horizontal handle is attached. The form is attested in the Athenian Agora between the 5th and 4th centuries BC.[19]

The fabrics are levigated and associated with Group 1, with the exception of **No. 18**, which with its frequent inclusions belongs to Group 4. The usual gloss completely covers the edge of the fragments; the only reserved area is the outer edge of the base on **No. 20**.

[16] Masiello 1992, 58–59.

[17] Masiello 1992, 59.

[18] *Agora* IV, Pl. 15, 421.

[19] *Agora* IV, 43.

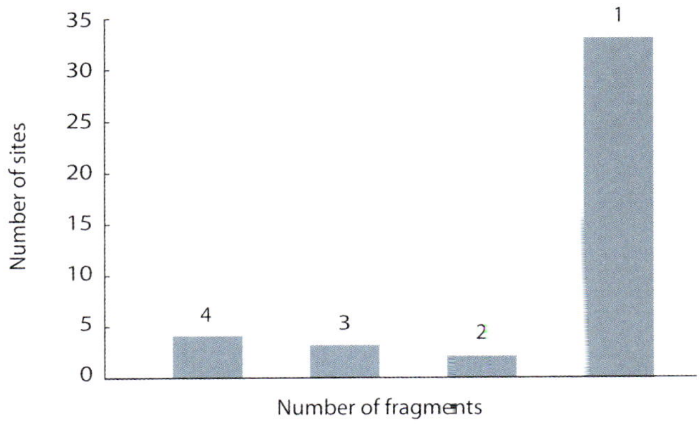

Figure 10.4 Distribution of lamp fragments per site.

This type of lamp is attested at Metaponto both in the kerameikos[20] and near the Hellenistic-period structures.[21] The survey examples can be dated typologically to the early 5th–3rd centuries BC, supporting the hypothesis that open-reservoir lamps circulated in Magna Grecia until the Early Hellenistic period.[22]

Cylindrical type

This type is characterized by a cylindrical reservoir, a wide shoulder, a rim that is sometimes raised, a splayed nozzle, a very slight base, and a vertical strap handle. No immediate parallel can be found in the repertoire of Greek types.[23] This type is represented by 3 fragments from the survey (*Table 10.1*). Although supported by comparisons, the attribution of **No. 23** (a base) to the cylindrical type is tentative, as the reservoir is not well-preserved. The entire profile of **No. 21** is preserved except for the base, which must have been flat or slightly concave, and the attachment of the characteristic vertical handle. A precise parallel can be found in the Locri type LAD2, which also has the same bevelling on the shoulder; the type has been associated specifically with colonial production in Magna Grecia because lamps from mainland Greece do not have the same morphological characteristics.[24] The fabrics that are well-levigated belong to Group 1; those with inclusions and voids belong to Group 3.

[20] D'Andria 1980a, 411, Figs. 56–57 (plain and black gloss).

[21] Kerameikos, deposit 1, plain and black gloss (D'Andria 1980a, 411, Nos. 224–26, Figs. 56–57); Hellenistic structures at contrada Sansone, plain (Lo Porto 1966, Pl. III, 6).

[22] Masiello 1992, 59.

[23] Masiello 1992, 65.

[24] Theseider Dupré 1992, 192–93.

Form	Date range, BC	Total NMI
Globular	375–150	31
Circular	475–200	5
Cylindrical	400–125	3
Apulian	350–250	3
Lids	320–270	2

Table 10.1 Lamp counts by date and form.

The cylindrical type is widespread in Apulia, where it is often associated with globular lamps in black gloss; it is, however, also noted among the finds at Metaponto.[25] Cylindrical lamps from the survey can be dated between the first half of the 4th and the early 2nd century BC.

Apulian type

Only three fragments can be associated with Apulian-type lamps, a production peculiar to southern Italy that does not have any immediate parallels among contemporary Hellenistic Greek lamps.[26] The distinguishing feature is the nozzle, with its pronounced anchor-like shape. The reservoir is barely convex and tapers toward the base; the disk is slightly rounded with a wide filling hole and vertical ring handle.[27] A wall fragment (**No. 24**) has the characteristic disk, which must have terminated in a thin cordon in relief around the filling hole, the barely tapering reservoir, and the handle. On the basis of the comparanda, this type is also associated with a concave base (**No. 25**). The fabrics all belong to Groups 3 and 7 (*Fig. 10.2*).

Apulian-type lamps are documented above all in Apulia (Peucezia, Messapia, Taranto), reaching Matera, and, along the Ionic coast, Metaponto,[28] Herakleia, and Thourioi; they are attested in the chora of Sybaris and in Campania.[29] From a chronological point of view, the lamps from the survey fit well within the 350–250 BC framework proposed for the Apulian type.[30]

[25] Hellenistic structure at contrada Sansone (Lo Porto 1966, Pl. III, 4); Pantanello necropolis, tomb 75-1, black gloss (Elliott 1998, 689); kerameikos, deposit 7, black gloss (D'Andria 1980a, 443, No. 394, Fig. 73); L. Giardino 1980a, Pl. 85, grey gloss.

[26] Masiello 1992, 66.

[27] Petitti 1997, 150.

[28] Castrum (Lo Porto 1966, 144, No. 1); Hellenistic structures at contrada Sansone (Pl. III, 1).

[29] Pavolini 1981, 141; Masiello 1992, 66–69.

[30] Pavolini 1981, 141.

Grey Ware

E. Vittoria

The single grey-ware lamp fragment (**No. 26**) from the survey precludes the creation of a typology. Its direct antecedent can be seen in the Apulian lamps that circulated between the second half of the 4th and the first half of the 3rd century BC. These were replaced by local forms, among which it is possible to find echoes of Hellenistic forms.

The local productions were exported, whereas the Apulian lamps were not.[31] The import of Hellenistic lamps into central Italy—already irregular—and into southern Italy became increasingly sporadic, favoring a greater production of local types associated with the drive toward export.[32] The rounded nozzles, still well-documented in the 3rd century BC,[33] changed gradually into the more elongated anchor or rostrum-shaped nozzle.[34] In general, this nozzle is associated with a body wall having a squat profile (less frequently cylindrical) with a concave disk foot and a vertical handle. It survives also in several mold-made wares.

The discovery of grey-ware lamps with anchor nozzles throughout southern Italy suggests that they were produced in the same centers where the fine table vessels were made.[35] The same situation appears to apply also in the non-Hellenized Italic area, at least beginning in the 3rd century BC, when this type of illumination became widespread; previously both local production and import of lamps were rare.[36] In fact, the lamps from the unpublished kiln deposit at Pantanello attest that even Metaponto produced its own sources of illumination.

The suggestion that grey-ware productions imitated metal vessels can be extended to lamps as well. Broneer emphasized the influence of bronze lamps on his type XIII on the basis of the nozzle shape, the fabric color, and the metallic appearance of the gloss.[37]

Lids

The survey recovered two plain lids (*Table 10.1*; **Nos. 27** and **28**). The examples, which were executed precisely, have wide rims; they are convex on the upper surface with a knob that is almost spherical or with a triangular section, and a cylindrical or trapezoidal flange. The lids explain the presence of rims with lid supports, which were noted on some lamps (**Nos. 1–3, 12,** and **15**). Both objects are nearly complete.

The fabrics are almost levigated with small inclusions and belong to Groups 1 and 6 (*Fig. 10.2*). This kind of object is uncommon in southern Italy but always associated with lamps.[38] The survey examples can be compared with those found at Metaponto itself as well as at Sybaris[39] and Pomarico.[40] A black-gloss lid found at Locri and dated to the 4th century BC demonstrates the use of these objects, although its shape is different from that of the Metapontine examples.[41]

---

[31] Pavolini 1981, 141–43; Masiello 1992, 69; 1994, 339.

[32] Pavolini 1981, 143, 279, n. 13; Masiello 1992, 65; 1994, 340.

[33] See *Agora* IV, 85–104, types 27–33; *Locri* IV, 200–01 and Pls. 56–57; Hellmann 1987, 5, Pl. 2, Nos. 14–15; Masiello 1992, Pl. XXIX, 1, and Pl. XXXI, 1.

[34] Prag 1992d, 210–11.

[35] Prag 1992d, 211.

[36] Pavolini 1981, 140, 143; 1988, 139–40; Masiello 1992, 67.

[37] *Corinth* IV.2, 54.

[38] Petitti 1997, 153.

[39] Petitti 1997, 153, n. 84; *Sibari* V, 486, No. 505, Figs. 436, 474. The dates for both items are not specified.

[40] Petitti 1997, Pl. 54, 31 (370–270 BC).

[41] Theseider Dupré 1992, Pl. LVII, 168.

Catalog of Lamps

Maximum height = h; maximum thickness = th; maximum diameter = max. diam.; base diameter = base diam.; rim diameter = rim diam.; maximum width = w; maximum length = l. All dimensions are in cm. Additional information about the sites can be found in the Gazetteer. Color was recorded with reference to the Munsell Soil Color Charts.

Globular

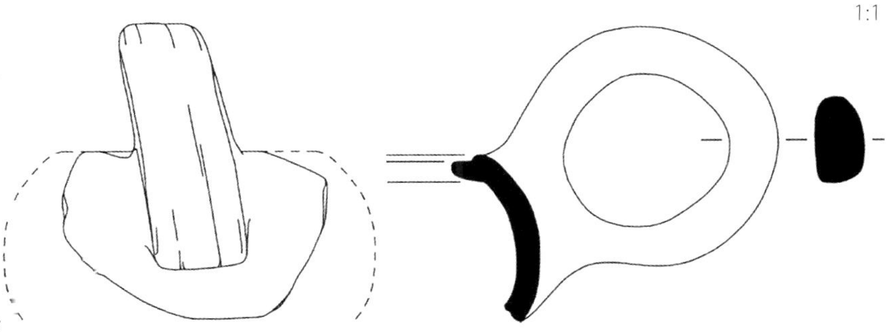

1. LA 125-01 Rim and handle
Dimensions: max. diam. 3.2 body th 0.3
Description: Rounded rim with possible groove for a lid, rounded vertical handle. Clay reddish-yellow (5YR 7/6), uneven, powdery, with white and red inclusions. Frequent voids. Black to reddish gloss outside.
Date: 325–second half of 3rd c. BC.
Comparisons: Petitti 1997, Pl. 53, 4 (Pomarico Vecchio, end of 4th–second half of 3rd c. BC).

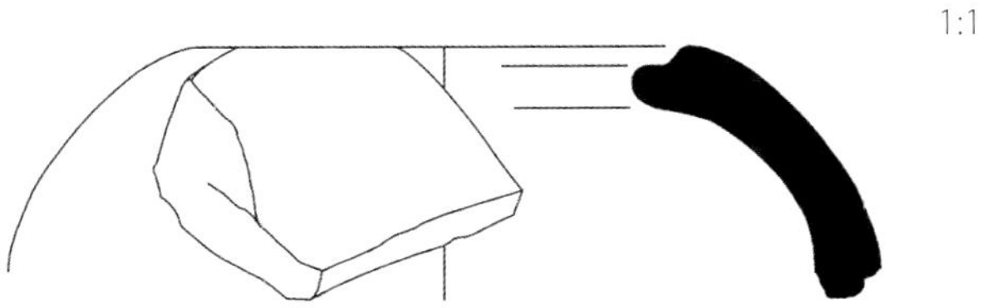

2. LA 266-44 Rim
Dimensions: max. diam. 3.0 h 2.1 th 0.5
Description: Rounded rim with a deep groove, possibly for a lid. Clay grey (10YR 6/1), uneven, very hard, levigated. Black gloss inside and outside.
Date: 325–250 BC.
Comparisons: *Laos* I, 79, Pl. 21, 89 (350–300 BC); Bailey 1975, 56, Pl. 17, Q86 (Athens, end of 4th–beginning of 3rd c. BC), closer to type 25 of the Athenian Agora (*Agora* IV, Pl. 10, 292).

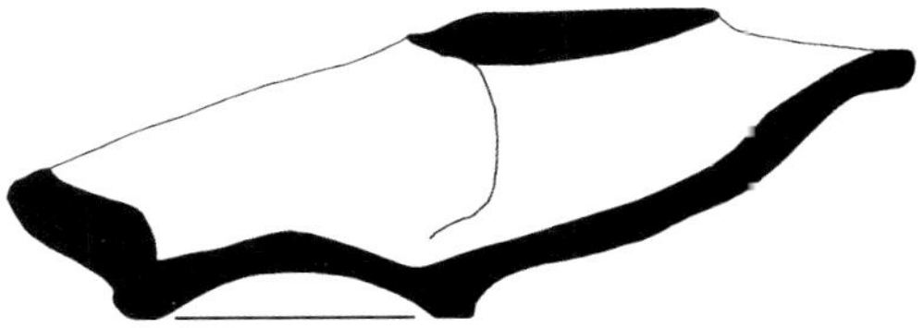

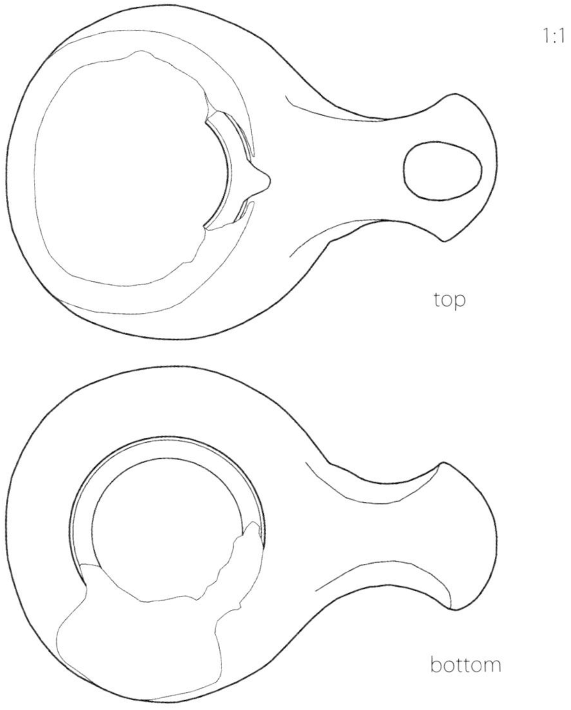

3. LA 320-84 Almost complete lamp, handle missing
Dimensions: h 2.25 base diam. 2.55 th 0.4
Description: Inward-sloping rim with a groove, possibly for a lid, globular squat reservoir, concave base, anchor-shaped nozzle. Clay pink (5YR 7/4), very hard, and levigated. Black gloss inside and outside.
Date: 300–150 BC.
Comparisons: For the reservoir, see Labellarte 1995, 142, 2.1.64; Giannotta 1997, 177, Fig. 52 (Vaste, Fondo Aia, TIII, beginning of 3rd c. BC); for the nozzle, Zaccaria Ruggiu 1980, 36, 41, No. 40 (southern Italy, 350–beginning of 1st c. BC); Prag 1992d 212, Fig. 94, Pl. XXA, 1652 (Botromagno-Gravina di Puglia, 2nd–mid-1st c. BC).

Globular cont.

4. LA 260-22 Base
Dimensions: h 1.45 base diam. 3.8 th 0.3
Description: Concave base. Clay reddish-yellow (5YR 7/6), powdery, levigated. Small voids. Black gloss inside and outside.
Date: 325–250 BC.

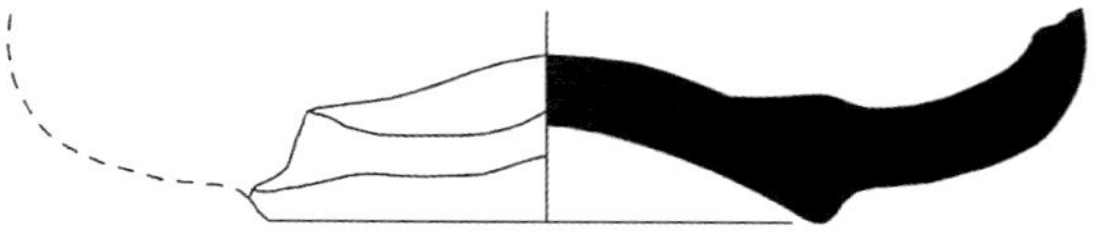

1:1

Comparisons: *Sibari* II, 485, No. 630, Figs. 450, 590 (Sybaris, 300–250 BC).

5. LA 260-23 Base
Dimensions: h 2.4 base diam. 3.4 th 0.3
Description: Disk slightly concave base. Clay light gray (10YR 7/1), uneven, hard, levigated, with sparse voids. Black gloss inside and out.
Date: 325–200 BC.
Comparisons: Petitti 1997, Pl. 53, 13 (Pomarico Vecchio, end of 4th–second half of 3rd c. BC); *Sibari* IV, 485, No. 169, Figs. 453, 485 (Sybaris, mid-3rd c. BC).

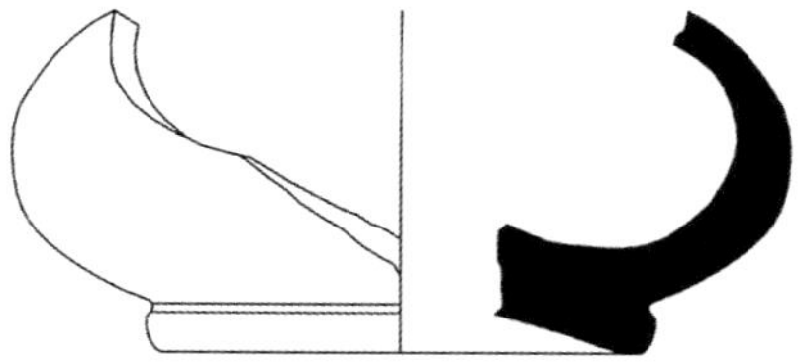

1:1

6. LA 320-85 Base
Dimensions: max. wall h 0.7 base diam. 4.7
Description: Concave base. Clay pinkish-gray (5YR 7/2), hard, levigated. Black to reddish gloss inside and out.
Date: 325–200 BC.
Comparisons: Petitti 1997, Pl. 53, 2 (Pomarico Vecchio, end of 4th–second half of 3rd c. BC).

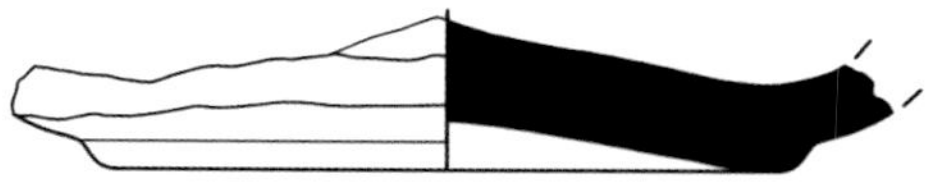

1:1

7. LA 343-12 Base
Dimensions: max. wall h 0.8 base diam. 3.1
Description: Concave base, marked by a slight groove. Clay reddish-yellow (5YR 6/6), uneven (grey nucleus), very hard, levigated, with sparse small voids. Black gloss inside and outside, but outer edge of base reserved.
Date: 350–250 BC.
Comparisons: Bailey 1975, 54, Pl. 17, Q 81 (Athens), closer to type 25A of the Athenian Agora (*Agora* IV, Pl. 9, 269 dated 350–275 BC).

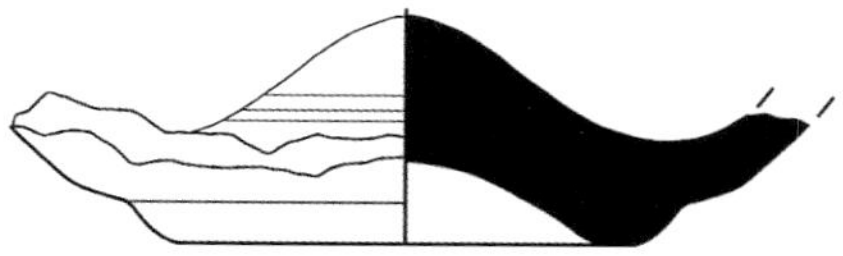

1:1

8. LA 506-34 Base
Dimensions: h 2.0 base diam. 3.5 th 0.4
Description: Disk concave base. Clay reddish-yellow (5YR 6/6) uneven, hard, with small white inclusions. Many voids. Black gloss inside and outside, but outer edge of base reserved.
Date: 325–200 BC.
Comparisons: Theseider Dupré 1992, 201, Pl. LVII, 159 (Locri, beginning of 3rd c. BC); Petitti 1997, Pl. 53, 11 (Pomarico Vecchio, end 4th–second half of 3rd c. BC).

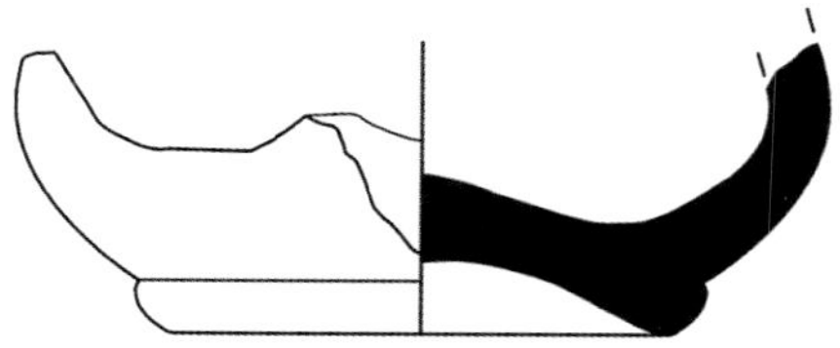

1:1

9. LA 558-03 Base
Dimensions: max. wall h 1.0 base diam. 2.6 th 0.4
Description: Almost flat base. Clay pinkish-gray (7.5YR 7/2), uneven, very hard, with many voids. Black gloss outside; 0.1 band reserved around base.
Date: 325–225 BC.
Comparisons: Possible globular type according to Lamboley 1996, 419, Fig. 48, 4 (Vaste, Fondo Pizzinaghe, 325–225 BC); see also D'Andria 1990, 162, 269 (Vaste, Fondo Ficorelli, mid-4th c. BC).

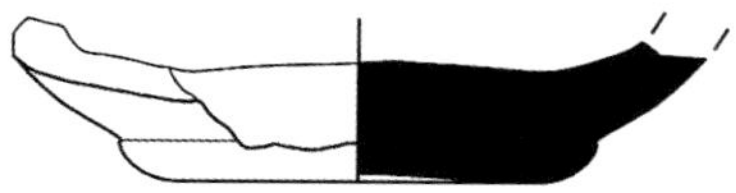

1:1

Globular cont.

1:1

10. LA 667-04 Base
Dimensions: h 1.4 base diam. 3.65 th 0.4
Description: Disk concave base. Clay reddish-yellow (5YR 7/6), powdery, with many voids. Black gloss inside and outside.
Date: 325–200 BC.
Comparisons: Theseider Dupré 1992, 200–201, Pl. LVII, 158 (Locri, end of 4th–beginning of 3rd c. BC); Petitti 1997, Pl. 53, 3 (Pomarico Vecchio, end of 4th–second half of 3rd c. BC).

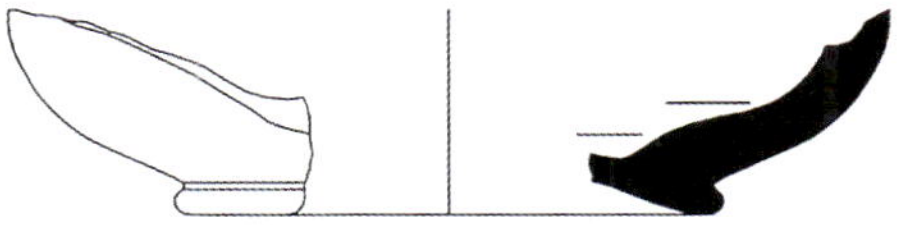

1:1

11. LA 783-01 Base
Dimensions: max. wall h 0.7 base diam. 3.7 th 0.4
Description: Disk concave base. Clay pink (7.5YR 7/4), uneven, very hard, with many black inclusions and voids. Black gloss inside only.
Date: 375–250 BC.
Comparisons: *Sibari* IV, 357, No. 159, Figs. 336, 374 (Sybaris, 375–250 BC).

1:1

12. LA 128-28 Nozzle
Dimensions: h 2.8 l 4.0 w 3.8 body th 0.3
Description: Short anchor-shaped nozzle; traces of possible support for a lid. The exterior surface of the reservoir is decorated with a pair of parallel vertical grooves on each side of the nozzle. Clay pink (5YR 7/4), very hard, and levigated. Black gloss inside and outside.
Date: 350–150 BC.
Comparisons: Possible globular type according to Petitti 1997, Pl. 54, 28 (Pomarico Vecchio, end of 4th–second half of 3rd c. BC); see also *Laos* I, 79, Pl. 21, 89 (350–300 BC), and Lo Porto 1966, 152, Pl. III, 3 (Metaponto, Sansone, 3rd–2nd c. BC).

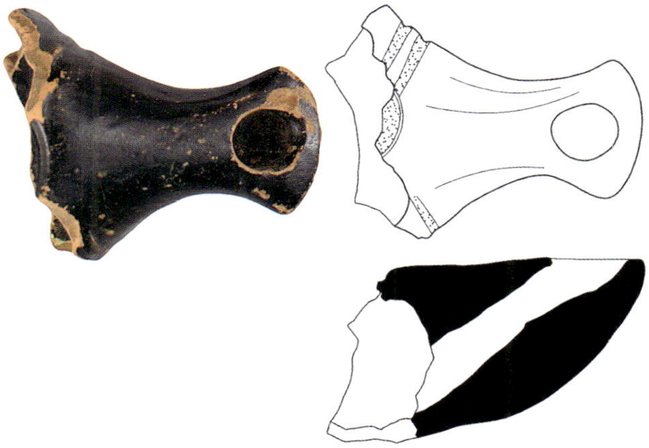

1:1

13. LA 354-23 Nozzle
Dimensions: l 4.4 w 2.9 th 0.6
Description: Short and rounded nozzle. Clay reddish-yellow (5YR 7/6), uneven, very hard, levigated, with rounded white and red inclusions. Black gloss inside and outside. Light traces of usage.
Date: 325–250 BC.
Comparisons: Possible globular type according to Masiello 1992, Pls. XXIX, 1 and XXXII, 5 (325–250 BC).

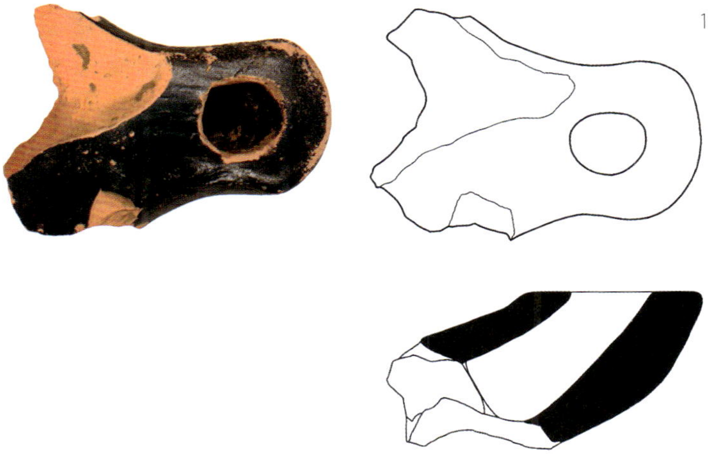

Globular cont.

1:1

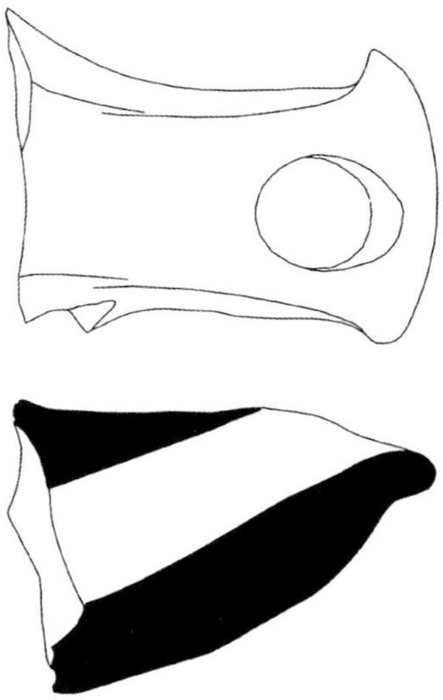

14. LA 476-01 Nozzle

Dimensions: l 4.4 w 3.3 th 1.0

Description: Long nozzle with tapered ends. Clay reddish-yellow (5YR 7/6), hard, with white, grey, and red inclusions. Black to reddish gloss inside and outside. Traces of usage present.

Date: 325–150 BC.

Comparisons: *Sibari* III, 225, No. 137, Figs. 232, 260 (Sybaris, 250–200 BC); Lo Porto 1966, 152, Pl. III, 3 (Metaponto, Sansone, 3rd–2nd c. BC).

1:1

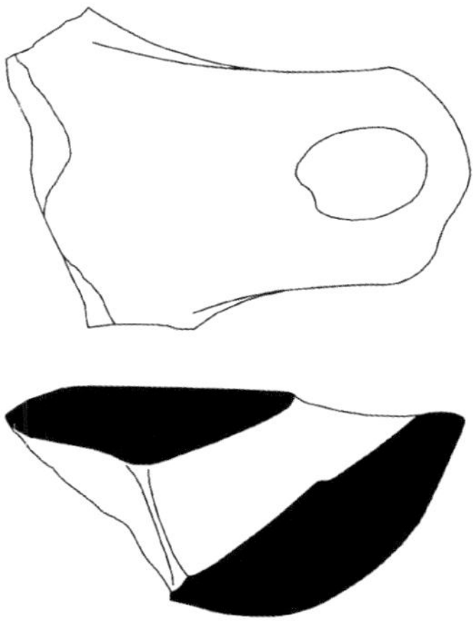

15. LA 503-11 Nozzle

Dimensions: h 2.25 l 4.3 w 3.1 th 0.4

Description: Short rounded nozzle; traces of possible support for a lid. Clay pink (5YR 7/4), uneven, very hard, and levigated. Black gloss inside and outside. Traces of usage present.

Date: 375–200 BC.

Comparisons: Possible globular type according to Petitti 1997, Pl. 54, 30 (Pomarico Vecchio); *Sibari* III, 228, No. 151, Figs. 233, 263 (Sybaris, 375–end of 3rd c. BC); Lo Porto 1966, 152, Pl. III, 2 (Metaponto, Sansone, 3rd–2nd c. BC).

Plain Ware

1:1

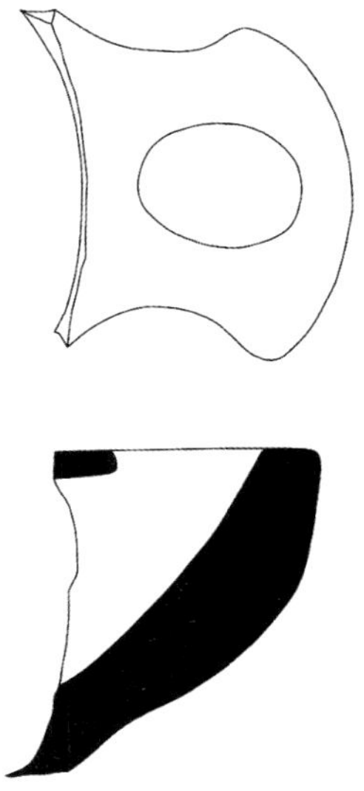

16. LA 506-31 Nozzle

Dimensions: h 3.0 l 3.9 w 4.6

Description: Rounded nozzle with barely pronounced tips, oval hole. Orange-red fabric (2.5YR 5/8), hard, levigated.

Date: 350–250 BC.

Comparisons: Masiello 1992, Pl. XXVIII, No. 4, and Pl. XXXII, No. 4 (325–250 BC). Petitti 1997, 149, Pl. 53, No. 7 (Pomarico, second half of 4th–first half of 3rd c. BC).

For the fragment from site 172, see: Masiello 1992, Pl. XXVIII, No. 4, and Pl. XXXII, No. 4 (325–250 BC). Pianu 1990, Pl. LVIII, n. 147.2 (Herakleia, 325–300 BC).

For the foot from site 312, see Petitti 1997, 149, Pl. 53, No. 13 (Pomarico, end of 4th–second half of 3rd c. BC). Prag 1983a, 362, No. 362 (Metaponto, end of 4th–2nd c. BC). (E. Vittoria)

Circular

1:1

17. LA 460-01 Rim
Dimensions: h 1.6 rim int. diam. 4.85 th 0.4
Description: Circular body with incurved rim. Clay pink (5YR 7/3), uneven, very hard, levigated. Black gloss inside and outside.
Date: 400–250 BC.
Comparisons: *Museo Taranto* II.1, 126–27, No. 51.15 (Laterza, end of 5th–4th c. BC); Lo Porto 1988–1989, 338, No. 14, Figs. 36, 40 (Montescaglioso, T2, first half of 4th c. BC); see also Petitti 1997, Pl. 54, 26 (Pomarico Vecchio, 250–200 BC).

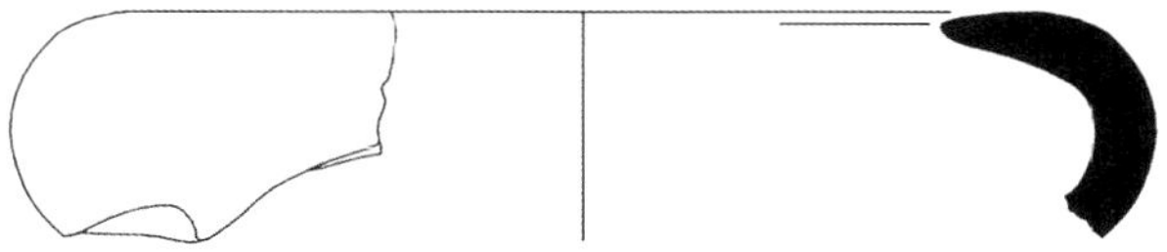

1:1

18. LA 193-07 Rim
Dimensions: h 2.5 rim int. diam. 4.5 th 0.6
Description: Cylindrical reservoir with vertical rim; horizontal handle with rounded section. Clay pink (7.5YR 7/4), uneven, very hard, with small white and black inclusions. Black gloss inside and outside.
Date: 475–450 BC.
Comparisons: *Kaulonia* I, Fig. 49, No. 346 (Kaulonia, 475–450 BC), compare type 20 of the Athenian Agora (*Agora* IV, Pl. 5, 154–55).

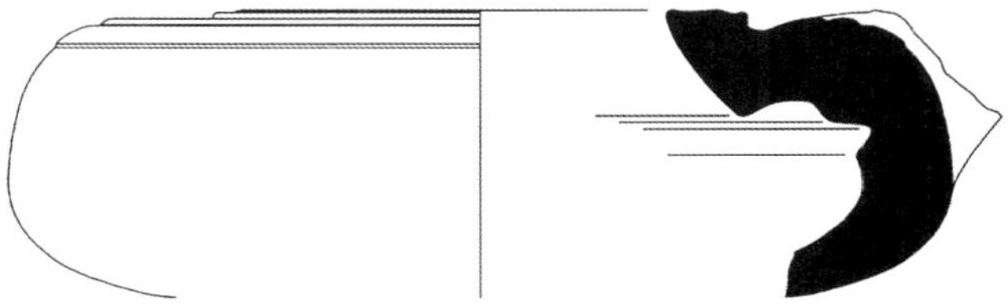

1:1

19. LA 260-30 Base
Dimensions: h 0.5 base diam. 6.4 th 0.4
Description: Slightly concave base. A slight groove marks the upper part of base. Clay pink (7.5YR 7/4), uneven, powdery and levigated, with sparse voids. Black gloss inside and outside.
Date: 400–350 BC.
Comparisons: Bailey 1975, 50, Pl. 15, Q 68 (Athens, 400–350 BC), compare type 23C of the Athenian Agora (*Agora* IV, Pl. 8, 229).

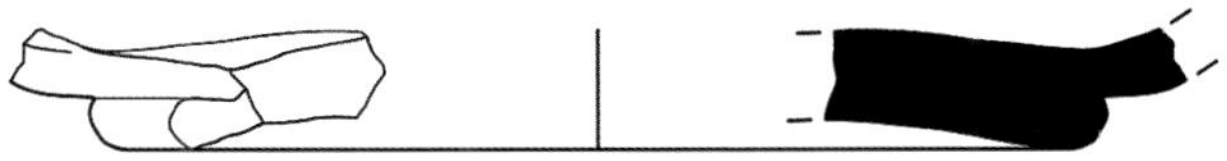

1:1

20. LA 501-04 Base
Dimensions: h 1.3 base diam. 6.0 th 0.8
Description: Disk base, marked by a slight groove. Clay pink (5YR 7/4), uneven, very hard, and levigated. Black gloss inside and outside; outer edge of base reserved.
Date: 350–200 BC.
Comparisons: D'Andria 1990, 118, No. 163 (Vaste, Fondo Melliche, 350–300 BC).

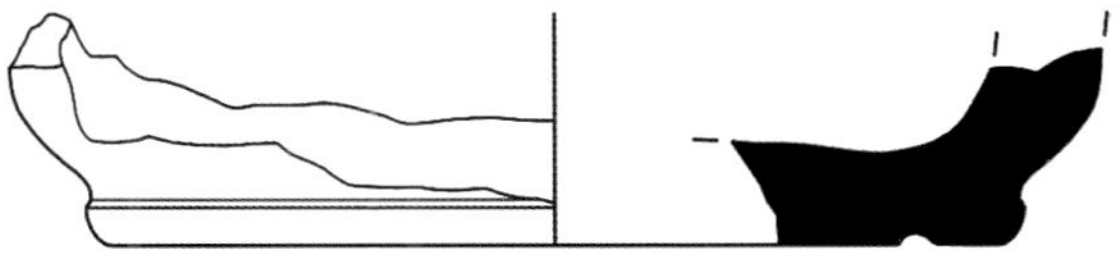

Cylindrical

1:1

21. LA 260-21 Shoulder and walls (handle and base missing)
Dimensions: h 2.9 rim diam. 2.0 th 0.3
Description: Vertical body, angular shoulder with narrow inward-sloping rim, groove on the shoulder, vertical handle not preserved. Clay pink (5YR 7/4), uneven, very hard, well-levigated, with sparse white and black inclusions; sparse small voids. Black gloss outside, partially reserved inner edge.
Date: 400–220 BC.
Comparisons: Theseider Dupré 1992, 199, Pls. 56, No. 150 (Locri, 400–350 BC); Elliott 1998, 689 (Pantanello necropolis, T75-1, end of 4th–beginning of 3rd c. BC).

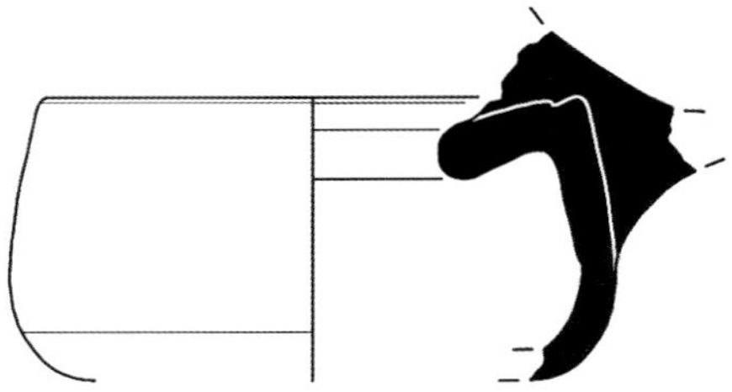

22. LA 043-04 Base
Dimensions: h 2.3 base diam. 3.9 th 0.4
Description: Disk concave base. Light reddish-brown clay (2.5YR 6/4), uneven, very hard and levigated, with white lithic inclusions and voids. Black gloss inside and outside.
Date: 300–250 BC.
Comparisons: Theseider Dupré 1992, 201, Pl. 57, 159 (Locri, beginning of 3rd c. BC).

1:1

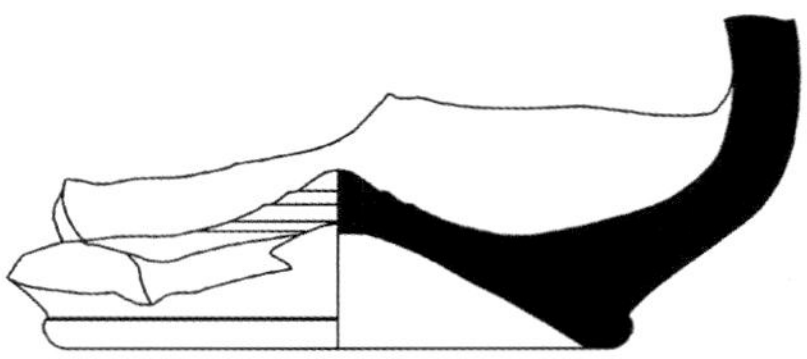

23. LA 057-16 Base
Dimensions: h 0.6 th 0.5
Description: Disk slightly concave base. Clay pinkish-gray (7.5YR 7/2), uneven, very hard, with small white lithic inclusions and voids. Black gloss inside and outside.
Date: 225–125 BC.
Comparisons: *Sibari* V, 419, No. 212, Figs. 375, 480 (Sybaris).

1:1

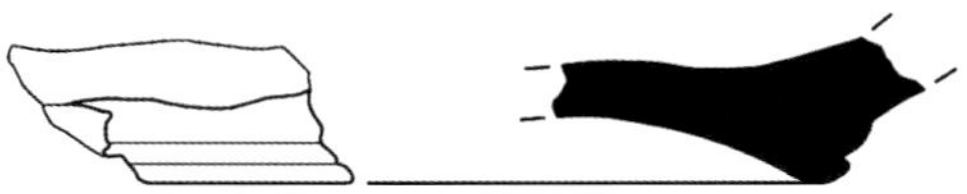

Apulian

1:1

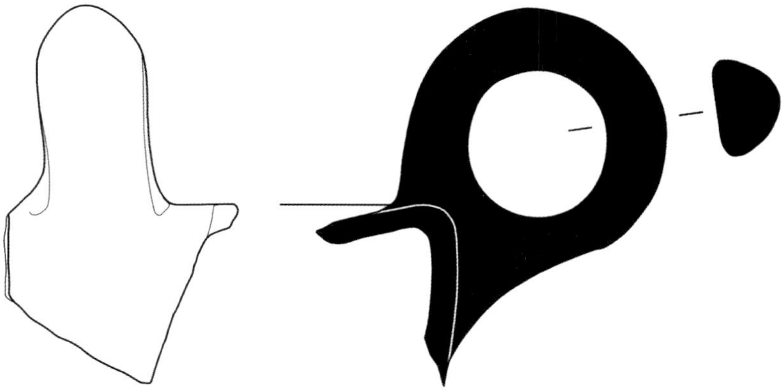

24. LA 096-03 Shoulder and handle
Dimensions: h 5.1 w 3.1 th 0.3
Description: Angular shoulder and inward-sloping rim, rounded vertical handle. Clay pink (5YR 7/4), uneven, hard with white inclusions. Black gloss inside and outside.
Date: 300–250 BC.
Comparisons: Petitti 1997, Pl. 53, 14 (Pomarico Vecchio, 300–250 BC); see also D'Andria 1990, 163, No. 270 (Vaste, Fondo Ficorelli, end of 4th c. BC); Labellarte 1995, 141, 2.1.63.

Apulian cont.

25. LA 121-01 Base

1:1

Dimensions: h 1.1 base diam. 3.7 th 0.3

Description: Slightly concave base. Clay reddish-yellow (7.5YR 7/6), uneven, very hard, with small white lithic inclusions and voids. Black gloss inside and outside.

Date: 350–250 BC.

Comparisons: Masiello 1992, Pls. XXXII, No. 2 (350–250 BC); Lo Porto 1988–1989, 351, No. 31, Figs. 46, 56 (Montescaglioso, T 5, second half of 4th c. BC); also D'Andria 1990, 163, No. 270 (Vaste, Fondo Ficorelli, end of 4th c. BC).

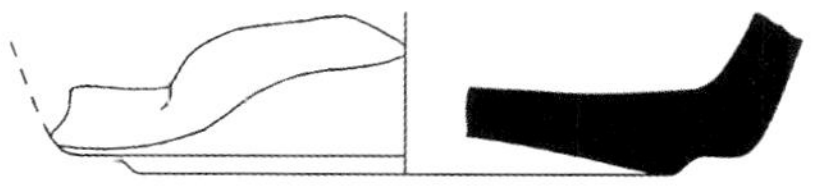

Grey Ware

26. LA 042-121P Nozzle

1:1

Dimensions: w 3.6 l 4.2

Description: Anchor-shaped nozzle with extremities barely turned upwards and slightly concave termination. Imperfect workmanship on the interior. Poor condition. Light grey fabric (7.5YR N7/), well-purified, compact, and soft. No traces of gloss.

Date: 200–50 BC.

Comparisons: The lamps from the Pantanello kiln deposit have the same form of the nozzle (Burgers 1991, Pls. 20–21); two examples from Metaponto are illustrated in L. Giardino 1980a, Pl. 85, Nos. 59–60, compared to Broneer's type XIII and to Howland's type 35, dated to the first half of the 2nd and last quarter of the 2nd–1st c. BC. In tomb 9 from Casa Ricotta in the Metapontino there was a grey-ware lamp like the survey piece (Lo Porto 1981b, 358, Fig. 74, No. 4; the tomb is dated to the middle of the 3rd c. BC, but the presence of the lamp suggests a lower date). There are similar lamps from Gravina (Prag 1992d, Fig. 97, No. 1680, type IIIc, attributed to period VIIIa, end of the 2nd–1st c. BC; Small et al. 1994, Fig. 15a, Nos. 44–45, from trench F202, dated 75±15 BC); the imperfect workmanship at Gravina is described as a peculiar aspect of the lamps in grey ware vis-à-vis those in black gloss (Prag 1992d, 211). (EV)

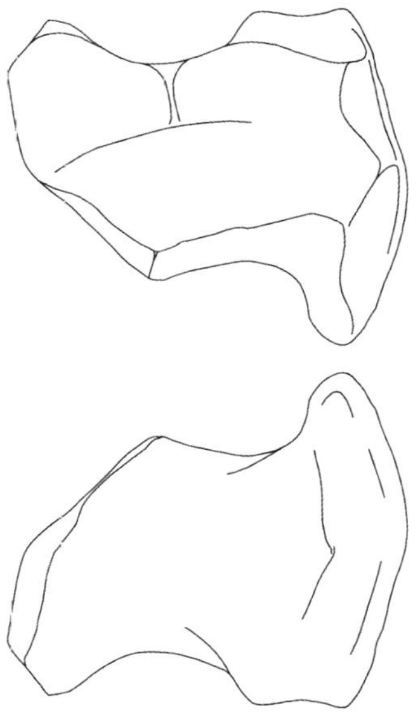

Lids

27. LA 341-45 Lid

1:1

Dimensions: h 2.4 rim diam. 3.15

Description: Small lid with trapezoidal full base, sloping brim and triangular knob. Clay pink (5YR 8/4), powdery, and levigated with common red and white inclusions.

Date: 320–270 BC.

Comparisons: Petitti 1997, Pls. 54, No. 31 (Pomarico Vecchio).

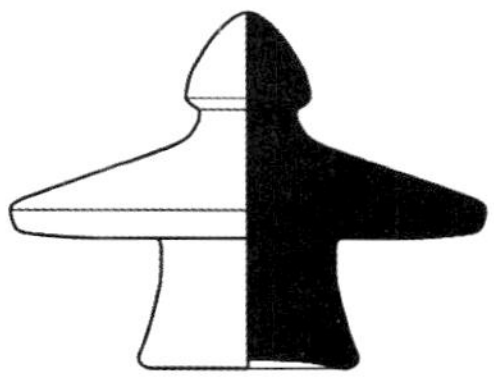

28. LA 638-01 Lid

1:1

Dimensions: h 1.4 rim diam. 2.3

Description: Small lid with cylindrical full base, sloping brim and triangular knob. Clay reddish-yellow (5YR 7/6), uneven, powdery, with white and black inclusions.

Date and Comparisons: see **No. 27**.

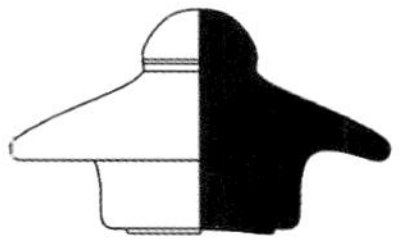

11
Greek and Roman Republican Cooking Wares

Ruth Smadar Gabrieli

Cooking wares were recovered from sites throughout the Metapontine chora, including farmhouses, necropoleis, and sanctuaries. The defining feature of these vessels is their coarse texture due to their high proportion of inclusions, and this provided a useful initial criterion for identification, followed by morphological attributes.[1] The cooking-ware forms from the survey have been divided into open shapes—pans and casseroles (lopades)—and closed-shape cooking pots (chytrai). Lids were also distinguished. The open shapes include lidded forms suitable for roasting and simmering (casseroles) along with shapes that are adapted for baking and frying (pans), and the specialized klibanoi, used as mobile ovens. One klibanos was identified with certainty in the survey material, and a few fragments tentatively, but they come from sites between the Basento and Cavone and will be included in future publication.

Cooking wares can be characterized as comprising a series of basic and long-lived shapes. This presents a major limitation to the use of cooking ware as an interpretive tool in a survey, as there is a lack of chronological sensitivity. The problematic nature of dating cooking-ware forms, challenging even in well-stratified excavations, is compounded in the fragmentary survey material, which usually lacks complete shape profiles.[2] For this study, the search for parallels was aimed, as much as possible, at sites in or near the Metapontine chora. Additional references can be made to major sites on the Greek mainland, given that the origin of the basic cooking-ware shapes employed by the Greek settlers in Magna Grecia was undoubtedly their homeland, and parallels from Athens and Corinth are therefore of some use for distinguishing the range of basic shapes present at survey sites in the Metapontine chora. On the other hand, cooking-ware forms also developed regionally; thus the development of local forms and their morphological details cannot always be expected to follow that of the basic parent shapes from Athens, for example.[3] Selected comparanda have been included if their dates were based on stratigraphic evidence from excavations in southern Italy. Generally two or three examples were considered sufficient. The attempt to further define the chronologies of cooking ware shapes on the basis of reliable chronological parameters proved unsuccessful: similar cooking-ware shapes were found in considerably dispersed chronological contexts at different sites even in close proximity.

Cooking-Ware Shapes

Together, open cooking-ware shapes account for over a third of all cooking-ware forms, with casseroles being much more numerous than pans: 177 examples of casseroles, or 30.4% of all forms, versus 37 pans, or 6.4% of all forms (*Table 11.1*). Comprising just below half (48.6%) of all cooking ware forms, cooking pots are far more frequent in the survey assemblages. Lids make up just below 15% of all cooking ware forms.

Pans

These are shallow cooking vessels with flat or slightly curved undersides. The forms from the survey can be divided broadly into four types:

1. Flanged pans, characterized by projecting rims which vary from up-turned (**No. 1**) to horizontal (**No. 2**) to down-turned (**No. 3**);
2. Pans with plain rims (**Nos. 4–5**);
3. Pans with thickened rims (**No. 6**);
4. Pans with flared rims (**No. 7**).

[1] For discussion of cooking wares at the different types of site identified in the chora—farmhouses, necropoleis/tombs, and sanctuaries—see Ch. 5. The proportion of inclusions typical of cooking-ware fabrics was determined by examination of casseroles and cooking pots using the charts published by Orton, Tyers and Vince (1993). Occasionally, non-cooking shapes were produced in similar fabrics to those used for cooking wares, consisting mainly of jugs and other liquid containers: see Ch. 9 on the plain/banded wares for the range of non-cooking ware forms represented in the survey.

[2] Forthcoming publications of the excavations at four farmhouses in the chora of Metaponto (Fattoria Fabrizio, Sant'Angelo Vecchio, Sant'Angelo Grieco, and Fattoria Stefan) and from the rural sanctuary and farmhouse at Pantanello will address the regional cooking-ware typologies and chronologies on the basis of substantial excavated assemblages containing relatively well-preserved forms.

[3] See comments by Sparkes and Talcott: *Agora* XII, 33.

Shape	Form	Count	% Forms
Pans (**Nos. 1–7**)	Flanged pan (**Nos. 1–3**)	7	1.2%
	Pan with plain rim (**Nos. 4–5**)	14	2.4%
	Pan with thickened rim (**No. 6**)	3	0.5%
	Pan with flared rim (**No. 7**)	1	0.2%
	Unidentified/unattributed	12	2.1%
	Subtotal pans	**37**	**6.4%**
Casseroles (lopades) (**Nos. 8–25**)	Shouldered casserole (**Nos. 8–10**)	26	4.5%
	Globular casserole (**Nos. 11–17**)	86	9.5%
	Casseroles with concave walls (**Nos. 18–24**)	55	14.8%
	Casserole with carinated shoulder (**No. 25**)	1	0.2%
	Unidentified/unattributed	9	1.5%
	Subtotal casseroles	**177**	**30.4%**
Cooking pots (chytrai) (**Nos. 26–37**)	Cooking pot with collared rim (**Nos. 26–27**)	2	0.3%
	Necked pots or jugs (**Nos. 28–29**)	68	11.7%
	Open-mouthed cooking pot with projecting rim (**Nos. 30–31**)	58	10.0%
	Lid-seated cooking pot (**Nos. 32–37**)	142	24.4%
	Unidentified/unattributed	13	2.2%
	Subtotal cooking pots	**283**	**48.6%**
Lids (**Nos. 38–39**)	**Subtotal lids**	**85**	**14.6%**
	All cooking-ware forms	**582**	**100%**
	Total body sherds	993	
	Total fragments	**1,575**	

Table 11.1 Quantification of cooking-ware shapes from the survey.

The pans do not have lid seats, though lids may have been placed over or within them. Pans and casseroles with rim diameter of up to 15 cm are considered small, rim diameter of 16–25 cm medium, and where the rim diameter is over 25 cm, the vessels are considered large. Of the 38 pans identified, diameter measurement was possible for nine; of these, two have a diameter of less than 16 cm (14 cm), while the rest are between 16 and 32 cm.

The shape is well suited for baking and frying. Differentiation of frying from baking pans is only possible if the single, straight handle of the frying pan survives. No examples of these are extant in the corpus of cooking wares from the survey.

Pans comprise a small component of the Metapontine cooking-ware repertoire (6.4% of all cooking-ware forms from the survey). The flanged pan is attested at the Pantanello necropolis in a tomb dated from the late 4th or early 3rd century BC, and in 3rd-century BC contexts at Pomarico Vecchio and Locri.[4] Pans with plain rims seem to have similar dates, attested in the Metapontine kerameikos in the first half of the 4th century BC.[5] The single example of a pan with a thickened rim is similar to an example from the Pantanello necropolis with a single, frying-pan style handle, dated to the late 4th or early 3rd century BC.[6] There is a single example of a pan with a flared rim, which preserves the edge of a pinched spout (**No. 7**).

[4] *Necropoleis*, 726, T 111-2, tomb dated 310–275 BC; *Pomarico Vecchio* I, 2.90, Pl. 82, No. 182; *Locri* II, Pl. XXXVII, No. 322.

[5] Quercia 2004, 188, Fig. 6.

[6] *Necropoleis*, 727, CD 89-4.

Casseroles (lopades)

Casseroles (lopades) are open, shallow cooking vessels that were a central component of the Greek cooking-ware repertoire. They are characterized by rounded bases and ever-present lid-supporting ledges, consistent with use for cooking or simmering in sauce. Casseroles and pans are the only shapes in the cooking-ware assemblage for which there are complete profiles. The casseroles are distinct from pans in being deeper and having a lid seat. The majority of casseroles have minimum rim diameters of 16 cm and maximum heights of 5.5 cm. Identification was based on body outline, handle shape, and rim profile, which resulted in the following groups:

1. Shouldered casseroles, with inclined rims and lid seats, which form a convex shoulder on the outer profile (**Nos. 8–10**);
2. Globular casseroles, with sagging bases and lid-seated rims varying from inclined to vertical (**Nos. 11–17**);
3. Casseroles with concave walls, in which the concave outline of the walls may be out-turned or vertical, and the turn to the base is often carinated (**Nos. 18–24**);
4. Casseroles with carinated shoulder, represented by a single example with a collared rim and lid seat (**No. 25**)

Two types of handles are associated with casseroles. The first is a high-swung handle with a round section, attached at the lip or just below it; the second is a handle hugging the wall and folded into an open rectangle that usually extends to the full height of the wall or just above the line of the rim. High-swung handles are characteristic of the shouldered casseroles, whereas wall-hugging handles are associated with the more upright shapes. At Locri the two types of handles appear to follow a chronological sequence. Casseroles with loop handles were found at Locri during the 5th and 4th centuries BC.[7] An intermediate shape, with handles rising only slightly above the rim close to the wall, but still with rounded section, has a minor presence in the first half of the 3rd century BC. Only the handles that fully hug the wall, which begin to appear in the mid-5th century BC,[8] continue through the 3rd century BC.[9] This sequence, however, is not necessarily applicable elsewhere: at Vaste, casseroles with loop handles were found in a context of the late 4th/early 3rd century BC,[10] while at Corinth casseroles with loop handles continued to be found in the destruction levels of the mid-2nd century BC, though with a different profile.[11]

Comparanda for the shouldered casseroles from the area of Metaponto is scarce, and date to the 4th or early 3rd century BC,[12] though the shape is of earlier origin.[13]

The globular casseroles are a popular type, extending over a long period. They are found in Gravina in contexts as early as the 6th century BC,[14] and continue there to the late 2nd to 1st century BC.[15] Globular casseroles are the most common type in the survey, with 86 examples accounting for just below 15% of all cooking ware forms. The casseroles with concave walls are the next most common type with 55 examples making up just below 10% of all cooking ware forms. The shouldered casseroles are much less frequent in the survey assemblages as a whole, with 26 examples accounting for just below 5% of all cooking ware forms. The casserole with carinated shoulder is rare, with a single example in the corpus of cooking wares from the survey.

Cooking pots (chytrai)

Broadly corresponding to the Greek chytra forms, cooking pots from the survey can be divided into five groups by the outlines of their necks and rims:

1. Cooking pots with collared rims. These lack a lid seat (**Nos. 26–27**).
2. Necked pots/jugs with projecting rims (**No. 28**).
3. Necked pots/jugs with flaring rim (**No. 29**).
4. Open-mouthed cooking pots with projecting rims (**Nos. 30–31**).
5. Lid-seated cooking pots (**Nos. 32–37**).

The cooking pots with collared rims (**Nos. 26–27**) are characterized by an almost vertical collared rim and a straight sloping shoulder. These are rare in the

[7] *Locri* II, 275, No. 313, Pl. XXXVI, Type C2.

[8] *Locri* II, 275–76, No. 314, Pl. XXXVII, Type C3a.

[9] *Locri* II, 275–78, Nos. 314–17, Pl. XXXVII, Types C3 and C4.

[10] Campagna 1995, 231, Fig. 9.1.

[11] *Corinth* VII.3, 124–26.

[12] Quercia 2004, 187 C4a and C4c, 186, Fig. 5 (first half of the 4th c. BC); Campagna 1995, 231, No. 61, Fig. 9.1 (cistern 121, in association with material of the late 4th–early 3rd c. BC).

[13] Bats 1988, 67.

[14] *Gravina* III.2, 130, No. 303; 131, Fig. 54.

[15] *Gravina* II, 179–80, Nos. 1356 and 1361, 328, Fig. 70; 180, No. 1363, 328, Fig. 70; 181–82, Nos. 1369–74, 329, fig. 71.

survey assemblages, with two examples, both shown in the catalog.

Much more common are necked cooking pots (**Nos. 28–29**), with 69 examples from the survey accounting for just below 12% of all cooking ware forms. The type represented by **No. 28** is characterized by a short distinct neck, sloping shoulder, and rim pulled outwards and often downwards; the neck can be straight, tapering, or concave. Where a handle is preserved, it is a vertical strap-handle with a double ridge along the center. Parallels from the Pantanello necropolis and Cozzo Presepe are termed "cooking pot" while at Gravina and Pomarico Vecchio the form is called a "necked jar."[16] Comparanda may date from as early as the late 6th or early 5th century BC (from Cozzo Presepe) to perhaps as late as the 1st century BC (from Gravina).[17] Rim diameter measurement was possible for 30 examples: 4 rims have a diameter of less than 10 cm, 7 greater than 15 cm. The overall diameter range is 8–19 cm. There are about 50% more examples of the type represented by **No. 28** than those represented by **No. 29**, which is characterized by the continuous curving line of the shoulder, neck, and rim, terminating in a pointed lip.[18] Sometimes there is a soft angle between the shoulder and the neck. Parallels from Gravina are described as "necked jars" and from Cozzo Presepe as "pots with flared rim."[19] Comparanda for this very basic and long-lived rim form appear to range in date from the first half of the 6th century BC (from Cozzo Presepe) to the late 2nd to 1st centuries BC (from Gravina).[20] Rim diameter measurement was possible for 20 examples; of these, 4 have a diameter of less than 10 cm, the rest between 10 and 15 cm.

The open-mouthed cooking pots with projecting rims (**Nos. 30–31**) have a long, narrow, sharply everted or horizontal rim; the lip is squared or segmented by a groove. Sometimes there is a short neck, and the shoulder, where preserved, is rounded. The 58 examples account for 10% of all cooking-ware forms. Rim diameter measurement was possible for 41; of these, 19 have a rim diameter greater than 16 cm (max. 26 cm), while the smallest measured diameter is 12 cm.

The lid-seated cooking pots (**Nos. 32–37**) follow the basic shape familiar from the Greek chytrai with parallels ranging in date from the 4th century BC through the Hellenistic period. They are the most frequent cooking-ware shape from the survey, with 142 examples accounting for just below 25% of all cooking-ware forms.

Lids

Lids are represented by 85 examples and account for just below 15% of all cooking-ware forms. Plain knobs like **No. 38** are very difficult to date. Lids are associated with both closed and open lid-seated forms, but can also be placed over the rim, or inside the rims/necks of vessels that do not have lid seats.

[16] *Necropoleis*, 726; *Cozzo Presepe*, 374, No. 443; *Pomarico Vecchio* I, 2.86; *Gravina* II, 337.

[17] *Cozzo Presepe*, 374, Fig. 145, No. 443; *Gravina* II, 337, Fig. 79, Nos. 1458 and 1460–61.

[18] This profile is suitable for jugs as well as for pots. See *Gravina* II, 183, No. 135, Fig. 76: 135.

[19] *Gravina* II, 333, No. 1403, "necked jar"; *Cozzo Presepe*, 374, No. 440, "pots with flared rim."

[20] *Cozzo Presepe*, 374, Fig. 145, No. 440; *Gravina* II, 333, Fig. 75, No. 1403.

Catalog of Greek and Roman Republican Cooking Ware

The cooking-ware catalog presents the reference sherds used for the identification of cooking-ware forms from the survey. Cooking wares are prefaced by the code Ck. All dimensions are in cm. Rim diameter indicated by "rim diam.," maximum height is indicated by "h," maximum width indicated by "w." The first three digits following the Ck abbreviation indicate the site number. An asterisk (*) after the site number indicates materials from the 1980 ICA pilot survey (sites numbered 1–44), as well as at the intensive survey sites and plots located south of the Basento River through the 2001 season (most of the sites numbered 562 and higher). Additional information about the sites can be found in the Gazetteer. The classification codes used during recording and for the survey data are given here, with some reclassification (of the CC1 codes in particular) and reordering to better fit the grouping of cooking wares by shape.

Flanged Pan (PD1)

These pans are very shallow with near-straight, outward-leaning walls and flat or near-flat undersides. The characteristic projecting rims, usually triangular in section, may have upper faces which are slightly rounded or concave.

1:3

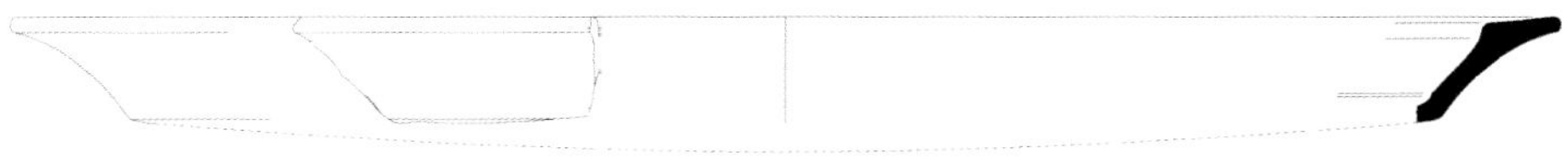

PD1
1. Ck 341-02 Pan with upturned flange
Dimensions: rim diam. 36.2 h 2.6 w 7.2
Comparisons: *Necropoleis*, 726 (T 111-2, 310–275 BC); *Pomarico Vecchio* I, 2.90, Pl. 82, No. 182 (second half of the 3rd c. BC); *Locri* II, Pl. XXXVII, No. 322 (with round horizontal handles, 3rd c. BC).

1:2

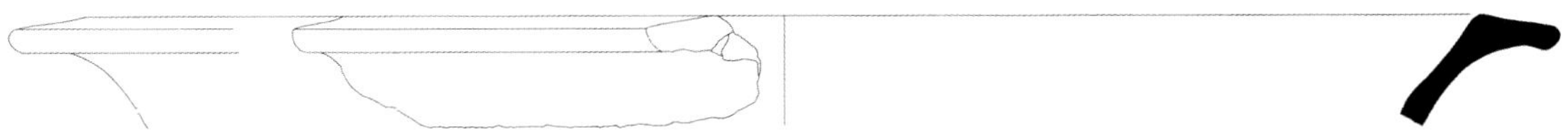

PD1
2. Ck 211-01 Pan with near-horizontal flange
Dimensions: rim diam. 29.0 h 2.1 w 8.8
Comparisons: See **No. 1**

1:2

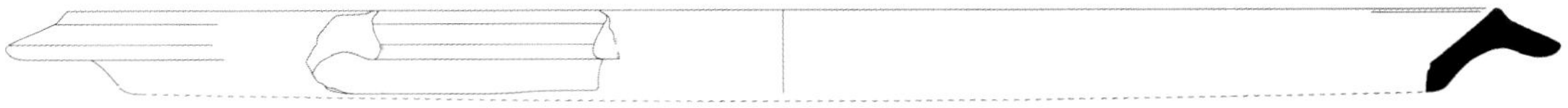

PD1
3. Ck 646-01* Shallow, flat-based pan with downturned flange
Dimensions: rim diam. 28.5 h 4.55 w 5.75
Comparisons: See **No. 1**

Pan with Plain Rim (PD2)

1:2

These pans are also shallow (though not always as shallow as **Nos. 1–3**), with a straight or slightly concave, outward-leaning wall, flat base, and diagonally flattened rim.

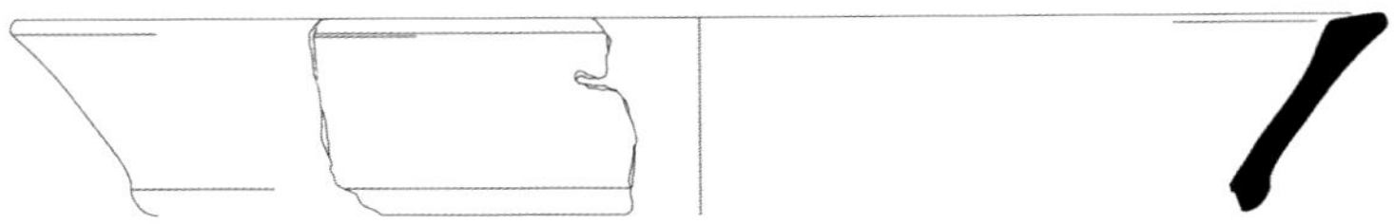

PD2

4. Ck 260-02 Pan with plain rim
Dimensions: rim diam. 19.2 h 2.85 w 4.55
Comparisons: Quercia 2004, 188, Fig. 6, D1a (first half of the 4th c. BC); *Eraclea lucana,* Fig. 50, Nos. 214–15 (375–270 BC).

1:2

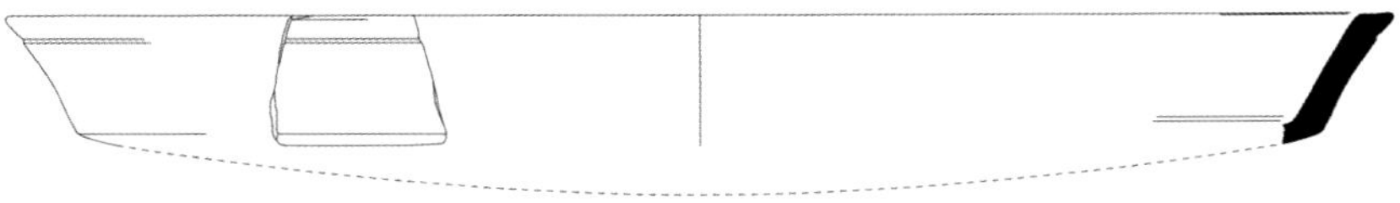

PD2

5. Ck 320-02 Pan with plain rim
Dimensions: rim diam. 21.8 h 2.1 w 2.7
Comparisons: See **No. 4**

Pan with Thickened Rim (PD4)

1:2

Very shallow pan with flat base and outward-leaning wall surmounted by a thickened and rounded rim; the lip is segmented by two grooves. On the basis of more complete parallels, this shape is associated with one long handle, as a frying pan, an example of which was present in the Pantanello necropolis.

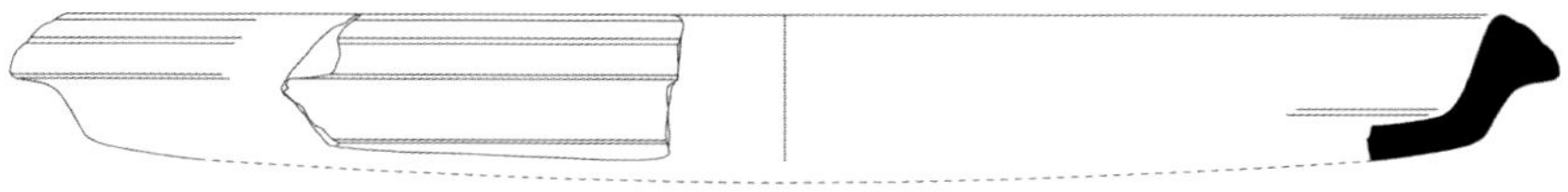

PD4

6. Ck 341-01 Pan with thickened rim
Dimensions: rim diam. 22 h 2.2 w 5.7
Comparisons: *Necropoleis*, 727 (CD 89-4, late 4th–early 3rd c. BC); *Locri* II, Pls. XXXVII and XLVIII, Nos. 323–24 (late 4th–early 3rd century BC); *Eraclea lucana* 173, No. 219, Fig. 50 (375–270 BC).

Pan with Flared Rim (PD5)

1:2

Shallow pan with concave wall, flat base, and flared rim. The form may have had a pinched spout, the edge of which seems to be preserved in the example presented here.

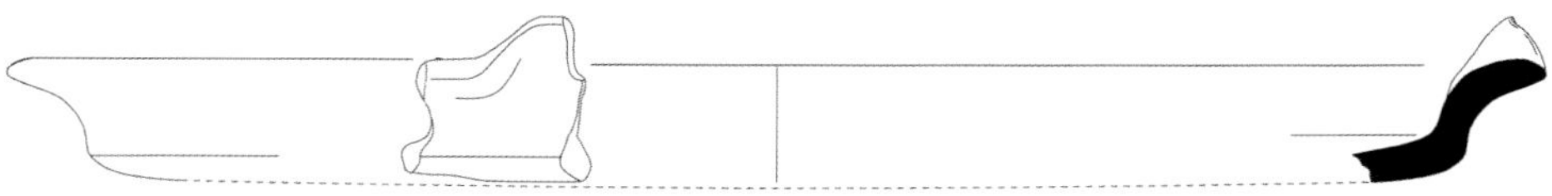

PD5

7. Ck 304-01 Pan with flared rim and terminus of vertical rim-mounted handle
Dimensions: rim diam. 24.5 h 2.7 w 3.0

Shouldered Casserole (CL)

1:2

These casseroles have sharply inclined rims with internal lid seats, which create a rounded shoulder with no distinct separation between the base and the wall, and loop handles rising high above the rim. The forked rim has three variants: the lid-supporting ledge is shorter than the rim (**No. 9**), equal in length to the rim (**No. 8**), or longer than the rim (**No. 10**). Examples from the survey are shallower than the other types of casseroles. Of the 26 vessels identified, diameter measurement was possible for 16; of these, 2 have a diameter of less than 16 cm (12 and 13 cm), while the remainder measure between 16 and 26 cm.

CL

8. Ck 549-01 Casserole rim with high-swung shoulder-mounted handle

Dimensions: rim diam. 19.0 h 2.9–6.6 w 5.9

Comparisons: Quercia 2004, 186, Fig. 5, C4a and C4c (first half of the 4th c. BC); Campagna 1995, 231, No. 61, Fig. 9, No. 1 (cistern 121, in association with material of the late 4th–early 3rd c. BC).

1:2

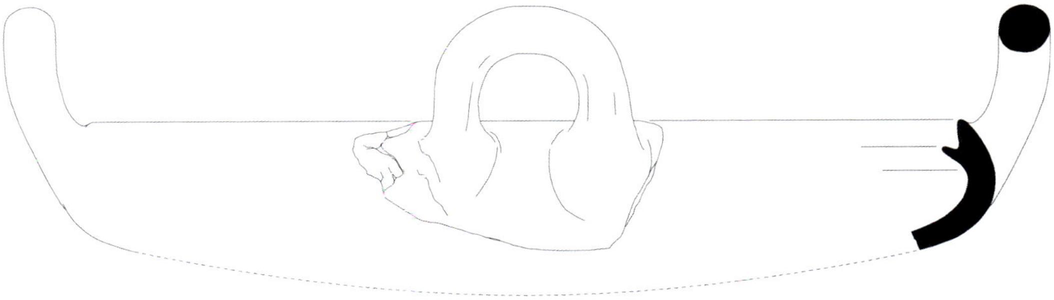

CL

9. Ck 182-01 Casserole rim with high-swung shoulder-mounted handle

Dimensions: rim diam. 22 h 3.6–6.7 w 8.3

Comparisons: See **No. 8**.

1:2

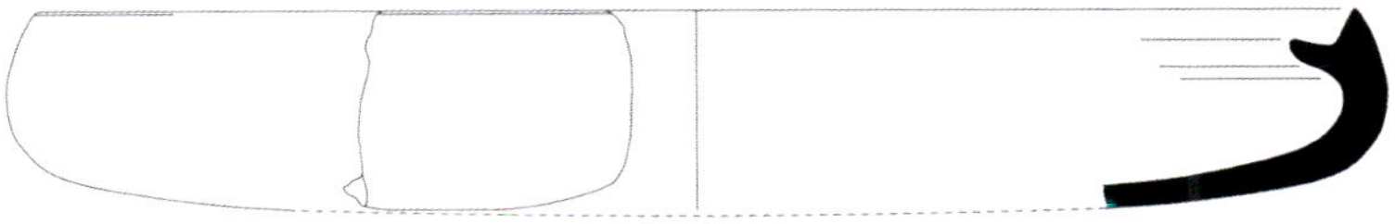

CL

10. Ck 322-02 Casserole rim to shallow rounded underside

Dimensions: rim diam. 18.8 h 2.9 w 4.1

Comparisons: See **No. 8**.

Globular Casserole (CG/CC)

1:2

The walls of these casseroles vary from inclined to near vertical. Associated bases are very slightly rounded. The handles, flattened in section, hug the wall and are folded into a rectangular outline. Of the 86 vessels identified, diameter measurement was possible for 59; of these, 8 have a diameter of less then 16 cm (11–15 cm), while the rest are between 16 and 30 cm.

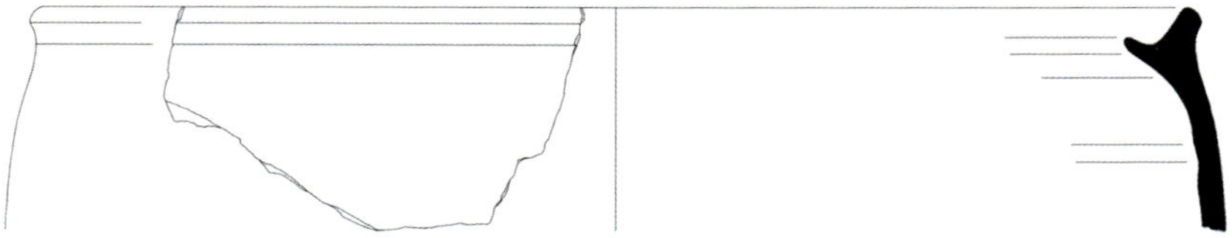

CG2b

11. Ck 296-01 Casserole with inclined, lid-seated rim
Dimensions: rim diam. 21.0 h 4.2 w 7.6
Comparisons: *Cozzo Presepe*, 377, Fig. 147, No. 464 (site B, phases I and III, late 4th–3rd c. BC).

1:2

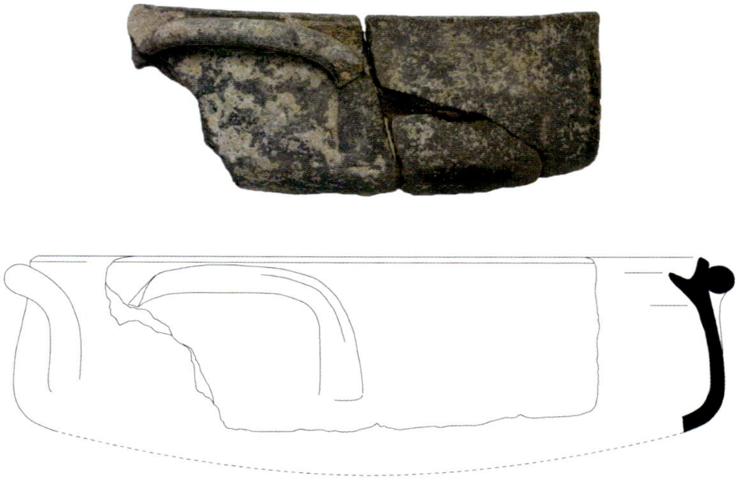

CG2b

12. Ck 465-02 Casserole with inclined lid-seated rim and wall-mounted handle
Dimensions: rim diam. 18.0 h 4.8 w 13.1
Comparisons: See **No. 11**.

1:2

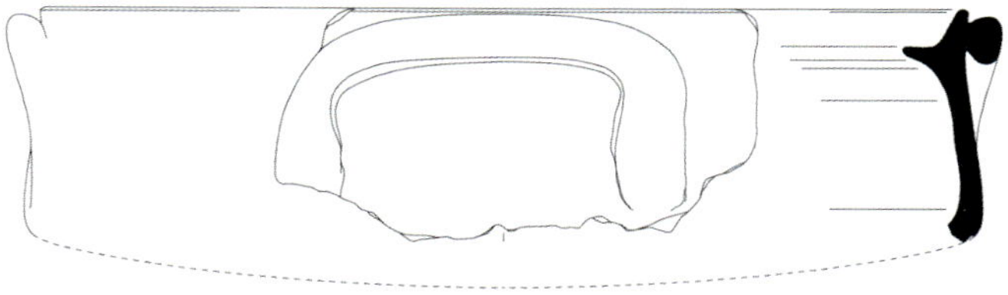

CC1b

13. Ck 358-03 Casserole with inclined lid-seated rim and wall-mounted handle
Dimensions: rim diam. 16.0 h 4.1 w 8.3
Comparisons: *Cozzo Presepe*, 377, Fig. 14,7 No. 462 (site A, phase VI, 4th–3rd c. BC); *Pomarico Vecchio* I, 2.89, Pl. 81, No. 173 (end of the 4th–3rd c. BC; note that the illustrated casserole has loop handles rising above the rim; in the text, however, the variant with hugging handles is described); *Locri* IV, Pl. LXXVI, No. 235 (first half of the 3rd c. BC); *Corinth* VII.3, Pl. 29, No. 671 (2nd c. BC).

Globular casserole cont.

1:2

CG2a

14. Ck 310-01 Casserole with everted rim

Dimensions: rim diam. 16.2 h 3.0 r 13.1

Comparisons: *Gravina* III.2, 130–31, Fig. 54, No. 303 (6th c. BC); *Gravina* II, 328, Fig. 70, Nos. 1356 and 1361 (site B, period VI, 4th–3rd c. BC; site DB, period VIIIa, late 2nd–1st c. BC), II.328, Fig. 70, No. 1363 (site DB, period VII, 3rd–early 2nd c. BC; site DC, period VIIIa, late 2nd–1st c. BC); *Cozzo Presepe*, 377, Fig. 147, No. 473 (site A, phase V, 3rd c. BC; site A, phase III, mid-6th or 4th c. BC); 377, Fig. 147, No. 476 (site E, phase III, late 4th–3rd c. BC).

1:2

CG1b

15. Ck 296-02 Casserole with near-vertical rim

Dimensions: rim diam. 22.0 h 3.8 w 3.6

Comparisons: *Gravina* II, 328, Fig. 70, No. 1359 (period VI, 4th–3rd c. BC; period VII, 3rd–early 2nd c. BC; period VIIIa, late 2nd–early 1st c. BC); *Gravina* III.2, 130, Fig. 54, No. 302 (common in the 5th–4th c. BC).

1:2

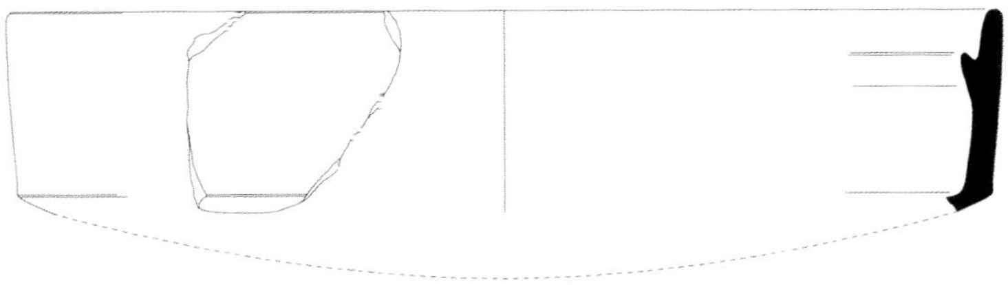

CC1a

16. Ck 365-02 Casserole with near-vertical rim

Dimensions: rim diam. 19.2 h 4.0 w 4.1

Comparisons: *Gravina* II, 328, Fig. 70, No. 1364 (period VIIIa, late 2nd–early 1st c. BC), II.329, Fig. 71, Nos. 1367–1368 (period VI, 4th–3rd c. BC; period VII, 3rd–early 2nd c. BC; period VIIIa, late 2nd–early 1st c. BC); *Cozzo Presepe*, 377, Fig. 147, Nos. 466 and 468 (sites B and E, phase III, late 4th–3rd c. BC; site B, phase V, 3rd c. BC).

1:2

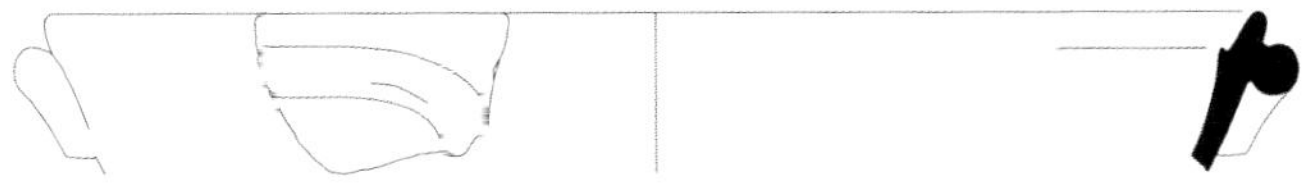

CG1a

17. Ck 266-02 Casserole rim with slight lid seat and wall-mounted handle

Dimensions: rim diam. 16.2 h 2.15 w 3 4

Comparisons: *Cozzo Presepe*, 377, Fig. 147, Nos. 469 and 477 (site E, phase III, late 4th–3rd c. BC); *Monte Sannace*, 258 (T 6, late 4th c. BC).

Casserole with Concave Walls (CC1)

The walls of these casseroles, which may lean outward or be nearly vertical, have a slightly concave outline. Of the 55 vessels identified, diameter measurement was possible for 35; of these, 9 have a diameter of less than 16 cm (12–15 cm), while 5 are greater than 20 cm (max. 30 cm).

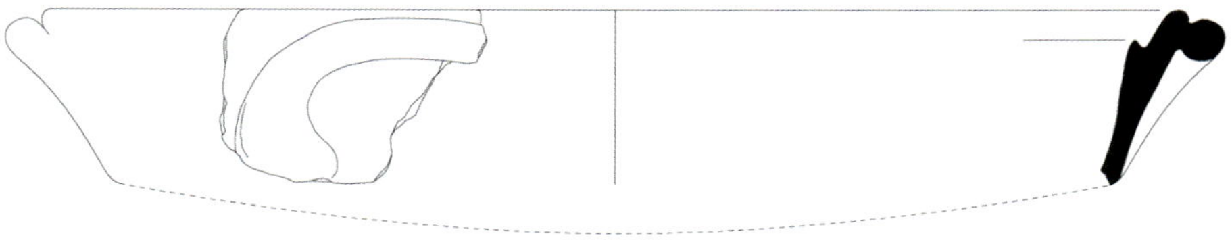

CC1a
18. Ck 147-01 Casserole with slightly convex wall and wall-mounted handle
Dimensions: rim diam. 22.5 h 3.5 w 5.6
Comparisons: See **No. 16**.

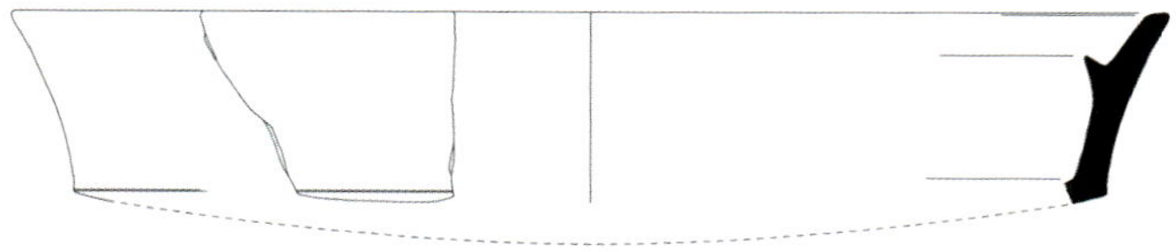

CC1c
19. Ck122-01 Casserole
Dimensions: rim diam. 17.4 h 3.0 w 3.9
Comparisons: *Gravina* II, 329, Fig. 71, No. 1366 (phase VIIIa, late 2nd–early 1st c. BC; phase VI, 4th–3rd c. BC); *Cozzo Presepe*, 377, Fig. 147, No. 467 (site B, phase V, 3rd c. BC).

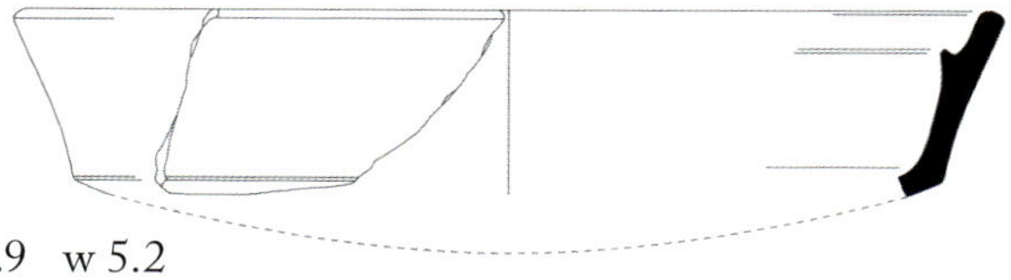

CC1d
20. Ck 320-01 Casserole
Dimensions: rim diam. 14.8 h 2.9 w 5.2
Comparisons: Quercia 2004, 184, Fig. 4, C2a (first half of the 4th c. BC); *Gravina* III.2, 130–31, Fig. 54, No. 304 (end of the 4th c. BC); *Gli ori* 486, No. 41 (T CXXXVIII, 150–125 BC).

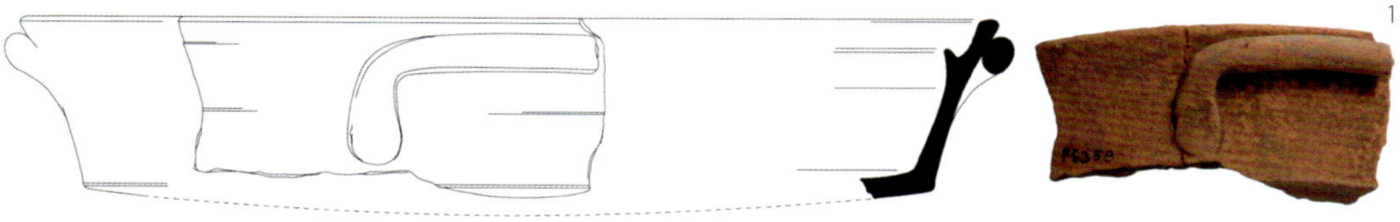

CC1e
21. Ck 358-04 Casserole with a lid seat and wall-mounted handle
Dimensions: rim diam. 19.5 h 3.7 w 8.5
Comparisons: See **No. 19**.

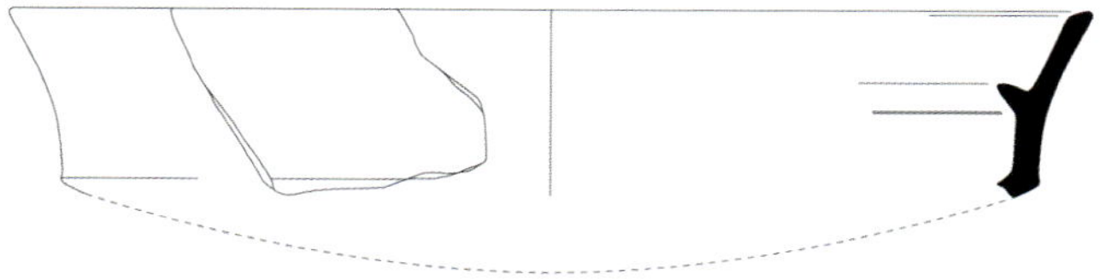

CC1c
22. Ck 353-02 Casserole with low lid seat
Dimensions: rim diam. 16.8 h 3.0 w 4.9
Comparisons: *Gravina* II, 329, Fig. 71, No. 1366 (phase VIIIa, late 2nd–early 1st c. BC; phase VI, 4th–3rd c. BC); *Cozzo Presepe*, 377, Fig. 147, No. 467 (site B, phase V, 3rd c. BC).

Casserole with concave walls cont.

1:2

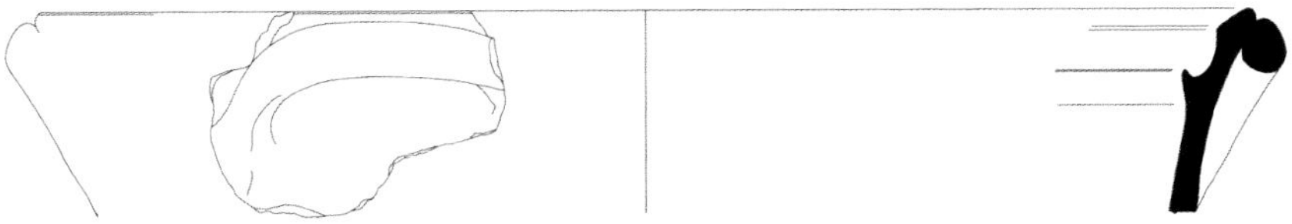

CC1f

23. Ck 154-02 Casserole with grooved lid seat and wall-mounted handle

Dimensions: rim diam. 21.0 h 3.6 w 5.1

1:2

CC1f

24. Ck 322-01 Casserole with grooved lid seat rim

Dimensions: rim diam. 25.3 h 4.5 w 4.9

Casserole with Carinated Shoulder (CR)

1:2

Represented by a single example from the survey, this casserole has a sharply carinated shoulder and a relatively thick rim with small lid seat. At Locri, carinated casseroles dated to the 5th–4th c. BC were found with associated loop handles; they are absent from 3rd-century contexts.

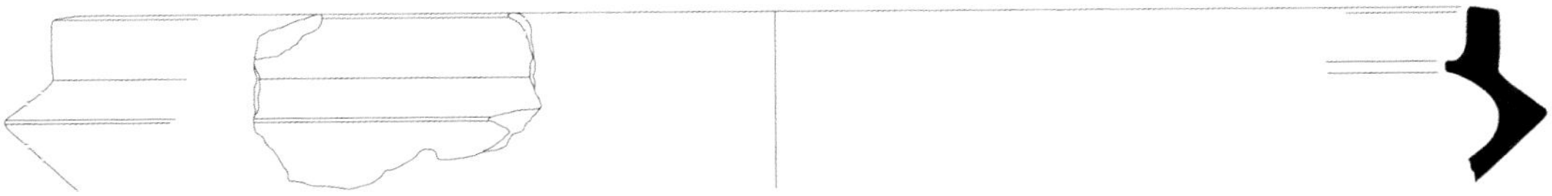

CR

25. Ck 482-01 Casserole with carinated shoulder

Dimensions: rim diam. 26.2 h 5.2 w 3.3

Comparisons: Quercia 2004, 186, Fig. 5, C3b (first half of the 4th c. BC).

Cooking Pot with Collared Rim (P4)

1:2

This form has a plain rim consisting of a more-or-less vertical collar extending from a near-straight, sloping shoulder.

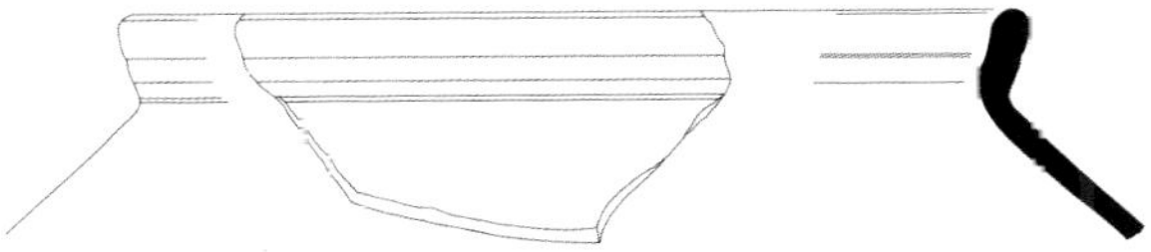

P4

26. Ck 269-01 Cooking pot with collared rim and straight, sloping shoulder

Dimensions: rim diam. 12.1 h 3.1 w 6.6

Cooking pot with collared rim cont.

1:2

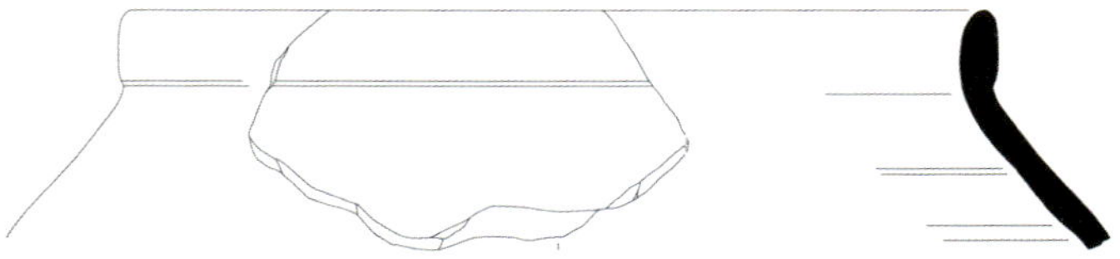

P4

27. Ck 309-01 Cooking pot with collared rim, groove at the junction with upper shoulder

Dimensions: rim diam. 14.0 h 4.0 w 7.0

Necked Cooking Pot with Projecting Rim (PJ1)

These pots have short necks with projecting triangular rims. Where a handle is preserved it is a vertical strap handle with a double ridge along the center.

1:2

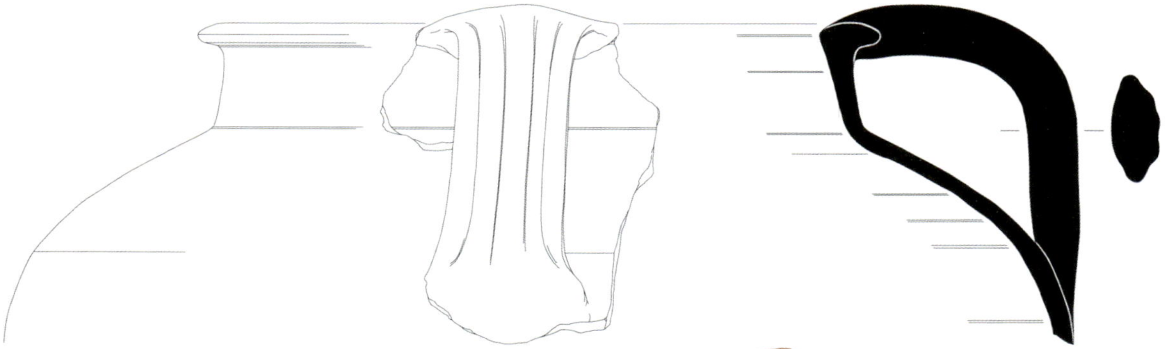

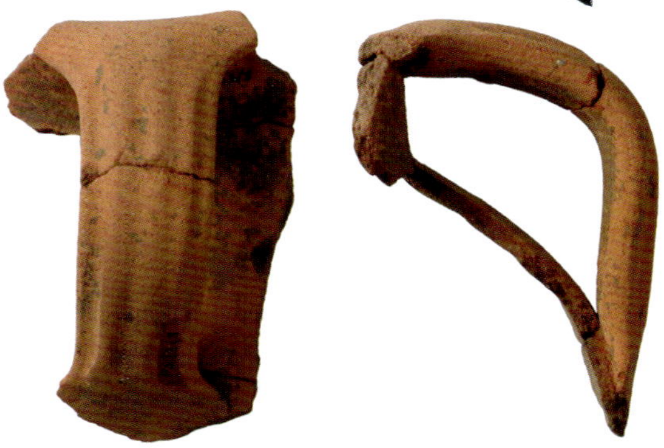

PJ1

28. Ck 113-02 Cooking pot rim to shoulder with vertical ribbed handle

Dimensions: rim diam. 18.1 h 9.3 w 7.4

Comparisons: *Necropoleis*, 726 (T 12-6, 300–275 BC); *Cozzo Presepe*, 374, Fig. 145, No. 443 (site A, phase VI, ca. 550–480 BC); *Pomarico Vecchio* I, 2.86, Pl. 78, No. 160 (levels of the second half of the 3rd c. BC); *Gravina* II, 337, Fig. 79, Nos. 1458 and 1460–1461 (period VI, 4th–3rd c. BC; period VIIIa, late 2nd–1st c. BC).

Necked Cooking Pot/Jug with Flaring Rim (PJ2)

1:2

The shoulder, neck, and rim of these pots form a continuous curve, terminating in a pointed lip. Sometimes there is a soft angle between the shoulder and the neck. Parallels are described either as necked jars (*Gravina* II) or pots with flared rims (*Cozzo Presepe*).

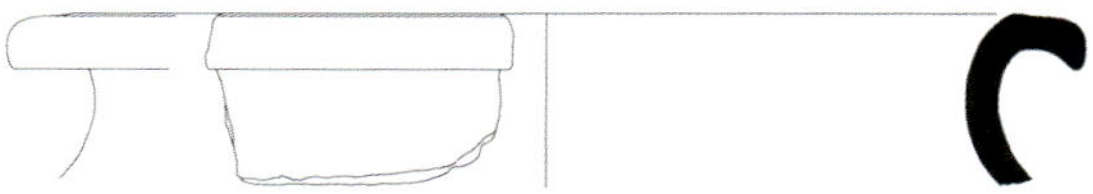

PJ2

29. Ck 260-03 Rim and neck of cooking pot with flaring rim

Dimensions: rim diam. 14.3 h 2.3 w 4.1

Comparisons: Quercia 2004, 179, Fig. 2, A1e (first half of the 4th c. BC); *Pomarico Vecchio* I, 2.87, Pl. 79, No. 161 ("jar," mid-3rd c. BC). One of the variants of the shape was found in Gravina, manufactured in well-levigated fabric (*Gravina* II, 183, No. 135, Fig. 76), and described as a jug, highlighting again the multi-functionality and/or ambivalence of shapes. See also *Gravina* II, 333, Fig. 75, No. 1403 (period VIIIa, late 2nd–1st c. BC); 337, Fig. 79, No. 1462 (period VI, 4th–3rd c. BC, with some intrusions); *Cozzo Presepe*, 374, Fig. 145, No. 440 (site A, phase IIIB, 600–550 BC; site A, phase V, mixed 6th and 4th c. BC; site A, phase VI, 4th–3rd c. BC, with some intrusions; site B, phase I, before the late 4th c. BC); *Corinth* VII.2, Pl. 82, Nos. An 291 and 303 (625–600 BC).

Open-Mouthed Cooking Pot with Projecting Rim (P2)

1:2

These cooking pots are characterized by a projecting rim with a central groove on the outer face.

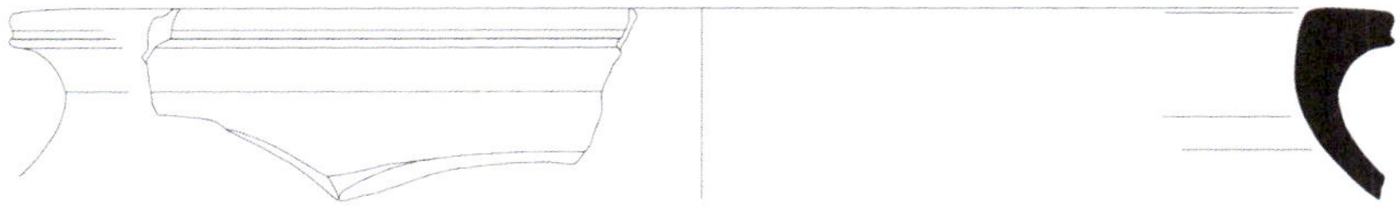

P2

30. Ck 358-02 Rim and shoulder of cooking pot with projecting rim
Dimensions: rim diam. 21.5 h 3.1 w 7.7
Comparisons: *Cozzo Presepe*, 374, Fig. 145, Nos. 447–49 (site E, phase IV, final occupation, 300–250 BC); *Pomarico Vecchio* I, 2.86, Pl. 78, No. 159, and 2.87, Pl. 79, No. 161 ("olla," second half of the 3rd c. BC); *Gravina* II, 334, Fig. 76, No. 1421 ("unlidded jars with bifid rims"—period VIIIa, late 2nd–1st c. BC).

1:2

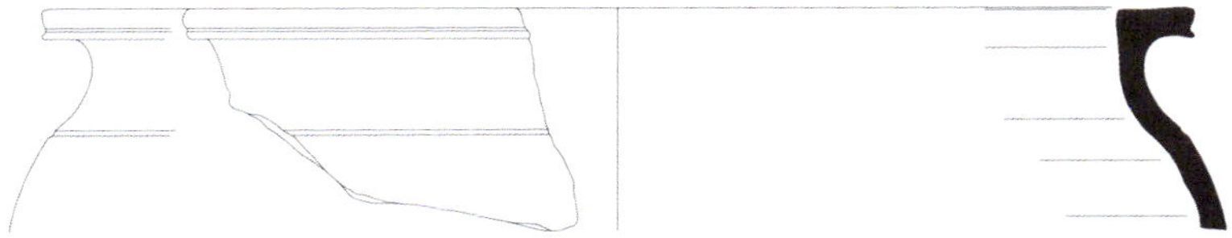

P2

31. Ck 113-04 Rim, neck, and shoulder of cooking pot with projecting rim
Dimensions: rim diam. 19.0 h 3.8 w 6.5
Comparisons: See **No. 30**.

Lid-Seated Cooking Pot (P1 and P3)

1:2

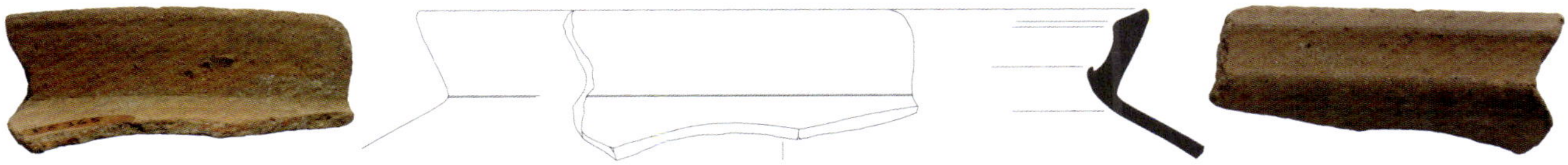

P1a

32. Ck 365-01 Lid-seated cooking pot with rounded shoulder
Dimensions: rim diam. 16.0 h 3.4 w 7.7
Comparisons: *Necropoleis*, 728 (CD 148, 400–350 BC); *Cozzo Presepe*, 376, Fig. 146, Nos. 454–55 (site E, final occupation, 300–250 BC); *Pomarico Vecchio* I, 2.87, Pl. 79, No. 165, 2.88, Pl. 80, Nos. 167–68; *Gravina* II, 335, Fig. 77, Nos. 1425, 1431, and 1435 (period VI, 4th–3rd c. BC; period VII, 3rd–early 2nd c. BC; period VIIIa, late 2nd–1st c. BC), II.336, Fig. 78, No. 1445 (period VIIIa, late 2nd–1st c. BC); *Corinth* VII.3, Pls. 27–28, Nos. 656 and 658 (300–146 BC).

1:2

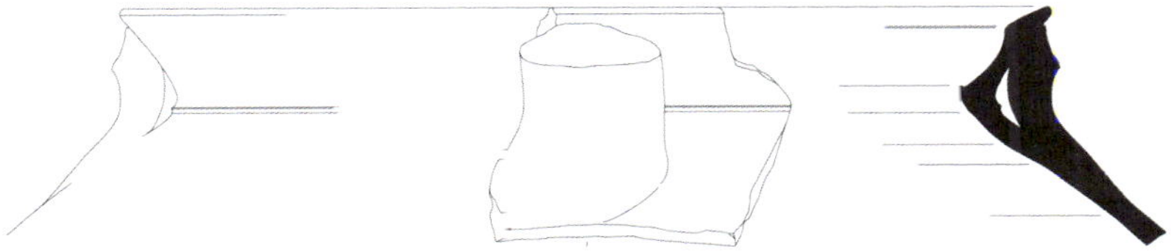

P1b

33. Ck 487-01 Lid-seated cooking pot with vertical handle
Dimensions: rim diam. 16.8 h 4.4 w 5.5
Comparisons: *Cozzo Presepe*, 376, Fig. 146, Nos. 451, 456 (site B, phase I, late 4th c. BC; phase V, 3rd century BC; site E, phase IVb final occupation, 300–250 BC); *Gravina* III.2, 130, Fig. 54, Nos. 310–11 (common in the 5th–4th c. BC); *Corinth* VII.3, Pl. 27, No. 651 (destruction phase, 146 BC).

Lid-seated cooking pot cont.

1:2

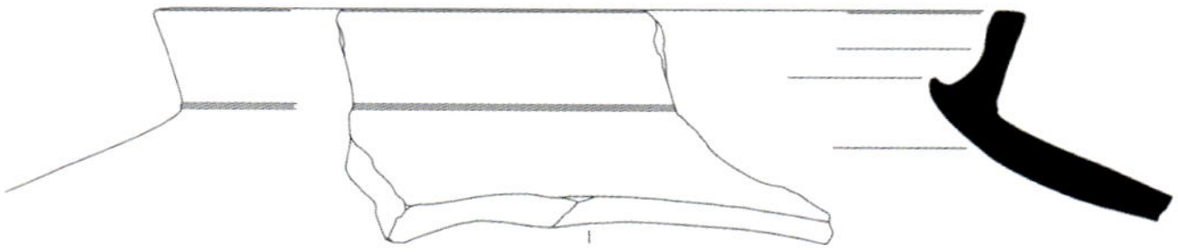

P1c

34. Ck 488-44 Lid-seated cooking pot with wide shoulder
Dimensions: rim diam. 13.0 h 3.6 w 7.4
Comparisons: Quercia 2004, 180, Fig. 3, B1 (first half of the 4th c. BC); *Locri* II, Pl. XXXVI, Nos. 301–02 (end of the 6th–3rd c. BC); *Corinth* VII.3, Pl. 28, No. 657 (300–146 BC).

1:2

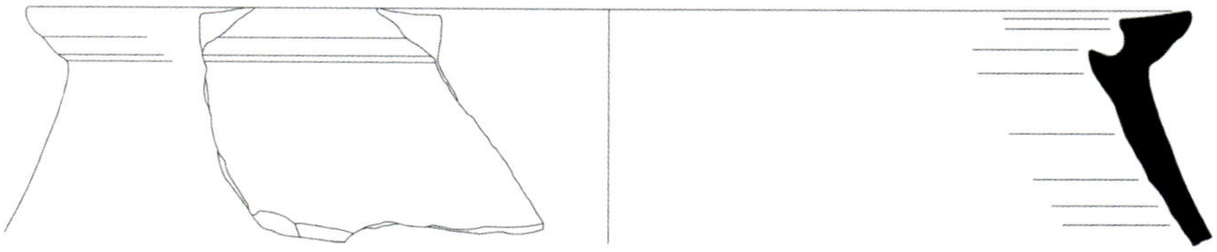

P1d

35. Ck 167-02 Lid-seated cooking pot with steep-sloping shoulder
Dimensions: rim diam. 22.0 h 4.5 w 6.5
Comparisons: *Necropoleis*, 727–28 (CD 147, 350–300 BC); *Cozzo Presepe*, 376, Fig. 146, No. 459 (site B, phase I, before the late 4th c. BC; site B, phase V, 3rd c. BC; site E, phase IVB, destruction), 377, Fig. 147, No. 471 (site E, phase IV, 300–250 BC); *Gravina* II, 328, Fig. 70, No. 1362 (period VIIIa, late 2nd–1st c. BC).

1:2

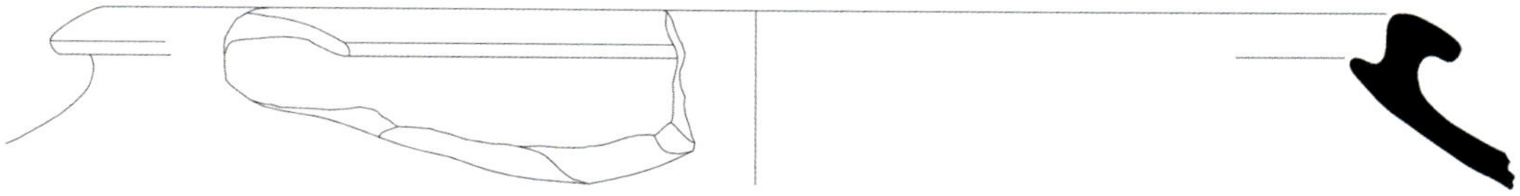

P3

36. Ck 394-01 Lid-seated cooking pot; the rim with an overhanging edge. This is the only rim-shape that is common to casseroles and cooking pots.
Dimensions: rim diam. 26.0 h 3.2 w 3.4
Comparisons: Very generally, *Pomarico Vecchio* I, 2.88, Pl. 80, Nos. 169–70 (second half of the 3rd c. BC).

1:2

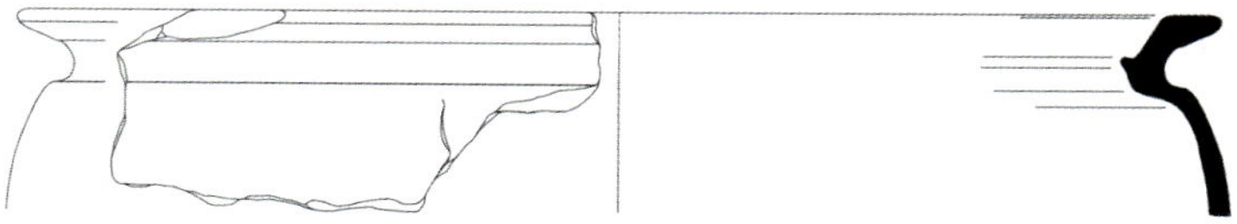

P3

37. Ck 113-05 Lid-seated cooking pot with carinated shoulder
Dimensions: rim diam. 18.5 h 3.5 w 7.5
Comparisons: None.

Lids (Li)

1:2

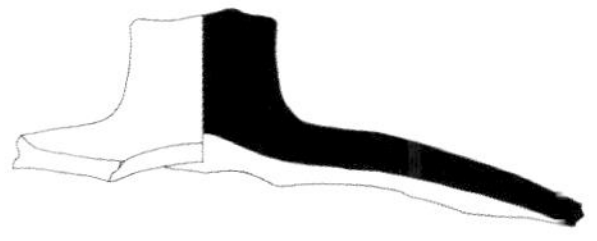

38. Ck 700-04 Lid knob
Dimensions: handle 2.0 h 3.0 w 7.6

1:2

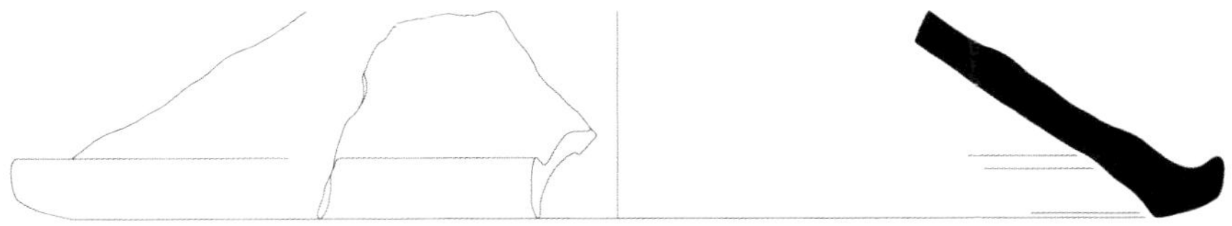

39. Ck 329-01 Lid with projecting rim
Dimensions: rim diam. 22.4 h 4.0 w 5.2
Comparisons: *Gravina* II, 340, Fig. 82, No. 1502.

Necked Pot/Jar (PJ3 = Jug J1 and J5b in Plain and Banded Ware)

These appear in coarse fabric similar to that used for cooking wares, with the general form following those of the plain and banded ware "Jug."

1:2

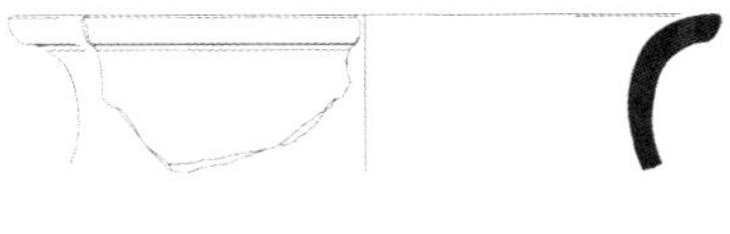

40. Ck 285-51
Dimensions: rim diam. 14.2 h 3.2 w 5.5

1:2

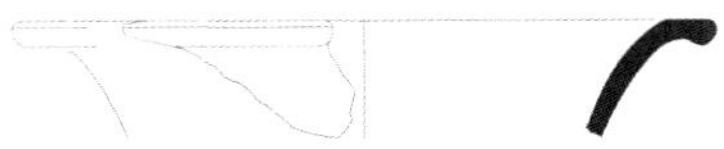

39. Ck 521-25
Dimensions: rim diam. 14.0 h 2.4 w 4.6

1:2

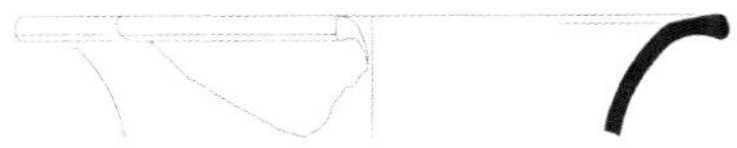

39. Ck 432-01
Dimensions: rim diam. 14.2 h 5.0 w 2.5
Comparisons: Ch. 9, J1 and J5b, **Nos. 111–116** and **125–126.** Quercia 2004, 177, 180–181 A1a–d, 179, Fig. 2 (first half of the 4th c. BC); *Necropoleis,* 726 (CD 289-51, associated with material of the late 7th–4th c. BC); *Gravina* II, 184, Nos. 1406–7 and 1409–10, 333, Fig. 75 (period VI, 4th–3rd c. BC; period VII, 3rd–early 2nd c. BC; period VIIIa, late 2nd–1st c. BC); *Cozzo Presepe,* 370, Nos. 406–7 and 410, 369, Fig. 139 (site A, phase IIIB, 600–550 BC; site B, phase III, late 4th–3rd c. BC).

12
Archaic to Late-Republican Transport Amphorae

Keith Swift

The corpus of transport amphora fragments from the ICA survey comprises 452 diagnostic pieces consisting of rims, bases, and handles. These provide evidence for the long-distance distribution of amphora-borne commodities and the inter-regional maritime networks with which Metaponto was engaged. They also attest the distribution of these commodities throughout the chora, beyond the confines of the asty and the immediate and most likely point of distribution from maritime networks—the urban area of Metaponto with its port installations. In this respect, amphorae are found throughout the survey area, and it seems likely that extra-regional transport amphorae were distributed much further inland to the limits of the chora. The amphorae from the survey also provide new evidence for the nature and production of local and regional amphora types—Metapontine type B amphorae in the Classical period, Metapontine Greco-Italic in the Early Hellenistic period, and Metapontine type amphorae in the Late Hellenistic/Late Republican period. In particular, transport amphora fragments from two production locations in the chora (sites 266 and 153) show new evidence for the production of late type B amphorae[1] of the Early Hellenistic period—the first definitive evidence for the long-suspected production of this amphora type in the region—and the production of Metapontine Greco-Italic amphorae.

Forms and Fabrics: Identification of Amphora Types from the Survey

The amphorae fragments were identified through a combination of their forms and their fabrics. These aspects are presented and discussed together, in broad chronological order, to reflect wider recent developments in amphora studies that have consistently shown that the study of forms alone is insufficient to attribute regional origin in most cases.

Attributions of many of the extra-regional types, particularly those which were produced in more than one area of the Mediterranean (e.g., Greco-Italic amphorae) are tentative; nonetheless these fabrics usually show clear differences from those produced in the Metapontine chora. The composition of the rocks and minerals from the Bradano Trough,[2] in which the Metapontine chora lies, varies little between the limestone plateaus of Apulia to the east and the schist massifs of Calabria to the southwest. It is therefore possible to identify only in general terms the regional Bradano Trough fabrics with their commonly occurring range of inclusions, and to be relatively secure in the attribution of many of these to a "local" source in the Metapontine chora. Comparisons with the fabrics of other wares produced in the Metapontine chora (black-gloss fine wares, plain wares, cooking wares) tend to confirm this, and usually most of the fabrics local to the Metapontine chora are readily distinguished in hand specimen.

An important caveat, however, is that fabrics produced elsewhere in the Bradano Trough—for example in the chora of Herakleia or the area just to the northwest of Taranto—are likely to be petrologically indistinguishable from those produced in the Metapontine chora. In short, any bulk amphora-borne commerce which occurred for example, within the chora, or between Herakleia and Metaponto, remains at present largely invisible. Given that the primary function of amphorae was transport, study of the distribution to Metaponto and throughout the Metapontine chora must take these geologically-imposed limitations into account when trying to infer economic interaction between the Metapontine chora, other areas of southern Italy, and other regions of the Mediterranean.

The presence of well-established and easily recognizable types with a known regional origin (such as Corinthian type A') alongside multi-regional types (for example, Greco-Italic amphorae and now type B) means that a certain amount of unevenness is inherent in the study of economic interactions.

[1] The term "type B" is used here to refer to amphorae which are similar in form to Corinthian type B amphorae. See discussion of the evidence for Metapontine production of this morphological type below.

[2] On the Bradano Trough, see Ch. 1, Figs. 1.1–4.

Period	Amphora type	Rim count		% Total rims	RBH
Archaic	SOS *7th c. BC*	2	40	1.1	2
	Corinthian type A *7th–early 5th c. BC*	14		7.4	51
	Archaic type B and related *Late 6th–early 5th c. BC*	24		12.8	29
Classical	Corinthian type A' *5th–4th c. BC*	21	80	11.2	72
	Type B *Early 5th–early 3rd c. BC*	58		30.9	69
	Metapontine type B *Late 4th/early 3rd c. BC*	1		0.5	2
Early Hellenistic	Hellenistic Corinthian type A' *3rd–2nd c. BC*	4	59	2.1	4
	Extra-regional Greco-Italic *4th–late 2nd c. BC*	19		10.1	25
	Metapontine Greco-Italic *Late 4th–3rd, probably into late 2nd c. BC*	36		19.1	42
Late Hellenistic/ Republican	Metapontine type (collared) *2nd–1st c. BC*	5	8	2.7	8
	Late extra-regional types *2nd–1st c. BC*	3		1.6	7
	Unidentified/ indeterminate	1	1	0.5	141
	Total	**188**		**100 %**	**452**

Table 12.1 Count of rims, percentages of total rims, and count of rims/bases/handles (RBH) by amphora type.

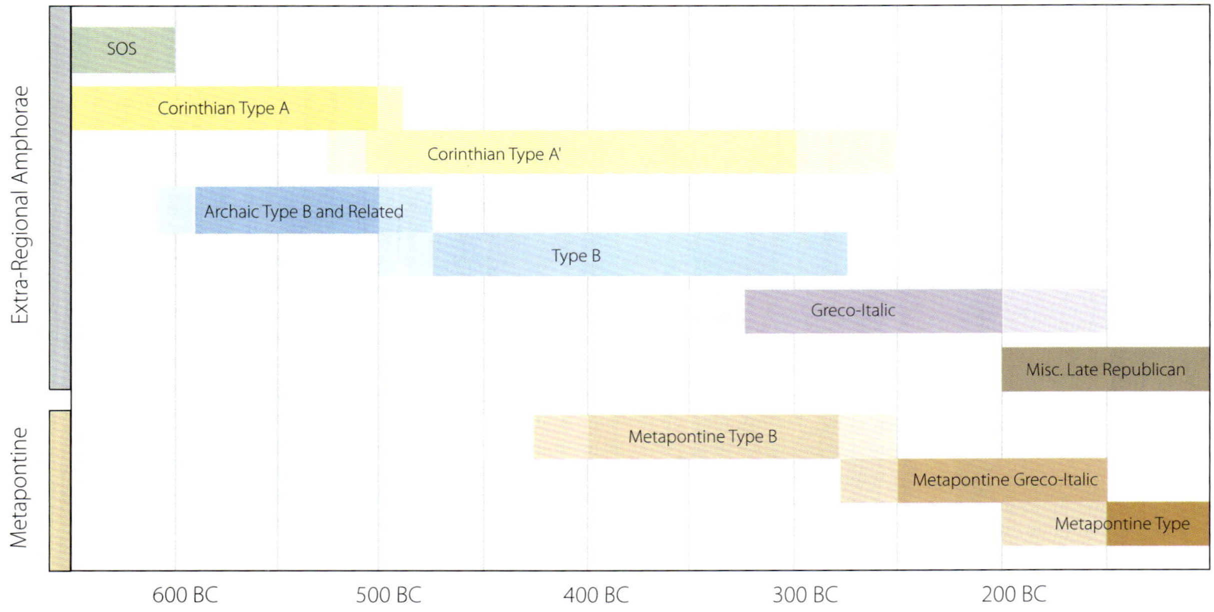

Figure 12.1 Chronological ranges for amphora types, 7th–early 1st c. BC.

Amphora Types from the Survey

Estimates of relative proportions of the various classes of amphorae can be made on the basis of the 188 rim fragments (*Table 12.1*). While handles and toes are identifiable for some classes, they tend to favor identification of certain amphora types over others, a bias that makes rims more reliable as an indicator of relative proportions of a wide range of different types.

Amphora types are discussed in chronological order from the Archaic to Late Hellenistic/Republican periods (*Fig. 12.1*), with a summary and overview prefacing discussions of the individual amphora types for each period. Since identification of amphora types relies on a combination of forms and fabrics, both are discussed together; a representative sample of both forms and fabrics is presented in the catalog for the common amphora types, along with key pieces.

Archaic period (pre-475 BC)

A total of 40 rims dateable to the Archaic period were identified in the survey assemblages. Two SOS rims dateable to the 7th century BC are among the earliest; there are 14 Corinthian type A amphorae that range in date from the 7th to the early 5th century BC. Of the amphorae dating to the Archaic period, Archaic type B and related amphorae predominate, accounting for 60% of all Archaic amphora rims, with many likely dating to the late 6th or early 5th centuries BC (*Fig. 12.1*).

Among the 14 Corinthian type A rim forms, a single example of a local/regional copy attests to southern Italian emulation of this widely distributed type. This appears to be the only local/regional fabric in the corpus of Archaic amphorae, which is otherwise made up of extra-regional types from both the east (Corinth) and west (southern Italy, perhaps also other regions).

The Corinthian type A and Archaic type B and related amphorae suggest that in the Archaic period the Metapontine chora received amphora-borne commodities both from central Greece and from the west, from at least as far away as southwestern Calabria and perhaps further afield. Although the eastern extent of their distribution remains to be established, the western types appear to have counter-circulated in relation to the Corinthian amphorae, at least along the southern Italian coast as far as Metaponto.

SOS amphorae (Nos. 1–2). Two rims of SOS amphorae are represented in the survey assemblages, both from sites identified as sanctuaries (site 270 and site 477) and both dateable to within the 7th century BC.[3] The example from site 270 (**No. 1**) has a rim profile falling within the Middle and Late I types in Johnston's classification, dated approximately 675–600 BC.[4] The black slip is almost completely worn and the fabric is consistent with an Attic origin. The example from site 477 (**No. 2**) has traces of black slip just visible on the surfaces, but the relatively poor preservation of the rim fragment means that it is attributable only in very general terms to the 7th century BC.

Corinthian Type A amphorae (Nos. 3–6). Produced in the area of Corinth in central Greece from the 7th to early 5th centuries BC, these amphorae had a geographically wide circulation in the Mediterranean and were distributed in southern Italy in significant and substantial quantities during the Archaic period. Earlier examples have distinctive forms with thick projecting rims and are hand formed rather than wheel thrown (**Nos. 3–5**). These features, combined with their distinctive fabrics (**No. 6**), which have been characterized petrographically by Whitbread, make them readily identifiable in hand specimen, even in fragmentary survey assemblages.[5]

The archaeological prominence of Corinthian type A amphorae in southern Italy may be explained in part by their secondary function in funerary contexts, either as grave markers or as grave goods. They are present at the necropolis in contrada Madonnelle at Policoro (ancient Siris/Herakleia) at a very early date along with other eastern Mediterranean amphora types; the Pantanello necropolis evinces later phases of distribution and use of this amphora type.[6] Indeed, as Koehler observed in the first key study of these amphorae, it was largely the evidence from Magna Grecia, including the Schirone necropolis near Policoro, that established the Corinthian type A amphorae as among the first transport amphora types produced in Greece.[7]

A general development in Corinthian amphora fabrics has been outlined by Koehler.[8] Earlier Corin-

[3] Identification, dating, and text for the SOS amphorae was by S. Savelli, translated and edited by K. Swift.

[4] Johnston and Jones 1978; Rizzo 1990, 16–17.

[5] Whitbread 1995.

[6] Personal observation of amphorae on public display in the Museo Nazionale della Siritide, Policoro, 2007; Morter and Leonard 1998, 731.

[7] Koehler (1992) cites Koehler 1981 and Adamesteanu 1974a, 112.

[8] Koehler 1992; Whitbread 1995, 278.

thian type A amphorae were produced in reddish-fired fabrics, often containing argillaceous rock fragments or "mudstone" (**No. 6**); later wheel-thrown Corinthian type A' amphorae were produced predominantly in the yellow Corinthian Neogene marl fabrics, with or without tempering of argillaceous rock fragments (see **No. 24**).[9] This lends an inherent chronological dimension to the study of Corinthian amphora fabrics and their production technologies.[10]

The final phase of development of the hand-formed Corinthian type A amphorae in the early 5th century BC shows a development towards sloping rim forms similar to those of Corinthian type A' amphorae, which began to be produced in this period: compare, e.g., the type A and A' rims from site 401 (**Nos. 5, 19–20**).[11] Later Corinthian type A hand-formed amphorae in fabrics similar to those used for Corinthian type A' are represented in the survey by a single rim and a single base likely to date to this period.

It is generally thought that the distribution of Corinthian type A amphorae beyond the Corinthia ceased around the middle of the 5th century BC, though production of this amphora type continued at Corinth until the end of the 4th century BC.[12] This has been used as a broad terminus ante quem for the distribution of Corinthian type A amphorae in the chora of Metaponto, and is generally supported by the corpus of 34 Corinthian amphorae from the Pantanello necropolis, where Corinthian type A' amphorae predominated in 5th-century funerary contexts.[13] Of the 14 Corinthian type A rim forms from the survey, all but one are of the earlier type, and the 5th-century example is massively outnumbered by the 21 rims of Corinthian type A' amphora rims (examples shown by **Nos. 19–24**).

Southern Italian Type A amphora (No. 7). There is new evidence from the Metaponto survey for the emulation of Corinthian type A amphora forms in southern Italy. A rim from site 349 is in a fabric that is not typically Corinthian and which may instead be southern Italian. A range of quartz-biotite rock fragment inclusions, along with individual, usually relatively fresh, grains of quartz, rare calcite, plagioclase feldspars, and biotite, are visible in a thin section of the fabric. These materials are present in the Bradano Trough and Calabria, and petrographically this fabric falls within the range of local/regional fabrics evident in other ceramic classes, including, for example, roof tiles (sampled from the survey assemblages) and impasto ware (sampled from the assemblages from the ICA excavations at Incoronata greca, located on the marine terraces above the Basento River). The fabric, however, is not typical of those produced in the Metapontine chora, and so a more distant southern Italian source seems likely, perhaps Calabria, as the fabric bears some similarities in terms of the range of inclusions to those thought to be from that area (see **No. 8** and discussions of Archaic type B amphorae below).

Given their ubiquity in the region and prominent secondary use in funerary contexts, it is perhaps not surprising that Corinthian type A amphorae were occasionally emulated by pottery producers in southern Italy. Indeed, it would be unusual if this had not been the case to some extent, and tends to suggest that Corinthian imports did not always satisfy local demand in southern Italy. This demand may well have been restricted to the secondary funerary function that many of these amphorae served, rather than their primary function as containers for the (maritime) transport of olive oil, as was no doubt the case with the genuine Corinthian imports. The extent and nature of local emulation of Corinthian type A amphorae in southern Italy remains to be established, but the single example among the 14 type A rims suggests that it was probably slight and/or sporadic.

Archaic Type B amphorae (Nos. 8–18). Aside from Corinthian type A amphorae, a range of other Archaic amphora forms of the late 6th and early 5th centuries BC are present in the survey corpus. The forms are represented by rounded to ovoid to lenticular rims associated with the relatively squat-necked forms typical of the Archaic and Early Classical periods. The broad rubric of "type B" has been used for these amphorae since these forms follow those of "Corinthian" type B amphorae, and can be readily distinguished from

[9] Whitbread (1995, 257, fabric class 1 and 2, respectively), following general observations by Koehler (1992).

[10] Vandiver and Koehler 1986; cf. reevaluation by Whitbread (1995, 284.)

[11] Whitbread (1995, 256) following Koehler 1979, 1981.

[12] Whitbread (1995, 256) following Koehler (1992), who observed that almost all of the Corinthian type A amphorae dating to the later 5th and 4th c. have been found at Corinth itself, but that there was a widespread distribution in the early 5th c. BC.

[13] Morter and Leonard 1998, 735.

the well-defined Corinthian type A amphorae.[14] Further classification on the basis of fragmentary rims is problematic, since some of the main types found in the western Mediterranean (e.g., Archaic "type B," "Ionio-Massaliote"/"Greco-Massaliote," and "Chian types") have similar basic forms that are difficult to distinguish on the basis of rim fragments.

The 24 Archaic type B and related amphora rims comprise a typologically and petrologically diverse group. The largest single group of fabrics is of likely Calabrian origin (**Nos. 8–13**), with 4 rims, 3 handles, and 2 toes (just over 16% of Archaic type B and approximately 10% of all Archaic-period amphorae). A single example of an Archaic type B rim is in a fabric microscopically consistent with type B amphora fabrics usually identified as Corinthian or Corcyrean (**No. 14**), making up ca. 4% of Archaic type B and 2.5% of all Archaic-period amphorae rims. A single 'Samian' type amphora rim (**No. 18**) stands out in the corpus of Archaic amphora rim forms, but the fabric may be southern Italian. Many of the remaining type B amphorae (18 rims) are also likely to be western Mediterranean or southern Italian in origin.

A range of types may have been produced at individual production centers, as was the case at Locri;[15] in the survey assemblages this situation may be reflected in the range of Archaic amphora rim forms in "Calabrian" fabrics, which have been characterized petrographically (see **No. 8** for fabric sample). On the current evidence, there does not appear to be a straightforward one-to-one relationship between morphological variations in rim forms and production centers in southern Italy. It is clear that further interregional studies of the fabrics of these amphorae are required for co-identification, formulation of regional groupings, and clarification of the typological complexities, which may largely be a consequence of multiregional production.

Differentiation and sourcing of regional groups through their fabrics is particularly important for the study of long-distance trade and economic interaction in the Archaic period. Several Archaic amphora forms similar to those represented in the survey have been found along the southern Italian coast from the Salentine (Apulian) to Calabrian peninsulas.[16] At present, the configuration of distribution networks is difficult to determine, given uncertainties about the sources of these amphora groups (for example, whether they come from the west or the east and were co-distributed or counter-circulated with Corinthian type A and early type A' amphorae).

In the corpus of Archaic type B and related amphorae from the survey, the most commonly occurring single group appears in fabrics that may be Calabrian (**No. 8**)—4 rims in these fabrics from a total corpus of 24. These fabrics are fired yellowish or off-white and are usually very porous, lending a distinctive texture to the matrix when viewed microscopically and in hand specimen. Petrographically, these fabrics are characterized by sand-sized inclusions of fine-grained quartz-biotite rock fragments (quartz-mica schist or similar), fresh biotite, and other weathered rock fragments consistent with the regional geology of Calabria.[17] Petrologically similar rock-fragment inclusions predominate in the fabrics of Late Roman Keay type 52 amphorae, which are known to have been produced in Calabria.[18]

A range of rim types are represented in these fabrics, from rounded and projecting rims (**Nos. 8–9**) to rims with ovoid or lenticular sections (**No. 10**) to lenticular sections. The fragmentary examples from the survey are insufficient to begin to formulate a typology or to identify discrete morphological classes. Archaic amphorae with a similar range of rim forms are known to have been produced in Calabria at Locri Epizephyrii.[19] Macroscopically, the fabrics appear similar to the descriptions of those identified as Locrian.[20] At present it seems safest to regard these imports as broadly Calabrian (i.e., produced from clays associated with the metamorphic solid geology of that region) rather than attempt an intraregional differentiation or attribution to a particular polis.[21]

[14] Whitbread 1995, 259. See in particular Savelli 2006, 116, and Fig. 35, for discussion of the problems of classification of Archaic western Mediterranean amphora forms. Later "Corinthian" type B forms and Greco-Italic amphorae in southern Italy in the 5th and 4th c. BC are discussed below.

[15] Barra Bagnasco 1992, 211–14.

[16] See in particular Auriemma 2004; Semeraro 1997 for Puglia; Barra Bagnasco 1992 for Calabria.

[17] *Carta Geologica*, Fogl. 222–55.

[18] Keay and Williams 2005.

[19] Barra Bagnasco 1992, Pls. 60–61.

[20] S. Savelli, pers. comm., 2008. A. Cavallo has observed that they are also macroscopically similar to plain-ware fabrics from this area (pers. comm., 2007).

[21] Forthcoming petrological study of the range of fabrics from the ICA survey of the territory of ancient Croton will deal more directly with the range of fabrics from Calabria and their relationships with those of the Bradano Trough and the Metapontine chora.

These Archaic amphora types attest the distribution of amphorae eastward along the southern Italian coast, circulating in the opposite direction to that of Corinthian type A and early type A' amphorae. The extent of their distribution eastward is still unclear—this might have encompassed the Adriatic and eastward across the Straits of Otranto to Corcyra and Corinth; when found in these areas or in other regions of the Mediterranean, they would presumably be attributed to Corinth.

The remaining Archaic type B and related amphora forms from the Metaponto survey are a macroscopically diverse group of fabrics with a wider range of rim forms of similar basic shapes to those represented in the "Calabrian" group of fabrics. A single example, with a rounded and projecting rim similar to some of the Calabrian forms, falls within the range of Whitbread's Corinthian type B class 2 fabric (**No. 14**).[22] While some of the fabrics show affinities with the "Calabrian" fabrics in the presence of quartz-biotite rock fragments along with quartz, biotite, and rare plagioclase feldspars, the rock fragments show a wider compositional range in contrast to the typical "Calabrian" amphora fabrics (**No. 8**). Consequently, these groups have been differentiated, since they are likely to derive from geographically or stratigraphically distinct clay sources. The inclusions are broadly consistent with the range observed in local/regional fabrics from the Bradano Trough. On the other hand, there is at present no unequivocal evidence for the production of Archaic type B and related amphorae in local/regional fabrics, at least in fabrics similar to those used for the production of Metapontine type B and Greco-Italic amphorae in the Late Classical and Early Hellenistic periods (**Nos. 26–39**). Reddish-firing fabrics are frequent in later type B amphorae of the 5th and 4th centuries BC and in some Greco-Italic amphorae are not represented in the corpus of Archaic-period amphorae from the survey. This may point to different sources of supply within or outside southern Italy, but it might also be a result of long-term developments in the kinds of sediments used for amphora production from the Late Archaic to Early Hellenistic periods.[23]

[22] Whitbread 1995, 276 and Pl. 5.14.

[23] Long-term developments in clay exploitation and related ceramic technologies are the subject of an ongoing ICA regional study of pottery fabrics from the Metapontino, the Bradano Trough, and adjacent regions of southern Italy.

A single example of a Samian-form type amphora form is represented in the survey (**No. 18**), with an almost squared rim and handles that have a rectangular section and angle sharply upwards from the termination with the neck. The form falls within the range of Samian amphorae of the late 6th into the early 5th century BC.[24] Petrographic analysis, however, suggests that it is closer to the range of southern Italian fabrics and is distinct from those typical of Samos, lacking the range of inclusions typical of Samian fabrics as characterized by Whitbread.[25] This example illustrates the problems of attribution and regional sourcing based on morphology. It appears that a range of amphora forms typical of other regions (particularly eastern Mediterranean types) were produced in southern Italy in the Archaic period, at least in Calabria and likely at other southern Italian production centers as well. Like the later type B amphorae of the 5th and 4th centuries BC, many or all of these Archaic amphora fabrics may come from southern Italy, in an arc extending from the eastern margins of the Bradano Trough to Sicily, or from elsewhere in the central or western Mediterranean. Aside from Corinthian amphorae, unequivocal examples of eastern Mediterranean fabrics, including the relatively distinctive fabrics characterized by Whitbread,[26] have not been identified in the survey assemblages, and this is supported by the absence of some of the more distinctive eastern Mediterranean/Aegean forms.

Classical period (ca. 475–325 BC)

Two types of amphora forms are attested during this period: Corinthian type A' amphorae (**Nos. 19–24**) and Classical-period type B amphorae, which, as in the Archaic period, occur in a range of fabrics and morphological variations (**Nos. 25–29**). The Classical-period type B amphorae appear to predominate over Corinthian type A' amphorae with 58 and 21 identified rims respectively (*Table 12.1*). The prevalence of type B amphorae in the assemblages (roughly 30% of all identified amphora rims from the Archaic Greek through Late Roman Republican periods) reflects in part their long span of production (5th–late 4th/early

[24] Whitbread 1995, 122.

[25] Whitbread 1995, 127 and Pl. 4.32, particularly in the absence of white mica and accessory inclusions of orthoclase, clinopyroxene, amphibole, and volcanic glass. Quartz-biotite schist is, however, present in the fabric characterized by Whitbread.

[26] Whitbread 1995.

3rd c. BC), the scale of their distribution in the central Mediterranean, and the frequency of sites of this date in the survey. Corinthian type A' amphorae had a span of production similar to the Post-Archaic type B amphorae (*Fig. 12.1*) and the assemblages include examples of rims of the earliest (early 5th c. BC, **No. 19**) and the latest, Hellenistic-period (early/mid-3rd c. BC, **No. 24**) forms.

Corinthian Type A' amphorae (Nos. 19–24). Circulated in the Metapontine chora in apparently substantial quantities, Corinthian type A' amphorae had an extremely wide distribution throughout the Mediterranean in the 5th century BC.[27] Corinthian type A' amphorae were produced in a different and very distinctive, typically Corinthian ceramic fabric made from Neogene marls of the Corinthia, either with or without argillaceous rock fragments.[28] Almost all of the Corinthian type A' forms from the survey were produced in the Corinthian fabric with argillaceous rock fragments.

Type A' amphorae predominated in the corpus of amphorae from the Pantanello necropolis where there were sufficient examples from the 5th century BC to assess the development of this form throughout the course of the century. The general conclusion reached was that, although broad developments in shape and morphological details (rims, toes, handles) could be identified, "to a large extent only a differentiation between those shapes likely to be from the earlier part of the 5th century versus those from the latter" was possible given the "considerable variations in the combinations of neck/handle, body, and base forms observed during the 5th century BC with no clear chronological trends in the occurrence of these combinations".[29] Given this demonstration of the limited applications of morphological details of type A' amphorae for chronological inferences, the fragmentary amphorae from the survey (consisting of isolated rims, bases, handles, and body sherds) have been assigned a general date based on the class.

Type B amphorae (Nos. 25–29). Type B amphorae were among the most common amphorae in the central Mediterranean in the later 5th and 4th centuries BC. Their ubiquity is reflected in the Metaponto survey corpus, where they are the most frequent amphora class (over 30% of all rims), though it should also be borne in mind that this period is well represented in the survey assemblages as a whole. Like earlier Archaic type B amphorae, the examples of this period from the survey make up a morphologically and petrologically diverse group. Defined as a single type, these amphorae used to be attributed to Corinth.[30] An amphora production site at the Corinthian colony of Corcyra provided the initial evidence that the production of these types may have been more complex and multiregional.[31] More recent evidence has significantly changed this picture: in addition to Corcyra, type B amphorae are now thought to have been produced elsewhere in the area of the Straits of Otranto and along the eastern Adriatic coast, including Illyria and Hvar (ancient Pharos), with production also likely in southern Adriatic Italy.[32] Production of type B amphorae is attested in southern Italy at Locri in Calabria.[33] Archaeometric analyses of type B amphora fabrics from Gela in Sicily point to a Calabrian origin.[34] More generally, analyses carried out by Desy and De Paepe on amphorae of type B forms from the Salento, dating to the late 5th/early 4th century BC, suggest a fabric that may be Calabrian or Sicilian in origin and another fabric that may be Sicilian or Liparian, expanding the areas of production still further westwards.[35] A general picture has emerged that shows production of type B amphorae in different regions, in one zone extending from Corcyra into the Adriatic, and in an arc from Puglia to southwestern Italy, perhaps as far as Sicily. As with the Archaic amphorae, the broad rubric "type B" is used in this study, permitting a range of geographical prefixes and probably incorporating a number of different regional and morphological variants.

Direct evidence for the Metapontine production of type B amphorae is provided by an overfired rim from an amphora production site in the survey area (**No. 28**, from site 266), with other examples in fabrics which appear to be regional (**Nos. 26, 27** and **29**). The

[27] Koehler 1981, 454–56.

[28] Corinthian type A' class 1 and 2 fabrics (Whitbread 1995, 270–73).

[29] Morter and Leonard 1998, 732.

[30] Koehler 1981, 1992; Vandiver and Koehler 1986.

[31] For a reevaluation of the evidence for a Corinthian attribution, see Whitbread 1995, 344.

[32] Summarized by Auriemma 2004, 150, with bibliography

[33] These are classified by Barra Bagnasco (1992, 214, Pls. 62–63 and 72–73) as *anfore con orlo ad echino*.

[34] Barone et al. 2002.

[35] Desy and De Paepe 1990, 215; cf. the discussion of the Apulian evidence in Auriemma 2004, 152.

rim form of **No. 28** is usually associated with the final stages in the morphological development of type B amphorae in the later 4th and early 3rd centuries BC, with a pressed figure eight rim, where handles were attached on either side, making the neck concave around the handle terminations.[36] There is a very similar rim from Oria in Puglia; the macroscopic fabric description is consistent with the Metapontine fabrics, presenting the tempting possibility that this may be a Metapontine export, but above all illustrating that much more study is required to identify, differentiate, and source these amphorae before they can be used for distribution studies.[37]

Other type B rims from the survey are in fabrics which are macroscopically and microscopically similar to those produced locally and perhaps more widely in the Bradano Trough (**Nos. 26–27**); production of late type B amphorae has been proposed at Herakleia.[38] At this stage of petrological study of regional fabrics there are no firm criteria to distinguish between the range of sediments used for pottery production in the area of Herakleia and that of the Metapontine chora, both located within the lower reaches of the Bradano Trough. It might be expected that the range of pottery fabrics from both areas would be petrologically similar. Production of type B amphorae at Herakleia and in the Metapontine chora at site 266 suggests that type B amphorae should be considered as a regional type and that neither production was an isolated phenomenon. Three examples of later type B amphorae were present in the Pantanello necropolis.[39] Of these, an example in a reddish fabric (T 190-8) appears macroscopically to be consistent with the range of fabrics produced in the Bradano Trough.[40]

Further study is required before a satisfactory attribution of provenance for these amphora groups can be made on the basis of either fabrics or forms. There is considerable typological confusion, which will continue until regional groups of forms can be differentiated by fabric analyses. The amphorae considered here are therefore grouped under the broad rubric of type B, with a number of regional subgroups and/or form types likely to be represented.[41] Given their ubiquity in the central Mediterranean and their potential as markers of economic interaction and maritime trade in the 5th to early 3rd centuries BC, the identification, characterization, and differentiation of type B amphora fabrics is a major research desideratum.

Early Hellenistic period (ca. 325–200 BC)

The Early Hellenistic period is characterized by the production of Metapontine Greco-Italic amphorae (**Nos. 30–39**), which appear to have succeeded the Metapontine type B amphorae of the Classical and very Early Hellenistic periods.

Identified by their characteristic range of triangular rims, Greco-Italic amphorae were the main types of amphorae produced in Magna Grecia and Sicily from the later 4th century until the late 2nd century BC.[42] Given the numerous regions of the Mediterranean now thought to have produced these amphorae, extending well beyond the triangle of Paestum-Akragas-Taras, it is hardly surprising that there is a great deal of typological variation and a very broad range in the fabrics. Greco-Italic amphorae were produced in the Metapontine chora, and the survey assemblages suggest that, in contrast to the Classical period, local/regional amphorae came to outnumber imported types in the Early Hellenistic period, based on the relative proportions of local/regional to extra-regional Greco-Italic amphorae: 65% and 35% respectively, a ratio of almost 2:1, from a total of 55 Greco-Italic amphora rims (*Table 12.1*). While this may suggest a decline in supply of extra-regional amphorae, it more likely points to a substantial increase in the production of local amphorae and their intraregional distribution throughout the chora. The frequency of Metapontine Greco-Italic amphorae relative to imports suggests that, compared to the preceding period, it was in the late 4th or early 3rd century that production of local and regional amphorae expanded in scale and scope.

The extra-regional Greco-Italic amphorae include two rims in fine pale fabrics containing volcanic inclusions that are probably Campanian, though perhaps Sicilian (**Nos. 41–42**). Either way, these attest to continued maritime distribution to the Metapontine

[36] Koehler 1992, Fig. 2c; Whitbread 1995, 259, and Pl. 5.5; cf. Morter and Leonard 1998, in particular T 190-1 and T 194-6.

[37] Semeraro 1997, 211, Fig. 189, No. 697.

[38] L. Giardino 2005, Pl. 36.

[39] Morter and Leonard 1998, 739-41 T 190-1, T 190-8, and T 194-6.

[40] Personal observation, 2007.

[41] Southern Italian types are often classified by rim shape, for example *a quarto di cerchio* (Lamboglia 1952), or *anfore con orlo ad echino* (Barra Bagnasco 1992; Brizzi 1999a, 299).

[42] Peacock and Williams 1986; Keay and Williams 2005. A terminal date of ca. 130 BC, proposed by Vandermersch (1994), is generally accepted.

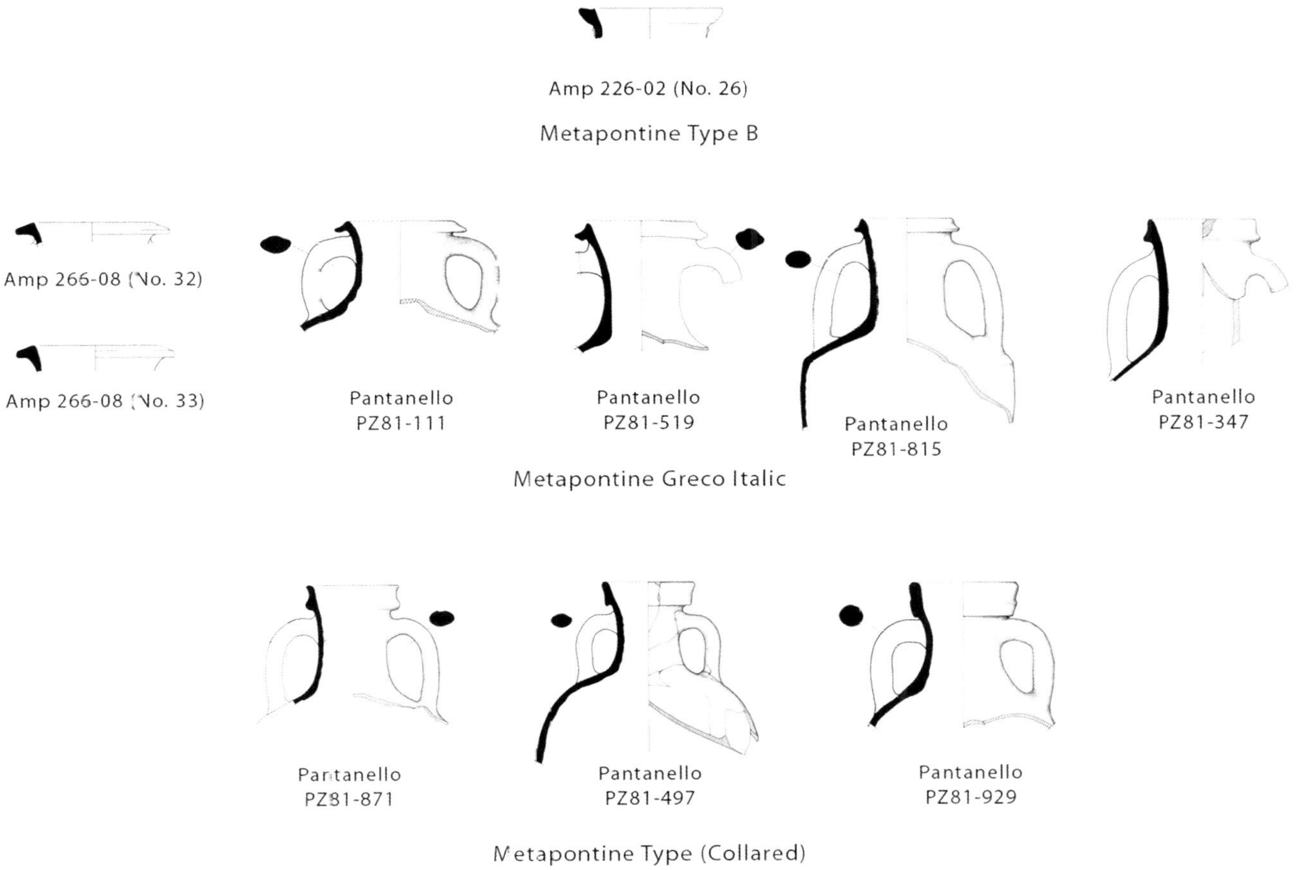

Figure 12.2 Proposed development of Metapontine amphorae from type B to Metapontine Greco-Italic to Metapontine type amphorae.

chora eastwards from the Straits of Messina during this period. During this time, Corinthian type A' amphorae also continued to be imported (**No. 24**), with four examples of the later rim type dating from the early/mid-3rd to the mid-2nd century BC.

Metapontine Greco-Italic Amphorae (Nos. 30–39). Metapontine Greco-Italic amphorae can be distinguished through their fabrics, and are known to have been produced at site 266 (**Nos. 34** and **35**), at the ceramic production site of Pantanello (*Fig. 12.2*) and probably also at site 153 (**No. 36**) in the chora.

These amphorae have been called Metapontine Greco-Italic amphorae on the basis of their rim morphologies, which follow the triangular projecting rims typical of Greco-Italic amphora types. Their shape history may, however, be more complex: a complete profile from a 3rd-century BC context in the stoa in the agora of Metaponto suggests that the prevailing influence on basic shape appears to have been Corinthian A' amphorae, with the handles showing particular affinities, in both their articulation and rounded cross sections (see **No. 58** from site 266).[43] A Late Corinthian A' amphora rim was present in the surface assemblage from site 266, showing that this amphora type was available for (selective) emulation at this production site. Although the rims have sloping upper surfaces, they are much more gracile and angular than late 4th-/early 3rd-century BC Corinthian A' amphorae, and in this respect they follow Greco-Italic rather than Corinthian forms. It therefore appears that there was some degree of morphological hybridization in the production of Metapontine Greco-Italic amphorae, involving the selection of different elements of other amphora classes—in the initial stages, at least, the same general body shape and handle articulation as Corinthian A' amphorae, but with rims more akin

[43] See the study of L. Giardino (2005, 409, Fig. 11), who cites numerous other, unpublished examples from Metaponto and observes that the body is heavier than the usual types produced in Magna Grecia during the 3rd c. BC.

to those of Greco-Italic amphorae produced elsewhere in southern Italy and in Sicily.

The distinctive hollow cylindrical toes with turned undersides (**Nos. 43–47**) show that the Metapontine amphorae did not conform to the usual Greco-Italic shapes in their morphological details and show a relatively restricted range of forms. These are probably the main type of toe for the Metapontine Greco-Italic amphorae; typical Greco-Italic hollow spike toes in Metapontine fabrics are conspicuous by their absence from the survey assemblages. Instead, there are numerous examples of Metapontine-type toes associated with Greco-Italic rims from the production site at Pantanello, and a complete 3rd-century BC example from the excavations of the stoa at Metaponto.[44] Further evidence from well-dated deposits is required before a more secure chronology can be proposed for this type, but production seems likely to have occurred in the late 4th and/or 3rd centuries BC at site 266, which would be broadly consistent with the rim forms within the general Greco-Italic amphora chronology. Like other Greco-Italic types, the Metapontine Greco-Italic amphorae may have continued to be produced into the later 2nd century BC, as is suggested by amphorae from a waste deposit associated with the Late Republican ceramic manufactory at Pantanello in the Metapontine chora.[45]

Extra-regional Greco-Italic amphorae (Nos. 40–42). Nineteen imported Greco-Italic amphora rims from the survey can be identified on the basis of their fabrics. Sourcing the range of Greco-Italic amphora fabrics from the survey must await a broader, interregional study of southern Italian amphora fabrics, drawing on a larger and less fragmentary corpus of amphorae, before these can be used for distribution studies. A southern Italian or Sicilian origin for **No. 40** seems likely, but other regions known to have produced Greco-Italic amphorae (for example, the Aegean) cannot at present be excluded.

Two likely Campanian Greco-Italic amphora rims are present in the survey assemblages (**Nos. 41–42**), with an extra-regional origin indicated by igneous inclusions in their fabrics. These suggest a source in a volcanic region, which may be consistent with an origin in the area of the Bay of Naples rather than other areas of igneous geology, such as Sicily, which are also known to have produced Greco-Italic amphorae.[46] If these are indeed Campanian, the two examples from the survey indicate some engagement with distribution networks involving Tyrrhenian Italy through the Straits of Messina and presumably the consumption of Campanian wine in the Metapontine chora in the 3rd century BC. The only examples of these microscopically distinctive fabrics are associated with these Greco-Italic amphorae, which fall within the range of Vandermersch's type MGS VI, conventionally dated ca. 260–210 BC.[47] If this chronology is followed, it points to the distribution of Campanian wine in the Metapontino during the mid- to late 3rd century BC; it has been suggested that the widespread distribution of Campanian Greco-Italic amphorae from the middle of the 3rd century BC was a consequence of Roman control of the Campanian wine trade.[48]

Hellenistic-period Corinthian Type A' amphorae (No. 24). Later Corinthian type A' amphorae of the 3rd century BC can be distinguished from those of the 5th and 4th centuries BC by their distinctive rims, with the projection increasingly sharply downturned to approach a collar.[49] The example presented here shows that the distribution of Corinthian amphorae in the Metapontine chora continued into at least the early 3rd century BC; the type is represented by 4 rims from the survey assemblages (from sites 247, 248, 249, and 266). A single example of this type was also present in the Pantanello necropolis.[50] Later Corinthian type A' amphorae were also frequent in the excavations of the stoa in the agora of Metaponto, with an apparent prevalence in the first half of the 3rd century.[51] The relationships between Corinthian type A' amphorae and the body and handle forms of early Metapontine Greco-Italic amphorae underscore the importance of these imports in the region in the later 4th century BC: it was the basic shape of the late 4th and early 3rd century BC Corinthian type A'

[44] L. Giardino 2005, 409, Fig. 11.

[45] Brehob 1983.

[46] Keay and Williams (2005) following Peacock and Williams (1986) for the identification of Campanian amphora fabrics associated with later Dressel 1 and 2–4 amphorae, and for the differences between these fabrics and those of Sicilian Greco-Italic amphorae. See Whitbread 1995, 239, for discussion of "black sand" amphora fabrics.

[47] Vandermersch 1994.

[48] Olcese 2004, 179.

[49] Koehler 1992, Fig. 2c.

[50] Morter and Leonard 1998, 739, T 103-3, dated ca. 300–275 BC based on context.

[51] L. Giardino 2005, 408, and Fig. 10.

amphorae that seems to have influenced the production of Metapontine amphorae in the 3rd century BC (though with different Greco-Italic type rims), and the affinities are particularly apparent in the handles with rounded sections represented in the survey at an amphora production site (**No. 58**, from site 266), where a Late Corinthian type A' amphora rim was also present (**No. 24**).

Late Hellenistic/Late Republican period (ca. 200–50 BC)
During this period, local or regional Metapontine amphorae appear to have continued to predominate over extra-regional types. Amphorae likely to date after the mid-2nd century BC are very poorly represented in the survey assemblages (8 in total compared to, for example, 80 rims dating to the Classical period, see *Table 12.1*), largely a consequence of the lower frequencies of sites of this date in the survey assemblages as a whole. Given the relatively small number of sites during this period, the relative importance of amphorae at the time is difficult to assess, but they probably continued to be at least a small component of the local and regional economies, as in the preceding periods.

Metapontine Type amphorae (Nos. 48–50). Metapontine type amphorae with their distinctive "collared" rim appear to have been produced in the region by the later 2nd century BC. Production is attested at Pantanello, postdating the mid-2nd century BC,[52] and in a 2nd-century BC deposit in the urban center of Metaponto.[53] Although only five examples of these distinctive rims are present in the survey, this should be seen within the context of the relatively small proportion of material of the 2nd and 1st centuries BC in the survey area as a whole (see discussion below of extra-regional amphorae).

The rims of Metapontine type amphorae developed from those of the Metapontine Greco-Italic amphorae, with a morphologically intermediate example represented in the survey (**No. 48**), a development which seems to have been accompanied in general by a shortening of the neck and handles. The precise chronology of this development is at present far from clear but both rim types are represented at the production site of Pantanello in deposits dating from the second half of the 2nd century BC (*Fig. 12.2*), which would be consistent with the general development of types from late Greco-Italics attested in other forms such as the Dressel 1 amphorae (cf. **No. 51**).

Extra-regional Amphorae of the 2nd–1st Centuries BC (No. 51). Extra-regional amphorae of the 2nd and 1st centuries BC are rare in the survey assemblages, represented by 3 rims in total, comprising less than 2% of all amphorae from the later 3rd to early 1st century BC.

A single Dressel 1C rim is present in the survey (**No. 51**). This form is thought to have developed from the Greco-Italic Will 1E type;[54] earlier versions have a flaring and very different development in rim form than that which occurred in Metapontine types. The date range for this type has been established as late 2nd to second quarter of the 1st century BC, following on from the late Greco-Italics after ca. 140/130 BC. This example may be earlier in the production range for this type.

Brindisi type amphorae may be represented in the survey corpus by a single poorly-preserved rim and a single handle. This type was produced in Apulia from the late 2nd to the late 1st century BC.[55] Microscopically the fabrics are consistent with characterizations of this type, with inclusions usually limited to quartz and limestone/calcium carbonates as a consequence of having been produced from sediments associated with the Puglian limestone plateau, but attribution on form and fabric is not certain.[56]

The paucity of extra-regional rims is accompanied by a general absence of relatively thick handles and substantial toes typical of Late Republican and Early Imperial transport amphorae. The apparent dearth of imports should, however, be seen against the low frequencies of all amphorae dated to this period, including Metapontine collared amphorae, and the lower frequencies of sites of this date in the survey assemblages as a whole. Imports are represented in the corpus of amphorae from excavated contexts at Pantanello, including Dressel 1 and 2–4 types, pointing to at least some degree of continuity in extra-regional trade during this period. The substantial and sustained distribution of amphorae of this period evidenced in other parts of the Mediterranean suggests that the very low frequencies of amphorae of this period in the survey are probably features of settlement organiza-

[52] For Pantanello, see Carter 1983a.

[53] L. Giardino 2005, 423, and Pl. 42.1.

[54] Keay and Williams 2005, with bibliography.

[55] Peacock and Williams 1986; Keay and Williams 2005.

[56] Keay and Williams 2005.

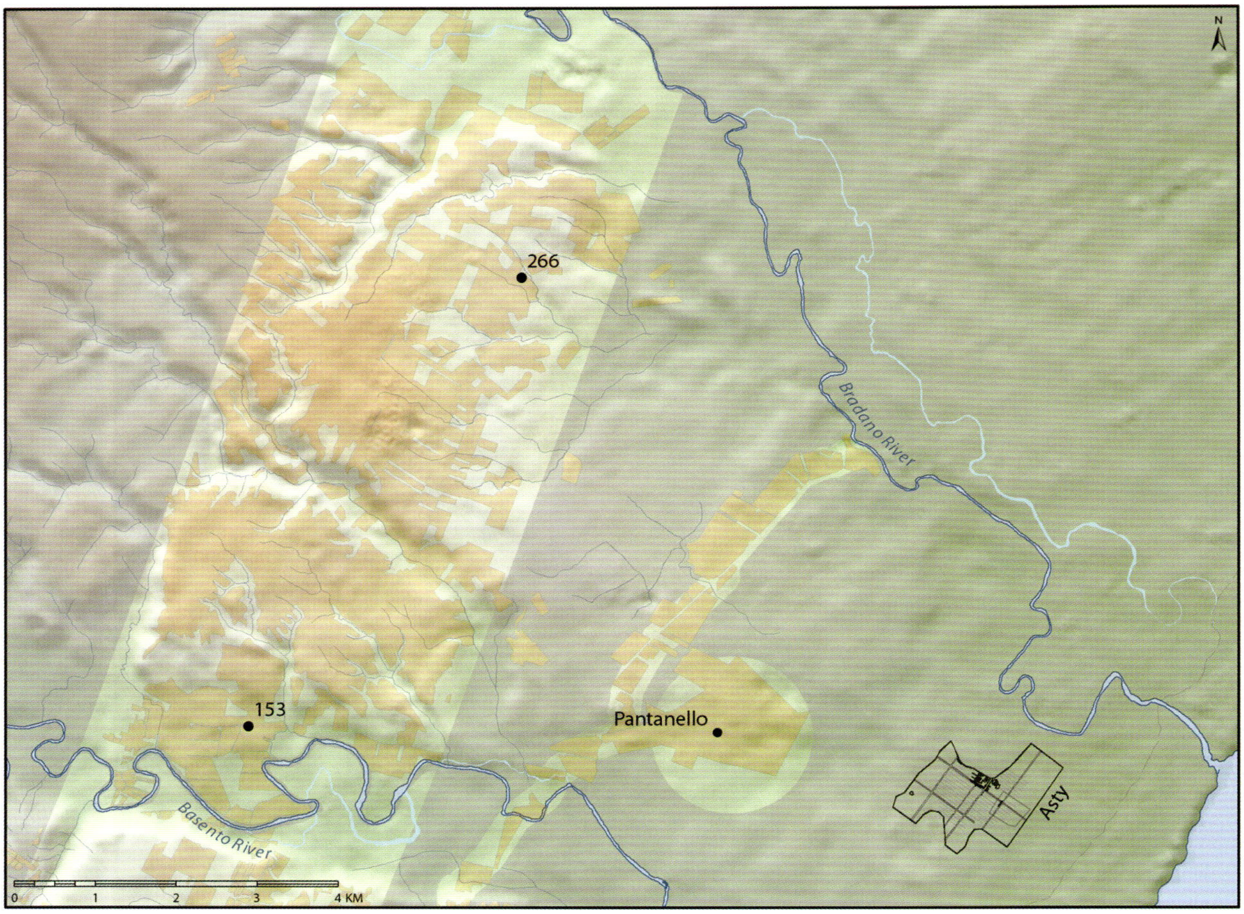

Figure 12.3 Identified amphora production sites in the chora: site 266 (Metapontine type B and Metapontine Greco-Italic amphorae); Pantanello (Metapontine Greco-Italic and Metapontine type amphorae); and site 153 (Metapontine Greco-Italic amphorae). (JT)

tion and the regional economy, rather than resulting from problems in supply connected with maritime distribution networks.

Amphora Production in the Metapontine Chora

The types produced locally in the Metapontine chora from the Classical to Late Republican periods show a sequence of development from Type B to Greco-Italic to Metapontine type amphorae (*Fig. 12.2*). The Metapontine type B amphorae followed the basic form of imported type B amphorae, which are ubiquitous in the central Mediterranean in the Classical period down to the early 3rd century BC. Site 266 provides evidence for the production of late type B amphorae dating to the late 4th or early 3rd century BC (**No. 28**) along with later Metapontine Greco-Italic amphorae (**Nos. 34–35**). Metapontine Greco-Italic amphorae were probably produced from the end of the production of type B amphorae in the course of the 3rd century BC and appear to have continued into the late 2nd century BC (*Fig. 12.1*) as suggested by their presence in the kiln deposit at Pantanello (*Fig. 12.2*), which is consistent with a more general terminal date for the production of Greco-Italic amphorae in other regions.[57] There is evidence for the production of this type at site 153 (**No. 36**), as well as at site 266 and Pantanello (*Fig. 12.3*). A morphological transition from Metapontine Greco-Italic to Metapontine type amphorae, with their distinctive collared rims, is apparent; an example of a morphologically intermediate form is represented in the survey (**No. 48**) along with other examples from the kiln deposit at Pantanello (*Fig. 12.2*).

Production at site 266

Production of Metapontine type B and Metapontine Greco-Italic amphorae is attested in the survey area at site 266 (**Nos. 27, 28, 34, 35** and **52–58**), along with

[57] Keay and Williams 2005.

production of plain wares. It also seems likely that terracottas were produced at this site: the majority of pieces are in a reddish fabric with a fine-grained matrix microscopically similar to those of the amphorae, but lack the fine sand-sized inclusions and are visually distinct from the majority of terracotta fabrics from other sites in the survey area (see Ch. 13).

Site 266 is located on a spur of the Pleistocene marine terrace, surrounded by gullies on either side (*Fig. 12.3*). As characterized by Abbott (Ch. 2), the underlying stratigraphic units consist of conglomerate rock rather than marine clays. These sedarenites are predominantly composed of siltstone, with "common limestone and some chert, vein quartz, metaquartzite, sandstone, and rare green plutonic rock" (see also Ch. 1). Thin sections of amphorae produced at site 266 show a relatively restricted range of inclusions, predominantly quartz (some with undulose extinctions indicating a degree of metamorphism, perhaps derived from the metaquartzite) with some limestone (or calcium carbonate precipitates). Some sandstone (mostly quartz) is evident in the larger inclusions in a sample of tile fabric from the site.

The solid geology underlying site 266 limits the range of sediments locally available for ceramic production to surface sediments (soils) and runoff sediments accumulated in the adjacent gullies and the small valley to the southeast, part of a network of channels draining into the Fosso Appio and, ultimately, the Bradano River. Levigation (refining the clay by pooling it to remove the coarse fraction) therefore seems a likely and necessary component for the production of ceramics from sediments in the immediate area of site 266. This may be consistent with the relatively fine-grained and compact matrices of the amphora, plain-ware, and terracotta fabrics that predominate in the survey assemblage. These fabrics are usually fired to a relatively intense reddish orange, suggesting the presence of iron oxides (as would be present in surface or basin sediments) and a relative absence of calcium carbonate in the matrices (usually resulting in lighter firing colors). The soils overlying the marine terraces are classified as noncalcareous, in contrast to lower deposits nearer the coastal plain and in the river valleys.[58]

Subsequent tempering with fine sand grains (predominantly quartz) seems likely. The tempering material was probably derived from well-sorted, sand-sized grains weathered from the conglomerates, where they are present in the cement between the larger pebble components. A tile sample from the site has inclusions similar to those present in the amphora fabrics, but larger, pointing to selection of the size fraction of the tempering material depending on intended function—some tile fabrics are not tempered, and some have coarse tempering, but all of the amphora fabrics from the site contain fine sand-sized grains.

The use of more calcareous clays is apparent in the range of tile fabrics from the site, and these generally have coarser-grained matrices. This suggests that a range of fabrics could be produced locally, and therefore that the restricted range of amphora fabrics probably represents the consistent application of clay-preparation practices for the production of this ceramic class. Such fine-grained fabrics were also used for the production of terracottas, although these lack the fine-sand tempering typical of the amphorae, probably associated with their coroplastic production.

The likely use of levigation/elutriation for the production of amphorae at site 266 points to a well-developed and relatively large-scale infrastructure for ceramic production, as might be expected for the production of amphorae. Levigation is labor intensive and generally avoided if suitable clay deposits are available locally (as was probably the case with production at sites located above marine clay deposits), but surface sediments are widely available for exploitation, including perhaps the initial selection of partially-sorted sediments in gullies or field runoff.[59] With the production of amphorae, plain wares, and (probably) also terracottas, site 266 bears all the hallmarks of an "estate" mode of production as defined by Peacock.[60]

Metapontine Greco-Italic amphorae in similar fabrics are present at sites throughout the survey area, from the margins of the Bradano valley to the Basento valley (e.g., sites 269 and 270, 383 and 153), and from the lower parts (sites 638 and 443) to the upper marine terrace to the northwest (site 309). Strong similarities between the fabrics of amphorae produced at site 266 and the fabrics of the Metapontine Greco-Italic amphorae from the survey suggest that they were pro-

[58] Characterized by < 0.5% carbonate content (*Carta pedologica*, 17 and 266–67).

[59] For discussion of the use of *terra rossa* sediments in Corinthian amphora production, see Whitbread 1995, 338. These sediments appear to have been used even though they did not contain a high proportion of clay components.

[60] Peacock 1982.

duced from similar sediments and/or from the application of similar ceramic-production technologies. This may indicate that the majority of Metapontine amphorae of the late 4th–2nd centuries BC from the survey area were produced from sediments similar to those on the marine terraces (surface/runoff sediments), rather than from the thick deposits of calcareous clays in the lower reaches of the survey area.[61]

Production at site 153

A waster of a carinated amphora shoulder from site 153 (**No. 59**) provides evidence for amphora production and is accompanied by a high-fired Metapontine Greco-Italic rim (**No. 36**). While the fabric of **No. 59** is macroscopically similar to those from site 266, it is overfired to the point of vitrification of the matrix, making it difficult to determine the nature of the raw material used to produce it. Site 153 is located at the northern margin of the Basento valley (*Fig. 12.3*), and the fabric was perhaps produced from fluvial/alluvial sediments, basin material that may be broadly similar to the surface and channel sediments on the marine terraces near site 266, particularly if similar clay manipulation processes were applied for its manufacture.

Study of the amphora fabrics from the survey suggests that it may be possible to differentiate between fabrics produced in different depositional zones within the survey area. The amphora fabrics from site 266 and site 153 are probably the result of a production process involving the refinement of source sediments using elutriation/levigation and subsequent alteration of the physical properties of the clays through tempering. While this has the effect of altering the fabrics significantly from the source material (in particular the granulometry of the non-plastic inclusions), the assemblage from site 266 and Metapontine amphorae from other sites in the survey present the possibility that particular production groups in the region may be differentiated not only by the types of source sediments, but also by the consistent application of these ceramic technologies for the production of amphorae and other ceramic classes in the Metapontine chora.

Summary

In the corpus of transport amphorae from the Metaponto survey, the two SOS amphorae are the earliest, dating to the 7th century BC. The more numerous Corinthian type A amphorae are the only other type which are likely to date before the 6th century BC. A possible southern Italian, perhaps local or regional, imitation of an Archaic Corinthian type A form is also represented. From the later 6th century BC, a morphologically and petrologically diverse group of Archaic amphorae, grouped together under the rubric of Archaic type B and related amphorae, comprise the majority of vessels of the Archaic period.

The range of fabrics and morphological variation of these amphorae suggest that a number of different types from different regions of the Mediterranean are represented; a group of fabrics likely to be western Calabrian in origin have been identified petrographically and other Archaic southern Italian or western Mediterranean amphorae are also likely to be represented, perhaps along with eastern Mediterranean types. Together, these types point to the importation of amphora-borne commodities from both east and west in the Archaic period.

Amphorae of the 5th and 4th centuries are characterized by type B and Corinthian type A' amphorae, with the ubiquitous "type B" amphorae *sensu lato* predominating, as elsewhere in the central Mediterranean during the Classical period. The corpus of type B amphorae comprises likely examples in southern Italian fabrics, including examples that are probably Calabrian in origin. A Metapontine type B amphora from a production site in the chora attests to some local production of type B amphorae, probably dating to the late 4th or early 3rd century BC, perhaps earlier.

It is from the Early Hellenistic period that Metapontine amphorae predominate over extra-regional amphorae in the survey assemblages. These Metapontine Greco-Italic amphorae point to a floruit of regional amphora production during this period. The Metapontine fabrics associated with these types are distinctive and are characterized petrographically to facilitate co-identification. Imports are represented by Greco-Italic amphorae including a single example of a likely Campanian Greco-Italic import, and by Late Corinthian type A' amphorae. In the Late Hellenistic/Late Republican period, Metapontine Collared amphorae predominate in the relatively small corpus, pointing to a continuation of local and regional amphora production, with imports largely restricted to southern Italian types, which were probably Apulian, though perhaps Adriatic Italian.

[61] See Ch. 2; also *Carta pedologica*, 17.

Catalog of Amphorae

This catalog includes a representative selection of amphorae and the key pieces from the ICA survey. Forms and fabrics of the main types are presented in chronological order. The last section of the catalog presents materials from two production sites in the Metapontine chora.

All dimensions for forms are in cm, for fabrics in mm. Rim diameter indicated by "rim diam.," extant height by "h," max width by "w," base diameter by "base diam." Color was recorded with reference to the Munsell Soil Color Charts. The first three digits following the Amp abbreviation indicate the site number, with fabric samples allocated a different series of codes for the reference collection. An asterisk (*) after the site number indicates materials from the 1980 ICA pilot survey (sites numbered 1–44), as well as at the intensive survey sites and plots located south of the Basento River through the 2001 season (most of the sites numbered 562 and higher), and in a single instance a fabric sample from the excavations of the Archaic sanctuary at Pantanello in the Metapontine chora (**No. 6**). See the Gazetteer for more information on sites.

Archaic Period

SOS amphorae (7th c. BC)

1:4

1. Amp 270-01 SOS amphora rim, Middle to Late I
Dimensions: rim diam. 16 h 3.8
Description: Ovoid, out-turned rim, with relatively sharply defined lip forming the highest point on the rim, a gently convex inner lip. Turned step on the upper neck. Traces of black slip, almost completely worn.

Fabric: Reddish yellow (7.5YR 7/6) with very fine inclusions of mica, fine white and brown grains.

1:4

2. Amp 477-01 SOS amphora rim
Dimensions: rim diam. 17 h 4.8
Description: Ovoid, out-turned rim, with a rounded lip. The upper neck is slightly concave on the inner wall and a turned step is extant on the upper exterior of the neck. Traces of black slip, almost completely worn.
Fabric: Very close to **No. 1**.

Corinthian type A amphorae (7th to early 5th c. BC)

1:4

3. Amp 254-01 Corinthian type A rim
Dimensions: rim diam. 20.0 h 4.3
Description: Irregular, hand-formed, thick, projecting rim with convex upper surface and concave outer face.
Fabric: Typical Corinthian type A amphora fabric with argillaceous rock fragments, as **No. 6**.

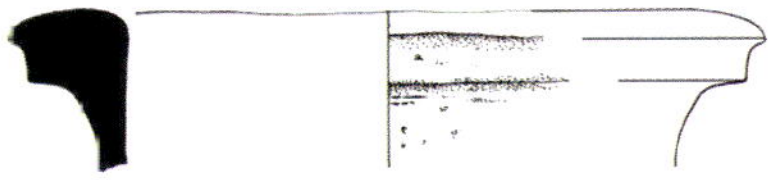

1:4

4. Amp 154-01 Corinthian type A rim
Dimensions: rim diam. 21.0 h 6.0
Description: Hand-formed, down-turned projecting rim and concave upper surface.
Fabric: Typical Corinthian type A amphora fabric with argillaceous rock fragments, as **No. 6**.

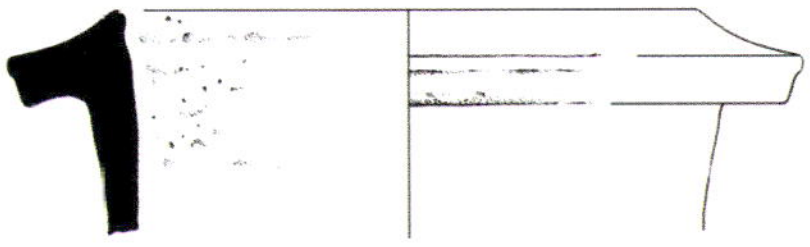

1:4

5. Amp 401-01 Corinthian type A rim
Dimensions: rim diam. 16.0 h 4.5
Description: Hand formed with relatively slender neck and sloping projecting rim. Probably an early 5th-c. form, morphologically similar to Corinthian type A amphorae of the same period, though produced by hand forming rather than the wheel throwing characteristic of Corinthian type A' amphorae. See **No. 19** for a wheel-thrown Corinthian type A' amphora rim, also from site 401, in a similar form.

Corinthian type A amphorae cont.

6. PZ-82-514P (*) Amphora fabric sample, Pantanello
Description: Typical Corinthian type A fabric from amphora found in Archaic sanctuary at Pantanello (see Carter 2006, 17), similar to **Nos. 3–5**.

A. Hand specimen: Scale 10 mm. Whitbread (1995, 338) suggests Corinthian type A amphorae were produced from Neogene marls of the Corinthia with an admixture of terra rossa sediments, resulting in the usual reddish-fired color. Terra rossa sediments are rich in iron oxides, in contrast to the typical Corinthian type A' fabrics, produced solely from the Neogene marl component, which yields a finer, yellowish fabric more typical of other Corinthian ceramics (see Corinthian type A' fabric, **No. 24**). The greyish component is probably a consequence of the addition of carbonaceous material, likely to have been present in the terra rossa soil. The angular, reddish inclusions are argillaceous rock fragments, i.e., petrified clays, typical tempering for Corinthian type A and A' class 1 amphora fabrics (Koehler 1981, 1992; Vandiver and Koehler 1986; Whitbread 1995, 268).

B. Microscopy: Corinthian type A amphora fabric sample, x10 magnification, width of field ca. 15.0 mm. Note the irregular fracture (largely a consequence of formation by hand, rather than by wheel throwing), mudstone inclusions, and admixed texture.

C. Microscopy: Corinthian type A amphora fabric sample, x20 magnification, width of field ca. 7.5 mm, showing the texture of the argillaceous rock fragments, which is close to the texture of the matrix, since these are essentially petrified clay.

D. Petrography: Corinthian type A amphora fabric sample in thin section, ppl (plain polarized light), x10 magnification, width of field ca. 15.0 mm. Elongated vughs (irregular voids) are visible in the section. The "flow" structure resulting from mixing and clay preparation and the boundaries of the angular argillaceous rock fragment inclusions can be seen more clearly than in hand specimen.

E. Petrography: Corinthian type A amphora fabric sample in thin section, xpl (cross-polarized light), x10 magnification, width of field ca. 15 mm.

A

B

C

D

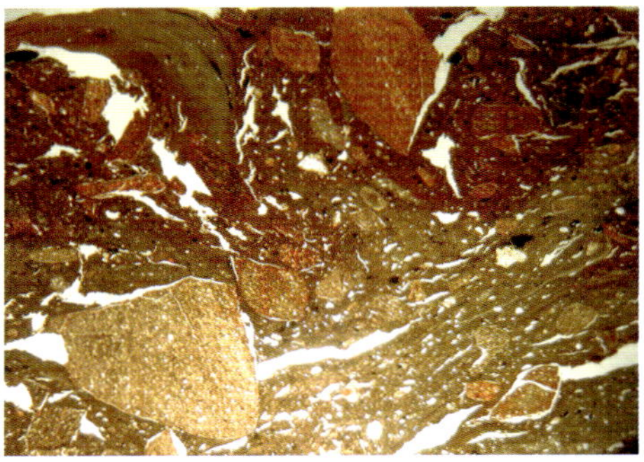

E

Type A amphora (southern Italian?) 1:4

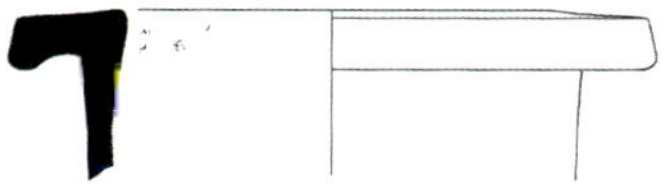

7. Amp 349-01 Type A rim (southern Italian?)
Dimensions: rim diam. 17.5 h 4.8
Description: Hand-formed, near-horizontal projecting rim, slightly concave outer face; cf. genuine Corinthian rim **No. 3**.
Fabric: Sample AMP007

A

A. Hand specimen: Cross section of rim, scale 10 mm. In hand specimen the fabric compares well to other types of Archaic amphorae; cf. in particular the fabric of the Archaic type B amphora **No. 17**, of likely southern Italian, perhaps southwest Calabrian origin, and the Classical-period type B amphora **No 25**.

B

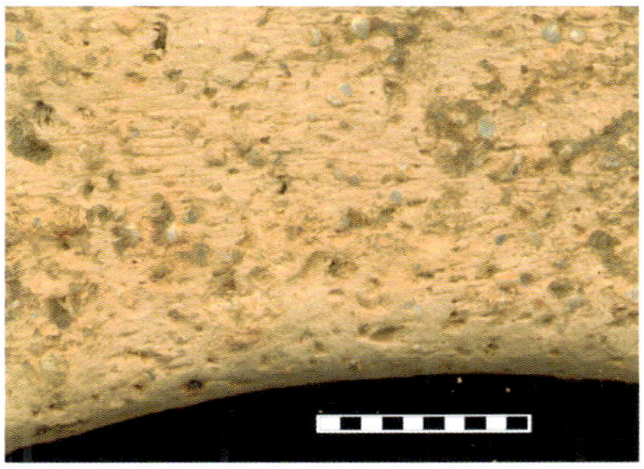

B. Hand specimen: Upper surface of rim, scale 10 mm. Note the superficial similarities in fabric to Corinthian type A amphorae, with a light-fired surface and angular rock fragment inclusions—microscopically and in thin section these can be seen to be very different from the argillaceous rock fragments typical of Corinthian fabrics.

C

C. Microscopy: Rim in section showing inclusions, x10, width of field ca. 15.0 mm

D

D. Petrography: Thin section, xpl, x30 magnification, width of field ca. 2.5 mm. A range of rock fragments are apparent, consistent with the range apparent in local/regional fabrics, which may be contrasted with the predominant quartz-biotite rock fragments in "Calabrian" fabrics (see **No. 8**). Also present are individual grains of monocrystalline quartz (white angular inclusions), polycrystalline quartz (two together in lower center right), plagioclase feldspar showing simple/albite twinning (center), and very fine biotite.

Archaic type B and related amphorae (6th–early 5th c. BC)

1:4

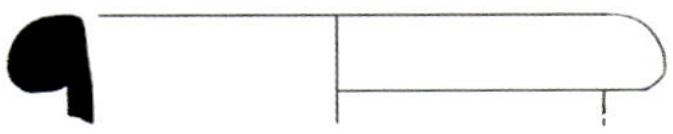

A

B

C

D

8. Amp 200-02 Archaic type B rim, Calabrian fabric
Dimensions: rim diam. 17.0 h 3.0
Description: Rounded, projecting rim; unfinished/weathered surfaces.
Fabric: Sample AMP093

A. Hand specimen: Ovoid rim in section, oriented vertically, scale 10 mm. The texture and color of the matrix are typical of the "Calabrian" group of fabrics from the survey (**Nos. 9–12**, with rim **No. 10** and handle **No. 12**). The five other examples of this fabric show some variation in the average size and proportions of rock-fragment inclusions. The angular rock fragments and color of the matrix are superficially similar to Corinthian type A' fabrics, but can be distinguished microscopically and are mineralogically different (see **Nos. 6** and **24**). The firing colors of these fabrics tend to be very pale brown (10YR 7–8/3–4) to white (10YR 8/2) (cf. Classical-period type B amphora fabric, **No. 25**), in contrast to typically yellowish Corinthian fabrics.

B and C. Microscopy: Ovoid rim in section, oriented vertically, x15 magnification, width of field ca. 10.0 mm and x30 magnification, width of field ca. 2.5 mm. Dark angular rock fragments are visible microscopically. Translucent, rounded grains are weathered quartz and/or plagioclase feldspars, identified petrographically (see below). Rare grains of mica are apparent on the surfaces but rarely visible in section. The component grains of the rock-fragment inclusion can be discerned, with the translucent component consisting of quartz, and the darker component consisting of weathered biotite.

D. Petrography: Ovoid rim in thin section, oriented vertically, xpl, x10 magnification, width of field ca. 15.0 mm. In xpl the rock fragments can be seen to consist of quartz and biotite. Larger, individual grains of these minerals are also apparent, presumably constituent grains of a coarser parent rock, but possibly derived from the clay source rather than associated with the finer-grained rock-fragment inclusions. Sparse, relatively fresh grains of plagioclase feldspar are present, exhibiting polysynthetic twinning and zoning in xpl, a typical and diagnostic feature of this very common mineral (MacKenzie and Adams 1994, 56). These do not appear to be associated with the quartz-biotite rock fragments and may have been present in the clay source along with the larger flakes of biotite and the quartz grains. Some of the quartz grains, angular and relatively fresh, show undulose extinctions indicating that the rock has been under stress; although this is a relatively common feature, it would be consistent with a metamorphosed rock (MacKenzie and Adams 1994, 48). The forming operation for the rim by folding or rolling is apparent in the preferred orientation of the inclusions and in the elongation of the voids, which are abundant, lending the fabric a relatively lightweight feel (e.g., compared with the more dense Corinthian type A' fabrics).

Archaic type B and related amphorae cont.

1:4

9. Amp 254-02 Archaic type B rim, Calabrian fabric
Dimensions: rim diam. 16.5 h 3.0
Description: Amphora with rounded projecting rim similar to **No. 8**, flattened upper surface.
Fabric: See **No. 8.**

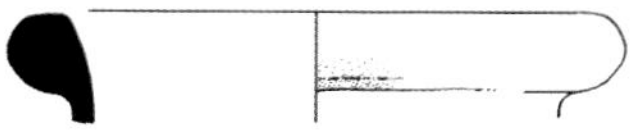

---

10. Amp 200-01 Archaic type B rim, Calabrian fabric
Dimensions: rim diam. 16.0 h 3.0
Description: Amphora rim with lenticular (elongated ovoid) section in similar fabric to **No. 8**. Although ovoid rims are usually distinguished typologically from rounded rims, the presence of both in nearly identical fabrics may point to a common origin and cautions against strict typological distinctions based on relatively minor variations in form. The rim profile ultimately depends on the degree of pressure applied during the folding or rolling of the rim (apparent in the orientation of the inclusions and elongation of voids in thin section as in **No. 8**) in what is essentially the same kind of forming operation.
Fabric: See **No. 8.**

1:4

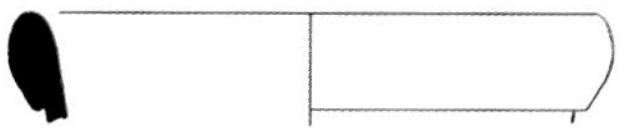

---

11. Amp 007-01 Amphora handle, Calabrian fabric
Dimensions: h 5.0 w 2.5
Description: Lower part of amphora handle with rounded section. Archaic-period amphorae generally had relatively squat necks with correspondingly short handles, tending to be more gracile than those typical of later 5th- and 4th-c. amphorae in both the western and eastern Mediterranean.
Fabric: As **No. 8**.

1:4

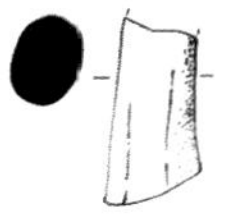

---

12. Amp 200-03 Amphora handle, Calabrian fabric
Dimensions: h 5.0 w 2.2
Description: Lower terminus of amphora handle with circular section.
Fabric: As **No. 8**.

1:4

1:4

13. Amp 074-01 Amphora toe, Calabrian fabric
Dimensions: base diam. 6.0 h 3.0
Description: Hollow toe in the same fabric as the rims and handles. Hollow toes are typical of a range of Archaic amphorae, including Corinthian type A and early type A' amphorae, early "Corinthian" type B amphorae, and related types produced in the western Mediterranean.
Fabric: Similar to **No. 8** in terms of matrix and finer inclusions, but lacks the larger rock fragment inclusions; it is closer to that of the Classical-period type B rim, **No. 25.**

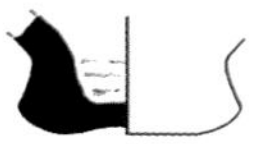

Archaic type B and related amphorae cont.

1:4

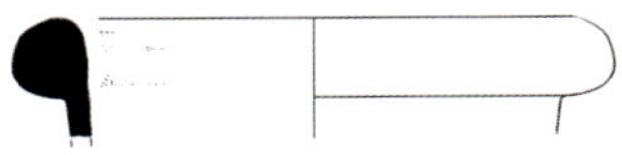

A

B

14. Amp 524-01 Archaic type B amphora rim
Dimensions: rim diam. 16.0 h 3.5
Description: Rolled rim. The form is similar to the Archaic type B amphorae in Calabrian fabrics (**Nos. 8** and **9**) and typologically similar to "Corinthian" type B amphorae (Koehler 1992, Fig. 3; cf. Auriemma 2004, 148 Fig. 4, for a "Corinthian" type B from Apulia).
Fabric: Sample AMP083

A. Hand specimen: Fine fabric characterized by frequent rounded voids (< 0.2 mm) and occasional larger voids (< 0.7 mm). Few inclusions are apparent in hand specimen, though frequent very fine biotite flakes are visible microscopically, particularly on the surfaces. In hand specimen and microscopically the fabric is similar to some later 5th- and 4th-c. Corinthian type B amphorae observed elsewhere and is consistent with Whitbread's Corinthian type B class 2 fabric (Whitbread 1995, 276 and Pl. 5.14).

B. Petrography: Archaic amphora in thin section, xpl, x10 magnification, width of field ca. 15.0 mm. Optically active micromass, appearing red (2.5YR 4/6) in xPl. The very fine-grained texture of the fabric is apparent in thin section, with inclusions consisting of abundant extremely fine biotite and larger grains of angular quartz, with less frequent plagioclase feldspars. The fabric also contains sparse larger rock fragments composed of fine quartz grains and laths of very fine biotite similar to the individual flakes visible throughout the fabric. Calcium carbonates (precipitates, limestones, etc.) are absent. Since the range of inclusions consists of individual grains of common minerals (quartz, plagioclase, biotite), it is difficult to infer a regional origin for this fabric, but it does not appear similar to local or regional pottery fabrics.

1:4

A

B

15. Amp 512-01 Archaic type B amphora rim (variant?)
Dimensions: rim diam. 17.0 h 3.5
Description: Ovoid rim with tooled plane at the junction with the neck, similar in form to **No. 17**.
Fabric: Sample AMP026

A. Hand specimen: Rim section. Folding of the rim is apparent in the fracture. The pinkish-white to pink matrix (7.5YR 8/2 and 7/4) is relatively porous, with frequent elongated vughs (< 2 mm) and abundant smaller rounded to lenticular voids. Occasional flakes of biotite ("gold mica") are apparent on the surfaces. Inclusions are frequent, with most < 0.5 mm.

B. Petrography: Rim in thin section, xpl, x15 magnification, width of field ca. 10.0 mm. Micromass appears reddish-brown in xpl, with abundant vesicules and vughs. Most inclusions are < 0.1 mm and include mono- and polycrystalline quartz, frequent plagioclase feldspars, and frequent quartz-biotite rock fragments, including undulose quartz with biotite laths. Very fine biotite is apparent in the matrix. Dark brown, optically inactive, iron-rich textural concentration features are also present, particularly prominent in pPl.

Archaic type B and related amphorae cont.

1:4

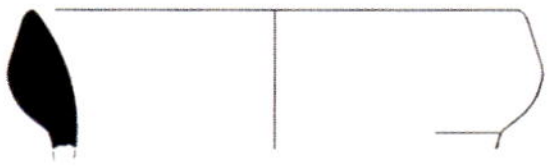

A

B

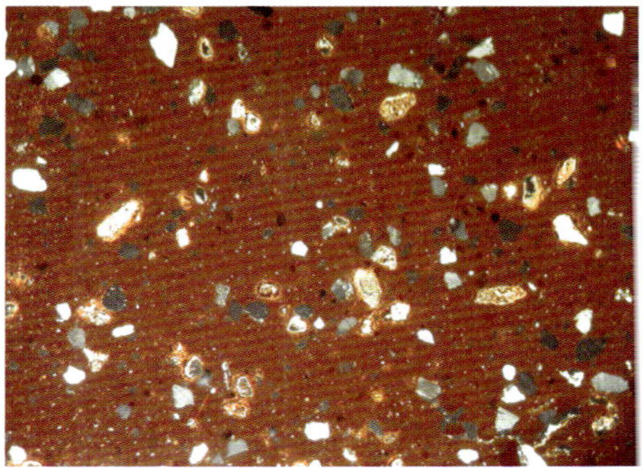

16. Amp 197-02 Archaic type B amphora rim (variant?)
Dimensions: rim diam. 14.5 h 4.0
Description: Lenticular rim with tapered lip.
Fabric: Sample AMP032

A. Hand specimen: Rim fragment in section, scale 10 mm. The light red matrix (2.5YR 6/6–8) is relatively porous, with frequent elongated vughs containing, and probably associated with, white powdery inclusions, presumably calcium carbonate precipitates. Inclusions are frequent, < 0.5 mm and mostly < 0.2 mm, with most appearing angular and grayish, identified petrographically as quartz, plagioclase, and (rarely) quartz-biotite rock fragments.

B. Petrography: xpl, x15 magnification, width of field ca. 10.0 mm. Micromass appears dark red (10R 3/6) in xPl. Inclusions are well sorted and apparently bimodal, with the grain size fraction 0.2–0.5 mm, perhaps added as sand temper (though natural sorting through depositional processes might also account for this feature). These are predominantly monocrystalline quartz and calcite/calcium carbonates lacking internal structure (possibly weathered calcite rock fragments—limestone, dolomite, etc.—decomposed during firing), frequent throughout the fabric, and they are visible in hand specimen as rounded to lenticular white, powdery inclusions. Rare, relatively fresh plagioclase feldspar grains are also present, as are quartz-biotite rock fragments with a fine-grained texture, along with common very fine biotite laths in the finer fraction. On the basis of the range of inclusions, particularly the quartz-biotite rock fragments, the fabric may be southern Italian.

1:4

A

B

17. Amp 366-01 Archaic type B amphora rim (variant?)
Dimensions: rim diam. 14.0 h 2.7
Description: Ovoid rim with tooled place at the junction with the neck, similar in form to **No. 15**.
Fabric: Fabric Sample AMP008. The fabric shows some strong similarities to those of the Calabrian amphorae, with a range of quartz-biotite rock fragments evident in thin section. Combined with textural differences in the matrix, this may indicate a distinct, though perhaps related, Calabrian fabric. On the other hand, the range of rock fragments is close to those observed in fabrics from the Bradano Trough, so a general southern Italian attribution seems safest. In particular, the fabric is close to the likely regional/southern Italian copy of a Corinthian type A form (**No. 7**).

A. Hand specimen: Scale 10 mm. Granular fabric with porous light brown matrix (7.5YR 6/4). Abundant angular inclusions, most < 0.3 mm.

B. Petrography: Rim fragment in thin section, xpl, x40 magnification, width of field ca. 3.5 mm. Inclusions of quartz and biotite rock fragments, polycrystalline quartz, sparse plagioclase feldspar, frequent very fine grains of quartz and biotite.

Archaic type B and related amphorae cont.

1:4

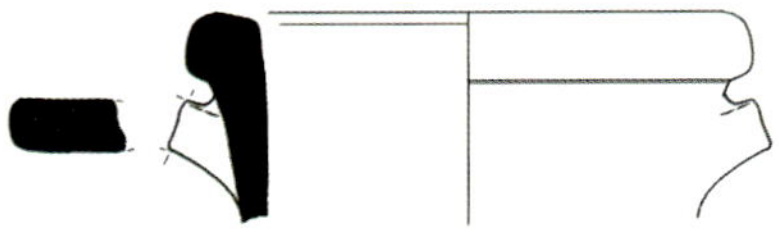

18. Amp 274-01 Amphora rim with squared profile, similar to Samian type amphorae
Dimensions: rim diam. 15.5 h 5.0
Description: Amphora rim with squared profile, tool-formed with well-defined junction at neck, in a similar form to Samian amphorae of the Archaic period.
Fabric: Fabric Sample AMP012

A

B

A. Hand specimen: Scale 10 mm. Hard-fired matrix, pink (7.5YR 7/4). Inclusions are frequent and relatively fine, < 0.5 mm.

B. Petrography: Rim fragment in section, xpl, magnification x40, width of field ca. 3.5 mm. Yellowish-red (5YR 4–5/6–8) micromass in xPl. Although the matrix is compact and hard-fired, this is an extremely porous fabric, with abundant vesicules < 0.5 mm and elongated vughs < 5.0 mm. Inclusions are predominantly mono- and polycrystalline quartz, frequently with undulose extinctions, and rock fragments of these along with biotite, similar to some of those observed in the other Archaic amphora fabrics. Angular, relatively fresh inclusions of plagioclase feldspars showing twinning and zoning are also present, with occasional flakes of biotite. The inclusions evident in thin section fall within the range of southern Italian fabrics as broadly defined in this study, though texturally and in hand specimen the fabric is distinctive and an eastern Mediterranean origin cannot be ruled out. This sample lacks the frequent white mica, orthoclase, rare amphibole, and clinopyroxene present in Samian fabrics characterized by Whitbread (1995, 127).

Classical–Early Hellenistic Periods

1:4

Corinthian type A' amphorae (early 5th–3rd c. BC)

19. Amp 401-02 Early 5th-c. BC Corinthian type A' rim
Dimensions: rim diam. 18.0 h 5.0
Description: Irregular wheel-thrown neck. The form compares well to that of a hand-formed Corinthian type A amphora from the same site (**No. 5**).
Fabric: As **No. 24**.

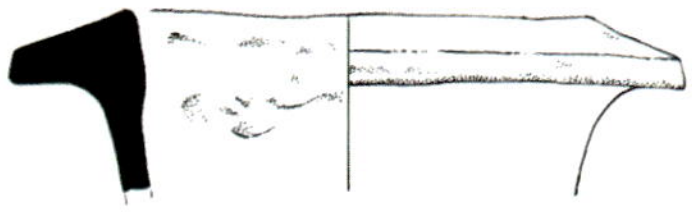

1:4

20. Amp 401-03 Corinthian type A' amphora rim
Dimensions: rim diam. 16.0 h 3.5
Description: Triangular rim with slightly convex upper surface and a turned plane on the outer face.
Fabric: As **No. 24** but lacking argillaceous rock fragment inclusions.

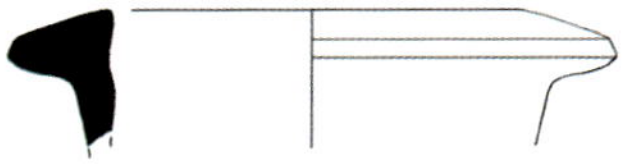

1:4

21. Amp 525-01 Corinthian type A' amphora rim
Dimensions: rim diam. 17.0 h 6.0
Description: Steeply sloping triangular rim.
Fabric: As **No. 24**.

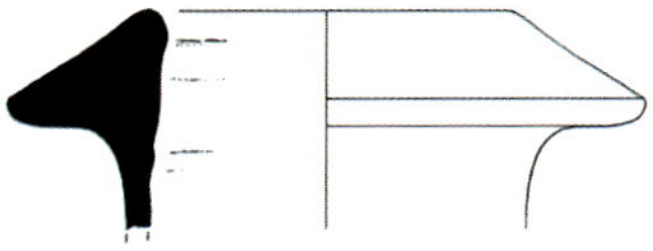

Corinthian type A' amphorae cont.

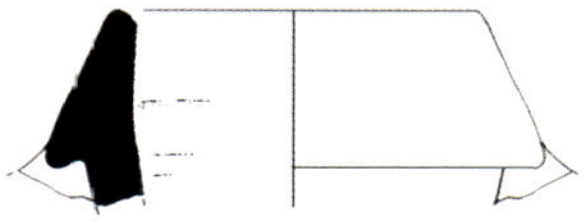

1:4

22. Amp 248-01 Corinthian type A' amphora rim
Dimensions: rim diam. 10.0 h 5.5
Description: Steeply sloping rim and part of the handle terminus merging with the lower part of the rim. The handle terminus is only partially preserved; the cross section (not extant) was probably ovoid or "pinched," as is usual for Corinthian type A' amphorae.
Fabric: As **No. 24** but lacking argillaceous rock fragment inclusions.

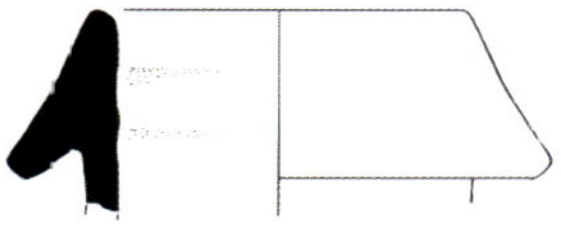

1:4

23. Amp 183-01 Corinthian type A' amphora rim
Dimensions: rim diam. 10.0 h 5.5
Description: Steeply sloping rim.
Fabric: As **No. 24** but lacking argillaceous rock fragment inclusions.

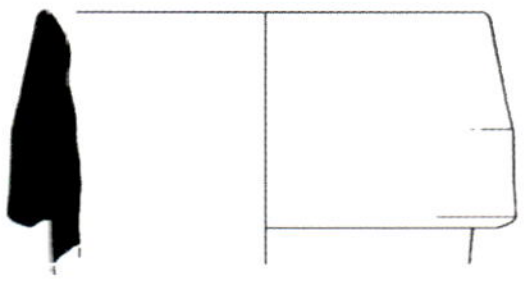

1:4

24. Amp 266-17 Later Corinthian type A' rim, 3rd-c. BC
Dimensions: rim diam. 12.0 h 7.0
Description: Later development sloping rim forming a collar. This form provides direct evidence for the distribution of Corinthian type A' amphorae at site 266 during the period of production of Metapontine amphorae, which emulate the handles and general body shape of Corinthian type A' amphorae, though with different rims.
Fabric: Fabric Sample AMP101.

A

B

C

A. Hand specimen: Corinthian type A' class 1 fabric from site 323, scale 10 mm, showing the typical Corinthian type A' amphora fabric. In contrast with the Corinthian type A fabrics (**No. 6**), the argillaceous rock fragments stand out clearly against the fine yellowish matrix produced from Neogene marl clays (Whitbread 1995, 344). This is the typical Corinthian fabric used to produce fine wares and other ceramic classes such as pithoi and louteria/basins (for an example of a Corinthian pithos from Incoronata greca, see Carter 2006a, 71, Fig. 2.23a). Corinthian fine wares were produced in similar fabrics, though lacking the tempering of argillaceous rock fragments. Almost all of the Corinthian type A' amphorae from the survey are in this fabric variant (Whitbread's Corinthian type A' class 1 fabrics); the remainder are in fabrics with similar matrices, either entirely lacking or containing rare inclusions of argillaceous rock fragments (Whitbread's type A class 2 fabrics; 1995, 270–73).

B. Microscopy: Corinthian type A' class 1 fabric x40 magnification, width of field ca. 3.5 mm. Note tempering of argillaceous rock fragments and distinctive splintery fracture of the fine-grained and relatively compact matrix.

C. Petrography: Corinthian type A' class 1 fabric in thin section, xpl, x15 magnification, width of field ca. 10.0 mm.

Type B amphorae (early 5th–early 3rd c. BC)

1:4

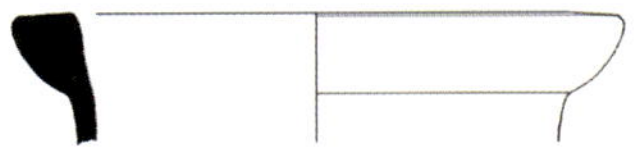

A

B

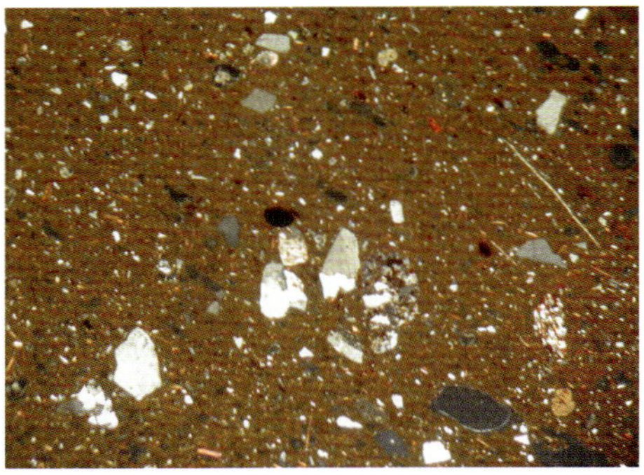

25. Amp 248-02 Type B rim, southern Italian (Calabrian?)
Dimensions: rim diam. 16.5 h 3.5
Description: Rim showing typical profile, with a slight concavity on the upper surface. For a Metapontine example with more pronounced concavity, typical of the late phase in the development of type B rim forms, see **No. 28**.
Fabric: Sample AMP095. The fabric does not appear to be local but bears some similarity to that of the type A amphora **No. 7**, which may be southern Italian. If this is indeed the case, the range of inclusions identified petrographically would suggest an origin in the Bradano Trough or, perhaps more likely, Calabria or eastern Sicily, but it may be eastern Mediterranean.

A. Hand specimen: Pinkish-white matrix (7.5YR 8/2) with frequent fine voids giving the fabric a lightweight feel. Inclusions < 1.0 mm, rounded to angular.

B. Petrography: Rim in thin section, xpl, x20 magnification, width of field ca. 7.0 mm. Abundant voids throughout the fabric. The larger inclusions are predominantly mono- and polycrystalline quartz, frequent biotite, and rare plagioclase feldspars, with rock fragments composed of polycrystalline quartz and biotite. Abundant finer inclusions consist mostly of quartz and abundant biotite, with occasional muscovite laths. Like that of **No. 3**, this fabric appears to fall within the range of Whitbread's type B class 1 amphorae, with the exception of the presence of quartz-biotite rock fragments (cf. Calabrian fabrics of Archaic type B amphorae, **No. 8**) and the apparent absence of microcline. The form and fabric are paralleled by a rim from the Pantanello excavations and a complete form from the Pantanello necropolis (Morter and Leonard 1998, T 256-1 and Fig. 17.4, 5th c. BC, not included in the petrographic analyses of the necropoleis amphorae; at that time type B amphorae were widely believed to be securely attributable to Corinth or Corcyra).

1:4

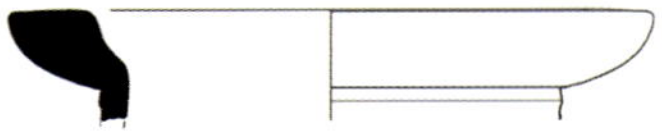

A

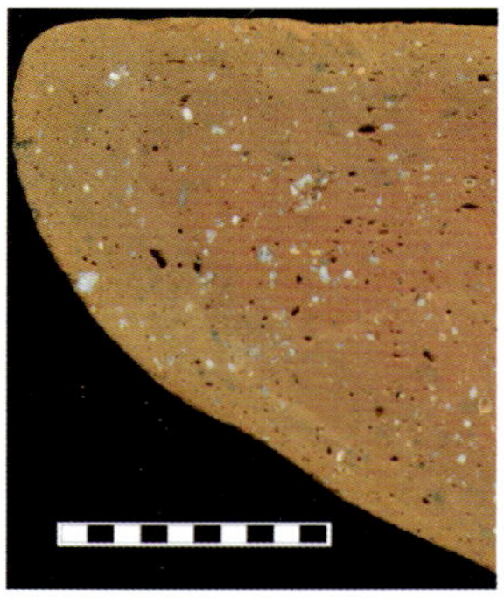

B

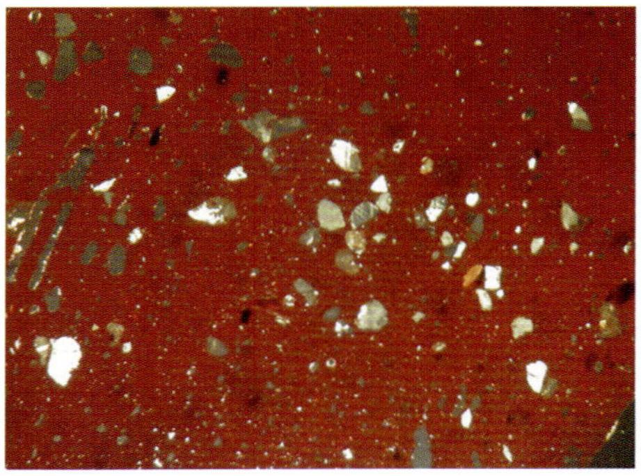

26. Amp 154-02 Metapontine type B amphora rim
Dimensions: rim diam. 17.0 h 3.0
Description: Rim with flattened upper surface and tooled junction with the neck.
Fabric: Sample AMP081. The fabric is typically and distinctively Metapontine, comparing well to those of Metapontine Greco-Italic amphorae from site 266 (**Nos. 27** and **28**) and Metapontine type amphorae.

A. Hand specimen: A very fine, hard-fired light red to red matrix (10R 6/6 to 4/8) with abundant inclusions, most < 0.2mm.

B. Petrography: Rim in thin section, xpl, x20 magnification, width of field ca. 7.0 mm. Although this fabric contains quartz, plagioclase feldspars, and very fine biotite, it can be distinguished from the usual local/regional fabrics by rare grains of muscovite, visible in thin section and more apparent on the surfaces and in hand specimen. This does not exclude beyond doubt a local/regional origin, however. Fine grains of microcline are also present (very rarely present in local tile fabrics). It may fall within Whitbread's type B class 1 fabric, which also contains muscovite (white mica) and microcline, though the relative proportions of the various mineral inclusions are not a particularly close match (Whitbread 1995, 275).

Type B Amphorae cont.

1:4

27. Amp 266-04 Type B rim (Metapontine fabric?)
Dimensions: rim diam. 16.0 h 4.5
Description: Type B amphora rim, slightly out-turned, approaching a triangular rim. The form is probably a morphological precursor to later Greco-Italic amphorae, though less pronounced than that of the Metapontine **No. 29**. From site 266, perhaps produced at that site.
Fabric: Sample AMP043. Metapontine, similar to, but slightly more granular than, that of **No. 26**.

A

B

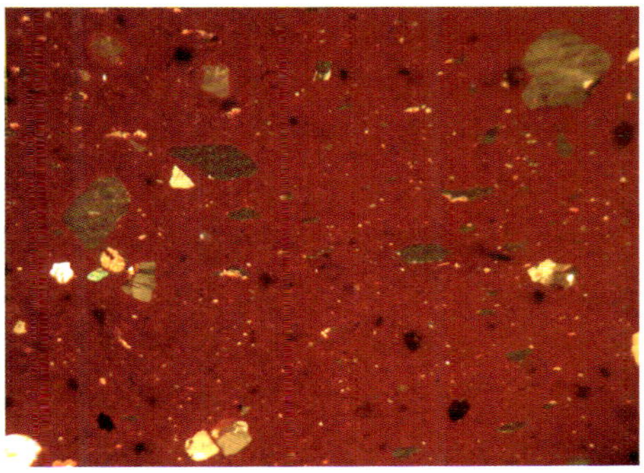

A. Hand specimen: Hard, red to light red matrix (2.5YR 5–6/6) with frequent fine-grain inclusions < 0.2 mm.
B. Petrography: Rim in thin section, xpl, x20 magnification, width of field ca. 7.0 mm. Microcline is sparse, as in the case with a tile sample from site 266. Very rare muscovite showing the presence of this mineral locally.

1:4

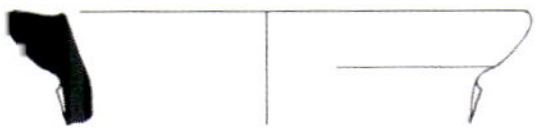

28. Amp 266-02 Late Metapontine type B rim, waster
Dimensions: rim diam. 14.0 h 3.0
Description: Overfired rim, not deformed. This piece provides direct evidence for the type's production in the Metapontino during its late phase of development. The rim's upper surface has a distinct concavity, indicative of a date in the late 4th/early 3rd c. BC (Koehler 1992, Fig. 2c; Whitbread 1995, 259, and Pl. 5.5; also Morter and Leonard 1998, T 190-1 and T 194-6). For its predecessor with less pronounced concavity, see **No. 25**.
Fabric: Sample AMP041. Although over-fired, the fabric is recognizable as Metapontine in texture and firing color (cf. texture and color range of overfired Metapontine Greco-Italic amphorae from same site, **Nos. 34** and **35**).

A

B

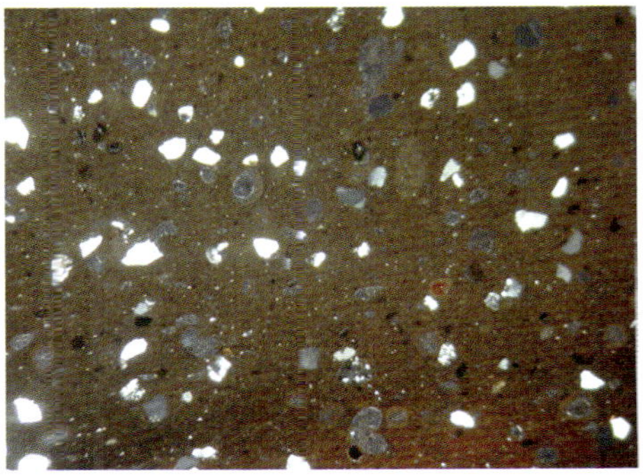

A. Hand specimen: Scale 10 mm. Misfired, fine-grained matrix, sintered to near vitrification, with frequent inclusions < 0.5 mm, most < 0.2 mm.

B. Petrography: x20, xpl, width of field ca. 7.5 mm. Predominant quartz, rare grains of plagioclase feldspars, and rock fragments consisting of quartz with some biotite.

Type B amphorae cont.

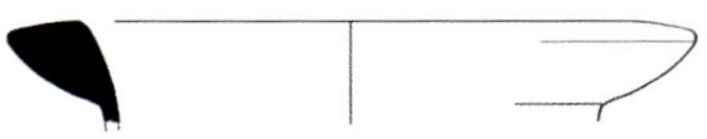

1:4

29. Amp 139-01 Metapontine type B amphora rim
Dimensions: rim diam. 18.5 h 3.0
Description: Rim showing a more pronounced development towards the typical triangular Greco-Italic rim forms than **No. 27**. These represent morphological precursors of the Metapontine Greco-Italic amphorae (**Nos. 30–39**).
Fabric: Sample AMP063, very close to **No. 27**, site 266.

A

B

A. Hand specimen: Red matrix (10R 4/6) with frequent voids and light red surfaces (2.5YR 6/6). Inclusions consist of relatively "fresh," angular grains of quartz and quartz-rich rock fragments, appearing microscopically similar to those in thin sections of regional fabrics. Rare grains of silver mica (presumably muscovite) are apparent on the surfaces, and in this respect the fabric is similar to that of **No. 27**.

B. Microscopy: x30 magnification, width of field ca. 2.0 mm, showing quartz-biotite rock fragment and polycrystalline quartz.

Metapontine Greco-Italic amphorae

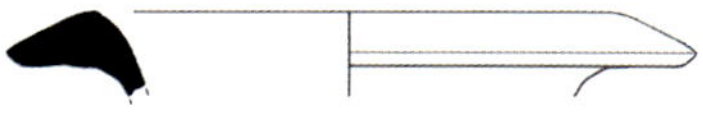

1:4

30. Amp 197-01 Metapontine Greco-Italic amphora rim
Dimensions: rim diam. 18.0 h 2.5
Description: Rim with triangular profile.
Fabric: Sample AMP031.

A

B

A. Hand specimen: Light reddish (10R 4/8 and 6/8) fabric with finely granular texture. Frequent gold and silver mica are apparent on the smoothed surfaces.

B. Microscopy: Rim section, x20 magnification, width of field ca. 7.5 mm. Abundant very fine grains of biotite and muscovite ("gold" and "silver" mica) along with quartz/feldspars visible on the surfaces. Inclusions in section are < 0.5 mm, mostly < 0.2 mm, consisting mostly of angular quartz/feldspar grains.

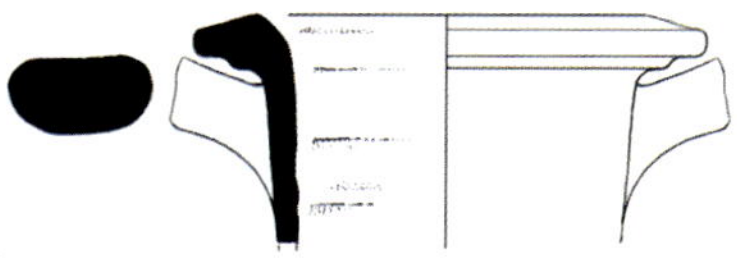

1:4

31. Amp 269-02 Metapontine Greco-Italic amphora rim, upper neck and handle termination
Dimensions: rim diam. 14 h 6.5
Description: Rim and upper neck with handle termination. Triangular rim with angular profile, tool-formed or trimmed, and slightly concave upper surface. Note the pronounced step on the underside; in this respect it is similar to the extra-regional, probably Campanian, Greco-Italic amphora **No. 41**, showing a degree of influence of the extra-regional types on Metapontine production. The handles are mounted slightly lower on the neck than was usually the case for the preceding late type B types; a feature even more pronounced on **No. 32**.
Fabric: Sample AMP020. Metapontine, falling within the range of **Nos. 34–36**.

Metapontine Greco-Italic amphorae cont.

32. Amp 057-02 Metapontine Greco-Italic amphora rim
Dimensions: rim diam. 14.5 h 5.5
Description: Rim with triangular profile, flared rim with concave inner face (for stopper?). As with **No. 31**, the handles are mounted lower on the neck than those of preceding type B amphorae, with the extant terminus on this example suggesting that the handles were positioned progressively lower and the upper parts of the handles were angled downwards, away from the line of the rim, as is the case with the succeeding Metapontine type amphorae (for more complete examples from the production site at Pantanello see *Fig. 12.2*).
Fabric: Sample AMP024. Metapontine, close to **No. 36**.

1:4

33. Amp 269-05 Metapontine Greco-Italic amphora rim
Dimensions: rim diam. 16.0 h 3.0
Description: Triangular rim with rounded profile, similar in basic shape to **No. 32** but less angular, particularly on the curved lower part of the rim, with a continuous curve at the junction with the neck. In contrast to **Nos. 31** and **32** there are no traces of tool forming or trimming, and this accounts for the curved profile of this example.
Fabric: Sample AMP022. Metapontine, close to **No. 36**.

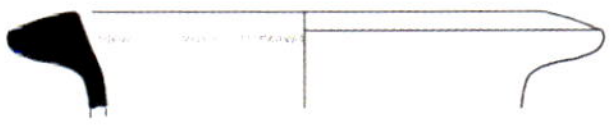

1:4

34. Amp 266-08 Metapontine Greco-Italic rim/handle
Dimensions: rim diam. 14.0 h 2.0
Description: Rim and handle terminus with over-fired surfaces. Tool-trimmed planes are visible on the rim's upper face.
Fabric: Sample AMP047. Typical Metapontine fabric.

1:4

A

B

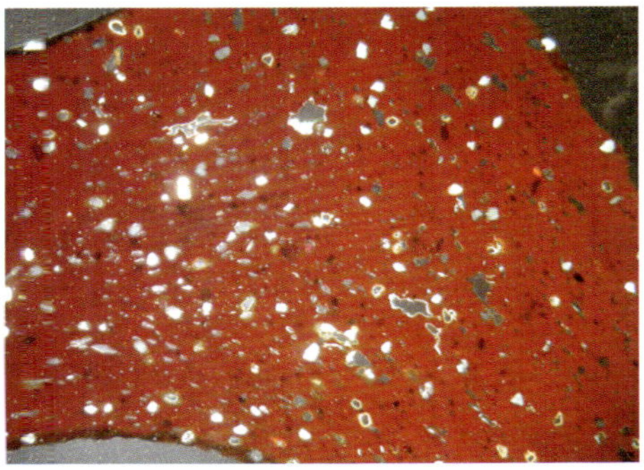

A. Hand specimen: Scale 10 mm. Hard, fine red matrix (10R 6/8) with overfired grey surfaces and frequent inclusions. Decomposed carbonate inclusions/voids with white haloes apparent in hand specimen.

B. Petrography: x10, xpl, width of field ca. 12.0 mm. Predominant quartz inclusions with common carbonate inclusions/voids. Rare biotite and very rare plagioclase feldspar. Vughs are frequent, probably a result of slight overfiring.

35. Amp 266-05 Metapontine Greco-Italic amphora rim
Dimensions: rim diam. 13.0 h 2.5
Description: Rim with overfired surfaces, similar to **No. 34**, with slightly thinner projecting rim.
Fabric: Sample AMP044. Typical Metapontine amphora fabric.

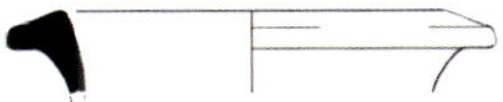

1:4

A

B

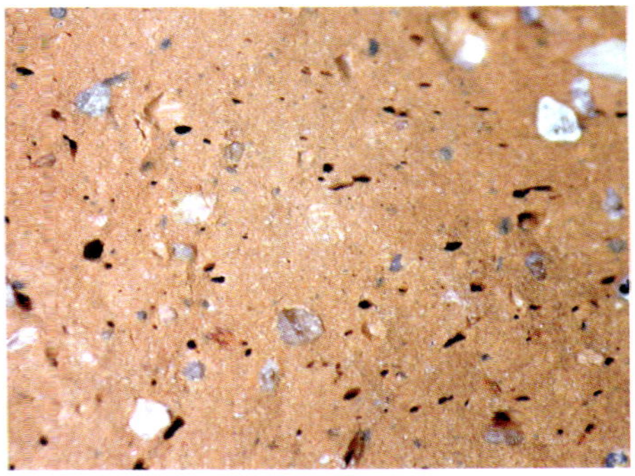

A. Hand specimen: Hard, fine red matrix (10R 6/6) with semi-conchoidal fracture.

B. Microscopy: x20 magnification, width of field ca. 7.5 mm, showing range of angular crystalline inclusions, most likely quartz, as seen petrographically in **No. 34**. Rounded to ovoid voids are frequent in this fine-grained fabric, probably the result of slight overfiring as with **No. 34**.

Metapontine Greco-Italic amphorae cont.

1:4

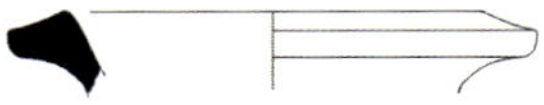

36. Amp 153-01 Metapontine Greco-Italic amphora rim
Dimensions: rim diam. 14.0 h 2.5
Description: Rim with pronounced concave upper face similar to **No. 37** and relatively thin projecting rim similar to **No. 35.** Regularly fired; an amphora waster **(No. 59)** comes from the same site.
Fabric: Sample AMP069. Typical Metapontine amphora fabric.

A. Hand specimen: Scale 10 mm. Hard, consistently fired, relatively fine red matrix (10R 5/8) containing frequent fine translucent inclusions, most < 0.2 mm, with common lenticular voids, most < 0.7 mm. The fine-grained texture of the matrix tends to result in a clean, semi-conchoidal fracture. (White surfaces visible in the photograph are post-depositional carbonate accretions.)

B. Microscopy: x20 magnification, width of field ca. 7.0 mm. The frequent angular inclusions with abundant fine lenticular voids are visible microscopically.

C. Microscopy: x30 magnification, width of field ca. 2.0 mm.

D. Petrography: Xpl, x20 magnification, width of field ca. 7.0 mm. Quartz and carbonate inclusions can be identified in thin section. The inclusions are relatively well sorted and indicate tempering with fine sand.

E. Petrography: Xpl, x40 magnification, width of field ca. 3.5 mm. Detail showing quartz inclusions and carbonate voids (with yellowish haloes). Occasional patches of iron-rich clay concentration features are apparent in xPl. These may have been present in the clay source, but they might also be consistent with levigation of the source sediments (see discussion of amphora production at site 266, below).

A

B

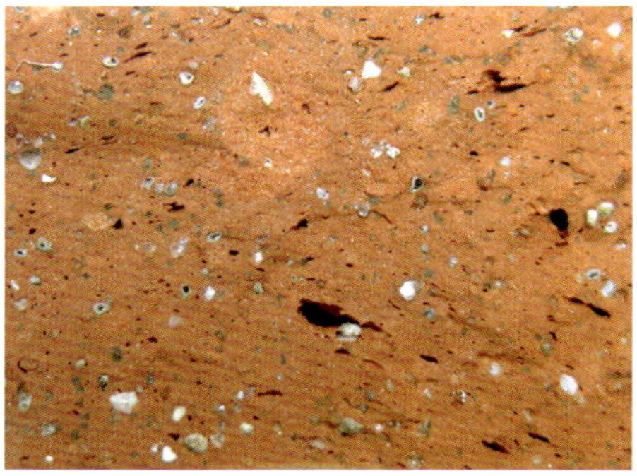

C

D

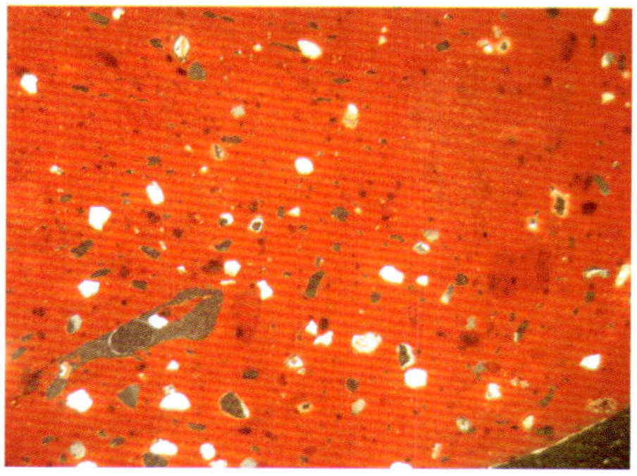

E

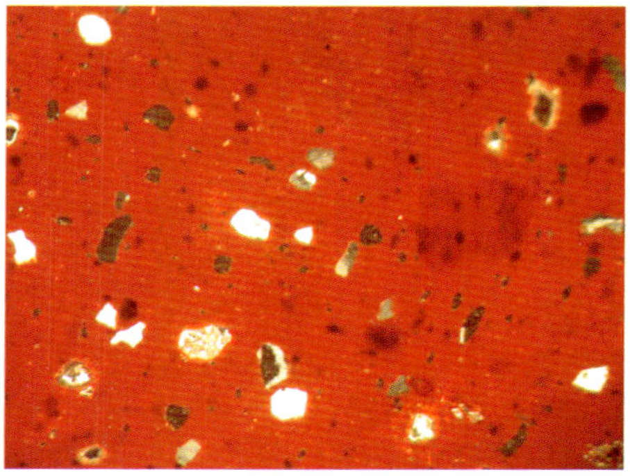

Metapontine Greco-Italic amphorae cont.

1:4

37. Amp 028-01 (*) Metapontine Greco-Italic rim
Dimensions: rim diam. 16.0 h 5.0
Description: Rim, neck, and handle terminus. Triangular rim with pronounced concavity on the upper face and rounded lower part. Handle is ovoid in section where attached to rim and projects from the neck and is angled slightly upwards towards the line of the rim.
Fabric: Metapontine, falling within range of **Nos. 34–36.**

1:4

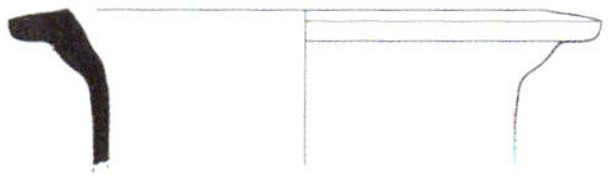

38. Amp 128-01 Metapontine Greco-Italic rim
Dimensions: rim diam. 16.0 h 4.5
Description: Rim and neck, relatively gracile projecting rim, distinctly flared with thickened upper neck, representing a variant of the more usual Metapontine Greco-Italic rim forms as show in **Nos. 34–36.**
Fabric: Metapontine, falling within range of **Nos. 34–36.**

1:4

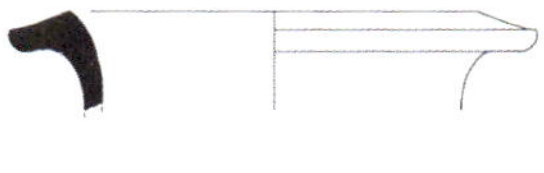

39. Amp 193-02 Metapontine Greco-Italic rim
Dimensions: rim diam. 14.0 h 2.5
Description: Rim and upper neck, slightly concave upper face of the projecting rim with sharply-formed outer edge and rounded lower projection.
Fabric: As **Nos. 32–34.**

Extra-local and extra-regional Greco-Italic amphorae

1:4

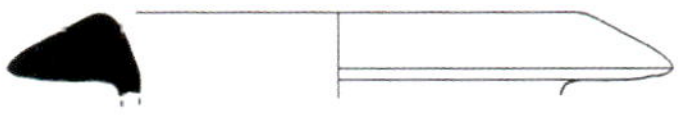

40. Amp 285-02 Greco-Italic rim (southern Italian?)
Dimensions: rim diam. 17.5 h 2.5
Description: Rim with pronounced triangular profile and sharply-defined outer edge, distinct from the Metapontine types, though bearing some similarity in basic form to that of **No. 30.**
Fabric: Sample AMP010. Area of origin unknown, may be southern Italian, but is not typical of fabrics produced locally.

A

B

A. Hand specimen: Finely granular fabric fired weak red to red (10R 5/3–5/6). Hand specimen is noticeably micaceous, particularly apparent on the surfaces, with muscovite flakes and biotite.

B. Petrography: Rim in thin section, xpl, x20 magnification, width of field ca. 7.5 mm. Petrography shows abundant very fine grains of quartz and biotite, frequent muscovite, and rare plagioclase feldspar inclusions. Quartz, plagioclase, and biotite are commonly occurring minerals that are not particularly diagnostic for attributing geographical or geological origin. This fabric may relate to some of the type B amphora fabrics with muscovite but is lacking the coarser fraction of inclusions.

Extra-local and extra-regional Greco-Italic amphorae cont.

1:4

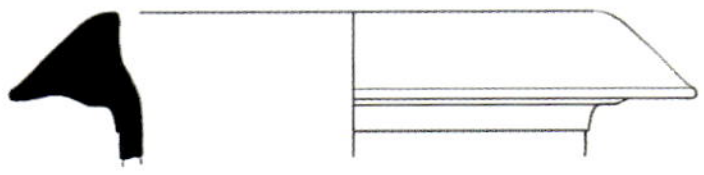

41. 154-03 Greco-Italic amphora rim (Campanian?)
Dimensions: rim diam. 18.0 h 4.0
Description: Rim of the Vandermersch type MGS VI. Near-vertical inner lip with concave inner face, presumably for a stopper. Sharply-sloping triangular projecting rim with pronounced lip on outer edge and tool-formed projection on the underside; see **No. 31** for the rim of a Metapontine amphora which shares this feature, though the profile of the rim is distinct from Metapontine rim forms. Slight step on the upper neck, produced from trimming of the lower part of the neck, a forming technique not attested by Metapontine amphorae.
Fabric: Sample AMP083.

A

A. Hand specimen: Finely granular fabric with abundant fine inclusions < 0.2 mm, fired to light red (2.5YR 6/6). The surfaces are salined, fired white, indicating the use of salt water for clay preparation and suggesting coastal manufacture using clay with a calcium carbonate content in excess of 10–15% (Peacock and Fulford 1984, 263).

B

B. Microscopy: x10 magnification, width of field 15.0 mm. Volcanic "black sand" inclusion evident (off-center), either volcanic rock or amphibole, both of which are evident in thin section.

C

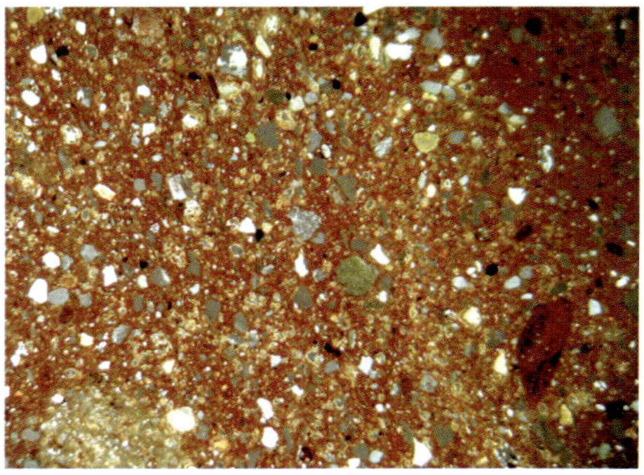

C. Petrography: xpl, x20 magnification, width of field ca. 7.5 mm. Along with plagioclase feldspar, rare amphibole inclusions and the presence of clinopyroxene, usually within the compositional range of augite (MacKenzie and Adams 1994, 36), may indicate a Campanian origin in the southern Tyrrhenian volcanic tract (Bay of Naples), rather than Sicily (Keay and Williams 2005; following Peacock and Williams 1986). Welded (presumably volcanic) rock fragments, with flow structure, are also present in the fabric, but their composition cannot be determined.

1:4

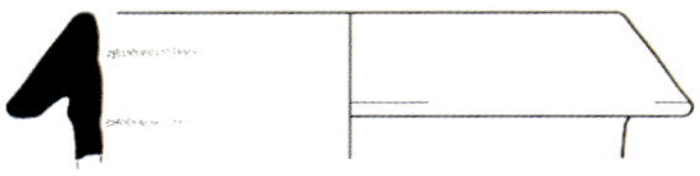

42. Amp 030-14 (*) Greco-Italic rim (Campanian?)
Dimensions: rim diam. 18.0 h 4.0
Description: Vandermersch type MGS VI amphora rim.
Fabric: As **No. 41**, though slightly finer-grained inclusions.

Metapontine amphora toes

1:4

43. Amp 269-04 Metapontine amphora toe
Dimensions: b 2.5 h 4.0
Description: Typical amphora toe in Metapontine fabric. Hollow toe with wheel-thrown and tool-formed button on the underside. This is the usual type of Metapontine toe, the range of which is shown by **Nos. 42–45.**
Fabric: Sample AMP021. As Nos. **34–36**.

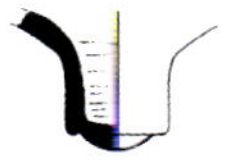

1:4

44. Amp 269-03 Metapontine amphora toe
Dimensions: b 2.5 h 3.5
Description: Typical amphora toe in Metapontine fabric. Hollow toe with wheel-thrown and tool-formed button on the underside. This is the usual type of Metapontine toe, the range of which is shown by **Nos. 43–45**.
Fabric: Sample AMP020. As **Nos. 34–36**.

1:4

45. Amp 057-03 Metapontine amphora toe
Dimensions: b 2.5 h 2.5
Description: Typical amphora toe in Metapontine fabric. Hollow toe with wheel-thrown and tool-formed button on the underside. This is the usual type of Metapontine toe, the range of which is shown by **Nos. 43–45.**
Fabric: Sample AMP024. As **Nos. 34–36**.

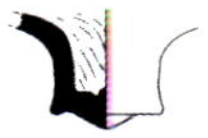

1:4

46. Amp 025-01 (*) Metapontine amphora toe
Dimensions: b 3.5 h 5.0
Description: Variant toe form in Metapontine fabric, with distinct tool-formed ridge around the base and a slightly larger button than is usual. It may represent a precursor to **Nos. 43–45**, and a development from earlier (type B?) toes such **No. 47**.
Fabric: As **Nos. 34–36**.

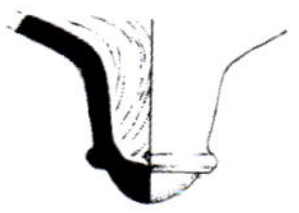

1:4

47. Amp 042-01 Metapontine amphora toe
Dimensions: b 4.0 h 4.5
Description: Amphora toe in Metapontine fabric. Presumably a precursor to the more typical Metapontine toes associated with Metapontine Greco-Italic and Metapontine type amphorae (**Nos. 43–46**). The form may have developed from the typical 'peg' toes of type B amphorae (Koehler reference), as part of a regional development towards the distinctively Metapontine amphora toes associated with the Metapontine Greco-Italics; this example may have belonged to a Metapontine type B amphora.
Fabric: Typical Metapontine fabric as **Nos. 34–36**.

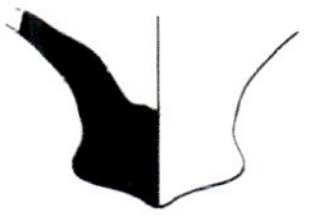

Late Hellenistic/Republican Roman Period

1:4

Metapontine type amphorae

48. Amp 030-01 Metapontine type amphora rim
Dimensions: rim diam. 18.0 h 4.5
Description: Intermediate Metapontine type amphora rim, showing development from the triangular rims of the Metapontine Greco-Italic amphorae (**Nos. 30–39**). See Figure 12.2 for the development and more complete examples from the Pantanello kiln deposit.
Fabric: Typical Metapontine fabric as **Nos. 34–36**.

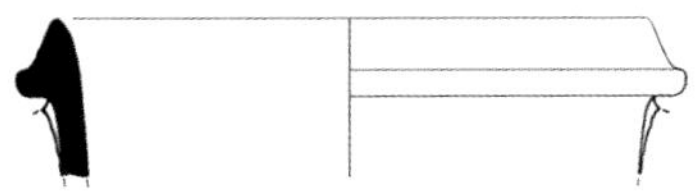

1:4

49. Amp 042-02 Metapontine type amphora rim
Dimensions: rim diam. 14.0 h 3.5
Description: Near-vertical collar with angular, tool-turned or -trimmed outer profile. This falls within the typical range of Metapontine type amphorae, of which there are many examples from the Late Republican kiln deposit at Pantanello (*Fig. 12.2*).
Fabric: Typical Metapontine fabric as **Nos. 34–36**.

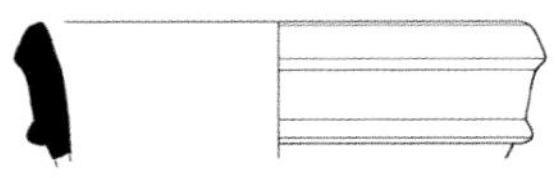

1:4

50. Amp 339-01 Variant Metapontine type amphora rim
Dimensions: rim diam. 17.0 h 4.0
Description: Out-turned rim with lower projection from the line of the wall, forming a collar. This presumably represents a further development of the collared Metapontine type amphorae (*Fig. 12.2*), similar to **No. 49**, and which are much more common than this variant.
Fabric: Sample AMP029. Typical Metapontine fabric as **Nos. 34–36**.

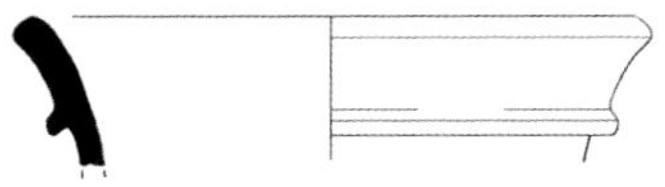

Dressel 1C Amphora

1:4

51. Amp 042-03 Dressel 1C amphora
Dimensions: rim diam. 16.5 h 8.5
Description: Collared rim of Dressel 1C amphora. From the same site as Metapontine type amphora rim **No. 50**.
Fabric: Pale, yellow fabric, not of the Campanian type.

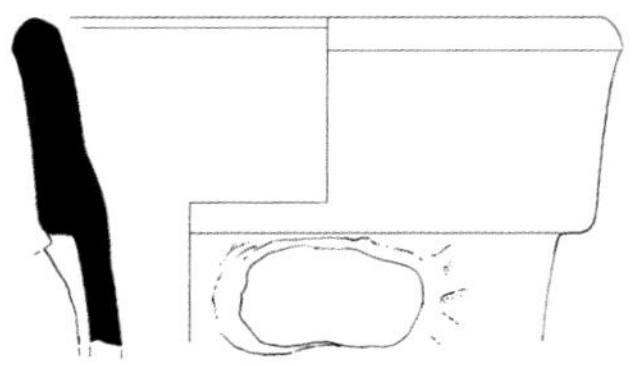

Amphora Production in the Metapontine Chora

This part of the catalog presents the evidence from two production sites in the Metapontine chora (site 266 and site 153). The pieces generally consist of over-fired fabrics comprising parts of forms which in themselves are not sufficiently diagnostic for inclusion in the main amphora catalog. For the Metapontine type B and Metapontine Greco-Italic amphora rims from site 266 see also **Nos. 28, 34,** and **35**, and for the Metapontine Greco-Italic amphora rim from site 153 see **No. 36**.

Amphora production at site 266

1:4

52. Amp 266-09 Amphora rim and handle waster
Dimensions: diam. > 14? h 7.5
Description: Amphora rim waster, strongly deformed. Bloated during firing and consequently thicker than the usual Metapontine amphorae. The handle terminus is ovoid in section, and the general articulation appears similar to Metapontine "Greco-Italic" amphorae from other sites in the survey, with rim forms similar to some of those from site 266.
Fabric: Sample AMP048. Waster, reduced semi-vitrified matrix with numerous voids and vacuoles.

A

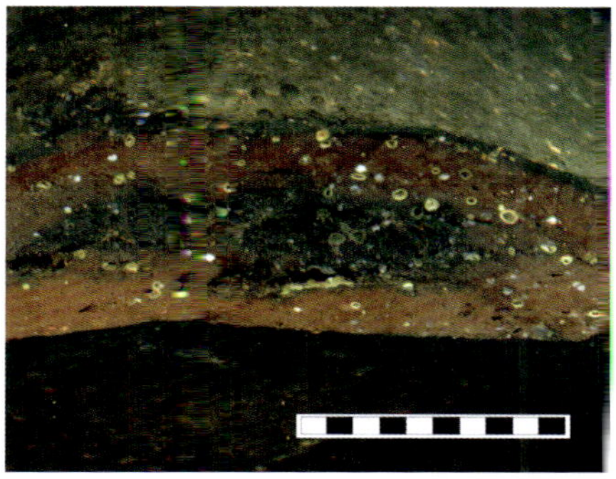

B

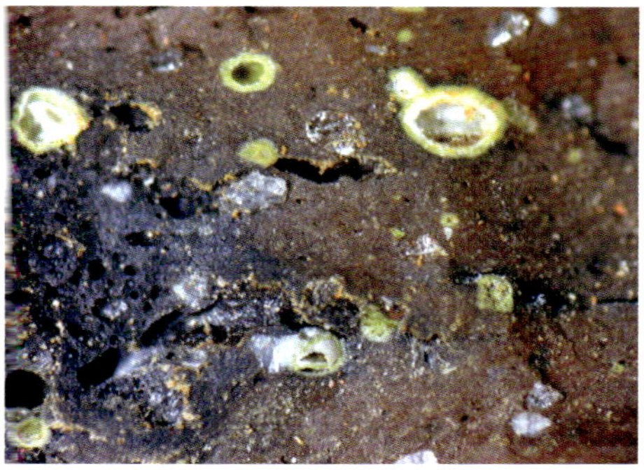

A. Hand specimen: Cross section of neck beneath handle, scale 10 mm, showing overfired surfaces and relatively fine-grained, over-sintered matrix. The well-rounded white inclusions were calcium carbonates.

B. Microscopy: Amphora waster fabric, x30 magnification, width of field ca. 2.5 mm. The over-sintered matrix, quartz grains, and erupted/decomposed calcium carbonate inclusions (rounded voids with yellowish white reaction haloes) are clearly visible microscopically. Carbonate inclusions appear to be more frequent in this fabric than is usual for Metapontine amphorae.

53. Amp 266-03 Amphora body sherd waster
Dimensions: h 4.5
Description: Metapontine fabric, misfired, with overfired surfaces, oxidized margins and reduced core.
Fabric: Metapontine, misfired.

1:4

A

B

A. Hand specimen: Misfired body sherd showing range of firing colors from red (10R 6/6) to very dark gray (2.5YR N3/). The gray core may result from carbonaceous material present in the clays and be consistent with the use of levigated surface sediments, but this is not certain given that this is an overfire.

B. Petrography: xpl, x20 magnification, width of field ca. 7.5 mm. A range of relatively fresh, angular inclusions similar to the type B and Greco-Italic amphorae from site 266 is apparent. Note the higher frequency of fine elongated voids in the grayish core—these may be the result of the incomplete burning off of carbonaceous material in the source clay.

Amphora production at site 266 cont.

1:4

54. Amp 266-19 Metapontine type B amphora handle
Dimensions: h 4.5
Description: Pinched, convex (figure-eight) neck, presumably associated with Metapontine type B amphora forms. For a rim waster of this type from site 266 see **No. 28**.
Fabric: Metapontine, falling within the range of **Nos. 34** and **35**, also from site 266.

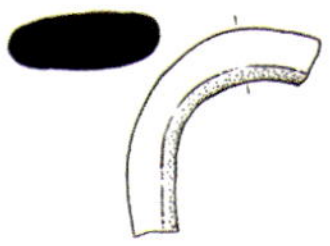

1:4

55. Amp 266-16 Metapontine amphora handle
Dimensions: h 5.5
Description: Upper part of strap handle.
Fabric: Metapontine, falling within the range of **Nos. 34** and **35**, also from site 266.

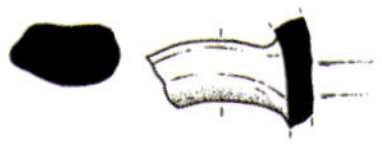

1:4

56. Amp 266-13 Metapontine amphora handle
Dimensions: h 3.0
Description: Amphora handle and terminus with the neck.
Fabric: Metapontine, falling within the range of **Nos. 34** and **35**, also from site 266.

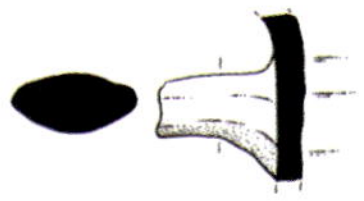

1:4

57. Amp 266-14 Metapontine amphora handle
Dimensions: h 4.0
Description: Upper amphora handle and terminus with neck. The length of the extant wall above the handle suggests a Metapontine Greco-Italic or a Metapontine type amphora rather than a Metapontine type B.

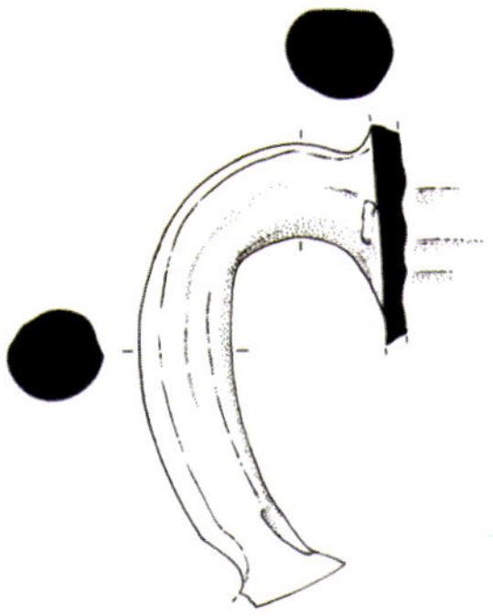

1:4

58. Amp 266-06 Metapontine amphora handle
Dimensions: h 14.0
Description: The articulation of the handle and circular section perhaps suggests inspiration from Corinthian type A' amphorae, particularly in the pinched upper terminus with the neck; in this respect, note the presence of a Late Corinthian type A' amphora rim from the same site (**No. 24**).
Fabric: Metapontine, falling within the range of **Nos. 34** and **35**, also from site 266.

Amphora production at site 153

1:4

A

59. Amp 153-02 Amphora shoulder waster
Description: Carinated amphora shoulder.
Fabric: Sample AMP070. Metapontine, overfired.

A. Hand specimen: Semi-vitrified matrix characterized by frequent voids from the decomposition of calcite or carbonate inclusions.

13
Terracotta Objects

Rebecca Miller Ammerman

The Metaponto field survey recovered 342 terracotta figurines, relief plaques, molds, and other molded objects from 73 different sites between the Bradano and Basento Rivers (*Fig. 13.1*).[1] All of these terracotta objects are preserved in an extremely fragmentary state, a circumstance to be expected in the collection of materials from agricultural land that is regularly plowed. As is characteristic of such surface collections, the majority of fragments indicate little more than the fact that they once belonged to an object that had been cast from a mold. A total of 160 fragments are either too small or not sufficiently distinctive in preserved detail to allow either dating or identification of the subject of the terracotta. Another 59 fragments may be recognized as the edge of a relief plaque or a portion of the drapery of a figurine, but again lack sufficient detail for dating or classification within a general typology.

Of the remaining 123 fragments, at least 90% may be dated and assigned to a given format, class, and subject of terracotta with a fair degree of certainty.[2] More than 70% can even be identified as belonging to a specific mold series or typology proposed in previous studies for a given group of molded terracottas. A "mold series" refers to a network of mechanically related molds and casts, often spanning several generations and including numerous variants, that derive from a single archetype or original model.[3] In this way, the original subject matter and general appearance of the complete terracotta to which the isolated fragment once belonged may frequently be ascertained.

Once identified, the fragment may then take its place within a larger network of relationships between such terracottas. This network reflects not only technological histories of manufacture and distribution of terracottas, but also patterns of ritual behavior and belief. Most of these terracottas were produced as visual expressions of religious concepts, often serving as votive offerings in both private and public settings.[4] Thus the small worn fragments of molded terracottas recovered from the plow zone in the chora of Metaponto, although perhaps unpromising at first sight, with patient study have much to tell about ancient life in the countryside of this Achaian city-state.

The accompanying catalog is organized in ascending order by site identification number. The terracottas are listed in chronological order for each site. This arrangement preserves the association between artifacts at each site. Entries for two molds and a molded attachment for a ceramic vessel or utilitarian object

[1] I would like to thank Professor Joseph C. Carter for his invitation to study this material, and Mary Malone for introducing me to the terracottas from the University of Texas excavations at Metaponto. I am also indebted to Dottor Antonio De Siena and the rest of the staff of the National Archaeological Museum at Metaponto for their gracious assistance during my visits to Metaponto. Special thanks are similarly due to various members of the ICA team who facilitated my work at various stages. These include not only those individuals on site at Metaponto—Rosetta Torraco and her family, Eloisa Vittoria, Cesare Raho, and Giambattista Sassi—but also those based at ICA at the University of Texas at Austin—Carol Cook, Chris Williams, Alberto Prieto, and Steve Thompson. I would finally like to thank three special students who assisted me at Metaponto in the summers of 2002 and 2003—Bryan Kraemer, Danielle Drimer, and Tara McCann.

[2] The dating of a terracotta is based primarily on stylistic criteria and, when possible, similarity to other terracottas from more closely dated archaeological contexts. Each terracotta presents a general format, such as figurine, relief plaque, bust, model of a fruit or animal, or utilitarian object. Terracottas of a given format may be further subdivided according to different groups that will be called classes. A class is usually determined by the subject of a terracotta. Figurines, for instance, may be divided into such classes as "Enthroned Goddess Figure" and "Standing Female Figure," while relief plaques may be assigned to such classes as "Reclining Banqueter" and "Satyr, Silenos, or Pan with Female Figure."

[3] Nicholls 1952; R. Ammerman 2002, 12–13. Often a given class of terracottas comprises several mold series that represent the same general subject matter, such as an enthroned goddess, but with differences of style, dress, attributes, and so on that distinguish one mold series from another. Typologies, based upon technically related groups of molded terracottas that derive from a common archetype, have been developed at Metaponto for the production of Archaic figurines from San Biagio alla Venella (Olbrich 1979, 18–25), of relief plaques portraying a reclining banqueter from the kerameikos (Signore 1996, 303–4, 348–49), and of relief plaques portraying a satyr, silenos, or Pan with a female figure from temple E (Postrioti 1996, 19–20). Postrioti, however, has not been able to follow technological criteria as strictly as others in developing a typology. Whenever a terracotta from the survey may be assigned to such a typology, references to appropriate comparanda are given in the catalog entry.

[4] For introductory discussion and additional bibliography on the ritual use of terracottas in sanctuary, funerary, and domestic contexts, see R. Ammerman 1990.

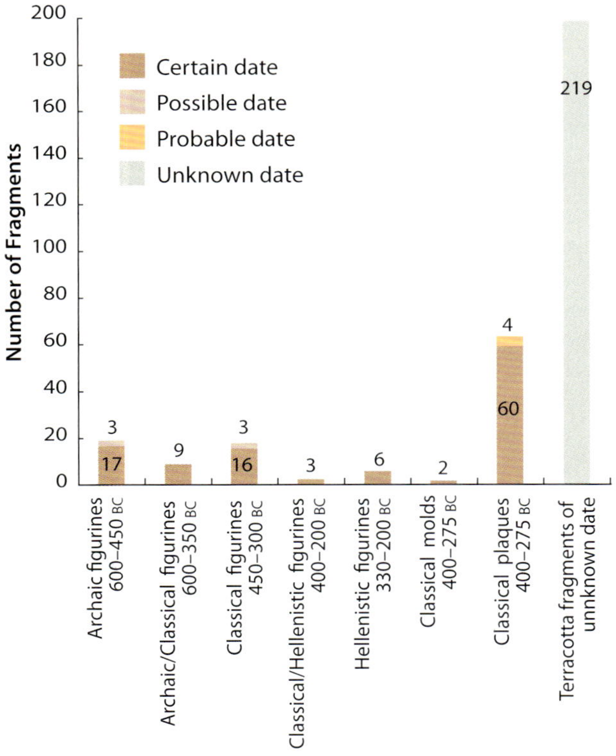

Figure 13.1 Chronological distribution of terracotta fragments.

are at the end of the catalog, along with an entry by Erminia Lapadula for a tile with molded decoration.

Technique, Style, and Imagery of Terracottas

It is useful to look beyond the group of terracottas found at an individual site to the entire collection of recovered terracottas as a whole. Of primary concern here will be the imagery presented by the terracottas in different periods. Stylistic traits as well as any unusual technological details will also be noted. The terracottas have been divided into three broad chronological groups—Archaic, Classical, and Hellenistic—based primarily upon stylistic considerations. The material will be discussed in chronological order according to the different subjects represented by the different classes of terracottas. In closing, the focus will shift to what the individual contexts of the terracottas found in the survey may reveal about ancient life in the chora of Metaponto.

The terracottas range in date from the second quarter of the 6th century to the 3rd century BC (*Fig. 13.1*).[5] A significant number of figurines belong to the 6th century BC. A majority of the terracottas are relief plaques dating to the later 4th and early 3rd centuries BC. Only a few terracottas may date as late as the second half of the 3rd century BC. This evidence thus coincides well with what the other survey and excavation materials indicate about the occupation of the chora of Metaponto. The consistent and widespread presence of Archaic settlement in the countryside, beginning towards the middle of the 6th century BC, appears to wane during the early decades of the 5th century BC, only to increase dramatically in the second half of the 4th century BC and then sink into a severe decline in the second half of the 3rd century BC.[6]

The terracotta figurines and relief plaques collected in the survey provide, moreover, a remarkably representative sample of the products of the coroplastic industry at Metaponto from the Archaic to Hellenistic periods. For the most part, the terracottas were made for use in ritual, either within the public setting of a sanctuary or within a private domestic context. They accordingly portray such subjects as divinities or cult participants. As is the case with votive terracotta figurines throughout the Greek world, female figures tend to be more prevalent than male. The relief plaques, on the other hand, present a narrative image of interacting characters, usually a male and a female figure.

Archaic terracottas (6th to early 5th century BC)

Figurines: seated and standing female figures

The oldest terracottas yet found at Metaponto were recovered from the urban sanctuary and from the rural spring sanctuary of San Biagio alla Venella.[7] They are small molded plaques representing a female figure of Daedalic style, and they date to the second half of the 7th century BC. Although the survey did not recover any terracotta of such an early date,[8] the

[5] There is one possible exception, a fragment from site 772 (see final catalog entry written by Lapadula).

[6] Carter 2001; 1985a, 480–82; 1981, 172; Carter and D'Annibale 1985, 151–53; Carter et al. 1985, 281, 298.

[7] From the urban sanctuary: Barberis 2004, 43–48, 203–5, Figs. 1–9, and for hand-modeled figurines: 125–27, 242–43, Figs. 218–23; Adamesteanu 1980, 48–51, Figs. 31–33 (from the area of *Sacellum C*, a small structure between temple A and temple B that is labeled *sacello* in Fig. 184), 137–38, Fig. 142a–f (from the area of temple B), 171–78, Figs. 185–89 (from the area of temple C). From the trench for enlargement of the east end of temple B and from the west side of temples A and B, respectively: De Siena 1980, 94, 96–97, Pl. 7, Nos. 45 and 54. From the area north of temple E: A. Bottini 1990, 565. From San Biagio: Olbrich 1979, 100–02, 194, Pl. 1 A1–A4, Pl. 47 C1.

[8] A Daedalic female head of the late 7th c. BC was recovered by the Soprintendenza per i Beni Archeologici per la Basilicata as an isolated find in the Saldone locality: Carter 1994a, 171, 178, Fig. 7, No. 4; 1998a, 253. A plaque representing a male and female figure in Daedalic style was recently found near Incoronata/San Teodoro at the Museo Archeologico Nazionale di Metaponto's site 68 along the track of the SNAM pipeline: Nava 2003a, 671, Pl. 36.2.

oldest terracottas that were found, **Nos. 11, 162**, and **210–11**, dating to the second quarter and middle of the 6th century BC, reflect the persistent influence of Daedalic workshop practices on Early Archaic terracottas. The figurines are still of small size and made from a solid plaque of clay. The human form is conceived as a flat graphic figure within a single plane rather than as a fully three-dimensional volume. The preferred subject is that of a standing or seated female figure wearing a polos and a simple belted peplos, sometimes with an overfold (*apoptygma*).

The hallmark of Metapontine terracotta production in the 6th century BC is a fairly large-scale figurine of a sometimes winged female figure who wears a tall, flaring polos and a peplos with *apoptygma* that is fastened by disk fibulae at her shoulders. She frequently carries an animal, such as a quadruped or fantastic sea creature.[9] These flamboyant terracottas are merely an elaborated extension of the coroplastic tradition of casting solid plaquelike figures from a frontal mold. A solidly cast plaque comprises the head and torso of the figure; this portion was subsequently attached either to a wheel-turned cylinder of clay that serves as the skirt of a standing figure, or to a folded slab of clay that represents the skirt and throne of a seated figure. Ornamental details—floral buds embellishing the polos, disk fibulae, wings, animals, and so on—are all hand-modeled additions to the plaques. The figurines belong to a large group of related mold series used to manufacture votive terracottas portraying a female deity at most Achaian city-states in southern Italy.[10] The figurines are present among the 6th-century BC dedications at the urban sanctuary at Metaponto, in the extramural votive deposits at Crucinia and Favale, and at the extra-urban sanctuaries at Tavole Palatine, San Biagio alla Venella, Pantanello, and Incoronata, and have been found even as far inland as Pomarico Vecchio.[11] Although the figurines have been interpreted at San Biagio, as well as elsewhere within the territory of Metaponto, as representations of Artemis,[12] they served as appropriate votive offerings to a range of deities,[13] and should thus be interpreted according to the individual context where each figurine was found. The survey recovered only a few fragments of this popular kind of votive figurine. These include a torso (**No. 163**) and a few hand-modeled additions of a disk, wing, and polos (**Nos. 234, 245**, and **232**).

By the middle of the 6th century BC, coroplastic workshops began to move away from the practice of filling a frontal mold entirely with clay to form a solid plaque. A thinner layer of clay instead is gently pressed into the concave curve of a deeper mold to produce a hollow figurine whose back is left open. Three female heads from the survey reflect this tendency toward a greater three-dimensionality of form and the production of hollow, open-backed figurines over the course of the second half of the 6th century BC. For example, two female heads (**Nos. 230–31**) are more volumetric than those of the earlier plaques, and their facial traits are fully Archaic in style. In another female head (**No. 236**) of the late 6th century BC, the shift to a three-dimensionally conceived figurine is almost complete: in terms of volume, the hollow head is deeper from back to front than it is broad from side to side. Other terracottas reflect the same developments with regard to the body of a figure. For example, **No. 233** still recalls the figurines of an enthroned goddess, described above, whose head and torso are a molded plaque attached to a curved slab of clay that serves as skirt and throne. This terracotta continues to portray a seated goddess whose peplos is fastened at her shoulders by disk fibulae, but the torso is cast from a deeper mold, hollow, and open at the back. It is rendered more three dimensionally than the flat, solid torsos of the contemporaneous but more widespread Achaian figurines.

This trend of increased three-dimensionality

[9] For a few examples from San Biagio alla Venella, see Olbrich 1979, 157–58, Pls. 29–30 A122–A124.

[10] For bibliography regarding terracottas belonging to this network of mold series from Kaulonia, Croton, Francavilla Marittima, Sybaris, Policoro, Taranto, and Satyrion on the Ionian coast and from Palinuro and Poseidonia-Paestum on the Tyrrhenian coast, see R. Ammerman 2002, 46–47 nn. 11–19; Barberis 2004, 81–82.

[11] For bibliography, see R. Ammerman 2002, 46 n. 10; Barberis 1995, n. 38, and 2004, 80–82, nn. 339–46. For an example from temple D: Adamesteanu/Mertens/De Siena 1975, 43, Fig. 78. For Favale Liseno 2004, 39–40, Pls. 6c, 7a–c. For recent recovery of such figurines from the sanctuary at Favale: Nava 2002, 732, Pl. 65.1. For others from a spring sanctuary (site 57 along the track of the SNAM pipeline) near Incoronata/San Teodoro, south of the Basento: Nava 2003a, 672–73, Pl. 37.2. For this type of terracotta from farmhouses in the chora: Barberis 1995, 18 n. 29, Fig. 1. Sporadic finds also were recovered at Masseria Avenia/Galotta on the right bank of the Basento near Pisticci: Sestieri 1940, 120–22, Fig. 53; Lo Porto 1983–1989, 303–7, Figs. 3–4. The presence of terracotta votives here has led some to postulate the presence of a rural sanctuary at Avenia: Adamesteanu 1967a, 25; Edlund 1987, 100–01; Carter 1994a, 176–77. The strength of the evidence for such an interpretation has been called into question by De Siena (2001a, 764), who elsewhere (1999a, 63) suggests, however, that there might have been a sanctuary at Avenia dedicated to Athena.

[12] Olbrich 1979, 70–86, 97–98. For additional discussion and bibliography, see Barberis 1995, 18–20; 2004, 172–76.

[13] For different deities to whom figurines were dedicated at different sites, see R. Ammerman 2002, 47.

continues in the portrayal of enthroned female deities throughout the first half of the 5th century BC. Fragmentary terracottas, **Nos. 243–44** and **138**, belong to an enthroned goddess (the first two fragments from site 477 belong possibly even to the same figurine) of Late Archaic stylistic inspiration. Although the figurines should be dated as late as the second quarter to the middle of the 5th century BC because of certain elements of Severe or Early Classical style that may be detected in better-preserved figurines belonging to this and other closely related mold series, the hieratic pose, costume, and ornamental nature of the drapery of the goddess, as seen in the omega fold of the apoptygma and the finely detailed folds of the underlying chiton in **No. 244**, hark back to a Late Archaic model. More complete casts belonging to this group of related mold series have been found at Taranto and Herakleia,[14] attesting to the popular nature of this terracotta image among Italiote city-states on the Ionian coast. Terracotta figurines closely related to the older 6th-century BC mold series of an enthroned goddess, **No. 231**, have also been found at Taranto.[15] This suggests that workshop practices at Taranto may have been responsible for the introduction to Metaponto of these more three-dimensional mold series, which eventually supplanted the popular terracottas made from solid plaques.

Protomai: female figures

In this context a small Archaic protome, **No. 229**, should be mentioned. The terracotta portrays a goddess wearing a polos and exhibits the same shift to the production of more three-dimensional terracottas in the second half of the 6th century BC as was noted above. It also attests the possible sharing of mold series and workshop practices between Metaponto and Herakleia, where an identical protome was found.[16]

Statuettes: standing female figure

Perhaps the most interesting Archaic terracotta recovered by the survey is a relatively large-scale statuette of what was probably a standing female figure, **No. 235**. The two joining fragments that belong to the statuette appear to preserve the pointed end of the woman's himation as it falls from her right arm along her side. The smooth, flattened forms that describe the woman's garments and the simplicity of their portrayal recall the figures represented in a procession on architectural terracotta revetments dating to the beginning of the 6th century BC, found at temple C in the urban center at Metaponto.[17] The statuette represents the efforts of local coroplasts to give three-dimensional expression to the human form on a large scale. Together with the architectural reliefs, it suggests that in the first half of the 6th century BC Metapontine craftsmen were beginning to develop a distinctive stylistic language of their own.

Summary

Regardless of style or manufacturing technique, all of the Archaic terracottas recovered by the survey portray female figures that are intended as representations of either a goddess or a worshipper. A polos (**Nos. 230, 232, 229, 72**) or a hieratically enthroned posture (**Nos. 211, 231**) is usually seen as a marker of divine status. A standing female lacking a distinctive headdress or attribute (**No. 210**) is best identified as a mortal votary. A more specific identity for any of these female figures cannot be proposed without detailed knowledge of the terracotta's original findspot. Even when such a figurine is recovered from a well-excavated context, identification of the divine personality may remain elusive. Terracotta figurines belonging to the same mold series were dedicated at most of the Archaic sanctuaries at Metaponto.[18] It is precisely their generic imagery that made the figurines excellent multipurpose ex-votos. Such a terracotta could serve equally well as an offering to a wide range of female deities. The same figurine, for instance, may have been intended by a dedicant as a representation of Artemis at the sanctuary at San Biagio alla Venella, but as a representation of Hera at

[14] For a figurine from Herakleia belonging to the same mold series as No. 244, see Neutsch 1981, 166, Pl. 27, No. 3. For similar figurines from Taranto: Iacobone 1988, 45, Pl. 35c; Higgins 1967, 90, Pl. 40b = Mollard-Besques 1963, 67, 134, Pl. 15, No. 3 = Mollard-Besques 1954, 123, Pl. 88 C247; Winter 1903, 110, No. 7; 122, No. 6; 123, No. 7. Photographs of two fragments of similar figurines, presumably from Favale at Metaponto, are published without discussion by Liseno 2004, Pls. 21a, 22b.

[15] Mendel 1908, 474, Pl. 14, No. 3; Breitenstein 1941, 23–24, Pl. 22, Nos. 227–28; Mollard-Besques 1954, 59, Pl. 41 B386; Iacobone 1988, 40, Pl. 31c.

[16] Rolley 1981,187, Pl. 38, No. 2.

[17] Adamesteanu 1980, 112–14, Figs. 112–13; Mertens-Horn 1992, 46–59, 110–17, Figs. 27–35 and 46, Pls. 8–15. Architectural friezes from the same or closely related mold series found at Francavilla Marittima and Policoro attest to close links between craftsmen at these Achaian city-states ca. 600 BC. For larger terracotta sculptures of Archaic korai from the urban sanctuary at Metaponto, see Adamesteanu/Mertens/De Siena 1975, 28–29, Figs. 12–17; Barberis 2004, 66, Fig. 58.

[18] De Siena 2001a, 762.

the sanctuary at Tavole Palatine. An image was thus probably endowed with specific meaning according to the intentions and needs of a dedicant on a case-by-case basis.

Classical terracottas (mid-5th to 4th century BC)
Figurines: enthroned goddess

The independent, creative spirit observed in several of the Archaic terracottas seems, for the most part, to be absent in coroplastic workshops at Metaponto during the 5th century BC. As noted above, figurines of an enthroned goddess (**Nos. 243–44** and **138**) that depend upon Late Archaic models but betray the influence of Early Classical or Severe style belong to a group of related mold series that are probably the products of Tarantine workshops. On the basis of stylistic considerations, the figurines should date to the second quarter and middle of the 5th century BC.[19]

Figurines: standing female figure with offerings

The most popular subject for figurines that date on stylistic grounds from the middle to the end of the 5th century BC at Metaponto is that of a standing woman holding such offerings as a sheep, goat, bird, deer, or basket of fruit and an oinochoe.[20] These terracottas also appear to derive from mold series originally fashioned in the 5th century BC at Taranto,[21] where multiple generations of casts from several of these mold series suggest their continued popularity into the 4th century BC.[22] At temple E in the urban sanctuary at Metaponto the figurines were present in a closed context dated to the end of the 4th or early 3rd century BC.[23] Again, the possible pitfalls of dating mold-made figurines solely on the basis of style should be kept in mind. Among the terracotta fragments recovered by the ICA survey, one (**No. 285**) preserves the image of a small sheep or goat nestled in the left arm of a standing woman,[24] while another fragment (**No. 12**) preserves a basket of fruit that the woman holds in her outstretched left hand. The first fragment belongs to a figurine of a peplophoros whose himation veils her head and shoulders. She cradles a sacrificial animal in her left arm while holding an oinochoe for libations in her right hand. Figurines from this or other closely related mold series have been found at Metaponto in the area of temples A, B, C, D, and E in the urban sanctuary,[25] at Sansone and the kerameikos within the urban center,[26] and in votive deposits west of the city wall at Crucinia and Favale,[27] as well as at a farmhouse in the chora.[28] The exact mold series to which the second fragment belongs cannot be ascertained, but the complete figurine doubtless resembled one of the numerous terracottas portraying a woman holding a basket of fruit in her left hand that have been found at most of the same sites as those where figurines holding a sacrificial animal described above occur.[29]

[19] The remarkable longevity, in terms of production and dedication, of a terracotta image of a hieratically enthroned female image, however, should be remembered. Figurines belonging to other, similar mold series that portray an enthroned goddess were found in a closed context dated to the end of the 4th or beginning of the 3rd c. BC at temple E in the urban sanctuary. These figurines may thus document the long-lived popularity of the basic subject of an enthroned goddess and underscore the perils of dating terracottas on stylistic criteria alone: see Postrioti 1996, 83–98, 139–40, Pls. 13–16a.

[20] For general discussion of this class of figurines and bibliography, see Barberis 1995, 22–24 nn. 64, 69–70; Postrioti 1996, 98–103, Pl. 16b–d; Liseno 2004, 40–50, 108–115.

[21] For similar figurines from Taranto, see Winter 1903, 110, Nos. 1–6, 8; Lo Porto 1964b, 251–52, Figs. 67–68; Iacobone 1988, 9–11, Pls. 6–10, 14–30; Lippolis 1995, 58–59, Pl. 18, Nos. 3–4 and Pl. 19, Nos. 1–2.

[22] For the long life of some of these mold series as well as the presence of casts from certain mold series in contexts of the third quarter of the 4th c. BC at Taranto, see Iacobone 1988, 10–11.

[23] Postrioti 1996, 98–103, 139–40, Pl. 16b–d.

[24] For figurines from the same or closely related mold series at Metaponto, see Lo Porto 1966, 162–64, Fig. 24, No. 1; Sestieri 1940–41, 99, Pl. 37, Fig. 16 upper right; Adamesteanu 1980, 270–73, Fig. 284a, Fig. 285a, e; Liseno 2004, 47–49, Pls. 17a–b, 18a–b. For similar figurines from Taranto, see Lo Porto 1964b, 251–52, Fig. 67; Winter 1903, 110, No. 4. From Timmari: Lo Porto 1991, 78–79, Pl. 25, Nos. 9–10.

[25] Urban sanctuary: Adamesteanu 1980, 93–94, Fig. 84a (temple A); 142–43, Fig. 146b (temple B), Postrioti 1996, 99, 101 (temple E); Calabria 2005, 78, Figs. 14–15 (temples C and D).

[26] Sansone: Pesce 1936, 445, No. 16 = Sestieri 1940–41, 99, Fig. 16 upper right; Lo Porto 1966, 162–64, Fig. 24, No. 1, Fig. 26, No. 2. Metapontine kerameikos: Signore 1996, 345, Fig. 70, No. 97.

[27] Crucinia: Lo Porto 1981b, 330, Fig. 39, No. 1, Fig. 40, No. 1. Favale: Adamesteanu 1980, 269–73, Fig. 284a–b, Fig. 285a, c–i; Liseno 2004, 47-49, Pls. 17a–b, 18a–b.

[28] Barberis 1995, 22, Fig. 5 upper left. Unfortunately, the farmhouse at which this fragment was found is not specified.

[29] For a good summary and additional bibliography regarding such figurines at Metaponto, see Meirano 1996, 76–77, 81–82, 85–86, 94, especially Figs. 23 and 30. It should also be noted that figurines belonging to a set of closely related Tarantine mold series that portray a woman holding a *kanoun* filled with fruit and sometimes also a crossed torch were found in the courtyard of the Mezzacapo caserma at Taranto in 1989, in an archaeological context closely dated to the last third of the 4th c. BC: Dell'Aglio 1996a, 72–73, No. 59. The image of a woman with offerings of fruit thus remained a popular subject for votive terracottas throughout the 4th c. BC. Although the ICA survey did not recover any terracottas from these 4th-c. BC mold series, excavations at several locations at Metaponto have found such figurines. For examples from the urban sanctuary, see Meirano 1996, 76 nn. 57–58 (temples B and C, and the large cloaca); 77, Fig. 12 (temple A). From a deposit northwest of the theater: Meirano 1996, 76 n. 59, Figs. 9–10. From Sant'Angelo Vecchio in the chora: Edlund 1979, Figs. 3, 34; 1984, 287, Fig. 7; 1988, 238–39. From farmhouses in the chora: Meirano 1996, 76 n. 58 (Pascucci-Torraca at Saldone); 77 n. 61 (Fattoria Stefan) = Barberis 1995, 15, 23, Fig. 6 lower center.

The survey recovered numerous terracotta fragments of drapery folds (**Nos. 9–10, 166, 212–13, 238–41**), as well as several feet and bases of figurines (**Nos. 3, 154, 177, 242**). The fragments lack sufficient detail to be securely associated with a given mold series, but many may well belong to a figurine portraying a standing woman holding an offering.[30]

Figurines: female heads

Two female heads illustrate more effectively than these small, battered fragments of drapery the graceful appearance that one might expect of the terracotta products of Metapontine workshops in the 5th century BC. **No. 335**, for instance, with its somewhat heavy and somber visage framed symmetrically by undulating tresses, reflects stylistic developments of Early Classical or Severe style. The head finds close parallels among Tarantine terracottas, including figurines of standing women holding offerings.[31] On the other hand, **No. 328** offers a delicate portrait of a young female wearing a low polos and veil that looks to sculptural models of the late 5th century BC.[32]

Figurines: Europa and the bull

Another unfortunately fragmentary terracotta, **No. 181**, presents a subject otherwise not previously attested among the Metapontine coroplastic repertoire: Europa and the Bull. Europa, sitting sidesaddle on the back of the bull, grasps the animal's horn with her left hand. The rounded zigzag and swallowtail drapery folds of her himation that fall from her outstretched arm over the bull's neck point to a date towards the middle of the 5th century BC, while the sensitive modeling of the bull's head and neck suggest a slightly later date in the second half of the 5th century BC. The subject is a common one among terracottas in Boeotia,[33] but not in Magna Grecia, where it does, however, find representation in monumental architectural sculpture.[34] At Metaponto, Europa appears on the back of a bull in a delicate sculpture that decorates the lid of a marble pyxis recently discovered in the tomb of a young woman dated from the middle to the third quarter of the 5th century BC.[35] The terracotta from the survey attests to an interest in the mythological subject on a more modest scale.

Figurines: summary

In terms of imagery, the Classical figurines at Metaponto continue to represent the same subjects found in the Archaic period: an enthroned goddess without specific attributes who was a suitable votive offering for a range of different deities and a standing female figure who probably represents a worshipper holding various animals and objects that were used in making sacrifices, offerings, and libations to the gods. Although the offerings held by the woman are appropriate for most deities and, accordingly, the figurines could serve as a gift at most cult places, some mold series represent a woman who holds objects that point to specific divine personalities. Figures who hold a deer have been interpreted by some as a reference, for instance, to Artemis,[36] while a mold series that is popular in the 4th century BC portrays a woman holding a crossed torch, a ritual instrument that is closely associated with the worship of Demeter and Kore.[37] Among the terracottas from the ICA survey, however, none appears to portray either of these two iconographic types.[38]

Relief plaques: reclining male banqueter

In the 4th century BC, as settlement continued to intensify in the chora of Metaponto, another format of terracotta, that of the relief plaque, became increasingly the votive offering of choice. One of the

[30] In certain cases, the fragments may belong as easily to a seated female figure or even a bust. The range of possibilities is discussed in the catalog entry for each piece.

[31] For closely related heads from Taranto, see Schneider-Herrmann 1971, 141–42, Fig. 5a–b, Fig. 7a–b. For similar heads from Taranto: Iacobone 1988, 10, 21, Pl. 8b, Pl. 10c (on standing women with offerings), 134, Pl. 127a–b. For similar head from Favale at Metaponto: Liseno 2004, 59, Pl. 25f.

[32] The terracotta does not, as far as I am aware, depend on Tarantine or other models. It resembles somewhat the head of a woman from a relief plaque portraying a reclining banqueter from proprietà Morlino at Metaponto: Nava 2002, 735, Pl. 66.1 lower right.

[33] Verhoogen 1956, 25, Fig. 17; Schmidt 1994, 70, Pl. 19, No. 87; Higgins 1954, 219, Pl. 112, No. 821. From Boeotia and Corinth: Winter 1903, 163, No. 6, 164, No. 5.

[34] It is, however, the subject of a Locrian relief plaque of the 5th c. BC and a terracotta mold, said to be from Taranto, dated to 330 BC: *LIMC* 4 (1988), 81, Nos. 89–90, s.v. "Europe I" (M. Robertson). For stone metopes from Selinous and Poseidonia-Paestum and architectural terracotta sculpture from Poseidonia-Paestum, *LIMC* 4 (1988), 81–83, Nos. 78–79, 113, s.v. "Europe I" (M. Robertson); *Greci in Occidente*, 674, No. 67.

[35] I thank A. De Siena for drawing my attention to this imported piece, possibly of Parian marble. Nava 2004, 961–962, Pl. 67b–c.

[36] For further discussion and bibliography, see Barberis 1995, 22–24.

[37] Barberis 1995, 24; Lippolis 1995, 58–59, Pls. 18, Nos. 4–19, Nos. 1–2; Iacobone 1988, 11, Pls. 17–20. See also n. 29 above.

[38] It is also worth noting that the survey did not recover any figurine of the so-called "Artemis Bendis" type popular in the 4th c. BC. For discussion and bibliography, see Barberis 1995, 25–27, Figs. 10–11; Lippolis 1995, 59–60, Pl. 19, Nos. 3–4; Iacobone 1988, 11, Pls. 11–12.

most popular classes of terracotta plaque portrays a reclining male banqueter holding a phiale or other wine-drinking vessel.[39] The mold technology for the plaques again depends upon the coroplastic industry at Taranto, where terracottas with this subject were made in the tens of thousands from the later 6th to 3rd century BC.[40] The male figure, who may be bearded or not, wears a simple polos in early examples of the reliefs. A festive wreath embellished with rosettes, palmettes, and a taenia or ribbon with ends that flutter down over his shoulders later replaces the polos. In the middle of the 5th century BC, a second figure, that of a woman seated at the end of the banqueter's couch, is added to the relief. By the beginning of the 4th century BC, the woman, either seated or sometimes standing at the end of the couch, appears regularly in the company of an infant or small boy who often scrambles out of her lap, stretching his arms out toward the reclining male. In some instances the heads of one or two horses are added to the image, which now begins to resemble a family portrait.

Interpretation of this long-lived, plentiful class of terracotta images at Taranto enjoys an extensive history of its own.[41] Some scholars have preferred to see a divine personality in the character of the reclining male figure and his companions. An identification of the figures as Dionysos-Hades, Kore, and Iacchos, for instance, has been suggested. Others have preferred to see the reclining symposiast as the male counterpart to the standing female, who holds animals and objects used for sacrifice and libation in the numerous contemporary figurines portraying a standing woman with offerings. Among those who see the terracottas as representations of a mortal, some have endowed the image with heroic, funerary, and chthonic overtones, while others stress elements that point to the political and military identity of the male figure as a citizen of the polis. In short, it seems that the complexity and evolutionary nature of the image entices the interpreter to seek semantic specificity, but ambiguity (perhaps intentional) regarding the identity of the figures portrayed in the reliefs ensures that such specificity remains elusive.

At Metaponto, terracottas derived from Tarantine mold series portraying a reclining banqueter have been widely found in the urban center as well as in the chora. They were the first kind of terracottas reported from Metaponto by dilettante archaeologists of the late 19th century.[42] Within the urban center, in addition to scattered finds from Masseria Sansone,[43] the reliefs have turned up in the castrum,[44] at temples A, B, C, and E in the urban sanctuary,[45] and in a votive deposit northwest of the theater that has been associated with the small temenos of Zeus.[46] Votive deposits outside the city wall at Crucinia[47] and Favale[48] have also yielded reliefs of a reclining male figure, and one example was found at a Hellenistic tomb near Casa Ricotta.[49] The votive reliefs are again present at the rural sanctuaries at Pantanello[50] and Sant'Angelo Vecchio,[51] as well as at several farmhouses.[52]

[39] Signore 1996, 304–40, 349–55; Barberis 1995, 27–28.

[40] For general discussion and bibliography, see Lippolis 1995, 51–53, Figs. 7–11; Iacobone 1988, 166–89; Signore 1996, 346–55; Barberis 1995, 27–28.

[41] Good summaries are found in Signore 1996, 351–55 and Lippolis 1995, 51–53.

[42] For example, Lacava (1891, 127) describes various terracottas of a banqueter with woman and child found in the Le Tegole zone within the urban center. Lo Porto (1966, 167–72, Figs. 28–32, Pls. 7–8) publishes 64 examples of Lacava's collection from Le Tegole. See also Letta 1972, 10–13, 61–108, Pls. 3, Nos. 1–20, No. 3. For other early references to the presence of this kind of terracotta relief at Metaponto, see Signore 1996, 349.

[43] Pesce 1936, 445, Fig. 6, Nos. 19–23; Sestieri 1940–41, 99, Figs. 14 and 16 (for woman seated at end of couch); Lo Porto 1966, 172–76, Figs. 32–34, 35, No. 2, 36, No. 2.

[44] Lo Porto 1966, 145–46, Fig. 5, Nos. 1–2.

[45] Temple A: Adamesteanu 1980, 93–96, Figs. 84b, 85e, 86d; possibly also 92–94, Figs. 81c, 84c–e. Temple B: Adamesteanu 1980, 145–46, Figs. 150b and possibly c; temple C: Adamesteanu 1980, 200, 202, Fig. 210a, c–e. The large cloaca: Meirano 1996, 86, Fig. 27. Temple E: Postrioti 1996, 105–6, Pl. 17a–b. Also, proprietà Di Taranto at the intersection of *plateia* C–C1: Adamesteanu 1980, 311, Fig. 329.

[46] Signore (1996, 353) mentions the terracottas, of which only a couple of examples have been published: Meirano 1996, 77, Fig. 14; *Greci in Occidente*, 708, No. 205.

[47] Lo Porto 1981b, 332, 334, Fig. 41, Nos. 1–3.

[48] Adamesteanu 1980, 273–76, Fig. 288a–b, d–h (possibly), Fig. 289c–d; Liseno 2004, 61, Pl. 25f.

[49] From investigations of a chamber tomb (No. 5) of the late 4th to early 3rd c. BC that had been previously robbed: Lo Porto 1981b, 351–52, Fig. 65, No. 1.

[50] More than 160 fragmentary terracottas of a reclining banqueter were found in the University of Texas excavations at the rural sanctuary: Carter 1994a, 170, 173, 194, 196, Table 7.1. Only one fragment of the woman seated at the end of the dining couch, which was instead found near the kiln site on the hill above the sanctuary, has been published: Carter 1983a, 442, 456, Fig. 32, No. 29.

[51] Edlund 1979, Fig. 32; 1984, 287; 1988, 238–39, Pl. 45, No. 2.

[52] Musillo and Pascucci-Torraca farmhouses in Saldone, Arezzo farmhouse in Lago del Lupo, Sciurti farmhouse at Marconia-Le Cesine, Risimini-Stefan farmhouse in Campagnolo, a farmhouse in Pagliarone near Montescaglioso: Barberis 1995, 15, 27–28, Fig. 12. Morlino farmstead near Pantanello: Nava 2002, 735, Pl. 66.1. While excavations at most farm sites recovered only from 1–4 fragments, 36 fragments of a reclining banqueter were found at the Risimini-Stefan farm site. Barberis (1995, 28 n. 99) also notes 15 fragments of a banqueter that are sporadic

An important group of relief plaques portraying a reclining banqueter was recovered in excavations of the kerameikos near the northern gate of the city.[53] Signore has developed a typology for these plaques, which were certainly produced in this industrial area of the urban center.[54] The terracottas were found in deposits in association with ceramics that date from the third quarter of the 4th to the beginning of the 3rd century BC.[55] The terracotta reliefs found in these deposits should accordingly date to the same period. Most of the reliefs of a reclining banqueter from Metaponto, as well as those from Taranto, have regrettably not been found in such closely dated archaeological contexts. As a result, stylistic and sometimes technical characteristics have served as the principal guides in dating the relief plaques. But, as is well known, terracottas cast from molds could be, and were, produced long after the original prototype of a mold series was fashioned. Thus a stylistic date refers to the date of the prototype, but not necessarily to the date of a specific cast. As closer attention is paid to the stratigraphic context of such terracottas, a more accurate picture of the history of the production and use of these reliefs may emerge. It would not be surprising if a slight downward shift in the dating of these reliefs results.[56]

The ICA survey recovered more than a dozen such figurines from the chora between the Bradano and Basento Rivers. A fragmentary mold used to produce the terracotta reliefs, **No. 340**, was also found. One terracotta, **No. 220**, portrays the banqueter wearing a polos. When scholars trace the development of the imagery of the reclining banqueter, the sober polos is usually identified as an older element that was replaced in later mold series by a festive wreath decorated with rosettes and palmettes. Reliefs that represent the banqueter wearing a polos are conventionally dated as deriving from a prototype of the 5th century BC. The presence of eight reliefs of a banqueter with a polos in the deposits from the kerameikos[57] suggests, however, that casts of this older image continued to be made at Metaponto throughout the 4th century BC.

Most of the fragments from the survey come from relief plaques belonging to a group of closely related mold series that portray the banqueter wearing an elaborate wreath: **Nos. 89, 221**, and **286**.[58] A woman whose himation veils her head (**Nos. 60, 87, 228**) sits demurely at the end of the dining couch. She holds in her left arm a child, who reaches to grasp the right hand of the reclining man (**No. 88**). The head of one or two horses (**Nos. 70** and **320**) has been added in the background between the man and woman in some mold series. These reliefs are the ones most commonly found at Metaponto and date to the 4th century BC. They were cast from molds that lacked sharply defined features and thus appear to be worn with age. In one variant, the skin of a lion or panther is draped over the dining couch. The paw of the feline pelt is preserved hanging over the side of the couch next to the foot of the woman in **No. 67**. The mold for a relief of a reclining banqueter that was recovered from site 404 (**No. 340**) preserves only the head of the child and a few drapery folds of the himation covering the legs of the banqueter. It is thus difficult to determine the exact mold series to which the fragment belongs.

In trying to understand the significance this imagery held at Metaponto, one is confronted by the wide variety of contexts in which such relief plaques have been found, ranging from urban and extra-urban sanctuaries, to tombs, to industrial areas and farmhouses. Until the votive materials from the Metapontine sanctuaries are fully published, it is difficult to assess the relative importance that the votive image of a reclining banqueter may have held in cult practice at a particular shrine, much less to determine the meaning of the imagery in private domestic or funerary ritual. It is worth stressing that the imagery of a votive terracotta from a single archaeological context needs to be evaluated not only within the entire assemblage of votive materials from the sanctuary where the terracotta was dedicated, but also against the assemblages of votive terracottas from all the sanctuaries of a given polity. Only through such comparative analyses can gains be made toward better understanding the significance a given image

surface finds from the chora. Nava (2002, 735) describes the number of plaques from Morlino as "considerevole."

[53] D'Andria 1980a, 418–23, Figs. 61–63; Signore 1996, 299–340, Figs. 1–58.

[54] Signore 1996, 304–40, Figs. 1–58, Nos. 1–84.

[55] Signore 1996, 299–302. Four other fragments of a reclining banqueter have also been found in a closed context dated from the end of the 4th to the beginning of the 3rd c. BC at temple E: Postrioti 1996, 105–6, 139–40, Pl. 17a–b.

[56] It is worth noting that the fragment of a reclining banqueter relief from the kiln site at Pantanello was found in a stratum in association with 4th-c. BC ceramics: Carter 1983a, 442, 446, 456, Fig. 32, No. 29. In addition, chamber tomb 5 near Casa Ricotta, which yielded another relief, dates from the late 4th to the early 3rd c. BC: Lo Porto 1981b, 351–52, Fig. 65, No. 1.

[57] Signore 1996, 304–8, Figs. 2–9.

[58] Signore's types III–VI: Signore 1996, 308–23, Figs. 10–37.

may have held for the individual who dedicated the object in one context versus dedicating the same image (or another image) in a different context. The varied proposals put forth in prior attempts to interpret the meaning of the banqueter reliefs at Taranto, mentioned above, should be kept in mind, but the analysis should be based on what the archaeological record reveals about the use of the reliefs at Metaponto. In this light, observations by Signore regarding the presence of banqueter reliefs in deposits associated with the temenos of Zeus (identified in part by a stele bearing the inscription *Dios agora*) in the urban center appear to be headed in a sensible direction.[59] She reads the image of the reclining banqueter within this specific context as one connected with the political sphere of the cult of Zeus in the agora. More work is needed on Metapontine terracottas, however, before the varied valences that the reclining banqueter may have taken in different ritual contexts at the city-state can begin to be sorted out.

Relief plaques: soldier or horseman

The survey recovered at least two terracottas (**Nos. 13** and **325**), belonging to a class of relief plaques that share several iconographical elements with the terracottas of a reclining male figure. These reliefs, which once again belong to mold series that were first fashioned by Tarantine craftsmen, portray a standing soldier or a horseman who usually carries a round shield and frequently wears the same festive wreath with a long taenia that falls over his shoulders as do the reclining banqueters.[60] Often the man is portrayed with a horse and, in some cases, he holds a kantharos or phiale for pouring a libation. The two fragments from the survey belong to a single mold series within this larger class of relief plaques. Complete casts of this mold series from Taranto portray a bearded man, standing in a hipshot pose, who wears a clamys that falls over his back and an elaborate headdress with rosettes and a taenia.[61] His right hand rests on his hip while his left raises a shield up to the level of his head. At his left a tripod and tall stele are depicted. A small boy, standing between the man and the tripod, gestures toward the older man with raised arms.

Two fragments preserving the head of a horse (**Nos. 70** and **320**) may belong either to a relief plaque of a reclining banqueter, as noted above, or possibly to a relief plaque in this class of terracotta portraying a horseman. In the latter case, the complete relief would have represented either an older, bearded soldier who stands in front of his horse holding a shield on his left arm and either the reins of the horse or a phiale for libation in his right hand, or a younger, beardless man wearing a pilos and holding a shield as he performs the cavalry maneuver of the apobates, jumping off and again onto his mount.[62] Again, too little of the original relief plaques is preserved to assign either head of a horse securely to one or the other class of terracotta.

Two other terracottas belong to a mold series that is otherwise undocumented but may be a variant within the class of terracottas portraying an equestrian soldier. One fragment, **No. 153**, was found at a site within the Bradano-Basento transect. The second, more complete fragment (TC 588-01) comes from a site lying between the Basento and Cavone Rivers; it will thus be published with the results of that survey project,[63] but it is nevertheless illustrated here next to **No. 153** for purposes of comparison. This new mold series portrays a beardless youth wearing a kausia, a cap generally associated with Macedonian youths beginning in the late 4th century BC.[64] The central button and soft, rounded rim of the cap are clearly indicated. In the more complete example, the ends of a taenia, which is rarely rendered with the kausia in terracotta figurines, fall from the head of the youth to his shoulders. In this fragment one can moreover make out the shadow of a tall conical shape that appears to lie underneath the kausia. It may be that this is the "ghost" of the pilos that youths are regularly portrayed wearing in those terracottas, discussed above, that represent a horseman. In such mold series from Taranto, for instance, the pilos is sometimes worn with a taenia and, on occasion, even a festive

[59] Signore 1996, 353. Postrioti (1996, 144) also attempts to interpret the four terracottas portraying a reclining banqueter found at temple E in the urban center within the particular context of this cult, which he believes focused on Dionysos Iacchos. He sees the reliefs as underscoring links that Dionysos had with Hades and the funerary sphere, which in turn allude to his generative powers.

[60] Lippolis 1995, 54, Pls. 12–13; Bartoccini 1936, 166–69, Figs. 74–79.

[61] Abruzzese Calabrese 1996, 201–2, Fig. 146; Lippolis 1995, 54, Pl. 12, No. 1; Iacobone 1988, 143, Pl. 133d–e; Winter 1903, 203, No. 12.

[62] For the standing horseman from Metaponto: Letta 1972, 110–11, Pl. 20, No. 4. From Taranto: Bartoccini 1936, 166–69, Figs. 76–77; Kingsley 1979, 204, Pl. 2, No. 6; Higgins 1954, 361, Pl. 181, No. 1327; Iacobone 1988, 121, Pl. 118c; Lippolis 1995, 54, Pl. 12, Nos. 2–4. For horseman jumping off his horse: Lippolis 1995, 54, Pl. 13, Nos. 1, 3–4.

[63] *Basento-Cavone* (forthcoming).

[64] D.M. Thompson 1963, 53–55.

wreath.[65] Perhaps the coroplast recut a mold from a mold series portraying an equestrian youth wearing a pilos with a taenia in order to update the image so that it would now portray a youth sporting the more fashionable Macedonian kausia.

Only a few relief plaques portraying a soldier or horseman have thus far been found at Metaponto. One, from the kerameikos, portrays a man who stands in front of his horse holding a kantharos in his right hand.[66] Two similar fragments are said to come possibly from the area of temple A in the urban sanctuary, while another is mentioned as having been found at Masseria Troia within the urban center.[67] A few isolated heads of a youth wearing a pilos have also been found within the urban center.[68] The heads are usually identified as representations of the Dioskouroi,[69] but without more of the terracotta preserved, it is difficult to determine whether the head belongs to one of the divine twins or to a youthful apobates rider.

The significance of the relief plaques portraying a soldier or horseman at Metaponto is not entirely clear, but a military and equestrian role for the male figure, whether of younger or more mature age, does seem to be stressed. The reliefs perhaps allude to the martial responsibilities of a male citizen, who is portrayed in a heroic guise. The recurring iconographic elements of the festive wreath and taenia, kantharos, phiale, shield, horse, and child whose gesture of outstretched arms visually links him to the adult male figure recall similar elements associated with terracotta representations of the reclining male banqueter. Both groups of terracotta relief plaques may refer to the idealized role of a citizen of the polis.

Relief plaques: satyr, silenos, or Pan with female figure

The terracotta most frequently found by the ICA survey is a relief plaque portraying a female figure, possibly a maenad, but more likely a nymph, in the company of a satyr, a silenos, or Pan. Some three-dozen fragments can be attributed to a group of four or five mold series representing this subject that are documented by better-preserved, more-complete reliefs,[70] coming mostly from temple E in the urban sanctuary. The fragments from such relief plaques represent more than one-tenth of all the terracottas recovered in the survey and almost one-third of those 123 fragments whose subject matter may be recognized. The most common of these reliefs depicts a female figure running to her right as she raises her himation in the gesture of anakalypsis with her right hand (**No. 203**) and holds a cornucopia in her left.[71] A paunchy, splayed-nosed, bearded male figure (**No. 269**), who in some variants of the mold series has the horns of Pan and caprine legs and hooves (**Nos. 33, 35, 92, 222**)[72] and in other variants is portrayed without horns, but with equine ears and booted human legs,[73] embraces the woman with his right arm and carries a volute krater in his left. In a second mold series, the position of the male and female figures (**No. 1**) is reversed as they appear to prance across the surface of the relief plaque, but otherwise their physical characteristics (including the variation between caprine and human legs) and the objects that they carry, such as the cornucopia (**Nos. 19–20**), are the same.[74] Certain variants of both mold series place the two dancing figures within the setting of a rocky grotto with clusters of grapes hanging over their heads (**No. 18**).[75] Less frequently attested mold series portray the two figures (**Nos. 4, 156**) in a more stationary pose,[76] while in one

---

[65] For youth identified as one of the Dioskouroi, with pilos and taenia: Besques 1986, 25, Pl. 21a–b, d. For bearded male with pilos worn over wreath and taenia: Lippolis 1995, 54, Pl. 12, No. 2.

[66] D'Andria 1980a, 422–23, Fig. 63, No. 281; Signore 1996, 340, Fig. 59, No. 85.

[67] Letta 1972, 10–13, 108, 110–11, Pl. 20, No. 4 (111 n. 337 for reference to terracotta from Le Tegole-Masseria Troia). For two reliefs that are identified as representations of one of the Dioskouroi, but which may belong to this class of soldier or horseman, see Lo Porto 1988c, 20, Fig. 15, Nos. 1–2.

[68] Pesce 1936, 445, No. 18 (cf. Winter 1903, 208 No. 6.); Sestieri 1940–41, 99, Fig. 14 lower, second from left; Sestieri 1940, 116, Fig. 55 lower left. From temple B in the urban sanctuary: Adamesteanu 1980, 145–46, Fig. 150d.

[69] A few relief plaques that belong to Tarantine mold series representing the Dioskouroi have been found at Metaponto—one fragment from the Sciurti farmhouse at Marconia-Le Cesine, two from the Musillo farmhouse at Saldone, and a single sporadic find from the chora; see Barberis 1995, 31. Dioskouroi relief plaques have also been found in the temenos of Zeus in the urban center; see De Siena 1998, 154.

[70] Postrioti 1996, 21–72, Pls. 1–9; Barberis 1995, 29 n. 110, Figs. 13–14; Tempesta 2001, 101, Fig. 106; De Siena 1999a, 67, 70, Fig. 45; Meirano 1996, 85, Fig. 24; Lo Porto 1988c, 22–24, Fig. 19; Letta 1972, 126–29, Pl. 26, Nos. 1–2; Prag 1983b, Fig. 150; Carter 1980a, 14–15, Fig. 11; Carter 2000, 90, Fig. 47.

[71] Postrioti 1996, 26–50, Pls. 1–3, types AIA–C.

[72] Postrioti 1996, 26–33, Pls. 1–2, types AIA–B. Also Letta 1972, 126–29, Pl. 26, No. 2; Carter 1980a, 14–15, Fig. 11; Carter 2000, 90, Fig. 47.

[73] Postrioti 1996, 33–50, Pl. 3, type AIC.

[74] Postrioti 1996, 50–59, Pls. 4–5, types AIIA–C. Also Tempesta 2001, 101, Fig. 106; De Siena 1999a, 67, 70, Fig. 45; Meirano 1996, 85, Fig. 24; Barberis 1995, 29, Fig. 14.

[75] Postrioti 1996, 32–33, 52–53, Pls. 2a–b, 4b, types AIB, AIIB; Prag 1983b, Fig. 150; Carter 1980a, 14–15, Fig. 11; Carter 2000, 90, Fig. 47.

[76] Postrioti 1996, 60–70, Pls. 6–8, types AIII–AIV. Also Barberis 1995, 29, Fig. 13.

mold series,[77] the female figure carries a situla instead of the cornucopia.

The greatest concentration of this engaging class of relief plaque comes from temple E in the urban sanctuary,[78] although the terracottas have been found also in the area of temples A and C and elsewhere within the urban center.[79] Outside the city walls, numerous fragments have been recovered at the extramural sanctuary at Favale,[80] as well as at the sanctuaries at Pantanello[81] and San Biagio alla Venella.[82] The reliefs are likewise known at several farmhouses,[83] including the kiln deposits at Sant'Angelo Vecchio [84] and are again present in a house at Cozzo Presepe[85] at the northwestern limits of the chora of Metaponto. The archaeological contexts of many of these relief plaques, especially the hundreds of plaques from temple E, suggest a date as late as the last quarter of the 4th to first quarter of the 3rd century BC for the terracottas.[86]

Unlike most Metapontine terracottas of the 5th and 4th centuries BC, these mold series seem to have been fashioned originally at Metaponto.[87] While terracottas similarly portraying a satyr or silenos with a maenad were made at Taranto,[88] they do not belong to the mold series represented by the relief plaques at Metaponto. Related terracottas portraying Pan and satyrs with a female figure have also been found at Herakleia and Francavilla Marittima.[89] The motivation for the creation and widespread production of this lively imagery at Metaponto is not entirely clear. They are the predominant votive terracotta at temple E and recur frequently, as the survey demonstrates, throughout the chora of Metaponto in both sanctuaries and domestic contexts. The horned figure with the legs of a goat may be identified as Pan and, by implication, his female companion as a nymph. The hornless figures of a silenos or satyr, however, point to the world of Dionysos and as such imply that the female figure may instead represent a maenad. Postrioti acknowledges that the reliefs may reflect the worship of Pan at Metaponto, but he is inclined to view the imagery as instead a reflection of Dionysiac concerns.[90] The rustic mythological creatures portrayed in the relief plaques make up the thiasos of Dionysos and express the concerns and hopes of the population regarding the agricultural fertility and abundance of the countryside.[91] The emphasis on the grape and wine, as reflected in the grape arbor under which the couple dances and the volute krater carried by the male figure, offers a visual counterpart for the increase in viticulture that is documented by an analysis of pollen at Metaponto for the mid-4th century BC.[92] The desire to propitiate divine powers that were responsible for the productivity

[77] Postrioti 1996, 70–71, Pl. 9a, type AV.

[78] Postrioti 1996, 21–72, Pls. 1–9; Meirano 1996, 85–86, Figs. 24, 26.

[79] Temple A: Letta 1972, 10–12, 127–29, Pl. 26, Nos. 1–[illegible]; Meirano 1996, 85 n. 108, Fig. 25, also mentioning numerous other fragments from the Sanctuary of Apollo area; Barberis 1995, 29 n. 110. Temple C: Adamesteanu 1980, 203–5, Fig. 213d–e. Elsewhere within the urban sanctuary: Adamesteanu 1980, 52, 54, Fig. 37; 94–96, Figs. 35f, 86b–c and f (according to Meirano [1996, 77 n. 64], this last piece is from the large cloaca along with three or four other fragments). Sporadic finds from the Sansone area in the urban center: Lo Porto 1988c, 22–25, Fig. 20, Nos. 1–6. Sporadic finds from Le Tegole-Masseria Troia: Letta 1972, 10–13, 127.

[80] Adamesteanu 1980, 275–76, Fig. 289e; possibly also 272–75, Figs. 286g, 288i; Liseno 2004, 64-66, Pl. 30a–d.

[81] More than 250 fragments from such relief plaques were found. For mention of plaques and two fragments that have been illustrated, see Carter 1994a, 170, 178, 194, 196–97, Fig. 7.18, Table 7.1; Carter 1998a, 245–46, Fig. 6b.

[82] Letta 1972, 14, 127; Barberis 1995, 29 n. 110; 34 n. 141; Meirano 1996, 85 n. 108. For mention of additional surface finds from the Venella area, see Barberis 1995, 16 n. 22; from a spring in proprietà Castellaneta, Barberis 1995, 29, possibly illustrated in Figs. 13–14.

[83] Arezzo farmhouse in Lago del Lupo, four fragments; Risimini-Stefan farmhouse in Campagnolo, 23 fragments; Incoronata-La Cappella, one fragment; Sciurti farmhouse at Marconia-Le Cesine, 42 fragments: Barberis 1995, 15, 29, Figs. 13–14. Also on Risimini-Stefan farmhouse: Meirano 1996, 77 n. 64, 86 n. 110, 93 n. 176. Also on Marconia-Le Cesine farm: Meirano 1996, 85 n. 108, 86 n. 110, 90 n. 142

[84] Edlund 1979, Fig. 31; 1984, 287; 1986, 119; 1988, 238; Carter 1980a, 14–15, Fig. 11; 2000, 90, Fig. 47.

[85] From site B in a house near a possible watchtower: *Cozzo Presepe*, 246; Prag 1983b, 379–81, Fig. 150. Also possible fragments from a votive deposit and others scattered over an area south of defensive walls: Prag 1983b, 381; Morel 1970a, 104, 112–13, Fig. 36.

[86] For the date of the terracottas from temple E, see Postrioti 1996, 139–40. For similar dates from Cozzo Presepe, see *Cozzo Presepe*, 246. Also possible fragments from a votive deposit dated 320–270 BC and scattered over an area south of the rampart: Prag 1983b, 381. Those from Sant'Angelo Vecchio are from kilns dated to the 4th c. BC: Edlund 1979, Fig. 31; 1984, 287; 1986, 119; 1988, 238.

[87] Postrioti 1996, 32.

[88] Wuilleumier 1939, Pl. 29, Nos. 2–4. For additional bibliography, see Barberis 1995, n. 110.

[89] Herakleia: Lo Porto 1988c, 22; Neutsch 1967, 167 n. 3, Pl. 31, No. 2 (reclining satyr with cornucopia). From Francavilla Marittima: Maaskant-Kleibrink 1993, 44–45; Stoop 1974–76, 130–39, Pls. 65–66. Also Ruvo: A. Levi 1926, 75, Fig. 64. Oria: Palumbo 1986, 104, No. 501/Pl. 15 and 130, No. 25/Pl. 44.

[90] Postrioti 1996, 141–47. He sees the cult practiced at temple E as dedicated to Dionysos Iacchos with links to Hades and Persephone.

[91] Postrioti (1996, 151–52) suggests that the cuffs of the satyr's boots are visual references to the *compedes*, or the heavy ankle chains of slavery, and the story of the silenos Marsyas, who represents the freedom granted to slaves. The imagery of the relief plaques, if viewed in this interpretive light, would then also express the social and political aspirations of the local population.

[92] Pollen analysis also shows an increase in olive production: Carter 1994b, 180.

of the land—regardless of their specific identities—accounts well for the prevalence of relief plaques with such imagery throughout the territory of Metaponto during the period that witnessed the most intensive occupation of the chora.

Protomai and busts: female figure

Because of its large scale, one fragment, **No. 150**, probably belongs to a protome or bust of a female figure of the 4th century BC. It preserves waved linear tresses that have been swept up on the right side of the head of a woman who wears a flaring polos, veil, and earrings. Given its fragmentary condition, whether the terracotta was an open-backed protome or a bust with a closed back cannot be ascertained. The piece resembles protomai, common at Herakleia, that portray a female figure holding a crossed torch and piglet associated with the worship of Demeter.[93] It may also be compared, however, to protomai and busts without attributes that offer a more generic imagery.[94]

Several other fragments recovered from the survey—**Nos. 9, 26, 213,** and **327**—may also belong to a Classical protome or bust, but again their fragmentary state precludes making a secure attribution. The fragments, which often preserve the shoulder of the figure, may possibly belong instead to a large-scale figurine representing a standing woman holding offerings (discussed above). The sharply raised ridges of drapery of the himation of **Nos. 9** and **327**, for instance, find parallels in a protome from Timmari as well as in figurines of a standing woman from Metaponto.[95] The same may be said of the terracotta, **No. 213**, of a female figure wearing a chiton with a buttoned sleeve.[96]

At Metaponto, protomai and busts are rare. When they do occur, they tend to serve as funerary offerings in tombs.[97] In such contexts, protomai and busts are often interpreted as chthonic and closely associated with Demeter and Persephone.[98] Although protomai and busts could serve as votive offerings to a number of different deities, those at Herakleia, mentioned above, that portray a female figure holding a crossed torch and piglet were indeed dedicated at a sanctuary where Demeter presided. The infrequent use of such terracottas at Metaponto, however, excludes any narrowly specific conclusions regarding their significance.

Hellenistic terracottas (late 4th to 3rd century BC)

Figurines: seated female figure

The survey recovered a fragment, **No. 97**, of the feet and hem of the skirt of a figurine portraying a seated female who, in more complete versions, is enveloped in a himation that frequently binds her bent right arm tightly to the front of her chest, allowing only the hand to protrude beyond the gathered upper edge of the mantle. Depending upon the mold series, she may also hold a tympanum, phiale, bird, or fruit. The terracotta from the ICA survey is too fragmentary to be assigned to a specific mold series of this group. Similar figurines come from Taranto but have lost their precise archaeological contexts, while others have been found at sanctuaries throughout Lucania.[99] The figurines, whose mold series may have been originally fashioned in Tarantine workshops, are dated primarily on stylistic grounds to the 4th century BC. At Metaponto, a few such figurines have been found in the urban center, while one was placed in a tomb dated 325–275 BC at the Pantanello necropolis.[100] At Herakleia the figurines have also been found

[93] Neutsch 1967, 189, Pl. 48, No. 4; Neutsch 1968, 776, Fig. 23. For another such protome from Metaponto, see Letta 1972, 113–14, type XII.1.

[94] For a similar, but smaller, protome from Montescaglioso, see Lo Porto 1973, 190, Pl. 38, No. 1. For a similar bust from Pomarico Vecchio: *Pomarico Vecchio* I, 1.220 and 2.150, Fig. 89. From Timmari: Lo Porto 1991, 85, Pl. 28, No. 21. From Satyrion: Lo Porto 2002, 16, Pl. 2 lower.

[95] Timmari: Adamesteanu 1974a, 221. For figurines of a standing female holding offerings, see Adamesteanu 1980, 271, 273, Fig. 285; Liseno 2004, 43–44, Pls. 10b, 11a; Lo Porto 1981a, 333, Fig. 40, No. 2. For figurine of possibly seated female with such drapery from a spring sanctuary (site 57 along the track of the SNAM pipeline) near Incoronata/San Teodoro south of the Basento: Nava 2003a, 673, Pl. 37.2 lower left.

[96] For a similar dress on protomai from Herakleia, see Neutsch 1968, 775, Fig. 22a; Otto 1996, 141, Pl. 1b. From Timmari: Adamesteanu 1974a, 220 lower left; Lo Porto 1991, 87–88, Pl. 29, No. 28. From Ferrandina: Lo Porto 1973, 204–5, Pl. 53, No. 1. For figurines of a standing woman with similar drapery, who sometimes holds a lamb, from Metaponto: Letta 1972, 119, Pl. 23, No. 4; Adamesteanu 1980, 270–73, Figs. 284a, 285c, 287; Liseno 2004, 47-49, Pls. 15b, 16a, 17b, 18a; Winter 1903, 110, No. 5.

[97] From the Pantanello necropolis, T 126 and 330, dated 420–380 BC: Malone 1998, 771–73, 780, 784, inv. nos. T 126-1, T 330-11, Fig. 19, No. 1. Sporadic find from the chora: Barberis 1995, 25, Fig. 9; Meirano 1996, 76 n. 57. For two other protomai, see Letta 1972, 113–14 type XII.1–2. For bust from temple C: Calabria 2005, 72, Fig. 4.

[98] R. Ammerman 2002, 290–91; Malone 1998, 772–73; Barra Bagnasco 1996a, 220–21, 1997b, 1.220–21.

[99] From Taranto, Egnatia, and southern Italy, now in European museum collections: Higgins 1954, 355, 363–65, Pl. 179, No. 1305, Pls. 186–87, Nos. 1336–45; Mollard-Besques 1954, 160–62, Pls. 104–5 C605–19. Others from Taranto, Ruvo, Egnatia, and so on: Winter 1903, 131–33. For their funerary and votive use at Taranto, see Abruzzese Calabrese 1996, 195. For a few of the many examples from Lucanian sanctuaries: Barra Bagnasco 1996a, 218–19; Lo Porto 1991, 108–28, Pls. 46–54 (Timmari); Adamesteanu et al. 1971, 83–84, Pl. 31, Nos. 54242, 54393 (Rossano di Vaglio); G. Greco 1982, 43, Pl. 22, Nos. 1–2 (Rivello); Letta 1972, 125 (Accettura); D'Anisi 2005, 170–71, Figs. 3–5 (Accettura).

[100] For examples probably from Le Tegole in the urban center, two from

in tombs, usually of women, dating to the later 4th through early 3rd century BC.[101]

Although certain aspects of the dress and rendering of the heads of the female figures recall Late Classical styles of the 4th century BC, the position of the arm bound tightly in the himation in several examples of this group of terracottas has suggested that similar arrangements used to portray standing draped female figures of Hellenistic style, especially evident in the "Tanagra-style" figurines, influenced the creation of this class of seated female figurines.[102] The presence of the figurines in tombs that can be closely dated to the later 4th and early 3rd centuries BC at Metaponto and nearby Herakleia suggests that the terracottas at these sites are contemporary in use with the better-known Tanagra-style figurines. For this reason it has seemed more appropriate to include discussion of this single fragment from the survey with those figurines of Hellenistic style and date.

Figurines: possibly dancing female figure

The head and torso of a young female figure is preserved by **No. 167**. Molds were used to cast both the front and the back of the graceful figurine. A wreath composed of hand-modeled ivy leaves with ribbons falling down her back has been added to her head. The high-belted chiton suggests that the terracotta portrays a youthful figure. Her pose is one of exuberant motion. Twisting her torso to her right, she raises her right leg and extends her arms outward away from her body. The animated figure may be represented as dancing, or, alternatively, her raised leg may indicate that she is seated. Her crown of ivy leaves suggests that she belongs to the sphere of Dionysos: perhaps the terracotta portrays a dancing maenad.

A few dancing female figures of Hellenistic style have been found at farm sites and in tombs at Metaponto.[103] A strikingly similar head with floral wreath has also been recovered from the Risimini-Stefan farmhouse.[104] The figure may instead be seated. If so, she finds a close parallel in a figurine of a woman seated on a swan from a tomb dated 325–275 BC at Taranto.[105]

Figurines: standing female figure

Of the few fragments of drapery that can be identified as belonging to a standing female figurine of Hellenistic style,[106] **No. 7** is particularly worthy of note. Only the right leg is preserved of what was undoubtedly an exquisitely graceful figurine. Complete terracottas of this mold series from Apulia (possibly Taranto) and Cyrenaica[107] portray a woman wearing a sleeveless chiton with a long, belted overfold whose right corner is visible on the upper thigh of the Metapontine figurine. She wears her mantle over both arms in the manner of a shawl. Its edge falls in a cascade of zigzagging folds along the side of her right leg.

At Metaponto, a few figurines of Hellenistic style portraying women have been found at the urban sanctuary and at the sanctuary at Tavole Palatine.[108] Others come from what are probably domestic contexts in the urban center,[109] and from farmhouses and industrial areas in the chora.[110] Most such figurines at Metaponto, however, have thus far been recovered from funerary contexts.[111] In fact, the meager number of Tanagra-style figurines that were dedicated as votive offerings at Metapontine sanctuaries stands in

Torre di Mare, and one from San Biagio: Letta 1972, 10–11, 124–25, Pl. 25, Nos. 2–3. For a closely dated example holding a tympanum from T 12 at the Pantanello necropolis: Malone 1998, 773, T 12-8. For other examples from Metaponto, but without specific context: Winter 1903, 131, No. 3A; 132, No. 1.

[101] Pianu 1990, 231–32, Pls. 25, No. 1; 26, No. 4; 28, No. 1; 40, No. 4; 41, No. 2; 55, No. 2; 71, No. 2. Also from tombs at Sant'Arcangelo-San Brancato: *Greci, Enotri, e Lucani*, 262–67, 3.40.23–25, 30 4[illegible]4.

[102] Letta 1972, 125.

[103] From the Sciurti farm site at Marconia-Le Cesine: Barberis 1995, 30, Fig. 16. From a Hellenistic chamber tomb of the end of the 4th c. BC in Orto di Crocinio (Crucinia), more than a dozen fragments of female figurines, many of which are dancing: Sestieri 1940, 55, 60, 66, 68–71, Fig. 17. One figurine from T 71, dated 325–275 BC, at the Pantanello necropolis: Malone 1998, 775, 779, inv. no. T 71-26.

[104] Barberis 1995, 30, Fig. 17.

[105] Graepler 1996, 244, 246, Fig. 195.

[106] Nos. 7, 178, 183.

[107] Apulia, possibly Taranto: Besques 1986, 38, Pl. 29a, c. Cyrenaica: Burn and Higgins 2001, 224, Pl. 114, No. 2706.

[108] Area of temple A: Sestieri 1940, 106, Fig. 45 left; Adamesteanu 1980, 95, Fig. 85a–c. Letta 1972, 10 n. 6. Temple B: Adamesteanu 1980, 145–47, Fig. 150e–f. Temple C: Adamesteanu 1980, 200, 202, Figs. 210i, 211a–b. Tavole Palatine: Sestieri 1940, 52–53, Fig. 1 = Sestieri 1940–41, 99, Fig. 15.

[109] From the castrum: Lo Porto 1966, 145–47, Fig. 4, Nos. 1–3. From the portico along the south side of plateia A–A1: Adamesteanu 1980, 300–01. From Casa Sansone: Pesce 1936, 443, No. 1.

[110] Sciurti at Marconia-Le Cesine: Barberis 1995, 30, Fig. 16. Risimini-Stefan, which may have had a terracotta workshop: Barberis 1995, 15, 30, Fig. 17. From deposit near kiln at Pantanello: Carter 1983a, 485, Fig. 59, No. 140.

[111] From a chamber tomb at Orto di Crocinio (Crucinia): Sestieri 1940, 55, 60, 66, 68–71, Fig. 17. From the Pantanello necropolis: Malone 1998, 774–79, 781–83. From chamber tombs at Casa Ricotta: Lo Porto 1981a, 352–53, Fig. 6c; 383, Fig. 111, No. 2. From tombs of mid- to late 3rd c. BC: De Siena 2007, 439, Pl. 12a; Osanna 2008, 928–29, Pl. 22b. Attributed to funerary contexts: Letta 1972, 11, 135–38, 140–41, Pl. 27, Nos. 1–4, Pl. 28, No. 2.

stark contrast to the abundance of such terracottas that have been recovered at other south Italian or Sicilian sites, such as Poseidonia-Paestum, Capua, Syracuse, or Morgantina in this period.[112] The increased use of terracotta figurines as grave offerings (especially in the tombs of children and women) is, however, common in the second half of the 4th and 3rd century BC throughout southern Italy.[113] Does the apparently infrequent use of Tanagra-style figurines as votive offerings at Metaponto reflect a local preference not to employ such terracottas in ritual practice,[114] or is it a symptom of the overall economic decline of the city-state that took place in the second half of the 3rd century BC?

Figurines: quadruped

A quadruped standing on a rectangular base is represented by **No. 98**. Only the base of the terracotta that is made from two longitudinal molds for each side of the animal and the front left leg and hoof of what may be a horse, bull, sheep, or goat are preserved. The fragment may belong to a simple model of an animal. Alternatively, it may belong to a figurine that portrayed a youth riding on the back of an animal, as in other terracottas found at or in the vicinity of Metaponto.[115] These terracottas show a youth sitting astride different animals including a goat, a sheep, a horse, and a bull. The figurines date to the later 4th and 3rd century BC and occasionally served as funerary offerings.

Archaeological Contexts of the Survey Terracottas

Background considerations

It is useful to consider the terracottas within the context of each individual survey site to see what may be inferred about the nature of the site and the activities performed there. In Magna Grecia, most terracotta figurines and relief plaques have been found within a sanctuary where they had been dedicated as votive offerings. The discussion above concerning the different classes of terracottas represented among the finds from the survey has shown, however, that many of the types of terracottas found at Metapontine sanctuaries have also been recovered from domestic and industrial settings. In addition, by the late 5th and increasingly in the 4th and 3rd centuries BC, terracottas were sometimes used in funerary ritual.

For example, relief plaques of a reclining banqueter have been found at urban (temples A, B, C, E, and temenos of Zeus),[116] extramural (Crucinia and Favale),[117] and rural (Pantanello)[118] sanctuaries, as well as at kilns in industrial areas (kerameikos and Sant'Angelo Vecchio),[119] where they were manufactured. Such reliefs also are present at several farmhouses in the chora where they may reflect domestic religious observances or industrial activity.[120] Reliefs have also been found just west of the city center in association with a chamber tomb of the late 4th to early 3rd century BC.[121] Again, the same wide range of contexts is documented at Metaponto for Classical

[112] For a general overview and additional bibliography, see Bell 1990; Abruzzese Calabrese 1996, 193–94; Barra Bagnasco 1996a, 221; R. Ammerman 2002, 145–47, with specific reference to Poseidonia-Paestum. For Capua: Baroni and Casolo 1990, passim. For Morgantina: Bell 1981, 51–67, 96–97, 174–86, 190–206.

[113] For such funerary practice at neighboring Taranto, see Graepler 1997, passim. At Herakleia, Pianu 1990, 230–31, 239–45, 247–48, 253–56.

[114] At Taranto, by the 2nd c. BC, Tanagra-style figurines are present only in graves, a situation that is seen as marking the end of the use of terracotta figurines as votive offerings in sanctuaries: Abruzzese Calabrese 1996, 195.

[115] Youth on goat from Casa Sansone: Pesce 1936, 445, Fig. 8, No. 25. From the kerameikos: Signore 1996, 345, Fig. 69, No. 69. Youth on horse, attributed to a funerary context: Letta 1972, 139–40, Pl. 28, No. 1. Youth on horse from a tomb at Montescaglioso: Lo Porto 1988–1989, 374–75, Fig. 84, Nos. 1–2. Youth on sheep and bull from Timmari: Lo Porto 1991, 147, Pl. 70, Nos. 182–83.

[116] Temple A: Adamesteanu 1980, 93–96, Figs. 84b, 85e, 86d; possibly also 92–94, Figs. 81d, 84c–e. Temple B: Adamesteanu 1980, 145–46, Fig. 150b and possibly c. Temple C: Adamesteanu 1980, 200, 202, Fig. 210a, c–e. Large cloaca in the urban sanctuary: Meirano 1996, 86, Fig. 27. Temple E: Postrioti 1996, 105–6, Pl. 17a–b. Temenos of Zeus near the theater: Signore 1996, 353; Meirano 1996, 77, Fig. 14.

[117] Crucinia: Lo Porto 1981b, 332, 334, Fig. 41, Nos. 1–3. Favale: Adamesteanu 1980, 273–76, Figs. 288a–b, d–h (possibly), 289c–d; Liseno 2004, 61, Pl. 26f.

[118] Mary Malone and I are preparing the publication of some 160 of these fragmentary relief plaques, mentioned by Carter (1987, 186). One fragment from the site of a kiln on the hill above the sanctuary has been published: Carter 1983a, 442, 456, Fig. 32, No. 29.

[119] Kerameikos: D'Andria 1980a, 418–23, Figs. 61–63; Signore 1996, 299–340, Figs. 1–58. Kilns at Sant'Angelo Vecchio: Edlund 1979, Fig. 32; 1984, 287; 1988, 238–39, Pl. 45, No. 2. For molds among terracotta fragments of a banqueter that suggest possible presence of a coroplastic workshop at Sansone within the urban center, see Lo Porto 1966, 157. For other evidence of possible workshops at Metaponto: Signore 1996, 353.

[120] Musillo and Pascucci-Torraca farmhouses in Saldone, Arezzo farmhouse in Lago del Lupo, Sciurti farmhouse at Marconia-Le Cesine, Risimini-Stefan farmhouse in Campagnolo, a farm site in Pagliarone, near Montescaglioso: Barberis 1995, 15, 27–28, Fig. 12. Morlino farmhouse near Pantanello: Nava 2002, 735, Pl. 66.1. While excavations at most farm sites recovered only one to four fragments, 36 fragments of a reclining banqueter were found at the Risimini-Stefan farm. The large number of various kinds of terracottas found at this site have suggested that in the 4th c. BC there was a workshop for making terracottas there: Barberis 1995, 15 n. 16 for additional bibliography.

[121] From tomb 5 near Casa Ricotta: Lo Porto 1981b, 351–52, Fig. 65, No. 1.

figurines of standing women holding offerings,[122] as well as for relief plaques representing a satyr, silenos, or Pan with a female figure.[123] These terracottas, too, are present not only at sanctuaries, but also in domestic structures and at kiln sites.

Although more figurines and relief plaques tend to be found in the excavation of a sanctuary, large numbers of terracottas are likewise found at farmhouses and industrial sites. For example, several hundred fragments of relief plaques portraying a satyr, silenos, or Pan with a female figure have been recovered from temple E and the Pantanello sanctuary,[124] but dozens of fragments of such plaques have similarly been found at farmhouses.[125] Excavations at the Kerameikos, moreover, yielded more than 80 fragments of relief plaques of a reclining banqueter.[126]

In other words, the presence of an isolated or even a handful of molded terracotta fragments at a site from the survey does not necessarily indicate the presence of a sanctuary. To assess the implications of the recovery of a figurine or a relief plaque, one must view the terracotta within the context of the other finds and any architectural structures with which it is associated at each site, as well as the topographic setting of the site itself.[127] In those cases where only a few terracottas have been recovered from a site, associated finds that are linked to food production and storage of agricultural produce, such as mortars and pestles, amphora, pithoi, and varied plain wares, might point to the use of the terracottas within a rural domestic context.[128] The presence of slag or numerous misfired vessels would suggest, on the other hand, industrial activities at the site. Other factors may indicate that an area was used as a burial ground.[129]

Each site has its own individual history, which may include successive periods in which different activities were carried out at the same location. It has been postulated, for instance, that at least one 4th-century BC farmhouse in the Metapontine chora was constructed in the area of an earlier Archaic sanctuary, the presence of which is asserted primarily on the basis of the recovery of terracotta figurines.[130] This is certainly what happened in the 3rd century BC at the sanctuary at San Biagio alla Venella.[131] Varied activities could also take place simultaneously at a single site. Kilns and clay pits used for the production of roof tiles, pottery, and occasionally terracotta figurines and relief plaques were sometimes located near working farmsteads as well as sanctuaries.[132] The artifacts collected from the surface of a given site in a survey may provide indications of such a long-term and varied history, but only excavation has the chance of obtaining more detailed clarification regarding the different chapters that make up the life story of an individual site. Even when such chapters can be written, the exact role that a given votive terracotta or group of terracottas played in that history may still need to be evaluated with caution.[133]

[122] See discussion on standing female figurines with offerings above, especially nn. 25–29.

[123] See discussion on the relief plaques above, especially nn. 78–85.

[124] More than 600 fragments were found at temple E: Postrioti 1996, 21–72, Pls. 1–9. More than 250 fragments were found at Pantanello; Mary Malone and I are preparing them for publication.

[125] Excavations at the Risimini-Stefan farmhouse, for example, yielded 23 fragments, while 42 fragments were collected from the area of what was probably a farm on the Sciurti property near Marconia: Barberis 1995, 15, 29, Figs. 13–14.

[126] D'Andria 1980a, 418–23, Figs. 61–63; Signore 1996, 3[illegible]–40, Figs. 1–58, Nos. 1–84.

[127] An earlier, preliminary analysis of the materials recovered or observed at 488 sites identified by the ICA survey indicated that 26[illegible] (55%) were farms, 133 (27%) necropoleis, 2 quarries, 15 kilns, and only 2 sanctuaries: Carter 1990, 408–9, Fig. 2; Carter 1987, 208. After further study, these earlier interpretations have, of course, been somewhat modified in this final publication of the Bradano-Basento transect of the survey.

[128] E.g., pestles have been found at site 358 (where the terracotta of Europa and the Bull was found), mortars at sites 638 (a relief of a satyr, silenos, or Pan and female figure) and 728 (indeterminate fragments), etc.

[129] Evidence for kilns has been found at sites 213, 507, and 521, although nothing suggests that the terracottas recovered from the three sites were being manufactured at these farmhouses. Several sites appear, on the other hand, to include tombs: sites 70, 132, 289, 380, 437, 520, and 638. A Hellenistic female figurine was found at site 132, and a relief of a satyr, silenos, or Pan was recovered from site 638, but most of the terracottas found at these sites are too fragmentary to determine their subject.

[130] For a farm at Lago del Lupo, proprietà Arezzo, where significantly more terracottas of Archaic date have been found than those of later date contemporaneous with the farmhouse, see Barberis 1995, 35–36. Farms dating from the 6th to 4th c. BC also lie near a sanctuary at Sant'Angelo Grieco: Edlund 1987, 100. For farms of the 6th and 5th c. BC built at Sant'Angelo Vecchio, close to what may have been a sanctuary, to judge from finds of 5th c. BC skyphoi and other cups as well as a votive inscription near a spring, see Edlund 1986; 1987, 99–100. For the presence of terracottas of the 5th and 4th c. BC at site 77 of the SNAM pipeline, which similarly enjoyed a long and varied history from the end of the 8th to the 1st c. BC, see Nava 2003a, 675.

[131] As was also the case at the spring sanctuary at Pizzica, a farmhouse was built near the earlier sanctuary at San Biagio in the 3rd c. BC: Nava 2001, 943.

[132] For clay pits at the Saldone-Musillo and Risimini-Stefan farms, see Barberis 1995, 17. For 4th-c. BC and later kilns at Sant'Angelo Vecchio, where there are earlier farmsteads, a small necropolis, as well as a sanctuary near a spring: Edlund 1988, 237–42, 1987, 99–100, 1986, 119–22. For kilns at the sanctuary of San Biagio alla Venella: Adamesteanu 1974a, 60–61; 1965, 130–31; Edlund 1984, 284; Carter 1994a, 178.

[133] Edlund (1984, 287), for instance, does not see the production of votive terracottas at the kilns excavated at Sant'Angelo Vecchio as evidence for the local presence of a sanctuary (which is instead documented by a

Site no.	Terracotta counts	Date of figurines				Date of plaques	Minature votives		Drinking vessels				Pateras	Jugs	Loom weights	Other votive objects
		Archaic: 6th c. to mid-5th c.	Classical: mid-5th c. to late 4th c.	Hellenistic: late 4th c. to 2nd c.	Unknown	4th c. to 1st quarter of 3rd c.	Archaic	Unknown date	Archaic	Classical	Hellenistic	Unknown date	Hellenistic	Archaic		
43	1					x		1								
101	1		x					1								
104	1					x		1								
128	1				x										1	
132	1			x				1								
154	1				x			4							1	
155	2		x					1							2	
183	1					x		1							1	1
193	9					x		1								3
221	2					x									2	
235	5				x			1								
247	1				x										1	
250	2				x			1							1	
266	48			x		x		2	x	x	x				2	
270	3		x					14	?							
309	3		x												6	
312	1					x		1								
322	1					x		2								
329	5					x									1	
334	4	x						24								
342	10		x	x							x		x		2	
352	1				x			2								
353	1		x												1	2
354	1			x				2							2	
358	1		x					1							1	
365	2			x		x									2	
375	1				x			1							1	
376	7					x									5	
394	8		x												2	
395	7					x		2								
397	10	x	x				85							x		
417	2					x									1	
449	4					x		2								
477	41	x					10							x	1	
488	15					x		1				1			1	
506	35		x			x		1							2	
516	2	x				x		2							1	
521	2					x		4							1	
523	1		x					4							1	
638	2					x		1								
726	1		x												1	

Table 13.1 Associations between terracottas and other possible votive objects. Sites with concentrations of terracottas highlighted in orange; additional sites with concentrations of miniature votive vessels are highlighted in blue.

Funerary rites seem, on occasion, to have employed certain kinds of terracottas at Metaponto. By the late 5th century BC, protomai, busts, and models of birds, for example, were sometimes placed in tombs.[134] By the later 4th century BC, Tanagra-style figurines of seated and standing women and models of animals might accompany the burial of (in most cases) a woman or a child.[135] In the absence of evidence indicating that a site is not a necropolis, the possibility remains that the few isolated fragments of Tanagra-style figurines and what may be a protome or bust from the ICA survey could possibly have played a religious role within a funerary setting as easily as within the context of a domestic or a public rural cult.[136]

Quantities of terracottas on survey sites

At most of the 73 survey sites where the presence of terracottas is documented, one or only a few fragments were recovered. As noted above, such a small quantity of figurines or plaques could have come as easily from a sanctuary as from a farmhouse, industrial area, or necropolis. Knowledge of the finds associated with the terracottas is thus critical for assessing the nature and history of the site as well as the potential function of the terracottas within that locale.

The majority of the sites are likely to be farmsteads. The terracottas, although collected usually in just a single fragment at any one site, nevertheless provide an extraordinary wealth of evidence for the practice of domestic cults within the chora of a Greek city-state. Future publication of the excavations of several such farmhouses in the territory of Metaponto should cast clearer light upon this fascinating topic. Of equal interest is the issue of the rural production of terracotta figurines and plaques at Metaponto. The presence of a mold (**No. 340**) for making a relief of a reclining banqueter at site 404, for instance, offers additional documentation for such industrial activity in the countryside. It is worth noting that misfired pottery was also found at site 404.

A few sites from the ICA survey have yielded, however, more than a handful of terracotta fragments and therefore should be examined to see if each group may have characteristics consonant with a rural sanctuary. Nine or more fragments have been recovered from seven sites, which are highlighted in Table 13.1 and noted on the map in Figure 13.2.

Site 477 is striking for the number of Archaic terracottas and their consistent imagery. The figurines and a single protome portray female figures who represent either deities, often enthroned, wearing a polos, or standing women. The character of the collection of terracottas from site 477 recalls the large corpus of Archaic female figurines recovered in excavations at the sanctuary of San Biagio alla Venella or the smaller number of Archaic terracottas from Tavole Palatine, Favale, and other sanctuaries within the chora.[137]

The terracotta fragments whose subject may be identified from sites 266 and 506 are predominantly relief plaques portraying a reclining banqueter with a woman and child, or a satyr, silenos, or Pan with a female figure. Exceptions are a seated female figure and quadruped (**Nos. 97–98**) from site 266 and a standing female holding the offering of a sheep or goat in her left arm (**No. 285**) from site 506. None of the three terracottas is inconsistent, however, in terms of its possible date with the relief plaques, which date from the 4th to first quarter of the 3rd century BC.[138] The preponderance of one or both of these two classes of

concentration of 5th-c. BC drinking cups in the area of the spring and an inscription). This stands in contrast to Carter (1994a, 173, Table 7.1), who sees the terracottas from the 4th c. BC kiln as evidence not only for the production of votive items to be used elsewhere, but also for the local presence of a sanctuary active at this time at Sant'Angelo Vecchio. Terracotta figurines, in association with miniature votive pottery of the late 6th c. BC in the kerameikos in the urban center, have likewise been interpreted as evidence for local cult activity, i.e., as a votive deposit dedicated to a divinity worshipped by the craftsmen within the industrial workshop area: see D'Andria 1980b, 119. Both the situation at Sant'Angelo Vecchio and that in the kerameikos illustrate the difficulties one may encounter in interpreting the presence of votive terracottas at a site where other archaeological finds document activities that are those regularly associated not with a sanctuary, but rather with industrial or agricultural work. One needs to consider what other (if any) criteria should be satisfied in order to stress the cultic value of terracottas in such contexts. Care should doubtless be exercised in weighing the evidence.

134 Malone 1998, 771–73, 778, 780, 784–85.

135 Malone 1998, 774–77, 778–79, 781–83. See also n. 111 above.

136 Again, the archaeological context and associated finds from each individual site must be evaluated for each of these pieces to see whether the terracotta is coming from a possible burial site or not. See n. 129 above for sites from the survey that may have burials. The Hellenistic Tanagra-style figurines that could possibly have served as grave offerings are **Nos. 7** (which comes from a site where a tomb is thought to be present), **178**, **183**, and **167**. The protome or bust is **No. 150**.

137 For brief summaries and additional bibliography, see Carter 1994a, 169–75, Table 7.1; Edlund 1987, 96–99; De Juliis 2001, 84–107. San Biagio alla Venella: Olbrich 1979, *passim*. Tavole Palatine: Lo Porto 1981a, 26, 35–37, Figs. 23–26; Galli 1926–28, 72–74, Figs. 10, 12, 14. Favale: Nava 2002, 732, Pls. 64.2, 65.1. For Archaic terracottas at a spring sanctuary near Incoronata/San Teodoro, south of the Basento (site 57 on the SNAM pipeline): Nava 2003a, 672–73, Pl. 37.2.

138 See discussion above regarding the 5th-c. stylistic date often ascribed to figurines of women holding offerings, such as **No. 285**, and the presence of figurines in contexts of the 4th to early 3rd c. BC.

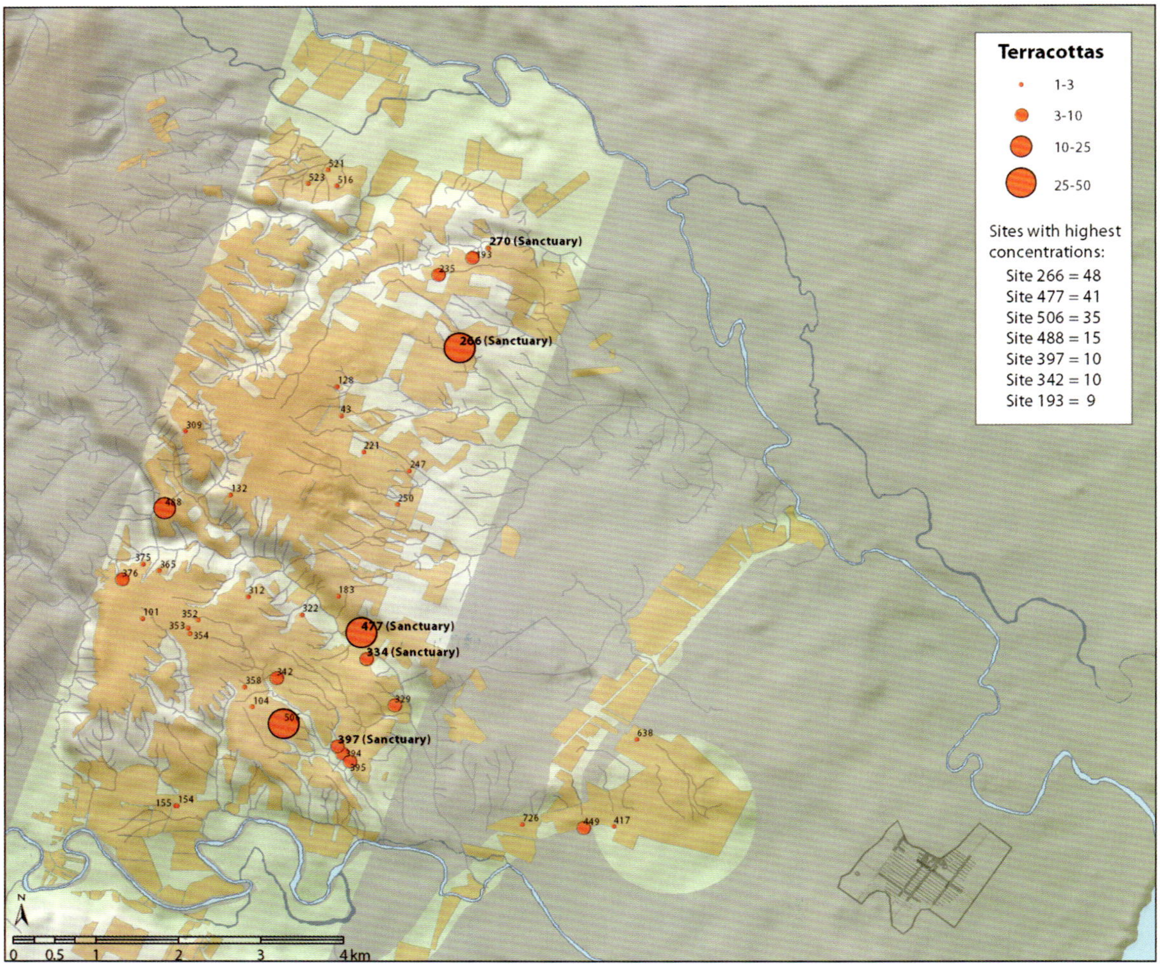

Figure 13.2 Sites with terracottas found in conjunction with other possible votive materials suggesting ritual activity. (JT)

relief plaques at the three survey sites falls in line with what were the most common votive terracottas of the 4th century BC at the spring sanctuary at Pantanello in the chora and at temple E in the urban center.[139]

Of the four sites where only 9 to 15 fragments were collected, those terracottas whose subject can be identified from sites 488 and 193 are all relief plaques portraying a satyr, silenos, or Pan with a female figure. Site 397, on the other hand, boasts some of the best-preserved and oldest terracottas found by the survey among the 10 pieces collected at this site: **Nos. 210–11** portray, respectively, a standing kore and an enthroned goddess. The other two fragments whose subject can be determined from site 397 likewise represent a female figure, but date to the 5th or 4th century BC. In contrast, only one Hellenistic female figurine (**No. 167**) and a fragment of Classical drapery (**No. 166**) can be identified among the 10 fragments found at site 342. The two terracottas are too small a sample to be indicative of the possible nature of the site.

Associated artifacts

The greater number of figurines and plaques at each of these sites (especially the first three, where 35 or more fragments were found) may indicate the presence of a sanctuary. But, as noted above, large numbers of terracottas have often been found at both farmhouses and industrial areas. It is advisable to examine the associated finds and topographical setting of these sites for additional evidence to help to assess more carefully

139 Pantanello: Carter 1994a, 170, 178, 194, 196–97, Fig. 7.18, Table 7.1. Temple E: Postrioti 1996, 140–42. The ICA survey also recovered 30 fragments of terracottas, mostly relief plaques portraying a satyr, silenos, or Pan with a female figure, from site 203. This material is included in the present catalog, but will be discussed in the projected volume that will publish the finds of the survey from sites farther inland near Cozzo Presepe. It is nevertheless worth noting that the large number of terracottas suggests that this may be the site of another rural sanctuary.

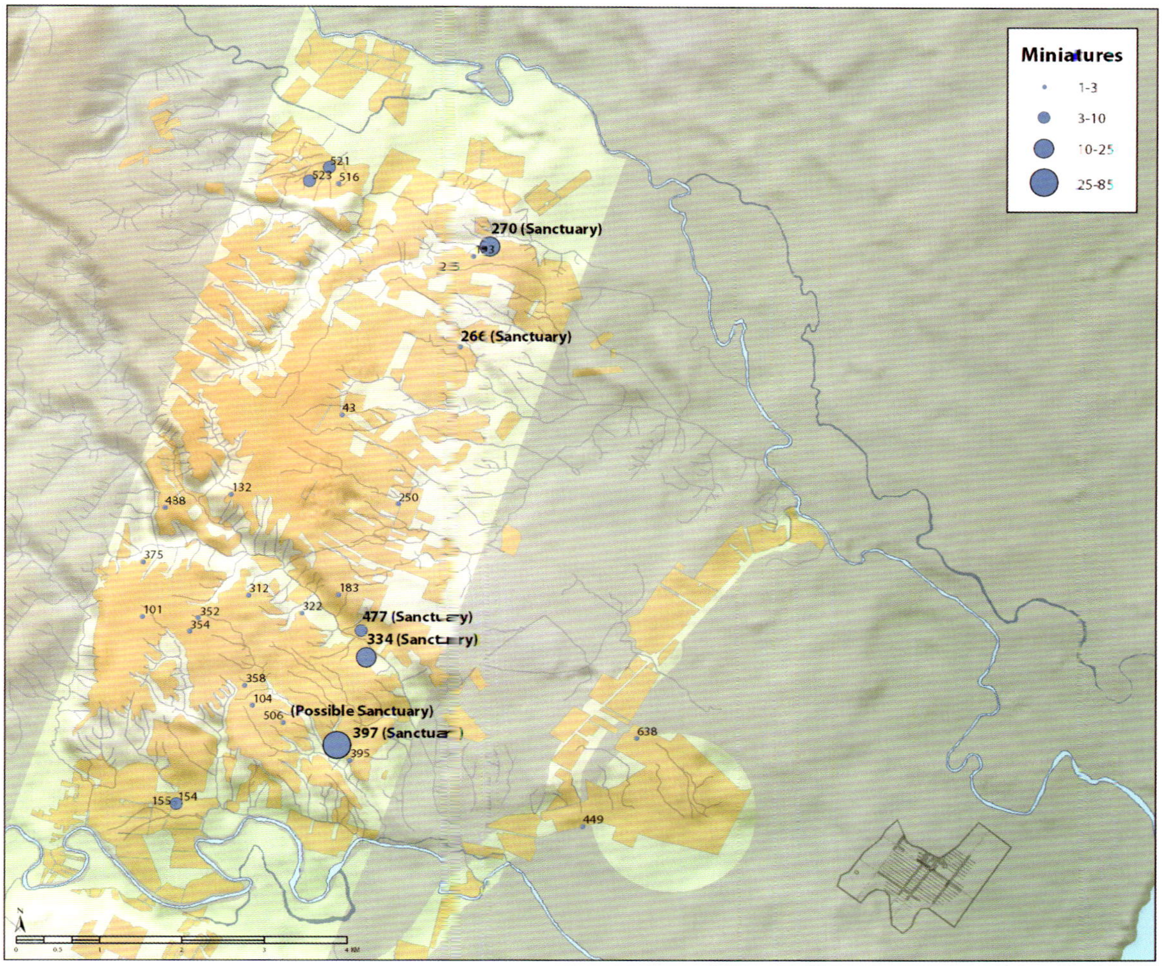

Figure 13.3 Sites with miniature vessels. (JT)

the reasons for the higher frequency of votive terracottas at each location.

Two types of ceramics have been commonly associated with ritual practice at Metapontine sanctuaries: perforated Ionic cups of the 6th century BC and miniature votive vessels that range in date from the 6th to the 4th century BC. Both types of pottery have likewise been found at farmhouses and necropoleis and were thus used also in domestic and funerary rites.[140]

In other words, as is the case with votive terracottas, the recovery of such pottery at a site, while it does probably reflect the performance of a ritual act, does not necessarily signal the presence of a sanctuary. A more global view of the archaeological evidence for a given site must be taken when evaluating the significance of the presence of such pottery in a particular setting.

Miniature votive vessels

Miniature votive vessels have been recovered from 29 sites where terracotta figurines or plaques were also found on the ICA survey (*Table 13.1, Fig. 13.3*).[141]

[140] For perforated Ionic cups at Pantanello sanctuary and necropolis, see Carter 1994a, 187–88, Fig. 7.14. For miniature votive vessels at the urban sanctuary: Adamesteanu 1980, 59, 63, Fig. 49; [illegible], Figs. 94–96c–d; 220–22, Figs. 228, 230. At temple E: Postrioti 1996, 146. At the Crucinia sanctuary: Lo Porto 1981b, 310–15. At Favale: Adamesteanu 1980, 282–87, Figs. 296–99; Nava 2002, 732. From San Biagio alla Venella: Olbrich 1976, 377; 1979, 16, 70. From the Pantanello sanctuary: Carter 1994a, 195, Fig. 7.19; Edlund 1987, [illegible]. From the spring sanctuary at site 57 along the SNAM pipeline near Incoronata/San Teodoro, south of the Basento: Nava 2003a, 673. From San Salvatore, site 29 of the SNAM excavations: Nava 2003a, [illegible], Pl. 40.2. From the necropolis at Casa Ricotta: Lo Porto 1966, 214–15, Fig. 69, No. 9. From the Pantanello necropolis: *Necropoleis*, 691, 72[illegible]–30, 828. In farm sites of the ICA survey miniature votives, along with louteria and thymiateria, constitute less than 5% of the typical assemblage of plain wares (see discussion of assemblages, Ch. 5). It is worth noting that cult activity has been suggested at the kerameikos at Metaponto on the basis of the discovery of miniature votive pottery associated with terracottas: see D'Andria 1980b, 119.

[141] The following sites have yielded miniature vessels as well as terracotta figurines or plaques: 43, 101, 104, 132, 154, 155, 183, 193, 235, 250, 266, 270, 312, 322, 334, 352, 354, 358, 375, 395, 397, 449, 477, 488, 506, 516,

The ceramics are present at most of the sites where larger numbers of terracottas were collected.

At sites 477 and 397, the miniature vessels are of Archaic date, as are the terracottas. At site 477 the early date, variety, and quantity of the terracottas all provide strong support for the identification of the site as a sanctuary. To this is added its location near a copious spring whose waters are still used in agricultural work today. As is well known, such springs were often the focus of cult activity in antiquity.[142] The presence of 10 miniature votive vessels among the artifacts collected at the site, while not clinching, does nevertheless add yet another supporting element to the picture that is developing of an Archaic rural sanctuary within the Metapontine chora. Although only 10 terracottas were found at site 397, the astonishing number of 85 miniature vessels (predominantly krateriskoi and a few pyxides) argues strongly for the identification of the site as the locale of another Archaic rural sanctuary.[143]

As at site 477, the relative abundance of 41 terracotta fragments argues strongly by itself for the identification of site 266 as a rural shrine where the votive practice of dedicating an image of a reclining banqueter or of Pan or a satyr with a female figure mirrors that at the Pantanello sanctuary in the 4th and early 3rd century BC.[144] The plentiful presence of pottery, including two miniature votive vessels and numerous drinking cups dating from Archaic to Hellenistic times, also supports such an identification. Again, the large number of terracotta plaques recovered from site 506, mostly from relief plaques of reclining banqueters or of Pan or a satyr with a female figure, provides solid ground for potentially identifying site 506 as that of a rural sanctuary. The presence of a single miniature vessel recovered from the site is consistent with such an interpretation.

The evidence for a shrine at site 488, on the other hand, is more tenuous. Only 15 terracotta fragments were collected there. Of these, the subject of just three fragments can be recognized: all belong to reliefs portraying a satyr, silenos, or Pan with a female figure. The single miniature vessel from the site adds only modest weight to any argument for identifying site 488 as a rural sanctuary. The large number of skyphoi and other drinking cups from the site may likewise reflect the performance of religious rites, but a mortar, pithoi, misfired pots, and possibly other artifacts argue for different kinds of activity at the site. Perhaps the terracotta plaques and miniature votive vessel reflect instead the practice of a domestic cult.

A similar situation presents itself at site 193, where only nine terracotta fragments were collected. Of the six fragments whose subject may be recognized, all belong to relief plaques portraying a satyr, silenos, or Pan with a female figure. The associated pottery from site 193, which includes a single miniature votive vessel, is plentiful and ranges in date from the 5th to 3rd century BC. Such an assemblage of material may point to the presence of a rural shrine, but more likely belongs to a farmstead. This may well be the case given the limited number of terracottas, all belonging to the same kind of plaque that, as we have seen, is commonly found in domestic and industrial contexts in the Metapontine chora.

It remains to consider site 342, where 10 terracottas were found but not in association with miniature votive pottery. The site's substantial quantity of pottery, ranging in date from the 5th to the 3rd century BC, does however include many Hellenistic paterai and skyphoi. It does not, regrettably, cast a light sufficiently bright to better our understanding of the two recognizable terracottas from site 342: a fragment of Classical drapery and a Hellenistic figurine possibly of a dancing woman. The terracottas by themselves, as noted above, constitute too small a sample from which to draw firm conclusions about the nature of the site where they were recovered.

Finally, it is useful to shift focus briefly to two sites where very few terracotta fragments were recovered, but where numerous votive miniature ceramics were found. The most interesting of these is site 334, with a date restricted to the 6th century BC. Here, two dozen miniature votive vessels were collected along with two Archaic terracottas portraying a seated female figure (**Nos. 162–63**). The site, with its association of miniature votive vessels and

521, 523, 638.

[142] At Metaponto, cult activity has been documented at various springs, including those at San Biagio alla Venella, Pantanello, Sant'Angelo Vecchio, a recently excavated Archaic sanctuary south of the Basento at Incoronata/San Teodoro (SNAM excavation site 57), and possibly site 77, a short distance farther to the north: Nava 2003a, 672–73/Pl. 37.2 and 675, respectively. For general discussion of the association of springs and religious rites, see Dilthey 1980; Edlund 1987, 60–61, 98–101; Barra Bagnasco 1999b; A. Russo 1999.

[143] This is the sanctuary site in the Avinella locality mentioned by Carter (1994a, 176).

[144] See nn. 50 and 81 above. Also Carter 1994a, 193–98.

Archaic votive terracottas portraying a female deity, appears to fit a pattern seen at other sanctuaries of Archaic date within the Metapontine chora: San Biagio alla Venella, the Archaic period of the Pantanello sanctuary, the spring sanctuary at SNAM site 57 near Incoronata/San Teodoro, and others to which may probably now be added sites 477 and 397 from the ICA survey.[145] At site 270 a figurine of an enthroned goddess of the 5th century BC was found in association with 14 miniature vessels as well as many Archaic drinking vessels. The assemblage thus offers tantalizing evidence for the presence of early cult activity also at this location.[146]

Of the 21 remaining sites where terracottas were found in association with miniature votive vases, usually only one or two terracottas and one or two ritual vessels are documented. It is worth noting that at seven sites, the votive vessels were found in association with fragments of a relief plaque portraying a satyr, silenos, or Pan with a female figure.[147] Although it is difficult to weigh the importance of this frequent association of artifacts when both classes of material are each present in such small numbers at a given site, the archaeological evidence may well reflect the domestic practice of a popular cult within the chora.

Other types of pottery and loom weights

No terracottas were found at the sites that yielded examples of such ritual vessels as perforated vessels (sites 113, 326, 341, and 390) and thymiateria (sites 730 and 742). An interesting correlation between the presence of terracottas, miniature votive vessels, and jugs among the cooking ware in the Archaic period is however worth noting. Vessels for pouring liquids (necked pots/jugs) predominate among the cooking-ware shapes found at sites 397 and 477.[148] The two sites also boast the presence of Archaic terracotta figurines and miniature votive vessels. They should probably be identified as sanctuaries. The presence of jugs at these sites, of which site 477 is also associated with a spring, may perhaps point to the performance of ritual libations.

[145] For a survey of the sanctuaries of the Archaic period in the urban center as well as the chora, see Carter 1994a, 161–77, 183–89; De Siena 2001a, 762–66. For sanctuary at SNAM site 57: Nava 2005a, 672–73.

[146] The presence of indigenous impasto pottery at the site is also intriguing for the history of the site.

[147] Sites 43, 104, 322, 395, 449, 516, and 638.

[148] See Ch. 11.

More than 100 loom weights have been recovered by the ICA survey.[149] While a loom weight normally reflects female industry at a farmhouse, it could also serve as a votive offering at a shrine where cult was practiced. A loom weight may thus be part of a larger assemblage of artifacts from either a domestic or a ritual setting. The association, in and of itself, of a terracotta with a loom weight at a given site from the survey thus illuminates neither the use to which the terracotta may originally have been put nor the character of the location where the objects were both found. Additional information about the archaeological context of these two kinds of artifacts is needed to understand better the nature of their shared findspot.

Conclusion

The presence of only a few terracotta figurines or relief plaques at a given site identified by the ICA survey does not by itself reveal the character of the site. Figurines and relief plaques representing most classes of terracottas manufactured at Metaponto have been found in the excavations of farmsteads, kilns, and sanctuaries within the chora. The recovery of a handful of terracottas at a site could potentially come from a domestic or industrial context as easily as from a sanctuary. Additional information gained from an analysis of associated artifacts, architectural features, and the topographic setting of a given site can, however, enable the nature of many of the sites from the survey to be identified. In addition, certain kinds of terracottas, such as protomai and busts as well as Tanagra-style figurines, were occasionally used in burial rites. Thus, the recovery of such a terracotta may indicate a funerary context. But again, support for a funerary interpretation is needed from associated finds at the site. A large quantity of terracotta figurines or relief plaques collected at a site, on the other hand, has a greater likelihood of indicating the presence of a sanctuary. When the terracottas are accompanied by other supporting evidence, such as votive pottery or the presence of a spring, one may be more certain they mark a locale that possessed a sacred character.

Of the 73 sites where terracottas were found, a vast majority were farmsteads. The presence of terracotta figurines and plaques in so many residential contexts is striking and unparalleled. It attests emphatically to the widespread practice of domestic cult in the Meta-

[149] See Ch. 14. Loom weights were found at 24 sites where terracottas also were recovered.

pontine countryside. In addition, the terracotta mold from site 404 provides complementary evidence for the manufacture of such images in the chora in order to supply the religious needs of the rural population.

An analysis of the different classes, imagery, dates, and quantities of terracottas found at each site in combination with associated artifacts, such as votive miniature pottery, and topographic setting suggests, however, that at least five—and possibly as many as six—sites identified by the ICA survey mark the location of rural sanctuaries. Sites 477, 397, 334, and 270 appear to be cult places established in the 6th century BC in honor of a female deity (potentially portrayed in the votive terracottas) and fall closely in line with what is known of other Archaic shrines in the Metapontine countryside. Sites 266 and possibly 506 are instead characteristic of rural sanctuaries at Metaponto of the 4th and early 3rd century BC, in which the predominant imagery of votive relief plaques portrays a satyr, silenos, or Pan with a female figure who may represent either a nymph or a maenad. The plaques depicting a spirited thiasos of rustic creatures who bear vessels filled with wine and cornucopiae brimming with fruit and cakes appear to have been originally fashioned by a Metapontine craftsman. The votive plaques, with their links to the world of Dionysos and the lesser deities of the countryside, evoke the festive blessings of fertility and abundance that such divinities could bestow upon the efforts of the inhabitants of the farmsteads that densely punctuated the territory of Metaponto in the 4th century BC. It should, in fact, come as no surprise that these same relief plaques are the most commonly recovered terracottas from the survey. Some of the sites, where only a few such plaques were found, may mark the location of yet another rural shrine, but most of the sites were no doubt farmhouses where, in a domestic setting, individual families propitiated these same divinities. This widespread popular devotion was apparently quite effective. The votive relief plaques belong to the period that archaeological research (much of it conducted in connection with the ICA survey and excavations) has shown to have experienced the most intensive and successful occupation and cultivation of the chora in the history of Metaponto.

Catalog of Terracottas

The terracottas are ordered first by site/inventory number and then chronologically. An asterisk (*) after the site number indicates materials from the 1980 ICA pilot survey (sites numbered 1–44), as well as at the intensive survey sites and plots located south of the Basento River through the 2001 season (most of the sites numbered 562 and higher). Additional information about the sites can be found in the Gazetteer.

Fragments that can be more securely identified according to class and often even mold series are listed first, while those that cannot are listed in a group at the end.

Comparisons are given for the relief plaques of a satyr, silenos, or Pan with a female figure to groups of closely related mold series distinguished by Postrioti (1996) as individual types, according to his scheme of typological classification.

Unless otherwise indicated, all catalog items consist of one fragment; none of the objects was found with completely intact sides. All dimensions are in cm. Maximum height is indicated by "h," maximum width by "w." Color of the fired fabric of the terracotta is described according to the Munsell Soil Color Charts.

Terracotta Figurines and Relief Plaques

1:1

1. TC 043-01 Relief plaque, probably satyr, silenos, or Pan with female figure (Postrioti type AII or AIII)
Dimensions: h 6.1
Fabric and technique: 7.5YR 7/6. Polos added by hand to molded visage.
Description: Abraded surface with incrustation. Head of female figure who turns slightly to her right and wears a broad polos. Centrally parted hair is pulled back under the polos in segmented waves. Longer tresses appear to be depicted falling along left side of neck. Edge of relief is preserved behind polos. The extension of clay from the right side of the polos probably linked the female figure to a silenos. Although the head shows a stylistic affinity with those of figurines that represent standing women bearing offerings, the slight turn of the head, the asymmetrical treatment of the locks of hair falling along the left side of her neck, and the extension of clay from the base of the polos on the right side of the head all suggest that the head belongs to the dancing female figure on a relief plaque with a silenos or a satyr.
Comparisons: Postrioti 1996, 51–52, 60–61, Pls. 4a, 6a. Similar stylistically to head of standing offering-bearer from Taranto: Iacobone 1988, 34, Pl. 26c.
Date: Last quarter of 4th to first quarter of 3rd c. BC.

2. TC 070-01 Unidentified type
Dimensions: h 3.6
Fabric and technique: 7.5YR 7/6, finely levigated. Hand-modeled, solid.
Description: Irregular cylinder of solid terracotta broken at both ends. One end has a circular depression and splays outward. The other end is closed in the center, while stumps of two flat strips of clay extend outward at each side. Incised line runs partially around cylinder from splayed end to middle of piece. Finely levigated fabric suggests an early date.
Date: Possibly 6th c. BC.

1:2

3. TC 101-01 Classical female figure of unidentified type
Dimensions: h 5.6
Fabric and technique: 5YR 7/4-6, fairly fine with small (< 1 mm) quartz inclusions.
Description: Left foot and hem of skirt rest on rounded base. The woman's toes are indicated. Flat vertical folds of drapery fall in arc over foot. The fragment is too small to allow one to determine the exact type of figurine or relief, but it may represent the foot of a standing woman bearing an offering, or that of an enthroned figure. It may also belong to the female seated at the end of the reclining banqueter's couch. In early examples of terracotta relief

3. TC 101-01 cont.

plaques with this imagery, the craftsman attached a cast, made separately from a mold of a seated woman, directly to a relief portraying a reclining male figure. This technical workshop practice accounts for the preservation of the rounded base under the foot of the female figure in such examples of reclining-banqueter reliefs.

Comparisons: For standing female: Lo Porto 1966, 162–63, Fig. 25, No. 3; Lo Porto 1981b, 331, 333, Fig. 40, No. 7; Adamesteanu 1980, 270, Fig. 284c. From Taranto: Iacobone 1988, 20, Pl. 10b; Lo Porto 1964, 251–52, Fig. 67. For enthroned female: Lo Porto 1966, 161–62, Fig. 23. From Taranto: Lippolis 1995, 60, Pl. 17, Nos. 1–2. For similar foot and base on relief of reclining banqueter: Lo Porto 1966, 172–73, Fig. 33, No. 1. From Taranto: Lippolis 1995, 51–53, Pl. 7, No. 4.

Date: Mid-5th to early 4th c. BC.

1:2

4. TC 104-01 Relief plaque of satyr, silenos, or Pan with female figure (probably Postrioti type AIIIB)

Dimensions: h 5.8

Fabric and technique: 5YR 7/6.

Description: Cornucopia held in left arm of female figure. Edge of relief is preserved along side of cornucopia and fruit.

Comparisons: Postrioti 1996, 64–65, Pl. 6c; Barberis 1995, 29, Fig. 13.

Date: Last quarter of 4th to first quarter of 3rd c. BC.

5. TC 128-01 Drapery fragment of unidentified type

Dimensions: h 5.2

Fabric and technique: 5YR 7/6.

Date: Unknown.

6. TC 128-02 Drapery fragment of unidentified type

Dimensions: h 2.1

Fabric and technique: 7.5YR 7/6.

Date: Unknown.

7. TC 132-01 Hellenistic figurine of standing female

Dimensions: h 10.4

Fabric and technique: 5YR 6-7/8. Back of figurine appears to have been formed from a smoothed slab of clay. Some retouch on drapery folds at hip.

Description: Surface incrustation. Right leg and drapery of standing female. She places her flexed right leg slightly forward, while her left leg probably carried her weight. She wears a chiton with a belted overfold (*apoptygma*) that falls to the level of her upper thigh. The front right corner of the overfold is shown bent back upon itself at the center of the right side of her thigh. Elaborately stacked folds of her mantle cascade down the right side of her leg to her knee. Modeling folds and catenaries render gracefully the volumes of the knee and leg. The hem of her skirt falls in an arc over a simple slipper with a pointed end. At the lower right side of the leg, portions of the clay slab that formed the back of the figurine and the strips of clay added by hand to the interior to strengthen the join between the front cast and the back slab are preserved. Foot and lower hem of garment preserve finished base. This finely crafted figurine finds close parallels in Cyrenaica as well as Puglia, possibly from Taranto. These figurines show a young woman wearing a sleeveless chiton with a V-shaped neckline and overfold belted under her breasts. She wears a himation wrapped like a shawl around both arms at the level of her wrists and hands. Her right arm is held akimbo as she places her right hand, enveloped in the himation, on the back of her right hip. Her left hand hangs at her side. The ends of the himation that have been bunched tightly together around her hands fall in decorative masses of zigzagging pleats and folds along her sides.

1:1

7. TC 132-01

Comparisons: From Cyrenaica: Burn and Higgins 2001, 224, Pl. 114, No. 2706. From Apulia, possibly Taranto: Besques 1986, 38, Pl. 29 a, c. For similar figurine from Taranto: *The Coroplast's Art*, 154, No. 41. For figurine similar only in general pose and *apoptygma*, from tomb dated from end of the 4th to early 3rd c. BC at Herakleia: Pianu 1990, 92–93, 231–32, Pl. 41, No. 2.

Date: Late 4th to 3rd c. BC.

1:2

8. TC 154-01 Drapery fragment of unidentified type
Dimensions: h 4.2
Fabric and technique: 5YR 7/8.
Date: Unknown.

1:2

9. TC 155-01 Protome or Classical figurine of female figure, unidentified type
Dimensions: h 9.7
Fabric and technique: 7.5YR 7/4.
Description: Two fragments preserve right shoulder and upper arm. Incrustation on surface. Himation falls in five rounded ridges of drapery over right shoulder of female figure. Edge of terracotta is preserved along back of shoulder and upper arm. The large scale suggests that it may belong to a protome. It does, however, find close stylistic parallels for the treatment of the himation folds in figurines, which are often of significant size, that portray a standing female who carries an offering.
Comparisons: For protome from Timmari: Adamesteanu 1974, 221. For standing female holding offerings: Adamesteanu 1980, 271, 273, Fig. 285l; Liseno 2004, 43–44, Pls. 10b, 11a; Lo Porto 1981b, 333, Fig. 40, No. 2. For possibly seated figure: Nava 2003a, [illegible], Pl. 37.2 lower left.
Date: Mid-5th to 4th c. BC.

1:2

10. TC 155-02 Early Classical figurine of female figure, unidentified type
Dimensions: h 6.1
Fabric and technique: 5YR 7/6.
Description: Drapery of *apoptygma* falls over right side of torso of female figure in stacked pleats that form an omega fold at their lower edge. The fragment is too small to determine whether it belongs to a figurine of a standing or seated figure. Its closest parallels are standing *peplophoroi* who carry a small quadruped that appears to be a lamb in their left arm, such as in **No. 285**.
Comparisons: For *peplophoros* holding sheep: Lo Porto 1966, 162–63, Fig. 24, No. 1; 1981a, 330–33, Fig. 39, No. 1, 40, No. 1. ; Liseno 2004, 48–49, Pls. 16a, 17b, 18a. For similar from Taranto: Herdejürgen 1971, 43, Pl. 28 a; Winter 1903a, 110, No. 6. For similar arrangements of *apoptygma* on seated figures from Taranto: Iacobone 1988, 45, Pl. 35b–c; Higgins 1967, 90, Pl. 40b = Mollard-Besques 1963, 67, 134, Pl. 15, No. 3 = Mollard-Besques 1954, 123, Pl. 88 C247; Winter 1903, 110, No. 7, 122, No. 6, 123, No. 7.
Date: Mid-5th c. BC.

1:1

11. TC 170-01 Head of Archaic figurine of female figure
Dimensions: h 3.1
Fabric and technique: 7.5YR 7/6. Mold of front of figurine filled solidly with clay. Back smoothed flat to form solid plaquelike figurine.
Description: Some incrustation on surface. Blocky, rectangular head of female figure wearing a polos. Rounded volume of hair protrudes from under polos running in a band across forehead. The eyes are rimmed, nose elongated, lips swollen, and ears indicated. The head may have belonged to either a standing or a seated female figure.
Comparisons: For similar heads on standing and seated figurines: Olbrich 1979, 104, 141, 166, 201–2, 244, Pls. 2 A8, 18 A84, 35 B7, 49 C11 Var. B, 67 C103; Adamesteanu 1980, 198–201, Figs. 207e–g, 209f.
Date: Second to third quarter of 6th c. BC.

1:2

12. TC 173-02 Classical figurine of female figure, unidentified type
Dimensions: h 5.1
Fabric and technique: 2.5YR 6/8.
Description: Top of fruit chipped. Left hand of female figure holds low, rounded basket of fruit out from her left side.

1:2

12. TC 173-02 cont.

Her fingers extend fully to the left edge of the basket. Two circular fruits at the basket rim are rendered in higher relief than the flatter background. Edge of figurine preserved along side of hand and upper right edge of basket. The fragment has no exact parallel among terracottas yet found in the Metapontino that portray either a standing or seated female figure who holds a basket of fruit at her left side.
Comparisons: Similar: Meirano 1996, 94, Fig. 30; Liseno 2004, 48-49, Pls. 16a, 17b, 18a.
Date: Second half of the 5th to mid-4th c. BC.

1:2

13. TC 173-01 Relief plaque of standing warrior
Dimensions: h 10.0
Fabric and technique: 5YR 7/6.
Description: Three fragments preserve right arm and right side of torso of male figure. Surface of arm chipped. Standing male wearing clamys clasped over his chest places right hand on hip. Drapery falls in vertical folds behind his bent right arm and muscular torso. Back edge of relief is preserved behind right arm. Complete examples of this mold series from Taranto portray a bearded male figure, standing in a hipshot pose, who raises a round shield on his left arm. He wears a festive wreath decorated with a taenia and rosettes and a clamys that falls behind his naked torso and legs to the level of his calves. A plain stele and tripod are depicted to his left, while a naked child stands between the man and the stele and reaches upward with outstretched arms towards the adult male figure.
Comparisons: See **No. 326**. From Taranto: Abruzzese Calabrese 1996, 201–2, Fig. 146; Lippolis 1995, 54, Pl. 12, No. 1; Iacobone 1988, 143, Pl. 133d–e; Winter 1903, 208, No. 12.
Date: 4th c. BC.

14. TC 173-03 Unidentified type
Dimensions: h 5.9
Fabric and technique: 5YR 6-7/8.
Description: Fabric quality suggests that it may belong to male torso TC 173-01.
Date: Unknown.

15. TC 173-04 Unidentified type
Dimensions: h 4.4
Fabric and technique: 5YR 7/8.
Description: One fragment preserves edge of plaque or figurine.
Date: Unknown.

16. TC 183-01 Relief plaque, probably reclining banqueter
Dimensions: h 4.7
Fabric and technique: 7.5YR 7/6.
Description: Dark incrustation covers surface. Right arm of female figure seated at end of couch. Edge of relief is preserved along right arm of female figure.
Comparisons: Similar: Signore 1996, 309–10, Figs. 10–12, 316–17, Figs. 23–25; Lo Porto 1966, 168–69, 171, Figs. 30, Nos. 1–31, No. 1, Pl. 8, Nos. 1–2.
Date: 4th c. BC.

17. TC 193-01 Relief plaque of satyr, silenos, or Pan with female figure (Postrioti type AIB)
Dimensions: h 6.5
Fabric and technique: 7.5YR 7/6.
Description: Traces of red paint preserved on surface. Upper left edge of plaque with overhanging rock of grotto and grape clusters.
Comparisons: Postrioti 1996, 32–33, Pl. 2a–b; Adamesteanu 1980, 275–76, Fig. 289e. From Cozzo Presepe: Prag 1983b, Fig. 150.
Date: Last quarter of 4th to first quarter of 3rd c. BC.

1:2

18. TC 193-02 Relief plaque of satyr, silenos, or Pan with female figure (Postrioti type AIB or AIIB)
Dimensions: h 8.0
Fabric and technique: 5YR 7/6-8.
Description: Two fragments (193-02, 193-06). Upper left curved plaque edge with overhanging rock of grotto and grape clusters. Stippled texture suggests rocky surface of background. A thin linear depression follows the curve of the edge of the relief demarcating the image's boundary.
Comparisons: Postrioti 1996, 32–33, 52–53, Pls. 2a–b, 4b; Adamesteanu 1980, 275–76, Fig. 289e. From Cozzo Presepe: Prag 1983b, Fig. 150.
Date: Last quarter of 4th to first quarter of 3rd c. BC.

1:2

19. TC 193-03 Relief plaque of satyr, silenos, or Pan with female figure (Postrioti type AIIC)
Dimensions: h 5.8
Fabric and technique: 5YR 7/6-8.
Description: Cornucopia in left arm of female and drapery of veil. Relief edge preserved at side of cornucopia.
Comparisons: Postrioti 1996, 53–54, Pl. 5c; Barberis 1995, 29, Fig. 14.
Date: Last quarter of 4th to first quarter of 3rd c. BC.

1:2

20. TC 193-05 Relief plaque of satyr, silenos, or Pan with female figure (Postrioti type AII)
Dimensions: h 7.5
Fabric and technique: 5YR 7/4.
Description: Cornucopia held by female figure. Her left hand grasps the lower part of the basket. This appears to be a variant similar to Postrioti type AIIB, where the figures are set into a background with a grotto and grape arbor. The fruit in the cornucopia has not been recut as in Postrioti's type AIIB. Instead, the pile of circular fruit, each piece with a central depression, is similar to that in such other variants as Postrioti type AIIA or AIIC.
Comparisons: Postrioti 1996, 50–54, Pls. 4a (type AIIA), 4b (type AIIB), 5c (type AIIC). Postrioti's type AIIA: Barberis 1995, 29, Fig. 14. Postrioti's type AIIC: Meirano 1996, 85, Figs. 24–25; Tempesta 2001, 101, Fig. 106; De Siena 1999a, 67, 70, Fig. 45.
Date: Last quarter of 4th to first quarter of 3rd c. BC.

21. TC 193-04 Relief plaque, probably satyr, silenos, or Pan with female figure (Postrioti type AIIC)
Dimensions: h 4.3
Fabric and technique: 10YR 7/4.
Description: One fragment, probably tufts from satyr tail.
Comparisons: Postrioti 1996, 53–54, Pl. 5b; Meirano 1996, 85, Fig. 24; Tempesta 2001, 101, Fig. 106; De Siena 1999a, 67, 70, Fig. 45.
Date: Last quarter of 4th to first quarter of 3rd c. BC.

22–25. TC 193-07–10 Unidentifiable fragments
Date: Unknown.

26. TC 200-01 Unidentified type
Dimensions: h 6.0
Fabric and technique: 5YR 7/4.
Description: Possibly hair fragment, female protome or bust.
Date: Unknown.

27. TC 201-02 Head, probably of Archaic or Classical figurine of female figure
Dimensions: h 2.5
Fabric and technique: 5YR 6/6.
Description: One fragment with lower portion of face and neck. The face appears to have been oval in shape with a rounded chin and straight-lipped, broad mouth, but the features are too fragmentary and worn to be able to date the terracotta closely.
Date: Second half of 6th to late 5th c. BC.

28. TC 201-01 Unidentified type
Dimensions: h 5.1
Fabric and technique: 10YR 7/3.
Description: Right forearm and hand placed over straight, broad drapery folds. Fabric differs from that of other Metapontine terracottas.
Date: 4th c. BC or later.

29. TC 201-03 Unidentified type
Dimensions: h 2.4
Fabric and technique: 5YR 6/8.
Description: Edge of plaque preserved.
Date: Unknown.

30. TC 203-01* Relief plaque of reclining banqueter
Dimensions: h 5.0
Fabric and technique: 7.5YR 7/6.
Description: Right foot and hem of dress of female seated at end of couch.
Comparisons: Similar: Signore 1996, 304, Fig. 2, 309–10, Figs. 10–12, 315, Fig. 23; Letta 1972, 97–98, Pl. 16, No. 1.
Date: 4th c. BC.

31. TC 203-02* Relief plaque of Pan with female figure (Postrioti type AIIA or AIIB)
Dimensions: h 4.6
Fabric and technique: 5YR 7/6.
Description: Left edge of relief with hoof of Pan.
Comparisons: Postrioti 1996, 52–53, Pl. 4b.
Date: Last quarter of 4th to first quarter of 3rd c. BC.

32. TC 203-03* Relief plaque of Pan with female figure (Postrioti type AIA, AIB or AIIA, AIIB)
Dimensions: h 3.2
Fabric and technique: 5YR 7/6-8.
Description: Cloven hoof of Pan.
Comparisons: Postrioti 1996, 26–27, 32–33, 52–53, Pls. 1a, 2a, 4b; Meirano 1996, 85–86, Fig. 26. From Sant'Angelo Vecchio: Carter 1980a, 14, Fig. 11.
Date: Last quarter of 4th to first quarter of 3rd c. BC.

1:2

33. TC 203-04* Relief plaque of Pan with female figure (Postrioti type AIA or AIB)
Dimensions: h 5.3
Fabric and technique: 5YR 6/8.
Description: Shaggy left leg of Pan. Edge of *nebris* ending in lion's paws falls along the left side of the satyr's leg.
Comparisons: Postrioti 1996, 26–27, 32–33, Pls. 1a–b, 2a–b; Meirano 1996, 85–86, Fig. 26.
Date: Last quarter of 4th to first quarter of 3rd c. BC.

34. TC 203-05* Relief plaque of satyr, silenos, or Pan with female figure (Postrioti type AIIA or AIIB)
Dimensions: h 6.2
Fabric and technique: 5YR 7/4-6.
Description: Left foot of female figure emerges from pleated folds of hem of garment.
Comparisons: Postrioti 1996, 52–53, Pl. 4b.
Date: Last quarter of 4th to first quarter of 3rd c. BC.

1:2

35. TC 203-08* Relief plaque of Pan with female figure (Postrioti type AIA or AIB)
Dimensions: h 5.4
Fabric and technique: 7.5YR 7/8.
Description: Shaggy left leg of Pan.
Comparisons: Postrioti 1996, 26–27, 32–33, Pls. 1a–b, 2a–b; Meirano 1996, 85–86, Fig. 26.
Date: Last quarter of 4th to first quarter of 3rd c. BC.

36–37. TC 203-06–07 Drapery fragments of unidentified types
Dimensions: h 4.5; h 3.1
Fabric and technique: 5YR 7/8; 7.5YR 7/8.
Date: Unknown.

38–59. TC 203-09–30* Unidentifiable relief fragments.
Date: Unknown.

1:1

60. TC 211-01 Relief plaque of reclining banqueter
Dimensions: h 3.6
Fabric and technique: 5YR 7/8.
Description: Some incrustation. Veiled head of female figure seated at end of couch. Hair is parted into two undulating masses swept up towards veil, which preserves edge of plaque.
Comparisons: Similar: Signore 1996, 309–10, Figs. 10–12, 315, 317, Figs. 23–25, 318, Fig. 27; Lo Porto 1966, 168–69, 171, Fig. 30, Nos. 1–31, Pl. 8, Nos. 1–2, 6; Sestieri 1940–41, 99, Fig. 16 upper left; Lo Porto 1981b, 332, 334, Fig. 41, No. 2; Adamesteanu 1980, 200, 202, Fig. 210c; Letta 1972, 94–95, Pl. 15, No. 2.
Date: 4th c. BC.

61–66. TC 211-02–07 Unidentifiable fragments
Date: Unknown.

1:2

67. TC 213-01 Relief plaque of reclining banqueter
Dimensions: h 8.7
Fabric and technique: 7.5YR 6/4.
Description: Traces of white slip on surface. Legs and drapery of female figure seated at end of couch. Paw of a lion's skin draped over the couch falls below left leg of female.
Comparisons: Signore 1996, 318–19, Figs. 27–29. For similar reliefs with lion's skin: Letta 1972, 77, Pl. 9, No. 3, 93, Pl. 14, No. 3.
Date: 4th c. BC.

68–69. TC 216-01–02 Unidentifiable fragments
Date: Unknown.

1:2

70. TC 221-02 Relief plaque, possibly reclining banqueter or standing horseman
Dimensions: h 5.2
Fabric and technique: 5YR 7/8.
Description: Left profile of horse with relief edge preserved at top. The terracotta is too fragmentary and worn to determine whether it belongs to a relief of a reclining banqueter, wherein a horse's head is sometimes depicted in the background between the banqueter and the woman seated on the dining couch, or to a relief of a naked male figure holding a shield standing heroically by his horse.
Comparisons: For fragments preserving only head of horse: Sestieri 1940–41, 99, Fig. 14; Schürmann 1989, 84, Pl. 48, No. 284. For horses with reclining banqueter: Signore 1996, 326, Fig. 41, 330–31, Figs. 46–47. For horse held by standing male figure: Letta 1972, 110–11, Pl. 20, No. 4. From Taranto: Bartoccini 1936, 166–69, Figs. 76–77; Kingsley 1979, 204, Pl. 2, No. 6; Higgins 1954, 361, Pl. 181, No. 1327; Lippolis 1995, 54, Pl. 12, Nos. 2–3. From S. Maria d'Anglona: Rüdiger 1967, 347, Fig. 20.
Date: 4th c. BC.

71. TC 221-01 Unidentified type
Description: Edge of relief preserved.
Date: Unknown.

1:1

72. TC 225-01 Head, Late Archaic figurine of female figure
Dimensions: h 4.8
Fabric and technique: 5YR 7/6. Mold of front of figurine filled solidly with clay. Back slightly hollowed out.
Description: Incrustation on surface. Badly battered and chipped. Figure wears a polos with a veil falling along the sides of her neck. Rounded volume of hair protrudes out from under polos, extending beyond the edge of the veil to frame forehead. The features of the rounded face are too worn to be distinguished. May belong to either a standing or seated female figure of Late Archaic style.
Comparisons: Somewhat similar: Olbrich 1979, 193, 250, Pls. 46 B63, 69 C117. On standing female: Adamesteanu 1980, 191–92, Fig. 201c. Stylistically similar, although not filled solidly with clay, is the head of a woman seated at end of a reclining banqueter's couch in which the female figure was originally derived from an independent figurine of an enthroned female: Lo Porto 1981b, 332, 334, Fig. 41, No. 1.
Date: Late 6th to first half of 5th c. BC.

73. TC 225-02 Unidentified type
Dimensions: h 4.1
Fabric and technique: 5YR 7/6. Slab attached to molded front of figurine to form back of solid figurine.
Date: Unknown.

74. TC 225-03 Unidentifiable fragment
Date: Unknown.

75. TC 228-01 Unidentified type
Dimensions: h 3.1
Fabric and technique: 5YR 6/4-6.
Description: Possibly arm, winged back of throne from left side of enthroned female. Back edge smoothed flat.
Comparisons: Postrioti 1996, 88–89, Pl. 15b–c.
Date: Possibly late 6th to mid 4th c. BC.

76–80. TC 235-01–05 Unidentifiable fragments
Date: Unknown.

81. TC 244-03 Relief plaque of satyr, silenos, or Pan with female figure (Postrioti type AIB or AIIB)
Dimensions: h 4.3
Fabric and technique: 5YR 7/6.
Description: Upper right edge of plaque with overhanging rock of grotto and grape clusters.
Comparisons: Postrioti 1996, 32–33, 52–53, Pls. 2a–b, 4b; Adamesteanu 1980, 275–76, Fig. 289e. From Cozzo Presepe: Prag 1983b, Fig. 150.
Date: Last quarter of 4th to first quarter of 3rd c. BC.

82. TC 244-02 Unidentified type
Dimensions: h 7.1
Fabric and technique: 5YR 7/4-6.
Description: Possibly right arm of Archaic or Classical figurine of female figure.
Comparisons: Classical figurine of standing female: Adamesteanu 1980, 270, Fig. 284a.
Date: Possibly 6th to 5th c. BC.

83. TC 244-01 Drapery fragment of unidentified type
Dimensions: h 4.7
Fabric and technique: 5YR 6-7/8.
Date: Unknown.

84–86. TC 247-01; 250-01–02 Unidentifiable fragments
Date: Unknown.

1:1

87. TC 266-01 Relief plaque of reclining banqueter
Dimensions: h 4.1
Fabric and technique: 5YR 7/6.
Description: Veiled head of female figure seated at end of couch. Hair is centrally parted into two undulating masses that are swept back towards veil. Rounded face has worn, indistinct features. Ball-shaped earring on left ear. Edge of relief is preserved above and along left side.
Comparisons: Edlund 1988, 238, Pl. 45, No. 2; Signore 1996, 320–21, Fig. 32.
Date: 4th c. BC.

1:1

88. TC 266-02 Relief plaque of reclining banqueter
Dimensions: h 4.0
Fabric and technique: 7.5YR 7/4.
Description: Lower right arm, hand, and lap of female seated at couch end. Thumb is distinguished from other fingers of her hand, which rests on side of her lap. The edge of the himation crosses over her lap, echoing the curve of her arm. Fragment includes right leg and feet of infant, who scrambles toward reclining male figure.
Comparisons: Signore 1996, 310, Figs. 11–12; Lo Porto 1966, 168–69, 171, Fig. 30, No. 1, and 31, Nos. 1, 3; Sestieri 1940–41, 99, Fig. 16 upper left; Letta 1972, 94–97, Pls. 15, Nos. 2, 4, and 16, No. 1.
Date: 4th c. BC.

1:1

89. TC 266-04 Relief plaque of reclining banqueter
Dimensions: h 4.4
Fabric and technique: 5YR 7/6.
Description: Incrustation on surface. Upper left side of head of bearded banqueter wearing wreath, decorated with rosettes and a taenia. The end of the taenia falls along his left shoulder. Relief edge is preserved along left side of rosette and taenia.
Comparisons: Signore 1996, 309–11, Figs. 10–11, 13–14; Lo Porto 1966, 171, Pl. 7, No. 8; Sestieri 1940–41, 99, Fig. 14; Adamesteanu 1980, 200, 202, Fig. 210d–e.
Date: 4th c. BC.

90. TC 266-05 Figurine of female, unidentified type
Dimensions: h 5.0
Fabric and technique: 7.5YR 7/6.
Description: Neck, left shoulder, and breast of female figure wearing garment with scooped neckline. Veil falls over shoulder. Edge of relief is preserved along shoulder. The fragment may possibly belong to the female figure who sits at the end of the couch of a reclining banqueter in older versions of such relief plaques, where the female figure derives from a mold series initially fashioned to depict an independent figurine of a seated female.
Comparisons: Lo Porto 1966, 166–67, 172–73, Fig. 28, No. 1, Fig. 33, No. 1.
Date: Second half of the 5th to early 4th c. BC.

91. TC 266-08 Relief plaque of satyr, silenos, or Pan with female figure (Postrioti type AIIA or AIIC)
Dimensions: h 7.3
Fabric and technique: 7.5YR 7/6.
Description: Incrustation on surface. Right hand of satyr grasps a volute krater at left edge of plaque. Fingers of right hand of female figure who embraces satyr may be present at other side of fragment. Edge of relief is preserved behind satyr's hand and volute krater.
Comparisons: Postrioti 1996, 51–54, Pls. 4a, 5a–b; Barberis 1995, 29, Fig. 14.
Date: Last quarter of 4th to first quarter of 3rd c. BC.

1:2

92. TC 266-10 Relief plaque of Pan with female figure (Postrioti type AIA or AIB)
Dimensions: h 7.3
Fabric and technique: 5YR 7/8.
Description: Surface incrustation. Shaggy lower right leg of Pan at joint where leg turns at a right angle toward hoof.
Comparisons: Postrioti 1996, 26–27, 32–33, Pls. 1a–b, 2a–b; Meirano 1996, 85–86, Fig. 26. From Sant'Angelo Vecchio: Carter 1980a, 14, Fig. 11.
Date: Last quarter of 4th to first quarter of 3rd c. BC.

1:2

93. TC 266-11 Relief plaque of Pan with female figure (Postrioti type AIA)
Dimensions: h 7.7
Fabric and technique: 7.5YR 7/8.
Description: Incrustation on surface. Lower right leg and finely pleated drapery of skirt of female figure. Tip of Pan's right hoof is preserved at lower right edge.
Comparisons: Postrioti 1996, 26–27, Pl. 1a–b; Meirano 1996, 85–86, Fig. 26. From Sant'Angelo Vecchio: Carter 1980a, 14, Fig. 11.
Date: Last quarter of 4th to first quarter of 3rd c. BC.

94. TC 266-12 Relief plaque of satyr, silenos, or Pan with female figure (Postrioti type AIB or AIIB)
Dimensions: h 4.7
Fabric and technique: 5YR 7/6.
Description: Edge of plaque represents overhanging rock of grotto.
Comparisons: Postrioti 1996, 32–33, 52–53, Pls. 2a–b, 4b; Adamesteanu 1980, 275–76, Fig. 289e. From Cozzo Presepe: Prag 1983b, Fig. 150.
Date: Last quarter of 4th to first quarter of 3rd c. BC.

95. TC 266-15 Relief plaque of satyr, silenos, or Pan with female figure (Postrioti type AIA, AIB, or AV)
Dimensions: h 4.4
Fabric and technique: 7.5YR 7/6.
Description: Drapery falls in fine vertical folds below breast of female figure.
Comparisons: Postrioti 1996, 26–27, 32–33, 70–71, Pls. 1a–b, 2a–b, 9a; Meirano 1996, 85–86, Fig. 26. From Cozzo Presepe: Prag 1983b, Fig. 150 left.
Date: last quarter of 4th to first quarter of 3rd c. BC.

96. TC 266-32 Relief plaque of satyr, silenos, or Pan with female figure (Postrioti type AIB or AIIB)
Dimensions: h 4.5
Fabric and technique: 7.5YR 7/6.
Description: Upper edge of plaque with pendant trefoil grape cluster.
Comparisons: Postrioti 1996, 32–33, 52–53, Pls. 2a–b, 4b; Adamesteanu 1980, 275–76, Fig. 289e. From Cozzo Presepe: Prag 1983b, Fig. 150.
Date: Last quarter of 4th to first quarter of 3rd c. BC.

1:1

97. TC 266-03 Late Classical or Early Hellenistic figurine of seated female figure
Dimensions: h 4.0
Fabric and technique: 7.5YR 7/6. Slab of clay added to molded cast of front of figure to form back.
Description: Incrustation on surface. Right foot wearing shoe and drapery hem of seated female figure. Although the exact mold series cannot be determined, the fragment belongs to a type common throughout Lucania, that portrays a seated female figure, wrapped in himation, holding a phiale, tympanum, swan, or basket with fruit.
Comparisons: From Pantanello necropolis, T 12, 325–275 BC: Malone 1998, 778, T 12-8. From tombs, usually of women, ranging in date from the last half of 4th to early 3rd c. BC, at Herakleia: Pianu 1990, 231–32, Pls. 25.1, 26.4, 28.1, 40.4, 41.2, 55.2, 71.2. From tombs at Sant'Arcangelo-San Brancato: *Greci, Enotri, e Lucani*, 262–67, 3.40.23–25, 30.41.14. From Timmari: Lo Porto 1991, 108–28, Pls. 46–58. From Egnatia or probably Taranto: Higgins 1954, 355, 363–65, Pls. 179, No. 1305, 186–87, Nos. 1336–45. From Accettura: D'Anisi 2005, 170–71, Figs. 3–5.
Date: Second half of 4th to early 3rd c. BC.

1:2

98. TC 266-07 Hellenistic figurine of quadruped, possibly ridden by a youth
Dimensions: h 4.7 w 7.3
Fabric and technique: 5YR 7/6. Each side of animal cast from a longitudinal mold. Slab of clay added as base between two side casts and slit by a knife in its center. Possible retouch to define front leg of animal.
Description: One fragment preserves front left leg of animal and front half of figurine base. Incrustation on surface. Front left hoof, fetlock, and leg of quadruped standing on a rectangular base. The fragment probably portrays a youth riding an animal such as a bull, horse, sheep, or goat.
Comparisons: For youth on horse: Letta 1972, 139–40, Pl. 28, No. 1. For youth on goat: Pesce 1936, 445, Fig. 8; also Signore 1996, 345, Fig. 69, No. 69. For youth on sheep and on bull from Timmari: Lo Porto 1991, 147, Pl. 70, Nos. 182–83. For horse with youth as rider from tomb at Montescaglioso: Lo Porto 1988–1989, 374–75, Fig. 84, Nos. 1–2.
Date: Second half of 4th to 3rd c. BC.

99. TC 266-09 Female figure of unidentified type
Dimensions: h 5.0
Fabric and technique: 7.5YR 7/6.
Description: One fragment, possibly drapery of himation pulled diagonally over underlying garment.
Date: Possibly 4th to early 3rd c. BC.

100. TC 266-06 Relief plaque of unidentified type
Dimensions: h 5.1
Fabric and technique: 7.5YR 7/6.
Description: Edge of relief is preserved.
Date: Possibly 4th c. BC.

101. TC 266-13 Relief plaque of unidentified type
Dimensions: h 9.1
Fabric and technique: 7.5YR 7/6.
Description: Incrustation on surface. Lower portion and edge of plaque.
Date: Possibly 4th c. BC.

102. TC 266-14 Relief plaque of unidentified type
Dimensions: h 6.1
Fabric and technique: 7.5YR 7/6.
Description: Edge of plaque with what is possibly an oval-shaped swag of drapery.
Date: Possibly 4th c. BC.

103–137. TC 266-16–31; 266-33–48; 269-01–03
Unidentifiable fragments
Date: Unknown.

1:2

138. TC 270-01 Severe-style figurine of seated female
Dimensions: h 7.2
Fabric and technique: 5YR 7/4-6.
Description: Drapery of chiton sleeve and himation falls in U-shaped loop beneath left forearm of enthroned female figure. Left side of throne and edge of figurine are preserved below drapery. The fragment is too small to be attributed to a specific mold series, but is similar to Severe-style figurines from Taranto and Herakleia representing an enthroned goddess that look back to Late Archaic models.
Comparisons: For similar figurines from Taranto: Lippolis 1982, 115, Pl. 31, No. 2; 1995, 60, Pl. 17, Nos. 1–2; Iacobone 1988, 45, Pl. 35a–c; Higgins 1967, 90, Pl. 40b = Mollard-Besques 1963, 67, 134, Pl. 15, No. 3 = Mollard-Besques 1954, 123, Pl. 88 C247; Higgins 1954, Pl. 172, No. 1253; Winter 1903, 122, No. 6. From Herakleia: Neutsch 1981, 166, Pl. 27, No. 3. For a fragment of the head and chest of an enthroned goddess figurine found at Metaponto that probably derives from one of the Tarantine mold series cited above: Pesce 1936, 444, Fig. 4.
Date: Second quarter to middle of 5th c. BC.

139–146. TC 270-02–03; 274-01–04; 288-01; 289-01
Unidentifiable fragments
Date: Unknown.

147. TC 296-02 Relief plaque of satyr, silenos, or Pan with female figure (Postrioti type AIB or AIIB)
Dimensions: h 3.9
Fabric and technique: 5YR 7-6/4.
Description: Upper edge of plaque with grape cluster. Bumpy, irregular form suggests rocky edge of grotto.
Comparisons: Postrioti 1996, 32–33, 52–53, Pls. 2a–b, 4b; Adamesteanu 1980, 275–76, Fig. 289e. From Cozzo Presepe: Prag 1983b, Fig. 150.
Date: Last quarter of 4th to first quarter of 3rd c. BC.

1:1

148. TC 296-01 Relief plaque of unidentified type
Dimensions: h 5.6
Fabric and technique: 7.5YR 6-5/4.
Description: Broken edge at top and lower diagonal of pilaster. Edge of relief plaque preserved around molding and upper vertical left side under rounded molding of pilaster. Pilaster crowned by rectangular horizontal molding supports an element that has broken away. Half-round molding outlines top and outer left edge of horizontal molding, suggesting that relief represents a crowning element that is more rounded than square. Irregularly broken fragment of relief extends beyond opposite side. Similar pilasters and stelai are represented in relief plaques portraying Zeus, horsemen, and warriors, but none resembles this example.
Comparisons: For relief plaques portraying Zeus and a stele: Lo Porto 1988, 16–18, Fig. 12, Nos. 4–5, Fig. 13, No. 4. For horseman: Lo Porto 1988 , 19–20, Fig. 15, No. 2. For warrior: Abruzzese Calabrese 1996, 201–2, Fig. 146; Lippolis 1995, 54, Pl. 12, No. 1.
Date: Possibly 4th c. BC.

1:1

149. TC 296-03 Relief plaque of unidentified type
Dimensions: h 5.8
Fabric and technique: 7.5YR 7/6.
Description: Flat slab is pierced by hole used for suspension of relief. Straight edge of plaque preserved on one side of fragment and marked by faint band. Similar holes are pierced in the plain pediment of a naiskos represented in relief plaques portraying the Dioskouroi from Taranto.
Comparisons: For Dioskouroi relief plaques from Taranto: Pirzio Biroli Stefanelli 1977, 320–22, 344–45, 347–48, Pls. 64–65, 80–81, 83; Lippolis 1995, 55–56, Pl. 14, Nos. 1–4.
Date: Possibly 4th c. BC.

1:2

150. TC 309-01 Protome or bust of female figure of unidentified type
Dimensions: h 10.4
Fabric and technique: 7.5YR 7/4-6.
Description: One fragment preserves upper right side of polos and hair. Hair on right side of forehead is swept up in individual groups of several linear, waved strands to base of a short polos, while tresses alongside of right cheek are swept back horizontally covering the ear. Traces of a veil and possibly an earring remain. The fragment cannot be securely attributed to a specific type. It bears a strong resemblance to certain protomes of a female holding a torch and piglet from Herakleia but also finds parallels in busts without specific attributes from other sites in the vicinity of Metaponto.
Comparisons: For similar protomai from Herakleia: Neutsch 1967, 189, Pl. 48, No. 4; 1968, 776, Fig. 23. For another such protome from Metaponto: Letta 1972, 113–14, type XII.1. For similar protome of female holding torch and piglet from Rubi, near Taranto: Higgins 1954, 348, Pl. 176, No. 1276. For similar protome, but of smaller scale from tomb 12 at Montescaglioso: Lo Porto 1973, 190, Pl. 38, No. 1. For similar bust from Pomarico Vecchio: *Pomarico Vecchio* I, 1.220 and 2.150, Fig. 89. From Timmari: Lo Porto 1991, 85, Pl. 28, No. 21. For somewhat similar bust, probably misclassified as a figurine, from Taranto: Iacobone 1988, 33–34, Pl. 27a. From Satyrion: Lo Porto 2002, 16, Pl. 2 lower.
Date: 4th c. BC

151–152. TC 309-02–03 Unidentifiable fragments
Date: Unknown.

1:2

TC 588-01

153. TC 312-01 Relief plaque of male wearing kausia
Dimensions: h 5.5
Fabric and technique: 5YR 7/8. Deeply impressed fingerprint on interior of figurine.
Description: Some incrustation on surface. Kausia with central button and rim is worn by youth whose brow is just visible under cap. Curved narrower upper rim of cap preserves the finished edge of the relief. This fragment belongs to the same mold series and generation as another terracotta fragment found by the ICA survey in 1992 between the Basento and Cavone Rivers (TC 588-01). It preserves the head of a youthful, beardless male figure wearing a kausia with a broad taenia, one end of which falls over his left shoulder. The head of the young man bears a striking similarity to heads of youthful and older bearded horsemen portrayed in Tarantine terracotta relief plaques. In some of these plaques, the bearded man, who stands in front of his horse with a phiale in his right hand and a shield in his left, wears a pilos with a taenia, the ends of which fall symmetrically over his shoulders. In other reliefs, youthful beardless horsemen, who ride their horses, also wear a pilos, but without a taenia. The outline of a similar pilos seems to appear in a ghostlike manner under the kausia of the Metapontine head. Perhaps the craftsman who made the mold used to cast this Metapontine terracotta did so by recutting the worn impression of a pilos in a mold taken from such a Tarantine horseman plaque into the more fashionably up-to-date kausia.
Comparisons: Unpublished terracotta from Basento–Cavone survey, TC 588-01. For older bearded horseman from Taranto: Lippolis 1995, 54, Pl. 12, Nos. 2, 4. For younger horsemen: Lippolis 1995, 54, Pl. 13, Nos. 1, 4.
Date: 4th c. BC.

154. TC 320-01 Classical figurine of female figure of unidentified type
Dimensions: h 3.2
Fabric and technique: 7.5YR 7/6.
Description: Left bare foot with toes indicated rests on curved base and may belong to a standing woman holding offerings.
Comparisons: For standing offering bearer: Letta 1972, 120–21, Pl. 24, No. 3; Adamesteanu 1980, 270, Fig. 284b. From Taranto: Winter 1903, 110, No. 4; Lo Porto 1981b, 331, 333, Fig. 40, No. 7.
Date: Possibly mid-5th to early 4th c. BC.

155. TC 320-02 Unidentifiable fragment
Date: Unknown.

1:1

156. TC 322-01 Relief plaque of satyr, silenos, or Pan with female figure (Postrioti type AIIIB)
Dimensions: h 6.0
Fabric and technique: 10YR 7/6. Color is unusual for Metaponto.
Description: Incrustation on surface. Head of female figure wearing a sphendone with waved tresses swept up at sides. Figure wears a veil and ball earrings. Blocky face has heavy eyelids and small, fleshy mouth.
Comparisons: Postrioti 1996, 64–65, Pl. 6c. Similar unpublished heads from Pantanello: inv. nos. PZ 77.666.02, PZ 77.946.01. For head with similar hairstyle but wearing a polos: Adamesteanu 1980, 143–44, Fig. 148c. For stylistically similar head of standing offering bearer from Timmari: Lo Porto 1991, 79–80, Pl. 25, No. 12.
Date: Last quarter of 4th to first quarter of 3rd c. BC.

157. TC 329-03 Relief plaque of reclining banqueter
Dimensions: h 5.2
Fabric and technique: 5YR 7-6/6
Description: End of lower cushion of couch under left arm of banqueter. Edge of relief is preserved behind cushion.
Comparisons: Similar: Lo Porto 1966, 167, 169, Fig. 29, No. 2, Fig. 31, No. 2.
Date: 4th c. BC.

158. TC 329-01 Probably plaque of satyr, silenos, or Pan with female figure (Postrioti type AIC)
Dimensions: h 4.5
Fabric and technique: 5YR 7/6.
Description: Cloaked hand is probably the left hand of a satyr, which is wrapped in his mantle and holds a volute krater. Edge of relief is preserved behind hand.
Comparisons: Postrioti 1996, 33, Pl. 3d.
Date: Last quarter of 4th to first quarter of 3rd c. BC.

159–161. TC 329-02; 329-04–05 Unidentifiable fragments
Fabric and technique: 5YR 7/6-8.
Date: Unknown.

1:1

162. TC 334-02 Archaic figurine of seated female figure
Dimensions: h 5.5
Fabric and technique: 5YR 7/6. Front of solid, plaquelike figurine cast from mold.
Description: Left side from shoulder to lap of seated female figure, whose left arm follows side of torso to rest in lap. Small, highly set left breast indicated. Figurine, which is little more than a thin curved slab of clay, is rendered in a simple fashion with little detail regarding dress. Edge of figurine preserved along side of arm.
Comparisons: Olbrich 1979, 165–67, Pl. 35 B7–B8; Lo Porto 1981b, 325–26, Fig. 34, No. 3; Sestieri 1940, 100–101, Fig. 39; Adamesteanu 1980, 188–89, Fig. 198a, 309, 311, Fig. 329; Barberis 2004, 93–94, Figs. 141–42; Nava 2002, 732, Pl. 44.2.
Date: Second to third quarter of 6th c. BC.

1:2

163. TC 334-01 Archaic figurine of seated female figure
Dimensions: h 5.6
Fabric and technique: 5YR 7/4-6. Smooth slab of clay added as back to molded front to form solidly filled figurine. Arms and skirt are hand-modeled additions.
Description: Some incrustation on surface. Summarily fashioned flat torso of female figure, originally placed on slab of clay bent into an upside down U-shape to form skirt and back of throne of figure. Arms extend outward. Only the apex of the curve of U-shaped slab and hand-modeled stubs of arms are preserved.
Comparisons: Similar: Olbrich 1979, 169, 181, Pls. 36 B12, 41 B41; Letta 1972, 34–35, Pl. 4, No. 4.
Date: Mid-6th c. BC.

164–165. TC 334-03–04 Unidentifiable fragments
Date: Unknown.

166. TC 342-02 Classical figurine of female figure of unidentified type
Dimensions: h 5.5
Fabric and technique: 5YR 7-6/6.
Description: Fine parallel folds of the *kolpos* fall between lower legs and over left knee of female figure. Similar drapery is found on a variety of seated and standing female figures.
Comparisons: For similar treatment of drapery on standing figure of Artemis: Lo Porto 1988, 9–10, Fig. 5, No. 2. On smaller-scale figurine of an enthroned goddess: Olbrich 1979, 191–92, Pl. 46 B59. On enthroned goddess from Taranto: Iacobone 1988, 45, Pl. 35a; Higgins 1967, 90, Pl. 40b = Mollard-Besques 1963, 67, 134, Pl. 15, No. 3 = Mollard-Besques 1954, 123, Pl. 88 C247. On seated female figure at end of banqueter's couch from Taranto that was taken from a separate mold of a seated woman: Iacobone 1988, 73, Pl. 63 b.
Date: Second quarter of 5th to early 4th c. BC.

1:1

167. TC 342-01 Hellenistic figurine of young female
Dimensions: h 6.7
Fabric and technique: 7.5YR 7/6. Molds used for both front and back of figurine; ivy leaves added by hand.
Description: The figure twists her torso to her right and raises her right leg forward, while lifting her left arm and stretching her right arm outward from her side. The animated pose suggests dancing, but the raised leg may indicate that the figure is instead seated. She wears a wreath of ivy with a ribbon falling over her back. Her sleeveless chiton has a V-shaped neckline and is belted under her breasts with a short *kolpos* or flounce falling just above her waist. Soft volumes of hair frame her rounded face, which presents worn, yet delicate, features. The ivy wreath links the figure to the world of Dionysos, suggesting that she represents a dancing maenad. Perhaps she was portrayed in the company of a satyr, such as in figurines from Taranto. The figure also closely resembles that of a girl seated on a swan in a figurine recovered from a Tarantine tomb.
Comparisons: For a head, possibly from the same mold series: Barberis 1995, 30, Fig. 17. For figurine of woman riding swan from tomb of last quarter of 4th to first quarter of 3rd c. BC at Taranto: Graepler 1996, 244, 246, Fig. 195. For youthful female raising right leg in active pose in company of satyr from tomb at Taranto: Palumbo 1986, 104, Pl. 15. For somewhat similar dancing figures from Metaponto: Barberis 1995, 30, Fig. 16; Malone 1998, 779, T 71–26; Sestieri 1940, 68–69, Fig. 17.
Date: Late 4th to early 3rd c. BC.

168–176. TC 342-03–10; 352-01 Unidentifiable fragments
Date: Unknown.

177. TC 353-01 Probably Classical figurine of female figure of unidentified type
Dimensions: h 7.5
Fabric and technique: 7.5YR 7/4-6.
Description: Left foot at edge of rounded base of figurine. The fragment may belong to a figurine of a standing woman representing various subjects.
Comparisons: For somewhat similar feet and base for standing female with offerings: Lo Porto 1966, 162–63, Fig. 25, No. 3; 1981b, 331, 333, Fig. 40, No. 7. For standing Artemis: Barberis 1995, 25, Fig. 10. From S. Maria d'Anglona: Rüdiger 1967, 349–50, Fig. 22d, 23c.
Date: Probably mid-5th to 4th c. BC.

1:2

178. TC 354-01 Hellenistic figurine of standing female figure of unidentified type
Dimensions: h 3.1
Fabric and technique: 5YR 6-7/8.
Technique: Thick walls of the figurine are unusual.
Description: Vertical, irregularly spaced drapery folds of chiton fall under mantle that crosses diagonally over the chiton.
Comparisons: For standing draped women of Hellenistic style from tombs dated 325–275 BC: Lo Porto 1981b, 352–53, Fig. 66; Malone 1998, 781–83, T 192.1–3. From Herakleia in tombs dated from last third of 4th to early 3rd c. BC: Pianu 1990, 79, 102–3, 231–32, Pl. 32, No. 1, Pl. 47, No. 2. From Taranto in tombs dated end of the 3rd to first quarter of 2nd c. BC: Graepler 1996, 244–45, Figs. 189–92.
Date: Late 4th to early 2nd c. BC.

179. TC 355-01 Plaque of satyr, silenos, or Pan with female figure (Postrioti type AIA or AIB)
Dimensions: h 5.4 w 6.2
Fabric and technique: 5YR 7/6.
Description: Two fragments (355-01 joins 355-02). Right shoulder, breast, and central torso of female figure. Strand of hair reaches edge of scooped neckline. Chiton falls in vertical folds from breast and catenaries between breasts.
Comparisons: Postrioti 1996, 26–27, 32–33, Pls. 1a–b, 2a–b; Meirano 1996, 85–86, Fig. 26; Adamesteanu 1980, 275–76, Fig. 289e. From Cozzo Presepe: Prag 1983b, Fig. 150 left.
Date: Last quarter of 4th to first quarter of 3rd c. BC.

180. TC 355-03
Description: Unidentifiable fragment
Date: Unknown.

1:1

1:1

181. TC 358-01 Figurine of Europa and Bull
Dimensions: h 5.3
Fabric and technique: 5YR 6/6. Right ear of bull added by hand.
Description: Right side of head, neck, and upper right shoulder of bull are represented in profile. Left arm of female figure reaches to her left to grasp horn of bull. Swallowtail drapery folds of woman's mantle fall from her left arm over neck of bull. Rounded protruding form behind shoulder of bull may represent the leg of the woman who rides the bull sidesaddle. Eye and soft folds of bull's skin are sensitively modeled. The figurine represents the abduction of Europa by Zeus disguised as a bull. The subject is common among figurines from Boeotia, which provide the closest parallels for this piece, but is also known in southern Italy and Sicily.
Comparisons: For similar figurines from Boeotia: Verhoogen 1956, 25, Fig. 17; Schmidt 1994, 70, Pl. 19, No. 87; Higgins 1954, 219, Pl. 112, No. 821. From Boeotia and Corinth: Winter 1903, 163, No. 6, 164, No. 5. For Europa and Bull on a relief plaque from Locri of the 5th c. BC and a terracotta mold, said to be from Taranto, of the later 4th c. BC: *LIMC* 4 (1988), 81, Nos. 89–90, s.v. "Europe I" (M. Robertson). For figurine of bull running in opposite direction from Tarsus: Besques 1972, 287, Pl. 358d, h. From Agrigento: A. Levi 1926, 179–80, Fig. 137.
Date: 5th c. BC.

182. TC 365-01 Relief plaque of reclining banqueter
Dimensions: h 3.3
Fabric and technique: 7.5YR 7/6.
Description: Rosette from headdress of banqueter. Five petals overlap in a spiral around central bead. Rosette preserves finished edge and back of relief.
Comparisons: Signore 1996, 309–11, Figs. 10–11, 13–15, 318, Fig. 28, 333, Fig. 49; Lo Porto 1966, 167–71, Figs. 29, No. 2, 30, No. 2, 31, No. 2, Pl. 7, Nos. 1–3, 5–8.
Date: 4th c. BC.

183. TC 365-02 Drapery fragment of unidentified type
Dimensions: h 5.8
Fabric and technique: 5YR 7/6. Hand modeled.
Description: Drapery swag from left side of Hellenistic figurine. Left side of drapery preserves finished edges of figurine.
Date: Late 4th to 3rd c. BC.

184. TC 375-01 Unidentifiable fragment
Date: Unknown

185. TC 376-01 Relief plaque of reclining banqueter
Dimensions: h 5.2
Fabric and technique: 2.5YR 6/8.
Description: Upper left side of headdress of banqueter. Wreath is decorated with a rosette of five petals and rises to possible central palmette or rosette, which is not preserved. Hair of male figure is centrally parted. Edge of relief is preserved along left side of headdress.
Comparisons: Signore 1996, 309–11, Figs. 10–11, 13–15, 318, Fig. 28, 333, Fig. 49; Pesce 1936, 445, Fig. 6; Sestieri 1940–41, 99, Fig. 14; Lo Porto 1966, 167–71, Fig. 29, No. 2, 30, No. 2, 31, No. 2, Pl. 7, Nos. 1–3, 5–8; Adamesteanu 1980, 200, 202, Fig. 210d–e.
Date: 4th c. BC.

186. TC 376-02 Relief plaque, possibly reclining banqueter
Dimensions: h 3.7
Fabric and technique: 5YR 7/8.
Description: Surface badly battered. Fragment of beardless face from top of nose to neck. Inner corner of right eye is preserved. The face may portray that of a youthful reclining banqueter.
Comparisons: Signore 1996, 307–8, 311, Figs. 7, 9, 15, 316, Fig. 24, 318, Fig. 23; Adamesteanu 1980, 200, 202, Fig. 210d–e.
Date: 4th c. BC.

187–194. TC 376-03–07; 380-01–03 Unidentifiable fragments, two from relief plaques
Date: Unknown.

1:2

195. TC 394-01 Drapery fragment of unidentified type
Dimensions: h 8.4
Fabric and technique: 7.5YR 7/4-6.
Description: Vertical drapery folds of irregular width belong possibly to a large-scale standing female figure.
Comparisons: Somewhat similar drapery on standing female figure: Adamesteanu 1980, 270, Fig. 284c.
Date: Possibly mid-5th c. BC.

196. TC 394-02 Drapery fragment of unidentified type
Dimensions: h 5.3
Fabric and technique: 5YR 7/6.
Description: Edge of relief or figurine is preserved next to softened zigzag fold.
Date: Possibly 5th to 4th c. BC.

197–202. TC 394-03–08 Unidentifiable fragments
Date: Unknown.

1:2

203. TC 395-01 Plaque of satyr, silenos, or Pan with female figure (Postrioti type AIB)
Dimensions: h 8.0
Fabric and technique: 5YR 7/6.
Description: Incrustation on surface. Raised right arm of female figure pulls veil away from her head. Broad zigzag and narrow vertical folds of veil fall below elbow and upper arm. Relief extended above hand and head to represent a rocky grotto with a grape arbor. Edge of relief is preserved along right side of figure's arm and her veil.
Comparisons: Postrioti 1996, 32–33, Pl. 2b; Letta 1972, 129, Pl. 26, No. 2. For female figure set more deeply within background setting of grotto and grape arbor: Adamesteanu 1980, 275–76, Fig. 289e. From Cozzo Presepe: Prag 1983b, Fig. 150 left.
Date: Last quarter of 4th to first quarter of 3rd c. BC.

204. TC 395-02 Plaque of satyr, silenos, or Pan with female figure (Postrioti type AIA, AIB, or AIC)
Dimensions: h 6.5
Fabric and technique: 5YR 7-6/4.
Description: Incrustation on surface. Right upper leg and side of lower torso of female figure who moves to her right. The parallel modeling folds of her chiton that encircle her thigh are faint.
Comparisons: Postrioti 1996, 26–27, 32–34, Pls. 1a–b, 2a, 3b; Meirano 1996, 85–86, Fig. 26.
Date: Last quarter of 4th to first quarter of 3rd c. BC.

205. TC 395-03 Relief plaque of Pan with female figure (Postrioti type AIA)
Dimensions: h 5.0
Fabric and technique: 5YR 7/6-8.
Description: Incrustation on surface. Left hand of Pan grasps a volute krater. A *nebris* wraps over his wrist and falls downward. Edge of relief is preserved behind satyr's hand and vessel.
Comparisons: Postrioti 1996, 26–27, Pl. 1a–b; Meirano 1996, 85–86, Fig. 26; Lo Porto 1988, 23, Fig. 20, No. 5.
Date: Last quarter of 4th to first quarter of 3rd c. BC.

206–209. TC 395-04–07 Unidentifiable fragments
Date: Unknown.

210. TC 397-01 Archaic figurine of standing female
Dimensions: h 11.3
Fabric and technique: 7.5YR 7/6-6/4. Mold of front of figurine filled solidly with clay. Back smoothed flat to form solid plaquelike figurine. Base indented upward. Incised lines mark moldings on back of lower base of figurine.
Description: One fragment preserves all but head of

1:1

210. TC 397-01 cont.

figure. Base chipped. Patches of incrustation on surface. Figure stands with weight evenly distributed on a low base. She places her left leg slightly forward and holds her hands at sides with thumbs outstretched. Her feet, which are bare with toes indicated, point to her left. A strand of hair falls over each shoulder to sides of breast in a long, undifferentiated, rectangular mass. She wears a belted peplos with rounded apoptygma falling to her waist while covering her upper arms in the manner of a short cape or poncho. Two incised lines mark moldings on back of base.
Comparisons: Olbrich 1979, 108–9, 142–43, Pls. 4 A19–A20, 19 A88 (this last figurine may be from same mold series, but of larger scale).
Date: Second to third quarter of 6th c. BC.

1:2

211. TC 397-04 Archaic figurine of seated female figure
Dimensions: h 5.9
Fabric and technique: 5YR 7/6-8.
Technique: Mold of front of figurine filled solidly with clay. Back smoothed flat to form solid plaque-like figurine.
Description: Lower right foot of throne and drapery and foot of enthroned goddess rest on base of figurine. Leg of throne ends in a lion's paw with three toes. Hem of smooth garment falls in an arc over bare right foot with straight, simple toes of enthroned figure. The depiction of the lion's paw of the throne leg within the same plane as the feet of the enthroned goddess and resting on a shared base finds no exact parallel among the terracottas recovered to date from Metaponto as well as from Herakleia. These are more three-dimensional figurines, in which the feet of the seated goddess extend beyond the feet of the throne. Plaquelike figurines from San Biagio alla Venella, Herakleia, and Francavilla Marittima near Sybaris may, however, represent a related version of the figurine in which much of the detail regarding the feet of the throne and feet of the goddess has been lost.
Comparisons: For plaquelike figurines with less well-defined detail, but with feet of throne and feet of goddess resting in same plane: Olbrich 1979, 167–68, 188–89, Pls. 35 B9–B10, 45 B56; Barberis 2004, 97, Fig. 149. From Herakleia: Neutsch 1981, 166, Pl. 26, No. 4. From Francavilla Marittima: Olbrich 1996, 182, 185, Pl. 2, No. 1. For more three-dimensional, hollow figurines of slightly later date, with feet of goddess extending beyond throne legs: Barberis 1995, 25, Fig. 8 right; Letta 1972, 57–58, Pl. 7, No. 2; Lo Porto 1966, 160–62, Fig. 21; Olbrich 1979, 192, Pl. 46 B60. From Herakleia: Neutsch 1981, 165–66, Pl. 25 = Olbrich 1996, 182, 185, Pl. 2, No. 2.
Date: Mid-6th c. BC.

212. TC 397-03 Classical female figurine, unidentified type
Dimensions: h 8.4
Fabric and technique: 7.5YR 7/4.
Description: Veil falls from head along right side of neck and over right shoulder in soft, irregularly stacked folds with a gently undulating zigzag fold along the edge of the veil. Part of right cheek and breast of figure are preserved. Edge of figurine is preserved behind veil at neck and shoulder. The fragment is too small to allow one to determine the exact type of figurine or relief to which it belongs. It may represent a standing woman bearing an offering, or a seated figure, or belong to the female standing at the end of the reclining banqueter's couch.
Comparisons: For figurines of standing women: Letta 1972, 120, Pl. 24, No. 2. From Taranto: Iacobone 1988, 20, 34, Pl. 10a–b, 27b; Winter 1903, 109, No. 8. For woman standing next to reclining banqueter from Taranto: Bartoccini 1936, 162–64, Figs. 64–65, 69.
Date: Second half of the 5th to mid-4th c. BC.

1:2

213. TC 397-02 Protome or Classical figurine of female figure of unidentified type
Dimensions: h 4.9
Fabric and technique: 5YR 7/8.
Description: Right shoulder and upper chest of female figure who wears a chiton with a scooped neckline that falls from shoulder in three vertical folds along right side of breast. Protruding linear volume of clay along top of shoulder may represent a transverse himation that falls along the back of the shoulder of figurines representing a standing female. The rendering also suggests that the open upper edge of the chiton is buttoned at intervals along the length of the arm, allowing the bare arm to be revealed in oval-shaped sections between buttons. Finished edge of terracotta is preserved along shoulder. The fragment may belong to a protome of a woman wearing a chiton in this fashion or to a figurine of a standing woman holding offerings.
Comparisons: For protomes from Herakleia: Neutsch 1968, 775, Fig. 22a; Otto 1996, 141, Pl. 1b. From Timmari: Adamesteanu 1974a, 220 lower left; Lo Porto 1991, 87–88, Pl. 29, No. 28. From a tomb at Ferrandina: Lo Porto 1973, 204–5, Pl. 53, No. 1. For figurines of a stand-

213. TC 397-02 cont.

ing woman with similar drapery, sometimes holding lamb: Letta 1972, 119, Pl. 23, No. 4; Adamesteanu 1980, 270–73, Figs. 284a, 285c, 287; Liseno 2004, 47–49, Pls. 15b, 16a, 17b, 18a; Winter 1903, 110, No. 5.
Date: Later 5th to end of 4th c. BC.

214–219. TC 397-05–10 Unidentifiable fragments
Date: Unknown.

1:2

220. TC 404-01 Relief plaque of reclining banqueter
Dimensions: h 7.2
Fabric and technique: 5YR 7/6.
Description: Bearded head and left chest of banqueter wearing polos with half-round base and taenia falling over shoulder. Hair is parted centrally into simple smooth masses. Edge of relief is preserved behind polos and shoulder of banqueter.
Comparisons: Lo Porto 1966, 172–73, Fig. 33, No. 3; Signore 1996, 304–7, Figs. 2, 4, 6.
Date: Second half of the 5th to 4th c. BC.

1:1

221. TC 417-01 Relief plaque of reclining banqueter
Dimensions: h 5.9
Fabric and technique: 7.5YR 7/7-6.
Description: Bearded head of male banqueter wearing wreath decorated with rosettes at center and sides. Hair of male figure is centrally parted into undulating smooth masses. Edge of relief is preserved along left side of headdress.
Comparisons: Signore 1996, 309–11, Figs. 10–11, 13–14; Pesce 1936, 445, Fig. 6; Sestieri 1940–41, 99, Fig. 14; Lo Porto 1966, 168–69, 171, Fig. 30, No. 2, Pl. 7, No. 6; Adamesteanu 1980, 200, 202, Fig. 210e.
Date: 4th c. BC.

1:2

222. TC 417-02 Relief plaque of Pan with female figure (Postrioti type AIA, AIB, AIIA, or AIIB)
Dimensions: h 5.9
Fabric and technique: 7.5YR 7/6.
Description: Shaggy left leg of Pan with genitalia preserved at upper edge of fragment. The piece is too fragmentary (as are likewise the best preserved examples of Postrioti's type AII) to allow one to determine whether the fragment belongs to type AI or type AII. It is not as fully rounded in volume as similar fragments from type AI, but flatter. It may be better assigned to type AII.
Comparisons: Postrioti 1996, 26–27, 32–33, 52–53, Pls. 1a–b, 2a–b, or 4a–b; Meirano 1996, 85–86, Fig. 26; Barberis 1995, 29, Fig. 14.
Date: Last quarter of 4th to first quarter of 3rd c. BC.

223. TC 437-01 Unidentifiable fragment
Date: Unknown.

224. TC 449-01 Relief plaque of Pan with female figure (Postrioti type AIA, AIB, AIIA, or AIIB)
Dimensions: h 5.2
Fabric and technique: 7.5YR 7-8/6.
Description: Shaggy leg of Pan. The piece is too fragmentary to determine whether the fragment belongs to Postrioti type AI or type AII.
Comparisons: Postrioti 1996, 26–27, 32–33, 52–53, Pls. 1a–b, 2a–b, or 4a–b; Meirano 1996, 85–86, Fig. 26; Barberis 1995, 29, Fig. 14.
Date: Last quarter of 4th to first quarter of 3rd c. BC.

225–227. TC 449-02–04 Unidentifiable fragments
Date: Unknown.

1:1

228. TC 453-01 Relief plaque of reclining banqueter
Dimensions: h 5.0
Fabric and technique: 7.5YR 7/6.
Description: Veiled head of female seated at end of couch. Hair is centrally parted into two smooth masses that are swept up towards veil. Round face has worn, indistinct features. Upper edge of relief is preserved above veil.
Comparisons: Signore 1996, 310–11, Figs. 11–12, 316–17, Figs. 24–25, 318, Fig. 27; Lo Porto 1966, 169, 171, Fig. 31, No. 1, Pl. 8 No. 1; Sestieri 1940–41, 99, Fig. 16 upper row; Adamesteanu 1980, 200, 202, Fig. 210c; Lo Porto 1981b, 332, 334, Fig. 41, No. 2; Letta 1972, 94–97, 99, Pls. 15, Nos. 2, 4, and 17, No. 1.
Date: 4th c. BC.

1:1

229. TC 477-14 Small Archaic protome of female figure
Dimensions: h 6.0
Fabric and technique: 7.5YR 7/6.
Description: One fragment, missing left side. Base of neck is chipped. Incrustation on surface. Right side of small protome representing a female head wearing a low polos. Smooth volumes of hair pulled back from central part towards ears. Long, oval-shaped face characterized by almond-shaped eyes, large nose, broad, fleshy lips, and rounded chin. Closely related protome found at Herakleia.
Comparisons: For protome, probably from same mold series, at Herakleia: Rolley 1981, 187, Pl. 38, No. 2. For stylistically similar heads: Adamesteanu 1980, 265–66, Fig. 281f. From Garaguso: Morel 1970b, 561, Fig. 28.
Date: Middle to late 6th c. BC.

1:1

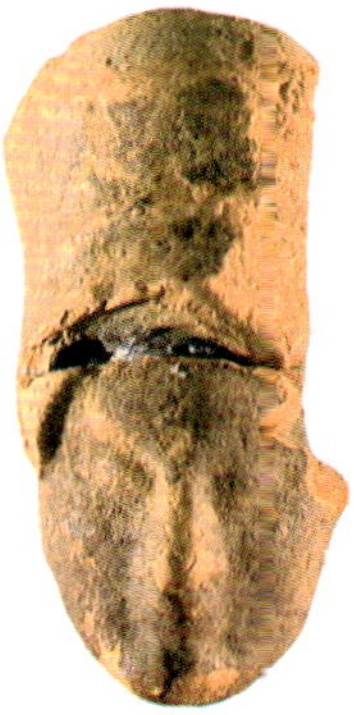

230. TC 477-12 Head of Archaic figurine of female figure
Dimensions: h 4.7
Fabric and technique: 7.5YR 7/6.
Description: Two fragments (TC 477-12 and TC 477-20) preserve face and part of polos. Chipped surface with dark incrustation. Head of female figure wearing a tall, conical polos with roundel marking its base. Smooth volume of hair frames oval-shaped face that has bulbous almond-shaped eyes, a large nose, and a small, fleshy mouth drawn into an Archaic smile. Similar head from San Biagio alla Venella belongs to standing female figure, albeit wearing a conical headdress. That the head may have originally belonged to a seated figure cannot be excluded.
Comparisons: Olbrich 1979, 140, 228, 231–32, Pls. 18 A82, 60 C67, 62 C76; Adamesteanu 1980, 88–89, Fig. 77h.
Date: Middle to third quarter of 6th c. BC.

1:1

231. TC 477-13 Head of Archaic figurine of female figure
Dimensions: h 3.5
Fabric and technique: 5YR 7/8.
Description: Some incrustation on surface. Lower right quarter of face of female figure. Smooth volume of hair frames face, which has bulbous almond-shaped eyes, a large nose, and a small, fleshy mouth. The head may belong to either a standing or a seated figure.
Comparisons: Similar, but of smaller scale: Olbrich 1979, 210, 228, 231, 228, Pls. 53 C31, 60 C67, 62 C76.
Date: Middle to third quarter of 6th c. BC.

232. TC 477-24 Head of Archaic figurine of female figure
Dimensions: h 7.0
Fabric and technique: 7.5YR 6/4-6. Polos is a hand-modeled addition to a molded head.
Description: Tall, flaring polos with rounded base worn by female figure. A wisp of hair is preserved beneath the polos. Top of polos has a finished edge.
Comparisons: Olbrich 1979, 311, 317–18, Pls. 94 C227, 97 C237.
Date: Second half of the 6th c. BC.

1:2

233. TC 477-06 Archaic figurine of seated female figure
Dimensions: h 9.3
Fabric and technique: 10YR 8/6. Disk included in mold of front of figurine.
Description: Incrustation on surface. Right shoulder, arm, and part of torso of seated female figure. Disk-shaped fibula pinned on shoulder. Hem of chiton sleeve marked on arm by raised surface. Edge of figurine preserved along shoulder and arm.
Comparisons: From Taranto: Mendel 1908, 474, Pl. 14, No. 3; Breitenstein 1941, 23–24, Pl. 22, Nos. 227-228; Mollard-Besques 1954, 59, Pl. 41 B386; Iacobone 1988, 40, Pl. 31c.
Date: Middle to late 6th c. BC.

1:2

234. TC 477-03 Probably Archaic figurine of seated female
Dimensions: l 7.7
Fabric and technique: 5YR 7/3. Lighter color than is common at Metaponto. Hand-modeled.
Description: One fragment preserves terminal disk of throne arm. Lower third of disk and areas of attachment to throne chipped away. Terminal disk probably belonging to throne armrest from an Archaic plank-style goddess.
Comparisons: Similar, but smaller: Olbrich 1979, 185–87, Pl. 44 B52–B54. From Taranto: Iacobone 1988, 39–40, Pl. 31a; Lippolis 1995, 60, Pl.16, No. 4. From Poseidonia-Paestum: R. Ammerman 2002, 45–51, 64, Pl. 12, No. 97 C.
Date: Middle to late 6th c. BC.

1:4

235. TC 477-07 Archaic figurine of standing female
Dimensions: h 17.6
Fabric and technique: 7.5YR 7-6/6. Thick, heavy layers of clay pressed into front mold. Smoothed interior edge and unoxidized interior suggest that back of the large-scale statuette was either hand-modeled slab of clay or also cast from a mold and then attached to cast from front mold.
Description: Two fragments preserve section of drapery (TC 477-07 joins TC 477-26). Drapery probably belongs to the right side of large-scale standing figure of unidentified type. Smooth, overlapping drapery folds run down the length of the piece. A fold, falling to a point, lies above other drapery folds and may represent the corner of a schematically rendered himation. Edge of molded front preserved inside. The large scale and unusual construction technique are noteworthy. Fabric similar to **Nos. 238–239** and possibly other fragments from site 477.
Comparisons: Rendering of drapery recalls stylistically female figures in architectural revetments from temple C at Metaponto: Adamesteanu 1980, 112–14, Figs. 112–13; Mertens-Horn 1992, 46–59, 110–17, Figs. 27–35, 46, Pls. 8–15. For larger terracotta statues of Archaic korai from the urban sanctuary: Adamesteanu/Mertens/De Siena 1975, 28–29, Figs. 12–17; Barberis 2004, 66, Fig. 58.
Date: Mid-6th c. BC.

1:2

236. TC 477-02 Head of Late Archaic figurine of female figure
Dimensions: h 7.2
Fabric and technique: 7.5YR 7/6.

236. TC 477-02 cont.

Description: Incrustation on surface of two fragments. Right side of head of Archaic female figure who wears a veil over her head. Smooth swelling volume of centrally parted hair frames side of face to level of ear. Facial features are indistinct because of worn surface of figurine. Back edge of figurine is preserved at middle portion of head. The head may belong to either a standing or seated female figure that reflects Ionian stylistic influence. Fabric similar to other terracotta fragments from site 477 (**Nos. 239–41**).
Comparisons: Similar to terracotta from Oria/Monte Papalucio: D'Andria 1990, 240, Fig. 28. Resembles head, perhaps mistakenly identified as part of a reclining banqueter plaque of 4th c. BC: Adamesteanu 1980, 93–94, Fig. 84e.
Date: End of 6th to early 5th c. BC.

1:2

237. TC 477-15 Probably Late Archaic figurine of standing female figure
Dimensions: h 5.6
Fabric and technique: 7.7YR 8-7/6.
Description: Drapery with omega fold that cuts back across center of torso of figure on shorter left side of the pleat while it continues further downward on right side of pleat and torso. The drapery probably belongs to the transverse himation of a standing kore figure. The omega fold describes the point at the right side of the torso where the long swag of the open front end of the himation, having passed under the left arm and crossed diagonally upward over the torso toward the right shoulder, turns sharply downward, falling in a swallowtail fold along the figure's right side.
Comparisons: Olbrich 1979, 124–25, Pl. 12 A47. Similar, but smaller: Lo Porto 1981b, 328, Fig. 36, No. 7. Similar from Taranto: Winter 1903, 109, No. 4.
Date: Late 6th to second quarter of 5th c. BC

238. TC 477-08 Late Archaic or Early Classical figurine of female figure of unidentified type
Dimensions: h 5.5
Fabric and technique: 7.5YR 6/6.
Description: Himation falls in soft swollen folds of drapery over left shoulder. Edge of figurine is preserved at neck and shoulder. The fragment is too small and lacks sufficient detail to determine the exact type of figurine to which it belongs and a corresponding date. The soft, full forms of the drapery folds suggest a date in the mid-5th c. BC; the fragment may represent a standing woman, possibly bearing an offering, or a seated figure. The similarity of its fabric to the fabric of other fragments from site 477 suggests it may, on the other hand, belong to the same statuette of the mid-6th c. BC, as **No. 235**, or to another terracotta to which also belong other fragments of drapery, such as **No. 239**, which are likewise characterized by the heavy quality of their fired fabric and swollen forms of their drapery.
Comparisons: For women holding offerings of mid-5th c. BC: Sestieri 1940–41, Pl. 37, Fig. 16 upper right; Lo Porto 1966, 162–63, Fig. 25, No. 1; Adamesteanu 1980, 271, 273, Fig. 285i-l; Lo Porto 1981b, 333, Fig. 40, No. 2. From Taranto: Iacobone 1988, 29, Pl. 22c–d; Winter 1903, 110, No. 5.
Date: Mid-6th to mid-5th c. BC.

239. TC 477-10 Late Archaic or Early Classical figurine of female figure of unidentified type
Dimensions: h 5.6
Fabric and technique: 7.5YR 7/6.
Description: Drapery fragment preserves what may be the undulating edge of non-angular zigzag fold with a series of soft vertical folds running between the meanders. The soft, swollen forms recall figurines of women holding offerings of the mid-5th c. BC. The unusually heavy fabric, however, is similar to other terracotta fragments recovered from site 477, which may be of Archaic date (**Nos. 236** and **240–41**).
Comparisons: See **No. 238**.
Date: Mid-6th to mid-5th c. BC.

1:2

240. TC 477-17 Late Archaic or Early Classical figurine of female figure of unidentified type
Dimensions: h 4.1
Fabric and technique: 7.5YR 7/6.
Description: Stacked pleat of drapery with softly meandering edge next to which two gently curved folds sweep away from stacked pleat. The fragment is too small to be assigned to an exact type of figurine. It may represent the upper part of a swallowtail fold of the transverse himation of a standing kore who pulls her chiton to her left side, creating gently curved folds to the side of the stacked pleat of the himation. It may alternatively represent an apoptygma on a seated or standing figure that falls in omega folds over both breasts, between which catenary

240. TC 477-17 cont.

folds are delineated. Fabric similar to other terracotta fragments from site 477, **Nos. 236**, **239**, and **241**, which may belong to same or closely related figurines.
Comparisons: For kore with transverse himation: Olbrich 1979, 124–25, Pl. 12 A47. From Taranto: Winter 1903, 109, No. 4. For omega folds falling from both breasts of seated figure from Taranto: Iacobone 1988, 45, Pl. 35 b. For similar arrangement on standing figure from Taranto: Winter 1903, 110, No. 6.
Date: Late 6th to first half of 5th c. BC.

1:1

241. TC 477-19 Late Archaic or Early Classical figurine of female figure of unidentified type
Dimensions: h 4.0
Fabric and technique: 7.5YR 6/6.
Description: Incrustation on surface. Drapery with an omega fold and vertical pleats running above upper curves of the fold. The drapery probably belongs to the apoptygma falling over the chest of a seated or standing female figure. The fragment is too small to be assigned to an exact type of figurine. Fabric similar to other terracotta fragments from site 477: **Nos. 236** and **239–40**.
Comparisons: For similar drapery on standing figures: Lo Porto 1966, 162–63, Fig. 24, No. 1; Lo Porto 1981b, 330–33, Figs. 39, No. 1, 40, No. 1. For somewhat similar drapery on enthroned figure: Postrioti 1996, 89, Pl. 15c. For possibly standing figure from Taranto: Mollard-Besques 1954, 60, Pl. 41 B394. For seated figures from Taranto: Winter 1903, 110 No. 7, 122, No. 6.
Date: Late 6th to mid-5th c. BC.

242. TC 477-23 Late Archaic or Early Classical figurine of female figure of unidentified type
Dimensions: h 4.9
Fabric and technique: 7.5YR 6/6.
Description: Tubular vertical folds of hem of skirt fall over right foot resting on a base. The foot lacks such details as the indication of toes. The foot may belong to an enthroned goddess or a standing figure.
Comparisons: See **No. 243**.
Date: Late 6th to mid-5th c. BC.

1:2

243. TC 477-11 Late Archaic or Early Classical figurine of female figure of unidentified type
Dimensions: h 5.3
Fabric and technique: 7.5YR 7/8.
Description: Chipped at right side. Incrustation on surface. Fairly flat, vertical folds of hem of skirt fall over feet that rest on rounded base. The feet lack such details as the indication of toes, and are placed evenly on the base. They may belong to an enthroned goddess or a standing figure. Similarity of fabric and scale suggest that the fragment may belong to the same figurine of an enthroned goddess as **No. 244**.
Comparisons: Enthroned goddess from Taranto: Lippolis 1995, 60, Pl. 17, No. 1; Higgins 1967, 90, Pl. 40b = Mollard-Besques 1963, 67, 134, Pl. 15, No. 3 = Mollard-Besques 1954, 123, Pl. 88 C247; Winter 1903, 122, No. 6. From Herakleia: Neutsch 1981, 166, Pl. 27, No. 3. Standing woman with offerings from Taranto: Iacobone 1988, 14, 20, Pls. 6c, 10b.
Date: Late 6th to mid-5th c. BC.

1:2

244. TC 477-16 Late Archaic or Early Classical figurine of seated female figure
Dimensions: h 7.2
Fabric and technique: 7.5YR 7/6.
Description: Incrustation on surface. Smooth upper garment falls in large omega fold in the manner of an apoptygma over the center of the torso of an enthroned goddess. The smooth upper garment also covers the upper arms of the figure in the manner of a cape or poncho. Its hem is visible at the right elbow. Fine vertical pleats of underlying chiton fall below upper garment on right arm and below apoptygma over torso into lap. A figurine from the same mold series was found at Herakleia. They are closely related to mold series of an enthroned goddess from Taranto that manifest elements of Early Classical or Severe style. Similarity of fabric and scale suggest that the fragment may belong to the same figurine as the feet resting on a base, **No. 243**.

244. TC 477-16 cont.

Comparisons: Presumably from Favale: Liseno 2004, Pls. 21c, 22a. From Herakleia: Neutsch 1981, 156, Pl. 27, No. 3. Similar from Taranto: Iacobone 1988, 45, Pl. 35c; Higgins 1967, 90, Pl. 40b = Mollard-Besques 1963, 67, 134, Pl. 15, No. 3 = Mollard-Besques 1954, 123, Pl. 88 C247; Winter 1903, 110, No. 7, 122, No. 6, 123, No. 7.
Date: Second quarter to middle of 5th c. BC.

245. TC 477-21 Unidentified type
Dimensions: h 4.7
Fabric and technique: 7.5YR 7/6. Hand modeled
Description: May be the wing or decorative disk of an Archaic figurine representing a female figure.
Comparisons: Olbrich 1979, 150–51, 156–57, Pls. 24 A106, 28 A121, 29 A122–A123.
Date: Possibly mid-6th c. BC.

246. TC 477-22 Base of unidentified type
Dimensions: h 11.3
Fabric and technique: 7.5YR 7/6. Slab pressed into a mold.
Description: Two fragments. Incrustation on surface. Larger rectangular side meets smooth strip at a right angle. A molded figure extended beyond the smooth strip, but has been broken away leaving only an irregular edge. Top of larger slab appears to present a finished edge. The fragments possibly preserve the side of a throne running along legs of a seated figure. The finished edge at the top of the larger slab is the horizontal side of the seat of the throne, which then turns down as a smooth strip (following knees and lower legs of the seated figure) to base. Such renderings of thrones are common among Archaic figurines of seated female figures from East Greece.
Comparisons: From Oria/Monte Papalucio: Palumbo 1986, 128, Pl. 40, No. 16. From Poseidonia-Paestum: R. Ammerman 2002, 94, Pl. 23, No. 196.
Date: Possibly second half of the 6th c. BC.

247. TC 477-09 Base of unidentified type
Dimensions: h 7.3
Fabric and technique: 5YR 7/6-8. Hand modeled from flat slabs of clay.
Description: Two slabs of clay meet at an angle, one slab extending slightly beyond the other. Vertical line inscribed down length of larger side. Base of slabs splays out to increase the size of the carrying surface. The fragment is possibly part of a relief plaque or the throne of a figurine of an Archaic goddess.
Date: Unknown.

248. TC 477-01 Two fragments. Incrustation on surface.
Dimensions: h 8.5
Fabric and technique: 5YR 7/8. Hand modeled from flat slab of clay.
Description: Unidentified type. Possibly part of relief plaque or throne of Archaic goddess.
Date: Unknown.

249. TC 477-04 Unidentified type; possibly part of relief plaque or throne of Archaic goddess
Dimensions: h 3.2
Fabric and technique: 7.5YR 8-7/6. Hand modeled from flat slab of clay.
Date: Unknown.

250. TC 477-05 Drapery of unidentified type
Dimensions: h 4.4
Fabric and technique: 7.5YR 7/6.
Date: Unknown.

251. TC 477-18 Drapery of unidentified type
Dimensions: h 4.4
Fabric and technique: 5YR 7/6.
Date: Unknown.

252. TC 477-25 Possibly drapery of unidentified type
Dimensions: h 6.7
Fabric and technique: 5YR 7/8.
Date: Unknown.

253. TC 477-27 Possibly drapery of unidentified type
Dimensions: h 9.4
Fabric and technique: 5YR 6/8.
Description: Incrustation on surface. Fragment appears to belong to a large-scale figurine.
Date: Unknown.

254–267. TC 477-28–41 Unidentifiable fragments.
Date: Unknown.

268. TC 485-01 Relief plaque of satyr, silenos, or Pan with female figure (Postrioti type AIA or AIC)
Dimensions: h 5.4
Fabric and technique: 7.5YR 7/6-8.
Description: Drapery of veil falls behind right upper leg of female figure, who moves to her right. Only the top of her thigh is preserved. The broad zigzag fold of the edge of her veil and the finer parallel catenary folds falling from her elbow are faintly marked. The edge of the relief is preserved behind the zigzag fold of her veil.
Comparisons: Postrioti 1996, 26–27, 33–34, Pls. 1a, 3a; Meirano 1996, 85–86, Fig. 26.
Date: Last quarter of 4th to first quarter of 3rd c. BC.

1:1

269. TC 488-02 Relief plaque of satyr, silenos, or Pan with female figure (Postrioti type AIA or AIB)
Dimensions: h 3.5
Fabric and technique: 5YR 7/8.
Description: Lower right side of satyr's or Pan's face with right side of broad nostril, moustache, mouth, and beard.
Comparisons: Postrioti 1996, 26–27, 32–33, Pls. 1a–b, 2a–b; Meirano 1996, 85–86, Fig. 26; Letta 1972, 129, Pl. 26, No. 2.
Date: Last quarter of 4th to first quarter of 3rd c. BC.

270. TC 488-04 Relief plaque of Pan with female figure (Postrioti type AIA)
Dimensions: h 4.0
Fabric and technique: 5YR 7/6.
Description: Right side of Pan's head. Short, curved horn rises above round wreath. Edge preserved at top of horn.
Comparisons: Postrioti 1996, 26–27, Pl. 1a–b; Meirano 1996, 85–86, Fig. 26.
Date: Last quarter of 4th to first quarter of 3rd c. BC.

271. TC 488-05 Drapery fragment of unidentified type
Dimensions: h 3.7
Fabric and technique: 5YR 7/6, finely levigated.
Description: Zigzag drapery folds fall next to vertical pleats.
Date: Unknown.

272–284. TC 488-03; 488-06–15; 493-01–02 Unidentifiable fragments.
Date: Unknown.

1:1

285. TC 506-04 Classical figurine of standing female figure holding quadruped
Dimensions: w: 7.4
Fabric and technique: 7.5YR 7/6.
Description: Head of animal chipped away. Small quadruped with short tail, best identified as a lamb, which a standing female holds across her torso in her left arm. Upper edge of draped left forearm preserved beneath animal, whose hind legs overlap front of arm and whose front legs bend forward beyond arm and hand of female. The complete figurine type depicts a standing woman who holds the sacrificial animal in her left arm and often carries an oinochoe for libation in her right hand.
Comparisons: Lo Porto 1966, 162–64, Fig. 24, No. 1; Sestieri 1940–41, 99, Pl. 37, Fig. 16 upper right; Adamesteanu 1980, 270–73, Figs. 284a, 285a, e; Liseno 2004, 47–49, Pls. 17a–b, 18a–b. For similar figurines from Taranto: Lo Porto 1964, 251–52, Fig. 67; Winter 1903, 110, No. 4. From Timmari: Lo Porto 1991, 78–79, Pl. 25, Nos. 9-10.
Date: Second half of 5th to late 4th c. BC.

1:1

286. TC 506-02 Relief plaque of reclining banqueter
Dimensions: h 6.0
Fabric and technique: 5YR 7/6.
Description: Upper portion of head of banqueter wearing wreath decorated with central palmette and rosettes with six petals at center and sides. Hair is centrally parted into undulating smooth masses framing the forehead. Facial features are worn and indistinct. Edge of relief is preserved along top of headdress.
Comparisons: Signore 1996, 309–11, Figs. 10–11, 13–15; 318, Figs. 27–28; Pesce 1936, 445, Fig. 6; Sestieri 1940–41, 99, Fig. 14; Lo Porto 1966, 168, Fig. 30, No. 2; Adamesteanu 1980, 200, 202, Fig. 210d–e; Letta 1972, 78, Pl. 9, No. 4.
Date: 4th c. BC.

287. TC 506-03 Relief plaque of reclining banqueter
Dimensions: h 5.1
Fabric and technique: 5YR 7/6.
Description: Lower right arm, hand, and lap of female figure seated at end of couch on which male banqueter reclines. Her thumb is distinguished from the other fingers of her hand, which rests on the side of her lap. Edge of relief is preserved along right arm of female figure.
Comparisons: Similar: Signore 1996, 304–305, Figs. 2–3, 309–10, 313, Figs. 10–12, 18, 322, Fig. 35, 323, Fig. 38; Lo Porto 1966, 169, Fig. 31, Nos. 1, 3; Sestieri 1940–41, 99, Fig. 16 upper left; Letta 1972, 94–97, Pl. 15, Nos. 2, 4, and Pl. 16, No. 1.
Date: 4th c. BC.

288. TC 506-01 Relief plaque of Pan with female figure (Postrioti type AIIB)
Dimensions: h 4.2
Fabric and technique: 5YR 6-7/8.
Description: Some incrustation on surface. Upper left side of head of Pan with lower part of curved horn rising above flattened double band of wreath encircling his head. Lids of left eye are clearly indicated beneath bulging brow. Pointed left ear overlaps band of headdress.
Comparisons: Postrioti 1996, 52–53, Pl. 4b.
Date: Last quarter of 4th to first quarter of 3rd c. BC.

289. TC 506-05 Plaque of satyr, silenos, or Pan with female figure (Postrioti type AIB)
Dimensions: h 8.5
Fabric and technique: 5YR 7/4-6.
Description: Some incrustation on surface. Drapery of veil of female figure falls behind her right upper leg and knee as she moves to her right. The broad zigzag fold of the edge of her veil, as well as the finer parallel catenary folds falling from her elbow, is faintly marked. The edge of the relief is preserved approximately 1 cm beyond the zigzag fold of her veil; a subtle vertical depression marks the division between the drapery folds and the plain border added to the relief plaque. This suggests that the fragment belongs to Postrioti's type AIIB, in which the figures are set within a grotto with a grape arbor.
Comparisons: Postrioti 1996, 32–33, Pl. 2b
Date: Last quarter of 4th to first quarter of 3rd c. BC.

290. TC 506-06 Relief plaque of satyr, silenos, or Pan with female figure (Postrioti type AIB or AIIB)
Dimensions: h 4.4
Fabric and technique: 5YR 7/4.
Description: Upper right curved edge of plaque with overhanging rock of grotto and trefoil grape clusters. Stippled texture is used to denote the grapes in the cluster. Right edge of relief is preserved beside the grapes.
Comparisons: Postrioti 1996, 32–33, 52–53, Pls. 2a–b, 4b. From Cozzo Presepe: Prag 1983b, Fig. 150.
Date: Last quarter of 4th to first quarter of 3rd c. BC.

291. TC 506-07 Plaque of satyr, silenos, or Pan with female figure (Postrioti type AIB)
Dimensions: h 7.2
Fabric and technique: 5YR 7-6/8.
Description: Incrustation on surface. Raised right arm of female figure pulls veil away from her head. Relief extends above hand of woman to represent a rocky grotto and grape arbor. Edge of relief is preserved along right side of arm.
Comparisons: Postrioti 1996, 32–33, Pl. 2b; Letta 1972, 129, Pl. 26, No. 2. For plaque in which female figure is set more deeply within background of grotto and grape arbor: Adamesteanu 1980, 275–76, Fig. 289e. From Cozzo Presepe: Prag 1983b, Fig. 150 left.
Date: Last quarter of 4th to first quarter of 3rd c. BC.

292. TC 506-08 Relief plaque of Pan with female figure (Postrioti type AIB)
Dimensions: h 4.1
Fabric and technique: 5YR 6-7/8.
Description: Left side of upper head of Pan with short, curved horn rising above rounded wreath.
Comparisons: Postrioti 1996, 32–33, Pl. 2a-b; Letta 1972, 129, Pl. 26, No. 2. From Cozzo Presepe: Prag 1983b, Fig. 150 left.
Date: Last quarter of 4th to first quarter of 3rd c. BC.

293–319. TC 506-09–35 Unidentifiable relief fragments
Date: Unknown.

1:1

320. TC 507-01 Possibly relief plaque of reclining banqueter or standing horseman
Dimensions: w 6.0
Fabric and technique: 5YR 6/6.
Description: Upper portion of left side of head of horse shown in profile. Raised brow over eye, ear, and four waved tufts of mane are portrayed. Edge of relief is preserved along top of horse's head. The terracotta is too

320. TC 507-01 cont.

fragmentary to allow one to determine whether it belongs to a relief of a reclining banqueter, in which the head of a horse is sometimes depicted in the background between the banqueter and the woman seated at the end of the dining couch, or to a relief of a naked male figure holding a shield and standing heroically in front of his horse.

Comparisons: For fragment with only head of horse: Sestieri 1940–41, 99, Fig. 14; Schürmann 1989, 84, Pl. 48, No. 284. For horse with reclining banqueter: Signore 1996, 326, Fig. 41, 330–31, Figs. 46–47. For horse held by standing male figure: Letta 1972, 110–11, Pl. 20, No. 4. From Taranto: Bartoccini 1936, 166–69, Figs. 76–77; Kingsley 1979, 204, Pl. 2, No. 6; Higgins 1954, 361, Pl. 181, No. 1327; Iacobone 1988, 121, Pl. 118c; Lippolis 1995, 54, Pl. 12, Nos. 2–3. From S. Maria d'Anglona: Rüdiger 1967, 347, Fig. 20.

Date: 4th c. BC.

321–322. TC 515-01–02 Unidentifiable fragments

Date: Unknown.

323. TC 516-01 Archaic figurine of female figure of unidentified type

Dimensions: h 3.9

Fabric and technique: 5YR 7/6. Front mold filled solidly with clay.

Description: One fragment preserves right arm from below shoulder to mid-forearm. Right arm of female figure with edge of sleeve indicated at elbow. Edge of thin, solid figurine preserved to right side of arm.

Date: Middle to late 6th c. BC.

324. TC 516-02 Relief plaque of Pan with female figure (Postrioti type AIA)

Dimensions: h 4.8

Fabric and technique: 7.5YR 7/4-6.

Description: Left leg of Pan from joint to middle of thin, lower portion of leg near hoof. Edge of relief is preserved behind leg.

Comparisons: Postrioti 1996, 26–27, Fig. 1a = Meirano 1996, 85–86, Fig. 26.

Date: Last quarter of 4th to first quarter of 3rd c. BC.

325. TC 520-01 Unidentified type

Dimensions: h 6.7

Fabric and technique: 5YR 6/6-8.

Description: Flat, heavily walled fragment with three rows of diagonally impressed designs.

Date: Unknown.

1:1

326. TC 521-02 Relief plaque of standing warrior

Dimensions: h 7.7

Fabric and technique: 7.5YR 7/4-6.

Description: Fragment preserves right hand placed on hip of standing male figure. Long, slender fingers are outspread, and the thumb, passing to the side of the hip, is not indicated. Drapery of clamys falls in vertical folds below little finger. Complete examples of this mold series from Taranto portray a bearded male figure, standing in a hipshot pose, who raises a round shield on his left arm. He wears a festive wreath decorated with a taenia and rosettes; a clamys falls behind his naked torso and legs to the level of his calves. A plain stele and tripod are depicted to the man's left, while a naked child stands between the man and the stele and reaches upward toward the adult male figure with outstretched arms.

Comparisons: See **No. 13**. From Taranto: Abruzzese Calabrese 1996, 201–2, Fig. 146; Lippolis 1995, 54, Pl. 12 No. 1; Iacobone 1988, 143, Pl. 133d–e.

Date: 4th c. BC.

327. TC 521-01 Unidentified type

Dimensions: h 5.3

Fabric and technique: 5YR 6/8.

Description: Fragment presents a curved profile and preserves three parallel ridges of clay that are wider over the curve of the fragment but narrow towards a single point as they move away from the curve. The fragment may represent the drapery folds of a himation drawn over the shoulder of a protome or female figurine as in **No. 9**.

Comparisons: See **No. 9**.

Date: Possibly 5th or 4th c. BC.

1:1

328. TC 523-01 Head of Classical figurine of female figure
Dimensions: h 4.2
Fabric and technique: 7.5YR 6/6.
Description: Upper left side of head chipped, end of nose worn away. Some surface incrustation. Figure wears a low polos and veil and turns her head slightly to her right. Centrally parted hair with stippled curls falls along sides of neck to shoulders. Delicately modeled features give a sweet expression to the oval-shaped face.
Comparisons: Nava 2002, 735, Pl. 66.1 lower right.
Date: Second half of 5th to early 4th c. BC.

329–332. TC 554-01–04 Unidentifiable fragments
Date: Unknown.

333. TC 638-01 Relief plaque of satyr, silenos, or Pan with female figure (Postrioti type AIB or AIIB)
Dimensions: h 4.2
Fabric and technique: 5YR 7/4-6.
Description: Pendant trefoil grape cluster, relief left side.
Comparisons: Postrioti 1996, 32–33, 52–53, Pls. 2a–b, 4b; Adamesteanu 1980, 275–76, Fig. 289e. From Cozzo Presepe: Prag 1983b, Fig. 150.
Date: Last quarter of 4th to first quarter of 3rd c. BC.

334. TC 638-02 Unidentifiable fragment
Date: Unknown.

1:1

335. TC 726-01 Head of Classical figurine of female figure
Dimensions: h 5.4
Fabric and technique: 10YR 8/6.
Description: One fragment preserves head and right side of neck. Surface battered and encrusted. Head of female of Early Classical style. Raised edge at top of head may represent a stephane. Undulating waves of hair cascade symmetrically from central part to ears and then fall in full masses along sides of neck. Individual strands are indicated by parallel lines that follow the waved pattern of the tresses. Face has large eyes, long nose, and full-lipped mouth. Finished edge is preserved at back of head. The head may have belonged to either a standing or seated female figure.
Published: *Ancient Territories* 1999, 17, Fig. 15.
Comparisons: Liseno 2004, 59, Pl. 25f. For closely related heads from Taranto: Schneider-Herrmann 1971, 141–42, Figs. 5a–b, 7a–b. For similar isolated heads from Taranto: Iacobone 1988, 134, Pl. 127a–b. For similar heads of standing women with offerings from Taranto: Iacobone 1988, 10, 21, Pls. 8b, 10c.
Date: Middle of 5th c. BC.

336–339. TC 728-01–02; 745-01; 774-01 Unidentifiable
Date: Unknown.

Molds and Miscellaneous Mold-made Objects

1:1

340. TC 404-02 Mold fragment for relief plaque of reclining banqueter with woman and child.
Dimensions: h 5.8
Fabric and technique: 10YR 8/3.
Description: The head of a child held by a female at her left side and the upper portion of the draped legs of a reclining male figure are preserved below the upper edge of the mold. The mold is too fragmentary to allow a secure attribution to a specific mold series.
Comparisons: Signore 1996, 308–10, 313, 318, Figs. 10–12, 18, 27.
Date: 4th c. BC.

1:2

341. TC 489-01 Mold fragment for antefix of Io.
Dimensions: h 12.3
Fabric and technique: 5YR 6/6-8.
Description: The fragment preserves the portion of the mold that portrays the nose, right eye, forehead, hair, and right horn of Io. The centrally parted tresses swirl into circular curls that frame the forehead. The mold is a variant of the Tarantine type of antefix portraying Io wearing a diadem, an adornment that seems to be absent in this mold.
Comparisons: Tempesta 1995, 44–46, Fig. 2; 2005, 83–84, Fig. 1; *LIMC* 5 (1990), 671–72, Nos. 91-93, s.v. "Io I" (N. Yalouris).
Date: First half of the 4th c. BC.

1:2

342. TC 016-01 Unidentified type, molded attachment probably to pottery vessel or other utilitarian object
Dimensions: h 4.7
Fabric and technique: 7.5YR 7/6, fairly fine with small calcareous inclusions (< 1 mm). One side is cast from a mold, while other is made from a slab of clay. Holes are pierced in back and base.
Description: One fragment, broken above loops. Surface chipped. Front presents molded design consisting of a floral-like motif with two, or possibly four, petals. Base is smoothed, with its center pierced by a hole approximately 1 cm deep. Rounded slab of clay attached as back is pierced by a hole for attachment to another element. Slab and molded front join at narrow center of object, but leave a hollow space where they separate toward top.
Date: 4th c. BC or later.

1:1

343. TC 772-01 Decorative terracotta, possibly tile
Dimensions: w 7.3 ; th 1.6
Fabric and technique: 2.5YR 6/8. Solid tile with molded decoration on one side.
Description: Circular tile with impressed decoration: outer band of irregularly arranged droplike impressions bordered by row of short vertical hatches with larger oval-shaped patterns filling center of tile. The impressed design alternatively resembles a face with oval-shaped eyes, nose, and brow outlined by ridges in center of tile, with short row of bangs under a broader mass of circular curls framing forehead.
Comparisons: Although the comparison is not definitive, the fragment has stylistic similarity to architectural terracottas recovered in indigenous contexts in Lucania and Campania. (Several of these appear in Orlandini 1972). The indigenous products of Lucania are sparsely published. The comparanda as well as the other material found at the site suggest a date in the 4th c. BC.
Date: Possibly 5th-4th c. BC.

14
Loom Weights

Lin Foxhall

This chapter examines the 119 loom weights collected in the survey areas between the Bradano and Basento Rivers, looking first at how loom weights were used in general, and then at what can be determined from their forms, use, and taphonomy. The types of loom weights found at Metaponto are then compared with those found elsewhere in the Greek world, to form the basis for investigating the chronological and spatial distribution of the Metapontine loom weights. An analysis of use patterns follows, with particular attention to the presence of some unusual stamps. The chapter concludes by considering the implications of the loom-weight data for our understanding of the chora of Metaponto. The examples illustrated in the select catalog are indicated in bold.

The Use and Significance of Loom Weights

Loom weights are clay weights of varying shapes, pierced at one end with one or two holes. Occasionally they were made from other materials such as stone or lead. Their use in classical antiquity to hold taut the warp (vertical) threads on a warp-weight loom is well documented from painted vases. Each loom weight was typically attached to a small section of the warp threads by tying a small bundle of the warp threads to either a loop of string threaded through the suspension hole(s) or a metal ring,[1] or sometimes by attaching them to a stick placed through the suspension hole.[2] At Corinth and occasionally elsewhere, stamps on conical loom weights dating to the 4th century BC depict a loom weight with a stick placed at an angle through the suspension hole.[3] Hence if loom weights have been used regularly, they often display specific wear patterns, notably (1) wear around the suspension hole from a thread, ring, or stick (each of these leaves a slightly different signature of wear), or (2) abrasion or chipping at the base where adjacent loom weights have bumped into each other while hanging from the loom. It also follows that the lighter the weight of the loom weights, the finer the cloth on the loom.[4] Although it is possible to increase the tension somewhat by tying fewer threads to each loom weight,[5] thereby using more loom weights per warp, this changes the texture and density of the finished fabric; if the thread is coarse and heavy, lightweight loom weights will not hold the warp sufficiently taut for weaving. Conversely, lightweight loom weights indicate that they were intended for the manufacture of very fine cloth.

Weaving, spinning, and textile manufacture were archetypally women's work in classical antiquity. The tools associated with these processes, including loom weights, can be linked both symbolically and practically with the presence of women in the landscape. The status of these women generally cannot be determined from the loom weights alone. However, within large assemblages of loom weights on Greek sites normally only a small minority are stamped. Many of these stamps appear to be made with personal seal rings of metal or with engraved seal gems. It is interesting that this phenomenon can be observed throughout the whole of the Greek world in the 5th and 4th centuries BC.[6] This could suggest that relatively prosperous women owned these particular loom weights and wished to identify them as their personal property, though of course other people, including slaves, may also have used them. This may also imply that women identified themselves closely with the tools of their trade, and that specific loom weights may thus have become material manifestations of personal identity and, to some extent, "fashion items."

Comparison of Loom-Weight Types and Chronology

The form of a loom weight is not in itself a reliable chronological indicator, since the specific details of

[1] Davidson/Thompson/Thompson 1943, 68, Fig. 30, shows what is probably a 4th c. BC example in the British Museum.

[2] *Corinth* XII, 147.

[3] *Corinth* XII, 154–55 and Fig. 25, 158 and Fig. 27.

[4] Mårtensson/Nosch/Andersson Strand 2009, 380–82.

[5] Cahill 2002, 453.

[6] Richter 1968, 149; *Corinth* XII, 153.

Type	Count		% Total
Disk	78	81	68%
Lenticular	3		
Pinched	6		5%
Pyramidal	32		27%
Total	**119**		**100%**

Table 14.1 Types of loom weight by count and percentage.

1:1

Figure 14.1 Fragment of a handmade loom weight (Lo 359-01).

loom-weight styles and the changes in fashion over time are to a considerable extent regionally specific. There are, however, some trends that are widely observable throughout the Greek world. Moreover, for such a mundane and homely item, loom weights move around more than one might expect.

The majority of Greek loom weights fall roughly into conical, pyramidal, and disk types. There is substantial stylistic variation within each type. In most places conical and pyramidal loom weights are the most common in the Archaic and Classical periods, whereas the disk and lenticular forms (on other sites more generally grouped together and labeled "discoid") are more characteristic of the Hellenistic period.

Conical loom weights

At Corinth the conical loom weight was always by far the most common type,[7] with examples ranging in date from the late 8th century to the 1st century BC. At Akragas they are also the most common Classical type,[8] but they do not outnumber other types to the same degree as at Corinth. In Athens conical Corinthian-style loom weights became popular in the 4th century BC. As well as local Attic imitations, there are a few examples of actual Corinthian imports,[9] made in Corinthian clay, found in the fill of the Third Period of the Pnyx (425–325 BC).[10] At Olynthos (late 5th to mid-4th c. BC) the conical form (type 8) shares its popularity with the pyramidal form (type 7).[11] Some examples of the conical form appear in Italy at Locri Epizephyrii.[12] It is therefore striking that among the Metaponto survey sites covered by this volume there are no conical loom weights. **No. 1**, the lone conical example from Metaponto found outside of the present study area in the transect between Basento and Cavone, appears to be an early type, most closely paralleled by examples of the late 8th–7th centuries BC from Corinth.[13]

Pyramidal loom weights

Pyramidal loom weights appear regularly all over the Greek world. In Athens they were the predominant form from the 7th to the early 4th centuries BC, thereafter vying with the conical form. At Poseidonia-Paestum the 4th-century BC deposit of loom weights from behind the linking wall adjacent to the oikos sanctuary consists almost entirely of pyramidal examples.[14] At Locri Epizephyrii they are also the most common type.[15] They are roughly equal in popularity to the conical weights at Olynthos and very common at Akragas (see above). They are regularly found at Corinth to the end of the 4th century, but not in the Hellenistic period,[16] although there are Hellenistic examples from Athens and elsewhere.

Although common at Metaponto, pyramidal loom weights of various kinds represent just over one-quarter (27%) of the total loom weights recovered in the survey (*Table 14.1*). All except three examples are narrow pyramids with a single hole at the top, generally 54–66 mm in height (only two examples are 70 mm or longer). The average weight of the fully preserved examples is 497 g (standard deviation: 96.80217). In Athens pyramidal loom weights such as these were often mold-made. While it is possible that some of the Metapontine examples were mold-made,

[7] *Corinth* XII, 147.
[8] De Miro 2000, 119.
[9] For example, *Corinth* XII, 84, No. 56.
[10] *Corinth* XII, 63.
[11] Cahill 2002, 4.50.
[12] Costamagna and Sabbione 1990, 118.
[13] *Corinth* XV.2, 269; *Corinth* XII, 148–50.
[14] Pedley and Torelli 1984, 375; Pedley and Sevink 1985, 59.
[15] Costamagna and Sabbione 1990, 118.
[16] *Corinth* XII, 162; *Corinth* XV.2, 268.

4:1

Figure 14.2 Stamp on pyramidal loom weight (Lo 204-01).

others appear to have been handmade, starting with a "sausage" of clay, as one broken specimen shows (*Fig. 14.1*). The fabric of most is quite fine, often resembling the fabric of black-gloss or plain ware. There are, however, two examples with coarser fabrics, **No. 2** resembling an amphora-type fabric, and **No. 3** resembling a very coarse pithos-type fabric. In Athens, Olynthos, and elsewhere, stamped examples of pyramidal loom weights are regularly found, but among the Metapontine examples only one bears an illegible round stamp on the underside of the base (*Fig. 14.2*).

The three anomalies among the pyramidal loom weights are all trapezoidal in shape. They are much heavier than the narrow pyramids. **No. 5** has a single hole through the long axis and is made of a fabric similar to both plain and cooking wares. It resembles late examples from Athens[17] and Locri Epizephyrii.[18] **No. 6.** has two holes at the top and is made of an unusual plain-ware–type fabric that contains sherd inclusions, similar in character to the fabric of **No. 19**, one of the disk loom weights. This might suggest that it is relatively late in date. The last of these examples, **No. 7**, is particularly interesting. It has two unevenly spaced holes at the top and is made of a coarse grey impasto-type fabric. A small impression visible between the holes may have been produced by a seed. This piece most closely resembles the indigenous loom weights characteristic of the Iron Age in southern Italy, but it is smaller than most of the Iron Age examples. Although Neolithic material appears at site 468, that chronology seems too early for this piece. A similar indigenous-type loom weight was recovered from the excavation of a Classical period (late 5th to early 4th c. BC) rural house at Bova Marina, Calabria (Umbro Greek site).

Disk loom weights are comparatively rare in Classical contexts. At Corinth there were only a few examples of disk weights. Davidson lists five,[19] and Stillwell lists four,[20] all from 4th-century BC excavated contexts. All of these were of the type with convex faces called lenticular at Metaponto. Similarly, at the Pnyx in Athens (425–325 BC) only four examples were found, three with the convex profiles of the "lenticular" type considered by Davidson to belong to the 5th century BC, and one (**No. 144**) with flat faces.[21] In Athens, the disk form becomes more common in the Hellenistic period.[22] The 4th-century BC deposit of loom weights from Poseidonia-Paestum (see above) contained two disk loom weights, in contrast to the 93 pyramidal examples. At Akragas, 62 disk loom weights were found, in contrast to the 210 conical and 160 pyramidal examples[23] in more or less Classical (5th–4th c. BC) contexts. The disk form is present at Locri Epizephyrii particularly in the 4th and 3rd centuries.[24] At Olynthos, disk loom weights do not seem to be present at all. In contrast, at the central Greek site of Halai, a "domestic" deposit of the late 2nd to early 1st century BC contained numerous loom weights, all of the flat-faced disk form.[25] Several of the examples from Halai were stamped using sheep or goat astragali.

As shown in Table 14.1, disk and "lenticular" loom weights are by far the most common types in the Metaponto study area, constituting 68% of the total number of loom weights. This balance seems odd compared with other parts of the Greek world. Also unusual is that the disk form with flat faces and two holes at the "top" (a form more regularly associated with Hellenistic contexts elsewhere) is far more common at Metaponto than the "lenticular" form with convex faces, although the flat-faced Metapontine examples are generally smaller than the Hellenistic examples found at Halai and Athens. However, given the relative paucity of securely dated Hellenistic material in most of the Metaponto survey corpus, these might better be associated with the later 4th/3rd-century phases of the sites on which they were found. This issue is explored further in the following section.

The fabric of the disk loom weights generally resembles plain-ware-type fabrics and is often hard and slightly sandy. There is, however, considerable

[17] Davidson/Thompson/Thompson 1943, 73–74, and 80, Fig. 32.

[18] Costamagna and Sabbione 1990, 118.

[19] *Corinth* XII, 162.

[20] *Corinth* XV.2, 271.

[21] Davidson/Thompson/Thompson 1943, 79, and 93, Fig. 41.

[22] Davidson/Thompson/Thompson 1943, 79.

[23] De Miro 2000, 119.

[24] Costamagna and Sabbione 1990, 118.

[25] Coleman et al. 1992, 276; O'Neill et al. 1999, 310, 312.

1:1

Figure 14.3 Loom weight with coil breaks (Lo 624-01).

Site type	Sites with loom weights	% Total
Farmhouse	75	93%
Tomb	5	6%
Farmhouse/tomb	(5)	—
Sanctuary	1	1%
Farmhouse/sanctuary	(1)	—
Total	**81**	**100%**

Table 14.2 Distribution of loom weights among site types.

variation: some examples approach cooking- (e.g., Lo 342-01) or amphora- (e.g., Lo 492-01) type fabrics, while others approach fine black-gloss fabrics (e.g., **No. 31**). It is clear that most of the flat-faced disk loom weights were carefully made by hand using the coil technique: on broken examples the coil breaks are often clearly visible (*Fig. 14.3*). Two examples of decorated mold-made disk loom weights survive: **No. 11** and **No. 34**; they are discussed in detail below. The lenticular disk weights were also made by hand, but sometimes perhaps less carefully. On **No. 43**, which has only a single hole at the top rather than the usual double holes, a dent in the center suggests that the maker began by intending to produce a spindle whorl but changed intention partway through the process.

With only two exceptions—pyramidal Lo 204-01 (*Fig. 14.2*) and pinched **No. 44**—all stamps in the Metaponto survey corpus appear on disk loom weights. As the "fashion" for stamping loom weights of all types elsewhere in the Greek world peaked between the mid-5th and the end of the 4th centuries BC, it is likely that many (or even most) of the Metapontine stamped loom weights date to this period. In most places stamped loom weights represent only a very small proportion of recovered assemblages.[26] However, among the disk loom weights at Metaponto, a surprisingly high 37% (29) were stamped. Issues raised by the stamps will be considered in more detail below. The complete disk and lenticular loom weights range in weight from 117 g (Lo 103-01) to 2041 g (Lo 156-01, a lenticular piece), but most cluster in the range of 1100–1600 g (with the average weight of whole examples 1283 g and standard deviation 360.1202). They are therefore much heavier than the narrow pyramidal group of loom weights and significantly heavier than the pinched loom weights.

[26] For example, *Corinth* XII, 153; Davidson/Thompson/Thompson 1943, 74.

The pinched loom weights that appear at Metaponto in small numbers (5% of the total corpus) are virtually unparalleled elsewhere. At Locri Epizephyrii similar "figure-eight" loom weights appear in the late 4th and 3rd centuries BC.[27] At Metaponto, these pinched loom weights are all handmade, and indeed they often appear to be casually made, having been squashed inward at the sides, but generally with very deliberate fingerprints impressed upon the two faces. **No. 45** appears to have been made by a child, to judge from the small size of the fingerprints. **No. 44** has been stamped using a ring. One example (Lo 413-01), originally identified as a pinched loom weight and similar in shape, is actually a spindle whorl. The fabric of most examples resembles plain-ware fabrics ranging from somewhat coarse to as fine as black-gloss fabrics; one example (Lo 734-01) approaches cooking-type fabric. Overall, the pinched weights are lighter than the disk weights but heavier than the pyramidal weights. The average weight of whole examples is 754 g. Although the sample is too small for the standard deviation (134.5232) to have any real statistical validity, the relative standardization of these apparently casually made loom weights is striking: with the exception of one example weighing 989 g, the rest are all close to 700 g in weight.

Distribution of Metapontine Loom Weights

Not surprisingly, a vast majority of the survey sites (75 of 81, 93%) on which loom weights were found are sites identified as farmhouses (*Table 14.2*). Only five sites (on which a total of six loom weights were found) were securely identified as Greek tombs (disk: sites 48, 116, 415; pinched and disk: site 300; pyramidal: site 253). Five sites appear to be Greek farmhouses with associated tombs nearby (disk: sites 531, 363; pinched: site 734; pyramidal: sites 521, 523), so it is impossible

[27] Costamagna and Sabbione 1990, 118.

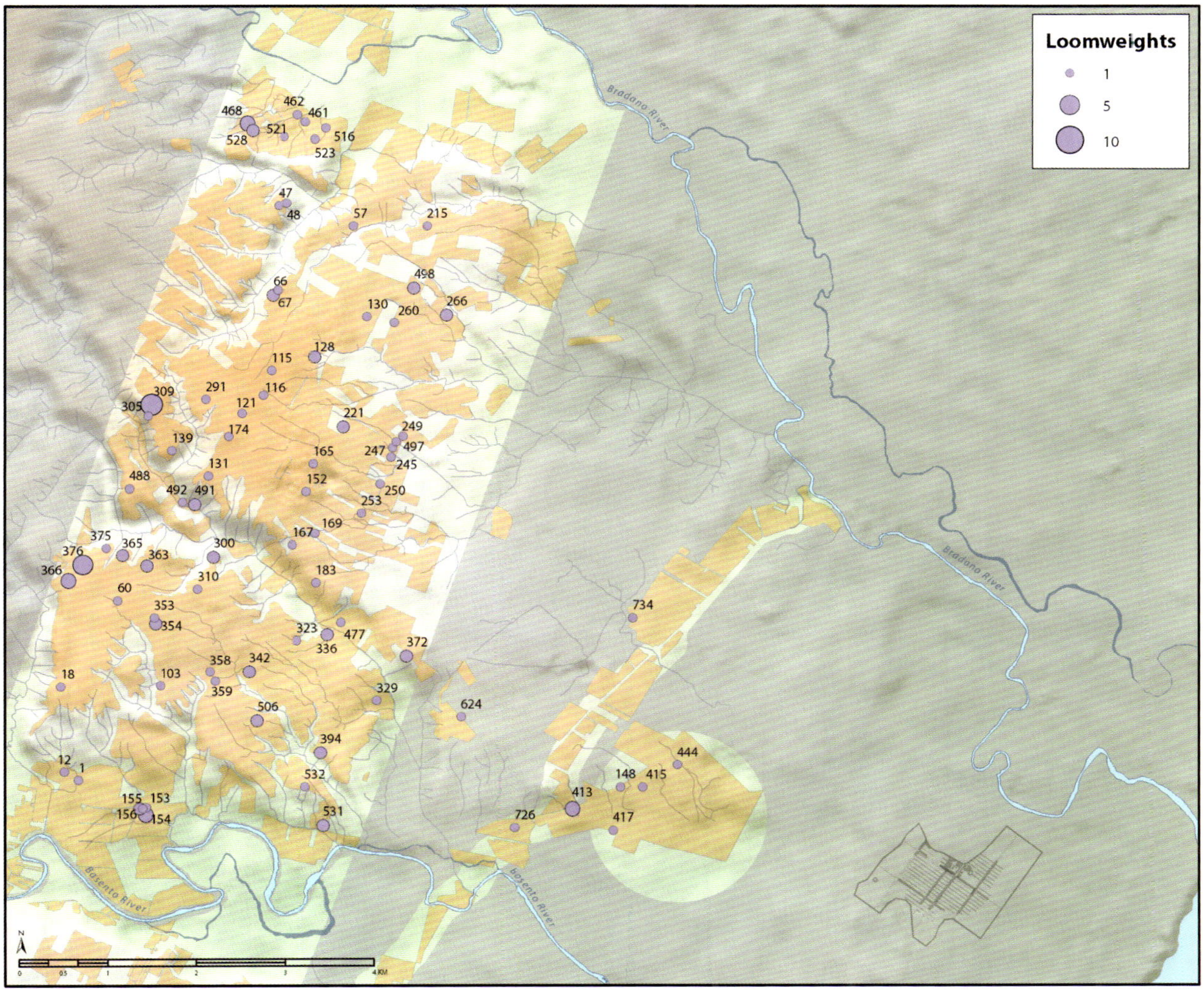

Figure 14.4 Map showing concentration of loom weights in the survey area. (JT)

to be certain whether the loom weights discovered on these sites were from the farmhouse or the tomb context—in most cases, the former is perhaps more likely. Only three loom weights were recovered at two sites securely identified as sanctuaries or farmhouse/sanctuaries (two disk: site 266; pyramidal: site 477). Hence, it seems safe to infer that loom weights are most likely to be associated primarily with domestic residence, offering evidence for the presence of resident women and their productive activities in the rural landscape. However, all types of loom weights are present in all of the principal types of contexts (farmhouse, tomb, and sanctuary).

Loom weights were used in sets for weaving, and the evidence of such "sets" found in excavated contexts suggests that these generally consisted of between 20 and 70 loom weights.[28] At Olynthos and elsewhere these sets of loom weights regularly contain a mixture of types (e.g., conical and pyramidal).[29]

Most of the Metaponto survey sites where loom weights have been found produced only a single weight (57 sites, 70%; *Fig. 14.4*). This must in large part be due to the incomplete nature of the evidence generated by surface survey, but it may also provide one indication that loom weights were valued as personal property by women and were not casually abandoned unless they had been broken beyond repair, or their abandonment was unavoidable.

On only 24 sites (30%) have two or more loom weights been found (*Fig. 14.5*). Of these, all except

[28] Davidson/Thompson/Thompson 1943, 69–70; Cahill 2002, 4.50–53.
[29] Cahill 2002, 4.51–52.

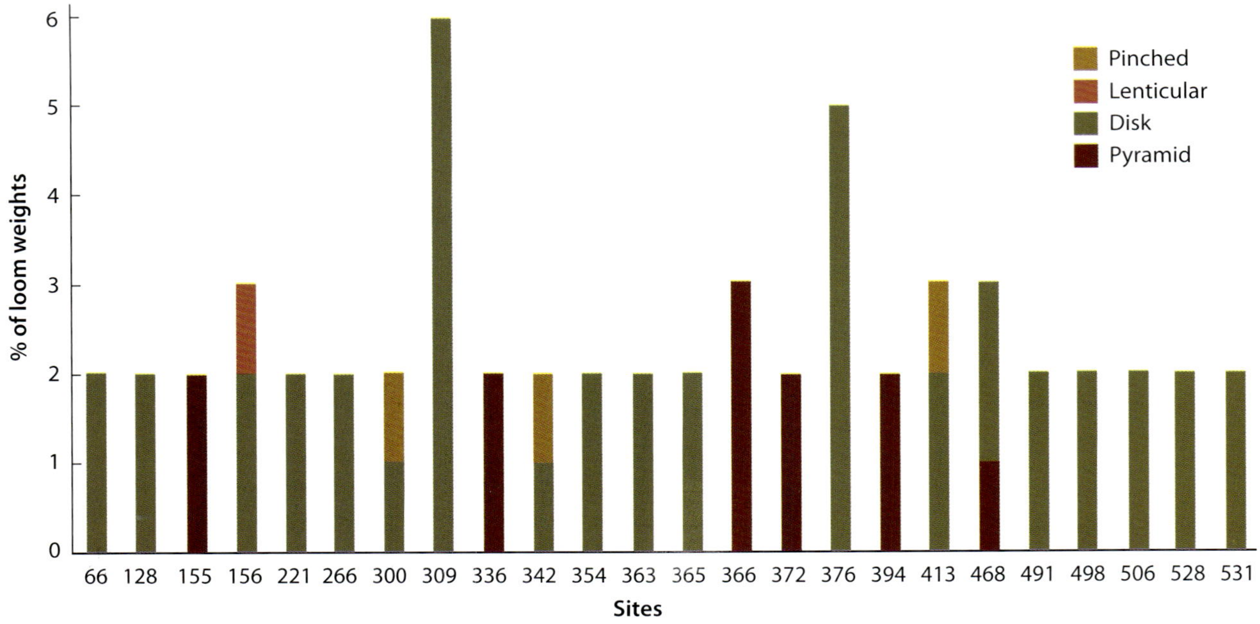

Figure 14.5. Distribution of types among sites with more than two loom weights.

sites 300 (Greek tombs) and 266 (Greek sanctuary) are farmhouse sites (although, as noted above, sites 363 and 531 are farmhouses with associated tombs). Most of these 24 sites have only 2 loom weights, both of the same type: 12 sites with disk only (sites 66, 128, 221, 266, 354, 363, 365, 491, 498, 506, 528, 531), and four sites with pyramidal only (sites 155, 336, 372, 394). There were only two exceptions: on sites 300 and 342 1 disk and 1 pinched loom weight were collected. Of the six sites with more than 2 loom weights, four have 3 loom weights: site 156 with 1 lenticular and 2 disk; site 366 with all pyramidal; site 413 with 2 disk, 1 pinched; and site 468 with 2 disk, 1 anomalous pyramidal (see above). Site 376 has 5 loom weights, all disk, and site 309 has 6, all disk.

It is clear that in the Metaponto survey area different types of loom weights appear only rarely on the same sites. It is hard to ascertain the significance of this observation since sites with single examples are the most common. It is nonetheless striking that, with the exception of site 468, where the pyramidal loom weight is anomalous and likely to be either prehistoric or indigenous in origin (see above), pyramidal- and disk-type loom weights are never present on the same site. Disk and pinched forms are combined on three sites (sites 200, 342, 413), and lenticular and disk forms appear together on one site (site 156), although these two types should probably be regarded as variations on the same form.

This evidence for the geographical segregation of pyramidal and disk types of loom weights, slight though it is, may suggest that the pyramidal types were generally earlier. Elsewhere in the Greek world, it is clear that, although there is considerable chronological overlap, the disk form begins later in date. Moreover, as noted above, weavers seem not to have minded mixed sets of loom weights in principle. In the Metaponto survey corpus, most loom weights appear on sites that are only Greek, with no Late Hellenistic occupation. By the Imperial period the warp-weighted loom had declined significantly in popularity, and it largely went out of use in the Roman world by the 2nd century AD.[30] Therefore, one would not expect to find loom weights much after the period designated as "Late Hellenistic" by the project. In the Metaponto survey corpus, no loom weights have been recovered from Late Hellenistic or later Roman sites that do not also have a Greek component.

A closer look at the findspots strengthens these impressions. Pyramidal loom weights appear at 19 sites in total. Of these, 16 (84%) have a Greek phase and no Roman presence. Only three of the sites (16%) where pyramidal loom weights were found have a Republican component in addition to the Greek. In contrast, disk loom weights appear at 56 sites in total; of these, 32 (57%) have only a Greek component, while 24 (43%) have both Greek and Republican compo-

[30] Davidson/Thompson/Thompson 1943, 70–71.

nents. This shows that the disk form of loom weight is considerably more likely to appear on sites occupied in periods following the Roman conquest of Lucania, while pyramidal loom weights are largely restricted to sites abandoned by the end of the colonial period.

Loom Weights and Personal Identities: Use Patterns and Stamps

Virtually all of the loom weights in the Metaponto survey corpus appear to have been used on the loom, and specific wear patterns are visible on some. Most often these take the form of wear or slight grooves at the top inner edge of the holes, or occasionally on the bottom inner edge (*Fig. 14.6*). Such a pattern suggests that Metapontine loom weights were attached to the warp threads with a piece of sturdy string rather than a stick or a metal ring. Although survey material is often battered, it is still possible to detect wear and chipping at the bottom of loom weights where they knocked against each other while hanging on the loom. Examples include **No. 17**, a disk that shows clear wear on the bottom half of each side, to the point that the shape is slightly distorted, and **No. 5**, a pyramidal loom weight that shows analogous wear and chipping near the base.

A number of the Metapontine disk loom weights are decorated or stamped. Unlike at Corinth[31] or Athens,[32] the kinds of abbreviated-name or monogram stamps probably representing maker's marks do not appear. Also absent are letters incised before firing, and graffiti, both relatively common elsewhere.[33] Only two weights from Metaponto preserve traces of slip or paint: **No. 34** and **No. 35**. As explained at the start of this section, the stamping of loom weights with personal seal rings was a practice widespread in the Greek world, although it is interesting that the proportion of stamped loom weights in the Metaponto survey corpus seems to be surprisingly high (37%) among the disk loom weights, the only form that is regularly stamped (with illegible stamps appearing on one pyramidal and one pinched example—see above).

The decoration of the two loom weights stamped with patterns covering the entire surface fits comfortably in the 4th century BC. Loom weight **No. 34** bearing a guilloche pattern and palmettes or lotus flowers is closely paralleled by the pyramidal weights with

Figure 14.6 Wear marks on loom weight (Lo 156-01).

stamped palmettes from the Pnyx; Davidson dates these to the 4th century BC on the basis of the similarity of the motifs to 4th-century pottery decoration.[34] The figure depicted in a conventional running pose on **No. 11** is very worn and faint, so it is difficult to identify the subject with certainty. The head is missing, but the pose might suggest a gorgon, a common motif on loom weights and seals. The similarity of the decorative technique to **No. 34** suggests a 4th-century date, and a disk loom weight decorated in the identical technique depicting a gorgon's face from Locri Epizephyrii is also suggestive.[35]

Of the 29 stamped or decorated loom weights nearly two-thirds (18, 64%) were stamped with personal seal rings or stamps. It is thus probably not unreasonable to suggest that personal ownership and identity played a significant role in the use and display of these items in the colonial Metapontine countryside. But the story becomes more complex when the stamps and decorations are examined more closely.

Most of the seal stamps fit comfortably with the 4th-century BC date suggested for the Metapontine disk loom weights. Where motifs are legible, they often have analogs elsewhere. So, for example, the standing figure of **No. 10** is paralleled by a stamp on a 4th/3rd-century disk loom weight from Akragas (1052),[36] and the general stylistic traits of the Metapontine figures are in accord with western Greek styles of the 4th century BC.[37] Similarly, the bird of **No. 9** corresponds in style to a dove on an unprovenanced seal in the British Museum[38] and to a cockerel on a 5th–4th-century BC bronze intaglio ring from Crete in the Fitzwilliam Museum.[39] Even where the motifs of the stamps are not legible, the shape of the stamp can help in dating. Many, e.g., **No. 14** and **No. 15**, are

[31] *Corinth* XII, 156–60.

[32] Davidson/Thompson/Thompson 1943, 77–78.

[33] *Corinth* XII, 150–52; Davidson/Thompson/Thompson 1943, 75–76.

[34] Davidson/Thompson/Thompson 1943, 75, Nos. 81–84.

[35] Costamagna and Sabbione 1990, 118.

[36] De Miro 2000, 210 and Fig. 133.

[37] Boardman 2001, 228–29.

[38] Walters 1926, 60, No. 560 and Pl. X.

[39] Henig 1994, 260, No. 557.

ovals typical of 4th-century rings;[40] **No. 29**, with an identifiable motif, is also of this shape. However, the shape of the stamp on the only stamped pinched loom weight is of a type that is generally somewhat earlier in date, common to the middle of the 5th century BC, although there are also some slightly later examples of this type of stamp.[41]

The two loom weights with letter stamps both marked X, **No. 8** and **No. 16**, also have 5th- and 4th-century BC parallels, notably at Athens,[42] but also at Akragas.[43] Davidson observed that in the case of Corinth the letter stamps used on loom weights (which here appear to be feminine names or fragments of them) never appear on other kinds of objects; single-letter stamps do not occur, but there are many instances of incised single letters.[44] It seems likely that in the case of the Metapontine examples these single-letter marks also signify personal ownership, and the two stamps are clearly different. Other stamped marks that are not letters or pictures, such as the various kinds of incised circles found on **No. 17**, **No. 21**, **No. 22**, **No. 28**, **No. 32**, **No. 33**, and **No. 37**, may also therefore be intended as distinguishing marks for purposes of identification and ownership. It is remotely possible that the same function was served by the barley-grain impression on **No 19**. This seems similar in inspiration to the practice of impressing loom weights with astragali at Halai.[45] The emblem on the coinage of Metaponto was an ear of barley.

Several stamps seem to be rare or unique. The large stamp depicting an ivy or smilax leaf seems to be unparalleled, but similar decorative motifs are well known on later 5th- and 4th-century BC pottery, so the stamp seems compatible with the 4th-century date of the loom weights. The footprint stamps of **No. 20** and **No. 30** are likely to be identical (see below). This motif is unknown on loom weights elsewhere, but 5th–4th century parallels as a seal stamp are found specifically in areas where there was a significant Phoenician presence, notably Cyprus[46] and Sardinia (Tharros).[47] It is interesting to speculate whether the presence of these footprint stamps in the Metapontine countryside might point to the perceived "Phoenician" identity of an individual or a family.

There appear to be instances in the Metaponto survey corpus of two interesting phenomena observed elsewhere: (1) loom weights bearing stamps that seem to be considerably older than the likely date of manufacture of the loom weights themselves,[48] and (2) loom weights appearing in contexts which are later than the date of the loom weights themselves.[49] A possible example of both phenomena may be **No. 27**, one of six loom weights found at site 309 (identified as a Greek farmhouse), the site with the greatest number in the survey area. All are disk in form, but **No. 27** is notably smaller and made from a finer fabric than the other five, which are similar to each other in fabric, dimensions, and probably originally in weight (*Table 14.3*).

There are as yet no close analogies for the remarkable stamp on **No. 27**. Its shape does not seem typical of the 5th–4th century BC. Although the stamp is worn, the manifestly orientalizing motif of the goddess behind a double-headed horse and the style of the work both seem more typical of the 6th century than of the 4th century BC. The most expedient interpretation of the loom weights at site 309 is that they constituted part of a "set" in use at (or shortly before) the time the house was abandoned. One, **No. 27**, was perhaps a weight that had been kept and passed on from another set, perhaps because it had some special significance. However, the weight itself is unlikely to be earlier than the late 5th century and more likely belongs somewhere in the 4th century, which means that the stamp used to make it was itself an heirloom of considerable age when the weight was made. The other loom weights, **Nos. 24–26**, probably belong somewhere in the second half of the 4th century and may well be 20 to 30 years (a generation?) later than **No. 27**.

In the case of the three identical stamps on **No. 39**, **No. 40**, and **No. 41**, it is also possible that the stamp is considerably older than the weights. I can find no close analogs for the rosette motif on these loom weights in the 5th–4th-century BC repertoire. On both seal rings and stones, as well as on stamped loom weights, rosette/star motifs seem to be rare in the Classical period. When they do occur, as for example at Akragas[50]

[40] Boardman 2001, 213–14, types VI and VIII.
[41] Boardman 2001, 212–13, type I.
[42] Davidson/Thompson/Thompson 1943, 74, 81, 83, Nos. 12–21 and Fig. 34.
[43] De Miro 2000, Fig. 132, Nos. 1125 and 696.
[44] *Corinth* XII, 150–52, 156–60.
[45] O'Neill et al. 1999, 312, Fig. 21.
[46] Richter 1968, 123, No. 481 (Oxford, Ashmolean 1896.1908.0.14).
[47] Walters 1926, 51, Nos. 430–31 and Pl. VII.
[48] Richter 1968, 143; Davidson/Thompson/Thompson 1943, 84, No. 56; *Corinth* XII, 152.
[49] *Corinth* XII, 146.
[50] De Miro 2000, 191, Fig. 133, No. 816.

Number	Diameter (cm)	Thickness (cm)	Weight (g)
Lo 309-01	7.2	1.8	1083
Lo 309-02 (fragment)	7.6	2.0	495
Lo 309-03	7.5	2.4	1549
Lo 309-04	7.9	2.3	1598
Lo 309-05	7.9	1.9	1350
Lo 309-06	6.2	1.7	747

Table 14.3 Loom weights from site 309.

and Athens,[51] the designs are much simpler than the rosette stamp at Metaponto. Rather worryingly, the closest parallels appear on seal stones of the 7th–6th centuries, a period in which the motif was common.[52] Again, there seems to be a serious possibility that all three of these loom weights were stamped with "heirloom" stamps, ones much older than the period of the manufacture of the weights themselves, which was most likely in the 4th century.

The appearance of identical stamps at different locations in the Metaponto survey area also raises interesting questions. There are two instances where this occurs: the footprint stamps **No. 20** and **No. 30**, and the rosette stamps of **No. 39, No. 40,** and **No. 41**. All of these sites have been identified as Greek farmhouses, but site 531 also has an associated Greek tomb and a Late Hellenistic occupation phase. In the cases of **No. 20** and **No. 30**, the two farmhouses lie about 3 km apart on either side of the Venella valley and appear to be contemporary in date. Sites 531 and 532 are relatively close together but appear to have slightly different chronological ranges of occupation, with site 531 extending slightly later in date.

From what is known more broadly of Classical Greek societies, it is easy to imagine a number of means by which loom weights might have migrated around the landscape. In both the cases where this happens in the Metapontino, the stamps are potentially special, in one case, **No. 20** and **No. 30**, because of an unusual motif that might have "ethnic," or at least familial, significance, and in the other case, **No. 39**, **No. 40**, and **No. 41**, because of the possible use of an heirloom stamp. These spatial distributions could have been the result of inheritance, movements of residence upon marriage, or even simply lending and borrowing or the giving of gifts. Whatever the underlying social practices that produced them, these two examples, along with the curious case of site 309, demonstrate that both loom weights themselves and the seals used to mark them were items of property that women considered sufficiently important to pass along to the next generation. The seals in particular appear to have remained in circulation for a very long time in some instances, and it seems most likely that they passed down the generations along female lines. This must surely underpin the suggestion that women's identities were closely linked to these objects, and that they may well have provided a material manifestation of familial links, particularly those between mothers, daughters, and sisters. In light of this it might not be too fanciful to suggest that the clear and deliberate fingerprints on the pinched loom weights might also have represented a similar expression of personal (indeed, very personal) identity.

Conclusions

The loom weights from Metaponto can tell us much about the social life of the countryside. They suggest that by the 5th and 4th centuries BC this was a settled landscape full of households with resident women. There is slight evidence to suggest that households including women were settled in the countryside as far back as the early 7th century. Loom-weight styles changed over time: it seems likely that the pyramidal loom weights belong largely to the 5th century BC and a bit earlier, the disk loom weights to the 4th century and a bit later, and the pinched loom weights in between, perhaps mid-5th to mid-4th century BC.

By the 4th century, a considerable number of these women must have been wealthy enough to own a personal seal ring in metal or with a semiprecious seal stone. These were used, among other things, to mark their loom weights. They seem to have considered their loom weights and weaving equipment sufficiently precious that they abandoned them reluctantly. Most sites produced only one loom weight, and only two spindle whorls were found, one at site 300 (Greek tombs, **No. 46**) and the other at site 413, a Greek farmhouse. On the other hand, there is clear evidence that both loom weights and seals moved through female connections and kinship lines. The stamps and other marks on the loom weights suggest that, as personal property, they played a key role in the personal and familial identities of the women of the chora of Metaponto.

[51] Davidson/Thompson/Thompson 1943, 81, 83, Fig. 34, Nos. 28–30.

[52] E. Brandt 1968, 37, Nos. 132–33 and Pl. 16, 40, No. 151 and Pl. 18.

Catalog of Loom Weights

All dimensions are in cm. Maximum height is indicated by "h," maximum diameter indicated by "diam.," maximum thickness indicated by "th." An asterisk (*) indicates materials from the 1980 ICA pilot survey (sites numbered 1–44), as well as at the intensive survey sites and plots located south of the Basento River through the 2001 season (most of the sites numbered 562 and higher). Additional information about the sites can be found in the Gazetteer.

Conical loom weights

1:2

1. Lo 819-01*
Dimensions: h 4.8
Weight: 471 g
Stamp: circular impresson on top.

Pyramidal loom weights

1:2

2. Lo 291-01
Dimensions: h 6.2
Weight: 437 g
Stamp: None.

1:2

3. Lo 329-01
Dimensions: h 5.6
Weight: 358 g
Stamp: None.

1:2

4. Lo 372-01
Dimensions: h 10.8
Weight: 3000 g
Stamp: None.

1:2

5. Lo 372-02
Dimensions: h 6.2
Weight: 493 g
Stamp: None.

Trapezoidal loom weights

1:2

6. Lo 394-01
Dimensions: h 5.5
Weight: 1231 g
Stamp: None.

1:2

7. Lo 468-01
Dimensions: h 6.2
Weight: 1033 g
Stamp: None.

Disk loom weights

1:2

1:1

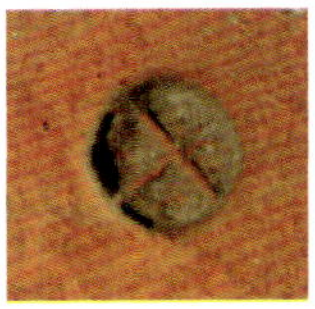

8. Lo 001-01
Dimensions: diam. 7.8 th 2.3
Weight: 1477 g
Stamp: Stamp (ring/seal?) of X (Greek letter chi) in circle.

1:2

1:1

9. Lo 032-01
Dimensions: diam. 7.0 th 2.0
Weight: 1136 g
Stamp: Ring stamp of bird facing right.

1:2

1:1

10. Lo 048-01
Dimensions: diam. 6.9 th 2.5
Weight: 1110 g
Stamp: Worn oval ring stamp of running figure, possibly a maenad, facing right with left leg straight and right leg bent.

1:2

11. Lo 060-01
Dimensions: diam. 7.8 th 2.0
Weight: 1321 g
Stamp: Stamp, covering whole surface, of a large figure in conventional running pose with head missing—a gorgon?

1:2

1:1

12. Lo 067-01
Dimensions: diam. 7.7 th 2.1
Weight: 1307 g
Stamp: Large stamp of heart-shaped ivy or smilax leaf.

13. Lo 128-02
Stamp: Illegible ring stamp.

1:2

1:1

14. Lo 131-01
Dimensions: diam. 7.6 th 2.2
Weight: 1406 g
Stamp: Elliptical ring stamp, almost illegible, of a basket?

Disk loom weights cont.

1:2

15. Lo 156-02
Dimensions: diam. 7.8 th 2.3
Weight: 1578 g
Stamp: Illegible.

1:2

1:1

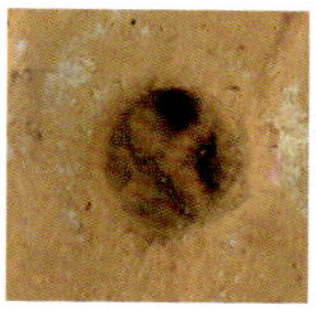

16. Lo 165-01
Dimensions: diam. 7.1 th 2.4
Weight: 1228 g
Stamp: Stamp (ring/seal?), just below holes on one side, of deeply impressed X (Greek letter chi) in circle.

1:2

1:1

17. Lo 169-01
Dimensions: diam. 7.4 th 2.2
Weight: 1335 g
Stamp: Small circle or "o" impressed on one side.

1:2

1:1

18. Lo 183-01
Dimensions: diam. 7.1 th 2.3
Weight: 1370 g
Stamp: One ring stamp on each side, one illegible, the other depicting a horse facing right.

1:2

1:1

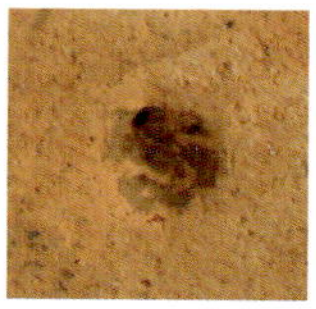

19. Lo 221-01
Dimensions: diam. 7.0 th 2.8
Weight: 1544 g
Stamp: Impression in center, perhaps from a grain of barley. Straw impressions on reverse.

1:2

1:1

20. Lo 221-02
Dimensions: diam. 7.0 th 2.1
Weight: 1187 g
Stamp: Footprint (probably ring) stamp. Similar fabric and identical stamp with Lo 358-01.

Disk loom weights cont.

21. Lo 266-01
Stamp: Impressed circle in center.

1:2

1:1

22. Lo 300-02
Dimensions: diam. 6.7 th 2.2
Weight: 1201 g
Stamp: Small circle impressed in center of one side.

1:2

23. Lo 309-01
Dimensions: diam. 7.2 th 1.8
Weight: 1083 g
Stamp: None.

1:2

24. Lo 309-03
Dimensions: diam. 7.5 th 2.4
Weight: 1549 g
Stamp: None.

1:2

25. Lo 309-04
Dimensions: diam. 7.9 th 2.3
Weight: 1598 g
Stamp: None.

1:2

26. Lo 309-05
Dimensions: diam. 7.9 th 1.9
Weight: 1350 g
Stamp: None.

1:2

27. Lo 309-06
Dimensions: diam. 6.2 th 1.7
Weight: 747 g
Stamp: Truncated oval seal stamp of goddess with pilon-type hat between the heads of a double-headed horse.

1:1

Disk loom weights cont.

1:2

1:1

28. Lo 354-01
Dimensions: diam. 6.8 th 2.4
Weight: 1365 g
Stamp: Impressed circle with raised center on one side.

1:2

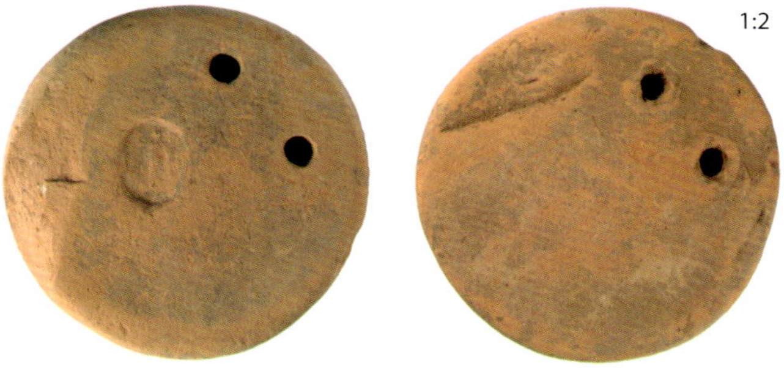

1:1

29. Lo 354-02
Dimensions: diam. 7.9 th 2.3
Weight: 1643 g
Stamp: Oval ring stamp of standing or running figure facing left(?).

1:2

1:1

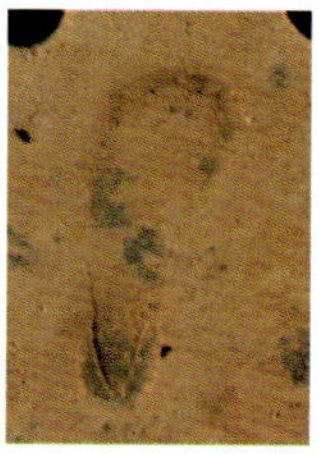

30. Lo 358-01
Dimensions: diam. 6.9 th 2.1
Weight: 1167 g
Stamp: Footprint (probably ring) stamp. Similar fabric and identical stamp with **No. 19.**

1:2

1:1

31. Lo 363-02
Dimensions: diam. 7.1 th 2.3
Weight: 1196
Stamp: Oval ring stamp, unclear, possibly seated or crouching figure?

1:2

1:1

32. Lo 376-01
Dimensions: diam. 7.7 th 2.1
Weight: 991 g
Stamp: Circular impression, identical to **No. 33.**

1:2

1:1

33. Lo 376-02
Dimensions: diam. 7.8 th 2.0
Weight: 1381 g
Stamp: Circular impression, identical to **No. 32.**

Disk loom weights cont.

1:2

1:1

34. Lo 376-03
Dimensions: diam. 7.8 th 2.2
Weight: 1656 g
Stamp: Pattern, stamped all over both sides, of running guilloche at top near holes, lotus-type flowers or palmettes with leafy tendrils above in center, leafy acanthus-like vegetation at sides. Traces of red paint preserved.

1:2

1:1

35. Lo 376-04
Dimensions: diam. 7.3 th 2.8
Weight: 1609 g
Stamp: Oval ring stamp, illegible. Traces of red paint preserved.

---

36. Lo 417-01
Stamp: Bottom of oval ring stamp (or footprint?) preserved on one side, largely illegible.

37. Lo 506-01
Stamp: Remains of impressed circle on one side.

---

no scale

38. Lo 506-02
Dimensions: diam. 7.9 th 2.0
Weight: 1265 g
Stamp: Part of ring stamp, surviving just below suspension holes, of seated woman facing right.

1:2

1:1

39. Lo 531-01
Dimensions: diam. 8.0 th 2.0
Weight: 878 g
Stamp: Round ring/seal stamp of star or rosette near edge; same position as, and probably identical with, stamps on **No. 40** and **No. 41**.

1:2

1:1

40. Lo 531-02
Dimensions: diam. 8.0 th 2.1
Weight: 1406 g
Stamp: Round ring/seal stamp of star or rosette near edge; same position as, and probably identical with, stamps on **No. 39** and **No. 41**.

1:2

1:1

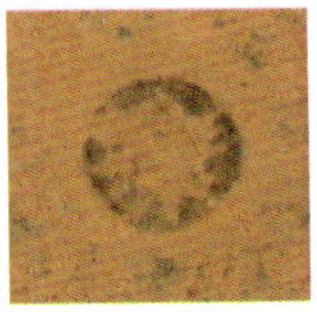

41. Lo 532-01
Dimensions: diam. 7.8 th 2.3
Weight: 1526 g
Stamp: Round ring/seal stamp of star or rosette near edge, same position as, and probably identical with, stamps on **No. 39** and **No. 40**.

Lenticular loom weights 1:2

42. Lo 156-01
Dimensions: diam. 8.2 th 3.0
Weight: 2041 g
Stamp: None.

1:2

43. Lo 310-01
Dimensions: diam. 5.7 th 5.0
Weight: 540 g
Stamp: None.

Pinched loom weights 1:2

44. Lo 300-01
Dimensions: h 6.1 th 2.7
Weight: 656 g
Stamp: Ring seal.

1:2

45. Lo 342-02
Dimensions: h 6.4 th 3.0
Weight: 708 g
Stamp: None, but small fingerprints suggest manufacture by child.

Spindle whorl 1:2

46. Lo 300-03
Dimensions: h 3.2 w 1.9
Weight: 199 g
Stamp: None.

15
Coins

Anna Rita Parente

Nine coins were recovered by the survey, including four from two separate tombs (**Nos. 1–4**), and three from the area between the Basento and Cavone (**Nos. 6–8**). Together they form a homogenous and coherent picture when compared to other such finds from the Hellenistic era in the Metapontino. As in most cases, they are examples in bronze from the mint of Metaponto, consisting of small coins—predominantly fiduciary in character—that were used in daily exchange.[1] These could easily be lost because of their small size and reduced value.

The introduction of coined bronze in the Greek world represented an important evolution, albeit more conceptual in nature than economic: coinage was assigned a nominal value that, in fact, greatly exceeded its real one. Thus overvalued, bronze coins became established as the preferred form of currency, substituting for fractional values in silver.

Production of the first bronze series began in the second half of the 5th century BC in Sicily and Magna Grecia, predating usage in mainland Greece, and was due to the influence of the local indigenous populations, who had long used bronze for weights.[2] In the course of the Hellenistic period (4th–3rd centuries BC), production of bronze increased considerably. The well-developed bimetallic system, with silver and bronze specimens in various sizes and weights, facilitated exchanges and commerce at every level in an economic environment that was already deeply monetized. The coining of gold remained occasional and was intended for large extraterritorial payments.

Evidence suggests that the Metapontine mint produced large quantities of bronze coins from the end of the 5th century BC until the Punic occupation.[3] After the defeat of Hannibal, Rome definitively consolidated her power in southern Italy, putting an end to local autonomy and therefore, with few exceptions, to local coinages.

Among the first Metapontine specimens, the heaviest ones (7.5–10.0 g) bear the legend "obolos", which indicates their nominal value. This name originally denoted the fraction in silver (1/6 of a drachma); the fact that it was adopted as well for bronze series opens new horizons for research on the (not always clear) relationship between the issues in the two metals. Soon afterwards, the bronze coinage of Metapontion, as at other Italiote mints, suffered a series of devaluations in weight, which culminated in the disappearance of the heaviest examples. The ascendancy of bronze in daily exchanges was undoubtedly favored also by the end of the silver issues, which at one time were produced abundantly and regularly at Metaponto. Save for a brief revival during the Hannibalic occupation, these were never again coined after the Pyrrhic War.

Proof of the definitive ascendancy of bronze comes also from the recovery of a significant number of hoards that contained only bronze examples from Metaponto. Among the earliest (the burial is dated around the middle of the 4th century BC) is a hoard of 33 coins discovered in the chora, in the Lama locality near Bernalda (IGCH 1935).[4] Another hoard, unfortunately not intact (only nine items were found) and dating a few decades later, was discovered in the territory of Pisticci.[5] During the 3rd century BC, the phenomenon of hoarding noticeably increased, probably as a reaction to the military activities that affected the area at that time. To this period date a large hoard of 1,042 coins in a vase found in contrada Sansone (IGCH 2038)[6] and a small hoard of 25 coins

[1] For similar discoveries, see Lo Porto (1966, 176 contrada Sansone), who nevertheless suggests that the four examples in question could be part of a hoard of silver coins (*IGCH* 1959); Lo Porto 1966, 185 (Metapontino). See also Johnston 1977, 383 (Cozzo Presepe); Lo Porto 1988–1989, 437 (Metapontino); Postrioti 1996, 131–33 (votive deposit of Temple E); Guardascione 2002, 304, 308 (Torre di Mare). Many more remain unpublished.

[2] On the question in general, see the synthesis in Price 1968.

[3] Johnston 1989, 121.

[4] Lo Porto 1966, 134–85; Johnston 1989, 123, 135.

[5] Lo Porto 1973, 181.

[6] Lo Porto 1966, 184.

found near the urban center.[7] There is less certainty regarding the find spots of two other hoards, now dispersed, that contained 650 (IGCH 2040) and 37 (IGCH 2039) Metapontine bronzes.

Other cases can be attributed to the same chronological period, wherein the bronze Metapontine currency was horded outside the polis of origin, along with other Italiote coins in silver, as with the hoards from Martina Franca (IGCH 2008) and Pisticci (IGCH 1996), or with bronze coins from mints in the Bruttian area, as at Strongoli (IGCH 2037). Extraterritorial circulation of Metapontine bronze is confirmed also by isolated discoveries in the indigenous chora, better documented in domestic contexts than in sanctuary areas.[8]

Among the most interesting features of the corpus of coins from the survey is the presence of two coins in each of two tombs (site 635 tomb 4 and site 736 tomb 20, see Ch. 29). Coins in funeral contexts are generally, but not always,[9] interpreted as a reflection of religious traditions linked to rituals of passage: according to a widespread belief, in order to cross the River Styx and enter the Underworld, the deceased had to pay a symbolic fare to the ferryman Charon. The evidence from the necropoleis of the Metapontino (Crucinia, Casa Ricotta, Pantanello[10]), which continues to accumulate, documents a practice that became widespread in the Hellenistic period (4th–3rd centuries BC) but in fact was never homogeneously and systematically adopted.[11] Use of this ritual seems to have depended on individual or family choice, not clearly defined socio-economic realities. The phenomenon in itself nevertheless embodies an economic interest as far as the idea of "payment," albeit symbolic, and was by then indissolubly linked to the concept of coinage.

[7] Lo Porto 1988–1989, 429–32.

[8] Siciliano and Sarcinelli 2004.

[9] For other possible interpretations of the meaning of coins in graves, see Cantilena 1995.

[10] See Parente 1999, 142–44, for information about these finds, with preceding bibliography. To this list should be added the most recent data from the excavations at the Pantanello necropolis (Johnston 1998, 830–33), as well as many other still-unpublished finds.

[11] Even the oldest examples, dated to the end of the 5th–first half of the 4th c. BC (e.g., **No. 1**. AE 736, T20-05), could have been deposited long after the moment of issue, after the many years of circulation indicated by their advanced wear.

Catalog of Coins

2:1

1. AE 736, T 20-05 Bronze
[MM Inv. 323.300]
Weight: 3.3 g
Diameter: 16 mm
Orientation: 270°
Obv: Male head facing right with pointed beard.
Rev: M (to left), E (to right). Barley ear.
Comparanda: Johnston 1989, 14; Rutter 2001, 1650.
Date: Last quarter of 5th to middle of the 4th c. BC.
(See Ch. 29 for tomb context)

2:1

2. AE 736, T 20-06 Bronze
[MM Inv. 323.301]
Weight: 3.9 g
Diameter: 16 mm
Orientation: 180°
Obv: Head of Demeter facing right, hair gathered under barley wreath.
Rev: Barley ear with leaf to right; above, thunderbolt.
Comparanda: Johnston 1989, 25; Rutter 2001, 1660.
Date: Second half of 4th to first quarter of the 3rd c. BC.
(See Ch. 29 for tomb context)

2:1

3. AE 635, T 04-02 Bronze
Weight: 0.9 g
Diameter: 11 mm
Orientation: 250°
Obv: Head of Apollo Karneios facing right.
Rev: M[ET]A (at left). Barley ear with leaf to right; above, fly.
Comparanda: Johnston 1989, 64; Rutter 2001, 1700.
Date: First quarter to middle of the 3rd c. BC.

2:1

4. AE 635–T04-08 Bronze
Weight: 1.3 g
Diameter: 12 mm
Orientation: 0°
Obv: Head of Apollo Karneios facing right.
Rev: M[ETA] (at left). Barley ear with leaf to right; above, bill-hook.
Comparanda: Johnston 1989, 44; Rutter 2001, 1679.
Date: First quarter to middle of the 3rd c. BC.

2:1

5. AE 506-01 Bronze
Weight: 0.7 g
Diameter: 10 mm
Orientation: 0°
Obv: Head of Apollo Karneios facing right.
Rev: Barley ear with leaf to right; above, unidentified symbol.
Comparanda: Johnston 1989, 44, 64; Rutter 2001, 1679, 1700.
Date: First quarter to middle of the 3rd c. BC.

2:1

6. AE 903-01 Bronze
Weight: 2.1 g
Diameter: 14 mm
Orientation: 0°
Obv: Head of Helios radiate facing forward.
Rev: Three barley grains radiating from center; M-E and race torch between grains.
Comparanda: Johnston 1989, 53; Rutter 2001, 1689.
Date: First quarter to middle of the 3rd c. BC.

2:1

7. AE 876-01 Bronze
Weight: 2.1 g
Diameter: 14 mm
Orientation: 0°
Obv: Head of Demeter facing right, wearing barley wreath and pendant earring.
Rev: Traces of letters at left. Barley ear with leaf to right.
Comparanda: Johnston 1989, 59; Rutter 2001, 1695.
Date: First quarter to middle of the 3rd c. BC.

2:1

8. AE 1007-01 Bronze
Weight: 2.9 g
Diameter: 16 mm
Obv: Illegible.
Rev: Barley ear.
Date: 3rd c. BC.

2:1

9. AE 339-01 Bronze
Weight: 4.1 g
Diameter: 17 mm
Orientation: 60°
Obv: Head of Demeter facing right with barley wreath.
Rev: Traces of letters at left; TIM (in the middle). Two barley ears, each with short curly leaf to right.
Comparanda: Johnston 1989, 72; Rutter 2001, 1708.
Date: Late 3rd c. BC.